Annual Review of Chinese Architectural Design Works

2009-2010

中 国 建 筑 设 计 作 品 年 鉴

○上

《中国建筑设计作品年鉴》编委会 编

華中科技大學出版社
http://www.hustpas.com

图书在版编目（CIP）数据

2009–2010中国建筑设计作品年鉴（上、下册）/《中国建筑设计作品年鉴》编委会 编.
—武汉：华中科技大学出版社，2010.11
ISBN 978-7-5609-6592-5

Ⅰ.①2… Ⅱ.①中… Ⅲ.①建筑设计–作品集–中国–2009–2010–年鉴 Ⅳ.①TU206-54

中国版本图书馆CIP数据核字（2010）第184116号

2009–2010中国建筑设计作品年鉴（上、下册）　　《中国建筑设计作品年鉴》编委会 编

出版发行：华中科技大学出版社（中国·武汉）
地　　址：武汉市武昌珞喻路1037号（邮编：430074）
出 版 人：阮海洪
责任编辑：姚　璐 余航帆
责任监印：秦　英
美术设计：郝泽星 彭　丹 张　琼 胡培瑜
印　　刷：北京信彩瑞禾印刷厂
开　　本：889 mm×1194 mm　1/12
印　　张：82（上：40，下：42）
字　　数：560千字
版　　次：2010年11月第1版
印　　次：2010年11月第1次印刷
书　　号：ISBN 978-7-5609-6592-5 / TU·951
定　　价：960.00元（上、下册）

投稿热线：(010)64155588-8009　hzjztg@163.com
销售电话：(022)60266190，(022)60266199（兼传真）
　　　　　(010)64155566（兼传真）
网　　址：www.hustpas.com　www.hustp.com

（凡购本书，如有缺页、脱页，请向本社发行部调换）

主　　管：中华人民共和国住房和城乡建设部
主　　办：中国建筑文化中心
编　　辑：《中国建筑设计作品年鉴》编委会
　　　　　北京主语空间文化发展有限公司
主　　编：陈建为
执行主编：肖　峰
编辑部主任：何建国
编　　辑：陈　石 刘笑艳 王　伟 冷娜英
　　　　　代　君 张　澜 时亦飞
编辑部地址：北京市海淀区三里河路13号
　　　　　中国建筑文化中心712室
邮　　编：100037
电　　话：+86-10-88151966/13910120811
传　　真：+86-10-88151958
网　　址：www.ccdu.com.cn
E-mail：zgjsmail@126.com

谨以此年鉴献给为中国建筑设计行业发展而辛勤付出的设计师们！

2009–2010
中国建筑设计作品年鉴

《2009-2010中国建筑设计作品年鉴》（以下简称《年鉴》）
收录了新中国成立以来，重点是近几年的1900多件优秀建筑设计作品，
这些不同类型、不同设计理念的作品基本反映了当前建筑设计行业多元化的发展状况和实践成果，
从中可以直观地领略到中国建筑设计行业融汇中西、贯通古今的设计思想。
对于各地建设主管部门和从事城市规划、建筑设计、景观设计、工程建设、房地产开发人员以及相关科研教育机构掌握建筑设计行业发展趋向、了解新的设计理念，《年鉴》会有很好的参考、借鉴作用。
《年鉴》在编辑过程中，得到了各地建设主管部门和众多设计院所、设计师的大力支持和帮助，
在此，谨向为《年鉴》提供帮助和支持的单位和个人表示衷心的感谢！
《年鉴》在国内公开发行，并委托中国图书进出口（集团）总公司在国外和港、澳、台地区发行。
《年鉴》编辑过程中，不免会有疏漏或错误之处，敬请读者指正，并提出宝贵意见，
以便我们不断提高《年鉴》的编辑水平，满足读者的需求。

《中国建筑设计作品年鉴》编辑部
2010年9月

Saving our environment is the most vital issue that humankind must address today leading into our fears that this millennium may be our last. An ecological approach to our businesses and design is ultimately about environmental integration. Ecodesign is designing for bio-integration.

建筑师在本世纪所担负最大的课题就是用生态设计的方法去解决我们人类社会所面临的环境问题。我们的终极目标就是实现建筑与环境的和谐共生。

杨经文

宗白华先生言，一切艺术综合于建筑，而礼乐诗歌舞剧之表演，亦与建筑背景协调成为一片美的生活。所以每一文化的强盛时代，莫不有伟大的建筑计划，以容纳和表现这一丰富之生命。先生此言发自半个多世纪以前，今日我们更应致力于这一和谐的美的意境早日来临。

吴良镛

希望"年鉴"成为全面总结和推动我国建筑创作发展的优秀典籍

何镜堂

贺《中国建筑设计年鉴》出版

作为中国建筑设计成就的记录，它将成为世界建筑史的辉煌篇章

程泰宁

愿《中国建筑设计作品年鉴》成为中国设计力量在世界设计之林崛起的推动力！

[illegible]

办年鉴，年年明鉴；推精品，精益求精；树大师，不遗余力。

[illegible]

为了中国建筑设计的崛起，为了本土建筑师的声望与力量，请有志同仁加入到"新地方主义"建筑设计运动中，也请《中国建筑设计作品年鉴》作为此方向的引导。

王小东

推陈出新

把祖国建设得更美好

[illegible]

希望"年鉴"成为高品质的盛世年鉴。

[illegible]建民

设计，说到底，都是一种选择：在喧嚣中选择宁静，在盲目中选择理性，在平庸中选择创意，在浮华中选择真诚。

[illegible]

希望本年鉴不只是建筑师或设计单位的"目录"，而是大家彼此观摩学习的园地。

潘冀

贺《中国建筑设计作品年鉴》出版发行

追求卓越

宁静致远

[illegible]

良师益友

梁应添

祝贺《中国建筑设计作品年鉴》发行

历史纪录

时代借鉴

刘景樑

2009–2010

中国建筑设计作品年鉴

建筑文化是一个最能感染其他的文化，而建筑设计，是一个最没有国界的产品。

中国建筑设计的发展，从过去20多年来，由抄袭西方，模仿西方，到最近两年来的探索中国建筑文化，可以说是逐渐的找到了“自我”。但我很希望，中国设计师们，不要重复義合团的“自大”。

的确，如何“古为今用”，“洋为中学”，进而“中为西传”，是我们设计人员，未来的重要工作。共勉之。

[illegible]

愿《中国建筑设计作品年鉴》在中国设计界更有权威性、代表性。

[illegible]

潜心研究市场需求
专注挖掘项目价值

认真解决建筑功能
努力创造优秀建筑

林怀文

立足澳门，背靠祖国，
面向世界，共创繁荣。

[illegible]

希望中国建筑师的职业水准能迅速提升；希望中国建筑师有更多更好的相互交流的平台；希望中国建筑师所面对的设计市场环境更开放、更公平。

[illegible]

佳作 前鉴
拙作 后鉴
新作 新鉴

AIM 陈晓宇

建筑设计能改变人的生活方式并影响人的生活质量，因此，科学设计是我们的责任。

[illegible]

追求品质
倡导原创
宏扬个性
力争创新

[illegible]

建筑文化源于生活，高于生活

愿《中国建筑设计作品年鉴》在中国建筑界树起中国建筑文化的大旗。

[illegible]

愿《中国建筑设计年鉴》成为东西方建筑文化交流的平台。

[illegible]

绘尽三江
笔涂四海

顺德建筑设计院
梁[illegible]书

回顾新中国60年的建筑设计历程，努力进取，迎接中国建筑创作更加美好的未来。

[illegible]

浓缩精华，展示精品
愿2009年“年鉴”成为建筑师的良师益友

[illegible]

Epigraphy Cmooittee 编委题词

特邀编委

（按姓氏首字母排序）

目录 上卷

目录 下卷

特邀编委索引 上卷
（按姓氏首字母排序）

设计师索引 上卷
（按所在机构首字母排序）

设计作品分类索引 上卷

设计作品分类索引 上卷

宾馆、酒店、餐厅、招待所

医院、医疗建筑

交通建筑

设计作品分类索引 上卷

园林、公园、规划设计

其他

Architectural

设计作品
Design Works

澳大利亚柏涛墨尔本建筑设计有限公司
PEDDLE THORP ARCHITECTS

澳大利亚柏涛建筑设计有限公司是澳大利亚最大的建筑设计公司之一。柏涛的历史可以追溯到1889年。一个多世纪以来，柏涛一直走在世界建筑设计行业的前列。

杰出、实用和经济是柏涛公司的设计原则；设计上的创新和技术上的更新是公司的设计宗旨；技术上的可靠和设计上的独特更是公司长期的声誉之所系。庞大的技术资源，广泛、丰富的经验，使柏涛公司能够承担各种规模、各种类型的区域规划设计和各类建筑的设计。

柏涛建筑设计集团除在澳大利亚几个大城市外，还在东南亚、欧洲和美国设有分支机构，设计业务遍布世界各地。柏涛墨尔本公司集中了一大批优秀的建筑师和相关专业人员，除了开展建筑设计业务之外，还在体育、医疗、住宅建筑领域设有专门的研究机构。主要作品有澳大利亚国家网球中心、澳大利亚墨尔本奥林匹克公园及自行车赛馆、ESSO澳洲总部、墨尔本水底世界水族馆、马来西亚运动中心以及众多建于澳洲本地与国外的酒店、商业大厦、政府大厦、写字楼工程、居住区建筑、医疗设施等。

1998年2月，柏涛墨尔本公司在中国成立办事处，随即设立柏涛亚洲公司，发展中国及周边地区的建筑设计业务，其设计业务范围包括城市规划、建筑设计及景观园林设计。经过澳中建筑师几年来的共同努力，亚洲公司如今已成功设计完成了许多令人瞩目的优秀工程项目，并多次获得中国权威机构颁发的奖项，同时与中国的政府、著名的开发企业及设计机构建立了良好、深入的合作关系。

柏涛墨尔本公司在中国的设计机构拥有众多高素质的中外建筑师，国际化的先进设计理念、本地化的优秀团队服务，使公司业务发展迅速。到目前为止，业务范围已覆盖了中国境内26个省、市、自治区，并在上海常设实力雄厚的设计机构。在澳洲本部的支持下，我们有能力在大规模的城市区域规划设计、大型公共建筑设计（包括办公楼、商业中心、酒店、教育行政文化设施、运动娱乐设施、医疗设施等）以及住宅规划设计、建筑设计及园林景观设计等方面，以独特的设计手法、先进的技术和丰富的经验，活跃在国际建筑设计舞台上，并始终如一地为客户提供一流的服务。

Founded in 1889, Peddle Thorp Pty Ltd. has been one of the largest architectural design firms in Australia. For over one hundred years, Peddle Thorp has always been at the forefront in the architectural world.

Pre-eminent, practical and efficient design works are the principles; design innovation and technical renovation are treated tenets; established reputation has been based on technical reliability and unique design. With superior technical resources, advanced computer skills and comprehensive experiences, Peddle Thorp has been specializing in regional planning and architectural design in various scales and categories.

Peddle Thorp Group has practices worldwide with offices in Australia, South East Asia, Europe and America. Supported by a number of outstanding architects and experts, Peddle Thorp provides professional design service for sports, health and various residential projects. Major works are Australian National Tennis Centre, Melbourne Olympic Park and Velodrome, ESSO Headquarters, Melbourne Underwater World Aquarium, Malaysia Sports Centre, and a series of hotels, retail projects, offices, residences, and health facilities both at home and abroad.

Peddle Thorp Asia has established business in 26 provinces, autonomous regions and cities throughout China and a branch office in Shanghai, to accommodate a quality team of both Foreign and Chinese architects, introducing an internationally advanced design philosophy. In collaboration with the Melbourne Office, Peddle Thorp Shenzhen is able to provide a wide range of professional planning, architectural design, residential design, and landscape design by utilizing a unique approach. Advanced skills and comprehensive experience commit Peddle Thorp to provide a specialized service for our clients worldwide.

澳大利亚柏涛墨尔本建筑设计有限公司亚洲分公司
PEDDLE THORP ARCHITECTS ASIA

地址：深圳市南山区华侨城生态广场A座302室
邮编：518053
电话：+86-755-2692 8866
传真：+86-755-2690 5186
邮箱：main@szpta.sina.net

Address: Room 302, Building A, Ecological Square, OCT, Shenzhen
P.C.: 518053
Tel: +86-755-2692 8866
Fax: +86-755-2690 5186
E-mail: main@szpta.sina.net
Http: //www.ptma.com.cn

1–2 阿里巴巴淘宝城 杭州

3–4 海口珠江新城（两个方案） 海口

1

4

2

3

5	中粮体验中心	天津
6	义乌金融商务区 01—09 地块	义乌
7	建发汇金国际	长沙
8	三亚水居巷	三亚

9–10 上海漕河泾新兴技术开发区科技绿洲三期 上海

11–13 弘生外滩中心 浙江舟山

14 绿地国际花都商业街 徐州

15—18 大连·星海湾壹号 大连

18

16

17

19

19	珠海中信湾	珠海
20	永庆盈滨岛	海南
21–23	新世界 四季山水	深圳

23

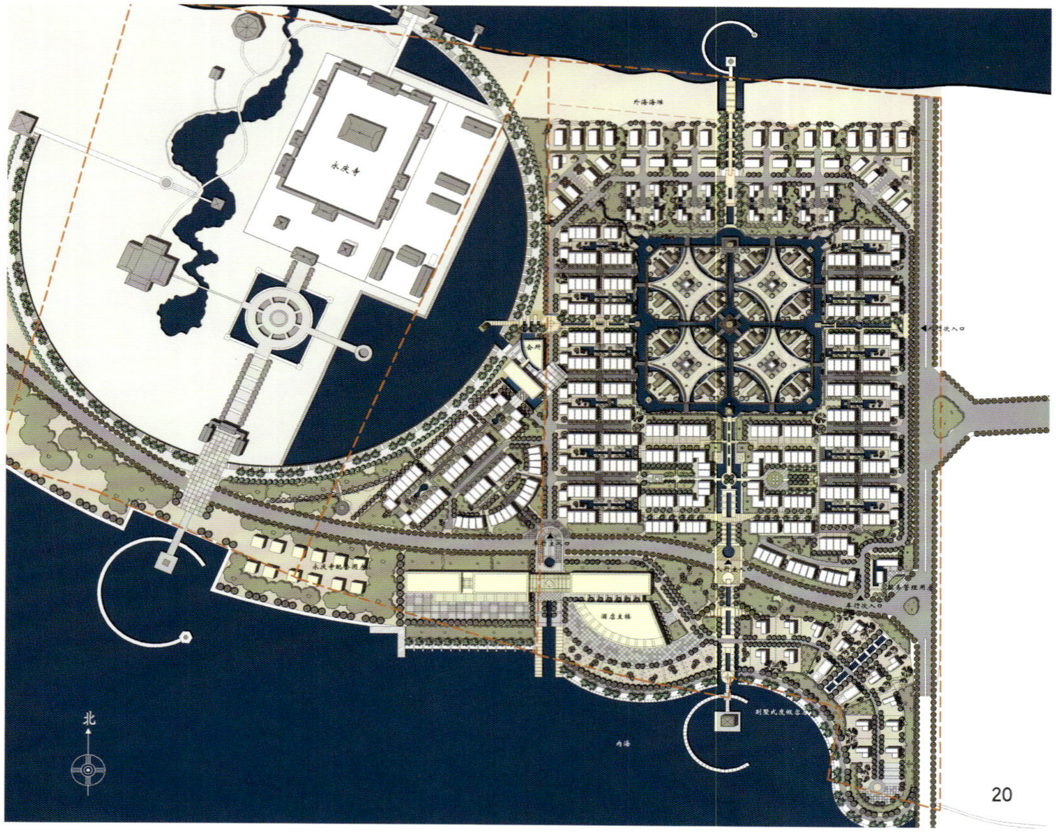

20

21

22

24	宝山褐石街坊	上海
25	松江新南商业项目	上海
26—28	中信山语湖	佛山

24

25

26

27

28

紫薇地产

西安紫薇永和坊

基地位置：陕西 西安
设计时间：2008 年
用地面积：23.4 万平方米
总建筑面积：48.2 万平方米
容积率：2.1

设计概述：

曲江是汉代宫苑，池畔杨柳成荫，行列如排衙，别号“柳衙”。永和坊，地处长安城外郭，体现了唐代长安城的包容教化与万家和谐。作为曲江池畔超大规模的复合型居住社区，丰富的产品和空间形态以及鲜明的文化主题特征，是取得市场认同度的核心要素。方案主题中的“唐庭”点出了建筑规划布局在有机规律中不拘一格、寻求变化的特点，“汉苑”则体现出景观空间设计在震撼效果中寄情于大自然、传承文化的特色，正如“永和坊上长安客，唐庭汉苑柳如烟。”

紫薇公园时光

基地位置：陕西 西安
设计时间：2009 年
用地面积：3.2 万平方米
总建筑面积：12.3 万平方米
容积率：3.2

设计概述：

装饰繁缛、尺度失真的常规高层豪宅只可以称之为奢华的消极空间，它不过是价格与面积的重复叠加，而公园时光力求创造的不是一种产品，而是一种生活方式。渴望自然是每个人内心的火花，环绕公园时光的城市大公园是每个人梦想的居住环境所在。公园时光的独特之处在于它源于别墅，优于别墅的空间形态和流线组织，以及它与自然环境的互动关系。公园时光的造型特征以一种极简主义的个性化姿态，突破了市场对豪宅产品认识的窠臼，用时尚、现代、清新的感觉引领新的潮流，在引导生活方式的同时，引导人们对居住建筑的审美文化。

中粮大悦城

基地位置：四川 成都
设计时间：2010 年
用地面积：5.5 万平方米
总建筑面积：14.6 万平方米
容积率：2.2

设计概述：

居住社区规划的原则在于基地的价值发掘与价值规划，并将基地中潜在的交通、景观、噪音、自然环境等因素明晰地判断并加以利用，充分提升居住社区的人文价值，为周边区域乃至整个城市规划起到积极的、可持续的拉动作用。成都中粮大悦城的设计目标就是通过与城市形成良好对话关系，以清晰优雅的英伦立面风格、恬静舒适的园林景观，来将其变成为该区域富有城市感、品质感和文化气息的高端物业标杆。

中粮锦云国际

基地位置：广东 深圳
设计时间：2010 年
用地面积：3.9 万平方米
总建筑面积：14.8 万平方米
容积率：4.0

设计概述：

住宅产品对于它人而言，不仅是人们栖息的空间，它还是人类的一种外化，是人类能力的象征，是人类艺术的形态、价值观念、伦理关系、社会关系的外化合反应。

住宅产品是个躯壳，而文明是它的灵魂，艺术是它的容貌，文明和艺术是住宅产品的价值所在。

住宅产品是文明和艺术的载体、结晶。住宅产品的价值，与其说是地段区位，不如说是环境；与其说是环境，不如说是文明、艺术的包容和享有程度。

沈阳港中旅国际小镇一期东组团

基地位置：辽宁 沈阳
设计时间：2007 年
用地面积：6.0 万平方米
总建筑面积：6.8 万平方米
容积率：0.9

设计概述：

港中旅国际小镇是港中旅进入沈阳市场的战略布局，一期东组团住宅区作为先期的市场试探，涵盖了从双拼别墅、联排别墅到花园洋房、电梯洋房、小高层住宅的一系列产品。地中海风格的温暖色调和层层叠叠的屋面与露台让建筑群为属于寒冷气候的当地带来了不同的生活氛围，方正实用的产品结合了当地居住习惯和先进的产品理念，同时欧陆风情的商业街和壮观的人工湖也为居住品质的提升创造了条件。主题化的居住空间是高尚生活的基础，也是契合旅游地产概念的表现形式，更是产品得以成功的关键。

沈阳港中旅国际小镇二期

基地位置：辽宁 沈阳
设计时间：2008 年
用地面积：16.4 万平方米
总建筑面积：12.0 万平方米
容积率：0.6

设计概述：

以双拼别墅和联排别墅为主的国际小镇二期既是一期组团空间上的延伸，也是产品的升级。方案充分利用不规则的用地属性，形成两个核心节点和一系列的景观通廊，并以此为基础划分出不同的产品组团。别墅结合户型平面的创新组织形式，强调独栋化的处理手法，让每个单位具有相对完整的外在形象，同时拼接而成的整体又具有宏大的气势。建筑风格顺应中旅国际小镇的总体风格，运用杂色红砖、深灰麻石、黑色屋顶以及标志性的细部语言体现北美建筑的独特性。该项目将成为沈阳主题高端住宅的典范。

远洋地产
SINO-OCEAN LAND 成就成长 相伴一生

大连远洋 TIMES

基地位置：辽宁 大连
设计时间：2010 年
用地面积：27.2 万平方米
总建筑面积：56.6 万平方米
容积率：2.1

设计概述：

项目位于大连经济开发区大连大学城内。项目定位为集高层住宅、别墅、公寓、会所、商业、学校为一体的大型中高端山地居住楼盘。

本项目以建筑、自然与人之和谐为主题。规划设计以环境内与外结合、建筑与环境结合、环境与功能结合、观赏性与实用性结合、环境与人性化结合为五大原则。设计尊重原有地形地貌特征，融建筑于绿化景观之中，以台地山地为特征，将社区构成从“都市”过渡到“自然”的序列，形成较好的建筑坡向与适宜建设的地形坡度，场地东西向约为 6%，东侧挡墙高度不大于 2 米，利用基地高差走势分离出住宅内庭院，打造更为独立的居住空间。

统一考虑项目的地形特点，形成立体景观系统，意在打造一个类似山地的生态城市精品社区。

蓝光BRC
—用 心 建 筑 生 活—

成都蓝光紫檀山

基地位置：四川 成都
设计时间：2008 年
用地面积：4.6 万平方米
总建筑面积：2.1 万平方米
容积率：0.4

设计概述：

作为一个低密度的、低容积率的高档住宅项目，项目设计从规划到建筑、景观、配套都遵循高品质，高起点的方向。无论是户型设计、社区的营造、公共空间的打造、还是建筑风格的融合和周边地块空间规划上的连续性，都遵循了新东方主义的精髓。作为"类独栋"双拼别墅的典范，每栋建筑的两户之间，仅通过露台相连，讲究空间上的"似连非断"，邻里间由内庭院水景绿化、分段毛石院墙、高大的树木和灌木丛等相隔离，既保持了居住的私密感，又强化了景观的整体性与共生性。

观岭国际社区二期

基地位置：四川 成都
设计时间：2010 年
用地面积：18.7 万平方米
总建筑面积：6.7 万平方米
容积率：0.3

设计概述：

本项目力求创造出沐浴在地中海阳光中的浪漫、悠闲、艺术、尊贵的圣塔芭芭拉贵族城堡的休闲与尊贵感，通过寻求差异化文化特质，引入小语种文化组群，为各组团均赋予的强烈鲜明的主题风格，形成差异化风情，以顶级生态资源及纯正人文资源为核心价值，营造国内顶级的精品高尔夫宽景低层住宅区。

福清核电生活区

基地位置：福建 福清
设计时间：2009 年
用地面积：16 万平方米
总建筑面积：41.6 万平方米
容积率：2.6

设计概述：

本项目位于福建省福清市龙江经济开发区观音埔村内，南靠观音山，环境优美，地势平坦，交通便捷，是多元复合的现代高档的福清核电职工的生活社区。设计主旨在于提供舒适的家居生活氛围及空间，为核电职工提供一个更加良好的精神家园。成为福清市新的居住物业标杆，并且为周边区域乃至整个城市规划起到积极的、可持续的拉动作用。

当"欧陆风"、"北美风"吹来的时候，更多的人开始怀念起从前的"中国人家"，中国特有的审美观和生活价值观被更多人所追求。

因此在规划中我们提出新"现代中式人文生态风情"作为本项目规划的主题思想。设计采用大方庄重的中式院落布局方式，以高贵典雅的现代中式建筑风格，让项目本身的属性与城市文脉以及福清核电特有的企业文化产生共鸣。设计致力于营造符合人的心理、保持传统文化的居住环境，使居者达到"人文精神上的回归"。

房北极星花园

基地位置：四川 成都
设计时间：2009 年
用地面积：1 万平方米
总建筑面积：8.5 万平方米
容积率：6.0

设计概述：

社区位于城市核心区域，融合商业、办公、酒店、居住、休闲等多种城市功能，各种元素有机结合，并成为商业、商务、居住活动的焦点，具有开放性，项目除了满足社区内生活需求之外，更服务于整个区域。

1. 广东深圳中信银行总部
2. 成都观岭商业样板区
3. 合肥领寓公寓
4. 西安港务局大厦
5. 湖北汤池温泉二期
6. 湖北黄石麒麟山庄
7. 广东惠州恒河金谷二期
8. 梅州世界客商中心
9. 江西泗阳紫金广场
10. 陕西西安名万丈八楠村
11. 黄石天虹金华庭
12. 广东惠州恒河大厦
13. 广东惠州天峰松子岭项目
14. 河南南阳尚庄
15. 湖北黄石中学
16. 江苏安芯数码港
17. 海螺集团芜湖两湖用地片区规划

10

13

15

11

14

16

17

徐州行政中心
连云港阳光国际大厦
深圳中心区现代国际大厦
深圳创维数字研究中心
成都多元国际总部一号
济南行政中心
上海兴业银行
河南洛阳美术馆
张家港建设局
四川成都蓝谷地商业
深圳南山中泰艺术公园
西安紫薇尚层
合肥江南九溪江南
鸿荣源深圳宝安西城上筑
西安紫薇风尚
广东惠州恒和大厦
连云港帝豪水榭花都
沈阳港中旅欧陆小镇
南宁荣合山水美地
汕尾万福黄金海岸
珠海华南名宇
珠海扬名海邑
东莞宏远江南雅筑
成都熊猫小镇
青岛红岛组团
西安华圣旅游区
三宝国际陶艺村
深圳宝能太古城
衡阳文华中心
……

CAPA

汉米敦 HMD | 上海汉米敦建筑设计有限公司

公司简介

上海汉米敦建筑设计有限公司，主要从事城市规划、建筑设计、景观设计等专业咨询服务，全方位参与到中国的城市化建设中。目前在上海及深圳分别设有分公司，主要从事的项目类型为城市规划及城市设计、城市综合体、商务办公楼、酒店、商业及房地产开发项目。其主要合伙人保罗·莱斯先生、徐石先生、李剑波先生均有着丰富的专业经验及中国项目的操作经历。汉米敦的业务宗旨是将先进的国际设计理念与中国独特的项目发展需求充分结合，为我们的客户创造更高的价值并为其实一更美好的期望。

Company Profile

HMD Design (Shanghai) is a professional consultant company on urban planning, architectural design and landscape design. HMD is dedicated to the urbanization development in China. It has offices both in Shanghai and Shenzhen. Its business covers urban planning and design, urban complex design, commercial building, hotel, business and real estate development. Our partners, Paul Rice, Xu Shi and Li Jianbo, all have years of professional experiences and projects experiences in China. The principle of HMD is to make much more value and create a prospective future for our clients by introducing advanced international design ideas to local project development.

上海
地址：上海市静安区常德路 800 号 9 号楼 6 楼
电话：+86-21-3255 3266
传真：+86-21-3255 3266*801
邮箱：sh@hmddesign-china.com
网址：www.hmddesign-china.com

北京
地址：北京市朝阳区建国路乙 118 号招商局京汇大厦 1002 室
电话：+86-10-8588 6080
传真：+86-10-8588 6963
邮箱：bj@hmddesign-china.com
网址：www.hmddesign-china.com

深圳
地址：深圳市福田区益田路 6009 号新世界中心 2410 室
电话：+86-755-8321 6996
传真：+86-755-2221 1820
邮箱：sz@hmddesing-china.com
网址：www.hmddesign-china.com

Shanghai
Address: 6th Floor of Building No.9, 800 Changde Road, ShangHai
Tel: +86-21-3255 3266
Fax: +86-21-3255 3266-801
E-mail: sh@hmddesign-china.com
Http: //www.hmddesign-china.com

Beijing
Address: Room 1002 of China Merchant Bureau Jinghui Mansion, No.188 of Jianguo Road B, Chaoyang District, Beijing
Tel: +86-10-8588 6080
Fax: +86-10-8588 6963
E-mail: bj@hmddesign-china.com
Http: //www.hmddesign-china.com

Shenzhen
Address: Room 2410 of New World Centre, No.6009 of Yitian Road, Futian District, Shenzhen
Tel: +86-755-8321 6996
Fax: +86-755-2221 1820
E-mail: sz@hmddesing-china.com
Http: //www.hmddesign-china.com

HMD

乌海市市委党校

设计时间：2009 年 12 月
基地位置：内蒙古 乌海
用地面积：100 000 平方米
建筑面积：47 516 平方米
容 积 率：≤0.4

Party School of the Wuhai Municipal Committee

Design Time: 2009.12
Base Location: Wuhai, Inner Mongdia
Site Area: 100,000 m^2
Building Area: 47,516 m^2
Plot Ratio: ≤0.4

规划理念

规划布局及设计理念：充分考虑利用高尔夫球场、黄河以及甘德尔山的景观资源，各个功能块按照不同的功能特点布置在最合理的位置，同时相互之间既不相互干扰，又可互动。

总体布局吸收中国古典皇家园林建筑的建筑特点，强调中轴线、空间序列及建筑与庭院的结合。采用轴线对称、三面围合、面向西侧黄河形成开放空间的布局，体现建筑群体的庄重大气。

Planning Concept

Layout and design concepts: We try to make the full use of Golf course, Yellow River and Gander Mountain's landscape resources, according to their different function characteristics, every functional block is distributed in the most reasonable position, and they don't impact each other and have good interaction.
The main layout absorbs architectural characteristics of Chinese classical imperial garden; it emphasizes good combination of central axis, special sequence and courtyard. The layout adopts open style through Axis symmetry, enclosed on three sides, facing to the Yellow River in the west, and it makes the building groups solemn and great.

设计理念

建筑设计借鉴了美国中西部经典的赖特草原风格，出檐深远的屋顶、宽大的露台和屋顶平台诠释了草原建筑样式的精髓，反应出人们对栖居于大地诗一般的建筑意境的追求。

建筑造型切合总体格局的气势，简洁、庄重，富有变化，在简朴中体现建筑的典雅气质。党校正入口的大台阶及柱廊处理，强调了中轴对称建筑的沉稳大气，酒店和会所正入口处大雨棚与挑高的大堂彰显建筑非凡气势和高贵品质。

Design Concept

Architectural design took examples from typical Wright Prairie Style of Midwest U.S., far-reaching eaves of the roof, wide balcony and roof garden interprets the essence of Prairie Style architecture, the harmony between heaven, earth, architecture and human, it also indicates the human's artistic conception towards the architecture and pursuit on poetical inhabitation.
The building form meets the overall pattern, which is simple, dignified, full of change and reflects elegance. The Party school' big steps and portico emphasize axis constructions' sedate and aroma, the hotel and club's large canopy and lobby highlight its magnificence and nobleness.

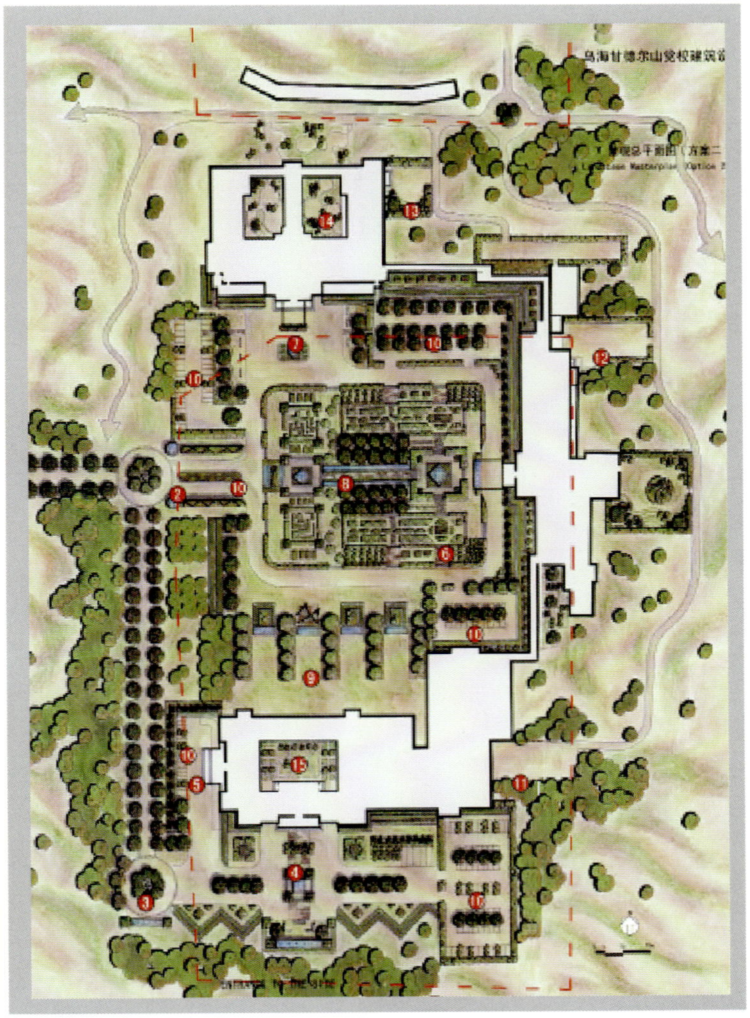

汉米敦

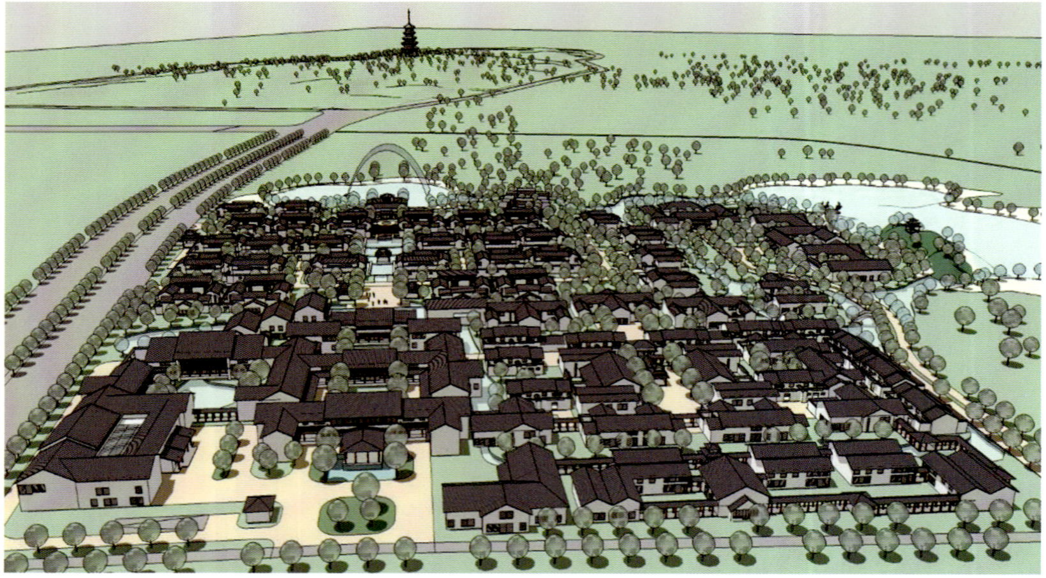

北京玉皇山贵宾馆

建设地点：北京
占地面积：400 000 平方米
建筑规模：90 000 平方米
设计时间：2010 年 1 月
项目类型：度假精品酒店
容 积 率：0.7
项目状态：深化设计中

Beijing Yuhuangshan Guesthouse

Construction Location: Beijing
Site Area: 400,000 m^2
Building Area: 90,000 m^2
Design Time: 2010.01
Project Type: boutique hotel resort
Plot Ratio: 0.7
Project status: under deepening design

设计理念

基地位于北京市海淀区，东临颐和园，西北临玉泉山。由于地理位置敏感，所以我们借鉴中国传统建筑的设计原理，通过项目中轴线的规划，使建筑与玉泉山玉峰塔相呼应，规划形态则呈扇形展开与颐和园取得关联。随着现代城市化进程的深入，城市中原有资源和构架不得不进行重新分配，著名的“燕京八景”的景观功能也渐渐退化，特别是“玉泉垂虹”一景更是随着地热温泉的过度利用而逐渐消亡，因此我们在轴线末端设计了一个“人工垂虹”的景观喷泉，重现历史典故的同时也提醒人们重新思考人与环境的深层关系。

Design Concept

The piece of land is located in Haiding District, Beijing. Due to the critical location, we are using Chinese traditional architecture theory, developing a fan-shaped layout which is connecting with the Summer Palace. With the development of urbanization, the original resources and structure of the city has to be distributed again. The famous scene “eight great sights of Yanjing” is no longer attractive and “Yuquan Chui Hong” has destroyed by over using thermal hot spring. But we have designed a fountain “artificial Chui Hong” from which we almost see the old scene and also remind people to think about the relationship between people and environment.

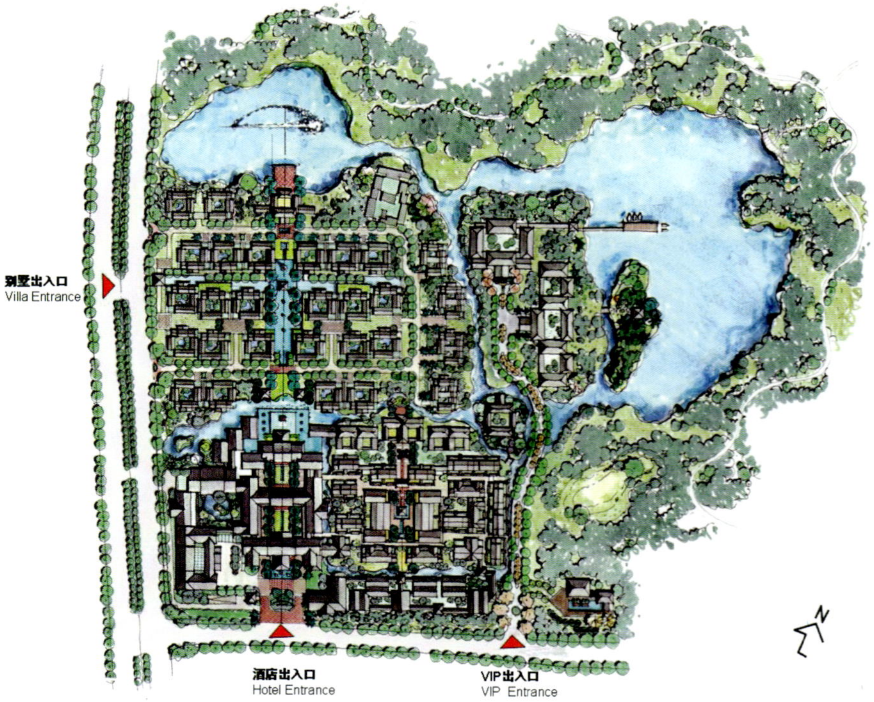

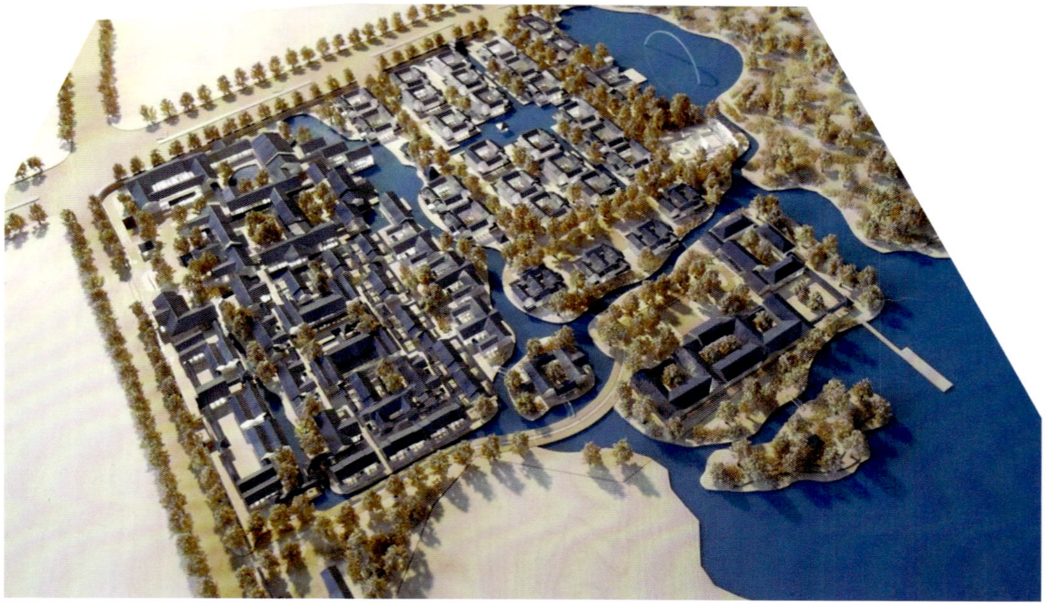

HMD

松江广富林总体规划方案设计

规划总用地面积：76.7万平方米
容积率：0.5—0.8

规划设计理念

项目充分利用地块独特的地域环境和城市特点，结合丰富的人文底蕴和历史背景，运用国际领先的设计理念，打造集居住、展览、休闲、娱乐于一体的文化社区，使项目形成亲切的地域特色，成为松江新城一大亮点。

总体规划的核心理念是营造一个复合型的特色区域，使广富林重新融入到城市生活和城市节奏中，并恢复它的生机与活力。

单体设计理念

设计延续和发展总体规划的构思，审慎处理不同类别的现状建筑。保护与利用相结合，分区域展现不同时期的建筑特色，以一种多元化、可持续的姿态融入新的城市规划当中。

建筑立面融合了现代时尚元素与传统建筑风格，体现江南传统建筑文化的特色，白墙黛瓦的简洁造型与和谐的色调，营造出一种新江南水乡风格。

项目整体的空间设计以中式风格为主基调，尽量维持原旧村建筑的体量、高度和规模，统一吸收了广富林旧村民居的建筑特征，突出院落形态。把开放空间与室内经营空间有效结合，形成和谐的、有地域感的弹性单元，同时，建筑组团之间在立面细节和材质上寻求差异，保持各自的细部特色，以适应产品多元化的需求。

建筑材料主要选用灰砖、青瓦、木材等传统建筑材料，恰当地配合钢、玻璃等现代建筑材料。建筑的主色基调以体现黑、白、灰为主，用现代的手法诠释传统的建筑文化。

汉米敦

成都泰信 10–3 号项目

建设地点：四川 成都
设计时间：2010 年 3 月
用地面积：16 437.77 平方米
建筑面积：200 000 平方米
容 积 率：≤ 12.0

Project No. 10-3 of Chengdu Taixin

Construction Location: Chengdu, Sichuan
Design Time: 2010.03
Site Area: 16,437.77m^2
Building Area: 200,000m^2
Plot Ratio: ≤ 12.0

设计理念

西部金融中心项目的设计创造了由三个塔楼形成的建筑雕塑群，240 米高的酒店和办公楼呈细长的锥状，精美的顶部由宽大的基层开始慢慢上升。塔楼顶部像一个灯笼，使这个项目成为城市的亮点。

虽然每幢塔楼都有自己的独特之处，但是从整体上看，他们形成了一种三角的形状，既各自独立又相互呼应，弧形的裙房地基线将弯曲的河岸与住宅楼有机结合起来，两个新广场将行人引至连接三个主塔楼和商业基地的新商业街。

Design Concept

The design of the Minyuon project creates a sculptural group of three towers. The main 240 m hotel and office tower is a slender tapering form that rises from a larger base to a delicate top. At the top of the tower is a lantern which illuminates the project in the city. Each of the three towers has its own distinct identity but are grouped to create a tripartite image. The sweeping horizontal lines of the podium base unify the curving forms of the bank and residential towers. Two new public squares are created to lead pedestrians into the new retail street which links the three main towers and the retail base together.

HMD

福星惠誉东澜岸建筑专业方案及扩初设计

设计时间：2010 年 2 月
建设地点：湖北 武汉
用地面积：232 680 平方米
建筑面积：461 200 平方米
容 积 率：2.0

Building professional scheme and development design of Donglan'an of Fuxinghuiyu Project

Design Time: 2010.02
Construction Location: Wuhan, Hubei
Site Area: 23,680 m^2
Building Area: 461,200 m^2
Plot Ratio: 2.0

武汉大学城

建设地点：湖北 武汉
建筑面积：460 000 平方米
项目类型：居住，商业城市综合体
设计时间：2009 年 12 月
项目状态：一期已开工建设

Wuhan College Town

Construction Location: Wuhan, Hubei
Building Area: 460,000 m^2
Project Types: residential and business urban complex
Design Time: December, 2009
Project status: Phase I of construction has started

设计理念

项目地块位于武汉市黄家湖大学城内，东临黄家湖，西临 G107 国道。我们通过研究当地日照，风，温湿度的情况对项目进行规划布置并进行建筑设计，对传统建筑形式进行了低碳化的改造，同时从城市角度构建黄家湖及 G107 国道两个城市开放空间的联系，创造最符合武汉当地气候特点和城市足迹的综合体。

Design Concept

The land of this project is located in Wuhan Huangjia Lake College Town. Through the research of local sunshine, wind, moisture and temperature conditions, we have done the layout and building design, low-carbon transformation on traditional architectures. At the same time, we tend to create a complex that conforms to local climate characteristics and Wuhan's developing footprint with building connections between Huang Jia Lake and G107 national highway.

Gensler

Gensler是一家全球性的建筑、设计、规划和战略咨询公司，专门从事各企业、机构和公共组织所拥有或使用的各种建筑及设施的设计咨询。我们提供全方位的建筑服务，从初始策划到设计、实施和管理。我们坚持客户至上的理念，深入了解客户目标和需求，力求通过我们的工作和服务显著增加客户的企业价值。

Gensler于1965年成立于美国旧金山，为了确保与客户密切互动，公司从一间办公室成长为一家拥有30间办公室和一个专业智囊团的大公司，人员超过2 500人。Gensler获得著名的美国《商业周刊》设计大奖评选出的多项荣誉，该奖项旨在评选出以战略性商业目的为导向的、创新设计解决方案。2000年，美国建筑师学会授予“Gensler年度最佳事务所”的荣誉称号，这是其授予合作事务所的最高奖项，并将Gensler列为“21世纪设计事务所典范”，《工程新闻记录Engineering News—Record》杂志及《世界建筑(World Architecture)》杂志将Gensler列为全球顶级建筑事务所，2006年，美国绿色建筑商会(US Green Building Council)授予Gensler“领袖奖”，另外，Gensler在酒店业、零售业等主要的专业领域中，均排名前五位。

早在20世纪80年代，Gensler就已涉足中国市场，20世纪90年代初，Gensler设立香港办事处以扩大亚洲市场，为国际客户提供服务。至90年代末，我们的业务已遍及中国深圳、北京、青岛、上海、西安等城市，Gensler积极参与大型规划、建筑设计、公司内部设计等项目。

2003年3月，Gensler在上海及北京设立了完全独立的办事处，Gensler在中国属于外商独资企业。我们于2004年取得执照，Gensler作为建筑和设计顾问开始经营业务。我们和本土设计研究院密切合作，他们负责确保项目符合建筑行业规范要求、准备施工文件、向当地政府部门提交图纸供审批。自2005年以来，Gensler上海办事处已经成功完成500多万平方米的总体规划项目和250多万平方米的建筑及室内设计项目，其中280万平方米已经于近期在北京、上海完成建设或正在建设中。

认识到中国在改善环境质量方面做出的巨大努力，Gensler将可持续设计作为其在亚洲工作中的意识目标。这一目标在所有层面中都得到了贯彻，例如，在地区规划上强调保护中国城市的历史建筑物，在建筑及装饰上强调可持续性发展，这种可持续性发展体现在注重建筑的遮阳功能、自然通风及制冷以及使用可重复使用的材料和产品等方面。

迄今为止，Gensler上海办事处已经完成了160多个项目，其中包括部分技术相当复杂的高难度项目。通过充分利用Gensler在全球的资源及专门技术，我们中国办事处有能力向每一位客户提供高水准、高质量的设计，并在客户的预算之内，按时完成。

Gensler is a global architecture, design, planning, and strategic consulting firm that specializes in a wide range of building and facilities owned or used by businesses, institutions, and public agencies. Our services engage the full building cycle from initial planning through design, implementation, and management. We focus on our clients, understand their goals and strategies, and seek to add substantial value to their enterprise through our work and services.

Gensler was founded in San Francisco in 1965. To ensure close interaction with its clients, the firm has grown from one office to a broad-based organization with 30 branch offices and a professional think-tank of over 3,100 staffs. Gensler is a multiple winner of the prestigious Business Week Design Award which recognizes innovative design solutions driven by strategic business purpose. In 2000, the American Institute of Architects named the company “Firm of the Year,” its highest award to a collaborative practice. The award recognized Gensler as “a model of a 21st Century design firm” .Engineering News-Record and World Architecture both rank the company as the world’s top architectural design firm. In 2006, we received the Leadership Award of the U.S. Green Building Council. We also rank among the top five in the majority of our specialist practices, including hospitality and retail.

Gensler entered the China market in the 1980’s. In the early 90’s we established a Hong Kong office to provide services for our international clients who were beginning to expand into the Asian market. By the late 90’s, we were actively engaged in large-scale planning, building design and corporate interiors projects in multiple cities including Shenzhen, Beijing, Qingdao, Shanghai and Xi'an.

In March 2003, Gensler opened an office in Shanghai and a satellite office in Beijing. Gensler works in China as a Wholly Foreign-Owned Enterprise (WFOE). Our license, issued in 2004, allows us to act as an architecture and design consultant. We work with a Local Design Institute that is responsible for building code compliance, construction documentation, and the submittal of drawings to local government agencies for approval. Since 2005, Gensler China has successfully worked on over 5,000,000 m^2 of master planning projects and over 2,500,000 m^2 of architecture and interior design projects, 2,800,000 m^2 of which we have recently completed or have been under construction in Shanghai and Beijing.

Recognizing the strong efforts underway in China to improve environmental quality, Gensler has emphasizedsustainable design as a conscious goal of its work in Asia. This goal has been pursued at all levels - from district planning, which aims to preserve the historic fabric of Chinese cities, to buildings and settings that incorporate sustainable features such as solar shading, natural ventilation and cooling, and the use of reusable materials and products.

With over 160 projects completed, our China portfolio has grown to include some of the firm’s most technically complex projects. By extending the full resources and expertise of Gensler Worldwide into Asia, our China offices can deliver to every client a high-standard design, within budget on time.

地址：上海市卢湾区湖滨路222号企业天地大厦908室
邮编：200021
电话：+86–21–61351900
传真：+86–21–61351999

Address: One Corporate Avenue, Unit 908 222 Hubin Road,Shanghai, PRC
P.C.: 200021
Tel: +86-21-61351900
Fax: +86-21-61351999

地址：北京世纪财富中心1号楼803室 朝阳区光华路5号
邮编：100020
电话：+86–10–65628900
传真：+86–10–85875108

Address: Room803 of Fortune Plaza Tower 1, No.5 of Guanghua Road, Chaoyang District
P.C.: 100020
Tel: +86-10-65628900
Fax: +86-10-85875108

上海中心大厦
项目类型：综合办公楼，酒店及商业零售
建设地点：上海
建筑高度：632米
占地面积：30 000平方米
建筑面积：380 000平方米
动工时间：2008年
竣工时间：2014年

Gensler设计的上海中心大厦有128层，总面积为380 000平方米，作为宏伟壮丽的都市天际线新标志，它奏响了陆家嘴金融贸易区摩天大厦的三重唱。这座定位为全球顶级金融中心的大厦高632米，设有甲级办公区、休闲娱乐设施、商业、展览中心、精品酒店及各种文化设施。

由于大厦主体透明，从地面上即可看见大厦垂直的城市邻里空间，在每个楼层都能欣赏到无与伦比的城市风景。大厦不仅拥有出众的外观，其造型更融入了最新的现代超级摩天大楼科技。其形态、材质和不对称性使风载降低了24%，大大节省了建筑用料与施工成本。

Gensler将双层表皮、风力发电机和由大厦顶部螺旋护墙组成的雨水收集利用系统等尖端的环保理念融入设计之中。由上海中心大厦建设发展有限公司开发的上海中心将于2014年竣工，预期届时将达到美国绿色建筑委员会（USGBC）及中国绿色建筑委员会的高水准认证标准。

Shanghai Tower
Project Type: Mixed-Use with Office,Hotel & Retail
Construction Location: Shanghai
Construction Height: 632 m
Site Area: 30,000 m^2
Building Area: 380,000 m^2
Construction Time: 2008
Completion Time: 2014

A new icon on the exciting Shanghai skyline, the Gensler designed, 128-storey, 380,000 m^2 Shanghai Tower completes a trio of "super-tall" buildings in Shanghai's Lujiazui financial district. Set to become a premiere global financial center, this 632-meter high tower contains Class-A office space, entertainment venues, retail, an exhibition center, boutique hotel, and cultural venues.

Shanghai Tower's undulating transparent facades are visible from all directions, offering excellent views of Shanghai from every floor. In addition to its beauty, the tower's shape advances the science and technology of modern super-tall buildings. The tower's taper and asymmetry all combine to reduce wind load on the building by 24%, offering considerable savings in both building materials and construction costs.

Gensler's design incorporates cutting-edge sustainable ideas, including a double-skin facade, wind turbines, and a spiraling parapet at the building's apex that serves as a rain collection system. Developed by the Shanghai Tower & Construction Co., Ltd. the tower is slated for completion in 2014 and is expected to achieve a high level of certification from the USGBC and China Green Buildings Committee.

上海陶氏化学中心

项目类型：研发实验室，办公区，培训中心，餐饮区，客户创新中心，安全室，保健中心服务：室内设计，指示标志设计，工程，管理

建设地点：上海

建筑面积：75 000平方米

竣工时间：2009年

Shanghai Dow Center

Project Type: R&D lab, office space, trainingcenter, cafeteria, customer innovation center, security room, fitness & health center

Services: Interior Design, Signage & Graphics, Construction Administration

Construction Location: Shanghai

Building Area: 75,000 m^2

Completion Time: 2009

为体现公司的环保责任感、创新精神和以人为本的理念，陶氏化学公司（Dow Chemical）找到Gensler为其座落在两幢平行而立的L形大厦中占地75 000平方米的办公室现有设计进行审改。

为了营造人文化的环境，Gensler将连接两边建筑物的中庭长廊重新设计成一个可联通不同工作区域的走廊，缤纷多姿，而又令人眼前一亮。突出的木材饰面、多彩多姿的板块和曲线形的墙壁使空间更为柔和舒适。位于中间的咖啡"岛"更便于沟通与交际，从而促进了员工间的相互交流。该项目还包括咖啡厅、健身中心、医疗中心和学习中心等功能区。

项目注册LEED银级认证。

To convey its environmental responsibility, innovation and emphasis on people, Dow Chemical looked to Gensler to overhaul an existing design for new office space in two parallel L-shaped buildings.

Creating a human scale, Gensler re-purposed a long atrium that linked the two buildings into a varied, inviting promenade connecting various workspaces. Wood accents, colored panels and curved walls soften the spaces. Centrally located coffee "islands" encourage connections and socialization, which in turn spur idea exchanges among employees. The project also includes a café and fitness, medical and learning centers.

This project is Registered LEED Silver.

上海浦东发展银行

项目类型：总部，办公室

数据中心，培训中心，酒店

建设地点：上海

占地面积：26 000平方米

建筑面积：27 000平方米

竣工时间：2003年

Shanghai Pudong Development Bank

Project Type: Headquarters, Offices, Data Center & Training Center with Hotel

Construction Location: Shanghai

Site Area: 26,000 m^2

Building Area: 27,500 m^2

Completion Time: 2003

上海浦东发展银行是中国历史最悠久的银行之一，项目希望建成21世纪的高科技总部大楼，成为公司和城市的标志性建筑。客户特别强调设计应反映团队氛围，融入各种元素，鼓励沟通和合作，在可持续性发展设计理念的指导下，探索使用多种建筑样式。

总面积为27 000平方米的大厦是银行的神经中枢，内设数据中心、培训和会议中心、餐厅、酒店和服务管理中心（包括110间酒店客房）、健身俱乐部、网球场、洗衣设施、商务中心功能设施等。

为了达到客户提出的项目目标，Gensler团队将中心的所有功能设施整合在U型格局的大厦中，这样，大厦中间景色优美的代表柔和、自然的内庭院，在大厦使用的现代材料和设计之间实现了平衡。大厦的可持续性特征体现在双层外立面、双层楼顶这样的设计及减少了外界温度对室内的影响，天窗和斜面墙壁用以增强自然采光，同时也使用了遮阳设备等。该项目成功地将高水准的工艺、环境保护理念与自然界的宁静结合起来，打造了一个方便员工互动，促进其思考、有益身心健康，能激发其创造性的高科技工作场所。

As one of the oldest banking institution in China, Shanghai Pudong Development Bank requested a model 21st century high-tech campus that would represent a landmark project for the company and the city. Specifically, the client wanted the design to reflect a community atmosphere with elements that would encourage collaboration and communication while exploring a variety of architectural forms derived from sustainable design concepts.

Serving as the bank's nerve center, the 27,000 m^2 complex houses its data center, a training and conference center, a dining area, and a hotel and service management center which provides 110 hotel rooms, a health club, a tennis court, laundry facilities and a business center.

To support the client's goals for the project, the team brought all of the center's functions together in a single "U"-shaped building. The shape places emphasis on the landscaped inner courtyard, which balances soft, natural elements with the building's modern materials and design. Sustainable features include a double-skin perimeter, a double roof to reduce heat gain, natural daylighting through skylight and sloped walls, and sun shading devices. The result successfully melds three elements – quality craftsmanship, environmental sensitivity, and the tranquility of the natural world – into a space that balances the demands of a high-tech workplace with the human need to interact, reflect, renew and create.

招商银行
项目类型：综合办公楼，校区
建设地点：上海
建筑类型：银行，数据中心
建筑面积：202 000平方米
项目状态：在建

国际性的金融机构需要国际性的人才，为了吸引业界精英的加盟，招商银行目前正建造总面积为202 000平方米的总部大厦，以提供客户及员工所需要的各种设施。

我们在建筑设计、室内设计及景观设计中力求彰显企业文化，使设计和谐统一、天衣无缝，正如招商银行服务多样的银行卡中心，竭诚为客户提供令其满意的舒适服务一样，我们的设计也力求做到形象统一，犹如一张发给客户的信用卡，而且也能时时处处提醒员工身处招商银行的大家庭之中。

简洁、优雅与宁静是本项目设计意向的三大核心概念。坚实可靠的石材象征着有力的基础，而清爽自然的布局则体现了招商银行合作无间的精神。

China Merchants Bank
Project Type: Building & Campuses
Construction Location: Shanghai
Building Type: Bank, Data Center, Training
Building Area: 202,000 m^2
Project Status: Under construction

The success of a world-class financial institution relies on its people.In order to attract the best talent in the industry, China Merchants Bank is developing a 202,000 m^2 headquarters campus that will provide every amenity desired for its clients and staffs salike.

The 200,000 m^2 campus is registered LEED Gold, and will house flexible workspace, credit card production and a processing center, along with data and training centers and a 4-star hotel. Reflecting the company's integrated approach to banking, architecture, interior and landscape design blend together as seamlessly as China Merchant Bank's varied card center departments, supporting the overall experience of the environment.

Simplicity, grace and tranquility are the three core concepts behind the design and serve to create a uniform image and identity. The Gensler team introduced a solution with a central atrium corridor that links the building's various functions, maximizing future flexibility and encouraging interaction between different departments. Stone and glass are used to create a feeling of transparency, integrating the design with its natural settings.

宝矿国际广场
项目类型：综合办公楼，酒店及商业零售
建设地点：上海
建筑高度：200米
占地面积：24 800平方米
建筑面积：182 800平方米
竣工时间：2009年

BM Plaza
Project Type: Mixed-Use with Office,Hotel & Retail
Construction Location: Shanghai
Construction Height: 200 m
Site Area: 24,800 m^2
Building Area: 182,800 m^2
Completion Time: 2009

宝矿国际广场位于上海市闸北区上海火车站不夜城附近。受客户委托，Gensler负责设计这一建筑面积为183 900平方米的现代高层建筑，其物业类型包括写字楼、酒店和零售业。Gensler根据宝矿国际广场项目特征，设计了3幢相对独立又彼此关联的建筑，包括一幢高层写字楼和两幢中等高度的酒店大楼，写字楼与酒店之间以商业裙房及会议中心等酒店设施相连接。

BM Plaza is located near the main railway station in the central area of the Zhabei District of Shanghai. Gensler was invited to design these 183,900 m^2 modern complex high-rise buildings that will house office space, a hotel, and retail. The project design includes 3 towers shaped by the nature of the program-a tall office tower, flanked by lower hotel towers. Retail and conference center facilities are located within the 2-story base that links the towers together.

于家堡金融区

建设地点：天津

建筑面积：430 000平方米

新于家堡世界金融区将屹立在中国新一轮金融改革的最前沿。如果天津的历史和金融经济发展中缺少了海河，则是缺少了城市商贸、发展以及历史基础。由于该项目组团位于海河岸边，因此从海河挖掘灵感则成为了项目规划设计的不二选择。这样不仅提供了一种与海河的实体联系，也提供了一种视觉联系，这种联系是该项目成败的关键。当人们从步行街向南进入园区内时，南部的商业空间如同一座灯塔招引着人们前去，海河及园区的视觉联系则吸引着人们向西移动，去享受沿河岸设置的绿化景观。园区内靠北的渡口则是另一个景点，与其他节点共同形成了一个环形流线，连接着城市中心。该通道充当着城市引擎的角色，组团内外的建筑及景点则是该引擎的发动器，给周边创造了具有活力，充满动感的空间。

Yujiapu Global Financial District

Construction Location: Tianjin

Building Area: 430,000 m^2

The new Yujiapu World Financial District promises to be at the forefront of economic reform in China. To speak about the history of finance and economic development in Tianjin without mentioning the Haihe River would ignore the city's foundation of trade, growth and history. The proposed development for Yujiapu borders along the edge of this important artery, and from it the project derives its inspiration.

The relationship of these three buildings to the river is important: they provide physical and visual connections to the river which are key to the project's success. As visitors move south from the pedestrian streets and cross into the park, they are beckoned by retail to the south. Visual connections to the river and park draw people westward to enjoy the green space along the river's edge. The ferry terminal in the park to the north also provides another node of interest that completes a circular path back into the city center. This path acts as an urban engine, fueled by architecture and points of interest within and outside the building cluster to generate energy and movement between neighborhoods.

万通历史陈列馆

项目类型：综合娱乐设施

建设地点：北京

建筑面积：650平方米

竣工时间：2009年

Vantone Museum

Type: mixed-use entertainment

Construction Location: Beijing

Building Area: 650 m^2

Completion Time: 2009

万通历史陈列馆位于北京近郊的万通新新家园，是公司的培训基地和文化中心。它提供了一个分享经验和交流公司价值观的平台，也成为了周围社区的文化活动中心。

设计的核心是采用两层通高的设计，创造出可兼作大厅、培训区、休息室、陈列馆和报告厅的多功能区。一层的其余空间用以展示公司的发展历史，另有一大型的电视屏幕滚动播放万通的最新项目信息。面积为650平方米的陈列馆还包括了档案室、VIP休息用餐室和备餐间的功能区域。

我们的设计灵感源于万通公司董事长对自然的热爱，并融入对中国传统园林文化的理解。选用适当的建筑材料和形式，充分利用自然采光，加强室外与室内的陈列馆之间的联系。折线形的隔断引导空间流向和自然采光，橄榄绿色的玻璃隔断营造出柔和的春夏景色，长凳沿弧形墙面设置，增添了园林山石的意向。

设计将陈列馆作为周围景观的延伸，从而形成更为广阔的空间氛围。尽管多数展品是永久性的，但我们营造出的空间氛围却是动态的、是充满活力且富于变化的。

The Vantone Museum, located in Vantone XinXin Village outside of Beijing, is a training and cultural center for the Vantone Corporation. It offers a place to share experiences and exchange ideas about the company's values, as well as provides a cultural hub for the surrounding community.

At the core of the design is a double-height entry space, a multi-function area that serves as a lobby, lounge, exhibit and lecture hall. The remainder of the ground floor displays company history while a huge video screen relays information on Vantone's latest projects. The 650 m^2 museum also includes flexible training rooms, a VIP lounge and private dining rooms.

Inspired by the Chairman's love of nature, the design concept is based on metaphors for the natural elements found in traditional Chinese gardens. Forms and materials channel natural light and create reflections, literally bringing the outdoors inside the museum. Partitions bend to direct the flow of the space and to guide natural light throughout, green tinted glass reference a soothing spring and summer landscape, bench seating curls around a curved wall, symbolizing rocks and mountainscape.

The museum's design flows as though it is an extension of the surrounding landscape, resulting in a space that feels larger and more prominent than expected. While many of the exhibits are permanent, the space is dynamic and alive, changing with the light throughout the day and year.

海口国际时尚创意岛

项目类型：概念规划设计
建设地点：海南 海口
规划面积：5.6平方千米

海口位于海南岛，北隔琼州海峡，与广东省雷州半岛相望。海口古称珠崖、琼崖或琼州，其较低的工业化程度，纯天然的绿色景观、美丽的海滩和洁净的空气，使其成为热门的旅游景点。

项目力求打造一处独特的生态旅游目的地，使其成为中国迅速增长的中产阶级群体的主要旅游目的地，使项目最终能成为全新的集休闲、文化、娱乐和体育活动为一体的完美体验地，同时使其成为该地区的示范性低碳排社区。

Haikou Fashion Island

Project Type: Concept masterplan
Construction Location: Haikou Hainan
Planning Area: 5.6 km^2

Haikou is located on Hainan Island, which is located in the South China Sea, separated from Guangdong's Leizhou Peninsula to the north by a shallow and narrow strait. Hainan Island was called the Pearl Cliffs Fine Jade Cliffs, and the Fine Jade Land. As Hainan Island is not heavily industrialized, its greenery, together with its beautiful beaches and clean air, make it a popular tourist resort.
The project envisioned to create a unique and vital premium tourist destination in China : Create a major tourist destination for the expanding middle-class group in China; Provide a multi-day destination centered on leisure, culture, entertainment and sports activities; Create a model low-carbon community development for the region.

世界贸易中心总体规划

项目类型：总体规划
建设地点：上海
建筑面积：1 962 600平方米

世贸中心城市设计总体规划旨在重新开发上海的这片城区，项目建成后，这片城区的面貌将发生翻天覆地的变化。该规划的目的是提供合理的构架，鼓励区域的协调发展，以实现提升上海作为主要金融中心及重要混合使用区的愿景。

在总体规划的编制中，独特且丰富的区域历史文化发挥了关键性的作用，包括通往旧城区的沿路高楼大厦的选址，以及从上海全景轮廓的角度出发，对整个区域定位进行考虑。对独特的历史街区进行选择性的保护是总体计划中的一个关键组成部分，这些历史街区将转变为混合用途的商业和文化分区。

The World Trade Center

Project Type: Master Plan
Construction Location: Shanghai
Building Area: 1,962,600 m^2

The World Trade Center Urban Design provides a vision for the redevelopment of an area of Shanghai that is poised to undergo a considerable transformation. The purpose of the plan is to provide a rational framework and to encourage a coordinated development for the area that supports the overall goal and vision of a modern Shanghai as a major financial center and a vital mixed use district.
The unique and rich site history played a key determinant role in shaping the master plan, including the location of high-rise towers to mark key gateways to the old City and to mark the whole district within the Shanghai skyline. A key component of the master plan is the selective preservation of unique historic neighborhoods that will be transformed into mixed use retail and cultural sub-districts.

思纳史密斯(集团)中国
SMITHGROUPcna
architecture engineering interiors planning

思纳史密斯（集团）中国

思纳史密斯集团是在美国、加拿大和中国同时注册的业务范围覆盖全球的设计和管理机构，也是世界公认的占有领先地位的设计公司之一。

思纳史密斯集团在全球共有13个分公司，正式员工已经超过3000名。集团在大型城市规划、机场车站、博物馆、图书馆、学院建筑、商业娱乐业等项目中获得过诸多良好的国际设计声誉。作品曾刊登在著名的《设计师》《世界建筑》《纽约时代周刊》《华盛顿邮刊》《洛杉矶时代周刊》等著名杂志的显著位置上。

思纳史密斯集团是当今建筑设计及工程领域中最能为客户提供全面的专业化和个性化服务的公司之一，业务范围包括建筑专业、工程专业、设备专业、室内设计、城市规划、前期策划、投资预算、高新建筑工程技术、环境分析、建筑能源分析、项目管理及设施管理等。在这些领域中，我们都有卓越的人才，我们的专业人员以独特的设计理念和专业化的服务在美国以及全世界都享有声誉。

集团的专业人员为自然环境和建筑环境提供规划、设计和管理的解决办法，我们强调的是如何提供不同的客户对市场的需求和对地方性的影响，我们以专业的知识和业主共同应对项目的各个问题，而不仅仅是设计的解决方案。

集团以大规模发展规划项目的实施和注重实效的运作方式而闻名，我们善于把城市规划、商业娱乐业以及展示性建筑的各种相关功能组合起来，创造和提倡快乐的消费设计模式，并将此理论运用到各种商业和规划项目上去。我们在居住区规划方面有着长足的经验，并一直在为美国的芝加哥、底特律和凤凰城等城市做长期的规划，我们的专业人员已经有效地帮助了很多成功的城市，并且帮他们使这些城市持续发展经济超过50年。集团着重于强调通过设计提高城市的综合水平，有效的建造和发展使我们规划的新社区实现了可持续性的发展。

对于大型的综合性项目，集团的优势也在与多个国际设计团队的合作中被充分体现出来，不同的文化背景、各种项目的经验组织起来形成强大的策划和设计源泉，最终使项目获得意想不到的成功。

思纳史密斯集团（中国）成立于2000年，是集团在中国的项目组织者和设计小组，是集团中国设计和管理体系的最高管理和决策者。2004年中国公司下属上海思纳建筑设计有限公司获得国家建设部的综合甲级设计和乙级城市规划、景观专项资质。2005年集团中国公司正式合资建立思纳北方公司和西南公司以及南昌泛美艺术职业学院等机构，2008年集团西南公司成都思纳史密斯又获得建设部建筑综合甲级资质证书。2010年集团注册商标正式获得中国注册商标许可。目前中国公司已有正式员工超过400名，是目前中国规模最大的外资设计公司之一。

现在思纳史密斯中国集团已成为可以在中国提供建设项目全过程服务的综合设计管理和咨询集团，我们相信事业是我们共同的事业，我们将全力以赴，努力工作、实现梦想。

CNASmithGroup is one of the dominant design companies in the world, which registered in United States, Canada and China.
CNASmithGroup has 13 branches all over the world with over 2000 outstanding employees. The Group has gained honors from global projects such as museum, library, commercial entertainment restore and etc. The approaches has been published on the famous magazines as *Designer*, *World Architecture*, *New York Times*, *Washington Post* and *Los Angles Times*.
CNASmithGroup is one of the most professional firms in the architectural design field that we could provide comprehensive and characterized service for our client. The services include: architectural design, engineering, equipment, interior design, urban planning, consultancy in the primary stage of project, investment budget, investment model, large scale renovation, site analysis, high-tech engineering, environment analysis, architectural resource analysis, project management and facility management etc.
Our professionals are famous for the unique concept and professional service in the United States and all around the world .
The experts in the group provide the planning, design and management solutions for natural or architectural environment. What we are always emphasizing is how to support different clients' influences to the place and market demands, as well as the corporation with clients.The scope of our work is: city/community, museum, entertainment, leisure facility, office building, research and manufacture center, educational buildings, aviation industry and transportation etc.
The Group is famous for operation and programmatic work of the large scale developing plan projects. We are good at integrating the urban planning ,commercial entertainment and displaying architectures to create or promote the consuming model .We have lots of experiences on the residential community planning. We are doing the planning all the way for Chicago ,Detroit and Phoenix. We help lots of successful cities to improve the economic developments for decades. The Group focuses on improving the living quality and efficiency construction by our design to finally realize the sustainable development of our new communities .
CNASmithGroup(China) is sent by CNASmithGroup of China in 2000 playing the roles of project organizer and project team all over China. We are currently one of the biggest foreign investment qualified architectural planning and landscape design firm in Shanghai with more than 400 employees.
In 2004 the CNASmithgroup design team in Shanghai has obtained Architectural Design Class A Qualification in Architecture Department and Planning Second Class A Qualification, Landscaping Professional Qualification. And in 2005 the north branch and south-west branch were established. In 2008 the Chengdu CNASMITH has obtained Architectural Design Class Qualification in Architecture Department.
Nowadays the CNASmithgroup(China) has become a comprehensive consultancy group which could provide the services of all stages. We believe the career is our shared career. We will go all out to realize our dream.

中国区域服务联系人
朱轶俊：中国区执行总裁，高级建筑师
张振涛：建筑设计院总经理，高级工程师
傅　威：西南公司总经理，注册建筑师
李　超：规划公司总经理，注册规划师
安庆东：创作公司总经理，注册建筑师
曹　洁：景观公司总经理，注册规划师
韦尔辉：室内公司总经理，高级室内设计师
胡　坚：光系统公司总经理，高级工程师

Team Principals(China)
Zhu Yijun: CEO of China, Senior Architect
Zhang Zhentao: Architectural Design Institution General Engineer
Fu Wei: General Manager of Chengdu Company, Registered Manager, Senior Engineer
Li Chao: General Manager of Planning Company, Registered Planner
An Qingdong: General Manager of Creation Company, Registered Architect
Cao Jie: General Manager of Landscaping Company, Registered Planner
Wei Erhui: General Manager of Interior Company, Senior Designer
Hujian: Lighting System Company General Manager, Senior Architect

中国公司
地址：上海市中山南路505弄老码头创意产业园12号楼
电话：+86-21-61526700
传真：+86-21-61526712
网址：www.cna-group.com

China office
Address: No.505 South Zhongshan Road, Building No.12, Shanghai
Tel: +86-21-61526700
Fax: +86-21-61526712
Http: www.cna-group.com

西南公司
地址：成都市航空路6号丰德国际广场D1座10楼
电话：+86-28-85265630
传真：+86-28－85265633
网址：www.cna-group.com

Southwest office
Address: No.6 Hangkong Road, Fengde Plaza D1, F10, Chengdu
Tel: +86-28-85265630
Fax: +86-28-61526712
Http: www.cna-group.com

北方公司
地址：天津市成都道46号
电话：+86-22-23390238
传真：+86-22-23126009
网址：www.cna-group.com

North Company
Address: No.46 Chengdu Avenue, Tianjin
Tel: +86-22-23390238
Fax: +86-22-23126009
Http: www.cna-group.com

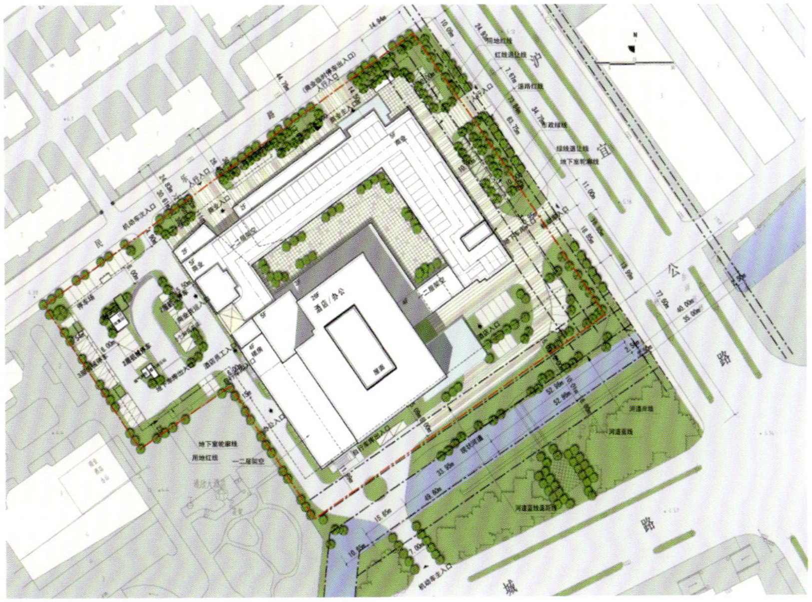

绿地嘉创国际商务广场方案设计

用地规模：13 700 平方米
建筑面积：55 000 平方米
容 积 率：4.0
项目地点：中国 上海
编制时间：2007 年 3 月

裙房商业部分通过室内中庭及放大的室外公共广场营造出舒适而丰富的空间。塔楼的形式强化了酒店的标志性形象，无论在其中间还是地块外，都呈现了不同角度的不同变化。

Greenland Jiachuang International Business Plaza

Site Area: 13,700 m^2
Building Area: 55,000 m^2
Plot Ratio: 4.0
Construction Location: Shanghai, China
Time: 2007.03

The podium creates a comfortable and flexible space through the atrium and exterior open square. The tower emphasizes the symbolic configuration of hotel .

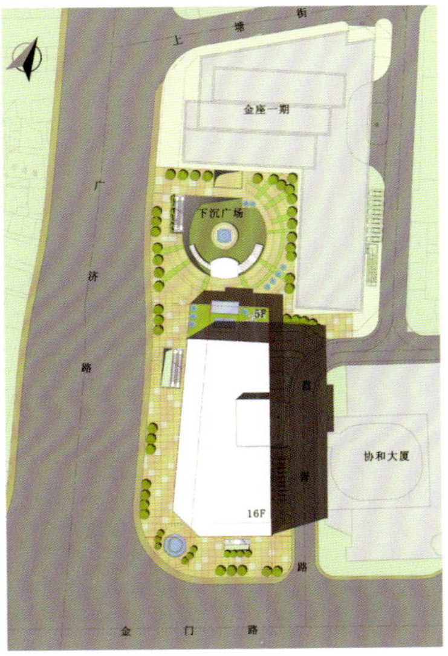

苏州石路金座项目方案设计

用地规模：5 100 万平方米
建筑面积：24 900 万平方米
容 积 率：4.9
项目地点：江苏 苏州
编制时间：2008 年 1 月

简洁的内部规划提供了空间的流动性和无限的分隔可能性，竖向交通核心筒辐射整个平面，并提供了无柱 4.8 米的层高创造出灵动的 Loft 空间。立面设计中重点突出端庄和严谨的感觉，在建筑顶部用共享的空中花园中庭，打破了建筑的呆板感。

Office Building Project Planning

Site Area: 5,100 m^2
Building Area: 24,900 m^2
Plot Ratio: 4.9
Construction Location: Suzhou, Jiangsu
Time: 2008.01

Simple interior planning provides the possibility for the limitless of movement and division. The circulation core radiates to the entire no-column plan. 4.8m floor to floor height creates a innervation LOFT space. The roof floor atrium is another active element for the building.

江苏省人民医院方案设计

用地规模：123 900 平方米
建筑面积：248 300 平方米
项目地点：江苏 南京
编制时间：2006 年 3 月
容积率：2.0

项目位于南京市鼓楼区西南部，规划在原有基础上进行增建，将其建成为集临床、教学、科研、预防于一体的大型现代化综合医院。由于基地位于南京城中的一处上下落差较大的坡地上，因此总体布局采用分散的"集中式"布局，依据设计师提出的"绿色医院""医疗大街"的核心理念，采用一条精心设计的医学时尚长廊将整个医院的新老建筑群完美地连接起来，以达到功能紧密联系、资源有效使用和节约能源的目的。

People's Hospital of Jiangsu Province

Site Area: 123,900 m^2
Building Area: 248,300 m^2
Construction Location: Nanjing, Jiangsu
Time: 2006.03
Plot Ratio: 2.0

The project locates on the southwest of Gulou district, Nanjing. It's a large scale comprehensive hospital with the functions of clinic, research and prevention. The planning strategy is dispersive "centralized "the corridor connects all departments, in order to get to the target of sustainable operation.

明宇金融广场项目规划及方案设计

基地面积：3 198 平方米
建筑面积：12 189 平方米
项目地点：四川 成都
容 积 率：14.5

项目以塔楼的形式将办公、酒店及其他功能垂直结合在一个单独的标志体中。项目建筑由 5 层的地下室，1–7 楼的入口大厅、会议及餐饮，8–31 楼的低区及高层甲级办公区，33–47 楼的五星级酒店等 4 个部分组成。功能布局设计强调"高效性"和"经济性"，可以适应五星级酒店标准间、甲级办公写字楼等多种使用形式，其功能位置分布、面积大小也可以随市场的需要弹性调整。

整体建筑意象的灵感来源于成都三星堆的出土文物——玉璋。玉璋在中国古代是一种极为重要的礼器。玉璋既是古人祭天、拜日、祈年等的礼器，又是权力和尊贵社会地位的象征。为了给城市空间造成富有冲击力的视觉震撼，形成具有象征意义的标志性焦点建筑，并呈现出一种平衡的和谐感，因此设计结合玉璋的文化延伸及体态形式，在平面和立体造型上大量运用弧线同时结合 200 米的高度形成舒展的体态。设计的塔楼不受裙楼阻隔，可直接连接到地面，高度和弧线的完美结合使本项目在这一地区大量的方正建筑群中脱颖而出，成为具有划时代意义的地标性建筑。

我们的主题塔楼与西边的新世纪塔楼形成和谐的半围合建筑形态，与南边的东大街形成一定的夹角，使主要的立面朝向市中心，朝向繁华的都市景观。上下等大的体量和略带弯曲面的形态极富有动感。区域规划，利用了项目上的矩形基地，并结合技术管理规定的退界要求，巧妙地在对角线上布置塔楼，这种橄榄形平面的斜角设置，不仅使自身具有完整的 180°景观视角，更使东大街的城市景观尽收眼底，同时这样的平面布局，基本达到全日照要求，使得这样高度的建筑对周边的建筑不产生任何的日照遮挡。

MingYu Financial Plaza Project Planning and Schematic Design

Site Area: 3,198 m^2
Building Area:12,189 m^2
Construction Location : Chengdu, Sichuan
Plot Ratio:14.5

This tower vertically integrates office, hotel and other functions into one single marking figure. Respectively, its four functional parts are 5 floors for basement, 1-7th floors for entrance hall, conference and dining, 8-31 floors for low-floor area connected with high grade A office, and 33-47 floors for five-star hotel. Its emphasis on functional layout is "efficiency" and "economy", which is suitable for multiple usages such as standard room of five-star hotel, grade A office building, etc., meanwhile, it can also be adjusted catering to the market flexibly in respect of its functional location distribution or area size.

In order to create an extraordinary visual shock for urban space, to form a landmark focus building with symbolic meaning, this project presents a balanced harmony for its best. Combined with the cultural deposits and body implication of Yu Zhang, arc is extensively used in plane and three-dimensional modeling, extending to a height of 200 meters to form a stretch of the entire figure. The tower connects to the ground without any blocking by the podium. Its perfect combination of height and arc, outweighs numerous square buildings in the region and thus gradually evolves into a magnificent, epoch-making landmark.

Our theme tower and the west new century tower create a harmonious semi-enclosed architectural form, which shapes an angle with the East Avenue in the south, thus the main façade is towards the city center, towards the bustling urban landscape. Dramatic and dynamic as it is, this tower has an equal shape from top to bottom, along with its surface form curved slightly. As to regional planning, it makes use of the rectangular base of the project, combines with requirements of setbacks at technical management stipulation, and also arranges tower in diagonal skillfully as well. Such oblique angle setting of olive plane, not only has a complete 180-degree landscape angle itself, covering the entire panoramic view of city landscape in the East Avenue, also plane and its self-condition are superior, basically up to full sunshine level, so that no any sunshine shade to the surrounding buildings due to a construction at such height.

沈阳音乐厅方案设计

用地规模：45 000 平方米
建筑面积：21 900 平方米
容积 率：1.15
项目地点：辽宁 沈阳
编制时间：2007 年 10 月

项目是位于沈阳市浑河以北的地标性建筑，是充满浑河精神和力度并突显优雅的时代产物，它将成为这座城市的大门。设计师将沈阳的历史文化和人民意志以及浑河本身作为这个项目的信息和资源进行解构，在平静中创造纪念的力量，推动的力量，启发人们去探究其自身曾经存在的历史真相。建筑物带着灵动的曲线骄傲地耸立，雕塑般的形态承载了令人印象深刻的"交响乐"，向人们展现其本身所代表的浪漫篇章。

Shenyang Concert Hall

Site Area: 45,000 m^2
Building Area: 21,900 m^2
Plot Ratio: 1.15
Construction Location: Shenyang, Liaoning
Time: 2007.10

As the land mark of north Hunhe in Shenyang, the project is outcome of spirit and power of the time. She would be the gate of city. The designer took advantage of the historical culture and people's spirit as the resource, create a memorable thinkable figure. The movable curves proudly stands there as a sculpture taken impressionable romantic chapters .

绿地昆山花桥企业总部方案设计

用地规模：66 700 平方米
建筑面积：99 100 平方米
容 积 率：1.5
项目地点：江苏 昆山
编制时间：2007 年 6 月

设计对商业、办公、居住、展示等几大功能进行功能整合，形成脉络清晰的建筑规划架构。整个规划可以看出设计思维从混合到清晰的演变。设计力图创造一个白昼黑夜着循环不息的活力之城，商业环抱着中心绿地，通过商业内街向基地中央的绿化渗透，创造亲水休闲购物环境，形成一种无导向性的休闲人流组织布局。

Kunshan Greenland Huaqiao Headquarters

Site Area: 66,700 m^2
Building Area: 99,100 m^2
Plot Ratio: 1.5
Construction Location : KunShan Jiangsu
Time: 2007.06

The integration of main function zones forms a clear structure and provides a progression of confusion to clearness. The active city is surrounded by central Greenland and the natural environment penetrates to commercial facilities as a guideline for pedestrian flow.

AIM GROUP INTERNATIONAL 国际设计集团

亚瑞建筑设计有限公司（中国甲级）AIM GROUP INTERNATIONAL （加拿大）

AIM 国际设计集团是在加拿大、香港注册的跨国品牌。近年来中国相关市场迅猛发展，因而，我们将集团总部设立在中国珠江三角洲的核心城市——广州。总部现有员工 200 多名，包括加拿大注册建筑师、香港著名艺术顾问。拥有 1 500 平方米办公写字楼。

集团分三个部门，一个子公司，分别是亚瑞建筑设计、铭唐室内设计、点子策划、中建南方建筑设计院（深圳）。经营范围有建筑设计、室内设计、平面设计、园林设计、策划、招商、销售。集团秉承设计全程服务，成就品牌价值的经营理念，业务足迹遍布国内外，如美国、澳大利亚、加拿大等国家，国内有北京、上海、香港、广州等各大城市。

亚瑞建筑设计在建筑和规划设计方面，以先进的国际项目管理方式，优质的服务质量和设计水平赢得了世界口碑。AIM 集团逐渐孕育出以开放、激情为主题的公司文化，以海纳百川之态吸纳了众多才华横溢的明星建筑师，团队精神深深植入公司文化中。AIM 集团一直将设计质量和客户满意度放在首位，公司伴随客户业主们的成功而迅速成长。

AIM International Design Group is a multinational brand registered by Canada and Hong Kong. In recent years, with the rapid development of related markets in China, we set up the group's headquarters in the core city of the Pearl River Delta in China---Guangzhou. Headquarters have more than 200 employees, including registered Canadian architects, Hong Kong's famous artistic advisers, owning 1,500 square meters office building.

The Group is divided into three departments and a subsidiary. They are: Yarui Architectural Design, Mingtang Interior Design, China South Architectural Design Institute (Shenzhen). Business scopes include: architectural design, interior design, graphic design, landscape design, planning, investment and sales. The Group upholds the business philosophy of whole course service of design and of the achievement brand value. Business footprint covers around the home and abroad, such as the United States, Australia, Canada and other foreign countries and domestic cities like Beijing, Shanghai, Hong Kong, and Guangzhou and so on.

In the aspect of architecture and planning design, Yarui Architectural Design has won the world reputation for advanced international project management, excellent service quality and design. AIM Group has gradually bred corporate culture for the subject of openness, passion and has absorbed a large number of talented star architects with the attitude of keeping an open mind about different opinions. Team spirit has ingrained into company culture. AIM Group has been putting the design quality and customer satisfaction in the first place. The company is developing rapidly with the success of our customers.

（中国）
广州体育西路 133 号天河大厦综合楼二楼
Tel: +86−20−38813815/38813785/38864882
Fax: +86−20−38811267/34387365（广美）

（香港）
Rm.306, Golden Gate Commercial Building,
136 Ausn Road, Tsimshatsui, Kowloon, Hong Kong
Tel: +852−24116631 Fax: +852−89497386

（加拿大）91 Glentworth Road, Toronto, Canada
Tel: +416−4919988 Fax: +416−5677803

AIM GROUP INTERNATIONAL 国际设计集团

亚瑞建筑设计有限公司（中国甲级）
AIM GROUP INTERNATIONAL （加拿大）

1–2 项目名称：东莞景天花园城
开发单位：联华国际集团
用地面积：22.6 万平方米
建筑面积：45.2 万平方米
设计时间：2008 年

3–4 项目名称：东莞道滘客运站
开发单位：耀盈房地产开发有限公司
用地面积：3 万平方米
建筑面积：8.3 万平方米
设计时间：2007 年

5 项目名称：东莞荷塘月色
开发单位：联华国际集团
用地面积：1.3 万平方米
建筑面积：6 万平方米
设计时间：2009 年

AIM GROUP INTERNATIONAL 亚瑞建筑设计有限公司（中国甲级）
国际设计集团 AIM GROUP INTERNATIONAL （加拿大）

6

7

6–7 项目名称：中信德方斯公寓
开发单位：中信（华南）集团
用地面积：4 万平方米
建筑面积：9.2 万平方米
设计时间：2006 年

8–9 项目名称：中信凯旋国际中心
开发单位：中信（华南）集团
用地面积：8 万平方米
建筑面积：30 万平方米
设计时间：2008 年

10–11 项目名称：中信商业中心
开发单位：中信（华南）集团
用地面积：3.5 万平方米
建筑面积：9.3 万平方米
设计时间：2008 年

8

9

11

AIM GROUP INTERNATIONAL 国际设计集团 亚瑞建筑设计有限公司（中国甲级） AIM GROUP INTERNATIONAL （加拿大）

12–15 项目名称：千山花园规划设计项目
开发单位：广晟逸涛万国房地产开发有限公司
用地面积：66.67 公顷
建筑面积：41 万平方米
设计时间：2009 年

16–17 项目名称：万国 14A 地块海韵兰庭
开发单位：广晟逸涛万国房地产开发有限公司
用地面积：3.4 万平方米
建筑面积：11.2 万平方米
设计时间：2009 年

AIM GROUP INTERNATIONAL 亚瑞建筑设计有限公司（中国甲级）
国际设计集团 AIM GROUP INTERNATIONAL （加拿大）

18．广州番禺汇珑新天地商业广场
19．清远碧江帝景（超高层）
20．广州从化名城御景绿洲
21．广州山河百货商业中心

罗根表演艺术中心（美国）
Logan Center for the Performing Arts(USA)

艾迪石 简介 AEDIS Introduction

美国 AEDIS（艾迪石）建筑与规划设计有限公司（以下简称艾迪石）位于加州旧金山区的硅谷首府圣荷西市，由创建于 20 世纪 50 年代的加利福尼亚知名建筑设计公司 PJHM 和 AWG 合并组成，项目范围涵盖了文教建筑、城市公建、商业和住宅建筑等，为客户提供从项目的前期策划、场地规划、建筑设计、工程管理到竣工验收的全套服务。迄今，艾迪石已多次荣获美国本土的诸多建筑设计奖项，并被多家权威专业杂志特别报道。

公司在近来的设计作品中充分应用了可持续建筑的设计理念，并以此获得了诸多荣誉和绿色建筑证书。

随着 BIM（建筑信息模型）在全球的建筑行业引领一场全新的变革，处在建筑设计领域前沿的艾迪石不仅掌握着这门最新的专业技术，同时又将这一优势应用到中国项目中，赢得了各方的好评，为我们的客户提供了具有具有当今国际水准的高品位、高品质的服务。

艾迪石（上海）建筑设计咨询有限公司
上海市 南京西路 920 号 南泰大厦 1601 室 邮编 200041
www.aedischina.com　　Tel:+86-21-61039698

圣荷西大学－视觉表演实训楼（美国）
San Jose City College(USA)

山景图书分馆（美国）
Hillview Branch Library(USA)

建筑和规划
ARCHITECTURE & PLANNING

南京金牛湖国际度假中心
Jinniu Lake International Resort, Nanjing

苏州穹窿山度假别墅酒店
Mountain Qionglong Resort Villa Hotel, Suzhou

建 筑 和 规 划
ARCHITECTURE & PLANNING

安徽芜湖师范学校
Wuhu Normal University, Anhui

特拉诺瓦剧场（美国）
Terra Nova Theatre (USA)

温州水上旅游服务基地
Wenzhou Aquatic Tourism Service Base

南京浦口科技工业园科学区
Pukou Scientific Industrial Park, Science Section, Nanjing

建筑信息模型（BIM）是指基于最先进的三维数字设计和工程软件所构建的“可视化”的数字建筑模型，被誉为 CAD 之后工程技术领域的又一次革命，把真实的建筑信息参数化、数字化以后形成一个模型，从设计师、工程师到施工单位和物业管理方，都可在建筑的全生命周期进行信息共享和改进。直观、高效是 BIM 的最大优点。艾迪石在 BIM 领域积累了十年以上的工作经验，以此大大提高了设计质量和服务品质，完成了令设计师与业主共同满意的项目。

上海丰益国际全球研发大楼
Wilmar International - Global R & D Building, Shanghai

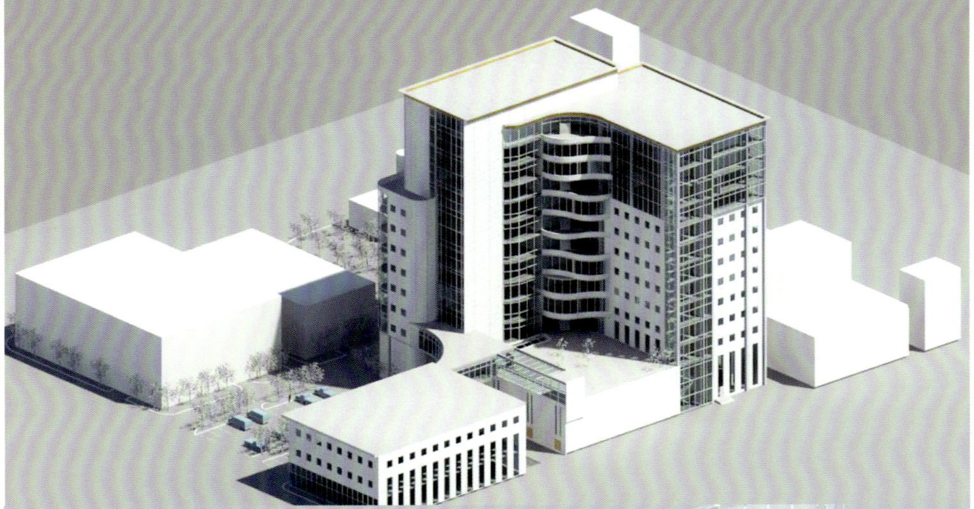

上海时代金融中心 -BIM
Shanghai Times Financial Center

室内设计和建筑信息模型（BIM）
INTERIOR Design & BUILDING INFORMATION MODELING

上海时代金融中心
The One Business Centhe Shanghai

苏州高新国际商务广场
Suzhou Gaoxin International Commercial Plaza

天津商务广场
Tianjin Commercial Plaza

Wei Li Design Studio
李卫设计工作室

李卫设计工作室成立于美国，麻省，波士顿，并随后在北京中央商务区成立中国分部——北京李卫设计工作室，以面向中国开放的巨大市场

李卫设计工作室相信“优秀的设计创造价值”，其愿景是“通过创作优秀的设计作品使社会与业主受益”。

工作室主要进行建筑设计、城市与规划设计，擅长设计教育建筑、公共建筑、商业建筑与医疗建筑，其主创人员均有在国内外丰富的设计实践经验。

北京李卫设计工作室创始人李卫先生曾任美国Perkins & Will波士顿设计总监和董事，以及英国Atkins中国设计总监和董事，具有丰富的实践经验。

Focus on China's huge open architecture design market and its desire of international design expertise, based on its founder's extensive architecture and business experience in China and United States, Wei L Design Studio was established at Boston, Massachusetts, US. Its China Branch at Beijing's Central Business District was opened right after.

Wei Li Design Studio believes "Good design and planning creates values". Wei Li Design Studio's mission is "Using creative design and planning ideas to benefit our clients and the society."

Wei Li Design Studio is an architecture design and planning practice focus on educational, public, commercial and HEALTHCARE projects. Its Key staffs have extensive experience in design / planning in domestic and overseas fields.

Wei Li Design Studio's past important project experience is based on its key member's professional experience in reputable design firms like Perkins and Will in US and Atkins in Britain.

中国工程

清华美院－打开的玉石

清华美院 模型

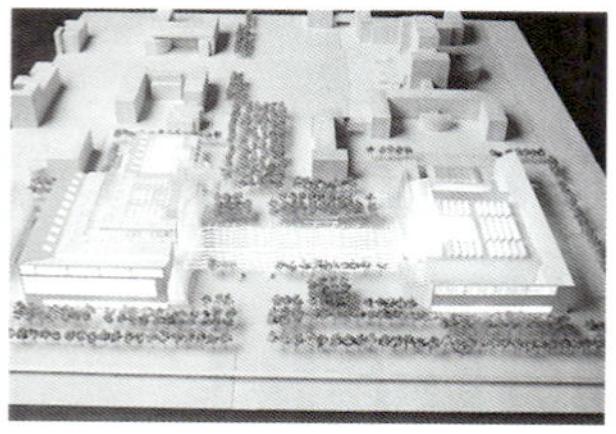

清华美院 实景照片

北京MAX空港企业园

外交学院山东外语学院

成都市中心商住综合体

美国工程

Columbus Center

888 Boylston St.

Telecommunication Center. NY

Gallaudet Univ. Conference Center

Boston Children's Hospital

Robertwood Johnson Children's Hospital

22 Mallard Drive Sharon, MA.02067, USA
T: 001.617.821.4559 F: 001.781.793.9611

中国北京市朝阳区光华路7号汉威大厦A座21层07号 邮编：100004
Tel: +86-10-65612162 Fax: +86-10-65612163
E-mail: weilione@yahoo.com

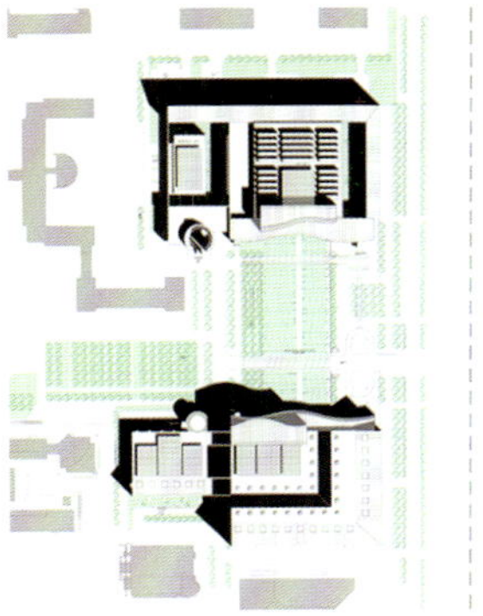

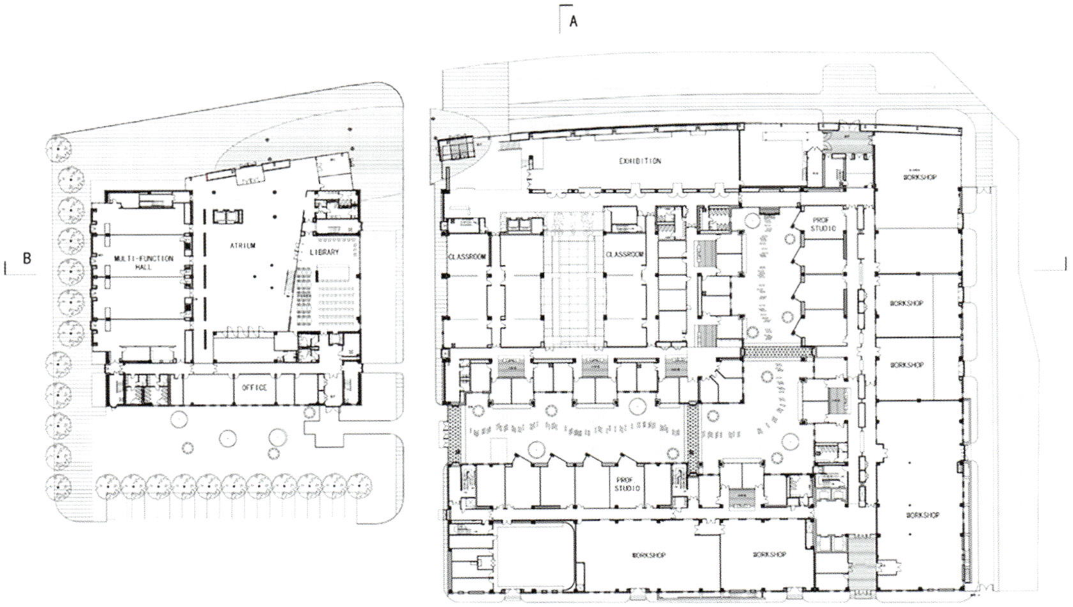

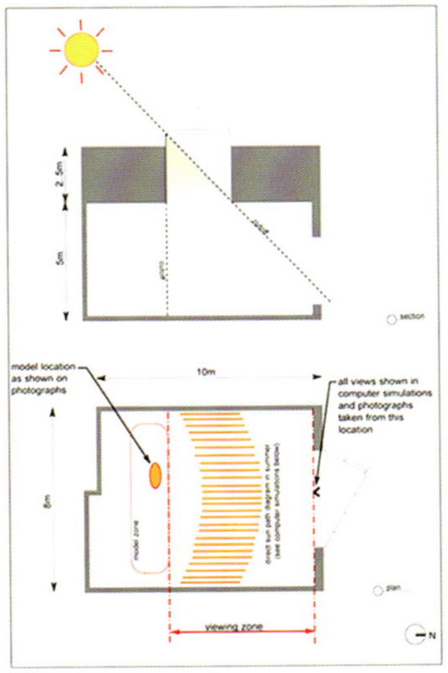

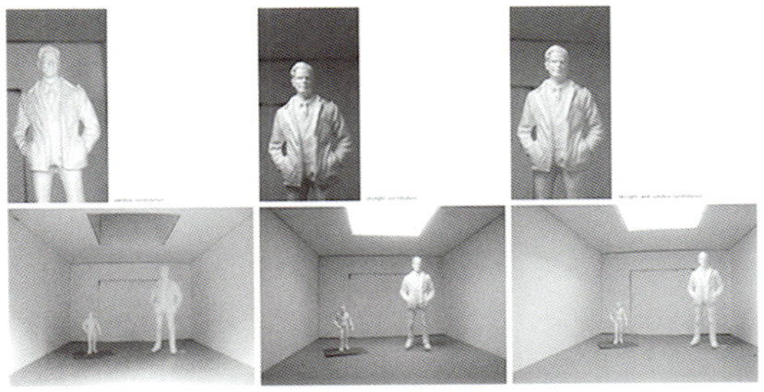

光学设计

清华大学美术学院

随着清华大学和中央工艺美术学院的合并，美术学院新楼的建设，成为清华大学向综合性一流大学迈进的关键一步。

该设计方案通过国际竞赛脱颖而出，于 2005 年秋投入使用，落成后，美术学院成为清华大学主楼前区的重要地标，它充分尊重周边景观环境，尤其是与邻近建筑的体量和色彩协调统一。

其浅红色有肌理的花岗岩外墙面突出了建筑典雅的气质，入口处通透明亮的展览大厅、尺度宜人的室外庭院和室内的研讨、休闲沙龙，为师生提供了舒适的交流场所，它们与教室，画室和工房等一起，形成富有生机的“艺术社区”。

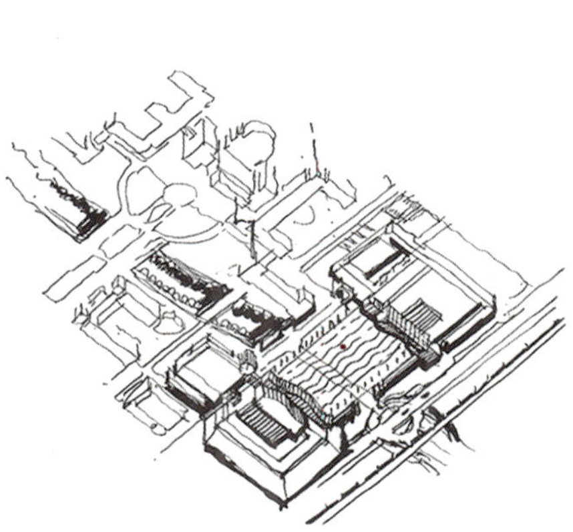

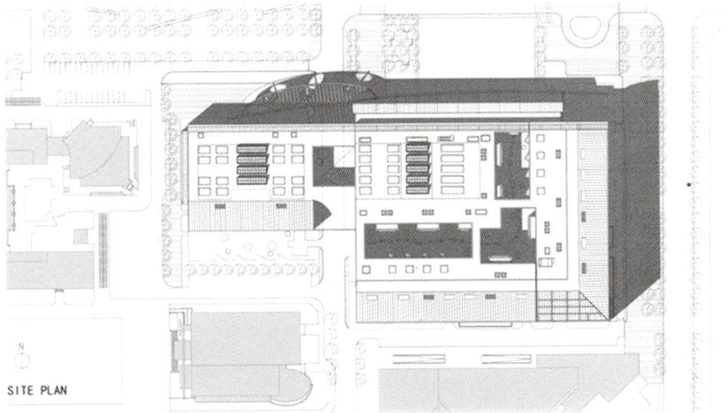

清华大学美术学院 2

1. 全景照片
2. 入口大堂
3. 入口雨篷
4. 展览厅

北京 Max 空港企业园

设计力求创造一个有生气、可变化的社区，随着现代工业、科技的不断发展，工业社区的发展与变化是不可避免的，由此，在本园区中，通过精心设计一个可以不断叠加的"企业独幢"母体，使其可以随着企业需求的变化而发展。在总图设计上，亦为未来的生长和变化预留了空间。

设计充分利用自然风与阳光，力求营造一个绿色园区。为了使每个员工都有良好的朝向与自然光，各个独栋采用了倾斜 30°角的总图布局，同时，在冬季风入园的西北边采用较为密集的布置，以遮挡寒冷的冬季风，而在夏季风入园的东南边，采用疏朗的布局，以导入凉爽的夏季风。各企业独栋的立面设计采用模数化的合成面板与玻璃幕墙系统，通过其不同的质感（纹理、光色）、色彩与透明度的设计体现了各个企业的不同特点。各企业独幢的入口则可据其性质（IT、化工等）进行个性化设计，为施工与未来的变化提供方便，模数化的外墙挂板使施工简便、有效率，同时为未来的变化提供了契机。

设计用适宜的建筑语言为园区创造视觉形象。工业化、简洁、高效的外墙模板系统符合空港工业区的个性，体块化的公共建筑与独幢建筑反映其高效的内部功能，通过不同的组合，营造各园区高效率、生机勃勃的氛围。

中国外交学院威海外语学校

中国外交学院威海外语学校总建筑面积 85 000 平方米，含教学楼、实验楼、图书馆、大礼堂、室内体育馆、办公区、食堂和学生生活用房，并配有野外夏令营、野外实验区和国际教育交流中心。校园设计依山面海，室内外空间与院落交相辉映，营造出怡人的教学气氛。

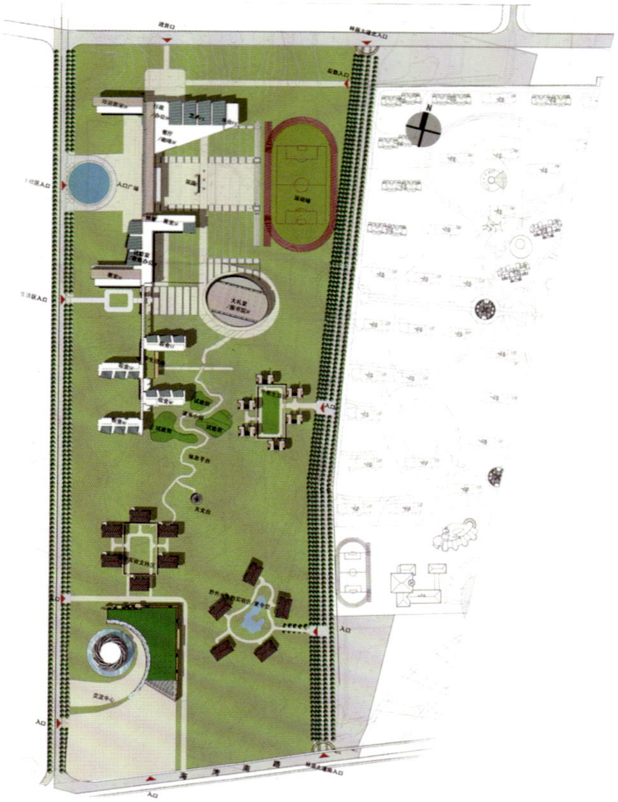

总平面图

北京师范大学新校区

中国传统的书院建筑给设计以很大的启发，书院一般坐落于寂静的山水之间，通过其外围的院落形成读书养性的氛围，寄情山水。北京师范大学新校区的规划充分分析了该地段“风水”，以中国书院为内核，以现代建筑为外围，成功地利用了阳光、夏季风和水系，抵挡了冬季风，从而形成了自己的院落体系，并为未来的发展预留了充分的空间。

在总平面的布局上，采用传统的中轴线与院风作为北京师范大学的内核区。通过沿中轴线分段布置入口区、纪念区、教研区和未来区这四个大的矩形院落，来体现自尊自强、严谨向上的气质，反映出北京师范大学“百年老校”的学术地位。其具体的空间尺度既考虑了建筑史上著名的美国佛吉尼亚大学中心院落的尺度，又融入了中国传统书院、杏坛等教育建筑的院落空间关系，营造出一个“诚信质朴”的精神内核。在传统的、严谨的“书院”风格中心区外，规划了一个自由的、流线型的，集科研、讨论、办公于一体的现代外延区，并以院落与中心教学区内核相连，同时将各院系教学管理部分设计成中国传统的“四合院”式，用其将教学内核与科研外延串联起来，形成空间三维的立体交叉互动模式。

当车流，行人沿校区外围道路行走时，会感受到一个流线的、崭新的、有气势的、前瞻性的现代化新校区。

成都市中心商住综合体

设计力求打造城市中心地段既现代时尚又优美宜人的现代都市立面，使其成为城市新景点。单体立面设计上，注重线与面的结合，注重立面与简洁流畅的体块相呼应，形成起伏的乐章。为进一步烘托体与面的转折与融合，色彩设计上选择在大气尊贵的灰紫色调基础上配以“温暖亲切”的棕红色，从而营造出现代城市时尚宜人的情调。在材料应用方面，则以简洁的透明玻璃、铝板、石板与瓷砖四种材料各取所需地塑造上述体面关系，达到近人处追求质感和肌理、远观处强调线和面的效果。

1-2 Columbus Center

该建筑建于波士顿，位于麻省 90 号高速公路边上的哥伦布中心区，是集商业、高端居住、豪华酒店于一体的现代建筑，建成后将成为波士顿的新地标。

3-4 888 Boston St. Office Building

项目位于波士顿会展中心与老图书馆之间的 888 号，为高级办公楼。建筑追求在比例与尺度上与周边环境的和谐，但在造型设计上追求前卫新颖，为有历史与文化内涵的街区增添了新元素。

5

8

6

5-7 Telecommunication Center NY

纽约哈德逊学院电讯中心设计在平面简练实用的同时，造型上更突出了其“电讯”的特征。

8-10 Gallaudet Univ. Conference Center

位于华盛顿的著名聋哑大学会议中心是集会议、酒店于一体的综合性建筑，设计上既体现了现代感，又能将其融入古老的校园环境。

9

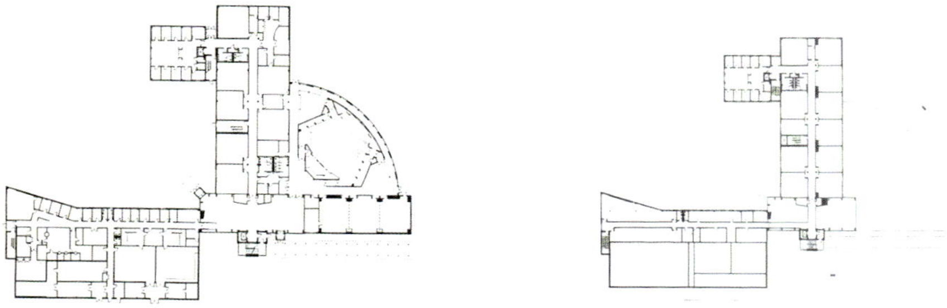
7

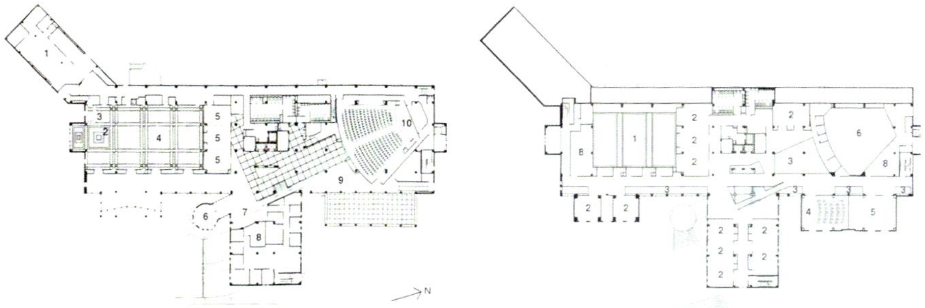
10

加拿大 OA

天津市霞光道 1 号大厦

大沽路风貌建筑

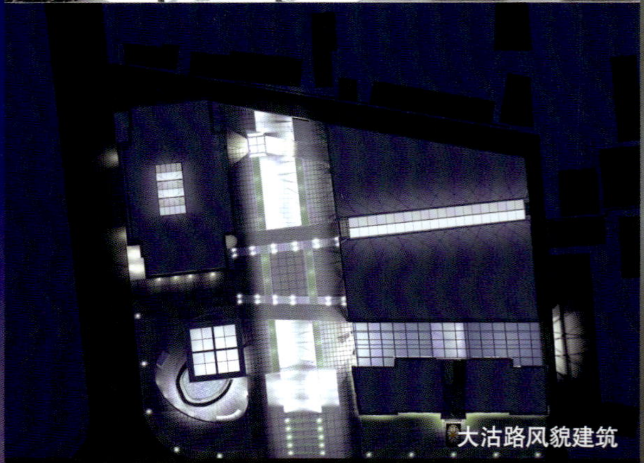

大沽路风貌建筑

天津梅江商务中心

天津梅江商务中心

京沪高速铁路天津南站

RCH DESIGN GROUP LTD.

天津梅江商务中心

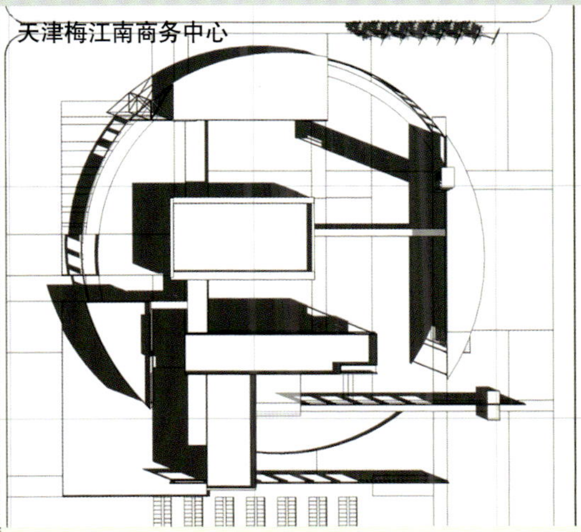

天津梅江南商务中心

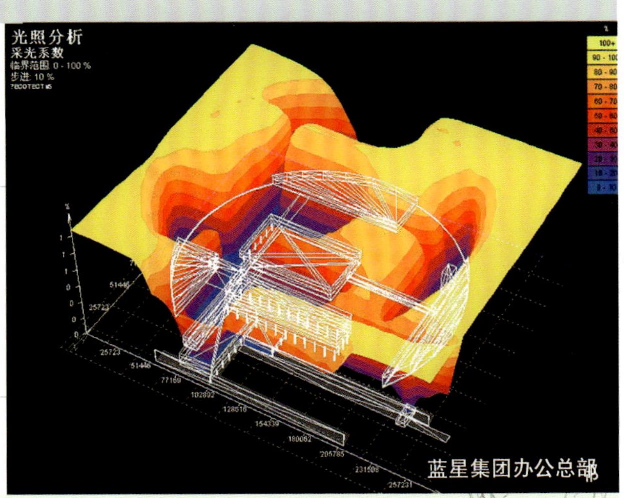

蓝星集团办公总部

蓝星集团办公总部

ORCH
DESIGN GROUP LTD.

地址：天津市南开区阳光公寓4-2-3F
电话/传真：86-022-23923720
邮箱：yuanchuangjianzhu@163.com

蓝星集团办公总部

北京-安联置业万寿路29号地
Wanshou Lu, Beijing

北京-首开璞
Shoukai Puti, Beijing

ADS建筑 规划设计事务所
ADS ARTEC DESIGN STUDIO architecture & urban planning

地址

北京代表处
海淀区车公庄西路甲19号华通大厦396室
中国 北京 100044　　邮箱：ads_china@163.com
电话：+86–10–68700548　　传真：+86–10–68700545

上海代表处
徐汇区肇嘉浜路738号欧江大厦23楼A室
中国 上海 200030　　邮箱：adssha@163.com
电话：+86–21–64660496　　传真：+86–21–64660497

Beijing office
Address: Room 396, Huatong Building, Chegongzhuang West Rd. No.19
Beijing, China 100044
Tel: +86-10-68700548　　Fax: +86-10-68700545
E-mail: ads_china@163.com

Shanghai office
Address: Room A, 23rd floor, Oujiang Building Zhaojiabang Rd. No.768
Shanghai, China 200030
Tel: +86-21-64660496　　Fax: +86-21-64660497
E-mail: adssha@163.com

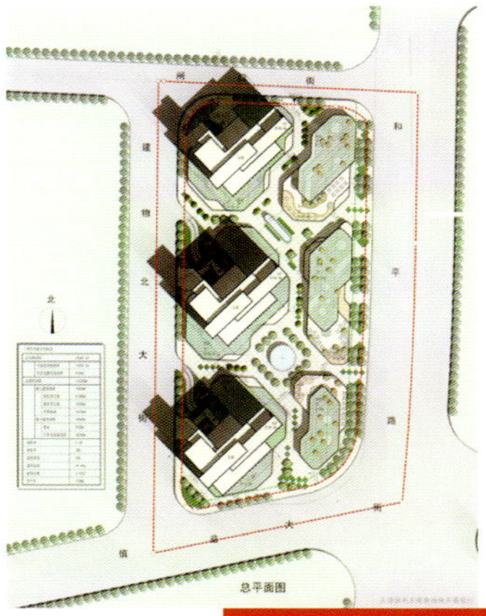

天津–保利香槟花园
Poly Xiangbin Garden, Tianjin

Artec Design Studio architecture & urban planning

简介

ADS建筑规划设计事务所创立于加拿大安大略省，公司在北京和上海两地设有办事处。近几年来，事务所在国内十几个城市参与了许多公共建筑与住宅项目的规划及设计工作，其中多个项目成为当地的代表性作品，并获得了众多优秀设计奖项。在设计过程中，不仅应用最先进的规划设计理念和方法，而且强调为业主提供国际水准的专业化全程服务理念。公司非常重视与业主间的充分沟通，尤其在概念设计阶段，只有在更深的层面上理解业主的需求和项目本身，才能做出既能充分体现客户意愿又能完整表达设计创意的方案。

Artec Design Studio (ADS) is an architecture & urban planning firm registered in Ontario, Canada. Its local representative offices are located in Beijing & Shanghai. It has been actively participating many public buildings and housing projects across China in recent years. Many of them are award-wining projects and become local landmark.In our design concept, we emphasize in-depth communication with our clients. Understanding the requirement and characteristic of each project enable us to address our clients' needs better while maintaining the artistic integrity of the design.

唐山–兆瑞广场
Zhaorui International Plaza, Tangshan

北京–首开苏活
Soho, Beijing

ADS建筑 规划设计事务所
ADS ARTEC DESIGN STUDIO architecture & urban planning

环渤海地区

01 保利 • 垄上别墅 （北京）
02 保利 • 西山林语 （北京）
03 慧城 • 君山别墅 （北京）
04 泰禾 • 红御别墅 （北京）
05 鲁商 • 蓝岸丽舍 （北京）
06 华润 • 风景翠园 （北京）
07 润泽 • 润泽庄园 （北京）
08 政泉置业 • 金泉广场 （北京）
09 保利 • 香槟花园 （北京）

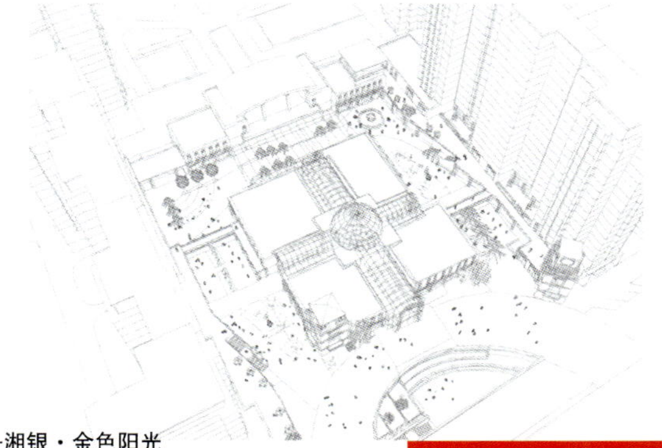

北京–保利・茉莉公馆
Poly JASMINE MANSION, Beijing

株洲–湘银・金色阳光
Golden Sunshine, Zhuzhou

Artec Design Studio architecture & urban planning

北京–润泽庄园会所
Runze Villas, Beijing

ADS建筑 规划设计事务所
ARTEC DESIGN STUDIO architecture & urban planning

公司近年主要项目一览表

长三角地区

24 保利 • 十二橡树庄园 （上海）
25 万科 • 提香别墅 （上海）
26 万科 • 新里程（上海）
27 招商 • 依云郡 （上海）
28 久事 • 西郊花园二期 （上海）
29 万里雅筑 （上海）

北京-泰禾红御
Thaihor Mansion, Beijing

北京-君山高尔夫会所
Mountain Duke Golf Villa, Beijing

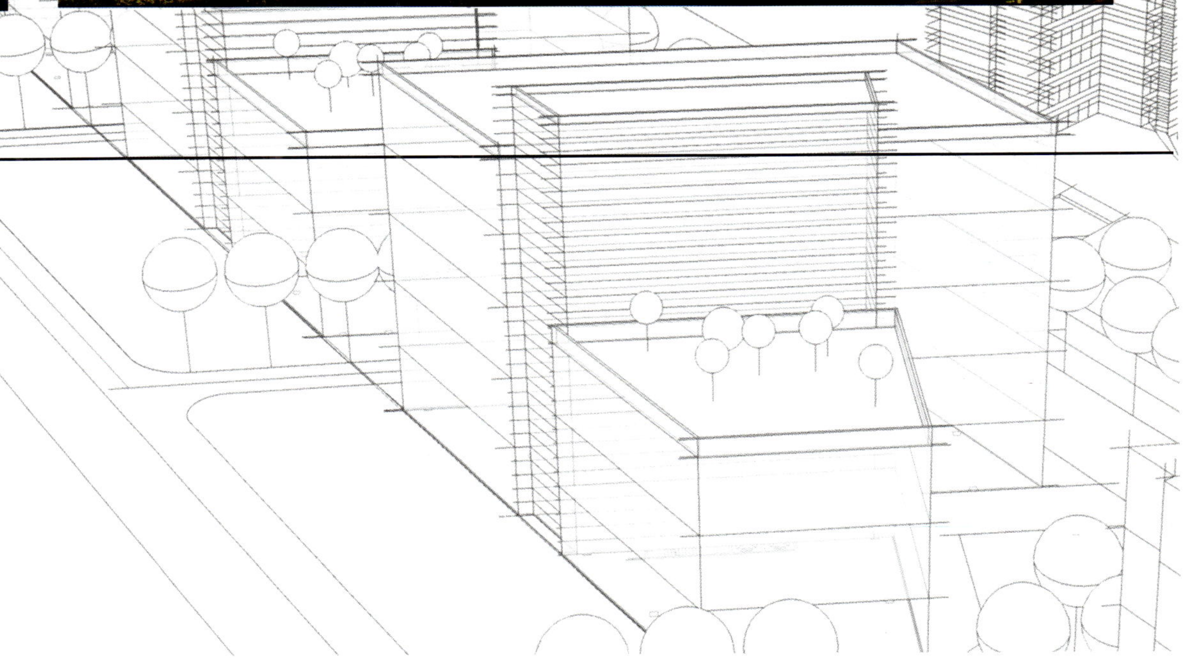

30 银都名墅 （上海）
31 浙电 • 九龙坞别墅 （富阳）
32 绿城 • 桃花源 （杭州）
33 地税局办公楼 （杭州）
34 在水一方 （苏州）
35 宝华 • 海滨庄园 （上虞）

北京-蓝岸丽舍
Blue Shoreline Villa, BeiJing

ADS建筑 规划设计事务所
ADS ARTEC DESIGN STUDIO architecture & urban planning

公司近年主要项目一览表

其他地区

36 上海城开 • 托斯卡纳 （长沙）
37 德普 • 美洲故事 （长沙）
38 泰禾 • 住宅项目 （福州）
39 新长城 • 龙湖花园 （郑州）
40 佳达利 • 阳宗海凹子山项目 （云南）
41 海悦天伦 （海口）

北京－君山高尔夫西区
Mountain Duke Golf Villa West Zone, Beijing

石家庄－合欢阳光城
Hehuan Sunny Town, Shijiazhuang

ADS建筑 规划设计事务所
ARTEC DESIGN STUDIO architecture & urban planning

公司近年主要项目一览表

海外项目

49 MNSK学院总体规划 （印度）
50 HSC MEDICAL CENTER W/SOMERSET SERVICE APARTMENT （吉隆坡）

Artec Design Studio architecture & urban planning

ÉTÉ 翌德国际设计机构

été lee et associés architectes urbanistes

■ 法国翌德国际设计机构　■ 上海翌德建筑规划设计有限公司

来自法国的翌德国际设计机构(ÉTÉ Lee et Associés architectes urbanistes)是城市规划、建筑与景观设计领域的实践先锋与思想先行者，机构项目遍布法国、西班牙、美国、韩国等国家。近10年来，设计机构在中国的50余座大中城市主持并完成了300余项设计任务。其中，上海2010年世博水门、上海市南外滩地区、上海市张家浜楔形绿地、上海芦潮港海滨国际花城、上海市外高桥商务别墅区、上海市金地·格林郡、成都市华新锦绣尚郡、宁波盛世天城，以及应用绿色建筑技术的北京鼎嘉恒苑(The House)等项目的建成和投入使用，充分体现了翌德国际设计机构(ÉTÉ)"适用·和谐·动人"的创作理念和对自然、文化、经济环境的高度责任感与洞察力，以及在新技术、新材料运用方面的领先地位。

迄今为止，机构荣膺国家、省部级奖励30余项，并先后获得了由联合国人居署、中国建筑学会颁发的集体与个人国际贡献嘉奖。翌德国际设计机构（ÉTÉ）拥有完善的项目管理体系，健康、活跃的企业文化和创作氛围，多年来，翌德国际致力于将欧洲先进的设计理念与中国的实际情况相结合，为中国的城市建设提供高质量的专业服务和可持续发展的技术方案。其国际化的专业协作团队将把对社会、经济、环境与时效的综合考虑融入创作过程，为每个项目带来超凡的品质和长远的效益。全方位的效益评估和缜密的设计思路，使翌德国际设计机构(ÉTÉ)成为众多地方政府、投资集团的理想合作伙伴。

été lee et associés Architectes Urbanistes,distributed in Paris, has first-class designers, engineers and planning consultants, and has an excellent reputation in planning, architecture and landscape designing areas. Many outstanding achievements have already been acquired by été lee et associés Architectes Urbanistes from the year of entering the market of mainland, China in 2000. Not only designed and implemented many projects in more than 50 big or medium-sized cities in China, été lee et associés Architectes Urbanistes also won more than 30 national, provincial and ministerial level awards as far, and the international contribution awards for group and individual. The same time, many projects were highly evaluated in industry after being put into use, such as Shanghai Waigaoqiao Business Villa, Shanghai Goldland Green County residential district, The House in Beijing, Shanghai TianYaoQiao Business Center, Ningxia Colorcast Center and Business Bank, etc. Major projects such as the general city plan for Qingpu district, urban design for the South Bund in Shanghai, Water Gate designing and implement for Shanghai Expo, and urban green land designing in Zhangjiabang, also won été lee et associés Architectes Urbanistes more good industrial and social reputation.
In the purpose of providing high-quality professional services and technical solutions for sustainable development for urban constructions in China, été lee et associés Architectes Urbanistes always working on the way to sufficiently combine the actual situation of China with the advanced design ideas of Europe. And a well co-operated professional team will always put their thinking about social, economy, environment and time into their creations of each of projects, for the excellent quality and a long-term benefit. Comprehensive assessment of the benefits and rigorous design ideas always make été lee et associés Architectes Urbanistes to be the ideal partner of many local governments and investment groups.

法　　人：Etienne MATHIEU，于洪滨
地　　址：上海市北京东路668号科技京城西楼23B
邮　　编：200001
联 系 人：邵尚来
电　　话：+86-21-53082309
传　　真：+86-21-53082776
网　　址：www.etelee.com
邮　　件：etelee@163.com

Legal Person: Etienne MATHIEU, Hongbin Yu
Address: West Building 23 B, Science & Technology Jing City, Beijing East Road No. 668, Shanghai
P. C.: 200001
Contact Person: Shanglai Shao
Tel: +86-21-53082309
Fax: +86-21-53082776
Http: //www.etelee.com
E-mail: etelee@163.com

翌德国际设计机构
été lee et associés architectes Urbanistes

上海2010世博会秦皇岛路站水门

项目地点：上海
建设时间：2009年
总用地面积：6.95公顷

上海市嘉定区农民集中安置小区规划暨农村建筑设计

项目地点：上海 嘉定
设计时间：2010年
规划用地：21.6公顷

南京金地板桥石闸湖商务度假中心

项目地点：江苏 南京
设计时间：2009年
技术指标：办公2 990平方米，酒店7 900平方米

江苏省常熟世茂新城项目4#、6#地块景观设计

项目地点：山东 济南
设计时间：2009年
用地面积：71 646.72平方米，绿地率：47%

四川都江堰紫荆城二期规划及建筑设计

项目地点：四川 都江堰
设计时间：2010年
总建筑面积：281 784平方米
容 积 率：1.98

山东济南云湖山墅景观设计

项目地点：山东 济南
设计时间：2009年
技术指标：景观设计面积8.32公顷

陕西省府谷中学、府谷职教中心建筑设计

项目地点：上海
设计时间：2009年
总建筑面积：271 063平方米

HOOP ARCHITECTS

美国HOOP建筑设计有限公司

HOOP ARCHITECTURAL DESIGN Co.,LTD.

城市设计 | 公共建筑 | 居住建筑

Urban Design | Public Facility | Residence

California U.S.A
AD:70 PACIFIC ST.ROOM615 CAMBRIDGE, MA02139 U.S.A
TEL: 001-617-452-5166
FAX: 001-617-452-5160
EMAIL: hproject@hoop-archi.com

中国 上海
地址：中国上海浦东新区世纪大道1589号长泰国际金融大厦908–911单元
AD:Rm908 Chamtime International Finance Center No.1589 Century Avenue Pudong Shanghai China
ZIP CODE: 200122
TEL: 0086-21-68590505
0086-21-58783137
FAX: 0086-21-58782763
E-mail: shhoop@vip.163.com
HTTP: //www.hoop-archi.com

中国 深圳
地址：中国深圳福田区保税区英达利科技数码园B栋101A室
AD:Rm101A,BlockB,Yingdali Technology Premises,Futian Futian Free Teade Zone,Shenzhen,China
ZIP CODE: 518038
TEL: 0086-755-83458622
FAX: 0086-755-83458633
E-mail: szhoop@vip.163.com

美国HOOP建筑设计有限公司1995年成立于美国加州，是美国加州具有较大影响力的建筑设计咨询公司。在美国、加拿大和东南亚等国家和地区留下了一批精心设计的作品。

美国HOOP建筑设计有限公司相继于2000–2003年在深圳、上海设立分公司，大力开拓中国市场，公司凭借强大的设计实力、先进的设计理念以及国内外设计师的精诚合作，取得了骄人的成绩。在中国上海、深圳、天津、浙江、江苏、山东等地参与了多个项目的规划及建筑设计工作，并多次获得国家级奖项和荣誉称号。

美国HOOP建筑设计有限公司强调设计理念的创新，以最优的策略解决项目中的审美问题、技术问题和经济问题。公司突出团队精神和员工的天然执业责任感，将客户的信赖视为企业的生命。全新的设计理念、高水准的服务水平以及不断自我超越的决心是我们高质量完成项目的保证。

Established in 1995 in California, U.S. HOOP Architectural Design Co.,Ltd. is an architectural design company with great influence in California, the United States. Over the years, it has accomplished a number of exquisite designs in areas such as the United States, Canada and Southeast Asia.

From 2000 to 2003, U.S. HOOP Architectural Design Co.,Ltd. has successively founded its branch companies in Shenzhen and Shanghai, vigorously developing China's market. With its strong capability, advanced concepts and sincere cooperation with domestic and international designers, HOOP has achieved remarkable success: it has participated in the planning and design for many projects in Shanghai, Shenzhen, province Tianjin, Zhejiang province, Jiangsu province and Shandong province, etc., receiving several national awards and honorary titles.

Emphasizing innovation in design concepts, U.S. HOOP Architectural Design Co.,Ltd. strives to solve the aesthetic issue, technical issue and economic issue in the project with the optimal strategy.

The company highlights outstanding team spirit and employees' natural professional responsibility, regarding the customer's trust as the life of the company. New design concept, high standard service and determination of continuous self-transcendence are our guarantee of completing projects with high qualities.

公司2009–2010年项目更新

01. 保利－南京保利仙林地块项目
02. 保利－武汉保利卓刀泉项目
03. 保利－南昌保利富大有堤项目
04. 万科－上海浦东大道项目
05. 绿地－常州世纪城项目
06. 华润－重庆“二十四城”
07. 大华－南京锦绣华城项目
08. 宁波城投－宁波古林地块项目
09. 苏宁环球－宜兴城东新区项目
10. 万达－武汉万达菱角湖项目
11. 万达－武汉万达积玉桥项目
12. 万达－淮安万达项目
13. 哈尔滨群力新区项目
……

2009-2010 Project Update

01. POLY - Poly Xianlin plot project in Nanjing
02. POLY - Poly Zhuodao Spring project in Wuhan
03. POLY - Poly Fudayoudi Project in Nanchang
04. Vanke - Pudong Road project in Shanghai
05. Lvdi - Century City project in Changzhou
06. Huarun - "24 City" project in Chongqing
07. Dahua - Fairview City in Nanjing
08. Ningbo City Investment –Gulin plot project in Ningbo
09. Suning Universal – Chengdong Zone project in Yixing
10. Wanda - Wanda Lingjiao Lake project in Wuhan
11. Wanda - Wanda Jiyu Bridge in Wuhan
12. Wanda - Wanda project in Huaian
13. Harbin Qunli Zone project
……

HOOP Architects
www.hoop-archi.com

保利南京仙林地块项目 项目地点：江苏 南京 项目规模：350 000 平方米 完成时间：2010 年 发 展 商：保利地产

南通“万濠华府”

项目地点：江苏 南通
项目规模：180 000 平方米
完成时间：2010 年
发 展 商：南通万通置业

宁波“天一公馆”

项目地点：浙江 宁波
项目规模：180 000 平方米
完成时间：2010 年
发 展 商：宁波城投置业

绿地常州世纪城　　项目地点：江苏 常州　　项目规模：750 000 平方米　　完成时间：2009 年　　发 展 商：绿地集团

1

1-2 华润“二十四城”

项目地点：重庆
项目规模：610 000 平方米
完成时间：2011 年
发 展 商：华润置地

3 武汉保利卓刀泉项目

项目地点：湖北 武汉
项目规模：280 000 平方米
完成时间：2012 年
发 展 商：保利地产

2

3

1

1 绿庭松江广场

项目地点：上海
项目规模：250 000 平方米
完成时间：2012 年
发 展 商：上海绿庭置业

2 沈阳万达铁西项目

项目地点：辽宁沈阳
项目规模：180 000 平方米
完成时间：2011 年
发 展 商：万达集团

3 武汉万达积玉桥项目

项目地点：湖北 武汉
项目规模：100 000 平方米
完成时间：2011 年
发 展 商：万达集团

2

3

1–2 苏州商贸城酒店　项目地点：江苏 苏州　项目规模：75 000 平方米　完成时间：2012 年　发 展 商：苏州市相城城市建设有限责任公司

3 苏州元和塘餐饮建筑　项目地点：江苏 苏州　项目规模：150 000 平方米　完成时间：2011 年　发 展 商：苏州市相城城市建设有限责任公司

J.A.O.Design International Architects & Planners Limited

美國龍安建築規劃設計顧問有限公司

美国龙安建筑规划设计顾问有限公司1984年成立于纽约，是由美籍华人、国际著名建筑规划专家饶及人先生创建的国际性建筑规划设计服务公司，作为在中国已成功经营了十余年的外籍设计公司，其规划与建筑作品遍及31个省、市、自治区的80多个城市，及中东、加勒比海等地区，并赢得111个知名设计奖项，同时自2004年起，连续6年获得“中国十大影响力设计单位”的殊荣，凭借自身优秀的设计能力和良好的市场口碑。公司规划设计的作品达300余个，为海内外投资者的开发提供了大量成功典范。美国龙安公司凭借扎实稳健的设计实力及先进的设计理念、丰富的经济学识、突出的工作业绩，在业界树立了良好的专业品质和企业形象，将“国际设计理念”引入中国的专业单位。

美国龙安建筑规划设计顾问有限公司作为美国龙安集团旗下最核心的品牌公司，始终将集团“以城市规划设计为切入点进行城市运营，为中国百姓提供世界级的低碳绿色建筑”的发展理念为己任，积极推动中国城市化发展的进程。

美国龙安公司拥有多种文化背景的专业设计人员，为客户提供专业服务。规划设计：城市规划（新市镇规划、旧区改建规划、旅游度假区规划、市政规划、CBD设计）及各专项规划。建筑设计：甲级写字楼、购物中心、住宅楼群、博物馆、会展中心、国际学校、综合大楼、五星级酒店等。景观设计：公园、居住区环境景观设计、游乐场、广场等。室内设计：星级酒店、营业厅、各商业场所、别墅、公寓、售楼处、会所等。迄今为止，城市规划总面积2 746平方千米　（约45个纽约曼哈顿岛）；住宅区规划设计总建筑面积约为1 785万平方米（约为17万个家庭房）；公建设计总建筑面积930万平方米（约相当于58个国家大剧院）；旧城改建街道总长度11 300米（相当于6条半法国香榭丽舍大道）；景观设计总面积776公顷（约2个半纽约曼哈顿中央公园面积）；室内设计总面积13万平方米（约430个样板间）。

2009年，美国龙安成功收购了建设部甲级设计单位北京众拓建筑工程设计有限责任公司（010518—sj），成为目前国内为数不多的具有甲级资质的外籍建筑设计企业之一。双方整合优势资源形成核心竞争力，强强联合进一步巩固了美国龙安在中国建筑规划设计领域内的领先地位。

与其他国际设计公司不同，美国龙安公司作为跨国经营的事务所，中西合璧的理念结合得最好，在中国获奖最多，成为中国最具影响力的国际设计事务所。由于充分了解与熟悉中国的体制与管理，并能和中国各城市主管领导进行无障碍沟通，美国龙安公司结合自身的国际资源优势与实际经济情况，同时通过集团总裁饶及人先生的充分整合，集中调动国际资金资源，介绍与带入投资，让项目得以落地执行。

提供专业服务，为客户量身定制配套顾问服务

美国龙安建筑规划设计顾问有限公司以“精品设计”“量身定制”为座右铭，针对不同种类的项目需求，向客户提供建设开发、投资可行性研究，以及项目投资、融资等配套顾问服务。

中西文化兼容并蓄，客户意愿完美体现

美国龙安建筑规划设计顾问有限公司着重于多重文化的结合，塑造“中西合璧”的理念，成功地将先进的设计思路与商贸哲学运用于各项规划和建筑设计项目中，使客户的意愿得到完美体现。公司凭借多年实战经验与丰富的社会资源为客户提供高品位、高回报率的专项服务。

中西交流的桥梁

美国龙安建筑规划设计顾问有限公司能将国际先进的规划设计理念引入，并通过资金板块、外商资源，使这些理念成为现实。对经过规划后，一些基础较好、前景较明朗的项目，可以协助或自主对国外招商，同时还引入境外投资机构包括亚洲开发银行（ADB）、世界银行（WD）的资金或引进大型金融机构参与项目，使整个项目从设计构思到实施完成达成完美的统一。美国纽约曼哈顿区驻中国商务代表处选择于美国龙安建筑规划设计顾问有限公司挂牌设立，即成为外界对其在中国业界和社会地位的肯定范例之一。

J.A.O Design International Architects & Planners Limited was established in 1984 in New York, providing internationally celebrated architecture and planning services. It was founded by Chinese American Mr. James Jao, a famous expertise in architecture and planning. As a certified foreign enterprise that has more than a decade years' running experience in China, its planning and construction works has been throughout more than 80 cities in 31 provinces, municipalities and autonomous regions, as well as the Middle East, the Caribbean and other regions, which have won 111 known design awards. Since 2004, it has been honored with "China Top Ten impact Power Design Unit "award for 6 years by virtue of its excellent design capability and good market reputation. The company has planned and designed more than 300 works which could provide a good many of success paragons for both domestic and foreign investment. With solid strength solid design and advanced design concepts, knowledge-rich economy, outstanding job performance, J.A.O establishes a good professional quality and corporate image in the industry, as a professional unit bringng "international design" into China.

As the company's core brand branch, J.A.O Design International Architects & Planners Limited, always holds the development ideas that "to take urban planning and design as a focal point for urban operations, to provide Chinese people with world-class low-carbon green buildings" as its mission, actively promoting the process of urbanization in China.

J.A.O employs professional designers with various cultural and national background. J.A.O Design provides professional services to its clients:
(1) Programming. Urban Planning: including new town planning, Urban reconstruction planning, city planning, tourism planning, Municipal planning, CBD design, and other specific plannings.
(2) Architecture Design: including Class A office buildings, shopping centers, residential buildings, museums, exhibition & convention centers, international schools, mixed-use apartments, and five star-rated hotels.
(3) Landscape Architecture: including gardens, parks, plazas, etc.
(4) Interior Design: Star-rated hotels, business hall, commercial establishments, villas, apartments, sales offices, clubs, etc.

So far, its city planning area has covered 2,746 square kilometers (equivalent to 45 Manhattan in New York); approximately 1,785 million square meters total floor area of residential planning and design (equivalent to 17 million households in housing); a total construction area of 9.3 million square meters in public designed building (equivalent to 58 National Grand Theatre); total length of the street of 11,300 meters in Urban reconstruction planning (equivalent to six and a half in France Champs Elysees); landscape design covering a total area of 776 hectares (about two and a half Manhattan's Central Park area in New York); interior design covering a total area of 130,000 square meters (about 430 showroom).

In 2009, J.A.O successively took acquisition of a Class A design unit, the Beijing Zhongtuo Construction Engineering Design Limited Liability Company (010518-sj), become one of the few foreign architectural design firms with Class A qualification. After integration of advantageous resources, the two parties formed a core competitiveness and powerful alliances to further strengthen the leadership status of J.A.O in Chinese architectural planning and design industry.

Other than international design firms, J.A.O as a multinational company sets up the best combination of Chinese and Western philosophy, winning the most awards in China, and becomes the most influential international design firm in China. Fully understanding and being familiar with China's system and management, along with the obstacle-free communication with competent leadership in all Chinese cities respectively, J.A.O combines international resources of it itself and the actual economic situation, relying on the fully integration under Group CEO, Mr. James Jao, and intensively mobilizes international financial resources to introduce and bring investment to enable the project to implement.

Providing customers with professional services with customized supporting consulting services

J.A.O takes "fine design" and "customized" as the motto. Based on projects demand of different types, it provides customers with supporting advisory services involving building development, investment feasibility studies, project investment and financing.

Inclusive Chinese and western culture, perfect embodiment of customers' wills

J.A.O focuses on the combination of amphibious culture, and shapes the concept of "when East meets West". It has successfully applied advanced design ideas and business philosophy to the planning and architectural design projects, so that customers' wills are perfectly embodied. By many years' practical experience and abundant social resources, it provides customers with high-grade and high rate of return special services.

Bridge for Chinese and Western communication

J.A.O can bring in the international advanced planning design concepts and make these ideas a reality through the funding plate and foreign resources. For those projects which have good base and bright perspective after planning, the company can help or independently attract foreign investment, and introduce foreign investment institutions including the funds of the Asian Development Bank (ADB), World Bank (WD). It can introduce large financial institutions involved in the project, so that the whole project achieves the perfect unity of completion from design concept to implementation. Business Representative Office of Manhattan, New York in China chose to be set up in J.A.O., which is one of the affirmation examples for its social status in the Chinese industry from public opinions.

中国总部（北京）

地址：北京市朝阳区霄云路26号鹏润大厦B座2601—2602室
邮编：100016
电话：+86—10—8458—4666
传真：+86—10—8458—4950
网址：www.jaodesign.com
邮箱：jaobj@jaodesign.com

China Headquarters (Beijing)

Address: Room 2601, of Eagle Plaza Building B, No.26 of Xiaoyun Rd. Chaoyang District, Beijing
P.C.:100016
Tel: +86-10-8458-4666
Fax: +86-10-8458-4950
Http//: www.jaodesign.com
E-mail: jaobj@jaodesign.com

密云开发区综合体项目
项目地点：北京
建设规模：160 000 平方米
服务范围：建筑概念设计
项目备注：业主直接委托

Complex Project in Miyun Development Zone, Beijing
Project Location: Beijing
Construction Scale: 160,000 m^2
Service Scope: Conceptual design
Remark: Commissioned design

该项目位于密云总部研发商务中心区内，总用地面积 15 公顷，总建筑面积近 18 万平方米，建筑总容积率 1.2，总绿化率达 42%。项目建成后主要为中心区提供商务性的服务功能，包括五星级商务酒店和会议中心、酒店式服务公寓及高端商务休闲型商业。整体风格定位为异域风情小镇，在品质上体现国际化、生态化、服务高端化的特色。

盐城市城南新区第一沟欧陆风情商业一条街
项目地点：江苏 盐城
建设规模：6.5 公顷
服务范围：修建性详细规划、建筑方案设计、景观方案设计、施工图设计
项目备注：业主直接委托　已建成

Yancheng food business street of Venice style in Jiangsu
Project Location: Yancheng, Jiangsu
Construction Scale: 6.5ha
Service Scope: Comprehensive planning, architectural, landscape design, construction drawing
Remark: Commissioned design/built

受盐城市政府委托，美国龙安建筑规划设计顾问有限公司对江西省盐城第一沟进行详细规划和建筑设计。本项目位于盐城城南新区，总规划面积 6.50 公顷，商业街全长约 700 米，宽大约 60 米。

项目背景

盐城市位于江苏省中部偏东地区，地势平坦，河渠纵横，交通发达。为了发挥盐城地理和经济优势，促进消费经济联动，增强城市活力，市政府决定对第一沟进行详细规划和建筑设计。

规划设计

美国龙安对此项目进行了 SWOT 分析，即通过对项目优势（Strengths）、劣势（Weaknesses）、机遇（Opportunities）和挑战（Threats）的分析，将第一沟规划定位为威尼斯风情餐饮商业一条街。规划设计理念为：1. 现代城市空间的营造；2. 人文生态城市环境的塑造；3. 品牌城市景观的打造。规划布局为“一轴、两心、四区”。建筑群将包括入口开放式广场、美食廊桥、健身休闲中心、数字影院及渔人码头。

建筑设计

建筑总体风格借鉴了欧陆风格，根据现代功能需要和当地气候及地形，审慎地采用特定的建筑设计手法，在营造威尼斯风情的设计中力图创新。建筑柱网为常规模数，开间柱距为 6 米、7 米等，门窗、檐口、柱头等取材于威尼斯建筑风格，但予以适度提炼、简化，使之成为具有现代功能的新建筑由此焕发新的生命。在本次设计中刻意设计并希望充分展现盐城沿河景观，通过柱廊及露台体现西式餐饮崇尚自然的情趣，探索并创造了具有现代意义的商业建筑空间。

通过点（跃马路西侧，入口广场，滨河路西侧渔人码头），线（沿河景观及建筑群）的结合，在有限的空间里创造异国风情的建筑环境及景观环境，给城市的休闲、餐饮、文化生活增添了色彩。第一沟威尼斯风情商业一条街将成为盐城市民休闲和外地游客体验异国风情的最好选择，也将成为盐城的旅游观光的品牌。

盐城先锋岛景观设计
项目地点：江苏 盐城
建设规模：10 公顷
服务范围：景观概念设计
项目备注：国际竞标中标

Yancheng Pioneer Island Landscape Design
Project Location: Yancheng, Jiangsu
Construction Scale: 10 ha
Service Scope: Landscape Concept Planning
Remark: Int'l competition/winning design

五色先锋，让生活更精彩

先锋岛位于盐城市区西部，由串场河、蟒蛇河、越河交汇成岛。本次景观设计总面积约 10 公顷。先锋岛自然环境优美，滨水景观资源丰富。岛上有省级文物保护单位——泰山庙，是新四军的重建军部旧址，历史人文资源底蕴深厚。该岛是盐城商业发源地，是盐城生态文化、红色文化、宗教文化及滨水文化的重要积淀区。

目标定位：现代生活及商业时尚的先锋

设计通过精雕细琢的细节，文化氛围浓郁的传统旅游区，生态宜人的滨水空间，活泼特色的商业空间等，将商业文明、生活文化与自然景观完美融合，处处充满创新和惊喜，为市民及旅游者创造出一个舒适迷人、富有亲和力的开放式生活及商业的城市。以唯美的园林景观空间作为背景迎接到访者，为城市开创一个融入自然景观的全新开放式空间，使得项目区真正成为引领新城市生活及商业时尚的先锋。

设计主题：五色先锋，让生活更精彩

在结合先锋岛先天资源条件及项目开发思路和先锋岛规划“文化、商业、娱乐、旅游、景观”五大功能定位并秉承“自然的、生态的、健康的、休闲的、文化的、商业的”思路基础之上，打造独特五色先锋城市，引领新城市生活时尚。

设计注重梳理城市文化脉络，统筹城市规划格局，寻找并确定城市文、理交汇点，形成先锋岛五色文化带。竖向分析与横向比较相结合，明确先锋岛功能结构。宏观把握微观分析，明确先锋岛亮点表现区。点、线、面结合，完善先锋岛整体城市特色。打造五色先锋城市，并以其为核心，个性扩散，引导苏、浙等滨水城市带，引领新城市生活时尚。

空间架构：

在对先锋岛上位规划及人文历史资源的解读分析基础之上，本次设计形成 158 空间架构，以此为骨架，指导细部景观设计工作。

1—— 一心：先锋广场景观核心。
5—— 五色：黄色国学文化、红色革命文化、橙色商业文化、绿色生态文化及蓝色滨水文化五色先锋。
8—— 八组团：内街道路、主干道路、商业内街、滨水休闲、传统文化、公寓、酒店、跨河大桥八大功能板块。

中国移动通信集团江西有限公司红角洲生产基地项目
项目地点：江西 南昌
建设规模：90 000 平方米
服务范围：建筑方案设计
项目备注：业主直接委托

China Mobile Production Base in Hongjiaozhou,Jiangxi Province
Project Location: Nanchang, Jiangxi
Construction Scale: 90,000 m^2
Service Scope: Architectural Design
Remark: Direct commission

本项目占地 6.67 公顷，总建筑面积 9 万平方米，其中地上 6 万平方米，地下 3 万平方米，项目周边交通方便，北面 100 米宽的学府大道可直达市区并直通高速路。

通过对中国移动 “正德厚生，臻于至善”的企业精神的解读，挖掘其“责任”和“卓越”的企业文化之核心内涵，本着绿色建筑及专业、实用的原则，项目将结合现代办公的理念进行设计。

一个江西的“j”（jiangxi）带出阡陌的良田：环保绿色建筑＋客家水塘＝创造一个属于江西的中国移动大楼。

建筑的主要特点：形如巨型桥梁般的建筑群联系紧密，功能集中。错落有致的建筑单体不仅利于自然通风，而且更提供了开阔的景观，看似复杂的单体叠加，来源于简单标准的框架结构。庭院的总体灵感来源于江西水田，它代表着生命与丰收，水田造型的花园设计与整齐不一的小路具有工业感和科技感。建筑中大小不一的庭院，令办公环境与周围的自然景观和谐统一。在建筑中如水源热泵、地源热泵、地面通风、太阳能泛光照明、雨水回收、屋顶绿化等环保、低能耗措施的运用将使建筑的运行费用大大降低。

宁波 BOBO 城
项目地点：浙江 宁波
建设规模：23 公顷
总建筑面积：450 000 平方米
服务范围：规划设计、方案设计、初步设计
项目备注：国际竞标第一名、已建成

Ningbo BoBo Residential Community
Project Location: Ningbo, Zhejiang Province
Construction Scale: 23 ha / 450,000 m^2
Service Scope: Residential Design, Scheme, Architectural,Design, Preliminary Design
Remark: 1st rank in International Competition / Completed

2003 年 7 月获得建设部 2003 年 “社区规划示范楼盘” 奖
2005 年 10 月获得第二届精瑞住宅科学技术奖居住建筑文化艺术优秀奖
2006 年 9 月 21 日荣获 “第 14 届中华建筑金石奖” 优良规划设计类国际组金石奖

“Award for Best Community Planning” issued by the Ministry of Construction in July 2003
The Ningbo BoBo Residential Community has won the Award for Excellence in Residential Design as part of the second Jingrui Residence Sci-Tech Awards in October 2005
On Sept 21, 2006, Ningbo BoBo Residential Community won the Bronze and Stone of Excellent Planning of International Team in the 14th China Architecture Inscriptions Award Competition.

宁波 BOBO 城项目位于宁波市江东区东北部、甬江南岸。项目由宁波市交通房地产有限公司开发，将建设成大规模的具有相应配套设施的居住社区。

设计师在充分领会 BOBO 族这一社会新阶层内涵的前提下，确定 BOBO 城的设计宗旨为严谨的规划结构与富有创意的空间交相辉映，前者象征了布尔乔亚对成功孜孜以求的精神，而后者则显示了追求生活品质的波希米亚那永不枯竭的创造之源。

BOBO 城的设计体现了新经济时代的社会精英阶层对成功和创意生活的双重追求，代表着崭新的价值理念。多功能、多层次、多形象、多种户型、多元景观的社区设计也表明了 BOBO 是一种精神追求而非僵化的程式语言。我们共同期待着 BOBO 城在历史悠久的宁波城的城市演进中，成为崛起的新兴阶层——BOBO 族的精神家园。

BOBO 城的设计是 “造城”，即：
建设一个社区，一个现代气息浓郁、智能化、配套完善的生态型社区，追求社会、经济、生态最佳效益；
营造一种生活，一种崇尚成功与创意的崭新理念，创造高品质的生活、交流、休闲空间；
塑造一种环境，既具现代特色又体现文脉、既丰富多彩又整体协调的景观环境；
树立一个样板，一个在 21 世纪社会、经济、科技背景下，体现现代生活形态的综合社区。

中国铁路馆位于上海浦西 D 片区，D11 街坊内，展馆距离轨道 8 号线西藏南路站约 580 米，距浦西世博轴约 100 米，D11 片区街坊由局门路、龙华东路以及 D 片区东侧边界围合而成。中国铁路企业馆位于街坊的西北角，高架平台北侧。

本展馆总用地面积 4 000 平方米，展馆总建筑面积 5 808.7 平方米，总高度 22 米，内部为一至三层，不设地下室。展馆为可拆除并可二次重建的建筑物，按建设方要求展会后全部建筑物要异地重建（仍在上海市区），故便于建设、方便拆除、大量材料可重复使用是本项目的重要特征。

中国铁路馆将围绕上海世博会 “城市，让生活更加美好” 和中国铁路馆 “和谐铁路的主题，创造美好生活新时空”。建筑的寓意是：有城市的地方，就有铁路。有了铁路的连接，城市才更有活力，而城市有了活力，人们的生活才会更美好。

建筑外观：不刻意追求风格，朴素、自然、精致，用朴素自然的材料与形式、恰当的比例关系、精致到位的细部处理，营造舒适气氛。

本工程为一类多层建筑，耐火等级一级，设计合理使用年限为 50 年，抗震烈度为七度，结构形式为钢框架结构。

2010 年上海世博会中国铁路馆
工程地点：上海
建设规模：5 800 平方米
服务范围：概念设计到初步设计
项目备注：已建成

The ministry of Railways Pavilion Design Shanghai Expo
Project Location: Shanghai
Construction Scale: 5,800 m^2
Service Scope: Conceptual, Design / Architectural, Design / Development, Design
Remark: Built

唐山南湖生态城总体概念规划
项目地点：河北 唐山
建设规模：10 500 公顷
服务范围：概念性总体规划设计及起步区城市设计
项目备注：国际方案征集获奖实施方案

Master conceptual planning of Tangshan Nanhu Ecological City
Project Location: Tangshan, Hebei
Construction Scale: 10,500 ha
Service Scope: Conceptual Master Plan and First Stage Urban Plan
Remark: Winning implementary scheme in the international design competition

唐山南湖生态城位于唐山市城市中心区南部，东至陡河，西至新京山铁路，南至唐津高速公路，北至南新道，总规划面积约105 平方千米（其中南湖生态城风景区 28 平方千米）。规划地块分属路南区、路北区、开平区和丰南区四个行政区域。

唐山市正面临着从工业城市向生态宜居的综合性城市转型期，本次规划旨在将南湖生态城建设成为老城区改造的亮点、资源型城市转型的典范、生态重建的旗帜，生态型示范新城，以生态修复、历史文化遗产挖掘、景观绿化、湖面拓展为契机，更新城市结构，着力打造休闲度假胜地、文化创意园区、国家城市湿地公园，推动景观地产发展。以高标准、生态、科学的理念规划南湖生态城，使其成为实施唐山市整体发展战略的重要支撑。

本次规划采用明智成长（Smart Growth）、TOD 模式（公共交通引导型开发）、循环经济等规划理念，以最大化利用资源和能源但不破环环境的增长方式，创造出可步行的综合使用社区、以行人为尺度标准的空间、有当地文化历史特色的场所，以公共交通为导向，引导城市开发，让便利、休闲的步行道系统与快速公共交通相结合，成为人们出行的优选方式，减少对私人汽车的依赖，避免交通堵塞、噪声及空气污染，营造静谧、休闲、幽雅、舒适的城市氛围。采用低开采、高利用、低排放的“资源—产品—再生资源”的反馈式流程替代传统的高开采、低利用、高排放的“资源—产品—污染排放”模式，使所有的物质和能源在这个不断进行的经济循环中得到合理和持久的利用，并把经济活动对自然环境的影响降低到尽可能小的程度。

本次规划还围绕生态策略、产业策略、城市形态策略、人文策略、交通策略，打造独特的南湖生态城。采用生态系统维护与恢复、空间管制分区与生态安全保障、生态资源保护和合理利用、宜居城市生态系统建设等方法，优化生态系统的保护和利用方式，充分利用现有生物多样性，发挥自然资源和潜力；利用绿色建筑与清洁能源，实现清洁生产，以减少废物源。根据南湖生态城建设发展面临的机遇和唐山未来经济发展所提出的要求，将南湖生态城打造成为京津冀地区服务产业带的重要组成部分、京津冀地区重要的蔬菜供应基地、循环经济理论支持下的生态工业园区、临港产业发展腹地、高新技术企业走出去参与竞争的服务平台、环京津休闲旅游产业带的重要组成部分、湿地生态景观集聚区等。以现状南湖水资源、绿地、湿地系统及规划绿地系统为依托，形成“蓝色雌凤凰”“绿色雄凤凰”，将自然生态与人工建筑紧密结合，设计和谐的健康型生态居住环境。通过文化活动空间的合理布局，在南湖生态城形成集中展示唐山面貌和人文精神的汇聚之地；通过南湖的生态空间产业化经营，完成从后院到前厅的转化。遵循生态性、多样性、人性化的交通原则，形成生态城环路和方格网相结合的布局形式，将道路系统分为高速路、主干路、次干路、支路四级；建设清洁能源公交与慢行体系相结合的绿色交通系统，实现人车分离、机非分离、动静分离，绿色出行成为市民首选，区内绿色出行比例达到 90%，营造环境宜人的交通空间的同时提高公共交通的可达性。

本次规划力争准确把握南湖地区发展的脉搏，全面分析资源、环境、生态等关乎生态城规划建设的关键要素，理性评估确定性与不确定性因素，科学、严谨、创新的提出富有针对性的方案，同时注意规划的前瞻性、指导性和实际操作性。我们相信，唐山南湖生态城在科学合理的规划指导下，定会创造出美好的明天。

首创武清高村驿站服务区概念规划设计
项目地点：天津 武清
建设规模：3 040 000 平方米
服务范围：概念规划设计
项目备注：业主直接委托

Initiative conceptual planning and design of the service area in Wuqing Gaocun Toll Station
Project Location: Wuqing, Tianjin
Construction Scale: 3,040,000 m^2
Service Scope: Conceptual plan
Remark: Commissioned design

昆明草海片区控制性详细规划设计

项目地点：云南 昆明
建设规模：873 公顷
服务范围：控制性详细规划设计
项目备注：业主直接委托

Regulatory Detailed Planning for Kunming Caohai Area

Project Location: Kunming, Yunnan
Construction Scale: 873 ha
Service Scope: Regulatory detailed planning
Remark: Commissioned design

项目位于滇池北滨的草海，南临茫茫滇池水，北望葱茏五华山，是滇池文化内涵的集中体现区。

规划以草海的开发建设为契机，利用草海现有的生态湿地、纵横交错的河流和独特的空间肌理及历史文脉，将自然山水引入城市公共生活，再现传统城市形态和历史性公共空间，为城市文化生活提供具有草海地区特色的物质载体，塑造"半城山水半城街"的城市形象。

作为连接东南亚的重要门户，昆明将越来越凸显其区域中心的地位，草海作为昆明城市花冠上一颗璀璨的明珠，必将成为立足中国西南部、面向东南亚和南亚开放的国际"游憩港"，成为昆明市重要的生态屏障和市民休闲游憩中心。

昆明五华区王家桥片区城市设计

项目地点：云南 昆明
建设规模：11.57 平方千米
服务范围：城市设计
项目备注：业主直接委托

Urban Planing for Wangjiaqiao Area, Wuhua Distrit in Kunming City, Yunnan Province

Project Location: Kunming, Yunnan
Construction Scale: 11.57 square kilometers
Service Scope: Urban design
Remark: Commissioned design

昆明市五华区王家桥，乃是藏风得水、生化万物的地方，是昆明西北地区重要的商贸中心和辐射西北区域的带动型城市片区，规划以"生态廊道与绿色网络"的生态格局为依托，形成以绿色生态为主要特征、现代服务业为主导方向的城市综合片区。

城市总体空间形态设计，致力于将王家桥的山水城市个性与走向现代化及国际化的共性融合，架构多重城市空间体系，形成主题鲜明、层次丰富、错落有序、城市空间形态，打造王家桥"山水辉映，城景交融；创新服务，活力之城"的城市品牌形象。

引入城市设计的科学理念，体现可操作的原则，对重点片区、核心地块进行详细设计，并有针对性的提出城市设计导则，指引城市开发建设。核心起步区是重点街区，设计考虑以轨道站为中心，对七个地块，包括商务办公、商业零售、酒店和公寓等配套设施的位置、体量、地下通道、出入口和交通流线等作出了详细设计。精心组织的地上和地下、人行和车行、下沉与地面、商业和办公等空间，对提升王家桥城市品质和土地价值起到了重要作用。

昆明 CBD 东风广场
项目地点：云南 昆明
建设规模：99 990 平方米
服务范围：前期策划、修建性详细规划
项目备注：业主直接委托

Kunming CBD Dongfeng Square
Project Location: Kunming, Yunnan
Construction Scale: 99,990 m^2
Service Scope: Earlier stage planning,regulatory planning
Remark: Commissioned design

昆明四季如春，气候宜人，有“春城”“花都”的美誉，并有丰富的自然资源和悠久的历史文化。昆明是拥有 600 万人口，6 万亿美元 GDP 的中国——东盟自由贸易区、全球第三大市场、未来中国吞吐量第二大的机场。陆路交通从“大后方”摇身变成“最前沿”，以及湄公河国际旅游航运的开通，所有这些迹象表明，昆明未来将成为中国西南地区国际大都市的资源和潜力。

因此我们认为昆明目前不是处在发展的转折点，而是处在一个机遇撬动发展的时期，需要各项建设来带动的时期，因此，项目设计的起点要高、眼光够远、配套要全、格局要大。在昆明 CBD 的这个核心地块，绝对需要一栋说明昆明未来发展的高层建筑，它对于一座城市来说是一种吸纳，是一种精神，是一张名片，结合上述分析，我们提出本项目方案构思的要点：

1. 超高层标志性建筑

唯一性、标志性、易识别性；

七边形的超高层塔楼在世界已知的超高层建筑中具有唯一性；

唯有彩云之南之“七彩”云南才能适配；

其七边对应自然浅淡色调的“七彩”；

技术、经济上的便利性、可行性；

超高层建筑受结构技术的制约，建筑形式受约束较多，应当从结构的内在逻辑出发设计建筑，本设计正是体现这一点。

对历史文化的传承，建筑从塔的意向呈现，体现登高以望远的诗般意境；

东方传统文化中供远眺的高塔形象深入人心，实际案例有上海金茂、台北 101 等，这方面未来设计可能会有探讨。

2. 大规模地下空间的开发

结合地铁系统

（1）建议与地铁商业空间整体开发本项目地下空间与地铁协同设计、整体开发。由此可以减少工程上的额外花费，便于组织地铁人流与地下商业空间人流的对接。

（2）处理好地铁商业空间与广场地下商业空间的衔接界面。本设计以室外或室内下沉式空间去衔接不同高度的地铁与广场地下商业空间。

以二层挑空的地下商业街组织地下二层高商业空间，而在地下一层有中心下沉水面回廊与室外联通，将沿河边的城市道路埋入地下，便利地下商业物流，避免地上人流与地下车流的交叉。

3. 保留、传承与创新的广场文化

历史积淀，文化宫富含几十年昆明人对城市的记忆，文化宫或将让位于新地标。

明城的卫星影像显示该城区高度密集，且缺乏规模稍大的开放公众空间。

设计上保留完整的休闲广场，营造长长的亲水平台，此举将带动崭新的广场文化。

重庆龙湖花园
项目地点：四川 重庆
建设规模：230 000 平方米
服务范围：方案、施工图
项目备注：获建设部优秀设计二等奖

Chongqing Longfor Garden
Project Location: Chongqing, Sichuan
Construction Scale: 230,000 m^2
Service Scope: Program/Construction plan
Remark: Second Prize of Excellent Design by the Ministry of Construction

大连希望大厦
项目地点：辽宁 大连
建设规模：90 000 平方米
服务范围：施工图

Dalian Hope Mansion
Project Location: Dalian, Liaoning
Construction Scale: 90,000 m^2
Design Time: 1997.5
Service Scope: Construction Plan

北京大西洋新城
项目地点：北京
建设规模：300 000 平方米
服务范围：施工图

Beijing New Atlantic City
Project Location: Beijing
Construction Scale: 300,000 m^2
Service Scope: Construction plan

北京昌平区回龙观龙泽苑东区
项目地点：北京
建设规模：230 000 平方米
服务范围：施工图

Eastern Zone of Long Zeyuan, Huilongguan, Changping District, Beijing
Project Location: Beijing
Construction Scale: 230,000 m^2
Service Scope: Construction plan

LSK国际建筑师事务所（L.S.K Design Group）

LSK国际建筑师事务所(L.S.K DesignGroup)是由加拿大LSKGROUP和中国建筑师共同创立的职业化的建筑设计事务所，在加拿大多伦多及中国上海分别注册。公司从事专业的设计服务，包括建筑咨询、规划设计、建筑设计、环境设计等全程服务。

LSK机构首席建筑师及其合伙人和多位国际、国内一流设计专业院校毕业生组成的精英设计团队充满活力，极富创造力，已完成了多项优秀设计作品，业绩斐然。LSK的成功源自不懈的创新追求，深刻的细节考虑和风格纯粹的设计理念。LSK能为客户提供从项目咨询、规划设计、建筑设计到景观设计的全方位服务。LSK的作品往往以独特的风格、创意和理念来表现设计的特色，并力求达到细节上的尽善尽美。LSK致力于精益求精的工作方法、现代科学的价值管理及组织流畅的专业团队来完成每一项设计作品。

人本现代功能主义是L.S.K在建筑设计上的孜孜追求。

“人本”：建筑设计依据使用对象的不同，时代背景的不同，将丰富的空间融入建筑设计，使建筑充满了时代信息并引领市场文化。

“现代”：以现代的、明快的手法与技术创作具有时代感的建筑。

“功能”：代表主流建筑设计思想，强调建筑立面形态的同时，忠实地反映其内在功能。

城市规划与社区设计——新城市主义+历史文脉+生态

小到居住社区，大到城市的规划设计都注重这三大理论的结合，并借助于公司多年丰富的实践经验，不断创造出独具特色、充满人性化的作品。

LSK Design Group is a professional architectural design company,co-founded By Canada LSK Group and Chinese architects. It's registered In Toronto, Canada and Shanghai, China. We engaged in professional design services, Including construction consultant, urban planning, architectural design and landscape design.

The Vice President of LSK Design Group, his partners and Chinese architects composed the design team. We are young & vibrant, very creative. We have completed a number of excellent design work and accessed a number of awards. The success of LSK from the relentless pursuit of innovation, Consider the details of the profound, purely aesthetic style design. We Provide customers with construction advice, urban planning, architectural design, landscape design. Our Design with unique style, creativity and ideas to the performance of characteristics, details to achieve perfection. We committed to improving the working methods of modern science and the value of the management and organization of the professional team fluent to establish its unique design concept.

Human Modern functionalism: Diligently in the pursuit of architectural design on the guiding ideology.

Urban Planning and Community Design: New Urbanism + Historical context + Ecological

Residential communities and cities, followed the three major planning and design theories, and depended on many years' practical practical experience and create a unique work is full of humanity.

L.S.K ARCHITECTS INTERNATIONAL CO.(SHANGHAI)

地址：上海市蒙自路169号智造局PARK² PACE
1号楼303、304、403、404室
电话：+86-21-53028213 53028214
传真：+86-21-53028214-12
邮编：200023
邮箱：lsk@lskgroup.com.cn
网址：www.lskgroup.com.cn

智造局 PARK²PACE　CONSTRUCTED
厂房改建项目

建设地点：上海
建筑面积：25 000平方米
占地面积：10 250平方米
竣工时间：2008年

浦江公馆 CONSTRUCTED

建设地点：上海
建筑面积：60 000平方米
占地面积：15 000平方米
竣工时间：2009年

■浦江公馆
■108创意广场

108创意广场 CONSTRUCTED

建设地点：上海
建筑面积：100 250平方米
占地面积：24 100平方米
竣工时间：2010年

L.S.K
ARCHITECTS INTERNATIONAL
CO.(SHANGHAI)

鞍山 · 爱家华府 CONSTRUCTING

建设地点：辽宁 鞍山
建筑面积：310 000平方米
占地面积：100 300平方米
设计时间：2010年

湖州 · 爱家上海花园洋房 CONSTRUCTING

建设地点：浙江 湖州
建筑面积：152 000平方米
占地面积：117 000平方米
设计时间：2009年

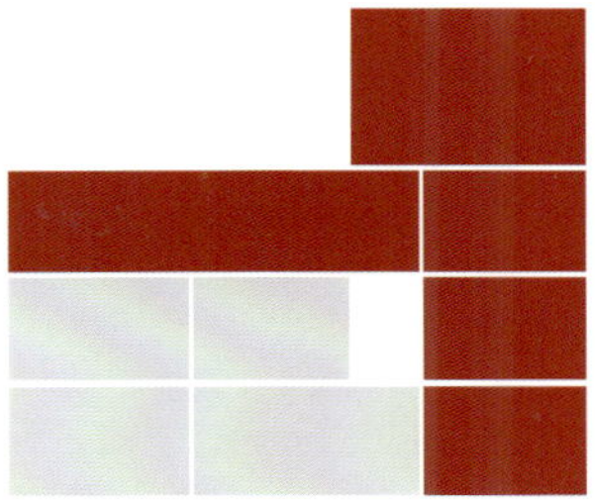

■鞍山 · 爱家华府
□湖州 · 爱家上海花园洋房

L.S.K
ARCHITECTS INTERNATIONAL
CO.(SHANGHAI)

东方圣杰罗 CONSTRUCTING

建设地点：江苏 南通
建筑面积：238 000平方米
占地面积：82 000平方米
设计时间：2010年

东方圣杰罗
德诚嘉元广场

德诚嘉元广场 CONSTRUCTING

建设地点：江苏 苏州
建筑面积：76 000平方米
占地面积：16 800平方米
竣工时间：2009年

南立面建筑渲染图

L.S.K
ARCHITECTS INTERNATIONAL
CO.(SHANGHAI)

嘉利华府商业街 CONSTRUCTING

建设地点：江苏 无锡
建筑面积：60 000平方米
占地面积：98 000平方米
设计时间：2009年

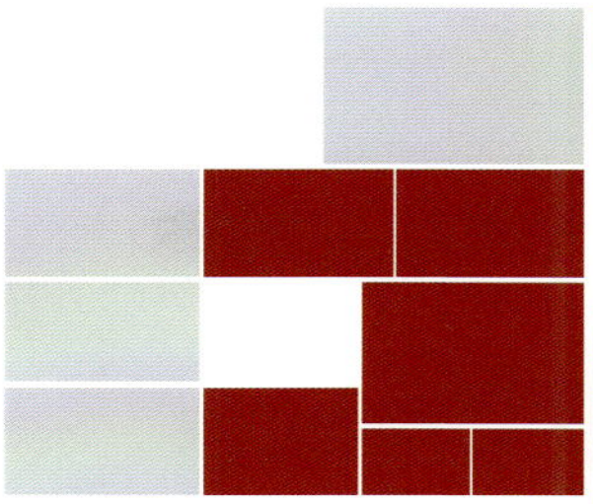

嘉利华府商业街
嘉利华府二期

嘉利华府二期 CONSTRUCTING

建设地点：江苏 无锡
建筑面积：100 000平方米
占地面积：154 000平方米
设计时间：2010年

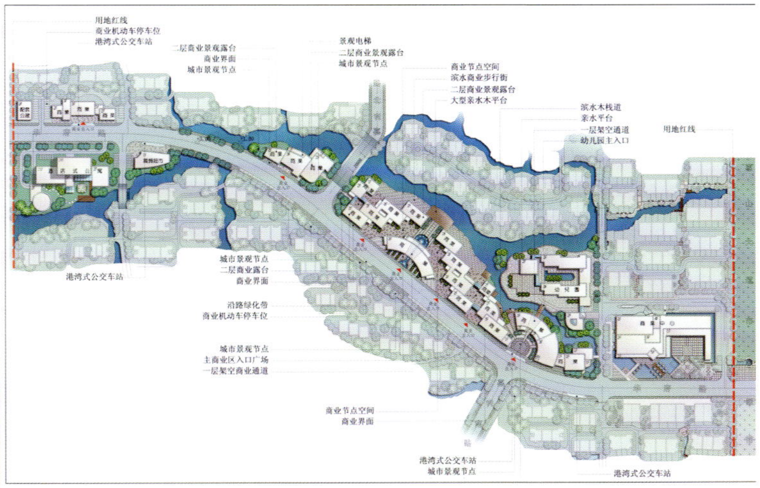

L.S.K
ARCHITECTS INTERNATIONAL
CO.(SHANGHAI)

龙王桥8号 CONSTRUCTING

建设地点：江苏 南通
建筑面积：223 000平方米
占地面积：86 000平方米
设计时间：2010年

龙王桥8号
嘉华新天地

嘉华新天地 CONSTRUCTING

建设地点：天津
建筑面积：120 000平方米
占地面积：120 000平方米
设计时间：2009年

L.S.K
ARCHITECTS INTERNATIONAL
CO.(SHANGHAI)

美国波士顿国际设计集团 Boston International Design Group

公司简介 / Company Introduction

公司总部位于美国麻省剑桥市，毗邻着著名的哈佛大学和麻省理工学院。公司的主要创始人及主要的设计人员均来自两所名校，因此波士顿国际设计有着浓郁的学院气氛。

波士顿国际设计集团包括
波士顿国际设计（美国麻省剑桥市）
美国先进建筑研究中心（美国印地安那州、拉费耶特）
波士顿国际设计集团（上海）公司（上海浦东）
波士顿国际设计集团（北京）办事处

波士顿国际设计一直致力于把项目的规划设计与业主的要求最大限度的保持一致。主要目的是和业主、合作者分享高标准的专业设计和学术成果，以平等的合作关系为业主提供杰出设计，并创造最终的经济效益。我们在世界各地的项目设计均尊重当地历史和文化，以价值和兴趣为基础，通过规划和设计，为增进当地城市生态环境做出实质的贡献。通过合作，我们的业主将获得建立在他们的需求和期望上的卓越的设计，这使得我们能够与业主建立长期的合作，甚至为有的业主服务了30多年，以确保每一座新建筑能够成为一座独特的、难忘的成功作品。
波士顿国际设计集团现设波士顿、上海两个公司，并在北京设有办事机构。波士顿国际设计总共有150名设计师，其中外籍设计师8名。我们在中国的项目主要集中在北京、上海、澳门、宁波、青岛、杭州、三亚等城市。

With the headquarter located in the city of cambridge in Massachusetts State of the USA and adjacent to the famous Harvard University and Massachusetts Institute of Technology, the major founders and principal designers of the company are from these two famous universities, therefore, Boston International Design Group is with great acdemic atmosphere.

Boston International Design Group is including
Boston International Design Group (Massachusetts, USA cambridge city street)
The U.S advanced architecture research center (Fayette County, Indiana)
Boston International Design Group(Shanghai) company (Pudong Shanghai)
Boston International Design Group Beijing office

Boston International Design Group is devoted to maximally interate the project planning design and the requirment of the owers during the cooperation. With the purpose of sharing the high-standard professional design and academic achievement with the owners and cooperated partners, we focus on creating the outstanding and remarkable design works for our lients, bring the ultimate econimic benefit in a fair cooperative relationship. In various regions of the world, we put emphasis on the local history and culture eased on the value and interest in our design projects. Through the plannig and designing, we made factual contribution in improving the urban ecologic environment and our clients obatin excellent design works which meets their needs and expectation. What we have achieved enables us to establish the long-term cooperation with the clients, some of whom have cooperated with us for more than 30 years. Whenever we accept a commission of project design,we will make study and negotiation with our clients to ensure that each of the new buildings will be a unique and long-lasting successful work..
WIth two companies in Boston and Shanghai, Boston International Design Group also has set up Beijing Office. There are 150 deisgners working in Boston International Design Group , including 8 overseas designers. The projects they involved covere Beijing, Shanghai, Macao, Ningbo, Qingdao, Hangzhou Sanya and other cities.

我们的设计是国际化的。我们的实践主要集中在以下几个方面:
1. 规划和城市设计
2. 多功能综合体
3. 酒店和度假村的建筑
4. 校园规划和教学设施
5. 生态节能住宅研究

Being an international design company, We manily concentrate on the following aspects:
1. City Planning/Urban Design
2. Mixed-use
3. Hotel/Resort
4. Campus Planning/Education
5. Sustainable Residential Study

公司联系方式 / Contact information

美国 马萨诸塞州
剑桥市考克德大街1000号（02138）| TEL：617–821–3188 | FAX：617–821–3188
中国 上海
上海市浦东新区福山路33号（200120）| TEL：+86–21–51327266 | FAX：+86–21–51327269

Massachusetts, USA
02138 No.1000 Concord Ave.Cambridge, Massachusetts, USA | TEL: 617-821-3188 | FAX: 617-821-3188
Shanghai China
No.33 Fushan Rd, Pudong, Shanghai, China(200120)| TEL: +86-21-51327266 | FAX: +86-21-5132726

BIDG
Boston International Design Group

三亚香水湾1号

建设地点：海南 三亚
建筑面积：400 200平方米
占地面积：920 000平方米
设计时间：2007年
竣工时间：2009年

项目介绍

规划突出海南陵水的地方特色及本身良好的自然环境资源优势，同时注重整体环境塑造，形成高品位、高质量的休闲度假环境，建筑风格追求朴实大方，以期达到保护和加强具有绿色生态特征的自然乡土环境。根据规划部门提供的设计依据和条件，按照建设单位的要求，项目定位为休闲式度假酒店。

我们一贯坚持对当地历史文化的尊重，跳出中国传统建筑白墙灰瓦的习用语言，探索能和现代建筑艺术与审美意识契合的精神层面上的表达，开辟了独特的中国建筑创作之路。

三亚香水湾1号，回归商周建筑特色，运用这种传统的建筑群组织方式，通过围合空间的改变，实现平面布局，强调空间的开放性与延续性，在独院高墙的怀抱中，院门、水池庭院、游廊、客厅、卧室等渐次铺排，达到对各个空间的体量大小、虚实、敞阖及方位照应关系的最佳拿捏程度。可以同时充分考虑三亚热带海洋性气候特点，合理采用开敞式走廊门厅设计，淡化室内外界限，使内外空间相互渗透贯通。建筑为错层设计，三面围合，一面看海，大多数户型可以足不出户，躺在自家游泳池里观海，开启一个阴凉舒适的生活环境。

在居住部分的设计中，充分显示中国传统居住建筑内向性空间的本质，空间流线组织，完美阐释步移景异的园林精神，使建筑融入自然景致，很好地映衬出原生地貌的自然美，达到中国传统哲学"大隐隐于市"的境界。置身其中，你会有一份富于中国意境的独特体会，在喧闹的都市中寻得一份淡然。

BIDG
Boston International Design Group

宁波郁家巷历史街区规划与保护设计

建设地点：浙江 宁波
占地面积：43 000平方米
设计时间：2006年
竣工时间：2010年

项目介绍

郁家巷历史街区地处宁波市城市核心区，如何解决老街区的保护与新商业环境的营造之间的矛盾是本项目的核心。

我们的功能定位是把郁家巷历史街区改造成为充满活力的城市中心区，创造宁波老城区新亮点：

(1)保留原有街区的城市肌理、建筑尺度和空间结构，把老城区融入现代生活，赋予新的商业功能；
(2)拆除部分老旧建筑，营造一个城市广场和公共中心，精心打造街区环境，营造高雅、浪漫的商业氛围；
(3)调整部分院落结构，让历史街区从私人化走向公众化；
(4)开创中国历史街区的改造尝试新模式。

设计完成的郁家巷，目标就是能够对宁波市的中心区文化功能进行完善和补充，能够加强和周边的地区的历史文化区和旧城改造区的联系，对城市的可持续发展，为余下的有保留价值的街区的改造提供值得参考的技术、经验和建议，提供一种切实可操作的开发模式。

BIDG
Boston International Design Group

上海露香园旧区改造

建设地点：上海
建筑面积：36 849平方米
占地面积：23 797平方米
设计时间：2009年
竣工时间：待定

项目介绍

作为上海中心城区黄浦区老城厢旧区改造的重要项目，露香园项目东至河南南路和旧仓街，南至方浜中路，西和北至人民路，整个项目将以住宅为主，包括配套设施及商业，项目以“打造上海老城厢特色的综合性旅游商业地产”为目标。

我们在设计上保持原有老城厢的城市特色，延续传统里弄建筑的空间形式、街道尺度、街道界面特征等，特别是体现露香园地区20世纪20年代左右所形成的建筑风貌特色。

改造项目继承和发扬上海19世纪末20世纪初所形成的海派建筑特征，延续中西交融的建筑氛围，体现露香园地区所承载的文化内涵。

彻底改变老城厢落后的城市建设水平，通过增加绿地和开放空间，完善交通和市政设施，保证完善的居住套型空间，提高居住质量，提升环境品质。

疏解中心城区的人口密度，降低对中心城区交通和公共服务设施的压力。

通过发挥本项目的区位优势、历史文化内涵优势，打造具有市场竞争力的高品质的物业产品，提升黄浦区的城市品位，再现黄浦区作为城市中心区的繁华景象。

BIDG
Boston International Design Group

北京国子监SPA酒店

建设地点：北京
地上面积：25 478平方米
占地面积：16 352平方米
设计时间：2008年
项目状态：在建

设计思路

北京国子监SPA酒店位于国子监历史建筑保护区，与国子监正门几乎相对。基地内现状建筑以新建仿古四合院和废旧厂房建筑为主。

在设计中我们延续国子监历史建筑规划布局的传统，强化中心轴线，以精心设计的新中轴串联不同功能建筑群体。景观设计处处体现中国园林造园艺术的精髓，达到“步移景异”的效果。突出“水”主题，把多种形式的水景精妙地组合在一起，形成“北国水园”。主入口空间以“重门”形式，体现从动态的、喧嚣的外部都市到静态的、幽静的内部酒店的空间过渡，建筑立面改造设计注重与区域环境的协调与融合，同时充满现代化与高档次的特点，以现代中式风格设计酒店客房，延续国子监历史与文化的脉络，突出中式文化主题。

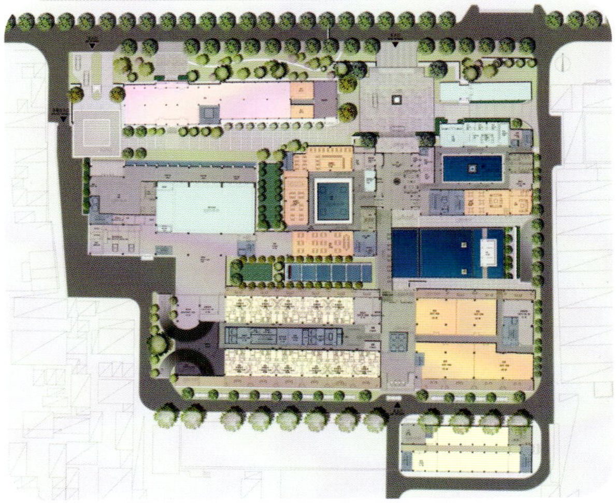

BIDG
Boston International Design Group

杭州湖滨国际名品街

建设地点：浙江 杭州
建筑面积：19 616平方米
占地面积：6 035平方米
设计时间：2009年
竣工时间：2010年

项目背景

杭州湖滨项目位于风景秀丽的西子湖畔，是市级商业中心的核心区块。设计从杭州市的商业规划与城市发展的角度出发，提出打造复合型、多元化的城市商业中心区的规划理念，达到提升湖滨区的商业品质与城市环境的目的。

在一期改造项目中，针对现状流线混乱、招商不佳的问题，我们提出了增加顶棚、改造入口、更新景观、优化立面等策略，同时结合区域商业的发展状况，提出品牌经营的规划战略，着力打造知名的高档消费品集中区域。

海南蜈支洲岛别墅项目

建 设 地 点：海南 三亚
总用地面积：4 000平方米
占 地 面 积：830平方米
设 计 时 间：2010年
竣 工 时 间：待定

项目背景

本案位于风景秀丽的蜈支洲岛，面临南海，景色独特。四周海域清澈透明，海水由浅入深，明丽动人，层次分明。对于本案，我们的设计理念是：

一、从空中俯瞰，犹如一堆树丛中的岩石，若隐若现。
二、从海上仰望，亦若生于山水间的奇石，高低错落。
三、在居室内远眺，无限海景尽收眼底，令人心旷神怡。
四、在入口往里，水院昌盛，一步一景，营造桃花源记之感。

英国UK.LA太平洋远景国际设计机构

U.K PACIFIC LONG-RANGE PLANNING & DEVELOPING DESIGN CONSULTANT LTD.

中国境内公司联系方式：
地址：南京市奥体大街128号奥体名座大厦F座10楼整层
电话：+86-25-84739678
传真：+86-25-87763798
网址：www.UKLADESIGN.com
邮件：UKLA2000@126.com

Contact Method (China):
Address: Fl 10, No. 2 Building, Aotimingzuo Mansion, 128, Aoti Street, Nanjing
Tel: +86-25-84739678
Fax: +86-25-87763798
Http: //www.UKLADESIGN.com
E-mail: UKLA2000@126.com

英国UK.LA太平洋远景国际设计机构（UK.LA PACIFIC LONG—RANGE PLANNING & DEVELOPPING DESIGN CONSULT LTD.）是一家英国专业设计公司，旗下设计机构遍布英国、中国香港、南京、上海和郑州。公司致力于城市与建筑的功能规划和空间设计，业务覆盖城市规划、建筑设计、景观设计、建设工程咨询等诸多领域。在设计的各个阶段，公司秉承“专业”“创新”的宗旨，不断在设计作品上精益求精。近年来在实践中所展示的创造能力、先锋的设计理念和不懈的探索精神得到了公众和学术界的广泛认可，是江苏规模最大的境外设计机构之一。

UK.LA的核心优势来源于其独特的、多专业、多文化和国际型的主设计师群体和设计团队。公司云集了一批不同年龄和不同背景经历的设计师，他们带来了创新的设计概念和手法、先进的设计管理经验和各种文化的精华。多种文化的融合和碰撞，形成了UK.LA独树一帜的创作风格，凭借其对自然、文化和经济环境的高度责任感和洞察力，UK.LA形成了“适用、和谐、创新”的原则，在设计理念和技术手段上不断追求更高的境界。

UK.LA的核心优势还来源于对卓越设计的不懈追求和坚韧探索，无论是在南京总建筑面积达2 300 000平方米的华欧国际友好城的规划中，还是在郑东新区意大利格拉姆集团的超高层双塔格拉姆国际中心的建筑设计中，公司都充分分析设计任务中每个要素的重要性和整体性，其根本宗旨是实现设计最优化，将城市设计、建筑设计、景观设计、生态设计交汇融合，自始至终将创造性、和谐性和超越性完美统一，最终提供给客户一个完整的卓越的设计方案。

世纪之交的中国，社会、经济、城市与建筑等各个方面均发生着巨变，资本、技术和思想领域都进行着前所未有的交流，城市空间和建筑作品必须适应这一潮流。在走向全球化的同时，必须发扬自身文化的特点，发掘本土文化的魅力。UK.LA的团队在结合世界先进理念的同时融入中国本土的建筑元素和语汇，UK.LA公司一直重视与开发商和政府建立和发展伙伴关系，准确把握市场脉搏，在业务不断扩展的过程中以创新的设计和优质的服务，赢得了众多客户的信赖与支持。

主要获奖情况

2009中国最具影响力境外设计机构
2009国际建筑设计创意企业金奖
2007亚洲建筑规划设计奥斯卡国际风尚大奖
2007中国建筑规划设计诚信百强品牌机构
2006江苏十强诚信品牌设计机构

UK.LA PACIFIC LONG-RANGE PLANNING & DEVELOPPING DESIGN CONSULT LTD. is a Britain-based professional design company with its subordinates spreading over UK, Hong Kong of China, Nanjing, Shanghai and Zhengzhou. The company is specialized in functional planning and spatial design of cities and buildings, and with its business covering urban planning, architectural design, landscape design, construction engineering consultancy, etc. In terms of design, the company sticks to the “expertise and innovation” philosophy, and keeps improving designs. In recent practices, the company has received broad recognition from the public and the academic circle for its creative capacity, pioneering design ideas and unremitting efforts in exploration. And its affiliate in Jiangsu is one of the largest overseas design institutions.

The core advantage of UK.LA originated in its multi-professional, multi-cultural and international chief-designer group and designer team. The company has a team of designers with different ages and design experiences, and they bring in innovative design ideas and tacts, experiences of advanced design management and diversified cultures. Integration of multiple cultures makes the unique creation style of UK.LA with high responsibility and far-reaching insight to the nature, cultures and economic environment.UK.LA has established its “Practical, harmonious and innovative” principle of creation, and the company has been pursuing greater achievements in design ideas and technical measures.

The core advantage of UK.LA also originated in unremitting efforts and diligent exploration. The company makes sufficient analysis on the importance and integrity of each factor of design, during both planning of Huaou International Friendship Town (located in Nanjing, covering a total building area of 2,300,000 m^2) and the architectural design for high-rise twin-tower Clam International Center of Clam Group of Italia. The company pursues optimal design by integrating urban design, architectural design, landscape design and ecological design, and provides customers with creative, harmonious and transcendent design services.

In the new century, China is undergoing huge changes in the society, economy, cities, architectures and other sectors, and as well as unprecedented exchanges in capital, technology and ideology. Therefore, urban space and architectural designs have to meet this trend. When going into the global, the company has to carry forward its cultural advantages and discover the charms of local cultures. In addition to combination with advanced ideas of the world, UK.LA also integrates local architectural elements and expressions into designs. UK.LA attaches great attention to establishment and development of partnership with the governments and developers and attach importance to the accurate market information. The company has received broad recognition and supports from customers for its innovative designs and high-quality services during business expansion.

Major Awards

The Most Influential Design Agencies Outbound, 2009
Gold Prize for the International Architectural Design Award, 2009
Asia Oscar International Fashion Creative Enterprise Architecture Planning and Design, 2007
China Top 100 Integrity Brand Institute of the Architectural Planning and Design, 2007
Jiangsu Top Ten Integrity Brand Design Institute, 2006

1–3 红太阳 · 乌江新城规划设计

建设地点：安徽 和县
项目业主：江苏红太阳房地产开发有限公司
总用地面积：约 700 万平方米
总建筑面积：约 538 万平方米

乌江新城项目位于安徽和县乌江镇，北靠乌江，东临长江被喻为“黄金水道”，是八百里皖江第一镇，处于南京“一小时都市圈”核心层。乌江是千年古镇，镇内有千年老街、3A 级国家风景区霸王祠等名胜。

作为上万亩的大型项目，整个规划将地块划分成三大功能片区即为生态居住区、商贸旅游区以及湿地观光区。规划结合安徽文化，恢复了江南水乡河街并行的传统格局，将老街改造成水街，形成一系列的水上旅游环形线路。这样一个通透的集旅游、观光、休闲、养生、度假、会务、交易、居住为一体的徽派社区，必将符合人们对中国传统文化的需求和在城市寻根的心理，同时也形成了原生态山水豪门的生活气势。

1-3 Planning and Design for Red Sun • Wujiang New Town

Construction Location: Hexian County, Anhui
Project Owner: Jiangsu Red Sun Real Estate Development Co., Ltd.
Total Site Area: about 7,000,000 m^2
Total Building Area: about 5,380 000 m^2

The project of Wujiang New Town is located in Wujiang Town, Hexian County, Anhui, adjacent to Wujiang River on the north, the "Golden Watercourse" - Yangtze River on the east, the No. 1 town along the 400 km Wanjiang River, located at the core layer of Nanjing's "1-hour Metropolitan Circle". Wujiang is an ancient town with one thousand years of history, boasting millennium streets and a great many places of interests such as Overlord Temple – a National 3A Scenic Area.

As a large-scale project covering a land of over 7 million square meters, the land is divided into three key functional areas in the overall plan, i.e. eco-residential area, trade and tourism area and wet-land sightseeing area. The planning combines the cultures of Anhui to restore the traditional pattern of southern Chinese riverside towns – with streets stretching along the river, and reconstruct the old streets into water streets to form a series of loop routes for cruising tourism. Such transparent Anhyui-style community will integrate tourism, sightseeing, recreation, health preservation, holiday, conference, trade and residence, and it will meet people's needs of traditional Chinese culture and the psychology of seeking the root of the town, and meanwhile, it makes a luxury life style in the primitive landscape.

5

1

4–7 海南儋州双联滨海住区规划设计

建设地点：海南 儋州
项目业主：海南儋州双联置业有限公司
总用地面积：101 万平方米
总建筑面积：88 万平方米

规划区位于海南儋州白马井镇，距海口市 135 千米，西临大海，具有得天独厚的景观优势。地块分为居住板块与度假板块两个独立的功能板块。

高档滨海社区与海上度假区这两个区域通过一条横贯东西的中央景观大道连接为一体，由东向西的序列依次为商业步行街——会所——景观大道——湿地公园——沙滩游乐区——内港游乐区——百米海景酒店。

在整个项目规划建设过程中，我们积极倡导生态节能和低碳环保的设计理念，推广生态技术工程的应用。力争使本项目成为儋州首席“低碳零排放示范社区”，在进一步保护自然生态环境系统的同时，提升居住品质和舒适度。

6

2

4-7 Planning and Design for Danzhou Shuanglian Coastal Residential Community, Hainan

Construction Location: Danzhou, Hainan
Project Owner: Hainan Danzhou Shuanglian Real Estate Co., Ltd.
Total Site Area: 1,010,000 m^2
Total Building Area: 880,000 m^2

The planed area is located in Baimajing Town, Danzhou, Hainan, 135 km from Haikou City, adjacent to the sea on the west, and boasting unique landscapes. The land is divided into two independent functional areas, i.e. residential area and holiday resort area.

The top-grade coastal residential community and the sea resort area are linked into a whole by the central sightseeing avenue stretching from the east to the west, the layout (from east to west) is: shopping street – clubs – sightseeing avenue – wetland park – beach recreation area – inner harbor carnie – 100m sea-view hotel.

During the planning and construction of the entire project, we actively advocate the ecological, energy-efficient, low-carbon and environment-friendly design concept, and promote the application of ecological engineering. Aspire to built the project into the first "low-carbon and zero-emission model community" in Danzhou, and improve the quality and comfort of residence besides further protecting the ecological and environmental system of the nature.

3

4

7

1–5 江苏新沂市窑湾古镇规划设计

项目业主：新沂市政府
建设地点：江苏 新沂
建筑规模：总占地面积：1000 000 平方米

窑湾古镇是目前苏北地区在京杭大运河滨水古镇中保存最完好的一个。她形成于春秋战国时期，明清时期达到鼎盛。古镇有发达的水系，分别是大运河、后河以及护城河，形成独特的半岛形态。

窑湾古镇滨水景观规划在京杭大运河 2013 年申报世界文化遗产的背景下展开，本次规划中充分考虑了窑湾古镇在京杭大运河遗产中的地位，在保护窑湾特有的“镇有前后河，城在两湖中”的滨水古镇形态的同时，打造中国“苏北第一运河古镇”。

我们在总平面规划中设计了三条水轴、四大功能区、十三组景观节点。

景观规划着力打造的窑湾新“十三景”，旨在将窑湾三条河的沿线打造成展示窑湾历史文化的重要景区，并通过景观步道紧密相联，形成窑湾穿越时空感的如翡翠项链般的沿河景观带。

1-5 Planning and Design of Yaowan Ancient Town, Xinyi, Jiangsu Province

Project Owner: Xinyi Municipal Government
Construction Location: Xinyi, Jiangsu
Construction Scale: Total Site Area: 1,000,000 m^2

Yaowan Ancient Town is the best preserved waterfront ancient towns along Beijing-Hangzhou Grand Canal in Northern Suzhou Area. It came into being in the Spring and Autumn and Warring States periods and riches to its greatest prosperity in Ming and Qing Dynasty. The Ancient Town has sophisticated river system, namely the Grand Canal, back river and the moat, forming a unique peninsula shape.

Yaowan Ancient Town waterfront landscape planning was carried out under the background of declaring for the World Historical and Cultural Heritage of 2013. In this planning the position of Yaowan Ancient Town in Beijing-Hangzhou Grand Canal Heritage—forging "First canal ancient town in Northern Suzhou" of China in the meantime of protecting Yaowan's unique shape of waterfront ancient town of "town with both front and back river, city between two lakes".

Our site planning has designed 3 water axles, 4 big functional areas and 13 groups of landscape nodes. The landscape planning puts forth efforts on forging Yaowan's new "Thirteen" sceneries, aiming at making the areas along the three rivers of Yaowan into significant scenic spot representing Yaowan's historical culture, and connected closely with landscape footpaths, forming Yaowan's emerald necklace through the veil of time.

3

4

1

2

5

6

7

8

6-8 Zhengzhou Tsinghua-Yijiangnan construction planning and design Construction

Construction Location: Zhengzhou, Henan
Construction Scale: Total Site Area: about more than 1,460,000 m^2
Total Building Area: 860,000 m^2

This project locates at the juncture of Huiji District of Zhengzhou and Guangwu Town of Xingyang, facing Yellow River Grand View in the east, near Hanba Two kings' City in the north, and having Guangwu Town Government 3 kilometers to the west. The transportation of the whole district is very convenient. The total site area of this district is 1.46 million square meters. In the middle of this district lies a long river traversing from west to east. In the district the plants are exuberant and distributes naturally, forming a great greenery condition. Three high-tension cables slants through the middle area. This project constructs ideal sustainable community and provides living space of higher leveled public health and better quality for the residents. At the meantime of creating good in and out door space state of the construction, it emphasizes creating good community humanity environment, adapting to the requirement of development of market economy and urban morphology, making the planning more flexible and adeptable. Making use of the mountain's momentum, incorporating with the terrain and river, Tsinghua Yijiangnan Subdistrict sets an ecological wetland of 320 thousand square meters within the boundary of 1.46 million square meters. Under the environment of lacking water in Zhengzhou, Tsinghua Yijiangnan pays attention to the utilization of natural resource, builds along the slopes and rises and falls according to the mountain terrain, which fuses the construction style and the natural landscape together and realizes the resort of "scenery in the house, house in the scenery". It is precious due to the location, to the river, and it owns the sky, the earth, the river and the scenery. According to the terrain, the subdistrict sets multiple residential states like waterfront independent villas, semi-detached villas, rowhouse villas, garden houses, medium height buildings, resort style apartments, wetland wooden house etc., and even manor style single island villas, which represents dignity.

6–8 郑州清华 · 忆江南建筑规划设计

建设地点：河南 郑州
建筑规模：总用地面积：1 460 000 平方米
总建筑面积：860 000 平方米

本案处于郑州惠济区与荥阳市广武镇交界处，东临黄河大观景区，北依汉霸二王城，西部 3 千米处为广武镇镇府。整个地块的交通十分便捷。地块总用地面积为 146.7 公顷。地块中部有一自西向东流的长河横贯东西，地块内植物茂盛，并且呈自然式分布，绿化条件优越。中间地区有三条高压线斜穿。

本案意在建设理想的可持续社区，为居民提供更高水准的公共式健身空间和更优质的生活空间。方案在创造良好的建筑室内外空间形态的同时，注重打造良好的社区人文环境，以适应市场经济和城市形态发展的需要，使规划具有弹性和应变能力。

清华 · 忆江南小区依山就势，结合地形及河流，在 146.7 公顷的范围内设置了 32 万平方米的生态湿地。在郑州水资源稀缺的环境下，清华 · 忆江南小区规划重视对自然资源的利用，建筑依坡而建，随山势起伏，将建筑风格与自然景观统一融合，实现了“景中有房，房中有景”的胜境。因地而贵，因水而贵，有天、有地、有水、有景。小区结合地形设置了滨水独立别墅、双拼别墅、连排别墅、花园洋房、小高层、度假酒店式公寓、湿地木屋等多种居住形态，更有庄园式独岛别墅，彰显尊贵。

1

2

3

1-3 南京红太阳旭日上城

建设地点：江苏 南京
项目业主：江苏红太阳集团
总用地面积：约 100 万平方米
总建筑面积：约 200 万平方米

南京红太阳旭日上城地处中国东南部交通要地，南京长江大桥北端。依托长江三角洲区域经济强劲发展的大背景，借助“南京市发展长江两岸，建设跨江发展的国际化大都市”的发展战略，将建成一个以购物、度假、旅游、休闲、娱乐、文化为一体的综合性、标志性商业城及一个国际化的高尚大型居住区。项目为南京市“跨江发展，两岸齐飞”的重要领航项目。项目规划中的旭日上城将成为中国、乃至全球城市综合开发的一大亮点。

规划区位于南京市（浦口区）南京长江大桥北侧，整个旭日豪庭项目的最东端地块。地块东西长约 581 米，南北长约 468 米，北邻柳州路东沿线，东依南浦路，南滨丽江河，基地西侧为规划市政道路。

本设计构想主要依据基地临水之特征，以中国江南水乡之风貌结合托斯卡纳人居之理念，以塑造一个具有住（居住）、游（休闲）、创（文化）特点的“生活理想国”，为南京市浦口区提供一个悠闲的、舒适的、崭新的居住环境，同时依据基地和城市文脉，旭日上城也将成为建筑与公共开放空间相结合的独特典范。

1-3 Nanjing Red Sun. Rising Sun Top Town

Construction Location: Nanjing, Jiangsu
Project Owner: Jiangsu Red Sun Group
Total Site Area: 1,000,005 m^2
Total Building Area: 2,000,000 m^2

The priject is Located at the traffic hub of Sortheast China ,THE north end of Nanjing Yangtze River Bridge. Under the background of the rapid development of Yangtze Delta Area and based on Nanjing's development strategy "To seek development on both banks of Yangtze River and build an international metropolis across Yangtze River", a comprehensive business town integrated with shopping, tourism, recreation, entertainment and cultural facilities and a large international modern residential community will be built. It's a pioneering project for Nanjing's "transYangtze development" strategy. The planned Rising Sun Top Town will become a spotlight of comprehensive urban development in China or even in the world.

The planned area is located at the north side of Yangtze River Bridge in Pukou District, Nanjing. Sun Rising Top Town sits on the east end of the land. The land extends 581 m from east to west and spans 468m from south to north, with east line of Liuzhou Road on the north, Nanpu Road on the east, Binli River on the south, and the planned municipal road on the west side.

The design, based on the geographic features (sitting on the riverside) and combining the style of South China water town and Tuscany's concept of residential buildings, aims at creating a "Utopian land of life" integrated with residential, tourism and cultural elements, and providing an easy, comfortable and fresh living place in Pukou District. Meanwhile, Sun Rising Top Town will become a model in respect of integration of architecture and public open space based on its location and the cultures of the city.

4

4 黄河生态科技博览园规划设计

建设地点：河南 郑州
项目业主：河南超达置业
总用地面积：560 万平方米

本项目地处黄河中下游南岸，位于郑州市惠济区东北部，隶属花园口镇，占地约 560 万平方米，西邻花园口景区，东临马渡，北有富景生态园，南部有规划建设中的 107 国道与大河路，周边生态旅游资源丰富，交通便利，与郑州新东区相连，是沿黄河文化产业带发展中较为有利的区域。

通过科学规划，设计将石桥建设为郑州“后花园明珠”，成为省内有特色的社会主义新农村建设示范区，同时也是沿黄河文化产业带中下游地区的中间纽带。

4 Planning and Design for Yellow River Ecological Technology Expo Park

Construction Location: Zhengzhou, Henan
Project Owner: Henan Chaoda Real Estate
Total Site Area: 5,600,005 m^2

This project is located on the south bank of the middle-lower reach of Yellow River, the northeast part of Zhengzhou Huiji District, subordinated to Huayuankou Town, covering a land of 5.6 million square meters, adjacent to Huayuankou Scenic Area on the west, adjacent to Madu Village on the east, and with Fujing Eco Park on the north; in the south part, there are the planned 107 National Road and Dahe Road which are under construction, there are abundant tourism resources on its periphery and with convenient transport facilities, and it's connected with the new East District of Zhengzhou. Therefore, it's an advantageous region in the development along Yellow River Cultural Industry Belt.

Through scientific planning, Shiqiao will be built into "Pearl of the Backyard" and a demonstration area for the new socialist countryside construction program with special characteristics, and it's also the intermediate link in the middle-lower region along Yellow Cultural Industry Belt.

1–4 郑东新区龙子湖湖心岛

建设地点：河南 郑州
项目业主：郑州市人民政府
总用地面积：990 000 平方米
总建筑面积：1 010 000 平方米

郑东新区龙子湖位于该区东部龙子湖区中部，是一个水面为 1.2 平方千米的人造湖，周围环绕 12 所高校和一块高校预留地。规划确定的 12 所高校建筑面积约 365.1 公顷，在校生人数约 14.48 万人。湖心岛为龙子湖中心岛，为科研中心和可共享资源用地。规划用地范围北至校园六路，南至校园七路，西起龙子湖纵贯一路，东止龙子湖纵贯三路，总用地面积约 1 平方千米。

湖心岛结合文化、休闲、娱乐、购物等设施，匹配大学城的特定功能，创造良好的科研环境，塑造绿色的教育园区；同时体现了郑东新区“共生城市”和“新城代谢城市”的规划概念。湖心岛的规划建设目的在于“以学兴城”“以城促学”的良性结合，提升城区整体活力，形成以大学为纽带，辐射周边地区，集教育功能、产业功能和生活服务为一体，文化氛围浓厚、科技产业发达、服务体系完善、优秀人才汇聚、生太环境良好的优化区域，最终建成一个发展的智力支持密集区、文化科技积累与传播的中心区、生态环境良好的新世纪现代化城区。

1-4 Chung Dong New Area Longzihu Lake Island

Construction Location: Zhengzhou, Henan
Project Owner: Zhengzhou Municipal People's Government
Total Site Area: 990,000 m^2
Total Building Area: 1,010,000 m^2

Located in Eastern Longzihu lake New District, the Dragon Lake is in central eastern area, a man-made lake covering 1.2 square kilometers, surrounded by 12 universities and a university set aside land. Plan identified 12 construction area of about 365.1 hectares University, in the number of about 14.48 million students. Lake island Longzihu center island, as research centers and land resources can be shared. Planned land area is in the north to campus bus No. 6 line and the south of campus No. 7 line, West from Longzihu No. 1, east to Longzihu longitudinal No.3, with a total land area of about 1 square kilometer.

Lake island has combined culture, leisure, entertainment, shopping and other facilities, to match the specific function of University City, create a good environment for scientific research, and shape the green zone for education; meanwhile it also reflects the planning concept of Chung Dong New Area, i.e. "Symbiotic City" and "Metabolic Metro City". the purpose of its planning and construction is lied on the virtuous combination of "let university strive the city", "Better city better education", to enhance the overall vitality of urban areas, formed by the University as a link, radiating surrounding areas, integrating educational, industrial and health services into one, optimizing to being with strong culture atmosphere, developed science and technology industry, sophisticated service system, outstanding talent gathering, and good regional environment. All in all, it is aimed to develop into a community intelligence-intensive support area, accumulation and spread center of culture and technology, and a modern city with well environment in the new century.

1

3

4

5

5-7 Donghu Hotel Planning and Design

Construction Location: Zhengzhou, Henan
Project Owner: Zhengzhou Jiuhua Real Estate Development Limited Company
Construction Scale: planning area of about more than 1,330,000 m^2
with about more than 460,000 m^2

This project is oriented as “healthy leisure and conference reception resort with typical natural environment and cultural characteristics of Central Plains, and excellent natural ecological environment, which can represent government image of Henan Province”. The sections are arranged as the following: multiple functional sections such as star level resort hotel, physical examination center, warm spring regimen, artistic culture exhibition, resort villa, Central Plains food street, ecological wetland, golf practicing, superaqueous tourism, ecological orchard, ball games, waterfront activities, wetland enjoyment, gardens greenery etc., which fully represents the district's comprehensive functions combining leisure and fitness, cultural entertainment, resort and tourism as a whole. The planning takes abundant green “surface” area like garden greenery, golf driving ranges, ecological wetland, ecological orchards etc. as the background, lying out construction monomers of delicate classical style with different functions as “spots”, with the coastwise footpaths and roundabout roads and “linear” paths, which connects each functioning organization parts, and makes this project form an enjoyable and pleasant resort.

6

7

5–7 东湖宾馆规划设计

建设地点：河南 郑州
项目业主：郑州九华房地产开发有限公司
建筑规模：规划用地约 1 330 000 平方米
其中水面约 460 000 平方米

本项目定位为“具有典型中原自然环境特征和文化特征，具有极佳自然生态环境，能够代表河南省的政务形象和生活品质的，健康休闲、会议接待的度假胜地”。地块规划分为星级度假酒店、体检中心、温泉养生、艺术文化展示、度假别墅、中原美食街、生态湿地、高尔夫练习场、水上游览、生态果园、球类运动、滨水活动、湿地观赏、园林绿化等多个功能区块，充分体现了该区集休闲健身、文化娱乐、度假旅游于一体的综合功能。规划以园林绿化、高尔夫练习场、生态湿地、生态果园等大量绿色“面”域为背景，以精致的古典建筑风格的不同功能的建筑单体为“点”缀，以水体沿岸步道及迂回弯曲的道路为“线”条，联系各个功能组成部分，使该案形成一个佳境宜人的旅游胜地。

1–2 徐州中山堂项目

建设地点：江苏 徐州
项目业主：徐州市规划局
总用地面积：约 5 400 平方米
总建筑面积：约 6 万余平方米

本项目位于徐州市，基地处于繁华的淮海东路与解放南路交叉口，东西临规划市政道路，交通极为便利。设计的出发点就是打造一个全新的中山堂娱乐影视中心，通过独特的建筑形体设计，树立起城市的标杆，进一步扩大其在徐州及周边城市的影响力，起到城市窗口的作用，展现出徐州经济和人文的巨大变化，同时也最大限度地挖掘城市土地地经济潜能。

3 郑州格拉姆国际中心

建设地点：河南 郑州
项目业主：意大利格拉姆财团 罗马市政府
建筑规模：12 万平方米

郑州格拉姆国际中心是受意大利格拉姆财团和罗马市政府委托，郑州唯——家外商独资开发企业独立开发的商业地产项目。该项目现为中原第一高楼。

意大利是欧洲文明的摇篮，河南是中国黄河文明的发源地，在我们的设计中，希望通过意大利和中国文化的交融与碰撞，建筑语言的融汇来塑造现代、高效，具有时代特色，体现中意两国文化的地标性建筑。

建筑造型层层收进，寓意“芝麻开花节节高”，顶部造型宛如钻石般熠熠生辉，成为郑州东部 CBD 中央商务区内一颗耀眼的明珠。

1-2 Zhongshantang Amusement Center in Xuzhou

Construction Location: Xuzhou, Jingsu
Project Wwner: Xuzhou Urban Planning bureau
Total Site Area: 5,400 m^2
Total Building Area: 60,000 m^2

Located in Xuzhou City, the base is located at busy intersection of Huaihai East Road and Jiefang South Road, neighboring municipal roads on east and west, so transportation is very convenient. Focused on creating a whole new Zhongshantang Amusement Center, the design tries to establish a benchmark through the unique building shape to further expand its influence in and around Xuzhou as the city window to reflect the enormous economic and cultural changes, as well as to maximize the economic potential of urban land.

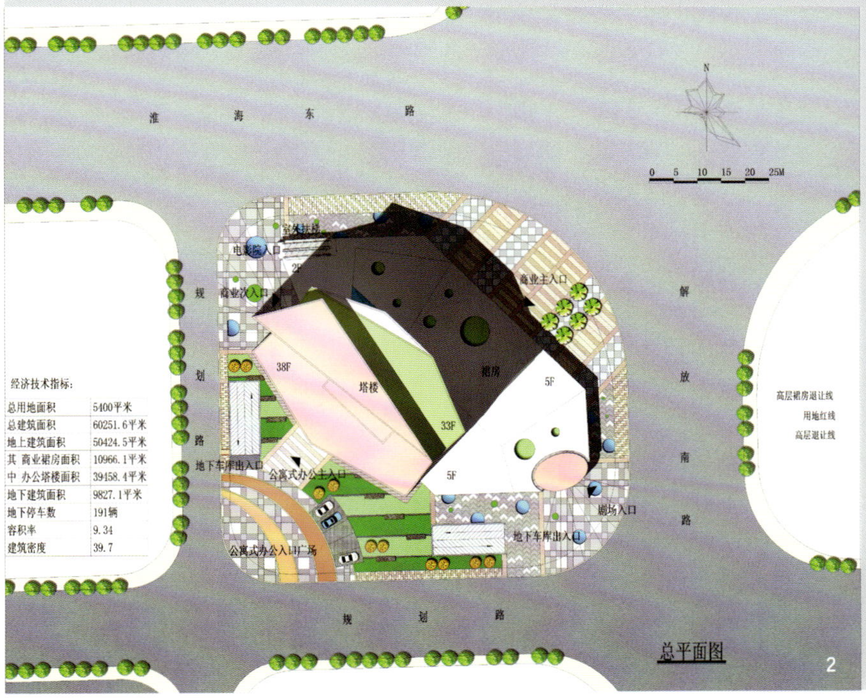

3 Zhengzhou Clam International Center

Construction Location: Zhengzhou, Henan
Project Owner: Clam Consortium of Italia, Government of Rome
Construction Scale: 120,000 m^2

Zhengzhou Clam International Center is the only one commercial real estate project developed by a foreign enterprise in Zhengzhou. The project is entrusted by Clam Consortium of Italia and Government of Rome, and it's currently the No. 1 high-rise in Central China.

Italia is the cradle of European civilization and Henan is the birthplace of Yellow River civilization of China. The design aims at Creating a modern and high efficient landmark building embodying Italian and Chinese cultures by integrating the architectrual languages of the two nations.

Each floor of the building recesses inward like “a gingili growing taller and taller”,the crests of the twin towers shine like two diamonds. The project has become a pearl of CBD in Zhengdong New Area.

1-4 伴山 · 香槟庄园

建设地点：江苏 南京
项目业主：南京万汇置业
总用地面积：59 589 平方米
总建筑面积：69 337 平方米

本项目位于江宁北区和西区的方山地区，邻近104国道，交通便捷，驱车到南京主城仅需20分钟。自然环境优美，是未来住宅规划的重点区域。项目总用地面积59 589.6平方米（不含绿化用地面积）。

本项目的规划及建筑设计，应确立以“创造社会精英人群的人居环境，开发竞争力强的产品”为原则。充分分析项目的地理位置、人文环境、气候特点、水文地质状况及景观环境等要素，关注目标客户群的文化素质层次、生活习惯等方面的需求，针对南京市房地产时常再规划、建筑的薄弱点，进行务实创新的规划、建筑、环境设计。以人本关怀为主导，结合法兰西风格的设计要素，借用文脉主义、隐喻主义、装饰主义来赋予建筑历史和传统要素，与现有环境融合，使其成为真正的经典建筑。

1-4 Banshan • Champagne Manor

Construction Location: Jiangsu Nanjing
Project Owner: Department of Nanjing Wanhui Propriety
Total Site Area: 59,589 m^2
Total Building Area: 69,337 m^2

Located at the northern and western side of Jiangning District Fanshan region, this project is close to 104 National Road, with convenient traffic, which just taking 20 minutes' drove to the main city of Nanjing. Beautiful natural environment is the focus of future residential planning area. The total land area is 59,589.6 m^2 (excluding the green land area).

The planning and architectural design of the project should be established under the principle of "Create living environment for social elites, develop competitive products". Being sufficiently analyzed on the project's geographical location, cultural environment, climate characteristics, hydrogeological, environmental and landscapes factors, it also pays attention to Cultural quality level of the target population, their life habits, and other needs. Based on the weak point of Nanjing real estate industry that re-plans and re-constructs from time to time, it carries out pragmatic and innovative planning, architecture, and environmental design. On the basis of humanity-oriented, combining the design elements of the French style, borrowing contextualism, allusionalism, ornamentalism to generate architecture with history and traditions, along with the integration of the existing environment, making it a truly classic construction.

1

2

3

4

5

5-8 Lianyungang Crystal Palace Architecture

Construction Location: Lianyungang, Jiangsu
Project Owner: Lianyungang Umikawa Properties Ltd.
Total Site Area: 22,467.8 m^2

The project base is located southwest of the planning East River new city of Lianyungang, China's well-known crystal capital. This project, East China Sea Crystal Palace, named after the Journey to the West as the case name, being a bond to cultural relics. In the design, the southwest corner of the building is presented in the typology of "Dragon" plane, and its facade is made of glass mosaic forms of polygons, implied crystal and "Dragon scale" metaphorical sense; in the middle of the its body we use the whole image of linking dragon head, dragon body, and dragon tail through, to generate the "Dragon gate" space pattern.

Dual enclosed structure is adopted to create an inner and outer ring layout, with a strong sense of direction and consumer guide, integrated business and leisure with the individual region cultural icon, like city landmarks in Shanghai "New World", Nanjing "1912", Hangzhou "West Lake" and so on. Modeling techniques and reasonable creative layout of the land contributes it a novel beautiful "pearl" in the East River New City of Lianyungang.

6

5-8 连云港水晶宫建筑规划设计

建设地点：江苏 连云港
项目业主：连云港海川置业有限公司
总用地面积：22 467.8 平方米

本项目基地位于规划中的东河新城西南侧，连云港为中国知名水晶之都，本项目以东海水晶宫为案名，以西游记传说为文化纽带。在设计过程中西南角的建筑以“龙”形平面的形态展示，其立面形式采用多边形的玻璃拼接，隐含水晶之喻意又有龙鳞附体的感觉。在中间部位我们采用了连接体将龙首、龙身、龙尾联系于一体，在空间形态上产生了“龙门”之意。

设计采用双围合结构，形成内外环布局，具有强烈的方向感和引导性，使休闲商业与各地区域文化相结合，成为富有城市特征的地标，类似于上海“新天地”、南京“1912”、杭州“西湖天地”等。现代的造型手法和合理的创意布局，将把地块打造成连云港东河新城新颖靓丽的“明珠”。

7

8

1 郑州一生缘 · 翡翠谷建筑规划设计

建设地点：河南 郑州
项目业主：郑州一生缘置业有限公司
总用地面积：117 万平方米
总建筑面积：77 万平方米

本案作为郑州远郊一个具有高度竞争力的项目，以其优美的景观为依托，在葱葱河畔，雕琢出一个个精致的组团。其最大的特点就是以隐映在河湾林丛之中，与自然融为一体。

翡翠谷融多种产品形态共存，有联排住宅、双拼住宅、花园洋房、坡地洋房、多层、小高层、高层以及酒店公寓等多种住宅形态，除此之外，还有商业街，假日酒店、温泉会所等多种商业形态。规划整体呈现两边高中间低的态势，全面体现自然水景生活，以舒适、自然、休闲作为样板标准，使舒适的生活记忆与休闲记忆永留在翡翠谷。

翡翠谷在注重空间设计的基础上，加强了对个人空间的重视和关注，保证了主体景观的突出与个性生活空间的延伸，力争做到整体景观与个人生活的融合统一。

1 Zhengzhou Yishengyuan • Jadeite Valley Architectural Planning and Design

Construction Location: Zhengzhou, Henan
Project Owner: Zhengzhou Yishengyuan Properties Ltd.
Total Site Area: 1,170,000 m^2
Total Building Area: 770,000 m^2

As a highly competitive project at the suburb of Zhengzhou City, this project relying on the beautiful landscape of river bank, carves out a set of exquisite composition. It is especially characterized by implicitly reflection in the bay, among the forest cluster, integrated with nature.

A variety of products co-exist in the Jadeite Valley, including townhouses, semi-detached house, garden houses, sloping houses, multi-layer, medium high-rise, high-rise, hotel apartments, etc.. In addition, there are also various business forms such as commercial street, holidays hotels, clubs, hot springs clubs. Generally, it presents a higher on both sides overall trend, explaining natural water life in an all-round way, running for comfortable, natural, recreational standard, so that only the comfortable and leisure memories are stayed in the Jadeite Valley forever.

This space-oriented Jadeite Valley enhances the emphasis and focus on personal space to ensure highlighting the main landscape and extending of individuality living space and also strives to achieve the integration and reunification between overall landscape and personal life in the best way.

1

2

2–4 清华 · 大溪地项目规划设计

建设地点：河南 郑州
项目业主：郑州清华园房地产开发有限公司
总用地面积：198 万平方米
总建筑面积：264 万平方米

本项目位于河南省郑州市与荥阳交界处，地处郑州西大门中原西路两侧，路况良好，交通出行十分便利，周边交通网络便捷。

规划设计了大型水上游乐设施、温泉浴场、酒店客房、餐饮娱乐以及大型商业等多种形式的产品，强力打造中原地区特色鲜明的"吃、住、游、购、娱、休"一条龙的特色休闲娱乐基地以及中原休闲娱乐顶级强势品牌。

该方案本着可持续发展，以人为本；社会环境、生态环境、建筑环境和谐统一；远近结合，保持弹性；坚持规划的可操作性为原则，力求打造缤纷、清新、气息浓郁、配套完善的都市近郊生活氛围，使这里成为城市精英圆梦的地方。

2-4 Tsinghua • Tahiti Project Planning and Design

Construction Location: Zhengzhou, Henan
Project Owner: Zhengzhou Tsinghua University Park Real Estate Development Co., Ltd.
Total Site Area: 1,980,000 m^2
Total Building Area: 2,640,000 m^2

This project is located in boundary of Zhengzhou City, Henan Province and Xingyang, near Xingyang side, where both sides are West Road of Zhongyuan, West Gate, Zhengzhou city. There are good road conditions, convenient traffic travel, expedite traffic network around.

Its planning has designed various forms of product function including large water amusement facilities, Spa pool, hotel rooms, dining, entertainment, Shopping Mall, etc., striving to create an one-stop recreation base featured at "food, housing, travel, shopping, entertainment, rest," and top entertainment brand with great influential in Central Plains.

This project is in line with sustainable development as well as people-oriented tenet; to realize the harmony among the social environment, ecology as well as built environment; to adjust as the immediate and long term development as well as maintaining flexibility; to adhere to the principle of manipulility; to create fun, fresh, rich flavor, and perfect supporting suburban life atmosphere. This will be the place where the dreams of urban elites come true.

3

4

James Wang Design Associates
北京杰地亚建筑咨询有限公司

地址：北京市海淀区中关村南大街35号（紫竹院公园西院）
邮编：100044
电话：+86-10-88510711
传真：+86-10-88510892
邮箱：jwda@vip.163.com

Address: No.35, Zhongguancun South Street, Haidian District, Beijing, China (west yard of Zizhuyuan Park)
P.C.: 100044
Tel: +86-10-88510711
Fax: +86-10-88510892
E-mail: jwda@vip.163.com

北京杰地亚建筑咨询有限公司原为美国的James Wang Design Associates公司成立于1987年，注册地是美国加利福尼亚州。该公司于1993年在中国设立事务部，进入中国市场，并以一流的设计与服务赢得了客户与市场的认可。2000年在北京成立北京杰地亚建筑咨询有限公司，为中国客户提供优质的设计服务。

公司的主要客户涉及金融、电信、住宅、商业、交通等多个行业，公司下设规划建筑设计部和室内装潢设计部两大部门，根据不同客户所特有的技术与地域特点，提供专业化的服务。

公司自1987年成立迄今，已为业界提供了大量杰出的建筑设计和区域规划的服务，作为一个建筑设计公司，我们能为业主在预定的时间、指定的设计预算内提供最理想的设计作品，同时，我们以此服务为荣，并且深信客户的成功就是我们通往成功之路的钥匙。事实证明，我们对客户的这种承若，使我们能与客户保持长期的合作关系。

在承接政府机构及私人企业项目方面有广泛的经验，所设计的范围涉及教育研究机构、办公楼、商业开发、医疗设施、住宅区、城市综合体、旅游度假中心的规划及建筑群体设计等。

业主对建筑师的选择必须基于对建筑设计公司负责人、主要专业人员及其员工在专业技术经验方面的信任。我们公司的成员包括了有创造天分的建筑师、规划师、室内设计师及施工管理人员。我们公司的建筑设计服务永远以客户的需求为先，公司每个成员都深切了解客户需求的重要性，并深信客户的成功和满意才是公司的成功。

公司创办人James Wang是美国伊利诺伊大学的建筑硕士，曾是美国建筑师协会会员。公司分别在洛杉矶、北京及大连设有办事机构，设计人员近100人。

目前公司已经完成了数百个规划建筑设计项目及室内装饰设计项目。公司采用最先进的各类设计软件，为客户提供高标准、高效率、高品质的设计服务。出色的设计成果，使公司得以闻名世界，公司先后在北京、上海、成都、广州、济南、沈阳、威海、秦皇岛、青岛、长沙、天津、大连、唐山等地设计建筑方案，并得到同行的普遍赞誉。

非凡的设计、富有创意的规划及专业技术上的知识，是我们设计哲学的基础。本公司的多元化设计经验，使我们每一个设计都具有独特的理念，所创造的空间不但实用，而且具有很高的艺术水平。

James Wang Design Associates was founded in 1987, registered in California, USA, China Affairs Department of James Wang Design Associates was established in 1993 to enter the Chinese market and it won the customers and market recognition by the first-level design and service. In 2000, Beijing James Wang Design Associates Architecture Consulting Co., Ltd was set up to provide top quality design services for Chinese customers.

Customers of James Wang Design Associates are mainly in financial services, telecommunications, residential, commercial, transport and other fields. According to specific technology and geographical characteristics of different customers, the company have architectural (planning)design and interior (decoration) design two majoy departments to provide specialized services.

Since its foundation in 1987, James Wang Design Associates has provided a large number of outstanding architectural design and regional planning services. As an architectural design company, what we are proud of is to provide the best design work for the owners in the scheduled time and specified design budget, and we believe that the success of customers is the key to our success. The fact that this commitment to customers led to long-term cooperative relations between customers and the company has been proved.

The company has a wide range of experiences in undertaking projects for government agencies and private sectors, and our design services cover from education research institutions, office buildings, commercial development, and medical facilities to residential districts and hotels.

The owners' choice to the architect must be based on the trust to leaders of architectural design company and major experts on professional technology and experience. The James Wang Design Associates team includes the creative talents of skilled architects, experienced planners, interior designers, specification writers and business support personnel. Our company always place customers' demands for the first, and each member is keenly aware of the importance of customers' needs, and convinces that the success and satisfaction of customer is the success of the company.

James Wang, the main leader of James Wang Design Associates, have his master degree in architecture in University of Illinois. The company has set up administrative bodies in Los Angeles, Beijing and Dalian. The number of the staff is nearly 100.

James Wang Design Associates has completed hundreds of projects for architectural design and interior design. The company provides customers with top-quality design services by using the most advanced design softwares. Many and outstanding design achievements of our company have been world-famous. Covered in Beijing, Shanghai, Chengdu, Guangzhou, Jinan, Shenyang, Weihai, Qianhuangdao, Qingdao, Changsha, Tianjin, Dalian and Tangshan, our project design has received universal praise from Chinese counterparts.

The extraordinary design, creative planning and professional technical knowledge are the foundation of our design philosophy. Every one of us has a unique design concept from the company's diverse experience in the design, and therefore, space created by us is not only practical, but also with high quality.

1–5 鄂尔多斯康巴什IEF国际公园

6–9 鲁能海口盈宾岛海蓝・福园

10

11

12

13

14

10–14 北京山水文园金海湖碧波花园

15

16

17

18

15–18 北京中建国际红杉溪谷

19

20

21

22

23

19–23 烟台 · 龙口松涛名苑

24

25

26

24–26 大连文达老虎滩综合楼

27–31 蓬莱登州府第

32–35 北京一渡新新小镇

36–38 烟台芝水

39 青岛檀宫

40–42 烟台三站（新区）商住区

43–45 海南盈滨岛别墅式酒店

46–49 乌海君正花园D户型别墅

50–52 武汉泛海

53–55 山水金海湖宾馆

56–59 厦门鑫塔

60–63 东四正和广场样板间

64–66 青岛海泉湾度假酒店

67–70 日坛国际戊区

71

72

73

74

75

76

77

78

79

80

81

82

83

71–72 北京民生金融中心泛海办公层

73–75 大兴龙熙温泉酒店

76–79 天津远洋临港路项目售楼处

80–83 西山一号院样板间

室 内 项 目

James Wang Design Associates

北京杰地亚建筑咨询有限公司

84

85

86

87

88

90

91

89

92

93

84–85 山水茅草屋　　86–87 东四正和广场商业部分　　88–89 内蒙古鄂尔多斯康城4期　　90–91 和裕亦庄开发区　　92–93 天津远洋城样板间

北京大学肖家河居住区 Xiaojiahe Residential, Peking University

长春昂展净月别墅
Angzhan Jingyue Villa, Changc

CONCORD CDG国际设计机构 城 市 区 域 规 划 建 筑 设 计 环 境 景 观 设 计
DESIGN GROUP

地址：北京市海淀区长春桥路 11 号万柳亿城中心 A 座 10 层 /13 层
电话：+86-10-58816545
+86-10-58816546
+86-10-58816547
传真：+86-10-58816156
邮箱：cdg@cdgcanada.com
网址：www.cdgcanada.com

万科鞍山惠斯勒小镇　Vanke whistler Town, Anshan

公司简介

CDG 国际设计机构是由多家境外设计事务所组成的专门针对中国市场的设计机构，2001 年在加拿大 BC 省（不列颠哥伦比亚省）首府维多利亚市注册成立。

CDG 国际设计机构利用其主要合伙人对中国市场的了解及其与长期合作客户的良好互信的伙伴关系，整合境外及本土的优秀设计资源，为不同要求的客户提供最优化的设计组合。

CDG 国际设计机构目前在北京设有办事处，有设计人员 80 余人。境外组合机构包括建筑设计及环境景观设计机构，服务范围涉及城市区域规划、建筑设计、环境景观设计。

CDG INTERNATIONAL DESIGN LTD. consists of several overseas design firms, all focusing on the Chinese market. It was registered in 2001 in Victoria BC, Canada.

The main partners of CDG are very familiar with the Chinese market, and they all have long established mutual trusting relations with a wide range of clients. The Group integrates excellent overseas and native design resources, providing the best design combination for each client with their differing demands.

At present there is a main branch of CDG in Beijing, including about eighty designers. All of the branches provide services in the fields of urban planning, architectural design and landscape design.

万科唐山南湖别墅 Vanke Nanhu Villa, Tangshan

CONCORD DESIGN GROUP CDG国际设计机构 城市区域规划 建筑设计 环境景观设计

公司近年主要项目一览表

综合区域规划／建筑设计

1. 万科烟台芝罘岛项目
2. 沈阳信达理想新城
3. 万科长春惠斯勒小镇
4. 万科天津张家窝片区规划
5. 万科天津士林苑
6. 营口海湾馨城
7. 廊坊华夏铂宫
8. 长春中东净月湾
9. 松原中东班芙小镇
10. 通化中东帕萨迪娜
11. 万科青岛城市花园
12. 万科天津金色雅筑
13. 万科鞍山惠斯勒小镇
14. 廊坊锦瑞尚城
15. 嘉铭天津中新生态城红树湾
16. 中铁置业沈阳人杰水岸
17. 万科长春 228 厂地块概念设计
18. 万科天津宁发项目概念设计
19. 万科沈阳金域兰湾
20. 万科青岛金色城品
21. 沧州大合庄旧城改造
22. 东莞同沙项目
23. 北京大学肖家河居住区
24. 北京阿凯笛亚别苑
25. 鞍山万科城
26. 北京密云中加会馆
27. 长春国信美邑
28. 万科长春净月 2 号地

万科天津士林苑　Vanke Shilinyuan, Tianjin

29. 长春吉林农大棚户区改造
30. 东莞香树丽舍居住区
31. 金隅万科城概念设计
32. 东莞锦盛年华居住区
33. 贵阳花溪大地之舞大剧院地块
34. 廊坊嘉轩凤凰城
35. 北京金隅小营西路项目
36. 东莞凯逸豪庭
37. 长春伊通河地王居住区
38. 北京海棠公社
39. 廊坊华夏水晶城
40. 东莞凯旋门江滨居住区
41. 东莞天骄峰景居住区
42. 北京首创京棉一厂改造
43. 北京首创吉普厂改造概念设计
44. 北京归谷园二期概念设计
45. 北京通州漷县镇中心概念规划
46. 东莞松山湖商住中心区规划
47. 东莞理想 0769 家园
48. 东莞海怡花园三期
49. 东莞金鳌湾江岸公寓
50. 北京中海高尔夫花园
51. 东莞盛世华南居住区
52. 北京鹿港嘉苑
53. 北京博雅西园

商业，公共建筑设计项目

1. 长春中东南部新城项目
2. 长春中东铁北项目
3. 通化滨江戴思酒店
4. 松原中东新天地 MALL& 帕萨迪娜
5. 北京国际青年创意中心概念设计
6. 廊坊东 115 项目
7. 通化规划大厦
8. 通化中东新天地 MALL
9. 大连正源华南餐饮城
10. 成都海峡两岸科技开发园中心商务区
11. 万科长春兰乔圣菲商业公寓
12. 张家口五一广场
13. 广饶商业中心区
14. 廊坊一号
15. 东莞横岗水库度假村
16. 北京顺义金街商业区
17. 东莞豪凯五星大酒店
18. 东莞松山湖锦绣山河城市中心
19. 东莞华南 MALL 生活城 E&F 区
20. 廊坊第六大街
21. 胜芳中心广场
22. 北京龙腾文化广场
23. 东莞樟木头沃尔玛广场
24. 宜昌长江帝景 6 号综合楼
25. 东莞亿兆厚街商业广场
26. 丹东边境经济合作 A 区—E 区
27. 北京彼岸青滩高尔夫会所
28. 贵阳半岛国际大酒店
29. 廊坊第八大街西区
30. 北京望京六佰本随便消费区
31. 北京后沙峪政府办公楼
32. 北京酒仙桥风情商业街
33. 东莞广华商贸中心区
34. 东莞万江区行政中心
35. 北京卷石天地
36. 华南 MALL

沈阳天琴湾 Tianqin Bay, Shenyang

松原中东班芙小镇 Banff Town, Zhongdong, Songyuan

CONCORD DESIGN GROUP CDG国际设计机构 城市区域规划 建筑设计 环境景观设计

别墅 / TOWNHOUSE

1. 万科唐山南湖别墅
2. 哈尔滨太平湖别墅
3. 沈阳天琴湾
4. 长春昂展净月别墅
5. 长春倚澜观邸C区
6. 沈阳岭墅
7. 北京阿凯笛亚庄园二期别墅
8. 万科长春潭溪别墅
9. 北戴河总部基地
10. 万科沈阳兰乔圣菲三期\四期\五期
11. 东莞松山湖锦绣山河二期别墅
12. 东莞紫园别墅
12. 东莞紫园别墅
13. 长春天安第一城三期
14. 北京中海安德鲁斯庄园
15. 北京阿凯笛亚庄园
16. 东莞松山湖锦绣山河一期别墅
17. 东莞江畔花园游艇别墅
18. 北京银湖别墅

营口海湾馨城　Bayshore Town, Yingkou

造型设计项目

1. 天津中天首府
2. 北京翰庭
3. 北京金隅七零九零
4. 潍坊白浪金沙商业广场
5. 北京福泰华城
6. 北京总部公寓
7. 北京半岛国际公寓
8. 北京金隅丽港城
9. 北京博雅园

景观设计项目

1. 万科唐山南湖别墅
2. 春倚澜观邸 C 区
3. 廊坊华夏铂宫
4. 天津中天首府
5. 北京阿凯笛亚别墅二期
6. 北京阿凯笛亚别苑
7. 万科长春潭溪别墅
8. 北戴河总部基地
9. 鞍山万科城
10. 长春国信美邑
11. 万科沈阳兰乔圣菲别墅三期＼四期
12. 北京酒仙桥风情商业街
13. 宜昌长江帝景
14. 北京阳光大厦广场

荣誉

CDG 国际设计机构荣获—2004 年中国建筑艺术奖
沈阳万科金域兰湾荣获—2006 年被美国新闻周刊评为“人类新七大奇迹之一”
国信美邑荣获—2007 年詹天佑大奖“住宅区”金奖
CDG 国际设计机构荣获—2007 亚洲房地产峰会“亚洲建筑规划设计区域十大品牌机构”奖
林世彤荣获—2007 亚洲房地产峰会“亚洲建筑规划设计区域十大创新人物”称号
中海安德鲁斯庄园荣获—2008 年中国人居典范建筑规划“设计方案规划”金奖
南 MALL 荣获—2008 年中国住交“中国名盘·优秀别墅”大奖
2009 年建设部中国建筑文化中心“中国最具影响力境外设计机构”奖

A&S 翰时国际建筑设计
INTERNATIONAL DESIGN

翰时国际简介

翰时（A&S）国际建筑设计咨询有限公司是由国内外建筑师共同创立的建筑设计公司。公司创立于美国亚特兰大，2002年在中国北京正式注册。公司的目标是把国际上先进的建筑设计与规划理念和技术介绍到中国，并按照国际标准，为业主提供高质量的设计服务，为新世纪中国城市与建筑的发展作出贡献。翰时（A&S）国际可在建筑设计、城市设计、室内设计、景观设计等各领域，为业主提供全方位的咨询服务。

A&S Introduction

A&S international Design is an international architecture design and consulting firm which was jointly established by the architects both at home and abroad. It was registered in 2002 respectively in Atlanta, USA and Beijing, China. A&S dedicates itself to create modern and functional construction with international standard, by integrating the international and advanced concepts and techniques with an architectural philosophy, in order to provide its worldwide clients with high standard design and consulting services. A&S is positioned to provide services in architectural design, urban planning, interior design, landscape design and etc.

翰时国际设计理念

翰时国际采用国际化的理念，专业化的设计，地域化的服务，强调对客户的理解与尊重。翰时国际的敬业精神、专业知识、实践经验、市场活力已经成为有建设任务的业主们强有力的帮手和顾问。

A&S Design philosophy

A&S provides an international design perspective and professional services with regional design fees. It excels at providing a high standard of service and respecting for its clients. A&S becomes its clients' powerful consultant and assistant with passion, vision, professional knowledge, advanced project management experience and marketing energies.

A&S 翰时国际建筑设计
INTERNATIONAL DESIGN

建筑设计项目类型

住宅建筑
办公建筑
医疗建筑
实验室建筑
商业建筑
教育建筑

公司主要服务内容

建筑规划
总体策划
总体规划
景观设计
工程／建筑投资估算分析

建筑设计

基地分析
方案设计
初步设计
施工图设计
室内设计
施工配合

Project Types

Residential Facility
Office Facility
Hospital Facility
Lab Facility
Commercial Facility
Educational Facility

Main Services

Architecture Design & Planning
Integrate Programming
Master Plan
Landscape Design
Engineering/Building Investment Analysis

Architecture Design

Site Analysis
Scheme Design
Preliminary Design
Construction Document
Interior Design
Site Cooperation

医疗实验室工艺流程设计

工艺流程设计
设备规划

工程设计

暖通空调工程
电气工程
给排水工程
消防工程

Medical /Lab Planning Consulting

Medical /Lab Planning
Equipment Planning

Engineering Design

HVAC
Electrical Engineering
Water Supply & Sewerage Engineering
Fire-protection Engineering

A&S 翰时国际建筑设计
INTERNATIONAL DESIGN

中国北京

北京市宣武区宣武门外大街10号，
庄胜广场中央办公楼北翼 1301室
邮编：100052
电话：+86(010)63109869
传真：+86(010)63109870
E-mail：office@as-arch.com
网址：www.as-arch.com

美国乔治亚洲

美国乔治亚洲德卢斯市
Wild Dunes 大街 1054号
邮编：30097
电话：+1(770)4971645
传真：+1(770)4979148
E-mail：office@as-arch.com
网址：www.as-arch.com

Beijing, China

Room 1301, Central Tower North Wing,
Junefield Plaza,
No.10, Xuanwumenwai Street,
Beijing, 100052 China
Tel: +86(010)63109869
Fax: +86(010)63109870
E-mail: office@as-arch.com
Http:// www.as-arch.com

Georgia, USA

1054 Wild Dunes Way,
Duluth, GA 30097
Tel: +1(770)4971645
Fax: +1(770)4979148
E-mail: office@as-arch.com
Http:// www.as-arch.com

公司主要负责人

余　立　董事长／总设计师
（荷兰籍建筑学博士）
陈娟娟　董事／总裁助理
张　兵　董事／副总设计师
张广亮　董事／副总设计师
林载舞　董事／副总建筑师

公司成员

90人

Directors

Li Yu Ph.D.
Juanjuan Chen
Bing Zhang
Guangliang Zhang
Zaiwu Lin

Number of Employees

90 persons

映時設計

TIME MIRROR international

美国 **Time Mirror** international 映时设计公司，以美国洛杉矶为基地，并积极在中国推展市场上进行推广，不断探求设计艺术与经营管理的最佳平衡点，力求使每个项目获得最大的成功。公司把不断为业主提供更好的设计服务作为服务的总目标，除在设计上求新外，更在项目管理上求精。公司永远站在客户的立场上去寻求最佳并有创意的解决方法。公司不仅有众多优秀的员工，更有一个稳定而坚实的领导结构，以使公司能有更多的机会提升设计与服务品质。Mr. Brian Liu 为公司的核心领导，在其他多位最具创意的合伙人的大力支持下，为每个项目配备最佳员工，最大程度地发挥设计水平，从而满足客户的需求。

创新是我们灵感的来源，作为设计师，倍感欣慰的是我们总能找到目标与价值的平衡点，使每个不同的客户类型都能享受到优质服务。多年来我们的设计项目和规模，也在不断地开拓发展。迄今为止，我们的设计已涉及各个领域，包括酒店、写字楼、住宅、商业等不同的设计项目。我们的设计人员是从众多既努力又富天赋的人才中精选出来的，这也是公司成功的源泉与基石。总之，TMI 设计公司的宗旨是在设计及服务上精益求精，为客户营造出有创意的设计作品。

America **Time Mirror** international Design Company, based in Los Angeles, USA, develops China market actively. The company constantly seeks the best balance between design art and operating management, trying to make each project successful. It makes the improving services continuously provided to the clients as the general goal of the company's services. It not only pursues innovation in design, but also pursues refinement in project management. It always seeks the best and creative solution from the stand point of the clients. The company not only has plenty of excellent employees, but also has a stable and strong leadership, which can make the company have more chances to improve design and services. Mr. Brian Liu, the core leader of the company, with the strong support of many other creative partners, provides the best staff for each project to put the designs which are made for the clients into full play. Innovation is the source of our inspiration.

As designers, what makes us gratified and content is the reaching of the balanced point between the object and value all the time, which can let every different client enjoy excellent service. For the past many years, our design project and scale have also continuously expanded and developed. So far, our design has already covered a lot of fields, including the design projects of hotel, office building, residence and business, etc. Our designers are chosen from a lot of talents who work hard and have gift, which is also the source and foundation of the success of the company. In a word, the aim of TMI Design Company is to seek for greater perfection in design and services and provide creative design works for the clients.

TIME MIRROR INTERNATIONAL
18566 EAST VANTAGE POINTE DRIVE
ROWLAND HEIGHT, CA 91748 USA
地址：北京市朝阳区建国路 93 号
万达广场 10 号楼 1909 室
电话：+86-10-5820-5860
传真：+86-10-5850-5760
邮箱：tmioffice@163.com
网址：www.timemirrordesign.com

TIME MIRROR INTERNATIONAL
18566 EAST VANTAGE POINTE DRIVE
ROWLAND HEIGHT, CA 91748 USA
Address: Room 1909, Building 10 Wanda Plaza,
Jianguo road No.93, Chaoyang District, Beijing
TEL: +86-10-5820-5860
FAX: +86-10-5850-5760
E-mail: tmioffice@163.com
Http: //www.timemirrordesign.com

1–2 清锦源，庄园别墅 样板间

建设地点：北京
设计面积：700 平方米

1-2 Chateau De La Vie Model Houses

Construction Location: Beijing
Planning Area: 700 m^2

1

2

3

4

3–4 天地一家餐饮会馆

建设地点：黑龙江 哈尔滨
设计面积：4 200 平方米

3-4 Between Heaven and Earth Dining Hall

Construction Location: Harbin, Heilongjiang
Planning Area: 4,200 m^2

5–6 凤凰城，庄园别墅 样板间

建设地点：河北 廊坊
设计面积：350 平方米

5-6 Phoenix Town Model Houses

Construction Location: Langfang, Hebei
Planning Area: 350 m^2

5

6

7

8

7–8 紫玉山庄别墅

建设地点：北京
设计面积：900 平方米

7-8 Purple Jade Villas

Location: Beijing
Planning Area: 900 m^2

9–10 和乔丽致酒店

建设地点：北京
设计面积：2 000 平方米

9-10 Chateau De Luze

Construction Location: Beijing
Planning Area: 2,000 m²

9

10

11

12

11–12 八号公馆俱乐部

建设地点：北京
设计面积：7 000 平方米

11-12 No.8 Clubhouse

Construction Location: Beijing
Planning Area: 7,000 m²

13

15

14

31-16 阿凯笛亚庄园 样板间

建设地点：北京
设计面积：350 平方米

13-16 Arcadia Villa Model Houses

Construction Location: Beijing
Planning Area: 350 m^2

16

LOOK architects

Architectural & Urban Design

LOOK Architects Pte Ltd
LOOK 建筑设计公司
18 Boon Lay Way
#09-135 TradeHub 21
Singapore 609966
Tel: +65-63168676
Fax: +65-63168690
Email: office@lookarchitects.com.sg

Founded in 1993, LOOK Architects is a design-intensive company committed to rigorous analysis and research to produce innovative, iconic buildings and urban design.

Founders Look Boon Gee and Ng Sor Hiang foster a collaborative studio environment to germinate and develop design ideas on a broad spectrum of works in the Southeast Asian region, making LOOK Architects extend new ground in various fields of design. Recipient of The President's Design Award – Designer of the Year 2009, Look Boon Gee's passion for his craft continues to lead the design practice towards exciting directions.

Versatility in handling projects of vastly different natures is evident in completed works, an array that includes private residences, apartment tower, university institution, community library, commercial headquarters, waterfront promenade and bridges. Embracing a design philosophy of the integrated whole, a high degree of inventiveness goes into construction detailing, ecological use of building material and planning of spatial relationships in relation to the environment.

Clients' personal and collective experiences are creatively interpreted to form perspectives that enrich the design approach of each project, and together with sustainable responses to the distinctive character of place, design enters the realm of eco-poetry, a lively dialogue that reverberates between nature and aspirations of man.

LOOK 建筑设计公司成立于 1993 年，公司重视设计实践，以严谨的分析和研究完成新颖的标志性建筑和城市设计。

公司创始人骆文义和黄素香以加强公司的合作环境为出发点，不断的萌发全新的设计理念，从而在东南亚地区形成了一个能全面、敏锐的开拓新领域的设计公司。作为 2009 年总统设计奖的获得者，骆文义以他对设计的热情，引领着公司迈向更加令人振奋的明天。

公司的多元化极大地体现在完成的多种不同类型的项目上，项目包括私人住宅、公寓大楼、大学机构、社区图书馆、商业总部、濒水步道和桥梁，以综合的设计理念和创新意识，来强化建筑细部，利用生态材料以及空间和环境的关系进行合理的规划设计。

客户个人和集体的经验，创造性地被阐释为设计的表现形式和观点，以丰富每个项目的设计理念并连同对地方特色的可持续性研究，开拓出一个人与环境和谐共生、生活气息鲜明的生态设计领域。

PRESIDENT*S DESIGN AWARD SINGAPORE
DESIGNER OF THE YEAR 2009

Http://www.lookarchitects.com

BISHAN PUBLIC LIBRARY

A 4-storey public library with basement and roof terrace. The prominent "pods" cantilevered from the facade serve as intimate niches for quiet contemplation. Using the treehouse as a source of reference, the library interior is animated by natural light filtering through colored glass panels, resembling dappled light coming through the tree canopy.

Site Area: 1,400 m^2
Building Area: 4,322 m^2
Construction Location: Singapore
Project Status:Completed in 2006

碧山公众图书馆

高四层的公众图书馆，包括地下室和屋顶露天阳台。图书馆立面突出的悬挑“盒子”给使用者提供了一个私密的静思空间，并使自然光线如同透过树冠照入树屋一般，使图书馆内部生动的表现出自然光穿过彩色玻璃后的所形成斑斓效果。

用地面积：1 400 平方米
建筑面积：4 322 平方米
建设地点：新加坡
项目状况：已完成（2006 年）

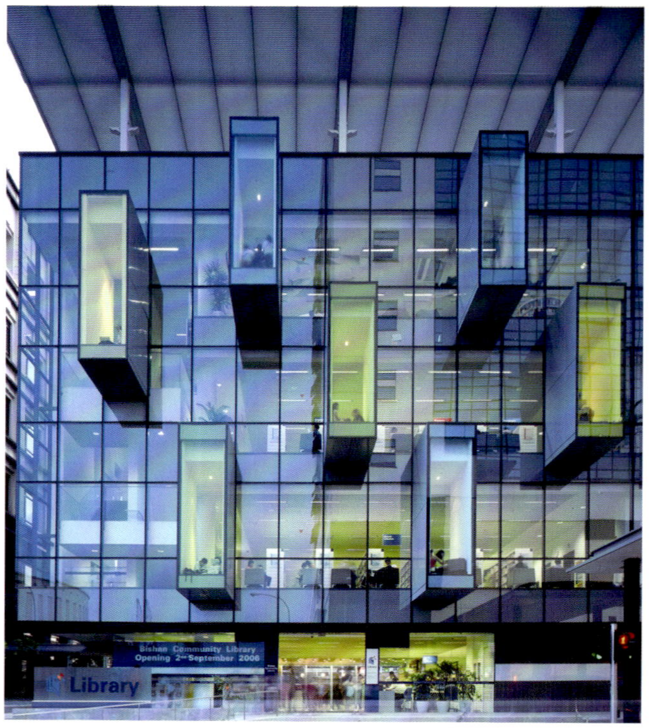

ALEXANDRA ARCH & FOREST WALK

Comprising a 80 metre long overhead pedestrian bridge over Alexandra Road and a 1.6 kilometre long elevated pedestrian walkway connecting to Telok Blangah Hill Park, Alexandra Arch & Forest Walk enables barrier-free public access into the natural heritage treasures of the Southern Ridges.

Site Area: 95,470 m^2
Construction Location: Southern Ridges, Singapore
Project Status: Completed in 2008

亚历山大拱桥和森林漫步

项目包括横跨亚历山大路的 80 米长的人行天桥和 1.6 千米长的悬空森林走道，并一直延绵到直落布兰雅公园。亚历山大拱桥和森林漫步走道可以使市民无障碍地进入南部山脊的自然遗产宝库。

用地面积：95 470 平方米
建设地点：南山脊 新加坡
项目状况：已建成（2008 年）

Luhu PEDESTRIAN BRIDGE

It's a design proposal for a 320-metre long pedestrian bridge at a 1,700-hectare new satellite town in Chengdu. The bridge forms part of a crucial groundwork to establish the circulation network in the new town. Luhu pedestrian bridge injects a sense of lyricism into the environment by introducing a weaving, sinuous way of passage through space.

Construction Location : Chengdu, China

麓湖步行桥

此设计方案位于成都南部的一个 1 700 公顷的卫星城市社区，320 米长的麓湖步行桥是社区交通网络的重要部分。通过蜿蜒曲折的空间设计，使麓湖步行桥为社区注入了新鲜的活力。

建设地点：成都 中国

SIGNAL HILL CONDOMINIUM

Design proposal for a 100 units of highrise luxury condominium development. The serrated profile of the 32 storey condominium tower allows for incorporation of narrow monsoon windows throughout the building perimeter as a passive environmental control device for wind catchment. Eco-ponds and contoured drainage courses covered with vegetation (bioswales) are introduced on the ground storey to manage surface runoff from the surrounding hilly terrain.

Site Area: 9,859 m^2
Building Area: 25,750 m^2
Construction Location: Kota Kinabalu, Malaysia

信号丘公寓

设计方案为 100 套高层豪华公寓的发展项目。32 层公寓大楼的锯齿形外观结合季风窗口，形成一个无源的风集控制装置。地面设计使用了生态池塘和植被覆盖的排水通道，以加强对地表径流的管理。

用地面积：9 859 平方米
建筑面积：25 750 平方米
建设地点：哥打京那巴鲁 马来西亚

3 BALMORAL

A 12 storey apartment block with semi-basement carpark in a residential enclave on the outskirts of the Orchard shopping belt. In response to the condition of ever-increasing density in the urban environment, the building mass is de-materialised by translucent perforated aluminium screen panels swathing cantilevered balconies on the facade. Rooftop gardens of duplex penthouse units offer panoramic views of nearby Goodwood Hill and the vibrant night lighting along Orchard Road. To enhance the public streetscape, sloping greenery instead of solid walls define the site boundary, creating an amiable, permeable interface between the private development and its surroundings.

Site Area: 2,283 m^2
Building Area: 3,652 m^2
Construction Location: Balmoral Road, Singapore

3 BALMORAL

本项目是位于乌节路购物区附近的一个 12 层公寓住宅楼，配有半地下室的停车场。针对不断增长的城市人口密度，我们设计的主要目标是探索以建筑材料“解材”建筑体，即在对建筑立面悬臂式阳台护栏的处理上，使用了半透明的穿孔铝板。从顶楼复式单位的天台花园可以鸟瞰 Goodwood Hill 及附近乌节路一带活跃的夜间全景。为了美化街景，我们设计了绿化斜坡，而不是实心的围墙，这样的设计更能强化一个亲切的以及使私人和周围环境相交流的氛围。

用地面积：2 283 平方米
建筑面积：3 652 平方米
建设地点：Balmoral Road 新加坡

BALCON EAST

It's a 5 storey high-density apartment block with basement carpark and communal swimming pool. There are 37 units, with apartment sizes varying from 45 sqm to 194 sqm to cater to a wide range of lifestyle choices. The sleek facade is accentuated by distinctive L-shaped balconies that allow all apartment units to enjoy privacy and unobstructed vistas.

Site Area: 2,587 m^2
Building Area: 3,621 m^2
Construction Location: Upper East Coast Road, Singapore

BALCON EAST

本项目为高密度的 5 层楼公寓住宅，配有底层停车场和公共游泳池。37 间套房，从 45 平方米到 194 平方米不等，能充分适应现代人不同的生活习惯和需求。流畅的外观设计更加凸显出独特的 L 形阳台，在保证隐私的同时，使每间套房都能享受宽阔的视野。

用地面积：2 587 平方米
建筑面积：3 621 平方米
建设地点：上东海岸街 新加坡

COBBLESTONE

加拿大考斯顿设计／上海考斯顿建筑规划设计咨询有限公司

Cobblestone Design Canada/ CobblestoneUrbanists + Architects Workshop Shanghai

Tel: +86-21-65073135 65078631 Fax: +86-65073135-8008 上海 200081 欧阳路681弄3号2F

公司简介

今天，信息技术日新月异并影响着人类生存环境的各个方面，作为营建人造景观的建筑及其相关领域毫无疑问地应当为此做出一份贡献，考斯顿正是清晰地认识到这一点，因此在设计中以城市主义为价值导则、以对城市生活和事件的深刻理解与关注作为设计起点去完成规划、城市设计、建筑和景观的设计。以城市专家、建筑师、景观建筑师和研究人员为研发设计组合的考斯顿团队致力于在信息多元化的今天，提供超越图纸空间的全方位的设计服务，对前期市场研发、产品定位、中期产品设计、后期适时应对市场变化的动态调整进行服务，孜孜不倦寻求设计的依据和价值，并努力平衡人的需求和所有外部压力之间的矛盾。重实效、敏锐、多元化、创造性和多选择的设计服务，以及对文化和生态环境的关注将持续贡献我们为之努力的当代人文景观建设。我们的专业服务涵盖商业办公、星级酒店、酒店式公寓、大型城市混合体、居住区、工业园区、城市街区的策划、规划、建筑设计及景观和室内设计全过程。

Profile

Nowadays, information and technology are changing with each passing day and impacting every aspect of the human environment. As an artificial scene, architecture associated with other design disciplinaries should undoubtedly respond to this change. Cobblestone comprehensively meets the issues of planning, urban design, architecture and landscape with clearly understanding for urban life and urban events. The Cobblestone team combined with urbanists, architects, landscape architects and analysts will be devoted in providing overall services beyond limited the paper space, which are pre-design study, project design, and post-design revision. We mediate between mankind and all external forces to create a balance. Pragmatic, sensitive, plural, innovative and optional design excellence to our culture and mankind environment would contribute to the contemporary human landscape. Our services include commercial, office, hotel, service apartment, urban complex, residential quarter, industrial park and urban block with programming, planning, architectural design, landscape and interior design.

CAUD

城市商业

项目：桃浦生产性服务业功能区常和路
地点：上海普陀区
用途：生产性服务业、现代服务业
建筑规模：60 000平方米
设计时间：2009年—2010年
Project: ShiFan-CreationTaopu, Shanghai
Logcation: Pudong district Shanghai
Use: Industrial Service Division\Service Commercial
Building Area: 60,000 m^2
Design Time: 2009-2010

项目：城市设计/景观设计，三水湾商业街 江苏 泰州
用途：商业
建筑规模：19 000平方米
设计/竣工时间：2009年/2010年
Project: Bay Sanshui Landscape and Architectural Facade Design, Taizhou, JIangsu Province
Use: Commercial
Building Area: 19,000 m^2
Design/Completion Time: 2009/2010

城市景观

创意地产

1. 项目：普陀区桃浦生产性服务业功能区F19地块
——桃浦MOHO，上海
用途：生产性服务业
建筑规模：38 000平方米
设计/建成时间：2009年/2011年
Project: MOHO Taopu, Shanghai
Use: Industrial Service Division
Building Area: 38,000 m^2
Design/Completion Date: 2009/2011

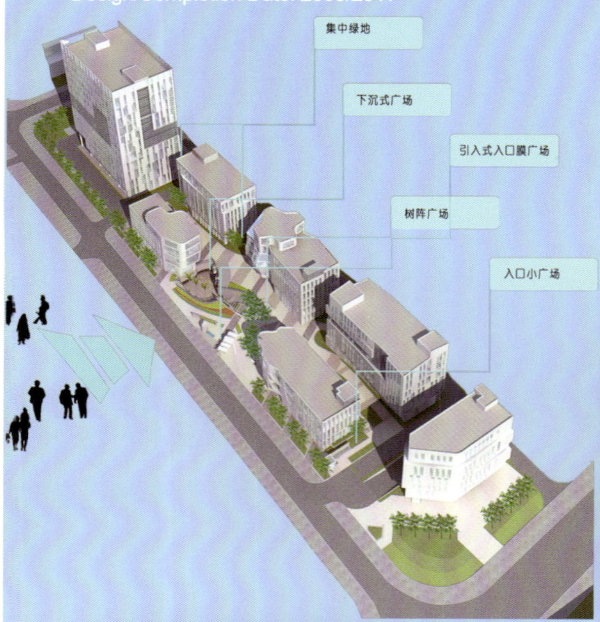

城市住宅

2. 项目：新江湾城05–2七星华府
地点：上海新江湾城
用途：公寓、别墅
建筑规模：25 000平方米
设计/建成时间：2008年—2009年/2010年
Project: 05-2 Seven Star Luxury Housing, Shanghai
Use: apartment, villa
Building Area: 25,000 m^2
Design/Completion Time: 2008-2009/2010

办公

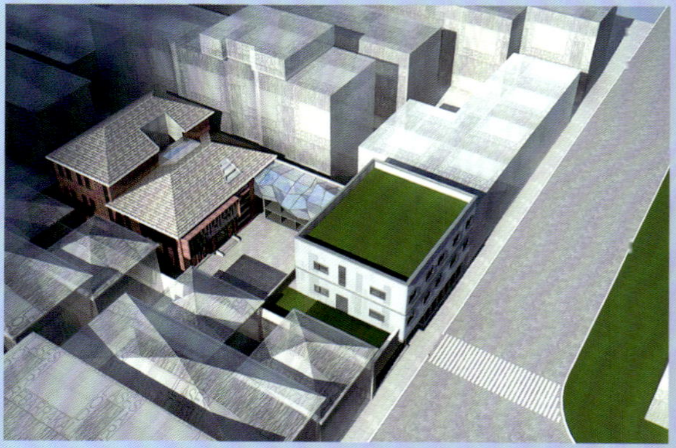

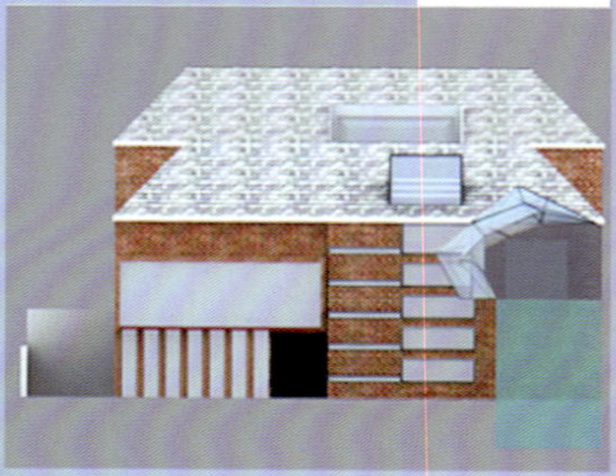

3. 项目地点：上海瑞金二路
用途：办公
建筑规模：2 840平方米
设计/建成时间：2009年—2010年/2010年
Location: Office Building, Ruijin No.2 Rd. Shanghai
Use: Office
Building Area: 2,840 m^2
Design/Completion Time: 2009-2010/2010

4.项目：新江湾城21—3、4公寓式办公
地点：上海
用途：公寓、办公
建筑规模：50 000平方米
设计/建成时间：2009年/2011年
Project: 21-3, 4 Service Office, New Jiangwan City, Shanghai
Use: Office, Apartment
Building Area: 50,000 m^2
Design/Completion Time: 2009/2010

办公

城市设计

5.项目地点：上海普陀区桃浦永登路景泰路
用途：城市空间改造更新重建
建筑规模：19 000平方米
设计时间：2009年
Project: Conceptual Urban Design Taopu, Shanghai
Use: Walkable Street
Building Areae: 19,000 m^2 Design Time: 2009

E-mail: shanghai@whiint.com
Tel: +86-21-62996297, 53930250
Fax: +86-21-62998336
Address:Rm.413, Chuangzhan Building, 928 Xikang Road, Jing'an District,Shanghai

E-mail: guangzhou@whiint.com
Tel: +86-20-38479329
Fax: +86-20-38477163
Address:Rm.611, Fuli Tianhe Business Mansion No.4 Hua Ting Rd, Tianhe District Guangzhou

WHI International is an Australian-based multidisciplinary design practice and a leader within the architectural field. We service the Asia-pacific region with offices and partners in Adelaide, Melbourne, Bangkok, Seattle, Guangzhou, Shanghai and Beijing. WHI International, in partnership with Thailand based P49 DEESIGN has now extended business throughout the Middle East and Europe.

WHI INTERNATIONAL provides innovative design services based on market needs. We employ designers with different cultural backgrounds from Australia, Europe, the U.S., Thailand, China and other Asian countries. This diverse cultural mix and our combined experiences make for a professional design practice, which has successfully provided designs for numerous projects.

Our design covers a broad scope from urban design and master planning of large scale urban areas to tourist zones and residential areas. We also specialise in architectural, interior, and landscape designs for high quality hotels, resorts, offices, retail, villas, residential buildings and public buildings.

WHI INTERNATIONAL 建筑设计集团是澳大利亚知名的综合性设计机构。多年来 WHI 一直走在亚太地区设计行业的前沿。

WHI 经过多年的发展，如今在亚太地区的阿德莱德、墨尔本、曼谷、西雅图、广州、上海、北京均建立了相关机构，包括曼谷的 P49 DEESIGN 在内，业绩已遍布亚太地区、中东、欧洲等地。

WHI 提供完整综合的创造性设计服务，设计范围包括规划、建筑、室内、景观、图文等专业。高素质的设计人才拥有不同的文化、地域背景，包括来自澳大利亚、欧洲、美国、泰国和中国等世界各地。

多元文化的氛围使我们的设计能针对不同的项目提出不同的创造性方案。

WHI 建筑设计集团专注于设计创意、优良服务和以市场为导向的设计定位，作品形态涵盖了城市规划、旅游度假区及住宅区规划、高档酒店、度假村、豪华住宅、别墅、商业、办公楼、综合性公共建筑等范围。

WHI International

1

澳大利亚 AUD 建筑设计公司（上海）
上海工程勘察设计有限公司（二所）

澳大利亚 AUD 设计有限公司（上海）是一家在澳洲注册的专业建筑、装潢、景观设计公司，是与上海工程勘察设计有限公司联合发展的设计单位。

公司承接有大中型住宅小区、商业广场、办公楼、各类学校等项目，在国内外有一定知名度。20 世纪 90 年代，上海的《解放日报》专访我公司的一篇文章中就将我公司比喻为“设计界跑出的一匹黑马”。

公司现有国家一级注册建筑工程师、国家一级注册结构工程师多人，并具有国家甲级设计资质。

公司愿与各业界共同创设未来新空间。

Australia AUD Architectural Design Co.,Ltd.(Shanghai) is a professional construction, decoration, landscape design company registered in ANZAC, which is a united company of Shanghai Engineering Survey and Design Co.,Ltd .

We have undertaken large and medium-sized residential area, commercial plazas, office buildings and other projects and we have received popularity at home and abroad. In 1990s Shanghai Liberation Daily reported our company in special interviews as a dark horse of designing industry.

There are so many national Class A registered architects and national Class A registened struct gineers in our company.

We would like to invite all elites hand in hand to our bright future.

地址：上海市武宁南路 318 号 2 楼
邮编：200042
电话：+86–21–62323145–201
传真：+86–21–62311424
邮箱：skdaud@21cn.com
网址：www.scsad2.com

Address: 2#, Wuning Southroac 318, Shanghai
P.C.: 200042
Tel: +86-21-62323145-201
Fax: +86-21-62311424
E-mail: skdaud@21cn.com
Http: //www.hfutadi.com.cn/

2

3

1–3 曹杨第五中学改造

地址：上海市普陀区
总用地面积：14 690 平方米
地上建筑面积：13 701 平方米
地下建筑面积：2 200 平方米
建筑占地面积：4 110 平方米
容积率：0.93

The reform project of Caoyang Fifth High School

Location: Putuo District, Shanghai Province
Gross Floor Area: 14,690 m^2
Building Area on the Ground: 13,701 m^2
Building Area Underground: 2,200 m^2
Building Area: 4,110 m^2
Plot Ratio: 0.93

4–5 上海市青浦区华新基地 1 号地块 B–1 建筑设计

地址：上海市青浦区
总用地面积：21 470 平方米
总建筑面积：62 946.63 平方米
地上建筑面积：57 512.48 平方米
地下建筑面积：5 434.15 平方米

The architectural design for Block No.1, B-1 of Huaxin Base, Qingpu District, Shanghai

Location: Qingpu District, Shanghai Province
Gross Floor Area: 21,470 m^2
Gross Building Area 62,946.63 m^2
Building Area on the Ground: 57,512.48 m^2
Building Area Underground: 5,434.15 m^2

4

5

6

7

8

9

6–9 安徽省池州商贸城建筑规划设计方案

地址：安徽省池州
总用地面积：32 554 平方米
地上总建筑面积：88 200 平方米
地下建筑面积：17 200 平方米
容积率：2.71

Architectural Planning and Design of Anhui Chizhou Trade City

Location: Chizhou, Anhui
Gross Floor Area: 32,554 m^2
Gross Building Area on the Ground: 88,200 m^2
Building Area Underground : 17,200 m^2
Plot Ratio: 2.71

10

11

12

13

14

10–12 新江湾城 21 号地块

地址：上海市
总用地面积：15 640 平方米
地上建筑面积：31 105 平方米
高层住宅：31 105 平方米
地下建筑面积：12 992 平方米
占地面积：1 725 平方米
容积率：1.99

Block 21# of New Jiangwan City

Location: Shanghai Province
Gross Floor Area: 15,640 m^2
Building Area on the Ground: 31,105 m^2
High-rise Residence: 31,105 m^2
Building Area Underground: 12,992 m^2
Site Area: 1,725 m^2
Plot Ratio: 1.99

13–14 鸿缘广场

地址：上海市普陀区真华路
总用地面积：12 800 平方米（其中红线内 10340 平方米）
容积率：3.325

Hongyuan Square

Location: Zhenhua Road, Putuo District, Shanghai
Gross Site Area: 12,800 m^2 (the areawithin the red line is 10,340 m^2)
Plot Ratio: 3.325

A N S
ARCHITECTS & PLANNERS

ANS国际建筑设计与顾问有限公司

中国总部：上海市静安区西苏州路 71 号 6 楼 200041 电话：+86-21-51697511 传真：+86-21-62668470 电子邮箱：info@ansarchitects.com

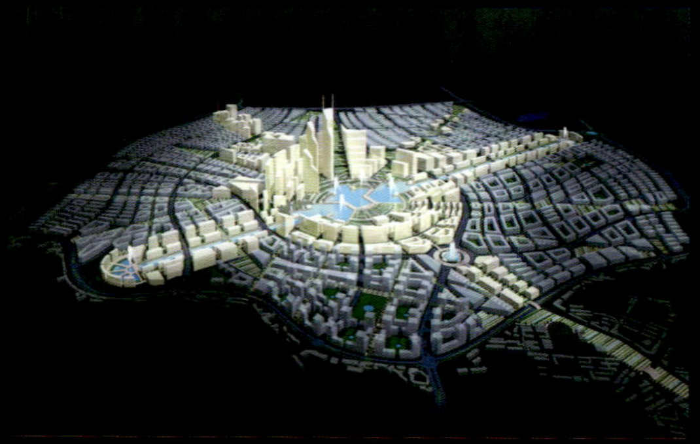

ANS 是一家起源于澳大利亚，并通过其在澳大利亚的合作伙伴进行全球推广及专业服务的国际建筑设计与顾问有限公司，于 1997 年进入中国市场并于上海建立上海爱恩斯建筑设计有限公司，并相继在武汉、天津、成都设立了分公司，其中还包括拥有甲级设计资质的北京彩恩建筑设计有限责任公司。公司主要业务包括：城市规划、建筑设计、室内设计、景观设计，同时提供房地产项目前期工程的咨询顾问服务。ANS 公司力图通过设计作品形式和功能的完美结合，成为生活和行为模式的创导者。公司的承诺是不断创造革新方案、国际设计标准并以客户的满意为最终结果，就像 ANS 所代表的含义——ANS WE R，我们一直致力于将国际化的视野融入本土化的充分理解从而给您带来最完美的解决方案。

ANS International Design & Consulting Pty. Ltd., is a multi-disciplinary international design company originally founded in Australia to offer expertise services in the fields of urban planning, architectural design, landscape design and interior design, as well as pre-project consultations during the early stages of a project through the global promotion of Australian partner. ANS established its first Chinese studio in Shanghai in 1997. Due to its success in its fields of expertise, ANS has since expanded its operations to include branch offices in Wuhan, Tianjing, Chengdu, and a Grade A architectural licensed studio in Beijing. ANS is dedicated to become a leading innovator and creator of the built environment with the goal of enhancing the qualities of life and lifestyle. We endeavor to exceed client expectations with our innovative designs in every project. ANS WE R=ANSWER. Our international designers' fresh and innovative perspectives coupled with our localized knowledge offer our clients THE ANSWER to their question.

www.ansarchitects.com

2010年上海世博会民营企业联合馆

CHINESE PRIVATE ENTERPRISE PAVILION, EXPO 2010 SHANGHAI

概念设计：日本 ADK
建筑设计：
ANS 国际建筑设计与顾问有限公司
项目位置：中国 上海
占地面积：5 570 平方米
建筑规模：5 850 平方米

展馆的建筑主体以"细胞"为设计理念，象征众多民营企业的无限活力以及蕴藏在其内部的能量。我们还为展馆起了一个很特别的名字，叫做"CosMoss 结构"，即"Corrugated Steel Cylinder Module Structure System"，这是一个与内部展示空间的连续性浑然一体的波形钢结构模块系统。通过世界前沿的展示手段和感人的故事构成，描述企业汇聚在一起所产生的"相乘效应"，为参观者展现了中国民营企业的"无限活力"。

外墙材料【智能膜】——挑战常规，勇于创新

展馆外墙材料由镀锌钢折板和镭射膜构成，这是对传统设计思路的一次极大挑战。大量的钢折板勾勒出建筑优美的外轮廓弧线，既是建筑的外围护结构，又给立面提供了丰富的肌理。这种设计构成在过去的建筑案例中从未出现过，无论是从效果还是耐久性方面，都提出了重要的课题。白天，随着外部环境条件的不同，外墙表面会呈现出不同的颜色，建筑仿佛细胞生命体般会表现出不同的表情与参观者进行交流。到了夜晚，在泛光照明和 LED 照明效果的映衬下，展馆外观又会呈现出另一番景象，流光溢彩的膜壁变化，美轮美奂。

屋面结构【靓丽顶】——结构复杂，美观震撼

展馆屋面由 42 个标高平面构成，从 19.4 米到 6 米不等，这是为了满足馆内展区的不同功能要求而设计的。整个建筑高低错落，结构形式相当复杂，丰富的层次变化在光线的映衬下带来了强烈的视觉冲击感，但同时也给技术人员带来了极大的挑战。相关人员经过多轮反复验证、讨论及多次实样制作，才确定了在如此复杂的屋面上同时解决保温、防水、防火以及俯瞰效果等问题的实施方案。民营企业联合馆还凭借其独特的屋顶设计被网易评为世博会九大奇特建筑之一。

等候区域【生命篷】——意兴盎然，流连忘返

展馆等候区域的顶部设有雨篷，从展馆入口呈波浪形向西侧延伸，与其说是等候区，不如说是内部展示内容在室外空间的延伸，是一个集生命、绿化、企业环保理念传达等于一体的特别展示，同时也是对以往传统观念中等候区定义的重大创新。地面上设有地下水槽，里面将出现具有生命的生物。在历届世博会中，很少出现将生命体融入到展示项目中的案例。中国民企联合馆的生命篷打破了这个规律，将具有生命力的生物汇集在一起。与此同时，该区域还将配置绿化、水泡柱以及下水槽的展示内容，使得每个部分彼此组合在一起，构成一个整体，向参观者讲述了一个关于生命的故事。

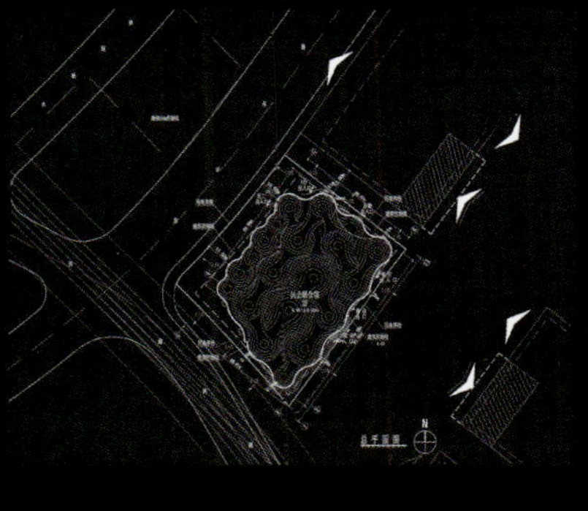

青岛小港湾海岸线规划及酒店设计 2008

XIAO GANG BAY COASTLINE & HOTEL MASTER PLAN, QINGDAO 2008

规划设计：建筑设计
设计团队：Joe Lau，袁德，吕钊
项目位置：中国 青岛
用地面积：15.4 公顷
总建筑面积：128 500 平方米

青岛小港湾位于胶州湾的东岸，城市老港区的南部，属青岛市西部老城区，与中港、火车站、中山路相邻，岸线长度约 2.8 千米。规划重新梳理了海岸线结构，设计以码头景观带串联起广场圆形舞台区域与主体建筑的两心，滨水的城市海岸景观轴和栈桥景观轴连动，形成人、海、城之间的整体空间，立体而有序地激发出滨水区域的空间活力。

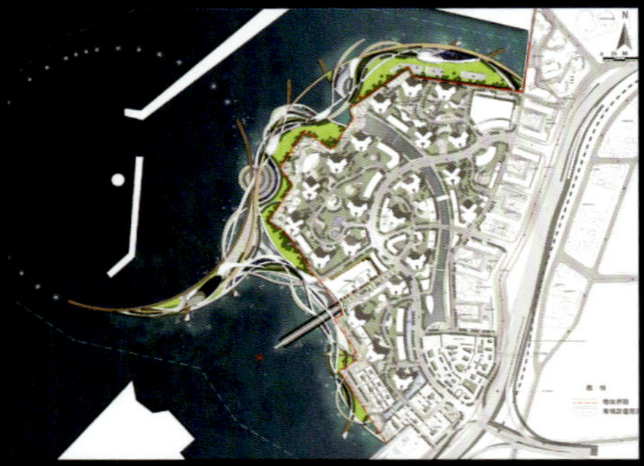

交通系统设计考虑采用回转式的方式（阻力最小的路径），因此设计确定了海滨长廊的形状与方向。对比现有的坚硬边缘，如柔软组织物一样的曲线散步道的形成将使海港更为出色。设计选用垂直交叉的网络来区分和连接行人道和车道，在静态和动态中，提供聚会场所和移动空间，更塑造了一个优质风景区。设计中再充分利用太阳能等自然资源，为废水过滤及循环再造提供能量支持。

酒店设计考虑将建筑建成为青岛最高的建筑，结构在 220 米以上当代城市的地标，它将成为青岛城市文明的新象征。主题酒店设计与区域海岸线规划融将为一体。酒店综合体提供了一系列的娱乐和商业设施，在公共通道和空间内，确保了酒店在商业上的可行性。酒店将包括大约 500 个客房，这些客房位于建筑大楼的下半部分，另有 250 个高档公寓客房位于建筑物上半部分，顶层景观平台将成为青岛高档星级酒店里程碑式的地标。设计充分利用海洋和海港优势，为酒店的客房提供两个截然不同的景观。酒店设施将包括主席台、零售、会议设施及广场、餐饮和商业。

品味青岛城市那自然天成的山水，油然而生一副水墨画卷的构想。眼前的小港湾墨兰摇曳盛开，片片飘逸的兰花叶勾勒出美丽的港湾之景，朵朵兰花宛如在港湾中散发出清新而幽雅的花香，充满韵味。建筑设计中西合璧，极富创意，海天一色的美丽图画随即展现在这东海之滨。

“不是生来偏爱兰，缘于神姿自天然。心逐碧草摇清风，嗅得幽香沁心田。”

FLA 梵瑞设计
www.fla-arch.com

城市设计	居住建筑	公共建筑	环境景观	工业建筑
金华白龙桥核心区城市设计	盐城澄达东景苑规划建筑设计	科技部上海培训中心建筑设计	上海高等法院景观设计	上海富士达电梯项目或厂区二期设计
盐城劝业场商业中心区规划设计	上海飞虹路地块规划建筑设计	盐城飞驰新天地建筑设计	盐城娱乐花园景观设计	上海正泰电器厂区设计
郑州高新区规划设计	上海川宝地块规划建筑设计	常州电脑城建筑设计	扬州新加坡花园景观设计	
盐城新洋经济区规划设计	扬州新加坡花园规划建筑设计	盐城市中医院建筑设计	大丰恒达国际城景观设计	
	盐城大地名居规划建筑设计	盐城市体校规划建筑设计	盐城机电学校景观设计	（以上为部分作品）

FLA 梵瑞设计是由加拿大 FORM LANDSCAPE ARCHITECTS 和上海梵瑞建筑设计咨询有限公司共同创办的全方位、多元化的设计公司，分别在加拿大多伦多和中国上海注册公司。公司是极富个性的新智力集团，拥有自成体系的设计理念、充满创意的工作方法和融洽的人际关系，并注重最大限度地激发设计人员的创作激情。公司拥有强大的设计团队，追求创新思维和前沿技术，具备国际视野与先进设计理念。

FLA 梵瑞设计秉承“客户满意、团队合作、尊重个人、追求卓越、注重结果”的企业价值观，以具有前瞻性的设计理念，经验丰富的人才优势为核心，关注及理解客户以及客户的业务，努力满足客户的需求，与客户结成战略性的伙伴关系，从而创造出具有建筑理想及专业精神且符合客户及市场需求的建筑作品。

梵瑞多年的耕耘和付出，使其在全国尤其江浙地带留下了众多经典的设计作品，赢得了社会各界的赞誉，其中重大项目的有上海交大闵行新校区规划、金华白龙桥核心区城市设计、盐城劝业场商业中心区建筑设计等，这些项目先后赢得多个设计大奖。作为中国知名的设计企业，FLA 借鉴世界范围内最新的设计理念和设计方法，坚持建筑创作的自主性，并将高品质的作品推向市场。

FLA (Fanrui Design) is a professional design and consulting firm, co-founded by the Form Landscape Architects (FLA) and Shanghai Fanrui Architectural Design Consulting Co., Ltd, and registered both in Toronto and Shanghai. This firm is a group of professionals. Their specialized excellence, systematical design concept, creative design methods and harmonious interpersonal relationship maximally inspire each other's creativity. This is a strong design team, pursuing innovation and cutting-edge technology, seeking an international vision and advanced design solutions.

FLA adheres to the value of “customer satisfaction, teamwork, respect for individual, pursuit of excellence, results orientation”. With forward-looking ideas and experienced personnel as its core, this firm shows concerns and understanding for customers and their business, endeavoring to satisfy customers' needs and forming a strategic partnership, creating architectural designs carrying with ideal and professional spirit which meet both of the customer's and the market's needs.

Through several years' efforts, FLA has produced many unique designs throughout China, especially in Jiangsu and Zhejiang Province, also got praises from all levels of society and won many design awards. The major projects include the planning for Minhang Campus of Shanghai Jiaotong University; the urban design for the core area in Jinhua; the architectural design for Quanyechang central business district in Yancheng, etc. As a well-known design firm, FLA absorbs the latest design approaches and methods all around the world, insists on creativity, initiative and innovation in architecture, and launches its high-quality designs into market.

上海梵瑞建筑设计咨询有限公司

网址：www.fla-arch.com
邮箱：fla@fla-arch.cn
地址：上海市杨浦区大连路1546号16楼F座
邮编：200093
电话：+86-21-5596 1791
传真：+86-21-5596 3977

Http: //www.fla-arch.com
E-mail: fla@fla-arch.cn
Address: 16F, No. 1546 Dalian Road, Yangpu District, Shanghai
P.C.: 200093
Tel: +86-21-5596 1791
Fax: +86-21-5596 3977

加拿大 FORM LANDSCAPE ARCHITECTS

地址：42 KERBAR ROAD, SCARBOROUGH, ON MIVIG3
电 话：416-278-5066　传真：416-817-1866

FLA 梵瑞设计
www.fla-arch.com

1

3

4

5

6

2

7

8

1–2 常州电脑城
建设地点：　江苏 常州
用地面积：　30 345 平方米
地上建筑面积：105 440 平方米
地下建筑面积：40 890 平方米
设计阶段：　规划及建筑方案

3–5 盐城劝业场商业中心（方案一）
建设地点：　江苏 盐城
用地面积：　52 170 平方米
地上建筑面积：204 400 平方米
地下建筑面积：54 500 平方米
设计阶段：　规划及建筑方案

6–8 盐城劝业场商业中心（方案二）
建设地点：　江苏 盐城
用地面积：　52 170 平方米
地上建筑面积：189 720 平方米
地下建筑面积：65 730 平方米
设计阶段：　规划及建筑方案

FLA 梵瑞设计

www.fla-arch.com

1

2

3

4

5

6

7

8

9

10

1–3　飞驰新天地
所在地：　江苏 盐城
用地面积：　15 550 平方米
地上建筑面积：59 090 平方米
地下建筑面积：13 000 平方米
项目阶段：　已建成

4–6　澄达东景苑
所在地：　江苏 盐城
用地面积：　88 789 平方米
地上建筑面积：137 827 平方米
地下建筑面积：38 940 平方米
项目阶段：　已建成

7–8　大地名居
所在地：　江苏 盐城
用地面积：　83 000 平方米
地上建筑面积：49 800 平方米
地下建筑面积：12 000 平方米
项目阶段：　施工中

9–10　昆山新天地（景观）
所在地：　江苏 昆山
用地面积：4 800 平方米
项目阶段：已建成

Kasian

Firm Overview

At Kasian, we see architecture and design as a powerful tool with which to influence the human experience. We harness the collective expertise and imagination of a global team of 400 innovative professionals to find feasible solutions to the complex questions asked by our clients. Our integrated, collaborative approach means that the needs of our clients are always met with exceptional places and sustainable environments for people everywhere. Kasian is one of the largest fully integrated architecture, planning, and interior design firms in Canada and we now have offices in Vancouver, Edmonton, Calgary, Toronto, Abu Dhabi, Dubai, Mumbai and Shanghai. Our award-winning projects range from the most advanced children's hospitals in the world to master plans for the most densely populated cities on the planet. Yet the same client focused approach that we started with drives our business today.

- *400 global staff with 80 registered architects, interior designers, and planners*
- *40 Awards won in the last 10 years*
- *Over 100 projects in China*
- *Consistently ranked among top 100 architecture firms in the world*

Market Sector

Each of our Practice Areas represents an area of expertise, where inspiration combines with experience: Health Care, Transportation, Civic, Academic, Science & Technology, Commercial, Master Planning, Interior Design, and Strategic Facilities Planning

Locations

Vancouver
Calgary
Edmonton
Toronto
Abu Dhabi
Dubai
Mumbai
Shanghai

KASIAN ARCHITECTURE
DESIGN CONSULTING (SHANGHAI) CO, LTD
Platinum Building,
233 TaiCang Road, Suite 2508
Shanghai, China 200020
www. kasian.com

1. Baoneng Asia City, China
2. Baoneng Asia City, China
3. Sun Yat Sen University, China
4. British Columbia Institute of Technology, Canada
5. Xiamen Design Competition, China

A. Mosiac City, United Arab Emirates
B. Canada Line - Airport Station, Canada
C. G + 50 Tower, United Arab Emirates
D. Kitsilano Reserve Master Plan, Canada

公司概况

在凯盛国际，我们坚信创造性思维的力量，这种创造力融合了我们的文化传承，我们与客户互动而产生的思想火花以及我们刻意追求的建筑空间意向。

凯盛国际成立于1985年，并迅速发展为北美建筑界首屈一指的设计公司。我们所获得的100多项专业嘉奖极大地验证了20多年来公司全体成员的敬业精神和专业素质。今天，它以横跨全球的8个办公室和多达400余人的专业服务团队无可置疑的成为加拿大规模最大的专业建筑设计公司。我们开放式的价值观念和彼此信任的合作机制多年来已经成为公司快速发展和应对挑战的制胜法宝，我们称之为“凯盛模式”。

我们的创作热情已经不仅仅停留在美学阶段，而是全面提升到理解客户的项目意图和需求，以及我们的作品如何影响人类社会的行为等更高层次上。我们更善于将这种热情与开放的心态、专业的追求以及工作的幽默结合起来，期望在一种放松而高效的工作环境中为客户提供最大的价值服务。

- 全球近400名员工，其中近80名为注册建筑师、室内设计师及规划师
- 过去10年中荣获近40项设计大奖
- 在中国有超过100项工程的设计经验
- 世界100强建筑设计企业

专业领域

作为一个大型建筑设计公司，我们的专业服务围绕着我们最擅长的项目知识和经验而为。这种对不同项目类型的深刻理解和丰富经验演变成为我们在以下专业领域的强大竞争力。我们在公共建筑、科教建筑、医疗建筑、交通建筑、都市开发、规划设计、室内设计以及空间策划方面都有独一无二的专业团队，随时能够为不同的项目提供无微不至的支持和服务。

全球办公室：
温哥华
卡尔加里
埃德蒙顿
多伦多
阿布达比
迪拜
孟买
上海

凯臣建筑设计咨询（上海）有限公司/
凯盛国际（中国）全资子公司
上海市卢湾区太仓路 233 号
新茂大厦2508座
www. kasian.com

kasian

▲ 深圳万科天琴湾 建设地点：广东深圳 设计时间：2007年6月—2008年1月 竣工时间：2009年6月

美國KAS設計有限公司
深圳凯斯筑景设计有限公司
KAS DESIGN COMPANY

KAS DESIGN COMPANY is a comprehensive architectural design company. Its business scope covers planning, architectural, landscape, interior design and architectural photography. Its headquarter is in Hawaii in the United States, the Shenzhen company is located in the Futian, Chegongmiao, Shenzhen.

Since the founding of KAS, it has been dedicated to planning and construction, architecture and landscape architecture and integration between the indoor and exploration. KAS thinks that a good work must be well understood over several aspects and, through the various fields of expertise together the way to achieve comprehensive, interdisciplinary, integrated design effect. KAS has been guided by honesty and reliable, excellence, innovative, visionary thinking and high sense of responsibility.

Since the establishment, KAS is steadily developing developments. The projects throughout the country, includes Vanke,Gemdale, Excellence Group, Poly,Fuli,Kingkey,FINANCIA STREET HOLDINGS,SPG,Galaxy, Fuchun Oriental, Pearl River Investment, HongRongYuan, Guangzhou Li Haiand other well-known real estate developers and involved in the construction phase of the programme to the coordination by many U.S., Britain, Australian and Japanese planning, architects, landscape, interior designers, and jointly create a high-quality, customer satisfaction, and with all sectors of the community recognized and appreciation for the work.

KAS DESIGN COMPANY INC.（凯斯设计公司）是一家综合性的建筑设计公司，其业务范围涵盖规划设计、建筑设计、景观设计、室内设计以及建筑摄影等。公司总部在美国夏威夷，深圳公司位于深圳市福田区车公庙。

KAS成立以来，一直致力于规划与建筑、建筑与景观、建筑与室内之间的融合与探索。KAS认为，一个好的作品，必须对上几个方面作出很好的理解，并通过将各个专业领域融合的方式，来实现整体性强、跨学科、一体化设计的效果。KAS一直遵循诚信可靠、精益求精、富有创意、有远见的设计思维并有着高度的责任感。

KAS成立以来，稳步发展，承接的项目遍布全国各地，与万科、金地、卓越、保利、富力、京基、金融街、盛高置地、星河、富春东方、珠江投资、鸿荣源、广州利海等著名的房地产开发商合作，并和多名美国、英国、澳大利亚及日本的规划师、建筑师、景观设计师、室内设计师共同参与了从方案到施工阶段的全过程，共同创造出了一个个高品质、让客户满意、并得到社会各界认可和赞赏的作品。

KAS DESIGN COMPANY
HEADQUARTERS
250 KAWAIHAE
HONOLULU,HAWAII 96825,USA
Tel: 001-808-927 8139

美国KAS设计有限公司
美国夏威夷檀香山Kawaihae大街250号
邮编：96825
电话：001–808–927 8139

SHENZHEN COMPANY
Room 11B, Yunsong Tower, Chegongmiao.
Futian,Shenzhen, Guangdong 518040, P.R.C
Tel: +86-755-8835 0900
Fax: +86-755-8835 0400
Http: //www.kas.cn
E-mail: kas@kas.cn

深圳凯斯筑景设计有限公司
深圳市福田区车公庙云松大厦11B
邮编：518040
电话：+86–755–8835 0900
传真：+86–755–8835 0400
网址：www.kas.cn
电子邮箱：kas@kas.cn

海口万科浪琴湾　建设地点：海南海口　设计时间：2007年3月—2007年12月
竣工时间：2009年5月

东莞丰泰观山碧水　建设地点：广东东莞　设计时间：2007年5月—2007年9月
竣工时间：2009年1月

KAS DESIGN COMPANY

▲ **南宁荣和中央公园** 建设地点：广西南宁 设计时间：2007年5月—2008年1月 竣工时间：2009年5月

▼ **常熟怡景湾** 建设地点：江苏常熟 设计时间：2007年1月—2007年12月 竣工时间：2009年10月

KAS DESIGN COMPANY

大连万科溪之谷

建设地点：辽宁大连　设计时间：2009年5月—2009年7月
竣工时间：2009年9月

深圳首地容御

建设地点：广东深圳　设计时间：2009年1月—2009年5月
竣工时间：2009年8月

KAS DESIGN COMPANY

H2 architecture design group.inc

地址：上海市浦东新区陆家嘴环路958号华能联合大厦36楼
邮编：200120
电话：+86–21–61623070
传真：+86–21–61623069

Address: Floor 36, Huaneng Union Tower, No.958 of Lujiazui Ring Road, Pudong District, Shanghai
P.C.: 200120
Tel: +86-21-61623070
Fax: +86-21-61623069

注册于美国加利福尼亚州洛杉矶的H2建筑设计公司（中国注册–意境（上海）建筑设计有限公司）是由一群杰出的设计师领导的面向国际的建筑设计公司。我们的原则是在坚持于成本效益的基础上为用户提供最高水平的设计，并努力使我们的每一个作品忠实于建筑的功能，环境，文脉和可持续性的发展。我们提供的服务包括以下几种。

Registered in Los Angeles, California, the United States (Sino-registered - Architectural Design Co.,Ltd. in Shanghai) this architecture design company is a group under the leadership of many outstanding designers serving for international architectural design. Insisting on cost-effective basis, our principle is to provide users with the highest level of design, and strive to make each work faithful to the function of the building, to environment, to culture, and to sustainable development as well. Our services include:

建筑设计 Architectural Design
城市设计 Urban Design
室内设计 Interior Design
规划设计 Master Planning
景观设计及主题公园 Landscape Design & Theme Park

开封西北湖酒店

建筑面积：24 020 平方米
设计说明：
项目应用抽象、分散的形态布局形式，酒店主体位于湖面中央区域，酒店周边均有湖面与外围隔离，基地西南侧设置高尔夫练习场，会所位于酒店西南和高尔夫练习场之间，浮于水面之上。

Kaifeng Northwest Lake Hotel

Building Area: 24,020 m^2
Design Description: The project is in rational and abstract dispersion layout, and the large part of hotel is located in the central region of the lake, which isolates the external environment off hotel by the lake. The southwest side of the base sets Golf Driving Range. Floating on the water, the club is located between the southwest part of the hotel and Golf Driving Range.

上海振龙酒店

项目说明：
振龙酒店总共有540个自然间和449间客房。酒店以其完善的硬件设施、超一流的服务，将为商务人士增添全新的体验。它将成为浦东地区超五星级的豪华商务休闲酒店。

Shanghai Zhen-Long Hotel

Project Description: Zhenlong Hotel owns 540 individual rooms and 449 guest rooms. With its perfect facilities, perfect service, the Hotel will provide a holy fresh experience with customers, striving for a super five-star luxury business and leisure hotel in Pudong area.

三亚富力湾万豪精品酒店

总建筑面积：16 200 平方米
设计说明：

酒店坐落在陵水县东部沿海地区，酒店的特色可以概括为"天然、豪华、新颖、奇特和广阔"。它后靠山、前面水，拥有便捷的交通设施和热带风光。

Fuli Bay Marriott Boutique Hotel, Sanya

Building Area: 16,200 m^2

Design Description: The hotel is located in the east of the coastal area of Lingshui County of Hainan, in feature of "wild, luxurious, novel, fancy and wide". It backs the mountain while faces to the sea, with convenient transportation facilities and many tropical scenes of fun, modern sports and recreation items.

西安威斯汀博物馆酒店

建筑面积：90 737 平方米
设计说明：

项目建成后将成为西安市档次最高、功能最全、规模最大的五星级酒店，将为曲江新区带来更全面、更高尚的经济文化发展和城市旅游发展。

Xi'an Westin Museum Hoteltel

Building Area: 90,737 m^2

Design Description: After completion, this project will be the highest grade, and the most fully featured five-star hotel in Xi'an, which will even promote the development of Qujiang New District with more comprehensive, more noble economy, culture and urban tourism as well.

苏州太湖精品酒店

建筑面积 28 450 平方米　　客房酒店：122 间　　豪华别墅：43 栋
设计说明：

项目人为本，步行优先，创造了度假村安全、便捷、舒适的环境，充分体现了动与静，风与林，山与水，绿化与铺地，建筑与公园，城市与自然的融合统一。

Taihu Boutique Hotel, Suzhou

Building Area: 28,450 m^2　　Guest rooms: 122　　Luxury Villas: 43

Design Description: Human-oriented, walking first, this project shoots at creating a resort round-the-clock which is safe, convenient and comfortable. It fully expresses the rich integration experience between the dynamic and static, wind and forest, mountain and river, afforestation and flooring, natural and artificial, exclusive and leisure, building and park, as well as city and natural environment relationships.

昆明抚仙湖高尔夫旅游度假村

规划面积：444.7 公顷
设计说明：

项目充分结合地形，并合理利用景观资源形成跨谷酒店这一独特创新点。规划中，设计师充分尊重现状，以地方文化元素贯穿全局，规划了以生命起源为主题的豪华酒店，以静养和居住为一体的养生园，以打造世界一流景观的高尔夫球场为聚焦点，以玉溪的民俗风情和抚仙湖的美丽传说给出行者送上一场非凡洗礼的盛宴。

Fuxian Lake Golf Resort, Kunming

Planning Area: 444.7 ha

Design Description: This project takes fully advantage of the terrain and rationally uses landscape resources to realize a unique innovation in the formation of cross-valley hotel. Its designer shows full respect for the status quo, lets local cultural elements through the entire project in its planning, and plans a luxury hotel themed by the origin of life, a resort integrated convalescence and living functions. It is focused as building world-class golf course landscape. Now it is waiting to offer a remarkable feast with Yuxi's legends and customs around its beautiful Fuxian lake for tourists all over the world.

公司简介 Aedas Profile

凯达是全球第二大建筑设计师事务所，凯达39所办事处为客户提供最合适的建筑设计方案。我们了解各地的风土人情，并将当地智慧融入国际级的建筑设计概念。凯达对保持生态平衡有一套整合的方针，这体现出我们不单重视建筑环境，同时对我们的合作伙伴和小区团体非常关注。

凯达为亚洲、中东、欧洲及美洲提供建筑设计、室内设计、总体规划、景观设计、城市规划与建筑咨询等顾问服务。我们对改善自然、建筑环境的承诺和坚持从我们的作品中体现出来，当中包括公共、教育、医疗、交通运输设施、商业大厦、住宅及工业大厦等。

Aedas, the world's second largest architectural firm, creates some of the best architectural design solutions from our 39 offices around the globe. We provide international expertise with innate knowledge and understanding of local cultures. Our designers are committed to contributing and lead in the societies we design for. We have a holistic approach to sustainability which shows our care not only for the built environment but also for the people we work with and the societies we live and work within.

Aedas global network provides consultancy services in architecture, interior design, master planning, landscape, urban design and building consultancy in Asia, the Middle East, Europe and the Americas. Our commitment and tenacity to improve the natural and built environment are reflected in our designs for all building types including civic, education, health, transport, commercial, residential and industrial buildings.

香港项目 Hong Kong Projects

1 中环新海滨城市设计研究
Central Reclamation Urban Design Study

2 天际万豪酒店
SkyCity Marriott Hotel

3 香港国际机场北卫星客运廊
North Satellite Concourse, Hong Kong International Airport

4 置地广场
The Landmark

5 港铁迪斯尼支线欣澳站
MTRC Sunny Bay Station, Disneyland Resort Line

1

2

亚洲 Asia

香港 澳门 北京 上海 成都 沈阳 新加坡
胡志明市 河内 新德里 孟买 阿拉木图 卡拉奇
Hong Kong Macau Beijing Shanghai Chengdu Shenyang Singapore Ho Chi Minh City Hanoi New Delhi Mumbai Almaty Karachi

欧洲 Europe

伦敦 伯明翰 布里斯托尔 爱丁堡 格拉斯哥
哈德兹菲 利兹 利物浦 曼彻斯特 什鲁斯伯里
贝尔格勒 基辅 莫斯科 华沙 卡托维兹 都灵
London Birmingham Bristol Edinburgh Glasgow Huddersfield Leeds Liverpool Manchester Shrewsbury Belgrade Kiev Moscow Warsaw Katowice Turin

美洲 America

纽约 华盛顿 洛杉矶 西雅图 圣保罗
New York Washington DC Los Angeles Seattle Sao Paulo

中东 Middle East

阿布扎比 巴林 迪拜 多哈 利雅得
Abu Dhabi Bahrain Dubai Doha Riyadh

3

Aedas 凯达

Aedas and KPF as Joint Design Architect
2008年优质建筑优异奖
4

Business Week / Architectural Record
2008中国环保大奖
5

中国项目 China Projects

6. 北京富力城公建项目
R&F City, Beijing

7. 苏州西交利物浦大学行政信息楼项目
Xi'an Jiaotong - Liverpool University Administration and Information Centre, Suzhou

8. 成都太升南路综合发展项目
South Tai Sheng Road Mixed-Use Development, Chengdu

9. 厦门港国际旅游客运码头
Xiamen International Cruise Terminal

10. 广州富力中心
R&F Centre, Guangzhou

11. 上海鹏利海景花园
TG Harbour View Apartments, Shanghai

12. 无锡恒隆广场
Centre 66, Wuxi

6

7

8

9

10

11

12

澳门及东南亚项目
Macau & Other Asia Projects

13. 印度格岗Raisina住宅发展总体规划
Raisina Residency Masterplan, Gurgaon, India

14. 新加坡313@somerset
313@somerset, Singapore

15. 越南胡志明市Park City
Park City, Ho Chi Minh City, Vietnam

16. 澳门威尼斯人度假村酒店
Venetian Macao-Resort-Hotel, Macau

17. 新加坡滨海湾金沙综合休闲度假村项目
The Marina Bay Sands™, Singapore

18. 澳门路氹金光大道四季酒店
Cotai Strip™ Four Seasons, Macau

13

14

15

16 (Aedas as Lead Consultant in Macau; HKS as Lead Consultant in the US)

17 Photo by Timothy Hursley
(Aedas as Lead Consultant & Executive Architect; Moshe Safdie as Lead Design Architect)

18 (Aedas as Lead Consultant & Executive Architect; Paul Steelman as Lead Design Architect)

1

2

1–2 天津经济技术开发区标志性建筑概念方案设计

建设地点：天津
项目类型：超高层公建
建筑面积：400 000平方米
占地面积：27 790平方米
设计时间：2005年
竣工时间：2012年
容 积 率：11

项目介绍

项目用地位于天津泰达经济开发区的滨海区域，总建筑面积为400 000平方米，占地面积为27 790平方米，基地东侧紧临开发区的城市景观轴，南侧和西侧是繁华的第一大街和新城西路，北侧是宽阔的CBD中心公园。

基地内呈“品”字形布置的三幢塔楼，分别是300多米和150多米高的商业综合楼及100多米高的政府检验检疫局大楼。商业综合楼集办公、酒店、公寓、会议、大型集中商业功能于一身，300多米高的1号塔楼更是开发区的第一高楼，是滨海区的地标性建筑。

Conceptual Design of Landmark Building in Tianjin Economic Technological Development Area

Construction Location: Tianjin
Type: high-rise building
Building Area: 400,000 m^2
Site Area: 27,790 m^2
Design Time: 2005
Completion Time: 2012
Plot Ratio: 11

Project profile

Located in waterfront of Tianjin Taida development district, this project has total constrcution area with 400,000 m^2, land using area with 27,790 m^2. East side is next to the landscape axis of development area, south and west side next to 1st street and Xin Cheng Xi Lu and north side next to CBD central park. Three towers are two commercial towers with more than 300 m height and 150 m height and one CCIB(China's Commodity Inspection & Quarantine Bureau) building. Commercial buildings have office, hotel, apartment, conference and shopping center etc. And the 1st tower with more than 300 m height is the highest building and landmark in this area.

3 成都希尔顿酒店

建设地点：四川 成都
项目类型：五星级酒店
建筑面积：200 000平方米
占地面积：29 304平方米
设计时间：2007年
竣工时间：2009年
容 积 率：6.8

项目介绍

成都希尔顿酒店阿特金斯上海分公司在成都的一项建筑综合体的设计任务，项目建成后将成为成都市地标建筑之一。

本项目由3个塔楼构成，将分别作为五星级酒店、公寓综合体以及办公楼使用，三位一体的形式组成了奇妙的对称组合。设计暗喻了将一个水晶体一分为三，同时创造了宏伟的门户形象，并在建筑的腹地创造了一处可供民众游憩的城市公共空间。

项目用地位于成都南部市中心通往新政府办公区及展览中心的主要道路上。阿特金斯的建筑师蒋力航介绍说：“在一块相当有限的用地内要达到甲方要求的建筑面积是非常困难的，我们考虑用一种单一的形式去塑造形体，虽然分为三部分但仍然作为一个建筑综合体。3个水晶体创造了一个独特而又引人注目的建筑形体，同时将基地的中心开放成为公共活动中心。”

Hilton Hotel, ChengDu

Construction Location: Chengdu, Sichuan
Type: 5-star hotel
Building Area: 200,000 m^2
Site Area: 29,304 m^2
Design Time: 2007
Completion Time: 2009
Plot Ratio: 6.8

Project profile

The project comprises three towers which contain a five star hotel, an apartment complex and an office tower arranged as a dramatic symmetrical group. The design alludes to the form of a crystal divided into three, creating a grand gateway and new public space at the heart of the complex. The project is located on the main thoroughfare that runs from central Chengdu to the new government and exhibition centre at the south of the city. Atkins Shanghai architect Lihang Jiang said, “The client had a very demanding programme that squeezed in a lot of accommodation. We wanted to create a landmark that separated the three main functions and yet suggested the complex to be a much greater single form. The tripartite form creates a unique and eye-catching shape yet opens up the centre of the site as a shared space for office workers, hotel guests and apartment residents”. The main entrance to the complex is marked by a crystalline canopy which hangs over the podium at the base of the hotel and apartment towers.

3

4

4–7 太原机场
建设地点：山西 太原
项目类型：机场
建筑面积：54 000平方米
设计时间：2005年
竣工时间：2008年

项目介绍

太原武宿机场是2008年北京奥运会的主要备降机场、国内干线机场。在建筑形体的设计上，根据用地的实际特点，采用45度斜向发展的方式，巧妙处理了主楼、走廊与现有航站楼三者之间的关系，创造了全新的室内外环境并由建筑功能特点衍生出3个极具山西地域特色的大院空间，同时内庭院也很自然地把国内、国际航班进行分流，使新航站楼成为一个会呼吸的生态建筑。建筑的立面设计上采用三角翼飞行器的造型，利用多片三角形屋面的交错起伏，形成动势，建筑的空间形体雄浑飘逸。飞翔的三角翼造型寓意为“起飞三晋”。太原机场将成为太原市新的地标性建筑。

Taiyuan Airport
Construction Location: Taiyuan, Shanxi
Type: Airport
Building Area: 54,000 m²
Design Time: 2005
Completion Time: 2008

Project profile

As the main backup airport for 2008 Olympic Game and a main domestic airport, Taiyuan airport creates a whole new internal & external environment and its terminal is an ecological building by using local charaters, dealing main building, finger corridor and current terminal relationship technically. Three Shanxi character yard-spaces derives from function charateres. At the mean time airport's internal yard has distributed the domestic and international naturally. The elevation deisgn uses triangle wings figure, which is dynamic and a symbol of "San Jin Taking off", becomes a lanmark of Taiyuan City.

5

6

7

8

9

10

8–10 滴水湖皇冠假日酒店
建设地点：上海
项目类型：五星级酒店
建筑面积：66 900平方米
占地面积：126 700平方米
设计时间：2007年
竣工时间：2010年
容 积 率：0.3

项目介绍

该项目在创作中强调“以人为本，环境为先”的原则，在满足使用的实用性、耐久性的同时，着重强调均好性，注重环境、节能、环保三大主题。现代建筑设计的节能措施在设计应用时，除需保证建筑的安全性、舒适性、智能化和生态环境等因素外，还应注重能源的有效使用，提高居住空间的品位，提升城市的品牌。

Holiday Inn Crown Plaza, DiShuiHu
Construction Location: Shanghai
Type: 5-star Hotel
Building Area: 66,900 m²
Site Area: 126,700 m²
Design Time: 2007
Completion Time: 2010
Plot Ratio: 0.3

Project profile

Using the humanistic perspective & primary environment as design principle, this design emphasizes the three themes which include environment,energy saving and environmental conservation. It also should satisfy the applicability and durability. Besides safety, comfortability and intelligent, it should consider more about energe efficiency and saving, also improve living taste and city branding.

gmp · 德国建筑师事务所 冯 · 格康，玛格及合作者

事务所介绍

麦哈德 · 冯 · 格康，福尔克温 · 玛格于1965年创立的建筑师联合事务所目前有六个合伙人。gmp拥有350多名员工，分布于十个不同的地区，在国内外积极地从事建筑设计工作。

gmp是少数进行全方位设计的建筑师事务所之一，其对建筑项目从方案设计到施工建造直至室内装修都会全面负责。

麦哈德 · 冯 · 格康，福尔克温 · 玛格与他们的合伙人过去40年中，于德国的各大城市进行了规划和建造设计工作。其涉及的项目包括私人别墅、旅馆、博物馆、剧场及音乐厅、办公楼、贸易中心、医院、科研教育设施、交通建筑、工业建筑和总体规划等方面。

国际上，gmp在创建之初是以机场建筑设计而闻名于世的。1975年由gmp设计的柏林特格尔机场作为第一座驶入式机场而被启用，这一创新设计理念在斯图加特机场及汉堡机场航站搂的出港和到港的设计建造中进一步得到了发展和完善。

受到世界广泛重视的不仅有已建成的作品如莱比锡新会展中心、意大利利米尼会展中心或2000年世界博览会基督教展厅，也有诸如柏林奥林匹克体育场改建加顶项目、科隆莱茵河能量体育馆和2006年竣工的“柏林中心火车站”的建筑设计。

在德国国内和国际性建筑设计竞赛中，gmp获奖430多次，其中一等奖230多项，另外因其创新性的建筑设计而获得了众多优秀奖。gmp至今共设计建成260多座建筑。

在中国已建成的有南宁和深圳的会议展览中心、上海浦东新区文献中心、北京中关村文化中心和佛山世纪莲体育中心。另外还有一系列项目仍在建设和设计中，其中包括重庆大剧院、青岛大剧院和北京的中国国家博物馆扩建工程。在2006年，位于越南河内的国际会展中心现已竣工，越南城市历史博物馆和越南内政部工程尚在建设中。

此外，在“斯图加特21”、“汉堡港口新城”和“布加勒斯特2001”的规划设计中以及可容纳80万居民的临港新城在内的众多中国城市规划设计项目，gmp也充分证明了其在城市规划方面的专业能力。

目前，gmp除了德国以外还在中国、印度、越南、南非、阿拉伯联合酋长国、意大利、西班牙、土耳其、罗马尼亚、波兰、拉脱维亚等国家和地区承担了建筑项目的设计任务。

The practice

The architectural practice von Gerkan, Marg and Partners was founded by Meinhard Von Gerkan and Volkwin Marg in 1965. Since its inception it has grown to include six additional partners and more than 350 employees in ten offices in Germany and abroad.

gmp is one of the few practices with a generalist position, which takes responsibility for a project from the design idea and its realization right through to the interior design.

In the past forty years Meinhard Von Gerkan, Volkwin Marg and their partners have planned and constructed buildings in most major German cities. They have designed small scale homes, hotels, museums, theaters and concert halls, office buildings, commercial centers and hospitals as well as research, educational and sports facilities, buildings for transportation, trade and industry buildings and master plans.

gmp became internationally known as a team of airport architects when, in 1975, the Berlin-Tegel Building with its drive-in airplane terminal for "stacked parking" was established. This innovative design has been further developed to transform the departure and arrival halls at Stuttgart and Hamburg airports.

They are internationally recognized not only for their projects like the New Trade Fair in Leipzig, the Rimini Exhibition Center or the Christ Pavilion for the EXPO 2000, but also for their designs. Key examples are the designs for the reconstruction and roofing of the Olympic Stadium in Berlin, the RheinEnergie Stadium in Cologne or Berlin Central Station, which was completed in 2006.

gmp has succeeded in winning more than 430 prizes in national and international competitions, among them more than 230 first prizes and a large number of awards for exemplary architecture. More than 260 buildings have been constructed by gmp worldwide.

Amongst others, the exhibition and congress centers in Nanning and Shenzhen, the Shanghai-Pudong museum and the Zhongguancun Cultural Center in Beijing have been realized in China. Numerous projects are under construction, for example the Chongqing Opera and the Chinese National Museum in Beijing. In Hanoi/Vietnam the National Conference Center has been built.

Von Gerkan, Marg and Partners have demonstrated town-planning competence with their visionary proposals for "Stuttgart 21" for the Deutsche Bahn, HafenCity Hamburg, and the urban master plan of Bucha-rest/Romania, as well as many urban projects in China, including the planning of Lingang New City for 800 000 inhabitants.

Aside from Germany and China, gmp is currently involved in projects in India, Vietnam, South Africa, the UAE, Italy, Spain, Turkey, Romania, Poland and Latvia, to name but a few.

中国区办公室联系方式

北京市东城区东直门南大街5号中青旅大厦1212室
邮编：100007
电话：+86-10-58156161
传真：+86-10-58156365
邮箱：Beijing@gmp-architekten.de

CYTS Plaza, No.1212 No.5 Dongzhimen South Avenue Dongcheng District 100007, Beijing P. R. China
Tel: +86-10-58156161
Fax: +86-10-58156365
E-mail: Beijing@gmp-architekten.de

上海市徐汇区汾阳路138号轻科大厦10楼
邮编：200031
电话：+86-21-54655151
传真：+86-21-54655131
邮箱：Shanghai@gmp-architekten.de

10th Floor QingKe Mansion No. 138 Fen Yang Road Xu Hui District 200031 Shanghai P. R. China
Tel: +86-21-54655151
Fax: +86-21-54655131
E-mail: Shanghai@gmp-architekten.de

深圳市福田区金田路4028号荣超经贸中心3706室
邮编：518035
电话：+86-755-82577766
传真：+86-755-82577799

Rm3706, 37/F, Landmark building, No.4028 Jintian Rd, Futian district Shenzhen,China 518035 Shenzhen P. R. China
Tel: +86-755-82577766
Fax:+86-755-82577799

Http: //www.gmp-architekten.de

1

2

1–5 北京中关村基督教海淀堂

设计竞赛：2002年一等奖
设　　计：曼哈德·冯·格康
　　　　　施蒂芬·舒茨
建筑面积：4 000平方米
建设时间：2005年—2007年

Christian Church, Haidian District, Beijing, China

Competition: 2002 - 1st prize
Design: Meinhard von Gerkan
Stephan Schütz
Building Area: 4,000 m²
Construction Time: 2005–2007

3

4

5

6–8 北京第三极

设计竞赛：2002年一等奖
设　　计：曼哈德·冯·格康
合 伙 人：施蒂芬·舒茨
建筑面积：85 000平方米
建设时间：2003年—2006年

Zhongguancun Cultural Center, Beijing

Competition: 2002 - 1st prize
Design: Meinhard von Gerkan
Partner: Stephan Schütz
Building Area: 85,000 m²
Construction Time: 2003–2006

6

7

8

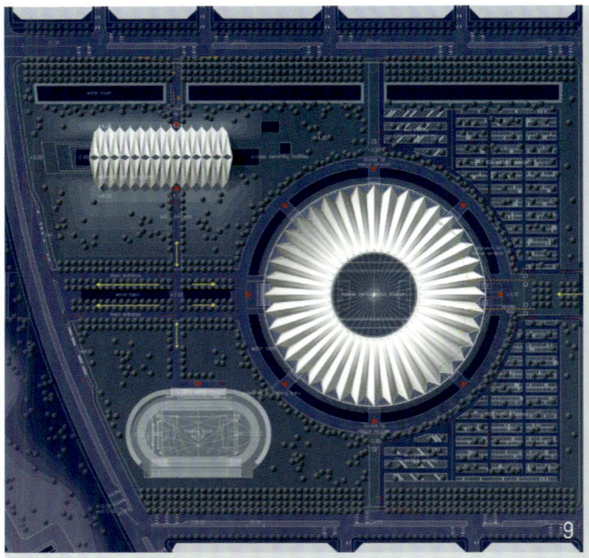
9

10

11

12

13

9–15 佛山世纪莲体育中心

设计竞赛：2003年二等奖
设　　计：曼哈德·冯·格康
合 伙 人：尼古劳斯·格茨
体 育 馆：36 000个座位
游 泳 馆：2 800个座位
建设时间：2004年—2006年

Century Lotus Sports Park with Stadium and Swimming Hall, Foshan

Competition: 2003 - 2nd Prize
Design: Volkwin Marg
Partner: Nikolaus Goetze
Seats Stadium: 36,000
Seats Swimming Hall: 2,800
Construction Time: 2004-2006

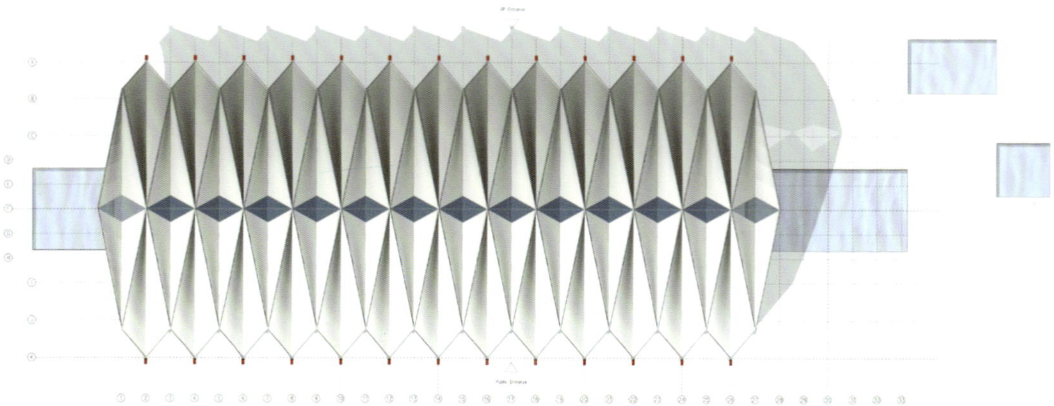
14

15

16

17

18

19

16–19 南宁国际会议展览中心

设计竞赛：1999年一等奖
设　　计：曼哈德·冯·格康
　　　　　尼古劳斯·格茨
建筑面积：130 000平方米
容　　积：1 200 000立方米
建设时间：1999年—2003年/2005年

Nanning International Convention & Exhibition Centre, China

International Competition: 1999 – 1st prize
Design: Meinhard Von Gerkan
Nikolaus Goetze
Building Area: 130,000 m²
Volume: 1 200 000 m³
Construction Time: 1999–2003/2005

20–21 青岛大剧院

设计竞赛：2004年一等奖
设　　计：曼哈德·冯·格康
　　　　　施蒂芬·舒茨
　　　　　尼古拉斯·帕姆仁克
建筑面积：80 000平方米
歌 剧 厅：1 600个座位
音 乐 厅：1 200个座位
多功能厅：400个座位
建设时间：2005年—2008年

Grand Theater, Qingdao

Competition: 2004 – 1st prize
Design: Meinhard von Gerkan
Stephan Schütz
Nicolas Pomränke
Building Area: 80,000 m²
Opera Hall: 1,600
Concert Hall: 1,200
Multifunctional Hall: 400
Construction Time: 2005–2008

20

21

GKK+ARCHITEKTEN

PROF. SWANTJE KÜHN
OLIVER KÜHN

德国 GKK 建筑师事务所

德国GKK建筑师设计事务所
斯万提・库恩 教授，奥利弗・库恩
Pariser Strasse 110719 Berlin/Germany
电话：0049–0–2830820（德国）
0086–18600033808（中国）
传真：0049–0–28308253
邮箱：info@gkk–architekten.de
网址：www.gkk–architekten.de

GKK+Architekten
Prof. Swantje Kuehn, Oliver Kuehn
Pariser Strasse 1
10719 Berlin / Germany
Tel: 0049-0-2830820 (DE)
0086-18600033808 (CN)
Fax: 0049-0-28308253
E-mail: info@gkk-architekten.de
Http: //www.gkk-architekten.de

我们的哲学

德国GKK国际建筑师设计事务所致力于追求生态的、明晰的能源自循环建筑。

多学科的协同工作与网络构架，使我们能在极短的时间内为业主量身定制出独具特色的解决方案。运用高新材料和先进技术使建筑具有独特的开放性和精巧性。

主要专长

引领最高工艺水平的建筑和设计
实现并高效完成建筑的创新概念
多学科协同工作的综合设计团队
融合当代文化艺术的一体化建筑

主要活动领域

建筑和设计
总体规划
详细规划和城市设计
再开发

通过理性创新和不断获得的建筑奖项的肯定，GKK成为了德国面向未来设计，明晰建筑构架的先锋代表。其设计及完成作品范围广泛，在国际上有着高度影响力，包括从住宅、办公楼高层塔楼到各类总部大厦，从航空业到重大交通枢纽站，再到展览馆、博物馆、歌剧院等各类重大建筑规划项目。

宣言

建筑是对特定场所的合适理想高效而精准的实现。

超越寻常的思考，我们致力于使用达到结果的最佳解决方法和创造令人惊讶的全新解决方案，在发现真正思想的同时，努力使每一个细节都趋于完美。积极与当代艺术对话，从中获取有效的灵感，赋予建筑自由创造性和积极思维。

从歌剧院到办公楼，从火车站到研发中心，我们为各种项目提供解决方案，并赋予其独特的创新潜能。

GKK+Architekten
Office Philosophy

GKK+Architekten pursue forward looking, transparent and energy-conscious architecture.

Multidisciplinary teamwork and networked thinking allow GKK to find several options tailored to different solutions in a very short time. Using state of the art materials and new technology their buildings emanate openness, precision and lightness.

Major expertise

Design excellence
Innovative building concepts
Multidisciplinary teamwork and process engineering
Integration of contemporary art in architecture

Main fields of activities

Architecture and design
General planning
Master planning and urban design
Redevelopment

With their innovative and winning architectural concepts GKK+Architekten are rewarded as major representatives of a future-oriented design, with transparent architecture.

The scope of their internationally well received work ranges from high-rise towers, headquarter buildings, buildings for the aviation industry and major railway stations further to exhibition buildings, museums and opera houses.

Statement

Architecture is the committed and precise translation of the right idea for a specific location.

Beyond formal considerations we focus on the best way to the result and the surprising new of solution. Once discovered the genuine idea will be present in every single detail.

Making architecture is creative free and positive thinking; the dialogue with contemporary art is a valid source for inspiration.

From operas to offices, from railway station to research centres we work on all projects, provided them with unique potential.

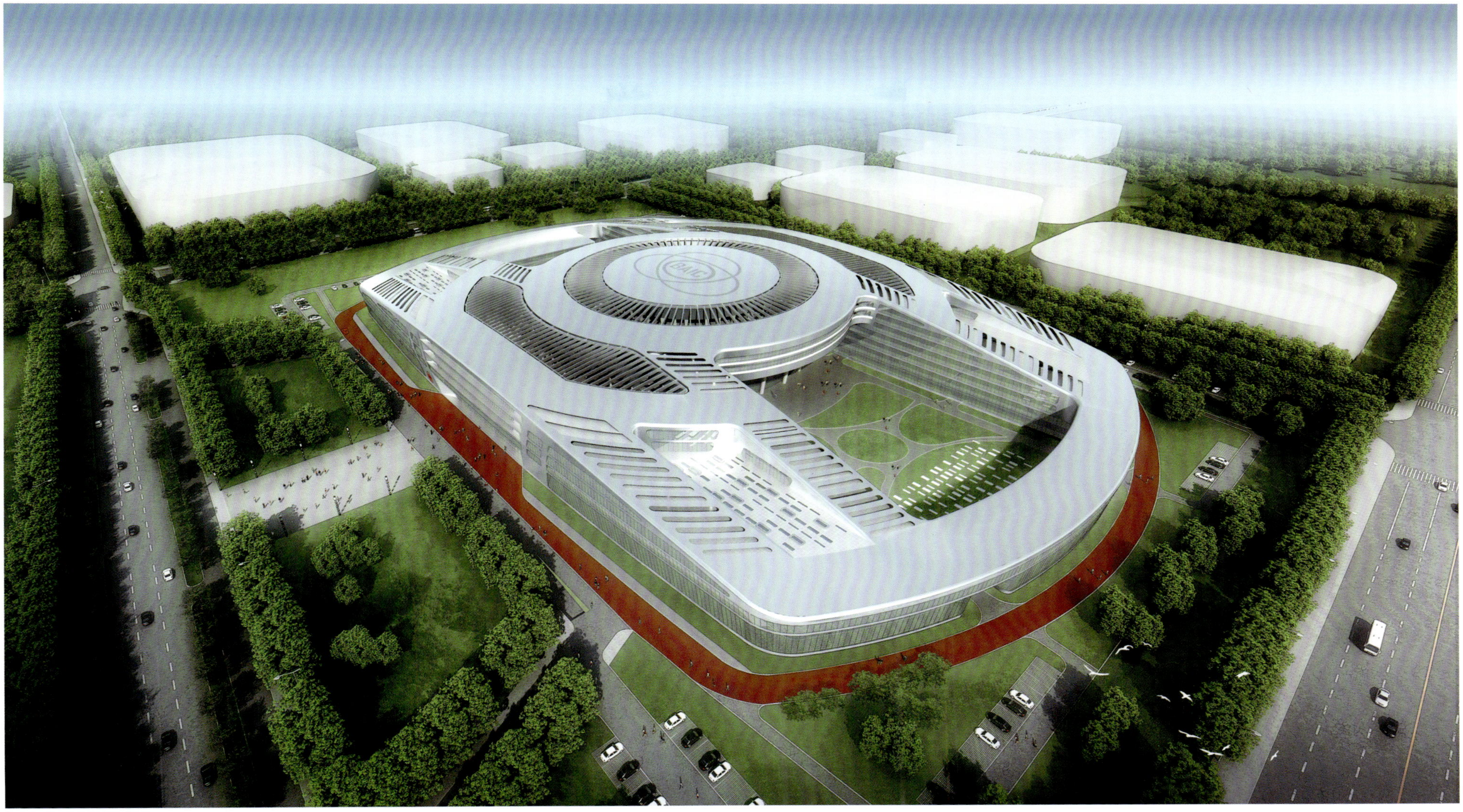

BAIC–北汽集团研发中心–总部大楼　北京

新的北汽集团总部大楼研发中心紧邻北京国际机场T3航站楼，为公司2 700名员工提供了全新的办公区域、试验中心、研发中心、试制等区域。流线形的建筑外形基于明晰的功能性概念而形成。访客和员工通过一个3层通高的具有强烈感染力的空间进入到建筑内部，左边设置了大型的会议中心，右边的餐厅提供超过2 700人的就餐环境。在入口大厅之后，在建筑物的中心部位是一个环形的向公众开放公共展览空间。

在这里，企业可以展示他们的最新产品，举办公司内部展览会或大型的重要接待会展。组装和制造领域的功能则安排建筑的北部。

所有办公空间均设置在楼上，并透过落地玻璃窗和中庭取得良好的照明和通风。建筑表皮为富有光泽的金属板，像一辆完美的汽车车身，使得整个建筑从远处便向人们展示了极强的标志性，同时也代表了北汽集团的企业核心竞争力。

BAIC – Beijing Automotive R & D Centre – Headquarters, Beijing

The new headquarters for BAIC is situated right next to Beijing International Airport and will provide new offices, lab spaces, design studios and manufacturing areas for more than 2,700 company members. The streamline building is based on a very clear and functional diagram. Visitors and employees enter the building from the south through a 3-storey impressive entrance hall, to the left large conference facilities are located, to the right a restaurant provides more than 2,700 menues per day. After the entrance hall and in the centre of the building a circular showroom is open to the public.

Here the company can exhibit their latest products, in-house fairs can be organized or large receptions and premiers can be arranged. Further to the north the assembly and manufacturing areas are arranged.

All offices are located in the upper floors and have good lighting conditions through large windows and glazed atria. The building skin is clad in shiny metal just like a beautiful bodywork of a car and makes the building an icon of the BAIC core competences visible for many thousands of people from afar.

广汽研究院化龙研发基地　广州 中国

GAC ENGINEERING HUALONG BASE, R&D Center, Guangzhou, China

随着2010年1月26日设计协议的签订，广汽研究院化龙研发基地成为GKK事务所在完成北汽研发中心总部之后，在中国实施的另一个重大项目。目前，广汽研究院（GAEI）在广州番禺，将建设约为15万平方米使用面积的化龙研发基地。

GAEI主要的研发任务将包括自有品牌的乘用汽车、先进的动力系统、汽车关键组件和部件，以及高新先进的汽车技术和新模式的汽车发展平台。该GAEI研发中心将设有总部和所有研发活动室，并将于2011年完成。此研发中心将为超过2 000名技术以及行政人员提供先进的办公及研发空间。

流线的造型和珍珠白色的金属表皮，以及大面积的玻璃幕墙，使得建筑本身具备了强有力的表现力，整个建筑从动态造型到细部处理都强烈的传达出了中国先进的汽车行业形象，企业技术精度和创造力。

With the signing of the contract on January 26, 2010, GKK+Architects has been awarded another major project in China, after the main BAIC corporate headquarters in Beijing. Now, the Guangzhou Automotive Engineering Institute (GAEI) is to build its new Automotive Technology Research & Development Centre(Hualong base) on the banks of the Pearl River in Panyu County, Guangzhou Province, with some 150,000 m² of net area.

The main R&D tasks conducted by the GAEI will include passenger vehicle platforms for own-brand cars, advanced powertrains, automotive key assemblies and components, as well as the development of high/advanced automotive technologies and new models. The GAEI R&D Centre will house both the headquarters and all R&D activities, and will be completed by 2011. It will provide office and work space for more than 2,000 engineers.

Streamlined and clad in pearl white enameled steel, with large seamless glass areas, the building will make a strong statement, and convey the image of advanced automotive expertise, technical precision and creativity – from its dynamic shape down to the last detail.

德国IBM新行政总部　埃灵根

IBM德国新行政总部位于斯图加特附近的IBM埃灵根产业园区，现已建成。

宽敞的庭院通往一个圆形中庭，并为所有客户提供空间导向，如小餐馆、单元会议和容纳600人的礼堂。

中庭玻璃电梯上部三层的办公空间，建筑师提出了简单明确的五角星图式，5个四层的办公翼围绕圆形的中庭展开，这样人们就可以通过中庭去到任何一个地方。工作场所是按照移动E空间的概念组织（借助笔记本、手机和随处可见的网络接口，实行移动化的IT工作方式，公司员工的工作地点可以进行随机分配，提高空间的使用效率），具有均好性和非领域性，装备也是灵活提供的。

从外形上看，IBM德国新总部大楼的金属外皮闪闪发光，依据观看角度，建筑会从银色变为深灰色，本身也具有减少空调能耗的功能。除此以外，建筑还通过中庭的烟囱效应和自然通风、光电转化设施、雨水回收利用等一系列措施进行节能，并提高能源利用效率。

New Executive Headquarters IBM Deutschland, Ehningen

IBM Germany's new Executive Headquarters at the company's campus in Ehningen, near Stuttgart, has been completed.

A spacious courtyard leads into a circular atrium, and on into all customer-oriented areas, such as the bistro, the conference units and the auditorium for up to 600 visitors.

Glass elevators in the atrium provide access to the three upper office stores, which are arranged in five bars on a base plate, radiating out from the atrium. Each workplace is organized according to the e-place concept, equivalent and non-territorial; the furnishing is flexible.

Bi-color strips in the façade bands change color depending on the position of the viewer, and make the building appear silvery with dark anthracite.

An energy concept with natural thermal flow uses the chimney effect of the space for cooling and ventilation of the atrium. A photovoltaic system on the roof produces electricity, while the sanitary facilities in new buildings are fed with rainwater from a cistern. Component cooling reduces the energy consumption for cooling and heating.

南德意志出版社总部　德国慕尼黑2008年10月竣工

Süddeutscher Verlag Head Office in Munich completed in October 2008

此建筑伫立在城市中心的入口，像一个被精致打磨的水晶，富有优雅比例的梯田中庭空间，连接了塔楼和边翼，为南德意志出版社总部大楼定义了独特的空间形制。

棱镜玻璃构成了整个建筑外表皮，并从不同的角度投影出城市和天空的面容，同时也为建筑增加了巨大的体积和重量感。

出版社总部位于一个由公园和绿色梯田围合的阳光充裕的广场，对面为其印刷工厂。底部两层建筑为餐厅、会议室、休息区，以及一个艺术空间和一个幼儿园。上部建筑为办公区，是为通讯和新闻领域的2 000工作人员提供了高品质的办公空间。在建筑顶部，是一座两层通高的大厅，作为重要访客的接待区。

此“绿色建筑”通过双层玻璃幕墙，分散幕墙通风、地热能和季节性蓄热，节省了约80%的主要能源消耗。同时，南德意志出版社总部大楼也成为德国第一个获得LEED的金奖认证的绿色建筑（由美国绿色建筑协会的认证），同时GKK也被授予了“最佳建筑师10”的奖项。

Standing at an important entrance to the city centre like a polished crystal, the Süddeutscher Verlag Building defines the space with its well proportioned terraced atrium, which connects the tower and a side wing. A prismatic glass shell sheaths the building. The façade's facets reflect the cityscape and sky; the huge volume appears weightless. Framed by a park and green terraces, the publishing house head office is on a sunny square, across from its own printing plant. The bottom two floors house the canteen, conference and lounge areas, an art space and a kindergarten. Above that are offices, communications areas and news rooms for 2,000 staffs. At the pinnacle is a two-storey hall, a reception area for guests.

The "green building" has a double façade, decentralised façade ventilation, geothermal energy and seasonal heat storage, for primary energy savings of approx. 80%. The SV high-rise in Munich is the first office building in Germany to be awarded a LEED certificate in gold (certified by the U. S. Green Building Council) and is also awarded with the Award "Best Architect 10".

Auer+Weber+Assoziierte
奥尔 + 韦伯 + 合伙人 建筑设计有限公司

公司简介 / Company Introduction

公司历史

奥尔 + 韦伯 + 合伙人建筑设计有限公司是 2006 年初由其前身在 1980 年成立的奥尔 + 韦伯建筑设计事务所改建而来。公司分别在斯图加特和慕尼黑两处设立工作室，拥有员工 90 ~ 100 人。

设计领域

公司侧重于为公共及私人业主规划设计新建和改建项目， 并日益加强自身作为设计总承包人与开发商及建筑公司之间的密切合作。近年来更多的是注重国际性项目的设计和建设。公司的设计宗旨在于针对各个项目不同的项目条件和要求，努力寻找最适合当地、最合理、最切实可行的设计方案。

设计范围

公司主要设计范围为办公管理建筑和文化性建筑，教育科研建筑和旅馆酒店和住宅建筑，以及体育交通建筑和城市的总体规划。

公司创始人

佛里茨 · 奥尔教授	德国建筑学硕士工程师，美国建筑学硕士，德国国家建筑师协会成员
卡罗 · 韦伯教授	德国建筑学硕士工程师，德国建筑师协会成员

公司主管合伙人

莫里茨 · 奥尔	德国建筑学硕士工程师，注册建筑师
飞利浦 · 奥尔	德国建筑学硕士工程师，注册建筑师
约恩 · 索茨	德国建筑学硕士工程师，注册建筑师，德国建筑师协会成员
阿西姆 · 索丁	德国建筑学硕士工程师，注册建筑师，德国建筑师协会成员
史蒂芬 · 苏克斯多夫	建筑学硕士工程师，注册建筑师

公司合伙人

马汀 · 科连坡	德国建筑学硕士工程师，注册建筑师
约格 · 穆勒	德国建筑学硕士工程师，注册建筑师
史蒂芬 · 尼泽	德国建筑学硕士工程师，注册建筑师
克里斯多夫 · 泰格	德国建筑学硕士工程师，注册建筑师
依罗娜 · 维茨莱茵	德国建筑学硕士工程师，注册建筑师

获得的重要奖项

2007 德国国家建筑奖 荣誉奖 德累斯顿市布鲁舍——泰阿森展览馆，德国
2006 德大罗——米露塞国际建筑奖 特别奖 林茨市太阳能城中心，奥地利
2005 Leaf 奖类别：最佳环境奖 林茨市太阳能城中心，奥地利
2004 Leaf 奖 类别：新建筑奖、全球奖 赛罗帕拉那天文学者宾馆，智利
2001 德国国家建筑奖 瑞克林豪森市鲁尔庆典演出大厅，德国

地址：Haussmannstrasse 103 A
70188 Stuttgart（斯图加特）
电话：+49-711-268404-0
传真：+49-711-268404-88
电邮：stuttgart@auer-weber.de
网页：www.auer-weber.de

地址：Sandstrasse 33
80335 Munich（慕尼黑）
电话：+49-89-381617-0
传真：+49-89-381617-38
电邮：muenchen@auer-weber.de

Office history

The architectural office Auer+Weber+Assoziierte originates from the Auer+Weber Architectural Group existing since 1980. With offices in Stuttgart and Munich,the company has 90~100 staffs.

Profile of activities

The company focuses on the conception and design of new buildings and alterations for public-sector institutions and private clients and gradually strengthen the close collaboration with developers and construction companies as the general contractor. Over the past few years, we have been increasingly involved in the design and realisation of international projects.The common characteristics of all the projects lie in the distinctive and coherent development of architecture which emerges from the specific brief and unique site conditions.

Design Scope

The design scope ranges from projects for administration and culture, education and research, hotels and residential buildings as well as sports facilities and transportation buildings to urban development and master plans.

Founding Partners

Prof. Fritz Auer, Dipl. Ing. Architect BDA
Prof. Carlo Weber, Dipl. Ing. Architect BDA

Managing Partners

Moritz Auer, Dipl. Ing. Architect
Philipp Auer, Dipl. Ing. Architect
Jörn Scholz, Dipl. Ing. Architect BDA
Achim Söding, Dipl. Ing. Architect BDA
Stephan Suxdorf, Dipl. Ing. Architect

Associates

Martin Klemp, Dipl. Ing. Architect
Jörg Müller, Dipl. Ing. Architect
Stefan Niese, Dipl. Ing. Architect
Christof Teige, Dipl. Ing. Architect
Ilona Werz-Rein, Dipl. Ing. Architect

Major awards since

2007 Deutscher Architekturpreis, Acclamation Exhibition Hall Brühlsche Terrasse Dresde, Germany
2006 Dedalo Minosse Prize, Special Prize,solarCity Centre Linz-Pichling, Austria
2005 Leaf Awards, Kategorie "Best Environmentally Sustainable Project", solarCity Centre Linz-Pichling, Austria
2004 Leaf Awards, Category "New Build" + "Overall",ESO Hotel on Cerro Paranal, Chile
2001 Deutscher Architekturpreis, Festival Hall Recklinghausen Germany.

Haussmannstrasse 103 A
D-70188 Stuttgart
Tel +49-711-268404-0
Fax +49-711-268404-88
stuttgart@auer-weber.de
www.auer-weber.de

Sandstrasse 33
D-80335 Munich
Tel +49-89-381617-0
Fax +49-89-381617-38
muenchen@auer-weber.de

上海辰山植物园

建设地址：上海
项目业主：上海辰山植物园建设管理委员会
委托时间：2006 年
竣工时间：2009 年
建筑面积：106 000 平方米
项目投资：6.4 亿元
摄 影 师：Jan Siefke

Shanghai Chenshan Arboretum

Construction Location: Shanghai
Project Owner: Shanghai Chenshan Arboretum Construction and Management Committee
Consigning Time: 2006
Completion Time: 2009
Building Area: 106,000 m^2
Project Investment: 640 million RMB
Photographer: Jan Siefke

在上海接近疗养区的中心地带，坐落着 13.34 公顷的大植物园。该园处在 70 米高的山上，周围环绕着带状花园。在这片绿洲之中建有各种重要的建筑物，例如入口建筑物、大暖棚以及一个研发中心等，共计有 75 000 平方米。它们将成为所有环状花园和周围景观的重要地标。通过其具有活力的建筑形式及其对混凝土和玻璃这两种材质交替使用手法，满足了与其景观和园林建筑相呼映的理念要求。

With a site area of 13.34 ha, Shanghai Chenshan Botanical Garden is located in the central area near the sanitarium area in Shanghai. It is situated on the hill of 70 meters, surrounded by linear parks. Various important architecture are built in this green space, such as architecture of the entrance, large greenhouses and a research center, with the total area of 75,000 m^2, they will become the important landmark of the linear parks and neighboring landscape. With the dynamic architecture patterns and the technique of alternate use of concrete and glass, it well meets the conceptual demand of correspondence in landscape and landscape architecture

Auer+Weber+Assoziierte

项目在不断延伸的景观中形成一个具有艺术气息的基地，它的力量来自植根于景观的建筑特征。四方形的基本模块有意识地叠加在起伏的地形之上，这些模块具有相同的单位面积。但不同的城市功能将根据其内容的重要性分别添加在街道、草地和广场系统之上。集会广场作为校园公共空间的重点，同博物馆一样，超越了单个基本模块的尺度。

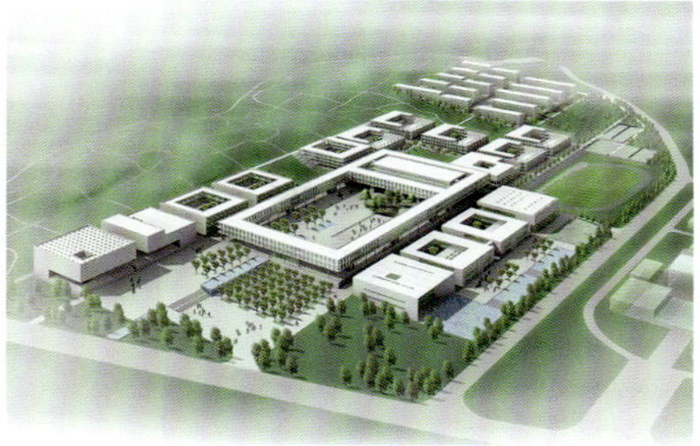

Within an expansive landscape, a place for the arts has emerged which draws its power from an architectural ensemble which has been implanted on the landscape. The flowing topography is expressed through the overlaying of an orthogonal grid layout.
The campus building components with their various functions, yet identical overall dimensions, are inserted in the system of streets, lanes and squares in accordance with their internal significance. Only the "agora" – the outdoor focus of the campus-city – and the museum extends beyond the dimensions of a single module.

沈阳鲁迅美术学院大连分院新校区

建设地址：辽宁 大连
合作单位：辽宁省建筑设计研究
委托时间：2004 年
竣工时间：2008 年
建筑面积：139 000 平方米
项目投资：360 000 000 元
摄 影 师：Jan Siefke

Shenyang Luxun Academy of Fine Arts, Dalian Branch New Campus

Construction Location: Dalian, Liaoning
Cooperation Organization: Liaoning Architectural Design and Research Institute
Consigning Time: 2004
Completion Time: 2008
Building Area: 139,000 m^2
Project Investment: 360 million RMB
Photographer: Jan Siefke

Auer+Weber+Assoziierte

Auer + Weber + Assoziierte

在英国公园附近的代表性地段上，1904 年建造的住宅楼将按照文物建筑标准被重新修复，此建筑的原住宅楼部分将根据建筑未来的办公用途进行改建，此外还将添加一个现代的新扩建部分。

比起原有的坐落在非常神圣的女王大街上的建筑，新的扩建部分就像是一个朴素的金属立方体，附属于原有建筑之上。扩建部分的建筑形式语言和材料运用有意识地与原有建筑形式语言保持一定距离，以此来尊重和强调原有建筑的独特性。

The residential building, built in prestigious surroundings near the Englischer Garten park in 1904, is being rebuilt with earnest regard to its listed status, converting it for future office needs and extending it with a modern building.

While the richly structured, existing building makes up the stately address in Königinstraße, the new one in the back garden is designed as a clearly subordinate, plain, metallic cube, deliberately contrasting the old building in stlye and material, thus respecting and enhancing its identity.

女皇街 23 号办公楼 /LfA 银行

建设地址：德国 慕尼黑
项目业主：巴伐利亚慕尼黑 LfA 银行，慕尼黑
委托时间：2006 年
竣工时间：2008 年
建筑面积：12 000 平方米（旧楼）
6 200 平方米（新楼）
摄 影 师：Roland Halbe, Stuttgart

23 Office Building, Queen Street / LfA Bank

Construction Location: Munich, Germany
Project Owner: Baravia Munich Lfa Bank,Munich
Consigning Time: 2006
Completion Time: 2008
Building Area: 12,000 m^2 (old building)
6,200 m^2 (new building)
Photographer: Roland Halbe, Stuttgart

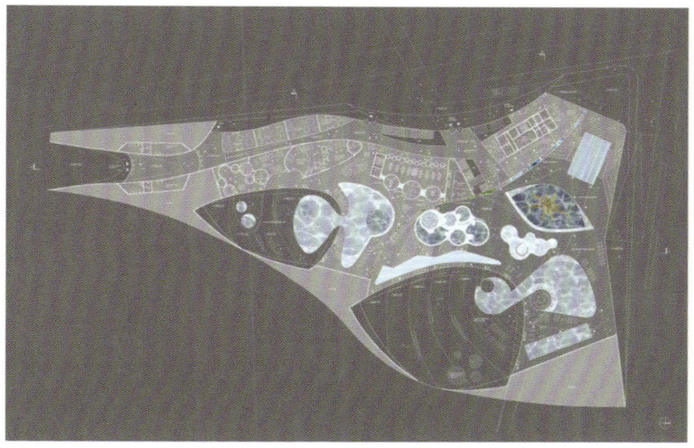

位于两个山间站之间的新建筑，由于其优越的地理位置，既具有连接功能又有大门式的特点，因此建筑连接桥梁连接山间站的两侧，使得人流可以不受车流影响，从而使建筑成为通往高处滑雪区的大门。

设计考虑到地块的地理位置和明显度，因此把建筑与周围自然的融合置于首位。建筑屋顶的设计，充分考虑到其与自然的结合在此规模的设计中所占的决定性的因素。

位于街道上的广场拓宽了此处的空间，建立了一个城市中心。这里几乎设立了所有重要的通往水中心、室内健身、宾馆和大场馆的出入口，并且邀请游人驻足观赏。

The prominent site between two peak stations assigns to the building the two functions of connection and gateway. A connecting bridge the buildings of two sides together combines, allows pedestrians to walk around unaffected by entering traffic and acts as a gateway on the journey up to the higher skiing areas.

The design acknowledges the prominence and visibility of the site. A considerate treatment to the enclosing wilderness is one of its crucial elements. The image of the "fifth façade", the landscape-like roofs, and their integration into the surroundings, is most influential to the design of this large complex.

A square on street level widens the space, makes up an urban centre, which almost all of the important access, of the aqua centre, indoor sports hall, hotel, and the big lobby and placed and wecome everyone.

科施威大山谷 · 法国

建设地址：Grandes Combes Courchevel 1550
项目业主：Commune de Saint–Bon Tarentaise 科施威
合作单位：Sarl Stidio Archm，特勒色维尔，法国
委托时间：2008 年
竣工时间：2011 年
建筑面积：82 422 平方米

Curswey Valley · France

Construction Location: Grandes Combes Courchevel 1550
Project Owner: Commune de Saint-Bon Tarentaise
Cooperation Organization: Sarl Stidio Archm, France
Consigning Time: 2008
Completion Time: 2011
Building Area: 82,422 m^2

Auer+Weber+Assoziierte

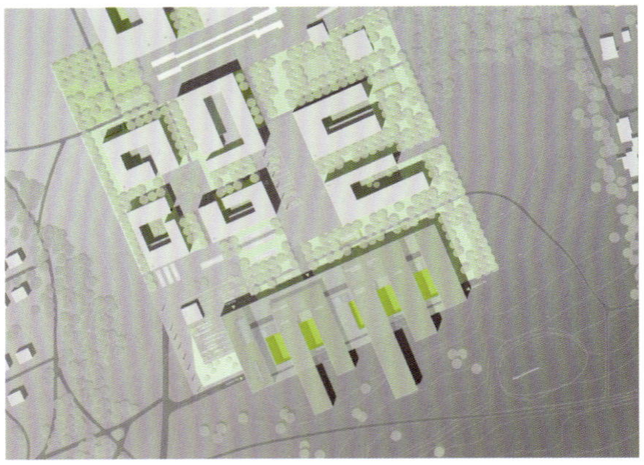

与比勒菲尔德大学紧密相连的北高校校区是北莱茵－威斯特法伦州提供给高校使用的一块引人注目的发展用地。

比勒菲尔德高等专科学院的新楼就坐落在该用地南侧 Gellershagner 溪谷处的自然景观地带。设计方案意在让建筑外立面形成未来校区的一面空间界限，同时把建筑面向自然景观的一面处理成多层次和开放的空间。高等专科学院的新楼将成为该高校新区的起点和"入口大厅"。

空间序列的设计方案中，"Hot Spots"的主题也将在交通流线设计上得以应用。

从校园中心广场走来，人们首先进入到的是两层架空的宽阔的入口大厅，交通主轴线以这里为基点向各个方向继续延伸展开。

In creating the northern university campus (Hochschulcampus Nord) the federal state of North Rhine-Westphalia has provided an exposed development area situated in immediate relation to the University of Bielefeld.

The new building in the university of applied sciences takes its place in the southern part of the northern university campus which has a landscape with the valley of a stream, the Gellershagener Bachtal. Toward the new campus, which is yet to be built, a clear spatial edge is set while the landscape is shaped in a more differentiated and open manner. The new building of the university is "prelude" and reception for the new development area.

The idea of a spatial succession of "hot spots" is continued in the concept of access to and within the building. Coming from the central campus square one enters a multi-storey lobby which makes a generous entrance from which the main campus axis is continued, generating a central access corridor for the building.

比勒菲尔德（Bielefeld）高等专科学院新楼

建设地址：Lange Lage 比勒菲尔德
项目业主：Bau—und Liegenschaftsbetrieb NRW, Morgenbreede 39, 33615 比勒菲尔德，
Jürgen Delkeskamp 先生
委托时间：2009 年
竣工时间：2012 年
建筑面积：92 000 平方米
项目投资：1.3 亿欧元

Bielefeld College New Building

Construction Location: Lange Lage, Bielefeld
Project Owner: Bau-und Liegenschaftsbetrieb NRW, Morgenbreede 39, 33615, Bielefeld
Mr. Jürgen Delkeskamp
Consigning Time: 2009
Completion Time: 2012
Building Area: 92,000 m^2
Project Investment: 130 million Euro

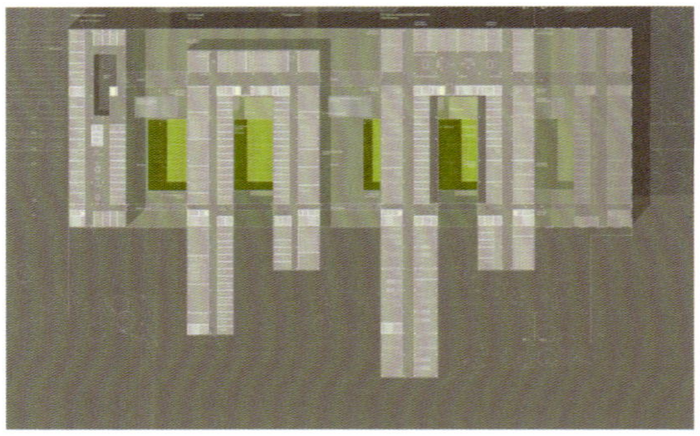

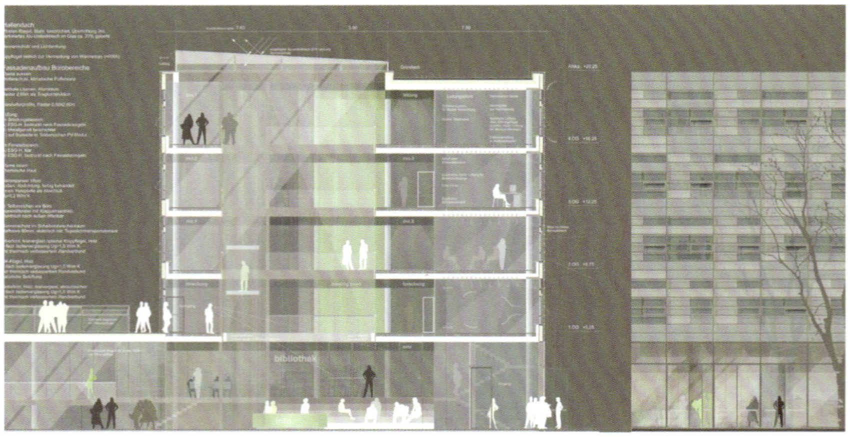

Auer+Weber+Assoziierte

韩德凯办公中心耸立在多瑙河畔，邻近千禧塔。建筑设计是与具有纵向优势的塔楼形成对比的横向设计，从而使其拥有了在多瑙河大桥旁不可替换的地位。

两个不同高度的玻璃大楼通过自由的建筑立面语言为正在发展的维也纳水畔设计书写了重要的一笔。

面向多瑙河的建筑大堂是整个办公建筑的核心。一份出色的建筑能源计划和充满智慧的建筑立面设计给予了这个玻璃大楼的使用者出色的工作环境。

Rivergate, the office centre on the Danube, placed in immediate vicinity of the Millennium Tower, with the crosswise design contrasted with the lengthwise tower building constitutes a strong accent within Waterfront Vienna.

Two buildings of different heights opening towards the Danube form a generously shaped atrium. A public walkway through the project helps to mark clearly the access to the building via the atrium. The free-form shell has become characteristic for the skyline on the Danube, it enhances the flexibility of use by varying building sizes and allows to react appropriately to building regulations.

Rivergate's presence and identity is established by its unmistakable appearance as well as the consistent combination of generosity, openness and flexibility.

河畔之门办公大楼，维也纳

建设地址：Handelskai 92，2002 维也纳，奥地利
项目业主：海德斯办公中心，房地产有限公司
委托时间：2007 年
竣工时间：2009 年
建筑面积：46 000 平方米
项目投资：1.15 亿欧元

Riverain Door Office Building, Vienna

Construction Location: Handelskai 92, 2002, Vienna, Austria
Project Owner: Handelskai Office Center, Real Estate Co., Ltd.
Consigning Time: 2007
Completion Time: 2009
Building Area: 46,000 m^2
Project Investment: 115 million Euro

Auer+Weber+Assoziierte

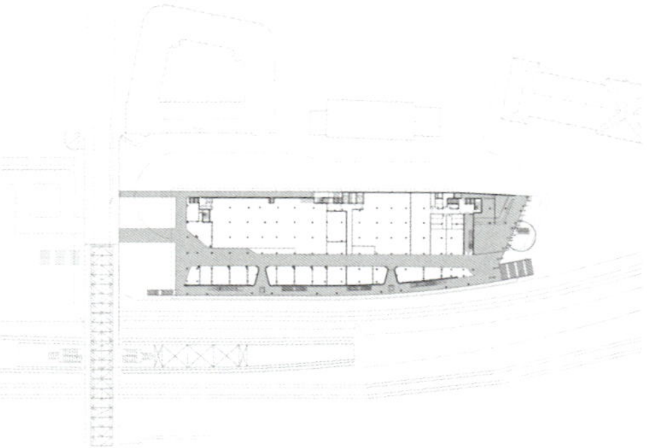

这栋建筑汇总了包括汽车站、办公、酒店与服务等多项功能，也为慕尼黑火车总站与帕辛站之间的铁道区域北侧勾勒出新的城市轮廓。在长途汽车到达与出发之处，汽车总站恰到好处地呼应了“交通”的动感，同时自身也成为了一个引人注目的商业基地。

具有集散层的“环景平台”通过天桥与自动扶梯与长途汽车始发站台、城市快速铁路及哈克桥紧密联系。候车大厅与其宽阔的站前广场，则朝向内城与火车总站。

The body of the building unites the functions of the terminal, office, hotel and service and constitutes the prelude to the new "city edge" north of the fields of tracks between Munich main station and the station of Pasing. At the point of both outset and destination of long-distance coach journeys, a building is being built which reflects the dynamic aspects of traffic and which makes an attractive place for commercial life.

A "promenade deck" serves as distribution level and is connected to the platforms of the central coach terminal and to the Hacker bridge by gangways and escalators. The terminal lounge opens up via a generous forecourt into the direction of the inner city and the main station.

慕尼黑长途汽车站中心

建设地址：Arnulfstra e/Hackerbr ü cke 慕尼黑
项目业主：HOCHTIEF，工程有限公司，慕尼黑东南分部
委托时间：2005 年
竣工时间：2009 年
建筑面积：25 000 平方米
摄 影 师：Roland Halbe，Stuttgart

Munich Coach Center

Construction Location: Arnulfstraße / Hackerbrücke, Munich
Project Owner: HOCHTIEF engineering Co., Ltd., Munich Southeast Branch
Consigning Time: 2005
Completion Time: 2009
Building Area: 25,000 m^2
Photographer: Roland Halbe, Stuttgart

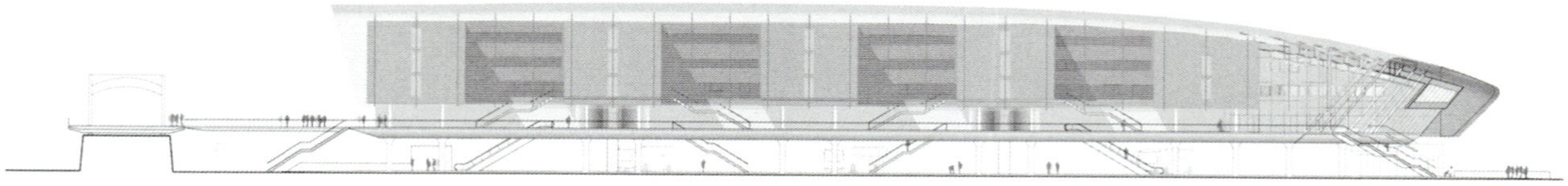

Auer+Weber+Assoziierte

欧洲南方天文台主办公楼扩建方案的空间秩序和建筑语言都与周边环境密切相关。方案既考虑了加兴科技园南部的绿化空间，同时也尊重了它的“亲密邻居”——现有办公主楼独特的造型和外向的性格。扩建的部分与主楼拥有着相似的建筑语汇，并与自然景观产生共鸣。主入口、接待处、公共空间等功能保留在主楼之中，扩建部分中主要分布着办公室、研讨厅和大会议厅等功能空间。

The new ESO (European Southern Observatory) headquarter in its spatial array and architectural language pays respect both to the green belt to the south of the Research Campus Garching and to the extrovert character of the existing main building in the immediate neighbourhood.
Taking up the main building´s formal language the new building develops a formal kinship and engages symbiotically with the landscape. Central entrance, reception and public spaces continue to be in the main building while the new building houses office and seminar spaces as well an an auditorium.

加兴欧洲南天文台主办公楼扩建，德国

建设地址：Karl–Schwarzschild–Strasse2，D–85748Garching 慕尼黑
项目业主：欧洲南方天文台
委托时间：2009 年
竣工时间：2012 年
建筑面积：12 200 平方米
项目投资：2 370 万欧元

Garching European South Astronomical Observatory Main Office Building Extension, Germany

Construction Location: Karl-Schwarzschild-Strasse 2, D-85748 Garching, Munich
Project Owner: European South Astronomical Obervatory
Consigning Time: 2009
Completion Time: 2012
Building Area: 12,200 m^2
Project Investment: 23.7 millon Euro

Auer+Weber+Assoziierte

公司简介 BACKGROUND

吕邓黎建筑师有限公司创立于1983年，公司以香港为基地，分公司分别设于上海和重庆。

公司的宗旨是坚持以创新精神与稳健周全的专业态度，服务不同顾客，满足不同的项目需求。多年来，吕邓黎积极参与香港的建设，他所设计的项目包括各类规模的住宅、办公楼、酒店、商场、厂舍、学校、公共房屋、宗教建筑生态环境等。通过长时间的考验，公司在业内建立了良好的信誉。

随着中国经济的迅速发展，吕邓黎20世纪80年代中期开始承接内地的设计项目，在各大都会如北京、上海、重庆、成都及广州等地，先后参于并完成了多个建筑及室内设计项目，尤其是在作为全国经济龙头的上海，公司已完成的项目包括瑞安广场、城市酒店、东方巴黎、东方剑桥、东方曼克顿及瑞虹新城等。以上项目都以优秀的质量完成，并获得多个奖赏，瑞安广场更荣获建筑工程鲁班奖的殊荣。

2002年吕邓黎建筑资询(上海)有限公司的成立，意味着公司会更积极地参与到内地和香港的建设。秉承公司的宗旨，公司愿与国内专业单位共同努力，为广大人民创造更美好的生活环境。

Lu Tang Lai Architects Ltd. originated in Hong Kong and started off as a small owner-managed design studio in 1983.

Realizing the rapid economic transformation and growth of China in the last decades of the twentieth century, Lu Tang Lai Architects Ltd. sought to contribute the development and redevelopment in China by actively participating in building and environmental design projects in the mainland as early as the mid eighties, focusing in Shanghai in the early nineties. To increase competitiveness and enhance services, the Shanghai and Chongqing offices were put into operation in 2002 and 2004 respectively.

Over the years, the practice maintains a steady growth in sizable architectural projects ranging from offices and commercial developments, hotels, service apartments, high-class residential developments, schools and institutional buildings to master planning in Hong Kong, Mainland China and S.E. Asia. Its reputation for providing personalized and quality services remains unchanged.

奖项 AWARDS

上海瑞安广场
Shui On Plaza, Shanghai
- 建筑鲁班奖
- 白玉兰优秀建筑奖

上海瑞虹新城一期
Rui Hong Xin Cheng Phase 1, Shangha
- 上海市优秀住宅评选优秀房型奖
- 上海市优秀住宅评选科技奖
- 上海市“四高”优秀小区
- 上海市优秀住宅评选优秀住宅银奖

上海瑞虹新城二期
Rui Hong Xin Cheng Phase 2, Shangha
- 詹天佑住宅大奖小区优秀规划设计奖
- 全国人居经典方案大赛住宅组环境金奖
- 最受欢迎楼盘综合金奖

上海东方曼克顿－世纪豪庭
Century Metropolis, Shanghai
- 上海市优秀住宅工程单体设计项目二等奖

广州西门口广场一，二期
Westmin Plaza Phase 1 & 2, Guangzhou
- CNBC Asia Pacific Commercial Property Awards 2009 5-star award for "Best Mixed Use Development China"

重庆市化龙桥雍江苑
The Riviera,Chongqing
- Residential Building 4-star Performance Award 2008

香港皇冠假日酒店
Crowne Plaza, Hong Kong
Asia Pacific Commercial Property Awards in Association with Bloomberg Television 2010
- The Best Hotel Construction & Design Hong Kong 5-star Award
- The Architecture(Leisure & Hospitality)Hong Kong 5-star Award
- Best Interior Design Hong Kong Highly Commended Award

香港骏逸峰
The Morrison, Hong Kong
- HKIS Property Marketing Award 2008 "The Best Environmental Design Awar

香港总办事处
HONG KONG HEAD OFFICE
香港柴湾永泰道 60 号柴湾工业城第一期 2001 室
2001, Chai Wan Industrial City Phase 160 Wing Tai Road,
Chai Wan, Hong Kong
Tel: +852-2521 0681
Fax: +852-2524 3319
E-mail: info@ltlarch.com
Http://www.ltlarch.com.hk

上海办事处
SHANGHAI OFFICE
中国上海市徐汇区，宜山路 407 号，筑园，南楼，303 室
邮编：200030
Room 303 South Block, Living Design Centre No.407 of Yi Shan Road, Shanghai 200030, China
Tel: +86-21-3363 0103
Fax: +86-21-3363 0120
E-mail: info.sh@ltlarch.com.cn

重庆办事处
CHONGQING OFFICE
中国重庆市青年路 3 号时代豪宛 A 座 3004 室
邮编：400010
Room 3004 Tower A, Timesquare No.3 of Young Road, Chongqing 400010, China
Tel: +86-23-6371 3592
Fax: +86-23-6371 3592
E-mail: info.cq@ltlarch.com.cn

1 上海瑞安广场
Shui On Plaza
Shanghai

2 上海城市酒店
City Hotel and Apartment
Shanghai

3 成都中汇广场一期及二期
Central Point Plaza
Phase 1&2
Chengdu

4 广州瑞安中心
The Centrepoint
Yuexiu, Guangzhou

5 广州西门口广场一期及二期
Westmin Plaza Phase 1&2
Zhong Shan Road
Guangzhou

6 成都中中环广场
Overseas Exchange Square
Chengdu

7 香港铜锣湾皇冠假日酒店
Crown Plaza
Leighton Road, Hong Kong

8 香港鸿图道工商项目
Industrial Building
Hung To Road, Hong Kong

9 香港大新金融中心
Dah Sing Financial Centre
Hong Kong

10 香港葵涌政府海关大楼
Kwai Chung Custom House
Hong Kong

11 香港丰田皇冠车行服务中心
Crown Motors Pre-Delivery Center
Yuen Long, Hong Kong

12 香港湾仔捷利中心
Jubilee Centre Phase 1&2
Fenwich Street, Hong Kong

13 广州恒宝华庭
Heng Bao Garden
Guangzhou

14 上海康城一期
Shanghai Cannes, Phase 1
Xin Zhuang, Shanghai

15–16 上海东方曼克
Century Metropolis
Xuhui, Shanghai

17–18 上海瑞虹新城一期
Rui Hong Xin Cheng, Phase 1
Shanghai

19 上海瑞虹新城二期
Rui Hong Xin Cheng, Phase 2
Shanghai

20 上海瑞虹新城三期
Rui Hong Xin Cheng, Phase 3
Shanghai

21–22 重庆化龙桥雍江苑
The Riviera
Hualongqiao, Chongqing

23 香港司徒拔道别墅项目
Residential Development
Stubbs Road, Hong Kong

24 香港赫兰道别墅项目
Residential Development
Headland Road, Hong Kong

25 香港蒲岗村道汇豪山
The Forest Hills
Po Kong Village Road, Hong Kong

26 香港湾仔道骏逸峰
The Morrison
Wanchai Road, Hong Kong

27 香港上水御景峰
8 Royal Green
Sheung Shui, Hong Kong

28 香港上水御皇庭
Royal Green
Sheung Shui, Hong Kong

Architectural

设计作品

Design Works

嘉柏建筑师事务所 于 2003 年创立。公司不愿再延续一般传统的设计，不愿再进行类似生产线的工作模式，坚持打开设计理念的新一页，集合一群充满活力和有志于创意设计的专业人才，在蜕变的环境中，以崭新的理念，敬业的精神，为客户提供独特的设计，力求确立卓越的建筑典范并负起对社会应有的专业责任。嘉柏成立至今已获得多家知名发展商对其独特设计的认同，并受委托参与多项不同类型的规划及建筑设计，包括万科房地产有限公司于深圳、佛山、东莞、珠海、成都、武汉、天津、大连等城市的项目，金地房地产有限公司于武汉、杭州的项目，招商房地产有限公司于深圳、天津的项目，厦门建发集团于厦门、福州、成都的项目。此外还有其他发展商如山东鲁能集团、重庆中安房地产开发有限公司、杭州坤和集团、广州侨鑫集团、惠州金融街置业有限公司、香港新鸿基地产和越南 Indochina Group 等。

Gravity Partnership was founded in 2003 with a vision to deliver innovation through non-conventional architectural designs and work processes to better serve society and the profession. This vision is shared by a group of enthusiastic and motivated professionals who are passionate and endeavors to create high quality design as wellas take every opportunity to explore new ways in achieving original and imaginative results. Recognized for its distinctive design, Gravity Partnership has been invited by prominent developers to participate in prestigious planning and architectural design commissions, including Vanke Real Estate Co. Ltd. projects in Shenzhen, Foshan, Dongguan, Zhuhai, Chengdu, Wuhan, Tianjin and Dalian; Gemdale Real Estate Co. Ltd. projects in Wuhan and Hangzhou, China Merchants Real Estate Co. Ltd. project in Shenzhen and Tianjin, Xiamen C&D Real Estate Co. Ltd. projects in Xiamen and Fuzhou; others such as Shandong Luneng, Chongqing Enrich Real Estate Development Co. Ltd., Hangzhou Canhigh Estate Group, Guangzhou Kingold Group, Huizhou Financial Street, Hong Kong Sun Hung Kai Properties and Indochina Group in Vietnam.

嘉 柏 建 筑 师 事 务 所

g r a v i t y partnership ltd

| architects & planners |

www.gravitypartnership.com

1–5 中国天津万科水晶城中心会所 Vanke Crystal City Sports and Recreation Facility, Tianjin, China

6–9 中国东莞万科运河东一号商业区 Vanke East Canal No.1 Commercial District, Dongguan, China

10

13

11

12

9

14

15

16

总监
Directors

设计总监
Principal - Design

余啸峰
Frank Yu
B Arch(Pratt Institute) New York

设计总监
Principal - Project

王克江
Claude Wong
Registered Architect
AA Dipl.
RIBA
HKIA
AP (List of Architects),
PRC Class 1 Registered Architect
Qualification

高级主任设计师
Senior Associates

方正道
Solomon Fong
M Arch (HKU)
LEED®AP

廖国安
Roy Liu
Registered Architect
RAIA
HKIA
AP (List of Architects)
PRC Class 1 Registered Architect
Qualification

主任设计师
Associate

汪皓
Wang Ho
Registered Architect
HKIA
PRC Class 1 Registered Architect
Qualification

联络
Contact

香港铜锣湾电气道 148 号 39 楼
Building No.39, 148 electric road
Causeway Bay, Hong Kong
t +(852) 3106 8711
f +(852) 3106 0754
e studio@gravitypartnership.com

www.gravitypartnership.com

10–13 天津万科水晶城北入口公建 Vanke Crystal City North Entrance Apartment, Tianjin, China

14–16 中国天津万科西青假日风景 Vanke Holiday Town, Tianjin, China

17–18 中国重庆中安翡翠湖一期 B Enrich Jade Lake Phase 1B, Chongqing, China

19–20 中国厦门建发爱琴海项目 C&D Aegean Sea, Xiamen, China

21–22 中国福州建发领域 C&D Domain, Fuzhou, China

23–25 中国深圳万科天琴湾 Vanke Vega Villa, Shenzhen, China

26–27 越南岘港 Montgomarie Links 项目 Montgomarie Links, Danang, Vietnam

28–29 中国深圳万科东海岸三期及四期 Vanke East Coast Phase 3&4, Shenzhen, China

30–31 中国成都新鸿基悦城 SHKP Jovo Town Phase 1, Chengdu, China

32–33 中国佛山万科顺德新城湾畔 Vanke The Paradiso, Shunde, Foshan, China

34–35 中国珠海市万科金域蓝湾 Vanke Paradiso, Zhuhai, China

36–38 中国东莞万科城市高尔夫花园 Vanke Golf Town, Dongguan, China

39–40 中国天津招商钻石山 China Merchants Diamond Hill, Tianjin, China

41–42 中国珠海市绿景凤凰山一号花园 Lujing Mount Phoenix Garden, Zhuhai, China

43–45 中国厦门海峡交流中心二期一号楼 Hai Xia Exchange Centre Phase II Tower 1, Xiamen, China

46–48 中国厦门夏新电子城研发办公楼 Amoi Research & Design Headquarter, Xiamen, China

49–51 中国厦门建发国际大厦 C&D International Tower, Xiamen, China

52–54 中国海南三亚鲁能美丽城 Luneng Sanya Beauty City, Hainan, China

55

56

59

60

57

58

61

62

65

63

64

66

67

55—58　越南河内 IPH 项目　Indochina Plaza Hanoi, Vietnam

59—61　中国成都东大街电视娱乐综合　East Street Cinematique & Entertainment Mall, Chengdu, China

62—63　中国佛山市顺德区中毅弘越中心　Zhongyi Hongyue Centre, Shunde, Foshan, China

64—67　中国深圳招商数码大厦　China Merchants Digital Building, Shenzhen, China

68

69

70

71

72

73

74

75

77

78

68–69 中国天津招商钻石山会所 China Merchants Diamond Hill Clubhouse, Tianjin, China

70–72 中国福州建发领域公建 C&D Domain, Fuzhou, China

73 香港中文大学教学楼（方案竞赛） Teaching Building at Chinese University of Hong Kong (Design competition)

74–75 香港中文大学晨兴书院 Morningside College, Chinese University of Hong Kong

77–78 中国成都四川航空广场 Sichuan Airline Plaza, Chengdu, China

79–82 中国佛山市北滘文化中心 Beijiao Cultural Centre, Foshan, China

83–85 中国广州侨鑫从化养生谷别墅及水疗中心
Kingold Cong Hua Regimen Gu Villa and Spa Centre, Guangzhou, China

86–88 中国北京鲁能优山美地 A 区二期 Luneng Yosemite Beijing, China

89–91 中国重庆新天地 Xingtiandi, Chongqing, China

91–93 中国佛山市跨 G325 国道乐从段人行天桥工程
Pedestrian Footbridge, G325 National Highway, Le Cong, Foshan, China

94–96 中国深圳梅沙 . 海与建筑 Seashore Promenade, Meisha, Shenzhen, China

97–98 中国重庆鲁能星城十三街区 13th Avenue, Luneng Star City, Chongqing, China

99–100 中国北京鲁能四惠桥商业项目 Sihuiqiao Development, Beijing, China

101–103 中国深圳招商光明科技企业项目（方案竞赛）
China Merchants Guangming Science and Technology Park, Shenzhen, China (Design competition)

嘉柏建筑师事务所

www.gravitypartnership.com

104

105

107

108

106

109

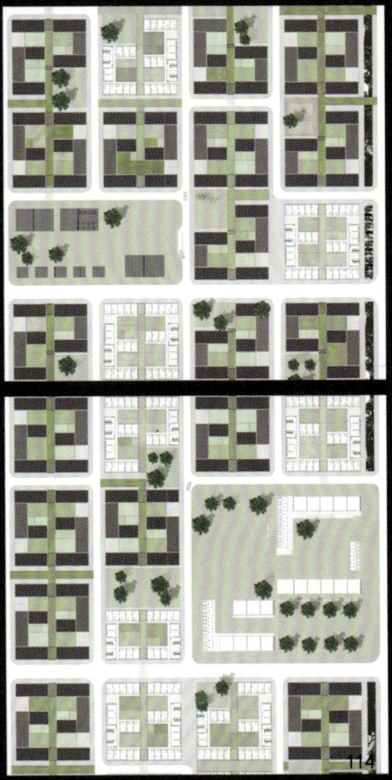
114

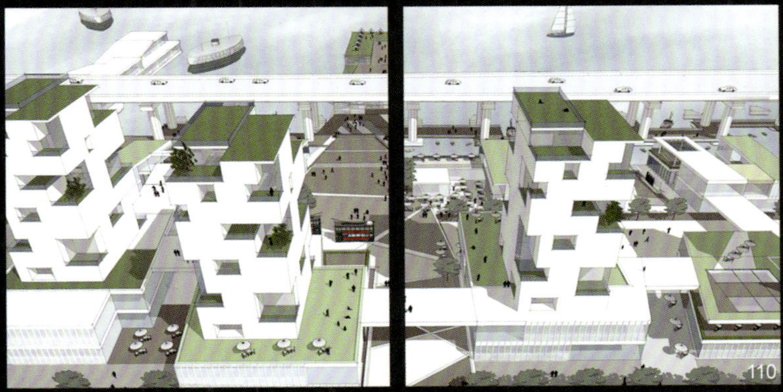
110

112

113

111

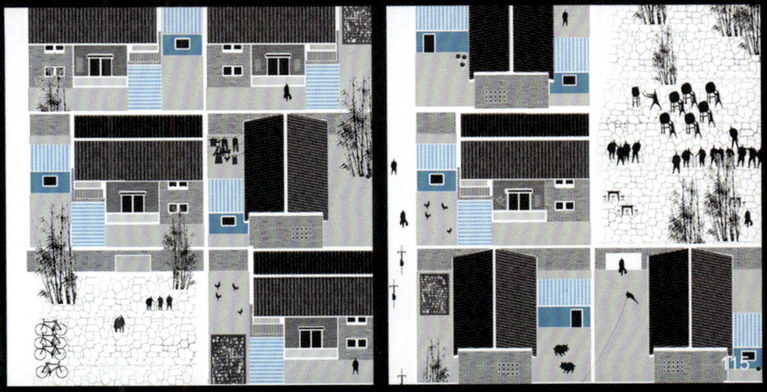
115

104–106 2010 上海世界博览会香港馆概念设计比赛（亚军）
Concept Design Competition for Hong Kong Pavilion, World Expo 2010 Shanghai (Second Prize)

107–109 嘉柏建筑师事务所办公室 Office of Gravity Partnership Limited

110–111 香港北角汀综合发展设计概念比赛（专业组冠军）
North Point Harbour Conceptual Design Competition, Hong Kong (Professional Champion)

112 香港 . 深圳城市 / 建筑双城双年展 2010（策展组）
Hong Kong & Shenzhen Bi-City Biennale of Urbanism / Architecture (Curatorial Team)

113–114 中国四川过渡安置房改善计划 Intermediate Relocation Housing, Sichuan, China

115 2008 万科青年建筑师设计大赛 2008 Vanke Young Architects Design Competition

ORIGINAL VISION

ORIGINAL PLANNING ORIGINAL ARCHITECTURE ORIGINAL INTERIORS

Hong Kong 22/F,88 Gloucester Road, Wanchai, HongKong
Tel:+852 2810 9797 Fax:+852 28109790

Phuket 393 Moo 1,Srisoontom Road, Chemgtalay Subdistrict
Thalang District,Phuket 83110, Thailand
Tel:+6676 270 755 Fax:+6676 270 757

DIRECTORS

Adrian McCarroll
B.Sc.,B.Arch(Hons),M.Sc,
HKIA,RIBA,
ARB(UK)

Stephen Gorton
BA(Hons)Arch,Dip Arch,ARB,
RIBA

Ken Leung
Bachelor of Built Environm
(Interior Design)

ASSOCIATE DIRECTORS

Jamie Jamieson
B.Arch,RIBA

Prasert Chanhom
B.Arch,M.Arch,ASA 3719

SELECTED CLIENTS

Asia Properties Asia Capacity Exchange **Asia Island Homes** Axiom Investment Management **Baker & McKenzie** Bovis Lend Lease **CastawayBay** Caf Piatti **Chesterton Petty** Cheung Kong Holdings Ciena Clifford Chance **Double Star Caf** FCC Cambodia **East Asia Properties** EliteModes Management **Frontline Clothing Ltd.**Hyson **IFM (Asia) Ltd**. Indochina Assets Ltd.Internet **Media House** Iregent **Jaspas** Lee Marie **Manhattan Holdings** Midas **Movenpick** Mundo Latino **Nopawong Construction Co.** La Rose Noire **Lotus Asset Management** Olivers **Penta Finance Asia Ltd.** PCCW **PCPD** Pepperonis **Pinnacle Asia** PT Padma Ohm **Raimon Land** Samsara **Salem Attitude** Sedgwick Richardson **Sergio Tacchini** Sino Land **Starcore Asia** Time Inc. **Time Life** Urban Entertainment **Wilson Parking** WSP **XS Media** Zac's **Zuellig**

COMPANY DATA

Established	创立年份	1992
HKIA member	香港建筑师学会会员	3
Professional	专业人员	17
Support	辅助人员	10

Creativity is not a linear process. It involves establishing then understanding a matrix of information and references which must be viewed inparallel to exploit the opportunities that lie within.

创作，从来不是直线进行的。在过程中，必须搜罗各式资讯及参考素材，然后深入剖析，并纵观彼此关系，方能发掘潜藏良机。

ORIGINALVISION

Many skills and disciplines are involved in creatingelegance and beauty in the building environment.Appropriate orchestration of that expertise is asessential as the spark of innovation

要缔造高雅优美的建筑环境，更需要运用不同技巧及专业知识。把智慧发挥得恰到好处，这其实和激发创意一样重要。

ORIGINAL VISION

山东莱州鸣玉泉
建筑面积：560 000平方米
建设地点：山东 莱州

Mingyuquan in Laizhou
Building Area: 560,000 m^2
Construction Location: Laizhou, Shandong

深圳皇家海湾公馆
建筑面积：56 280平方米
建设地点：广东 深圳

Shenzhen Royal Golf Mansion
Building Area: 56,280 m^2
Construction Location: Shenzhen, Guangdong

海峡两岸文化交流中心
建筑面积：92 688平方米
建设地点：北京

Center for Cultural Exchanges across the Taiwan Strait
Building Area: 92,688 m^2
Construction Location: Beijing

香港华天国际建筑与城市设计有限公司
HONGKONG WITEN INTERNATIONAL LTD.

香港华天国际建筑与城市设计有限公司(HONGKONG WITEN INTERNATIONAL LTD.)成立于2000年6月2日，公司拥有对工作精益求精并富有团队合作精神的专业人员，追求卓越的设计的设计品质，为顾客提供城市规划设计、建筑工程设计和房地产发展咨询等服务。公司于2008年7月被香港名牌理事会授予常务理事会员单位。

Founded on June 2, 2000, Hong Kong Witen International Ltd adheres to quality first and strives for excellence by the spirit of seeking perfection and professional staff with strong team-spirit. The company engages in urban planning design, construction engineering design and real estate consultation and other services. In July., 2008, the company was authorized to be executive member unit of the council by Hong Kong International Brands Council.

深圳清湖创意产业园
建筑面积：90 000平方米
建设地点：广东 深圳

Qinghu Creative Industry Park in Shenzhen
Building Area: 90,000 m^2
Construction Location: Shenzhen, Guangdong

湖南永州鸿星帝景湾
建筑面积：169 000平方米
建设地点：湖南 永州

Hongxingdijingwan in Yongzhou
Building Area: 169,000 m^2
Construction Location: Yongzhou, Hunan

湖南永州创宏大厦一期
建筑面积：69 000平方米
建设地点：湖南 永州

Hunan Yongzhou Chuanghong Plaza Phase 1
Building Area: 69,000 m^2
Construction Location: Yongzhou, Hunan

湖南宁远国际新城
建筑面积：86 517.13平方米
建设地点：湖南 永州

Ningyuan International New City in Hunan
Building Area: 86 517.13 m^2
Construction Location: Yongzhou, Hunan

广东惠阳紫金华府
建筑面积：87 000平方米
建设地点：广东 惠阳

Huiyang Zijin mansion in Guangdong
Building Area: 87,000 m^2
Construction Location: Huiyang, Guangdong

山东博兴豪门庄园
建筑面积：102 000平方米
建设地点：山东 博兴

Boxing Noble Manor in Shandong
Building Area: 102,000 m^2
Construction Location: Boxing, Shandong

遵义喇叭河小区
建筑面积：156 360平方米
建设地点：贵州 遵义

Labahe Community in Zunyi
Building Area: 156,360 m^2
Construction Location: Zunyi, Guizhou

香港公司：香港九龙弥敦道625号雅兰中心二期15楼1508室
电　话：852-67662969　30766658
传　真：852-30626606
Company Address: Room 1508.Floor 15, Phase II of Grand Tower.625 Nathan Road, Kowloon, Hong Kong
Tel: 852-67662969 / 30766658
Fax: 852-30626606

南宁公司：广西南宁市星湖路南二里6号4楼
电　话：+86-711-4504668
传　真：+86-711-4504668
Nanning Company: Floor 4, No.6 of Nanerli, Xinghu Road, Nanning, Guangxi
Tel: +86-711-4504668
Fax: +86-711-4504668

深圳公司：深圳市南山区马家龙艺园路田厦IC产业园3014室
电　话：+86-755-26470085
传　真：+86-755-86604392
邮　箱：witen999@foxmail.com
Shenzhen Company: Room 3014 of IC Industry Park of Tin Sha, Majialongyiyuan Road, Nanshan District, Shenzhen
Tel: +86-755-26470085
Fax: +86-755-86604392
E-mail: witen999@foxmail.com

烟台公司：山东省烟台开发区万寿山路5号宝威科研楼677室
电　话：15853592277　13828809897
Yantai Company: Room 677 of Baowei Research Building, No.5 of Yantai Development Zone, Shandong
Tel: 15853592277 13828809897

山东烟台炉上村
建筑面积：419 000平方米
建设地点：山东 烟台

Lushang Village in Yantai, Shandong
Building Area: 419,000 m^2
Construction Location: Yantai, Shandong

惠州阳光城市花园
建筑面积：111 788.82平方米
建设地点：惠州 惠阳

Sunshine City Garden in Huizhou
Building Area: 111,788.82 m^2
Construction Location: Huiyang, Huizhou

长春范家店项目规划
建筑面积：980 000平方米
建设地点：吉林 长春

Project Planning for Changchun Fanjiadian
Building Area: 980,000 m^2
Construction Location: Changchun, Jilin

蔡田田建築師事務所
Joy Choi Arquitecta

地址：澳门南湾大马路 325 号昌辉大厦一楼 A 和 C 座
Address: Avenida Praia Grande No.325, Edif. Cheong Fai 1ºA e 1ºC, Macau
Tel: 28752667　Fax: 28752544　E-mail: joyc@macau.ctm.net

氹氹仔岛圣善国际学校重建计划
Reconstruction plan of Shengshan International School in Bingzai Island

公司资料 Company Data

成立年份	Established	2001
专业人员	Professional	15
协助人员	Support	5

主要专业人员表 Key Staff

蔡田田	Joy, Choi Tin Tin	澳门注册建筑师，主理建筑师 Principal Architect
蔡子钰	Dominic, Choi Chi Iok	澳门注册建筑师，英国注册建筑师，设计总监 Design Director, Partner
高武洲	Joseph, Wu-zhou Gao	澳门注册建筑师，联营董事 Associate
梁志伟	Tony , Leung Chi Wai	澳门注册机电工程师，高级项目经理 Senior E/M Engineer, Senior Project Manager

圣善学校现况
Shengshan International School current status

澳门城市发展的历史悠久，旧城散布着不同类别的文物建筑、历史建筑和传统民居，根据澳门政府的资料显示，澳门超过三十年楼龄的建筑中属单一业权登记的物业及土地超过 60%，倘若再加上未曾登记但长期被使用的土地和物业，那么这类土地资源的再利用，将在整体旧城区物业发展的比例中占有举足轻重的作用，足以改变或保留整个旧城的风貌。

近年来，本事务所有机会在业主的委托下参与多项旧城区特色建筑的改造和重建工作，建筑计划的探讨重点在于建筑同历史环境的协调和建筑单体的风格体现。

这些项目的摸索过程比较典型，反映出政府对建筑环境的关注，具有条例标准的平衡、使用者在功能和建筑风格上的要求和配合、设计者的思考历程等多重意义。

Over a long decade of urban development in Macau, the Old Towns nurtured different types of heritages; some of the historic buildings, traditional business are still under preserved. According to the Macau SAR's record, more than 60% of lands and properties which are over thirty years old are registered as single ownership, whereas a high portion of the other 40% are declared to be not registered and have long been occupied illegally. These premises and lands have a strong pivotal potential for an Old Town Redevelopment and Conservation.

In the recent years, Joy Choi Arquitecta has been granted the opportunity to participate on the Design Research for several Old Towns in Macau. In respect to the past and present, our study models and analysis revealed the history of the local heritage in coordination with the environment of the presence.

For projects as such often leads to a contradictions between the government regulations, site constraints and the users' requirements; our practice aims to maintain a healthy balance of these aspects in every project.

澳门旧城区高楼街及西坑街住宅建筑计划
Rua do Padre António and residence area of Sai Street architecture plan

氹仔旧城区多座单栋楼宇重建
Reconstruction of single buildings in Old town of Bingzai

大原建筑设计咨询（上海）有限公司
Dayuan Architecture Design Consulting Co.,Ltd

李祖原联合建筑师事务所

李祖原联合建筑师事务所是李祖原先生于1978年创立的，是一个只有6个人及业务个案范围限于台北市的事务所，现已成长为一个有着强烈建筑及规划设计使命感，公司员工超过200人，且设计作品遍及世界的工作组织。事务所以东方文化为主轴，以生命建筑为设计理念，结合生态与现代科技，提供高创意性、高整合性、高技术性的建筑设计服务，以创造人类优质生活环境为目标，并秉持〝以人为本〞的精神与员工共同成长。李祖原联合建筑师事务所是合伙制的事务所，有李祖原先生及王重平先生等6位合伙人，总部设于台北，由吴文圳总经理负责。

大原建筑设计咨询（上海）有限公司

大原建筑设计咨询（上海）有限公司是李祖原联合建筑师事务所在大陆地区的分公司。分别设于北京、上海，由王声翔总经理总负责。

大原建筑设计咨询（上海）有限公司业务涉及东北、华北、西北、西南、华南、华东6个地区，对城市规划、各类商业规划、大型购物商场设计、星级酒店设计，办公大楼设计以及大型住宅群设计都有着相当丰富的项目经验。公司谨慎控制案件人员的成长比例和业务导向，以专业原创的模式培养出了一个在整体规划、建筑设计、室内设计以及营建管理等领域都非常优秀的建筑设计团队。

本公司认为建筑师必须直接参与设计项目早期的计划阶段，与业主一同工作，并统合各个相关项目的工作小组，协助确立目标，界定发展方向，提供在设计发展中设计的准则，并透过预见的各种可能性，满足业主对项目功能的需求。建筑师综合考虑基地状况、美学要求、预算限制、能源要求以及开发执行过程，并找到最适宜的解决方式。

C.Y.Lee established C.Y.Lee and Partners in 1978. It is an architectural firm share held by Mr. C.Y. Lee, Mr. C.P.Wang and other 4 partners. Design projects concentrated in Taipei, CYL has grown to an organization of over 200 staffs with architectural and planning commissions through the world. C.Y.Lee & Partners Currently CYL Provides Planning, Architectural and Interior Design Services in Asia and North American Area.

Dayuan Architecture Design Consulting, the filiale of C.Y.LEE & Partners, is established in 1997 in Mainland. Recently it has expanded office to Beijing, Shanghai, which are in charge by Mr. James Wang.

Through its history, CYL has developed specialized expertise in a wide diversity of building types. A most important factor in the firm's progress has been the ability to manage the Urban planning and Architectural planning, Architectural design and Interior design, and construction process for projects of any size or scope, along with its record of delivering projects on time within budget. A commitment to excellence in design and complete familiarity with current problem-solving techniques and building methods enables CYL to meet any challenge in the field of architectural designing, or planning.

台北 Taipei
李祖原联合建筑师事务所
C.Y. LEE & PARTNERS Architects/Planners
10595 台北市敦化北路 6 号 3 楼
3F., No.6 of Tunhua N. Rd., Taipei 10595, Taiwan
Tel: 886-2-2781-1666 ext. 503
Fax: 886-2-2781-2002
E-mail: info@mail.cylee.com
Http: //www.cylee.com

上海 Shanghai
大原建筑设计咨询(上海)有限公司
DAYUAN Architecture Design Consulting Co.,Ltd.
上海市虹桥路1452号古北国际财富中心302室 200336
Rm 302, Gubei International Fortune Center, No.1452 of Hongqiao Road, Shanghai 200336
Tel: +86-21-32095999
Fax: +86-21-32098860
E-mail: cylsh@mail.cyleecn.com
Http: //www.cylee.com.cn

北京 Beijing
大原建筑设计咨询(上海)有限公司(北京分公司)
DAYUAN Architecture Design Consulting Co.,Ltd. (Beijing office)
北京市朝阳区建国路91号金地中心711−715室 100022
Rm 711-715, Tower A, Gemdale Plaza, No.91 of Jianguo Rd, Beijing 100022
Tel: +86-10-65693068
Fax: +86-10-65693069
E-mail: cylbj@mail.cyleecn.com

上海世博台湾馆
Taiwan Pavilion

蕴含深厚内涵及意义的台湾馆：

山水心灯——自然·心灵·城市

当代城市文明历经了17世纪法国政治革命、19世纪共产资本主义经济革命和20世纪科学至上时代的科技革命之后，形成了在高技术、高规则的形式规范下结合科技至上的中心理念，最终使得当代城市文明成为21世纪环保、反核、地球危机与金融危机的重大议题。

这些新时代重大议题，使我们认为唯有彰显“生活革命”的全新思维，并提出“自然·心灵”作为“生活革命”的两大核心要素，方能有建构21世纪人类新文明的可能性。2010上海世博会主题“城市，让生活更美好”，更是一个可以催化21世纪“生活革命”的最佳实践舞台，这就是2010上海世博会台湾馆的核心设计思维。

西方文化以二元论对立辩证的观点形成文化的深层结构，并开展出强调工具理性的科技文明。东方文化则以“天、人、地”三元结构合一的观点，作为文化的深层基因，并开展出强调生命智慧与和谐共生的东方文明。我们认为，21世纪“生活革命”当从城市构建开始实践，并引进“天、人、地”三元结构合一的东方生命哲学，应结合当代科技技术，创造一个强调“自然·心灵”的新城市文明，以平衡当代过度强调科技消费所造成的城市种种不和谐发展缺陷。

2010上海世博会台湾馆“山水心灯”，在外部主体形式上，呈现出多层表达的多媒体展演。“山水心灯”透过如同透过心灯般透明率为60%的建筑外膜和高流明LED球体的“台湾之心”，形成虚实互生的动态展演，呈现出万蝶飞舞、海洋守护、台湾花布、自然永续与民间艺术的多种影像，向世界传递台湾多元自然的生命图像，并展现建筑日夜不同的景象。

在建筑的外部形式上，台湾馆以“山水剧场”作为守护“心灯”的建筑基台，以玉山、太平洋为隐喻，以台湾艺术表演为主角，彰显“未来城市”中“回归自然”的设计理念和实现生活和谐的理想。

台北101国际金融大楼 Taipei 101

业　　主：台北金融大楼股份有限公司

功　　能：办公大楼 / 购物商场

服务范围：设计、室内设计、监造

地　　点：台北市　信义路 / 市府路 / 松智路

结　　构：钢骨

完工时间：1998年1月—2004年10月

基地面积：30 277平方米

建筑面积：370 000平方米

楼 层 数：B5～101层

备　　注：世界第二高楼 508米

Client : Hung Kuo Group

Category : Office / Shopping Mall

Service : Design / Interior / Supervision

Location : Hsin-Yi District, Taipei

Completion Time: 1998-2004

Site Area : 30,277 m^2

Total Area : 370,000 m^2

Floor : B5-101F

Remarks : 2nd tallest building in the world 508 m

北京西门子总部 Siemens Center

业　　主：西门子中国有限公司
空间性质：办公楼
服务范围：设计 / 扩初
完工年份：2008年
坐落位置：北京
土地面积：44 400平方米
建筑面积：60 406平方米
楼 层 数：B1–28层　150米
备　　注：国际竞标第一

Client : Siemens Group
Category : Office
Service : Design
Category : Office
Service : Design
Completion Time : 2008
Location : Chaoyang, Beijing, China
Site Area : 44,400 m^2
Total Area : 60,406 m^2
Floor : B1-28F
Remarks : Int'l Competition 1st Prize

外滩延伸段（中山东二路）建筑立面改造案

自新中国诞生以来，外滩建筑群以新的风格与面貌向南发展。从延安东路至十六铺地段的建筑群体以现代功能主义为设计规划原则，其作为老外滩的“过渡期”，是本次改造及规划的重点地段。此地段已建的七栋建筑均是体现现代建筑造型与立面风格的作品。在过渡外滩历史风貌与空间的战略定位的基础上，建筑设计必须以改造和整治作为重新发展的动力源泉。

本案规划的重点在于以下几个方面。

风貌的延续

外滩的风貌是世界建筑古典艺术多元化的综合表现，风貌的延续必须体现古典元素意向、历史呼应以及都市动态与静态相结合的整体品质。

1. 对原有建筑顶部加以改进，使其形式更完整，以便加强建筑使用性。
2. 秉承老外滩建筑造型内在的DNA。
3. 修正原有建筑立面比例关系，使其整体更符合古典风格；调整微观比例，使其配合外滩尺度。
4. 借鉴老外滩建筑元素，通过细部处理使其与老建筑相协调。
5. 采用与老外滩外墙相同或相似色彩、质感的材料。

空间的延续

以历史外滩的空间构成元素为延续的主体，从天空线的延续、地面基度形体序列、墙面发展的空间架构以及结合水面的空间形态等多方位的角度入手，使其成为外滩空间的生命延续。

活动的延续

空间与风貌的延续是群落的发展主轴，根据外滩在都市发展的角色扮演和都市战略定位，建筑设计以商贸旅游为动态的延续主题，重新构筑本地区的功能发展和开发定位的核心动力。

法门寺合十舍利塔 Famen Temple

空间性质：宗寺规划

服务范围：规划／设计

设计年份：2003年

坐落位置：中国 西安

占地面积：200 000平方米

建筑面积：105 709平方米

建筑高度：147米

备　　注：荣获2007年度人居规划建筑经典奖

Category : Religious Structure

Services : Planning / Design

Design Time : 2003

Location : Xi'an, China

Site Area : 200,000 m^2

Total Area : 105,709 m^2

Height : 147 m

上海长风汇银金融中心 **Shanghai Huiyin Center**

空间性质：办公 / 商业

服务范围：规划 / 设计 / 营建管理

设计年份：2005年

坐落位置：上海

土地面积：54 766平方米

建筑面积：187 460平方米

Category : Office / Commercial

Services : Planning / Design / Construction

Design Time : 2005

Location : Shanghai, China

Site Area : 54,766 m^2

Total Area : 187,460 m^2

张文明建筑师事务所

W.M.CHANG ASSOCIATES. ARCHITECTS & PLANNERS

地址/Address：台湾 80661 高雄市民权二路 6 号 16F–2
16F–2，No.6，MingChiuan 2nd Rd.KAOSHIUNG
电话/Tel：886–7–3315896
传真/Fax：886–7–3390853
网页/Http//：www.wmchang.com.tw
电邮/E–mai：yj5896@ms23.hinet.net

内地主要设计作品

哈 尔 滨　顺迈集团 昆仑商城
哈 尔 滨　顺迈集团 荣耀宝座
哈 尔 滨　顺迈集团 丽景天地
哈 尔 滨　顺迈集团 宏景天地
哈 尔 滨　顺迈集团 世纪广场
哈 尔 滨　顺迈集团 荣耀世纪
哈 尔 滨　顺迈集团 得意居
上海昆山　阳明房地产 阳明山庄
哈 尔 滨　华风房地产 华风世贸广场
大　　连　顺迈集团 星海国宝
大　　连　顺迈集团 星海新天地
北　　京　顺迈集团 顺迈金钻
大　　连　国泰盛达集团 帝柏湾 立面 · 景观 · 灯光

事务所有着专业的建筑素养，可为业主提供土地开发的策略与规划、设计。理论与实践相结合，塑造符合业主需求及周边环境的优良作品，以建筑产业的发展方向和加入 WTO 后对建筑界的冲击为出发点，立足台湾、放眼大陆是事务所未来的发展方向，本所早在 1994 年即随业主西进，目前于大陆各地区已有完成的建筑个案，并致力于开拓新市场，期望能在此过程中建立更完善的制度与坚实的专业技能，提供更完整的服务。

本事务所以专业的方式为业主提供集合住宅、办公大楼、百货商场、饭店、休憩中心、高科技厂办、校园规划、校舍设计等相关信息及规划设计，并以高效率的作业提供最佳的设计方案。

Our institute provides customers variety services such as land development consulting, planning with professional experience. We combine academic research and design practice to create excellent environment design works which can satisfy customer's needs. Due to the construction industry's developing trend and the mainland has entered WTO, we should expand our business to the mainland while bassed in Taiwan. As early as 1994, we had entered the mainland market and at now have already finished several projects. We hope we can refine our service system and professional skills during such course.

Our service category includes multi-family residential facilities, office buildings, shopping centers, hotels, resort facilities, high-tech, factories, campus planning and design...... and so on. We can serve kinds of customers with excellent design works of high efficiency.

中盛地产 颐和香榭

建设地点：辽宁 大连
建筑用途：住宅、商业
业　　主：大连 大和中盛
建 筑 师：张文明
占地面积：63 550 平方米
设计时间：2006 年—2008 年
完工时间：2009 年

Construction Location: Dalian, Liaoning
Building Purpose: residence, commercial
Project Owner: DA/ WA Zhongsheng, Dalian
Architect: Wenming Zhang
Site Area: 63,550 m^2
Design Time: 2006-2008
CompletionTime: 2009

都市设计（Urban design）是介于都市计划和建筑设计之间的一个设计桥梁，对都市景观和居住环境有着重大的影响，因此是设计者在建筑设计之前需要特别关注的一个课题。

在“颐和香榭”这个项目中，业主对现有的都市涵构（Urban context）、户外空间的尺度，展现了最大的尊重和配合，为都市街道赋予了新的创意和景观。

Being the bridge of urban planning and architecture design, urban design has great influence in the urban landscape and living environment.Thus it is a subject the designers most concern before architecture design procedure.
In this project of Yihexiangshe, the project owner's respect and coordination to the urban context and outdoor space measurement, showing their utmost sincerity and tolerance, providing innovative ideas and landscape to the historical record of Longevity street at full stretch.

中盛地产 颐和香榭

在建筑外观和细部处理上设计师采用了内敛和低调的设计手法体现了项目的高雅和尊贵。

在大连这个中国沿海最美丽的城市里，“颐和香榭”展现出的是一种低调中带有奢华的美。

From the exterior of the building to the art deco style, we adopt the reserved and low-pitched design technique matching the style of the project, revealing its corresponding elegance and honor.

In Dalian, this most beautiful costal city in China, Yihexiangxie is the model of unique low profile design aesthetics with a sense of luxury.

太普建设 太普－尚海

太普建设 太普－尚海

建设地点：台湾 高雄
建筑用途：集合住宅
业 主：太普建设
建 筑 师：张文明
占地面积：589.88 平方米
设计时间：2006 年—2007 年
完工时间：2009 年

Construction Location: Kaohsiung, Taiwan
Building Purpose: Collective Housing
Project Owner: Type Construction
Architect: Wenming Zhang
Site Area: 589.88 m^2
Completion Time: 2006-2008
Complete Time: 2009

区位分析

基地使用面积为 1 364 平方米，共规划了 15 层 31 户（包含 2 户临路透天住宅）的住宅建筑，此项目位于苓雅区成功一路附近，距离百货商圈大约有 3 分钟的车程。建筑周边因临近码头港区，交通流量不大，并与南高雄市中心相连交通极为便利，且因与邻接的高雄港区码头仅有 200 ~ 300 米，因此主体建筑配置面向新光码头（水岸公园），拥有约 180° 以上的海景视野。

Location Analysis

The site area is 1,364 m^2, it is a 15 storeyed building with 31 households (including 2 wayside households) this project is located near the No.1 Chengong Road in Lingya District, it only takes 3 minutes to get the shopping mall. Being close to the port area and with less traffic flow, the transportation to the downtown of Kaohsiung is convenient. As it is only 200~300 meters away from the Kaohsiung port area wharf, and with main building facing the Singuang Ferry Wharf (The Lake King Park), it provides excellent 180 degree sea view.

太普建设 太普－尚海

规划构想

建筑配置取其最佳视野角度设计建造，符合基本物理环境条件朝南向的配置方式，局部采用深窗形式处理，降低了阳光的直射时间。朝南向的夏季风，有效地缓和室内空间的热辐射，使单位空间拥有更优化的物理条件，项目邻接开阔的 20 米宽的规划道路（目前尚未开辟，暂时为空地绿化带）使建筑外围拥有更多空间。设计采用 L 形架构延长了表面积，增大了对外开口的范围，让户主（客厅）能拥有两面的视野角度，收纳更多港景（海景），建筑第一线的视野条件具有无可取代的价值。

Planning Conception

The layout of buildings presents in a best viewing angle to meet the configuration that ensures basic physical environment conditions towards the south direction, meanwhile to match the local form of deep windows, thereby direct sunlight time will be cutback, and the southward summer monsoon will be reconciled, so as to effectively alleviate the thermal radiation of interior space, further more claims a more optimal physical conditions for individual spatial units; adjacent to the 20-meter-wide open road in plan (not open up yet, temporary as open space green belt), to create more breathing space for the building, by adopting L-type structure to extend the surface area thereby increasing the scope of external openings, thus our two primary spaces (living room)can own two horizon angles of facade to incorporate more harbor view (sea view) simultaneously, while there is still irreplaceable value in the first line of vision.

太普建设 太普－尚海

量体计划

建筑形态造型简单，合适的动线与尺度才是一个住宅方案应有的设计目标，简单的实体体现出具韵律感与空间感的空间。建筑设计增加遮阳与挡水的机能属性，利用力度强的垂直向板墙做切割，使单纯的体块形态表现出很强力道。同时建筑设计选用深色材质（黑色小口砖）强化建筑的线条，在厚实的体块中搭配简洁的玻璃阳台。

窗刻意做出较高的开口高度与宽度，此案设计的目标是为了尽量符合人的使用需求，并优化环境，建筑设计以简约的线条与架构，清楚地定位出此建筑所表达的低调、奢华的意象。

Accordingly Build Up Plan

Its architectural style applies the simple undecorated space vocabulary, and its residential space primarily focuses on functional allocation, while appropriate scale and dynamic lines should be the motivation of a residential design, based on rational configuration to integrate the appropriate open division for functional plane. Simple entity leaves deep windows virtual space with sense of rhythm and space, meanwhile this form can improve the function properties of shade and water retaining. This building simply makes use of vertical wall with strong strength, to cut off the joint-surface of the material, and pull out the guideline for spatial vision, so that the pure block pattern can be able to show a strong force, at the same time, the use of dark material (black little dimension brick) to enhance the strength of texture.

Its thick block is matched with a simple style of glass balcony, appropriately brings in lightweight construction to the heavy feeling in external environment, at the same time it also converses the outside architectural surface in texture.

On the configuration space, the main view window facing the sea deliberately pulls out a higher height and width of the opening, which presents a different opening atmosphere, also express properties of its main space. This program is designed to meet the optimized human needs and environmental conditions, combined with simple lines and structure, thereby clearly defines a low-key luxury image this building intended to convey.

(C.J Chen Architects) 陈昭榕建筑师事务所

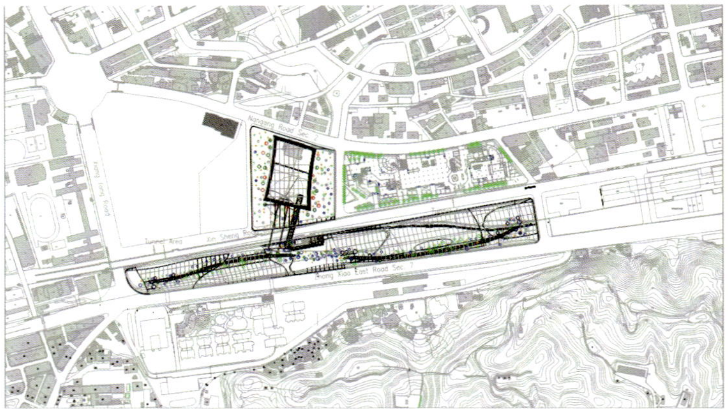

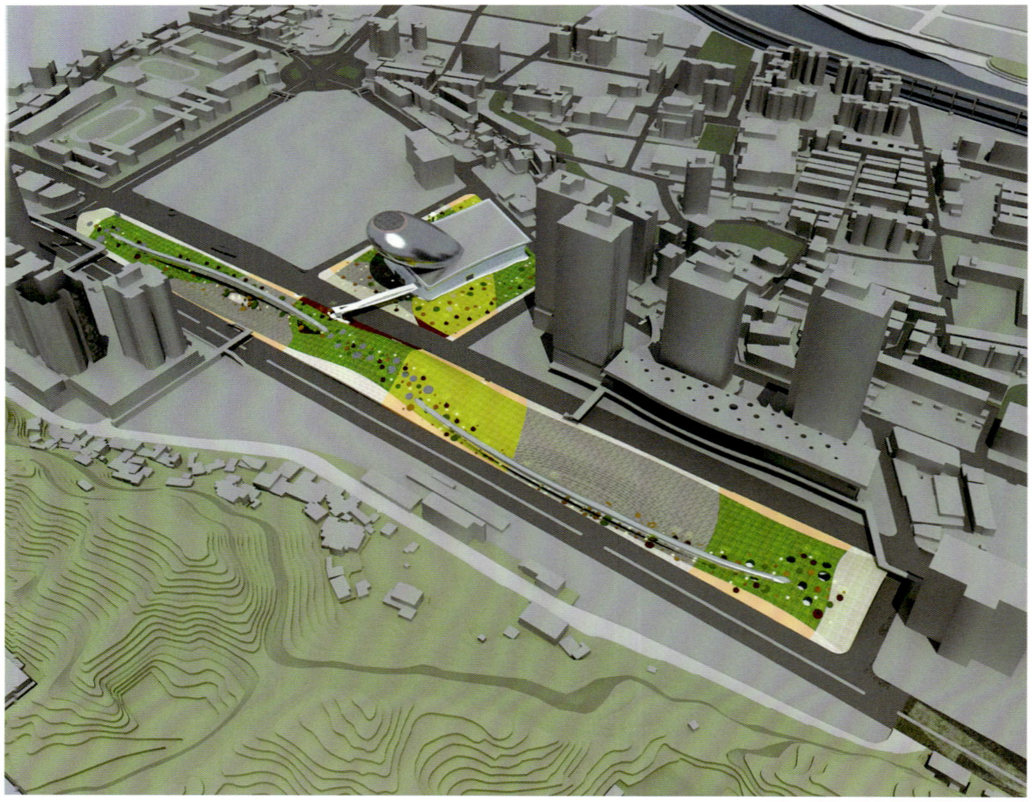

陈昭榕建筑师事务所（C.J Chen Architects）于 2006 年 1 月在台北成立。事务所主要的服务项目包含私人住宅、商业、办公，室内设计，混合使用内容以及相关的公共建筑等。除了实质的景观设计、建筑设计、室内设计服务外，事务所也提供概念设计，以前卫的设计思潮配合数字科技，从而创造出独特的空间感。

2007 年，为了提升事务所的设计质量与水平，陈昭榕建筑师前往纽约哥伦比亚建筑研究所进修，汲取建筑设计上的形而上的养分。事务所的经营理念以满足顾客的需求为目标，并在设计过程中寻找创意设计的空间，通过团队的合作将空间体现出来。

C.J Chen Architects is a 20-person international design firm offering services in architecture and interiors. Founded in 2006, we are committed to forgoing a greater connection between the built environment and the quality of life, and creating thoughtful and responsible architecture or" design product", which is tied to a true story of place.

Focused on the unique and compelling retail and mixed-use destinations worldwide, the cohesive of sensibility of our practice is expressed in a customer-centric approach, an emphasis on communication and teamwork, and a dynamic and positive office culture.

地址：台湾台北 110 基隆路二段 51 号 3 楼之 7
电话：886–2–27392075
传真：886–2–27392327
网址：www.cjchenarchitects.com

Address: 7th floor,3building,No51,110jilong road 2nd, Taipei, Taiwan
Tel: 886-2-27392075
Fax: 886-2-27392327
Http: //www.cjchenarchitects.com

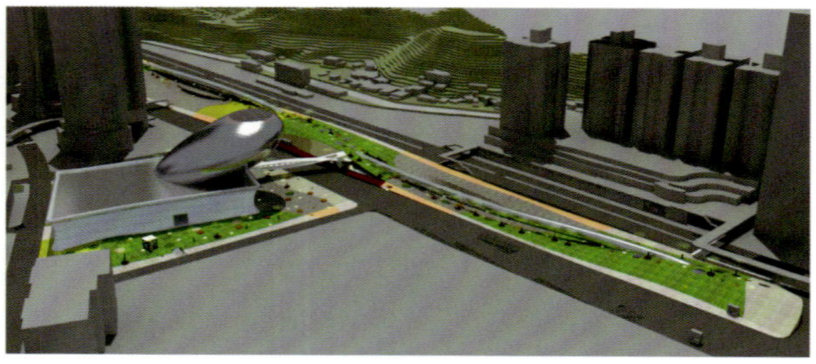

北部流行音乐中心

设计单位：陈昭榕建筑师事务所
建 筑 师：陈昭榕建筑师
地板面积：7.65 公顷
主要材料：ETFE 膜构 ＋ 薄壳构造
建设地点：台湾 台北
总工程造价：3 500 000 000 新台币
完工时间：2012 年 9 月

Northern Centre for Popular Music

Designer: C.J Chen Architects
Architect: C.J Chen
Floor Area: 7.65 hectares
Main Material: ETFE membrane structure + Shell Construction
Location: Taipei, Taiwan
Total Project Cost: 3,500,000,000 NTD
Completion Time: 2012.09

天一制药厂迁厂暨创意文化馆

设计单位：陈昭榕建筑师事务所
建 筑 师：陈昭榕建筑师
面　　积：17 000 平方米
主要材料：膜构造 +SC
建设地点：台湾 台南
总工程造价：800 000 000 新台币
完工时间：2011 年 12 月

Relocation of Tianyi Pharmaceutical and Creative Cultural Center

Designer: C.J Chen Architects
Architect: C.J Chen
Area: 17,000 square meters
Main Material: Membrane structure + SC
Construction Location: Tainan, Taiwan
Total Project Cost: 800,000,000 NTD
Completion Time: 2011.12

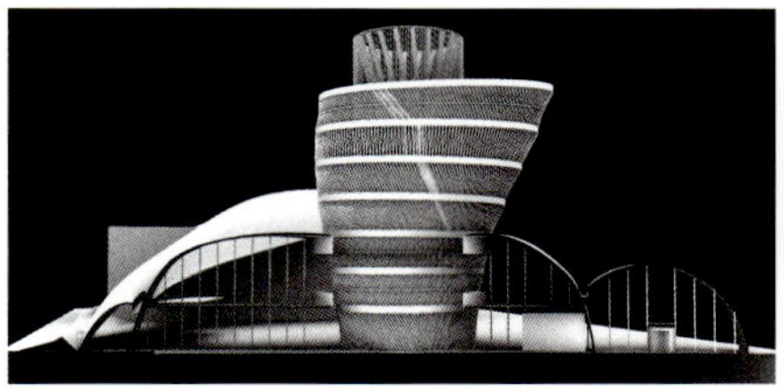

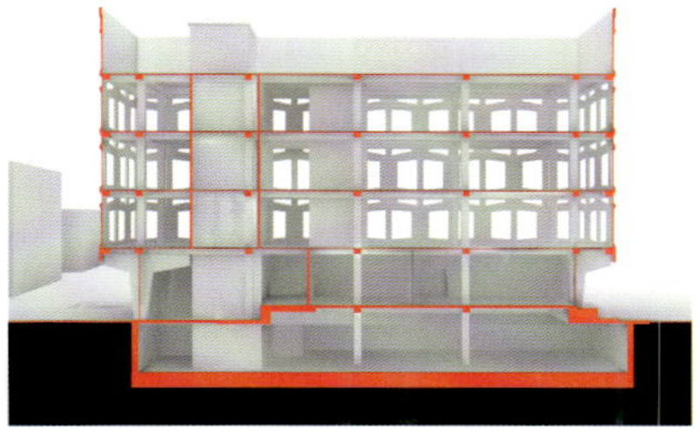
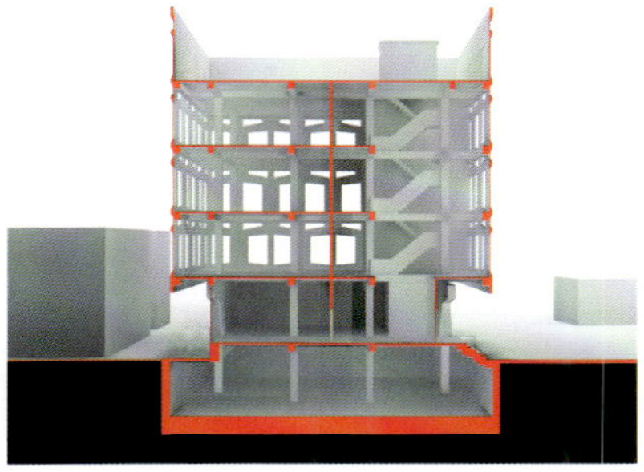

中央印制厂整钞中心大楼

设计单位：陈昭榕建筑师事务所
建 筑 师：陈昭榕建筑师
地板面积：2 900 平方米 ±10 平方米（地下一层，地上四层及屋顶突出层）
主要材料：预制混凝土板配合柱梁模矩系统
建设地点：台湾 台北
总工程造价：82 000 000 新台币
完工时间：2011 年 3 月

Central Building of Banknote Arrangement of Central Printing Plant

Designer: C.J Chen Architects
Architect: C.J Chen
Floor Area: 2,900 m^2 ± 10 m^2(one layer underground, four layers above ground and roof prominent layer)
Main Material: pre-cast concrete plate coordinated with beam-column model moment system
Construction Location: Taipei, Taiwan
Total Project Cost: 82,000,000 NTD
Completion Time: 2011.03

441 号室内设计

设计单位：陈昭榕建筑师事务所
建 筑 师：陈昭榕建筑师
地板面积：200 平方米 ±10 平方米（15 层 +16 层）
主要材料：室内装修常用材料
建设地点：台湾 台北
总工程造价：5 000 000 新台币
完工时间：2011 年 4 月

NO.441 Interior Design

Designer: C.J Chen Architects
Architect: C.J Chen
Floor Area: 200 m^2 ± 10 (15F+16F)
Main Material: Commonly used materials of interior decoration
Construction Location: Taipei, Taiwan
Total Project Cost: 5,000,000 NTD
Completion Time: 2011.04

万邦建筑师事务所
WANBANG ARCHITECTS & ASSOCIATES

万邦建筑师事务所成立于 1974 年，擅长体育设施设计如棒球场、体育场、体育馆、游泳馆、休闲运动中心等多元功能用途空间的建筑设计公司，已有 35 年以上的专业经验。

目标放眼于："世界村全球化的观点，体育事业已是超越疆界、肤色、人种及国籍的代表性活动，亦是衡量国家、社会、经济进步发展的重要指标"。

本公司期待持续发展"自然、人文、科技、技术"的整合，以设计出深广兼俱、适用于未来的体育场馆为激励。

Founded in 1974, WANBANG ARCHITECTS & ASSOCIATES has 35 years' experience in the design for the multi-functional space, sports facilities for example, the baseball field, stadium, gymnasium, swimming pool and recreational sports center, etc. Our orientation is the globalization of the world village. Sports have become the representative activity which is beyond borders, racial colors, ethnic origins and nationalities, which is also a key indicator in measuring the national, social and economic progress and development.

Our company anticipates to develop continuously the integration of "nature, humanity, science and technology", taking as its incentive the production of gymnasium designs both with width and depth, and the applicability for the future use.

地址：台北市中正区罗斯福路二段 50 号 9F(R.O.C)
电话：886–2–2394–8135
传真：886–2–2394–8139
邮箱：wanbang@ms16.hinet.net
法人：庄辉和

Address: 9F (ROC), No. 50, Sec. 2, Roosevelt Road, Zhongzheng District, Taipei City
Tel: 886-2-23948135
Fax: 886-2-23948139
E-mail: wanbang@ms16.hinet.net
Legal Person: Huihe Zhuang

高雄县澄清湖棒球场

建设地点：台湾 高雄
时　　间：1995 年—1999 年
建筑面积：35 863 平方米
摄　　影：潘瑞琮 / 齐柏林

设计说明

高雄县澄清湖棒球场坐落于南台湾风光明媚的澄清湖畔，为台湾观众容量最大的棒球场，这座"有人性的棒球场"，无论在观众视野、空间动线、球员休息还是草坪平整、排水设施、夜间照明均达国际标准，是一流的比赛场地。

ChengChing Lake Baseball Field

Construction Location: Kaohsiung, Taiwan
Time: 1995-1999
Building Area: 35,863 m^2
Photography: Ruicong Pan / Bolin Qi

Design Description

Located in the waterside of the scenic ChengChing Lake in the southern Taiwan, ChengChing Lake Baseball Field is of the largest audience capacity in Taiwan. This "baseball field of humanism" is a first-rate competition venue satisfying the international standards in terms of the audience perspective, the space dynamic line, the players' rest place, the lawn formation, drainage facilities and the nocturnal lighting.

花莲县立棒球场

建设地点：台湾 花莲
时　　间：1999 年—2001 年
楼地板面积：7 869 平方米
摄　　影：郑锦铭

设计说明

花莲县立棒球场坐落于好山好水的"台湾后花园"，周边天然景观丰富，建筑规划塑造为具有花莲缩影的地域特色设计并导入各种环境艺术，融入地景风貌。屋顶膜构造型与沙婆当山曲径相连，展现出天人合一的境界。

HuaLine Baseball Field

Construction Location: Hualien, Taiwan
Time: 1999-2001
Floor Area: 7,869 m^2
Photography: Jinming Zheng

Design Description

Located in the scenic "back garden of Taiwan", HuaLine Baseball Field is surrounded by a rich natural landscape. Its architectural planning applies the geographical features of a miniature Hualine, and introduces various environmental arts. It impacts on the low-altitude environment, merging into the landscape feature. The membrane structure of the roof connects with the winding paths in the mountain, exhibiting the harmony between man and nature.

新竹县立体育馆暨游泳馆

建设地点：台湾 新竹
时　　间：2004 年—2007 年
楼地板面积：52 516 平方米

设计说明

新竹县立体育馆暨游泳馆采用分栋式设计规划，外观造型设计结合新竹地域客家精神的内涵，体育馆主体与游泳馆屋顶立面的比例是从建筑的整体性入手而设计的。特将两栋立面设计以流线型延伸，相映成趣。本场馆是 2009 台北夏季听障奥运会游泳与水上运动竞赛的正式比赛场地。

Hsinchu Municipal Gymnasium and the swimming pool

Construction Location: Hsinchu, Taiwan
Time: 2004-2007
Floor Area: 52,516 m^2

Design Description

Hsinchu Municipal Gymnasium also as the swimming pool adopts the design of "seperate design". The modeling of the external image combines with the Hakka spirit of the Hsinchu region and the consideration in the proportion as well as it is based on the integrity of the main body of the gymnasium and the roof façade of the swimming pool. Thus the facades of these two constructions are in streamline extension, echoing with each other. This gymnasium has been selected into the official venues for the swimming and water sports competitions of the 2009 Taipei Summer Deaflympics.

HFUT Institute of Architectural Design

合肥工业大学建筑设计研究院

地址：安徽省合肥市屯溪路 193 号
邮编：230009
电话：+86-551-2901598
传真：+86-551-2901598
网址：www.hfutadi.com.cn

Address: T'un-hsi Road 193, Hefei, Anhui Province
P.C.: 230009
Tel: +86-551-2901598
Fax: +86-551-2901598
Http: //www.hfutadi.com.cn

合肥工业大学建筑设计研究院成立于1979年，隶属于国家教育部，持有多项甲级资质证书，包括建筑工程设计甲级、工程咨询甲级、工程总承包甲级、建筑智能化系统工程设计甲级、工程勘察设计岩土工程甲级、房屋建筑工程监理甲级以及市政工程设计乙级、电力工程设计乙级和国家二级保密资质。

研究院现有正、副教授级高级工程师116名、工程师62名，其中国家一级注册建筑师23名、国家特许一级注册建筑师1名，国家一级注册结构师26名、国家特许一级注册结构工程师1名、国家注册咨询工程师（投资）9人，取得国家注册公用设备（给排水）工程师资格的有4人，取得国家注册公用设备（暖通空调）工程师资格的有4人，取得国家注册电气工程师资格的有3人，取得国家注册规划师资格的有8人，取得国家注册土木（岩土）工程师资格的有11人，国家注册监理工程师50人和相当数量的硕士、博士研究生，同时还有安徽省工程勘察设计大师5人。

1999年单位被建设部确认为全国76家骨干建筑设计单位之一。

为强化质量意识和提高设计水平，研究院经常对全体员工进行质量教育和技术培训，强调员工严格执行国家和地方有关的强制性规范、标准的重要性。精心设计，努力满足用户需求，防止和杜绝不合格的设计作品出现，保证合同的履行，定期开展工程回访活动，不断改进服务质量，对工程全过程进行跟踪服务，对工程全面负责。1992年研究院通过全面质量管理达标验收，2008年通过中质协质量管理、环境管理和职业健康安全三项体系通过认证，取得质量管理、环境管理和职业健康安全三项体系证书。

研究院本着“为社会提供一流的建筑产品与服务”的精神，力求为社会提供更多、更好的高质量建筑产品。

Established in 1979, Architectural Design & Research Institute, Hefei University of Technology is subordinate to Ministry of Education. This institute has been granted many Class A qualifications, including Class A qualification for architectural design, for consulting engineering, for EPC, for architectural intelligence system engineering design, for geotechnical engineering investigation, for building construction supervision. It has also some Class B certificates like Class B municipal engineering design, Class B electrical engineering design, Class B security institute.

At present, there are altogether 116 full professors, associate professors, and senior engineers, and 62 engineers. Among them, 23 are PRC class I registered architects, 1 national franchise class I registered architects, 26 national class I registered structural engineers, 1 national franchise class I registered structure engineers, 9 national registered consulting (investment) engineers, 4 national registered public utilities (water) engineers, 4 national registered public utilities (HVAC) engineers, 3 national registered electric engineers, 8 national registered planners, 11 national registered civil engineers, 50 national registered supervision engineers. There are a number of doctors and masters and 5 provincial survey and designing masters in this institute.

In 1999, this institute was rated as one of 76 main architectural design units by Ministry of Construction.

To strengthen quality awareness and enhance the design, this institute often provides quality education and technical training for staffs. In the process of education and training, this institutes attaches importance to related national and provincial norms and mandatory standards and underlines significances of well-designed products. It makes efforts to satisfy needs of its clients, prevent and put an end to unqualified products, and guarantee contract performance. Meanwhile, it conducts project revisits at regular intervals to improve services and it shoulders overall responsibility for projects and offers all-round and follow-up services. In 1992, it obtained quality management certification. In 2008, it passed QAC quality management certification for quality control, environmental management, and occupational health and safety and got QMS certificates.

Taking “providing society with quality construction products and superior services” as its goal, HFUT Institute of Architectural Design will constantly supply more high-quality construction products for society.

1 合肥工业大学建筑技术研发中心

建设地点：安徽 合肥
建筑规模：19 221 平方米
设计时间：2004 年 12 月
获奖情况：2009 年度全国优秀工程勘察设计行业奖三等奖
教育部优秀工程勘察设计二等奖
安徽省优秀工程勘察设计行业奖一等奖

R&D Centre of HFUT Institute of Architectural Design

Construction Location: Hefei, Anhui
Construction Scale: 19,221 m²
Design Time: 2004.12
Awards: The third prize of State Outstanding Engineering Investigation and Design, 2009
The second prize of Outstanding Engineering Investigation and Design issued by Ministry of Education
The first prize of Outstanding Engineering Investigation and Design in Anhui Province

2

2 安徽广播影视职业技术学院新校区规划

建设地点：安徽 合肥
建筑规模：385 300 平方米
设计时间：2008 年 6 月

Planning and design for the new campus of Anhui Broadcasting Film and Television College

Construction Location: Hefei, Anhui
Construction Scale: 385,300 m²
Design Time: 2008.06

3 合肥工业大学翡翠湖校区图书馆

建设地点：安徽 合肥
建筑规模：32 500 平方米
设计时间：2001 年 10 月
获奖情况：2005 年获得安徽省优秀设计一等奖
2006 年获得建设部优秀设计三等奖

Library in Emerald Lakes Campus of Hefei University of Technology

Construction Location: Hefei, Anhui
Construction Scale: 32,500 m²
Design Time: 2001.10
Awards: The First Prize of Outstanding Design in Anhui Province, 2005
The Third Prize of Outstanding Design issued by Ministry of Construction

4 合肥工业大学学科群实验楼

建设地点：安徽 合肥
建筑规模：70 000 平方米
设计时间：2004 年 2 月
获奖情况：2007 年度教育部优秀工程勘察设计三等奖
2008 年度全国优秀工程勘察设计行业奖三等奖

Experiment Building in Academic building complex of Hefei University of Technology

Construction Location: Hefei, Anhui
Construction Scale: 70,000m²
Design Time: 2004.02
Awards: The Third Prize of Outstanding Engineering Investigation and Design issued by Ministry of Education, 2007
The Third Prize of State Outstanding Engineering Investigation and Design, 2008

3

5

5 中科院合肥物质科学研究院研究生教育基地

建设地点：安徽 合肥
建筑规模：18 000 平方米
设计时间：2005 年 12 月
获奖情况：2008 年度安徽省优秀工程勘察设计行业奖二等奖

Education base for graduate students of Hefei Institutes of Physical Science

Construction Location: Hefei, Anhui
Construction Scale: 18,000 m²
Design Time: 2005.12
Awards: The second prize of Outstanding Engineering Investigation and Design in Anhui Province, 2008

6 黄山市人大政协办公大楼

建设地点：安徽 黄山
建筑规模：12 000 平方米
设计时间：2005 年 2 月
获奖情况：2009 年度安徽省优秀工程勘察设计行业奖三等奖

Office block of Huangshan Municipal People's Congress and People's Political Consultative

Construction Location: Mount Huang, Anhui
Construction Scale: 12,000 m²
Design Time: 2005.02
Awards: The third prize of Outstanding Engineering Investigation and Design in Anhui Province, 2009

7 合肥学林雅苑住宅小区

建设地点：安徽 合肥
建筑规模：286 548 平方米
设计时间：2004 年 10 月
获奖情况：2008 年中国勘察设计协会优秀工程勘察设计（住宅与住宅小区）三等奖

Xuelin Yayuan Residence Community, Hefei

Construction Location: Hefei, Anhui
Construction Scale: 286,548 m²
Design Time: 2004.10.20
Awards: The third prize (residence and residence community) of State Outstanding Engineering Investigation and Design, 2008

8 安徽省教育学院新校区艺术楼

建设地点：安徽 合肥
建筑规模：14 700 平方米
设计时间：2004 年 9 月
获奖情况：2008 年度安徽省优秀工程勘察设计行业奖一等奖

Art Building in new campus of Anhui Education College

Construction Location: Hefei, Anhui
Construction Scale: 14,700 m²
Design Time: 2004.09
Awards: The first prize of Outstanding Engineering Investigation and Design in Anhui Province

6

7

8

9

9 桐城玉屏山商务生态休闲区

建设地点：安徽 桐城
用地面积：78.74 公顷
建筑规模：121 320 平方米
设计时间：2009 年 11 月

Mount Yuping Business and Zoology Leisure Area, Tongcheng

Construction Location: Tongcheng, Anhui
Site Area: 78.74 ha
Construction Scale: 121,320 m²
Design Time: 2009.11

10 安徽师范大学新校区一期院系教学楼

建设地点：安徽 芜湖
建筑规模：44 827 平方米
设计时间：2004 年 5 月
获奖情况：2007 年度教育部优秀工程勘察设计三等奖
2008 年度安徽省优秀工程勘察设计行业奖一等奖

Teaching Building in new campus of Anhui Normal University, Phrase One

Construction Location: Hefei, Anhui
Construction Scale: 44,827 m²
Design Time: 2004.05
Awards: The third prize of Outstanding Engineering Investigation and Design issued by Ministry of Education, 2007
The first prize of Outstanding Engineering Investigation and Design in Anhui Province, 2008

10

11

11 圣大休博园概念性规划

建设地点：安徽 巢湖
规划用地：180 公顷
设计时间：2009 年

Conceptual design for Shengda Xiubo Garden

Construction Location: Lake Chao, Anhui
Construction Scale: 180 ha
Design Time: 2009

12 合肥望湖城望湖云居小区

建设地点：安徽 合肥
建筑面积：154 213 平方米
设计时间：2006 年 6 月
获奖情况：2009 年度教育部优秀工程勘察设计三等奖

Wanghuyunju in Wanghu Residence Community, Annhui

Construction Location: Hefei, Anhui
Building Area: 154,213 m²
Design Time: 2006.06
Awards: The third prize of Outstanding Engineering Investigation and Design issued by Ministry of Education, 2009

12

Architectural

设计作品
Design Works

黄山市建筑设计研究院
Huangshan Architectural Design and Research Institute

黄山市建筑设计研究院成立于1958年，是拥有国家建设部批准的建筑工程设计甲级，工程造价咨询、工程建设监理、招投标代理、岩土工程勘察乙级，市政工程设计丙级资质的综合勘察设计单位，内设院长室、总工办、综合部、技术经营部、徽派建筑设计研究所、设计一所、设计二所、市政设计所、岩土工程所、造价咨询部、东莞设计部几大部门。其下还有装饰工程公司、控股黄山市建设监理有限公司、蓝鹰图文公司三个公司。

我院专业设置齐全、技术实力雄厚，专业技术人员占所有员工94%，其中国家一级注册建筑师5人、一级注册结构工程师5人、二级注册建筑师6人、注册监理工程师22人、注册造价工程师9人、注册咨询工程师（投资）4人、注册岩土工程师1人、注册设备工程师2人。院内有完善的计算机网络应用及管理系统，是ISO 9001：2000 质量达标认证单位。我院业务遍布黄山市三区四县及福建、山东、江西、浙江、上海等省市。我院为国家城市建设和徽派建筑设计研究作出了较大贡献。

近年来我院获奖项目主要有"安徽省徽州文化博物馆"荣获省2009年度优秀建筑工程设计一等奖；"水墨宏村"项目获全国工商联地产商会第四届"精瑞住宅科学技术奖、建筑文化金奖"，并被列入国家科技支撑计划，获科研经费资助；"黄山管委会、股份公司综合办公楼"获2008年度省优秀建筑工程设计一等奖、部三等奖。我院获奖的项目还有徽商故里大酒店（省二等奖）、黄山狮林饭店（建设部鼓励奖、省二等奖、结构设计专项省三等奖）、黄山翡翠宾馆（省二等奖）、黄山市示范幼儿园（建设部三等奖、省二等奖）、黄山市委市政府大楼暖通专项设计（省三等奖），黄山机场侯机楼、旌德梓山宾馆、黄山市老干部活动中心、昱阳山庄、歙县北岸中学教学楼（省三等奖）及屯溪一中教研楼获建筑结构专项鼓励奖。

黄山市建筑设计研究院先后承担国家建筑标准设计"地方传统建筑（徽州地区）"（GJBT–652）和安徽省工程建设标准设计"坡屋面"（DBJT11–115）图集的编制工作，为安徽省"新农村建设农民建房通用图集"提供两套入选方案，并一直致力于新工艺推广、新技术应用工作，为城市建设、社会发展作出了应有的贡献，多次受到省、市级表彰，荣获2005年度"全省建设系统"十五"发展创新工程先进集体""科技兴市先进集体"、黄山市地税A级纳税诚信单位、"文明单位"等荣誉称号，我们的设计院和监理公司都是省级"重合同守信用"单位，社会知名度不断提升。我院秉承"诚信、务实、优质、高效"的经营宗旨，充分发挥人才、技术、管理优势，一如既往地为社会奉献精品项目设计。

Architectural Design and Research Institute, Huangshan City was established in 1958. It is a comprehensive survey design unit which has a Class A qualification of architectural engineering design approved by the Ministry of Construction, Class B qualification of engineering cost consulting, engineering construction supervision, bidding agent, and geotechnical engineering investigation, Class C qualification of municipal engineering design. The institute consists of Dean Office, Overall Engineering Office, Comprehensive Department, Department of Technology Management, Architectural Design Institute of Anhui School, No.1 Design Institute, No.2 Design Institute, Municipal Design Institute, Geotechnical Engineering Institute, Cost Consulting Department, Dongguan Design Department. Decoration Engineering Company is its subsidiary and it holds shares of construction supervision Co., Ltd, Huangshan City and Blue Eagle Graph-text Company.

The institute has complete specialty provision and strong technical strength. Among all the staff, the professional and technical staff account for 94%, of which there are 5 registered architects, 5 Class I registered structural engineers, 6 secondary registered architects, 22 registered supervision engineers, 9 registered cost engineers, 4 registered consulting engineers (investment), 1 registered geotechnical engineer, 2 registered equipment engineer. Having a perfect application and management system of computer network, the institute is a certification unit up to the standard of ISO9001: 2000 quality. Business covers three districts and four counties in Huangshan City and Fujian, Shandong, Jiangxi, Zhejiang, Shanghai and other cities. It has made a great contribution to national urban construction as well as architectural study and design of Anhui school.

In recent years, winning projects are: "Cultural Museum in Huizhou, Anhui" which won the First Prize for 2009 Outstanding Provincial Engineering Design; "Shuimo Hongcun Village" was awarded the 4th "Science and Technology Award of Fine Housing" and "Construction Culture Award" given by the Estate Commerce of National Federation of Chamber. This project is listed in the support programs of national science and technology and got financial aid of research funds; "Huangshan Management Committee and Office Building of Shares Company " won the First Prize for 2008 Provincial Excellent Construction Engineering Design, and the Third Prize of the Ministry of Construction, Huishang International Hotel (the Provincial Second Prize), Huangshan Lion Forest Hotel (the Encouragement Prize of the Ministry of Construction, the Provincial Second Prize, the Third Prize for Provincial Structure Design Special), Huangshan Jade Hotel (the Provincial Second Prize), Huangshan Demonstration Kindergarten (the Third Prize of Ministry of Construction, the Provincial Second Prize), the Special Design of HVAC of Huangshan Municipal Government Building (the Provincial Third Prize); Terminal Building of Huangshan Airport, Zishan Hotel in Jingde, Huangshan Exchange Center for Old Cadres, Yuyang Villa, Teaching Building of North Shore School, Shexian County (the Provincial Third Prize); the Encouragement Prize for Building Structure Special of Teaching and Research Building of Tunxi No.1 Middle School.

Architectural Design and Research Institute, Huangshan City has successively undertaken compiling work of Atlas for national architecture standard design---"local traditional architecture (Huizhou region)" (GJBT-652) and Anhui construction standard design---"sloping roof"(DBJT11-115). It offers two sets of selected programs for "General Drawing Album for Farmers' housing of New Rural Construction" in Anhui Province. Meanwhile, it has been dedicated to the promotion of new technology, to the application of new technology, which makes its due contribution to urban construction and social development. It has been awarded many times by province and city. The institute won the " the Advanced Collective for Innovative Engineering of the Tenth Five-year-project of Provincial Building System" in 2005, "Advanced Collective for Vitalizing the City by Means of Science and Education", "Credit unit of A class Tax Payment for Land Tax in Huangshan, "Vivilized Unit", etc. The design institute and supervision company are provincial "Strictly abiding by contracts and keeping promise" units. Social popularity and fine reputation are enhanced gradually .

Upholding the business aim "Integrity, being pragmatic, high quality, efficiency", Huangshan Architectural Design and Research Institute will give full play to advantages of human resources, technology, management and keep creating design boutiques to serve the society and to achieve better results.

地址：黄山市屯溪区齐云大道35–1号
邮编：245000
法人代表：洪祖根
电话：+86-559-2513998
传真：+86-559-2517387
邮箱：2513998@vip.sina.com
网址：www.hsadi.com

Address: NO.35-1 of Will Awenue,Tunxi District, Huangshan, Anhui
P.C.: 245000
Legal Representative: Zugen Hong
Tel: +86-559-2513998
Fax: +86-559-2517387
E-mail: 2513998@vip.sina.com
Http: //www.hsadi.com

黄山市建筑设计研究院办公楼

建设地点：安徽 黄山
时间：2008
规模：10 000平方米

Office Building of Architectural Design and Research Institute, Huangshan City

Construction Location: Huangshan, Anhui
Construction Time: 2008
Construction Scale: 10,000 m^2

安徽省徽文化博物馆

建设地点：安徽 黄山
建设时间：2005 年
建设规模：16 000 m^2
获奖情况：安徽省优秀建筑工程设计一等奖

Cultural Museum in Huizhou, Anhui
Construction Location: Huangshan, Anhui
Construction Time: 2005
Construction Scale: 16,000 Square meters
Awards: the First Prize for Outstanding Engineering Design, Anhui

水墨宏村

建设地点：安徽 黄山
建设时间：2007 年
建设规模：一期 15 000 平方米
获奖情况：科技部集约型住宅推广奖，全国工商联地产商会第四届"精瑞住宅科学技术奖、建筑文化金奖"

Shuimo Hongcun Village
Construction Site: Huangshan, Anhui
Construction Time: 2007
Construction Scale: 15,000 m^2 of Phase 1
Awards: Promotion Award for Intensive Houses of Ministry of Science and Technology, the 4th "Science and Technology Award of Fine Housing" and "Construction Culture Award" given by the Estate Commerce of National Federation of Chamber"

徽商故里大酒店

建设地点：安徽 黄山
建设时间：2005 年
建设规模：8 000 平方米
获奖情况：安徽省优秀建筑工程设计二等奖

Huishang International Hotel
Construction Location: Huangshan, Anhui
Construction Time: 2005
Construction Scale: 8,000 m^2
Awards: the Second Prize for Outstanding Engineering Design, Anhui

山东省滨州市中海一号星接待中心

建设地点：山东 滨州
建设时间：2005 年
建设规模：5 000 平方米

Reception center of China Shipping Star-1, Binzhou City, Shandong
Construction Location: Binzhou, Shangdong
Construction Time: 2005
Construction Scale: 5,000 m^2

中国建筑设计研究院

CHINA ARCHITECTURE DESIGN & RESEARCH GROUP

中国建筑设计研究院是以原建设部建筑设计院、中国建筑技术研究院为母体，吸纳中国市政工程华北设计研究院、建设部城市建设研究院为所属单位，于 2000 年 4 月组建成为科技型中央企业，隶属于国务院国有资产监督管理委员会。我院现有职工 3 000 余人，其中有工程院院士 2 人、全国工程勘察设计大师 6 人，经国务院批准享受政府津贴的专家近百人，国家一级注册建筑师、结构工程师和其他注册人员近 600 人，高、中级设计师和研究人员近 2 000 人。2000 年 11 月通过 ISO 9001 国际质量体系认证。

中国建筑设计研究院是改革开放后第一批进入国际建筑市场并较早获得对外经营权的设计院之一。主营业务包括建筑工程设计与咨询、城市与小城镇规划、市政工程综合设计、城市燃气与工业煤气设计、污水与垃圾处理、道路桥梁设计、建筑智能化及系统工程设计、建筑标准设计、景观艺术设计、工程监理、工程总承包、房地产开发和居住工程、建筑标准、科技信息技术、建筑历史、建筑经济等研究以及与上述业务相关的科研工作。附属业务主要有技术培训、技术服务、技术开发、技术转让等。基本形成了以建筑设计为主业，以城市建设规划、城市市政、建筑标准、建设信息、工程咨询、室内装饰、园林绿化、住宅研发、房地产为重点业务的集团化产业构架。

中国建筑设计研究院是国家建筑行业中综合实力强、科技含量高、辐射范围广、具有全国影响力和国际竞争力的建筑设计与科研企业，秉承"科学管理、竭诚服务、精心设计、质量第一"的执业宗旨，充分利用设计与科研、人力与技术等优势，积极探索和开拓更广阔的业务市场，力争将研究院打造成为拥有自身知名品牌的世界一流的科技型设计企业。

China Architecture Design & Research Group (CAG), a merger of the Architecture Design Institute Ministry of Construction and the China Building Technology Development Center, is a large high-tech state enterprise grouped in April 2000. The North China Municipal Engineering Design & Research Institute and the China Urban Construction & Research Institute as its solely owned branch institutions were incorporated into the Group. The Group is affiliated to the State-owned Assets Supervision and Administration Commission of the State Council.

The Group boasts of over 3,000 employees, and among them there are 2 academicians of Chinese Academy of Engineering, 6 national design masters, nearly 100 experts who enjoy the government allowances by the State Council, nearly 600 class-A registered architects/structure engineers and other registered personals, around 2,000 senior and medium titled designers and researchers. In November 2,000, the Group passed the ISO9001 certification exam.

CAG is among the first group who has set up its reputation in the international building construction market after the open policy was carried out in China and it is among the early group of design institutes that have obtained permission of managing foreign trade and import/export qualification from the concerning authorities. Its main business scope covers, architectural design and consultant, urban and town planning, municipal engineering comprehensive design, urban gas and industrial gas design, sewage and garbage disposal, road and bridge design, building intelligence and systems design, building standardization design, landscape design, project supervision, project turn-key contracting, real estate development and residential projects, research works concerning building standardization, sci-tech information, architecture history, building economy and other above mentioned items. The related business covers technical training, service, development, transfer, etc. The Group has now set up its business structure that takes architectural design as its fundament and the urban planning, municipal engineering, building standardization, construction information, project consultant, interior decoration, landscape and gardening, housing research/development and real estate development as its key businesses.

CAG is now an architecture design and scientific research enterprise that is strong in capabilities, has high scientific contents in its works, covers a wide range in business scopes and has a strong national influence and international competitive edge in the construction sector of the country. In pursuit of its practice principle of "best management, best service, best design and best quality", CAG will make full use of its advantages in scientific research, design, technologies and human resources, and will actively explore and expand its business fields. It will strive to become a first-class science and technology-oriented design enterprise with its own brand name and international competitive edge in the near future.

国家体育场
The National Stadium

建设地点：北京
用地面积：24 134 平方米
建筑面积：63 390 平方米
设计时间：2001 年
竣工时间：2005 年
方案设计：法国 AREP 建筑设计公司、中国建筑设计研究院

Construction Location: Beijing
Site Area: 24,134 m^2
Building Area: 63,390 m^2
Design Time: 2001
Completion Time: 2005
Scheme Design: France AREP Architectural Design Company and China Architecture Design & Research Group

国家体育场是 2008 年北京第 29 届奥运会的主体育场。奥运会期间，国家体育场的观众坐席约为 91 000 个，其中临时坐席约 11 000 个（可在赛后拆除）。奥运会的开幕式、闭幕式、田径比赛和男子足球决赛以及残奥会的开闭幕式和田径比赛在这里举行。奥运会后，国家体育场的观众坐席约 80 000 个，可承担特殊重大比赛、各类常规赛事以及文艺演出、团体活动、商业展示会等非竞赛活动，并可提供运动、休闲、健身和商业等综合性服务。

The National Stadium is the main stadium of the 29th Olympiad in 2008. Its seating capacity amounts to 91,000, including 11,000 temporary seats. The venue hosted the opening and closing ceremonies of the Beijing Olympic Games and Paralympic Games, the track and field competitions, and the football finals. After the Olympics, the stadium became a large-scale sports and entertainment facility for the residents of Beijing as an architectural landmark and Olympic legacy.

首都博物馆
Capital Museum

项目地点：北京
用地面积：204 100 平方米
建筑面积：258 055 平方米
建筑高度：67.21 米
设计时间：2002 年
竣工时间：2008 年
方案设计：瑞士 HdeM 建筑设计事务所、中国建筑设计研究院

Project Location: Beijing
Site Area: 204,100 m^2
Building Area: 258,055 m^2
Construction Height: 67.21 m
Design Time: 2002
Completion Time: 2008
Scheme Design: Switzerland HdeM Architectural Design Firm and China Architecture Design & Research Group

首都博物馆新馆是长安街上的标志性建筑之一，设计以“绿色首博、科技首博、人文首博”为目标。建筑简洁的矩形平面与北京城的格局相协调，非对称的建筑形体则呼应了建筑所在的街角空间。建筑中所运用的青铜、木材、陶砖等传统材料代表老北京的建筑历史，先进的建造技术则体现了现代北京的新特点。设计打破了传统博物馆封闭沉闷的感觉，为市民营造开放、明亮的文化休闲场所，并通过采用屋顶太阳能电池板、虹吸式雨水系统、预应力混凝土和钢管混凝土、大型钢屋盖和金属吊顶、先进的消防安防和智能化系统等措施，全面提升了建筑的科技含量。

The New Capital Museum is a remarkable building of Chang'an Street. A simple rectangular plan is well responding to the urban structure of the Beijing, and a dissymmetrical arrangement of the volumes is responding to the location of it by a street corner. The traditional materials of the building with bronze, wood, and brick are deriving from the tradition of Beijing, while the advanced techniques of the building are echoing the modernity of the city. In the project, an open atmosphere and a bright interior environment are introduced instead of the depressive looks of the traditional museums. In addition, many new techniques are used in the building, including the roof solar panel, siphon drainage system, prestressed concrete, concrete filled steel tube, large roof structure, metal false ceiling and the advanced intelligent building system.

北京西环广场
Xihuan Plaza, Beijing

建设地点：北京
用地面积：59 900 平方米
建筑面积：294 800 平方米
设计时间：2000 年
竣工时间：2005 年
方案设计：法国 AREP 建筑设计公司、中国建筑设计研究院

Construction Location: Beijing
Site Area: 59,900 m^2
Building Area: 294,800 m^2
Design Time: 2000
Completion Time: 2005
Scheme Design: France AREP Architectural Design Company and China Architecture Design & Research Group

项目的建成为北京提供了新的城市轮廓线，塔楼建筑的南北立面和屋顶既有创新又继承了北京传统的建筑形式。室内空间是设计的核心，设计尽可能争取充足的自然光，裙房内的大型中庭使日光可以直接照射到商业中心六层各空间。塔楼的立面设计作了特别的处理，将复杂的使用功能统一在相对简单的建筑体形内，方便使用。外立面先进的施工技术保证了建筑耐久性和施工速度，建筑的立面材料主要为 Low-E 中空玻璃。裙房中庭部分和整个西立面均采用了玻璃幕墙，黑色密封干挂花岗岩石板则用于实墙。夜间灯光的设计使三座塔楼看上去犹如三座灯塔映照在北京的夜色之中，裙房中庭和交通系统的换乘大厅的灯光也显现着无限的活力。

In this project, an innovative figure is introduced to respect the traditional architectural characters of Beijing old city, and creates a new skyline for the developing city. The main focus on the design is to gain more nature daylight for interior space. The complicated functions are reasonably arranged in a single body, which is convenient for the use of space. The construction techniques are well chosen to improve the durability and the construction speed. The distinct contrast of elevation effects is created by the Low-E glass curtains of the office towers and the dark granites of skirt building. In the evening, the lighting effects make the office buildings look like three light towers shining in the dark sky of the old city, while the lights glittering in the atrium of skirt building also respect the activity of this traffic center.

北京市建筑设计研究院
BEIJING INSTITUTE OF ARCHITECTURAL DESIGN

北京市建筑设计研究院（以下简称为BIAD），是与中华人民共和国同龄的大型国有建筑设计咨询机构。其业务范围包括城市规划、投资策划、大型公共建筑设计、民用建筑设计、室内装饰设计、园林景观设计、建筑智能化系统工程设计、工程概预算编制、工程监理、工程总承包等领域。

伴随着新中国的社会发展，BIAD在北京完成了许多重大项目的设计工作。如20世纪50年代象征新中国形象的人民大会堂、国家博物馆、民族文化宫、工人体育场，60、70年代体现我国自主科技实力的北京工人体育馆、首都体育馆和北京饭店工程；80年代体现改革开放的中国国际展览中心和第11届亚运会场馆；90年代完善国际大都市功能的北京首都机场2号航站楼、北京西站、北京植物园温室、首都图书馆、西单文化广场。进入21世纪，BIAD完成了包括2008年部分奥运工程、北京首都机场3号航站楼、中国石油大厦、全国人大机关办公楼等标志性建筑的设计工作。

几年来，BIAD在全国范围内设计完成了一大批城市标志性建筑，如海南博鳌亚洲论坛、浙江中国美术学院、烟台市体育中心体育场、秦皇岛体育馆、重庆人民大厦、新疆体育中心、绵阳市政府办公楼、河南省体育中心体育场、西安国际金贸中心、三亚喜来登酒店等项目。

BIAD与许多国家的著名设计公司保持着良好的合作关系，合作设计了包括中国工商银行总行办公楼、北京顺义国际学校、深圳文化中心、国家大剧院、东方广场、LG大厦、中国电影博物馆、中国科技馆、凯晨世贸中心等工程。

改革开放以来，BIAD积极参与全国各地的经济建设，推动房地产业的健康发展，每年完成大量的民用建筑设计项目，累计完成设计面积超过1.5亿平方米，其中恒基中心、国际金融大厦、海淀剧院、投资广场、北京新闻文化中心、信远大厦等项目已成为北京城市新的地标建筑，而望京新城、北京现代城、万科星园、北京橘郡、万科西山庭院、深蓝华亭、颐源居、远洋山水、朗琴园、回龙观居住区均对房地产市场产生了较大影响，成为明星楼盘。

BIAD始终将设计精品工程和保持技术领先作为自己的责任，从1977–2008年，设计作品获国家奖53项，获建设部奖194项，获北京市奖475项；科研成果获国家科技进步奖44项，获部科技进步奖68项，获北京市科技进步奖133项，获詹天佑土木工程奖5项。在多年的工作过程中，BIAD逐步形成了一个优秀的设计团队，并集中了一大批优秀的建筑师和各个专业的工程师，其中有工程院院士1名、国家设计大师9名、突出贡献专家10名、高级建筑师和高级工程师438人、建筑师和工程师361人、一级注册建筑师185人、一级注册结构工程师124人、注册造价工程师9人。

BIAD自成立以来的59年中，始终致力于向社会提供高品质的设计服务，在行业中享有极高声誉，并逐渐形成了"建筑服务社会"的企业核心理念。面临新世纪设计行业的严峻挑战，全院1800名员工深刻认识到，只有进一步强化市场意识，切实尊重业主需求，通过整合全院乃至国内外优势资源，逐步提升设计作品的完成度，才能体现国家大型设计机构的综合竞争实力。

开放的BIAD愿与社会各界进行广泛合作，为创造人类美好的生活环境而不懈努力。

The Beijing Institute of Architectural Design (BIAD) is a large-scale state-owned architectural design and consulting institute established in October 1949, following the founding of the People's Republic of China. Every year BIAD undertakes many architectural design projects, and its total design area now amounts to over 150 million square meters.

BIAD's business scope includes: city planning, investment planning, large-scale public building design, civil engineering. Interior Design. Landscape Design. Design of intelligent systems for buildings, Project Estimating, Project Administration, and Project Contracting.

In the 1950s, BIAD designed the symbolic buildings that represented the image of the New China, including, the Great Hall of the People, the National Museum, the Cultural Palace of the Nationalities, and the Worker's Stadium.

In the 60s and 70s, BIAD's projects reflected the independent scientific and technological strength of China, with projects such as the Beijing Worker's Gymnasium, the Capital Stadium and the Beijing Hotel.

In the 80s, BIAD's designs displayed China's achievements in its reform and opening-up process, for example, the China International Exhibition Center and the gymnasiums and stadiums for the 11th Asian Games.

In the 90s, BIAD's projects contributed to Beijing's function as an international metropolis with projects such as the Beijing Capital International Airport Terminal 2, the Beijing West Railway Station, the Conservatory of Beijing Botanical Garden, the Capital Library and Xidan Cultural Square. Other landmark commercial, office, and cultural buildings by BIAD included the Beijing Henderson Center, the International Financial Mansion, Haidian Theater, the China Academy of Art, Investment Plaza, Beijing Media and Cultural Center and Xinyuan Mansion. Some of BIAD's residential building complexes have exerted considerable influences on the real estate market, such as, Wangjing New City, Beijing SOHO New Town, Vanke Star Garden, Beijing Orange County, Vanke West Mountain Courtyard, Deep Blue Houses, Yiyuan Houses, Lar Valley, and Huilongguan Residential Area.

In the 21st century, BIAD has dedicated itself to some of the architectural projects for the 2008 Olympic Games and a number of landmark buildings in Beijing, including Beijing Capital International Airport Terminal 3, China National Petroleum Mansion and the office building for the Standing Committee of the National People's Congress.

Since China's reform and opening-up, BIAD has been actively taking part in economic construction in other cities around China, making contributions to the healthy development of the real estate industry. BIAD has designed many landmark buildings nationwide, including Hainan Boao Forum for Asia, Yantai Sports Center Stadium, Qinhuangdao Stadium, Chongqing People's Mansion, Xinjiang Sports Center, the Mianyang municipal government offices, Henan Sports Center Stadium, Xi'an International Finance and Trade Center, and the Sanya Sheraton Hotel.

BIAD has established good cooperative relations with many well-known foreign design offices. Some joint designs include the Head Office of Industrial and Commercial Bank of China, Beijing Shunyi International School, Shenzhen Cultural Center, China National Grand Theater, the Oriental Plaza, LG Mansion, China Movie Museum, chemsunny Plaza,Beijing and the China Science and Technology Museum.

BIAD has always devoted itself to top-quality design and technological leadership. From 1977 to 2008, BIAD was awarded 53 national prizes for its designs, 194 awards from the Construction Ministry and 475 awards from the Beijing Municipal Government. Its scientific research has won 44 national scientific and technological awards, 68 from the Construction Ministry, and 133 from the Beijing Municipality. BIAD has been a five-time winner of the Zhan Tian You Construction Award.

Over the years, BIAD has built up an outstanding design team with a large number of excellent architects and engineers specialized in various fields, including a member of Chinese Academy of Engineering, 9 national design masters, 10 experts with outstanding contributions, 438 senior architects and senior engineers, 361 architects and engineers, 185 Class 1 certified architects, 124 Class 1 certified structural engineers, and 9 certified cost engineers.

Over the 59 years since its establishment, BIAD has devoted itself to providing good quality design services, thus enjoying a good reputation in the industry. BIAD has also gradually developed its key corporate concept "Architecture serves society". In face of the severe challenges to the design industry at this time, the 1800 staff members of BIAD have realized that it can fully display its comprehensive competitiveness as a large-scale design institute only if it: strengthens its market awareness, earnestly respects the needs of clients, and increases its design quality by integrating the resources of the whole institute and even those at home and abroad.

BIAD is willing to join hands with all circles of society, making sustained efforts to create a wonderful living environment for all human beings.

地址：北京市南礼士路62号
邮编：100045
电话：+86-10-88043999
传真：+86-10-68034041
网址：www.biad.com.cn

Address: No.62 of South Lishi Rd., Beijing
P.C.: 100045
Tel: +86-10-88043999
Fax: +86-10-68034041
Http: // www.biad.com.cn

国家大剧院

建设地点：北京
建筑面积：219 400平方米
占地面积：120 000平方米
设计时间：2000年—2002年
竣工时间：2007年
合作单位：法国安德鲁

National Grand Theater

Construction Location: Beijing
Building Area: 219,400 m^2
Site Area: 120,000 m^2
Design Time: 2000—2002
Completion Time: 2007
Cooperation: Andrew, France

首都机场T3航站楼

建设地点：北京
建筑面积：902 000平方米
占地面积：13 260 000平方米
设计时间：2003年12月—2005年2月
竣工时间：2007年12月

Capital International Airport Terminal T3

Construction Location: Beijing
Building Area: 902,000 m^2
Site Area: 13,260,000 m^2
Design Time: 2003. 12—2005. 02
Completion Time: 2007.12

国际金融大厦

建设地点：北京
建筑面积：100 000平方米
占地面积：18 000平方米
设计时间：1996年9月
竣工时间：1998年

International Financial Building

Construction Location: Beijing
Building Area: 100,000 m^2
Site Area: 18,000 m^2
Design Time: 1996. 09
Completion Time: 1998

北京植物园展览温室

建设地点：北京
建筑面积：7 250平方米
占地面积：55 000平方米
设计时间：1999年6月
竣工时间：2000年

Exhibition Greenhouse of Beijing Botanical Gardens

Construction Location: Beijing
Building Area: 7,250 m^2
Site Area: 55,000 m^2
Design Time: 1999. 06
Completion Time: 2000

国家奥林匹克体育中心

建设地点：北京
建筑面积：110 000平方米
占地面积：660 000平方米
设计时间：1986年
竣工时间：1989年

National Olympic Sports Center

Construction Location: Beijing
Building Area: 110,000 m^2
Site Area: 660,000 m^2
Design Time: 1986
Completion Time: 1989

中国国际展览中心2–5号馆

建设地点：北京
建筑面积：75 000平方米
占地面积：150 000平方米
竣工时间：1985年

China International Exhibition Center Hall No. 2–5

Construction Location: Beijing
Building Area: 75,000 m^2
Site Area: 150,000 m^2
Completion Time: 1985

万科第五园

建设地点：广东 深圳
建筑面积：125 000平方米
占地面积：130 000平方米
设计时间：2004年7月—2005年7月
竣工时间：2005年7月

Vanke Park V

Construction Location: Shenzhen, Guangdong
Building Area: 125,000 m^2
Site Area: 130,000 m^2
Design time: 2004. 07–2005. 07
Completion time: 2005. 07

人民大会堂

建设地点：北京
建筑面积：171 800平方米
占地面积：150 000平方米
设计时间：1958年
竣工时间：1959年

Great Hall of the People

Construction Location: Beijing
Building Area: 171,800 m^2
Site Area: 150,000 m^2
Design Time: 1958
Completion Time: 1959

中国历史博物馆和中国革命博物馆

建设地点：北京
建筑面积：65 152平方米
设计时间：1958年
竣工时间：1959年

National Museum of Chinese History & Museum of the Chinese Revolution

Construction Location: Beijing
Building Area: 65,152 m^2
Design Time: 1958
Completion Time: 1959

BIAD

毛主席纪念堂

建设地点：北京
建筑面积：28 225平方米
占地面积：57 200平方米
设计时间：1976年
竣工时间：1977年

Chairman Mao Memorial Hall

Construction Location: Beijing
Building Area: 28,225 m^2
Site Area: 57,200 m^2
Design Time: 1976
Completion Time: 1977

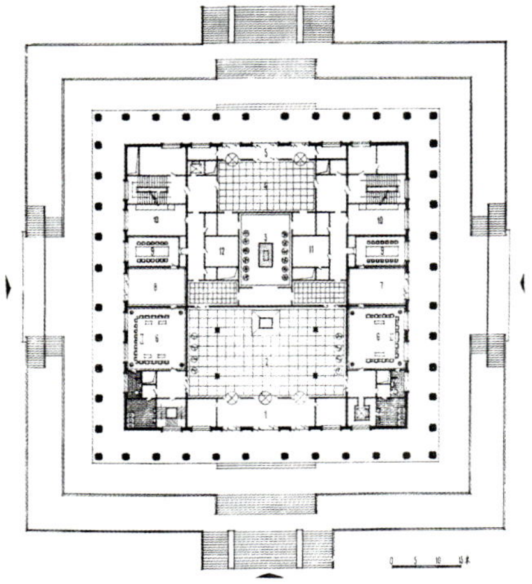

北京白林建筑设计咨询有限公司

Bailin Architectural Design and Consultant Corporation Ltd., Beijing

公司简介
Company profile

公司自2002年创立以来，共完成80多项设计项目，项目涉及城市设计、城市中心区改造、居住区规划、大型公共建筑、大学校园规划、环境景观设计、造型艺术设计等众多领域。合作单位有北京（市）建筑设计研究院、中国（城市）规划设计研究院、清华大学建筑学院以及国外的川崎清、宗本顺三、山本理显、安藤忠雄、岸和郎设计事务所等。公司涉及的项目遍及全国各地包括北京、天津、广州、深圳、西安、太原、石河子、常州、赣州、青岛、运城、吴江、泰安、唐山等地。

办所理念
“研究型，国际化，追求卓越，培育人才”。

设计理念
追求“思想性，艺术性，前瞻性”。

设计方法论
“建筑是思想的容器”；“以人为本，以环境为源，以科技为手段”。

学术目标
探索中国的建筑现代化与中国文化的融合。
探索中国人的价值观与建筑的结合点。
为中国的建筑学术事业做出贡献。

服务态度
超越自我，超越甲方，为甲方谋求最根本的利益。
深入挖掘建筑的思想性、文化性、艺术性、精神性内涵。
坚持“先做人、后做事、再做建筑”的做人做事原则。

地址：北京市西城区西外大街1号西环广场T2 19层C1
电话：+86-10-58301336/7
传真：+86-10-58301336/7转606
邮箱：bailin.bailin@263.net
网址：www.bailindesign.com

Address: Xicheng District, West Acenue, West Central Plaza T2 19-C1, Beijing
Tel: +86-10-58301336/7
Fax: +86-10-58301336/7-606
E-mail: bailin.bailin@263.net
Http://www.bailindesign.com

Since the foundation in 2002, the company has completed a total of over 80 design projects, including urban design, urban center reconstruction, residential area planning, large-scale public construction, university campus planning, environmental landscape design, plastic art design and other fields. The company cooperates with Beijing Institute of Architectural Design, China Academy of Urban Planning and Design, School of Architecture, Tsinghua University and international design firms, such as Kawasaki Kiyoshi, Munemoto Junzo, Riken Yamamoto, Tadao Ando and Waro Kishi, etc. Its projects have covered throughout the country (Beijing, Tianjin, Guangzhou, Shenzhen, Xi'an, Taiyuan, Shihezi, Changzhou, Ganzhou, Qingdao, Yuncheng, Wujiang, Taian and Tangshan, etc.)

Aim
"Research", "Globalization", "Pursuit of Excellence," "Cultivation of Talents"

Design Concept
Pursuit of "In-Depth Thinking, Artistic Spirit and Inspiring Prospect"

Design Methodology
"Architecture is the container of the thinking." "People as the base, environment as the source, science and technology as the means".

Academic Goals
To explore the intergration of the modernization of architecture in China and the Chinese culture.
To explore the joint of the Chinese values and architecture.
To make contribution to the architectural academic career of China.

Attitude for Service
Surpass ourselves, surpass Party A, and seek for the fundamental interests of Party A.
Go deep into the ideological content, cultural content, artistic quality and spiritual connotation of architecture.
Persist in the principle to conduct ourselves and our business of “firstly conduct ourselves, secondly conduct our business, and then the architecture”.

沱牌集团科技研发中心办公楼
建设地点：四川 遂宁
建筑面积：31 400 平方米
设计时间：2009 年

什么是建筑？
建筑是空间的哲学，
使用物质、空间表达思想、精神的容器。

酒是什么？
酒，是注入了思想和文化的溶液。

什么是国际竞标？
国际竞标的本质是东西方思想的碰撞和较量。
竞标单位（美国 HOK、澳大利亚 SPW、德国 SOL、日本.ASSOCLATES、北京白林建筑）

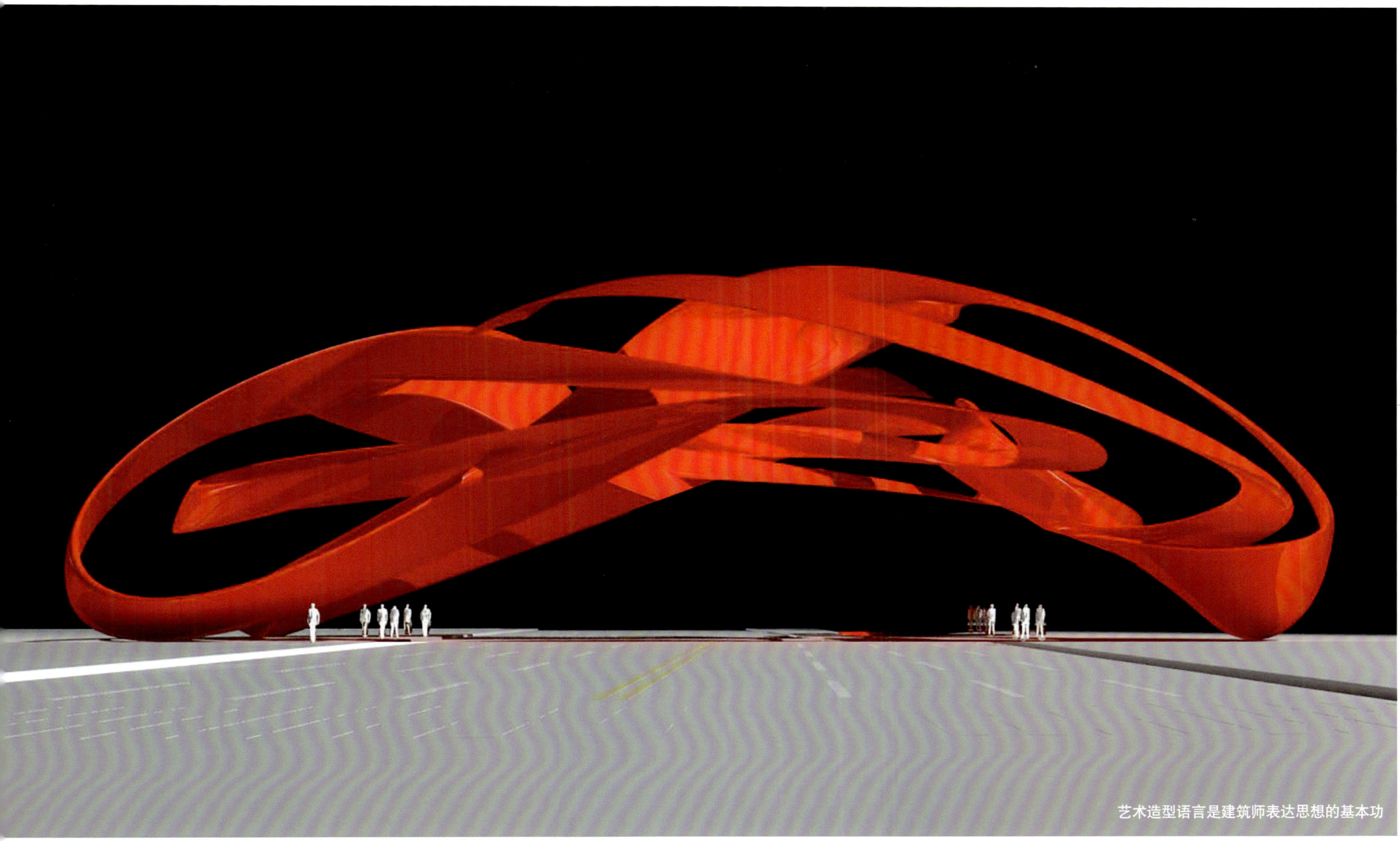

艺术造型语言是建筑师表达思想的基本功

吴江市东太湖大道及周边区域城市设计（国际竞标第一名）
建设地点：江苏 吴江
规划面积：5 900 000 平方米
设计时间：2008 年

太湖湾广场（旅游度假区北大门）
建设地点：江苏 常州
规划面积：87 319.7 平方米
竣工时间：2008 年

唐山南湖之门
建设地点：河北 唐山
长／高：99.9 米／30 米
设计时间：2010 年

太湖旅游度假区西大门
建设地点：江苏 常州
建筑高度：16.7 米
竣工时间：2009 年

深圳前海地区概念规划国际竞标
建设地点：广东 深圳
规划面积：18 040 000 平方米
设计时间：2010 年

1．深港同城一体，共建世界都市。
2．理顺周边关系，描绘发展策略。
3．布局框架结构，构建整体形象。
4．紧凑城市空间，合理路网结构。
5．系统组织交通，绿色健康出行。
6．商业商务公交，相互支撑体系。
7．运用山水资源，打造滨海景观。
8．地下空中结合，构筑立体城市。
9．倡导低碳行动，思考落实行动。
10．活用一国两制，推进开发策略。

1.Integrate Shenzhen and Hong Kong; build a global city.
2.Harmonize peripheral relations; portray future blueprint.
3.Deploy a frame pattern; build up an overall image.
4.Refine the urban space; rationalize road network structure.
5.Systemize traffic organization; encourage green travel.
6.Combine commerce; business and common transport; establish a mutually supporting system.
7.Exploit natural resources; build a coastal landscape.
8.Link with underground; create a vertical city.
9.Promote low-carbon operation; set up an eco city.
10.Utilize "One Country, Two Systems" flexibly; push forward development strategy.

这个规划究竟贯穿了一种什么思想呢？
贯穿了一种以人为本，人与自然、人与社会、城市与城市和谐相处，“彼此相融，天人合一”的哲学思想。

深圳前海地区概念规划国际咨询
建设地点：广东 深圳
规划面积：18 040 000 平方米
设计时间：2010 年

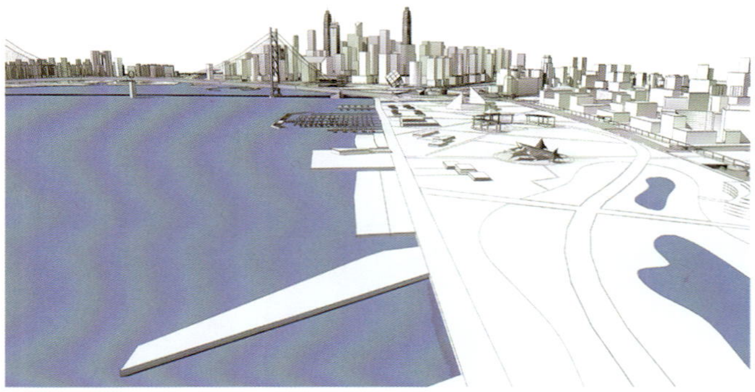

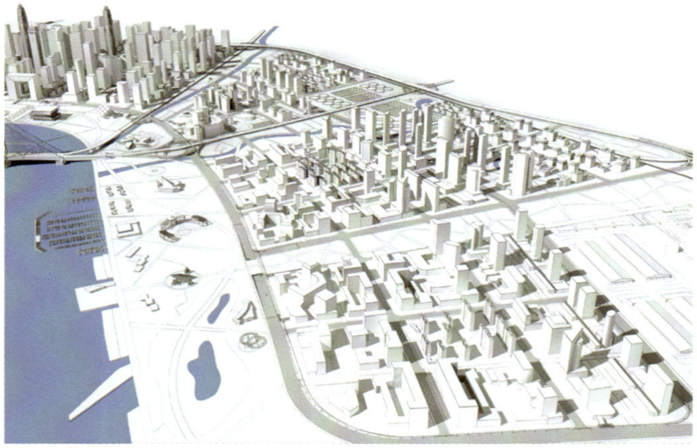

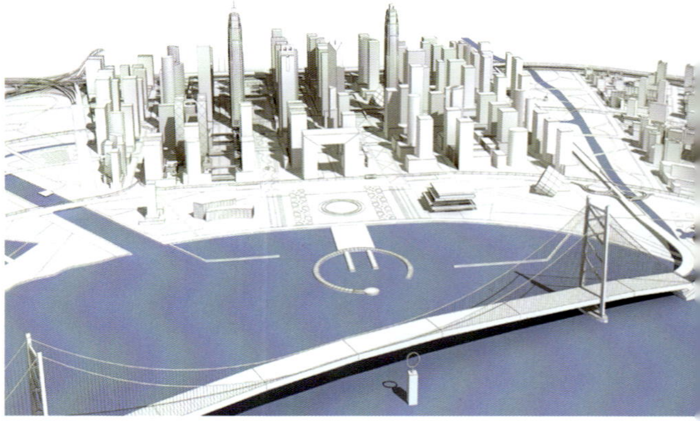

中科院建筑设计研究院有限公司前身为中国科学院北京建筑设计研究院，成立于1951年，公司拥有建筑行业建筑工程甲级、市政公用行业（热力）甲级资质。目前公司有员工共409人，其中工程技术人员336人（高级以上职称75人、中级技术人员76人、一级注册建筑师26人、一级注册结构工程师20人、水暖电热力一级注册工程师28人）。除总部外，公司在广东东莞、浙江杭州、辽宁沈阳、四川成都、河南郑州和上海设有分公司，此外还设有综合一所、环艺设计所、景观设计所、惠中设计所4个专业设计所，以及综合二所、综合三所、环境工程所、热力工作室、照明工作室、景观工作室、公共艺术工作室等专业设计机构。

公司能够完成大型文化、办公、商业、医疗、体育类建筑设计、城市规划设计、居住区规划和各类住宅设计、历史街区保护规划及设计、环境景观及室内装修设计、结构改造加固等，尤其擅长科研教育类建筑规划及设计，包括科研基地规划、科研办公建筑设计、学校规划及设计等。同时能够进行项目策划、项目管理、工程造价咨询等工作。

公司创作设计了一批具有社会影响力的建筑精品，如中科院文献情报中心、国家开发银行、中国人民银行重点库、中国驻埃塞俄比亚大使馆、中国驻贝宁大使馆、中科院研究生院教学楼、中科院苏州纳米技术与纳米仿生研究所、中科院生态中心实验楼、中科院青岛生物能源与过程研究所、中科院烟台海岸带可持续发展研究所、中科院动物所、中科院电子所总装备楼、中科院计算所、北京正负电子对撞机工程、国家天文台、石油部物探局计算中心及亿次银河机机房工程、北京基督教丰台堂和朝阳堂工程、上海万科城市花园、北京万科星园（二三期）、天津万科假日风景、北京阜内大街历史文化保护区的保护规划等。公司曾获得国家级奖励10项，省部级奖励60余项。

1998年到2006年间（2007年未举办首都汇报展），在第5届到第13届首都建筑设计汇报展中，我公司11个项目获得14个奖项，其中6个为方案设计最高奖。我公司设计的“中国科学院文献情报中心”荣获全国优秀工程设计最高奖项“金奖”。2006年公司被地产界评为“北京地产十佳建筑设计机构”，获得万科集团最佳设计合作伙伴奖，并成为沿海集团的策略联盟指定设计合作伙伴。

2009年，公司负责设计的中国科学院文献情报中心(中国国家科学图书馆)、九寨沟国际大酒店均获得“中国建筑学会建国60周年建筑创作大奖”； 国家动物博物馆及中科院动物研究所科研实验楼、标本楼获北京市第十四届优秀工程设计公共建筑设计一等奖；北京万科紫台家园获北京市第十四届优秀工程设计住宅建筑设计三等奖；北京万科四季花城获2009詹天佑大奖优秀住宅小区金奖；郑州龙福国际获2009全国人居经典建筑金奖。此外，为表彰公司在北京奥运会残奥会环境建设工作中做出的突出贡献，北京市“2008”奥运环境建设指挥部为我公司颁发了荣誉证书。

公司先后与欧美、日本、澳大利亚等国家和地区的有关专家机构进行了广泛的学术交流或设计合作，并与法国安东尼·贝叙建筑设计公司、美国Perkins Eastman建筑设计事务所、意大利罗马大学签订了长期合作协议，具有丰富的国际合作经验，并赢得了良好的声誉。公司设计专家团队相信物理环境对生活工作及学习质量有重大的影响，而设计良好的空间则会使人能更好的对环境进行感受与使用。公司专业化的团队同时也密切关注擅长的专业领域的发展趋势，以保持在行业创新中的领先地位。企业秉承“精心设计，诚信守约，追求精品，锐意创新”的经营理念，充分利用实力和优势为业主提供最优的服务。

中科院建筑设计研究院

Institute of Architecture Design and Research, Chinese Academy of Sciences

Architecture Design & Research Institute Co., Chinese Academy of Sciences (CAS) grew out of Beijing Institute of Architectural Design & Research (BIAD), CAS. Founded in 1951, this company ranks Class-A in construction works and in heating power of municipal public utilities. At present, the staff number of this company is 409 altogether, 336 of whom are engineering and technical personal. Among them 75 people are qualified with senior professional title, 76 are intermediate technicians, 26 get PRC Class-A Registered Architect Qualification, 20 are PRC Class-A Registered Structure Engineers, and 28 are PRC Class-A Registered Water and Electric Heating Engineers. The company has also established professional designing institutions such as Comprehensive Institutes No.1, No.2 and No.3, Environment and Art Institute, Heating Power Studio, Lighting Studio, Landscape Studio, Public Arts Studio and so on.

The business scope of this company covers architectural design of large-scale cultural buildings, office premises, hospital and medical buildings and sports venues. This company also works on urban studies, residential district planning, all kinds of residential designs, protection planning and programming of historic district, landscaping and inner decoration designs, and reform and reinforcement of construction. This company boasts the excellent architectural drafting and designing of research and education buildings such as planning of science facilities, design of scientific research buildings, and planning and design of school buildings, Meanwhile, project planning, project management and cost consultation of construction project are involved in the business scope.

This company has designed and completed an array of high-quality construction projects which have immense social influence. Those brilliant construction projects include Library of Chinese Academy of Sciences, China Development Bank, key locations of the People's Bank of China, Chinese Embassy in Ethiopia, Chinese Embassy in Benin, the academic building of Graduate School, CAS, Suzhou Institute of Nano-tech and Nano-bionics, CAS, laboratory block of Biotechnology Center, CAS, Qingdao Institute of Biomass Energy and Bioprocess Technology, CAS, Yantai Institute of Coastal Zone Research for Sustainable Development, CAS, Institute of Zoology, CAS, general equipment building of Institute of Electron, CAS, Institute of Computing Technology, CAS, Beijing electron-positron collider project, Nation Astronomical Observatory, computing center of Geophysical Prospecting Bureau, Petroleum Department, Fengtai church and Chaoyang church, Beijing Christian Council, Shanghai Vanke City Garden, Beijing Wankexingyuan(Phase and Phase), Tianjin Vanke Holidays Scenery, Protection planning of FUCHENGMEN Inner Street conservation of historic sites and so on. This company has obtained 10 state-level awards and over 60 province-level encouragement awards.

From the 5th to 13th capital planning exhibitions on the architectural design (from 1998 to 2006), the 11 project accomplished by this company had won 14 awards, 6 of which were the highest conceptual design awards. Buildings of Library of Chinese Academy of Sciences designed by this company won gold medal, the highest prize of National Excellent Projects. In 2006, it was named as one of ‘Beijing Top Ten Planning Architectural Design Organizations’ by the real estate industry, was honored as ‘Best Designing Partner of Vanke’ and became the unique authorized designing partner of strategic alliance of Coastal Group.

In 2009, the designs of buildings of Library of Chinese Academy of Sciences and Jiuzhaigou Valley International Hotel won Architectural Creation Prize held by Architectural Society of China in celebration of the 60th anniversary of the founding of PRC. The designs of The National Zoological Museum of China, the research activities and laboratory building and the specimen building of Institute of Zoology, CAS were awarded Meritorious Winners of the 14th Beijing Excellence Engineering and Public Building Designs. The design of Beijing Vanke Four Season Flower City won gold award of outstanding residential areas of 2009 Zhan Tianyou Prize. The designs of buildings of Zhengzhou Longfu International Group won 2004 National Construction Award Habitat classic program. In addition, 2008 Beijing Olympic Environment Construction Command issued this company certificates to award it which have made outstanding contributions in environment development during the Paralympics in Beijing in 2008.

This company has successively conducted extensive academic exchange and designs cooperative with the relevant exports and professional institutions in Hong Kong and Taiwan, Europe, the USA, Japan, Australia and other countries and regions. It signed a large long-term cooperation with Anthony Architects in France , Perkins Eastman in US and Univ Roma La Sapienza Italy. Through the cooperation, this company has gained extensive experience and well-deserved reputation. Specialist team in this company believes that physical environment tremendously influence people's working and studying life, so the beautified exterior space is very important. The team, in the meantime, pays close attention to the present situation and the trend in this field and attempts to continue to rank among the leading position through innovation. This company will stick to operation principles of supplying with well-designed and competitive products, valuing honesty and integrity, upholding the spirit of innovation to provide the best service.

1

1–2 光大国际中心

设计/建成时间：2004 年 /2007 年

建造地点：北京

建筑面积：120 000 平方米

1-2 Everbright International Center

Design / Completion Time: 2004/2007

Construction Location: Beijing

Building Area: 120,000 m^2

2

3–5 北京林业大学教学楼

设计/建成时间：2008 年 /2010 年

建造地点：北京

建筑面积：50 000 平方米

3-5 Teaching Building of Beijing Forestry University

Design / Completion Time: 2008/2010

Construction Location: Beijing

Building Area: 50,000 m^2

4

5

6–8 中国科学院动物研究所
设计/建成时间：2002 年 /2004 年
建造地点：北京
建筑面积：42 900 平方米
获北京市优秀工程建筑设计一等奖（2009）

6-8 Institute of Zoology, Chinese Academy of Sciences
Design / Complete Time: 2002/2004
Construction Location: Beijing
Building Area: 42,900 m^2
The First prize of Outstanding Architectural Design in Beijing. (2009)

9–11 北川央企办公楼
设计/建成时间：2009 年 /2010 年
建造地点：四川　绵阳
建筑面积：50 000 平方米

9-11 Beichuan Central Enterprise Office Building
Design / Complete Time: 2009/2010
Construction Location: Mianyang, Sichuan
Building Area: 50,000 m^2

12–15 中国科学院地球化学研究所金阳新所园区

建设地点：贵州 贵阳
建筑面积：140 000 平方米（其中一期 54 000 平方米，二期 86 000 平方米）
总用地面积：9.7 公顷
建筑高度：60 米
设计时间：2008 年—2009 年
竣工时间：2012 年

12-15 Institute of Geochemistry Chinese Academy of Sciences in Jinyang

Construction Location: Guiyang, Guizhou
Building Area: 140,000 m^2 (54,000 m^2, 86,000 m^2)
Site Area: 9.7 ha
Construction Height: 60 m
Design Time: 2008-2009
Completion Time: 2012

12

13

生态西部的科研园区

——记中国科学院地球化学研究所金阳新所园区

一、科研建筑群落的设计思路

传统生活方式所造就的空间群落，形成了各具特色的群落空间，建筑设计继承和发扬这种独特的建筑群落形态，在不同空间中采用各具特色的建筑语汇，形成区域内形态各异、丰富多彩的建筑群落。科研园区基本由科研办公区、配套服务区和生活区三部分组成，各个区域形成各具特色的建筑群落空间。

二、规划理念及设计手法

1．关注绿色、生态、低碳的设计理念

1）多层建筑优于高层建筑，整体能耗低。

2）规划布局与建筑物的自然通风。

3）单体节能保温处理通过通风屋面、屋顶花园、遮阳构造、体型系数进行处理。

4）环境空间处理采用嵌入式内庭园（全围合、半围合），来营造小环境、改善提高环境品质。

5）雨水收集利用系统

2．科研建筑单体模式

1）交通组织：回廊式。双廊式。单廊式。

2）平面特点：平面各异，少有标准层。功能布局的唯一性。单元式组合。空间的模糊性。

3）立面特点：理性、秩序感强。风格朴素、外观典雅。功能决定形式。

3．构成与变异

运用基本单元构成逻辑，在群体设计中有所变异，组成新的秩序和层次。

三、设计出发点

1．因地制宜

从（贵州）地域文化、地源特征出发来着手设计。

2．因物制型

从科研园区的功能出发着手设计、从国家级科研院所的形象定位着手设计。

3．因势利导

因势利导就是顺着事物发展的趋势加以引导，科学研究具有不确定性和长期性，可持续发展的科研园区，按照国际现代化研究与研发机构的标准进行规划、设计和建设，满足研究所自身发展和区域科学研究、技术开发、成果转化、人才培养等需求，提供一处安全、高效、舒适、环保的智能化生态园区，这是我们设计的长远目标。

由此就形成了第三个设计出发点，即可持续发展、前瞻性的设计。

Scientific Research Park of Ecological West

——Jinyang New Zone, Institute of Geochemistry, Chinese Academy of Sciences

1.The design concept of the scientific research building community

Community created by the traditional way of life forms all kinds of distinctive space. The architectural design inherits and carries forward this unique community pattern, adopting different space architectural vocabularies with various characteristics and producing a building community varying in shape and rich in color.

Basically the scientific research park composes three parts: the research office area, the service area and the living area, each of which forms a building community space with distinctive features.

2.The planning concepts and design methods

1. Concerns for the green, ecology and low-carbon

1) the multi-layer construction is superior to high-rise buildings, for the overall power consumption is much lower

2) the layout planning and the natural ventilation of buildings

3) the single energy-saving thermal insulation—roof ventilation, roof garden, shade structure and shape factor

4) the treatment of the environmental space—the embedded gardens (enclosed and semi-enclosed), creating a small environment and improving the environmental quality

5) the rainwater harvese system

2. The single mode for scientific research buildings

1) the traffic organization: corridor-style; double-gallery-style; single-gallery-style

2) the plane features: various planes and few standard layers; uniqueness of the function layout; modular combination; fuzzy space

3) the facade features: rational and a strong sense of order; a modest style and an elegant appearance; function-determined forms

3, Composition and variation

The design adopts the basic unit as the logic, varying in the group design and forming a new order and hierarchy.

3. The starting points of the design

1. The design is carried out according to local conditions

The design takes (Guizhou) culture and regional characteristics as its starting point.

2. The form is determined according to the material

The design takes the functions of the scientific research park and the image of the national research institute as its starting point.

3 To improve the occasion

To improve the occasion is to follow the development trend and couple with guidance. Because scientific research is uncertain and always lasts a long term, a scientific research park of sustainable development has to satisfy: its planning, design and construction need to be done in accordance with modern international standards of research and development institution, meeting the requirements of its own development, the regional scientific research, the technological development, the achievement application as well as the personnel training, producing an intelligent ecological park which is safe, efficient, comfortable and environmentally friendly. This is the long-term goal of our design.

Thus is formed the third starting point of the design: sustainable development and forward looking.

15

14

16

17

涿州市文化活动中心作为代表千年历史古城的一座现代化综合性文化建筑，其设计既关注悠久的城市历史文化，又具有鲜明的时代特征。

方案构思从涿州民间手工艺——花灯出发，融合音乐的元素，通过不断地推敲与演化，形成颇具特色的建筑形式。文化中心整体形象取意于打开的灯笼，由剧场、音乐厅、影城、文化宫及群艺馆前厅外侧的弧形玻璃幕墙组成一个巨大的“圆形灯笼”造型。另外，由青少年宫、小剧场及市民广场的灯塔组成散落在其周边的小灯笼，各灯笼之间通过S形曲线形体相联系，宛若飘落于千年古城的音符。在市民广场耸立的景观灯塔更是将音符与灯笼这两个主题强化出来，灯塔造型是一只金色长笛，外罩巨大的倒锥形玻璃幕墙，取意金笛奏响时代最强音符，同时以通透现代的形体与城内的辽代双塔遥相辉映，谱写出一曲历史与现代共奏交响曲的时空乐章。

Zhuozhou Cultural Activity Center is a modern integrated cultural building in the ancient city of long history, the design not only concerns with the city's history and culture, but also shows distinct characteristics of the times. The program concept derives from the lantern design, Zhuozhou folk handicraft, combining with musical elements, and producing a quite distinctive architectural form through constant scrutiny and the evolution. The overall format of the Cultural Centre takes the image of an open lantern, and a huge "round lantern" shape is formed by the theater, the concert hall, the studios, the cultural center and the curved glass curtain-wall outside of the lobby of the art center. In addition, at its surroundings scatter some small lanterns formed by the Youth Palace, the small theater and the lighthouse of the public square, the interval of which are linked by the S-shaped curve, just like notes falling within the ancient city. The landscape lighthouse stands in the public square strengthens the two themes of note and lantern: the lighthouse assumes the shape of a golden flute, covered by a immense glass curtain-wall of inverted cone form, embodying the strongest note of times played by the golden flute. At the same time, its transparent body echoes with the towers of Liao Dynasty, composing a space-time movement with the resonance of history and the modern time.

18

16–18 涿州市文化活动中心

建设地点：河北 涿州
建筑面积：47 400 平方米
总用地面积：6.46 公顷
建筑高度：24 米
设计时间：2009 年

16-18 Center of Cultural activities in Zhuozhou

Construction Location: Zhuozhou, Hebei
Building Area: 47,400 m^2
Site Area: 6.46 ha
Construction Height: 24 m
Design Time: 2009

19

20

19–22 沈阳万科城一期

建设地点：辽宁 沈阳
总用地面积：7.31 公顷
总建筑面积：109 600 平方米（其中住宅建筑面积 93 200 平方米）
容积率：1.67
绿化率：30.1%

19-22 Shenyang Vanke City, Period I

Location: Shenyang, Liaoning
Gross Land Area: 7.31 hs
Gross Building Area: 109,600 m^2 (including residential building area 93,200 m^2)
Plot Ratio: 1.67
Greening Rate: 30.1%

规划设计中综合运用城市设计的现代手法，为居民创造一系列收放有序、功能与氛围统一的有机空间序列。整个社区形成高低错落、结构清晰、且具有独特风貌、和谐形态及明显的标识性的特点。规划中高度重视城市界面和空间的设计，强调沿城市界面高层住宅空间造型的处理，通过虚实间隔和适当的高度及立面变化，避免了板式建筑造成的单一城市界面和空间。对城市空间与形态的组合做出了良好回应。

沈阳万科城一期 2007 年竣工后获得了一致好评，荣获沈阳市优质住宅小区称号，沈阳市新貌杯和 2007 詹天佑大奖优秀住宅小区金奖、2008 年荣获国家科技奖"精瑞住宅科学技术奖"——住宅规划设计优秀奖。在专家评审会上对本项目的评价是该项目的实施将使沈阳乃至东北地域的住宅设计理念和设计手段有了一次质的提高，必将成为沈阳市城市建设和房地产市场的新亮点。现在随着沈阳万科城逐期推进，使该项目所在地域的城市建设品质和层次都得以显著提高。

The planning comprehensively adopts modern method for urban design, creating for the residents a series of orderly, organic space sequences with a unification of function and atmosphere. The whole community, alternate in height, has an obvious structure, possessing all the virtues of unique feature, harmonious formation and distinct identification. The planning attaches great importance to the design of urban interface and space, emphasizing the treatment of the high-rise residential space along the urban interface. It avoids the monotonous urban interface and space caused by the plate construction by intervals of actual situations and appropriate changes of height and elevation, presenting a sound response to the urban form and space.

The first period construction of Vanke City recerved favorable comments after its completion in 2007, won the title of high-quality residential area in Shenyang, the gold award of Shenyang's New Cup and the 2007 ZHANTIANYOU Awards for Outstanding Residential Quarter, and in 2008 won Excellence Award for Residential Design of the National Science and Technology Award "Glitter domestic science and technology Award". The evaluation of the project on the expert review meeting is: The implementation of this project will make a qualitative improvement in the residential design concept and design method in Shenyang as well as the whole Northeast area, and it will surely become the new highlight in the urban construction and real estate market in Shenyang, the new phenomenon. Now with the further construction of Vanke City, the quality and level of the urban construction of the entire region have received significant improvement.

22

21

23

24

23–25 中科院高能所北京正负电子对撞机改造

北京正负电子对撞机工程是在邓小平同志提出"中国必须在世界高科技领域占有一席之地"的战略思想指导下在党和政府亲切关怀下兴建的重大科学工程之一。它一举填补了我国高能物理实验和同步辐射实验的空白，依靠其先进的加速器和谱仪及所取得的科研成果，实现了小平同志的宿愿。北京正负电子对撞机工程 1990 年获得了国家科技进步奖特等奖及中科院科技进步特等奖。

北京正负电子对撞机工程由多项单体组成，位于北京石景山区玉泉路高能所园区内，总建筑面积 57 500 平方米，本次改造主要为对撞实验大厅的改造及辅助部分土建方面的改造。改造内容包括建筑物外立面改造、室内装修改造、原有空调、暖气系统维修、更换，新增分体空调、计算机房气体灭火系统、综合布线及楼宇管理系统以及卫生间给排水系统和电气系统配套改造等。

改造设计中依据"以人为本"的原则和时代的要求，对建筑物进行改造，力图创造符合高能所研究使用功能及科研内涵的一流科研实验建筑. 坚持"经济、适用、美观"的原则，力求减少对原建筑结构的改动。在保证科研功能的前提下，对具有展示效果的建筑公共部分予以重视，充分注重建筑内部功能的完善，使之赶上中等发达国家水平，与北京正负电子对撞机的定位相匹配，并为国家大科学工程 BEPC Ⅱ，打下良好的基础。

23-25 Reform of Beijing Electron Positron Collider in High Energy Institute, Chinese Academy of Sciences

Beijing Electron Positron Collider is one of the major scientific projects constructed under the strategic thinking and guidance of Comrade Deng Xiaoping: "China needs high-tech place in the world". Relying on this advanced accelerators and the gorerament spectrometers as well as the scientific research results achieved, It fills up the blank of our high-energy physical experiments and synchrotron radiation experiments, realizing the long-cherished wish of Comrade Xiaoping. The Beijing Electron Positron Collider project in 1990 won the Grand Prize of National Science and Technology Progress Award and the Grand Prize for Science and Technology Progress of Chinese Academy of science and technology.

The Beijing Electron Positron Collider project composes a number of monomers, located within High Energy Institute Park in Yuquan Road, Shijingshan District in Beijing, the gross floor area of which is 57,500 square meters. The reform is mainly concerned with the transformation of the collision experimental hall and the civil aspects of the auxiliary parts. The reform project include the transformation of the building facades and interior decoration; the maintenance and replacement of the original air conditioning and heating system, installing new split air-conditioning; installing gas fire-extinguishing system for the computer room; installing integral cabling and building management system, and the supporting transformation of the feeding/draining system of the bathroom and the electrical system.

Basing on the principle of "people orierted" and the requirement of the times, the reform design carries out transformation to the building, attempting to generate a first-class scientific experimental building satisfying the utilitarian function and research implication of High Energy Institute. Adhering to the principle of "economic, applicable, beautiful", this project strives to reduce changes to the original building structure. Under the premise of ensuring research capabilities, it attaches importance to the public part of visibility and pays full attention to the sound function of the internal construction. The project aims to catch up with middle-level developed countries, match with the position of Beijing electron positron collider, and lay a good foundation for National Science and Engineering BEPC Ⅱ .

25

26

27

26–30 郑州金钻五星大酒店

项目用地位于河南省郑州市郑汴产业新区内。项目总建筑面积 130 331 平方米，其中酒店面积 57 955 平方米、商业面积 17 812 平方米、写字楼面积 54 564 平方米。
建筑群体由三部分体量构成：超高层五星级酒店位于用地西北端，成为区域标志性建筑；主题商业百货面向东侧拥有最大的商业界面；5A 写字楼布局为“一”字形，成为酒店和商业的建筑背景。建筑群体立面设计着重突出建筑时代感，体现郑州大都市新的精神面貌。超高层五星级酒店建筑顶部造型设计采用切削手法，形成独具特色的，象征财富的钻石形切面造型，使建筑具有较强的标志性，进而成为郑汴产业新区内的新地标。

Diamond Five Star Hotel, Zhengzhou

The project site is located in Zhengbian new industry zone in Zhengzhou City, Henan Province. Total construction area is 130,331 m^2, of which the hotel area is 57,955 m^2, commercial area is 17,812 m^2, and office area is 54,564 m^2.
The building group consists of volumes of three parts: super high-rise five star hotel located in the northwest end of site forms a regional mark, commanding the overall situation; the theme commercial store faces the east with the largest commercial interface; the layout of 5A office is like the shape "—" , becoming the construction background for the hotel and commercial store. The facades of construction group highlight the sense of times of buildings, reflecting the new spirit of Zhengzhou city. The top of super high-rise five star hotel is designed by cutting technique, forming a unique diamond-type section which symbolizes wealth. All these make the building a strong symbol and become a new landmark for Zhengbian new industry zone.

30

29

28

世纪安泰建筑工程设计有限公司
SJAT Architecture & Engineering Design Co.,Ltd.

北京世纪安泰建筑工程设计有限公司(简称SJAT)，成立于1994年，隶属中国建筑学会，持有国家颁发的建筑工程甲级设计资质证书，是具有法人资格的工程设计、咨询企业。

公司的主要业务包括建筑设计、城市规划设计、环境艺术设计、居住区规划设计、古建园林设计、室内装饰及设计、建设项目可行性研究、建筑工程开发咨询服务等。

公司自2000年组建以来共完成公共建筑设计逾500万平方米、居住建筑设计逾1 000万平方米，并在全国范围内完成了大量城市设计和规划设计工作，均得到了社会及业主的好评，并有多项作品荣获国际大奖。

公司现有专业技术人员126人、国家一级注册建筑师10名、国家一级注册结构师13名、高级职称60人、中级职称36人，公司专业配置合理，具备同时承担多项大型建筑工程设计的能力。

公司建立了以项目管理为中心，以专业管理为基础的设计质量、进度管理控制体系。标准体系运行有效，实现公司的动态管理，并于2007年8月取得了ISO 9001质量管理体系认证证书。

公司荣誉

2007—2010年，连续四次，世纪安泰在中国地产年度风云榜评选中被评为“中国地产十佳建筑设计机构”。

2009—2010年，世纪安泰被评为北京领袖建筑设计机构。

2010年6月，世纪安泰获得由中国商业联合会颁发的中国商业地产优秀设计团队奖。

Established in 1994, Beijing Century Aetna Building Engineering Limited (SJAT) is under the Architectural Society of China, holding the Class A architectural design qualification certificate issued by the State, which is also the engineering design as well as consulting firm with legal personality qualification.

Its principal businesses include: architectural design, urban planning and design, environmental design, residential area planning and design, ancient architecture landscape design, interior decoration and design, feasibility study on construction project, construction development consulting services, etc.

Since its establishment in 2000, has completed the design of more than 500 million square meters public buildings and more than 1,000 million square meters of residential buildings, and completed a large number of urban design and planning design work across the country, which has won a wide range of praise by the community and owners, along with a number of internationally awarded works.

The Company currently has 126 professional and technical personnel, 10 national Class-A registered architects, 13 national Class-A registered structural engineers, 60 senior engineers, 36 staffs with intermediate titles, all under reasonable professional configuration, who are capable to undertake multiple large-scale construction design works.

Setting the project management as the center, professional management as the basis, the company runs on a design quality, schedule management control system, and the standard system runs effectively to achieve the dynamic management, and in August 2007, it has obtained the ISO9001 quality management system certification.

Honor of the company

2007-2010, SJAT (Aetna century) was named "Top Ten China Real Estate Agencies for Architectural Design" for four times successively in the China Real Estate Annual Billboard Award, 2009-2010, SJAT has been awarded as Beijing leading architectural design agency.

June 2010, SJAT was awarded as the Outstanding Design Team among China Commercial Real Estate issued by the China Chamber of Commerce.

地址：北京市西城区车公庄大街9号 五栋大楼A3-5层
邮编：100044
电话：+86-10-88395430 88395440 88393624 88393694
传真：+86-10-88395729
邮箱：sjat@sina.com
网址：www.sjat.com.cn

Address: 5F. A3 of 5 Buildings, No.9, Chegongzhuang Street, Xicheng District, Beijing
P.C.: 100044
Tel: +86-10-88395430 88395440 88393624 88393694
Fax: +86-10-88395729
E-mail: sjat@sina.com
Http: //www.sjat.com.cn

北京环球·国际马戏城

设计时间：2009年7月

建设地点：北京

用地面积：100.5公顷

建筑规模：26 000平方米

项目建成后可容纳观众约8 000人（其中包括500个VIP席位），每年将为观众带来300场以上的精彩演出。北京环球城 国际马戏城是打造以马戏为主题的文化休闲游乐中心，落成后将成为北京市首家专业的马戏演出场馆。

Beijing Universal City • International Circus City

Design Time: 2009.07

Construction Location: Beijing

Site Area: 1,500 acres

Construction Scale: 26,000 m^2

The project will have a audience capacity of about 8,000 people (including 500 VIP seats), and each year it can present the audience more than 300 performances. Beijing Universal City • International Circus City aims at building a circus-themed cultural recreational center, which will become the premiere professional circus venues in Beijing.

邮件综合处理中心

建设地点：北京

用地面积：289 411.342平方米

邮件综合处理中心（一期）是中国邮政集团公司重点工程项目。项目的建设将极大地提高网络运行效率，降低运行成本，加快邮件传递时限，极大提高邮政服务水平，促进地方经济快速、持续发展。建成后的邮件综合处理中心，将成为亚洲最大的邮件处理基地，大幅度提升了中国的邮政处理能力。

Integrated Mail Processing Center

Construction Location: Beijing

Site Area: 289,411.342 m^2

Integrated Mail Processing Center (Phase I) is a key project of China Post Group. The project will greatly improve network efficiency, reduce operating costs, speed up mail delivery, so as to greatly enhance the postal service, and promote local economy rapid and sustainable development. When completed, it will be the base for Asia's largest mail processing and significantly enhance the postal ability in China.

伯朗峰国际中心

建设地点：北京

用地面积：19 730.65平方米

建筑面积：100 721平方米

项目由京客隆超市大卖场和五星级酒店组成，既满足了京客隆超市的大卖场空间需要，又满足了酒店、客房的功能需要，兼顾地上与地下的空间使用需求。项目对城市景观及街道空间起到了开放与融合的作用。

2008年5月，联合国人居署授勋由世纪安泰设计的伯朗峰国际中心为"2008联合国人居署迪拜最佳国际范例（中国）推动大奖"。

Brown Peak International Centre

Construction Location: Beijing

Site Area: 19,730.65 m^2

Building Area: 100,721 m^2

This project is comprised by Jingkelong hypermarkets and a five-star Hotel. It not only satisfies the space need of Jingkelong Supermarket, also meet the unified column grid needs of hotel rooms feature; cater to both ground and underground space requirements. It plays a role in opening up and integration in respect to the street space on the urban landscape.

In May 2008, UN-Habitat honored "Brown Peak International Centre" designed by SJAT with "2008 UN-HABITAT best international model of Dubai (China) promote award".

山水文园

建设地点：北京

占地面积：61.18公顷

建筑面积：970 000平方米

本项目通过对环境的设计与营造，创造出宜居的生活空间。项目十分重视环境的营造，以景观的设计与实施，实现每户在景观、朝向、日照时间方面的均好性，实现了建筑节能与环保的目标。

荣获的主要国际奖项

联合国环境署颁发的"国际生态安全示范小区"

亚洲环境协会颁发的"规划设计奖"

全球花园城市与花园社区评选"金奖"

Landscape Culture Garden

Construction Location: Beijing

Site Area: 61.18 ha

Building Area: 970,000 m^2

Through the design and decoration to the environment, it creates a livable living space, with focusing on design and implementation of environment and vision landscape. It aims to realize an equalization in the aspects such as landscape, orientation, sunshine lasting time for each household, to fulfill the ideal energy conservation and environmental building.

Major International Awards

"International Ecological Security Model Community" issued by the UNEP

"Plan and Design Award" by Asia Environmental Association

"Gold Award" in Global Garden Cities and Garden Community Competition

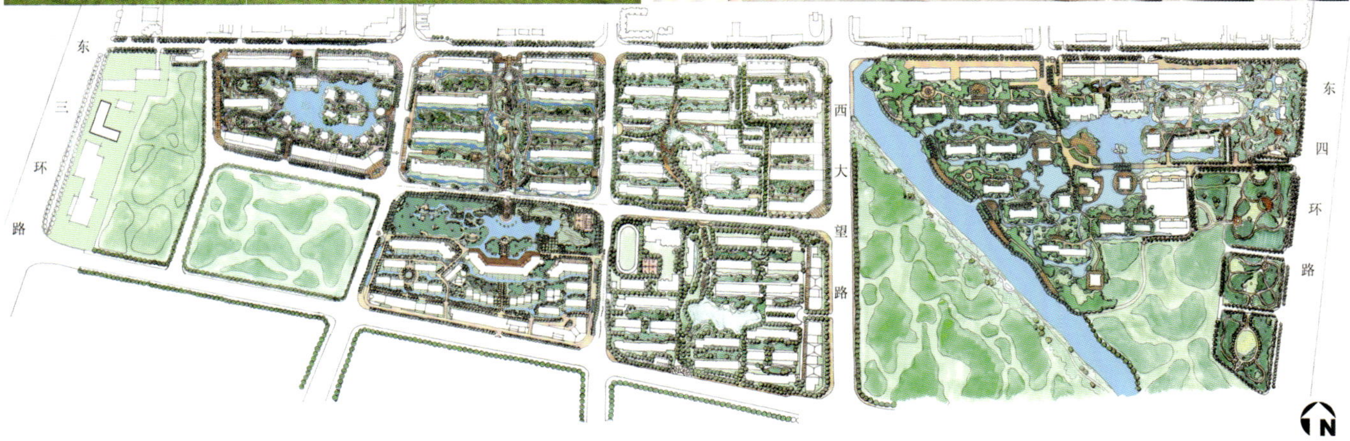

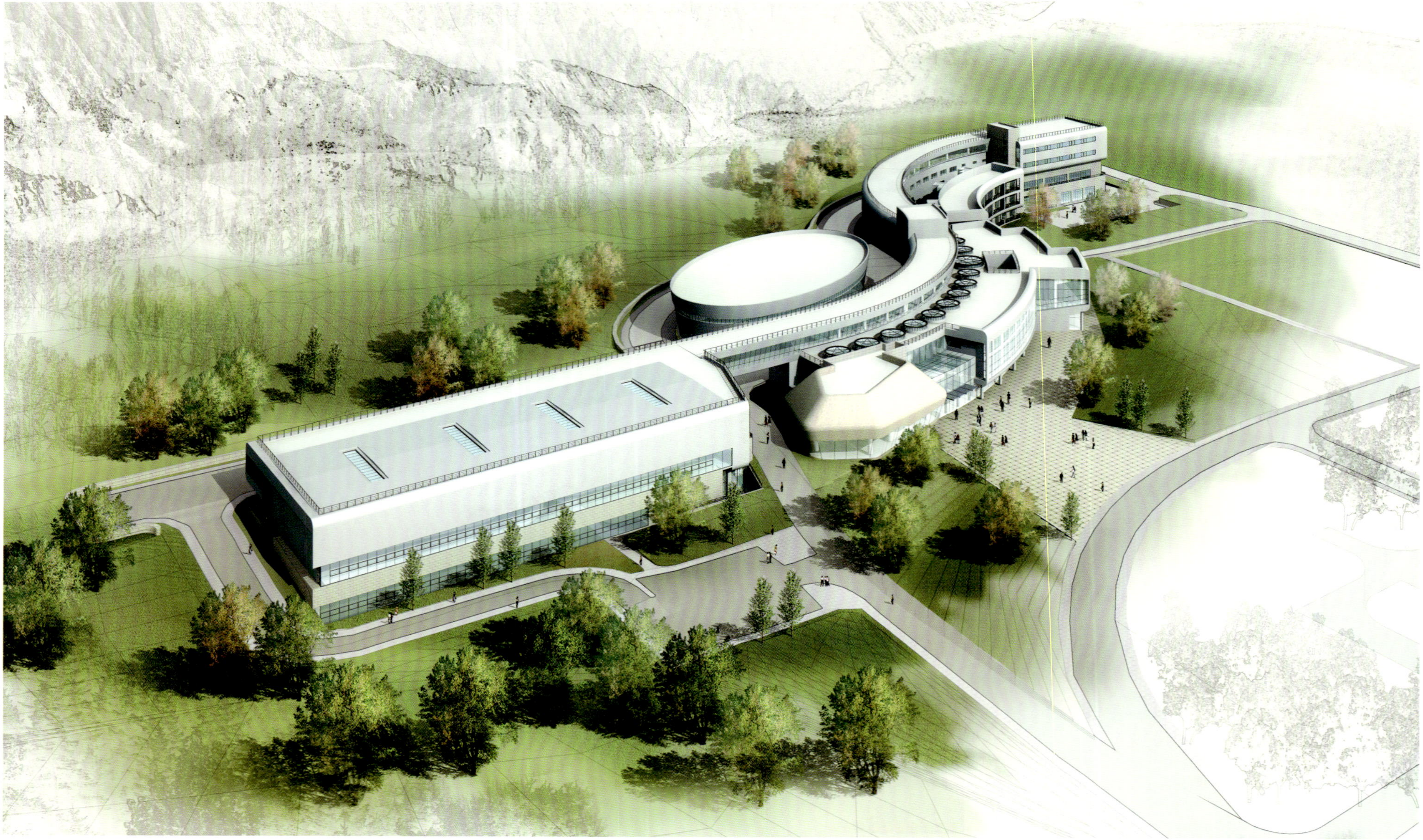

国家地震紧急救援训练基地

建设地点：北京

占地面积：13公顷

项目是一所专业化、规模化、现代化、国际化的国家地震紧急救援训练基地，可以模拟多种救灾环境和场所，开展搜索、营救、紧急处置、指挥等内容的培训。本项目的建设实施有助于提高我国地震紧急救援专业队伍和志愿者的整体素质和救援能力。

National Earthquake rescue training center

Construction Location: Beijing

Site Area: 13 ha

This project is a national earthquake rescue training center characterized by specialization, scale, modernization and internationalization. It can simulate a variety of environmental and disaster-relief sites, and conduct the training such as search, rescue, emergency treatment, and commanding. It helps to improve the country's earthquake emergency relief team of professionals and volunteers, the overall quality and rescue capabilities.

梧桐大道

建设地点：河北 唐山
占地面积：18 799平方米
建筑面积：491 100平方米

本项目的建筑风格以简约的现代风格为主，设计结合阳台、飘窗、楼梯间等建筑构件元素的归纳与整理，强化建筑的细节设计与建筑的韵律。温暖的色彩体现居住建筑"家"的感受。项目成为唐山高端住宅产品中的新标志。设计全方位营造高级住宅甚至豪宅的建筑氛围，带动全区品质的提升。

Indus Boulevard

Construction Location: Tangshan, Hebei
Site Area: 18,799 m^2
Building Area: 491,100 m^2

Its Architectural style is primary in simple, modern style, combined with arrangement and induction to architectural elements like balcony, bay window, staircase, etc., strengthening the building design and construction details along with the rhythm itself. Warm colors imply that a residential buildings should be full of "home" atmosphere. This project has become the new landmark among high-end residential products in Tangshan. It creates a variety of experiences in senior housing as far as luxury mansions to promote the region's quality .

亚胜·铂第公馆

建设地点：北京
占地面积：7 557.65平方米
建筑面积：53 792平方米

项目在狭小用地上解决了商业、金融、餐饮、酒店等不同功能的交互关系。立面设计采用将外墙的分隔方式进行错动变化的设计方法，从而体现前卫感与现代感，传达出新北京在新时期的特点。

Yasheng • Bodi Hotel

Construction Location: Beijing
Site Area: 7,557.65 m^2
Building Area: 53,792 m^2

Given to its narrow space, this project solves the interaction issue among different functions including business, finance, catering, hotels, etc. As to the facade design, the separation way of exterior wall has been changed with dislocation, which both convey a slightly avant-garde feature, and also showcase on behalf of the new era Beijing.

诺杰建筑设计事务所

ROGGEO DESIGN ASSOCIATES INC.

公司简介

关于 ROGGEO

诺杰自 1993 年 11 月 18 日在加拿大多伦多创始以来，在全世界范围内参与过许多重大项目的咨询服务，业务范围遍布加拿大、美国、日本、欧洲、中国。近几年，诺杰公司凭借多年国内外丰富的设计经验参与了多项国内重大项目。公司凭借众多的设计精英团队，严谨的执行力，国际惯例思维，在高速发展的国内市场环境中，创造出一套完整的、综合的，独特的建筑工程解决方案。

诺杰中国区总部设在北京，在成都设有分支机构，上海、天津、深圳设有合作机构，为遍布各地的客户提供高品质的市场研究、商业策划、规划及方案设计、工程设计、工程管理等方面的专业咨询服务。诺杰高度本地化运作方式和中、英双语设计团队的优势，使其和客户之间的沟通更加顺畅，信息反馈更加便捷，从而大大提高了设计效率及质量水准。

诺杰承诺用智慧、诚信和专业服务来满足客户的需求和愿望，用国际化的水准灵活应对市场挑战，尊重客户，尊重团队，在和谐的氛围中打造最完美的设计作品。

联系方式

法　人：余茂林 Roger M. YU
中国总部：北京诺杰建筑咨询有限公司
地　址：北京市朝阳区东三环北路霞光里 18 号佳程广场 A 座 16D 单元
邮　编：100027
电　话：+86-10-84400606
传　真：+86-10-84400062
E-mail：info@roggeo.com
网　址：www.roggeo.com

河北石家庄西美第五大道

西美第五大道项目位于河北省石家庄市裕华区，集高端住宅、商业、社区配套为一体。项目区域内建筑被规划为六栋高层住宅和沿街二层商业裙房，商业裙房引入高端零售专卖店，高层住宅在满足当地日照要求前提下采用点形式均匀布置，商业沿东、北、西三侧布置，与高层住宅一起自然围合出小区内部景观庭院。

诺杰建筑设计事务所采用独特的手法将项目打造成舒适的城市花园，并通过总体环境的布置、建筑底商围合布局以及多彩的建筑语言凸显强烈的时代都市感。该项目百米高的高端住宅隐喻“石门”的建筑构成主义风格，简洁、现代，已成为石家庄槐安路上的新地标。

规划用地：4.19 公顷　　总建筑面积：18.15 万平方米
容 积 率：4.3

上海浦东软件园1、2期
PUDONG SOFTWARE PARK, PHASE 1、2

建筑地点：上海 浦东
建筑用途：软件开发
用地面积：90 000平方米
建筑面积：125 300平方米
高　　度：22.5米
层　　数：3～5层

Construction Location: Pudong, Shanghai
Construction Purpose: Software Development
Site Area: 90,000 m^2
Building Area:125,300 m^2
Construction Height: 22.5 m
Total Floors: 3-5

中国电子工程设计院集团创建于1953年，是直属于国家开发投资公司的全国大型设计机构，拥有建筑工程综合甲级设计资质。院直属设计机构建筑设计研究工作室成立于2009年10月，由院总建筑师、上海世博会沙特馆总设计师王振军主持，工作室业务范围涵盖建筑与工程设计的全部内容，并在超大型城市综合体、高科技园、机场塔台和博览建筑设计等领域取得了出色的成绩。

基于多年打造的品牌优势、服务各类顾客的丰富经验、业内首屈一指的人才团队、日益完善的经营和服务体系，我们致力于为顾客提供最专业的优质服务。

Founded in 1953, with Class A building integrated design qualification, China Electronics Engineering Design Institute Group is a national large design agency directly under the State Development Investment Corporation. Its architectural design research studio, also directly under the institute was established in October 2009, and hosted by its Chief Architect and the General Designer of Shanghai World Expo, Saudi Arabia Pavilion. Its studio design business spans all architectural and engineering projects, and has achieved remarkable outcome in super urban complex, high-tech park, airport control tower and the Expo building design and other fields.

Based on years of the brand advantage, various types of customer service experience, the industry's premier talent team, as well as the increasingly sophisticated management and service system, we are committed to providing clients with the most professional and platinum service.

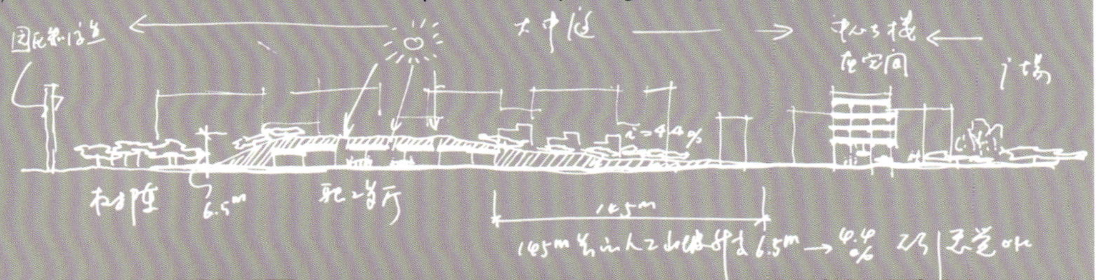

中国电子工程设计院王振军工作室（建筑设计研究工作室）
CEEDI-ADR ARCHITECTURE RESEARCH STUDIO

上海浦东软件园3期
PUDONG SOFTWARE PARK, PHASE 3

建筑地点：上海
建筑用途：软件开发
用地面积：463 000平方米
建筑面积：379 916平方米
容 积 率：0.82

Construction Location: Shanghai
Construction Purpose: Software Development
Site Area: 463,000 m²
Building Area: 379,916 m²
Plot Ratio: 0.82

首都机场东区塔台
CAPITAL AIRPORT EASTAREA NEW AVIATION CONTROL TOWER

建筑地点：北京机场
建筑用途：导航、办公
用地面积：13 334平方米
建筑面积：2 647平方米
高　　度：79.6米
层　　数：13层
结构类型：筒体

Construction Location: Beijing
Construction Purpose: Aviation Control, Office
Site Area: 13,334 m²
Building Area: 2,647 m²
Construction Height: 79.6 m
Total Floors: 13
Structure: Tube Structure

海南三亚山海天大酒店（五星级）
SANYA SHANHAITIAN GRAND HOTEL(5 STAR)

建筑地点：海南 三亚
建筑用途：酒店、会议
用地面积：21 849平方米
建筑面积：27 202平方米
高　　度：26.5米
层　　数：7+1层
结构类型：框架

Construction Location: Sanya, Hainan
Construction Purpose: Hotel, Conference Center
Site Area: 21, 849 m²
Building Area: 27,202 m²
Construction Height: 26.5 m
Total Floors: 7+1
Structure: Frame Structure

1 2 3

1 上海浦东软件园3期
2 海南三亚山海天酒店
3 首都机场东区塔台

上海世博会沙特国家馆是唯一由中国设计单位独立设计完成的外国自建馆。该设计用一种轻松的建筑语言描绘了阿拉伯神话中的月亮船由富饶的阿拉伯半岛漂浮至东方时尚港口的场景。极具动感和未来感的体量漂浮在地表之上，与屋顶的树木一起为周边创造出大量的阴影。室内外参观流线围绕中庭环形布置，进出人流有序分离。阿拉伯风情的场地景观、开敞的表演及展示平台、全球独创的全景立体融入式参观方式、视野极佳的屋顶花园，为中外游客奉献了一场阿拉伯文化的饕餮盛宴。

The Saudi Arabia Pavilion at Shanghai World Expo 2010 is the sole foreign-built pavilion whose design is independently performed by a Chinese design firm. The design depicts a scene of a moon boat in the Arabian mythology floating to the fashionable oriental port from the affluent Arabia using an easy architectural language. The extremely dynamic and futuristic massing stands adrift on the surface, creating a large area of shade around the pavilion and the vicinity together with the date palm trees on the roof. The indoor and outdoor visitors' circulations are arranged around the atrium, separating the entering visitor flow from the quitting one. The Arabian-flavored site landscape, the open performance and exhibition platform, the globally unique visiting pattern of panoramic 3D melting and the roof garden with a nice view all help provide the visitors from home and abroad with a magnificent feast of Arabian culture.

清華大學建筑设计研究院

清华大学建筑设计研究院成立于1958年，为国家甲级建筑设计院。依托于清华大学深厚广博的学术、科研和教学资源，并作为建筑学院、土水学院等院系教学、科研和实践的基地。我院十分重视学术研究与科技成果的转化，规划设计水平在国内名列前茅。

清华大学建筑设计研究院现有工程设计人员300余人，其中国家一级注册建筑师74名，一级注册结构工程师55名，高级专业技术人员占60%以上，人才密集、专业齐全，人员素质高、技术力量雄厚。院设有六个建筑工程设计所、四个由国家科学院和工程院院士以及国家设计大师领衔的工作室、两个建筑专业设计所、工程软件室、文物与历史建筑保护设计研究所、绿色建筑工程设计所、工程咨询与建筑策划所、工程设计文件审查所等。作为国内久负盛名的综合设计研究院之一，我院业务领域涵盖各类公共与民用建筑工程设计、城市设计、居住区规划与住宅设计、古建筑保护及复原、景观园林、室内设计、检测加固、前期科研和建筑策划研究以及工程咨询。

近年来，由我院设计并建成的工程已获得国家级、省部级优秀设计奖达六十余项，位居全国甲级院前列。代表作有北京菊儿胡同居住区、北京大学图书馆、清华大学图书馆、北京天桥剧场、清华大学设计中心楼、中国美术馆改造装修工程、雷峰塔复原工程、中央美院迁建工程、世界大学生运动会游泳跳水馆、海淀社区服务中心等一批脍炙人口的精品之作，均获得很高的社会评价。我院在生态建筑设计方面也做了大量的研究工作，清华大学设计中心楼和超低能耗办公楼都是绿色办公建筑的有益实践。此外，在一系列的国内外竞赛和招投标中，我院设计的国家大剧院、北京国际会展体育中心等项目均取得佳绩，尤其在2008年北京奥运会场馆工程设计投标中，我院一举中标三个项目，2008年北京奥运会射击馆、飞碟靶场以及柔道跆拳道馆，这些场馆现已完成设计任务并建成投入使用。

科研方面，我院自主研发了TUS工程设计软件，并通过国家鉴定。学术方面，在2001年受全国注册建筑师管委会的委托，与建筑学院合作编纂完成了全国注册建筑师继续教育教材之三《建筑设计的生态策略》、之四《建筑设计与经济》。

清华大学建筑设计研究院积极发展国际间的交流与合作，已与法国、美国、日本、德国、澳大利亚、韩国以及中国香港等十几个国家和地区的高等学校、科研机构、设计单位和业主有着良好的合作关系。现已获得中国（CNAB）和英国（UKAS）质量体系认证证书。

成立至今，清华大学建筑设计研究院始终严把质量关，秉承"精心设计、创作精品、超越自我、创建一流"的奋斗目标，热诚地为国内外社会各界提供优质的设计和服务。我们的队伍是年轻的、充满活力的，如果说建筑是一座城市的文化标签，我们的建筑师将用流畅的线条勾勒它，灵魂的笔触描绘它，迸发的激情演绎它，目的只有一个——让世界更加美好。

Tsinghua University Architectural Design and Research Institute, was established in 1958. Relying on professional and wide-range academic research and instructional resources, it has become an architectural design and research institute with national Class A architectural design certificate. Meanwhile, it is a teaching, scientific research and practice base for School of Architecture, School of Civil Engineering and other departments of Tsinghua University. This institute lists transformation from academic researches to scientific research achievements on its top priority and its design capacity ranks among the most successful candidates in PRC.

At present, the engineering and technical staff number of this institute is about 300 altogether. Among them 74 get PRC Class I Registered Architect Qualification and 55 are PRC Class 1 Registered Structure Engineers and senior professionals and technicians account for over 60%. So there are intensive hi-qualified talents and advanced technical forces in this institute. It consists of 6 building engineering design offices, 4 studios led by members of NAS and NAE and national design masters, 2 professional architecture design offices, an engineering software office, a design and research office for historical relic and building protection, a green building engineering design office, an engineering consulting and architectural programming office, a review office for engineering design documents and so on. As one of the prestigious comprehensive design and research offices, the business scope of this institute covers various architectural design of public and civil architectures, urban designs, residential planning and designs, conservation and restoration of historic buildings, landscape architectures, interior designs, building examinations and reinforcements, preliminary scientific researches and architectural programming researches, engineering consulting, etc.

In recent years, architectures designed and built by this institute has won more than 60 national and provincial designing awards and this institute is among the best national Class A architectural design and research institutes. Representative works of this institute include Beijing Chrysanthemums Alley residential area, Peking University Library, National Tsinghua University Library, TQ Theater, Tsinghua University Design Center Building, renovation and decoration project of National Art Museum of China, Lui Fung Pagoda restoration project, Central Academy of Art removal and building project, Universidad Swimming and Diving Natatorium, Hai-tien Community Service Center and other famous and popular competitive products which get high social appraisals. This institute has also done a lot of research in ecological building designs. Tsinghua University Design Center Building and office buildings of ultra-low energy consumption are the beneficial practice of green office buildings. In addition, National Grand Theatre, Beijing International Conference and Exhibition Center and other architectures designed by this institute do very well in domestic and international competitions and biddings. In design bids for Beijing Olympic Venues, this institute won 3 projects, i.e. Beijing Shooting Range Hall, Beijing Shooting Range Clay Target Field, and Judo and Taekwondo Museum for 2008 Beijing Olympic Games, which are finished and put into service.

As for scientific research, TUS design software is a product of independent innovation of this institute and it has been verified by the state. In the sphere of learning, National Registered Architect Management Committee authorize this institute to edit teaching materials for further education of certified architects with School of Architecture. Ecological Tactic of Architectural Design and Architectural Design and Economy are 2 textbooks edited by this institute.

Tsinghua University Architectural Design and Research Institute, has been developing international communication and cooperation actively and it has established good cooperation with universities, research institutes, design units, and owners in France, America, Japan, Germany, Australia, South Korea, Hong Kong and other countries and regions. It has got QMS certificates issued by CNAB and UKAS.

Tsinghua University Architectural Design and Research Institute, has always strictly controlled the quality since its establishment. Taking 'creating competitive products through elaborating designs and being the top-notch through surpassing ourselves' as its service tenet, this institute provide domestic and international customers with excellent designs and services. Besides, this team is a youthful and vigorous one. If architectures are cultural icons of a city, architects of this institute are depicting them with smooth strokes with incorporeal strokes and with their passion. To make our world a better place is their goal.

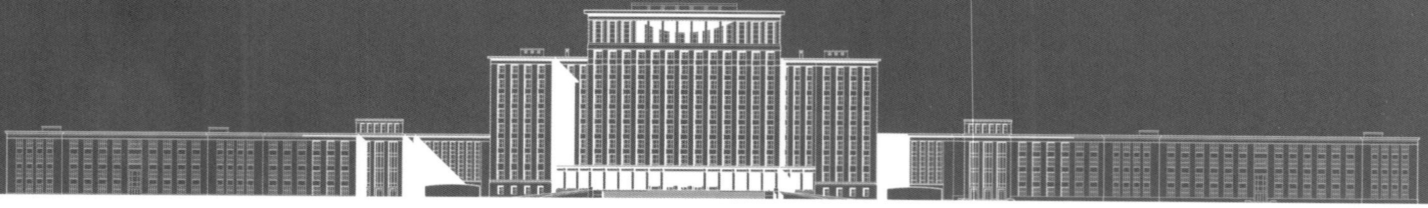

中国　北京　100084
Beijing　100084　P.R.China
电话/Tel：+86-10-62782581
传真/Fax：+86-10-62784727
网址/Http：//www.thad.com.cn
邮箱/E-mail：jzsjy@mail.tsinghua.edu.cn
法人/Corporate：庄惟敏

清华大学建筑设计研究院
ARCHITECTURAL DESIGN AND RESEARCH INSTITUTE OF TSINGHUA UNIVERSITY

清华大学医学院

总 用 地 面 积：23 800平方米
总 建 筑 面 积：46 500平方米
设计起止时间：2004年10月—2005年3月
竣 工 时 间：2006年10月

清华大学医学院位于清华大学老校园内，与其相邻的许多建筑都已被列为国家级文物保护单位。本设计着重于将空间体量、材料尺度等与周边环境协调，同时积极塑造具有文化品质的校园环境。

该建筑为一座现代化的实验室建筑，其中设有900兆核磁共振实验室、SPF级实验动物中心、冷冻尸库、解剖室等实验设施。

设计中把确保实验人员的安全及身心健康放在首位，通过可控的人工环境满足不同的实验要求。在安排好复杂技术功能的前提下，创造出诸多不同尺度的交往空间，以促进学术交流，同时，考虑到医学科学兼具道德、伦理等方面的社会责任，通过一些纪念性的空间形象、铭文塑像等，力求对使用者形成潜移默化的影响。

School of Medicine, Tsinghua University

Gross Land Area: 23,800 m^2
Gross Building Area: 46,500 m^2
Starting and Ending Time of Design: 2004.10-2005.03
Completion Time: 2006.10

School of Medicine, Tsinghua University locates in the old campus of Tsinghua University. It is surrounded by many old buildings which have been designated as national cultural relic units. By using appropriate spatial entity and material scale, architects intend to let the school harmonize with the external environment, and actively shape a campus with high quantity of culture.

As a modern laboratory building, it has 900 MHz Nuclear Magnetic Resonance Laboratory, Experimental Animal Center of SPF degree, Corpse Freezer and Anatomy theatre etc.

The design fully considers the mental and physical health and safety of laboratory personnel. The controllable artificial environment can meet the requirements of different kinds of experiments. On the premise of exercise of complex technological function, architects create communication space in different scales to promote academic communication. Meanwhile, architects take social responsibility, such as moral and ethnic, into consideration. Through some memorial spatial image, inscriptions and statues, architects exert a subtle influence on users.

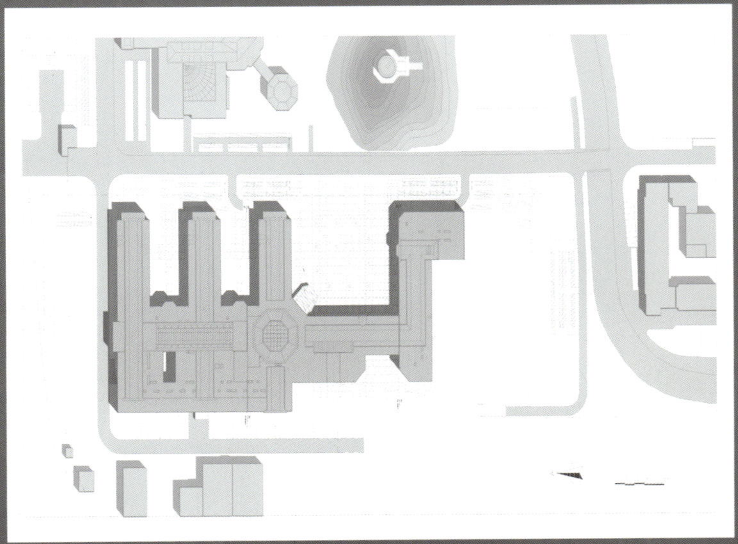

清华科技园科技大厦

总 用 地 面 积：16.5公顷
总 建 筑 面 积：188 027平方米
设计起止时间：2002年10月—2004年1月
竣 工 时 间：2005年7月

清华科技园位于北京中关村科技园、清华大学、北京大学及中科院的核心地带，是全国著名的高校科技产业园区，由科技创新、研发、孵化、国际会议等功能建筑组成，总建筑面积逾40万平方米。科技大厦是科技园的标志性建筑，高110米的四栋主楼统领科技园建筑群。大厦二层绿化广场与顶层绿化会所相呼应，体现绿色节能、科技与自然相结合的设计理念。

The Technological Mansion in Tsinghua Science Park

Gross Land Area: 16.5 ha
Gross Building Area: 188,027 m^2
Starting and Ending Time of Design: 2002.10-2004.01
Completion Time: 2005.07

Tsinghua Science Park is located to the nuclear region among Beijing Zhongguancun Science Park, Tsinghua University, Peking University and Chinese Academic of Sciences, is a famous Science Park in higher education institutions around China. In Tsinghua Science Park, different buildings exert different functions including technological innovation, research and development, incubation, international conference. The gross building area is over 400,000 m^2. The Technological Mansion as a landmark building, is composed by four main buildings,each of them is 110 meters,which also the leader of the building group among the Tsinghua Science Park.

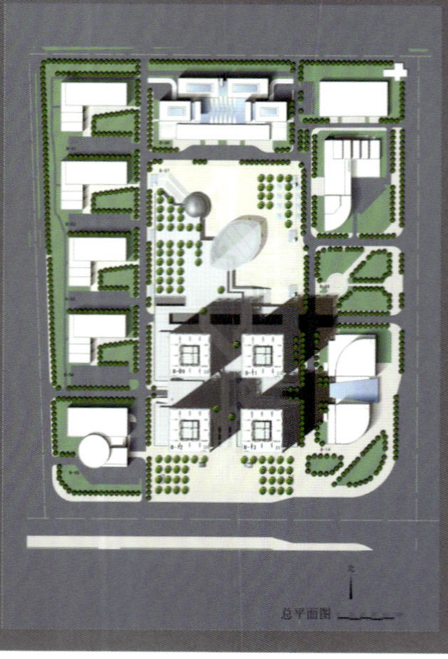

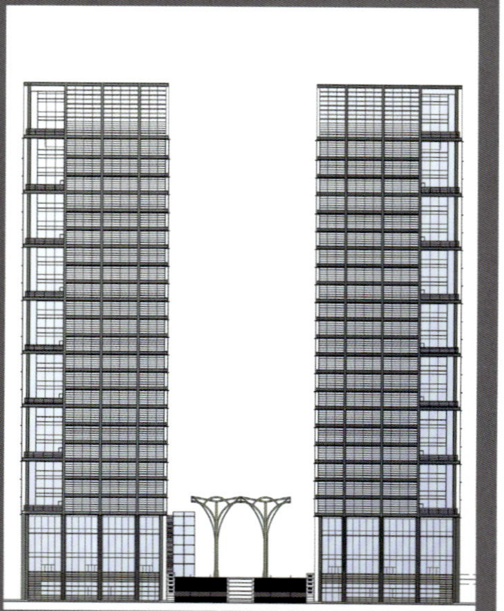

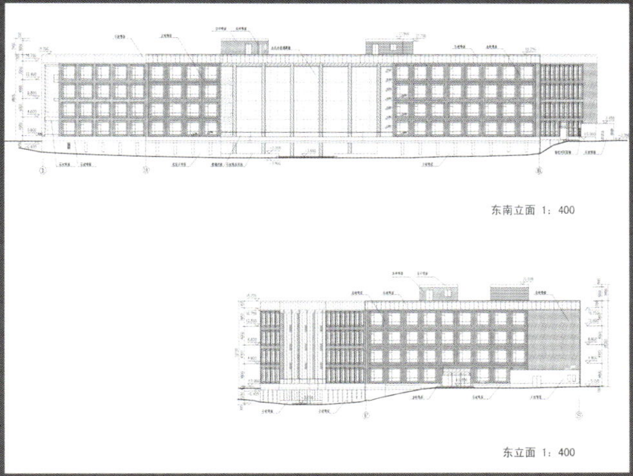

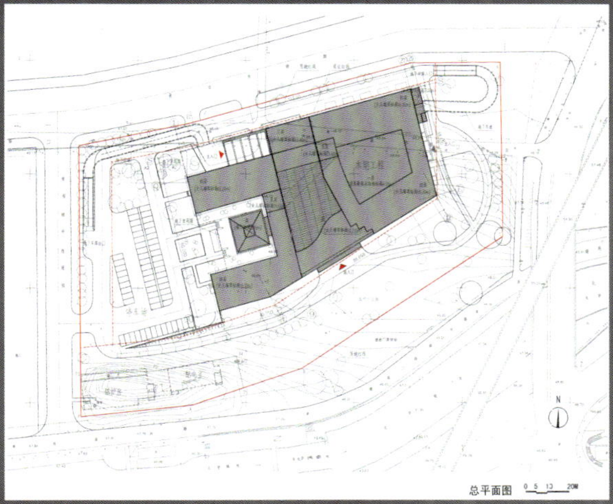

中国工程院综合办公楼

总用地面积：20 612平方米　　总建筑面积：22 883平方米
设计起止时间：2003年7月—2004年11月
竣工时间：2006年12月

中国工程院位于北京德胜门西北侧，北侧为德胜尚城写字楼，南侧为开阔的城市绿化带。

建筑的基地位置以及使用功能的特殊性，决定了其设计有别于普通办公楼有其自身的独特之处。设计中力图解决两个问题：其一是相目处于敏感的地段环境，如何处理建筑与城市、环境的相互关系；其二是中国工程院作为中国的最高学术机构，建筑如何帮助中国工程院塑造自身性格形象。设计中，建筑界面沿地段红线而建，以维护城市街区的完整性，建筑体量完整简洁，符合德外大街的车行尺度，同时立面采用匀质的正方形窗洞，饰以灰色面砖，在色彩和造型上与德胜门箭楼呼应协调，做到“和而不同”。中国工程院作为中国的最高学术机构，建筑在造型的处理上借用了新古典主义的风格，体现端庄典雅的内涵，同时避免了普通办公楼的商业形象以及政府部门大气的形象。

The Comprehensive Office Building of Chinese Academy of Engineering

Gross Land Area: 20,612 m^2　　Gross Building Area: 22,883 m^2
Starting and Ending Time of Design: 2003.07-2004.11
Completion Time: 2006.12

Chinese Academy of Engineering locates in the northwest of Beijing Desheng Gate. The north of it is Desheng Shancheng Office Building, while south a broad green belt.

Compared with the ordinary office building, it has special location and applicational function. This design strives to solve two problems. Firstly, as locating in a sensitive district environment, this building should deal with the relationship between urban and environment. Secondly, as the Chinese highest academic institution, the building should shape its image accordingly. In order to keep district complete, building entity integrated and simple, its interface is built along the building line, conforming the traffic scale of Dewai road. The façade is decorated with gray face brick and homogeneous square windows to harmonize Jian Tower in Desheng Gate, no matter from color or shape. As the highest academic institution, this building followed neoclassic style, embodies elegance, and avoids commercial manner of ordinary office building as well as grand manner of government building.

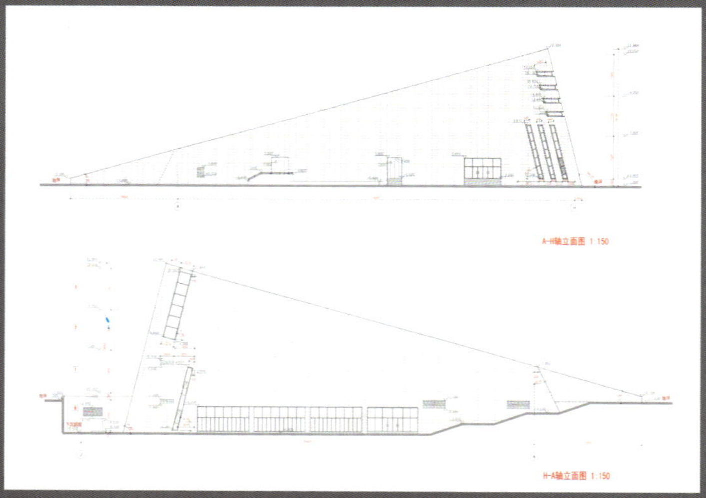

成都金沙遗址博物馆文物陈列馆

总 用 地 面 积：195 288平方米
总 建 筑 面 积：20 190平方米
设计起止时间：2005年1月—2005年10月
竣　工　时　间：2007年4月

本工程由文物陈列馆、文物保护中心及配套的地下车库等服务设施共同构成，同时还与遗迹馆共同构成金沙遗址公园。

斜台体造型的文物陈列馆以抽象的造型，创造出破土而出的动态，体现了文物出土时的特征，延续了历史，沟通了现在。洞石外表面表现出厚重的文化气息，屋面的方格网象征了考古挖掘的探方形态。

内部空间设计打破了传统博物馆将展区、公共空间割裂的模式，而将各种功能区有机融合，考虑公共空间的多功能使用，将陈列和人的活动相结合，集陈列、教化、收藏及娱乐为一体，使博物馆成为成都重要的文化殿堂，成为高品位的城市客厅。

Cultural Relics Exhibition Hall of Chengdu Jinsha Site Museum

Gross Land Area: 195,288 m^2
Gross Building Area: 20,190 m^2
Starting and Ending Time of Design: 2005.01-2005.10
Completion Time: 2007.04

This project is composed by Cultural Relics Exhibition Hall, Cultural Relics Protection Center and under ground garage. Together with Remains Exhibition Hall, it constitutes into Jinsha Site Museum.

Cultural Relics Exhibition Hall shaped into abstract sloping bench looks like something is breaking out of the earth, which represents cultural relics are unearthed. The overall design continues history and links up present. The surface of rock wall is rich in culture, while the grid network roof symbolizes the plot dug in archaeological site.

The design of internal space is different from traditional museum model which separate the exhibition hall from public space. As all kinds of functional zones merge organically, the public space has various functions. Associating exhibition and people's activities means that exhibition, teaching, collection and entertainment combine into one. This museum will become an important cultural palace in Chengdu, a high grade urban living room.

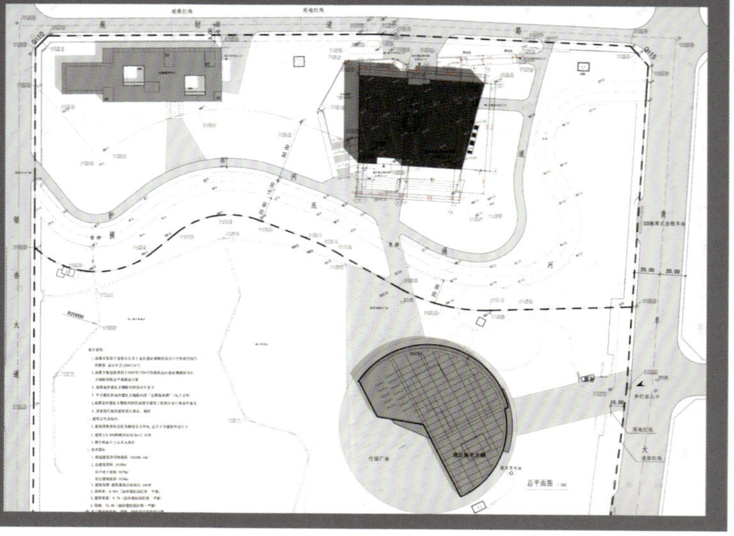

丹东市第一医院（一期主体工程）

总 用 地 面 积：653 000平方米
总 建 筑 面 积：29 941平方米
设计起止时间：2005年4月—2006年5月
竣 工 时 间：2008年6月

创造先进模式

丹东医院的设计围绕“双轴”理念展开，设计改变传统医院单轴通道的设计模式，两条轴线贯串医院，成为联系各部门的交通空间，两轴之间设置核心医技单位，服务一期及未来的二期。

双轴理念给医院建筑的生长带来了极大便利，一二期的衔接将更加灵活，二期可以根据不同需要在西轴不同位置介入而不影响整个医疗构架的完整性。

轴线的宽度可以根据功能和人流加以调整，一期东轴由南向北逐渐变窄，局部结合庭院扩大为等候及停留空间；二期西轴同样可根据功能进行收放处理，将东西双轴打造成极富特色的医院空间。

打造人文医院

丹东医院建筑设计着力改变医院固有的白色或者暖灰色调外装的形象，以有着四种色差的红色面砖按一定比例拼贴使其成为主要建筑饰面，配合玻璃、金属，力求打造一种具有人文气质的医疗建筑。

医院的外立面设计对红砖进行了多样的处理：裙楼的红白体块搭配，病房楼的红砖跳跃布置，建筑洞口等细部的排砖设计，体现了建筑沉稳而又精致的现代气息。

Dandong First Hospital (Phase I of Main Project)

Gross Land Area: 653,000 m^2
Gross Building Area: 29,941 m^2
Starting and Ending Time of Design: 2005.04-2006.05
Completion Time: 2008.06

Creating Advanced Model

The design of Dandong First Hospital follows the concept of “two axes”. This design changes the traditional hospital model where a single alleyway is built. Two axes running through the hospital becomes a traffic channel of each section. Core medical technology institution is built between two axes, which are beneficial to both phase Ⅰ and Ⅱ construction.

The concept of “two axes” brings construction huge convenience. Firstly, two phases could connect more flexible. Secondly, the construction of phase Ⅱ could be placed at any place of west axle according to different requirement, which would not break the integrity of the hospital.

The width of axle can be adjusted according to different functions and the number of people. The east axle of phase I gets narrow from south to north. In partial area, space is left for future expanding. Accordingly, the west axle can be adjusted. Architects intend to build west and east axes into a characteristic space in hospital.

Creating a Humanistic Hospital

Compared with the traditional hospital which is colored by white or warm grey, Dandong Hospital is covered by four kinds of red face brick in specific proportion. Together with glass and metal, architects strive to build a hospital in humanistic manner.

Red bricks are arranged differently on the external surface of hospital: the skirt building is covered by red and white bricks; the red bricks of sickroom building are like leaping notes; the detail part, such as the mouth of the cave, is paved with red bricks.

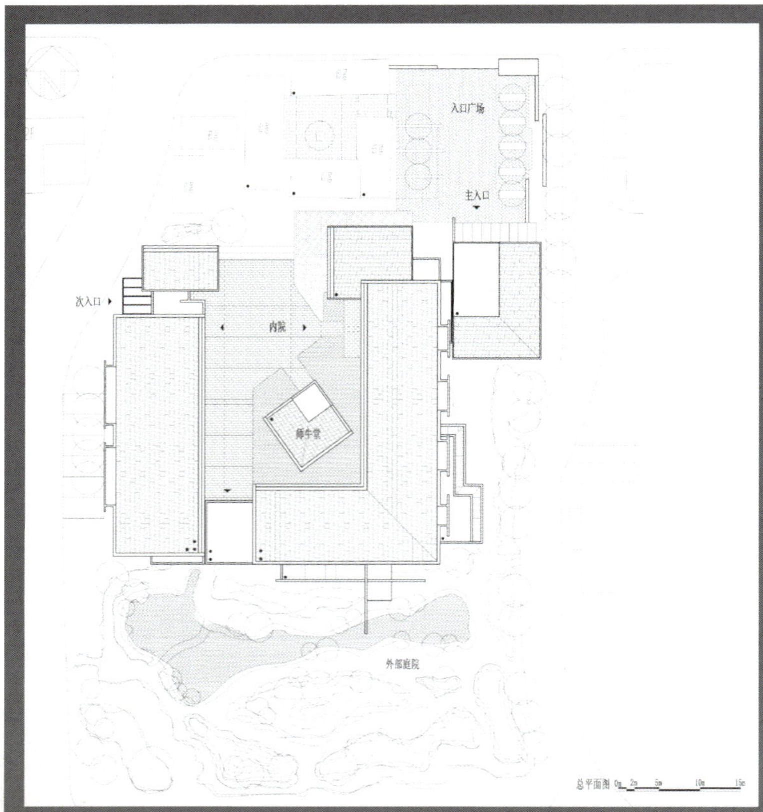

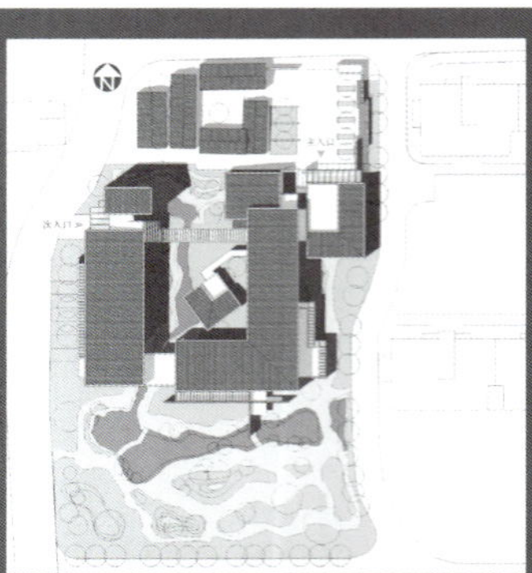

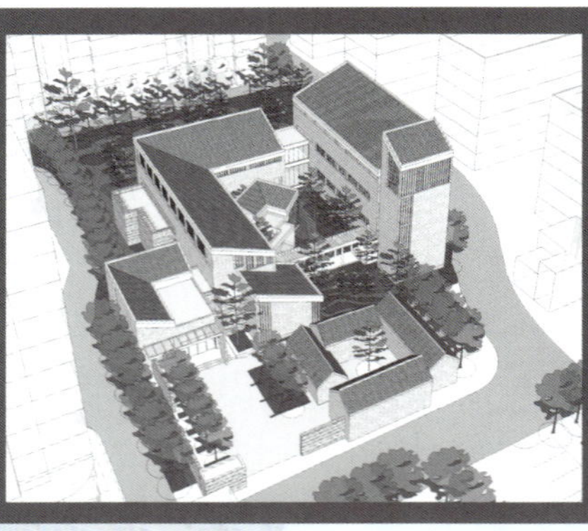

李可染艺术馆

总 用 地 面 积：4 232平方米
总 建 筑 面 积：2 835平方米
设计起止时间：2005年8月—2005年12月
竣 工 时 间：2007年11月

李可染艺术馆的建筑创作立意于将国画大师李可染先生绘画中展现的艺术精神融入建筑设计思维的价值观中，延续紧邻的李可染旧居中传统建筑所蕴含的脱离建筑表面形式的本质精髓，给旧居的存在形成恰当的生存空间和文化土壤，在摆脱形式的束缚中求得对传统建筑神似的传承，并在建筑表达方式上大胆突破。

艺术馆的规划设计是旧居传统民居空间的延伸，总平面设计沿革了旧居院落化的空间格局。建筑追求水墨山水画的意境，建筑外墙采用了产自宜兴的紫砂灰砖，屋顶采用深灰色瓦，特意突出建筑厚重的感觉。师牛堂在整个艺术馆的空间构成中起到画眼的作用，位置处于新馆院落围合的中央位置，遥望旧居，建筑特意偏斜角度位置，与主体似而不同的形式，以及运用木材而形成独特温润的质感，形成了所谓“拙者巧之极，奇者正之华”的建筑美感。

李可染艺术馆是一座专业美术馆，建筑设计体现小而精的设计原则，两层四个展厅呈线性联系，连贯流畅，展厅之间设置休息及过渡空间，十分便于举办各种规模的展览，也便于观众观展。展厅采用无柱空间，根据地形实现不同进深的展厅空间，二层的坡屋面形成高低不同的展墙，利于展出各种尺度的画作。通过侧向和顶向的反射遮阳窗为所有展厅实现自然通风及自然采光，十分利于节能。

Likeran Art Museum

Gross Land Area: 4,232 m^2
Gross Building Area: 2,835 m^2
Starting and Ending Time of Design: 2008.08-2005.12
Completion Time: 2007.11

Mr. Li is a master in traditional Chinese painting. The design of Likeran Art Museum is inspired by artistic spirit of Mr. Li's paintings. The design of Mr. Li's former residence embodies the quintessence of traditional architecture. Likeran Art Museum reflects the cultrual factors and provides living space for the former residence. Meanwhile, it gets rid of the fetter of forms and inherits similarities in spirit. So the design of Likeran Art Museum can be regarded as a breakthrough in architectural presentations.

The design of Likeran Art Museum is spatial extension of the former traditional residence. General layout design inherits and develops spatial pattern of country yard of the former residence. The conception of design seeks artistic perfect of traditional Chinese landscape paintings. Yixing purple sands are used in the outer wall of this architecture and the roof is made up of dark grey tiles, which delivers a sense of massiness. Shiniu Hall plays the role of centre in the whole art museum. It is located in the center of courtyard of this art museum and has some distance from the former residence. The deliberated deflect angle seems different from the main body and application of wood expresses the unique and gentle texture. This design could be summarized by a line of a poem that is "Careless design is perfect. Strangeness is a shinning point".

Likeran Art Museum is a professional art gallery. The design of which embodies design philosophy of being small but excellent. The four exhibition halls in the two-storey architecture are connected with each other coherently and smoothly. There is relaxation space located between two halls. Different scales of exhibitions can be held here and the design also facilitates views for visitors. The construction of large exhibition space is without pillars. In line with the various landscapes, depths of exhibition halls are different. Sloping roof on the second floor forms exhibition walls of different heights, which suit different scales of paintings. Sunshade windows on the roof and side direction walls provide ways of nature ventilation and daylighting, which is also a way of energy conservation.

中冶京诚 北京京诚华宇建筑设计研究院有限公司
CERI ARCHITECTS & ENGINEERS

地址：北京市宣武区白广路 4 号
邮编：100053
电话：+86-10-83587418/83587419
传真：+86-10-83587418
E-mail：ceria@ceria.com.cn
Http：//www.ceria.net.cn

北京京诚华宇建筑设计研究院有限公司（简称京诚华宇）于 2007 年正式从中冶京诚工程技术有限公司（简称京诚公司）改制分立而成立。京诚公司是由原中冶集团北京钢铁设计研究总院进行分立式整体改制后设立的有限公司。新成立的京诚华宇由原京诚公司建筑设计研究院原班人马组成，从事建筑工程设计、施工图审查及设计咨询业务，是京诚公司下属的独立法人有限责任公司。

改制前的北京钢铁设计研究总院已经经历了五十多年的历程，取得了辉煌的业绩。京诚公司是经国家认证的综合甲级大型工业与民用建筑工程设计单位，是第一家部属通过国际 ISO 9001 质量体系认证的设计单位。在 1992 年全国首次设计勘察单位的百强评比中，荣登“百强之首”，2007 年与 2008 年在全国设计勘察单位评比中列第一位。

京诚华宇现总人数近 200 人，其中中高级及以上职称 120 余人，中冶集团高级技术专家 1 人，美国注册建筑师 1 人，香港注册建筑师 1 人，英国注册结构工程师 5 人，香港注册结构工程师 3 人，注册建筑师、注册规划师、注册结构工程师、注册岩土工程师、注册电气工程师、注册公用设备工程师共 67 人。

京诚华宇在超高层设计、大型公建设计、钢结构设计、地基基础工程、抗震加固设计等领域具有领先优势，在施工图审查及设计咨询方面具有丰富的实践经验。

京诚华宇是国家钢结构设计规范主编和管理单位。

京诚华宇将坚持“以诚为本、以信为先、不断求索、勇于争先”的企业精神，继往开来，不断创新，为打造国际一流的建筑设计咨询公司而努力奋斗。

CERI ARCHITECTS & ENGINEERS(CERIA for short)separated from Capital Engineering & Research Incorporation Limited(CERI for short)in 2007. CERI is a limited company by primary MCC Beijing Central Engineering and Research Incorporation of Iron and Steel Industry(CERIS for short) after separate reforming.The newborn CERIA is composed of the original staff of the primary CERI Design Research Institude and is engaged in architecture engineering design,construction drawing review and design consultation. it is a Limited liability company of independent legal person affiliated to CERI.

CERIA has been experienced fifty years. And achieved spectacular achievement before the reforming. CERI is an A Grade large industrial and civil architecture engineering design unit certificated by the authority, and also the first design unit passed through the international ISO 9001 quality framework certification. it had been granted the first place of the “top 100”during the “top 100 comparison” among design survey units first time in china in 1992, had been in top one winner among design survey units in china in 2007 and 2008.

CERIA has nearly 200 employees, 120 employers are senior engineers.1 MCC senior expert, 1 USA registered architect, 1 HK registered architect, 5 UK registered structural engineers, 3 HK registered structural engineers, 67 engineers are A grade registered architects, registered planning engineers, grade A registered structural engineers, registered geotechnical engineer, registered electric engineers, registered public facility engineers.

CERIA has leading advantages on such areas, super high structure, public buildings, steel structure, foundation engineering, earthquake-resistant strengthening, etc., and has various experiences on design documents examination and design consultation.

CERIA is the supervising and managing unit of national code for design of steel structures.

CERIA Will insist that “Honest is essential and credit is first, And Innovation-keep learning and exploration ” the entrepreneurial spirit, inherits, innovates unceasingly. And work hard for the making of an international first-class architectural design Consultant firm.

超高层建筑 Super High Rise Buildi

1

2

3

1 中国国际贸易中心三期 A 阶段工程

工程规模：30 万平方米　主塔楼 330 米高
结构形式：型钢混凝土筒中筒结构
业态：商业、写字楼、香格里拉酒店
合作单位：美国 SOM、王董国际有限公司、奥雅纳工程顾问、柏诚（亚洲）
设计时间：2006 年
建设地点：北京
获奖：第十三届首都城市规划建筑设计汇报展优秀设计方案二等奖

Stage A of phases III of the project of China International Trade Centre

Project Scale: 300,000 m^2, 330 m high main tower
Structural Style: tube in tube structure of steel reinforced concrete
Type of Operation: commercial building, office building and Shangri-La Hotel
Cooperation Partners: America SOM, Wong & Tung International Limited, Ove Arup & Partners (Hong Kong) Ltd and Parsons Brinckerhoff (Asia)
Design Time: 2006
Construction Location: Beijing
Award: Second prize of the excellent design of the 13th capital city planning architectural design exhibition

2–3 中国国际贸易中心一、二期工程

一期：43 万平方米　塔楼 156 米高
二期：13 万平方米　塔楼 156 米高
结构形式：一期筒中筒钢结构，二期框架筒体混合结构
业态：商业、写字楼、酒店、会展
合作单位：一期：美国 Sobel/Roth　日本日建设计
二期：王欧阳有限公司
设计/完工时间：一期：1987 年/1990 年
二期：1997 年/1999 年
建设地点：北京

Phases I, II of the project of China International Trade Centre

Phase I: 430,000 m^2, 156 m high tower
Phase II: 130,000 m^2, 156 m high tower
Structural Style: Phase I: tube in tube steel structure
Phase II: Mixing barrel framework structure
Type of Operation: commercial building, office building, hotel and exhibition
Cooperation Partners: Phase I: America Sobel/Roth, Japan Nikken Sekkei
Phase II: Wang Ouyang Ltd.
Design/Completion Time: Phase I:1987/1990. Phase II: 1997/1999
Construction Location: Beijing

4

5

4 北京财富中心三期

建设面积：17.59 万平方米　主塔楼 260 米高
结构形式：型钢混凝土筒中筒结构
业态：商业、办公楼
合作单位：美国 KPF、奥雅纳工程顾问、柏诚（亚洲）
设计时间：2006 年
建设地点：北京

Phases III of Beijing Fortune Center

Project Scale: 175,900 m^2, 260 m high main tower
Structural Style: tube in tube structure of steel reinforced concrete
Type of Operation: commercial building and office building
Cooperation Partners: America KPF, Ove Arup & Partners Hong Kong Limited and Parsons Brinckerhoff (Asia)
Design Time: 2006
Construction Location: Beijing

5 烟台世茂海湾一号

工程规模：35 万平方米　主塔楼 323 米高　高档公寓主塔楼 195 米高结构形式：型钢混凝土筒中筒结构
业态：写字楼、希尔顿酒店、高档公寓
合作单位：王董国际有限公司、柏诚（亚洲）
设计时间：2007 年
建设地点：山东 烟台

Shimao No.1 Harbor Yantai

Project Scale: 350,000 m^2, the main tower is 323 m,the major flat tower is 195 m
Structural Style: Steel Reinforced Concrete tube in tube structure
Type of Operation: office building, hiltonhotel, high-end flat
Cooperative Company: Wong&Tung International,Parsons Brincker hoff(Asia)
Design Time: 2007
Construction Location: Yantai, Shandong

超高层建筑 Super High Rise Buildings

钢结构 Steel Structure

6 长春国际金融中心

工程规模：28.5万平方米　主塔楼 250 米高
结构形式：钢结构
业态：商业、写字楼、五星级酒店、公寓
设计时间：2009 年
建设地点：吉林 长春

Changchun International Financial Centre

Project Scale: 285,000 m^2, 250 m high main tower
Structural Style: steel structure
Type of Operation:commercial building, office building, five-star hotel and apartment
Design Time: 2009
Construction Location: Changchun, Jilin

7 天安门钢结构观礼台

工程规模：0.66万平方米
结构形式：钢结构
设计／竣工时间：2009 年 /2009 年
建设地点：北京

Changchun International Financial Centre

Project Scale: 6,600 m^2
Structural Style: steel structure
Design/Completion Time: 2009/2009
Construction Location: Beijing

6

7

8

9

规划设计 Overall Programming

8 秦皇岛卢龙县三里庄安置房工程

工程规模：47.7 万平方米
业态：住宅
设计时间：2010 年
建设地点：河北 秦皇岛

Relocating housing project of Sanlizhuang, Lulong county, Qinhuangdao

Project Scale: 477,000 m^2
Type of Operation: Residence
Design Time: 2010
Construction Location: Qinhuangdao, Hebei

9 中国钢研涿州基地南区

工程规模：25 万平方米
业态：住宅 厂区
设计时间：2010 年 4 月
建设地点：河北 涿州

The South of China Iron and Steel Base in Zhuozhou

Project Scale: 250,000 m^2
Type of Operation: Residence Factory
Design Time: 2010.04
Construction Location: Zhuozhou, Hebei

10

11

12

13

15

14

16

10 人民大会堂宴会厅改造

设计／完工时间：1997 年／1999 年
建设地点：北京

Renovation for Grand Banquet Hall in Great Hall of the People

Design/Completion Time: 1997/1999
Construction Location: Beijing china

11 人民大会堂万人大礼堂改造

工程规模：17 万平方米
设计／完工时间：1998 年／2000 年
建设地点：北京

Renovation for People's Congress Hall in Great Hall of the People

Project Scale: 170,000 m^2
Design/Completion Time: 1998/2000
Construction Location: Beijing

12–13 北京火神庙国际商业中心

工程规模：34.7 万平方米
业态：商业、公寓式写字楼、电影院、室内冰场
设计时间：2007 年
建设地点：北京

Beijing Huoshen Temple International Business Centre

Project Scale: 347,000 m^2
Type of Operation: commercial building, apartment-style office building, cinema and indoor rink
Design Time: 2007
Construction Location: Beijing

14 青岛胶南红树林酒店

工程规模：65 万平方米　主楼 155 米高
业态：室内运河商街、写字楼、五星级度假酒店、电影院会展中心、豪华温泉馆、室内水上乐园、室内冰场、生态餐厅、美术馆、儿童游乐城
合作单位：ZNA
设计时间：2009 年
建设地点：山东　青岛

Jiaonan Mang rove Tree Hotel in Qingdao

Project Scale: 650,000 m^2, mian building is 155 m
Type of Operation: indoor chanel commercial street, office buildings, Five-Star Resort Hotel, Cinema Exhibition Center, Luxury Spa Resorts, Indoor Waterpark, indoor icerink, ecological restaurant, gallery, Children's entertainment city
Cooperative Company:ZNA
Design Time: 2009
Construction Location: Qingdao, Shandong

15 中冶京诚工程技术有限公司亦庄办公区

工程规模：9 万平方米
业态：写字楼、研发楼、员工宿舍、体育场
设计／完工时间：2004 年／2006 年
建设地点：北京
获奖：2006 年冶金行业部级优秀设计一等奖

Yizhuang office area of Capital Engineering & Research Incorporation Limited

Project Scale: 90,000 m^2
Type of Operation: office building, research and development building, staff dormitory and stadium
Design/Completion Time: 2004/2006
Construction Location: Beijing
Award: First prize of Excellent Design of the Construction Ministry of metallurgy industry of year 2006

16 中冶京诚工程技术有限公司亦庄第二办公区

工程规模：22 万平方米
业态 写字楼、研发楼、员工宿舍、礼堂、游泳馆、网球馆
建设地点：北京

Yizhuang No.2 office area of Capital Engineering & Research Incorporation LimitedIncorporation

Project Scale: 220,000 m^2
Type of Operation: office building, research and development building, staff dormitory, assembly hall, natatorium and indoor tennis stadium
Design Time:2009
Construction Location: Beijing

17

18

住宅建筑 Residential Buildings

17 南京市河西莲花地块经济适用房和中低价房

建设面积：55.4 万平方米
设计时间：2009 年
建设地点：江苏 南京

Economically affordable housing and houses with low-medium prices of Hexi Lotus plot, Nanjing

Project Scale: 554,000 m^2
Design Time: 2009
Construction Location: Nangjing,Jiangsu

18 新奥蓝城

建设面积：48 万平方米
设计／完工时间：2006 年／2007 年
建设地点：吉林 长春

New Aulan Town

Project Scale: 480,000 m^2
Design/Completion Time: 2006/2007
Construction Location: Changchun,Jilin

工业建筑 Industrial Buildings

19 安泰科技北京空港新材料产业基地

工程规模：14 万平方米
业态：高科技园区
设计／完工时间：2003 年／2005 年
建设地点：北京
获奖：2006 年冶金行业部级优秀设计二等奖

Beijing Konggang New Materials Industry Base of Antai Advanced Technology & Materials Co., Ltd.

Project Scale: 140,000 m^2
Type of Operation: high-tech park
Design/Completion Time: 2003/2005
Construction Location: Beijing
Award: Second prize of Excellent Design of the Construction Ministry of metallurgy industry of year 2006

19

20

20 安泰科技北京永丰新材料产业园

工程规模：10 万平方米
业态：高科技园区
设计／完工时间：2003 年／2005 年
建设地点：北京
获奖：2006 年冶金行业部级优秀设计二等奖

Beijing Yongfeng New Materials industrial Park of Antai Advanced Technology & Materials Co., Ltd.

Project Scale: 100,000 m^2
Type of Operation: high-tech park
Design/Completion Time: 2003/2005
Construction Location: Beijing
Award: Second prize of E xcellent Design of the Construction Ministry of metallurgy industry of year 2006

21

22

21 安钢集团职工总医院

建设地点：河南 安阳

Employees General Hospital of AnGang Group

Construction Location: Anyang, Henan

22 鞍山市烧伤整形医院

工程规模：7.8 万平方米
设计时间：2009 年
建设地点：辽宁 鞍山

Anshan Burn Orthopaedic Hospital

Project Scale: 78,000 m^2
Design Time: 2009
Construction Location: Anshan, Liaoning

医疗建筑 Medical Buildings

文体建筑 Culture And Sport Buildings

23 北京第十四中学

工程规模：3.5 万平方米
设计／完工时间：2006 年／2009 年
建设地点：北京

Beijing No.14 Middle School

Project Scale: 35,000 m^2
Design/Completion Time: 2006/2009
Construction Location: Beijing

24 北京宣武体育馆二期游泳馆、网球馆

工程规模：1.53 万平方米
设计时间：2006 年
建设地点：北京

Phase II Natatorium and Indoor Tennis Stadium of Beijing Xuanwu Gymnasium

Project Scale: 15,300 m^2
Design Time: 2006
Construction Location: Beijing

23

24

室内设计 Interior Buildings

25–26 中国人民公安大学礼堂改扩建

工程规模：2 400 座位
设计／完工时间：2001 年／2003 年
建设地点：北京
获奖：2006 年冶金行业部级优秀设计一等奖

Redevelopment and expansion for Chinese People's Public Security University Assembly Hall

Project Scale: 2,400 seats
Design/Completion Time: 2001/2003
Construction Location: Beijing
Awards: won the first excellent design prize of ministry grade metallurgical industry

25

26

景观设计 Landscape Buildings

27–31 京汉香城

工程规模：22 万平方米
设计时间：2008 年
建设地点：河北 香河

Jinghanxiangcheng

Project Scale: 220,000 m^2
Design Time: 2008
Construction Location: Xianghe, Hebei

32 A 派公寓

工程规模：6.5 万平方米
设计／完工时间：2004 年／2006 年
建设地点：北京

A-pie Apartment

Project Scale: 65,000 m^2
Design/Completion Time: 2004/2006
Construction Location: Beijing

27

28

29

30

31

32

CSA 中天伟业 中天伟业（北京）建筑设计集团（甲级）

做建筑设计专家 CHINA SKY (BEIJING) ARCHITETURE GROUP

中天伟业(北京)建筑设计集团成立于1995年,是经由建设部部属国有企业、社团企业改制重组而成的股份制企业。具有建设部颁发的建筑工程设计甲级资质以及工程监理资质。集团注册资金 1 000 万元。中天伟业现拥有 150 余名工程设计人员，其中一级注册建筑师、注册规划师、一级注册结构工程师、注册设备工程师及注册电气工程师数十名，并已通过中国质量认证中心 ISO 9001 质量体系认证。

中天伟业专业齐全，人才济济，不仅具有原国有企业雄厚的技术实力和高度的社会信誉，同时还拥有国内外优秀建筑设计企业所具有的核心竞争力，以及独特的客户意识和优良的运营管理模式。

中天伟业坚持"质量第一、信誉第一、服务第一"的原则和"为客户创造价值"的理念，广泛涉及建筑创作领域，为客户提供优质服务。从区域规划到城市设计，从办公楼宇、星级酒店到商业地产，从文教建筑、医疗建筑到体育建筑，从高档别墅到居住建筑，从旧城改造到仿古建筑，再从室内装饰设计到环境景观园林设计，中天伟业无不创作出优秀的设计精品，在业内和市场上建立了良好的口碑。中天伟业于 2009 年荣获中国银行北京分行年度（2009—2012）设计单位。

中天伟业凭借合理的组织结构、科学的管理方法、雄厚的技术实力以及先进的服务模式，为客户提供包括区域规划与城市设计、建筑与工程设计、室内装饰设计、景观园林设计、结构设计、给排水、采暖空调、电气照明、楼宇自控、弱电智能设计、房地产前期策划、工程项目的可行性研究、咨询，项目管理、工程监理等全方位的服务。

中天伟业以北京为基础,业务辐射全国,同时为拓展业务需要,在乌鲁木齐、兰州已设立分公司，并陆续将在上海、天津、重庆、沈阳、山东等地设立办事处或分公司。中天伟业的目标是立足北京，面向全国，走向世界，同时，中天伟业亦与世界著名的美国雅马萨奇设计公司开展紧密和广泛的合作，努力设计出中国最好的建筑。

作为低碳绿色建筑设计的先行者，中天伟业长期致力于绿色建筑的设计和探索，成绩卓越，2010 年被《中国企业报》评为"中国建筑设计绿先锋"。

China Sky (Beijing) Architecture Group, founded in 1995, is a corporate enterprise restructured and recombined from the state-owned enterprise and community enterprise affiliated to Ministry of Construction. It owns the Class A qualification of constructional engineering design issued by Ministry of Construction as well as qualification of construction supervision. The registered capital of the group is 10 million yuan. There are more than 150 engineering design staffs, among which tens of Class I registered architects, registered planners, Class I registered structural engineers, registered facility engineers and registered electrical engineers. It has passed quality system certification of China Quality Certification Center ISO9001.

China Sky has got complete majors and plenty of talents. It possesses not only solid technical strength and high social reputation of the former state-owned enterprise, but also the core competence of the excellent constructional design enterprises home and abroad, as well as unique client consciousness and eminent operation management mode.

China Sky sticks to the principle of "Quality first, reputation first and service first" as well as the concept of "create value for courtesy". It involves widely the domain of architectural creation and provides the clients withhigh-quality service. It has created excellent design works from regional planning to urban design, from office buildings and starred hotels to commercial properties, from cultural and academic constructions, medical constructions to sports constructions, from top-grade villas to residential architectures, from old town reformation to pseudo-classic architectures, from indoor decorative design to environmental landscape design and gained great public praises in the industry and in the market. In 2009 it has won the award of Annual (2009-2012) Design Unit of Beijing Branch of China Bank.

By virtue of reasonable organizing structure, scientific managing method, solid technical strength and advanced service mode, China Sky provides the clients with all-around services as the following: regional planning and urban design, constructional and engineering design, indoor decorative design, landscape design, structural design, water supply and drainage, heating and air conditioning, electric lightening, building automation, weak current intelligent design, real estate pre-job planning, feasibility research and consultation of engineering projects, project management and construction supervision etc.

Based in Beijing, China Sky exposes its business all around the nation. For the purpose of extending business, it has established branches in Urumqi and Lanzhou, and will successively establish agencies or branches in Shanghai, Tianjin, Chongqing, Shenyang and Shandong etc. It has got the goal of keeping a foothold in Beijing, facing the whole nation, and stepping into the world. Meanwhile it has carried out intimate and wide cooperation with world-famous Yamasaki Design Company of USA, endeavoring to design the best architectures in China.

As a pioneer of low carbon green constructional design, China Sky devotes itself to long-term design and exploration of green architecture and has got outstanding achievements. In 2010 it has been rated by "China Enterprise News" as "Green Pioneer of China Constructional Design".

地址：北京市海淀区车公庄西路乙 19 号华通大厦 B 座 9 层
邮编：100048
电话：+86-10-88018011/8012/8569
传真：+86-10-88018011/8012/8569 转 202
邮箱：csabj@163.com
网址：www.csabj.com

Address: 9th floor of Huatong Mansion Building B, Yi No.19 of Chegongzhuang
West Road, Haidian District, Beijing
P.C. : 100048
Tel: +86-10-88018011/8012/8569
Fax: +86-10-88018011/8012/8569 transferring to 202
E-mail: csabj@163.com
Http: //www.csabj.com

CSA
中天伟业

超高建筑

河北唐山曹妃甸工业园五星级国际酒店

首钢集团河北迁安办公中心

重庆钓鱼嘴半岛标志性双子座

昆明永升广场

昆明南亚论坛项目标志性双子座

中材集团大厦

深圳生命保险大厦

CSA
中天伟业

城市规划

海南陵水东方棕榈滩规划

河南南阳天鹅湖农业观[illegible]

河北安国关汉卿
文化旅游产业园规划

天津静海新城规划

重庆钓鱼嘴半岛中央休闲商务区规划

北京国际传媒大道新闻出版城

昆明南亚论坛项目城市设计

CSA
中天伟业

公共建筑

北京通用时代国际中心办公楼

北京曙光计算机科研楼

天津曙光计算机产业基地

北京雅宝城商业中心

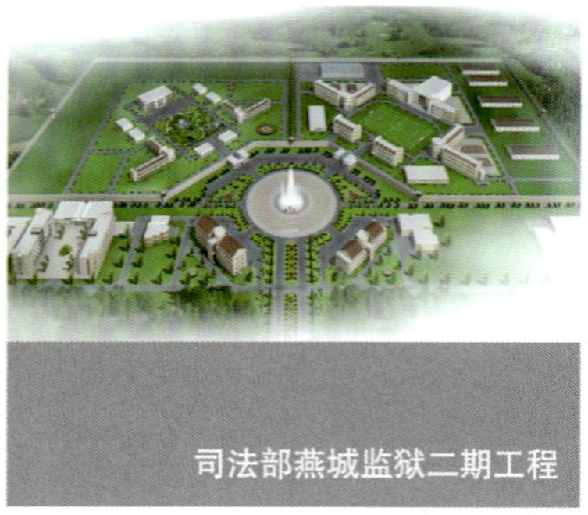
司法部燕城监狱二期工程

海南三亚美丽之冠七星级酒店

山东泰安银座城市广场

辽宁建平县医院门诊及病房楼

内蒙古通辽科尔沁迎宾馆

北京邮件综合处理中心

北京一号地震憾艺都艺术园

河北三河汇福国际健康中心

中行北京分行使馆支行

山东东营银座城市广场

北京翠微百货清河店

辽宁清原县医院病房楼

神华可再生能源展示中心

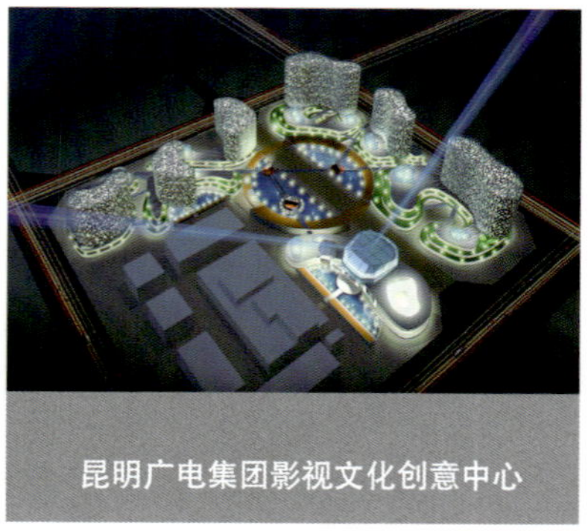
昆明广电集团影视文化创意中心

新疆哈密基地吐哈石油基地科技馆

CSA
中天伟业

室内设计

内蒙古通辽科尔沁迎宾馆

国家食品药品监督管理局新办公楼

国家保密局办公楼

国家无线电监测中心办公楼

居住建筑

北京通用时代国际中心高级公寓

山东淄博巴黎春天二期

北京中广·宜景湾项目

北京样本高尚社区

新疆库尔勒香梨置业和合家园

河北唐山滦南鼎盛花园

北京强佑清河新城

北京雁栖半岛别墅项目

鑫苑中国郑州中央华府项目

北京第六大洲项目

新疆阿克苏多浪人家住宅小区

青岛鲁商置业西海岸住宅小

清华大学八家教师住宅区

北京东一时区项目

北京圣世一品项目

辽宁宏运集团朝阳凤凰新城二期

新疆新农世纪城

北京通典铭居项目

SOHO中国广渠门项目

益田集团长春净月湖别墅项目

北京金碧湖畔二期

首钢地产南戴河鸥洲项目

北京翠湖花园别墅项目

北京清华城市规划设计研究院
Beijing Tsinghua Urban Planning & Design Institute

北京清城华筑建筑设计研究院有限公司
THCA Architectural Design & Consulting Institute
(北京清华城市规划设计研究院建筑分院)
Beijing Tsinghua Urban Planning & Design Institute Architectural Department

地址：北京市海淀区中关村东路8号东升大厦A座901室
邮编：100083
电话：+86-10-82526161
传真：+86-10-82526867
邮箱：thca@thupdi.com
网址：www.thupdi.com

Address: Rm 90, Tower A, Dongsheng Plaza, 8 Zhongguancun EastRoad Haidian District, Beijing, P.R, China
P.C.: 100083
Tel: +86-10-82526161
Fax: +86-10-82526867
E-mail: thca@thupdi.com
Http: //www.thupdi.com

北京清城华筑建筑设计研究院有限公司（北京清华城市规划设计研究院建筑分院），为清华城市规划设计研究院下属22个院所之一，是以建筑设计为主的专业设计和研究机构，执有建设部颁发的甲级建筑设计资质证书，公司注册资金为500万元。建筑分院和清华规划院各所之间形成了城市总体规划、详细规划、建筑设计、景观设计、市政、交通、能源等专业配置完整的规划、建筑、景观、市政设计体系，在城市规划与建筑设计领域具有技术领先，多学科、多专业交叉融合的技术优势。

设计理念：绿色、数字建筑的创新实践

分院以绿色建筑、数字建筑为核心设计理念，采用最新设计方法和技术，将城市规划、建筑设计、景观设计、绿色设计交汇融合，追求最高设计品质及艺术境界。

绿色建筑

我们主张积极的被动式绿色设计，强调通过建筑的被动式设计创造出可持续发展的建筑与环境，而不仅仅依靠技术手段。

数字建筑

建筑数字技术的普及应用，使建筑形式面临一场新的革命，我们始终以最新的设计手段，以国际化的视野，倾力打造数字建筑新体验。

我们的工作模式：多专业、多学科协同工作的高效模式

建筑分院以专业化的工作模式对设计流程进行高效率的科学管理，严格控制每一设计阶段的设计质量，充分发挥团队合作的力量和优势，务必使每一个作品达到尽善尽美。

分院与规划院内的其他各专业保持密切深入的合作，充分发挥各专业的技术优势，形成高水准、专业完备的项目服务团队，这种所际之间共同完成项目的工作模式，保证了项目在不同学科内的高质量和学科之间的连续性，实现了设计的整体性，跨学科性，一体化设计的目标。

分院同时把市场即客户的需求作为设计的目标，既注重项目所在地的经济效益和社会条件，又以一流的设计方法和技术，在建筑、结构、能源、环保、智能化等方面向客户提供全方位的优质服务。

我们的团队：国内最具潜力、最新视野、最具前瞻性的设计团队

绿色建筑设计团队

分院集合了一批充满活力和有志于创新设计的专业人才，其核心优势来源于他们是一群多专业、多文化，具有国际化视野的设计师，他们以绿色、数字建筑设计为主旨，以创新的设计手段和手法，形成了独树一帜的创作风格。

结构设计优化团队

结构优化是本着3R(Reduce、Reuse、Recycle)原则，对建筑的结构及材料进行合理设计及优化。我们赞成开发商为了经济效益，以少的代价，满足必要的安全要求，而我们也正在追求以同样少的代价，实现更高的安全标准。

适老住宅设计团队

我们一直关注中国社会老龄化问题，专注于适老化设计研究工作，积极响应政府居家养老的政策倡导，着力推进在常规商品房住宅小区内进行适老化商品住宅的设计营造。

绿色建筑推广团队

绿色建筑推广中心是一个绿色建筑技术专业咨询机构、“四节一环保”绿色建筑科技应用推广平台，我们致力成为中国最优的绿色建筑领域专业咨询顾问团队。

我们的优势：实力强大，潜力无限

建筑分院依托清华大学建筑学院的教学、科研平台，能迅速将教学及科研学术成果的新理念、新技术，通过项目应用到工程实践当中。同时分院依托规划院二十多个专业所的多专业、多学科、多领域的技术支撑，在建筑设计领域具有得天独厚的技术优势。

多学科技术的融合，形成了建筑分院独特的技术特点和优势，对设计的原创性和创新性的追求，使分院时刻处于积极的创作状态，凭借对中国地域文化的高度责任感和洞察力，分院始终能以崭新的理念、敬业的精神，为客户提供高素质及独特的设计，务求确立卓越的建筑典范，并担负起对社会应有的专业责任。

我们的项目：类型丰富，遍布各地

分院业务范围广泛，涵盖文化教育、体育场馆、写字楼、酒店、商业、居住区等各类大型公共及居住建筑设计，并提供项目前期策划、可行性研究、绿色建筑、结构优化、适老住宅等设计咨询业务。

建筑分院业务发展迅速，相继设计了一批具有时代特点的各类建筑，工程遍布北京、天津、吉林、辽宁、河北、内蒙古、新疆、河南、浙江、江苏、四川、西藏、海南、山东、湖北等20多个省市自治区，受到地方政府及业主的好评。

THCA Architectural Design & Consulting Institute is one Branch out of 22 of Beijing Tsinghua Urban Planning & Design Institute. The institute is based on architectural design and professional design of Construction. The THCA Branch has the Class A architectural design qualification certificate issued by the Ministry of Construction. THCA has registered the capital of 500 million yuan. The branch of the Planning Institute of Tsinghua University are formed between the urban master planning, detailed planning, architectural design, landscape design, municipal, transportation, energy and other professional chain complete the planning, architecture, landscape, municipal design system. First place in the urban planning and architecture design field of technology in many disciplines, the professional field of technology convergence.

Designing Methods: Innovation practice basis of Green & Digital Architecture

THCA Branch Focuses on the green buildings, the figures for the building design philosophy, using the latest design and technology, design the city planning and architecture, the green design interchange highest quality of arts.

Green Building

We advocate constructive passive and green design, stressing on creating sustainable development of architecture and environment by passive building design, not just rely on the technique skills.

Digital Design

Digital technology and popularization of the application form is facing the building of a new revolution, we always design means to an international perspective, vigorous build numbers of buildings as new experiences.

Our Working Mode: Working Efficiently with multiple proffessions and technologies

THCA specialized work on the design process for the efficient and scientific management and strictly controls the design phase of the quality, full play and team work force, making every work to achieve perfection.

THCA has further cooperation within the institute. And making full use of professional technical advantages of a high level and professional services to complete the project team, the international common operational mode of the project, project on different subjects in the high quality and consistency among subject to comprehensive, integrated science, the target of the designs.

THCA splits the market and customers demand as a design of project, importance of the economic and social conditions, and design technical buildings, structure and energy and environmental protection, intelligent and to provide one-stop services.

Our Team: The most potential, latest view and and leading team

The green building team

Gathered a lot of active and innovative professional personnel from the core of the advantages more professional, culture, an international outlook of the main designer group, with green, digital designing themes, to design means and methods of the unique artistic style.

Structure Design Optimizing Team

Structure optimizing team are designing as 3R which is Reduce、Reuse and Recycle aply to the architectural structure and the material be reasonable design. We agree with developers in economic returns, and a price to meet the necessary security, we are being pursued at the same low price, for a better safety.

Elder Housing Designing Team

We have been concerned about the problem of Chinese society getting older, focusing on the aging of the design of research work. Responding to the government of the old-age household policy initiatives and efforts in promoting conventional commercial residential building district in the aging of the product design.

Green Building Expanding Team

Green buildings expanding centre is a green building technical and specialized advisory body, the "four savings one green" green buildings to promote science and technology application platform, it is becoming the best of green field of the building of professional consultant teams of China.

Our Advantages: Great Strength and Infinite Potential

THCA is supplemented by Tsingha University teaching and researching. It will put academic results of the new concept, and new technology, new approved project inro the practice. While supporting THCA, the Urban & Planning institute has more than twenty professional, and more discipline in the field of technical support in the construction sector is with advantaged technology.

Subject of technology convergence of THCA unique technological features and advantages for the design of the original and innovative and positive actions are ready all the times in the form of chinese culture, geography is a high degree of responsibility and insightful, THCA always in the new concept, and hardworking spirit, to provide high-quality and unique design to its brilliant building model, and take in the professional social responsibilities.

Our Projects: Multiple types, Crossing the country

THCA has a wide range of business included literacy education, sports buildings, office buildings, enterprises and hotels, commercials, residential areas of major public and living building design, and offering the design of planning feasibility study, green, architectural and structural optimization and the old house and the design consultancy services.

THCA business are developed rapidly and cover a number of time features of construction and engineering in Beijing, Tianjin, Jilin, Liaoning, Hebei and Inner Mongolia, Xinjiang, Henan, Zhejiang, Jiangsu, Sichuan, Tibet, Hubei, Hainan and Shandong for more than 20 provinces, well reputated by the local government and the developers.

开放创新、精益求精、勇于拼搏、追求卓越
Open Innovation, Accuracy Perfection, Dare Struggling, Pursue Excellence

顺义文化中心

项目规模：用地面积：55 894平方米

建筑面积：65 893.9平方米

设计说明

规划设计：绿色"生态走廊"的设计，将文化、健康、绿色融为一体，成为整个文化中心的灵魂。

建筑设计：通过每层之间的错动不仅创造了丰富的室内外空间，而且使立面灵活有层次，表皮肌理充分展示并夸张的表现出中国石文化的特点，打造出顺义别具一格的中心建筑群。

Culture Center of Shunyi

Project Scale: Site Area: 55,894 m^2

Building Area: 65,893.9 m^2

Design Brief

Planning Design: the green "Ecological corridor" which combining culture, healthy and green as one, becomes the soul of the whole culture center.

Architecture Design: by making every floors different shapes, creating more active spaces, using ancient Chinese symbols of characters, the unique architectures group of Shunyi will be established.

义马体育中心

项目规模：用地面积：295 000平方米

建筑面积：149 166平方米

体育馆及训练馆4 212座

游泳馆2 000座

体育场33 000座

设计说明

规划设计充分考虑地形因素，梳理、归纳了坡度起伏特征，将建筑融入其中，实现建筑与景观的完美融合。

建筑设计以煤矿为出发点，由地质演变的层次感过渡到建筑层次，然后又如同给建筑施加力的作用一般，使建筑扭曲起伏，充满动势。

Stadium of Yima

Project Scale: Site Area: 295,000 m^2

Building Area : 149,166 m^2

Training center 4,212 seats

Natatorium 2,000 seats

Stadium 33,000 seats

Design Brief

This project combines with the building and landscape, clears out the level differences, and makes the architecture melt into the site. The architectural idea was coals, from the levels of coals deeap underground transforming into the architecture façade, makeing the building full of movements.

江苏省常熟市江南文化中心

项目规模：用地面积：43 000平方米

建筑面积：21 000平方米

地上建筑面积：20 446.73平方米

地下建筑面积：553.27平方米

设计说明

项目位于江苏省常熟市主城区东部，在环境设计方面充分考虑本项目地处城市绿地空间的特质，实现了"虞城蓓蕾、城市园林"的设计理念，在满足建筑功能需求的同时，也活跃了广场的活动空间，成为了市民广场的景观。景观建筑化、建筑景观化是这个项目的又一特征。建筑设计吸取了传统建筑尤其是江南园林建筑的精华要素，结合现代建筑的语言，形成了表达常熟深厚历史文化积淀以及时代感的建筑形式。建筑意向契合了这个城市地段空间开放、宽阔的特质。

Jiangnan Culture Center in Changshu, Jiangsu province

Project Scale: Site Area: 43,000 m^2

Building Area : 21,000 m^2

Upper Ground Area: 20,446.73 m^2

Under Ground Area: 553.27 m^2

Design Brief

This project locates at the east part of the Changshu downtown, Jiangsu Province. In the environmental design part, considering of the urban green space where this project lays , we adopt the urban green space design idea. It not only meets the architectural function needs but also provides an active dynamic space, creating a landscape view of the people's square. The landscape is the architecture and the architecture turns to be the landscape, which is another design feature of this project. Taking the prime of the traditional architecture, especially from the Jiangnan garden architecture and integrating the modern architecture technique ,it expresses the profound historical and culture accumulation and the sense of times in Changshu. The architecture disposition is conform to the space characteristic of being open and vast in urban city.

顺义马坡体育中心

项目规模：用地面积：27 697平方米

建筑面积：152 422平方米

设计说明

以体育为主题的文化公园；

由多个室内外体育场所装点起来的，形成绿地、山丘、水系相互对话的坡地景观公园；

高端体育赛事与大众休闲体验的兼顾；

冰球特色休闲与大众娱乐消费结合；

体育娱乐项目多元化，场馆实现一馆多用；

采用先进的结构模式，形成场馆的建筑特色，打造城市地标性建筑；

运用先进的建筑技术，创造舒适节能的活动空间，以人为本，回归自然。

Mapo Stadium of Shunyi

Project Scale: Site Area: 276,697 m^2

Building Area: 152,422 m^2

Design Brief

Central park of sports theme.

Landscape park decorated with multiple spots with lands, hills and water talking to each other.

Considerations of sports competition and informal matches.

Ski ball entertainment combining with people common costs.

Sports entertainment venues for diversity, a multipurpose use.

Advanced structural form, making it a unique stadium and a signature of the city.

Advanced architecture technologies, creating a comfort and energy efficient space.

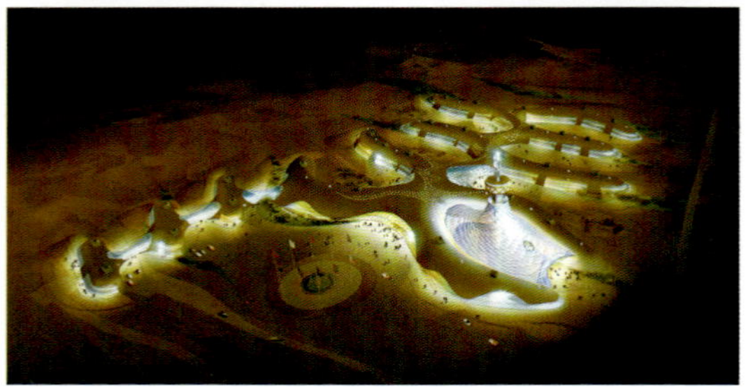

库布其沙漠论坛酒店

项目规模：建筑面积：50 000平方米

Desert Forum Hotel of Kubuqi

Project Scale: Building Area : 50,000 m²

设计说明

项目为库布其沙漠论坛会议中心酒店，酒店定位为五星级酒店。方案模拟沙漠自然形态，取意于风吹起的沙丘。建筑蜿蜒起伏，与周边环境相融相生，气势磅礴。

Design Brief

The desert forum hotel of Kubuqi was set as five starred hotel. This hotel was imitating the natural shape of the desert, with the meaning of the wind blowing up the sands and combining the nearby environments.

鄂尔多斯成吉思汗大饭店

项目规模：用地面积：200 096平方米
建筑面积：87 857平方米
地上建筑面积：82 723平方米
地下建筑面积：5 134平方米

Genghis Khan Hotel in Ordos

Project Scale: Site Area: 200,096 m²
Building Area: 87,857 m²
82,723 m² overground and
5,134 m² underground

设计说明

建筑位于内蒙鄂尔多斯市伊金霍洛旗甘德利草原上。该建筑单体设计主要体现内蒙古草原文化的特点，融合独特的民族气息。在立面造型设计上我们运用了内蒙古建筑独特的元素，同时借鉴了内蒙古已有建筑的形式，在体现中国传统观念“天圆地方”的同时，清晰展现建筑结构形式，使建筑造型与结构形式完美统一；将内蒙古草原文化中圣洁、纯净的蓝色、白色、红色运用在建筑立面上，赋予建筑独特的文化气息，将内蒙古草原文化中独特流畅的纹样抽象刻画于建筑细部。

Designing Brief

The building is located on Gandri Prairie, Yijinhuoluo Country, Inner Mongolia. The design of this single building, which is accordant with the ethical style, mainly reflects the grassland culture of Mongolia. The facade design uses unique factors and former shape of buildings in Mongolia for reference, in order to clearly exhibit the framework of the construction at the same time with presenting Chinese traditional idea of "round heaven and square earth", so as to achieve a perfect combination of design mode and form of structure. Besides, the building has blue, white and red used on the facade, which represent sanctity and pureness in the Mongolia culture, and has special fluent veins abstractly carved in the details as well.

设计说明

建筑位于浙江台州市新区主要街道的交汇处。主楼面向十字路口呈L形布置，与路口形成一个相互咬合的围合空间，使建筑成为城市中的景观节点，并成为城市空间的一个区域性地标。立面设计力图在繁杂喧闹的城市背景下创造出一个清新典雅的建筑形象，具有装饰意味的铝合金建筑外表皮体现了天台历史悠久的文化传承和艺术品位，同时经过仔细推敲设计而成的外表皮系统又蕴含着一些传统的建筑元素，突出了天台市在佛教与道教发展历史中的重要地位，从而完成了传统与现代的完美结合。

Designing Brief

Building was located in the new districts of Taizhou main street in zhejiang. In the face of the type that is one at each other with a bite out of space, into the building points and give city a regional space. The facade is designed to the noisy cities on a complex background to create a clear and elegant decor. The image of the aluminium building outside surface of the day's long history of the cultural and artistic taste. Also been reviewed by the skin system and with some traditional architecture elements, which gives prominence of Tiantai city and development in the history of the importance that accomplishes a perfect combination of traditional and modern.

台州金字雷迪森国际大酒店

项目规模：建筑面积：49 850平方米

Golden Character of Leidison international Hotel inTaizhou

Project Scale: Building Area : 49,850 m²

曹妃甸新区市民服务中心及行政服务区

项目规模：用地面积：396 900平方米
建筑面积：100 000平方米

设计说明

营造富有中国传统特色的城市格局；

塑造充满活力、人性化的城市空间；

打造“城市地毯”上的城市客厅；

构造自然、绿色、可持续发展的城市系统。

The City Hall and Administration Service Center of Caofeidian District

Project Scale: Site Area: 396,900 m^2
Building Area: 100,000 m^2

Design Brief

To make a rich traditional Chinese feature of urbanism

Vibrant city and humanized space

Creating “City Carpet” aboving City parlor

Constructing natural, green and sustainable city system

豆各庄艾泽斯盖高科技园

项目规模：用地面积：18 320平方米
建筑面积：15 000平方米

设计说明

以办公建筑为主导体现低碳、绿色、技术的设计思路，实现园区整体的和谐，统一。将办公功能与辅助功能分区整合，实现实用功能与管线布局的合理搭配，节约能源。

Assky High-Tech Region

Project Scale: Site Area : 18,320 m^2
Building Area : 15,000 m^2

Design Brief

Office building shows as low-carb, green and technology design ideas, making the whole region as one. Separating the office building region, makes the real function and pipelines reasonable, in order to save energies.

天津塘沽西部新城社区服务中心

项目规模：用地面积：6 000平方米
建筑面积：6 750平方米

设计说明

以绿色、生态、生长、展示为平台，提出形态、功能、生态科技三位一体的处理办法。“蔓藤”是设计构思的主题，沿螺旋线向上生长建筑形式，象征着人对自然环境积极、自信的态度。

Service Center of the New West City in Tanggu, Tianjin

Project Scale: Site Area: 6,000 m^2
Building Area : 6,750 m^2

Design Brief

Using Green, Ecology, Growing and Expressing as the platform, coming up with the Shape, Function and Ecological Techniques as one. “The Plant” as the main idea of the scheme, growing up follows a sparrow path, symbolizes the confidence of the humanity to the environments.

北京大兴区北臧村镇九年一贯制学校

项目规模：用地面积：80 097平方米
建筑面积：32 000平方米
小学共有39班，中学共有27班

设计说明

以“书”为主要创意主题，体现学校特有的文化特征，突出为以下三大特点：

整体化设计：用连廊将所有建筑单体相联系，增强校园的整体感。

通过对建筑第五立面的处理，将书的形态演绎于建筑的造型设计之中，体现了“书院”的独特文化氛围。

独特的建筑立面：数字化的分格，将窗口模糊在色彩之中，给人以丰富的表情。

Nine Years in One School of Beizang Village in Daxing District of Beijing

Project Scale: Site Area: 80,097 m^2
Building Area: 32,000 m^2
Junior School: 39 Classes; Senior School: 27 Classes.

Designing Brief

"The Book" as the main idea of the scheme, shows the characteristics of school culture, with the most outstanding features.

Integrating Design: Integrating every single buildings with the connection corridors. Expressing the image of the books by creating the 5th elevation, coming up with the characteristic of a cultural space.

Unique façade: Split the shapes of windows within the façade making people feel its abundant expression.

四川省绵竹市土门镇中心小学及幼儿园、土门卫生院

项目规模：土门中心小学及幼儿园
用地面积：18 991平方米
建筑面积：10 091.44平方米，
其中小学总建筑面积：7 901.48平方米
幼儿园总建筑面积：2 189.96平方米
小学18班，幼儿园9班
土门卫生院
用地面积：11391.55平方米
建筑面积：5331.04平方米

设计说明

规划设计中，注重挖掘地方传统建筑空间策略和工艺技术，并思考在新的功能要求和现实条件限制下对其进行新的阐释；注重以简单适宜的设计手段贯彻可持续发展的理念；设计中还注重考虑时间和成本的因素，在保证建筑质量和追求建筑艺术价值的前提下，尽可能的缩短工期、降低造价。

Central Primary Schools, Kindergartens and Healthy Center in Tumen, Mianzhu City of Sichuan Province

Project Scale: **Central primary schools and kindergartens**
Site Area: 18,991 m^2
Building Area: 10,091.44 m^2
Primary school area: 7,901.48 m^2
Kindergartens area: 2,289.96 m^2
Primary 18 Classes Kindergarten 9 Classes
Healthy Center of Tumen
Site Area: 11 391.55 m^2
Building Area: 5 331.04 m^2

Designing Brief

In the project planning and design, we lay stress on the exploration of local traditional architectural space strategy and technique skills, considering the new interpretation under the new function demand and the practical condition. We implement the simple and suitable design method in the sustainable development idea. Ensuring the construction quality and architecture artistic value, we taking account of the time and coast factors in the design process, making effort in reducing the construction time and cost.

三河市新天地花园

项目规模：用地面积：308 670平方米
建筑面积：1 039 077方米

设计说明

以“水”为创意主线；
生态思想贯穿始终；
深入思考和解析“宜居”理念；
营造舒适的生活环境；
谋求人工环境与自然环境的和谐共生。

The New Garden of Sanhe City

Project Scale: Site Area: 308,670 m^2
Building Area: 1039,077 m^2

Design Brief

Using "water" as the main idea of the scheme
Ecological Design all the way through the project
Deep thought and analyzing "Suitable Living" theory
Creating comfortable living spaces
Combining natural and artificial environments together

新疆畜牧厅集资建房项目

项目规模：用地面积：88 000平方米
建筑面积：250 000平方米
总 户 数：1 849户

设计说明

项目为集资建房项目，规划容积率较高，业主为政府公务员，人群素质及居住品质较高。根据需求，规划与建筑形态追求“健康、生活、绿色宜居”，力求创造最佳的高品质生活环境，使其成为特色的新型社区。

Capital Formation of Constructing Buildings, The Department of Animal Husbandry of Xinjiang

Project Scale: Site Area: 88,000 m^2
Building Area: 250,000 m^2
Total Households: 1,849

Design Brief

It is a collaboration house building project with high bulk rate; the owners are government officials with high cultural and living quality demand. According to their requirement, we emphasize the healthy, living, green and livable design and architectural patterns, striving for creating the best high quality living environment and a new characteristic community.

Architectural

设计作品
Design Works

北京炎黄联合国际工程设计有限公司

Beijing Yanhuang United International Engineering Co.,Ltd.

北京炎黄联合国际工程设计有限公司是一家股份制设计公司，拥有建筑工程设计甲级资质及城市规划乙级资质。总部设于北京中关村南大街甲 18 号北京国际大厦 B 座 9 层及 12 层，设计人员近 200 人。公司为适应国家城市化进程和与日俱增的设计任务的需要，正在稳步发展壮大中，公司在天津、南京、合肥、厦门等地还设有分公司。

公司由我国资深建筑师刘永梁先生于 20 世纪 80 年代末创建，至今已有 20 余年的历史。2003 年公司改成股份制，改称为北京炎黄联合建筑设计有限公司。2010 年 3 月，与中外建建筑设计有限公司第一分公司合并重组，更名为北京炎黄联合国际工程设计有限公司（简称炎黄国际）。从此，这家创立了多年的设计企业开始了全新的历史时期，炎黄国际将以前所未有的活力和生命力去迎接新的挑战，为实现“建设全国一流设计公司”的宏伟目标而努力奋斗。

炎黄国际的业务范围包括各种建筑工程设计、室内装修设计、园林景观设计、城市规划和风景区旅游规划等多个领域，尤其是在高尔夫社区、酒店及商业建筑的设计上拥有自己专业的设计团队。公司专门设有高尔夫社区规划院和商业设计院，完成和正在设计的项目从住宅、别墅、公寓、学校、医院、商场、购物中心、体育馆到写字楼、宾馆酒店、度假中心、高尔夫社区，几乎囊括了所有民用建筑设计的领域，项目所在地点除北京外，从最北端的伊春、呼伦贝尔、哈尔滨到最南面的昆明、贵阳、三亚，遍及全国各地。公司还设有海外设计部，中东、北非的一些设计项目也正在运行中。

炎黄国际技术力量雄厚，专业配备齐全。它拥有一支业务精堪、经验丰富、骨干稳定、充满活力的设计团队。公司的宗旨是：用先进的设计理念和技术，用精益求精、一丝不苟的工作态度，为客户提供国际化、专业化、品牌化、本地化的设计服务，并为国家的经济建设和人民工作、生活环境的改善做出自己应有的贡献。

Beijing Yanhuang United International Engineering Co., Ltd. is a joint-stock agency, with Class A qualification in construction design and Class B qualification in urban planning. Headquartered in Beijing Zhongguancun South Street No.18, story 9 and 12, Block B, Beijing International Building, it owns nearly 200 designers. Catering to the national urbanization and the increasing design tasks, nowadays it is in steadily growing, besides, there are also a series of branch offices of the company in Tianjin, Nanjing, Hefei, Xiamen and other cities.

Created in the late 80s by a Chinese senior architect Mr. Liu, the company has been through 20 years of history. Since joint-stocked in 2003, it renamed into the Beijing Yanhuang Joint Architectural Design Co., Ltd. In March 2010, it conducts merge and restructuring with No.1 Branch of Beijing CCI Architectural Design Co., Ltd., and renamed as Beijing Yanhuang United International Engineering Co., Ltd. (Yanhuang International for short). thence, founded more than 20 years, this design company began a new history period of unprecedented vigor and vitality to meet new challenges, to strive for the grand goal of "building the first-class design company in China".

Yanhuang International serves for a variety of fields including building design, interior design, landscape design, urban planning and tourism planning in scenic areas, and its team particularly expertise in the design of golf community, hotels and commercial buildings. The company specializes in Golf Communities Planning and Business Design Institute, programs completed and under construction including almost all areas of civil construction like residence, villas, apartments, schools, hospitals, shopping malls, shopping centers, stadiums to office buildings, hotels, resorts, golf communities. Apart from Beijing, its programs locate from the northern end of Yichun, Hulunbeier, Harbin to the southern end Kunming, Guiyang, Sanya, throughout the country. The company also has overseas design departments, like some of the design programs in the Middle East, and North Africa are also running.

Strong technical force, equipped with complete professional, Yanhuang international has a professionally worthy, experienced, core stability, and dynamic design team. The company's mission is: to use advanced design concepts and technology, with excellence and meticulous work attitude, providing customers with international, professional, branded, localized design services, and make its due contribution for the country's economic construction and for the improvement of people's work and life environment.

地址： 北京市海淀区中关村南大街甲 18 号北京国际 B 座 9 层及 12 层
邮编： 100081
法人： 冈钢
电话： +86–10–62117746
传真： +86–10–62122704
邮箱： yanhuangguoji@vip.163.com
网址： www.yuie.com.cn

Address: Beijing International Building, Block B 9th Floor and 12th Floor,
No.18 of Zhongguancun South Street, Haidian District, Beijing
P.C.: 100081
Corporate: Gang Gang
Tel : +86-10-62117746
Fax : +86-10-62122704
E-mail: yanhuangguoji@vip.163.com
Http: //www.yuie.com.cn

北京炎黄联合国际工程设计有限公司
Beijing Yanhuang United International Engineering Co.,Ltd.

全国政协综合楼及恒奥中心写字楼

该项目位于北京市西城区，西二环以东，阜成门内大街以南的位置。是北京金融街区域东侧的最后一个项目。该建设用地是一个完整的街坊，此街坊北距阜内大街 115 米，距西二环约 500 米。建设用地面积约为 15 317 平方米，总建筑面积为 112 608 平方米。

建筑总体布局采用围合式，合理的将建筑划分为五个相对独立但又相互联系的部分，将较大的体量有序化处理，使其便于使用和管理，同时也将整个建筑充分的展示在外界面前。我们通过细腻的细部设计化解了建筑体量上的厚重感，丰富而不凌乱，并具有节奏感。建筑群整体感和体积感较强。高、低错落的建筑采用现代的设计手法，使用新的建筑材料，通过三种不同的建筑语言对其进行描绘，使其特点突出。建筑立面光影丰富而有韵律，变化中又有统一，既体现了其位于商业、金融业区段应有的现代建筑的特点，又传达出一种与古城环境相协调的文化气息。

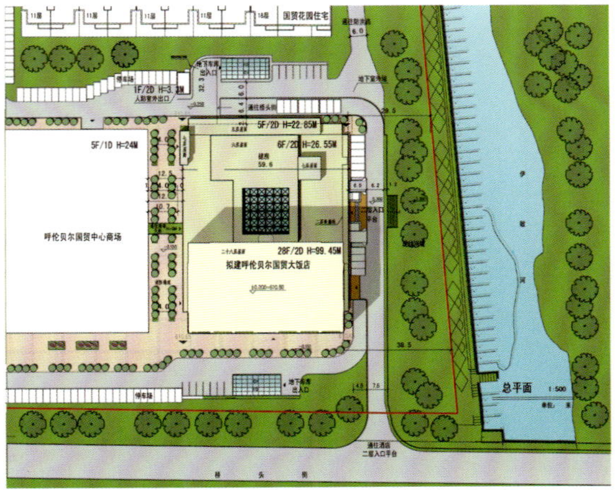

呼伦贝尔国贸大饭店

建设地点：内蒙古　呼伦贝尔
建设单位：呼伦贝尔炎黄世纪投资置业有限公司
总建筑面积：77 566.09 平方米
地上面积：63 889.12 平方米
地下面积：13 676.97 平方米
建筑层数：地上 28 层，地下 3 层

呼伦贝尔国贸大饭店位于该市中心中央桥西北处，建筑南面是主要大街，西接商业中心，东临美丽的伊敏河。此处交通方便、视野开阔、闹中取静，是建设酒店的绝佳地段。

国贸大饭店地下 3 层，地上 28 层，高 99.45 米，是该市目前最高的建筑。总建筑面积为 77 566 平方米。6 层以下为酒店，7 ~ 28 层为公寓，两部分功能各有单独出入口和垂直交通，既互不干扰又能互相连通，可以共享餐厅、商店、健身场馆等各项公共设施。

由于该市气候寒冷，为了节能，我们将平面布置的尽量方整，建筑立面不用玻璃幕墙，外墙除下面二层裙房采用石材外，全部采用浅色涂料，使建筑在呼伦贝尔特有的蓝天白云的映衬下，在清澈的伊敏河边，显得十分端庄大气。

北京炎黄联合国际工程设计有限公司
Beijing Yanhuang United International Engineering Co.,Ltd.

三亚美丽之冠七星级酒店

项目用地概述

本项目建设用地位于海南省三亚市区。其用地东接城市干道凤凰路，西依流经城区的临春河，南侧是举办“世界小姐”的竞赛场馆——“美丽之冠”文化会展中心，北侧是三亚市的名花公园。项目建设用地为112 072．27平方米。

规划设计概念方案说明

1．总平面地形的再造

在良好建筑布局的基础上，通过对地形的再造使室外形成逐渐升高的地势，如同“山状”的建筑物，有序的布置在“山”上，形成层次分明，具有特色的被绿色树林覆盖的酒店楼群。

“山”内拟建酒店配套的功能用房及附属用房，设置大面积城市服务性商业经营及旅游休闲娱乐场所，并设置为上述功能服务的大型停车库。

2．建筑设计理念、设计构思

我们设计的主题很明确——世界级标志性建筑

度假天堂 · 三亚

中国的三亚 · 世界的三亚

面向全球 · 七星级酒店

成为世界标志性建筑之一

设计构思

在三亚有着世界珍稀的植物——红树林。本项目用地内临春河畔就拥有规模极大的红树林，用地南侧则是曾承办过第53届世界小姐选美大赛的“美丽之冠”。自然环境和人文环境突出的优势都将促使本项目成为世界级别的建筑集群。

我们将建筑以大树的形态呈现出来，让它们静静的“长”在河边，与成群的白鹭为伴，与动人的红树林相映成趣，让人们童话般的梦想成为现实。就住在“大树”里吧！它的“果实”就成为了你的房间，它的“树肩”就是你最爱的泳池。你能在树根里吃饭，你可以站在树梢上看海……

其实我们没有设计，“大树”本应该生长在这里！

世界将因这棵“树”而精彩！

3．建筑单体外形的塑造

我们设计的目标是要建设世界一流的、美丽的、三亚的、服务设施齐全的、综合性的标志性建筑集群。让前来享受三亚的阳光、空气、海水、沙滩的人们过目不忘，分享建筑如同城市雕塑而传达出的愉悦之感，使人们对三亚的美好印象得到进一步的提升。

在多方案论证和探讨之后，我们隆重提出、绿色而美丽的“大树”方案，这个方案能够充分反映人们追求自然，追求环保的先进理念，同时还反映了三亚独特的地区特色。

北京炎黄联合国际工程设计有限公司
Beijing Yanhuang United International Engineering Co., Ltd.

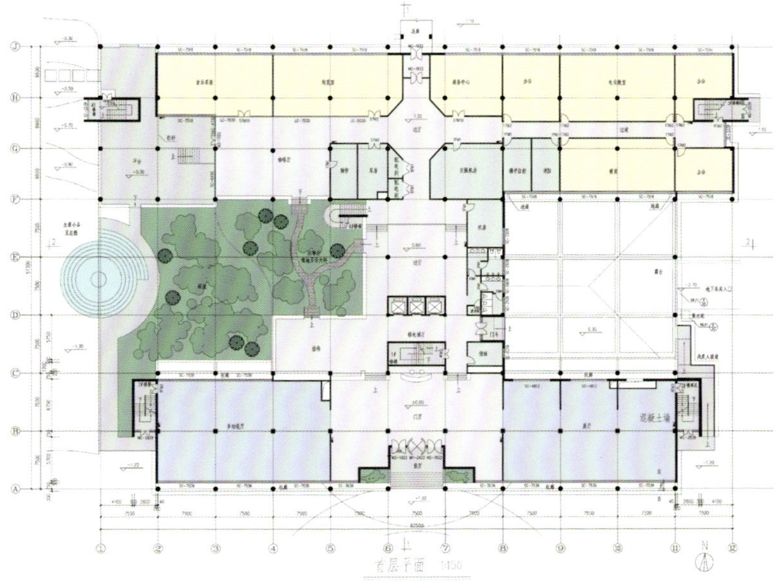

亿阳集团大连基地工程

设计时间：2000 年
总建筑面积：25 400 平方米

亿阳集团大连高新园位于大连西部经济开发区，靠山面海，环境优美。园区被一条道路一分为二，北部为研发办公及住宿，南部为厂房展示及销售。该项目最后只建成了北部的研发楼及住宿部分，建成后又转给了中石油下属的一家公司所用。

该项目地处半山坡，建筑随地形布置，分成南北两组，南侧为研发及办公，北面为宿舍、客房、餐厅，附设有带游泳池的健身中心，两组建筑在其二层有走廊相连。建筑外观简洁朴素，除入口部分和共享大厅西侧用了玻璃幕墙外，其余外墙全部为普通涂料，但由于在建筑的比例、尺度、色调、空间变化、绿化景观及细部处理等方面精心设计，使整组建筑显得新颖、活泼、充满活力，具有科技建筑的现代感。

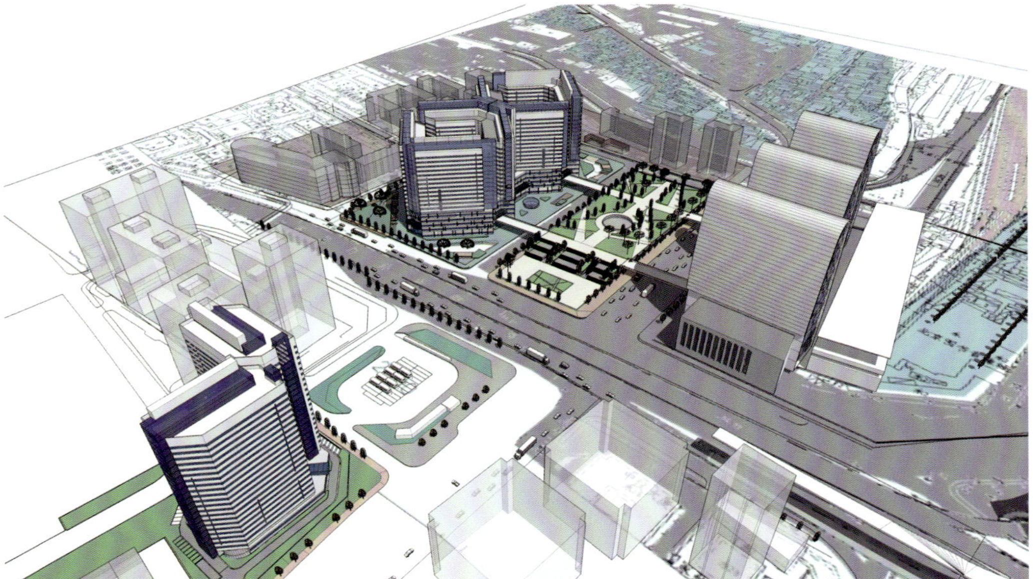

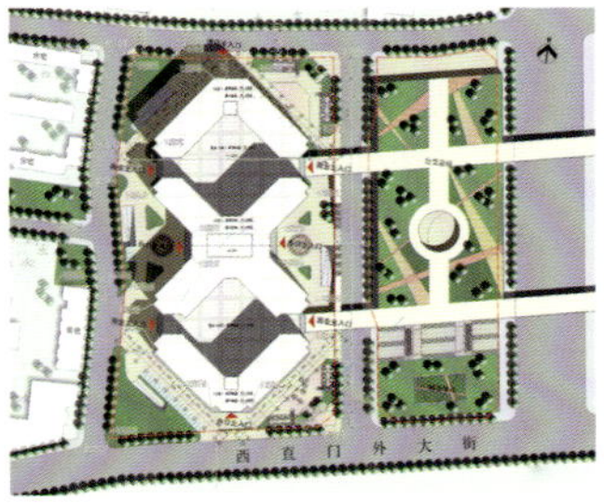

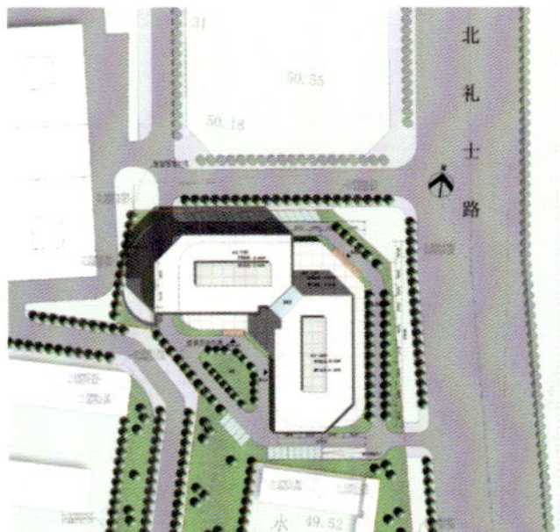

经济技术指标

2 号地：规划用地面积：50 537 平方米
建设用地面积：28 866 平方米
市政代征地面积：21 671 平方米
总建筑面积：288 026 平方米
地上建筑面积：199 474 平方米
地下建筑面积：88 552 平方米
建筑基底面积：10 189 平方米
容积率：6.91
建筑密度：34.4%
空地率：35.6%
建筑高度：27.9 ～ 94.4 米
停车位：1388 辆。（其中地下停车：1308 辆，地上停车位：80 辆）
绿化率：28%

7 号地：规划用地面积：20 769 平方米
建设用地面积：10 569 平方米
市政代征地面积：10 200 平方米
总建筑面积：105 010 平方米
地上建筑面积：78 210 平方米
地下建筑面积：26 800 平方米
容积率：7.43
建筑密度：35%
建筑高度：91.5 米

德宝二期危改 2 号、7 号地控制性详细规则调整方案

德宝二期危改 2 号地位于北京西直门外大街北侧，东临西直门交通枢纽西环广场，西至德宝饭店，北至北展路。规划建设用地 1.66 公顷，市政代征地 1.99 公顷，容积率 3.0，建筑控高为 45 米，地上可建房 5 万平方米。7 号地位于北京西外大街南侧城市广场以南，东起北礼士路，西至华正大厦及团结大院东墙，南至西外南路。占地 0.73 公顷，容积率 4.5，地上可建房 3.3 万平方米。

西直门地区作为北京市的一个重要的集商业、办公、交通、服务为一体的城市节点，是北京市中关村德胜科技园和西城区旅游商务功能街区的重要组成部分，具备良好的开发前景，结合该区域良好的民众认可度，综合各种经济、社会资源，根据西城区的发展战略，拟在该地区建立一个面向国际的旅游商务区。

这次**控制性详细规则**调整方案，力求通过规划设计和土地整合提高建设容量，节约城市土地及市政资源，实现区域价值，我们认为这一**控制性详细规则**调整方案是具有可实现性的。

地址：北京海淀区增光路 45 号综合楼 601
电话：+86–10–88560230
传真：+86–10–88560310
网址：www.aadoffice.com
Address: Room 601 of Comprehensive Building, No.45 of Zengguang Road, Haidian District, Beijing, China.
Tel: +86-10-88560230
Fax: +86-10-88560310
Http: //www.aadoffice.com

AA·都市建筑设计

金台夕照站站厅

双井站站厅

国贸站出入口

金台夕照出入口

国贸站站厅

Archi And design office

AA·都市建筑设计由刘弘（日本注册建筑师，加拿大安大略省建筑师协会会员）2003 年成立于北京，为外资设计事务所。

从成立初期承接的北京总部基地规划设计到近期的地铁系列空间设计，AA·都市建筑设计都把独特的设计理念和有效的设计手法相结合，通过融合规划、建筑、室内、景观的一体化设计，为人们提供具有时代精神、高品质的空间场所。

Archi And Design Office was founded in Beijing 2003 by Liu Hong who is a registered architect in Japan and the member of Canada Ontario Association of Architects.

From the early days of undertaking Beijing Advanced Business Park Master Planning to the recent series of subway space design, Archi And Design Office combined the unique concept and effective design method together through the integrated design of planning, architecture, interior and landscape design which focus on providing a modern and high-quality space for people.

大兴韩园子站综合体

崇文门站出入口

次渠站站厅

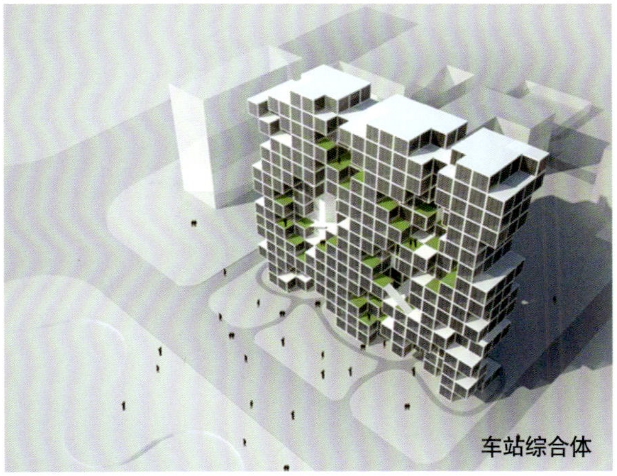
车站综合体

车站综合体

金台夕照站出入口

地铁都市

北京地铁 5 号线出入口设计总数共 51 个	竣工时间：2007 年
北京地铁 10 号线出入口设计总数共 68 个	竣工时间：2008 年
北京地铁 10 号线车站公共空间概念设计共 20 座	竣工时间：2008 年
北京地铁亦庄线车站公共空间概念设计共 13 座	竣工时间：2010 年

Subway City

The total number of Beijing Subway Line 5 entrance building design projects is 51
Completion: 2007

The total number of Beijing Subway Line 10 entrance building design projects is 68
Completion: 2008

The conceptual design projects of public space for Beijing Subway Line 10 is 20
Completion: 2008

The conceptual design projects of public space for Yizhuang Subway Station is 13
Completion: 2010

OAD
Office for Architecture & Design
欧安地

1

2

美国 OAD 欧安地建筑设计事务所

● **背景**　美国 OAD 欧安地建筑设计事务所是成立于美国新泽西州的国际化建筑师事务所，主要合伙人由美国注册建筑师及美籍华人建筑师组成，他们具有多年在美国规划、建筑及室内设计领域的工作经验，曾获美国建筑师协会 AIA 奖项和建筑师国际论坛 ATF 建筑事业奖。

公司凭借国际化的设计理念及对中国建筑技术的熟知，OAD 多次在国际及国内设计竞赛获得第一名。自 2003 年设立北京分公司至今，OAD 在北京、天津、成都等城市已经完成了涉及高级居住区、城市综合体、商业地产、酒店建筑、文化教育建筑等不同类型及规模的多项出色作品。其中，北京富力丹麦小镇、成都花水湾度假酒店项目分均得 2009 年联合国人居奖。

OAD 服务的客户包括富力集团、世贸集团、保利集团、中铁二局、大商集团等中国著名的房地产公司，同时 OAD 也服务于北京交通大学、中科院等中国的科研教育机构。

● **设计哲学**　OAD 强调从城市和环境的角度理解建筑，以创新作为设计的灵魂，提倡生态、人性化的空间设计理念，积极面对每项工程的特殊的限定条件及要求的挑战，提供具有创造性的卓越的解决方案。

● **使命**　OAD 的使命是提供高品质的设计和专业服务，以协助客户实现他们的目标，创造更美好的生活环境。

主要设计师

李颖悟
国家一级注册建筑师
美国新泽西理工大学建筑学硕士
清华大学建筑学学士

沈辉辉
美国注册建筑师
美国华盛顿大学建筑学硕士
清华大学建筑学学士

Helder Santos
美国新泽西理工大学 NJIT 建筑学学士、硕士
美国注册建筑师

中国：北京市朝阳区建国路 89 号华贸中心 15 号楼 501 室
邮编：100025
Tel: +86-10-65305785/5786/5753
Fax: +86-10-65305752
Http: //www.oadgroup.com
E-mail: office@oadgroup.com

3

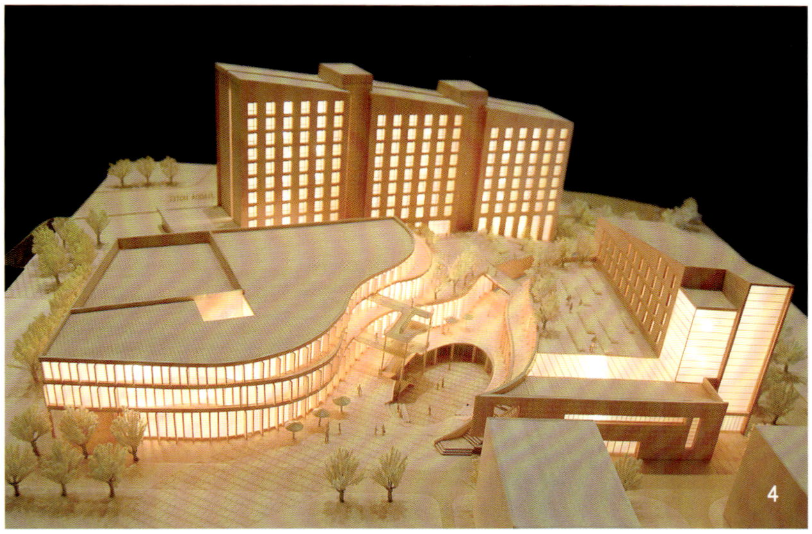

4

5

1. 方恒国际中心
2. 成都大邑五星级酒店
3. 成都花水湾度假小镇度假酒店及会议中心
4. 北京交通大学学生活动服务中心
5. 成都百伦酒店式公寓
6. 中国科学院微电子中心研究楼
7. 富丽丹麦小镇别墅区

6

7

北京昂众设计事务所
ANG ATELIER

昂众设计事务所（ANG ATELIER）是一个年轻、充满活力、对设计创作充满激情的团队。团队的成员都具有极佳的专业精神和创新精神，并有着很强的服务意识，团队由规划、建筑、环境艺术、园林、结构、水、电等多个专业的具有丰富经验的设计人员组成，是从事规划、建筑、景观设计的综合事务所，能够从整体上把控每一个项目。团队从概念方案到立项实施全程跟踪，力求把每个项目真正落到实处，由精至美。

昂众设计多次与融科智地房地产开发有限公司、无锡融创地产有限公司、大同阳光嘉业房地产发展有限责任公司、志高实业集团、天津现代投资有限责任公司、天津房地产发展（集团）有限公司、山东山海天城建集团、华远地产、光大集团等地产公司合作，并与中国规划设计研究院、中国建筑设计研究院、何显毅建筑工程师楼（中国）有限公司等国内外著名设计机构共同完成从规划、建筑到景观的设计工作。公司以景观设计见长并在多种景观类型上均有良好业绩，业务涉及风景旅游度假区景观设计、城市公园景观设计、河道景观设计、街道景观设计、城市广场景观设计、商业办公建筑景观设计、大型综合居住区景观设计（新古典风格、美式风格、新中式风格等）、高档低密度社区（高档别墅、联排别墅）景观设计等。

团队精神

时尚生活、激情创作、崇尚自然、精美实践。

团队口号

用智慧融化激情，用精美打造作品。

设计理念

以遵循自然与人的和谐为前提，运用大道无为的手法，创造出文化、时尚、精致、前卫的作品。

（1）生态性原则：强调设计遵从自然。自然的空间格局，自然的生态过程。
（2）文化性原则：强调人文景观的再创造。
（3）科技性原则：借助科技的手段和先进的设备，创造高科技的新景观特色。
（4）地方性原则：充分利用地方材料，符合本地区的风格进行设计。
（5）相宜性原则：建立人与空间，人与水和植物，人与人，植物与水等的亲切互利的关系。
（6）经济性原则：强调用经济、实用、节约这一原则，构筑亲切宜人的空间。
（7）现代性原则：体现时代特征、构思新颖的现代社区景观。
（8）国际化原则：按照国际化标准提供各种标准和服务。

地址：北京市西城区车公庄大街 9 号院五栋大楼 B 座 1 单元 1002 室
邮编：100044　　联系人：徐刚
电话：+86–10–88312656　　传真：+86–10–88312856
邮箱：ang_bj@126.com　　网址：www.angbj.com.cn

This team is youthful and vigorous, and its members are professional and innovative. They have enthusiasm towards design and strong consciousness for providing service. They major in various fields including design, building, environment art, park design, building structure, water resources and electric engineer, so they compose a comprehensive institute whose business covers planning, building and landscape design. This institute takes a project as a whole, and follows the process from conceptual design to project determination. When they finally put into practice, they endeavor to put forward best projects and elaborate works in beauty.

ANG ATELIER has launched cooperation with many companies including Rongke Zhidi Real Estate Development Co., Ltd., Wuxi Rongchuang Real Estate Co., Ltd., Datong Yangguang Jiaye Real Estate Development Co., Ltd., Zhigao Industry Group Corp, Tianjin Xiandai Investment Co., Ltd., Tianjin Real Estate Development Group Co., Ltd., Shangdong Shanhaitian Building Group Co., Ltd., Huayuan Real Estate and Guangda Group Corp. This company, together with many well-known design companies, at home and abroad, including China Planning & Design Research Institute, China Building & Design Research Institute and Hexianyi Institute of Architects Co., Ltd. (China Branch) has completed designing, building and landscaping for many projects. This company is good at landscape design. It has made great achievements on various landscape designs in various places including resort districts, urban parks, rivers, streets, urban squares, commercial office buildings, large residential site (in New Classic, American and New Chinese styles) and low-density residential site in high-quality (top grade villa and town house)

Team Spirit

Life is fashion. Creation is Passion. Nature is foundation. Work is elaboration.

Team Slogans

We will use wisdom to melt passion. Our goal is to bring you an exquisite masterpiece.

Design Concept

Based on the harmonious coexistence between nature and human, we will do everything to protect nature and create fashionable, exquisite works closely related with culture.

(1) Ecologic Concept: Natural is a key point. We will try our best to make good use of natural space and respect the ecologic process of nature.
(2) Cultural Concept: We emphasize the recreation of man-made landscape.
(3) Scientific Concept: We will apply scientific methods and advanced equipments to create high-tech landscape with new character.
(4) Regional Concept: We will make good use of new material to create a style in conformity with regional culture.
(5) Harmonious Concept: We will establish a mutually beneficial relationship between human and space, human and water, plant, human and human and plant and water etc.
(6) Economic Concept: We emphasize that we should do things economically and practically, through proper material, we try to create nice and pleasant space.
(7) Modern Concept: Special-designed landscape will embody modern spirit.
(8) International Concept: We will provide all kinds of services according to international standards.

Address: Room 1002, Unit 1, Tower B ,Building 5, No.9 of Chegongzhuang Street, Xicheng District, Beijing
Post code: 100044　　Contact: Gang Xu
Tel: +86-10-88312656　　Fax: +86-10-88312856
Email: ang_bj@126.com　　Http: //www.angbj.com.cn

1

2

1–2
委托方：无锡融创地产有限公司
项目名称：天鹅湖 B1&B2 景观设计
项目规模：景观 63 000 平方米 +75 000 平方米
项目类型：高层住宅、双拼及联排别墅、商业街
项目地点：江苏 无锡
项目特色：回归自然的亲和主义

Consignor: Rongchuang Real Estate in Wuxi
Project Name: Landscape design of Swan Lake B1 & B2
Scale Area: 63,000 m^2 + 75,000 m^2 landscape
Project Types: High-rise residence, semi-detached house, townhouse and commercial streets
Construction Location: Wuxi, Jiangsu
Project Feature: The affinity brought by returning to nature

3–4
委托方： 华远地产有限公司
项目名称：静林湾景观设计
项目规模：景观面积 25 000 平方米
项目类型：TOWNHOUSE
项目地点：中国 北京
项目特色：原生林、溪、滩相映成趣的古朴气质

Consignor: Huayuan Real Estate
Project Name: Environmental landscape design of Jinglinwan
Scale Area: 25,000 m^2 landscape
Project Types: townhouse
Construction Location: Beijing, China
Project Featrue: The original manner set off by natural forest, brook and beach

5–6
委托方： 融科智地房地产开发有限公司
烟台开发区产业联合发展有限公司
项目名称：烟台融科林语景观设计
项目规模：景观 45 000 平方米
项目类型：高层住宅
项目地点：山东 烟台
项目特色：精致的新古典自然风格

Consignor: Raycom Real Estate Development Co.,Ltd.
Yantai Development Zone United Development Co., Ltd.
Project Name: Landscape project of Rongkelinyu in Yantai
Scale Area: 45,000 m^2 landscape
Project Types: High-rise residence
Construction Location: Yantai Shangdong
Project Featrue: Elaborate new classic style

7–8
委托方： 天津房地产发展（集团）有限公司
项目名称：天房晴川家园景观设计
项目规模：景观面积 56 000 平方米
项目类型：多层和高层住宅
项目地点：中国 天津
项目特色：清新雅致、舒适浪漫的英式园林

Consignor: Tianjin Reality Development(Group)Co., Ltd.
Project Name: Landscape project of Tianfangqingchuan garden
Scale Area: 56,000 m^2 landscape
Project Types: Multi-storey and high-Rise residence
Construction Location: Tianjin, China
Project Feature: Elegant and fresh, comfortable and romantic parks in British style

9
委托方： 大同市阳光嘉业房地产开发有限责任公司
项目名称：山西大同凯德世家项目景观设计
项目规模：景观 80 000 平方米
项目类型：高层住宅
项目地点：山西 大同
项目特色：用新古典主义的手法描绘一幅森林溪谷、花海跌瀑、山水印象的图画

Consignor : Yangguangjiaye real estate in Datong
Project Name: Landscape project of Dekaishijia in Datong, Shanxi
Scale Area: 80,000 m^2 landscape
Project Types: High-rise residence
Construction Location: Datong, Shanxi
Project Feature: A picture of forest, brook, valley, flowers, waterfall, mountain and rives dawn by a new classic method

10–11
委托方： 泰安志高实业集团有限责任公司
项目名称：泰安志高国际项目景观设计
项目规模：景观 72 000 平方米
项目类型：高层住宅及商业建筑
项目地点：山东 泰安
项目特色：溪林谷地，北欧印象

Consignor: Zhigao group in Taian
Project Name: Zhigao group international landscape projects
Scale Area: 72,000 m^2 landscape
Project Types: High-rise residence, commercial buildings
Construction Location: Tai'an, Shandong
Project Feature: The valley with brook and forest in west European style

奥兰 archland ARCHITECTS AND ENGINEERS

土地规划 | 城市设计 | 建筑设计 | 室内设计 | 项目设计管理

奥兰是一家具有国家甲级工程设计资质的综合性建筑工程设计事务所，公司拥有近百名国内外设计精英，建筑工程作品遍布各地。公司业务范围涵盖办公、工业、公建、教学、酒店、居住、商业、室内、体育、医疗等建筑类别，可提供从项目概念策划、规划设计、建筑方案、施工图及其他配套专业设计到项目管理的全程一体化的设计服务。

Archland is a well-established architectural design and engineering company consisting of over 100 international and domestic world-class designers with architeetural design works all over the world. Archland design capabilities include but are not limited to municipal, industrial, commercial, residential, institutional and other facilities. Archland offers comprehensive services including conceptual planning, design, on site service, and other construction professional service through all phases of project management.

地址：北京市海淀区首体南路 38 号创景大厦 4 层
邮编：100037
电话：+86–10–58892158
传真：+86–10–58892160
邮箱：market@archland.cn market@onstarc.com
网址：www.archland.cn，Http：//www.onstarc.com

项目名称：厦门原石滩国际社区
项目地点：福建 龙海
占地面积：100 公顷
设计类型：规划设计、酒店设计、居住区设计

土地规划 | 城市设计 | 建筑设计 | 室内设计 | 项目设计管理

archland
ARCHITECTS AND ENGINEERS

项目名称：定鼎城
项目地点：河南 洛阳
占地面积：63.63 公顷
建筑面积：1 400 700 平方米
设计类型：住宅区

项目名称：平湖三港新城
项目地点：浙江 平湖
占地面积：90 公顷
设计类型：新城规划设计

土地规划 | 城市设计 | 建筑设计 | 室内设计 | 项目设计管理

archland
ARCHITECTS AND ENGINEERS

项目名称：深圳市中心区水晶岛
项目地点：深圳
占地面积：45 公顷
设计类型：城市公共空间设计

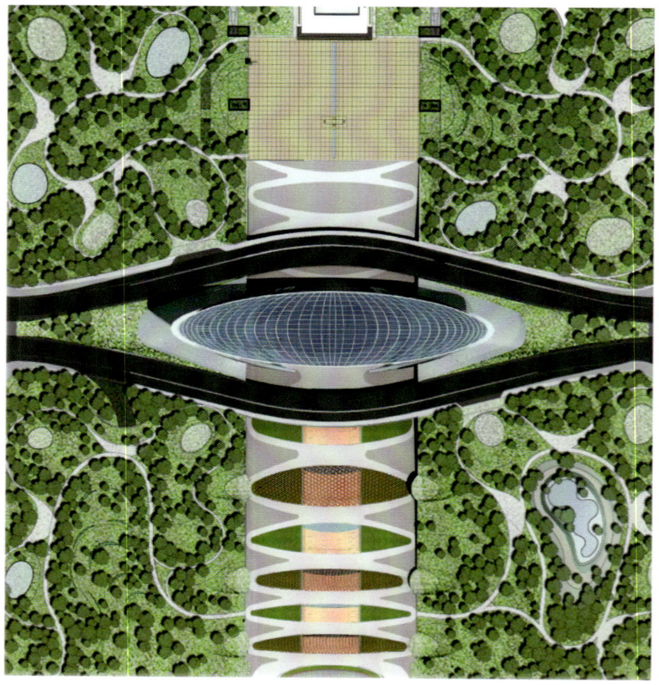

土地规划 | 城市设计 | 建筑设计 | 室内设计 | 项目设计管理

archland
ARCHITECTS AND ENGINEERS

项目名称：金陵紫清湖温泉度假区
项目地点：江苏 南京
占地面积：80 公顷
建筑面积：3 700 平方米
设计类型：文化建筑、博物馆设计

项目名称：日照阳光海岸广场
项目地点：山东 日照
占地面积：78 公顷
设计类型：城市公共空间

archland
ARCHITECTS AND ENGINEERS

土地规划 | 城市设计 | 建筑设计 | 室内设计 | 项目设计管理

项目名称：西安秦岭老年公寓
项目地点：陕西 西安
占地面积：6.67 公顷
建筑面积：15 000 平方米
设计类型：居住区

项目名称：南昌国际温泉城
项目地点：江西 南昌
占地面积：100 公顷
建筑面积：44 328 平方米
设计类型：温泉度假酒店

archland ARCHITECTS AND ENGINEERS

土地规划 | 城市设计 | 建筑设计 | 室内设计 | 项目设计管理

项目名称：丰台旧厂房改造——1978 文化艺术产业园
项目地点：北京
用地面积：17 133 平方米
建筑面积：17 114 平方米
设计类型：文化艺术产业园、旧场房改造

项目名称：1949（盈科商务会所）
项目地点：北京
占地面积：3 910 平方米
建筑面积：1 804 平方米
设计类型：高档商务会所、旧场房改造

北京城景联合建筑设计机构

Beijing Landscape Joint Architectural Design Agency

地址：北京市东城区新中西街2号新中大厦11A
电话：+86-10-65006971
传真：+86-10-65074578
邮箱：Zhao.yang20080224@163.com

Address: No.11A, Xinzhong Mansion, No.2 Xinzhong West Street, Dongcheng District, Beijing
Tel: +86-10-65006971
Fax: +86-10-65074578
E-mail: Zhao.yang20080224@163.com

北京城景联合建筑设计机构中国区共有设计师30余人，建筑设计核心团队之间有多年的项目配合经验，曾以全新的房地产设计理念和优秀的设计意识服务于万科、旭辉、亿城、昂展投资、同景等知名品牌开发商，长期的合作积累了丰富的设计经验和市场营销意识，并获得国内多项设计大奖。同时，公司作为国际联合设计机构，在拥有国外先进设计理念的同时又做到了从方案设计到单专业扩出设计的一体化和标准化建设，为客户提供了更优质的服务。

业务范围

城市设计、建筑设计、房地产前期策划咨询、景观设计、室内设计。

工作范围

概念设计、方案设计、施工图设计、技术规范、设计策划咨询、工程管理。

服务标准

能敏锐感知客户的需求；
协助客户追求卓越的市场业绩；
鼓励创新，致力于专业领域的探索；
不断提升面对复杂问题能寻找多种解决方案的专业能力；
力求综合各种专业知识，让其产生最现实的社会效益；
一直在为客户创造超越客户期望的产品和服务而努力。

设计理念

设计体现出地域性文化以及生态延续的概念，每一个作品就是一个宣言，为城市建筑的发展提出自己的独特见解。

权威媒体曾赞扬他设计的建筑象从地上长出来一样，并认为他的设计风格高雅，细腻，与环境十分协调。

Beijing Landscape Joint Architectural Design Agency China Branch has more than 30 designers in all. Its core teams of architectural design have many years' experience in project coordination. With its brand-new design concept in real estate and excellent design consciousness, this company has provided services for many famous brand developers, such as Vanke, CIFI, Yeland, Angzhan Investment and Verakin, etc. Through the long-term cooperation, it has accumulated rich experience in the design and marketing consciousness, and has won many design awards at home and abroad. Meanwhile, as an international joint design agency, the company does not only render its design teams the international advanced design concept, but also achieves the integration and standardization from the program design to the single professional expansion design, providing better services for the customers.

Business Scope

Urban design; architectural design; real-estate pre-planning consultation; landscape design; interior design.

Working Scope

Conceptual design; program design; construction drawing design; technical specification; design and planning consulting; project management.

Service Standards

Be sensitive to the customers' demand;
Help customers pursue the excellence of market performance;
Encourage innovation and devote to professional exploration;
Continuously improve the professional ability in seeking various solutions in the face of complicated problems;
Strive to integrate all kinds of professional knowledge in order to produce the most realistic social benefits;
Make constant efforts to provide the customers with products and services exceeding their expectations.

Design Concept

The design reflects the regional culture and ecological continuation; each design is like a manifesto, presenting a unique understanding for the development of urban architecture.

The media authority comments that buildings of his design seem like growing out of ground; his design is elegant and exquisite in style, and coordinates perfectly with the surrounding environment.

南昌万科润园

Vanke Run Yuan in Nanchang

北京 8 哩岛

用地面积：66 486.47 平方米
容 积 率：1.6
建筑限高：30 米
建筑面积：143 062.55 平方米
地上面积：106 378.35 平方米
地下面积：36 684.2 平方米

Beijing Bali Island

Site Area: 66,486.47 m^2
Plot Ratio: 1.6
Height Limit: 30 m
Building Area: 143,062.55 m^2
Floor Area on the Ground: 106,378.35 m^2
Floor Area Underground: 36,684.2 m^2

项目特点

项目用地属于泛 CDB 板块项目，所以业主定位为通州本区域业主和 CBD 的白领人员，因此将立面风格定位为现代风格，并采用现代 arcdec 风格。项目毗邻温榆河，而项目又有限高要求，所以我们把沿河高层定义为“灯塔”意向。为了克服形体的弊端，我们参考了大量竖线条的案例，决定将建筑定位为现代哥特式风格。

Characteristics of the project

The project site belongs to the general CDB plate, so the owners are located on those in Tongzhou region and the white-collar staffs of CBD. The façade design is positioned on the modern style, but it also has to be situational, so I think the modern arcdec style is appropriate. This project is adjacent to Wenyu River and has height limit, thus we define the high-rise building along the river as the "lighthouse" image. In order to overcome the squatness in form, we referred to a large number of vertical-line cases, and eventually decided to design it with the modern Gothic style.

方案改造前

方案改造后

方案改造后

方案改造后

旭辉大兴 10 号地块立面改造

原方案的前
1. 屋顶死板缺乏层次。
2. 中间段落单薄，缺乏分量感。
3. 阳台等充满情趣的地方处理简陋。
4. 缺乏细节，没有品质感。

原方案的后
1. 对屋顶层次重新输理，使建筑舒展而气势磅礴，体现大宅感。
2. 顶部层次错落有致，不再呆板。
3. 对开窗重新梳理，增强建筑的厚重感。
4. 对人在阳台的行为重新梳理，使阳台进深变化有致。栏板虚实结合。

Facade Reconstruction of CIFI Block 10 #, Xuhui Daxing

Before the original program
1.The roof looks rigid and lacks hierarchy;
2.The middle part looks flimsy and lacks a sense of weight;
3.The treatment to places of fun, such as the balcony, is crude;
4.The project is deficient in detail treatment, possessing no sense of quality.

After the original program
1.Reprocess the roof hierarchy. Thus the building stretches itself and demonstrates its magnificence, reflecting a sense of mansion;
2.The building top is no longer rigid, exhibiting a well-organized sense of level;
3. Reprocess the windowing and display a sense of weight;
4. Reconsider people's behavior on the balcony and show an orderly change in the depth of the balcony.
The treatment of the fence adopts a “virtual-actual combination”.

北京工业大学建筑勘察设计院

地址：北京市朝阳区平乐园 100 号
邮编：100124
法人：王奎仁
电话：+86–10–67392327
传真：+86–10–67391808
邮件：bgdsjy@yahoo.com.cn

北京工业大学建筑勘察设计院是集建筑设计、勘察设计、科研、咨询于一体，拥有建筑设计甲级和勘察设计乙级资质的设计院。设计院走产、学、研一体化发展的道路，通过不断努力开创进取，逐步形成了自己的技术特色。近年来在学校、科研楼、图书馆、博物馆等文化建筑以及体育建筑设计上多有建树并受到用户好评。

Institute of Architectural and Exploratory Design of Beijing University of Technology, is a comprehensive institute of architectural and exploratory design, research and consultation, with class A qualification of architectural design and B-level qualification of exploratory design. The design institute has its own technical characteristics by efforts on the integration developing road of production, study and research. In recent years, there are achievements of cultural building design and gymnasium design, such as school, library, and museum, which always got favorable comments from the users.

龙潭公园中心岛露天文化广场

用地面积：3 800.00 平方米
建筑面积：3 236.67 平方米

该项目是为迎接北京 2008 年奥运会所建的室外剧场的代表项目之一，项目坐落于风光秀丽的北京龙潭公园内中心岛上，具有可开闭式活动屋顶，具备了可以举办露天和室内文体活动的双重功能，项目建成时该柔性（膜）结构屋顶开合结构在北京尚属首例。

华北水利水电学院新校区一期学生生活区

用地面积：77 496.05 平方米
建筑面积：95 717.65 平方米

本项目位于郑州市郑东新区大学城，一期学生生活区为 8 500 名本科生使用，主要建设内容包括学生宿舍、食堂、浴室、学生活动中心等。本项目的特色是较好的组织了建筑的组群关系，将食堂、学生活动中心作为生活区的中心景点，利用屋顶、天桥、钟楼、平台等建筑手段突出了建筑组群的标志性特征。

北京工业大学第三教学楼

用地面积：14 025.00 平方米
建筑面积：21 259.51 平方米

项目平面布局采用三个相同的建筑组群为排列方式，既有韵律又便于人流相对独立的疏散。首层平面设一条贯穿建筑东西的内部街道，内外空间相互交融。建筑形象处理上力求做到简洁、完整，突出建筑的空间美，避免不必要的琐碎和做作，外部空间变化丰富，虚实对比强烈。

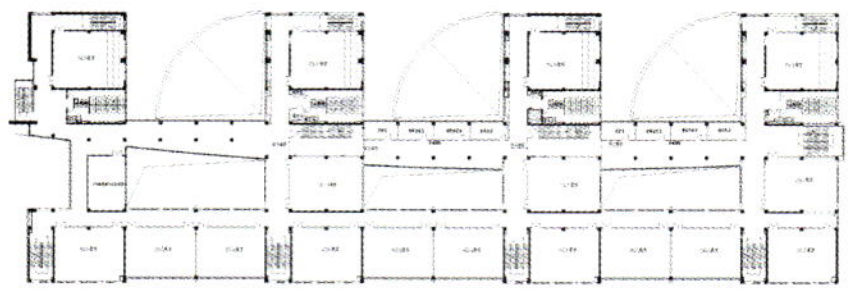

建筑主要平面图

北京市东城区史家小学通州分校

用地面积：61 001.03 平方米
建筑面积：30 629.30 平方米

该项目是一所全新的48班小学，建设内容包括教学楼、体育馆、游泳馆、食堂、宿舍楼、运动场等全套设施。设计中采用了地源热泵、智能节能控制等先进的节能环保措施。建筑布局紧凑明晰，风格简洁明快，通过对楼梯间、主入口、大门、过渡空间的强化处理，突出了活泼的符合小学特点的建筑氛围。

石景山区图书馆

用地面积：5 402.80 平方米
建筑面积：9 044.62 平方米

石景山区图书馆位于一个转角地块，用地紧张，规划条件限制也比较大，它的斜对面是一个公园。设计通过建筑体块的穿插和虚实对比突出了建筑的文化特性，主入口强调了与公园的对话。室内空间处理上通过带状中庭空间，给各阅览室引入自然采光，创造了一个明亮、宁静、通透的图书馆建筑。

HENO
慧诺国际建筑咨询有限公司

Taipu Academic Medical Research Institute

北京泰普医疗科研楼（图 1 2 4 7）

北京泰普医疗单位是以“两癌”筛查为主的科研医疗机构，受其委托，我公司承接了该单位老楼建筑及外立面改造工作，该项目将于 2011 年竣工。

Taipu Academic Medical Research institute focus on doing the medical research of cancer cases. Commisioned by the client, HENO focus on making a brilliant design to promote the rebuild of the construction. The construction will be finished in 2011.

1 | 2
3 | 4

Kangbashi Police Office

鄂尔多斯康巴什公安大楼（图 5 6）

慧诺国际通过竞标的方式，以第一的成绩赢得了鄂尔多斯康巴士公安大楼项目。项目建筑占地为 2.8 公顷，建筑规模为 41 900 平方米，容积率为 1.48。

HENO won the bid of Kangbashi Police Office project. The building covers 2.8 hector, total floor area is 41,900 m^2, and plot ratio is 1.48.

慧诺国际是一家集城市规划、建筑设计、景观设计、平面设计于一体的综合设计单位，本部注册地在澳大利亚。作为首次进入中国的中方独资、外方技术入股企业，慧诺国际在中国要真正做到的是，引入先进的设计理念，高质量的设计水准和优质的设计服务到中国规划建筑市场，为国内的规划与建筑设计提供一流的设计服务。慧诺人将用自身的设计智慧向我们的服务方，呈献高水平、高质量的设计作品。

地址：北京市海淀区三虎桥南路 17 号院 6 号楼 211 室
电话：+86-10-88415541
传真：+86-10-88415541-11
邮箱：mail@henodesign.com   henodesign@yeah.net

ONE
慧诺国际建筑咨询有限公司

Kangbashi Comprehensive Welfare House

鄂尔多斯综合福利院（图 3）

慧诺国际通过竞标的方式，以第一的成绩赢得了鄂尔多斯综合福利院项目。项目建筑占地为 2 公顷，建筑规模为 13 500 平方米（含 3 000 平方米地下室），容积率为 0.5。

HENO won the bid of Kangbashi Comprehensive Welfare House project . The building covers 2 hector, total floor area is 13,500 m^2, and plot ratio is 0.5.

5 | 6
7 | 8

Xueye Lake Tourism Real-estate Planing

雪野湖旅游地产规划（图 8）

慧诺国际在大型城市设计与旅游地产方面同样具有强大的优势。受居易集团委托，承接了山东莱芜雪野湖旅游地产规划项目。该项目占地 233 公顷，建设总量为 700 000 平方米，容积率为 0.3。

HENO has great experience on doing central area urban design and tourism real-estate planning. Commisioned by Juyi Group, client HENO is responsible for the project of Xueye Lake Tourism Real-estate Planning. The project covers 233 ha, total floor area is 700, 000 m^2, and plot ratio is 0.3.

HENO International co. ltd. Which domain reserved in Australia, is a comprehensive design branch combined urban planning, architecture design, landscape design and VI and CI design. HENO International, which was setup by Chinese and collaborated with foreign technical support, brings latest design theories, high standard design quality and best design services to Chinese market to promise to our clients, HENO is your best choice!

Address: Room 211, Building 6, Sanhu qiao nan road No.17, Haidian District
Tel: +86-10-88415541
Fax: +86-10-88415541-11
E-mail: mail@henodesign.com　　henodesign@yeah.net

北京建院约翰马丁国际建筑设计有限公司

John Martin International Architecture Design Co.,Ltd of BIAD

法人：李海南
地址：北京市南礼士路66号
邮编：100045
电话：+86-10-68080265
传真：+86-10-68086058

Legal person: Hainan Li
Address: No. 66 Nan Lishi Road, Beijing
P.C.: 100045
Tel: +86-10-68080265
Fax: +86-10-68086058

内蒙古盛乐别墅区

北京建院约翰马丁国际建筑设计有限公司(原北京金田建筑设计有限公司 Origional BEIJING GOLDEN EARTH ARCHITECTURAL DESIGN)是由北京市建筑设计研究院控股，经中华人民共和国商务部批准，具备建设部与北京市颁发的甲级设计资质证书的中外合作企业。公司成立于1994年，工程设计证书编号：010929-sj。

随着企业改革的不断深入，北京建院约翰马丁国际建筑设计有限公司逐渐由事业单位转变为事业单位企业化管理单位。在这50年中，北京建院约翰马丁国际建筑设计有限公司始终致力于向社会提供高品质的设计服务，在全国民用设计单位中名列前茅，在行业中享有极高的声誉，并逐渐形成了“建筑服务社会的企业核心理念”。为了应对新世纪之初设计行业面临的严峻挑战，通过整合全院乃至国内外优秀设计资源，逐步提升设计作品的完成度，北京市设计研究院决定将北京建院约翰马丁国际建筑设计有限公司在2008年初再一次整合，并作为BIAD的改革先锋将单位推向市场。

公司经营范围有：承接与组织国内外各种建筑工程设计、室内外装饰工程设计、城市规划及小区规划；开展与建筑设计、室内外装饰设计、小区规划等有关的咨询业务与技术服务。

公司的宗旨是通过在中国及世界各地的经营业务，以优质的设计服务和信誉广泛地开拓与建筑设计、室内设计、城市规划相关的业务项目，促进中外建筑技术交流，发挥各自优势，致力于建筑创作、人才培训、探索设计方法和经营机制的改革。

公司实行董事会领导下的总经理负责制，董事会为公司最高权力机构。公司设总经理一人，副总经理一人，由总经理召集并主持总经理会议，讨论执行董事会决议，共同负责公司业务。

公司拥有老中青结合的基本设计队伍，其中高级建筑师、高级工程师占48%（其中10人为教授级），建筑、结构、设备、电气等专业配套人才完善。除基本队伍之外，公司还设有由国际建筑师学会建筑评论家委员会（CICA-UIA）委员刘开济、中国建筑学会高层建筑抗震专业委员会主任委员、美国地震工程研究会会员、中国工程设计大师胡庆昌等7位享有国务院突出贡献专家特殊津贴的资深专家组成的总工顾问专家组，指导并亲自介入重点工程设计。视工程项目的需要，北京市建筑设计研究院会派出专业人员服务于项目。

公司以总经理、总建筑师及各专业总工程师组成的技术委员会为核心，严格执行三审制及设计人员岗位责任制，形成有效的质量保证体系，以保证工程设计的质量。

公司成立以来，工程设计作品获部、市级优秀设计规划及方案奖10余项。有18位基本设计人员因设计质量优秀23次获得建设部或北京市级奖励，如北潞春绿色生态小区于2001年1月获北京市第七界首都建筑设计汇报展住宅区规划设计优秀方案奖，同时获得建设部规划设计、建筑设计二项金奖。刘琪市长在1999年北京市计划会议上说：“这种住宅小区应该是北京的发展趋势。”

公司重新整合后，先后完成的设计项目类型有医疗建筑类、居住区规划及居住建筑类、博览建筑类、公共建筑类、学校建筑类、体育建筑类、规划类。

John Martin International Architecture Design Co.,Ltd of BIAD. is a joint venture held by Beijing Institute of Architectural Design (BIAD), approved by the Ministry of Commerce of the People's Republic of China, owing Class-A Design Quality Certificate conferred by the Ministry of Construction and Beijing government. It was founded in 1994, and its certificate number is 010929-sj.

As the further process of economic reform, John Martin international architecture design Co.,Ltd of BIAD. gradually transformed from government institution to enterprisingly management institution. During the past 50 years, it has always been making efforts to provide high-quality services, and has always been among the top of residential design all over the country, earning high reputation in the industry, and at the same time forming a core idea of "Architecture services the society". By synthesizing excellent resources in the whole institution, and even in and outside the country, BIAD decided to reorganize John Martin international architecture design Co.,Ltd of BIAD for another time at the beginning of 2008, and pushed it into the market as its key company.

The company's scope of business includes undertaking and organizing various architectural engineering design, indoor and outdoor decoration engineering design, city planning & community planning, and other consultant and technological services related to architectural design, indoor and outdoor decoration, and community planning etc.

The company's tenet is through its operation in China and all over the world, to widely develop projects associated with architectural design, indoor design and city planning with excellent service and reputation, and to facilitate international communication of architectural technology, making full use of each side's advantages. The company has committed itself to creating architectural work and training professionals, at the same time exploring reform of design method and management mechanism.

The company is using president responsibility system led by the board of directors, which plays as the highest authority. There is one president and one vice president, and the former one summons and host the president meeting to discuss about executing decisions of the board, and take responsibility for all the business.

The company owns a basic design team combining staff of different ages and from majors of architecture, construction, equipment, electric and other necessary ones, 48% of which are senior architects and senior engineer (10 are of profession level). Besides, the company has a council anticipating key projects design composed of 7 experts who enjoy the "Outstanding Contribution" special allowance from the State Department, including CICA-UIA member LIU Kaiji, Chinese Architectural Association High-rise Building Anti-seismic Division director member, American Earthquake Engineering Association member, and Chinese master designer HU Qingchang and so on. BIAD may also send specialists for some projects according to details of the case.

The company has the president, chief architect and chief engineers from different specialties as its core, and forms an effective quality guarantee system strictly executing three-examining system and designer responsibility system to ensure the quality of engineering design.

Since the company has been founded, its design works have won more than 10 departmental and municipal excellent planning and design awards. 18 basic designers have gain outstanding design prizes from the Ministry of Construction or Beijing City for 23 times. For example, Beiluchun Residential Quarter has been conferred the Excellent Residential Planning Prize in Beijing 7th Capital Architectural Exhibition, and also two golden prizes of planning design and architectural design by the Ministry of Construction. LIU Qi, the mayor of Beijing, mentioned in the Beijing City Planning Meeting 1999 that "This kind of community should be the development trend of Beijing."

After recognizing, the company has successively completed design projects including medical building, residential planning and residential building, exposition building, public buildings, school building, sports building and various planning.

晋国博物馆

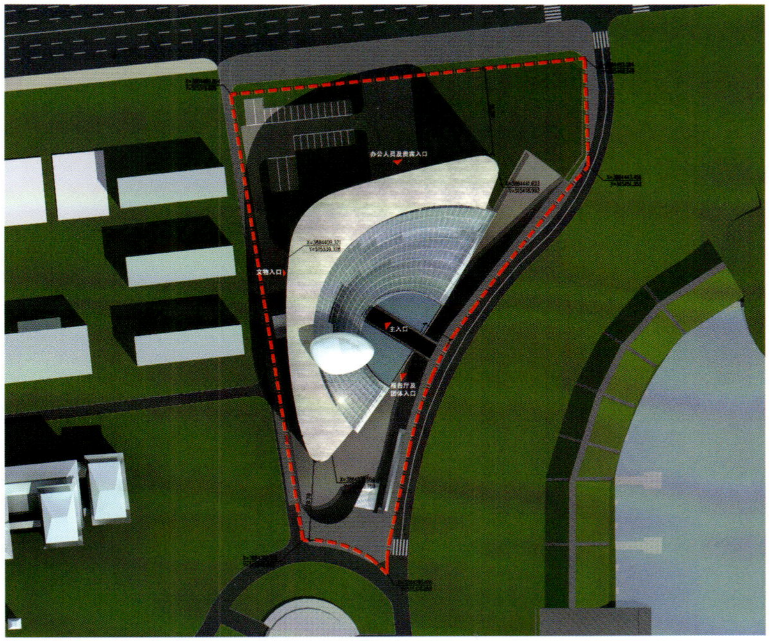

滕州博物馆
建设地点：山东 滕州
占地面积：13 830平方米
建筑面积：17 880平方米

Tengzhou Museum
Construction Location: Tengzhou, Shandong
Site Area: 13,830 m^2
Building Area: 17,880 m^2

成都盛邦超高层商业综合体
建设地点：四川 成都
占地面积：22 555平方米
建筑面积：3 480 000平方米

Shengbang High-rise Commercial-Residential Building
Construction Location: Chengdu, Sichuan
Site Area: 22,555 m^2
Building Area: 3,480,000 m^2

上海世博会中国铁路馆
建设地点：上海
建筑面积：5 200平方米
建筑层数：2层

Railway Pavilion of the World Exposition in Shanghai
Construction Location: Shanghai
Building Area: 5,200 m^2
Building Storey: two storeys

LD 朗东国际建筑设计有限公司
LANDON INTERNATIONAL ARCHITECTS DESIGN CO., LTD.

历史

朗东创立于2000年10月的深圳，
2004年在北京设立分部，
2006年总部迁至北京并设立深圳分部，
2009年在越南河内成立朗东国际建筑设计公司。

团队

公司现有员工60人，专业包括建筑、结构、机电等，是专业齐全、配套完善的服务体系。

业务

朗东的业务范围涵盖了城市设计、居住区规划与居住建筑设计、公共建筑设计、园林景观设计、室内设计及结构、机电等工程设计。

理念：品德、品质、品位、品牌

朗东追求根植于本土文化，以国际先进的理念和技术实现的设计创新，强调设计质量与服务并重，为客户提升最优质的价值，与客户实现共赢

History

In October of 2000, Landon was founded in Shenzhen;
in 2004, Landon Beijing branch was established.
In 2006, the headquarter moved to Beijing and the Shenzhen branch was set up.
In 2009 Landon International Architecture Design was established in Viet Nam-Hanoi

Team

There are more than 60 staffs in this working team, most of them specialized in architecture, structure, mechanical electronics, building up an excellent service system with a full range of specialties.

Business

Landon's business scope covers urban design, residencial area planning and residence architecture design, public architecture deisign, landscape architecture design, interior design, enginnering design of mechanical electronics and so on.

Philosophy: character, quality, taste, brand

Taking root in the local culture, realizing the design innovation through international forefront ideas and techonology, stressing on the design and service quality, to provide the best value for the clients and achieve the win- win cooperation.

北京朗东建筑设计咨询有限公司
地址：北京市海淀区西三环北路 72 号世纪经贸大厦 B 座 2600
电话：+86–10–51798330
传真：+86–10–51798330–801

Beijing Langdong architecture design consulting co., ltd.
Address: 2600, Block B, Millennium Plaza.No.72, Xisanhuan Bei Road, Haidian District, Beijing
Tel: +86-10-51798330
Fax: +86-10-51798330-801

深圳市朗东设计咨询有限公司
地址：深圳市福田区景田路擎天华庭擎天阁 32A
电话：+86–755–83544377
传真：+86–755–83285144

Shenzhen langdong design consulting co., ltd.
Address: 32A Qingtiange, Qingtian Huating, Jingtian Road, Futian Disctrict, ShenZhen
Tel: +86-755-83544377
Fax: +86-755-83285144

朗东国际建筑设计有限公司
地址：越南河内堆多郡阮志清路 91 号 M3–M4 栋
电话：0084–462751225
传真：0084–462751226
邮箱：landon_zp@163.com
网址：www.LD–A.cn

Landon International Architecture Design Co., Ltd.
Address: MS-M4 (91 Nguyễn Chí Thanh, Quận Đống Đa, Tp.Hà Nội, Việt Nam
Tel: 0084-462751225
Fax: 0084-462751226
Email: landon_zp@163.com
Http: //www.LD-A.cn

2

1–2 残疾人综合活动中心

建设地点：北京
建设规模：3 200 平方米
设计／竣工时间：2009 年／在建

1-2 Activity center for disabled persons

Construction Location: Beijing
Construction Scale: 3,200 m^2
Design / Completion Time: 2009/ under construction

3–4 中宏 · 新界

建设地点：河北 固安
建设规模：477 955 平方米
设计 / 竣工时间：2009 年 / 在建

3-4 Zhonghong • Xinjie

Construction Location: Gu'an, Hebei
Construction Scale: 477,955 m^2
Design / Completion Time: 2009/ under construction

3

4

5

6

7

8

5–6 河北省—迁西国家地质博物馆

建设地点：河北 唐山
建设规模：3 900 平方米
设计 / 竣工时间：2010 年 / 在建

5-6 Qianxi National Geological Museum in Hebei

Construction Location: Tangshan, Hebei
Construction Scale: 3,900 m^2
Design / Completion Time: 2010/ under construction

7–8 坤鼎国际企业花园规划设计方案

建设地点：北京
建设规模：212 665 平方米
设计 / 竣工时间：2010 年 / 在建

7-8 Kunding international enterprise garden master planning scheme

Construction Location: Beijing
Construction Scale: 212,665 m^2
Design / Completion Time: 2010/ under construction

9–10 通州文化中心

建设地点：北京
建设规模：28 000 平方米
设计 / 竣工时间：2009 年 / 在建

9-10 Tongzhou culture center

Construction Location: Beijing
Construction Scale: 28,000 m^2
Design / Completion Time: 2009/ under construction

9

10

都市意匠 URBAN DESIGN & TECHNOLOGY

都市意匠城镇规划设计（北京）中心
Urban Design and Technology Limited

都市意匠城镇规划设计（北京）中心（简称都市意匠）成立于2007年，是北京市海聚工程唯一的设计企业。创建人朱冰，为哈佛大学设计学博士，北京市首批引进的海外高层次人才，北京市特聘专家。

目前都市意匠团队由资深城市设计师、规划师、景观设计师、建筑师、工程师、生态规划师和经济规划师构成。业务范围涵盖城镇发展战略性规划、前期策划、总体概念规划、城市设计、旅游度假区规划、景观规划、修建性详细规划、建筑设计、平面与视觉识别设计等领域。

成立至今，都市意匠已完成近30项委托项目，其中包括国家重点工程——黄河流域最大的淡水湖泊湿地内蒙古乌梁素海生态产业园的产业规划、旅游规划以及重点地区的城市设计；中澳合作开发的天津蓟县长城文化休闲功能区总体概念规划；国家创意文化企业改革五大重点项目之一的江苏创意文化产业基地的城市设计和中华戏曲博览园的详细概念设计等项目。

此外，都市意匠与世界知名的咨询公司共同合作完成了国家郑州经济开发区汽车产业园的发展规划、上海莘庄商务区的可行性研究报告，以及三亚海棠湾C6片区的发展规划等报告。

项目的进行过程中，都市意匠都与业主、合作单位形成了战略合作伙伴关系，并秉承“专业精神、创新思维和团队工作”的核心理念，以“发现问题、解决问题（Problem–solving）”的工作方法，对城市规划的决策、开发、保护以及提倡可持续发展、生态敏感和社会完整性的城市规划与设计，提出了具有前瞻性和极强操作性的解决方案，并强调人居、生态、经济发展和个性体验的重要性。

都市意匠的设计师将国际一流的规划设计理念与当地环境和发展相融合，经常以独到的见解、打破常规的解决方法和个性化的创意方案带给业主以惊喜。都市意匠所追求的创作风格可概括为三点：
第一，追求人工景观和自然环境的完美融合，以设计和新技术最大限度地保持环境的原貌；第二，追求自然简单的问题解决方法和设计风格；第三，追求设计与功能的完美结合。 最后一点来源于都市意匠对中国建筑文化的理解。

设计的语言是国际的，人文的背景是中国的。都市意匠这一群对规划和设计怀着至诚热情的专业工作者，希望在历史赋予的机会面前，以国际化的视野和创新思维，为中国的城市化发展贡献力量。

都市意匠们相信环境可以改变人们的生活，为了更美好的家园，都市意匠和所有的业主一起努力着……

网址：www.urban–dtek.com
电话：+86–10–65525201
传真：+86–10–65525203

都市意匠 URBAN DESIGN & TECHNOLOGY

中关村数字电视产业园发展战略规划

项目业主：北京牡丹电子集团
项目类型：发展战略规划
项目位置：北京
用地规模：6公顷
完成日期：2010年

项目受托于北京牡丹电子集团，在集团企业转型、国家数字实验室落户于中关村数字电视产业园以及世界新型产业发展和北京建设世界城市的大背景下，为牡丹电子集团和中关村数字电视产业园进行新的战略部署、产业调整和发展空间规划。

发展报告针对牡丹集团的发展历程、技术以及产业布局特点，从产业国际宏观发展环境、中国发展趋势、北京未来建设、园区开发拓展等几个层面对中关村数字电视产业园的发展重新定位，使其成为建筑风格新颖、功能复合、设施齐全的企业总部基地，配合北京牡丹电子集团新的产业战略部署，成就老牌国企的产业复兴之路。

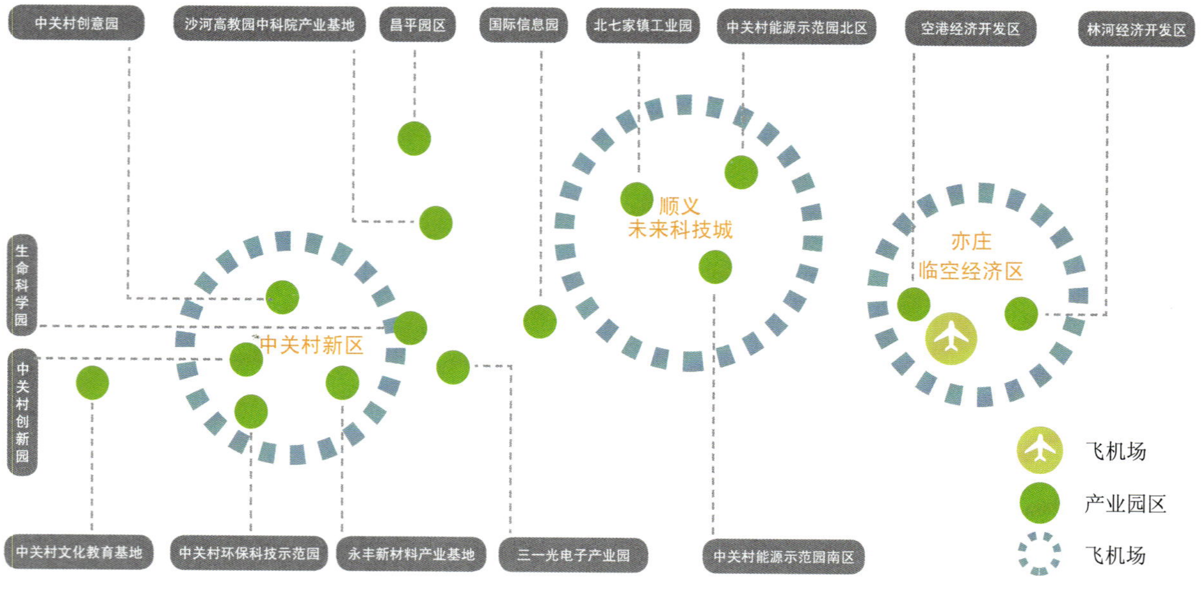

长城文化休闲功能区总体概念规划

项目业主：蓟县规划局
项目类型：概念性旅游规划、城市设计
项目位置：天津 蓟县
用地规模：205.21平方千米
完成日期：2010年2月

《长城文化休闲功能区总体概念规划》通过对蓟县北部山区、黄崖关长城以南地区现状资源分析和未来发展潜力的研究，提出以下营镇改造为中心，以中国特色养生为品牌，塑造世界知名的康体养生度假区。

设计方案采用生态优先的规划方法，通过对基地详细的调研并加以GIS科学的分析，找到发展、保护和生态之间的契合点，从而妥善解决新农村规划与管理、产业结构调整与转换、生态环境保护与修复、地域特色挖掘与保护、旅游资源整合与管理等问题，寻求一条保护与发展的新道路。

内蒙古乌梁素海生态产业园区总体概念规划

项目业主：内蒙古乌梁素海水生态产业园管委会
项目类型：总体概念规划、生态治理与保护规划、产业规划、概念性旅游规划
项目位置：内蒙古 巴彦淖尔
用地规模：760.5平方千米
完成日期：2009年12月

本次规划希望通过对乌梁素海水生态系统的综合治理，有效地改善乌梁素海水环境状况，并带动地区经济的可持续性发展。在这个背景下，乌梁素海水生态产业园区概念规划利用科学的手段、前瞻性的设计理念，针对园区及其周边地区现状提出了包括土地整合、经济产业规划、生态治理、旅游开发、新城镇开发在内的综合性建议，以帮助产业园制定乌梁素海开发和保护的政策、规划和设计。

乌梁素海红柳湾湾区规划及城市设计

加强乌梁素海旅游区的入口门户建设，改造湖区岸线，城市空间沿滨水发展，形成两区一带的发展格局。通过道路与滨水岸线景观的有机结合，塑造出富有层次的滨水岸线。

乌梁素海坝头旅游小镇规划及城市设计

坝头拟发展为特色风貌旅游小镇，以滨水幽静的环境、草原背景和浓郁的民俗风情为背景，创建城镇和湖区之间的空间联系以及和草原环境的融合；开发滨水活动，依托草原氛围，加强旅游设施建设，塑造悠闲舒适的休闲度假环境，从而带动当地旅游产业和其它产业的发展。

乌梁素海旅游区VI设计

以湖区平面为框架，以鸟、芦苇为意向展示一个天鹅自由的在乌梁素海的天空翱翔，水鸟在芦苇丛中生活的景象，充分体现了乌梁素海绿色生态的自然环境。翅膀形状的海岸，寓意乌梁素海是鸟儿的天堂。

江苏创意文化产业基地城市设计

项目业主：江苏演艺文化产业有限公司
项目类型：概念性旅游规划、城市设计
项目位置：江苏 南京
用地规模：4.56平方千米
完成日期：2008年10月

本案是以创意文化为主题思想，生态环境为组织框架的大型生态创意文化产业园区，在上轮控规的基础上对土地进一步细分，空间架构进一步深化，并全面思考城市映像的要素构成。

合理统筹土地，尊重并充分利用地形地貌，加强改善各区域山水环境品质，挖掘内在的潜力，提升土地价值。针对项目发展特点进行特色分区，强调优质的生态资源与用地规划的有机结合。

通过综合性的开发方法，江苏创意文化产业基地将被打造成包含产业、文化、研发、居住、商业、休闲娱乐等功能于一体的活力社区。

中华戏曲博览园概念规划

项目业主：江苏演艺文化产业有限公司
项目类型：概念性旅游规划、城市设计
项目位置：江苏 南京
用地规模：4.56平方千米
完成日期：2008年10月

博览园以中国各地的特色古镇、民居为设计蓝本，设计出不同种类的中华戏曲的实景舞台。

在深刻理解基地自然条件的基础上，巧妙利用已开发的用地、自然地形和水系，确定可开发用地范围、生态保护范围及旅游景观道路的组织。并在此基础上进行主要景区的详细设计。根据不同的地形地貌，塑造景区特色。以戏楼、戏台为中心，创造不同的小镇及村落空间，情景再现当地戏曲的生活场景。通过不同的景观要素确定开放空间的主题，从而创造具有鲜明特色、绿色生态的戏曲博览园。

对国内外传统民居街区进行深入的民俗研究，找到传统民居和民俗文化的重要组成要素，并将其融入到与地形条件、生态环境相匹配的地块中，用以再现传统民居特色，传承民俗文化。

文登葛家镇湖上田园概念性详细规划

项目业主：威海三元置地有限公司
项目类型：概念性详细规划
项目位置：山东 文登
用地规模：45公顷
完成日期：2009年5月

湖上田园是一个旅游加地产的综合项目，通过旅游来聚集人气，加强地产品牌，从而形成一定规模的地产经济效益，以此来塑造地区形象。因此，我们的设计宗旨是在尊重现有生态环境、基地地形、地貌的基础上，充分解读“湖上田园”的概念。

首先，在尊重生态自然的基础上，利用GIS对基地进行客观、科学的分析，以达到充分了解地形的目的；其次，对基地周边项目市场进行研究，以确定项目发展的趋势和配比及类型；最后，利用基地分析和市场研究的结果，结合基地独特的景观空间进行设计。

仁和文化休闲度假广场详细设计

项目业主：天津万事兴房地产集团
项目类型：修建性详细规划
项目位置：天津 蓟县
用地规模：44.42公顷
完成日期：2010年

基地存在着北侧的山体遭到破坏、交通季节性拥堵、铁路严重分割基地等问题，我们针对基地现状存在的问题提出如下解决方案：

利用高层住宅建筑遮挡由于挖矿而造成的被破坏的山体，从而形成新的天际线；利用设置停车场解决基地周边季节性交通拥堵问题；利用基地内部被荒废掩埋的水系打造景观；利用货运铁路后退的空间打造中心绿地；利用高大的酒店综合体提升基地的形象，同时将基地内部空间与外部嘈杂的空间加以划分，达到动静分区的效果。

蓟县兴华洋街详细设计

项目业主：天津万事兴房地产集团
项目类型：建筑改造、街巷改造
项目位置：天津 蓟县
建筑用地面积：4 800平方米
完成日期：2010年

商业萎靡不振的兴华商阜随着时间的推移建筑也黯然失色，因此，我们需要寻求新的方式重塑兴华商阜的城市形象和功能。

我们将其立面加以动感时尚的元素，将两条建筑通道打造成酒吧街——兴华洋街，使其白天成为蓟县休闲活动、浪漫时尚的新地标，晚间成为蓟县朋友相聚，情人浪漫邂逅的梦幻舞台。

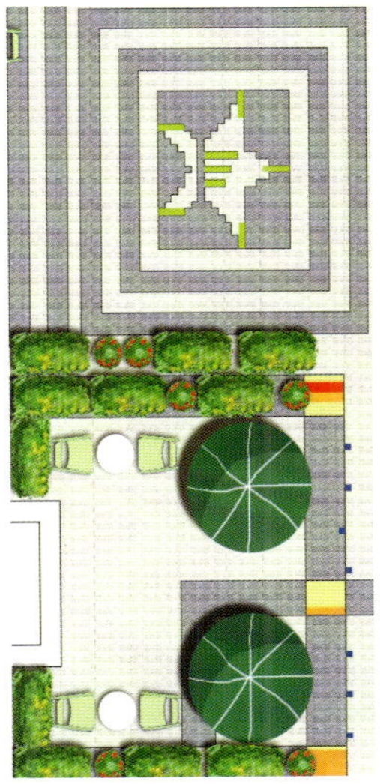

王暐建筑工作室 中广电广播电影电视设计研究院

WANGWEI Architecture Studio

地址：北京西城区南礼士路13号中广电广播电影电视设计研究院404室
Address: Room 404, Zhongguangdian Radio, Film and Television Design Institute, South Lishi Rd.No.13, Xicheng District, Beijing
电话/Tel: 13501197969 传真/Fax: +86-10-68026482
邮箱/E-mail: wangwei@dsarft.com 网址/Http: // www.dsarft.com

王暐，教授级高级建筑师，国家一级注册建筑师，曾获得清华大学优秀硕士论文奖、国家广播电影电视总局十大杰出青年等荣誉称号。2002年9月组建水木建筑策划研究部，主要从事建筑项目的策划研究及概念设计。2003年6月，本部门作为中广电广播电影电视设计研究院创作团队的一员，参加了公安部办公楼的投标设计，在众多的竞争对手中脱颖而出，获得设计权。王暐建筑工作室于2004年3月组建，开始从事建筑项目的全程设计与服务。

Introduction

Being the professor-level senior architect and PRC class I registered architect, Wei Wang, won Best Master's Thesis Award in Qinghua University, and the Top Ten Excellent Youth honor awarded by SARFT(the State Administration of Radio Fimand Television). In September of 2002 he founded the Water and Wood architecture planning and researching institute , mainly offering service in the project researching and the conceptual design. As a member of the creative team of SARFT design and researching institute, the institute participated in the bidding design of Ministry of Public Security office building in June 2003 and finaly wonthe bid, exceeding other competitors with the exquisite works , In March 2004, Wangwei Architecture Studio was established and started the overall architectural design works and service.

理念

倡导朴素自然的建筑思想和生命哲学，并称之为“水木精神”。老子曰：“上善若水，水善利万物而不争。”；“天下莫柔弱于水，攻坚强者莫之能胜，以其无以易之。”。水以其清澈、明净、慈柔、随缘的特质表达了无为而治的生命观；木以其丰华、殷实、坚韧，有机的特质表达了和谐发展的建筑观。水、木不仅代表了一种哲学思想，同时也反映了一种美学观念。水和木既是一种品性，也是一种方法；既是建筑设计的范式，也是生命行为的准则。本工作室将这种对建筑与生命的理解与思考，深刻融入建筑设计的全过程，并依此重新诠释建筑、社会和人之间的关系。

Conception

We advocate simple and natural construction thought and life philosophy ,and names it as "spirit of water and wood". Lao Tzu said, "the highest motive is to be like water. Water is essential to all life, yet it does not demand a fee" "In the world there is nothing more submissive and weaker than water. Yet for attacking that is hard and strong, nothing can surpass it." Water,with the quality of clarity, lucidness, mercy and flexibility, demonstrates the life philosophy of "to govern by doing nothing that goes against nature", and the "water and wood" indicates the architectural conception of harmonious development, through its flourishing, solid and tough characteristis. "water and wood" is not only a term of of philosophy but also an aesthetic concept, It's a character and a measure. It is the mode of architectural design and principle of life behavior. That is why we put in application of all the understanding and thoughts into the design works, and portray the new relationship among architecture, society and human.

公安部办公大楼

业绩

作为专业设计室，王暐建筑工作室参加创作和进行专业设计的项目有：

Achievement

As a professional design studio, we have participated the creation and specialized designing of the following projects:

1. 公安部办公楼（中标实施）
Ministry of Public Security Office Duilding (under construction)
2. 中影集团电影生产基地（中标实施）
China Film Group Corporation Film Production Base (under construction)
3. 郑州市广播电视中心（中标实施）
Zhengzhou Radio and Television Center (under construction)
4. 合肥广播电视中心（中标实施）
Hefei Radio and Television Center (under construction)
5. 毛里求斯广播电视中心（中标实施）
Mauritius Radio and Television Center (under construction)

郑州市广电中心

中国妇女儿童活动中心二期

公安部大院整体规划改造工程

公安部招待所及警察技能用房工程

曹妃甸渤海明珠电视塔

民政部办公楼

人民公安报社装修改造工程

公安部秦城绿化基地工程

日都（北京）高新建筑科技设计咨询有限公司
有限会社AGC都市设计

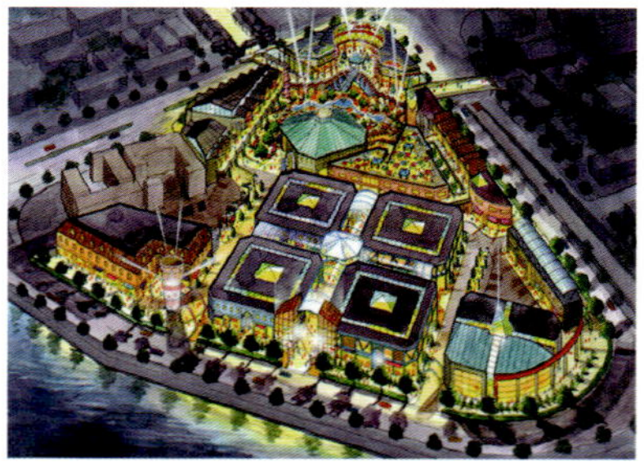

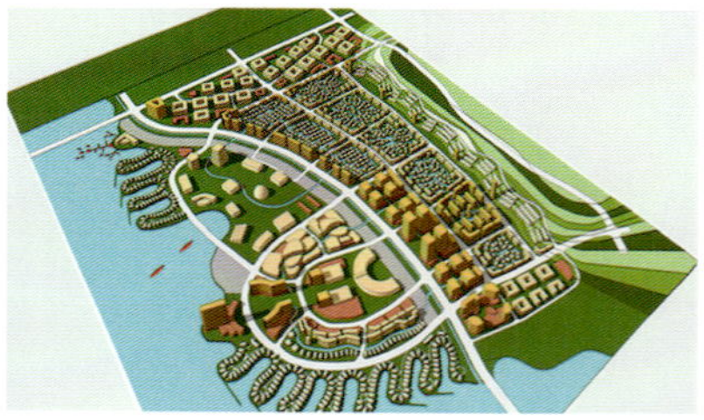

日都（北京）高新建筑科技设计咨询有限公司于2004年4月由日本有限会社AGC都市设计与海归学者共同组建。公司承担的策划、规划及设计项目北京证券广场（AGC设计）曾获1994年北京市民投票十佳建筑设计第五名，沈阳航天三菱办公楼曾获2007年东北三省金奖工程，北京金台夕照会馆曾获2008年全国最佳主题酒店。

公司主要承担以下几方面的工作：

大型项目的策划与规划

公建、商业、住宅的建筑设计

室内、外环境景观的设计

相关联业务

尤其在温泉设计方面有所专长，设计结合国情，并加入了日式温泉的理念与技术。1997年设计的广东清新温矿泉度假区（AGC设计）为中国温泉设计提供了新的思路。公司随后又承担了福州青云山御温泉、贵阳天邑温泉度假区、海南康乐园日式温泉、成都青城白鹭洲温泉岛、大连安波温泉新乐园等温泉项目的策划与设计，积累了丰富的专业经验。目前公司正着手北京南面的大千森林温泉的策划。

In Apr.2004, Ridu(Beijing)High Architectural Technology Consulting Company was co-founded by AGC Urban Design INC and returned scholars. Beijing Securities Square (designed by AGC) ranked the fifth in Top Ten Excellent Architectural Design voted by Beijing citizens in 1994, and Office Building of Shenyang Aerospace Mitsubishi was awarded the Golden Project of Three Northeast Provinces in 2007, as well as Beijing King-Talent Hotel was awarded the best national theme hotel in 2008.

Mainly responsible for:

Planning design for large scale projects

Architectural design for public buildings, business buildings and residence

Interior design and exterior environment landscape design

The relevant business

Especially on hot spring design, Japanese style hot spring combined with our national conditions was introduced into China; Qingxin Spa Resort in Guangdong (design by AGC) in 1997 has brought great changes in China's hot spring, and since then he has hosted a series of planning design for hot spring projects including Qingyun Mountain Royal Hot Spring in Fuzhou, Tianyi Hot Spring Holiday Resort in Guiyang, Japanese-style Spa in Hong Lok Yuen, Hainan, Bailuzhou Spa Island in Qingcheng, Chengdu and Anbo Spa New Paradise in Dalian. Therefore, it has accumulated rich professional experience. Currently it is busy with planning design for Daqian Forest Hot Spring in southern Beijing.

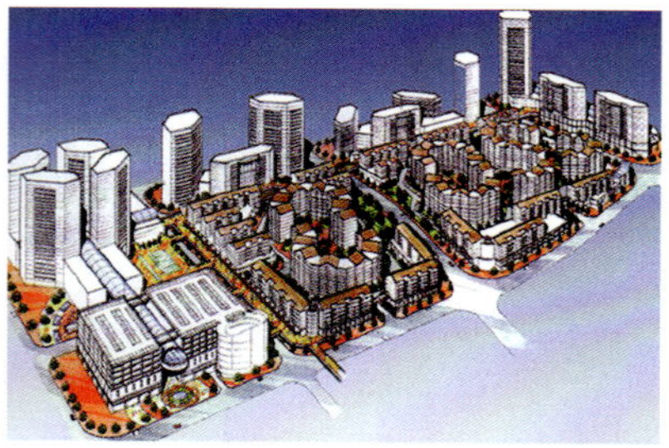

日都（北京）高新建筑科技设计咨询有限公司
有限会社AGC都市设计
地址：北京丰台区航丰路9号颐安鑫鼎科技大厦709室
电话：+86-10-83682627　传真：+86-10-83682673
网址：www.riduad.com　邮箱：rd@riduad.com
日本国群马县前桥市大手町 2-5-10,2F
邮箱：agc_urban@hotmail.com

Ridu(Beijing)High Architectural Technology Consulting Co., Ltd
AGC Urban Design INC.
Address: Room 709 in Yi'an Xinding High-Tech Mansion, No.9 of Hangfeng Road, Fengtai District, Beijing
Tel: +86-10-83682627　Fax: +86-10-83682673
Http: //www.riduad.com　E-mail: rd@riduad.com
2F, 2-5-10, Otemachi, Maebashi city, Gunma, Japan
E-mail: agc_urban@hotmail.com

北京金台夕照会馆
建筑面积：20 000平方米
获奖情况：获选2008年全国最佳主题酒店
合 作 方：北京都林国际工程设计咨询有限公司

Beijing King-Talent Hotel
Building Area: 20,000 m^2
Awards: the best national theme hotel in 2008
Partner: Beijing DBLANT Design International LLC

1

3

4

5

6

日都（北京）高新建筑科技设计咨询有限公司

1–5 福州青云山御温泉
用地面积：110 000平方米
建筑用途：度假/休闲

6–9 贵阳天邑温泉度假区
用地面积：120 000平方米
建筑用途：度假/休闲

Qingyun Mountain Royal Hot Spring in Fuzhou
Site Area: 110,000 m^2
Building Purpose: for holiday or leisure

Tianyi Hot Spring Holiday Resort in Guiyang
Site Area: 120,000 m^2
Building Purpose: for holiday or leisure

8

日都（北京）高新建筑科技设计咨询有限公司

1–2 沈阳航天三菱汽车发动机厂办公区
用地面积：30 000平方米
办 公 楼：10 000平方米
食　　堂：6 000平方米
宿　　舍：3 000平方米

Office working area of Shenyang Aerospace Mitsubishi Motors Engine plant
Site Area: 30,000 m^2
Office Building: 10,000 m^2
Dining Hall: 6,000 m^2
Living Quarters: 3,000 m^2

3–4 中山市蓝波湾豪庭二期
建筑面积：83 000平方米

Stage Ⅱ for Lanbowan Noble Residence in Zhongshan city
Building Area: 83,000 m^2

5 桂林桃花江旅游区规划Ⅱ
用地面积：1 090 000平方米

Planning for Peach River Tourism Region in Guilin
Site Area: 1,090,000 m^2

方略建筑设计有限责任公司

Strategy Architectural Design Co.,Ltd.

方略建筑设计有限责任公司前身是中国九三学社全资下属机构——九三建筑设计研究院，2002年经改制在国家工商总局注册登记成立有限责任公司，具有国家甲级设计资质（证书编号010509-sj）。

方略设计有限责任公司实行与国际化接轨的现代设计企业管理模式，以各建筑师工作室为核心，其中，业内知名的建筑师工作室——九三组是一个具有国际化背景的建筑师工作室，在这里，来自德国，加拿大等国家的设计师与本土设计师一同工作，具备了一流国际水准的设计理念和设计质量。同时，在内部的管理上，强调贯彻全面的技术质量保证体系和营销服务体系，以此形成了企业鲜明的市场化，细节服务型的管理理念。

多年的发展，方略设计公司已形成了以北京总部为中心，全国多点分院布局的发展格局，其设计业务涉及全国各大城市的大型酒店及综合商业设施，高档办公写字楼、居住小区、医院及学校设施、工业建筑等诸多方面。

方略设计尤以大中型商业及公建设施，大型写字楼、学校及医疗设施等规划方案设计和施工图设计见长。在设计创作中，方略始终坚持技术设计和服务意识的结合，形成了一种以技术服务和营销服务为核心价值观的企业营销服务体系。在这一体系指导下的服务意识和设计生产行为使方略设计能够在市场价值、开发商理念和技术设计手段的结合部分发挥重要的衔接作用，创造了价值、理念、市场共赢的结合点，使设计服务达到了一个新的层面。

方略设计的作品追求品质感和价值感的创作风格，展现着市场前瞻意识和不俗的建筑创新力。其代表作品如北京国际、北京大成国际中心、北京朝外MEN等大型城市建筑，被公认为极具城市雕塑感和地标性的建筑，体现了一种诠释明快、内蕴深厚、大气实用的设计理念。而如北京印象、长安新城、北京元大都等住宅项目更体现了国际化先进理念与中国实际相结合的特点，表达出对完美精致、舒适实用、生态环保的城市住宅的执着探求，这些都展现了方略建筑智库的特质和修养。

多年的耕耘创作，方略设计博得了客户及社会的认同与尊重，获得了较高的市场口碑与美誉度。

The Strategy Architectural Design Company Limited used to be part of the Jiusan Society—Jiusan Architectural Design Institute. After its neform in 2002, it'sregistered to establish the limited company with national Grade A architectural design certification (No.010509-sj).

The Strategy Architectural Design is under the international enterprise management mode, build the leading specialties take the chief architect to establish individual studio as the core strength. Jiusan Team of The Strategy Architectural Design Co.Ltd based on an international architectural background, here are German and Canadian architects to ensure the design concept and quality are up to the first class of world wide. At the same time, the office emphasis on carrying out the overall technical quality certification system and marketing service system to form the distinctive marketization and detailed service management.

After several development, the Strategy Architectural Design Co.Ltd established its headquarter in Beijing and a few branches in other cities of China. The scope of its services extended to large scale commercial, hotels and urban hybrids, high-class office building, residential community, hospital and educational institute all over China.

The Strategy Architectural Design Co.Ltd major in large scale public, office and commercial buildings. During the design process, architects always sustain to integrate the technical design and service together. Under the this specific system, Strategy Architectural Design use this unique conception to create values, theories, and market success as well as enhance its design service up to a new extent.

The Strategy Architectural Design's projects emphasis on the quality and stylish values, they reveal its bold and innovative tastes to the current architectural market. The representative project sare: Beijing Logo, Beijing Da Cheng International Center, Beijing Chaowai Men etc. These large scale public building are conceived Chang'an the symbolic landmark, they are refresh, sophisticated and practical. As for Beijing Image, Chang'an New Town, Beijing Yuan Da Du etc, they all designed under the international modern concept and articulated with Chinese current market trend, they seem comfortable and sustainable. All of these projects display the wisdom and standards of Strategy Architectural Design Co.Ltd.

地址：北京市海淀区阜成路115号北京印象1号楼1门5层
法人：张诣
电话：+86-10-88116413
传真：+86-10-88114178
邮箱：fanglue_arch@263.net
网址：www.fla.com.cn

Address: 5th Floor, 1ST Door, Building 1,115 Fucheng Road,
Haidian District, Beijing impression
Corporate: Yi Zhang
Tel: +86-10-88116413
Fax: +86-10-88114178
E-mail: fanglue_arch@263.net
Http: //www.fla.com.cn

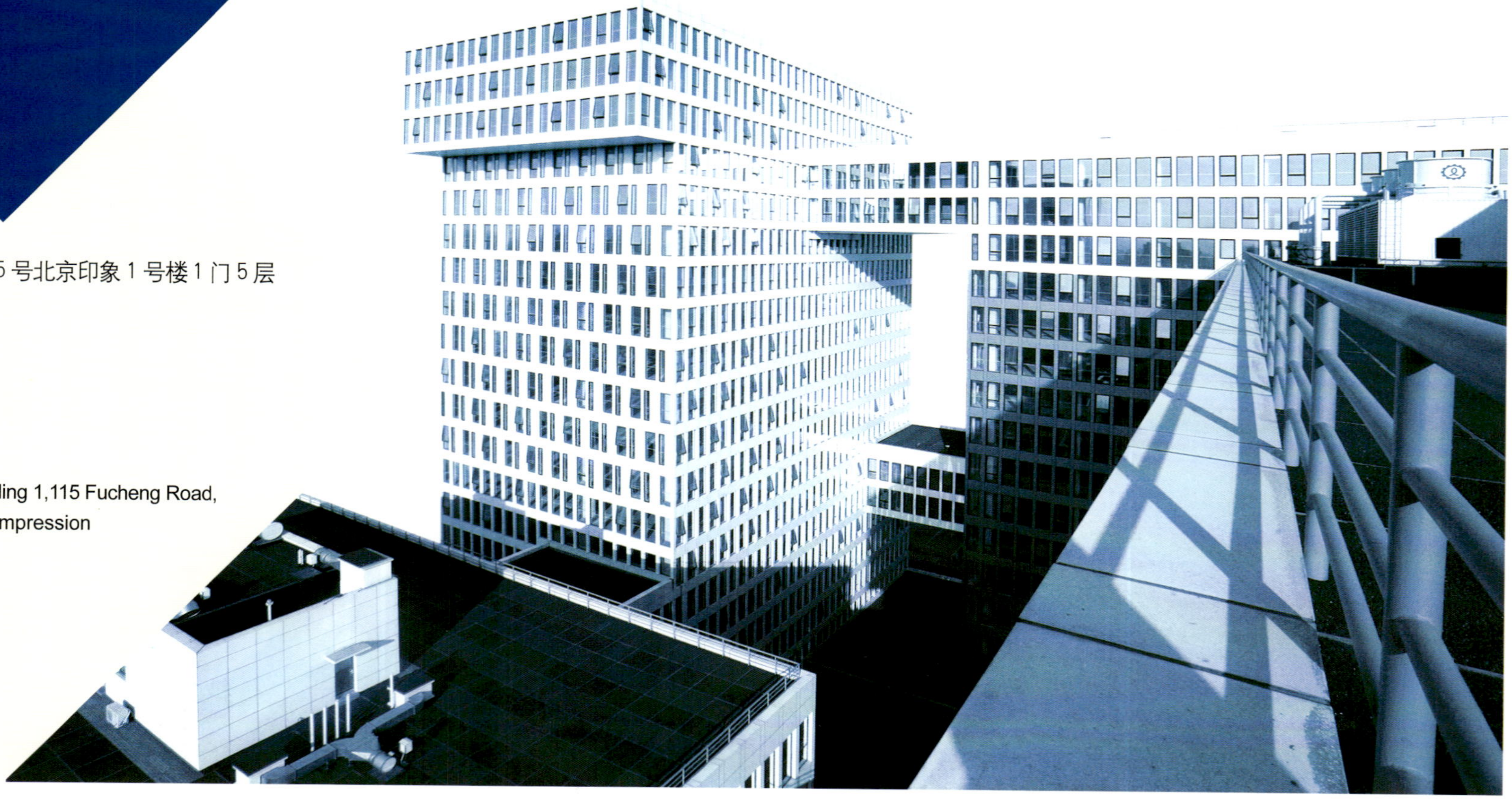

北京国际
建筑地点：北京
总用地面积：12 428 平方米
总建筑面积：124 000 万平方米
容 积 率：6.69
建筑高度：85 米
建筑层数：地下：4 层　地上：3 ~ 22 层
项目状况：完工已投入使用

Beijing International
Construction Location: Beijing
Total Site Area: 12,428 m^2
Total Building Area: 124,000 m^2
Plot Ratio: 6.69
Construction Height: 85 m
Building Storeys: underground: 4 storeys; overground: 3~22 storeys
Project status: completion and being used

本建筑以强烈雕塑感的造型，表现办公建筑简练、高雅、清朗的气质。立面设计两种不同形式的窗户互相交融、穿插，所有幕墙板均以一种尺寸排布，看似纷杂的立面实则具有完美的规律性、逻辑性，既简单又丰富、既活泼又统一，使本建筑具有超凡脱俗的个性。

The building embodies conciseness, elegancy, clearness of office buildings by its modeling with strong sense of sculpture. Windows with two different forms which are vertically designed mutually intersperse. All curtain wall plates are arranged in one size. The facades which look complex are actually with perfect regularity and logic. They are simple and plentiful, both lively and unified to make the building detached.

北京印象
建筑地点：北京
总用地面积：47 000 平方米
总建筑面积：170 000 平方米
容 积 率：2.77
建筑高度：12 ~ 44 米
建筑层数：地下：1 层　地上：4 ~ 15 层
项目状况：完工已投入使用

Beijing Image
Construction Location: Beijing
Total Site Area: 47,000 m^2
Total Building Area: 170,000 m^2
Plot Ratio: 2.77
Construction Height: 12~44m
Building Storeys: underground: 1 storey; overground: 4~15 storeys
Project status: completion and being used

建筑设计的特点之一是门厅，走进四层高的带天窗的阳光中庭，进入眼帘的是阳光下的空中廊桥、栏杆上悬挂的植物花卉、木地板上放置的桌椅，这些无不吸引人们驻足，由此增进了邻里交往，使其充满生活的气息。在装修品质上，着重设备的耐久性。门窗、墙体、防水层、保温层等重要构件都选用高档产品，确保其对安全性、密闭性、保温隔热性品质的要求。施工工艺的质量追求德国标准，以期达到现代工业体系下的高标准。

One of the architectural design features is the hall. Walking into the four-storey sun court with skylights, you will see the sky bridges under the sun, plants and flowers hanged on railings, wooden tables and chairs placed on the floor. They invariably attract people to stop. They are full of the smack of everyday life and enhance the contacts among neighborhoods.
In the decoration quality, the focus is device durability. Important components such as doors, windows, walls, waterproofing layer, and insulation layer are selected from top quality products to ensure quality requirements of security, impermeability and heat insulation. The quality of construction techniques is pursuing Germany standards, hoping to achieve the high standard under modern industrial system.

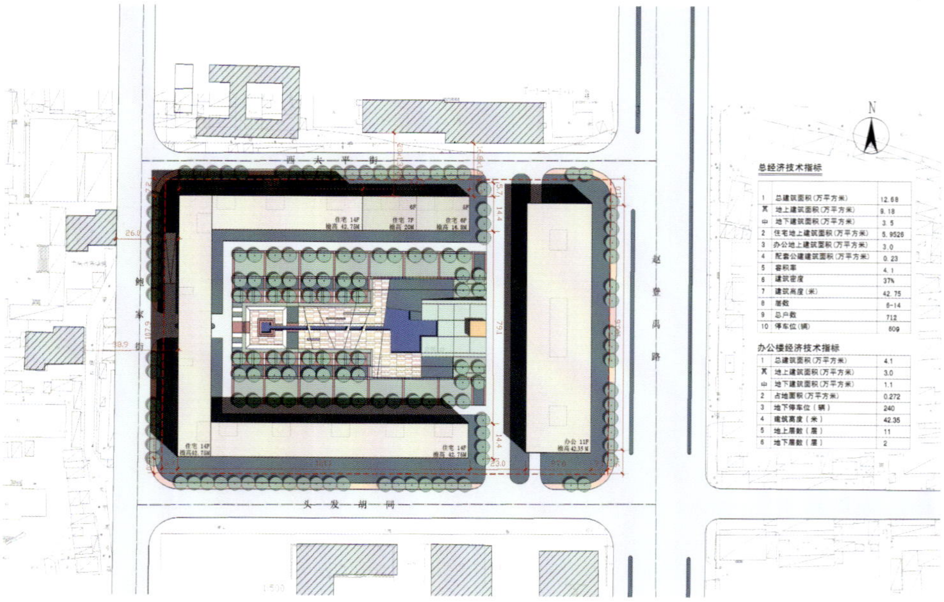

北京尊府
建筑地点：北京
总用地面积：22 400 平方米
总建筑面积：128 000 平方米
容 积 率：3.95
建筑高度：40 米
建筑层数：地下：3 层　地上：5 ~ 14 层
项目状况：完工已投入使用

Beijing Zunfu
Construction Location: Beijing
Total Site Area: 22,400 m^2
Total Building Area: 128,000 m^2
Plot Ratio: 3.95
Construction Height: 40 m
Building Storeys: underground: 3 storeys; overground: 5~14 storeys
Project status: completed and being in use

朝外 MEN
建筑地点：北京
总用地面积：13 680 万平方米
总建筑面积：107 300 万平方米
容 积 率：5.77
建筑高度：99.90 米
建筑层数：地下：3 层　地上：8 ~ 26 层
项目状况：完工已投入使用

Chaowai MEN
Construction Location: Beijing
Total Site Area: 13,680 m^2
Total Building Area: 107,300 m^2
Plot Ratio: 5.77
Construction Height: 99.90 m
Building Storeys: underground: 3 storeys; overground: 8~26 storeys
Project Status: completed and being in use

大成郡
建筑地点：北京
总用地面积：204 800 平方米
总建筑面积：350 000 平方米
容 积 率：1.87
建筑高度：20.55 ～ 34.50 米
建筑层数：8 ～ 12 层
项目状况：在建

Dacheng County
Construction Location: Beijing
Total Site Area: 204,800 m^2
Total Building Area: 350,000 m^2
Plot Ratio: 1.87
Construction Height: 20.55~34.50 m
Building Storeys: 8~12 storeys
Project Status: under construction

金隅大成时代中心
建筑地点：北京
总用地面积：16 600 平方米
总建筑面积：174 000 万平方米
容 积 率：6.28
建筑高度：27 ～ 80 米
项目状况：完工已投入使用

Jinyu Dacheng Times Tower
Construction Location: Beijing
Total Site Area: 16,600 m^2
Total Building Area: 174,000 m^2
Plot Ratio: 6.28
Construction Height: 27~80 m
Project Status: completed and being in use

本方案是对高密度、高容积率的大型综合体设计的一次成功探索。在解决多功能、多业态以及满足各专业规范的同时，最大可能的提升高级公寓的居住品质，使之具有良好的景观效果和闹中取静的舒适度。

The program is a successful exploration for large complex design of high-density, high plot ratio. In the process of solving the multiple functions, multiple formats and complex contradictions, it meets the professional standards, while enhancing the living quality of apartment as possible as it can, so that it has a good scenery and comfort of serenity amid chaos.

东莞理工学院学生宿舍区

建筑地点：广东 东莞
总用地面积：157 000 平方米
总建筑面积：113 000 平方米
其中学生宿舍：98 000 平方米
学生食堂：15 000 平方米
建筑高度：17.43 ～ 24.03 米
建筑层数：5 ～ 7 层
项目状况：完工已投入使用

Student dormitory of Dongguan Institute of Technology

Construction Location: Dongguan, Guangdong
Total Site Area: 157,000 m^2
Total Building Area: 113,000 m^2
Student Dormitory: 98,000 m^2
Cafeteria: 15,000 m^2
Construction Location: 17.43~24.03 m
Building Storeys: 8~7 storeys
Project Status: completed and being in use

大成国际中心

建筑地点：北京
总用地面积：4.308 万平方米
总建筑面积：23.04 万平方米
容 积 率：3.365
建筑高度：99.90 米
项目状况：完工已投入使用

Dacheng International Centre

Construction Location: Beijing
Total Site Area: 43,080 m^2
Total Building Area: 230,400 m^2
Plot Ratio: 3.365
Construction Height: 99.90 m
Project status: completed and being in use

建筑造型的基本指导思想是利用对建筑的造型处理，弱化建筑在体量上庞大的感觉，使建筑单体与建筑群体之间层次分明、和谐统一，形成一个有机的整体。

The basic guiding idea of architectural modeling is to weaken the sense of the large body of building by dealing with architectural modeling, so that gradations between single building and building groups are distinct, forming an organic whole with harmony and unity.

TCBDRI

北京市古代建築設計研究所

BEIJING TRADITIONAL CHINESE ARCHITECTURAL DESIGN AND RESEARCH INSTITUTE

北京市古代建筑设计研究所初创于1980年3月，是全国成立最早、实力雄厚的古建筑专业设计研究单位，具有文物古建筑保护修缮设计甲级资质和建筑设计乙级资质。

北京市古代建筑设计研究所的主要业务为文物古建筑保护修缮设计和传统建筑设计等，古建筑技术研究并出版、发表各类古建筑专业学术技术著作。

单位拥有国内一流的古建筑技术专家，并在古建筑研究、设计、教学、办刊、编制规范、培养人才等方面成绩显著，在行业内颇具影响力，是古建园林行业的学术带头人。

自成立以来单位已完成建筑设计项目700余项，始终坚持古建筑传统技术及理论研究，撰写了《中国古建筑木作营造技术》《中国古建筑瓦石营法》《中国清代官式建筑彩画技术》《中国建筑彩画选》和行业标准《古建筑修建工程质量检验评定标准》《古建筑修建工程施工与验收规范》《北京四合院建筑要素图集》等，对我国古建园林设计、施工、教学研究发挥重要作用。设计研究所所还是全国唯一的古建筑学术技术刊物《古建园林技术》杂志的创办单位，为我国传统建筑文化的继承和发展作出了重要贡献。

20多年来，该所还在创办大学古建筑专业、开展古建筑科普教育、培养古建筑技术人才方面做了大量工作。1985年单位的古建专家为北京市房地产职工大学(国家教委批准的成人高校)古建筑工程专业撰写了古建筑木作、瓦作、油漆彩绘等专业课教材并在校兼职任教，首创了我国第一个大学古建筑专业。现在，这些教材已被清华大学、天津大学、北京林业大学、北京建筑工程学院、东南大学、华南理工大学等十几所高等院校的建筑系用作专业课教材和教学参考书。

地址：北京市东城区安德里北街甲 20号紫萱园写字楼三层
邮编：100011
电话：+86—10—84126198
传真：+86—10—84126698
邮箱：gj_cad@163.com
网址：www.ctad—bj.com

Address: 3rd floor, office building of Zixuan garden, No.20, North Andeli Street, Dongcheng District, Beijing
P.C.: 100011
Tel: +86-10-84126198
Fax: +86-10-84126698
Email: gj_cad@163.com
Http: //www.ctad-bj.com

Initially established in March 1980, Beijing Traditional Chinese Architectural Design And Research Institute is the earliest ancient architecture design institute in China with high capability, which has class A qualification of ancient architecture renovation design and class B qualification of architecture design. Their research area covers technique research of ancient architectures, reparation design of cultural properties and traditional architecture design. They also have published lots of works on various kinds of ancient architecture design.

With national first class ancient architecture technical experts, the institute has made outstanding accomplishment in ancient architecture research and design, education, publication, making the code of conduct and cultivating the talents. With their influence in the profession, they have become the academic leader of the classical architecture profession.

Since its establishment, they have conducted more than 700 projects, maintaining the traditional technical and theoretical research of ancient architecture. For these years, they have composed many relative works including carpentry project building technique of Chinese ancient architecture, tile and stone construction technique, official building color painting technique in Qing Dynasty, selection of Chinese building color painting, as well as professional standards, such as quality evaluation standard of ancient architecture construction, construction and inspection criterion for ancient architecture project, graphics collection of basic elements of Chinese traditional rectangular courtyard in Beijing, which play an important role in the Chinese ancient architecture design, construction and education area. It is also the sponsor of Traditional Chinese Architecture and Gardens, making significant contribution to the succession and development of traditional architecture culture.

For more than 20 years, they have made great effort in establishing architecture subject in universities, promoting the ancient architecture education and cultivating the architectural talents. In 1985, they edited the text books and took the teaching works of carpentry project building of ancient architecture tile works, oil painting in ancient architecture engineering major of Beijingshi zhigong daxue(Adult Vocational College proved by Ministry of Education) Now these books has been used as the text book and reference books in more than 10 universities such as Tsinghua university, Tianjin university, Beijing Forestry University, Beijing University of Civil Engineering and Architecture, Southeast University, South China University of Technology and so on.

内蒙古通辽孝庄园达尔罕王府

建设地点：内蒙古 通辽
占地面积：71 199.05平方米
工程主持：马炳坚
参与人员：张郁、李小龙、王颖、姜家慎、王剡、杨国莉、张惠芳

设计说明

达尔罕王府原为清代蒙古王府，它是清代孝庄皇太后的出生地，这里曾经出过四个皇后，乾隆皇帝还曾经亲临王府驻跸，政治地位十分显赫，后被毁。本次建设为重建。

达尔罕王府为通辽市特色旅游项目孝庄园旅游区的重点工程。整座王府有大小建筑66幢，总建筑面积12 761.85平方米，为全木结构，规模宏大、气势雄伟，是自清中后期至今的二百余年中建设的第一座大型王府建筑。建成后将辟为王府博物馆。

Darhan Prince Residence, Tongliao Xiaozhuang Estate, Inner Mongolia

Construction Location: Tongliao, Inner Mongolia
Site Area: 71,199.05 m^2
Conducted by: Bingjian Ma
Other design participants: Yu Zhang, Xiaolong Li, Ying Wang, Jiashen Jiang ,Yan Wang, Guoli Yang, Huifang Zhang

Design Introduction

Darhan Prince Residence was formerly a palace of Mongolia at Qing Dynasty, which is the birthplace of Xiao Zhuang, the Empress Dowager of Qing Dynasty. There have born four Queens early and late, where Emperor Qianlong once had visited the palace as lodging, so it was in a very prominent political position. Later it was destroyed. This time it is under reconstruction.

Darhan Prince Residence is one of the key projects for the special tourism at Xiaozhuang tourist areas in Tongliao city. The whole palace has 66 buildings in size, with a total construction area of 12,761.85 m^2, all in wooden structure, which is heroic in scale, and magnificent in air. It is the first large-scale palace buildings construction among 200 years since the late-mid-Qing Dynasty hitherto. This Darhan Prince Residence will be opened up as a Palace Museum after completed.

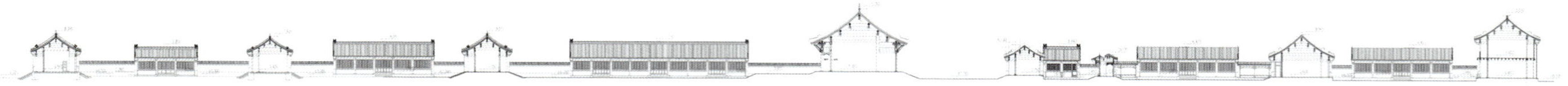

BEIJING TRADITIONAL CHINESE ARCHITECTURAL DESIGN AND RESEARCH INSTITUTE
北京市古代建築設計研究所

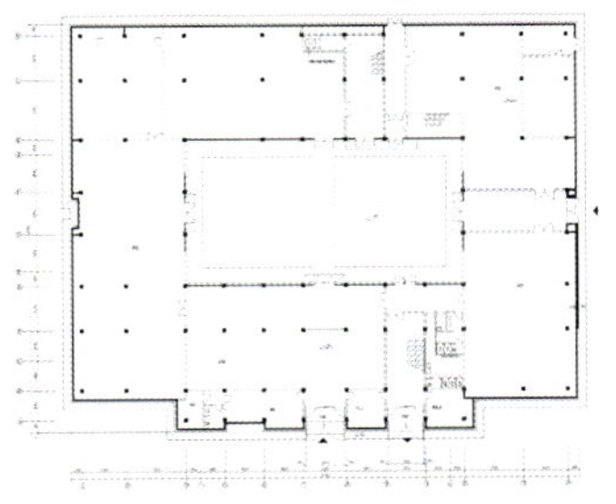

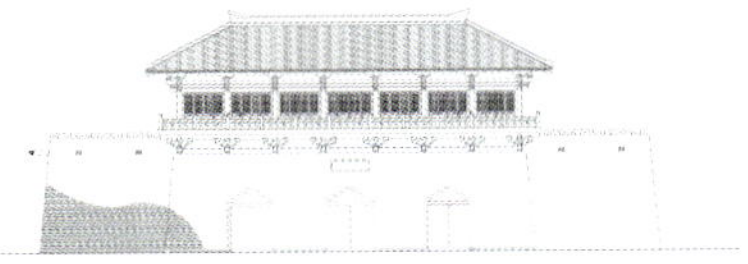

河北邯郸邺城博物馆

建设地点：河北 邯郸
用地面积：25 253平方米
建筑面积：5 228平方米
工程主持：徐新伟
参与人员：王大欣、周启玲、李栋、单成福

设计说明

邺城博物馆为综合性公共建筑，具有介绍古邺城历史与文化，展示邺城出土文物等功能，同时也是邺城考古工作站，建于河北临漳县王村以北。

博物馆为主体二层、局部三层的汉代风格城楼式建筑，主楼为高度23.10米的钢筋混凝土框架结构。

Handan Yecheng Museum in Hebei Province

Construction Location: Handan, Hebei
Site Area: 25,253 m^2
Building Area: 5,228 m^2
Conducted by: Xinwei Xu
Other design participants: Daxin Wang, Qiling Zhou, Dong Li, Chengfu Shan

Design Introduction

Handan Yecheng Museum is a comprehensive public architecture, with the function of demonstrating the history and culture of Yecheng in the past, gathering and collecting the unearthed artifacts; it is also the working station of archaeology. The museum is a gate tower architecture of Han Dynasty style with a two-storey main body and 3-storey part. The main building is 23.10 m in height and the architecture of is reinforced concrete frame structure.

北京大学国家发展研究中心

建设地点：北京
规划用地：19 500平方米
建筑面积：22 093.5平方米
工程主持：张丽鲜、
参与人员：李大雪、张郁、李小龙、谢成涛

设计说明

本项目是在北京大学中国经济研究中心基础上新组建的一个以综合性社会科学研究为主的科研教学机构。

项目本着尊重历史遗迹、保持传统风格、合理利用空间、优化功能分区、合理利用资源，及“可持续发展”的指导思想，以古典园林为优势，以传统风貌建筑为特点，把研究院建成一座外部传统古典，内部设施先进、功能齐全、配套合理、高效综合，具有浓郁中国特色的研究院。

National School of Development at Peking University

Construction Location: Beijing
Planning Site Area: 19,500 m^2
Building Area: 22,093.5 m^2
The Project is Conducted by: Lixian Zhang
Other design participants: Daxue Li, Yu Zhang, Xiaolong Li, Chengtao Xie

Design introduction

The project is a comprehensive science research education institute, which is based on China Center for Economic Research.

According to the guideline of respecting the historical remains, maintaining the traditional style, fully utilizing the space, optimizing function division, well allocating the recourses and the Sustainable development ideas, this project keeps the classical landscape architecture and traditional style and features in design. The exterior of the building is of classical traditional style, inside, there are advanced facilities with complete and efficient function. It is a research institute with distinct Chinese character.

TCBDRI

BEIJING TRADITIONAL CHINESE ARCHITECTURAL DESIGN AND RESEARCH INSTITUTE

北京市古代建築設計研究所

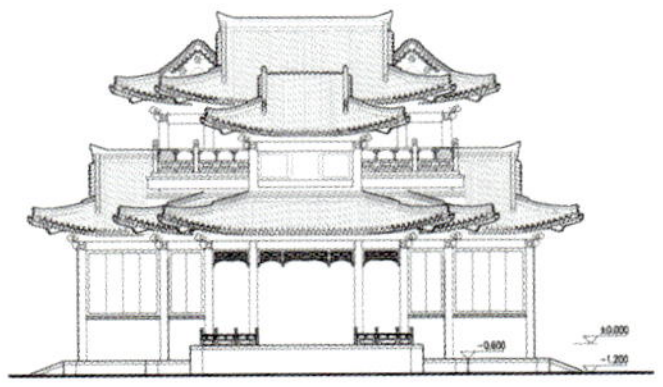

济南大明湖风景名胜区“超然致远”

建设地点：山东 济南
占地面积：3.06公顷
建筑面积：5 160平方米
工程主持：张越
参与人员：王思萌、李小龙、任一帆

设计说明

超然致远组团采用北方传统建筑形式青砖灰瓦、玉阶丹楹。地上建筑一到二层，坐落于高台之上。平面采用合院式布局，中轴对称，室内外采用全隔扇形式分隔，既保留对称格局又有所变化。

总体采用双轴线、多节点布局，形成多条游览路线。形成以博艺堂和牌楼为节点的东西主轴；以超然楼和戏台为节点的南北主轴，文化广场是两主轴的交汇。保留贺胜25号民居院落，将其补充完整，作为服务用房使用，既满足了功能需要又保留了历史建筑。

Tranquility- Scenic spot of Daming Lake in Jinan

Construction Location: Jinan, Shandong
Site Area: 3.06 ha
Building Area: 5,160 m^2
The Project is Conducted by: Yue Zhang
Other design participants: Simeng Wang, Xiaolong Li, Yifan Ren

Project introduction

Boyi Hall adopts the traditional architectural forms in the North China – decorated with black bricks, gray tiles, jade steps, and red principle columns. The floor construction, from first storey to the second, locates above the high platform. Its flat is in courtyard layout, with axial symmetry, whose indoor and outdoor separated with an entire sliding door, thus not only symmetrical pattern but also with changes in details

Overall it adopts a layout of dual-axis and multi-node, hereby forming a number of tour routes. East and west axis with Boyi Hall and the Memorial Archway as the node; North-South axis with Chaoran Tower and Stage as its node; Culture Square is the intersection of the above two axis. It Kept and repaired the No.25 residents' courtyard of Hesheng, making it as the service house, meeting the functional demand and in the meantime, well preserve the historic architecture.

TCBDRI

合道设计集团

HORDOR DESIGN GROUP

厦门合道工程设计集团是由有着40多年历史的原厦门市建筑设计院改制发展而来的。1960年厦门市建筑设计院成立，2003年经改制后成立厦门市建筑设计院有限公司，2007年8月厦门市建筑设计院有限公司及下属十余家参与控股公司，本着"道同而融合，循道而成业"的事业愿景，正式成立"厦门合道工程设计集团（以下简称为'合道设计集团'）"。

合道设计集团具有建设部颁发的建筑工程设计甲级资质，国家发改委颁发的工程咨询甲级资质，以及城市规划、市政工程（景观）、建筑智能化系统工程设计等资质。

合道设计集团拥有规范的现代企业制度、科学有效的管理机制和灵活的市场应变能力，集团立足设计主业，专注打造设计品牌，业务涉及建筑规划、建筑设计、景观设计、施工图审查、智能化设计、工程咨询及经济咨询、房地产开发与策划等领域，客户遍及北京、上海、深圳、福州、武汉、大连、青岛，南昌、昆明、成都、济南等地，迄今已与美国、加拿大、意大利、日本、英国、德国、西班牙、澳大利亚、新加坡、中国香港、中国台湾等国家和地区的设计机构建立广泛的合作关系。此外，集团始终致力于人才成长平台的研究和打造，培育了一支功底扎实、思想敏锐、敬业守纪的专业技术队伍，其中教授级高级工程师、高级工程师、工程师人数占公司人员总数的44%以上，至今已有超过50个工程项目获得部省级优秀设计奖项，赢得中国建筑设计行业分会华东联席会"2001—2005年度改革与发展最具创意奖"、中国勘察设计协会"2006年优秀勘察设计企业"、福建省勘察设计协会"2006年度优秀勘察设计单位"、"2007中国十大民营建设设计企业"等荣誉称号以及2007—2008中国民用建筑设计市场排名总榜第24名。

汗水与荣誉，变革与坚守，50年风雨历程，合道工程设计集团历久弥新。2010年的今天，他们一如既往地关注着您的生活，构筑着城市的蓝图，他们正以专业的精神打造百年合道品牌！

Xiamen Hordor Engineering Design Group is developed from the original Xiamen Architectural Design Institute which has 40 years of history. In 1960, Xiamen Architectural Design Institute was established; in 2003, Xiamen Architecture Design Institute Co., Ltd. was established by reform. In August 2007, Xiamen Architectural Design Institute Co., Ltd. and it's more than ten subordinate holding companies, in the career vision of "integrating because of the same target, perusing the target to establish the career", officially founded "Xiamen Hordor Engineering Design Group (hereinafter referred as 'Hordor Design Group')".

Hordor Design Group possesses Grade A qualification of architectural engineering design issued by the Ministry of Construction, Class A qualification of engineering consulting issued by the National Development and Reform Commission, as well as the qualifications for urban planning, municipal engineering (landscape), building intelligent systems engineering design and so on.

Hordor Design Group has the standardized modern enterprise system, scientific and effective management mechanism and flexible market adaptability. The Group takes design as the main business, and focuses on brand building. Its business involves building planning, architectural design, landscape design, construction diagrams checkup, intelligent design, engineering consulting and financial consulting, real estate development and planning and other areas. They have customers in Beijing, Shanghai, Shenzhen, Fuzhou, Wuhan, Dalian, Qingdao, Nanchang, Kunming, Chengdu, Jinan and other places, and so far they have established the broad cooperation with the design institutions in the United States, Canada, Italy, Japan, Britain, Germany, Spain, Australia, Singapore, Hong Kong, China Taiwan and other countries and regions. In addition, the Group is committed to research and create the talent development platform, and they have nurtured a disciplined and dedicated professional team with solid foundation and perceptive thought, including professor senior engineers, senior engineers, and engineers which represents above 44% of the total member of company. So far there are more than 50 projects received the provincial excellent design awards. The group has been awarded many titles of honor, such as "2001-2005 Most Creative Award for Reform and Development" of China Architectural Design Industry Branch East China Ally, "2006 Best Engineering design enterprise" of China Exploration & Design Association, "2006 Best Engineering Design Unit" of Fujian Survey and Design Institute, "2007 China Top Ten Private construction design enterprises", and so on. The group ranked the No.24th in 2007-2008 overall civil architectural design market in China.

With sweat and honor, reform and insistence, and 50-year ups and downs, Hordor Engineering Design Group stands the test of time. In 2010's today, they continue concerning your life and building a blueprint for the city, and they are building a century Hordor brand with the professional spirit.

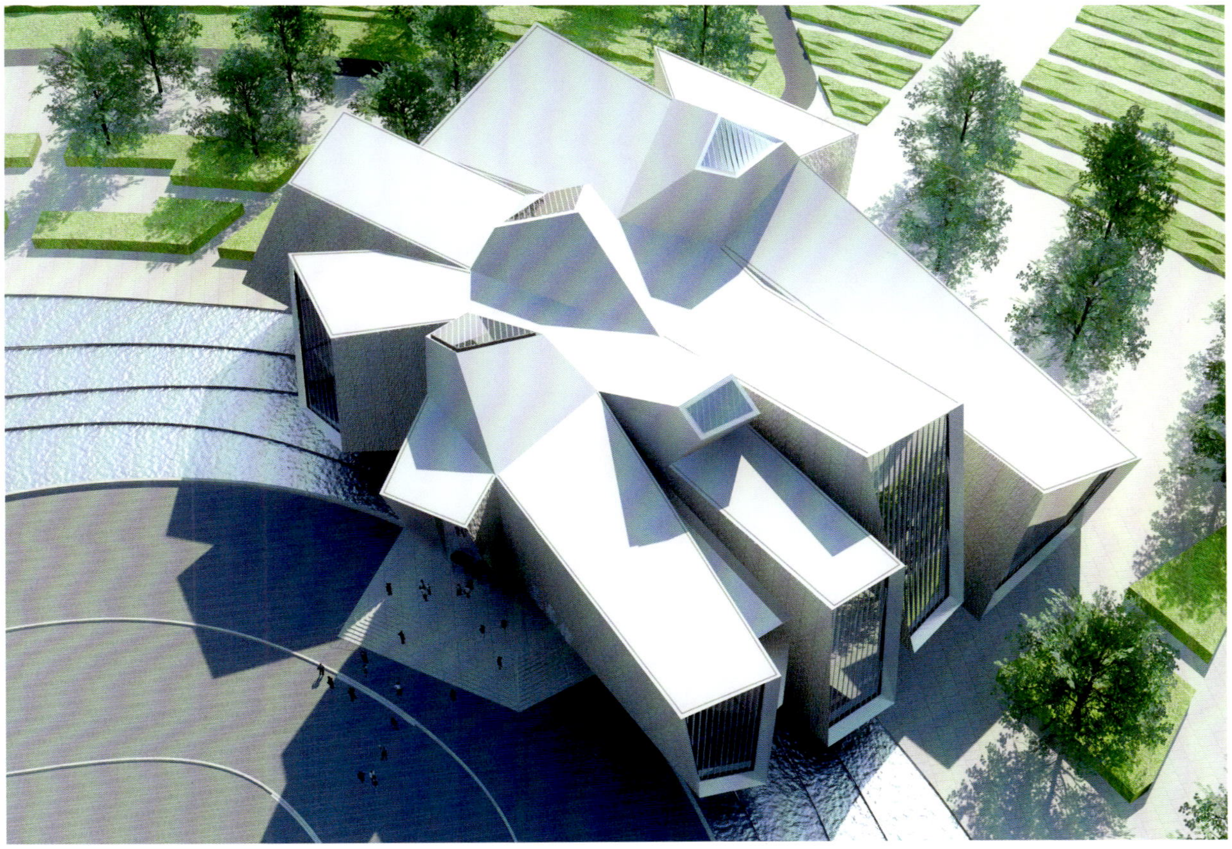

曲靖市文化中心

建设地点：云南 曲靖
设计单位：厦门合道工程设计集团有限公司、ATELIER ALTER 工作室
建设时间：待建
建筑规模：76 000 平方米

曲靖市文化中心包含了博物馆、美术馆、图书馆、科技馆、规划展示馆等五馆。本设计遵循了“一水滴三江、一脉隔两盘、一线串五珠”的区域规划理念，将“水舞朱雀、绿连三江，纵星拱月、天圆地方，水梯映照、绿波隐伏”的三大规划概念作了淋漓尽致的表达。五个文化场馆如同五颗明珠环水而居，围合出中心的文化广场。入云的梯田与连绵的山峦是云南的土地记忆设计中，设计中用这两个媒介联系了五个各具特色又相互辉映的单体建筑，在它们之间形成一个大的场地基调－梯田与水系的结合，解决了建筑与广场之间的过渡问题。

博物馆 上重下轻的造型使建筑产生漂浮感中，犹如4亿年前曲靖江河中的“鬼鱼”神游，通过悬挑的手法，表现博物馆雄壮的体量，加强建筑的雕塑感。建筑造型中的筋骨取法于爨碑书法，完整雄浑、曲折扭动，力求使其内含爨碑书法的遒劲、刚健、拙朴之美。

美术馆 建筑立意为艺术之花，一朵半开待放的石头之花，设计构思取材云南的石林壮景及海峰湿地。未完全放射张开的体量关系更增加了各体量要向外撑开生长的张力，设计手法取意海内外闻名的曲靖版画的刀法，鬼斧神工、浑然天成。

图书馆 取意书山，书山有路勤为径，设计构思取材于中国古书的折页形态结构。建筑的折页形态让人联想起中国古书的折叠结构，建筑体块组合形态也形似立起来的藏书。起翘的屋顶是中国古代建筑元素的抽象化表现，是中国古代书院记忆的传承。

科技馆 立意来源于古代中国朴素的科学观念，所谓科技乃仰观天文，俯察地理，究天人之际。科技馆将顺应环形中心广场的简单弧形建筑体量，化整为三。上、中、下三个体块是中国古代朴素科学观中天、地、人三界的隐喻，三个层次的弧形体块旋转错动，产生了建筑的层次感，是梯田层叠结构的原型。

规划馆 建筑的U形体量既内敛聚气又呈现展开双臂、拥抱世界的态势，建筑由南向北逐渐翘起，雄浑的体量形成向上腾飞的动势，寓意城市的发展。建筑形体兼具爨碑书法文字的飞跃动势，以及石林的敦厚稳健，展现了曲靖的文化历史，使建筑具有很强的整体感、雕塑感。

Qujing Culture Center

Construction Location: Qujing, Yunnan
Design Institute: Xiamen Horbor Design Group Co., Ltd., ATELIER ALTER Studio
Construction Time: to be built
Construction Scale: 76,000 m^2

Qujing Culture Center includes five halls which are museum, gallery, library, science museum, planning exhibition hall. The design follows the regional planning concept of "one drop of water in the three rivers, a pulse across two trays and one line strings five pearls", and the three plan concepts of "birds dance with water, trees connect three rivers; stars around the moon, sky surround the earth; water ladder map the trees, green wave concealed" are vividly expressed. Five cultural sites are like five pearls ringing around the water and they enclose a central cultural square. Terrace terracing into the cloud and waving mountains are the land marks of Yunnan. With these two land marks as the media linking the five single buildings with different characteristics, a large venue tone is formed between them. The combination of terraces and water serves as the transition between the building and plaza.

Museum: top-heavy shape creates a sense of floating to architecture, like the "ghost fish" of 400 million years ago swimming in Qujing rivers. By cantilever approach, the majestic body mass of the museum is performed, which strengthens the sense of sculpture to the architecture. The skeleton of the building adapts the form of calligraphy Cuan tablet, complete and forceful, twisting twists and turns. The designers seek to express the vigorous, energetic and plain beauty of Cuan tablet calligraphy.

Gallery: the conception of the building is the flower of the arts--a half-blossomed stone flower. The design concept is drawn from the scenery of Yunnan Stone Forest and the Sea Peak wetlands. The shape of the incomplete-opening radiation increases the outward growth tension of every part. The design intend to take the famous Qujing engraving techniques—natural superlative craftsmanship.

Library: the design takes the book mountain as the concept—there is no royal road to learning. The design idea is drawn from the leaflet shape of ancient Chinese book. The folding form of the architecture reminds people of the folded structure of ancient Chinese books. The building block combination shape also resembles the standing collected book. The raise roof is the abstract of Chinese ancient buildings element, and is the heritage of ancient Chinese academy library memory.

Science Museum: the concept originates from the simple science concept of ancient China. The so-called science and technology is to search up the astronomy and observe the geography, and seek the relationship between Heaven and Man. The Science Museum will follow a simple circular arc-shape of the culture center. The building dimension is divided into three, the upper, the middle and the lower. These three individual pieces are the metaphors of sky, earth and man in ancient Chinese plain science. The rotating dislocation of the three levels of arc dimensions results in the gradation of the construction, which is the prototype of terraces cascading structure.

Planning Center: the U-shape dimension of the building both possesses the introvert character of gathering and the opening state of embracing the world. The building gradually rises from south to north. The forceful dimension of the building forms the upward momentum, which have the meaning of city's development. The building shape both have the leap move potential of the calligraphy Cuan monument and the honest sound of the Stone Forest, which displays the overall sense of Qujing cultural history, and strong sculpture sense.

厦门湖里万达广场

建设地点：厦门
设计单位：厦门合道工程设计集团有限公司 万达设计研究院
建设年代：建设中
建筑规模：约 530 000 平方米（其中大型商业广场 MALL，建筑面积为 200 600 平方米）

Wanda Plaza, Huli, Xiamen

Construction Location: Xiamen
Designer: Xiamen Hordor Engineering Design Group Co., Ltd. and Wanda Design Institute
Construction Time: under construction
Construction Scale: approximately 530,000 m^2 (the floor area of the large-scale commercial plaza MALL is 200,600 m^2)

大型商业广场 MALL

商业广场 MALL 主体楼高 5 层。其功能为百货、影院、超市、数码、电器等商城。地下一层主要为超市、地下停车场和设备用房，地下二层为停车库。建筑主要以自动扶梯、电梯实现垂直交通，楼梯做为消防疏散。

就整个万达广场而言，无论从建筑体量，还是从内部空间的角度去衡量，空间的设计都是整个商业广场的重点，为了让将购物街室内外空间做得更加丰富，设计过程中针对这一点作下列几点考虑：

挑空 建筑室内大量运用了挑空设计，目的是希望内部的空间层次更加丰富、趣味。同时透过挑空的设计，空间在视觉上会显得较为宽敞，各楼层关系会变得更加密切。其中条形挑空区作为动区街道的延伸，圆形挑空多为入口，广场休息空间。

光线 如何让自然光充满空间，同时兼顾到厦门的气候特点，一直是我们在设计过程中努力追求的目标。在部分 mall street 上空及内部广场上空配置采光罩，采光罩下部采用型材管密布排列，有效控制进入商场的光线量，内衬复合板将光线隔离在外；上部以能阻挡能耗的镀膜玻璃为材料，借此减少能耗的流失量。Mall 中部广场和挑空部分阳光可以直接照射，使整个商业购物环境通透明亮。

通道 商业广场通道的设计目的除了向提供消费者通畅无阻的走道外，它还必须提供其他使用功能，才能让购物中心内部气氛变得更加活跃、舒适。通道上空条形挑空，视觉上时购物街变得更加宽敞明亮，这样它不只单纯扮演通道角色而已，也可以提供足够的空间让消费者驻足橱窗前欣赏橱窗内部的摆设，也可以轻松的倚靠栏杆，一面聊天，一面感受楼层间的空间变化。

户外广场 万达广场除建筑体量外，留设了相当宽敞的开放空间，在商业街入口前设置户外商业广场来聚合人流。除了内部空间的塑造外，外部空间的规划也由基地本身所留的大面积空地而变得更加重要。尤其，对于如何将内部空间与外部空间的活动、动线紧密的连接在一起，这也变成万达广场规划设计的另一个设计重点。

The large-scale commercial plaza MALL

Space: The main building of the commercial plaza MALL is five-storey layers in height and it is equipped with functions such as the department store, cinema, supermarket, digital and electrical appliance shops, etc. The first storey underground is mainly used for the parking lot and equipment room of the supermarket. The second storey underground is the parking garage; the escalator and elevator are used for the vertical transportation while the stairs for the fire evacuation.

For the whole Wanda Plaza, space is the focus of the commercial plaza in terms of both the architectural body mass and the internal space. In order to enrich the indoor and outdoor space of the shopping street, the design takes the following points into consideration:

Aerial Structure: The interior design of the building makes an extensive use of the aerial structure, aiming to enrich the internal spatial level. This structure also produces a more spacious visual effect, bringing the relationship between each storey even closer. The strip-shaped aerial structure is an extension of the dynamic streets, while the circular one is used mainly for the entrance and the plaza rest space.

Light: The design strives to pursue this goal: how to make the space full of natural light while taking into account the climatic characteristics of Xiamen. The overhead of partial mall streets and the indoor plaza adopts the dome skylight, under which the shaped tubes are in close configuration. This can effectively control the amount of light into the mall, and the ceramic-lined composite plate isolates the light outside of the building; the upper part takes use of the coated glass which can prevent the energy consumption, aiming to reduce the energy loss. The central plaza and aerial structure of the mall are exposed to direct sunlight, which makes the general shopping environment bright and transparent.

Channel: The design of the commercial plaza channel is to provide the customers with more functions apart from footpaths without obstruction, thus to make the internal atmosphere of the shopping center more active and comfortable. The strip-shaped aerial structure above the channel produces a brighter visual effect for the shopping street. Therefore, it is not simply a channel, but also can provide enough space for the customers to enjoy the internal decoration of the shopping windows, or leaning on the railing, chat leisurely while observing the spatial change of the building.

Outdoor Plaza: In addition to the building body mass, Wanda Plaza also sets a relatively spacious open space, aggregating the human flow by an outdoor plaza in front of the entrance of the commercial streets. Apart form the internal space shaping, the planning of the external space becomes even more important because of the large open space left by the base. How to bring the activity and the dynamic line of the internal and external space into a close relationship is another focus in the planning and design of Wanda Plaza.

宝龙城市广场

长沙建发汇金国际

建设地点：湖南 长沙
设计单位：厦门合道设计集团有限公司 澳大利亚柏涛墨尔本建筑事务所
建设时间：设计中
建筑规模：130 000 平方米

Kenfair•Huijin International, Changsha

Construction Location: Changsha, Hunan
Designer: Xiamen Hordor Engineering Design Group Co., Ltd. Peddle Thorp (Melbourne) Architects, Australia
Construction Time: in design
Construction Scale: 130,000 m²

本项目办公楼拟按照国际甲级写字楼的标准设计，建设成为长沙乃至湖南顶级办公楼。办公楼内部空间按模块化、景观化、生态化的设计宗旨，营造人性化的办公环境。外部形象以简洁、现代、稳重为原则，使建筑成为区域的标志。

平面设计上办公楼平面方正、实用，控制了核心筒及不必要的交通面积，提高了使用率，空间实用率高，办公区注意考虑面宽、进深的尺度及柱位的布置，标准层按大空间设计，布局灵活，由四个标准办公模块组成，可灵活分配给一到四家公司，每个办公模块均能很好的享受空中花园的景观资源，形成人性化的办公场所，避免了常规高层办公楼的封闭、单调、冷漠的弊端。

由于L形平面的互相咬合，咬合连接处为空中花园，电梯厅可自然采光通风，并拥有享受外部城市景观的极佳视野。而由于空中花园的设置，每个办公单元都具有自然通风的条件，充分体现了生态节能的设计理念，有效的改善了办公空间的使用率。SOHO 为兼有酒店式公寓及写字楼功能的时尚建筑。平面布局灵活，单元划分更加细致。裙楼部分配置私人健身会所、室内游泳池、会议、餐饮等配套功能，使其满足甲级写字楼及 SOHO 的配套功能。

In this project, the office building is to be built in accordance with the international Grade-A standards for office buildings, striving to become first-rate office building in Changsha, even in Hunan Its inner space is designed according to the purpose of modularity, landscaping and ecology, endeavoring to form a people-orientated office environment. Its external image is following the principle of simplicity, modernity and stability, which becomes the symbol of the whole region.

Graphic Design: The plane of the office building, which is square and pragmatic, controls the core walls and the unnecessary traffic areas, enhances the utility rate and the practical efficiency. The office district takes into consideration the width, the depth scale and the layout of column position. The standard layer, flexible in layout, is designed according to the standards of large space. It composes four standard office modules, which can be flexibly allocated to 1–4 companies. Each office module can fairly enjoy the landscape resources of the hanging garden, forming a people-orientated office place and avoiding the disadvantages of occlusion, monotony and coldness of the conventional high-rise office buildings.

As the L-shaped planes interlock with each other, the hanging garden is placed at their conjunction. The elevator hall enjoys the natural lighting and ventilation, as well as an excellent view for the external urban landscape. Because of the location of the hanging garden, each office unit meets the requirements of natural ventilation, fully embodying the design concept of ecological energy-saving and effectively improving the utility rate of the office space. SOHO is a fashion building which combines the function of hotel apartments and the office building together. Its plane layout is flexible and the unit division is even more meticulous. The podium is equipped with supporting facilities such as private fitness club, indoor swimming pool, conference room and restaurants, etc., satisfying the requirements of both the Grade-A office building and the SOHO.

厦门国际物流中心

建设地点：厦门
设计单位：厦门合道工程设计集团有限公司 株式会社日本设计
建设时间：建成
建筑规模：156 000 平方米

厦门国际物流中心聚集了世界一流的技术与信息。将海关、办公、行政三项功能有机融合为一体，其设计风格将引领未来建筑的潮流，并充分的体现生态环境理念，打造具有厦门地方特色的现代建筑。

设计风格

屋顶 南侧的低层部分和北侧的高层部分都被连贯的大屋顶所覆盖。从基地南面看，南北建筑立面重叠在一起，优美的屋顶曲线使人联想起闽南民居的独特风景。

灰空间 五栋建筑物之间由回廊连接。回廊带有屋顶但不设封闭的墙壁，在每一栋建筑物的一层部分将外墙后退，形成骑楼式空间。厦门四季气候温暖、这样一些半室外的灰空间，既为人们提供了舒适的休息交流场所，又为建筑内部引入了自然的通风和采光。

庭院 建筑之间的庭院内营造了丰富的绿色空间和水景，这里是建筑使用者交流和休息的核心空间。

红砖 整个设施的标志性部分——海关联检中心，大楼外墙采用具有闽南特色的红土砖，而其他建筑采用与之形成对比的米白色，通过建筑色彩来表达传承传统文化的重要性。

石材 行政办公楼和商务写字楼部分的主要外墙材料采用福建产的花岗岩石材，利用磨面和毛面的不同材质肌理进行组合。

高窗 为了适应厦门高温多雨的环境，传统闽南建筑在门与梁之间设置有装饰窗。本方案借鉴这一传统建筑要素，设计了可用于采光和通风的高窗。

Xiamen International Logistics Center

Construction Location: Xiamen
Designer: Xiamen Hordor Engineering Design Group Co., Ltd. and Japan Design Co., Ltd.
Construction Time: completed
Construction Scale: 156,000 m²

Xiamen International Logistics Center assembles world-class technology and information. Its design style which integrates functions of the Customs, the office and the administrative section into an organic whole, will lead the future architectural trend. It also fully manifests the concept of ecological environment and creates a modern architecture of Xiamen local characteristics.

Design Style

Roof

The lower-rise building in the south and the high-rise building in the north are both covered by the consistent roof. Viewed from the south of the base, the facades of the north and south buildings overlap with each other, and the graceful roof curve is reminiscent of the unique beauty of South Fujian residence.

Gray space

The five buildings are connected by a corridor which possesses a roof but no enclosed walls.

For each building the external wall will retreat at the first floor, forming a canopy space. Xiamen enjoys a warm weather all around the year, so this semi-outdoor gray space not only provides people with a comfortable place for rest and communication, but also facilitates the natural lighting and ventilation for the internal building.

Courtyard

The courtyard between the buildings creates a wealth of green space and water landscape, forming a core space for the residents to communicate and rest.

Red Brick

The landmark building of the whole facility—"Customs Combined Detection Center" adopts the red brick as the material for its external wall, which demonstrates the architectural characteristics of South Fujian. In contrast, other buildings adopt the color of beige, projecting the importance of tradition.

Stone Material

The main material for the external walls of the administrative office building and business building are the granite stone produced in Fujian; the design combines these stones according to the different material texture in grinding face and rough face.

High Windows

To adjust to the high temperature and rainy weather of Xiamen, the traditional building in South Fujian installs decorative windows between the door and the beam. The project adopts this traditional architectural element and designs high windows for lighting and ventilation.

崴豪酒店

建设地点：厦门
设计单位：厦门合道工程设计集团有限公司　西班牙 Melvin Villaroel Architect
建设时间：待建
建筑规模：34 000 平方米
工程性质：五星级旅馆类建筑

1. 酒店大堂

大堂是酒店形象塑造、气氛渲染的最为主要的部分，本酒店设计了二层透空的大堂，此外，大堂与咖啡厅等连成一体，形成流动感很强的开放空间，使大堂开敞气派、高贵典雅的气氛得到有力表现。

2. 生态庭院

这是与城市公园相连的景观中庭，利用原有地形高差塑造观景露台，休闲院落结合叠水泳池的设计来营造温馨、舒适的酒店庭院，成为本建筑的一大亮点。

3. 外观选型

本设计的特点是追求浓烈的西班牙建筑风格，采用独特的造型元素，创造独特的视觉感受。

厦门是一个风景优美的旅游城市，其城市形象风光秀美，建筑淳和优雅。本设计以西班牙式的建筑形象，着重于从城市设计的角度出发，力求创造和谐共荣的城市建筑形象。酒店采用简洁的体量、和谐的对比来塑造形象，有特色而不突兀。大进深的阳台，上下错落，大尺度的折角造型加上墙体上韵律化的开洞，以及富有西班牙地方特色的窗棂形式，一方面能达到城市范围内的和谐，一方面又不会被淹没在城市森林之中，使之有自己的特色。

Wei Hao Hotel

Construction Location: Xiamen
Designer: Xiamen Hordor Engineering Design Group Co., Ltd. and Spain Melvin Villaroel Architect
Construction Time: to be constructed
Construction Scale: 34,000 m^2
Project Attribute: construction of the five-star hotel type

1. Hotel Lobby

The lobby is the most important part of the hotel in terms of image building and atmosphere creation. Therefore, this hotel designs a permeable lobby of two storys. In addition, the lobby forms an open space of strong fluidity by connecting with the café house, fully demonstrating an atmosphere of grandness and elegancy.

2. Ecological Courtyard

The landscape atrium connecting with the urban park, the viewing terrace molded by utilizing the original altitude difference and the warm, comfortable hotel courtyard created by the combination of the leisure courtyard and the cascading swimming pool, all these become the lightspot of this construction.

3. Selection of the External Appearance

The feature of this design is to create a unique visual experience by adopting special styling elements in pursuit of a strong Spanish flavor.

Xiamen is a tourism city of beautiful landscape and its urban image is exquisite in scenery, simple and elegant in construction. Adopting the Spanish-style architectural image and focusing on the perspective of urban design, this project design strives to create an urban architectural image of harmony and common prosperity. The hotel builds its image through the simple body mass and the harmonious contrast, which is distinctive and not abrupt. The balcony of great depth which scatters up and down, the large-scale corners up and the open holes of regular rhythm on the wall, as well as the window lattice full of Spanish regional characteristics, are able to achieve harmony within the city limits; on the other hand these distinctive characteristics will save the hotel from being drowned in the urban forest.

温德姆和平大酒店

建设地点：厦门
设计单位：厦门合道工程设计集团有限公司
建设时间：已建成
建筑规模：87 014.674 平方米
建筑高度：98.9 米

Wyndham Peace Hotel

Location: Xiamen
Designer: Xiamen Hordor Engineering Design Group Co., Ltd.
Construction Time: completed
Construction Scale: 87,014.674 m^2
Construction Height: 98.9 m

功能布局

地下一层为车库及平战结合的六级人防工程及酒店后勤配套用房。地上一至四层为酒店配套用房，其中一层设有酒店大堂、商场和酒店后勤办公空间，二层为桑拿、西餐厅、咖啡厅和酒吧等，三层为多功能厅、中餐厅等，四层为会议室等，五至十八层南侧为酒店式公寓，北侧为花园，十九至二十五层为客房，二十六层为贵宾房和总统套房，客房套数为 324 套，酒店式公寓为 126 套，合计 450 套。

Functional Layout

The basement is equipped with the garage, the six-grade civil defense project with a combination of ordinary and military application and the logistic supporting room of the hotel. One to four floors of the building on the ground are the supporting rooms: the first floor is equipped with the hotel lobby, shopping malls and logistic office; the second floor the sauna, western restaurant, café house and bars, etc.; the third floor the multi-purpose hall, Chinese restaurant, etc. and the forth floor is conference room. The south side of the fifth to eighteenth floor of the building is for the hotel apartments, and the north is the garden. The nineteenth to twenty-fifth floor are for the guest rooms and the twenty-sixth are the VIP rooms and presidential suites. There are 324 guest rooms, 126 hotel apartments with the total number of 450.

阿罗海城市广场

建设地点：厦门
设计单位：厦门合道工程设计集团有限公司
美国那郭达·刘联合建筑师事务所
（LAGUARD.LOWARCHITECTS）
建设时间：建设中
建筑规模：100 000 平方米

本案为2～3层的集大型超市、百货、餐饮、娱乐、停车于一体的多层商业综合建筑群。总平面为哑铃状布局，分成南北两区，两区之间主广场及两区次广场将内湖景观引入建筑入口，体现独特的滨水购物氛围。两区又分别以数个小广场为中心的单元，均以回廊环绕，相互分隔又相互串联，步移景异，给消费者提供不同的视觉效果和充满变化的空间体验。

本案风格以热带风情为理念。沿湖设置有供市民活动的休闲设施及小岛。主广场提供集中的商业、休闲活动场地。内街上设有一连串的袖珍广场，设置不同的活动场所，给消费者创造出自然、温馨的环境。具有现代感的膜结构、钢结构等遮阳设施提高了开放式街道的步行舒适度，也减轻了厦门气候带给人的炎热感。沿湖岸线的区域设置休闲餐饮区，将内街和滨水步行道连接在一起，体现本案休闲的特色。

Aluohai Urban Plaza

Construction Location: Xiamen
Designer: Xiamen Hordor Engineering Design Group Co., Ltd.
LAGUARD•LOWARCHITECTS
Construction Time: under construction
Construction Scale: 100,000 m^2

This project is a commercial complex building group of 2–3 storeyed integrating the large-scale supermarket, department store, restaurants, entertainment and parking garage. The general plan is of the dumbbell-shaped layout, which can be divided into two districts: the northern and southern district. The main square between these two districts and the minor squares within them introduce the inner lake landscape into the building entrance, reflecting a unique waterfront shopping atmosphere. These two districts respectively form several small-square-centered units which are all surrounded by corridors. The units separate with each other while lay in a serial connection, presenting new landscapes at different angles, thus providing consumers with different visual effects and various spatial experiences.

The design for this project adopts the concept of tropical coquette. Along the lake are supplied with recreational facilities and islands for the citizens to exercise. The main square provides a place for commercial and recreational activities. The inner streets set a series of pocket squares and different activity places, creating a natural, warm environment for the consumers. The shading devices which posses a sense of modernity, such as the membrane structure and the steel structure, improve the walking comfort of the open streets and lower the sweating heat of Xiamen. Along the lake bank there are leisure dining areas which connect together the inner streets and the waterfront walking streets, reflecting the relaxation feature of this project.

即墨宝龙城市广场

建设地点：山东 即墨
设计单位：厦门合道工程设计集团有限公司
建设时间：建设中
建筑规模：约 580 000 平方米

建筑定位

由于项目处于地理条件优越且开发规模较大的用地，因此项目拟建为即墨市集大型商业广场MALL、休闲商业街、酒店式公寓及高档景观住宅于一身的城区综合大盘，规划从如下几方面对产品进行定位设计：

1. 商业业态

商业在塑造新型的社区与地区中心中扮演着至关重要的角色，同时商业空间对城市其他公共空间的影响也无所不在，并且正重新唤起人们对商业所扮演的都市中心角色的关注。商业的成功，使它并不仅仅作为社区或地区中心的一个配套，而其本身所带来的与都市生活之间的相容性、延展性、灵活性及流动性特质，也正在成为都市中不可忽视的部分。本案的规划设计正试图在住宅小区及大型商业MALL、休闲商业街三者的空间塑造及活力激发中找到一个最好的平衡点，来构筑地区和谐的生活链。

2. 高档景观住宅

由于项目南侧有墨水河这一自然景观资源可利用，规划设计以高档景观住宅定位为引领，居住本身即代表了舒适、便利及相应的尊贵气韵。与水共生，带有城市空间的开放与共享气息，这就成为景观尊贵豪宅的必要特质。同时作为商业区的住宅，它最大的价值还在于既可达到与城市空间、国际气息的最大融合，同时又可以享受私家领域的惬意生活，使居住环境宁静而不失繁华。

Jimo Powerlong Urban Plaza

Construction Location: Jimo, Shandong
Designer: Xiamen Hordor Engineering Design Group Co., Ltd.
Construction Time: under construction
Construction Scale: approximately 580,000 m^2

Construction Orientation

Because of the superior geographical condition and the large-scale development land, this project endeavors to build into an urban integrated zone of Jimo which combines the large-scale commercial plaza, shopping malls, leisure commercial streets, hotel apartments and high-end landscape residential community together. The planning positions this project in terms of the following aspects:

1. Commercial Form

Business plays a crucial role in shaping the new community and the regional center. Meanwhile the ubiquitous intervention of the commercial space into other urban public spaces is reigniting people's attention to the role business plays in shaping the urban center. The commercial success makes it not only a supporting component of the community or the regional center; it is becoming a pole which is not insignificant in city because of its compatibility, extensibility, flexibility and mobility with the urban life. The planning and design of this project strives to find a good balance between the spatial molding and vigor stimulating of the residential community, the large-scale commercial mall and the leisure commercial streets, creating a living chain of regional harmony.

2. High-end Landscape Residential Community

Because the south side of the project is adjacent to Moshui River, the planning and design take advantage of this natural landscape resource and position the project as a high-end landscape residential community which represents comfort, convenience and the corresponding nobility. Co-living with water and enjoying an open, sharing atmosphere of the urban space have become the essential features of the landscape mansion of nobility and luxury. At the same time as a residential community located in the business district, its greatest value lies in not only the maximum integration with the urban space and the international flavor, but also the possibility of enjoying one's private domain and having a life of tranquility and prosperity.

盐城宝龙城市广场

建设地点：江苏 盐城
设计单位：厦门合道工程设计集团有限公司
建设时间：建设中
建筑规模：约 490 000 平方米

设计原则

统一原则

商业部分强化商业街的完整性与统一性，强化主力店与商业街的联系，以风格统一的环境设计来规划整个项目。公寓部分着力创造一个布局合理、设施完善、生活方便、利于管理、环境优美且具有时代精神的高档社区。

以人为本的原则

贯彻以人为本的原则，设计充分体现对人的关怀，力求创造轻松、舒适、独具特色的休闲购物环境。

文化原则

努力提高文化品位，提炼经典商街的符号，以此为载体，精心设计小品与绿化，并通过空间变化、建筑语言形式等增加设计的文化情趣。

精致的原则

从整体到细部设计都遵循现代、典雅、精致的原则，并延续一定的文脉及本土特色，以环境自身的特色体现商业氛围。

Yancheng Baolong Urban Plaza

Construction Location: Yancheng, Jiangsu
Designer: Xiamen Hordor Engineering Design Group Co., Ltd.
Construction Time: under construction
Construction Scale: approximately 490,000 m^2

Design Principle

The principle of integrity:

The commercial component emphasizes the integrity and unity of the commercial street, and strengthens its relationship with the main-force shops, designing the whole project through a uniform environment. The apartment component endeavors to create a high-quality community with beautiful environment which is reasonable in layout, perfect in facility, convenient in life, easy in management and with the spirit of times.

The principle of humanism:

Upholding the people-oriented principle and fully reflecting the concern for people, the design strives to create an easy, comfortable and unique leisure shopping environment.

The principle of culture:

This design endeavors to improve the cultural quality, refine the signs and carriers of the classic commercial street, meticulously design the sketches and greening, and enhance the cultural flavor through space change and architectural language, etc.

The principle of refinement:

From the whole to the detail, the design follows the principle of modernity, elegancy and refinement, extends to a certain degree the contextual and local features, and reflects the commercial atmosphere by the characteristics of the environment.

李沧宝龙城市广场

建设地点：青岛
设计单位：厦门合道工程设计集团有限公司
建设时间：建设中
建筑规模：约 360 000 平方米

1. 城市文脉的思想

开放性：

商业及公建形象的建立及辐射作用使项目确立了在城市空间环境中的格局；公寓区的空间环境和绿化景观以及配套服务功能对城市的开放与渗透。

包容性：商业建筑与公寓的各种建筑形式和环境形态的相互包容的及有机组织，使项目拥有多相互样性、多层次的环境特征，增强商业与居住生活的趣味性、丰富性。

流动性：

建筑布局顺应周边条件变化，商业街蜿蜒曲折，以围合的方式布置，使空间和视觉联系形成"流动感"，景观设计结合保留的水系给人以丰富的环境体验。

2. 整合设计思想

整合设计思想是从建筑设计、城市设计到经济、文化、社会、生态的统一考虑与创新。

设计形成过程中充分尊重基地的原有地形，地貌，减少对地表生态结构的破坏。同时设计中充分利用地貌资源作为小区的重要景观要素。以大片的绿化景观和高处良好的视线渗透，形成居住区核心的公共活动空间，保留水系形成的丰富的休闲商业环境。设计同时充分尊重城市的区域环境，使项目的商业区和公寓区与城市的空间和景观融为一体。

Licang Baolong Urban Plaza

Construction Location: Qingdao
Designer: Xiamen Hordor Engineering Design Group Co., Ltd.
Construction Time: under construction
Construction Scale: approximately 360,000 m^2

1. The principle of urban context

Openness: The image of the commercial public building and its radiation function establish this project's pattern in the urban spatial environment. The space environment and greening landscape as well as the accessoty services of the apartment area are of openness and permeability to the city.

Compatibility: the compatibility and organic combination among the various architectural forms and environmental modes of the commercial public building and apartment impart to the project a diverse and multi-level environmental feature, enhancing the flavor and richness of the commercial and residential life.

Fluidity: The construction layout conforms to the changes of the surrounding condition and the enclosed form of the zigzagging commercial streets produces "a sense of fluidity" by relating the space with vision. The landscape design reserves and integrates with the water system, creating a rich environmental experience.

2. The principle of integrated design

The integrated design principle is the unified consideration and innovation from the architectural design, the urban design to economy, culture, society and ecology.

The formation of the design fully respects the original topography of the base, reducing damage to the surface ecological structure of the land surface. Meanwhile the design takes advantage of the geological resource as an important landscape element in the community and forms the residential core of residential community—the public space by large tracts of green landscape and good sight view from the high place. The reservation of the water system creates a rich leisure commercial environment. In addition, the design shows full respect to the city's regional environment, integrating the commercial public building and apartment area with the urban space and landscape.

规划结构及组织

本项目被贯穿南北的青年路自然划分为 2 号地块和 5 号地块两块用地。2 号地块为滨水休闲商业街，5 号地块东北部为一大型购物中心，南部是由高层住宅、多层和花园洋房组成的居住区，二者之间自然形成一条商业街，利用高差和建筑体量关系均匀串联起各主要公共空间和各个商住单元，体现了较好的整体性和均好性。

1. 规划结构　根据社区的总体布局形态，在遵循原有城市规划要求的前提下，结合城市干道及围绕小区的道路自然连接成为一个有机的整体。东西向水平轴与南北向的垂直轴构成了本社区的主体构架，小区结构明晰、简洁，充分显示了社区的高尚品质。

2. 景观与商业、住宅布局相结合　地块的布局东侧和北侧以商业为主，西南侧以住宅为主，板式平面的组合角度，使住户拥有更好采光面和观水视野，5 号地块东侧为 6 层情景洋房及宝龙复式住宅，东侧滨水区域为 2 -3 层的商业经营体，良好的景观形成了公众乐而往之的休闲性空间。西高东低的布局丰富了小区的天际线及沿街景观，小区空间灵活，富有动感。

3. 园林风格　地段显著的景观特点是东侧的古黄河运河景观带，总体的意象是充分利用周边的景观资源，在社区内通过情景洋房的生态性和植物的生机勃勃、绿意盎然，综合景观小品与建筑，组合成为"绿化厅廊"的园林风格。

4. 绿化贯穿　利用步行道、车行道的路边绿化及入口绿化广场及底层架空绿化将绿地连在一起，使各个单元围合的中心绿地相互渗透贯通，形成点、线、面相结合的绿化网络体系，最终形成绿地、景观小品、建筑艺术形态相融合的精品社区格局。

Planning Structure and Organization

This project is naturally divided into Block 2 and Block 5 by Qinnian Road running from south to north. Block 2 is Waterfront Leisure Commercial Street. Block 5 has a large-scale shopping center in the northeast, and the residential community composing high-rise residence, multi-storey and garden houses in the south, thus, a commercial street is naturally formed between these two blocks, which equally connects each main public space and each commercial & residential unit by the height difference and the body mass, displaying a good sense of integrity and sharing.

1. Planning Structure: According to the overall layout of the community and in compliance with original requirements of the urban planning, the design produces an organic integrity by the natural connection between the urban main roads and the roads around the community. The east-west horizontal axis and the north-south vertical axis form the main framework of this residential community, which is clear, simple and fully demonstrates its high quality.

2. Integration of landscape and commercial & residential layout: The east and north side of the block layout are business-oriented, while the southwest focuses on residence. The combination angle of the plate plane provides households with a better lighting surface and water vision. The east of Block 5 is 6-storey landscape houses and Powerlong duplex apartments and the east of Waterfront District is the business operation body of 2-3 floors. The fine landscape forms a leisure space which is the residents' popular resort. The west-high east-low layout enriches the skyline and the landscape along the streets, producing a flexible and dynamic residential space.

3. Garden Style: A significant scenery feature of the block is the canal landscape belt of ancient Yellow River. The overall concept is to take full advantage of the surrounding scenic resources and combine the facilities of landscape and the architecture into a garden style of "Green Gallery" through the ecological features of landscape houses and the rich, vibrant vegetation.

4. Greening Penetration: The design connects the green space into integrity by utilizing the roadside greening along the walking lane, traffic lane, the green entrance plaza and the elevated green at the bottom floor. The mutual infiltration and penetration of the central green spaces enclosed by each unit produce a network system with a combination of points, lines and planes, and eventually forms a boutique community structure integrating greening, facilities of landscape and the architectural artistic form.

宿迁宝龙城市广场

建设地点：江苏　宿迁
设计单位：厦门合道工程设计集团有限公司
建设时间：建设中
建筑规模：约 400 000 平方米

Suqian Powerlong Urban Plaza

Construction Location: Suqian, Jiangsu
Designer: Xiamen Hordor Engineering Design Group Co., Ltd.
Construction Time: under construction
Construction Scale: approximately 400,000 m^2

国贸．天琴湾

建设地点：厦门 五缘湾
设计单位：厦门合道工程设计集团有限公司
建设时间：建设中
建筑规模：260 000 平方米

ITG • Horizon Dominion

Construction Location: Wuyuan Bay, Xiamen
Designer: Xiamen Hordor Engineering Design Group Co., Ltd.
Construction Time: under construction
Construction Scale: 260,000 m^2

小区内建筑主要分为多层住宅与高层住宅两种建筑类型，我们在多层住宅立面处理上运用了古典建筑语言与现代元素的混搭。在 1 ～ 3 层，以古典形式为主，而在第 4 层，穿插有大片的玻璃，使建筑的顶部更加通透和轻盈，加强建筑的滨海特色。值得一提的是，在 1 ～ 3 层的古典立面上，我们放弃了过多的装饰细节的堆砌，而主要靠形体的组织来丰富建筑效果，使建筑在呈现古典风格的同时，又具有现代气息，避免给人过于繁琐和复古的印象。高层住宅立面以 Art-Deco 风格为主，追求明确、平衡和形体的简洁美，通过变形、简化、几何化将古典装饰转变为摩登装饰，将时尚和传统有机的结合，在表现时尚的同时又充满怀旧的情调。设计在高层建筑的底层部分使用局部石材装饰，使整个小区在近人的尺度上能够相互协调。

Buildings within the community are mainly of two architectural styles: the multi-storey building and the high-rise building. In the façade treatment of the multi-storey building, we adopt the mix & match of the classical architectural language and the modern elements. The first to third floor mainly follows the classical style; the forth floor is interspersed with large areas of glass, making the roof lighter and more transparent, thus enhancing the buildings' coastal features. It is worth mentioning that for the classical facade of the first to third floor, we abandon the excess piling of decorative details, but mainly depend on the form organization to enrich the architectural effect. Thus while the buildings demonstrate a classical style, they also possess a flavor of modernity, avoiding the impression of over complexity and retrospection. The facade of the high-rise building mainly adopts the Art-Deco style in the pursuit of clarity, balance and simplicity. The classical decoration has been converted into the modern counterpart by means of transformation, simplification and geometry. The vogue and the tradition achieve an organic combination, expressing fashion in a nostalgic mood. The bottom floor of the high-rise building is partially ornamented by the stone material, striving to coordinate the whole community in a close dimension.

万科．金域华府

建设地点：厦门 杏林
设计单位：厦门合道工程设计集团有限公司
建设时间：设计中
建筑规模：110 000 平方米

Vanke•King Metropolis

Construction Location: Xinglin, Xiamen
Designer: Xiamen Hordor Engineering Design Group Co., Ltd.
Construction Time: in design
Construction Scale: 110,000 m^2

在这个位于厦门杏林湾的住宅区的设计中，我们希望营造一个理想的居住场所，提炼一种更富文化情结、更加人性化的场所精神，创造出好的生活环境。我们希望在这里，居住区不只是无名栖息的所在，更是一种优雅而完美的艺术形式，进而唤醒人们对海湾城市海景生活的向往。我们希望能通过设计规划来充分挖掘本地块的价值，在保证开发的商业效益的同时，为城市新区的建设有所贡献，并对杏林新区居民住房消费理念的提升起到一定的引导作用。

贯彻以人为本的设计思想，以建设功能布局合理，配套及设施完善，建筑造型新颖的居住小区为目标，注重对地域性、文化性的挖掘，空间环境的渗透，使小区既具有现代气息又不缺传统特色。

建筑的单体设计上，平面布置动静有序、洁污分流、功能齐全适用，平面紧凑合理，既保证卧室、客厅有良好的朝向与通风，又保证厨房能够自然通风、采光良好且餐厅、卫生间都有阳光照射进来。

外观造型上运用新古典主义建筑风格，并通过适宜的尺度处理力，求创造一种典雅、宁静、安享的人居环境。

During the design for this residential area located in Xinglin Bay, we hope to create an ideal place to live, refine a place spirit of more cultural complex and humanism, and produce a better living environment. We hope that here, the residence is not just an unknown habitat, but an artistic form of elegancy and perfection, thus evoking people's yearning for the seaside life of coastal cities; we hope to fully exploit the value of the land through the design and planning, while ensuring commercial benefit of the development, contributing to the construction of the new urban district and playing a guiding role in enhancing the housing consumption concept of the residents in Xinglin New District.

Upholding a people-oriented design concept and aiming to build a residential community which is reasonable in functional layout, sound in supporting facility and novel in architectural form, the design focuses on the exploitation of regionality and culture, as well as the penetration of the space environment, endowing the community a modern flavor while reserving the traditional characteristics.

Unit Design: The plane layout is orderly in static, dynamic and sound in functional application. It also separates the clean and the dirty. The plane is compact and reasonable, which ensures good orientation and ventilation for not only the bedroom and living room, but also the kitchen, providing enough light for the dinning room and the bathroom. The appearance form adopts the neo-classical architectural style and endeavors to create a living environment of elegancy, tranquility and comfort through an appropriate scale processing.

水晶湖郡 1 号地块

建设地点：厦门 杏林湾
设计单位：厦门合道工程设计集团有限公司
设计时间：设计中
建筑规模：265 000 平方米

Block 1 of Crystal Lake County

Construction Location: Xinglin Bay, Xiamen
Designer: Xiamen Hordor Engineering Design Group Co., Ltd.
Design Time: in design
Construction Scale: 265,000 m^2

平衡传统与现代的理性空间，以简约、洗练、清新且具有时代感的建筑语言，使整体形象具有英式及澳大利亚建筑风格的表现力，让居住者情感回归自然和宁静，营造高尚的人文社区形象。

本项目最大的优势在于每户都拥有宽阔良好的景观面，因而高层的立面设计采用通透的落地玻璃，大开窗，以大飘板作为建筑的主要造型元素，即统一了整体细部格调，又使光影富于变幻——玻璃上天空的倒影和飘板在不同时段投射的阴影，与建筑的其他材质形成虚实效果及不同质感的对比。与高层相接的商业建筑采用现代英国风格，而低层与多层住宅建筑则采用了传统的英伦风格，与高层的现代风格，有对比、有呼应，更有着千丝万缕的联系，给人无限的遐想空间。

The design balances the rational space between the tradition and modernity. Through a simple, condense and fresh architectural language with a sense of times, it imparts to the overall image an architectural expression of British and Australian style, helping the residents feel to return to nature and tranquility by creating a noble community image of humanism.

The greatest advantage of this project lies in that each household has a fine and wide landscape view, thus the facade of the high-rise building adopts the transparent floor glass and large windows. Utilizing the large floating plate as the main modeling element, the design not only unifies the overall detail style, but also produces light of infinite changes—the sky reflection of the glass and shadow projection of the floating panel at different time periods together with other building materials, produce an effect of falsity or truth and a contrast in different textures. The commercial building connecting with the high-rise adopts the modern British style, while the low-rise and the multi-storey residences assume the traditional British style, echoing and contrasting with the former. Their inextricable relation leaves people an unlimited reverie space.

福州市规划设计研究院

People's Republic Of China
Fuzhou Planning Design and Research Institute

福州市规划设计研究院是国家首批综合性甲级勘测规划设计研究单位之一。全院有员工590人，其中国家注册规划师、建筑师、结构师、造价师、咨询师、工程师335人，高级工程师80 人，教授级高级工程师27人。业务部门有勘测院、规划设计所、市政设计所、建筑设计所、综合设计所、环境工程所、工程经济所、园林设计所、交通设计研究所、公路设计所、古代建筑所、城市设计及景观所、工程项目管理中心、工程服务中心、半导体照明应用研究所。研究院还通过了ISO 9001：2000质量管理体系认证。

研究院具有城市规划、市政公用行业（含道路、桥隧、给水、排水、风景园林、环境卫生、公共 交通、热力工程、载人索道）、建筑工程、工程咨询、工程勘察（综合类）、工程测量、地理信息系统工程、摄影测量与遥感、文物保护工程勘察设计监理、公路行业（公路）甲级资质，土地规划、旅游规划设计、大地测量、地籍测绘、行政区域界线测绘、地图编制、房产测绘乙级资质。

研究院先后完成了福州、福清、长乐、石嘴山、武威等城市的多项总体规划、专业规划项目如：福州“三坊七巷”历史文化街区保护工程规划、设计，鼓屏路至八一七北路城市设计、茶亭街城市设计、闽江至乌龙江两江四岸城市设计；福州、泉州、漳州、广州、宁德等城市主干道、大型桥梁、隧 道工程；福州、漳州、南平、山西、江苏、云南等城市污水处理厂及厂外管网设计；福州屏•山镇海楼、君临•盛世茶亭、闽江学院、职业技术学院、福州文体中心、福州市委市政府办公楼、金山工业区、福州大洋百货等建筑设计；日本那霸“福州园”、西藏林芝“福建园”、福州闽江公园（南、北）、福州新动物园、西湖公园、漳州、宁波、宜昌、广州、河南、南 昌、武汉等城市重要园林景观、绿地规划等500多项大型规划、设计及大量工程勘察、测量、地理信息系统开发应用。项目遍及福建、北京、上海、广东、浙江、江苏、安徽、江西、山东、湖南、湖北、河南、贵州、重 庆、宁夏、甘肃、云南、西藏、青海、四川及日本那霸市等。

As the first batch of national comprehensive research institute with Class A Certificate of Survey and Planning design, Fuzhou Planning Design & Research Institute has a total of 590 staffs, of which there are 80 senior engineers, 27 professor-level senior engineers, and 335 professionals including national registered planners, architects, structure engineers, cost engineers, consultants and engineers.
Business Department: Survey Institute, Planning Design Institute, Municipal Design Institute, Architectural Design Institute, Integrated Design Institute, Environmental Engineering Institute, Engineering Economy Institute, Landscape Design Institute, Transport Design Research Institute, Highway Design Institute, Ancient Architecture Institute, Urban Design & Landscape Institute, Engineering Project Management Center, Engineering Service Center and Institute of Semiconductor Lighting Applications;Management Standard: ISO9001:2000 Quality Management System.

Class A Certificate: urban planning, municipal public industries (including road, bridge and tunnel, water supply, drainage, landscape architecture, environmental health, public transportation, thermal engineering and manned cableway), architectural engineering, engineering consultant, engineering survey (Comprehensive), engineering survey, geographic information system engineering, photogrammetry and remote sensing, exploration and design supervision for preservation project of cultural relics, and highway transport industry(highway);
Class B Certificate: land planning, tourism planning design, geodetic survey, cadastral surveying and mapping, administrative district boundary surveying and mapping, map compilation and real estate mapping.

It has completed many master planning and professional planning in Fuzhou, Fuqing, Shizuishan, Weiwu and so on; protective project planning design for historical cultural block of "Three Districts and Seven Lanes" in Fuzhou, urban design from Guping Road to Bayiqi North Road, urban design for Chating Street, urban design for two rivers and four banks from Minjiang River to Wulong River, main roads, large scale bridges and tunnels in Fuzhou, Quanzhou, Zhangzhou, Guangzhou, Nngde and so on, design for sewage treatment plant and pipe network outside the plant in Fuzhou, Zhangzhou, Nanping, Shanxi, Jiangsu, Yunnan, and so on; architectural design for Pingshan Zhenhai Tower in Fuzhou, Junlin•Shengshi Tea Pavilion, Minjiang College, College of Vocational and Technical, Fuzhou Recreational and Sports Center, Fuzhou Municipal office building, Jinshan Industrial Zone, and Fuzhou Grand Ocean and so on; "Fuzhou Park" in Naha, Japan, "Fujian Park" in Lizhi, Tibet, Minjiang Park (south and north) in Fuzhou, Fuzhou new zoo, Xihu Park, planning for important garden landscape and Greenland in Zhangzhou, Ningbo, Yichang, Guangzhou, Henan, Nanchang, Wuhan and more than 500 large scale planning design and a large number of engineering exploration and survey as well as development and application of Geographic Information System.

Projects spread Fujian, Beijing, Shanghai, Guangdong, Zhejiang, Jiangsu, Jiangxi, Anhui, Shandong, Hunan, Hubei, Henan, Guizhou, Chongqing, Ningxia, Gansu, Yunnan, Tibet, Qinghai, Sichuan, Naha in Japan and so on.

地　　址：福州市湖东路188号　　邮　　编：350003
联系电话：+86—591—87807936　　传　　真：+86—591—87826264
邮　　箱：ghy@fzghy.com　　网　　址：www.fzghy.com

Add: Hudong Road No.188, Fuzhou　P.C.: 350003
Tel:+86-591-87807936　Fax: +86-591-87826264
E-mail: ghy@fzghy.com　Http://www.fzghy.com

福州南后街

南后街是中国十大名街之一，它沿南北方向，贯穿整个三坊七巷地区，是集中展示三坊七巷乃至整个福州旧城风貌的重要窗口。南后街北至杨桥路，南至安泰河，全长615米，沿街两侧进深6–30米不等。街内有三处文物保护单位及多处保护建筑和历史建筑、老字号等，文化遗产非常丰富。南后街保护性改造以传统商业为主，除传统木结构建筑外，还有近代典型的砖石建筑，这些建筑形成较为丰富的、多元化的沿街立面效果。设计强调小规模、连续化的沿街商业建筑格局，以促成生动、有趣的街道空间，并适当加入集中式商场、庭院式商业空间，丰富商业业态，塑造具有强烈域特征的历史文化街道。

福州市规划设计研究院　地　址：福州市湖东路188号　邮　编：350003　电　话：+86-591-87807936　传　真：+86-591-87826264　邮　箱：ghy@fzghy.com　网　址：www.fzghy.com

福州茶亭街地下空间

项目设计充分尊重、理解茶亭街总体规划，并保持整条茶亭街高品质、高水准的整体连贯性与有机统一性，充分发挥该项目在八一七路商业中心的纽带与核心作用，为提升与强化八一七路商务功能，塑造“海峡第一街”作出积极有效的贡献。

设计运用城市、建筑、区域交通三位一体的设计理念，实现地面与地下的功能及交通一体化，构建独具特色的中央商务休闲核心区。贯彻商业与休闲、文化与景观有机结合的设计原则，坚持以人为本，充分保留茶亭河及茶亭公园自然资源与人文特色，营造设施完善、功能齐全的景观空间，力求建构人工与自然、传统与现代和谐融合的、极具地域特色的都市中央商业核心区。

设计充分挖掘基地环境要素，将城市开发空间、茶亭公园、地铁群众路站、茶亭水街、8号和9号地块大型商业空间有机地融为一体，创造出极富当代休闲购物、时尚体验、生态、人文游历的城市中心区的活力场所。

在群众路口、洋头口设置大型城市下沉广场空间，沿八一七路东侧设置下沉水街，下沉广场设置动态自动扶梯、台地花园，下沉水街设置坡地林区和大面积静水池与高大瀑布相结合，形成生动时尚的当代城市开放景观。南北两端的下沉广场通过带状下沉水街形成了既富自然生态特征，又具有强烈时代人文品质的都市活力空间，与城市公园（茶亭公园）融为一体，将为福州市民提供一处极具魅力与个性的城市大型开放空间。

福州市图书馆乌山新馆

环境设计

尊重基地周边现有环境，合理组织交通，与周边公园及各景区连为一体。保留基地内部较大树木，依地势进行规划设计，并借助周边自然人文景观，营造清幽、文化氛围浓厚的环境。

院落空间

规划图书馆的院落空间包括主院落和次院落两种空间形态。

主院落：由南北向主要建筑及其东西向侧房围合而成，其空间形态相对方正，空间较大，在院落四角进行绿化，院落中间可设置方形水池或绿地。

次院落：由两列主要建筑山墙及侧房围合而成，其空间形态较为细长，空间较小，在院落四角结合台阶进行绿化，院落中间为砖石铺地。

建筑风格

新图书馆是具有地域文化特色的现代建筑，其建筑风格特点是采用新材料、新技术、新手法来诠释地域传统文化。

福州市规划设计研究院　地　址：福州市湖东路188号　邮　编：350003　电　话：+86-591-87807936　传　真：+86-591-87826264　邮　箱：ghy@fzghy.com　网　址：www.fzghy.com

福州镇海楼

镇海楼重建意义

历史上的功能意义

明洪武年间始建镇海楼是作为福州府城的城楼的样楼，除在军事上作为城墙上的敌楼外，还起到进入闽江口岸的航标作用，也成为福州西湖的外八景之一——“样楼观海”。

在福州古城城市总体格局中的意义

福州古城城市总体格局是“三山两塔一条轴”，于山白塔、乌山乌塔、屏山镇海楼，三大景观建筑与三山构筑了福州城的特色城市总体格局，位于城南的二山二塔似城市的门阙，而屏山镇海楼则是城市中轴线的端点，是全城的标志。

建筑规模尺寸

建筑高21.25米，宽43.5米，深23.9米，面阔七间加复阶周匝。建筑占地面积2 159平方米，总建筑面积4 218平方米。

建筑特征

屋顶举折较缓，檐口出挑深远的屋面形式，均施以鹊尾及角鱼，并将正脊升起，形成优美的屋面曲线。墙身结合现代使用功能及古典建筑形式，虚实结合，门窗采用较为简洁、大气的玻璃门窗。在建筑物最重要的构件即承托出檐的构件处理上，采用清式大式作法，台基采用简洁的石栏杆围合。

福州市规划设计研究院　地　址：福州市湖东路188号　邮　编：350003　电　话：+86-591-87807936　传　真：+86-591-87826264　邮　箱：ghy@fzghy.com　网　址：www.fzghy.com

福州安泰河水街

安泰河水街位于福州市三坊七巷保护区范围内，东起南后街，西至雅道巷路口，含金斗桥地块等。水街与南后街相连接，沿安泰河两侧布局，传承了福州传统商业建筑的精髓，充分体现了福州“水乡”的文化特色。

福州天宁阁

烟台山天宁阁位于福州历史文化中轴线的最南端，烟台山历史风貌区发源于宋、元时期，形成并繁荣于清朝末年和民国时期。它记述了福州城市城进水退、不断南进的空间格局发展过程，是福州城市历史文化中轴线和“三山两塔一条轴”的历史风貌特色的重要组成部分。

平面尺寸大小确定

（1）山顶平台尺度与阁楼尺度关系根据现场踏勘结果，确定楼阁尺寸为15米×15米的大小比较合适。

（2）平面形式确定

不同于镇海楼位于中轴线起点与北端点，烟台阁处于中轴线南端并延中轴线向南发展，同时位居两条城市主轴线交汇点，因此，采用方形平面呼应其所处位置，契合环境特征，成为自然的选择。

（3）四角配楼

以“一主四从”的组合平面形成组群关系，表达其应有的体量。

福州市规划设计研究院　地　址：福州市湖东路188号　邮　编：350003　电　话：+86-591-87807936　传　真：+86-591-87826264　邮　箱：ghy@fzghy.com　网　址：www.fzghy.com

兰州市城市建设设计院

实力雄厚。兰州市城市建筑设计院拥有核心精良技术，自1959年成立以来，已发展成为以兰州为核心的区域性综合甲级设计院，具有建筑全行业（含人防）、市政全行业、工程勘察、工程测量、工程咨询、工程监理等15个专业甲级资质，工程代建一级资质和公路、城市防洪、招标代理、施工图审查乙级资质及规划丙级资质。主要承接道路、桥梁、城市给排水、防洪、公共交通、大中型工业与民用建筑、风景园林、环境卫生、二程勘察、工程测量、工程咨询、工程代建等业务。通过了 ISO 9001：2000 国际质量体系认证。

精英荟萃。我院现有职员工276人，其中，专业技术人员240人，高级、正高级职称67人，一级注册建筑（结构）工程师等各类注册人员65人，享受政府特殊津贴专家及省、市学术带头人52人，是一个具有务实、创新精神的高效服务的专业团队。院内设有三个建筑设计所、市政工程设计所、设备设计所、经济室、测量队、地勘队。

业绩卓著。我院的勘察设计成果遍及西北地区及全国部分省市，在青海、内蒙、宁夏等省区及白银、天水、平凉、临夏、甘南、河西等省内周边地区广泛开展业务。在广东珠海、广西北海、浙江台州、甘肃榆中等地都设有分院、分部和分公司。以城市基础设施建设勘察设计为龙头，勘测、建筑专业等取得了长足的发展，完成了省人民医院急门诊楼、国芳国际曦华源小区、兰州市孙家台经济适用房小区、兰州水车博览园等一大批建筑工程设计，完成各类大中型勘测设计项目6 000余项，并承担了兰州市南、北滨河路、城关黄河大桥、雁滩黄河大桥、黄河河道规划、南河道综合整治、三大出口等大部分城市基础设施工程设计任务。有近百项勘察设计项目获得国家及部、省级优秀工程设计、优秀勘察项目奖，黄河兰州段河道规划获国家银质奖。

创新求发展，和谐促繁荣。我院精神文明和企业文化建设成果卓著，积淀了“团结、务实、创新、奉献”的企业精神，被评为全国建设系统党风廉政建设工作先进单位，全国建设系统精神文明建设工作先进单位、甘肃省文明单位、兰州市文明单位、兰州市先进基层党组织，同时还通过全国勘察设计诚信单位评估并在网上公示。

站在新的起点上，我们将向着新的目标，以国际竞争的眼光来制定发展战略，整合有效资源，用5～10年时间，发展成为西北领先，国内一流，集设计、研发、咨询、生产实践为一体的，社会敬仰、同行尊敬、员工爱戴的学习型、知识型、集团化的区域性咨询设计企业。

Lanzhou Urban Construction Design Institute has great ability in design and excellent core technclogy. Since its founding in1959, taking Lanzhou as the core, the institute has developed into a Class A design regional comprehensive institute. It has 15 professional qualifications including qualification of the whole construction industry (including air defense), all municipal fields, engineering exploration, engineering survey, engineering consulting, engineering supervision and so on. It has also got a Grade A qualification in project authorized construction, and Level B qualification in road, city flood control, tender agency, construction drawing review and Class C qualification in planning. The institute is mainly responsible for constructions of roads, bridges, urban drainage, flood control, and public transport, medium and large industrial and civil construction, landscaping, sanitation, engineering exploration, engineering survey, engineering consulting, project construction and other tasks. It has passed the international quality system cerification ISO 9001: 2000.

The institute has gathered lots of elites. Of its 276 employees, it has 240 professional and technical staffs, 67 that have senior and senior titles, and 65 that registered First-level Registration Construction (structure) Engineer. It is a professional and efficient service team that has a practical and innovation spirit, because there are 52 people who enjoy the government special allowance, and are the academic leaders of the provinces and municipalities. It has three architecture institutes, municipal engineering designinstitute, equipment design instituteeconomic studio, survey team and geological exploration team.

It has made outstanding achievements with successful designs and explorations throughout the Northwest and parts of provinces and cities nationwide. The institute has extensive businesses in provinces and districts like Qinghai, Inner Mongolia, Ningxia, and inside some provinces like Baiyin, Tianshui, Pingliang, Linxia, Gannan, Hexi and so on. It has branch, division and branch offices in Zhuhai (Guangdong), North Sea (Guangxi), Taizhou (Yuzhong), Yuzhong (Gansu) and other places. taking the urban infrastructure exploration and design as the first priority, exploration, architecture and so on have also made significant development. It finished a large number of construction projects like the emergency clinic building for People's Hospital of the province; Guofang international Xin Huayuan community; economic and usable Housing for Sun Jiatai, Lanzhou city; Lanzhou Water Tankers Expo and so on. It completed survey and design of 6,000 large and medium projects, undertook most of the urban infrastructure engineering design tasks in South of Lanzhou City, , North Binhe Road, Chengguan Yellow River Bridge, the Yellow River Yantan Bridge, the Yellow River Planning, South River Restoration, and the three major exports; many times it has been awarded excellent design for exploration and planning by the state, Ministry and provinces. It got a silver medal for the planning of Yellow River section in Lanzhou.

Innovation strives the development, and harmony promoted prosperity. The enterprise has an outstanding output in construction of spirit and cultural civilization, and accumulates a spirit of "unity, pragmatic, innovation and dedication". Therefore, it was named as advance unit of cultivating an incorrupt spirit for the party members in national construction system, civilized unit of Gansu province, civilized unit of Lanzhou city, advanced grass-roots party organizations of Lanzhou, and it also passed the assessment of national exploration and design integrity unit, which publicized in the internet.

Standing at a new starting point, we will go towards the new target. Formulating development strategies from the international competition point of view and integrating resources effectively, within 5-10 years, we plan to become the Northwest's leading domestic first-class institute that integrates design, research and creation, consulting, and production and practices, and to become a learning, knowledge-based, group-based consulting and design company that respected by the public and our peers, and supported and adored by our employees.

Lanzhou Urban Construction Design Institute

地址：甘肃省兰州市南滨河东路875号
电话：+86–931–8465166
传真：+86–931–8465106
网址：www.lzcjy.com.cn

Address: 875 South Binhe Road, Lanzhou, Gansu Province
Tel：+86-931-8465166
Fax: +86-931-8465106
Http: //www. lzcjy. com. cn

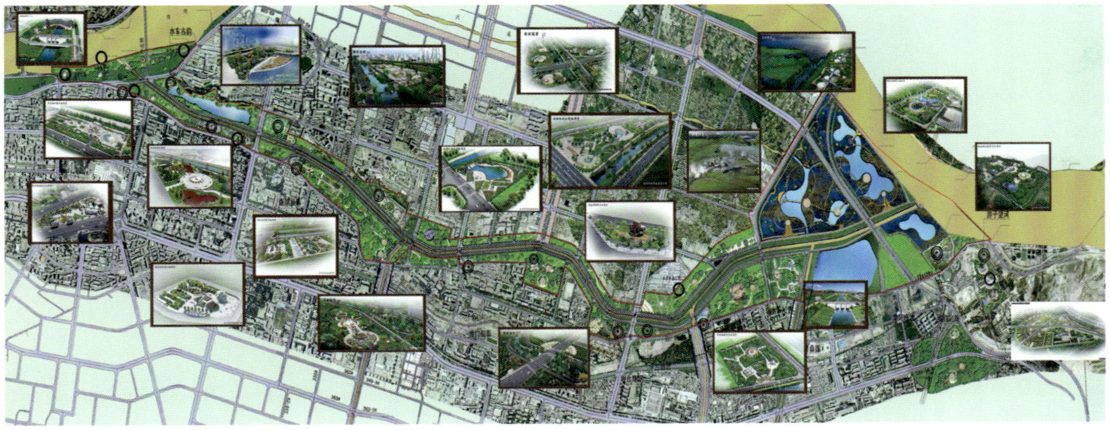

兰州市雁滩南河道综合治理工程西起雁滩滩尖子水车博览园，东至雁儿湾草地公园东南角，河道长 8 240 米。规划范围在 608 号路以西，以河道中心线南北各 80～150 米，608 号路以东，河道中心线南北各 100～200 米段，出口处结合湿地公园的建设宽度扩展到 1 660 米，同时还包括与之相交的五里铺排洪沟、烂泥沟、鱼儿沟各以中线两侧 80 米处进行控制性详细规则。规划用地面积 439.30 公顷。

总体规划结构是“一河两岸三维流”，“四区八园十六景”。

南河道景观造型分三个层次。

第一个层次——以“水”为母体，以“人”为根本。南河道的设计无疑是以水为设计主题的，其中蕴含着水能带给人的欢乐与愉悦。

第二个层次——以“历史、环境、文化”三方面为主题，建设中充分挖掘南河文化、地域文化，体现水和城的关系，塑造独特的环境景观。

第三个层次——“水景圈、绿地圈、文化圈”三圈一线（河道）的景观。沿河两岸，建设绿化长廊，让绿色植物发挥城市空气清滤器的功能；以河水为依托，构成恢复自然生态的绿化带，两边的道路建成具有浓厚文化氛围的交通线，从系列的园林景观和雕塑小品中，使人们领略兰州市城市文化的芬芳，陶冶人们乐观向上的精神情操。

景观规划以南河道为轴，形成链式结构。在河道出入口和城市主干道交叉口作为重要节点处理。以“平沙落雁”开始，依次建设“雁荡南河”“轻舟逍游”“南河晚渡”“柳林含烟”“南湖凝翠”“芦堤飘絮”“怀古幽思”“游子望河”为主题的景观园林。两侧带状河岸以南河文化为背景，或以雕塑、或以石刻展示丰富的文化资源，给以无限的想象空间，使游人和市民在休闲中接受文化的熏陶。

景观规划注重文化建设，充分发掘和诠释南河（教场河）文化素材，提炼主要元素进行文化建设，结合南河道已有的敦煌研究院、读者、画院、图书馆等文化建筑规划主题鲜明的文化林、园。建设中充分展现水车文化（祖宗车、十轮水车），渡口文化、教场文化、祖师楼遗址、马家窑文化遗址以及雁滩夹河滩、十八滩等地貌文化，将文化融入一草一木、一路一石之中，表达形式多采用雕塑、堆石、石刻、文化墙等。

治理后，新生的南河道，将是一片“蝉鸣莺啼目昆虫，鸟语花香绿映红；小桥流水俏码头，悠悠南河荡轻舟；芦堤飘絮雁儿湾，怀古幽思十八滩；黄河东去不复返，回首眺望金城观”的诗意画面。

The Comprehensive Control Project of south Yantan River, Lanzhou city begins at Yantan Jianzi Water Tanker Expo Park and ends in Grass-Land Garden of Yan'erwan, with a total length of 8,240 m. Planning area: it is in the west of No. 608 Road, centering the River, 80 meters and 150 meters from south to north respectively, east of No. 608 Road, 100 to 200 meters from south to north respectively. At the exit, it takes the construction width of the wetland garden into consideration extending to 1,660 m width. At the same time, where the wulipu flood discharge trench, ditch mud, and Yu'er ditch intersect with it, the distance is controlled within 80 meters beside the River. Planned land area is 439.30 hectares.

Overall planning structure: a three-dimensional river flow, four Districts, 8 Parks and 16 landscapes.

South River Landscape modeling is classified into three levels.

The first level – making "water" as mother body; taking "people" as fundamental. The South River is no doubt based on the content of water, with water, containing the joy and pleasure like water.

The second level – taking "history, environment, and cultural" as theme of the area. In the construction, it digs deeply the South River culture and regional culture, reflecting the relationship between water and the city. Therefore, it creates a unique environmental landscape.

The third level – integrating "waterscape circle, green circle, and culture circle" into one line, this makes a great landscape. Along the river, it constructs green corridors, so that green plants can play the role of urban air clean filter. Basing on the waters, it constitutes a restoration of the natural ecology of Green Belt. Both sides of the roads are built into the lines of communication with a strong cultural atmosphere. From the series of gardens and sculptures, people enjoy the fragrance of Lanzhou urban culture, cultivating their optimistic sentiments.

Landscape planning's axis is south of the river, forming a chain structure. Seriously, it treats the river entrance and main road intersection as an important node. From “Swans on the Shoal”, it builds "the Wild Goose swimming in the South River", "Sailing little boat", "night crossing Nanhe", "fogging and the Willows," "South Lake Verdant", "catkin floating dam", "nostalgia meditate", and " River-looking wanderer " respectively, and takes them as the theme of landscape gardens. on both sides of river banks, its background is Nanhe River's culture, which displays sculptures and rock carvings that reflect its rich cultural resources, providing ample room for imagination to people and giving visitors and citizens an opportunity to expose to the culture at the time of leisure.

Landscape planning highlights culture construction. The project fully digs and interprets cultural elements of Nanhe River (Jiao Chang River), and extracts the main ingredients to undergo cultural construction. It also plans a cultural garden with a sharp characteristic by combining the Dunhuang Research Institute, readers, Painting, libraries and other cultural architectures that have long been possessed by Nanhe River district. In the process of construction, it fully shows Tanker Culture (Ancestor Cars, 10-wheel water tankers), Dukou culture, Jiaochang culture, ruins of Shizu building, cultural site of Ma Jiayao, sandbank between Yantan Rivers, Shiba Beach and other geological culture landscapes. In this way, it integrates culture into trees and grasses, stones and roads, which displays in the form of Sculpture, rubble, stone, cultural wall and so on.

By bringing the new Nanhe River under control, a romantic poem can be cited to describe the beautiful landscape “Cicadas and orioles are singing and insects are within people's view, birds' songs, good smell of flowers and beautiful colors of nature compose a harmonious picture; With small bridge, rivers and pretty port, a lovely boat is floating leisurely on the river Nanhe; In the sky of Reed dam and the Yan'er bay catkins are flying around slowly, on the Shiba Beach one can easily fall into thought about the old times; The Yellow river has gone east forever, looking back one can only get a view of the Golden City.”

平凉行政中心

总占地面积：155 676.4 平方米
硬质铺装面积：124 605.55 平方米
水域面积：13 089.9 平方米
绿地面积：19 815.92 平方米
绿化覆盖面积：16 952 平方米
主要建筑面积：6 597.55 平方米
地下面积：12 356.53 平方米
其中地下车库面积：7 524.18 平方米

Pingliang Administrative Center

Gross Area: 155,676.4 m^2 approximately 155,600 m^2
Rigid Pavement Area: 124,605.55 m^2
Water Area: 13,089.9 m^2
Green Area: 19,815.92 m^2
Green Aoverage Area: 16,952 m^2
Main Building Area: 6,597.55 m^2
Underground Floor Area: 12,356.53 m^2
Including Underground Garage Area: 7,524.18 m^2

设计思想及理念

科技导入，引领时尚

本地段作为平凉市的新区和新的行政中心，将集中体现平凉市的城市形象，所以在城市设计中多采用新技术、新材料。建筑风格以高新技术为导向，创造简约的造型，精美的细部，提高结构和饰材的高科技含量，重视节能环保及生态建筑的推广。建筑色彩以白色、金属银色、浅蓝色、浅黄色及天然花岗岩石材本色等明快、淡雅的色调为主，形成新区新颖、简洁的整体建筑风格，倡导简约、高效、文明、时尚的城市文化。

依托文化，塑造典范

平凉崆峒山是"道"的发源地，皇帝问道广成子奠定了平凉成为"道源圣地"的地位，设计中"道"作为新区行政中心及绿地公园的主题。在文化渊源上兼有丝绸之路文化、泾河文明和民俗风情于一身，规划中通过绿化小品、雕塑、浮雕等形式，表现平凉的悠久历史和灿烂文化，赋予新区各个空间场所以丰富的文化内涵，塑造城市环境与艺术的典范。

以人为本，生态名城

泾河从平凉市区穿城而过，这是平凉独特的景观资源优势。新区中心位于泾河南岸水域环绕之中，因此要求本区域的建筑设计均向生态规划、绿色建筑设计方向发展。在绿地公园设计中注重序列性、公共性、开放性，在建筑设计中注重亲水性、近水性，以此使平凉成为极具地域特色的生态旅游名城。

Design ideas and concepts

1. Import technology, lead the fashion

This plot is a new area of Pingliang City and a new Administrative center. It will embody the image of Pingliang City, so in urban designing, it employs a lot new technologies and new materials. The architectural style is with high technology-oriented school, and simple shapes are created. The building has fine details, increases the proportion of high-tech in the structure and decoration materials. And it also takes energy-saving, environmental protection, and promotion of the eco-constructions into consideration. Building colors are mainly bright, elegant ones such as white, metallic silver, light blue, light yellow, natural granite stone's natural color and so on, which forms novel, compact style of the overall architecture. It advocates a simple, efficient, civilized, modern urban culture.

2. Relying on the culture, create classics

Pingliang Kongtong Mountain is the birthplace of Daoism. The emperor asked Guangchengzi about Daoism, which makes Pingliang the "bethel of Daoism". The administration takes "Daoism" as its center and the theme of the green-land garden. In terms of cultural origin, it inherits Silk Road culture, Jinghe River civilization and folk customs. In planning, green pieces, sculpture, relief and other forms of displays fully unfold its long history and splendid culture. It brings rich cultural contents to various space of the new district and creates a model of urban environment and art.

3. People-oriented, eco-city

Jinghe River goes through Pingliang city, which is a unique resource advantage of Pingliang. The center of the new area is located in the district that surrounded by waters of Jinghe River. It is required that in the design of buildings, it must head toward eco-planning and green building design; in the design of grass-land gardens, it must highlight sequence, public, and open; also, in the design of buildings, it must pay attention to the quality of water-friendly and water-accessibility, so that Pingliang will be a famous eco-tourism city with a remarkable regional characteristic.

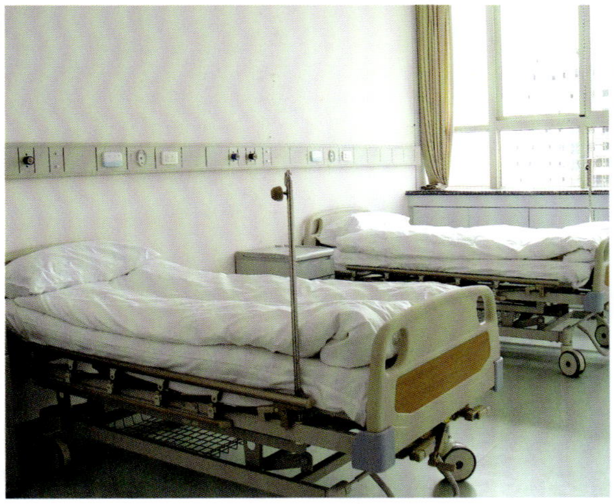

甘肃省人民医院门急诊楼

建设地点：甘肃 兰州

设计起止年月：2001 年 3 月 –2002 年 6 月

竣工验收时间：2004 年 6 月

项目类型：公共建筑

结构类型：框架 – 剪力墙结构

Emergency Clinic Building for People's Hospital of Gansu province

Construction Location: Lanzhou, Gansu

Design Beginning And End Years: 2001.03-2002 .06

Time of Final Acceptance: 2004.06

Project Type: public buildings

Type of Construction: frame - shear wall structure

项目建筑面积 37 012 平方米，建筑总高度 94.8 米。地下二层，地上二十二层， 裙楼五层。地下室为车库和设备用房，1 ~ 5 层为门急诊用房，6 ~ 8 层为检验试验科室，9 ~ 19 层为住院部，共有病床 374 个，20 ~ 22 层为设备用房。设计概算为 1.27 亿元，竣工决算为 1.30 亿元。

主要特点

甘肃省人民医院门、急诊大楼的设计，打破单体医院的传统设计手法，妥善解决了各个出入口及内部流线的组合，使门诊、急诊、住院、医技等功能部门容纳于一幢综合性建筑中，实现了节约用地的目的，以病人为中心的设计目标，在建筑布局和造型上充分体现热爱生命、尊重生命的主体形象。儿科诊室设有专用的活动场地，增加了绿化用地和患者户外活动的空间，从而使患者的就医条件和医生的工作环境得以改善。该工程的平面布置是本工程的设计重点和要点，设计中充分利用水平和垂直交通将门诊、急诊、检验试验室、住院病房、设备用房有机的结合，使该大楼在功能上满足三类甲等综合医院的要求，这也是建筑设计的创新点之一。由于该工程结构采用全现浇框架——剪力墙结构，钢筋混凝土外墙的低热阻使得外墙保温成为重中之重。设计中外墙采用了粘贴 50 毫米厚聚苯乙烯泡沫塑料板，单组分聚合物砂浆外保温系统；屋面采用 50 毫米厚挤塑聚苯乙烯泡沫塑料板保温层；外窗采用塑钢窗。这些节能措施的实施，使该工程的建筑节能指标达到了《民用建筑节能设计标准甘肃省节能实施细则》的要求。

Construction area: 37,012 m^2; total building height: 94.8 m. There are two stories underground, 22 levels over ground and 5 levels for podium buildings. The basement is used for the garage and equipment storage, 1 to 5 floor for outpatient and emergency department, 6 to 8 layers for the inspection and test sections, 9 ~ 19 layers for inpatient with a total beds of 374, and 20 to 22 layers for the equipment room. The design budget is about 127 million yuan, and the budget needed for completed the project is 130 million yuan.

Main features

The design of Gansu Emergency Clinic Building for People's Hospital employs an unconventional method, properly resolving the combination of various entrances and exits and inner flow lines. In this way, it saves land, integrating outpatient, emergency, hospital, medical technology and other functional departments in a comprehensive building. With a patient-oriented goal, this design shows the figure of the love of life and the respect to life in the building's layout and shape. The pediatrics room has an exclusive sport space. By increasing the area of green land and the space of patient outdoor activities, this design improves patients' medical conditions and doctors' working environment. The layout of this project is the focus and emphasis of the design. One of this design's breaking points is to make the function of the building met requirements of general hospital's three Class A types. Because this project employs a frame - shear wall structure, the Lower thermal resistance of reinforced concrete outside wall makes the outside wall thermal insulation to take full advantage. In this design, the outside walls use a paste of 50 thickness polystyrene board and a one-part polymer mortar exterior insulation system; the roofs use 50 thick extruded polystyrene foam board insulation; the outside windows employs PVC windows. By all these energy-saving measures, the project's building energy-saving level has achieved the requirements of the civil building energy saving design standards implementation rules in Gansu province.

兰州水车博览园位于兰州百里黄河风情线滨河东路黄河南岸，东连中立桥码头、体育公园，西接亲水平台、兰州港、中山桥、白塔山公园等景点。兰州水车博览园由水车园、水车广场、文化广场三部分组成，是一个展现水车文化的主题公园。兰州水车大观园规划用地从黄河大桥到中立桥的河滩地段，总长 1 200 米左右，三分之二地段平均宽度 60 米，为狭长形的空间。总用地面积为 55 900 平方米，总建筑面积为 1 865 平方米。

如今，水车博览园的水车已与黄河南岸的自然景色融为一体，构成了兰州独特的滨河风光，水车转动时嘎嘎的木轮声音、倒挽河流的气势及粗犷雄浑的形象，令人赞叹不已。远望如云，严冬时节大地银妆素裹，岸边水车如冰雕玉琢的精美工艺品，别有一番清寂之美。夏季园内车轮林立，景象十分壮观，同时水车园还成为人们避暑纳凉的好地方。"

兰州水车博览园被命名为国家 4A 级旅游景点，水车之都于是成了兰州的代名词。

Lanzhou Water Tanker Expo Park is located in southern bank of Yellow River, east Binhe Road of Yellow River style line. In the east, it connects harbor of Zhongli bridge and Sports park. In the west, it neighbors Qinshui platform, Lanzhou harbor, Baita theme park and other landscapes. Lanzhou Water Tanker is composed of three parts: Water Tanker yard, Water Tanker square and Culture square. It is theme park that displays Water Tanker culture. Lanzhou Water Tanker Grand View Garden plans to use an area that begins from Yellow River Bridge to the river Beach of Zhongli Bridge. Its total length is about 1,200 m, and two thirds of it has an average wide of 60 meters, a long and narrow space. The total land area: 55,900 m^2 ; total construction area: 1,865 m^2 .

Now the waterwheels in the waterwheel exhibition park and the scenery of the south bank of Yellow river blended, creating an unique riverside landscape in Lanzhou. The sound of the running wheels , the momentum of refluence and the rough image arouse Profound fondness from visitors. In the winter time, while ground is all white the waterwheel at the bank of river becomes the utmost delicate arts and crafts, giving a sense of peace and tranquility . In the summer time, A forest of waterwheels comes to be a magnificent sight at the riverbank, turing to be a great summer resort for people to spend summer holiday.

Lanzhou Water Tanker Expo is Park thus has been named the National 4A Grade Tourist Attractions. Consequently, the waterwheel city has become a synonym of Lanzhou.

孙家台小区

"孙家台"小区位于兰州市七里河区西津路以南，雷坛河西侧，北临兰新铁路线，西靠华林坪，东临雷坛河，总用地面积约14.0329公顷，其中建设用地10.9649公顷，其他用地3.0680公顷。建筑用地南高北低，西高东低，呈不规则地形，南北长约538米，北端最宽处320米，南端最窄处139米。

设计理念

经过对基地、周边环境及城市发展的整体考虑后再进行深入分析与研究，并与建设单位进行反复沟通，充分尊重建设方的需求并分析市场定位，最终确定本方案并从以下几个方面着手，来构建本小区的基本规划设计理念。

合理准确的定位

本小区位于兰州市中心区西关什字的黄金地带，随着西关什字公交枢纽站西迁至红门子，该地块居民出行将更为方便快捷。目前该地块为兰州市主要的城乡结合部之一，为改善现状环境，充分利用城区内有限的土地资源，解决公交枢纽站西迁的拆迁还建及响应国家积极倡导的开发经济适用住宅小区的重大举措，因而本小区主要解决城建拆迁还建用房建设问题并以工薪阶层为主要销售对象为主．虽然建筑户型面积不是很大，但是要求校区的环境建设一流，设施配套齐全，彻底改善工薪阶层的生活和居住环境。

创造独具特色的小区环境

充分挖掘现有条件，利用好用地内部和外部的各种环境设施，不仅有利于保护环境，利于小区与周边环境相融合，还会给小区居民创造别具特色的独特环境与氛围。通过现场多次勘探，该场地内现有的梨树历史悠久，东侧的雷坛河通过整治也将成为该小区的独特景观，场地内复杂的地形也将成为小区造景的有利条件。本方案将充分利用好这些独特的资源环境，设计出独一无二的生态园林景观．

塑造独具魅力和文化内涵的小区品牌

为给该小区赋予更深远的文化内涵，提升小区的整体品位，小区规划在设计时充分挖掘兰州地方历史文化，利用雷坛河（古名阿干河）流域的文化资源和场地内的兴远古寺，通过丰富多样的表现手法将文化融入到小区环境建设和设计理念中，为古寺异地复原重建，起到提升品牌知名度的效果．

Sunjiatai community

Sunjiatai community is located in south of Xijin Road, Qilihe District, Lanzhou city, west bank of Leitanhe River, north of Lanxin railway, west of Hualinping and east od Leitanhe River. The total land area is 14.0329 hectares, of it, construction land 10.9649 hectares, and other use of land 3.0680 hectares. The land is higher in the south and west, and lower in the north and east. It's a irregular terrain, about 538 meters from north to south. The widest part in the north part is 320 meters and the narrowest part I the south is 139 meters.

Design Concept

After a deep analysis and study to the base, the surrounding environment and a overall consideration to the urban development, and rounds of re-negotiations with the construction unit, we respect the construction party's need and analyze the market positioning. This project develops the basic design concept of the community from the flowing aspects:

Reasonable and accurate positioning

This community locates in the golden zone of Xiguanshizi, center of Lanzhou city. With the Transit Hub of Xiguanshizi moved west to Hongmenzi, residents here will enjoy a more convenient transport. At present, this plot is mainly one part of Lanzhou city and countryside integration. To improve the current environment, make full use of the limited land resources, solve the westward move of the transportation hub and its re-construction, and respond actively to the national advocate measures of developing affordable housing, this community mainly targets at solving problems in pulling down and re-construction of urban buildings, and its target customers are mainly working-class people. Although the chloroplast area is not large, the environment required is excellent, and relative facilities are complete. The living environment is improved greatly.

The creation of a unique residential environment

This project fully exploits the existing conditions, and makes good use of environmental facilities in- and outside of the land. In this way, not only is the environment being protected and fits well to the surroundings, but also creates a special environment and atmosphere for residents of the community. After several times of exploration, we found that pear trees in this area has a long history, a splendid landscape will be formed after the Leitanhe River in the east being controlled, and the complex terrain of this area will also be an advantage for the creation of landscape. This project will make full use of these special environmental resources and design a unique eco-park landscape.

Creating a community brand with unique charm and culture content

To provide the community with far-reaching culture content and enhance its overall quality, when planning the community, it went deeply to search the local history cultures of Lanzhou, employing cultural recourses of Leitanhe River (ancient name Agan Rive) and Xingyuan ancient temple inside the land. It integrates all the elements mentioned into community environment construction and design concept by various performing methods, promoting the re-construction and restoration of the ancient temple, and creates a renowned brand.

白银市城市建设设计院

法人代表：马智同
联系电话：+86-943-8229856
+86-943-8221966

成立于1986年初的白银市城市建设设计院，是白银地区具有工程设计乙级，工程勘察、测量、城市规划编制、工程监理丙级等资质的综合性设计院。多年来，设计院在困境中求生存，在竞争中谋发展，练内功、强素质、抓管理、树形象，以"一发展三提高"为目标，以"顾客至上，挚诚服务"为服务宗旨，使得设计院的各项事业得到蓬勃发展。历年的设计生产实践，使设计院的机构设置、专业配备更为合理，技术力量不断加强，设计经验进一步丰富，设计产品日臻精美，后期服务热情周到，这些大大提高了设计院的整体水平和社会知名度。院里有设计所、勘察公司、测量队、规划所、造价所等专业部门和办公室以及晒印室等服务部门，截至2009年底，设计院共有员工43人，其中专业设计人员36人，具有高级职称者12人，中级职称者21人，初级职称者3人，注册工程师11人，其中一级注册工程师7人，二级注册工程师4人。设计院设有建筑、结构、给排水、暖通、电气、测量、总图、效果图、预算等专业，对外可承接20层住宅建筑、高度50米以内的公共建筑设计、30米跨度的工业厂房设计、岩土工程勘察、工程测量、城市规划、工程概预算等业务。2004年设计院通过了ISO 9001质量管理体系认证，使设计管理工作更加科学化、规范化、制度化，提高了市场竞争能力。截止2009年设计院底完成各类建筑设计项目720项，共340万平方米，工程投资总额达28.8亿元，其中白银高科技园创业大厦、白银西区管委会办公大楼、白银四龙农业示范园工程、白银市委党校综合楼、靖远县人民法院办公楼、平川中恒商场、平川颐康园住宅小区、白银市大地开发公司惠民住宅小区、白银颐华苑住宅小区、白银区强湾乡、武川乡移民搬迁工程、陇南天润嘉园（6万平方米）等中高层建筑设计及规划得到了社会上较高的评价。在获取社会效益的同时，近几年间，设计院设计产值稳定增长，职工收入逐年提高，固定资产由当初的不足10万元增至100多万元，累计完成产值2 620万元，保持了年均28%的增长率。

设计院历经坎坷，勇于应对挑战、转变观念、创新机制、深化管理、锐意改革、更新服务理念、增强竞争能力，以精诚团结、不懈奋斗的精神，打基础、创名牌，如今已成为白银地区设计行业的龙头企业。

靖远县新城区同盛雅苑小区

平川区城市棚户区改造（七泵房片区）

陇南市金泰房地产开发公司天润嘉园

白银市福门数码港

白银市福利院规划建筑

靖煤体育馆

白银市党校教学楼

广西华蓝设计（集团）有限公司

Guangxi Hualan Design & Consulting Co., Ltd.

（原广西建筑综合设计研究院）

广西华蓝集团是以广西华蓝设计（集团）有限公司作为母公司，与其控参股的13家公司共同组建成立的企业集团。

广西华蓝集团拥有员工1 960余人，以工程设计与咨询为主，同时开展相关业务，多元化经营。集团业务涵盖建筑工程设计与施工图审查、城市规划编制、市政与环保工程设计、轻纺与建材行业设计、建筑智能化系统设计与施工、建筑装饰工程设计与施工、园林景观工程设计与施工、工程咨询、工程监理、招标代理与造价咨询，以及岩土工程的勘察、设计、施工与检测等。集团拥有以上业务的各类资质32项，其中甲级资质15项。同时集团在资本运营、房地产开发、视觉文化艺术设计、围棋活动组织策划等领域也开展了业务。

广西华蓝设计（集团）有限公司是由始建于1953年的原广西建筑综合设计研究院改制而成的现代科技型企业，于2007年1月1日正式成立，是广西规模最大的国家综合甲级建筑设计单位。公司拥有建筑设计院、规划设计院、市政设计院、项目管理与咨询院4个专业设计院，1个研究院和7个分公司，在深圳、广州、北京、福州、海口等地设立分支机构。公司与美国、日本、加拿大、德国、法国、比利时、丹麦、澳大利亚等国开展业务合作与交流并积极开拓非洲和东南亚各国市场。

Guangxi Hualan Group is an enterprise group which owns 14 members, including the parent company -- Guangxi Hualan Design & Consulting Co., Ltd. and its 13 holding companies.

The services of Guangxi Hualan Group cover a vast range including building design, construction drawing reviewing, urban planning, municipal and environmental protection project design, light industrial and building material project design, intelligent building system design, interior and exterior design and construction, landscape design and construction, project consultation, project supervision, project bidding agency, budget estimate, exploration, geological engineering design and construction. Beside traditional services, the group enters capital management, real-estate, image design and Weiqi competition, etc. Guangxi Hualan Group has more than 1,960 employees and holds 32 qualification certificates, among which 15 are Class A.

Guangxi Hualan Design & Consulting Co. Ltd. is a large science and technology based enterprise that was reformed from Guangxi Institute of Architectural Design and Research, the largest state issued Grade A architectural & engineering comprehensive design institute in Guangxi founded in 1953. It is established in 2007, there are four professional design institutes, one research institute and seven branches in the company. The four professional design institutes are architectural design institute, planning design institute, municipal design institute, and project management& consulting institute. The company not only covers its business all over China, but also cooperates in project design as well as academic exchanges with many overseas architectural and planning firms in U.S., Japan, Canada, Germany, France, Belgium, Denmark and Australia, etc. Moreover, the company is exploring and expanding its business in Africa and Southeast Asia and has opened an office in Hanoi, Vietnam.

地址：广西南宁市华东路39号
法人：雷翔
电话：+86–771–2438149
传真：+86–771–2435011
电子邮件：hl@gxhl.com
网址：www.gxhl.com

Address: No.39 of Huadong Road, Nanning City, Guangxi Province
Legal Person: Xiang Lei
Tel: +86-771-2438149
Fax: +86-771-2435011
E-mail: hl@gxhl.com
Http: //www.gxhl.com

南宁国际会展中心（合作设计）

建设地点：广西 南宁
建筑面积：151 370 平方米
竣工时间：2003 年（一期）、2005 年（二期）
合作设计：德国 GMP 设计事务所

获奖情况

2005 年广西优秀设计一等奖
2005 年第五届詹天佑土木工程大奖
建国六十周年中国建筑学会建筑创作大奖

Nanning International Exhibition Centre (cooperative design)

Construction Location: Nanning, Guangxi
Building Area: 151,370 m²
Completion Time: 2003 (Phase I); 2005(Phase II)
Cooperation Design: German GMP Design Studio

Awards

The First Prize of Excellent Design in Guangxi Province, 2005
The 5th China Civil Engineering Zhantianyou Prize, 2005
The Outstanding Design Issued by Architect Commission on the 60th anniversary of the founding of PRC

设计特点

建筑依山就势，布置有会议中心、展览中心大厅和露天展厅。建筑主体正面为圆形多功能大厅和竖立其上的膜结构穹顶，12 片折板式白色膜结构在蓝天下像一朵盛开的朱槿花一样格外耀眼。展厅两侧，两层楼高的修长有序的柱列，赋予整个建筑以强烈的韵律感。展厅结构新颖，造型独特的钢屋架飞架于中曲大厅和各个展厅之上，贯通整个展览中心，将展览建筑的特征充分地表达出来。

Design Features

This exhibition centre is built in total harmony with the hill terrain, including conference center, exhibition center and an outdoor exhibition hall. A round multi-purpose hall with a membrane structure dome is in front of main body. Twelve white folded membrane plates are shaped like a China rose which is very distinctive under the blue sky. On both sides of the exhibition hall are ordered colonnade of two-story high, which makes the whole building rhythmical. The exhibition hall has a unique shape and a novel structure. The steel roof truss is over the center hall and other exhibition halls, fully showing characteristics of architecture.

广西·南宁荔园山庄

建设地点：广西 南宁
建筑面积：100 130 平方米
竣工时间：2004 年

Guangxi • Nanning Liyuan Villa

Construction Location: Nanning,Guangxi
Building Area: 100,130 m²
Completion Time: 2004

获奖情况

2005 年度广西优秀工程设计奖一等奖
2005 年度建设部优秀建筑设计奖三等奖
2005 年度全国人居经典建筑规划设计方案竞赛建筑与环境双金奖
第四届中国建筑学会建筑创作佳作奖
建国六十周年中国建筑学会建筑创作大奖

Awards

The First Prize of Excellent Engineering Design in Guangxi Province, 2005
The Third Prize of Excellent Construction Design issued by the Ministry of Construction, 2005
The State Classical Design of Double Gold Metal for Architecture and Environment,2005
The 4th Excellent Awards of Architectural Creation issued by Architect Commission
The Outstanding Design Issued by Architect Commission on the 60th anniversary of the founding of PRC

设计特点

作为中国东盟博览会的主要接待基地，荔园山庄是一个集政府高级别接待设施、商务别墅、国际会议中心、高尔夫球场为一体的综合项目。

设计师在建筑风格的创作中，充分利用场地自然资源的亲和力和包容性，结合对地域与时代的理解，将别墅式接待中心与会议中心这两类不同的建筑相融会并协调于大环境之中，体现出"现代建筑地域性、乡土建筑时代性"的设计理念。

Design Features

As an important base for the China-ASEAN Expo reception, Liyuan Villa is an integrated project including high-level reception facilities, business villa, international convention center and golf course.

As to the architectural style, architects have taken advantages of attractiveness and inclusiveness of nature. Combining both territoriality and contemporary spirit, architects integrate the villa-style reception building and convention center into natural environment, embodying design ideas of the regional spirit of the modern architecture and the modernity of regional architecture.

北海冠岭山庄

建设地点：广西 北海
建筑面积：77 200 平方米
项目状态：在建

设计特点

项目整体划分为三大功能区：酒店会议区、贵宾接待区、生态观光区。考虑到市民休闲及旅游观光的需要，通过道路交通规划有效地解决了项目建设和市民享用冠头岭风景区的矛盾。三大组团之间相互独立、互不干扰，便于经营管理。

Beihai Crown Ridge Resort

Construction Location: Beihai, Guangxi
Building Area: 77,200 m²
Project Status: In Construction

Design Features

This project can be overall divided in 3 functional zones: hotel conference area, VIP service area, and eco-tourism area. Taking the residents' need for sight-seeing into account, architect effectively solves problems between project construction and beauty spot enjoying by traffic planning. The three functional zones are independent and not interfering with each other, which does good to operating management.

广西体育中心主体育场

建设地点：广西 南宁
建筑面积：123 618 平方米
项目状态：在建

The main stadium of Guangxi Sports Center

Construction Location: Nanning, Guangxi
Building Area: 123,618 m²
Project Status: In Construction

设计特点

主体育场建筑从体育的内涵出发力求体现出体育建筑应有的力度、动感和尺度，同时又表现出南方建筑的轻巧和飘逸。设计上从南宁绿城的概念中提炼出“绿叶”的理念，同时，又从绣球和花朵的“花瓣”中获得启发，结合使用功能的需求，利用东、西两片类似“绿叶”的罩棚的上翘和下翘所产生的曲线，展现出建筑极强的运动感。二片“绿叶”般的造型，分别向外和向内扭动，从不同的角度展现了建筑的“轻盈”与“飘逸”。

Design Features

This project is related to connotations of sports. The main stadium strives to reflect the strength, dynamic and dimension of sport building. Meanwhile, the stadium shows such characteristics as simplicity and elegance of south architecture. The design highlights the concept of green leaves from the idea of green city. Besides, architects take inspirations from petals of hydrangea and other flowers. The two pieces of coverings which look like curve of green leaves give people a sense of movement.

广西体育中心网球中心

建设地点：广西 南宁
建筑面积：10 305.9 平方米
项目状态：在建

设计特点

建筑延续了广西体育中心总体构思，设计将其隐喻的广西体育中心建筑序列中的“种子”。网球中心小巧可爱的圆形造型，宛如一个露出地面的种子，设计巧妙的“破”，更生动地表达了种子破土而生的瞬间。

Tennis Center of Guangxi Sports Center

Construction Location: Nanning, Guangxi
Building Area: 10,305. 9 m²
Project Status: In Construction

Design Features

This tennis center contributes to the overall conception and design idea of Guangxi Sport Center. It represents “seeds” in the sequence of Guangxi sports center construction. The small and exquisite round shape of tennis center is like a seed just breaking through the soil, which is the embodiment of birth of a new life.

广西体育中心游泳馆

建设地点：广西 南宁
建筑面积：31 500 平方米
项目状态：在建

Natatorium of Guangxi Sports Center

Construction Location: Nanning, Guangxi
Building Area: 31,500 m²
Project Status: In Construction

设计特点

建筑延续了广西体育中心的总体构思，设计将其隐喻为广西体育中心建筑序列中的“叶芽”。建筑如同错动的叶片展现出初生叶芽所饱含的生命力与张力，彰显朝气活力、坚韧不拔、积极进取的体育精神。

Design Features

Natatorium of Guangxi Sports Center contributes to the overall conception and design idea of Guangxi Sports Center. It symbolizes “leaf buds” extending from the Guangxi Sport Center. The crossed “leaves” not only fully display the vitality embodied in “leaf buds”, but represent the youthful, enterprising, persistent spirit related with sports.

广西体育中心体育馆

建设地点：广西 南宁
建筑面积：35 817.9 平方米
项目状态：在建

Stadium of Guangxi Sports Center

Construction Location: Nanning, Guangxi
Building Area: 35,817.9 m²
Project Status: In Construction

设计特点

建筑延续了广西体育中心的总体构思，设计将其隐喻为广西体育中心建筑序列中的“嫩叶”。建筑外形纯净而舒展，虽然还带着少许的羞涩，但其内在的张力正如同旺盛的生命力一般迸发，表达了体育和运动带来的生命力和活力。

Design Features

Stadium of Guangxi Sports Center contributes to the overall conception and design idea of Guangxi Sport Center. It symbolizes “tender leaf” extending from the Guangxi Sports Center. Appearance, pure, even with a little shy, but implied tension as exuberant vitality is bursting, sports and sports bring vitality and vigor.

广西民族博物馆

建设地点：广西 南宁
建筑面积：28 360 平方米
竣工时间：2009 年

设计特点

项目是一座以征集、收藏、保护、研究、展示、演示广西 12 个民族繁衍生存、融合发展的文化与艺术遗存、物证、典籍等物质与非物质遗产为主的国内一流的专题民族博物馆。将壮族古老的铜鼓文化与现代建筑艺术结合，并以现代的建筑新型材料体现壮民族文化中朴实、热情的民俗特点。

Guangxi National Museum

Construction Location: Nanning ,Guangxi
Building Area: 28,360 m²
Completion Time: 2009

Design Features

The purpose of this museum is to collect, protect research and display the heritage of 12 nationalities in Guangxi. The museum is first-class theme museum at home. It collects both the tangible and intangible heritage, such as the cultural remains, artistic material evidence and books. This design compromises the merits of drum culture of Zhuang nationality and art of modern architecture. The new materials are used to manifest the artless and enthusiastic personality of Zhuang people.

广西城市建设规划展示馆

建设地点：广西 南宁
建筑面积：48 562.9 平方米
项目状态：在建

Guangxi Urban Planning Exhibition Hall

Construction Location: Nanning, Guangxi
Building Area: 48,562.9 m²
Project Status: In Construction

设计特点

项目是第一个省级的城市规划展示馆。设计以独特的创意打破惯用的建筑设计手法，营造一座具有无限遐想空间和承载无限含义的建筑，寻找用建筑表现城市，用建筑回忆城市的另一种可能。设计结合项目的特点，用一种诙谐的叙事手法去阐述“城市的起源—城市的发展—城市的印象—城市的目标”的发展过程。

Design Features

It is the first provincial urban planning exhibition hall. This unique creative hall brings people boundless imagination and carries infinite meaning. The designer intends to use architecture to express urban, use architecture to memorize urban. Through a humor narrative, this design expresses the origin, development, image and objective of the urban.

广西美术馆

建设地点：广西 南宁
建筑面积：36 000 平方米
项目状态：在建

Guangxi Art Museum

Construction Location: Nanning, Guangxi
Building Area: 36,000 m²
Project Status: In Construction

设计特点

项目建成后将成为中国西部省份最大的美术馆，成为中国面向东南亚区域性的美术研究中心和美术产业基地。

以简单之美、基本之术作为设计主题，建筑构成了最自然的写生素材，每个角度都是一副画。人们用画笔描绘美术馆，走近美术，走进美术殿堂。以最基本的素描技法为理念，化繁为简，将变化多端的美术流派包罗其中，这种形式的美术馆建筑前所未有。

Design Features

This museum will become the biggest museum in West China, a research center and an art industry base in Southeast Asian.
Architects take simplicity as the theme, use nature as material. You can see different pictures from different angles. The museum seems to be drawn by paintbrush. If you enter it, you will feel you are in the palace of art. By the means of fundamental sketch technique, architects compromise various artistic schools, simplifying everything. This is initiative.

广西建设大厦

建设地点：广西 南宁
建筑面积：38 777 平方米
竣工时间：2003 年

获奖情况

2005 年度广西优秀工程设计一等奖
2009 年度全国优秀工程勘察设计行业奖建筑工程三等奖

Guangxi Construction Building

Construction Location: Nanning, Guangxi
Building Area: 38,777 m²
Completion Time: 2003

Awards

The First Prize of Excellent Engineering Design in Guangxi Province, 2005
It wins the construction engineering third prize in the national outstanding engineering survey and design industry.

设计特点

本建筑是集四星级酒店、娱乐、办公于一体的多功能高层综合楼。

建筑的主楼立面上重复使用灰、白色墙体、线条相间的设计手法，使建筑有节奏地产生韵律，充满逻辑的连贯性，同时还在现代感极强的立面上注入了一些经过提炼的古典风格的细部元素。这次建筑设计实践是对现代新古典主义的一次有益尝试，营造出具有岭南建筑特质，具有现代感、细部细腻充满变化、富于个性且与环境完美融合的建筑形象。

Design Features

The building is a multifunctional, high-rise and comprehensive building in which the four-star hotel, entertainment and office are integrated together. The main building façade walls are painted gray and white. The lines are appeared repeatedly and produced rhythmically. Full of logical consistency, the façade which is extremely rich in modern sense and which is full of constancy is infused with some refined classical detail. This architectural design practice is a useful attempt of the neoclassical in order to create a building image which is full of Linngnan architectural features and modern sense, has subtle change in detail, full of personality and integrated with environment.

南宁华蓝弈园

建筑地点：广西 南宁
建筑面积：10 000 平方米
项目状态：在建

设计特点

项目含棋牌、会议、时尚发布及餐饮等功能，属文化体育建筑。

建筑利用架空等手法形成具有中国传统风格的立体庭院，体现了中国传统文化内敛含蓄的性格。一至五层布置跨空间的大柱，交通核沿建筑四角布置，以满足多种经营需求，既能开展棋类比赛、培训、会议也能开展餐饮、接待、婚礼等活动。六层布置南宁传统风格的园林式棋院。

Nanning Hualan Chess Park

Construction Location: Nanning, Guangxi
Building Area: 10,000 m²
Project Status: In Construction

Design Features

The construction is a project of traditional Chinese three-dimensional garden, using techniques of building on stilts and so on to manifest characteristics of being modest and reserved in traditional Chinese culture. Longspan structure is used from the 1st to the 5th floors and distribution cores are arranged in line with four corners of the building, which meets demands for diversifying functions such as holding chess games, trainings, conferences, catering, receptions and wedding ceremonies. The 6th floor is the traditional Nanning chess club.

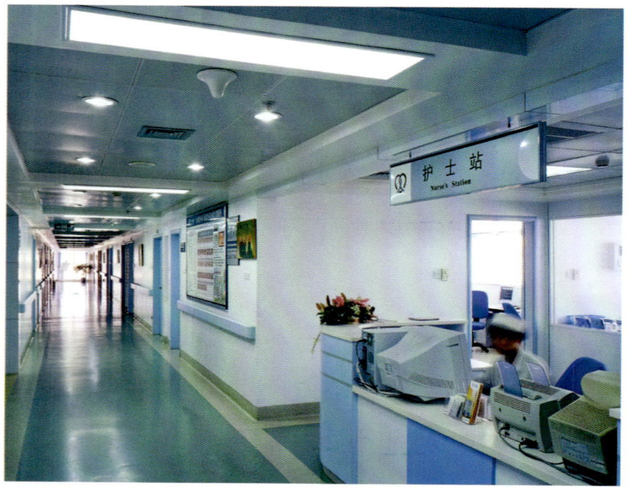

广西区人民医院门诊综合楼

建设地点：广西 南宁
建筑面积：56 548 平方米
竣工时间：2003 年

获奖情况

2005 年度广西优秀工程设计奖一等奖
2005 年度建设部优秀勘察设计奖三等奖

设计特点

广西壮族自治区人民医院门诊综合楼是为了适应现代化医院要求，提高医院工作效率，加强管理，改善医院当前门诊和内科病房不足及陈旧的状况，为病患提供一个更为舒适的就医环境而筹划建设的。医院的门诊量为每日 4 000 人，总病床数约 650 床。该工程为一类建筑，防火等级为 1 级，抗震烈度为 6 度，设 6 级人防。

Outpatient Complex Building of the People's Hospital of Guangxi Zhuang Autonomous Region

Construction Location: Nanning, Guangxi
Building Area: 56,548 m²
Completion Time: 2003

Awards

The First Prize of Excellent Engineering Design in Guangxi Province, 2005
The Third Prize of the Annual Outstanding Investagation and Design issued by Ministry of Construction, 2005

Design Features

Outpatient Complex Building of the People's Hospital of Guangxi Zhuang is designed and built to adapt to meet the needs of modern hospitals, to improve the hospital efficiency, to strengthen management, to improve the present inadequacy and obsolete situation of the outpatient and medicine wards, and to provide a more comfortable and ideal environmental for patients. The outpatient amount is 4000 outpatients/day, and the total hospital beds are about 650. This is a Category-I building, and the fire-protection rating is degree and the seismic fortification intensity I is 6 degree and has six degree of civil defense.

广西南宁市二中东校区

建设地点：广西 南宁
建筑面积：200 000 平方米
竣工时间：2009 年

设计特点

项目总平面设计吸取中国山水园林城市理念，因地制宜，依山据势，按功能合理布置各建筑单体。校区功能分区清晰，功能建筑互不干扰，同时合理解决消防疏散。项目充分体现了南方建筑的特点空、透、露、平台天井，良好的自然通风、采光的舒适、环境优雅，由此改善了校区内环境小气候，节约项目投资的同时满足节能需求。

The East Campus of Guangxi Nanning No.2 iddle School

Construction Location: Nanning, Guangxi
Building Area: 200,000 m²
Project Status: In Construction
Completion Time: 2009

Design Features

This graphic design contributes the whole concept on building scenery and garden city. Suiting to local conditions, this campus is built in total harmony with the hill terrain. Buildings are located separately according to different foundations, which solve the problem about fire-fighting and emergency evacuation. The whole design fully embodies the characteristic of south china architecture: empty, transparent and exposed. It allows natural light and wind into patios and small yards, which will not only improve the environment of the campus and save investment but also meet the requirement of energy conservation.

广西南宁市人大代表活动中心

建设地点：广西 南宁
建筑面积：478 987 平方米
竣工时间：2004 年

获奖情况

2008 年度全国优秀勘察设计行业奖建筑工程三等奖
2007 年度广西优秀工程设计一等奖

设计特点

活动中心是一座可容纳 1500 个座位的甲等剧场，建筑体现庄严、现代、生态及民族的特性，挖掘广西少数民族文化元素及构图形式，结合现代建筑的理念和政府建筑的设计特征，创造出了具有地域性和标志性的建筑。

The Activity Centre for NPC Member of Nanning, Guangxi

Construction Location: Nanning, Guangxi
Building Area: 478,987 m²
Project Status: In Construction
Completion Time: 2004

Awards

The Third Prize of State Excellent Investigation and Design on Architectural Engineering, 2008
The First Prize of Excellent Engineering Design in Guangxi Province, 2007

Design Features

It is a grade A theater which can accommodate 1,500 persons. The building is grandeur and modern, embodying characters about ecology and nationality. Architects integrate ethnic culture into composition, combine features of government building with modern architectural concept and create a regional and landmark building.

广西柳州市人民医院

建设地点：广西 柳州
建筑面积：158 900 平方米
竣工时间：2009 年

设计特点

项目中最具标志性的建筑属门诊住院楼，其建筑布局来源于对"桥"和"医疗港"的理解，半圆形呈放射状布置的门诊大楼，以环状公共交通系统连接着每一个门诊单元。门诊的每个医疗单元之间，设置有良好自然采光和通风的内部庭院，既满足现代化医院方便、快捷的要求，也营造了一个舒适的医疗环境。建筑立面造型风格以简约、明快的现代风格为主，力求实现个性与共性相和谐。

Liuzhou People's Hospital in Guangxi

Construction Location: Liuzhou, Guangxi
Building Area: 158,900 m²
Complete Time: 2009

Design Features

The Landmark Architecture of this project is the in-patient building, the idea of the architecture layout comes from the understanding of bridge and medical treatment harbor, the circular and radical-formed-out-patient building and the ring-fromed transportaion system link each medical department. With the inner yard of good daylight and ventiation, it not only meets the requirment of convenience and rapidity but also creats a comfortable medical treatment environment. The elevation design focus on simple and bright modern style, pursuing the harmony of universality and individuality.

雅克设计机构
APCE DESIGN GROUP

ARCHITECTURE • PLANNING • CONSULTANTS • ENGINEERING

城乡规划甲级　　建筑设计甲级

雅克设计机构创办于1992年，由雅克设计有限公司、海南雅克城市规划设计有限公司、雅克设计(中国)有限公司为拓展市场、迎接入世挑战于2002年联合组建的设计集团，其成员公司均具有建设部核发的建筑工程设计、城乡规划编制甲级证书。海南雅克城市规划设计有限公司还是全国第一家被授予规划甲级资质的民营设计企业。雅克设计机构于2003年12月通过ISO 9001：2000质量管理体系认证。机构在海南、上海设立区域业务中心，在杭州、厦门、重庆、南昌、长沙、宁波、南京、成都、广州、深圳等地设立分公司、办事处以及联盟单位。

雅克设计机构是中国最早以城市设计为主业的设计公司，主要业务领域包括城市总体规划、城市中心区规划、旧城改造、滨水地区规划、城市广场及景观规划、街道及商业区设计、社区及居住区规划、旅游度假区规划、公共建筑设计、酒店建筑设计、住宅建筑设计、市政设计、室内设计、园林设计及工程项目可行性研究等。18年来累计完成各类设计委托2500余项，赢得了客户的广泛赞誉。

雅克设计机构拥有500余位具有职业素养、创新精神和务实经验的专业人士。雅克秉承"设计创造价值"的公司理念，对每一项委托都从构思、技术、资源、经济及实施等方面精心研究，为达到目标并肩合作、全力以赴，共同实现客户与设计师的理想和荣耀。

APCE Design Agency was originally founded in 1992. In 2002, APCE Design Co., Ltd., Hainan APCE Urban Planning and Design Co., Ltd. and APCE Design (China) Co., Ltd. established APCE Design Group with an aim to expand the market and to greet the entry into WTO. Its member companies have obtained the Class-A Certificate of architectural engineering design and urban & rural planning issued by the Ministry of Construction. Hainan APCE Urban Planning and Design Co., Ltd. is the first private design firm which was granted the Class-A Qualification of Planning in China. In December 2003 this agency received the certification of ISO9001:2000 Quality Management System. APCE Agency has respectively established its regional business center in Hainan and Shanghai; it has also set up branches, offices or union units in Hangzhou, Xiamen, Chongqing, Nanchang, Changsha, Ningbo, Nanjing, Chengdu, Guangzhou and Shenzhen, etc.

APCE Design Agency is the earliest design company in China which specializes in urban planning. Its business scope mainly involves: the urban master planning, the urban center planning, urban renewal, the waterside district planning, the planning of urban plaza and landscape, the design of streets and commercial areas, the planning of community and residential district, the planning of holiday resort, design of public buildings, architectural design of hotels, domestic building design, municipal design, interior design, landscape design and project feasibility studies, etc. For 18 years, this agency has completed more than 2,500 project designs of all types, receiving wide praises from the clients and the society.

More than 500 organizations with professional quality, innovative spirit and practical experience of professionals. Adhering to the "design value" of the company's philosophy of each commission are from the concept, technology, resources, economy and other aspects of the implementation of careful study to achieve the goals work together to joint efforts, customers and designers to achieve the ideals and glory.

APCE Design Agency assembles more than 500 professionals with professional quality, innovative spirit and practical experience. Adhering to the philosophy of "Design Creating Value", it carries out careful studies on each commission in terms of the conception, technology, resource, economy and implementation, etc. Employees of APCE Agency cooperate with each other and make a concerted effort, striving to achieve the ideals and glory for both the clients and the designers.

法人：侯百镇
地址：海南省海口市龙华区玉沙路19号城中城A座二楼
电话：+86-898-68546187
传真：+86-898-68546186
邮箱：apce@apce.com.cn
网址：www.apce.com.cn

Legal person: Baizhen Hou
Address: 2nd floor of Chengzhongcheng A, No.19 of Yusha Road in Longhua District, in Haikou, Hainan Province
Tel: +86-898-68546187
Fax: +86-898-68546186
E-mail: apce@apce.com.cn
Http: //www.apce.com.cn

绿色技术概览
建筑能效模拟技术
室外风环境模拟技术
室内光环境模拟技术
室内热环境模拟技术
室内声环境模拟技术
太阳能光（热）伏系统
日光采集系统
雨水收集利用系统
建筑遮阳及导光技术
毛细管空调系统
可再生材料及循环利用材料技术
绿化滴灌技术

1. ASLOC成型墙体
2. 隔热断桥
3. 溶液调湿新风机
4. 地板辐射空调
5. LED照明
6. 透水地面
7. 遮阴率控制
8. 地下水涵养
9. 中水回用
10. 零照明能耗地下室
11. 烟尘过滤装置
12. 光导管
13. 变频技术
14. 节能管理系统
15. 绿化隔离墙
16. 高斯笼
17. 太阳能热水
18. 太阳能空调
19. 立体绿化
20. 新风预处理
21. 渗灌
22. 防自然通风口
23. 贴膜防辐射
24. 中空玻璃
25. 遮阳板
26. 导光板
27. 生态补偿
28. 太阳能光热利用
29. 可再生材料收集
30. 屋面绿化
31. 垃圾分类收集
32. 碳排放控制
33. 空气龄控制
34. 负氧含量控制
35. 雨水收集
36. 径流控制
37. 热环境控制
38. 可循环利用材料

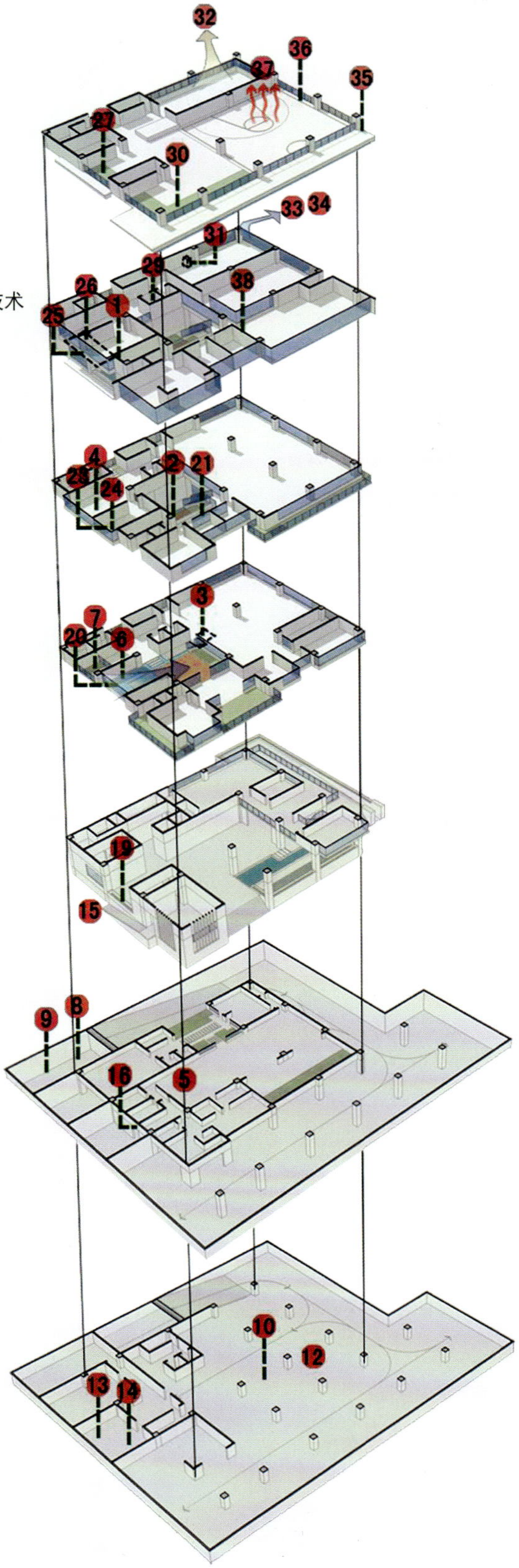

雅克设计中心
项目地点：海南 海口
建筑面积：7 400平方米
设计时间：2009年
主　　创：侯彦良、赵勇

项目为海南省第一个按三星级标准设计的绿色建筑，方案通过了建设主管部门组织的专家评审。设计遵循人文、绿色、开放、创新的理念，结合海南热带气候特点，进行大胆、有效的创新，在建筑全生命周期内，最大限度地节约了资源（节能、节地、节水、节材）。贯彻以人为本的思想，注重健康、适宜和高效空间的营造；注重与自然和谐共生；注重可再生能源的利用，减少碳排放。同时，公司在资源节约、环境友好、建筑可持续发展等方面也进行了有效的探索。

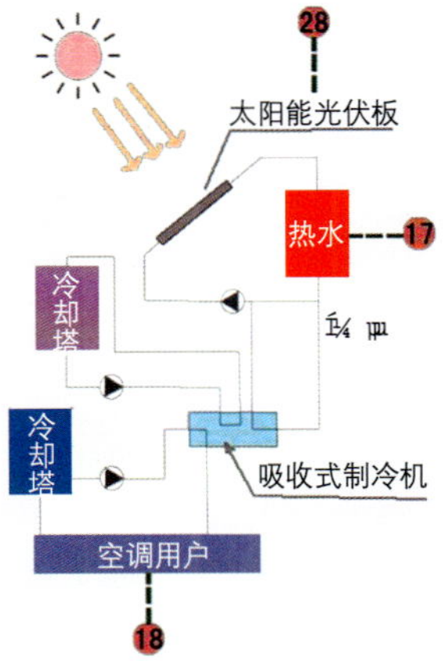

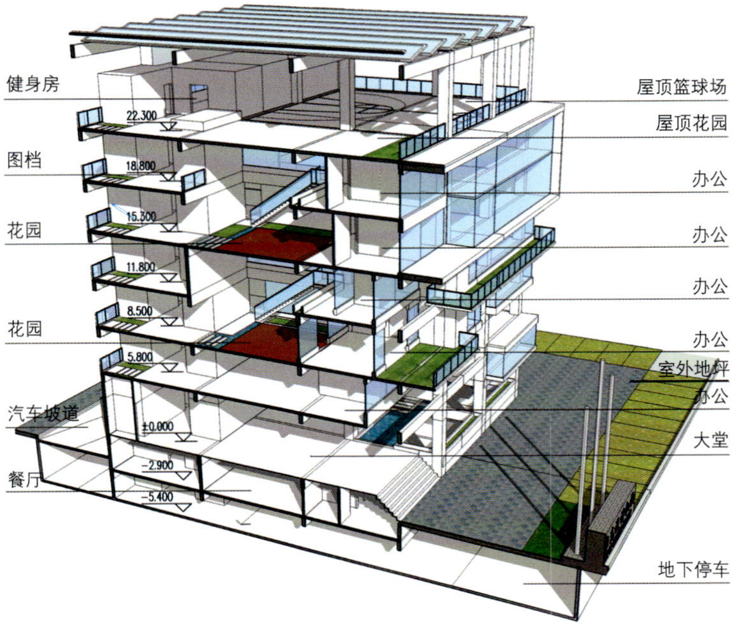

APCE APCE

1

1

冠豸山温泉酒店及度假村

项目地点：福建 龙岩

建筑面积：60 000平方米

设计时间：2005–2007年

主　　创：游涛

冠豸山温泉度假村以涉外五星级酒店标准配置为主体，拥有近六万平方米的温泉泡池休闲区。项目由会议中心、客房中心、总统套房及一百多幢山地度假屋等建筑群组成。

冠豸山度假村以客家传统文化为主要设计元素，依托自然山势，形成跌落式建筑组合空间序列，建筑多采用双坡或单坡式屋面形式，质朴而富于新意，坐落在高低错落的自然坡地上，为度假营造独特的空间氛围。

2

2

昆山曼氏商业街

项目地点：江苏 昆山

建筑面积：30 000平方米

设计时间：2008年

主　　创：王忠好

设计注重对商业建筑空间的合理规划，以提升其商业价值，合理实现城市从资源优势向经济优势转化的方式，并创造出丰富、舒适、有品位的现代商业空间。

建筑采用既典雅又不失现代感的设计风格，充分利用现代技术，和丰富的建筑符号，创造大气、细致、丰富、完整的建筑群体形象。建筑细部设计采用现代简洁的建筑语汇，通过艺术的构造和构成，力求充分展现建筑的时代性，弘扬传统建筑文化的精髓，为城市形象增添光彩，在建设"具有生活情调"的商业休闲区，同时充分体现自然水体景观。在沿青阳港一侧的商铺均设大面积露台，既和绕中心广场一周的连廊形成呼应，又形成了良好的商业氛围，便于水景与商业空间的相互渗透。

APCE

3

兴隆太阳河温泉度假中心

项目地点：海南 万宁
建筑面积：63 000平方米
设计时间：2010年
主　　创：傅培儒

本项目酒店、会所区建筑采用东南亚热带休闲建筑风格，以精致的浅坡屋顶、独特的造型及清新淡雅的色彩，营造一种简约大方的空间氛围，体现地域的热带气候特色并具有开放性和外向性特点。

酒店式公寓、别墅则以大面积开窗、坡屋面、众多的阳台以及中明度的色调形成通透、简洁、流畅的热带休闲建筑风格。

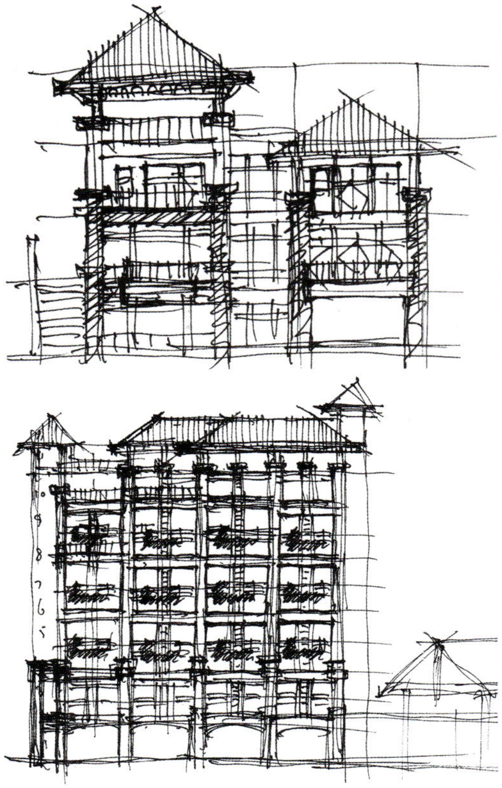

3

4

万平嘉园二期（3）

项目地点：海南 保亭
建筑面积：70 000平方米
设计时间：2010年
主　　创：周忠云

项目为酒店式公寓与住宅，是黎族苗族岭南特色建筑风格的重生，营造了一种热带度假风情的特色景观。

4

保定市建筑设计院

Baoding Architectural Design Institute

保定市建筑设计院始建于1952年7月，是河北省成立最早的建筑设计单位之一，具有建筑工程勘察、设计、监理甲级资质证书和施工图设计文件审查建筑工程一类、市政工程二类资质证书及建筑行业人防设计乙级、市政公用行业（道路）设计乙级、招标代理乙级、咨询丙级资质证书等。

我院现有职工191名，具有中高级职称人员105名，其中国家一级注册建筑师9名，国家一级注册结构工程师22名，国家注册监理工程师17名，注册岩土工程师7名，国家注册咨询师5名、注册造价工程师2名，注册设备师8名。

近年来，我院完成居住建筑、教育建筑、医疗建筑、工业建筑等设计项目近千项，建筑面积近500万平方米。业务范围遍布河北、北京、山西、山东、甘肃、湖北、青海、海南和内蒙古等地。获国家优质工程银奖1项、河北省工程勘察设计优秀奖20项，建设部推广应用新技术示范工程2项。我院连续多年被评为"重合同守信用单位""勘察设计先进集体""建设系统信誉、信用AAA单位""优秀勘察设计院""国家质量奖优质奖"等荣誉称号。

Founded in July 1952, our institute is one of the earliest architectural design units established in Hebei Province, possessing qualifications such as Class-A Certificate in construction survey, design and supervision; Grade I Construction Engineering Certificate and Grade II Municipal Engineering Certificate in the assessment of construction drawing design; Class-B Certificate in the design of civil air defense basement and in the design of city public utilities (road), Class-B Certificate in bidding agency and Class-C in consultation, etc.

Our institute has 191 employees in total and 105 of them have senior/intermediate titles, including: 9 Class-A national registered architects, 22 Class-A national registered structural engineers, 17 national registered supervision engineers, 7 registered geotechnical engineer, 5 national registered counselors, 2 registered cost engineers and 8 registered equipment engineers.

In recent years, our institute completed nearly one thousand design projects of residential, educational, medical and industrial projects, the gross floor area of which is close to five million square meters. Our business covers Hebei, Beijing, Shanxi, Shandong, Gansu, Hubei, Qinghai, Hainan and Inner Mongolia, etc. We are awarded one silver medal of national excellent construction, twenty excellent awards of Hebei Engineering Survey & Design and two demonstration projects of Ministry of Construction which promotes the application of new technology. Our institute has received titles of honor for many years, such as "the unit which fulfills contacts and keeps promises"; "Advanced Unit of Survey & Design"; "Credit AAA Unit of the architectural system"; "Excellent Survey & Design Unit" and "High Quality Award of National Quality Award", etc.

1–2 天威综合科技楼

本项目是与HARRY LIM ARCHITECTS的合作项目，位于河北省保定市朝阳北大街与保满路的东南角，占地约1.67公顷，建筑面积为70 940平方米，地上26层，地下3层，是集办公、科研、餐饮、住宿为一体的综合性建筑。建筑物主体布置在地块中部偏东北一侧。这样布置一方面使其尽量保持与道路交叉口有一定的距离，以缓冲其对该区域所造成的压迫感；另一方面在建筑的南侧主入口处建造入口广场，利用该广场丰富朝阳北大街的街景。在立面造型方面，三棱柱的形体在不同角度看到的面积不同，给人以"横看成岭侧成峰"的视觉感受，并且南侧立面紫色非晶硅玻璃和其他立面材料形成对比，丰富了视觉效果。而大厦底部柔和的平面形状以及周边绿化弱化了主体三角形带来的尖锐感。

This project is a cooperation project with HARRY LIM ARCHITECTS, which is located in the southeast corner of Chaoyang North Avenue and Baoman Road, Baoding, Hebei Province. Its land area is about 1.67 hectares, floor area 70,940 m^2, and there are twenty-six floors on the ground and three floors under ground. This building is an integrated construction which combines functions of office, scientific research, catering and accommodation together. Its main body is placed in the northeast side of the land center. On one hand, the design endeavors to keep the distance from the road intersection, buffering the pressure to this region; on the other hand, it creates an entrance plaza at the main entrance on the south side of the construction, which enhances the street landscape of Chaoyang North Avenue. For the facade configuration, the tri-prism shape displays different proportions at different angles, creating a visual sensation of "walk clean around the hill". The purple amorphous silicon glass of the south facade forms a contrast with other facade materials, enriching the visual effects. The plane shape at the bottom of the building and the surrounding greening soften the sharpness of the triangle-shaped main body.

1

2

3

3 水岸花城

水岸花城小区位于涿州市拒马河以北、107 国道以西，场地平整。

规划用地面积 20.387 公顷，规划总建筑面积 65.71 万平方米，容积率为 2.8，绿地率为 42%。

项目包括 19 栋 26 ~ 33 层的高层住宅、6 栋多层住宅及小学、幼儿园、会所等配套设施。

现已完成建筑面积约 40 万平方米。

Zhuozhou Waterfront Flower City

Waterfront Flower City Housing Estate is located in north of Zhuozhou Juma River, west of State Road 107, and the ground is grading. Planning land area is 20.387 hectares, with a programming construction area of 657,100 m^2, floor area ratio 2.8, and green rate 42%. It includes 19 high-rise residential buildings of 26-33 storeys, six other facilities such as multi-storey residential and primary schools, kindergartens, and clubs. Now, it has finished about 400,000 m^2 of the overall building area.

4 华科房地产非常空间

本工程为产权式公寓建筑，地下一层，地上二十层。其中地下一层为预留商业和设备用房。

地上一、二层为商业，三至十八层公寓楼。

工程总建筑面积：18 528 平方米。

建筑造型简洁明快，清新大方，通过开窗的上下错位变化赋予建筑物以动感和个性。

Special Space of Huake Real Estate

This project is an extended stay building, with a basement and 20 levels on the ground. The basement is the reserved commercial area and equipment room. The first and the second floor are the commercial sectors; the third to eighteenth are the apartments . The total construction area is18,528 m^2 . The modeling of the building is concise and lively, refreshing and in fine taste, and the change of upper and lower dislocation of window-opening brings the building dynamic and

4

5

5 北方家具产业基地行政服务中心

项目名称：北方家具产业基地行政服务中心
建设单位：河北建设集团中诚土地整理开发有限公司
建设地点：河北 香河
项目概况：行政服务中心及公寓楼及沿街商业楼占地面积 3.7 公顷，行政服务中心建筑面积 30 000 平方米，地上二十层，地下一层，主要功能为行政办公和电子信息服务中心。

Project Name: Administrative Service Center of Northern Furniture Industrial Base
Construction Unit: Loyal Land Administrative and Development Co.,Ltd. of Hebei Construction Group
Construction Location: Xianghe, Heibei
Project Description: area of Administrative Service Center and commercial building along the street: 3.7 ha. Building area of Administrative Service Center 30,000 m^2, 20 floors over ground, one floor underground, and it mainly serves as the center of administrative office and electron information service.

机械工业第六设计研究院

No.6 Institute of Project Planning & Research of Machinery Industry

机械工业第六设计研究院（简称中机六院）创建于1951年，是国家大型综合设计研究院，全国勘察设计行业综合实力百强单位，隶属中央大型企业集团——中国机械工业集团有限公司。

研究院现有7个工程所、3个分院、3个子公司、1800多名员工，其中中国工程院院士1人、中国工程设计大师2人、英国皇家特许建筑设备注册工程师协会荣誉资深会员1人、享受政府特殊津贴专家28人、研究员级高级工程师110 人、高级工程师386人、各类国家注册工程师439人。

研究院拥有工程咨询设计甲级资质8个，乙级资质7个，以及压力容器、压力管道设计资质共17个，具有工程监理综合资质，工程总承包甲级资质，建筑企业施工资质二级，具有国家商务部颁发的对外工程承包经营资格证书。

业务涵盖工业、民用、市政和环境工程领域的咨询、设计、监理、管理、代建和总承包。建院以来，研究院共完成工程项目10000余项，主编、参编国家和行业标准、规范21项。荣获国家科技发明二等奖1项，中国土木工程创新最高奖詹天佑奖1项，鲁班奖6项，国家科技进步及优秀工程设计金、银、铜奖25项，省部级奖300余项。

工业工程涵盖机床工具、无机非金属材料、重矿机械、轻工烟草、机车车辆、石化机械、风电机械、工程机械、轻纺机械、通用机械、农用机械、电工电器、仪器仪表、标准件、汽车及汽车零部件、军工等16大类机械行业。

民用工程涵盖大型公建、会展、文化、体育、交通、办公、商业、金融、医疗、教育、宾馆、酒店、住宅以及城镇、景观规划等。

市政与环境工程涵盖城市集中供热、城市供水、污水、工业废水治理、生活垃圾处理及危险固体废物处理等。

中机六院是国内机床工具和无机非金属材料两个行业唯一的专业设计院，是烟草、铸造、重矿、工程机械、民用建筑五大行业设计强院，在大型工厂和园区规划、企业生产流程再造、高难度结构、暖通空调、工业除尘、信息智能化、绿色工业建筑、市政和环境工程等许多方面具有国内一流的工程技术。

我院秉承"服务是立院之本、创新是兴院之道、人才是强院之基"的理念，竭诚以一流的技术、一流的队伍、一流的管理为国内外客户提供工程建设领域的全过程、全方位服务！

打造中国机械工业工程设计第一强院

No.6 Institute of Project Planning & Research of Machinery Industry (Abbreviated as SIPPR) is a large scale national comprehensive design institute and one of the top 100 design institutes in China survey & design industry subordinated to central enterprise group—China National Machinery Industry Corporation which was established in 1951.

SIPPR has 7 engineering institutes, 3 branches, 3 subsidiary companies and over 1,800 staff, of whom there are 1 academician of Chinese Engineering Academy, 2 Chinese engineering design masters, 28 experts enjoying special allowance from the government, 110 senior engineers at researcher level, 386 senior engineers and 439 national certified engineers of various fields.

Besides, SIPPR has 8 Class A qualifications, 7 Grade B qualifications, 17 pressure vessel & other qualifications, comprehensive supervision qualification, Grade A general contract qualification, Grade B architecture construction qualification and direct foreign business right.

Business scope of SIPPR includes consultation, design, supervision, management, construction agent & general contract of industrial, civil, municipal & environmental protection projects. It has finished more than 10,000 projects since its establishment. What's more, SIPPR attended the compilation of 21 national, local & industry standards and won the Second Prize of National Sci-tech Invention, Zhan Tianyou Prize---the highest prize of China civil engineering projects, Lu Ban Prizes for 6 times, 25 items of national gold, silver & copper prizes & more than 300 items of ministerial & provincial prizes.

Industry project includes such 16 machinery industries as machine tool, inorganic nonmetal material, light industry, tobacco, rolling stock, petroleum machinery, wind power machinery, engineering machinery, textile machinery, general machinery, agricultural machinery, electrical engineering and electric apparatus, instruments and meters, standard parts, automobile and auto parts and military etc.

Civil projects includes large scale public buildings, exhibition center, cultural buildings, sports buildings, transportation buildings, office buildings, commercial buildings, financial buildings, hospital buildings, educational buildings, hotels, restaurants, residential buildings, urban & landscape planning etc.

Municipal projects include urban central heat supply, municipal water supply & drainage, industrial sewage treatment, domestic refuse treatment and hazardous solid waste treatment etc.

SIPPR is the only professional machine tool & inorganic nonmetal material design institute in China and strong design institute of tobacco, casting, heavy mining machinery, engineering machinery & civil projects with advanced technology in large scale plant & industrial park planning, enterprise flow reconstruction, high-difficulty structure, industrial dust remove, information intelligence, green industrial buildings, municipal and environmental protection projects.

SIPPR sticks to the concept of "Service is the base of existence, innovation is the way to prosper and talents is the foundation to make SIPPR stronger " and sincerely looks forward to cooperating with friends both home and abroad by our powerful comprehensive strength, rich design experience, advanced design concept and best service.

To build the strongest institute for China's mechanical industrial engineering design

地址（Address）：郑州市中原中路191号/No.191，Middle Zhongyuan Road，Zhengzhou，China
邮编（P.C.）：450007
电话（Tel）：+86-371-6760 6088/6005 6087　+86-371-6760 6016
传真（Fax）：+86-371-6763 9571/6762 8091
网址（Http）：www.sippr.com.cn

工业项目
Industrial Projects

沈阳机床（集团）数控产业园
Numeric Control Industrial Park of Shenyang Machine Tool (Group) Co., Ltd. (SMTCL)

该项目是沈阳机床（集团）有限责任公司发展数控机床及成套装备整体改造项目。该项目是中机六院承担“设计、监理、项目管理”业务的典型代表，总投资 22 亿，占地 76 万平方米，建筑面积 58 万平方米，被誉为“东北振兴的样板和典范”“中国最大的数控机床装备制造基地”。

SIPPR undertakes design, supervision & project management of this project. Its total investment is 2,200 million with site space of 760,000 m^2 and building area of 580,000 m^2. It's regarded as "model and sample factory in machine tool industry" and "the biggest intelligence and net factory".

大连重工·起重集团重组改造工程
The Reform and Restructuring of the DHI.DCW Group

该项目包括泉水、中革、旅顺、双 D 港四个生产基地，占地面积 105 万平方米，建筑面积 45 万平方米，建设投资 37 亿元。大连重工、起重集团强强联合的战略性重组的成功实施，对全国重机行业大型骨干企业重组以及东北老工业基地改造具有重要的示范意义。该项目荣获机械工业科技进步三等奖、机械工业优秀工程咨询成果二等奖。

This project includes such four manufacturing bases as Quanshui, Zhongge, Lushun & Double-D Port with the site space 1,050,000 m^2, building area 40,000 m^2 and total investment 3.7 billion RMB. It plays important role in the regroup & transformation of Northeast China industrial base and won the 3rd Prize of Sci-tech Improvement of Machinery Industry and the 2nd Prize of Excellent Consultation Achievement of Machinery Industry.

宝鸡石油机械有限公司搬迁改造项目
Removing & Transforming Project of Baoji Petroleum Machinery Co., Ltd.

该项目占地 94.8 万平方米，建筑面积 38.5 万平方米，建设投资 30 亿元，建设内容包括渭滨新厂区建设及原有北厂区改造两部分，建成后将成为我国最大的石油钻采装备制造基地和石油钻采装备研发制造中心，全球最大的陆地钻机和泥浆泵研发制造基地。

This project is 948,000 m^2 of site area, 385,000 m^2 of building area and total investment 3 billion RMB. The construction mainly includes Weibin New Plant & transformation of original north plant. When it's finished, it will be the largest petroleum drilling equipment manufacturing base & research & manufacturing center in China and largest land rig & slurry pumps researcher, developer & manufacturer in the world.

中煤张家口煤矿机械有限责任公司煤矿装备产业园
Coal Mine Equipment Industrial Park of China Coal ZhangJiaKou Coal Mining Machinery Co., Ltd.

该项目占地 171 万平方米，一期建筑面积 36 万平方米，建设投资 21 亿元。企业通过搬迁改造，实现企业流程再造和产品升级，建成后将成为国内最大的煤机装备产业园。

This project is 1,710,000 m^2 of site area, 360,000 m^2 of building area and total investment 2.1 billion RMB. The company will be the largest coal mine machinery industrial park by the removing, transformation, production process reconstruction & product up-gradation.

国电联合动力（保定）技术有限公司一、二、三期建设项目

Construction project of Phase I,II,III of Guodian Joint Power (Baoding) Technology Co., Ltd.

该项目由国电集团公司投资建设，位于保定市国家高新技术产业园区，占地 20.6 万平方米，建筑面积 7.1 万平方米，由总装车间、叶片生产车间、打磨车间、电控车间、原料仓库、综合楼、公用站房等组成。国电联合动力技术有限公司年产 800 台套风力发电整机装备和 800 台套电控及风机叶片，是中国第一、亚洲最大的风电设备生产基地。

This project locates in Baoding National High-tech Industrial Park and is invested by China Power Group Company with the site area of 206,000 m^2, 71,000 m^2 of building area and annual output 800 wind power generation whole-set equipments and 800 electric control & fan blades. It's consisted of assembly workshop, blade production workshop, polishing workshop, electric control workshop, raw material warehouse, comprehensive building & public station and is the largest wind power equipment manufacturing base in China & Asia.

沈阳铸锻工业园

Shenyang Casting & Forging Industrial Garden

该项目占地面积 70 万平方米，建筑面积 31 万平方米，年产铸锻件 30 万吨，由沈阳重型机械集团公司铸锻公司、沈阳鼓风机集团铸造公司、沈阳矿山集团铸锻公司等近十家企业组成，是中国已建成的最大的铸锻工业园，已成为国内第一个集中建设铸锻企业的典范，对国内城市中铸锻件生产企业的搬迁改造重组起到示范作用。

This project is 700,000 m^2 of site area, 310,000 m^2 of building area and annual output 300,000 t of casting & forging products. It's consisted of about 10 companies, such as Shenyang Heavy-duty Machinery Group Corporation Casting & Forging Company, Shenyang Fan Group Casting Company, and Shenyang Mine Group Casting & Forging Company etc. It's the largest casting & forging industrial park finished in China and first successful model of centralized-built casting & forging company which plays an important role in removing, transformation & recombination of casting & forging products.

中国北车集团大连机车旅顺基地建设项目

Construction project of LuShun Base of Dalian Locomotive and Rolling Stock Co., Ltd. CNR. Group

该项目占地 188.6 万平方米，建筑面积 73.5 万平方米，建设投资 50 亿元，年生产各类机车 1000 台、城市轨道车辆 1000 辆、中高速柴油机 1000 台，将成为世界级、国内最大、技术最先进的轨道交通装备和通用动力机械研发和制造基地。

This project is 1,886,000 m^2 of site area , 735,000 m^2 of building area, with total investment of more than 5 billion Yuan and annual output of 1000 all kinds of rolling stocks, 1000 urban rail cars and 1000 middle & high speed diesel engines. When it's finished, it will be the world level and the largest & most advanced railway transportation equipments & general motive power machine research & manufacturing base in China.

郑州煤矿机械集团高端液压支架生产基地建设项目

High-end Hydraulic Support Manufacturing Base of Zhengzhou Coal Mine Machinery Group

本项目占地 32 万平方米，建筑面积 24 万平方米，建设投资 11.4 亿元，年产各类高端液压支架 15 000 架。该项目的实施进一步巩固郑煤机集团在国内煤机行业的领先地位，形成亚洲最大、世界一流的液压支架研发、生产基地。

This project is 320,000 m^2 of site area, 240,000 m^2 of building area, with total investment of 1.14 billion Yuan and annual output of 15,000 all kinds of hydraulic supports. This project makes Zhengzhou Coal Mine Machinery Group the largest & most advanced hydraulic supports research, development & production base in Asia & even in the world.

将军集团济南卷烟厂易地技术改造项目

Technical Innovation Project of General Group Jinan Cigarette Factory in other place

该项目占地约 7 万平方米，建筑面积 47.4 万平方米，建设投资 20 亿元，是国家烟草专卖局批准立项的我国烟草行业第一家占地 66.7 公顷、年产 100 万箱企业的易地技术改造项目，项目荣获机械工业科技进步二等奖。

This is the 1st technical innovation project of a tobacco company with the site area of about 70,000 m^2, building area of 474,000 m^2, total investment about 2,000 million Yuan and the annual output of 1 million boxes approved by State Tobacco Monopoly Bureau. It won the 2nd Prize of Sci-tech Improvement of Machinery Industry.

杭州卷烟厂“十一五”易地技术改造项目

Technical Innovation Project of the 11th Five-Year Plan for Hangzhou Cigarette Factory in other place

该项目位于杭州市西湖区科技经济工业园，占地 66.7 公顷，年产卷烟 100 万箱以上，总投资 20.6 亿元，其建筑面积近 19 万平方米的联合工房是国内烟草行业第一家按照最高标准“三星级”设计的“绿色工房”，对我国烟草行业建设“环境友好型、资源节约型”企业具有重要的示范作用。

Site area of this project is about 405,000 m^2, total investment RMB 2,060 million Yuan, total building area 620,000 m^2 and annual production capacity is more than 1 million boxes of cigarette. United workshop of this project with the building area 190,000 m^2 is a three-star green workshop which adopts such over ten technologies as roof greening, waste regeneration, after heat recycling & utilization and rain water utilization etc.

湖北中烟工业园区总体规划及武汉卷烟厂易地技改项目

General Planning of Hubei China Cigarette Industrial Garden and Technical Innovation Project of Wuhan Cigarette Factory

该项目建设用地面积 115.5 万公顷，一期工程总建筑面积 38.8 万平方米，建设投资 50 亿元，年产卷烟 150 万箱，是迄今为止我国烟草行业最大建设规模的易地技术改造项目。

Site area of this project is 1,150,000 m^2, building area of the Phase I is 388,000 m^2, total investment is 5 billon about 5,000 million Yuan and annual output is 1.5 million boxes of tobacco. It's the largest removing & technical innovation project in domestic tobacco industry.

C民用项目 Civil Projects

郑州白庄城中村改造—甲天下西湖新城

Zhengzhou Baizhuang Urban Village Reconstruction ——Jiatianxia Xihuxincheng Residential District

时间/Time: 2006年
地点/Location: 郑州(Zhengzhou)
建筑面积/Building Area: 613 000平方米

大观 · 国际住宅小区

Daguan International Residential District

时间/Time: 2009年
地点/Location: 郑州 (Zhengzhou)
建筑面积/Building Area: 330 000平方米

北京原乡美利坚别墅

Beijing Yuanxiang USA Villa

时间/Time: 2006年
地点/Location: 北京(Beijing)

民用项目 Civil Projects

郑州国际会展中心
Zhengzhou International Conference & Exhibition Center

郑州国际会展中心由中机六院和日本黑川纪章建筑都市事务所联合设计，是一个集展览、会议、交流、信息发布、观光等于一体的多功能、现代化、智能化的特大型公共建筑，项目总投资23亿元，建筑面积226 000平方米，展览中心展厅跨度102米，拥有3 560个国际标准展位。会议中心由可容纳5 000人的多功能厅、1 200人的国际会议厅、两个400人的会议厅和十几个中小型会议室等组成，无线局域网覆盖会展室内外所有空间，30余个智能化子系统高度集成。

中机六院在设计中大胆突破、勇于创新，创下了国内会展中心八项设计之最：7万平方米的钛锌金属屋面，60 000平方米原浆混凝土外墙饰面，360度旋转电子大屏幕，大面积室外透水混凝土，34 000平方米无柱展览大厅，10余米高的超重型活动隔断，单件重34吨的铸钢件节点，30米大跨度预应力梁板结构等。

该项目荣获中国建筑工程鲁班奖和第七届中国土木工程詹天佑奖。

Zhengzhou International Conference & Exhibition Center is an oversize multi-functional, modern and intelligence public building combined with the functions of exhibition, conference, communication, information and visiting. Its total investment is 2,300 million Yuan, building area is 226,000 m^2, span of the exhibition hall is 102 m and international standard exhibition booths number is 3,560. The Conference Center is consisted of a 5,000-people multi-functional hall, 1,200-people international conference hall, two 400-people meeting halls and more than 10 middle or small-sized meeting rooms. Besides, wireless LAN integrated by more than 30 subsystems covering all the space in and out of exhibition hall.

SIPPR creates eight "the first" in the design of Zhengzhou International Exhibition & Conference Center, they are: 70,000 m^2 titanium zincum metal roof, 60,000 m^2 virgin pulp concrete exterior wall finish, 360° rotary electric screen, large area exterior pervious concrete, 34,000 m^2 exhibition hall without column, over 10 m overweight movable partition, cast steel node with the single weight of 34 t and 30 m long-span pre-stressing beam & slab structure. All these mentioned above reflect the comprehensive design ability of SIPPR in oversize public building design.

This project won Lu Ban Prize & 7th Zhan Tianyou Prize for China Civil Engineering.

重庆龙头寺片区保利花园
Chongqing Longtousi Area Baoli Garden

时间(Time): 2008年
地点(Place): 重庆(Chongqing)
建筑面积(Building Area): 19 300平方米

青岛颐和・国际
Qingdao Yihe International Mansion

时间/Time: 2004年
地点/Location: 青岛 (Qingdao)
建筑面积/Building Area: 120 000平方米

金科国际大厦
Jinke International Mansion

时间/Time: 2009年
地点/Location: 郑州(Zhengzhou)
建筑面积/Building Area: 75 000平方米

国家棉花及纺织服装产品质检中心综合楼
Comprehensive Building of National Cotton, Textile & Clothes Product Quality Control Center

时间/Time: 2007年
地点/Location: 郑州 (Zhengzhou)
建筑面积/Building area: 42 000平方米

河南省建筑设计研究院有限公司

Architectural Design Research Institute Company Limited of Henan Province

河南省建筑设计研究院有限公司是由原事业单位河南省建筑设计研究院整体改制成立的民营科技型企业，2007年6月1日完成工商注册登记。原河南省建筑设计研究院成立于1953年，是河南省最早成立的国家甲级综合勘察设计单位。

公司拥有建筑工程设计、工程咨询、工程监理、工程造价咨询等甲级资质证书，工程勘察综合类甲级资质证书，以及市政行业（热力、给水、排水、道路、桥梁）设计、建筑行业人防工程设计、商物粮行业（冷藏库）工程设计、工程招标代理等乙级资质证书；公司还拥有建筑工程施工图设计文件审查一类资质证书，特种专业工程专业承包资质证书，地基与基础工程及建筑装修与装饰工程专业承包二级资质证书，建筑工程质量司法鉴定资质证书以及建设工程项目管理河南省试点企业。

公司现有职工308人，专业技术人员占80%以上，其中教授级高级工程师5人、高级专业技术人员67名、中级专业技术人员94名，被授予有突出贡献专家称号的有2人，6人享受国务院特殊津贴。公司共有国家一级注册建筑师26名、国家一级注册结构工程师38名、国家二级注册建筑师15名、国家二级注册结构工程师5名，注册土木（岩土）工程师14名，水、暖、电、化工专业注册工程师37人，注册监理工程师14人，注册造价工程师6人，注册咨询（投资）工程师10人。公司下设五个综合设计分院及十个分支机构，并在上海、珠海等城市设有分院。

公司业务范围包括工程勘察、建筑工程设计、人防工程、商物粮行业（冷藏库）、市政行业（热力、给水、排水、道路、桥梁）工程设计、工程咨询、土工试验和工程测量、工程监理、基础工程处理、工程造价咨询、工程招标代理、工程项目代建、建筑工程施工图审查、特种专业工程专业承包、地基与基础工程及建筑装修与装饰工程专业承包、建筑工程质量司法鉴定等。

五十多年来，公司立足中原，面向全国，放眼世界，先后完成数以百计的大型公共民用建筑（涵盖办公建筑、商业建筑、旅游建筑、交通运输用房、科教文卫建筑）设计，住宅小区规划及设计，大、中型工业厂房以及城市供排水等工程的建筑设计与工程勘察任务。

公司重视建筑创作，推进技术进步，讲求社会信誉，创作成果遍及全国26个省市及全球10余个国家和地区。20世纪90年代以来，有128项勘察设计和科研成果荣获国家、省（部）级的优秀设计、优秀勘察、科技进步奖。近年主要作品有河南博物院、河南国际博览中心、北京河南大厦、郑州豫港大厦、郑州邮政综合生产楼、河南电信综合楼、河南省政府办公室楼、河南省高级人民法院审判厅、建业金水花园、明辉城市花园、郑州火车站广场、上海影视乐园、河南大学教学组群等。其中河南博物院主馆设计，获得了“国际建筑师协会第20届大会——当代中国建筑艺术奖”，中国百年经典建筑奖，2005年度部级优秀建筑工程设计项目二等奖，“上海现代设计杯”全国首届城市标志性、纪念性建筑设计三等奖，郑州邮政综合生产楼荣获2005年度国家优质工程银奖。

公司于2002年通过GB/T19001–2000–IS90001：2000标准质量管理体系认证。2008年通过GB/T19001–2000–IS90001：2000质量管理体系、GB/T24001–2004–ISO14001：2004环境管理体系、GB/T28001–2001职业健康安全管理体系三标一体化管理体系认证。我们将长期遵循“以人为本、质量第一、安全环保、务实创新”的管理方针，始终致力于向社会提供高品质的设计服务，使公司在行业中享有极高的声誉。

河南省建筑设计研究院有限公司全体员工贯彻科学发展观，坚持推进改革，以更科学的手段加强质量管理，力气以更加热诚的态度为社会各界提供最优质的设计服务，向社会奉献更多的优秀建筑。在新的征途上，全公司员工将同心协力，不断开拓创新，再绘神州新蓝图。

Architectural Design Research Institute Company Limited of Henan Province is founded as a Civilian-run technological enterprise on June 1st, 2007, entirely transformed from Henan Architectural Design Academy, which is a State-owned institute and at the same time the first national class A comprehensive survey and design institute in Henan Province.

The company owes Class A certificates of construction project design, project counselling, project supervision, project cost consulting, and project prospection, also class B certificates of public utility industry design (like heating, water supply, drainage, road and bridge), civil air defense works design, grain-and-goods industry project design (like cold store), project bidding agency and so on. Besides, the company is endowed various certificates, including censor of construction project shop drawing, contract of engineering industry, professional contract of foundation works and architectural decoration project, judicial expertise of construction project quality, and construction project management pilot enterprise of Henan Province.

Currently the company has 308 employees in total, more than 80% of whom are professional technicians. Among them there are 5 senior engineers, 67 senior professional technicians, 94 junior professional technicians, 2 with the title of professor making outstanding contribution, 6 with special allowance by the State Council. Also there are 26 national Class 1 registered architects while 15 Class 2 ones,38 national first Class 1 registered structural engineers while 5 Class 1 ones, 14 national registered civil engineers (geo-engineers), 37 registered worker major in water, heating, electric and chemic, 14 registered manager engineers, 6 registered cost engineers, 10 registered consultant (Invest) engineers. The company has set up five comprehensive designing branches and ten offshoots, and affiliates in Shanghai, Zhuhai and other cities as well.

Our designs covers a great number of fields like project prospection, construction project design, design of civil air project, grain-and-goods industry project(cold store), public utility industry (heating, water supply, drainage, road and bridge), project counselling, geotextile experiment, project measurement, project supervision, foundation project management, project cost counseling, project bidding agency, censor of construction project shop drawing, construction of engineering project, professional contract of foundation works and architectural decoration project, judicial expertise of construction project quality and so forth.

For more than five decades, the company has been basing itself upon the central plains, facing the whole country, as far as Western Asian, Eastern Europe and Africa. We have successively accomplished hundreds of tasks: large-scale public civil building design (including office building, commercial building, tourist building, transport building, and buildings of science, education, culture or health care), residential district planning and design, large and middle scale industrial workshops, urban water supply and drainage project.

We always emphasize the architectural invention and credit standing, and promote technology progress. We have extended our works to 26 provinces / cities all over the country and a dozen of other countries and areas. Our 128 works won outstanding design, outstanding prospection and science technology progress reward at national or provincial level since 1990s. Our recent works are as following: Henan Museum, Henan International Expo Center, Henan Mansion in Beijing, Yugang Mansion in Zhengzhou, Zhengzhou Postal Incorporation Production Building, Henan Telecommunication Incorporation Building, office building of Henan Province's government, Court-room of Henan Higher People's Court, Jinshui Park in Jianye, Minghui City Park, Square of Zhengzhou Railway Station, Shanghai Film & TV Park, group of teaching buildings in Henan University etc. The design of Henan Museum's main building won the reward of Modern Chinese Architecture Art in the 20th conference of International Architect Association, as well as the Chinese Classical Architecture Reward of the Century, the second prize of Outstanding Engineering Design Award of Construction Ministry in 2005, and the third prize of First National Urban Symbolic / Memorial Building Design Award. Zhengzhou Postal Incorporation Production Building won the silver prize of National High-quality Project Award in 2005.

We get GB/T19001-2000-IS90001 proved and 2000 Standard Quality Management System proved in 2002, then get GB/T19001-2000-IS90001 proved, 2000 Quality Management System proved, GB/T24001-2004-ISO14001 proved, 2004 Environment Management System proved, GB/T28001-2001 Occupational Safety and Health Management System proved in 2008. Keep following the guidelines for management which are 'People oriented; quality in the first place; safe and green; practical and creative', we've been trying hard to provide the society high-quality design service and the company have built up a reputation in this industry.

Pursuing scientific development view, staff of Architectural Design Research Institute Company Limited of Henan Province will persist in improving, enhancing quality management with more scientific methods, providing the clients circles with best design service more warmly, and contributing more excellent architectures to the society. On the way to the future, we will all be of one mind, focus on exploration and innovation, and help to make a new blueprint of our country.

项目名称：河南博物院主馆
设计时间：1992年10月—1994年7月
建成投产时间：1998年5月1日
建筑规模（层数、建筑面积）：地上4层、地下1层
建筑面积：55 467平方米
结构类型：框架结构
2008年2月获得国家级优秀工程设计铜奖
2006年9月获得建设部优秀工程设计二等奖
主要设计单位：河南省建筑设计研究院有限公司
协作单位：东南大学建筑研究所

Project Name: Main Building of Henan Museum
Design Time: 1992.10 - 1994.7
Completion Time: 1998.5.1
Construction Scale (number of storeys, floor space): 4 storeys above ground, 1 storey under ground
Building Area: 55,467 m^2
Construction Type: Frame construction
Won the Bronze Prize of National Outstanding Engineering Design Award in February, 2008
Won the Second Prize of Outstanding Engineering Design Award of Construction Ministry in September, 2006
Major Designing Institution: Architectural Design Research Institute Company Limited of Henan Province
Associate Institution: Building Research Institute of Southeast University

项目名称：郑州旅游旅馆（河南国际饭店）
设计时间：1978年9月—1979年10月
建成投产时间：1981年
建设规模（层数、建筑面积）：主体13层、裙房1层，主体高度49米
总建筑面积：14 750平方米
结构类型：框架
1984年获得国家级优秀工程设计表扬奖
1984年获得建设部优秀工程设计二等奖
主要设计单位：河南省建筑设计研究院

Project Name: Zhengzhou Tourist Hotel (Henan International Hotel)
Design Time: 1978.9 - 1979.10
Completion Time: 1981
Construction Scale (number of storeys, floor space): 13 storeys and 49-meter high of the main building, 1 storey of annex
Gross Building Area: 14,750 m^2
Construction Type: Frame construction
Won the recognizing prize of National Outstanding Engineering Design Award in 1984
Won the Second Prize of Outstanding Engineering Design Award of Construction Ministry in 1984
Major Designing Institution: Architectural Design Research Institute of Henan Province

项目名称：郑州高新技术开发区管委会办公楼
设计时间：1995年10月—1996年12月
建成投产时间：1998年
建筑规模（层数、建筑面积）：主体4层，主体高度16米
总建筑面积：9 580平方米
结构类型：框架
2000年获得建设部优秀工程设计表扬奖
主要设计单位：河南省建筑设计研究院
协作单位：天津大学

Project Name: Zhengzhou New and Hi-Tech Development Zone Management Committee Office Building
Design Time: 1995.10 - 1996.12
Completion Time: 1998
Construction Scale(number of storeys, floor space): 4 storeys and 16-meter high of the main building
Gross Building Area: 9,580 m^2
Construction Type: Frame construction
Won the recognizing prize of Outstanding Engineering Design Award of Construction Ministry in 2000
Major designing institution: Architectural Design Research Institute of Henan Province
Associate Institution: Tianjin University

项目名称：中原国际博览中心主展馆
设计时间：1991年1月—1992年8月
建成投产时间：1994年
建筑规模（层数、建筑面积）：主体2层、附房四层主体高度21米
总建筑面积：19 457平方米
结构类型：框架及拱形网架
1995年获得建设部优秀工程设计表扬奖
主要设计单位：河南省建筑设计研究院

Project Name: Main Exhibition Hall of Central Plains International Expo Center
Design Time: 1991.1 - 1992.8
Completion Time: 1994
Construction Scale (number of storeys, floor space): 2 storeys and 21-meter high of the main building, 4 storeys of annex
Gross Building Area:19,457 m^2
Construction Type: Frame and vaulted space truss construction
Won the recognizing prize of Outstanding Engineering Design Award of Construction Ministry in 1995
Major designing Institution: Architectural Design Research Institute of Henan Province

河南圣地建筑景观设计有限公司

河南圣地建筑景观设计有限公司是一家集建筑、景观、规划设计及其他相关工程设计、策划为一体的设计单位。公司拥有十几年行业经验，在国内具有较高的知名度和权威性，对项目的整体规划、旅游规划、建筑设计、运营策划，具有熟练的专业操控水准和较高的前瞻性与行业带动性。公司在建筑、规划、园林、结构、水电、装饰、环艺等方面有固定专业人才30多人、专业顾问6名。近年来，已经完成河南、安徽、甘肃、山西、陕西、湖北、河北、山东等诸多省份三十多项大、中、小型项目的开发设计工作。同时公司还特别注重对不同地域文化、历史、民俗、象征等方面的思路整合。

Henan Shendi architectural Landscape Design Co., Ltd. is a design insitute that provides services of architecture design, landscape design, master planning, and other relative construction deisign and planning. With more than ten years of professional experience, they have good reputation and authority in China. Their professional manipulative level and good prospective in master planning, tourism planning, architecture deisign, operation planning area established their leading role in the industry. They have more than 30 professional talents in architecture, planning, landscape architecture, construction, water and electricity, decoration and environment art design, including 6 professional consultants. In recent years, they have finished over 30 small, medium and large projects in Henan, Anhui, Gansu, Shanxi, Hubei, Hebei, Shandong Provinces. In the design process, they put strong emphasis on the intergration of culture, history and custom of different areas.

地址：河南省郑州市经三路32号财富广场1号楼14层C室
电话：+86-371-65350762
+86-371-65350782
传真：+86-371-65350763
+86-13592662087（刘世洛）
邮箱：xsd808@163.com

Address: Room C, floor 14th, Building No.1, fortune suare, No. 32 Jingsan Rd, Zhengzhou, Henan
Tel: +86-371-65350762
+86-371-65350782
Fax: +86-371-65350763
+86-13592662087 (Shiluo Liu)
E-mail: xsd808@163.com

郑东新区宏信商业步行街

建设地点：河南 郑州
设计时间：2008年10月
设计师：刘世洛 郎研华

Hongxin pedestrian shopping street in new - east zone of Zhengzhou

Construction Location: Zhengzhou, Henan
Design Time: 2008.10
Designer: Shiluo Liu, Yanhua Lang

番城遗址公园

Fan town site park

建设地点：河南 信阳
占地面积：550亩
设计时间：2010年2月
设 计 师：刘世洛、张丹、杜刚成

Construction Location: Xinyang, Henan
Site Area: 550 acre
Design Time: 2010. 02
Designer: Shiluo Liu, Dan Zhang, Gangcheng Du

天宝阁综合文化建筑

Tianbaoge comprehensive cultural architecture

建设地点：山西 晋中
建筑面积：24 000平方米
设计时间：2009年11月
设 计 师：刘世洛 高光磊 张代令

Construction Location: Jinzhong, Shanxi
Building Area: 24,000 m^2
Design Time: 2009. 11
Designer: Shiluo Liu, Guanglei Gao, Dailing Zhang

宝天曼景区规划

建设地点：河南 南阳　　占地面积：650亩　　设计时间：2010年1月
设 计 师：张丹、张代令、李小霞

Planning of Baotianman scenic spot

Construction Location: Nanyang, Henan　　Site Area: 650 acre　　Design Time: 2010. 01
Designer: Dan Zhang, Dailing Zhang, Xiaoxia Li

FANGZHOUARCHITECTS

哈尔滨方舟建筑设计有限公司

Introduction to Harbin Ark Design Co., Ltd.

法人：刘远孝
地址：黑龙江省哈尔滨市道外区红旗大街 991 号
电话：+86-451-87858515
传真：+86-451-87858504
邮件：fz2001@vip.163.com
网址：www.fzjz.net

Legal Persen: Yuanxiao Liu
Address: No.991, Hongqi Street Daowai District, Harbin, Heilongjiang Province
Tel:+86-451-87858515
Fax:+86-451-87858504
E-mail: fz2001@vip.163.com
Http: //www.fzjz.net

哈尔滨方舟建筑设计有限公司坐落于美丽的冰城哈尔滨，是我国建筑设计行业改革大潮中涌现出来的一支新军。多年来，方舟人凭借卓越的团队智慧及拼搏精神始终立于本地区建筑设计行业的不败之地，企业经历了由小到大、由弱到强的深刻变化过程，发展为拥有建筑、规划、景观、结构、给排水、暖通、空调、电气、工程咨询等专业的强大设计企业。目前公司拥有员工 128 人，一级注册建筑师、一级注册结构工程师、注册设备工程师与高级工程师 48 人。成立近 10 年来，为社会、为城市贡献了许许多多精美的设计作品，得到了社会各界的广泛赞誉。许多作品在省优、部优以及全国民营设计企业华彩奖评选中多次获奖。今天的方舟依靠自身的实力，已经成为了黑龙江省内建筑设计行业的顶尖设计机构，充分占有了本地区的设计市场。

我们在《年鉴》中推出的一小部分作品，仅仅反映我们企业的点滴成果，目的在于与全国乃至境外的优秀设计机构创造更多的切磋、交流机会，从而使我们的设计水平有更大的提高。

Harbin Ark Design Co., Ltd. located in the beautiful ice city of Harbin, is a new army emerged in the reform tide of domestic architectural design industry. Over the years, the Ark people has been holding an invincible position in the regional architectural design industry by outstanding virtues of team cooperation and fighting spirit, making the company experience a process of profound change: from small to large, from weak to strong. Now Ark has become a powerful design company with professions including construction, planning, landscape, structure, water supply and drainage, heating, air conditioning, electric and engineering consulting, etc. Currently the company has 128 employees, 48 of whom are Class-A registered architects, Class-A registered structural engineers, registered equipment engineers and senior engineers. Over the past decade since its foundation, Ark has contributed lots of fine designs for the society and city, receiving wide praise from all works of life and winning many prizes in the province, ministry as well as the Nightlife Awards of National Private Design Enterprise. Today's Ark, by its own strength, has become a top institution in the architectural design industry in Heilongjiang province, receiving full recognition from all social circles.

We will present a small part of our design work in the "Yearbook", which merely exhibit a few achievements of Ark, with an aim to create more opportunities for mutual learning and exchanges with national, even international excellent design agencies, thus greatly enhancing our design performance.

哈尔滨工程大学教学办公主楼群体
设计时间：2001 年 7 月
竣工时间：2003 年 12 月
建筑面积：74 200 平方米
建筑层数：10 层

The main building group for teaching and office in Harbin Engineering University
Design Time: 2001.07
Completion Time: 2003.12
Building Area: 74,200 m^2
Building Storey: 10

大庆奥林匹克公园体育中心
设计时间：2009 年 8 月
竣工时间：建设中
建筑面积：80 000 平方米

Sports Centre of Daqing Olympic Park
Design Time: 2009.08
Completion Time: Under construction
Building Area: 80,000 m^2

哈尔滨市第一医院门诊病房综合大楼
设计时间：2008 年 11 月
竣工时间：建设中
建筑面积：87 773 平方米
建筑层数：22 层

The building complex of out-patient wards in Harbin First Hospital
Design Time: 2008.11
Completion Time: Under construction
Building Area: 87,773 m^2
Building Storey: 22

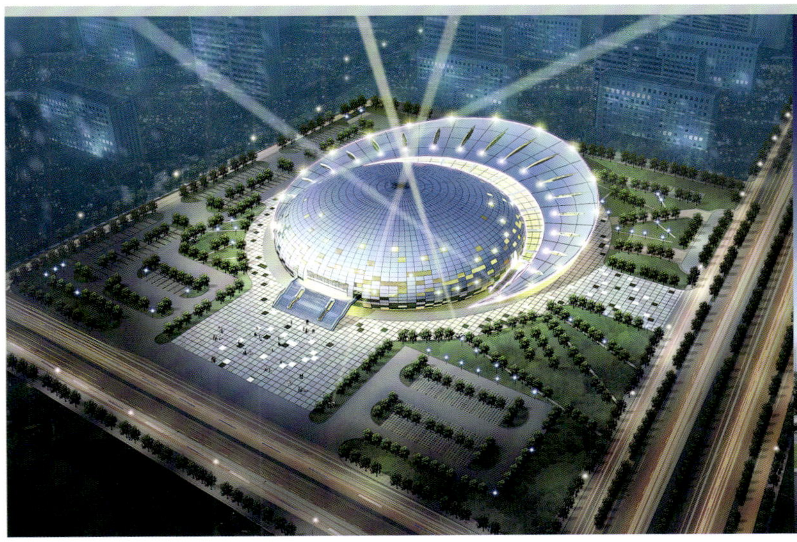

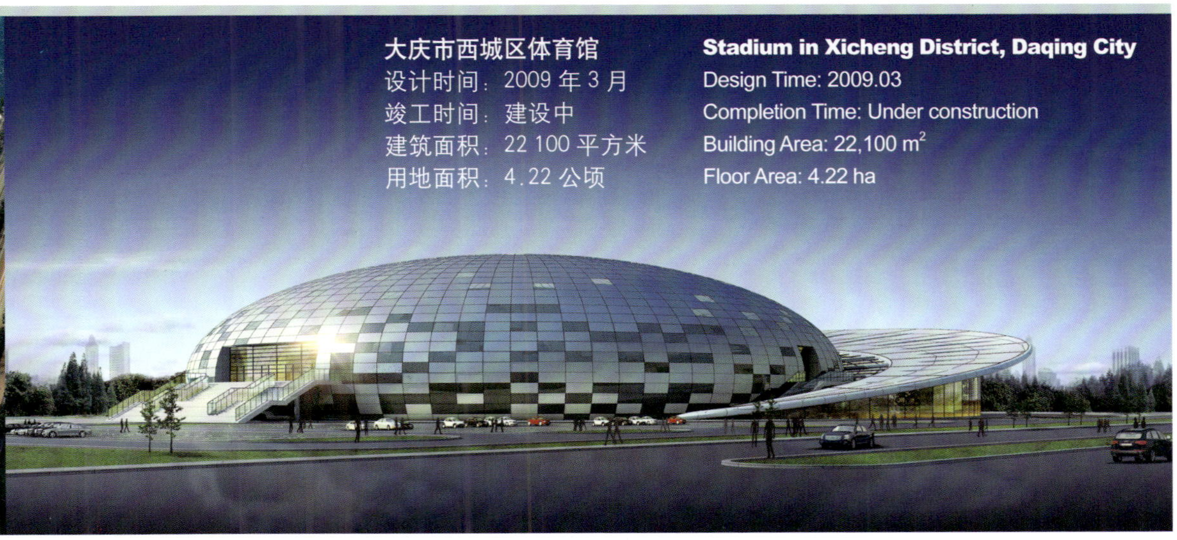

大庆市西城区体育馆
设计时间：2009 年 3 月
竣工时间：建设中
建筑面积：22 100 平方米
用地面积：4.22 公顷

Stadium in Xicheng District, Daqing City
Design Time: 2009.03
Completion Time: Under construction
Building Area: 22,100 m^2
Floor Area: 4.22 ha

齐齐哈尔中医院
设计时间：2009 年 8 月
竣工时间：建设中
建筑面积：93 860 平方米
建筑层数：15 层

Chinese medicine hospital in Qiqihar
Design Time: 2009.08
Completion Time: Under construction
Building Area: 93,860 m^2
Building Storey: 15

立汇－美罗湾居住区售楼中心
设计时间：2006 年 5 月
竣工时间：2007 年 3 月
建筑面积：1 300 平方米
建筑层数：2 层

The sales center of Lihui – Meiluowan residential quarter
Design Time: 2006.05
Completion Time: 2007.03
Building Area: 1,300 m^2
Building Storey: 2

哈尔滨理工大学教学主楼
设计时间：2004 年 7 月
竣工时间：2009 年 7 月
建筑面积：145 000 平方米
建筑层数：22 层

The main teaching building of Harbin Institute of Technology
Design Time: 2004.07
Completion Time: 2009.07
Building Area: 145,000 m^2
Building Storey: 22

绿地世纪城牡丹江塞纳丽舍会所
设计时间：2006 年 8 月
竣工时间：2007 年 7 月
建筑面积：2 100 平方米
建筑层数：4 层

Mudanjiang SeNaLiShe Club in Greenland Century City
Design Time: 2006.08
Completion Time: 2007.07
Building Area: 2, 100 m^2
Building Layer: 4

哈尔滨荣耀上城居住小区
设计时间：2007 年 7 月
竣工时间：2009 年 7 月
建筑面积：187 300 平方米
用地面积：3.76 公顷
容 积 率：4.98
建筑层数：32 层

The Glory of City Residential Quarter in Harbin
Design Time: 2007.07
Completion Time: 2009.07
Building Area: 187,300 m^2
Floor Area: 3.76 ha
Plot Ratio: 4.98
Building Storey: 32

哈尔滨东升江畔居住小区
设计时间：2006 年 4 月
竣工时间：2008 年 7 月
建筑面积：178 000 平方米
用地面积：2.13 公顷
容 积 率：6.10
建筑层数：33 层

Dongsheng River-bank Residential Quarter in Harbin
Design Time: 2006.04
Completion Time: 2008.07
Building Area: 178,000 m^2
Floor Area: 2.13 ha
Plot Ratio: 6.10
Building Storey: 32

中南建筑设计院
Central-South Architectural Design Institute

院况简介：

中南建筑设计院创建于1952年，是国家最早组建的六个大区综合性建筑设计院之一；现约有员工660人，各类专业技术人员约570人。50多年来，中南建筑设计院坚持以精心设计、优质服务为宗旨，先后在全国各地及世界18个国家、地区完成了6000余项工程设计，创造出一批享有盛誉的建筑设计作品。

业务范围包括建筑工程设计、商物粮行业（冷藏库）工程设计、市政公用行业（给水排水）工程设计、工程勘察专业类岩土工程、建筑智能化系统工程设计、建筑工程总承包、工程建设管理、工程咨询。

Profile:

Established in 1952, Central-South Architectural Design Institute fell into one of the first six regional comprehensive architectural design institutes in China, currently with about 660 employees and about 570 technical staffs in different professional fields. Over the past 50 years, the institute has finished more than 6,000 items of engineering design in the whole country as well as over 18 countries and regions around the world by sticking to the principle of "elaborate design and quality services", creating a large number of architectural design works with high reputation.

Business scope:

Architectural engineering design; engineering design for commodity, goods and food supplies (refrigerated warehouse), engineering design of urban utilities (water supply and discharge), geotechnical engineering in the field of engineering survey, engineering design of building intelligent system, architectural engineering general contract, engineering construction management, engineering consultation.

法定代表人：张柏青
地址：武汉市中南二路10号
邮编：430071
网址：www.csadi.net

Legal representative: Boqing Zhang
Address: No.10 of Zhongnan No.2 Road, Wuhan City
P.C.: 430071
Http: //www.csadi.net

延安火车站

建设地点：陕西 延安
建筑面积：23 987平方米
用地面积：40 309平方平方米
容 积 率：0.60
建筑高度：30.6米
层　　数：地上2层，地下1层
设计时间：2006年5月
竣工时间：2007年4月

项目介绍

在建筑设计中概括提取了窑洞的典型特征，将其作为建筑造型的基本元素延安精神的象征，体现延安车站的地域性文化特征；另外，将提炼后汉代门阙造型作为建筑的主入口。在车站功能设计方面，打破传统条块分割的车站格局，营造出具有视觉导引系统的开放式共享候车空间。延安站的进站大厅和候车大厅均设置连通上下两层的中庭，这里没有传统形式的如同迷宫般的候车室，只有低矮玻璃隔断划分出不同的候车和服务区域，明朗宽敞的室内空间，阳光透过半透明的屋顶天窗，形成华丽的光影效果，旅客来到在这里很容易就对整个车站的格局一目了然。

站台雨棚设计为无站台柱式雨棚，分为两个跨度，其跨度分别为54米和20.5米。雨棚垂直轨道方向采用两跨的"T"字形张弦钢梁结构；张弦钢梁结构两根弧形的上弦矩形主管为压弯构件，下弦为张紧的拉索，腹杆为平行排列的受压圆管，这样的雨棚设计富有韵律，恰似琴弦，又似张满的弓。建成后的雨篷轻巧通透，富有张力。

Yan'an Railway Station

Location: Yan'an Shanxi
Building Area: 23,987 m²
Site Area: 40,309 m²
Plot Ratio: 0.60
Building Height: 30.6 m
Storey Number: 2 storeys above the ground and 1 storey under ground
Design Time: May 2006
Completion Time: April 2007

Project Introduction

In building's image design, the typical element-cave dewelling has been adopted for serving as basic subject in shaping and a symbol of Yan'an Spirit as well as representing regional culture. Besides, shape of a Han Dynasty-style portico has been borrowed in designing the main entrance of the building. In functional design, the conventional fragmentation arrangement has been discarded and an open waiting space has been built for a visual guiding effect. Patios connecting the upper storey and lower storey have been designed in the entrance hall and the waiting hall. In this station, the conventional waiting rooms resembling a labyrinth have disappeared and only low glass partitions have been employed to separate different service and waiting zones. Bright and broad indoor space together with the translucent roof scuttle makes the sunlight generate a magnificent effect of shade and shadow. Passengers can command the whole layout of this station at this point.

Non-column platform awning has been adopted in design, which consists of two bays with spans of 54 m and 20.5 m respectively. In the direction of awning vertical to the track, T-shaped chord-tension steel girder structure with two bays has been used, upper-chord rectangular main of which is a compressive and flexural member, lower chord of which is a tense guy cable and web member is made of compressive circular pipe with parallel arrangement, resembling a music wire and a full-blown bow. The completed awning is featured with light quality, transparency and sense of tension.

武汉协和医院外科病房大楼

建设地点：湖北 武汉
建筑面积：74 117平方米
用地面积：8 628平方米
容 积 率：8.59
建筑高度：144.7米
层　　数：地上32层，地下2层
设计时间：2001年12月
竣工时间：2006年9月

项目介绍

平面设计遵循“以病人为本”的设计指导原则，针对任务书的要求和外科病房的功能特点，将标准护理单元平面设计成舒展的弧形平面，所有病房皆朝南向，最大限度获得自然通风和采光，使每位病人都能享受和煦的阳光、清新的空气和开阔的视野，以利病人身体的康复。护士站采取半岛模式，居于建筑平面的中央，使护士能给病人提供最快捷、方便的服务。

建筑的立面设计针对基地位于两条城市道路交叉口的情况，采用弧形的建筑外型，起到了空间延续与过渡的重要作用。柔和舒展的弧形建筑形象充分展现医院建筑亲切宜人的性格特征。主楼顶部的大屋面处理赋予了大楼强烈的性格特征感与可识别性。平滑曲面上凹槽与挑板形成丰富的光影效果，赋予大楼细腻的建筑表情。完整简洁的形体加上精致的细部处理，使该大楼成为该区域的标志性建筑。

Surgery Sickroom Building of Wuhan Union Hospital

Location: Wuhan, Hubei
Building Area: 74,117 m²
Site Area: 8,628 m²
Plot Ratio: 8.59
Building Height: 144.7 m
Storey Number: 32 storeys above the ground and 2 storeys underground
Design Time: Dec. 2001
Completion Time: Sept. 2006

Project Introduction

Graphic design of this building has fully reflected the “patient-oriented” design guiding principle: based on the requirements of the Assignment and functional characteristics of the surgery sickroom, the plane of standard nursing unit has been designed as an extended arc-shaped one and all sickrooms face south in order to maximize natural ventilation and daylighting, so that all patients can enjoy sunshine, clean air and a broad view to facilitate body’s recovery. The nurse station has been designed as a shape of peninsula in the center of the plane so as to enable the nurses to offer the most convenient and efficient service.
According to the situation that the base is located at the crossing of two urban roads, the arc-shaped form has been applied in elevation design to play a key role in spatial continuity and transition. This mild and extended arc-shaped building fully reflects a sense of kindness and comfort generated by hospital buildings. Large roofing on the top of the main building makes the building unique and identical. Groove on the smooth curve together with cantilever slab has formed various effects of shade and shadow, which endows the building with an exquisite representation. Completed and simple building shape with elaborate detail treatment enables it to serve as a unique landmark in this area.

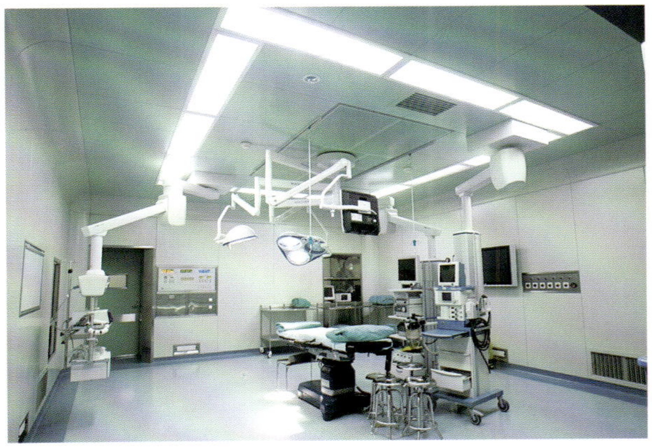

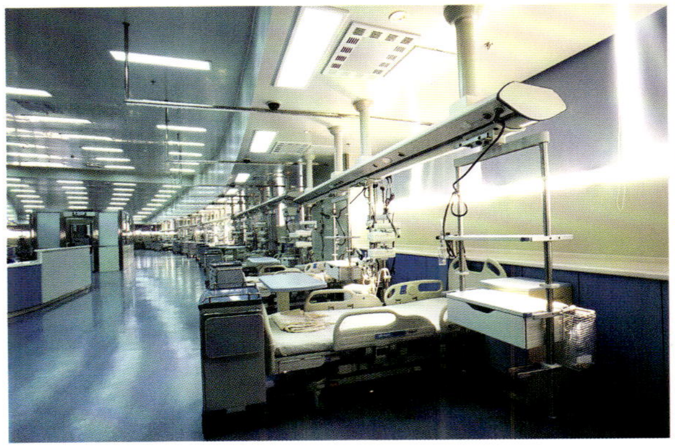

湖北省博物馆
建设地点：湖北 武汉
建筑面积：36 832平方米
用地面积：63 127平方米
容 积 率：0.52
建筑高度：37.9米
层　　数：地上4层，地下1层
设计时间：2003年8月
竣工时间：2007年9月

项目介绍

湖北省博物馆以曾侯乙编钟、郧县人头盖骨、越王勾践剑等一批国家顶级文物的收藏和20余万件珍贵历史文物的巨大藏量，成为国家财政重点扶持的省级博物馆。

湖北省博物馆的建筑平面布局采用一主两翼的构图，与主入口门楼共同形成“楚宫双阙对阳台”的完整格局。以左翼的楚文化馆为序馆，以居中的综合馆为核心馆，以右翼的编钟演奏馆为结束馆，流线清晰，展示内容丰富完整。文物库房和主要辅助功能设于综合馆底层，与陈列区相对独立，文物调度与管理便捷。

建筑造型结合楚式高台建筑的特点加以创新，主体建筑屋面以传统的坡顶建筑为基础，在覆斗形的建筑体量上作减法处理，结合各层的景观外廊，形成具有整体感和独特个性的建筑样式。精益求精的细部设计，从建筑的基座、台阶、栏杆、墙身、门窗、檐口、脊饰到室外环境的各个方面，均精心推敲，力求达到古朴而具有新意、简洁又耐人寻味的视觉效果。

Hubei Provincial Museum
Location: Wuhan, Hubei
Building Area: 36,832 m²
Site Area: 63,127 m²
Plot Ratio: 0.52
Building Height: 37.9 m
Storey Number: 4 storeys above the ground and 1 storey underground
Design Time: Aug. 2003
Completion Time: Sept. 2007

Project Introduction

Hubei Provincial Museum is a province-level museum highly supported by the State fund with the vast collection of a group of top-grade cultural relics nationwide such as Zenghou Yi Chime Bell, Yun-county Man Skull and Sword of “Yue's King Goujian” as well as over 200,000 pieces of valuable historical and cultural relics.

The layout is composed of a main building and both flanks, which combines with portico of main entrance to form a full pattern of “double Chu-style palaces facing a terrace in the sunshine”. The Chu Culture Exhibition Hall on the left is the beginning, the Comprehensive Hall in the middle the core and the Chime Bell Playing Hall the end, thus a smooth and explicit route displays those substantial contents. Cultural relics storeroom and main auxiliary functions have been arranged at the bottom storey of Comprehensive Hall, relatively independent from the display zones for facilitating deployment and management of cultural relics.

This building's shape is an innovation from the Chu-style high-platform buildings. Based on the traditional sloping roof buildings, roof of the main building is constructed through subtraction from dimensions of capsized bucket-shape roof. Thus, the roof together with the landscape side corridors will form an integral and distinguished building style. Elaborate detailed design is embodied in base, step, handrail, wall, door and window, eaves, ridge decorations and outdoor environment for reaching an effect of primitive simplicity, vividness and meaningfulness.

宁波国际汽车城
建设地点：浙江 宁波
建筑面积：100 727平方米
用地面积：47 080平方米
容 积 率：1.70
建筑高度：59.7米
层　　数：地上15层，地下1层
设计时间：2004年4月
竣工时间：2007年3月
项目介绍

建筑裙房部分围绕着内院布置有汽车展厅，首层入口处设有贯通4层裙房的圆形大门厅，气势宏大，并在空间上形成室外广场和东西展厅的过渡，同时也将室外广场、室内展厅和内院连接成一体，构成一个完整的空间序列。5～15层为办公部分，采用中走廊式布局，建筑布局合理，结构经济。

在建筑立面设计上，力求突出其商业性，使建筑物本身就体现出汽车城的特征。裙房以流线型的金属顶盖结构，结合大片的通透玻璃，创造了一个象征性的建筑形象。巨大的建筑体量，亲切的细部造型，既给人以强烈的视觉冲击，又给人以亲和力和吸引力，充分体现其商业建筑的性格特征。

采用金属、玻璃等现代建筑元素，塑造一个具有“汽车”内涵，能体现汽车工业及高新科学技术发展的建筑产品，创造一个既有活跃商业氛围又具有舒适性的人性化的商业、展览、办公空间。

Ningbo International Automobile City
Location: Ningbo, Zhejiang
Building Area: 100,727 m²
Site Area: 47,080 m²
Plot Ratio: 1.70
Building Height: 59.7 m
Storey Number: 15 storeys above the ground and 1 storey underground
Design Time: April 2004
Completion Time: March 2007
Project Introduction

The skirt building surrounds the inner courtyard with automobile exhibition hall in it. At the ground-floor entrance, a circular and spacious porch has been arranged for connecting the skirt building and serving as a transitional space for outdoor square and exhibition halls on eastern and western flanks. Thus, the outdoor square, indoor exhibition halls and inner courtyard have joined together as a whole, forming a completed spatial sequence. The part from the 5th floor to the 15th floor has been utilized as offices with middle corridors, demonstrating a reasonable and cost effective structure arrangement.

In elevation design, commercial function of this building has been spotlighted and the building itself emits a smell of automobile city. The skirt building boasts a streamlined metal roofing structure along with large pieces of transparent glass, generating a symbolic architectural image. Colossal building mass and friendly detailed construction not only causes a strong visual impact, but also sheds a touch of affinity and appeal, which fully embodies the characteristics of commercial buildings.

Such modern building elements as metal and glass help to provide an architectural work indicating “automobile” connotation and reflecting automotive technique and high-technology. Therefore, this building has created a cozy and humanistic space with active commercial atmosphere for business, exhibition and office affairs.

湖北中江建筑设计院

公司简介

湖北中江建筑设计院有限公司具有甲级建筑设计、乙级岩土工程勘察和丙级市政设计等多个资质。公司广纳贤才，优化组合，聚集了业内众多专业技术人才，组成了一支优秀的团队，公司有着积淀二十多年的业界勘察设计经验，并且运用运用行业成功的管理模式，致力于提供规划设计、建筑设计、景观设计、市政设计、工程勘察及项目咨询策划等专业服务。

Aedas Profile

Hubei Zhongjiang Architectural Design Institute Co., Ltd. has obtained qualifications such as Class-A Architectural Design, Class-B Geotechnical Engineering Investigation and Class-C Municipal Design, etc. Through wide recruitment and optimal combination, it assembles many professional and technical talents in the industry and forms them into an excellent team. Drawing on 20 years' experience in survey and design and applying the successful management model, this company is committed to providing professional services in planning and design, architectural design, landscape design, municipal design, engineering survey and project advisory & planning, etc.

地址：武汉市洪山区图书城南文秀街 8 号天宇创意大厦
邮编：430070
电话：+86-27-67126712 67126708
传真：+86-27-87384539
邮箱：zjjz_688@163.com
网址：www.hbzjjz.cn

Address: Tianyu Creative Building, No.8 Wenxiu Street, South of Book City, Hongshan District, Wuhan City
P.C.: 430070
Tel : +86-27-67126712 67126708
Fax :+86-27-87384539
E-mail: zjjz_688@163.com
Http://www.hbzjjz.cn

项目 /Projects

1 蕲春会展中心
2 卓尔食品工业园
3 汉口北国际商品交易中心
4 武汉优抚医院

1

2

3

4

项目 /**Projects**

5 武汉客厅

6 河南平顶山外国语学校

7 遗爱湖八仙嘴规划

8 鑫瑞 小城故事

9 萍乡政务中心

5

6

7

8

9

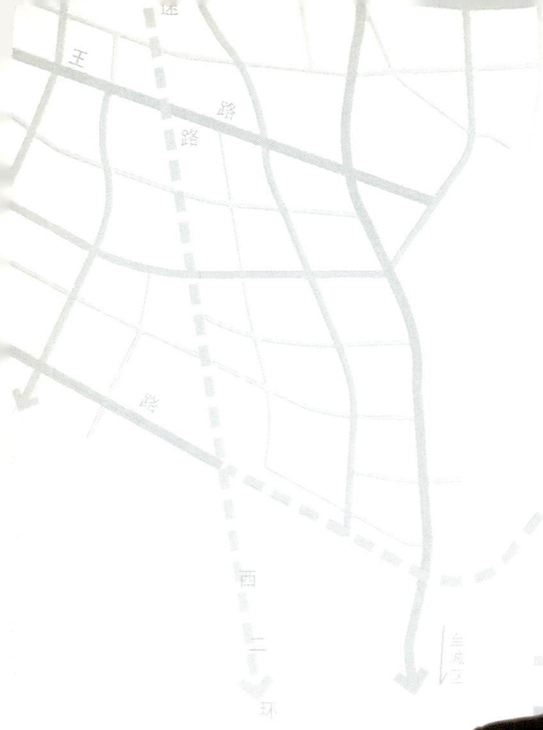

Editor,Designer

编委、设计师

澳大利亚柏涛墨尔本建筑设计有限公司
PEDDLE THORP ARCHITECTS

王漓峰
Lifeng Wang

王漓峰

董事长
首席代表

教育背景
1980—1984年　中国上海同济大学建筑学士
1990—1991年　澳大利亚皇家墨尔本理工学院建筑学硕士

资　　历
1997年至今
首席代表　　澳大利亚柏涛墨尔本建筑设计有限公司
董事长　　澳大利亚柏涛（墨尔本）建筑设计亚洲公司

在信息、理念、创意、文化急速传播的时代，各种创意、构思需要一个巨大的平台去融合、研究和实施。特别是在中国“城市化”进程的巨大舞台上，澳大利亚柏涛公司和具有本土特征的柏涛咨询（深圳）有限公司正以它的国际化视野和理念以及设计手法和本土化的服务特征，在澳大利亚和中国两国之间扮演着文化交流、创意交流、技术交流和经验交流的重要角色。

澳大利亚柏涛公司成立于1889年，在一个多世纪的时间里，柏涛公司一直是澳大利亚城市规划、建筑设计队伍中的主力军，是澳大利亚建筑设计领域中最有影响力的设计公司之一，特别是从20世纪80年代至今，柏涛公司在城市规划、区域规划、大型体育综合设施、商业设施、公共设施、文化设施、教育设施以及大型住宅区规划、景观设计等领域创造了一个又一个辉煌，如澳大利亚国家网球中心、ABC广播公司总部、墨尔本水上运动中心等都体现了柏涛公司在创意、技术以及项目实施上的卓越才能和贡献。

1998年初，中国的“城市化”进程开始加速，又适逢中国的设计市场向海外开放，澳大利亚柏涛（墨尔本）公司抓住了这个历史契机，在中国深圳设立了办事处。2001年又设立了柏涛咨询（深圳）有限公司，为中国的设计业务提供了一个直接沟通交流的平台，并将符合中国建筑行业的实际情况和法规规范好国外先进的设计理念完整地应用于中国项目中，为客户提供了面对面沟通、交流、碰撞的服务平台，最大程度地实现了本土化与国际化的接轨。在过去的十年中，柏涛公司的作品遍布中国的大江南北，创作设计了400多个项目，其所设计的作品屡屡中标，30多个项目获得多个国家级和国际大奖。

柏涛公司十年的发展见证了中国城市化进程的标志性阶段，并在这个进程中扮演了非常重要的角色。从1996年深圳万科城市花园、1999年深圳湾蔚蓝海岸、2001年华侨城波托菲诺纯水岸、2003年万科第五园到2005年中信红树湾，柏涛公司的这些作品标志了中国大规模住宅开发的各个不同阶段，柏涛公司亦成为中国住宅区规划设计最有影响力的公司之一。

柏涛公司把它进入中国的前十年的重点放在了在中国城市化进程中起主导和推动作用的住宅规划设计领域，我们认为，对中国城市的自然生态、社会生态和人文生态产生巨大影响的中国“城市化”运动的急剧发展将迅速改变中国社会，文化的架构，冲击中国社会、生活和文化价值观。“住宅”作为社会最大需求以它强大的力量改变着城市的功能、文脉和肌理，改变着中国城市的空间和内容。如何运用国外大型居住区规划设计中对人、自然、生态、社区空间在物质层面和精神层面多方面研究的实例和经验，对这次城市化运动中的大型住宅开发提供负责任、策略性的设计咨询服务，使中国的住宅开发能够沿着正确的道路走下去，是我们的责任和义务。我们希望通过我们的国际化视野和本土实践的经验，用理性、负责的理念和态度为中国急剧扩张的城市版图提供战略上的发展纲要和理论上的方向指导，保持城市文脉的连续性和发展的可持续性。

十年的实践使柏涛公司在中国住宅发展的过程中，起到了开疆拓壤、引领时尚的先锋作用，这使柏涛成为最具影响力的建筑公司之一。同时，柏涛公司在商业、文化、体育建筑领域也取得了很大的成功，阿布扎比国家银行、天津泰达体育场、上海体育文化中心和上海外滩十六铺码头等项目的投标，显示了柏涛公司在这些领域的创造力和实力，也证实了柏涛在这些领域具有很大的发展潜力。

我们在中国的十年实践，始终贯彻柏涛公司一直倡导的设计哲学：每一个建筑都是新的挑战，我们以对环境、文脉和地域文化的仔细研究，对客户要求的积极反应和专业诠释，对新技术新材料的积极应用，在设计理念上大胆创新，演绎激情，以最佳的解决办法，使我们的作品达到提升人类精神面貌、丰富建筑文化内涵的境界，从而实现人类的自身价值、社会价值和文化价值。

柏涛的十年是我们努力用国际化的经验和视野指导本土实践的十年，亦是我们自我发现、自我完善、自我实现的十年。回顾柏涛十年来在中国的几百个作品，其中都贯穿了以下几个重要的词汇：

- 尊　重：我们尊重自然，尊重地脉文化，尊重地域的文化和习惯，尊重客户的理念和理想。
- 激　情：我们充满激情追求每一个作品的创意和完美。
- 激发力：我们和合作伙伴及客户互相激发，共同创作。
- 创造力：我们充满了创作的热情，追求每一个作品的创意和个性。
- 融合力：我们努力把设计中许多相互对立的因素和谐地组合起来，以达到一种突破社会和空间界限的效果。
- 协调力：我们努力在设计理念的完美和客户的需求中寻找平衡，以达到“双赢”的效果。

我们倡导的是：理性、逻辑的思维分析，创意、灵感的迸发，有效、可实施的技术应用。柏涛公司始终坚持“主创核心”团队精神来保证设计作品品质的一贯性和连续性，我们提倡的是互相尊重、各取所长、互相包容、共同进取的团队精神，以及多元化的企业文化。柏涛公司成为境外设计师，海外设计师、本土设计师在一起互相学习、博采众长、共同研究、超越自我的平台。十年的历程，充分证明了“尊重、包容、创新”的精神是我们不断前进的保证。

当我们重新审视这十年来的作品，我们看到了国外先进设计理念和本土实践相碰撞的火花，看到了创意和灵感在大时代洪流中的闪现和飞扬，看到了迸发的激情和理性的思考。这些作品是柏涛公司在中国十年时间的缩影，它表明了在一个社会文化发生巨大变革的大时代背景下，我们做了什么，我们能做什么，我们将做什么。我们的目标是：以建筑师的社会责任感和敬业精神，以科学、理性的态度，用心去创作今后的每一个作品。

CAPA

蔡 明

Ming Cai

蔡 明

董事长、总建筑师、深圳市规划局建筑设计专业工作组专家

职业经历

1994年毕业于天津大学建筑学院，师从中国科学院院士及设计大师彭一刚先生。

1994–2003年工作于香港华艺设计顾问有限公司，任副总建筑师、高级建筑师、规划设计研究中心主任，兼中国海外集团住宅开发中心负责人，期间从事主持大量房地产开发规划设计专题研究并完成系列论文集。

现任美国开朴建筑设计顾问（深圳）有限公司董事长，总建筑师。

主要获奖作品

北京中国建筑文化中心

专业负责人。68 000平方米。展览、办公、会议。

获建设部银奖。

深圳福田区区委政府办公综合体

140 000平方米。行政办公、会议。

获深圳市年度优秀工程设计一等奖。

深圳创维数字研究中心

69 000平方米。集团办公、展示、研发。

获中建总公司年度优秀方案设计一等奖。

南昌金域名都

360 000平方米，集住宅、商业、办公为一体的城市综合体。

获建设部人居经典规划设计大奖。

南海怡翠花园

380 000平方米，纯高档住宅，南海第一大盘。

获广东省年度优秀规划设计二等奖。

安徽合肥科园九溪江南

280 000平方米。

获2006年度亚洲最佳人居环境社区奖

西安紫薇尚层

获2006年建设部90/70中小套型设计二等奖

2007年获亚洲建筑师协会颁布“亚洲建筑规划设计区域十大创新人物”称号

主讲专题

蔡明经常受国内外权威机构邀请，参加多种高规格的专题演讲及交流沙龙。

2006年中国（北京）第二届中外建筑师执业交流大会专题演讲

2007年香港建筑交流大会“高层高密度住宅舒适度研究及实践”

北京大学城市与环境学院21世纪房地产总裁黄埔班特聘讲师

多次参加行业杂志、专业机构、高校的建筑设计的专题讲座。

Ming Cai

Chairman of the board, general architect

Expert of Architecture and Design Expert Group in Shenzhen Urban Planning Bureau

Experience

In 1994, graduated from Academy of Architecture, Tianjin University, was apprentice of Mr. Peng yigang who is the academician of Chinese academy of sciences and design master.

1994-2003, as the vice general architect, advanced architect and director of Design & Plan Research Center in HuaYi Designing Consultants(Hong Kong) Co., Ltd. Meanwhile, worked as the charger of Residential Development Center in China Overseas Corporation, and during that period of time, mainly worked on the planning and design studies of a large number of real estate development, and completed series collection. Now working as the Chairman of the board and general architect in CAPA Architecture Designing Consultant (Shenzhen) Co., Ltd. USA

Major Winning Prize

China Architectural Culture Center, Beijing (being the chief manager, 68 000 m², Exhibition, Office, Conference, branch site of the 20th World Architects Conference in 1999, be awarded the silver medal of National Ministry of Construction)

Complex office building of Shenzhen Futian District Government (140 000 m², Administrative office, conference. Be awarded the first prize of Shenzhen Annual Excellent engineering design)

Digital Researching Center of Shenzhen Skyworth Company (69 000 m², office working, Exhibition, Researching and developing, be awarded the first prize for annual excellent designing plan of China State Construction Corporation)

Golden Area•Noble City, Nanchang (360 000 m², purely high-grade residence, integrating residence, commerce, and office into a urban complex building, be awarded the classical residence designing prize of Ministry of Construction)

Nanhai Green Garden (380 000 m², purely high-grade residence, the first magnificent building of Nanhai, be awarded the second prize of Guangdong Annual Excellent Design and Plan)

JiuXi JiangNan, Hefei KeYuan, Anhui(280 000 m², be awarded the prize for the best Asian residential environment district in 2006)

ZiWei Nobel Garden, Xi'an (be awarded the second prize for 90/70 small and medium-size housing design of ministry of construction in 2006)

In 2007, be awarded the honorary title for "the Top-ten Innovative Figures of Asian Architecture Designing and Planning Region"

Thematic Lecture

Invited by domestic and foreign authoritative organizations, often attends kinds of High-level Thematic Lectures and Sharon

Gave a lecture in the second session China (Beijing) of Chinese and foreign architects exchange conference in 2006

Attended the Hong Kong Architecture Exchange Conference on high-rise and high-density residential comfort Research and Practice in 2007

Be the lecturer specially engaged in the HuangPu class for the 21th century real estate president in Peking University Academy of Urban and environment.

Participated in kinds of thematic lectures on the magazine, specific institute, and architectural design of universities for many times

思纳史密斯(集团)中国

SMITHGROUP cna

architecture engineering interiors planning

朱轶俊

Yijun Zhu

朱轶俊

中国区执行总裁 高级建筑师

教育 & 经验

上海华东建筑设计研究院，设计师，1990年

上海联合工程咨询有限公司，设计师，1993年

上海欧亚工程咨询有限公司，主任设计师，1993年

上海协兴置业发展有限公司，总经理，建筑总监，1996年

凯悦酒店管理投资集团亚洲区，设计总监，1997年

美国史密斯集团 上海代表处，首席代表，2006年

上海思纳史密斯建筑设计咨询有限公司，执行总裁，2006年

上海思纳建筑设计有限公司，董事长，2006年

思纳史密斯（集团）中国，执行总裁，2006年至今

专业简介

朱先生在项目规划统筹、城市规划设计、酒店业、展览业及零售业的设计等项目中都具有丰富而广博的经验，这些项目包括的从室内设计到几十万平方米的综合建筑以及相关项目的策略规划问题的研究等。到目前为止，由朱先生亲自领导的设计项目已超过550万平方米，至今朱先生仍然坚持不懈地在他的工作上努力专研并使其达到更高的水平。

作为一位管理者，朱先生也是其他两个建筑开发和咨询公司的董事长和管理者之一，这也充分显示出他卓越的管理才能。他有效地整合国内外优秀的设计人才，并始终使公司在各类项目的竞争中总获得优势。

Yijun Zhu

Regional Executive officer in China Senior Architect

Education & Experience

Architect, Shanghai East China Architectural Design Institution,1990

Architect, Shanghai Alliance Engineering Consultant Company, 1993

Shanghai Ouya Engineering Consultant Company, Principal Architectural Designer, 1994

General Manager, Shanghai Xiexing Property Development Co.,Ltd

Design Director,1996

Hyatt Hotel Management and Investment Group (Asia),Design Director ,1997

SmithGroup Shanghai Office, Chief Representative, 2006

Shanghai CNA. Smith Architectural Design Consultant Company, Ltd,CEO,2006

Shanghai CNA Architectural Design Company ,Ltd , General Director ,2006

CNA Smith (Group) China, CEO ,2006 till now

Professional Summary

Mr.Zhu has encyclopedically experiences on programming various kinds of projects such as urban planning, hotel, exhibition and commercial. From interior design to hundred of thousands square meters complex buildings, the service scope also covers the plan strategy and problem research .Till now, the total area of projects led by him is more than 5,500,000 square meters. And he will never stop exerting himself to reach a greater success in the future...

Mr.Zhu is also the leader of other two architectural development and consultant companies. His excellent talent on management helps incessantly in efficiently organizing excellent staff to work for various competitions and getting lots of achievements.

汉米敦 HMD

徐 石
Shi Xu

徐石

董事总经理　高级建筑师
国籍：新加坡　出生日期：1968年

学历及资格
1990年天津大学建筑系学士
2005年上海交通大学／多伦多大学 EMBA
2008年清华大学 EMBA

语言
汉语－母语
英语－流利

曾经工作的国家和地区
中国、新加坡、中国香港

个人履历

徐先生具有近二十年的规划及建筑设计的工作经历。曾先后工作于中国及新加坡的设计公司。在城市设计、酒店、商业办公等类型的项目上积累了丰富的经验。2000年加入阿特金斯中国，分别任职于中国香港、深圳及上海公司，主要负责设计及业务管理。目前徐先生为汉米敦建筑设计公司总裁，全面负责公司在中国的建筑设计业务。

现任工作
2009年至今　汉米敦中国　董事总经理

曾任工作
2007–2009年　阿特金斯中国　副董事长
2005–2007年　阿特金斯中国　建筑景观设计部　董事
2003–2004年　阿特金斯中国　城市规划设计部　助理董事
2000–2002年　阿特金斯中国　城市规划设计部　建筑部经理
1998–2000年　PWD Consultant Pte Ltd（Singapore）准建筑师
1996–1997年　LWC Pte Ltd（Singapore）　建筑师
1990–1995年　中国建筑总公司东北建筑设计研究院（中国）助理建筑师／建筑师

近期完成的重大项目
中国项目
- 乌海五星级酒店及国宾馆
- 乌海东外滩坝上公园
- 天津泰达 MSD 城市商务中心发展项目
- 上海佘山世茂酒店
- 乌海滨河五星级酒店
- 上海松江新区泰晤士小镇
- 北京永泰五星级酒店
- 廊坊开发区五星级酒店
- 沈阳五里河高级酒店式公寓
- 苏州建屋发展8号地块英式居住区
- 沈阳方迪山庄别墅区
- 上海佘山银湖别墅二、三期概念性总体规划
- 上海松江新城交通枢纽区概念性规划及初步方案设计
- 上海松江大学城北区示范区规划设计
- 浙江医科大学附属第一医院新门诊大楼方案设计，

国外项目
- 新加坡德明中学外籍学生公寓方案设计及施工图设计
- 新加坡 New National Heart Center 方案设计
- 新加坡 Tampinis Polyclinic 方案设计

Shi Xu

Managing Director, Senior Architect
Nationality: Singapore　Date of birth: 1968

Education and Qualifications
Bachelor of Architecture in Tianjin University, 1990
EMBA of Shanghai Jiaotong University / University of Toronto, 2005
EMBA of Tsinghua University, 2008

Language
Language - mother tongue
English - fluent

Countries and regions where he has worked
China, Singapore, Hong Kong China

Profile
Mr. Xu has nearly two decades' working experience in architectural planning and design and has been working in the design companies in both China and Singapore. He has accumulated a wealth of experience in urban design and in projects such as hotels, commercial-office building, etc. Joining Atkins•China in 2000, Mr. Xu has been working in Hong Kong, Shenzhen and Shanghai, mainly responsible for the design and business management. Currently Mr. Xu is the president of HMD Architectural Design Company, taking overall responsibility for the architectural design business in China.

Current Position
MD In HMD China since 2009

Previous Position
- 2007-2009 Atkins•China Vice Chairman
- 2005-2007 Atkins•China Architectural Landscape Design Department Director
- 2003-2004 Atkins•China Urban Planning and Design Department Assistant Director
- 2000-2002 Atkins•China Urban Planning and Design Department Construction Manager
- 1998-2000 PWD Consultant Pte Ltd (Singapore) Associate Architect
- 1996-1997 LWC Pte Ltd (Singapore) Architect
- 1990-1995 Northeast Architectural Design & Research Institute of China Construction Corporation (China) Assistant Architect / Architect

Major projects completed recently
Projects in China
- Wuhai Five-star Hotel and Guesthouse
- Bashang Park in East Bund, Wuhai
- Development project of Tianjin Teda MSD City Business Center
- Shimao Hotel in Sheshan, Shanghai
- Wuhai Binhai Five-star Hotel
- Thames town in Songjiang New City, Shanghai
- Beijing Yongtai Five-star Hotel
- Five-star Hotel of Langfang Development Zone
- Senior Hotel Apartments in Wulihe, Shenyang
  British residential area of Suzhou Housing Development Block 8
- Fundi Villa Area in Shenyang
- Conceptual master planning for Project Phase I&II of Silver Lake Villa, Sheshan, Shanghai
- Conceptual planning and preliminary design for the transportation hub area of Songjiang New City, Shanghai
- Planning and design for the Demonstration Zone in North District of Songjiang University City, Shanghai
- Design for the new outpatient building of the First Affiliated Hospital of Zhejiang Medical University

Projects Aboard
- Project design and construction design for Foreign Students' Apartments of Singapore Deming Middle School
- Design for New National Heart Center in Singapore
- Design for Tampinis Polyclinic in Singapore

Gensler

夏 军
Jun Xia

夏 军

副总裁
亚洲地区设计总监
1991年加入Gensler

背景资料

科罗拉多大学丹佛分校	城市设计建筑专业	硕士
上海同济大学	建筑专业	学士

夏军是一位天才设计师，他擅长于在建筑的可施工性、经济性和美观性基础上对建筑进行物理、空间和行为方面的分析。他诠释设计理念的能力在项目初期能使客户对项目有更为清晰和深入的理解，并促使项目成功。夏军出生并成长于上海，曾在北京中国城市规划和设计院工作，参与了总体规划、城市设计以及住宅、零售和商业办公楼的建筑设计。在美国，夏军担任过数家著名公司的主设计师工作，这些设计公司有Gallup，Invesco，MCI及Sun Microsystems。在全球，他已参与了诸多重要建筑项目的设计工作，包括预计成为世界第二高楼的上海中心。

项目经验

综合用途建筑

上海中心	上海	中国	380 000平方米
上海21世纪大厦	上海	中国	110 800平方米
北京传媒生活大厦	北京	中国	251 000平方米
宝矿国际广场	上海	中国	182 800平方米
长安国际中心	北京	中国	195 000平方米
长沙规划设计	长沙	中国	700 000平方米
朝阳广场	北京	中国	223 200平方米
东海中心	深圳	中国	415 000平方米
呼家楼建筑综合体	北京	中国	326 500平方米

机场

		规模
Jefco机场航站楼	科罗拉多州	2 500平方米
米德兰国际机场	米德兰，德克萨斯州	18 500平方米
威尔罗杰斯世界机场	俄克拉荷马州	49 000平方米

教育设施

Auraria 高等教育中心	丹佛 科罗拉多	13 300平方米
中医药大学	上海 中国	135 000平方米

酒店

上海衡山路12号	上海	中国	42 000平方米
北京饭店	北京	中国	246 000平方米
三亚四季酒店	三亚	中国	
海南度假酒店	海口	中国	20 000平方米
金茂深圳JW 万豪酒店	深圳	中国	44 000平方米
佘山酒店	上海	中国	8 600平方米
王宝和大酒店	上海	中国	1 000平方米

大堂/公共空间

东银中心	上海	中国	83 800平方米

规划设计

北京通州Maxon	北京	中国	106公顷
滨江71号地规划设计	杭州	中国	158 800平方米
长沙规划设计	长沙	中国	700 000平方米
成都住宅规划设计	成都	中国	124公顷
中国电信数据中心，	上海	中国	
崇明岛规划设计	上海	中国	43公顷
重庆寸滩港	重庆	中国	829 500平方米
大华大场规划设计	上海	中国	196 000平方米
大连软件园	大连	中国	100 000平方米
董家渡规划设计	上海	中国	
武汉复地	武汉	中国	527 900平方米
福州路规划设计	南昌	中国	
福州中地规划设计	福州	中国	424 500平方米
银河湾	沈阳	中国	950 400平方米

Jun Xia

Principal
Regional Design Directorr
Joined Gensler in 1991

Background
Master of Architecture in Urban Design, University of Colorado at Denver
Bachelor of Architecture, Tong Ji University, Shanghai

Jun Xia is an exceptional designer who thrives on analyzing the physical, spatial, and behavioral aspects of a building within the framework of its constructability, economy, and aesthetics. His ability to realize design concepts has given his clients a clearer, deeper understanding of their projects at an early stage, allowing for more design collaboration, leading to award winning projects. Mr.xia joined Gensler in 1991, was appointed Principal in 1999 and became Design Director of the Shanghai Office in 2004.
In the United States, Mr.xia has worked with several leading companies. Internationally, he has worked on some of the most visionary projects to be developed in the past 20 years, including the Shanghai Tower, expected to be the world's second tallest building.

Project Experience

Mixed-Use

Shanghai Tower, Shanghai, China	380,000 m^2
21st Century Tower, Shanghai, China	110,800 m^2
Beijing Massmedia Life Plaza, Beijing, China	251,000 m^2
BM Plaza, Shanghai, China	182,800 m^2
Chang'an International Center, Beijing, China	195,000 m^2
Changsha Master Plan, Changsha, China	700,000 m^2
Chaoyang Square, Beijing, China	223,200 m^2
Donghai Center, Shenzhen, China	415,000 m^2
Hujialou Complex, Beijing, China	326,500 m^2

Airports	size
Jefco Airport Terminal, Broomield, CO	2,500 m^2
Midland International Airport, Midland, TX	18,500 m^2
Will Rogers World Airport, Oklahoma City, OK	49,000 m^2

Educational Facilities

Auraria Higher Education Center, Denver, CO	13,300 m^2
Traditional Chinese Medicine University, Shanghai, China	135,000 m^2

Hotels

12 Hengshan Road, Shanghai, China	42,000 m^2
Beijing Hotel, Beijing, China	42,000 m^2
Four Seasons Hotel, Sanya, China	
Hainan Resort, Haikou, China	20,000 m^2
JW Marriott Hotel, Shenzhen, China	44,000 m^2
Sheshan Hotel, Shanghai, China	8,600 m^2
Wangbaohe Hotel, Shanghai, China	1,000 m^2

Lobby/Public Spaces

DongYin Center, Shanghai, China	83,800 m^2

Master Planning

Beijing Tongzhou Maxon, Beijing, China	106 ha
Binjiang Plot 71 Master Plan, Hangzhou, China	158,800 m^2
Changsha Master Plan, Changsha, China	700,000 m^2
Chen Du Residential Master Plan, Chendu, China	124 ha
China Telecom Data Center, Shanghai, China	
Chongmin Island Master Plan, China	43 ha
Chongqing Cun Tan Port, Chongqing, China	829,500 m^2
Dahua Dachang Master Plan, Shanghai, China	196,000 m^2
Dalian Software Park, Dalian, China	100,000 m^2
Dong Jia Du Master Plan, Shanghai, China	
Forte Wuhan, Wuhan, China	527,900 m^2
Fuzhou Master Plan, Nanchang, China	
Fuzhou ZhongDi Master Plan, Fuzhou, China	424,500 m^2
Galaxy Bay, Shenyang, China	950,400 m^2

AIM GROUP INTERNATIONAL 国际设计集团 亚瑞建筑设计有限公司（中国甲级） AIM GROUP INTERNATIONAL （加拿大）

陈晓宇
Calvin Chen

陈晓宇

职务
AIM 国际设计集团（加拿大）首席建筑师、总裁

教育背景
2004 年　加拿大注册建筑师
2000 年　加拿大皇家建筑师学会会员
1997 年　建筑学硕士

工作经历
1998–2004 年　KNEIDER ARCHITECTS
2004–2006 年　JULIAN JACOBS ARCHITECTS
2006 年至今　AIM 国际设计集团

主要作品
加拿大森林山（Forest Hill）私人豪宅
加拿大多伦多圣天玛菲（St.Timothy）小学
加拿大多伦多卡门社区活动中心和图书馆
加拿大多伦多艾斯美社区活动中心
加拿大多伦多温哥华积架（捷豹 Jaguar）车行
加拿大多伦多富人山冰球馆
加拿大伦敦积架、富豪、路虎车行
东莞铂尔曼酒店（五星级）
广州大一山庄（中国 2009 十大豪宅之首）
从化檀宫温泉酒店
东莞联华花园城二、三期
东莞道滘客运站及商住小区
恒大成都恒大绿洲
保利福州滨江地块规划
中信东莞德方斯
中信东莞商业广场
中信凯旋国际
佛奥阳光花园
龙华星座尚筑
龙华尚墅名门
联华东莞星河传说・荷塘月色
英德仁兴广场
广州沙河百货
番禺百业 Mall
从化名城御景绿洲

Calvin Chen

Position
Chief Architect / CEO of AIM GROUP INTERNATIONAL (Canada)

Education Background:
2004　Registered architect of Canada
2000　Member of Royal Canadian Institute of Architects
1997　Master of Architecture

Work experience
1998 -2004　KNEIDER ARCHITECTS
2004 -2006　JULIAN JACOBS ARCHITECTS
2006- present　AIM GROUP INTERNATIONAL

Major Designs
Private mansion in Forest Hill, Canada
St.Timothy Primary School in Toronto, Canada
Carmen Community Activity Center and Library in Toronto, Canada
Iceman Community Activity Center, Toronto, Canada
Jaguar Dealer of Vancouver, Toronto, Canada
Hockey Hall of Rich Mountain in Toronto, Canada
London Jaguar, Volvo and Land Rover dealers of Canada
Dongguan Pullman Hotel (five star)
Guangzhou Dayi Villa (2009 China Top Ten Luxury Villas)
Conghua Tangong Spa Hotel
Lianhua Garden City Phase II & III , Dongguan
Daojiao Bus Terminal and the Commercial & Residential District, Dongguan
Hengda Hengda Oasis, Chengdu
Planning for Poly Riverside Block, Fuzhou
CITIC La Defense, Dongguan
CITIC Business Plaza, Dongguan
CITIC Triumph International
Foao Sunshine Garden
Hualong Constellation Construction
Hualong Nice Life & City
Lianhua Galaxy Legend • Moonlit Lotus Pond, Dongguan
Ren Xing Square, Yingde City
Guangzhou Shahe Department Stores
Panyu Baiye Mall
Famous City Scenic Oasis, Conghua

ÉTÉ 翌德国际设计机构
été lee et associés architectes urbanistes
■ 法国翌德国际设计机构 ■ 上海翌德建筑规划设计有限公司

于一凡
Yifan Yu

于一凡

上海翌德建筑规划设计公司董事
法国ETE国际设计机构董事
巴黎建筑与城市规划研究院研究员

于一凡2003年毕业于法国国立高等社会科学院（Ecole des Hautes Etudes en Sciences Sociales, France）建筑与景观专业。获得博士学位；1996年毕业于上海同济大学城市规划与设计专业，获得硕士学位。

近十年来，于一凡致力于城市规划设计、建筑与景观设计，主持完成规划与建筑设计项目近百项，获得中国省、部级优秀设计奖五项。在中、英、法学术刊物上发表文章二十余篇，主编中文著作一部，参与编纂法文专著一部，参与和主持多项中、法城市课题研究并热心推动中、法建筑与城市文化方面的交流，在巴黎协助组织了多次宣传中国建筑文化的专题展览。

Yifan Yu

Director of été lee et associés Architectes Urbanistes Shanghai
Director ofété lee Architectes et Urbaniste
A Fellow at Academy of Paris Architecture and Urban Planning

After obtaining the bachelor's degree of urban planning and design in Shanghai Tongji University in 1996. Yang graduated in Ecole des Hautes Etudes en Sciences Sociales, France with doctor's degree of architecture and landscape in 2003.
For over 10 years, Yu has been engaged in urban planning and design, architecture and landscape design works. Yu has conducted more than one hundred architecture design projects, which won five national, provincial and ministry level excellent design awards. Yu also published more than twenty articles in the academic journals of China, Britain and France, editing a Chinese book and participating in the edition of a monograph in French. Yu has conducted and participated in many China and France subject research activities and promoted the culture exchange of architecture and urban culture among the two countries. In these years, Yu has organized several subject exhibition of Chinese architectural culture.

李继军
Jijun Li

李继军

李继军，上海翌德建筑规划设计公司董事
上海同济城市规划设计研究院规划设计所长
高级规划师、建筑师

李继军1999年毕业于上海同济大学城市规划与设计专业，获得硕士学位。1992年获得建筑学学士学位。

十余年来，李继军致力于城市规划设计、建筑与景观设计，主持完成规划与建筑设计项目超过百项，获得中国省、部级优秀设计奖六项，对于建筑设计、景观设计与城市规划有着独特的见解与研究经验，在各种学术刊物上发表文章十余篇，参与和主持多项大、中城市课题研究，在中国诸多城市均有具备时代气息与影响力的代表作，并积极参与城市问题研究，作为若干城市的政府专家顾问，在城市总体规划、城市设计、居住区规划设计等领域积累了大量的经验。

Jijun Li

Director of été lee et associés Architectes Urbanistes Shanghai
Head of Planning and Design institute in Shanghai Tongji Urban Planning & Design Institute
Senior planner, architect

After obtaining the bachelor's degree of architecture in 1992.Li graduated from urban planning and design department in Shanghai Tongji University with the master degree.

For over 10 years, Jijun Li has devoted in urban planning and design, architecture and landscape design works. He has conducted more than one hundred planning and architectural design projects, which won six national, provincial and ministry level excellent design awards. With the unique understanding and research experience in landscape design and urban planning, Li published over ten articles in various academic journals and conducted and participated in many subject research of large and medium cities, creating numerous projects which is of the current atmosphere influence. Being the government expert consultant in several cities, Li is active in the research of urban issues, accumulating rich experiences in urban design, residence quarter planning and design.

J.A.O.Design International Architects & Planners Limited
美國龍安建築規劃設計顧問有限公司

饶及人
James C.Jao

饶及人

饶及人(James Jao)教授，美国龙安集团行政总裁、著名美籍建筑规划专家，1990–1994年任美国纽约规划局委员/局长，是美国纽约市360年历史上第一位亚裔局级官员，也是纽约市最年轻的首位有建筑师执照的局级官员。饶及人先生在纽约市城市规划领域中提出了平衡发展（Balance Development）的规划理论，并且对曼哈顿中国城的改造规划研究、皇后区法拉盛新中国城规划调整等做出了杰出贡献，在1993年12月7日荣获纽约州州长、纽约市市长的嘉奖令，曼哈顿区长将当日宣布为"曼哈顿区饶及人日"。

饶及人先生目前担任美国龙安集团行政总裁，集团由饶及人在1984年创立于纽约，目前在中国主要从事规划，设计，资金，地产投资，以及慈善五个板块的业务。龙安集团是一个"以城市规划为切入点从事城市运营，为中国百姓提供世界一流的绿色低碳建筑"为主业的集团公司。他同时担任集团旗下：北京龙安国际规划建筑设计咨询公司董事长；北京众拓建筑设计工程设计有限公司（建设部甲级010518–sj）董事长；翰诺城市（北京）投资顾问有限公司董事长；艾普克中国投资有限公司董事长；美国龙安建筑规划设计顾问有限公司设计总指导，并受联合国人居署署长安娜女士（Anna Tibaijuka）的委托出任联合国人居署人居年会(UN Habitat World Urban Campaign Steering Committee)程序委员（2009–2012）。在中国还出任国家外专局美籍规划建筑专家、全国工商联房地产设计联盟第二任C.E.O.（首席执行官）、中国20多个省市规划与经济发展顾问、纽约市亨特大学土地研究所、中国市长培训中心、清华大学EMBA班、南昌大学兼任教授、苏州大学研究生导师、全国市长培训中心特邀教授等社会职务，同时也是《中国建设报》《中国房地产报》《法制晚报》等主流媒体的学术顾问，以及《中国日报》（China Daily）特约专栏作家。

著书立说，饶氏箴言影响中国规划界：

在中国期间，饶及人先生被中国官员、业界人士、主流媒体称作最了解中国国情的外国建筑规划专家，他特别关心并全力研究中国的城市发展问题，陆续发表了近百篇重要文章，阐述了大量的城市规划理论并提出许多权威、具体且可操作的建议，他经常为中国主管城市的领导培训最新规划理念，被称为"城市医生"。饶及人先生利用他优秀的中英文双语能力，从20世纪80年代起就在美国的许多中文媒体上介绍纽约市的规划法规，是首位将许多英文规划专有名词翻译成中文的人。饶及人先生在搜狐焦点网与搜房网的"饶及人博客"分别成为这两个专业网站最具影响力博客第一人和地产博客红人。在房地产搜房网2009年第六届房地产网络人气榜票中，饶及人先生的博客被网民选为"最具影响力房地产博客"。他同时也被搜狐焦点网选为搜狐焦点"2009年度房地产意见领袖"。

为了让中国的城市官员能够了解与运用国际先进的规划理念，让中国老百姓的生活日益国际化，饶及人著有《十年规划中国情》（32.8万字），于2005年6月由新华出版社出版，他将西方的规划理念用小故事的方式说明。同时在2007年8月，由城市出版社出版的《寻找中国城市魂》（38万字）一书，让中国城市都能够找到自己的定位，同时倡导与推动绿色建筑在中国的发展，以及城市更新和中国文化在现代规划与建筑设计中的体现。为了培养老百姓对规划的兴趣，中国铁道出版社出版的《从美国梦到中国梦，城市医生——饶及人》已于2009年10月下旬在全国发行。这三本书都是深深影响中国领导层与规划界的重要城市建设理念的专业书籍。饶及人先生利用他的国际资源以及将国际设计理念落地的模式，为积极推动中国生态城市和绿色建筑的专业技术发展做了很多工作，这些工作得到了中国住房部和房地产业界等社会专业机构的充分肯定，并授予他很多的荣誉。

他所带领的龙安设计公司已经在中国做过2746平方千米的城市规划项目、设计过1 785万平方米的住宅项目、930万平方米的公建项目、776公顷的景观项目，共计获得111个国内外设计大奖。中国城市出版社在2008年将他所负责与带领的设计作品整理成了《西为"中"用——美国龙安项目选集》。

2009年美国龙安力鼎之作——世博会铁路馆在众多设计方案中脱颖而出，成为2010年中国世博会铁道部展览馆最终中标方案。

饶及人先生获奖过许多荣誉，在最近三年中，比较有代表性的荣誉包括：

2010年1月在"和谐中国2009年度影响力人物"颁奖盛典上被授予"2009年十大影响力人物"；2009年12月在"2009年中国城市建设与管理高峰论坛暨中国城市建设与管理60年表彰活动中"被中国城市经济学会、《建筑》杂志社授予"中国城市规划60年十大杰出人物"荣誉；2009年11月年饶及人先生被浙江省青田县委、县政府授予青田侨界"六十杰""十大创业精英"荣誉；于2008年获得中国住房和城乡建设部科学技术委员会及中国房地产协会联合颁发的"2008中国责任地产年度人物"奖，以及中国住房和城乡建设部技术委员会及中国国际城市化发展战略研究委员会共同颁发的"2008 中国城市化进程10大贡献力人物"年度重要奖项。

同时饶及人先生在"谁将影响下一个三十年（2008–2038）赢在未来"活动中荣获（2008）年度"赢在未来，商界领军人物100强"，并同期入选《安家iHome杂志》二十位"改革开放三十年影响中国地产人物"。

2007年美国龙安公司成为全国工商联房地产商会首批认证的"全国绿色生态建筑示范项目签约规划设计机构"，同年在庆祝内蒙古自治区成立60周年的活动中，美国龙安公司的内蒙古呼和浩特多元文化特色景观街项目获得"最佳建筑工程设计奖"，饶及人先生荣获"市长特别奖"的荣誉。

2006年在中国建设报、中国楼市主办的"影响中国商务建筑"推介活动中，饶及人先生荣获"影响中国十大公建设计师"称号。

饶及人先生投身城市规划与建筑设计领域已三十余年。

饶及人先生1980年毕业于纽约普拉特大学(PRATT)建筑系，1984年毕业于纽约佩斯大学（PACE）MBA，1990年获得普拉特大学"百年校庆校友奖"，1994年获得普拉特大学颁发的"杰出建筑系校友奖"。

饶及人先生长期从事城市规划和建筑设计工作，是全美注册建筑师，他持有美国纽约州、新泽西州、康州、宾州、乔州、加州、麻省注册开业建筑师资质。1988年被美国建筑师协会选为"杰出少数族裔建筑师"，1989年成为美国十大青年建筑师。

J.A.O.Design International Architects & Planners Limited
美國龍安建築規劃設計顧問有限公司

饶及人
James C. Jao

James C. Jao

NCARB, NABAR, SAFEA, AIA, RA

Mr. James C. Jao, (Rao Jiren, in Chinese) who is referred to by the Chinese government as the "Urban Doctor", has successfully established and developed his architecture, planning, and land development firm over the past 25 years. Besides being a senior consultant to many government sectors, he is also the Land Use Law Expert to the Chinese State Council. He is highly respected in the private industry and banking sectors, and in the halls of government in China. He has authored two important planning and land use books in China: "A Decade of Planning in China" (Xinhua Press 2005), and "Searching the Souls of Chinese Cities" (China Urban Press 2007). A book about himself, "The Urban Doctor- James Jao", will be published by the Rail Road Press in Oct. 2009, which will be sold in over 2000 train station book stores in China. He has delivered several hundred lectures and published over a hundred important papers on land use, urban planning, and real estate development in China, which have made a great impact on city planning and real estate policy after China's reform and opening up policy. He is demonstratively the pivotal player in the decision process for many planned and built major Chinese realty projects.

J.A.O. Design International Ltd is a design firm founded by him in New York circa 1984. It entered the Chinese market in 2001. Insofar, the firm has performed planning and architectural design services in over 50 cities with a total area of 2750 square kilometers (about 6 times the area of Manhattan Island), plus 13,000,000 square meters of new residential buildings, 15,000,000 square meters of commercial offices, hotels, civic centers, shopping malls, and mixed use developments. His projects have been awarded over 90 distinguished awards in China and abroad.

The cities and governmental entities where he serves as a senior advisor include Beijing CBD, Tongzhou District, Tianjin Yujiapu CBD, Tianjin Hongqiao District, Gaolihu District, Tangshan City, Hohhot City, Kunming City, Gongqing City, Wenzhou City, Shanghai City, Jiangxi Province, Sichuan Province, Liaoning Province, etc. He is the only person engaged by 26 cities, municipalities, and provinces in China with regards to their land use, economic development consulting, project planning and land development.

Now he is the CEO of US Long On Group (www.longongroup.com), and the founding president of J.A.O. Design International (www.jaodesign.com), a leading planning and architectural design firm headquarter in Beijing, with offices in Hong Kong, mainland China and United States. In addition, Mr. Jao is the founding President of the China Planning and Development Institute (CPDI, www.cpdi.org), a non-profit organization that encourages the sustainable development of China's urban centers through partnership between the U.S. and China. He is also the C.E.O. of China Real Estate Design League (CREDL) (www.sjlm.net.cn), the largest professional design organization within P.R.C. In July 2009, United Nations Under-Secretary General Anna Tibaijuka, who is also the Executive Director of UN Habitat, has appointed him as a member of the Steering Committee of the World Urban Campaign for the period of 2009-2013.

In Dec. 2008, he was honored by the Central Chinese Government and the People's Congress with distinguished awards recognizing his influence in the real estate industry, business community, urbanization, and eco-development with the "2008 China Responsible Real Estate Person of the Year Award".

Mr. Jao is an Honorary Citizen of the People's Republic of China, and the ONLY Certified Foreign Expert in Architecture, Planning and Regional Economic Development by the State Administration of Foreign Experts Affairs (SAFEA). He is a certified architect by the US National Council of Architectural Registration Boards (NCARB), and is a registered architect in the States of New York, California, New Jersey, Connecticut, Pennsylvania, Georgia, and Massachusetts. He is also certified by the Chinese National Administrative Board of Registered Architects (NABAR) as a Class A architect.

In recognition of his in-depth knowledge of zoning, city planning, architecture, development, and Chinese American relations, New York City Mayor David N. Dinkins, New York State Governor Mario Guomo, Manhattan Borough Presidents Ruth Messinger and Virginia Fields have honored Mr. Jao over the years.

James C. Jao is a US Citizen who was born in Taiwan in 1957 and grew up in New York. He received his Bachelor of Architecture degree from New York Pratt Institute in 1980. In 1984 Mr. Jao received his MBA from Pace University in New York. In 1994, he was elected by his alma mater Pratt Institute and received the "Alumni Achievement Award in Architecture".

美国HOOP建筑设计有限公司
HOOP ARCHITECTURAL DESIGN CONSULTANTS,INC

欧阳颖
Ying Ouyang

欧阳颖

一级注册建筑师

毕业于天津大学，获建筑学学士学位，
重庆建筑大学，获建筑学硕士学位
现任美国HOOP建筑设计咨询有限公司　首席设计师

主持设计项目
宁波雅戈尔东湖花园二期（500 000平方米）
浙江慈溪慈土储10号地块
雅戈尔东钱湖别墅设计（120 000平方米）
内蒙古呼和浩特天骄公馆设计（378 000平方米）
山东烟台海港北路城市设计（366 000平方米）

Ying Ouyang

First Grade Registered Architect

Graduated as a Bachelor of Architecture from Tianjing University and a Master of Architecture from Chongqing Jian zhu University one after another.
Currently hold the post of the chief designer of the HOOP Architectural Design Consultant INC. of America.

Designs Projects
The Second Phase of Ningbo Youngor East Lake Garden (occupying an area of 500,000 m^2)
The No.10 Plot of Zhejiang Cixi Land Reserve
The Design of Ningbo Youngor Dongqian Lake Villas (occupying an area of 120,000 m^2)
The Design of Inner Mongolia Hohhot Hero Mansion (occupying an area of 378,000 m^2)
Shandong Yantai Haigang North Road City Design (occupying an area of 366,000 m^2)

龚　俊
Jun Gong

龚　俊

一级注册建筑师

毕业于重庆建筑大学，获建筑学学士学位
先后就职于新加坡杰盟建筑设计公司、新加坡新仕建筑设计公司
现任美国HOOP建筑设计咨询有限公司　首席设计师

主持设计项目
上　海　仁恒河滨花园
上　海　保利"海上五月花"居住区
上　海　绿庭四季生态别墅
上　海　新时代富嘉花园
宁　波　世茂宁波北仑居住区
哈尔滨　万达商业广场
南　京　万达商业广场

Jun Gong

First Grade Registered Architect

Graduated as a bachelor of architecture from Chongqing Jianzhu University.
Once worked in Jgp Architecture Pte Ltd. of Singapore and Xinshi Architecture Design Ltd. of Singapore one after another.
Currently hold the post of the chief designer of the HOOP Architectural Design Consultant INC. of America.

Designs Projects
Shanghai - Renheng Hebin Garden
Shanghai - Poly "Mayflower On the Sea" Residential Area
Shanghai - Luting Four-season Ecological Villas
Shanghai - New Age Fujia Garden
Ningbo - Shimao Ningbo Beilun Residential Area
Harbin - Wanda Commercial Plaza
Nanjing - Nanjing Wanda Commercial Plaza

美国波士顿国际设计集团
Boston International Design Group

朱儁夫
Junfu Zhu

朱儁夫

总裁、董事、首席设计师

1992—1997年就读于清华大学建筑学院。1997年进入美国麻省理工学院（MIT）学习，2000年毕业于麻省理工学院建筑与城市设计专业，并获硕士学位。2000年开始任职于波士顿的斯塔宾建筑事务所。2003年年底被任命为高级副总裁，合伙人，是当时美国大型事务所中罕见的年轻合伙人。2004年9月，朱儁夫先生在波士顿创建波士顿国际设计集团（BIDG），同年年底到上海设立波士顿国际设计集团上海公司。波士顿国际设计集团现有设计人员100多人，在波士顿、上海、北京设有三个机构，主要从事重要城市历史街区的保护、更新与发展，校园可持续规划及校园建筑设计，高端物业及商业设计等。朱儁夫先生现担任美国波士顿国际设计集团总裁、首席设计师。

朱儁夫先生近年来一直从事可持续发展的城市规划及建筑设计工作。朱儁夫先生主要的国际设计项目包括：拉斯维加斯的威尼斯酒店、澳门氹仔岛整体规划、波士顿空权计划、德国德累斯顿市城市中心城区再发展计划、波士顿麻州总医院及监狱酒店改造方案、亚特兰大艾莫瑞大学表演中心设计等。

近期，在中国由朱儁夫先生主持设计的作品包括北京钟鼓楼区域保护与设计、北京雍和宫国子监区域规划研究及米高梅MGM酒店设计、北京隆福寺历史街区改造规划、北京南锣鼓巷及地铁周边地区规划及建筑设计、北京运河北段及南段历史保护规划、清华大学核心区规划、清华大学主楼东区建筑立面改造、杭州西湖商业整体规划及具体分期设计、宁波月湖西区保护规划及设计、宁波月湖盛园郁家巷地区整体设计、宁波莲桥街区规划改造及设计、上海黄浦区露香园区块保护设计、三亚香水湾一号、北京红楼梦主题文化园规划等。朱儁夫先生受北京市规划委员会和教育委员会委托，主持研究了"北京市高等学校规模研究"。朱儁夫先生设计的作品曾多次荣获国内外大奖，其中威尼斯酒店二期工程获2004年拉斯维加斯建协金奖。朱儁夫先生主持编写的《世界名校校园研究》等将在近期出版。朱儁夫先生曾在麻省理工学院（MIT）、清华大学建筑学院、芝加哥大学、美国内华达大学、韩国汉城大学湖西学院、浙江大学等高等院校作专题报告。

Junfu Zhu

President, Director, Chief Designer

Zhu Junfu studied at Architecture Institute of Tsinghua University from 1992 to 1997. In 1997 he was admitted to the Massachusetts Institute of Technology (MIT) and graduated with the Master's Degree of Architecture and Urban Design in 2000. Since 2000 Zhu began to work in the Stubbing Architectural Firm in Boston. At the end of 2003, he was named the senior vice president and a partner, which is rare in major U.S. firms at such a young age. In September 2004, Mr. Zhu established the Boston International Design Group (BIDG) in Boston and at the end of the same year founded the Shanghai branch of Boston International Design Group in Shanghai. BIDG now has more than 100 designers and three bodies in Boston, Shanghai and Beijing. BIDG mainly engaged in the protection, renewal and development of the historic district in major cities, campus' sustainable planning and architectural design, high-end property and commercial design, etc. Mr. Zhu currently serves as the president and chief designer of Boston International Design Group.

In recent years Mr. Zhu has been engaged in the urban planning and architectural design of sustainable development. The major international design projects in which he participated include: Venetian Hotel in Las Vegas, the overall planning of Taipa Island in Macau, Boston air rights plan, Dresden City Center redevelopment plan in Germany, Boston, the reform program of Massachusetts General Hospital and the prison hotel in Boston, Emory University, the design for the performance center of Emory University in Atlanta, etc.

The recent designs in China chaired by the Mr. Zhu include: the protection and design for Beijing Bell and Drum Tower area; the regional planning for Imperial College District in Beijing Lama Temple and MGM Hotel's Design; the transformation planning of the historic district in Beijing Longfu Temple; the planning and architectural design for South Luogu Lane in Beijing and the metro surrounding region; the historic preservation planning for the north and south section of Jade River in Beijing; the core area planning of Tsinghua University; the building facade reconstruction for the east district of the main building in Tsinghua University; the overall planning and the design of specific stages for West Lake business district in Hangzhou, the protection planning and design for the western district of Moon Lake in Ningbo; the overall design for Yujia Lane District of Moon Lake•Shengyuan in Ningbo; the transformation planning and design for Lotus Bridge District in Ningbo; the protection design for Luxiang Park of Huangpu District in Shanghai; Sanya Perfume Bay No.1 and the planning for the Honglou Dream theme cultural park in Beijing, etc. Under the commission of Beijing Municipal Planning Commission and Board of Education, Mr. Zhu conducted the study of "the Scale of Colleges and Universities in Beijing." His designs have won many domestic and international awards, in which the second phase project of Venetian Hotel won the Gold Medal by Las Vegas Building Society in 2004. The book "The Study of World Famous Campus" compiled by Mr. Zhu will be published in the near future. In addition, he has made special reports in many institutions of higher education, such as Massachusetts Institute of Technology (MIT), Architecture Department of Tsinghua University, Chicago University, U.S. Nevada University, Husi College of Seoul National University in South Korea, Zhejiang University, etc.

英国UK.LA太平洋远景国际设计机构
U.K PACIFIC LONG-RANGE PLANNING & DEVELOPING DESIGN CONSULTANT LTD.

张 凯
Kai Zhang

张 凯

华中科技大学建筑学硕士学位
东南大学建筑学博士学位
英国UK.LA太平洋远景国际设计机构大中华区总裁
南京环洋远景建筑规划设计顾问有限公司董事长
南京泛奥建筑规划设计顾问有限公司董事长
郑州九华房地产开发有限公司董事
先后被评为:
中国房地产协会专家委员会委员
亚洲建筑规划师杰出100强
中国建筑规划设计突出贡献人物
江苏建筑规划十大影响力人物
亚洲建筑规划设计风云人物
法国建筑师协会会员 (20090618)
中房商协城市规划与建筑设计专业委员会委员

主要作品

1. 城市规划部分
A.苏州古城一、二街坊保护性规划 (76公顷)
B.郑州龙子湖湖心岛城市设计 (99公顷)
C.新沂窑湾古镇规划 (101公顷)
D.弘业国际无锡锡东大道城市设计 (56公顷)

2. 公共建筑部分
A.郑州意大利格拉姆中心 (12万平方米超高层)
B.南京金宁广场 (18万平方米综合建筑群)
C.连云港甬连再生资源加工园区 (100万平方米科技园区)
D.南京名爵汽车广场 (26万平方米)
E.徐州中山堂改造项目 (6万平方米高层)
F.南京金港创业园 (76万平方米)

3. 居住社区部分
A.南京华欧国际友好城 (230万平方米)
B.南京旭日上城建筑设计 (202万平方米)
C.郑州清华·忆江南 (86万平方米)
D.海南儋州双联滨海住区 (88万平方米)
E.南京爱涛漪水园·爱涛天成 (40万平方米)
F.郑州清华大溪地 (260万平方米)
G.红太阳乌江新城 (530万平方米)
H.郑州中南海知音 (50万平方米)

4. 景观设计部分
A.郑州西南绕城高速景观设计 (全长52千米)
B.南京河西江东路景观设计 (总长16.8千米)
C.南京徐庄软件园景观设计 (设计面积12公顷)
D.信阳百花园景观设计 (设计面积20公顷)

Kai Zhang

BA of Architecture in Huazhong University of Science and Technology
Ph.D of Architecture in Southeast University
President of UK Pacific Long-Range Planning & Developing Design Consultant Ltd.
President of Nanjing Huanyang Long-Range Architecture Planning and Design Consultant Co., Ltd.
President of Nanjing Fanao Architecture Planning and Design Consultant Co., Ltd.
Director of Jiuhua Real Estate Co., Ltd. in Zhengzhou
He has honorbly joined the Committee of China Real Estate Association
He Ranks among the Ten Most Influential Figures in Jiangsu architecture Planning Area.
Top Ten Persons of Asia Architecture Planning and Design
He is the member of France Association of Architects (20090618)
Member of China Property Developer Association Urban Planning and Architectural Design Committee.

Main works

1. Urban planning
A. protection planning for Suzhou ancient city neighborhood I II (76 ha)
B. The mid-lake island urban design for Zhengzhou Longzihu lake island (99 ha)
C. planning of the Yaowan town in Xinyi(101 ha)
D.Urban designing of Hongye International Xidong avenue in Wuxi

2. Public architecture
A. Zhengzhou Italy Gram Centre (120,000 m^2, extra-highbuilding)
B. Jinning square in Nanjing(building complex of 180,000 m^2)
C. Yonglian renewable resource processing industrial park in Lianyungang (industrial park og 1,000,000 m^2)
D. MG car plaza in Nanjing (260,000 m^2)
E. Rconstruction of Zhongshan hall in Xuzhou (60,000 m^2)
F. Nanjing Jingang Pioneering Park(760,000 m^2)

3. Residence community
A. Nanjing, Hua'ou International friendly city (2,300,000 m^2)
B. Architecture design of Rising Sun Top Town in Nanjing(2,020,000 m^2)
C. Tsinghua yijiangnan in Zhengzhou (860,000 m^2)
D. Shuanglian seaside residence in Danzhou , Hainan(880,000 m^2)
E. Aitaoyishui Park, Aitaotiancheng(400,000 m^2)
F. Tstinghua Tahiti in Zhengzhou(2,600,000 m^2)
G. Red Sun Wujiang Metro (5,300,000 m^2)
H. Associate of Zhongnanhai(500,000 m^2)

4. Landscape design
A. Landscape deisgn of the loop high way in the south west of Zhengzhou (5,300,000 m^2)
B. Landscape design of Jiangdong road in Hexi, Nanjing(total length 16.8 km)
C. landscape design of Xuzhuang Software Park in Nanjing(design area:12 ha)
D. Landscape design of Spring Garden in Xinyang(design area: 20 ha)

James Wang Design Associates
北京杰地亚建筑咨询有限公司

王振楣
James Wang

王振楣

职　　务：美国JWDA建筑师事务所总裁
注册资格：美国国家注册建筑师
　　　　　美国建筑协会（AIA）会员

学　　历

1974年　　　台湾淡江大学建筑学士
1976–1977年　芝加哥伊利诺理工大学建筑设计研究院研究生
1977–1978年　美国伊利诺大学建筑设计研究院建筑硕士

James Wang工作的方针与原则

与客户密切沟通以充分了解客户所需并进行详尽的分析，提供针对各种困难的解决方案，最终为客户提供使用功能与经济效益合理、美观大方的优秀设计。

James Wang的工作成就得到了多方面的认可，在美国多次荣获美国建筑师协会（AIA）、加利福尼亚州政府所授予的奖励与奖状，在中国获得全国多个优秀设计奖，在美国及国际设计竞赛中，公司的设计方案曾多次获得首奖及其他殊荣。

公司的主持人建筑师James Wang在美国伊利诺大学获得建筑硕士，并在芝加哥伊利诺理工学院进修建筑设计，任美国建筑师协会会员（AIA）。

James Wang

Position: President of American Firms of JWDA Architects
Qualification of registration: American Registered Architect
member of American Institute of Architects(AIA)

Educational background

In 1974, obtained Bachelor's Degree in Taiwan Tamkang University, majoring in Architecture
From 1976 to 1977, studied as graduate student in Architecture and Design of Illinois Institute of Technology, Chicago
From 1977 to 1978, obtained Master's Degree in Architecture and Design of University of Illinois

Working policy and principle of James Wang

Deeply communicate with customers to know what they really want, and then make a detailed analysis to provide solutions to all kinds of difficulties, ultimately provide customers with a nice outstanding design with reasonable cost-effectiveness using function.

James Wang's achievements have been recognized by many aspects. He has been awarded for many times in America by American Institute of Architects(AIA), American Planning Association (APA), ASLA and state government of California. Meanwhile, he has been also awarded National Excellent Design and Beijing Excellent Design. In American and the international design competitions, designs of company have won awards and the first prize for many times.

James Wang has obtained master degree of architecture in University of Illinois and has further studied architecture and design in Illinois Institute of Technology. Now, he is member of American Institute of Architects(AIA).

LOOK architects
Architecture & Urban Design

骆文义
Look Boon Gee

骆文义

Look建筑设计公司创办人及董事总经理，新加坡

骆文义，Look建筑设计公司的创始人，可能很多人会简单的误以为他是一个现代主义者，然而，在研究了他近10年的非凡作品后，展现给我们的却是一个坚持不懈挑战设计现状的创新者。

他的作品的主要特点是综合性，你无法简单的将他们分类，以及不断的创新和每个项目的新奇设计。他不断地在众多领域搜索新的突破，这主要体现在他对尖端建筑技术，可持续生态设计和文化保护的独特见解上。

他对美学的追求体现在将所有的难题、限制、理想与机遇完美的融为一体，释放出项目的全部潜能。依靠开放的心态和无与伦比的活力，骆先生以他天赋的才华，明确的解析人和场所的关系，以得出作品的独特解决方案。

骆先生坚信对现代建筑和人文的追求可以创造出一个更美好的世界。在建筑设计领域的关键路口，他是专业人士的典范，并开始发掘出亚洲建筑的真实潜力。

LOOK BOON GEE

Founder and Managing Director of LOOK ARCHITECTS, SINGAPORE

The architecture of Look Boon Gee, founder of LOOK Architects, may beguile many to simply term him as a modernist. However, on closer examinations of an extraordinary portfolio of works accumulated over a short period of slightly over 10 years, it reveals a craftsman, economist, ecologist and innovator at heart that is consistently challenging the status quo of architecture.

The main characteristics of his works is that they defy any straight jacketing into any simple categorization, other than the continual ability to innovate and surprise with each new project. It is an architecture that constantly searches for new breakthroughs in multitude of fields; be it in cutting edge building technology, sustainable ecological designs or conservation interpretations.

Instead of being hampered by the pursuit of a stylistic aesthetic, his architecture seeks its beauty in a well resolved piece of architectural puzzle where all the problems, constraints, ideals and opportunities clicked in place to release the full potential of the project. Adopting an open mindset and unrivalled vigor in approaching very piece of work, Look Boon Gee has a gifted flair to identify and dissect the necessities and aspirations of the people and place involved, to arrive at a unique solution for each piece of work.

The precise reading of the forces at work often materializes itself in the finished project resulting in architecture of clarity that calms and inspires at the same time. Look Boon Gee's relentless pursuit of architecture based on the humanistic belief that it can be used to create a better world. This sets him as an exemplar for his fellow professionals at the critical crossroads of the development of architecture; we are only beginning to uncover a truly Asia n Architecture.

Aedas
凯达

纪达夫
Keith Griffiths

纪达夫

亚洲及中东区主席
凯达环球有限公司

1975年 建筑学硕士（剑桥大学）
1978年 建筑文凭（剑桥大学）
1979年 英国注册建筑师
1979年 英国皇家建筑师学会会员
1985年 香港注册建筑师
1986年 香港建筑师学会会员
2001年 香港建筑师学会资深会员

纪达夫先生为凯达环球集团董事局与亚洲、中东区主席。他曾于英国剑桥大学圣约翰学院修读建筑，从事建筑设计工作有30年。1985年，纪先生在香港创立自己的建筑设计师事务所，并带领事务所发展成为今日的凯达。在纪先生领导下，凯达发展迅速，现已跻身成为全球第二大建筑设计师事务所，并在全球设立39间办事处。纪先生拥有出众的领导才能和创新思维，其优秀的设计受到公认。纪达夫先生致力于推广卓越的可持续设计与文化融合的理念，并将其奉为凯达设计宗旨。除此之外，纪先生着力推动多项内部活动如国际设计工作坊、设计方案检讨、员工培训及教育外展课程等，带领凯达及员工融入众多不同的社区团体、文化及国家，他提倡通过建筑设计改善社区的面貌。纪先生在亚洲、及中东有多项地标项目，其中包括中国广州白天鹅宾馆的扩建及翻新工程、位于迪拜的达10万平方米的办公楼及商场综合发展项目、中东迪拜Boulevard Plaza、中国长沙新河三角洲总体规划、香港国际机场二号客运大楼达10万平方米的扩建项目，以及香港中环中心区包括置地广场、遮打大厦、历山大厦的翻新重建项目。

Keith Griffiths

Chairman – Asia and the Middle East
Aedas Ltd.

M Arch (Cambridge) 1975
Dip Arch (Cambridge) 1978
Registered Architect (UK) 1979
RIBA 1979
Registered Architect (HK) 1985
HKIA 1986
FHKIA 2001

Keith is the Chairman of Aedas' global board and of Asia and the Middle East. He studied Architecture at St John's College, Cambridge University and has practiced architecture for three decades. Keith formed his architectural practice in Hong Kong in 1985 and led the company into the creation of Aedas in 2001. He has overseen the phenomenal growth of Aedas to become the world's second largest architectural practice with 39 offices worldwide. Keith has a well recognized reputation for outstanding leadership and innovation. He is a recognized and sought after designer who promotes sustainable design excellence and cultural integration as the keystones of Aedas' philosophy. Keith facilitates international design workshops, reviews, staff trainings and educational outreach programs, and is an instigator and champion of Aedas' integration into many communities, cultures and countries the company works with. He provides a driving force in Aedas' work to improve our environment through our built form and its integration into the fabric of our communities. Keith has been leading the design for a number of landmark projects in Asia and the Middle East. To highlight a few, those includes the redevelopment of the White Swan Hotel, Guangzhou; Boulevard Plaza, Dubai; Xinhe Delta Masterplan; Terminal 2, Hong Kong International Airport: 100,000 sq.m extension at Hong Kong International Airport; The regeneration project for The Landmark, Chater House, Alexandra House and Prince's Building in Central, Hong Kong.

GKK+ARCHITEKTEN

PROF. SWANTJE KÜHN
OLIVER KÜHN

斯万提・库恩
Swantje Kuehn

奥利弗・库恩
Oliver Kuehn

奥利弗・库恩

奥利弗・库恩先生出生于建筑世家，是家族中的第四代建筑师。他曾就读于德国慕尼黑工业大学建筑系，并在最短的时间内完成学业，后在理查德　罗杰斯伦敦事务所开始了他的职业生涯。作为项目建筑师的他，完成了如国际机场，高层办公楼等诸多项目。

奥利弗的重要兴趣在于大型尺度的项目，并赋予项目本身最大可能的技术，功能以及审美水平。他既是一个全球化主义者，同时又是一个独特的完美主义者，这也使得他能够超越项目本身的社会、文化和环境问题的要求，通过分析和发展，为每个项目提出最佳设计方案。他频繁的国际曝光，使得他与世界各国的合作伙伴，合作建筑师有着广泛的项目合作，这也使他对不同的人以及文化率持有开放的观念和态度。

为了加强自己作为GKK建筑师事务所总裁的业务技能，奥利弗在瑞士就读了企业管理，并以最佳毕业生的荣誉完成其学业。奥利弗同时也担任了一系列大学的客座教授，并已获得多项国家和国际级别的设计大奖。

Oliver Kuehn

Oliver Kühn is an architect in the fourth generation. He studied architecture at the Munich University of Technology, and graduated in the shortest possible time. He began his professional career in the office of Richard Rogers Architects in London, and worked as a project architect for many international projects, including airports and office high-rises.

Oliver's key interest is in large-scale projects which he develops at the highest possible technological, functional and esthetic level. He is both a globalist and a distinct perfectionist, which enables him to look beyond the immediate project requirements to social, cultural and environmental issues, and then to analyze and develop the best technical design solution for each project.

His strong international exposure allows him to cooperate worldwide with his partners and co-architects, and makes him open minded to different cultures, and to other people's views.

To strengthen his business skills as the CEO of GKK Architects, Oliver studied business administration in Switzerland and graduated at the top of his class. Oliver is a guest speaker sought after at a number of universities, and has been awarded several national and international design awards.

斯万提・库恩

斯万提・库恩女士是建筑与室内设计的大学教授，她的设计风格受到了其在芝加哥伊利诺斯州韦斯恩大学的当代艺术教育的强烈影响。毕业之后，她又就读于德国慕尼黑工业大学建筑系，并将其对艺术的追求与德国工程技术的理性与逻辑相互结合。其最早的工作经验完成于非洲，并作为设计建筑师在理查德罗杰斯伦敦事务所工作了数年。

她作为德国戴特蒙德应用科学大学的斯万提教授的同时，也是GKK建筑师事务所的合伙人之一，并担任德国汽车制造业的内饰设计的市长级顾问。

从2005年起，奥利弗和斯万提开始深入中国。他们广泛而深入地走访了许多中国城市，并开展了一系列的学术讲座、展览以及设计工作。

奥利弗・库恩先生与斯万提・库恩女生已结婚20多年，并有3个孩子。

Swantje Kuehn

Swantje Kühn is a university professor for architecture and interior design. Her design approach is strongly influenced by her education in contemporary arts at Illinois Weslyan University in Chicago. After graduation, she studied architecture at the Munich University of Technology, and was able to combine her artistic approach with the rational and logical approach of German engineering. Her first work experience was in East Africa, where she worked for several years later she joined Richard Rogers Architects as a design architect for and worked on a number of different projects in London.

Although a professor at the University of Applied Science in Detmold, Swantje is also a partner and director of GKK Architects, as well as a design adviser for a mayor German automaker in the development of auto interior components.

Since 2005, Oliver and Swantje have been engaged in China. They have travelled widely in the country, and have given several series of lectures, followed by exhibitions and design workshops.

Oliver and Swantje Kühn have been married for twenty years and have three children.

LU TANG LAI ARCHITECTS LTD.
L T L 吕 邓 黎 建 筑 师 有 限 公 司

邓文杰
Tang Man Kit, Joseph

邓文杰
董事

学术资格
1979年 香港大学建筑学（一级荣誉）文学士
1981年 香港大学建筑学士（优异）

专业资格／公职
1985年 香港建筑师学会会员
1986年 英国皇家建筑师学会会员
1987年 香港屋宇机构认可人士（第一类别）
1990年 香港注册建筑师
2002年 香港建筑师学会资深会员
2004年 中国一级注册建筑师资格
2006年 香港大学建筑系助理教授
2008年至今 建筑师事务所商会秘书长
2010年至今 公共事务论坛（论坛）成员

奖项
1979年 香港大学何福建筑奖
1979–1981年 香港大学怡和奖学金
1981年 世界建筑师联会之建筑系学生国际设计比赛
法国建筑师学会奖项
法国都市规划师学会奖项
匈牙利布达佩斯科技大学奖项
1983年 香港专业教育学院（屯门）设计项目比赛（一等奖）
1985年 九龙公园设计项目比赛（一等奖）
1987年 香港科技大学设计项目比赛（二等奖）

Tang Man Kit, Joseph
Director

Academic Qualification
1979 Bachelor of Arts in Architectural Studies
University of Hong Kong (Awarded with First Class Honours)
1981 Bachelor of Architecture
University of Hong Kong (Awarded with Distinction)

Professional Qualification / Appointment
1985 Member of The Hong Kong Institute of Architects
1986 Member of The Royal Institute of British Architects
1987 Inclusion in the registrar of Authorized Person List 1 of Buildings Department
1990 Registered Architect of The Architects Registration Board, Hong Kong
2002 Fellow of The Hong Kong Institute of Architects
2004 PRC Class I Registered Architect Qualification
2006 Assistant Professor, Dept. of Architecture University of Hong Kong
2008 - Present Honorary Secretary of The Association of Architectural Practice
2010 - Present Public Affairs Forum (the Forum) - Member

Awards
1979 Ho Fook Prize in Architecture (HKU)
1979 - 81 Jardine Henry Lo Scholarship (HKU)
1981 UIA International Design Competition for Students of Architecture
Prize from Societe Francaise des Architects,France
Prize from D'instit et urbanizme, France
Prize from the Technical University of Budapest, Hungary
1983 Design Competition
Tuen Mun Technical Institute (First Prize)
1985 Design Competition - Kowloon Park (Co-Winner)
1987 Design Competition - Hong Kong University of Science and Technology

* 任职于关善明建筑师事务所有限公司 Work under Simon Kwan & Associates Ltd.

余啸峰 FRANK YU

余啸峰 设计总监

现任香港嘉柏建筑师事务所设计总监。生于香港，12 岁前赴美接受教育，1986 年毕业于美国纽约 Pratt Institute 建筑学院，随即到当时美国最大的建筑师事务所——美国纽约 Ellerbe Becket 任规划设计师，并取得多项建筑设计大奖，其后一直参与不同的著名项目及比赛，屡获奖项。2000 年决定回到香港发展，为了实践对建筑设计的理想及抱负，于 2003 年与资深建筑师王克江先生创立了嘉柏建筑师事务所，并集合一群志同道合的专业人才，共同努力于提供高质素、具创意及独特的设计理念，致力于为现代建筑设计打开新的一页。

教育经历

1986 年毕业于美国纽约 Pratt Institute 建筑学院

工作经历

2003 年至今　香港嘉柏建筑师事务所有限公司设计总监
2000—2003 年　香港许李严设计亚洲有限公司副董事及设计总监
1990—1994 年　美国加州洛杉矶 B.T.A. 有限公司设计总监及亚洲办事处常务董事
1987—1990 年　美国纽约 Ellerbe Becket 设计总监

FRANK YU Principal - Design

Design Principal of Gravity Partnership Limited. Born in Hong Kong, Frank received his education in the United States from the age of 12. He studied architecture at Pratt Institute, New York, and after graduating in 1986, he was invited to join the largest architectural firm in the US – Ellerbe Beckett New York as a Design Architect. He has worked on various renowned projects and international design competitions, and has won numerous awards. In 2000, Frank decided to return to Hong Kong and founded Gravity Partnership Limited with Claude Wong in 2003. Together with a group of dedicated designers, who have the same enthusiasm for non conventional design ideas and working methodologies, they look to create high quality and inspired designs and starting a new chapter in modern architecture.

Academic Background

1982-1986　Bachelor of Architecture - Pratt Institute, New York, USA

Professional Experience

2003-Present　Gravity Partnership Ltd. HKSAR, Principal
2000-2003　RDL Asia Ltd. HKSAR, Design Principal / Deputy Director
1990-1994　B.T.A. Ltd. (USA) Design Principal, (Asia office) Managing Director
1987-1990　Ellerbe Becket (New York office), Project Designer
1987-1990　Ellerbe Becket

王克江 项目总监

1984 年毕业于英国伦敦建筑协会 Architectural Association。香港注册建筑师，香港屋宇处认可人士及中华人民共和国一级注册建筑师资格。期间曾任香港特别行政区建筑署高级建筑师、规矩建筑师有限公司总监、许李严设计亚洲有限公司副董事。

教育经历

1984 年毕业于英国伦敦建筑协会 Architectural Association

工作经历

2003 年至今　香港嘉柏建筑师事务所有限公司技术总监
2001—2002 年　香港许李严设计亚洲有限公司副董事
1992—2001 年　香港规矩建筑师有限公司总监
1984—1992 年　香港特别行政区建筑署高级建筑师

专业资格

香港注册建筑师及认可人士
香港建筑师学会会员
英国皇家建筑师学会会员
中华人民共和国一级注册建筑师资格

CLAUDE WONG Principal - Project

Graduated from the Architectural Association in the United Kingdom in 1984, Claude practices as Registered Architect and Authorized Person in Hong Kong; and received the PRC Class A Registered Architect Qualification. Professional experiences include Architectural Services Department, Cartesian Architects Limited and RDL Asia Limited.

Academic Background

1979-1984　AA Diploma, Architectural Association, London, United Kingdom

Professional Experience

2003-Present　Gravity Partnership Ltd. HKSAR, Principal
2001-2002　RDL Asia Ltd. HKSAR, Deputy Director
1992-2001　Cartesian Architects Ltd. HKSAR, Director
1984-1992　Architectural Services Department, HKSAR, Architect

Professional Qualification

Registered Architect, Authorized Person (List of Architects), HKSAR
Member of Hong Kong Institute of Architects
Chartered Member of Royal Institute of British Architects
PRC Class 1 Registered Architect Qualification

嘉 柏 建 筑 师 事 务 所
gravity partnership ltd
architects & planners
www.gravitypartnership.com

张文明建筑师事务所

W.M.CHANG ASSOCIATES. ARCHITECTS&PLANNERS

张文明
W.M.CHANG

学历

1983年 中原大学建筑学士
1985年 淡江大学建筑硕士

经历

张文明建筑师事务所负责人
正修科技大学建筑系讲师
王西村建筑师事务所
苏成基建筑师事务所
嘉义东洋建设设计部

获奖作品

1989年 金狮奖
尖美建设 温哥华
1990年 金狮奖
皇家建设 台南名家
1991年 金狮奖
百立建设 首府宫廷
1993年 金狮奖
国城建设 高雄站前广场
美屋奖
尖美建设 尖美百货商业广场
1994年 美屋奖
乡田建设 璞园瑷园瑾园
1995年 优良建筑师作品
乡田建设 瑷园
1996年 建筑师杂志奖佳作
乡田建设 璞园
园冶奖
鼎宇建设 达尔文
中国建筑鲁班奖
哈尔滨 昆仑商城长春小区
1997年 中国建筑鲁班奖
哈尔滨 荣耀宝座
1998年 园冶奖
尖美建设 尖美市
中国建筑鲁班奖
哈尔滨 丽景天地
哈尔滨 美景天地
1999年 园冶奖
王象建设 梵赛荷
中国建筑鲁班奖
哈尔滨 荣耀世纪
2000年 园冶奖
允堂建设 原山主人
太普建设 美树苑
大连市政府最佳建筑规划设计奖
大连 星海国宝
2001年 园冶奖
福懋建设 丽庄
2002年 园冶奖
福懋建设 采云天
2003年 园冶奖
福懋建设 采华天
2004年 园冶奖
太普建设 自由自在
南统建设 湖畔城堡
2005年 园冶奖
串本建设 空间
太普建设 美术北一街
福懋建设 采扬天
2006年 园冶奖
福懋建设 采翔天
福懋建设 采荷天
京城建设 亚历山大
京城建设 莫扎特
东南吉建设 温莎城堡
2007年 园冶奖
太普建设 太普风格
太普建设 MoMA
福懋建设 梦世代
福懋建设 梦翔
鼎宇建设 美术园邸B案
2008年 园冶奖
福华建设 美术观止
家亿建设 崇明绿邑

Education

Zhong Yuan University, B. Arch (1983)
Dan Jiang University, M. Arch (1985)

Experience

Principal, W.M. Chang Associates, Architects & Planners
Zheng Xiu Technology University, Guest Design lecturer
Architect, S.C. Wang Architect's & Associates
Architect, C.C. Su Architect's & Associates
Architect, Dong Yang Development Group Co., Ltd.

Awards

Golden Lion Design Award
1989 'Vancouver', High-Rise Residential
Jian Mei Development Group Co. Ltd.
1990 'Asia Business Center', Commercial High-Rise
Yun Tai Development Group Co. Ltd.
1991 'Grand Palace', High-Rise Residential
Bai Li Development Group Co. Ltd.
1993 'Kaoshiung Station Plaza', Commercial High-Rise
Guo Cheng Development Group Co. Ltd.
1993 Aesthetic Excellence Award
'Jian Mei Commercial Department Tower', Commercial High-Rise
Jian Mei Development Group Ltd.
1994 Aesthetic Excellence Award
'Pu, Ai, Jin Gardens', Villa Complex
Xiang Tian Development Group Co. Ltd.
1995 Taiwan National Distinguished Architecture Award
'Ai Gardens', Villa Complex/Xiang Tian Development Group Co. Ltd.
1996 'Chinese Architect' Journal Excellence Award
'Pu Gardens', Villa Complex/Xiang Tian Development Group Co. Ltd.
Landscape Excellence Award
'Darwin', Residential House/Ding Yu Development Group Co. Ltd.
China National 'Ruban' Architectural Excellence Award
'Chang Chun Suburb, Kun Lun Commercial District', Residential Complexes, Harbin
1997 China National 'Ruban' Architectural Excellence Award
'Rong Yao Bao Zuo', Residential Complexes, Harbin
1998 Landscape Excellence Award
'Jian Mei City', High-Rise Residentia/Jian Mei Development Group Co. Ltd.
China National 'Ruban' Architectural Excellence Award
'Li Jing Tian Di', Residential Complexes, Harbin
'Mei Jing Tian Di', Residential Complexes, Harbin
1999 Landscape Excellence Award
'Fan Sai He', High-Rise Residential/Wang Xiang Development Group Co. Ltd.
China National 'Ruban' Architectural Excellence Award
'Rong Yao Shi Ji', Residential Complexes, Harbin
2000 Landscape Excellence Award
'Yuan Shan Zhu Ren', High-Rise Residential/Yun Tang Development Group Co. Ltd.
'Mei Shu Yuan', High-Rise Residential/Tai Pu Development Group Co. Ltd.
Da Lian City Council, Design Excellence Award
'Xing Hai Guo Bao', Residential Complexes, Da Lian
2001 Landscape Excellence Award
'Li Zhuang', High-Rise Residential/Fu Mao Development Group Co. Ltd.
2002 Landscape Excellence Award
'Cai Yun Tian', High-Rise Residential/Fu Mao Development Group Co. Ltd.
2003 Landscape Excellence Award
'Cai Hua Tian', High-Rise Residential/Fu Mao Development Group Co. Ltd.
2004 Landscape Excellence Award
'Zi You Zi Zai', High-Rise Residential/Tai Pu Development Group Co. Ltd.
'Hu Pan Cheng Bao', High-Rise Residential/Nan Tong Development Group Co. Ltd.
2005 Landscape Excellence Award
'Kong Jian', Chuan Ben Development Group Co. Ltd.
'Mei Shu Bei Yi Jie', /Tai Pu Development Group Co. Ltd.
'Cai Yang Tian', / Fu Mao Development Group Co. Ltd.
2006 Landscape Excellence Award
'Cai Shiang Tian', / Fu Mao Development Group Co. Ltd.
'Cai He Tian', / Fu Mao Development Group Co. Ltd.
'Ya Li Shan Da ', /Jjing Cheng Development Group Co. Ltd.
'Mo Ja Te ', /Jjing Cheng Development Group Co. Ltd.
'Wen Sha Cheng Bau ', /Dung Nan Ji Development Group Co. Ltd.
2007 Landscape Excellence Award
'Tai Pu Feng Ge', /Tai Pu Development Group Co. Ltd.
'Mo MA', /Tai Pu Development Group Co. Ltd.
'Meng Shr Dai', / Fu Mao Development Group Co. Ltd.
'Meng Shiang', / Fu Mao Development Group Co. Ltd.
'Mei Shu Yuan Di (B)', / Ding Yu Development Group Co. Ltd.
2008 Landscape Excellence Award
'Mei Shu Guan Zhi ', / Fu Hua Development Group Co. Ltd.
'Chong Ming Lu Yi', / Jia Yi Development Group Co. Ltd.

HFUT Institute of Architectural Design
合肥工业大学建筑设计研究院

张云海
Yunhai Zhang
Hou Baizhen

张云海

安徽省土木建筑学会理事
建筑专业委员会副主任委员
安徽省教育基本建设学会理事
合肥市城市规划行业协会理事
合肥市房地产业协会专家委员会委员

国家一级注册建筑师
国家注册咨询工程师
香港建筑师学会会员资格
安徽省工程勘察设计大师
合肥工业大学建筑设计研究院副院长
合肥工业大学建筑与艺术学院副院长

主要作品
合肥工业大学建筑技术研发中心
荣获 全国优秀工程勘察设计三等奖
教育部优秀工程勘察设计二等奖
安徽省优秀工程勘察设计一等奖
合肥工业大学南校区学科群实验楼
荣获 全国优秀工程勘察设计三等奖
安徽省优秀工程勘察设计二等奖
寿州宾馆
荣获 安徽省优秀工程勘察设计二等奖
中科院合肥物质科学研究院研究生教育基地
荣获 安徽省优秀工程勘察设计二等奖
淮南市山南新区规划展示厅
荣获 安徽省优秀工程勘察设计二等奖
合肥工业大学逸夫科教楼
荣获 教育部优秀工程勘察设计三等奖
合肥工业大学机械实验楼
荣获 机械工业优秀工程设计三等奖
合肥六中体育馆
荣获 安徽省优秀工程勘察设计三等奖
安庆师范学院龙山校区教学主楼
荣获 度安徽省优秀工程勘察设计三等奖
安庆师范学院龙山校区行政办公楼
荣获 安徽省优秀工程勘察设计三等奖
芜湖长江大桥收费站
荣获 安徽省优秀工程勘察设计专项工程三等奖

Yunhai Zhang

Director of Anhui Civil Engineering and Architecture Institute
Member and vice president of Architectural Profession Committee
Institute director of Capital Construction of Education of Anhui province
Association director of City Planning Industry of Hefei
Member of Real Estate Association Experts Committee of Hefei

National First-class Registered Architect
National Registered Consulting Engineer
Membership of Hong Kong Institute of Architects
Engineering survey and design master of Anhui
Assistant dean of Architectural Design Academy of Hefei University of Technology
Assistant dean of Architecture &Art College of Hefei University of Technology

Representative Works
Building Technology R & D Center of Hefei University of Technology
Awards: The third Prize of national outstanding engineering survey and design
The second prize of the outstanding engineering survey and design of the ministry of education
The first prize the outstanding survey and design in Anhui province
The north campus Laboratory Buildings of subjects of Hefei University of Technology
Awards: The Third Prize of national outstanding engineering survey and design
The second prize of the outstanding engineering survey and design in Anhui province
Shouzhou hotel
Awards:The second prize of outstanding engineering survey and design in Anhui province
Award: The postgraduate education base for Heifei Institutes of Physical Science, Chinese Academy of Sciences
Award: The second prize of the outstanding engineering survey and design in Anhui province
The Panning Exhibition Hall of the Shannan new district,Huainan city
Award: The second prize of outstanding engineering survey and design in Anhui province
The Yifu Building of Hefei University of Technology
Award: The third prize of the outstanding engineering survey and design of the ministry of education
Mechanical Laboratory Building of Hefei University of Technology
Award: The third prize of the mechanical engineering design
The Gymnasium of Hefei No. Six Senior High School
Award: The third prize of outstanding engineering survey and design in Anhui province
The main building of Longshan campus, Anqing teahers college
Award: The third prize of outstanding engineering survey and design in Anhui province
The administration building of Longshan campus, Anqing teachers college
Award: The third prize of outstanding engineering survey and design in Anhui province
Wuhu Yangtze Rive Bridge toll station
Award: The third prize of outstanding special engineering survey and design in Anhui province

唐望松
Wangsong Tang
Hou Baizhen

唐望松

总建筑师
高级建筑师
国家一级注册建筑师
同济大学工学学士
合肥工业大学硕士生导师
香港建筑师学会会员资格
安徽省土木建筑学会建筑专业学术委员会委员
合肥市房地产业协会专家委员会委员

主要作品
安徽师范大学新校区院系教学楼
荣获 安徽省优秀工程勘察设计行业奖一等奖
教育部优秀建筑设计三等奖
南京林业大学教学六楼（生物技术大楼）
荣获 江苏省第十二届优秀工程设计三等奖
徐州师范大学泉山校区实验楼
荣获 江苏省城乡建设系统优秀勘察设计三等奖
南京审计学院体育馆
荣获 南京市优秀工程设计三等奖
迈向21世纪全省住宅设计方案竞赛
荣获 安徽省优秀工程勘察设计一等奖
中国药科大学研究生实验楼
中国药科大学国际会议中心
安徽电子信息职业技术学院新校区规划
安徽电子信息职业技术学院信息中心
安徽环境监测科研应急指挥综合楼
安徽银瑞林国际大酒店
安徽六安新都会

主要论著
《现代博物馆建筑设计理念》
《创意——现代博物馆建筑设计的灵魂》
《从分析环境入手》
《塑造有性格特征、时代感和个性的建筑》

Wangsong Tang

chief architect
Senior Architect
national Class Ⅰ Registered Architect
BE of Tongji University
MA student adviser in Hefei University of Technology
Membership of the Hong Kong Institute of Architects
Member of Architectural Academic committee of Anhui Civil Engineering Institute of Anhui province
Committee member of Experts of Hefei Real Estate Association

Representative Works
New campus department teaching buildings of of Anhui Normal University
Awards: The first prize of outstanding engineering survey and design industry in Anhui province
The third prize of outstanding architectural design of ministry department
The teaching building(The Biotechnology Building) of Nanjing Forestry University
Award: the third prize of the twelfth outstanding engineering design in Jiangsu province
The laboratory building of Quanshan campus, Xuzhou Normal University
Award: The third prize of outstanding survey and design of the Urban and Rural Construction System
The gymnasium of Nanjing Audit College
Award: the third prize of outstanding engineering design in Nanjing
The residential design competition throughout the province towards the 21st century
Award: The first prize of the outstanding engineering survey and design in Anhui province
The laboratory building for graduate students, China Pharmaceutical University
The International Convention Center of China Pharmaceutical University
The new campus planning of Anhui Vocational College of Electronics & Information Technology
The information center of Anhui Vocational college of Electronics & Information Technology
The emergency command comprehensive building of Anhui environmental monitoring research
Anhui Inruilin Hotel
Maxmall of liuan city, Anhui province

Representative Works
Architectural Design Conception of Modern Museum
Originality-The Soul of Architectural Design for Modern Museum
From the Analysis of Environment
Molding architecture with the personality traits, sense of the time and individuality

北京炎黄联合国际工程设计有限公司
Beijing Yanhuang United International Engineering Co.,Ltd.

杨正光
Yeung Ching Kwong

杨正光

1966年毕业于东南大学建筑系，教授级高工，国家一级注册建筑师。
1967—1993年在中国科学院建筑设计研究院工作，曾任主任工程师，室主任等。
1987—1992年受聘清华大学建筑系兼职副教授，授建筑设计课。
1994—1995年创建人和建筑设计所，任法人代表，总经理。
1995年到中国对外建设总公司设计咨询部工作，是北京中外建建筑设计有限公司创建人之一，曾任总建筑师，董事，董事长。退休后，于2010年又担任北京炎黄联合国际工程设计有限公司工作任董事、总工程师。
40余年来，杨正光一直从事建筑创作，并在设计第一线工作，具有丰富的建筑设计经验和实践知识。设计作品涵盖科研、教育、电信、医院、剧场、宾馆等公建、古建及地产项目，曾参与建筑资料集及相关建筑规范的编写，其主持和参加设计的的获奖项目有京都信苑饭店、中国科学院外国专家招待所、望京电信局、西单电话局、碧水庄园、山水宜家、西城晶华、碛口大酒店等。

Yeung Ching Kwong

In 1966, graduated from Department of Architecture, Southeast University, professorial senior engineers, registered architect at the national level.
1967 -1993 worked at Architectural Design Institute of Chinese Academy of Sciences, formerly as chief engineer, director and so on.
In 1987 -1992 employed as part-time Associate Professor at the Department of Architecture of Tsinghua University, building design course.
In 1994 -1995, found Renhe Architectural Design Agency, legal representative, general manager.
1995 worked at CCI Architectural Co., Ltd., at design and consultation Department. He is the founder and former chief architect, director, chairman of Beijing CCI Architectural Design Co., Ltd, after retiring in 2010, reworking at Beijing Yanhuang United International Engineering Design Co., Ltd as Director and Chief Engineer.
For 40 years, Yeung Ching Kwong has been engaged in the front lines of architectural design, owning a wealth of architectural design experience and practical knowledge. Design works cover research, education, telecommunications, hospitals, theaters, hotels and other public buildings, ancient buildings and real estate projectsd. He has also participated in the editing and writing of architectural data collection and related building codes, many projects he conducted and designed has won prizes , including Beijing Xinyuan Hotel, Chinese Academy of Sciences Foreign Experts Hotel, Wangjing Telecommunication Bureau, Xidan Telephone Office, Bishui Housing, Shanshuiyijia, Xicheng Jinghua, Qikou Hotel and so on.

BIAD 北京市建筑设计研究院
BEIJING INSTITUTE OF ARCHITECTURAL DESIGN

朱小地
Xiaodi Zhu

朱小地

院长 总建筑师
国务院政府特殊津贴专家

清华大学建筑专业学士
1988年进入北京市建筑设计研究院工作
2003年以来任北京市建筑设计研究院总建筑师
2002年荣获首都劳动奖章
2004年荣获北京市五四奖章
2005年荣获北京市先进工作者称号
2005年荣获全国先进工作者荣誉称号
1995年获得中国建筑学会“青年建筑师”奖
2000年入选国务院政府特殊津贴专家

中国建筑学会常务理事
中国建筑学会建筑师分会理事
中国建筑学会建筑师分会理论与创作专业委员会副主任委员
北京地方志学会理事
北京市建筑业联合会副会长
北京节能环保促进会常务理事
北京市青年企业家协会理事

代表作品
北京市第四聋哑学校
海南寰岛（泰得）大酒店
1998年建设部部级优秀工程设计奖：建筑设计三等奖
海南博鳌金海岸大酒店
2002年全国第十届优秀工程设计奖：建筑设计银奖
南天大厦
2001年全国第十届优秀工程设计奖：建筑设计铜奖
SOHO现代城
2001年全国第十届优秀工程设计奖：建筑设计铜奖

光大花园住宅区
万科西山庭院规划
盈创大厦

Xiaodi Zhu

President, Chief Architect
experts with special government allowances by State Council

Bachelor of Architecture, Tsinghua University
In 1988, served at Beijing Architectural Design and Research Institute
Since 2003, was appointed as Chief Architect of Beijing Architectural Design and Research Institute
In 2002, won the Capital Labor Medal
In 2004, won the Youth 5.4 Medal of Beijing
In 2005, was awarded the title of Beijing Advanced Worker
In 2005, was awarded the honorary title of National Advanced Worker
in 1995, was awarded Young Architectural Guru by China Architectural Society
In 2000, was elected into the experts with special government allowances by State Council

Standing Director of Architectural Society of China
Executive director of Architects branch of Architectural Society of China
Vice committee director of Professional Theory and Creation Committee of Architects branch of Architectural Society of China
Director of Beijing Local History Society
Vice president of the Beijing Municipal Federation of Construction
Standing director of Beijing Energy Saving and Environmental Protection Association,
Director of Beijing Youth Entrepreneurs Association

Representative Works
Beijing No.4 School for deaf-mute
Hainan Huandao (Ted) Hotel
Ministerial Construction Award for Outstanding Engineering Design in 1998: third prize for Architectural Design
Hainan Boao Golden Coast Hotel
the Tenth National Excellent Engineering Design Award in 2002: Silver Award for Architectural Design
Nantian Mansion
the Tenth National Excellent Engineering Design Award in 2001: Bronze Award for Architectural Design
SOHO New Town
the Tenth National Excellent Engineering Design Award in 2001: Bronze Award for Architectural Design

Everbright Garden Estate
Planning for Vanke Xishan Garden
Yingchong Building

北京白林建筑设计咨询有限公司

Bailin Architectural Design and Consultant Corporation Ltd., Beijing

白 林
Lin Bai

白 林

1957年9月出生
1985年日本京都艺术短期大学毕业
1992年日本京都府立大学硕士学位
1996年日本京都大学工学博士学位（建筑学专业）
1994年至1996年京都大学建筑系执教（助教）
1996年至1998年京都环境建筑研究所和（株式会社）日本设计工作
1998年至2000年在清华大学建筑学院做博士后研究
2000年至今任北方交通大学教授（前建筑系主任）
2002年至今任北京白林建筑设计咨询有限公司主持建筑师

发表论著

在《日本建筑学会研究论文集》《清华大学学报》《世界建筑》等期刊杂志发表论文30余篇，出版了《中国传统民居》（在日出版）《安藤忠雄论建筑》等6部专著。博士论文《中国传统住宅空间与生活行为对应关系研究》（日文）、博士后研究报告《建筑设计方法的理论研究》受到广泛关注。

近期代表作

深圳前海地区概念规划国际咨询
泰山环山路景观带城市设计
常州市花园街城市设计
常州市武进区遥观镇剑湖区块概念规划
吴江市东太湖大道及周边区域城市设计
天津小白楼地区概念规划
四川沱牌科技研究中心规划设计
晋商银行总部大厦方案设计
太湖湾广场及北大门设计
太湖湾旅游度假区西入口区域总体规划与西大门设计
太湖湾庙堂山宕口度假公寓
新疆石河子大学校园规划
青岛奇石博物馆
唐山南湖大门设计

Lin Bai

Born in September 1957
In 1985, graduated from Kyoto Art Junior College, Japan
In 1992, Master of Kyoto Prefectural University, Japan
In 1996, Doctor of Engineering of Kyoto University, Japan (majoring in architecture)
From 1994 to 1996, Faculty of Architecture, Kyoto University (Assistant Professor)
From 1996 to 1998, Kyoto Institute of Environmental Architecture & Corporation (Designer in Japan)
From 1998 to 2000, School of Architecture of Tsinghua University (Postdoctoral Research)
From 2000 to present, Beijing Jiaotong University (Professor, formerly Head of the Department of Architecture)
From 2002 to present, Bailin Architectural Design and Consultant Corporation Ltd., Beijing (Chief Architect)

Publications

Over 30 papers published in "Proceedings of Architectural Institute of Japan ", "Journal of Tsinghua University" and "World Architecture", etc.; published 6 monographs, such as "Traditional Chinese House" (in Japan)" and "Tadao Ando on Architecture". Doctoral dissertation: "Study of the Relationship between Residential Space of Traditional Chinese House and Living Behavior" (in Japanese), post-doctoral research: "Study of the Theory of Architectural Design"

Recent Representative Works

Conceptual planning and international advisory for Qianhai Region, Shenzhen
Urban design for landscape of Huanshan Road, Taishan Mountain
Urban design for Garden Street, Changzhou City
Conceptual planning for Jianhu Lake block, Yaoguang Town, Wujin District, Changzhou City
Urban design for East Taihu (Lake) Boulevard and the surrounding area, Wujiang City
Conceptual planning for Xiaobailou Area in Tianjin
Planning and design for Tuopai Technology Research Center in Sichuan Province
Design for the Headquarters building of Jinshang Bank
Design for Taihu Bay Plaza and North Gate
Master planning for the west entrance area of Taihu Bay Resort and design for West Gate
Dangkou Resort Apartment of Miaotangshan Mountain, Taihu Bay
Planning for the campus of Shihezi University, Xinjiang
Qingdao Rocks Museum
Design for the Gate to Nanhu Lake, Tangshan

世纪安泰建筑工程设计有限公司
SJAT Architecture & Engineering Design Co.,Ltd.

王光新
Guangxin Wang

王光新

执行董事、总建筑师

1989年 毕业于北京建筑工程学院建筑系建筑学专业，并留校任教至今

1992年 北京建筑工程学院深圳设计部工作，从事建筑设计、城市规划、风景园林设计、室内设计及经营管理工作

1994年 返回北京后，继续从事建筑教育、建筑工程设计、城市规划、居住区规划工作，并担任多家房地产公司的设计顾问

2000年至今 担任北京世纪安泰建筑工程设计有限公司执行董事、总建筑师

部分代表作品

北京华澳中心写字楼
北京汇置通大厦写字楼
紫竹花园高级公寓
国展英特公寓
北京银创小区御景山项目
翡翠城一期别墅
青岛市北区政府综合培训中心
华城住宅小区
西贸中心
桂林盛景中心
大理新大陆科苑
山水文园E区
北京联盟天地广场
新天地住宅小区
唐山市机场新区项目
邮件综合处理中心
安阳新区示范引领区商业中心

Guangxin Wang

Executive Director, Chief Architect

In 1989, he was graduated from Beijing Institute of Architecture, Department of Architecture, and has been teaching there hitherto.

In 1992, he worked at design department Beijing Institute of Architecture in Shenzhen, engaged in architectural design, urban planning, landscape design, interior design and business management.

Returned to Beijing in 1994, he continue to engage in architectural education, building design, urban planning, residential planning, and shouldered as the design consultant at several real estate companies.

Since 2000, he has been served as Executive Director & Chief Architect of Beijing as Century Aetna Building Engineering Co., Ltd.

Representative Works

Beijing Chine et Macao Center Office Building
Beijing Huizhitong Office Building
Zizhu Garden Service Flat
State Fair Intel Apartments
Yu Jingshan Project Yinchuang District in Beijing
Emerald City Villa Phase I
Comprehensive Training Centre of North District Government in Qingdao
Huacheng Residential
West Trade Center
Guilin Shengjing Center
Dali New World Keyuan
Landscape Culture Garden Area E
World Square, Beijing Union
Xintiandi Residential
New District Project of Tangshan City Airport
Integrated Mail Processing Center
Business Center of Anyang new area for model leading

MCC 中冶京诚 北京京诚华宇建筑设计研究院有限公司
CERI ARCHITECTS & ENGINEERS

万晓峰
Xiaofeng Wan

万晓峰

职务
北京京诚华宇建筑设计研究院有限公司副总经理、总建筑师

简历
1966 年 7 月出生
1992 年毕于西安建筑科技大学　建筑学硕士
导师为著名建筑教育及理论家刘宝仲先生

资质
国家一级注册建筑师
国家注册城市规划师
香港建筑师学会会员
教授级高级工程师

工作经历
在中冶京诚工程技术有限公司（原北京钢铁设计研究总院民用设计院）工作期间，在多项大型公共建筑、超高层建筑、城市规划、居住区规划的方案投标和施工图工作中担任主持和领导工作，多项作品在行业内评比中获奖。

主要作品

武警特警学院	94 800 平方米
秦皇岛外国语职业学院	472 000 平方米
北京第十四中学广外新校址	35 000 平方米
宣武区体育中心游泳馆／网球馆	5 300 平方米
中冶京诚工程技术有限公司新办公区	90 000 平方米
营口沿海产业基地城市规划展览馆	4 000 平方米
武警水电指挥中心	40 000 平方米
长春“新奥蓝城”商品住宅区规划与设计	480 000 平方米
“京汉香河”商品住宅规划与设计	220 000 平方米
南京河西莲花经济适用房及中低价住宅规划与设计	920 000 平方米
南京秦淮区下码头地块危旧房改造	90 000 平方米
河南安阳“世纪城”城市结合体	40 000 平方米
邯郸市农村路住宅小区	260 000 平方米
焦作市人大会堂及会议中心	15 000 平方米
安泰科技永丰产业基地	100 000 平方米
金自天正丰台产业基地	50 000 平方米
北京火神庙国际商业中心	360 000 平方米
营口北海新区回迁安置小区	700 000 平方米
青岛胶南红树林酒店 旅游综合体	850 000 平方米
北京财富中心二期写字楼	176 000 平方米

Xiaofeng Wan

Position
Deputy General Manager and Chief Architect of CERI ARCHITECTS & ENGINEERS

Brief introduction
Mr. Wan was born, in July 1966
Graduated from Xi'an University of Architecture and Technology in 1992
He received Master's Degree of Architecture under the instruction of Liu Baozhong, a famous educator and theorist in Architecture.

Qualifications
National Class a registered architect
National Registered Urban Planner
Member of Hong Kong Institute of Architects
Professor level Senior Engineer

Performance
During working in the Capital Engineering & Research Incorporation Limited (former Beijing CERIS Civil Designing Institute), Mr. Wan serves as the project director and leader in many bidding proposals and construction drawings of the large-scale public building, mega-high-rise building, urban planning and residential district planning. A number of his designs have won prizes in the professional competitions.

Main works

Armed Special Police Academy	94,800 m^2
Qinhuangdao Vocational Colleege of Foreign Language	472,000 m^2
The new school site of Beijing No.14 middle school in Guangwai	35,000 m^2
Swimming Hall and tennis stadium of the Sports Center in Xuanwu District	15,300 m^2
New office block for Capital Engineering & Research Incorporation Limited	90,000 m^2
Urban Planning Exhibition Hall of Yingkou Coastal Industrial Base	4,000 m^2
Armed Resources Command Center	40,000 m^2
Planning and design for the commercial residential buildings"New Aolan city" of Changchun	480,000 m^2
Planning and design for the commercial residential buildings"Jing Han Xianghe"	220,000 m^2
Planning and design for the economically affordable housing and middle and low-cost housing in Hexi lotus region in Nanjing	920,000 m^2
Reconstruction of old and dilapidated buildings in Xiamatou block, Qinhuai District, Nanjing	90,000 m^2
The urban complex of "Century City" in Anyang, Henan province	40,000 m^2
Village Road residence community of Handan city	260,000 m^2
Great Hall of the NPC and Conference Centre of Jiaozuo city	15,000 m^2
Antai Technology Yongfeng Industrial Base	100,000 m^2
Aritime Fengtai Industrial Base	50,000 m^2
International business center of Temple of Fire, Beijing	360,000 m^2
Resettlement Residential Area of Beihai New District, Yingkou	700,000 m^2
The tourism complex of Mangrove Tree Hotel, Jiaonan, Qingdao	850,000 m^2
The office building (Phase II) of Beijing Fortune Plaza	176,000 m^2

CSA 中天伟业
做建筑设计专家

薛世勇
Shiyong Xue

薛世勇

中天伟业（北京）建筑设计集团董事长 总裁

沈阳建筑大学工学学士
清华大学经管学院EMBA
高级工程师

执业经历
1984–1995年 建设部建筑设计院（现中国建筑设计研究院）
1995年至今 中天伟业（北京）建筑设计集团

在建设部建筑设计院工作的11年中，与数名设计大师共同合作，设计出了众多的优秀作品。因工作出色，曾获中直机关"全国优秀青年知识分子"称号、建设部"优秀青年知识分子"称号、建设部建筑设计院优秀党员，深圳蛇口优秀团干部。

于1995年创立中天伟业（北京）建筑设计集团。目前集团具有建设部颁发的建筑工程甲级资质和工程监理资质。集团注册资金1000万，拥有150余名员工。

由于集团在绿色建筑设计领域的领先技术和设计探索，2010年被《中国企业报》誉为中国绿色建筑设计先锋，受到全国百家媒体的热捧。

主要作品
中国美术馆第一次改造（与戴念慈大师合作）
辽沈战役纪念馆（与戴念慈大师合作）
西安阿房宫宾馆（与崔恺大师合作）
深圳东方大厦
内蒙古通辽宾馆
北京通用时代国际中心
北京雅宝城商业中心

Shiyong Xue

China Sky (Beijing) Architecture Group President

Shenyang Jianzhu University bachelor of engineering
EMBA of Tsinghua SEM
Senior architect

Working experience
1984-1995 Architecture Design Institute, Ministry of Construction (China Architecture Design & Research Group)
From 1995 China Sky (Beijing) Architecture Group

With 11 years of architecture design working experience in Ministry of Construction, Shiyong Xue designed numerous excellent works in cooperation with design masters. For his outstanding achievement, he was awarded the title of National Outstanding Young Intellectual and excellent Young Intellectual of Ministry of Construction, excellent party member of Architecture Design Institute, Ministry of Construction and Model Communist Youth in Shekou of Shenzhen.
In 1995, he established China Sky(Beijing) Architecture Group, which has class A architectural engineering qualification issued by Ministry of Construction and Project supervision qualification. The group is of 10,000,000 registered capitals and over 150 staffs.
Due to the advanced technology and design exploration the group adopted in green architecture design area, China Sky (Beijing) Architecture Group was entitled as the pioneer of Chinese green architecture design by China enterprise news, gaining praises from a great many mass medias.

Major works
The first innovation of National Art Museum of China (cooperated with master Nianci Dai)
Memorial Hall of Liaoshen Campaign (cooperated with master Nianci Dai)
Hyatt Regency Hotel Xi'an (cooperated with Master Cui Kai)
Oriental Mansion in Shenzhen
Tongliao Hotel in Inner Mongolia
GM International Center in Beijing
Yabaocheng Commercial Center in Beijing

北京清城华筑建筑设计研究院有限公司
THCA Architectural Design & Consulting Institute
（北京清华城市规划设计研究院建筑分院）
Beijing Tsinghua Urban Planning & Design Institute Architectural Department

袁 昕
xin Yuan

袁 昕

北京清华城市规划设计研究院 副院长
北京清城华筑建筑设计研究院有限公司 院长
清华大学工学博士
美国麻省理工学院访问学者
高级工程师
国家注册规划师

工作业绩
北京国子监历史文化保护区保护规划研究
北京中关村西区详细规划
北京国际展览体育中心规划设计
北京奥林匹克公园规划设计
北京中轴线城市设计
北京朝阜大街城市设计
天津市西青区杨柳青镇控制性详细规划
天津市东丽区城乡空间发展战略研究
南宁市相思湖新区规划设计
汶川灾后重建总体规划
南宁市城市总体规划（2004–2020年）
银川市城市总体规划（2005–2020年）
内蒙古呼包鄂城镇群规划等

Xin Yuan

Beijing Tsinghua Urban Planning & Design Institute Co.,Ltd. Vice President
Beijing Qingchenghuazhu Architectural Design & Research Institute Co.,Ltd President
Doctor of Technical Science in Tsinghua University
Visiting scholar of Massachusetts Institute of Technology
Senior engineer
National registered planner

Research Projects
Beijing Guozijian historical and cultural preservation planning research
Regulatory plan of the west part of Zhongguancun area in Beijing
Planning and design of Beijing International Exhibition & Sport Center
Planning and design of Olympic Park
Urban planning for the axis of Beijing city
Urban planning of Chaofu street
Regulatory plan of the Yangliuqing Town ,Xiqing district in Tianjin
Research on the urban and rural spacial development strategy of Dongli District in Tianjin
Planning and design of the Xiangsi Lake new district in Nanning
Planning of the Wenchuan post-disaster reconstruction
Urban planning of the Nanning City (2004-2020)
Urban planning of the Yinchan City (2005-2020)
Planning of the town group of Hohhot, Baotou, Erdos in Inner Mongolian

广西华蓝设计（集团）有限公司
Guangxi Hualan Design & Consulting Group
（原广西建筑综合设计研究院）

蒋伯宁
Boning Jiang

蒋伯宁

广西华蓝设计（集团）有限公司总建筑师。1968年毕业于华南工学院建筑学专业，教授级高级建筑师，国家一级注册建筑师，中国建筑学会理事、广西建筑学会副理事长、香港建筑师学会法定会员，建设部历史文化名城专家委员会委员，中国建筑学会资深会员，广西十佳建筑设计师。

主要建筑作品

广西·南宁荔园山庄（建国六十周年中国建筑学会建筑创作大奖、建设部优秀建筑设计三等奖、第四届中国建筑学会建筑创作佳作奖、全国人居经典建筑规划设计方案竞赛 建筑、环境双金奖、广西优秀工程设计一等奖）、南宁国际会展中心一期工程（建国六十周年中国建筑学会建筑创作大奖、第五届詹天佑土木工程大奖、广西优秀工程设计一等奖）、梧州市骑楼城保护与利用规划（建设部优秀规划设计二等奖、广西优秀城乡规划设计一等奖）、广西区人民医院门诊综合楼（设部优秀勘察设计三等奖、广西优秀工程设计一等奖）、南宁明园五号楼（建设部优秀勘察设计奖三等奖、广西优秀工程设计一等奖）、广西医科大一附院内科大楼（全国优秀工程勘察设计行业奖建筑工程三等奖、广西优秀工程设计一等奖、全国医院建筑优秀设计奖三等奖）、广西人民大会堂（广西优秀工程设计一等奖）、广西南宁西园饭店十号楼（建设部表扬奖、广西优秀工程设计一等奖）、广西电力综合楼（建设部优秀设计表扬奖、广西优秀工程设计一等奖）、百里柳江景观控制性规划（广西优秀规划设计二等奖）。

主要著作

《现代建筑技术的艺术表现》

发表数十篇的论文，主要有《现代建筑地域性、乡土建筑时代性：南宁荔园山庄建筑与环境设计》《还自然于本色：记广西民族博物馆中标方案创作》《重历史文脉，塑造有地域文化特色的建筑个性——南宁明园五号楼设计》《延续、重塑与共生——浅谈建筑设计重的环境塑造》《新需求、新课题、新探索——现代化大型集中式综合医院的建筑设计研究》等。

Boning Jiang

Boning Jiang is the chief architect of Guangxi Hualan Design & Consulting Group. He graduated from South China Science and Technology College, majoring in architecture. Being a senior architect with professorship and Class A registered architect qualification, he is also the director and senior member of Architectural Society of China, Vice Director of Guangxi architect association ,member of Hong Kong Architect's Academy and famous historic and cultural city expert committee of MOHURD,ranking among the top ten excellent architects .

Major works

Liyuan Villa in Nanning, Guangxi(it has won the Architecture Design Award of Architectural Society of China on the 60th Anniversary of Founding of PRC, the Third Prize of Excellent Architecture Design of Ministry of Construction, Excellent Works Awards of the Fourth Architecture Design Competition of Architectural Society of China, double Gold Prizes of Architecture and Environment Design in Architecural Planning Deisgn Proposal Competition of national classic inhabits housing.,the First Prize of Excellent Project Design of Guanxi Province. Phase Ⅰ Project of Nanning International Convention and Exhibition Centre (Architecture Design Award of Architectural Society of China of the Sixtieth Anniversary of the founding, the Fifth Tianyou Zhan Civil Engineering Award, the First Prize for Excellent Engineering Design in Guangxi), Planning for protection and utilization of Qilou town in Wuzhou City (the Second Prize for Outstanding Planning Design for Ministry of Construction, the First Prize of Excellent Urban and Rural Planning and Design in Guangxi), Outpatient Complex Building of People's Hospital in Guangxi Region (the third prize for Excellent Survey and Design of Ministry of Construction, the First Prize for Excellent Engineering Design in Guangxi), Building 5 of Nanning Ming Park (the Third Prize for Excellent Survey and Design of Ministry of Construction, the First Prize for Excellent Engineering Design in Guangxi), Internal Medicine building of the first Affiliated Hospital of Guangxi Medical University (the Third Prize of Construction Engineering for the Survey and Design Award of national excellent engineering, the First Prize for Excellent Engineering Design in Guangxi, the Third Prize of Excellent Design Award of the National Hospital Buildings), the Great Hall of Guangxi (the First Prize for Excellent Engineering Design in Guangxi), Building 10 of Guangxi Nanning West Park Hotel (Recognition Awards of Ministry of Construction, the First Prize for Excellent Engineering Design in Guangxi), Guangxi Power Plaza (Recognition Awards of Ministry of Construction for outstanding design, the First Prize for Excellent Engineering Design in Guangxi), Baili Liujiang landscape control planning (the Second Prize of Excellent Planning and Design in Guangxi).

Major works

Art of Modern Architectural Technique

He has published dozens of papers including: *Region of Modern Architecture and theTimes of Vernacular Architecture: Architecture and Environmental Design of Nanning Liyuan Villa*; *Returning Characters to Nature: Bid Winning Program Creation of Guangxi National Museum*; *Value Historical Context, Create Building Features with Regional Cultural Characteristics – the Design of Nanning Ming Park Building 5*; *Extension, Remodeling and Coexistence – Discussing the Environment Creation of Architectural Design*; *New Needs, New Issues, New Exploration – Architectural Design Research of Modern Large Centralized general hospital*, etc.

日都（北京）高新建筑科技设计咨询有限公司
有限会社AGC都市设计

陈俊菁
Junjing Chen

陈俊菁

日本注册一级建筑师
1984年 获清华大学（建筑系）工学学士学位
1992年 获东京艺术大学美学修士（硕士）学位

在日本就职于安井建筑设计事务所，任主任建筑师。主持设计了伊藤洋华堂商场、光村印刷大楼、中央自动车大楼、琦玉中央医院、东洋信托银行支行、国民金融公库支行、东川口站等众多项目。现任AGC都市设计代表取缔役所长、日都（北京）总经理、安井事务所北京代表。

在中国主持设计的北京证券广场曾获1994年北京市民投票十佳建筑设计之第五位

沈阳航天三菱办公楼曾获2007年东三省金奖工程

北京金台夕照会馆曾获2008年全国最佳主题酒店

归国数年中在珠海新城、天津南市、北京北七家全国小城镇商务中心、东升华庭水乡城、长沙橘郡、桂林桃花江旅游区、重庆北碚十里温泉城、珠海动漫硅谷等十余项大型项目的策划与规划中游刃有余。

尤其在温泉设计方面颇具专长，设计结合国情并引入日式温泉的理念与技术。1997年主持设计的广东清新温矿泉度假区为中国温泉设计提供新的思路，随后主持设计的福州青云山御温泉、贵阳天邑温泉等先后开业，获得客户的好评。现正着手策划与设计北京南面的大千森林温泉度假区。

Junjing Chen

Japan Class I registered architect

In 1984, received Bachelor's Degree in Engineer in Department Architecture, Tsinghua University

In 1992, received Master's Degree in Arts Aesthetics in Tokyo University

In Japan he worked as director architect in Yasui Architectural Design Firm, and hosted for designing a large number of projects including Ito-Yokado shopping mall, Mitsumura printing building, Central Motor Building, Saitama Central Hospital, Toyo Trust Bank Branch, Branch of the National Finance Corporation and East Kawaguchi Station.

Currently he works as director of AGC Urban Design, General Manager of Ridu (Beijing) and Beijing representative of Yasui Architectural Design Firm.

In China, he designed many projects, among which Beijing Securities Square ranked the fifth in Top Ten excellent architectural design voted by Beijing citizen in 1994

Office Building of Shenyang Aerospace Mitsubishi was awarded the Golden Project of Three Northeast Provinces in 2007

Beijing King-Talent Hotel was awarded the best national theme hotel in 2008

Since he returned to China, he has been very skilled in planning design for large-scale projects such as New Zhuhai City, Tianjin South City, Beijing, Beiqijia National small cities and towns Business Center in Beijing, Dongsheng Huating River City, Orange County in Changsha, Peach River tourism region in Guilin, Shili Spa City in Beipei District in Chongqing and Cartoon Silicon Valley in Zhuhai.

Especially on hot spring design, he introduced concept and technology of Japanese-style spa combining the national conditions.Qingxin spa resort in Guangdong, design by him in 1997, has brought great changes in China's hot spring, and then Qingyun Mountain Royal Hot Spring in Fuzhou and Tianyi Hot Spring in Guiyang set up business in succession, and were greatly appreciated by customers. Now he is busy with planning design for Daqian forest hot spring in southern Beijing.

雅克设计机构
APCE DESIGN GROUP
ARCHITECTURE · PLANNING · CONSULTANTS · ENGINEERING

侯百镇
Baizhen Hou

侯百镇

研究员
国家一级注册建筑师
国家注册城市规划师
香港建筑师学会会员
1989年获重庆大学建筑工程学院建筑设计硕士学位
2005年获同济大学城市规划博士学位
2007年获新加坡国立大学工商管理专业硕士（EMBA）学位

现任
雅克设计机构董事、总经理
《规划师》杂志社社长
海南省土木建筑学会秘书长
海南省四届人大常委会环境资源工作委员会委员

20年来，侯先生在建筑设计、城市规划领域不断研究、探索和实践，先后发表了《关于城市设计问题的思考与实践》《海口旧城中心区历史文化价值和传统商业特色保护》《城市转型：周期、战略与模式》《建立在统计分析基础上的商业步行街规划设计模式》《转型与城市及发展》等十余篇专业论文和研究报告，并主编出版《雅克印记》。

其主持的海口市机场中心区城市设计、三亚城乡统筹概念规划、海口市滨海高层区城市设计、海口市城市总体规划纲要、桂林正阳步行街规划设计、湖州太湖度假区规划等项目多次荣获国际、全国、省级竞赛大奖。

1992年创建雅克设计机构，成为全国第一家被授予城乡规划和建筑设计甲级资质的民营设计企业。机构主要业务领域包括城市总体规划、城市中心区规划、旧城改造、滨水地区规划、城市广场及景观规划、街道及商业区设计、社区及居住区规划、旅游度假区规划、公共建筑设计、酒店建筑设计、住宅建筑设计、市政设计、室内设计、园林设计及工程项目可行性研究等。机构拥有500余位具有职业素养、创新精神和务实经验的专业人士，秉承“设计创造价值”的公司理念，对每一项委托都从构思、技术、资源、经济及实施等方面精心研究，为达到目标并肩合作，全力以赴，十七年来累计完成各类设计委托2100余项，赢得了客户和社会的广泛赞誉。

Baizhen Hou

Researcher
National Class 1 registered architect
National registered urban planner
Member of Hong Kong Institute of Architects (HKIA)
In 1989, received Master of Architectural Design in Institute of Civil Engineering and Architecture, Chongqing University
In 2005, received Doctor's Degree on Urban Planning at Tongji University
In 2007, received Master of Business Administration (EMBA) in National University of Singapore

Current position
Director and general manager of APCE Design Group
President of "Planner" Magazine
Secretary general of Hainan Civil Engineering Institute
Member of the Environmental Resources Committee, the 4th Hainan Standing Committee

Having researched, practised and explored on architectural design and urban planning for 20 years, he has in succession published more than ten professional papers and research reports, including "Reflection on the Practice of Urban Design", "the Conservation of History Cultural Value and Traditional Commercial Features in Old City's Central Area in Haikou", "City Transition: Cycle, Strategy and Model", "Plan and Design Mode for Pedestrian Street on the Basis of Statistic Analysis" and "Transformation and Urban Development". He has also edited and published "APCE Impression".
Projects hosted by him such as urban design for central area of Haikou Airport, concept plan for urban and rural coordinating in Sanya, urban design for Binhai High-rise area in Haikou, general planning framework for Haikou city, planning design for Zhengyang pedestrian street in Guilin and Taihu Lake Holiday Village in Hukou have been awarded prizes in international, national and provincial level.
Founded in 1992, APCE Design Group was qualified with class-A urban and rural plan as well as architectural design certification as the first private enterprise in our country. Its lines of business cover a variety of fields including general urban planning, urban central area planning, urban renewal, waterfront planning, urban square and landscape planning, design for streets and commercial areas, community and residential area planning, tourism resort planning, public buildings design, hotels architectural design, residential architectural design, urban design, interior design, garden design and project feasibility research. The group has more than 500 professional personnel with good work ethic, innovative spirit and practical experience. Adhering to the enterprise culture "Design Creates Value", the group focuses on each commission on aspects of idea, technology, resource, economy and implementation, and makes every effort to achieve the goal. For 17 years the group has completed more than 2,100 design commissions of all types, and has won a wide range of praises from clients and the society.

方略建筑设计有限责任公司
Strategy Architectural Design Co.Ltd

张　诣
Yi Zhang

张　诣

1987年	毕业于重庆建筑大学建筑学专业
1987–1988年	中国建筑工业出版社
1989–1992年	中国建筑学会设计所
1992–1993年	利尔企业集团公司
1994–1999年	中建科产业有限公司
1999至今	方略建筑设计有限责任公司（原九三建筑设计研究院）

作品和荣誉

1987年获英国皇家艺术学院世界大学生设计竞赛“New in the Old”入选奖
1992年获联合国人居中心、教科文组织举办的世界注册建筑师设计竞赛“CINTUS–II”“亚洲奖”
国家小康示范工程《锦城园》居住小区项目甲方代表
国家小康示范工程《龙湖花园》居住小区项目甲方代表
北京“长安新城”居住小区规划策划
武汉“锦湖花园”居住小区规划策划
北京“北京印象”居住小区规划策划
北京香江商业广场策划
广州广州纺织中心策划
《世界建筑》杂志社特聘编委

Yi Zhang

1987, graduated from School of Architecture, Chongqing University
1987-1988, China Building Industry Press
1989-1992, Design of the Architectural Society of China
1992-1993, Lier Enterprise Group company
1994-1999, China Construction Science and Technology Industrial Co., Ltd
1999 – Present, Strategy Architectural Design Co., Ltd. (former Jiusan Architectural Design and Research Institute)

Works and honors

•1987, won "New in the Old" Symbol Of Excellence in World University Design Competition of the Britain Royal College of Art
•1992, won "CINTUS-II", "Asia Award" in the registered architect design competition of the world organized by the UN Habitat, UNESCO
•Party A representative of *Glorious Garden* residential area project of state well-off demonstration project
•Party A representative of "*Dragon Lake Garden*" Residential area Project of State Well-off Demonstration Project State Well demonstration project
•Urban Planning and Architectural Design for Beijing "Chang'an New Town" residential area
•Urban Planning and Architectural Design for Wuhan "Kumho garden" residential area
•Urban Planning and Architectural Design for Beijing "Beijing Impression" residential area
•Urban Planning and Architectural Design for Beijing "Hong Kong Commercial Plaza" planning
•Urban Planning and Architectural Design for Guangzhou "Guangzhou Textile Center" planning
• *International Architecture* invited editorial board member

北京市古代建築設計研究所
TCBDRI
BEIJING TRADITIONAL CHINESE ARCHITECTURAL DESIGN AND RESEARCH INSTITUTE

马炳坚
Bingjian Ma

马炳坚

著名古建筑专家
注册建筑师　高级工程师

现任职务

北京市古代建筑设计研究所所长
《古建园林技术》杂志主编
中国建筑学会建筑史学分会学术委员
中国紫禁城学会常务理事
中国民族建筑研究会副会长
中国民族建筑风格艺术专业委员会主任委员
中国文物学会传统建筑园林委员会常务理事

马炳坚从事中国古建筑施工、设计、研究、教学、办刊40余年，业绩显著，著作颇丰。

1983年与同仁发起创办《古建园林技术》杂志并在其中担任重要工作。

1985年参与创办大学古建筑工程专业并担任教学工作同时在其中担任重要工作。

1990年以后重点转入古建筑工程设计，为继承、弘扬中华传统建筑文化做出了突出贡献。

他所设计的古建园林工程遍布海内外，深受各界人士好评。曾参与天安门城楼翻建和天坛祈年殿大修工程，以及北海、西苑（中南海）等皇家园林的修建工程。1984年受北京市政府委托，随团赴美国进行建造华盛顿中国城牌楼的现场考察，回国后进行建造华盛顿牌楼设计，并于1986年赴美参加牌楼施工。

其代表著作《中国古建筑木作营造技术》被海内外学者誉为"近代对中国古建筑最有分量的书"，"是从事祖国古建筑维修保护工作者必读之书"，并多次获奖；著作《北京四合院建筑》，从北京所经历的建国初期所有制变更、由于发展生产需要而对历史建筑的扰动、文化大革命极左思潮对传统建筑及其文化的浩劫，以及改革开放初期城市开发建设对胡同四合院的破坏说起，论述了20世纪90年代初以来北京四合院的历史价值、文化价值、艺术价值，并采取一系列措施加强了对北京四合院及其历史文化保护区的保护工作。他在《北京规划建设》杂志上发表的论文：《对三眼井地区保护、更新、改造规划的几点意见》《北京四合院保护理念及政策方法问题的探讨》以及《北京四合院工程十滥》等，对北京四合院保护建设提出了非常中肯而实际的意见。

Bingjian Ma

Prominent ancient architecture expert
Registered architect　Senior engineer

Post

President of Chinese ancient architecture design research institute
Chief Editor of Traditional Chinese Architecture and Gardens magazine
Director of Chinese Society of Architectural History Academic committee
Director of the Forbidden City academy
Vice chairman of National Architecture Institute of China
Director of China National Architectural Style Art Commission
Director of wenwuchina traditional architecture landscape committee

Bingjian Ma has engaged in ancient architecture engineering, deisign, research, teaching and publication for over 40 years, made great achievement and contributed many professional works.

In 1983 he established the Traditional Chinese Architecture and Gardens magazine with other colleagues and undertook the major teaching works.

1985 He helped with the foundation of ancient architecture engineering course .

From 1990 he mainly worked on the design of ancient architecture projects, made prominent contribution to succession and promotion of Chinese traditional architecture culture.

His ancient architecture works spread all over the world and is much popularized. He has participated in the renovation projects of Tiananmen gate tower and the Hall of prayer for good harvest in Temple of heaven, and the projects of royal gardens such as Beihai, Xiyuan(Zhongnanhai). In 1984, authorized by Beijing government, he went to the U.S. for site inspection of Chinese decorated archway construction in Washington D.C, after designed it in China; he went to the U.S. in 1986 for construction.

His representative works carpentry project building technique of Chinese ancient architecture is mentioned as the most significant Chinese ancient architecture book in modern times, and a book that all the people who works for the reparation and protection of Chinese ancient architecture have to read, winning many awards. His book Chinese traditional rectangular courtyard in Beijing talked about the disturbance and catastrophe that the Great Cultural Revolution, the ultra-left political movements and other social activities brought to the traditional architecture and the rectangular courtyards in Beijing. He analyzed the historical value, culture value, artistic value of the rectangular courtyards and the measures took to for protection and preservation from 1990s. The thesis he published in Beijing City Planning & Construction Review include: Suggestions for protection , renovation , reconstruction of Sanyanjing district, discussion of Protection ideas and policies for the rectangular courtyards in Beijing and Ten misuse in the rectangular courtyards projects in Beijing. In these articles, he offered very earnest and practical advice in Beijing rectangular courtyards construction.

合道设计集团

HORDOR DESIGN GROUP

黄　琰

Yan Huang

黄　琰

副董事长　副总经理　总建筑师
厦门城市规划委员会委员
福建省勘察设计协会常务理事
福建省土木建筑学会建筑师分会副会长
厦门市规划学会副会长
厦门市鼓浪屿风貌建筑保护委员会委员
厦门市规划局规划项目评审专家库成员
入选厦门市第六批拔尖人才

学习经历

1986–1990年　同济大学　建筑系本科
　　工学学士学立，厦门大学MBA研习班
2004年至今　同济大学　建筑与规划学院　工程硕士

工作经历

1990–2003年　厦门市建筑设计院　建筑师
2003–2007年　厦门市建筑设计院有限公司　总建筑师
2007年至今　厦门合道工程设计集团有限公司　副董事长
　　副总经理　总建筑师

主要工程经验

政府、学校、会展
厦门国际会展中心（竣工）2002年省优一等奖
厦门国家会计学院（竣工）2006年省优一等奖、2005年省土建"第一届建筑创作优秀奖"
福州大学新校区（竣工）2005年省土木建筑学会"第一届建筑创作佳作奖"
集美大学嘉庚教学楼群2006年省土建"第二届建筑创作佳作奖"
福建省漳州行政中心（实施方案）
福建省漳州会展中心（一等奖方案）
福建省龙海交通大楼2003年省优三等奖
厦门市海沧区行政服务中心（竣工）2006年省优二等奖
厦门市思明区检察院（竣工）
厦门市高级技工学校新校区（在建）
江西省赣江职业技术学院新校区国际交流中心
江西省贵溪市人民法院审判庭办公楼
江西省贵溪市人民检察院技侦大楼
云南省曲靖市文化中心
商业、办公、酒店
厦门市中山路商业街金同成（竣工）2003年省优三等奖
厦门市中山路商业街达意商业城（竣工）
厦门市鼓浪屿三友百货（竣工）
厦门大西洋海景城（竣工）
厦门港务大厦（在建）
厦门航空港酒店（在建）
风貌建筑保护
厦门市鼓浪屿[海天堂构]修复改造　2008年省土木学会"第三届建筑创作奖优秀奖"
厦门市鼓浪屿[林氏府]修复改造
住宅
厦门市金利·兰湾
厦门市海晟·维多利亚
厦门市联发·五缘湾一号
莆田市万辉世纪城
龙岩市水木莲花

主要学术活动

厦门市土木学会与台湾建筑师公会作海峡两岸学术交流
受美国SWA景观规划公司邀请，作《海沧新景城景观规划》研讨WA（世界建筑）苏州论坛—博物馆与城市
厦门房地产协会高尔夫社区论坛—厦门高尔夫现象
福建省土木学会建筑师分会年会作《项目一体化运作》专题发言，参加福建省土木学会建筑师分会年会优秀建筑创作奖评选

Yan Huang

Vice Chairman of the Board　Deputy General Manager　Chief Architect
Committee member of Xiamen City Planning Board
Standing director of Fujian Provincial Survey and Design Association
Vice President of Architects Branch of Fujian Provincial Institute of Civil Engineering
Vice President of Xiamen Planning Institute
Member of Xiamen Gulangyu Building Style Protection Committee
Planning project evaluation expert of Xiamen Municipal Planning Bureau
Selected as the sixth top-notch talent in Xiamen

Education background

1986-1990　Department of Architecture in Tongji University
　　Bachelor of Engineering, MBA Course in Xiamen University
Since 2004,　School of Architecture and Planning in Tongji University, Master of Engineering

Work experiences

1990-2003　Architect of Xiamen Municipal Architectural Design Institute
2003-2007　Chief Architect of Xiamen Municipal Architectural Design Institute Co., Ltd.
Since 2007,　Vice Chairman, Deputy General Manager, Chief Architect of Xiamen Hordor Design Group Co., Ltd.

Main projects and awards

Government, schools, exhibition halls
Xiamen International Exhibition Centre (completed) which was awarded the first prize of the province in 2002
Xiamen National Accounting Institute (completed) which was awarded the first prize of the province in 2006 and, "the first Architectural Award of Excellence" of provincial civil engineering in 2005
Fuzhou University new campus (completed) which was awarded "the first Architectural Works Award" of Provincial Civil Engineering Institute in 2005
Jimei University Jiageng Teaching Buildings which was awarded "the Second Architectural Works Award of Excellent" of provincial civil engineering in 2006
Fujian Zhangzhou Administrative Center (implementation plan)
Fujian Zhangzhou Exhibition Center (first prize program)
Fujian Longhai Transportation Building, which was awarded the provincial superior third prize in 2003
Xiamen Haicang District Administrative Service Center (completed) which was awarded the provincial superior second prize in 2006
Xiamen Siming District Procuratorate (completed)
New Campus of Xiamen Senior Technician School (under construction)
International Exchange center of Ganjiang Vocational and Technical College New Campus in Jiangxi Province
Trial Division office building of Guixi People's Court in Jiangxi Province
Technical Investigation Building of Guixi People's Procuratorate in Jiangxi Province
Cultural Center of Qujing in Yunnan Province
Commercial buildings, office buildings, hotels
Jin Tongcheng building in Xiamen Zhongshan Road Commercial Street in Xiamen (completed) which was awarded the provincial superior third prize in 2003
Dayi Commercial center of Zhongshan Road Commercial Street in Xiamen (completed)
Sanyou department store in Gulangyu,Xiamen (completed)
Xiamen Atlantic Seascape City (completed)
Xiamen Port administrative affairs Tower (under construction)
Xiamen Airport Hotel (under construction)
Style building protection
Renovating and rebuilding of Xiamen Gulangyu [Sea Paradise Structure] which was awarded "the Third Architecture Design Award for Excellence" of province civil construction society in 2008
Xiamen Gulangyu [Lin's House] fixed renovating
Housing
Xiamen [Kim Lee • Blue Bay]
Xiamen [Haisheng • Victoria]
Xiamen [Lianfa• No. 1 of Five edge Bay]
Putian [Wanhui Century City]
Longyan [Watering Lotus]

Main Academic Activities

Cross-Strait Academic Exchange of Xiamen Civil Engineering Society and Taiwan Architects Association
Invited by U.S. SWA Landscape planning company; made a speech on "Haicang New View City Landscape Planning" on WA (World Architecture) Suzhou Forum – Museum and the City
Real Estate Institute of Xiamen Golf Community Forum -- Xiamen golf phenomenon
Given a speech on "Integrated operation of the project" on the annual meeting of Architects Branch of Fujian Provincial Institute of Civil Engineering and was selected to take part in the election of Outstanding Architectural Design Award of Architects Branch of Fujian Provincial Institute of Civil Engineering

张建新
Jianxin Zhang

张建新

男　汉族　中共党员

毕业于兰州铁道学院工民建专业
1974年5月参加工作 高级工程师
现任兰州城市建设设计院院长，党委书记

积极参加社会活动和学术活动，首批入选兰州市跨世纪学术带头人“222工程”，市级专业技术拔尖人才。
被聘为甘肃省工程力学、计算力学、实验力学三专业联合委员会委员。
兼任甘肃省科技专家委员会委员；
省土木建筑学会建筑结构学术委员会委员；
省抗震防害协会常务理事；
中国勘察设计协会建筑分会理事；
中国勘察设计协会市政分会理事；
西北民族大学兼职教授。

在专业技术方面努力实践、大胆创新，具有扎实的理论功底和较强的项目组织管理能力，积累了丰富的工作经验，尤其在组织、主持重点项目、高层建筑设计及推广新技术、新材料、资源综合利用、科技进步、生产经营方面为我院的发展作出了巨大的贡献。

自张先生担任院级领导以来，狠抓生产经营，拓展市场空间，推行品牌战略，推进改企建制，全面加强企业文化建设和精神文明建设，各项工作取得了显著成绩。在他的领导和组织下，兰州市城市建设设计院的勘察设计甲级资质已达到15个，经济效益不断增长，精神文明建设也收获了累累硕果，张先生为设计院物质文明和精神文明建设作出了突出贡献。

Jianxin Zhang

Male　Han nationality　Party members

He graduated from Lanzhou Railway Institute, majored in Industrial and Civil Construction projects
In May, 1974 he began to work as a senior engineer, and currently
he is the president of Lanzhou Urban Construction Design Institute and party secretary

As one of the first selected cross-century academic leaders of “222 project” in Lanzhou city and municipal professional top-notch talents, he has always actively participated in social and academic activities. He is employed as the Joint Committee members of engineering mechanics, computational mechanics, and experimental mechanics.
And he also holds concurrent posts which include member of Expert Committee on Science and Technology
Gansu Province member of Civil Engineering Academic Committee of the Provincial Institute of Building Structures
The executive director of provincial Anti-seismic and Damage Prevention Association
Branch director of China Construction Exploration and Design association
Director of the municipal branch of China Exploration and Design association
Part-time professor of Northwest University for Nationalities.
In technical terms, he has a down-to-earth attitude to practice, make bold innovations, a solid theoretical foundation and strong capacity in project organization and management, accumulating plenty of work experience. And he has made great achievements for the development of our institute, especially in organization and presiding over key projects, high-rise building design, promoting new technology and new materials, spreading the comprehensive utilization of resources, pushing forward scientific and technological progresses, and improving production and management.
Since assumed the leading position of our institution, he has focused on production and management, expanding market space, implementing brand strategy, and promoting enterprise reform. By comprehensively strengthening the building of enterprise culture and spiritual civilization construction, every aspect of the work has made significant achievements. Under his leadership and management, the Grade A in exploration and design has reached 15 in our institution. At the same time, economic benefits keep growing, and spiritual civilization also obtained fruitful results. He did make remarkable contributions to the material and spiritual civilization for our institution.

机械工业第六设计研究院
NO.6 INSTITUTE OF PROJECT PLANNING & RESEARCH OF MACHINERY INDUSTRY

赵景孔
Jingkong Zhao

赵景孔

男，汉族，1953年8月生，1971年2月参加工作，1973年10月加入中国共产党，1977年2月毕业于南京工学院，就职于机械工业第六设计研究院（以下简称中机六院），2002年毕业于中央党校研究生班。历任技术员、工程师、高级工程师、室主任、副院长、党支部书记、院党委委员等职，1992年5月任副院长，1993年9月任院长，1994年4月任党委书记。原机械工业部中青年专家，享受国家政府特殊津贴的研究员级高级工程师，国家注册咨询工程师，国家一级注册建造师，荣获河南省劳动模范，河南省勘察设计行业先进管理者、优秀院长，中国勘察设计优秀企业家等诸多荣誉称号。

赵景孔同志现任中机六院院长兼党委书记、中国机械工业勘察设计协会理事、中国机械工程学会环保分会副主任委员、河南省勘察设计协会常务理事、河南省二十一世纪城市发展战略规划研究中心特约研究员、合肥工业大学兼职教授。
主要学术成果
1. 承担的上海机床公司利用世界银行贷款技术改造项目获1994年机械工业部科技进步三等奖，1995年机械工业部优秀工程勘察设计二等奖；
2. 承担的银川机床铸造厂硅铁扩建二期工程获1992年国家优秀工程设计银质奖；
3. 承担的上海机床厂利用银行贷款技术改造项目获1993机械工业部科技进步二等奖、1995年机械工业部优秀工程勘察设计一等奖；
4. 在我国技术刊物上发表《铸造工程总承包的发展趋势》等多篇铸造技术研究和行业发展趋势研究论文。

Jingkong Zhao

Jingkong Zhao: male, born in Aug. 1953. He began to work in Feb. 1971, joined the Communist Party in Oct. 1973, and graduated from Nanjing College of Technology in Feb. 1977 & from Party School of the CPC Central Committee as master in 2002. He has worked successively as technician, engineer, senior engineer, section chief, vice president and party branch secretary, and worked as Vice President of No.6 Institute of Project Planning & Research of Machinery Industry in May 1992, President in Sep. 1993 and Secretary of the Party committee in Apr. 1994. Besides, he is the expert of the former Ministry of Machinery Industry and senior engineer at researcher level enjoying special allowance from State Council, national certified consultation engineer, national Class I certified construction engineer, Labor Model of Henan Province, outstanding president & administrator of design & research association of Henan Province and Outstanding Entrepreneur of China Survey & Design Industry.
Mr. Zhao is the President & Secretary of the Party Committee of No.6 Institute of Project Planning & Research of Machinery Industry, senior engineer at researcher level, national certified consultation engineer, national Class I certified construction engineer, director of Survey & Design Association of China Machinery Industry, vice director commissioner of Environmental Protection Committee, standing director of Henan Survey & Design Association, Researcher of Henan 21st Century City Development Tragedy Strategy Research Center & part-time professor of Hefei University of Technology.
Main Academic Achievements:
1. Technical Innovation Project of Shanghai Machine Tool Company by World Bank Loan won 3rd Prize of Sci-tech Improvement of Machinery Industry in 1994 and 2nd Prize of Excellent Project Survey & Design of Machinery Industry in 1995;
2. Silicoferrite Expansion Project Phase II of Yinchuan Machine Tool Casting Company won Silver Prize of National Excellent Project Design in 1992;
3. Technical Innovation Project of Shanghai Machine Tool Company by World Bank Loan won 2nd Prize of Scien-tech Improvement of Machinery Industry in 1993 and 1st Prize of Excellent Project Survey & Design of Machinery Industry;
4. Has published many research essays on casting technology research on national technical magazines, such as Development Trend of General Contract of Casting Projects etc.

汉米敦 HMD

李剑波
Jianbo Li

李剑波

资深设计师　董事
国籍：中国　出生日期：1969 年

学历及资格
1991 年湖南大学建筑系学士
中国一级注册建筑师

语言
汉语 – 母语　英语 – 流利　粤语 – 流利

曾经工作的国家和地区
中国、英国

个人履历
李剑波具有 20 年的建筑设计及规划设计的工作经验：作为资深规划师，李剑波在城市设计、大型综合居住社区及城市综合体的规划设计上具有丰富的经验和能力；作为资深建筑师及项目经理，在居住、办公、酒店、教育、文化等各类建筑设计及项目管理方面也有着丰富的经验和能力；作为建筑部门经理，也同时具备较强的团队组织及业务管理能力。

现任工作
2009年至今　汉米敦中国　资深设计师／董事

曾任工作
2008–2009 年　中深瑞城规划建筑设计有限公司
设计总监／首席设计师
2004–2007 年　阿特金斯顾问（深圳）公司
建筑部经理／主任建筑师
2001–2004 年　阿特金斯顾问（深圳）公司
高级建筑师及项目经理
1991–2000 年　广东省中山市建筑院设计有限公司
合伙人及主任建筑师

Jianbo Li

Senior designer　Director
Nationality: China　Date of birth: 1969

Education and Qualifications
1991 Bachelor of Architecture in Hunan University
Class-A Registered Architect of China

Language
Language - mother tongue　English - fluent
Cantonese - fluent

Countries and regions where he has worked
China, United Kingdom

Profile
Mr. Li has nearly two decades' working experience in architectural planning and design. As a senior designer, he has accumulated a wealth of experience and ability in urban design large-scale integrated residential communities and the planning and design for urban complex. As a senior architect and project manager, he is specialized in the construction design and project management in the fields of residential quarters, office buildings, hotels, educational and cultural constructions, etc. As the manager of the construction sector, he also has a strong capacity in team organization and business management.

Current position
Senior designer / Director in Hanmi Dun•China since 2009

Previous position
2008-2009 Zhongshen Ruicheng Planning and Architectural Design Co., Ltd. Design Director / Chief Designer
2004-2007 Atkins Consultant Company (Shenzhen) Construction Manager / Director Architect
2001-2004 Atkins Consultant Company (Shenzhen) Senior Architect / Project Manager
1991-2000 Guangdong Zhongshan Architectural Design Institute Co., Ltd. Partner and Director Architect

Paul Rice

Paul Rice

设计总监
国籍：英国　出生日期：1965 年

学历及资格
英国 Glasgow 大学建筑设计学士
英国 Bath 大学施工经济与管理硕士
英国皇家建筑师学会现任会员
英国、新加坡注册建筑师

语言
英语 – 母语

曾经工作的国家
中国、英国、新加坡、马来西亚、约旦、印度尼西亚

个人履历
具有近 20 年建筑及规划设计工作经验，在城市设计、居住区规划、公共建筑、居住建筑设计方面积累了丰富的经验，同时具备很强的项目协调与管理能力。

现任工作：2009 年至今　汉米敦中国　设计总监

曾任工作
2003–2009 年　阿特金斯中国　设计董事
1999–2000 年　新加坡 ARCHETECTS VISTA PTE. LTD 高级建筑师
1998–1999 年　伦敦 PRP ARCHITECTS 项目建筑师
1996–1998 年　伦敦 PORPHYRIOS ARCHITECTS 项目建筑师
1991–1996 年　新加坡 ARCHETECTS VISTA 建筑师
1990–1991 年　英国 GLASGOW AUKETT ASSOC. ARCHITECTS
助理建筑师

Paul Rice

Design Director
Nationality: United Kingdom　Date of birth: 1965

Education and Qualifications
Bachelor of Architectural Design in Glasgow University, UK
Master of Construction Economics and Management in Bath University, UK
Current member of the Royal Institute of Architects
Registered architect in both the United Kingdom and Singapore

Language
English - mother tongue

Countries and regions where he has worked:
China, UK, Singapore, Malaysia, Jordan, Indonesia

Profile
Paul has nearly two decades' working experience in architectural planning and design. He has accumulated a wealth of experience in urban design, residential planning, public building and residential design, etc., and has a strong capacity in project coordination and management.

Current position
Design Director in Hanmi Dun•China since 2009

Previous position
2003-2009 Atkins•China Design Director
1999-2000 ARCHETECTS VISTA PTE. LTD (Singapore) Senior Architect
1998-1999 PRP ARCHITECTS (London) Project Architect
1996-1998 PORPHYRIOS ARCHITECTS (London) Project Architect
1991-1996 ARCHETECTS VISTA (Singapore) Architect
1990-1991 GLASGOW AUKETT ASSOC. ARCHITECTS (UK) Assistant Architect

Gensler

Benjamin Ward

Benjamin Ward

副总裁
设计总监

Benjamin是加利福尼亚注册建筑师，拥有超过15年的设计经验。来中国之前曾先后在纽约、巴黎和旧金山工作，参与从商业中心到城市规划等各种项目。Benjamin自加入Gensler后，已参与了三座大型机场、一个商业中心改造以及麦迪逊广场花园扩建等项目。作为建筑设计和总体规划项目的设计总监，Benjamin处理各项工程均细致到位，在总体规划项目中也总以独特的视角进行构思规划，同时，他也非常擅长于使用设计软件进行构图创意，运用三维技术更好地理解空间规划和设计方向。此外，Benjamin曾被《建筑实录》杂志评为全球建筑行业杰出青年设计师之一。

Benjamin Ward

Principal
Design Director

Benjamin Ward is a licensed California architect who has worked in New York, Paris, and San Francisco prior to arriving in China. He has over 15 years of varied design experiences, from retail architecture to urban planning. Since joining Gensler, he has worked on three international airports, a major retail revitalization, an addition for Madison Square Garden, and is a Design Director overseeing architecture and master planning projects in China.
Benjamin tackles every project with dedication and a unique master-planning angle that is essential to urban design. He also specializes in designing in the virtual world, using 3-D applications to better understand the space and design direction. In 2005, Benjamin was featured in Architectural Record as one of the industry's top young professionals.

Reja Bakh

Reja Bakh

副总裁
设计总监

Reja Bakh于2006年加入Gensler上海办事处，担任建筑设计总监。凭借着其卓越的职业生涯，包括在纽约HLW 建筑设计公司（HLW International）和在理查德迈·耶建筑设计事务所（Richard Meier）的17年工作经验，Bakh先生完成了一系列多样化的住宅和校园建筑设计作品。Bakh先生一直对科学与技术之间的关系，以及自然物理法则如何产生需求从而在技术上与建筑形成互动有着浓厚的兴趣，在他的设计中，他积极探索这种关系，以此创造出满足客户需要，并且与周围环境相协调的设计，实现建筑与自然环境之间的无缝衔接与统一。

Reja Bakh

Principal
Design Director

Reja Bakh joined the Gensler Shanghai office as a architectural design Director for. With a distinguished career, including over 17 years' experience in New York working with HLW International and Richard Meier, Mr. Bakh has developed a strong, diverse portfolio in both residential, hospitality, mixed-use and campus design.
Mr. Bakh has always been fascinated by the relationship between science and technology, and how the physical laws of nature produce a need to respond technologically with architecture. In his work, he explores this relationship to create designs that not only meet client needs, but also respond to their surroundings for a seamless integration between the built and the natural environment.

Christopher Chan

Christopher Chan

设计总监

Christopher Chan拥有超过20年的美国国内及国际项目设计和施工经验。他参与的项目遍及加利福尼亚、洛杉矶、休斯敦、西雅图和纽约。Christopher最近的工作重心转向亚洲，包括新加坡、中国及中国香港和中国台湾。虽然他一直致力于开拓不同的市场，但仍继续参与办公大楼、五星级酒店、大型综合体开发等高品质项目的设计工作。他擅长严密的设计研究和技术整合，为基地量身定制兼具独特设计和高市场价值的项目。

Christopher Chan

Design Director

Christopher Chan has over twenty years' of experience in design and construction both domestically and internationally. His projects can be found domestically in California, Los Angeles, Houston, Seattle, and New York City. Christopher's most recent work has been focused in Asia, spanning Hong Kong, Singapore, and Taiwan, with an emphasis in China. While he has been working across these different markets, Christopher continues to be involved in high-quality projects from high-rises offices, 5-star hotels, to large-scale, and mixed-use developments. His work centers on rigorous design investigation and strong technical integration to produce site-specific projects of distinctive design and high market value.

思纳史密斯(集团)中国
SMITHGROUPcna
architecture engineering interiors planning

李 超
Chao Li

李 超

规划公司总经理　注册规划师

教育 & 经验
加拿大C3城市规划与景观设计事务所，项目负责人，2001–2005年
加拿大LUSO建筑设计事务所，1999–2001年
上海城市规划设计研究院，1989–1999年
同济大学，城市规划，1989年

Chao Li

General Manager of Planning Co.　Registered Planner

Education & Experiences
He worked in Canada C3 Urban Planning and Landscape Design Office from 2001 to 2005.
Position: Project Principal
He worked in Canada LUSO Architectural Design Office from 1999 to 2001.
He worked in Urban Planning & Design Institute, Shanghai from 1989 to 1999.
He graduated from Department of Urban Planning, Tongji University in 1989.

傅 威
Wei Fu

傅 威

西南公司总经理　注册建筑师

教育
重庆建筑工程学院（现重庆大学），建筑学士，1991年
国家一级注册建筑师、高级建筑师

职业经历
四川省建筑设计院设计资深设计师，1991–2002年
四川省建筑设计院副总建筑师，2002–2006年
成都市建设委员会建筑专家组成员，2002年至今
建筑设计一所所长2005–2006年
思纳史密斯成都设计中心总经理、总建筑师，2006年至今

Wei Fu

General Manager of Southwest Co.　Registered Architect

Education
He graduated from Chongqing Architecture & Engineering College (Chongqing University for now) and received bachelor degree in 1991
PRC Class I Registered Architect Qualification, Senior Architect

Working Experiences
As a Senior Architect in Sichuan Architectural Design Institute, 1991-2002
As Chief Architect in Sichuan Architectural Design Institute, 2002-2006
As a member of Expert Group of Chhengdu Architectural Commission, 2002-Present
As the head of NO.1 Branch office of Architectural Design Institute, 2005-2006
General Manager and Chief Architect of Chengdu Cna Smith Design Center, 2006-Present

安庆东
Qingdong An

安庆东

创作公司总经理　注册建筑师

教育 & 经验
天津大学，建筑学硕士 1994年
上海现代建筑设计集团上海建筑设计研究院有限公司一所，主任建筑师 1994—2004年
上海交大安地建筑设计有限责任公司，总建筑师 2004—2006年
上海尊邦建筑设计有限公司，总建筑师及合伙人 2006—2007年
思纳史密斯集团（中国）上海思纳建筑设计有限公司，副总建筑师 2007年至今

Qingdong An

General Manager of Creation Co.　Registered Architect

Education & Experiences
Tianjin University,　Master of Architecture　1994
No.1 Branch office of, Shanghai Architectural Design Institute Co., Ltd., Shanghai Modern Architectural Design Group, Chief Resident Architect　1994-2004
Andi Architectural Design Co., Ltd. of Shanghai University of Communications, Chief Architect　2004-2006
Shanghai Zunbang Architectural Design Co., Ltd., Chief Architect and Partner 2006-2007
Shanghai CNA Architectural Design Company ,Ltd , CNA Smith (Group) China, Vice President of Architects　2007 - present

曹 洁
Jie Cao

曹 洁

景观公司总经理　注册规划师

教育 & 经验
上海市同济大学建筑城规学院规划系风景园林专业，学士，1986—1990年
上海市城市规划设计研究院，项目负责人，中级规划师，1990—2001年
上海市同济大学建筑城规学院规划系城市规划，硕士，1996—1999年
加拿大C3城市规划与景观设计事务所，技术主管，高级规划师，注册规划师，2001—2006年
思纳史密斯（集团）上海思纳史密斯规划和景观设计有限公司，规划总监，2006年至今

Jie Cao

General Manager of Landscape Co.　Registered Planner

Education & Experiences
Landscape Architecture of Planning Department, College of Architecture and Urban Planning of Tongji University, Shanghai, BA　1986-1990
Shanghai Urban Planning and Design Institute, the project leader, Mid-level planner 1990-2001
Urban Planning of Planning Department, College of Architecture and Urban Planning of Tongji University, Shanghai,Master　1996 and-1999
Canada C3 Urban Planning and Landscape Design Firm, technical director, senior planner, registered planner　2001-2006
Shanghai CNA Architectural Design Company ,Ltd , CNA Smith (Group) China,Planning Director　2006 - present

ÉTÉ 翌德国际设计机构
été lee et associés architectes urbanistes
■ 法国翌德国际设计机构 ■ 上海翌德建筑规划设计有限公司

王 渝
Yu Wang

王 渝

上海翌德建筑规划设计公司建筑师
建筑设计部主管

王渝2002年毕业于苏州城建环保学院建筑学专业，2002年加入上海翌德公司工作至今。

5年来，他从一名普通的设计助理做起，凭着饱满的热情和对设计的无限热爱，参与和负责设计项目数十项，尤其是在专业地产项目上积累了丰富的经验，同时兼顾部门行政管理工作，逐渐成为单位骨干。

Yu Wang

Architect of été lee et associés Architectes Urbanistes Shanghai
Director of architecture design department

After graduation from Frchitecture Department in Suzhou Institute of Urban Construction and Environmental Protection in 2002, Wang has worked in été lee et associés Architectes Urbanistes Shanghai.

Starting from a design assistant, during the five years, Wang has been working hard with great enthusiasm and interest. He has taken charge of over ten design projects, accumulating a wealth of experience in real estate projects. In the mean time, Li also undertakes administrative management works in the department, becoming the backbone member of the institute.

陈 民
Min Chen

陈 民

上海翌德建筑规划设计公司副总建筑师
1992年毕业于西北建筑工程学院

从业以来，他一直从事建筑设计、城市规划设计、房地产营销、室内设计工作，曾经负责或参与上海、南京、合肥、重庆、成都、无锡、苏州、长春、大连等城市的多项重大设计项目。工作学习期间，努力探索与建筑设计有关的房地产市场营销策划、室内装饰等相关行业、领域间的联动关系。

Min Chen

Vice Chief Architect of été lee et associés Architectes Urbanistes Shanghai
Graduated from Northwestern Institute of Architectural Engineering in 1992

Since he began to work, he has been engaged in the works of architecture design, urban planning and design, real estate marketing, interior design. He took charge of and participated in many significant design projects in big cities such as Shanghai, Nanjing, Hefei, Chongqing, Chengdu, Wuxi, Suzhou, Changchun, and Dalian. During his working and studying days, he made great effort in the exploration of the relationship in real estate marketing planning, interior decoration and other areas related to architecture design.

彭 松
Song Peng

彭 松

上海翌德建筑规划设计公司设计师

2004年毕业于东南大学，获建筑学硕士学位。2001年毕业于华中科技大学，获建筑学学士学位。

从业以来，一直从事城市规划设计与建筑设计工作，曾经负责或参与南京、武汉、上海、开封等城市的多项重大设计项目，以城市物质空间形态为主要研究方向，发表过三篇专题论文，参与编写空间研究专著一部。

Song Peng

Designer ofété lee et associés Architectes Urbanistes Shanghai

After gained the bachelor's degree of architecture in 2001 in Huazhong University of Science and Technology, Peng got his architecture master's degree in Southeast University in 2004 .

Since he began to work, he has been engaged in works of urban planning and architecture design. He has conducted and participated in many important design projects in Nanjing, Shanghai, Kaifeng cities. With the focus in the research of urban space form, he published three thematic thesises and participated in the edition of a monograph in space research.

杨大龙
Dalong Yang

杨大龙

毕业于西安建筑科技大学
上海翌德建筑规划设计公司建筑师

杨大龙从事建筑设计工作近10年，项目类型涉及各个层面。

他以创造适合人使用、融合于城市及可持续性建筑为工作理念，本着严谨务实、锐意创新的工作态度，多年来杨大龙在建筑设计领域积累了大量实践经验。

他善于挖掘建筑本质、把握建筑发展最新动态，结合项目本身性质为业主提供最优化方案。

通过多年对建筑设计本源的研究，已形成自己的建筑涉及语言，并将细节等的设计落实于真正的建筑设计实践中。

Dalong Yang

Graduated from Xi'an University of Architecture and Technology
Architect of été lee et associés Architectes Urbanistes Shanghai

Yang has been engaged in architecture design works for over 10 years, his projects cover all levels.

He is holding the working concept of creating the sustainable development architecture which suits people and cities. Yang is disciplined and innovative with down to earth attitude, accumulated great experience in architecture design field in these years.

Being efficient at seizing the essence of the architecture and the new trend of architecture development, Yang provides the optimization design method according to the project properties.

Through many years research of architecture origin, Yang has established his own architectural design language and implement the design of details in the architectural design projects.

美国HOOP建筑设计有限公司
HOOP ARCHITECTURAL DESIGN CONSULTANTS,INC

赵 恺
Kai Zhao

赵 恺

一级注册建筑师

毕业于华中科技大学，获建筑学硕士学位
先后就职于中船第九设计研究院、新加坡新仕建筑设计公司
现任美国HOOP建筑设计咨询有限公司 副总建筑师

主持设计项目
杭州保利东湾居住区
上海南汇区委党校
上海阳光欧洲城
湖北武汉爱家国际华城
辽宁鞍山爱家花园
南通南通奥特莱斯商业中心

Kai Zhao

First Grade Registered Architect

Graduated as a Master of Architecture from Huazhong University of Science and Technology
Once worked in the Ninth Design Research Institute for Shipbuilding Industry of China and Xinshi Architecture Design Ltd. of Singapore one after another
Currently hold the post of deputy chief architect of the HOOP Architectural Design Consultant INC. of America

Designs Projects
Hangzhou, Poly dongwanResidential Area
Shanghai, The Communist Party Institute of Nanhui District Committee
Shanghai, Sunshine Euro Town
Wuhan, Hubei Province Aijia International Hua Town
Anshan, Liaoning Province Anshan Aijia GardenNantong City
Nantong Outlets Commercial Center

王晓泉
Xiaoquan Wang

王晓泉

一级注册建筑师

毕业于天津大学，获建筑学学士学位
现任美国HOOP建筑设计咨询有限公司 副总建筑师

主持设计项目
宁波都市森林（265 000平方米）
宁波核心居住区一号地块（270 000平方米）
湖州升华俞家漾住宅小区（410 000平方米）
慈溪东外环居住小区规划（用地900 000平方米）
湖州南浔城市设计（用地4 000 000平方米）
深圳市仙桐体育公园（40 000平方米）

Xiaoquan Wang

First Grade Registered Architect

Graduated as a Bachelor of Architecture from Tianjin University
Currently hold the post of the deputy chief architect of the HOOP Architectural Design Consultant INC. of America

Designs Projects
Ningbo Metropolitan Forest (occupying an area of 265,000 m^2)
The No.1 Plot (occupying an area of 270,000 m^2) of Ningbo Core Residential Area,
Huzhou Shenghua Yujiayang Residential Community (occupying an area of 410,000 m^2)
The Planning of Cixi East Outer Ring Residential Community (occupying an area of 900,000 m^2)
Huzhou Nanxun City Design (occupying an area of 4,000,000 m^2)
Shenzhen Xiantong Sports Park (occupying an area of 40,000 m^2) and other known projects

ALVARO LEONARDO

Alvaro Leonardo

西班牙注册建筑师

毕业于西班牙FACULTY OF ARCHITECTUFE OF VALENCIA，建筑学硕士
现任美国HOOP建筑设计咨询有限公司首席设计师，

主持设计项目
BERUT CENTRAL DISTRICT(黎巴嫩)(166 000平方米)
上海芦潮港旅游度假区（52 000平方千米）
武汉万达积玉桥项目（180 000平方米）
上海绿庭松江广场（250 000平方米）

Alvaro Leonardo

Spanish Registered Architect

Alvaro Leonardo graduated from Faculty of Architecture of Vzlencia in Spain, receiving Master's Degree of Architecture
Currently he is the chief designer of U.S. HOOP Architectural Design Consulting Co., Ltd.

Designs Projects
"BERUT CENTRAL DISTRICT" (Lebanon) (166,000 m^2)
"Shanghai Luchaogang Resort" (52,000 k m^2)
"Wuhan Wanda Jiyu Bridge "(180,000 m^2)
"Shanghai Green Court Songjiang Square "(250,000 m^2)

J.A.O.Design International Architects & Planners Limited
美國龍安建築規劃設計顧問有限公司

商广敏
Guangmin Shang

商广敏

从事设计工作40多年。亲自主持和参加过许多工业与民用建筑的建筑与规划设计工作。

主要项目

工业建筑曾做过各种类型工业厂房，包括许多恒温恒湿的精密车间。民用建筑做过各种住宅及楼盘规划设计，多功能剧场、商业、体育及科研等各类建筑设计。比如北京虎坊桥工人俱乐部大修改造工程、北京前门大街东侧历史文化街区保护规划设计、北京怀柔体育中心、北京良乡西路园小区、广东东莞厚街商贸中心、广州气象卫星地面站、天津塘沽开发区中信天津工业发展公司办公楼及标准厂房设计、天津环美家具有限公司等项目。另外还参与了苏州国际教育园区规划设计、浙江乐清市乐城区概念规划设计、浙江温州瓯江南岸景观设计、新疆库尔勒市中心建筑规划与设计等项目。

Guangmin Shang

He engaged in designing for 40 years, organized and participated in many architectural designs for industrial and civil use.

Major Projects

He has completed many projects for factory building including many workshops in constant temperature and constant humidity and for civil use including residential buildings, multi-functional theaters, commercial, sports and research centers. For example, Overhaul and modification to Beijing Hufangqiao Worker's Club, protection design of Historical Cultural Block on the east side of Beijing Front Door Street, Beijing Huairou Sports Center, Beijing Liangxiang Xiluyuan District, Guangdong Dongwan Houjie Commercial Center, Guangzhou Satellite Station Office building and standard workshops for Zhongxin, Tianjin Industrial Development Company in Tianjing Development District, Tanggu Tianjin Huanmei Furniture Co., Ltd. Besides, he participated in overall planning for Suzhou International and Educational Park, conceptual design of Lecheng District, Leqing in Zhejiang Province, landscape planning on the south shore of Oujiang river, Wenzhou in Zhejiang Province, and architectural design and planning for Korla Shi in Xingjiang Autonomous Region.

黄立敏
Limin Huang

黄立敏

规划部总经理，国家注册规划师

1994年毕业于内蒙古工业大学交通土建工程专业，擅长城市总体规划，深谙“四因二应”的规划设计理念，即“因山而起，因水而洄，因地制宜，因人而异；回应历史，呼应未来。”从事项目负责人工作8年。

Limin Huang

General Manager of Planning Department , national registered planner

Graduated from Inner Mongolia University of Technology, majored in Tansportation and Civil Engineering. Huang excels in urban planning and has a good understanding of the" four relations and two correspondences" planing and design idea , that is " the relation of the water, mountains, region and people the correspondence of history and future."Undertaking the post of Project Manager for 8 years.

罗伊·爱德华特里耶
Roy Edward Tellier

罗伊·爱德华特里耶

国际建筑部经理、集团业务发展经理

1954年6月26日出生在美国纽约，在长岛长大。曾在科罗拉多大学学习医学，自然科学和数学，受过越战时期空军交通管制、航空摄影测量和土地测量培训的老兵。之前在美国有三家公司，其中两家是测量与工程方面，另一家为软件设计和支持的公司。2001年从纽约州立大学获得专业土地测量许可，有20多年土地测量，土地规划，土地法，全球定位系统（GPS），地理信息系统（GIS）和制图经验。

2005年来到北京，并投入饶及人先生的龙安集团旗下，美国龙安建筑规划设计顾问有限公司工作。第一年担任地理信息系统部经理，主要制作和改造地形三维模型，以取得平衡土方量并进行水文研究。2007年进入市场部，帮助引入外国设计师及设计理念，后升任国际建筑部负责人至今。

Roy Edward Tellier

Principal of International Architecture Department, Manager of Group Business Development

Born in New York City in June 26, 1954 and raised on Long Island. Studied medicine, science and mathematics at the University of Colorado located in Boulder, Colorado. In the era of Vietnam veteran trained by the United States Air Force in air traffic control, aerial photogrammetry and land surveying. Former owner of three companies at various times in the United States, two were involved in Surveying and Engineering and one was a software design and supporting firm. Obtained Professional Land Surveying Lentificate from the State University of New York in 2001, with over 20 years' of experience in land surveying, land planning, land law, global positioning systems (GPS), geographic information systems (GIS) and cartography.

Moved to Beijing, China in 2005 and worked with James C. Jao at J.A.O Design International Architects & Planners Limited. and the Long On Group. For the first year he was the Manager of the Geographical Information Systems Department where we concentrated on generating three-dimensional models of geography and improvements to produce balanced earthwork quantities and hydrographic studies. In 2007 became a part of the Marketing Department J.A.O Design International Architects & Planners Limited to help introduce and promote the use of foreign designers and concepts to our projects in China. He was eventually promoted to Principal-his current position.

王一铁
Yitie Wang

王一铁

建筑部副总经理，硕士研究生

已完成大量规划、建筑设计项目，参与和负责部分房地产项目策划、研发、设计、运营、管理等工作。对当前市场下的客户价值管理、绿色设计等有专项研究。

Yitie Wang

Vice Manager in Construction Department, Master Degree

He has completed many planning and architectural designs, and participated and organized planning, researching, operating and managing for real estate projects. He also has special researches on clients' management value and green designs.

J.A.O.Design International Architects & Planners Limited
美國龍安建築規劃設計顧問有限公司

丁海茹
Hairu Ding

丁海茹

规划部经理，城市规划硕士研究生

参与规划项目数十项，熟悉各种类型的项目规划，理解、掌握国家对城市发展的策略、规范、法规，积极参与将西方先进理念落户中国的实践，所参与的项目多项获得国际及国内的奖项荣誉。

Hairu Ding

Manager in Planning Division, Master Degree in Urban Planning
She has participated in many projects, so she is familiar with all kinds of projects and Chinese policies and standards for urban development. She combined advanced theories in western countries with practices in China. There are many projects she participated that won awards at home and abroad.

商贝尔
Sean Bell

商贝尔

资深建筑设计师，美国注册建筑师，美国绿色建筑评估师

Sean Bell

Senior Design Architect, LEED B.S.A architecture

Jason Balinbin

Jason Balinbin

资深建筑师，美国注册建筑师，美国建筑师协会，美国全国建筑师注册公会及美国绿建筑协会成员。1975年在西雅图大学获得美术学学士，1977年从加利福尼亚州立大学获得美术学硕士，1980年从加利福尼亚大学获得建筑硕士学位。

Jason Balinbin

Senior Design Architect, Registered architect; AIA, NCARB, LEED Green Associate
Seattle University, Bachelor of Arts, Fine Arts in 1975; University of California, Master of Arts, Fine Arts in 1977; University of California at Los Angeles, Master of Architecture in 1980.

丹尼斯·尼古拉耶夫
Dennis Nikolaev

丹尼斯·尼古拉耶夫

首席建筑师，美国建筑师协会成员

毕业于美国弗吉尼亚理工大学，建筑学专业学士，拥有美国加州建筑师许可证。

Dennis Nikolaev

Chief Architect, Membership of U.S. Association of Architects
He graducated from Virginia Tech University, with Bachelor's degree of architecture and architectural design certificate of

刘赞杰
Zanjie Liu

刘赞杰

建筑部副总经理

参与或负责的项目数十个，项目涉及住宅、公建等多种类型，部分已建成。

Zanjie Liu

Vice President of Construction

He involved in or responsible Department for ten projects including dwelling houses, public buildings and many other types. Some of his projects have been completed.

英国UK.LA太平洋远景国际设计机构
U.K PACIFIC LONG-RANGE PLANNING & DEVELOPING DESIGN CONSULTANT LTD.

申丽萍
Liping Shen

申丽萍

武汉大学城市规划学士
华中科技大学建筑学硕士学位
东南大学建筑学博士学位
南京大学——美国康奈尔大学EMBA
英国UK.LA太平洋远景国际设计机构大中华区董事长
南京环洋远景建筑规划设计顾问有限公司总经理
南京泛奥建筑规划设计顾问有限公司总经理
郑州九华房地产开发有限公司董事
先后被评为：
中国房地产协会专家委员会委员
亚洲杰出建筑规划师
法国建筑师协会会员（20090619）
中房商协城市规划与建筑设计专业委员会委员

主要作品

1．城市规划部分
A．苏州古城一、二号街坊保护性规划（76公顷）
B．郑州龙子湖湖心岛城市设计（99公顷）
C．新沂窑湾古镇规划（101公顷）
D．平顶山湛河两岸城市设计（80公顷）
2．公共建筑部分
A．郑州意大利格拉姆中心（12万平米超高层）
B．南京金宁广场（18万平米综合建筑群）
C．连云港甬连再生资源加工园区（100万平米科技园区）
D．连云港水晶宫（5万平方米）
E．郑州东湖国宾馆（133.4公顷国宾馆）
3．居住社区部分
A．南京华欧国际友好城（230万平方米）
B．南京旭日上城建筑设计（202万平方米）
C．海南儋州双联滨海住区（88万平方米）
D．郑州清华大溪地（260万平方米）
E．红太阳乌江新城（530万平方米）
4．景观设计部分
A．郑州西南绕城高速景观设计（全长52千米）
B．南京河西江东路景观设计（总长16.8千米）
C．南京徐庄软件园景观设计（设计面积12公顷）
D．信阳百花园景观设计（设计面积20公顷）

Liping Shen

Wuhan University, Bachelor of Urban Planning
Huazhong University of Science, Master of Architecture
Southeast University, Architecture Ph.D.
Nanjing University, - Cornell University EMBA
Chairman of British UK.LA Pacific Vision International Design agencies in Greater China
General Manager of Nanjing Pan Ocean Long Range Architectural Planning and Design Consultants Co., Ltd.
General Manager of Nanjing Fan'ao Architectural Planning Consultants Co., Ltd.
Manager of Zhengzhou Jiuhua Real Estate Development Co., Ltd.
Once named as
Member of Expert Committee of China Real Estate Association
Asian Outstanding Planners
Membership of French Architects Association (20090619)
Member of China Housing Business Association of Urban Planning and Architectural Design Committee successively

Main works

1. Urban planning
A. protection plan for Suzhou ancient city neighborhood I II (76 ha)
B. the mid-lake island urban design for Zhengzhou Longzihu lake island (99 ha)
C. Xinyi yaowan Town Planning (101 ha)
D. Urban Design for Cham river, Pingdingshan (80 ha)
2. Public buildings
A. Zhengzhou Italy Gram Centre (120 000 m^2 of high-rise)
B. Nanjing Jin Ning Square (180 000 m^2 of building complex)
C. Lianyungang Ningbo Processing Zone with renewable resources (1 000 000 m^2 of Science and Technology Park)
D. Lianyungang Crystal Palace (50 000 m^2)
E. Zhengzhou East Lake Guesthouse (133.4ha Guesthouse)
3. Living community
A. Nanjing, Hua'ou International friendly city (2 300 000 m^2)
B. Nanjing Xuri Shangcheng architectural design (2 020 000 m^2)
C. Hainan Danzhou doublehouse coastal settlements (880 000 m^2)
D. Zhengzhou Tsinghua Tahiti (2 600 000 m^2)
E. Red Sun Wujiang Metro (5 300 000 m^2)
4. Landscape design
A. landscape design for high-speed way in Zhengzhou Southwest Loop (total length 52 km)
B. Landscape Design for Jiangdong Road, Nanjing Hexi (total length 16.8 km)
C. Landscape Design Nanjing Xu Village Software (design area 12 ha)
D. Landscape Design for Xinyang Flowers Garden (design area 20 ha)

崔晨旻
Chenmin Cui

崔晨旻

UKLA设计部经理
毕业于天津城市建设学院

主要作品

1．平顶山湛河沿岸景观改造（全长8千米）
2．英地置业农科路A地块居住项目规划设计（38.4万平方米）
3．郑州一生缘　翡翠谷综合旅游度假项目规划设计（77万平方米）
4．盱眙帝景国际居住项目规划设计（23.7万平方米）
5．新沂　新北花苑居住项目规划设计（21万平方米）
6．淮安国华　御翠园居住项目规划设计（20万平方米）
7．淮安水渡口广场地块综合商业居住项目规划设计（26万平方米）
8．洛阳定鼎路北侧地块居住项目规划设计（73万平方米）
9．涟水　水岸城邦居住项目规划设计（28.7万平方米）
10．南京市雨花区正欣和通信研发中心建筑规划设计（7.5万平方米）

Chenmin Cui

Design manager of UKLA
Graduated from the Tianjin Institute of Urban Construction

Main works

1. Transforming the landscape along the Cham River, Pingdingshan (length 8 km)
2, planning and design for Yingdi Properties, Agricultural Road, block A housing project (384 000 m^2)
3 Zhengzhou Yishengyuan • Jadeite Valley Integrated tourism resort project planning and design (770 000 m^2)
4, planning and design for Xuyi Royal View international living project (237 000 m^2)
5, Planning and Design for Xinyi • New North Garden Residential Project (210 000 m^2)
6, planning and design for Huai'An Guohua • Regency residence project (200 000 m^2)
7, planning and design for Huai'An water Ferry Plaza block integrated business housing project (260 000 m^2)
8 Planning and Design for north of Luoyang Dingding Lu Project housing block (730 000 m^2)
9, Lianshui • waterfront city planning housing project (287 000 m^2)
10, Nanjing Yuhua District Zhengxinhe Communication R & D Center Building Planning and Design (75 000 m^2)

张俊鹏
Junpeng Zhang

张俊鹏（澳大利亚籍）

郑州分公司主任

参与项目

1．郑州意大利格拉姆中心（12万平方米超高层）
2．郑州清华·忆江南（86万平方米）
3．郑州中南海之音（50万平方米）
4．郑州聚龙城（10万平方米）
5．郑东新区龙子湖湖心岛城市设计（99公顷）
6．郑州经三名筑（12万平方米）
7．洛阳香榭里黎明（9.8万平方米）
8．信阳百花园规划（设计面积20公顷）

Junpeng Zhang (Australian)

Director of Zhengzhou Branch

Projects involved

1, Zhengzhou Italy Gram Centre (120 000 m^2 of high-rise)
2, Zhengzhou Tsinghua Yi Jiangnan (860 000 m^2)
3, Zhengzhou Zhongnanhai Voice (500 000 m^2)
4 , Zhengzhou Julong City (100 000 m^2)
5, The mid-lake island urban design for Zhengzhou Longzihu lake island (99 ha)
6, Zhengzhou, Jingsanmingzhu (120 000 m^2)
7, Luoyang Champs Dawn (98 000 m^2)
8, planning for Xinyang Flowers Garden (design area 20 ha)

James Wang Design Associates
北京杰地亚建筑咨询有限公司

于华清
Huaqing Yu

于华清

学历：山东建筑工程学院　本科
　　　湖南大学结构工程专业　硕士学位
职务：常务副总经理
主要作品
橘郡
优龙别墅
天津金湾大厦
塞纳维拉
永定华亭
乐成中心
东方华府
圆梦园休闲园
天津团泊湖
烟台圣安湾

Huaqing Yu

Educational Background:
Graduated from Shandong Jianzhu University, and obtained Bachelor's Degree
Graduated from Hunan University, majored in structure engineering, and obtained Master's Degree
Position: Executive Deputy General Manager
Main works
Orange County Villa
Youlong Villa
Jinwan Building in Tianjin
Sai Na Wei La Villa
Yongdinghuating Apartments
Lecheng Center
Dongfanghuafu Building
Yuanmengyuan Leisure Garden
Tuanbohu Villa in Tianjin
Sheng'anwan in Yantai

李昊斐
Haofei Li

李昊斐

学历：内蒙古科技大学建筑学　本科
职务：建筑设计部经理
主要作品
济南大地锐城商业综合体
上海长甲龙湖别墅
北京天恒别墅山
北京金海湖刘家沟别墅区
秦皇岛夏宫别墅区
威海华瀚四季花园住宅区
烟台南山马山寨海景酒店
杭州富阳复地综合商业中心
大连红旗镇张家二期住宅区
北京银河湾售楼处

Haofei Li

Educational Background
Graduated from Inner Mongolia University of Science & Technology, majored in Architecture, and obtained Bachelor's Degree
Position: General Manager of Department of Architectural Design
Main works
Dadiruicheng Business Complex-building in Jinan
Changjialonghu Villa in Shanghai
Tianheng Villa Area in Beijing
Liujiagou Villa Area at Jinhaihu where is in Beijing
Xiagong Villa Area in Qinhuangdao
Huahan four season graden Residential Area in Weihai
Mashanzhai Seaviwe Hotel at Nanshan where is in Yantai
Fudi complex business center at Fuyang where is in Hangzhou
Second phase of Zhangjia Residential Area at Hongqi town where is in Dalian
Yinhewan Sales Offices in Beijing

刘文博
Wenbo Liu

刘文博

学历：武汉科技大学城建学院　本科
职务：建筑设计部项目经理
主要作品
华南MALL
沈阳棋盘山商业中心
大连金石滩商业区
福建金银岛大酒店
八达岭温泉度假酒店
东隆别墅
江畔花园别墅
山水宜家
优山美地别墅
罗山绿洲
东方威尼斯
9号国际公寓
于洪新城

Wenbo Liu

Educational Background:
College of Urban Construction, WUST, obtained Bachelor's Degree
Position: Project Manager of Department of Architectural Design
Main works
Huanan Mall
Qipanshan Business Center
Jinshi beach Business Area
Gold and Silver Island Hotel
Badaling Hot Spring Resort Hotel
Donglong Villa
River Bank Garden Villa
Shanshuiyijia Villa
Yosemite Villa
Luoshan Oasis Villa
Qrient Venice Villa
No.9 International Apartments
Yuhong New City

James Wang Design Associates
北京杰地亚建筑咨询有限公司

陈学松
ALEX HSUEH-SUNG CHEN

王世刚
Shigang Wang

付 斌
Bin Fu

陈学松

学历：1982年　台湾中原大学土木工程学士
1986年　美国加州州立大学佛伦斯诺分校（CSUF）土木工程硕士（结构专业）
1989年　美国加州大学洛杉矶分校（UCLA）建筑硕士
职务：副总经理

主要作品
美国加州Pomona市12单位仓库办公室
美国加州Bastro市Quality Inn酒店
美国加州Palm Desert市Candlewood Hotel酒店
美国加州Rosemead市Big Island住商综合大楼
美国加州Rosemead市Garvey Plaza住商综合大楼
美国加州San Gabriel市李兴华珠宝城商场
美国加州Rosemead市Del Mar住商综合大楼
台湾台北市信义区南山人寿总部大楼
台湾台北开南中学教室大楼
台湾台南东帝士厂房办公大楼
台湾澎湖关务大楼及乘客码头大楼
湖南张家界天门山入口景区整体规划设计
天津宝坻创业园老年社区及住宅区整体规划设计
东莞华南MALL
天津世贸中心
长沙上河村商业中心

ALEX HSUEH-SUNG CHEN

Educational Background:
1982 Bachelor of Civil Engineering in Chung Yuan Christian University, Taiwan
1986 Master of Civil Engineering (major in structure) in California State University, Fresno, U.S.A
1989 Master of Architecture in California State University, Los Angeles, U.S.A
Position: Deputy General Manager

Main works
12-unit depository office in Pomona, California, U.S.A
Quality Inn in Bastro, California, U.S.A
Candlewood Hotel in Palm Desert, California, U.S.A
Big Island commercial-residential building in Rosemead, California, U.S.A
Garvey Plaza commercial-residential building in Rosemead, California, U.S.A
Xinghua Li Jewelry Market in San Gabriel, California, U.S.A
Del Mar Commercial-residential Building in Rosemead, California, U.S.A
Main building of Nanshan Life Insurance in Xinyi District, Taipei, Taiwan
Teaching building of Kainan High School in Taipei, Taiwan
Office Building of Tuntex factory in Tainan, Taiwan
Customs Building and passenger dock building in Penghu, Taiwan
Entrance scenery spot planning and design of Tianmen Mountain, Zhangjiajie, Hunan Province
Old-age community and residential quarter planning and design of Baodi Business Area in Tianjin
Huanan Mall in Dongguan
World Trade Center in Tianjin
Shanghe Village Commercial Center in Changsha

王世刚

学历：沈阳鲁迅美术学院　　本科
职务：副总经理

主要作品
北京尊府
山水文园
太阳公元
山水铂宫
中轴国际
日坛国际广场
北京世贸天阶
泛海国际居住区
9号公寓
沈阳华锐·世界城

Shigang Wang

Educational Background:
Graduated from Luxun Academy of Fine Arts in Shenyang, obtained Bachelor's Degree
Position: Deputy General Manager

Main works
Zunfu in Beijing
Shan Shui Wen Yuan
Taiyanggongyuan Building
Shan Shui Bo Gong
Zhongzhouguoji Building
Ritan International Square
Shimaotianjie in Beijing
Fanhai International Residential Area
No.9 Apartments
Huarui World City in Shenyang

付 斌

学历：天津大学建筑学院　　本科
英国诺丁汉大学　　硕士学位
注册资格：国家一级注册建筑师
职务：副总经理

主要作品
北京9号国际公寓
上海环球翡翠湾花园
上海长甲龙湖别墅
天津卡梅尔
华南MALL
日坛国际广场
青岛海泉湾度假城
北京山水文园E区别墅
中关村新城
成都逸源香舍别墅

Bin Fu

Educational Background:
Academy of Architecture Tianjin University, obtained Bachelor's Degree on Architecture
University of Nottingham, obtained Master's Degree on Architecture
Qualification of registration: National 1st Class Registered Architect
Position: Deputy General Manager

Main works
No.9 International Apartments in Beijing
Global Feicuiwan Garden in Shanghai
Changjialonghu Villa Area in Shanghai
Caramel in Tianjin
Huanan Mall
Ritan International Square
Haiquanwan Resort in Qingdao
E Regional of Shanshuiwenyuan Villa in Beijing
New City of Zhongguancun in Beijing
Yiyuanxiangshe Villa in Chengdu

James Wang Design Associates
北京杰地亚建筑咨询有限公司

刘晓强
Xiaoqiang Liu

陈永钢
Yonggang Chen

赵　波
Bo Zhao

刘晓强

学历：沈阳建筑大学建筑系　　本科
职务：副总经理
主要作品
沈阳　棋盘山欧陆游小镇
沈阳　华府天地购物中心
中南MALL
乐城国际公寓
八达岭温泉酒店
北京亦庄中央公馆
海口华联酒店
善美商业街
小平岛中心广场3号楼
北京世贸天阶

Xiaoqiang Liu

Educational Background:
Graduated from Shenyang Jianzhu University, majoring in architecture, obtained Bachelor's Degree
Position: Deputy General Manager
Main works
Ouluyou small town at Qipan mountain where is in Shenyang
Huafu Heaven and Earth shopping mall in Shenyang
Middle-south Mall
Lecheng International Apartments
Badaling Hot Spring Hotel
Yizhuang Central Kung-kuan in Beijing
Hualian Hotel in Haikou
Shanmei Business Street
No.3 Building on center square at Xiaoping Island
Shimaotianjie Shopping Mall in Beijing

陈永钢

学历：天津建筑工程学院
　　　建筑室内外装饰工程专业　本科
主要作品
乐成国际
9号国际公寓
光彩国际公寓
温哥华森林别墅
东莞阳光山庄别墅
世贸天阶商场
沈阳华府天地
北京华联创新培训中心
禧福汇样板间
9号公寓会所
世贸天阶SPA–VIP区
北京尊府
内蒙乌海君正花园样板间
厦门鑫塔项目样板间
北京山水金海湖宾馆

Yonggang Chen

Educational Background:
Bachelor of Interior and Exterior Architectural Decoration Engineering, Tianjin Construction Engineering Institute
Main works
Lecheng International
9# International Apartment
Guangcai International Apartment
Vancouver Forest Villa
Sunshine Villa in Dongguan
Shimaotianjie Shopping Mall
Huafu Shopping Mall in Shenyang
Creating Training Center of Beijing Hualian
Sample Rooms of Xifuhui
9# Apartment Club
SPA-VIP District of Shimaotianjie Mall
Zunfu Building in Beijing
Junzheng Park Sample Rooms in Wuhai, Inner Mongolia
Sample Rooms of Xingta Project in Xiamen
Gimhae Landscape Hotel in Beijing

赵　波

学历：沈阳航空工业学院　　本科
职务：室内设计部经理
主要作品
中国人民银行营业管理部办公大楼
理想国际大厦酒店
北京市1号楼
星明湖度假村
山西红金龙商务会馆
金海湖宾馆
乐成国际交流中心
奥中基业会所
春天百货
安邦国际金融中心

Bo Zhao

Educational Background:
Graduated from Shenyang Institute of Aeronautical Engineering
Position: General Manager of Department of Indoor Design
Main works
The Management Building of ThePeople's Bank of China
Ideal International Dajing Hotel
No.1 Building in Beijing
Xingminghu Resort
Hongjinlong Business Hall in Shanxi
Jinhaihu Hotel
Lecheng International Communication
Aozhong Foundation Club
Spring Shopping Mall
Anbang International Financial Center

加拿大考斯顿设计
Cobblestone Design Canada

上海考斯顿建筑规划设计咨询有限公司
Cobblestone Urbanists & Architects Workshop Shanghai

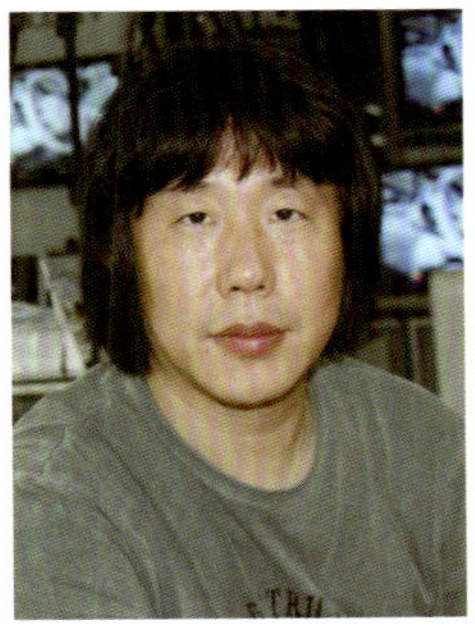

刘廷杰
Tingjie (Peter) Liu

刘廷杰
设计总监
加拿大马尼托巴大学专业建筑硕士
加拿大皇家建筑学会会员
中国注册建筑师、高级建筑师
20年中加两地经验
曾任职于加拿大温哥华IBI Group，温哥华 DGBK事务所，温尼泊Raymond Wan事务所，深圳大学设计院等。

获奖作品
- 2000年4月　入围OTIS国际建筑学校住宅设计竞赛
- 2001年6月　入围加拿大不列巅省建筑学会青年建筑师作品展
- 2004年　主创设计Multi–Tenant Facility of Smart Park in University of Manitoba
获2004年度加拿大总督奖提名（Ray Wan Architects）
- 2006年12月　中国创新90中小套型住宅设计竞赛命题类鼓励奖
- 2008年12月　黄石“东风路地块”中国人居典范建筑规划设计方案竞赛金奖
- 2009年10月　上海松江“新弘国际”获建筑学会中国人居奖综合设计大奖
- 2009年10月　上海新江湾城“05–2地块”　获建筑学会中国人居奖规划和方案设计大奖

Tingjie (Peter) Liu

Design Principal
Professional Master's Degree in Architecture, University of Manitoba, Canada
Member of Royal Architectural Institute of Canada
Registered Architect and Senior Architect in China
20 Years Experience in China and Canada
Worked in Vancouver IBI Group of Canada, Vancouver DGBK firm, Winnipeg Raymond Wan firm, Design Institute of Shenzhen University, etc.

Prize-winning works
- April, 2000, qualified for OTIS Residential Design Competition of International Architectural School
- June 6-22, qualified for Exhibition of Young Architects in Academy of Architecture of Province of Britain, Canada
- In 2004, leading the design of Multi-Tenant Facility of Smart Park in University of Manitoba, and won the nomination of Canada Governor General's Award, (Ray Wan Architects)
- December 2006, won Consolation Prize of the topic-assigned class in the Design Competition of China Innovation 90 Small and Medium Dwelling Size Residences
- December 2008, won the gold award in the Competition of China Dwelling Model Construction Planning and Design Programs by "Dongfeng Road Plot", Huangshi
- October 2009, won the Comprehensive Design Award in China Dwelling Award of Academy of Architecture by "Xinhong International" in Songjiang, Shanghai
- October 2009, won the Planning and Program Design Award in China Dwelling Award of Academy of Architecture by "05-2 Plot" in Xinjiangwancheng, Shanghai

FLA 上海梵瑞建筑设计咨询有限公司
FORM LANDSCAPE ARCHITECTS CO.,LTD

朱　兵
Bing Zhu

朱　兵
国家一级注册建筑师
设计总监

学历

南京工业大学	建筑学学士
同济大学	建筑学硕士

执业经历

1994–2000年	国内贸易部无锡科学研究设计院	建筑师
2000–2002年	日本MAO一级建筑士事务所	副总建筑师
2003–2010年	上海梵瑞建筑设计咨询有限公司	设计总监

主要项目

上海交通大学闵行东校区	2000年
扬州新加坡花园	2002年
盐城市娱乐花园	2003年
浙江金华白龙桥新区	2003年
上海飞虹路地块规划	2004年
常州电脑城商业中心	2005年
浙江永康五金总部中心	2007年
盐城市劝业场商业区	2009年
盐城市飞驰新天地高层综合楼	2008年
盐城市澄达东景苑高档住宅区	2008年

Bing Zhu
National Class 1 Registered Architect
Design Director

Education
Bachelor of Architecture, Nanjing University of Technology
Master of Architecture, Tongji University

Practicing Experience
1994-2000 Architect, Wuxi Scientific Research & Design Institute of Domestic Trade Department
2000-2002 Deputy Chief Architect, Japanese MAO Level One Architecture & Engineering office.
2003-2010 Design Director, Shanghai Fanrui (FLA) Architectural Design Consulting Co. Ltd.

SELECTED PROJECT EXPERIENCE
2000 East part of Minhang Campus planning, Shanghai Jiaotong University, Shanghai
2002 Singapore Garden planning and architectural design, Yangzhou
2003 Yule Garden redidential design, Yancheng
2003 Bailongqiao new district in Jinhua, Zhejiang Province
2004 Planning and design of Feihong Road in Shanghai
2005 Design for Changzhou PC Mall
2007 Design for Yongkang Hardware Headquarters in Zhejiang
2009 Design for Quanyechang business district in Yancheng
2008 Design for high-layered comprehensive building of Feichi New World in Yancheng
2008 Planning and design for the high-end residential of Chengda East View Court in Yancheng

格雷·海瑞斯　高级顾问

亚历山大·聪尼斯　高级顾问

A&S 翰时国际建筑设计 INTERNATIONAL DESIGN

余立 董事长／总设计师／博士

陈娟娟 董事／副总裁

张兵 董事／副总设计师

张广亮 董事／副总设计师

林载舞 董事／国家一级注册建筑师

Aedas
凯达

罗秉枢
David Roberts

罗秉枢

行政总裁
凯达环球有限公司

1986年 环境设计学文学士（TSIT）
1988年 建筑学（TSIT）
1992年 英国皇家建筑师学会会员
1992年 香港注册建筑师
1992年 香港政府认可人士（建筑师）
2000年 英国仲裁学会会员
2003年 项目管理协会附属会员
2006年 香港建筑师学会资深会员
2007年 澳洲皇家建筑师学会资深会员

罗秉枢先生在香港从事建筑设计逾24年，拥有建筑师资格达16年，尤擅长为香港大型复杂和高速度要求的项目作处理与报批事务。他在香港，内地，亚洲地区以及中东地区的项目经验丰富，曾参与不同类型的建设项目，包括办公楼、酒店、零售商场、文娱设施、工业发展项目、基础设施建设及医疗保健设施等。他对基础设施建设项目管理尤感兴趣，目前正担任香港国际机场二号客运大楼及最新北卫星客运廊建设项目的项目董事。

罗秉枢领导公司进军中东市场，推动公司业务踏上新的台阶，建树良多。2004年，他承接了中东多份大型工程合约，包括迪拜Marina港口Ocean Heights住宅项目、迪拜Emaar Towers办公楼及阿布达比Dancing Towers综合发展项目。

David Roberts

Chief Executive Officer
Aedas Ltd.

BA (Env Des) (TSIT) 1986
B Arch (TSIT) 1988
RIBA 1992
Registered Architect (HK) 1992
Authorized Person (List of Architects) (HK) 1992
CIArb 2000
AMAPM 2003
FHKIA 2006
FRAIA 2007

David has over 24 years of experience in Hong Kong architecture of which 16 years have been as a practicing Authorized Person. He is particularly experienced in gaining statutory approvals on complex and fast track projects in Hong Kong. David has also developed considerable expertise in Hong Kong, China, the Asia region and the Middle East upon a wide spectrum of projects including office, hotel, retail, recreational, industrial, infrastructure and healthcare projects.

As Aedas Asia CEO, David has often been at the forefront in establishing new links in international markets including China, North and South Asia, India and the Middle East and more recently in Europe and America. Also as a Global Board Director, his forward-looking business strategies have consistently created numerous opportunities for Aedas in various sectors and markets, not only generating significant commissions and he also strategically forges strategic alliances with Government Agencies, Developers and Engineering partners around the world.

David continues to be involved in some of the larger projects by Aedas with an emphasis on rail and air transportation and infrastructure. He believes good design should be a given with any Aedas project; the real challenge is now very much in designing ever more sustainable buildings.

麦志杰
Frank McGoldrick

麦志杰

董事总经理
凯达环球有限公司

1985年 建筑学荣誉文学士
1988年 建筑学文凭
1989年 建筑学硕士
1990年 英国皇家建筑师学会会员
1991年 香港建筑师学会会员
1991年 英国注册建筑师
1991年 香港注册建筑师

麦志杰先生为凯达亚洲区董事总经理及环球集团董事。自1993年加入凯达后，麦先生承接及管理了多个交通运输设施项目，包括港口客运码头、航运大楼、铁路站及车厂等。从2003年开始，麦志杰先生领导威尼斯人澳门路·金光大道项目，为凯达的娱乐设施业务谱写新章。而后，他成功协助凯达新加坡办事处赢取新加坡滨海湾金沙项目（综合休闲度假村）的合约，将凯达发展娱乐设施的专业经验带至新加坡。新加坡滨海湾金沙项目面积为600万平方米，设有坐拥2500个客房的五星级酒店、会议展览中心、两个各设2 000个座位的剧院、娱乐场及艺术科学博物馆。

Frank McGoldrick

Managing Director
Aedas Ltd.

BA (Hons) 1985
Dip. Arch 1988
MA 1989
RIBA 1990
HKIA 1991
Registered Architect (UK & HK) 1991

Frank is the Asia Managing Director and Global Board Director. He joined Aedas in 1993 and has built up a diverse transport portfolio including seaport terminals, air freight terminals and rail stations and depots. Frank has been instrumental in shaping Aedas' new chapter in the gaming industry with the Venetian Macao at Cotai Strip® since 2003. Later on, he assisted the Singapore office to further extending Aedas' expertise in the sector by winning the contract of the Marina Bay Sands which is a 6,000,000 sq.f. project comprising a 2500-room five-star hotel, a convention and exhibition centre; two 2000-seat theatres, a casino and an Arts and Science Museum.

韦业启
Ken Wai

韦业启

董事总经理（中国）
凯达环球有限公司

1983年 建筑学（荣誉）学士
1990年 建筑学文凭（优异级）
1993年 英国皇家建筑师学会会员
1995年 香港建筑师学会会员
1996年 香港注册建筑师
2004年 中华人民共和国一级注册建筑师资格

韦业启先生为凯达中国区董事总经理，负责项目设计及管理，并专注发展大型团体设施、基础设施建设、商业及住宅项目。韦先生曾参与全球各地多个极具代表性建筑物的设计工作。目前韦先生正担任香港国际机场的香港天际万豪酒店和圣保罗男女中学重建工程的总设计董事，同时作为上海半岛酒店、广州富力中心办公大楼、北京富力城公建项目（包括拥有539套客房的万丽酒店、办公楼及商场）等项目的设计及项目董事，他透彻了解每个项目的目标，成功将新颖理念付绪于行动。他优越的设计天份亦显见于其得奖之作，包含兴建客运码头大楼及三十三层高的办公楼即厦门港国际旅游客运码头。

Ken Wai

Managing Director (PRC)
Aedas Ltd.

BA (Hons) 1983
Dip Arch (Dist) 1990
RIBA 1993
HKIA 1995
Registered Architect (HK) 1996
PRC Class I Registered Architect Qualification 2004

Ken Wai is the Managing Director of PRC, who is responsible for project design and management, specializing in large-scale institutional, infrastructure, commercial and residential projects. His design effort has been widely involved in various iconic buildings around the world. Currently, he is the Lead Design Director for the development of SkyCity Marriott Hotel at Hong Kong International Airport as well as the regeneration of St. Paul Co-educational College. Also as the Design & Project Director for the Peninsula Hotel in Shanghai, the R&F Centre office in Guangzhou and the R&F City commercial complex consisting of a 539-room Renaissance Hotel, an office and retail components in Beijing, Ken has energetically put forward his innovative ideas into execution with the thorough understanding of project objectives. He further demonstrates his creative talent in his award-winning work – Xiamen International Cruise Terminal in progress consisting of a terminal building and a 33-storey office tower.

Aedas
凯达

施家殷
Kyran Sze

施家殷

执行董事
凯达环球有限公司

1988年 建筑设计理学士（澳洲新南威尔斯大学）
1990年 建筑荣誉学士（澳洲新南威尔斯大学）
1993年 工程管理学硕士（澳洲新南威尔斯大学）
1993年 澳洲特许建筑师
1993年 澳洲皇家建筑师学会（附属）会员
1994年 香港建筑师学会会员
1994年 香港注册建筑师
1994年 香港政府认可人士（建筑师）
1995年 英国皇家建筑师学会会员
2004年 中华人民共和国一级注册建筑师资格
2005年 亚太经合建筑师

施家殷先生为凯达执行董事及亚洲区董事局成员。施先生拥有超过20年建筑设计经验，其大量的客户及政府官员网络对凯达在大中华区设立与发展业务有极大帮助。在施先生的领导下，凯达成功取得多个大型项目建设，包括位于香港的获奖项目播道书院与港铁欣澳站。施先生亦主力为公司开拓在中国内地的业务。

Kyran Sze

Executive Director
Aedas Ltd.

BSc (Design Studies) (UNSW) 1988
B Arch (Hons) (UNSW) 1990
M Proj Mgt (UNSW) 1993
Chartered Architect (Aust) 1993
RAIA (Associate) 1993
HKIA 1994
Registered Architect (HK) 1994
Authorized Person (HK) 1994
RIBA 1995
PRC Class I Registered Architect Qualification 2004
APEC Architect 2005

Kyran is the Execntive Director and Asia Board Member of Aedas. He has over 20 years experience as an architect, whose extensive network of contacts with clients and government officials is significant to the set up and development of Aedas' business in the Greater China. He has been successful in securing a number of major projects which includes the award-winning Evangel College and Sunny Bay Station in Hong Kong. Kyran also has particular responsibility for leading the expansion of Aedas into Mainland China market.

安得宝
Andrew Bromberg

安得宝

国际设计董事 / 执行董事
凯达环球有限公司

1990年 环境设计学（特别荣誉）学士
1994年 建筑学（荣誉）硕士

安得宝先生拥有超过15年的建筑设计经验，其中8年在香港发展。他擅长各类型项目，能构思复杂而又切实可行的设计概念，因而备受赞扬。力求创新加上具有远见，令安先生分布世界各地的设计项目获奖无数。最近，安先生设计的Pentominium住宅发展项目刚于Cityscape中东房地产大奖赛中赢得其第6个设计大奖。安先生于中国的代表项目有北京北苑北辰居住区B5区公共建筑设计，位于广州琶州的新建综合发展项目以及惠州皇冠假日酒店。其新加坡市政及文化中心项目，占地62 000平方米，将于2012年完成，届时将成为该地区的新地标。

Andrew Bromberg

International Design Director / Executive Director
Aedas Ltd.

B.Env. Design (Special Hons) 1990
M Arch (Hons) 1994

Andrew has over 15 years of experience in architecture, of which 8 years in Hong Kong. He is recognized for his ability to conceptualize and realize complex designs within a wide range of projects. Striving for innovation and vision, Andrew has led the design on numerous award-winning projects throughout the world. Recently, his design for the Pentominium Tower in Dubai has won its sixth design award at Cityscape Middle East Real Estate Awards. Andrew's significant projects in China include the North Star Mixed-use Development in Beijing, a new exhibition mixed-use development in Pazhou Guangzhou and the Crowne Plaza Hotel in Huizhou. The Singapore Civic and Culture Centre, a 62,000 sq.m development due to complete in 2012, is going to be a new architectural icon in the region.

郑育良
Stephen Cheng

郑育良

执行董事
凯达环球有限公司

1982年 设计学学士
1984年 建筑（荣誉）学士
1987年 澳洲皇家建筑师学会会员
1988年 香港建筑师学会会员
1989年 英国皇家建筑师学会会员
1990年 香港政府认可人士（建筑师）
1991年 香港注册建筑师
2004年 中华人民共和国一级注册建筑师资格

郑育良先生拥有丰富的大型建设项目经验，由深入探讨有关政府部门各类规定的可行性研究开始，直至施工完成，期间每个执行项目的过程，郑先生都亲自参与。目前，郑先生正领导建筑团队，负责海洋公园等多项重建工程，同时担任广深港高速铁路西九龙总站的项目董事，为项目提供初步设计。

Stephen Cheng

Executive Director
Aedas Ltd.

B Design Studies 1982
B Arch (Hons) 1984
RAIA 1987
HKIA 1988
RIBA 1989
Authorized Person (List of Architects) (HK) 1990
Registered Architect (HK) 1991
PRC Class I Registered Architect Qualification 2004

Stephen Cheng has extensive experiences in large-scale developments in which he is responsible for execution of projects starting from feasibility study stage of making detail investigation into restrictions imposed by relevant government departments till completion of construction stage. Stephen is leading an architectural team for several redevelopment packages for the Hong Kong Ocean Park Redevelopment. He's also one of the project directors for the preliminary design works on the Guangzhou-Shenzhen-Hong Kong Express Rail Link West Kowloon Terminus.

Aedas 凯达

王敬智
Kenneth KC Wong

江立文
Max Connop

陈梁材
Chan Leung Choi

毕保罗
Paul Blazek

王敬智

执行董事
凯达环球有限公司

1990年 建筑文学士（香港大学）
1992年 建筑学学士（香港大学）
1995年 香港建筑师学会会员
1995年 香港注册建筑师
1998年 哲学硕士（香港中文大学）
2004年 中华人民共和国一级注册建筑师资格

王敬智先生的建筑设计曾赢过不少公开设计比赛的奖项，为凯达赢取了多份发展项目合约，其中大部分为北京、上海、成都、武汉及广州等城市黄金地段的大型综合及高档住宅发展设计项目。王先生以新颖的建筑设计为中国各大城市塑造新貌，而他亦以创新的手法，处理总体规划上的各样难题，成就多项新式而富创意的建筑。王先生的代表项目有北京富力城49公顷的综合发展项目及总体规划设计，中国长沙星沙镇发展项目，成都THE LOFT大型住宅发展项目，成都石羊片区一级住宅及商业发展项目，中国徐州新城区大龙湖总体规划设计，以及越南胡志明市Park City第二期住宅发展项目。

江立文

执行董事
凯达环球有限公司

1989年 建筑学（荣誉）理学士
1992年 建筑文凭
1998年 建筑管理学专业证书
1998年 英国注册建筑师

江立文为凯达执行董事，拥有逾15年的业内经验，其中9年集中在香港发展。这些年来，他主力设计了多个大型基础设施建设和商业发展项目，负责项目中概念构思至施工等多个阶段，而且，他也是一位追求革新的环保建筑、建筑可持续发展顾问，致力协助凯达为环保建筑屡创新猷，推动公司成就多项领导业界的可持续发展项目。他负责的港铁欣澳站便荣获香港建筑师学会优异奖、环保建筑专业议会之环保建筑大奖和“Architectural Record／Business Week”绿色大奖。

陈梁材

董事总经理（澳门）
凯达环球有限公司

1986年 建筑学文学士（一级荣誉）
1988年 建筑学学士（优异级）
1990年 香港建筑师学会会员
1990年 英国皇家建筑师学会会员
1991年 香港注册建筑师

陈梁材先生现为凯达澳门董事总经理，从事建筑设计接近20年，在各个领域拥有丰富经验，其中澳门金沙娱乐场为其代表作之一，项目由2003年初开展，于2004年第一季竣工，顺利地在高速下完成。其出众的项目管理能力，令陈先生成为澳门威尼斯人一度假村一酒店此大型项目的项目董事之一。此项目占地1 000万平方米，为亚洲区内最具规模的主题度假区及娱乐场。发展项目包括8 500平方米的商场、设有15 000个座位的展览中心、设有2 000个座位的表演场馆、50万平方米的赌场及近3 000个套房的酒店。

毕保罗

董事总经理
凯达城市规划及园景设计有限公司

1989年 建筑学学士
1993年 景观建筑学硕士
1999年 注册建筑师（纽约）
2009年「领先能源与环境设计」认证专业人员

毕保罗先生曾于美及亚洲等地从事城市规划及园林景观设计，拥有逾20年专业经验。他曾设计和管理多个大型建设和园林景观设计项目，并拥有独特的园林设计触觉，往往能在项目中就建筑与景观部分作出平衡和协调。毕先生尤其擅长概念设计，他的草图及视像化技术精湛，这有助于他向客户阐释设计内容，促进项目成功。毕先生现正参与越南胡志明市Park City大型住宅项目的总体规划，他的经验亦覆盖如医疗设施、体育及文娱设施、博物馆及其他大型文化活动场所、大型商场及房屋项目、主题公园及其他多个住宅项目。

Kenneth KC Wong

Executive Director
Aedas Ltd.

BA (HKU) 1990
BArch (HKU) 1992
HKIA 1995
Registered Architect (HK) 1995
MPhil (CUHK) 1998
PRC Class I Registered Architect Qualification 2004

Kenneth is highly design-focused and has been responsible for procuring work through winning design competition. Most of his projects are large-scale mixed-use and super deluxe residential development, occupying the best locations in the cities of Beijing, Shanghai, Chengdu, Wuhan and Guangzhou. His vision is to create a new urban landscape in the major cities of China, and his unconventional approach in solving master planning problems produces a lot of exciting and innovative projects. His significant projects include Beijing R&F City masterplan, Xingsha residential masterplan in Changsha, THE LOFT in Chengdu, a residential & commercial development at Shiyangpian District, Chengdu, Da Long Hu masterplan consultancy services in Xin Cheng District, Xuzhou, and Park City Phase 2 Residential Development in Ho Chi Minh City, Vietnam.

Max Connop

Executive Director
Aedas Ltd.

BSc (Hons) 1989
DipArch 1992
Cert. Prof. Practice and Mgt 1998
Registered Architect (UK) 1998

Max Connop is an Executive Director of Aedas and he has over 15 years of professional experience, of which 9 of those years have been spent in Hong Kong. Over the years, Max has been working predominantly on large infrastructure and commercial projects from conception to construction. He is also a progressive green building/sustainability adviser, and has significantly influenced Aedas' industry-leading sustainability initiatives. His Sunny Bay MTRCL Station project received an HKIA Merit Award, Council Green Building Award, and Architectural Record Business Week's Sustainable Design Award.

Chan Leung Choi

Managing Director (Macau)
Aedas Ltd.

BA (Arch Studies) (1st Hon) 1986
B Arch (Dist) 1988
HKIA 1990
RIBA 1990
Registered Architect 1991

LC is the Managing Director of Aedas Macau, who has been practicing for almost 20 years with numerous significant experiences across various sectors. One of his most significant works is the Venetian Sands Macao. This fast-track project commenced in early 2003 and was completed in the first quarter of 2004. His outstanding project management skills make him one of the Project Directors of the mega-scale Venetian-Macao- Resort-Hotel.
The 10 million sq.f. project is the largest themed-resort & casino development in Asia, which includes a 8,500 sq.f. retail space, a 15,000-seat Exhibition Centre, a 2,000-seat Zaia Theatre, a 500,000 sq.f. Casino, and a 3,000-suite Hotel.

Paul Blazek

Managing Director
Aedas Urban Design & Landscape Ltd.

B Arch 1989
M Landscape Arch 1993
Registered Architect (NY) 1999
LEED AP 2009

Paul has over 20 years of professional experience on a wide range of projects in the U.S., Europe and Asia. He has designed and managed numerous large and complex building and landscape projects. Paul has a unique perspective as a landscape designer, focusing on the relationship between building and landscape. He is particularly skilled in concept design. His strong sketching and visualization skills make him particularly effective when working with client groups. Paul is currently working on Park City in Ho Chi Minh City, Vietnam. He has been involving in a wide range of work including healthcare, sports and recreation, museums and other major cultural facilities, large-scale retail and housing projects, theme parks, as well as numerous residential projects.

Aedas
凯达

麦伟德
Logan MacWatt

费基利
Gregory Farrell

白礼达
Nenad Bratic

狄美师
Dan Meis

麦伟德

董事总经理
凯达室内设计有限公司

1975年 建筑学（荣誉）文学士
1988年 MCSD

麦伟德先生拥有逾27年室内设计经验，其中20年在亚洲发展。在凯达室内设计，他负责监察香港、澳门、中国、新加坡及中东的国际业务运作。麦先生在室内设计的各个范畴皆拥有丰富的经验，尤其是为零售及企业作设计。麦先生擅长透过弹性空间规划及灵活多变的设计概念迎合各类型的办公室环境为企业降低营运成本。麦先生近期的项目客户包括东亚银行、香港交易所、香港机场管理局总部办事处、毕马威会计师事务所、巴林世贸中心、中东迪拜AL MAS DMCC Tower、仁和春天广场、上海半岛酒店、德基广场、中东迪拜国际金融中心Lighthouse大楼、高德置地广场、上海九百城市广场、香港国际机场北卫星客运廊，以及其整体室内设计及环境改善研究。

Logan MacWatt

Managing Director
Aedas Interiors Ltd.

BA (Hons) 1975
MCSD 1988

Logan has been a practicing Interior Designer for over 27 years of which 20 years have been spent in Asia. He oversees the international operations of Aedas Interiors throughout Hong Kong, Macau, PRC, Singapore and the Middle East. Logan has extensive experience in all sectors of Interior Design with a current emphasis on retail and corporate office environments. Logan is passionate about creating flexible plannings and design concepts to keep pace with fast and ever-changing design directions. His recent projects include Bank of East Asia, Hong Kong Stock Exchange, Hong Kong Airport Authority Headquarters, KPMG, Bahrain World Trade Centre, Al Mas DMCC Tower, Renhe Spring Retail Centre, Peninsula Shanghai Hotel, Deji Plaza, DIFC Lighthouse Tower, GT Landmark, Join Buy Retail Centre, North Satellite Concourse and the overall ambience enhancement study of the Hong Kong International Airport.

费基利

董事
凯达室内设计有限公司

费基利先生拥有逾20年设计及管理经验，擅长酒店、休闲设施及不同类型的住宅设计，他曾参与不同地区的建设项目，包括澳大利亚、新西兰、东南亚、中东及欧洲，项目经验丰富。费先生加入凯达后，主要负责酒店及休闲设施室内设计项目，最近负责的项目有中国北京朗廷酒店、印度浦那朗廷酒店、香港旺角朗廷酒店、中国上海波特曼丽嘉酒店、俄罗斯秋明丽笙SAS大酒店、阿曼Blue City－Anantara酒店、迪拜Island Global Yacht Marina游艇俱乐部、越南藩切一拥有150房间之度假别墅，以及位於阿联酋及俄罗斯之Jumeirah酒店。

Gregory Farrell

Director
Aedas Interiors Ltd.

Greg has over 20 years of design and management experience specializing predominantly in hotel, hospitality and multi-residential design. He has extensive experience throughout Australia and New Zealand as well as South East Asia, the Middle East and Europe. Greg is responsible for hotel and hospitality interior design at Aedas Interiors. His recent projects include Langham Place Hotels Beijing, Pune and Mongkok, Ritz-Carlton Portman Shanghai Hotel, Radisson SAS Hotel Russia, Anantara Resort Oman, Island Global Yacht Marina Clubhouse at Dubai, a 150-room resort in Phan Thiet Vietnam and Jumeirah Hotels in the UAE and Russia.

白礼达

董事
凯达室内设计有限公司

1983年 Dip Ing Arch（Grade 10）
1985年 SAS，SAY

白礼达先生拥有逾25年的业内经验，其中超过20年专职负责多个位于南亚区的建筑及室内设计项目。他熟习各类型的设计手法及深化设计和制图等范畴，他的经验充份展示在其负责的度假村、酒店、高档低层住宅别墅、办公楼、商场以及教育设施等。白先生负责澳门威尼斯人项目的室内装修设计部分，除此之外他还曾设计多个位于香港浅水湾及寿臣山的豪宅项目，以及中国北京东方广场住宅部分等。

Nenad Bratic

Director
Aedas Interiors Ltd

Dipl Ing Arch (Grade 10) 1983
SAS, SAY 1985

Nenad Bratic has over 25 years of experience, which over 20 was in South Asian architecture and interior design. He is particularly experienced in design, design development and documentation in a wide spectrum of hospitality projects including integrated resorts, hotels, luxury low-rise residential villas as well as office, retail and educational projects. Nenad was a team leader for fit-out component of Venetian Macao. Nenad has also completed several residential projects in Hong Kong, which include the luxury houses at Repulse Bay Road and Shouson Hill and the Beijing Oriental Plaza residential development.

狄美师

董事总经理（洛杉矶）
凯达环球有限公司

1985年 建筑学士
1987年 注册建筑师（美国－伊利诺州）
2000年 注册建筑师（美国－宾夕法尼亚州）
2006年 美国建筑师学会资深会员

狄美师先生从事建筑设计逾25年，现正担任凯达洛杉矶办事处董事总经理，被誉为美国最具影响力的体育及娱乐设施设计师。在NBBJ任职期间，狄先生创立了体育设施专责部门，以满足各种不同的相关需求。他担任设计的洛杉矶Staples Center被誉为空前未有的最佳运动场建筑，其本人更在2001年成为《时代杂志》专访"全球体坛一百位最杰出的创意人"之一。他的作品曾两度获得商业周刊、建筑实录奖项，以肯定建筑设计对客户发展的重要性。狄先生亦是唯一两度获得《商业体坛杂志》四十位四十岁以下最具影响力体坛领导人殊荣的建筑师。他现正为占地百万平方米的巴林萨基尔城总体规划项目担任总设计师。

Dan Meis

Managing Director (Los Angeles)
Aedas Ltd.

B Arch 1985
Registered Architect (US – Illinois) 1987
Registered Architect (US – Pennsylvania) 2000
FAIA 2006

Widely regarded as one of the American most influential designers of sports and entertainment facilities, Dan Meis is the Managing Director of Aedas Los Angeles with over 25 years of experience. As the co-founder of NBBJ Sports and Entertainment, he established the sports division responding to various facility needs. Dan's design for Los Angeles' Staples Center has been heralded as the 'greatest arena ever built', and in 2001 he was featured in Time Magazine as one of the '100 Top Innovators in the World of Sports'. His work has twice been awarded the prestigious Business Week/Architectural Record Award, recognizing the value of design to a client's business, and he is the only architect twice recognized as one of Sports Business Journal's '40 under 40 Most Influential Sports Executives'. Dan is currently working on several projects including Sakhir City Master Plan in Bahrain, a development over 1,000,000 sq.m. in which he is the leading designer.

Aedas 凯达

温子先
Andy Wen

温子先

设计董事
凯达环球建筑设计咨询（北京）有限公司

1987年 建筑学文学士
1990年 建筑学硕士
1999年 建筑学博士

温子先先生从事建筑设计逾20年，曾参与不同类型的建设项目，包括总体规划、住宅大楼、公寓楼、别墅、五星级酒店、广场及办公楼等。他曾分别在美国、中国台湾和中国内地工作，拥有庞大的客户网络，从国际大型地产发展商及连锁酒店集团，到小型住宅物业公司等都能广泛涵盖。除了擅长各类型建设项目外，温先生的室内设计天份同样出众，他曾为不同项目肩负起室内设计重任，包括办公室、大学、图书馆、私人企业、零售商铺、餐厅及公寓等，并成就了十六个原创家具、照明及配件系列，展现出形形色色的艺术，由摩登到现代亚洲风情都应有尽有。温先生最近负责的项目有南京河西奥体新城苏宁广场、徐州苏宁广场、鼓楼科技园及苏州西交利物浦大学信息楼。

Andy Wen

Design Director
Aedas (Beijing) Ltd

BA Arch 1987
M Arch 1990
PH.D Arch 1999

Andy has been working as a Design Architect for over 20 years with significant experience across sectors. He was involved in numerous projects including master plans, residential towers, apartment complexes, villas, five-star hotels, plazas and offices. Andy has established an extensive client network in the United States and China, ranging from huge international real estate developers and hotel chains to small-scale property operators. Not only is Andy an excellent architect, his interior service endowment has further diversifying his presence in the industry. While allowing end-users more stylish yet comfortable interiors in offices, universities, libraries, corporate, retailers, restaurants, and apartments, Andy has worked out over 16 lines of designer furniture, lighting and accessories in styles, ranging from modern to Asian contemporary. He has recently contributed to the projects of Suning Hexi Nanjing Plaza; Xuzhou Suning Plaza, Gulou Science Park and Xi-an Jiao Tong - Liverpool University, Suzhou.

陈英琪
Angela Chan

陈英琪

董事（中国）
凯达环球有限公司

1988年 建筑学文学士
1991年 建筑学士
1993年 香港建筑师学会会员
1994年 注册建筑师（香港）
1995年 香港政府认可人士（建筑师）
2002年 英国仲裁学会会员
2004年 香港规划师学会会员
2006年 中华人民共和国一级注册建筑师资格
2009年「领先能源与环境设计」认证专业人员

陈英琪小姐于2008年加入凯达时任职主任，于去年晋升为高级主任，现为公司董事（中国）。效力凯达前，她曾成立非常空间设计有限公司，并出任设计董事。2001至2004年间，陈小姐在保良局担任产业及工程部主管，负责监督部门运作，其间，她曾为保良局各种规模的学校、社会服务中心及其他公司物业工程提供专业的意见及指导。在多年的建筑设计中，陈小姐专心致志，在于多个领域内取得令人注目的成绩，包括城市设计、商业项目及室内精装等的设计及方案实施工作，更荣获有不少奖项。此外，她亦积累相当丰富的项目管理经验，与客户和顾问紧密合作，承接多项综合发展住宅的大型建设。目前，她正代表凯达负责中国徐州湖畔发展及成都太升南路综合发展项目。

Angela Chan

Angela Chan
Director (PRC)
Aedas Ltd.

BA (AS) 1988
BArch 1991
HKIA 1993
Registered Architect (HK) 1994
AP 1995
CIArb 2002
HKIP 2004
PRC Class 1 Registered Architect 2006
LEED AP 2009

Angela Chan was an Associate of Aedas Ltd in 2008, and was promoted to Senior Associate last year, and now is a Director (PRC). Prior to working at Aedas, Angela had been leading Varispace Design Ltd/ HS Design Int'l Ltd, her own practice as a Design Director for 4 years. From 2001 to 2004, she had served at the Po Leung Kuk as the Head of Property overseeing the Property and Works Department. During her tenure, she had provided professional consultations and guidance over constructions in different scales including schools, social service centres and other company's properties. Over these years of devoted service, Angela has made her footprints remarkable with numerous awards won in urban design, commercial developments and interior fit-outs. With her extensive experiences as a project manager, she has been working closely with a good network of clients and consultants which leads to the success in various large-scale mixed-use residential developments. Currently at Aedas, she is working on Xuzhou Lakeview development, Chengdu Tai Sheng Nan Road mixed-use development in China.

陈旭晴
Jean Chen

陈旭晴

董事
凯达环球建筑设计咨询（北京）有限公司

1993年 建筑学学士（同济大学）
2006年 工商管理硕士（香港大学）

陈旭晴于2009年底加入凯达环球，现出任董事一职，并常驻凯达环球上海的分公司。上海同济大学毕业后，她前往新加坡展开其专业生涯，在文化及公共建筑方面积累了宝贵经验。其间，她参与建设了闻名遐迩的新加坡艺术中心、新加坡淡马锡理工学院三期扩建项目，以及位于那辛路的高级住宅项目，其天份和努力备受客户推崇。

2000年，她回到上海为著名的新天地项目担任高级项目建筑师。完成该项目后，她受邀于地产发展商，任职创作总监，先后与世界著名建筑事务所合作，负责策划并管理位于上海闵行区，面积达10平方千米的浦江森利半岛。

陈小姐擅于将其丰富的国际经验及对中国国情的独到见解融合，并转化为创新的建筑设计。她不时对中国城市的演变提出意见，在从事城市化发展的过程中，她相信自己已在平衡技术与艺术之间取得正确方向，并有效回应不同层面的经济需求。

Jean Chen

Director
Aedas (Beijing) Ltd

B Arch 1993 (Tongji University)
MBA 2006 (HKU)

Jean Chen has joined Aedas since 2009 and is now a Director based in Shanghai. After graduated from Tong Ji University, Shanghai, she had embarked on her profession in Singapore where valuable experience in cultural and civic architecture was accumulated. Meanwhile, she had significantly contributed to some of the region's famous buildings including the reputed Singapore Art Centre, Temasek Polytechnic Phase III extension and the high-end residence on Nassim Road. The synergy of her talent and effort has brought her considerable compliments from the clients.

Jean returned to Shanghai in 2000 as the Senior Project Architect for the well-known XinTian Di development. Afterwards, she was invited to joining local developers as a Design Director collaborating with world-renowned architectural firms, which enabled her to plan and manage the master planning of the 10sq.km. PuJiang Forest Peninsular in Ming Hang District.

Jean is good at interlacing international experience with her unique understanding of China's circumstances into innovative designs. To offer insightful input to the dynamic developments of China's cities, she believes she has been on the right track of achieving urbanization with the well balance of technology and art responding to economic needs in different levels.

张居健
William Cheung

张居健

董事（中国）
凯达环球有限公司

1990年 建筑文学士
1993年 建筑理硕士（建筑管理及经济学）
1995年 建筑学深造文凭
2000年 香港建筑师学会会员
2000年 香港注册建筑师
2001年 香港政府认可人士（建筑师）

张居健先生在凯达任职董事（中国）。他从事建筑设计逾15年，其中4年受聘于香港数间大型地产发展商，专职负责项目管理。从概念设计阶段至施工完成，期间每个执行项目，他都亲力参与，在管理及设计住宅方面经验尤其丰富。加入凯达后，张先生投入参与不少重要发展项目，包括上海浦东高级住宅（占地46万平方米，设有240幢独立式别墅、150幢半独立式别墅、会所及国际学校），以及香港地铁公司东涌站东涌市地段第5号住宅。最近，他亦为内地成都石羊片总体规划担任项目建筑师，负责一、二期的发展。目前一期设计已经完成，二期则于顺利发展当中。

William Cheung

Director (PRC)
Aedas Ltd.

BAAS 1990
MSc Arch (Construction Economics & Management) 1993
Post Dip Arch 1995
HKIA 2000
Registered Architect (HK) 2000
AP 2001

William is a Director of Aedas Ltd, who has over 15 years of experience of which 4 years as a project manager at some major developers in Hong Kong, and in particular familiar with managing and designing residential development from conception to completion. Since joining Aedas, William has devoted himself to several significant developments including the luxurious Putong residence in Shanghai which covers 460,000 sq.m consisting of more than 240 villas, 150 semi-detached houses, clubhouses and international schools, and Planning Area 19, TCTL 5 MTRC Tung Chung Station Development. Furthermore, he has been recently involved in Shiyanpian District Masterplan in Chengdu as the Project Architect responsible for both Phase One and Two, in which the former has already been completed and the later is currently underway.

Aedas
凯达

徐子苹
Zi Ping, Xu

柳景康
Cary Lau

林静衡
Christine Lam

祈礼庭
David Clayton

徐子苹

董事（中国）
凯达环球有限公司

1990年 建筑学学士（天津大学）
2004年 建筑学博士（香港大学）
2002年 中华人民共和国一级注册建筑师

徐子苹小姐于凯达任职董事（中国），在内地和香港拥有十多年专业经验，商业设计方面的造诣尤其深厚。她非常熟悉国内建筑设计条例，并擅长概念设计、细部设计及项目管理，令不少杰出建筑得以落成，范围包括商场、办公楼、工厂、高层及低层住宅及学校。她在研究方面的付出亦备受肯定，曾参与工作包括香港屋宇署绿色建筑评估法CEPAS（环保效能表现全面评审计划）的研究。

柳景康

董事
凯达环球有限公司

1995年 建筑（荣誉）学士
1998年 建筑硕士（成绩优异）
2001年 香港建筑师学会会员
2002年 香港注册建筑师

柳景康先生于1998年加入凯达，现出任董事一职。柳先生的设计才华出众，屡获设计殊荣，其中凭香港教育学院赛马会小学设计，获颁2002年香港建筑师学会全年建筑大奖；2004年获“弘扬抗疫精神建筑景观”设计比赛优异奖；2002年又获迈向多元化校园建筑设计比赛冠军。作为一位极富创意和洞察力的建筑师，他往往能够应客户的要求，构思别具时代特色和极富魅力的建筑。

柳先生最近的作品包括香港将军澳第65区香港基督教播道会联会播道书院直资中小学校舍的获奖设计，此设计竞赛由香港建筑师学会赞助，香港特别行政区教育统筹局推行，柳先生的设计方案经多番遴选后脱颖而出，并已落实执行，完成兴建。

林静衡

董事
凯达环球有限公司

1995年 建筑学荣誉学士
1998年 建筑学硕士（成绩优异）
2000年 香港建筑师学会会员
2000年 香港注册建筑师

林静衡小姐的设计作品以优雅著称，其对客户需要的透彻了解以及设想周到的构思亦是其设计的强大后盾。林小姐拥有出众的设计才华，在多个项目中均担任重要的设计角色，其作品充份反映她对特定项目在文化及知识层面上的深刻反思。林小姐的客户网络覆盖香港企业及国际发展商，无论新旧设计，林小姐为客户悉心孕育的作品始终杰出如一，足以证明其作品持续的优秀品质及完整性。林小姐现正率领团队为中国无锡恒隆广场作设计、制图及方案实施。此项目面积255 000平方米，是当地的地标建筑。

祈礼庭

董事
凯达环球有限公司

1997年 （荣誉）理学士
1999年 建筑学士（成绩优异）
2001年 英国皇家建筑师学会会员
2001年 英国注册建筑师

祈礼庭先生毕业于英国加的夫威尔士建筑学院，并以优异的成绩取得建筑学士学位，同时荣获Alwyn Lloyd 奖，成为该年度最佳建筑学士学生。祈先生拥有丰富的国际经验，曾参与全球多个地区的建设，包括欧洲、俄罗斯、香港、新加坡、韩国和中国。从项目的可行性研究、概念设计，到工地现场的协调工作等，他都参与其中。最近，祈先生刚完成香港中环一甲级办公楼及裙楼翻新工程，目前正领导团队参与中国无锡恒隆广场综合发展项目，包括设计、制图及方案实施。此项目面积255 000平方米，包含高档商场和两幢甲级办公楼。

Zi Ping, Xu

Director (PRC)
Aedas Ltd.

BEng Dept of Arch 1990 (Tianjin U, China)
PhD, Dept of Arch 2004 (HKU, HK)
PRC Class I Registered Architect Qualification 2002

Zi Ping, Xu is a Director of Aedas Ltd. with more than a decade of professional experience in both mainland China and Hong Kong, especially on retail design. With comprehensive knowledge on mainland building regulations, she has made perfect use of her creative talent in concept design, detail design and project management actualizing outstanding shopping malls, offices, industrial estates, high-and low-rise residences and schools. Zi Ping's effort has been further recognized in the research field with regard to Comprehensive Environmental Performance Assessment Scheme for Buildings, a green building assessment tool study for Buildings Department in Hong Kong.

Cary Lau

Director
Aedas Ltd.

BA (Arch Studies) (Hons) 1995
M Arch (Dist) 1998
HKIA 2001
Registered Architect (HK) 2002

Cary Lau joined Aedas in 1998 and is now a Director of Aedas in Hong Kong. Cary has great design talents and has received series of design awards from time to time including the HKIEd Jockey Club Primary School (HKIA Medal of the Year) 2002, Fighting SARS Memorial Architectural Scene Design Competition (Honorable Mention) 2004, In Search of Diversity in School Design (Champion) 2002. As an architect striving for innovation and vision, Cary is capable of producing modern and inspired architecture which client has always asked for.

His recent significant works include the award winning design from a competition sponsored by HKIA and promoted by the Education and Manpower Bureau HKSAR:
Evangel College - A "Through-train" Primary cum Secondary School, which has been selected and implemented amongst a wide selection of schemes.

Christine Lam

Director
Aedas Ltd.

BA (Arch Studies) (Hons) 1995
MArch (Dist) 1998
HKIA 2000
Registered Architect (HK) 2000

Christine's work is characterized by a supplicated elegance that is underpinned by a thorough understanding of the brief as well as the holistic inputs of design effort. Her instinctive flair for design has placed her at the forefront of projects. She has gained recognition for stylistically innovative design work whose context responds to cultural and /or intellectual concerns. Her clients have ranged from Hong Kong based corporations to international developers. The repeated work that Christine nurtures with her clients beats testimony to the quality and 'completeness' that consistently infuse her designs. Christine is now leading a team to design, document and realize Wuxi Hang Lung Plaza, a 255,000 sq.m mixed-use development in China, which will be an icon in the region.

David Clayton

Director
Aedas Ltd.

BSc (Hons) 1997
B Arch (Dist) 1999
RIBA 2001
Registered Architect (UK) 2001

David studied at the Welsh School of Architecture, Cardiff in the UK where he gained a distinction in Bachelor of Architecture. He also won the Alwyn Lloyd prize as the Best Architecture Student in his year. David's involvement has ranged from front-end feasibility and concept design through to project co-ordination on site. David has international experience having worked on projects in Europe, Russia, Hong Kong, Singapore, Korea and China. Recently, David has completed a Grade A Office and Podium renovation in Central Hong Kong. He is currently co-leading a team to design, document and realize Wuxi Hang Lung Plaza, a 255,000 sq.m mixed-use development in China. The scheme comprises prestige retail with two Grade A office towers.

Aedas
凯达

梁世豪
Edward SH Leung

李静雯
Helena Lee

叶敏翘
Irene Ip

黄政朗
Jerome CL Wong

梁世豪

董事
凯达环球有限公司

建筑学荣誉学士
建筑学硕士（香港大学）
国际建筑管理理学硕士
1996年 香港建筑师学会会员
1996年 香港注册建筑师
1997年 香港政府认可人士（建筑师）
1999年 英国仲裁学会会员
2003年 ISO 14001:1996 环境保证内部审核员
2009年 「领先能源与环境设计」认证专业人员

梁世豪先生在凯达任职董事，拥有逾10年香港建筑设计经验、8年认可人士资格。他曾参与不同类型的建设项目，包括基础设施建设、办公楼、酒店、零售商场、高科技设施及住宅项目等，尤其擅长为本港大型复杂和讲求速度的建筑项目处理入则报批。他曾为香港国际机场的翔天廊项目和海天客运码头项目担任项目建筑师，前者为1 000 000平方尺的扩建工程，连接到多元化模式交通枢纽、购物中心及机管局总部大楼。后者则提供跨境空海交通转驳服务，及延长客运大楼的旅客捷运系统至翔天廊。

李静雯

董事
凯达环球有限公司

1997年 建筑学学士
1997年 美国建筑师学会会员
2003年 美国注册建筑师
2004年 香港建筑师学会会员

李静雯小姐于2005年加入凯达任职高级建筑师，同年晋升为主任，并分别于2006年及2009年升为高级主任及董事。

李小姐主力处理亚洲及中东多个大型项目，包括新加坡市政及文化大楼、北京北苑北辰居住区B5区公建、广州琶洲综合发展项目、迪拜Boulevard Plaza甲级办公楼及DAMAC Ocean Heights Ⅰ & Ⅱ，以及阿布比Empire Tower。最近，她亦积极参与香港高速铁路即以西九龙为总站，连系香港、深圳、广州的大型交通建设。

叶敏翘

董事
凯达环球有限公司

1986年 建筑理学士（McGill）
1987年 建筑学士（McGill）
1989年 香港建筑师学会会员
1990年 英国皇家建筑师学会会员
1991年 香港注册建筑师
1992年 澳洲皇家建筑师学会会员
2001年 ISO 9001:2000 品质保证内部审核员
2003年 ISO 14001:1996 环境保证内部审核员
2006年 项目管理协会附属会员

叶敏翘小姐于2000年加入凯达，于2007年升任董事。
叶小姐目前正负责着港铁沙中线新国际邮件中心和香港国际机场国泰航空货运大楼项目，并在项目中担任项目董事，同时她亦负责为凯达在香港国际机场的其他工程项目联系相关政府部门和机管局，以进行报建审批工作。她同时正参与香港规划署委托的中环新海滨城市设计研究。

黄政朗

董事
凯达环球有限公司

1997年 建筑学文学士
2000年 建筑学硕士
2003年 香港建筑师学会会员
2003年 注册建筑师（香港）

黄政朗先生在凯达任职董事，从事香港及中国内地的建筑项目逾8年，善于为复杂和高速度要求的项目送审报批及联系相关的政府部门，对建筑细部设计、施工图设计、合同及施工管理等也十分熟悉。他曾参与香港多个范畴的建筑项目，包括零售、商业、教育、酒店及多用途建设等，积累了相当丰富的经验。

Edward SH Leung

Director
Aedas Ltd.

BA (AS) Hons
M Arch (HKU)
MSc (Int'l Construction Mgt)
HKIA 1996
Registered Architect (HK) 1996
AP (HK) 1997
MCIArb 1999
ISO 14001:1996 Internal Auditor, 2003
LEED AP 2009

Edward Leung is a Director of Aedas with over 10 years of experience in Hong Kong, of which 8 years as a practicing Authorized Person. Upon a wide spectrum of developments covering infrastructure, offices, hotels, retail and residence, he is particularly skillful in gaining statutory approvals on complex and fast track projects in Hong Kong. As a project architect, Edward had dedicated his talent to the design and construction of SkyPlaza (a 1,000,000 sq. ft. extension to a multi-modal transport hub, a retail centre and the Airport Authority headquarter building) and SkyPier (a sea-to-air cross-border terminal with an extension of the PTB Automatic People Mover to SkyPlaza), which are both located at Hong Kong International Airport.

Helena Lee

Director
Aedas Ltd

BArch 1997
AIA 1997
Registered Architect (US) 2003
HKIA 2004

Helena joined Aedas in 2005 as Senior Architect and was promoted to Associate in the same year, then to Senior Associate in 2006, and to Director in 2009.

Helena is particularly experienced in managing large-scale establishments in Asia and the Middle East including Singapore Civic & Cultural Complex in Singapore, North Star mixed-use development in Beijing, Pazhou mixed-use development in Guangzhou, the Boulevard Plaza and DAMAC Ocean Heights I & II in Dubai, and Empire Tower in Abu Dhabi. Moreover, she has been recently devoted to constructing the mega express railway terminus at West Kowloon linking up across Hong Kong, Shenzhen and Guangzhou in the future.

Irene Ip

Director
Aedas Ltd.

BSc Arch (McGill) 1986
B Arch (McGill) 1987
HKIA 1989
RIBA 1990
Registered Architect (HK) 1991
RAIA 1992
ISO 9001:2000 Internal Auditor 2001
ISO 14001:1996 Internal Auditor 2003
APM Associate Member 2006

Irene Ip joined Aedas in 2000, and was promoted to Director in 2007. She is currently working on the new International Mail Centre in Shatin-to-Central Link and Cathy Pacific Cargo Terminal at Hong Kong International Airport as a Project Director. In liaisons with the relevant government departments and the Airport Authority, She is also intensively involved with the consultant team for the procurement of tender and the on-site constructions for other airport constructions in Hong Kong. Moreover, she has been engaged in the Urban Design Study for the New Central Harbour Waterfront project assigned by the Planning Department.

Jerome CL Wong

Director
Aedas Ltd.

BA Arch Studies 1997
M Arch 2000
HKIA 2003
Registered Architect (HK) 2003

Jerome is a Director of Aedas with 8 years of experience in Hong Kong's and the Mainland China's architecture, who is specialized in architectural detailing, construction documentation, contract and construction management, as well as liaising with government departments on statutory processing for complex and fast track projects. His effort has been covering a wide spectrum of sectors including retail, commercial, education, hotel and mixed-use development in Hong Kong.

Aedas 凯达

莫志豪
Joseph Mok

莫志豪

董事
凯达环球有限公司

1988年 建筑学学士
1999年 美国全国注册建筑师委员会会员
1996年 美国建筑师学会会员

莫志豪先生从事建筑设计达16年，现为凯达董事。他曾参与香港、内地、美国、中东及欧洲等地多个大型建设项目。加入凯达前，莫先生曾任地铁公司建筑经理，亦曾于香港数间建筑师事务所任职。除了住宅、商厦、铁路、民用及社区建设及教育建筑等领域拥有丰富经验外，他在总体规划和室内设计方面亦表现出色，并备受推崇。

Joseph Mok

Director
Aedas Ltd.

B Arch 1988
NCARB 1999
AIA 1996

Joseph Mok is a Director of Aedas, who has over 16 years of experience in international project design and development throughout Hong Kong, PRC, United States, the Middle East and Europe. Prior to joining Aedas, he was the Architectural Manager of Mass Transit Railway Corporation and had once been with some other local architectural practices. While having a wealth of experience in residence, commercial, railway, civic & community and education, he has also marked some well-respected footprints in masterplanning as well as interior design.

Karma Barfungpa

Karma Barfungpa

董事
凯达环球有限公司

1993 建筑学士
1993 印度建筑协会
2001 英国皇家建筑师学会会员

Karma于1998年加入凯达，现为公司董事。Karma拥有逾13年的全面的建筑设计经验，主力香港建设，从项目启动到施工，期间各个阶段的工作，他都驾轻就熟。他曾参与的项目遍布亚洲各地，其中以中国香港、中国内地、蒙古、印度、中东等为主，涉及的类别亦一分多元化，包括商场、商厦、宗教设施、政府机关及交通建设。他除了对建筑环境感兴趣外，亦热衷于图纸与文件编制和施工细部，能以敏锐的触觉和闲熟的技巧处理相关的问题，另外对国际建筑标准和规范也有充分的掌握。

Karma Barfungpa

Director
Aedas Ltd.

B Arch 1993
The Council of Architecture, India 1993
RIBA 2001

Karma joined Aedas in 1998 and is currently a Director of the company. Karma has a well rounded solid 13 years of architectural experience, mainly in Hong Kong. His experiences have been garnered from project inception stages onwards to the construction stages. Geographically Karma has worked extensively in Asia with particular experiences in Hong Kong, China, Mongolia, India, Middle East, etc. He has also diversified his expertise in retail, commercial, religious, institutional and transport sectors. Apart from his professional interest for the built environment, he has a particular interest in documentation and construction detailing, and has achieved a unique industry understanding and a fluency in the necessary skills. He is also well versed in various international building standards and codes.

Malcolm E Sage

Malcolm E Sage

董事
凯达环球有限公司

1981年 建筑文凭
1983年 英国皇家建筑师学会会员
1985年 香港建筑师学会会员
1985年 香港政府认可人士（建筑师）

Malcolm Sage于2005年加入凯达，现任董事。他在亚洲区已积累了逾25年的经验，曾参与中国香港、中国澳门、中国内地、印尼及巴布亚新几内亚等地多个不同类型的建设项目。他起初在英国工作，期间4年不时穿梭英、亚两地。他曾与多间知名的国际级设计顾问公司合作，领导多个建设项目，从独立住宅，到大型商业项目及高要求的总体规划发展都一一包罗。Malcolm灵活多变、多才多艺，在拟备项目发展纲要及施工管理方面都很擅长。他不但富有商业触觉，更擅于沟通及领导团队解决不同的问题。其参与的项目有大型地产发展项目、基础设施建设、著名公共建设及室内设计项目。

Malcolm E Sage

Director
Aedas Ltd.

Dip Arch 1981
RIBA 1983
HKIA 1985
Authorized Person (List of Architects) 1985

Malcolm is a Director of Aedas Ltd. He has been with Aedas since 2005 and has over 25 years of professional experience in Asia on a broad variety of projects in Hong Kong, Macau, China, Indonesia and Papua New Guinea. Before moving to Asia, Malcolm had worked in the UK for 4 years and traveled frequently between the two. Malcolm has worked at Director level with highly recognized international design consultants on a diverse range of commissions varying in size from individual houses to the largest commercial developments and demanding Master plans. Equally at ease with formulating briefs for clients or managing construction contracts Malcolm's skills are versatile and multi-faceted.

Mark Pollard

Mark Pollard

董事
凯达环球有限公司

1992年（荣誉）文学士
1995年 建筑（荣誉） 学士
2005年 工商管理硕士
1996年 英国皇家建筑师学会会员
1996年 英国注册建筑师（ARB）

Mark于2006年加入凯达任职高级主任，于2007年晋升为董事。

Mark从事建筑设计逾11年，是一位全面的项目建筑师，他多方面的才能从他出色的管理技巧和沟通协调事务的优秀表现上可见一斑。他曾为中国香港和东南亚地区多个大型且高速度要求的项目进行深化设计、送审报批、施工图制作、施工监控及管理等，在设计创作、项目执行和结构系统使用等方面均有丰富的经验。

此外，Mark在管理和领导建筑团队方面亦同样拥有丰富的经验和十足的信心，而且他奉行减少监控的原则，尤其在电脑辅助设计的范畴，图纸均由电脑制作，准确度高，这也相应地减少干预监督的需要。另外，他对设计及建筑的过程均有透彻的了解，并能以精密分析，决定每个项目所需的元素和建筑方法。

Mark Pollard

Director
Aedas Ltd.

BA (Hon) 1992
BArch (Hon) 1995
MBA 2005
RIBA 1996
ARB (UK) 1996
Registered Architect (UK) 1996

Mark joined Aedas as a Senior Associate in 2006, and was promoted to Director in 2007.

Mark has over 11 years of experience in architecture, and has seen himself as a versatile Project Architect with good management, communication and co-ordination skills. He has extensive experience in creating, implementing and utilizing structured systems for the control of detailed design and submission and administration of contracts regarding some large fast track multi-use developments in Hong Kong and Southeast Asia.

Mark is experienced and confident in leading the architectural team with the minimum of supervision, from a CAD perspective, which can get the correct product from the computer. He also has a thorough understanding on design and construction process and a strong analytical skill in determining the elements and construction methods required for each projects.

LU TANG LAI ARCHITECTS LTD.
吕邓黎建筑师有限公司

黎绍坚
Lai Siu Kin,Rembert

黎绍坚
董事

学术资格
1979年 香港大学建筑学（荣誉）文学士
1981年 香港大学建筑学士

专业资格/公职
1983年 香港建筑师学会会员
1984年 英国皇家建筑师学会会员
1984年 澳洲皇家建筑师学会会员
1984年 香港政府屋宇署认可人士（第一类别）
1990年 香港注册建筑师
2002年 香港建筑师学会资深会员
2003—2007年 香港建筑师学会建筑物条例委员会会员
2003年至今 上诉审裁团（建筑物）成员
2004—2008年 承建商注册事务委员（主席/后补主席）
2005年 中国一级注册建筑师资格
2006—2009年 专业测评委员会副主席
2009年 香港中文大学建筑系客座副教授
2010年至今 专业测评委员会主席
2010年至今 公共事务论坛（论坛）成员

Lai Siu Kin, Rembert
Director

Academic Qualification
1979 Bachelor of Arts in Architectural Studies
University of Hong Kong (Honours)
1981 Bachelor of Architecture - University of Hong Kong

Professional Qualification / Appointment
1983 Member of The Hong Kong Institute of Architects
1984 Member of The Royal Institute of British Architects
1984 Member of The Royal Australian Institute of Architects
1984 Inclusion in the registrar of Authorized Person
List 1 of Buildings Department
1990 Registered Architect of The Architects Registration Board, Hong Kong
2002 Fellow of The Hong Kong Institute of Architects
2003 - 07 Member of BD Committee Panel of H.K.I.A.
2003 - Present Member of Appeal Tribunal Panel set up under Buildings Ordinance Section 48
2004 – 08 Chairman / Alternate Chairman of Contractors Registration Committee
Panel set up under Buildings Ordinance Section 8(1)
2005 PRC Class I Registered Architect Qualification
2006 – 09 Deputy Chairman of Professional Assessment Committee of H.K.I.A.
2009 Adjunct Associate Professor School of Architecture The Chinese University of Hong Kong
2010 - Present Chairman of Professional Assessment Committee
2010 - Present Public Affairs Forum (the Forum) - Member

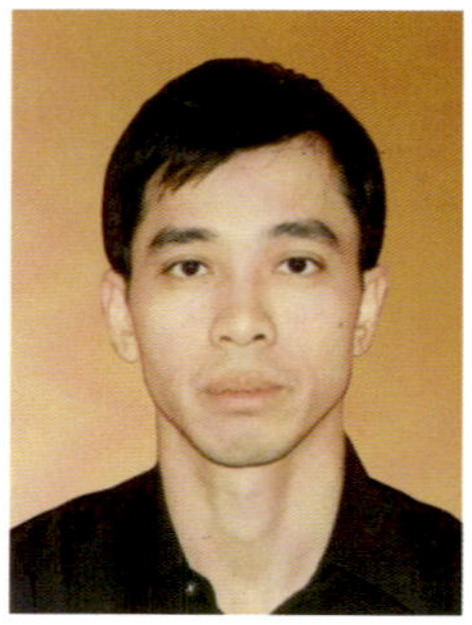

郭嘉辉
Kwok Ka Fai

郭嘉辉
董事

学术资格
1987年 香港大学建筑学（荣誉）文学士
1989年 香港大学建筑学士

专业资格
1991年 香港建筑师学会会员
1992年 香港注册建筑师
2004年 中国一级注册建筑师资格

Kwok Ka Fai
Director

Academic Qualification
1987 Bachelor of Arts in Architectural Studies
University of Hong Kong (Honours)
1989 Bachelor of Architecture - University of Hong Kong

Professional Qualification
1991 Member of The Hong Kong Institute of Architects
1992 Registered Architect of The Architects Registration Board,Hong Kong
2004 PRC Class I Registered Architect Qualification

蔡振贤
Choi Chun Yin,Kenny

蔡振贤
联席董事

学术资格
1989年 香港大学建筑学（荣誉）文学士
1992年 香港大学建筑学士

专业资格
1994年 香港建筑师学会会士
1995年 香港注册建筑师

Choi Chun Yin, Kenny
Associate Director

Academic Qualification
1989 Bachelor of Arts in Architectural Studies (Honours)
University of Hong Kong
1992 Bachelor of Architecture
University of Hong Kong

Professional Qualification
1994 Member of The Hong Kong Institute of Architects
1995 Registered Architect of The Architects Registration Board,Hong Kong

何 婷
Tain Mhaira Kirkwood

何 婷

学术资格
1986年 爱丁堡大学（荣誉）文硕士
1987年 爱丁堡大学建筑文凭
专业资格

1988年 英国注册建筑师
1989年 英国皇家建筑师学会会士
2004年 香港建筑师学会会士
2004年 香港注册建筑师

Tain Mhaira Kirkwood - Associate

Academic Qualification
1986 Master of Arts (Honours)
University of Edinburgh
1987 Diploma in Architecture
University of Edinburgh

Professional Qualification
1988 Registered Architect, UK
1989 Member of The Royal Institute of British Architects
2004 Member of The Hong Kong Institute of Architects
2004 Registered Architect of The Architects Registration Board,Hong Kong

张文明建筑师事务所
W.M.CHANG ASSOCIATES.
ARCHITECTS&PLANNERS

梁宇元
Y.Y.LIANG

学历

1984年 成功大学建筑学士
1988年 成功大学建筑硕士
东南大学博士班研究

经历

张文明建筑师事务所建筑师
正修科技大学建筑系讲师
高雄市民学苑室内设计班讲师
高雄市民学苑庭园景观设计班讲师
台湾省建筑师公会城乡景观风貌改造委员
台湾省建筑师公会学术委员

主要作品

高雄尖美建设 尖美市 集合住宅
台东大政建设 知本箱根 集合住宅
台南南纺建设 健康大地 集合住宅
高雄京城建设 京城世界 集合住宅
屏东尖美建设 东山河 集合住宅
屏东尖美建设 尖美国际观光饭店
高雄瑞堂建设 原山主人 集合住宅
哈尔滨顺迈集团 昆仑商城 商服
哈尔滨顺迈集团 丽景天地·美景天地 商服及住宅
哈尔滨顺迈集团 荣耀宝座·荣耀世纪 商服及住宅
哈尔滨顺迈集团 宏景天地·世纪广场 商服及住宅
哈尔滨顺迈集团 得意居 商服及住宅
上海昆山阳明建设 阳明山庄·俱乐部 会馆及住宅
大连顺迈集团 星海国宝 商服及住宅
大连顺迈集团 星海金钻 商服楼
哈尔滨顺迈集团 华风世贸广场 商服办公及酒店
哈尔滨顺迈集团 大都会米兰D.C 商服及住宅
大连国泰盛达集团 帝柏湾

Education

Cheng Kung University, B. Arch(1984)
Cheng Kung University, M. Arch(1988)
Southeast University, D.Arch(researching)

Experience

Architect, W.M.CHANG Associates Architects & Planners
Instructor, CHENG SHIU University
Instructor, Interior Design Class of Kaohsiung Citizen lyceum
Instructor, Landscape Design Class of Kaohsiung Citizen lyceum
'City & Country Feature Improve' Committeeman, Taiwan Architects Association
Committeeman, Taiwan Architects Association

Major Works

" Jian Mei Shi ", Residential /Jian Mei Development Group Co. Ltd, Kaohsiung
" Zhi Ben Xiang Gen ", Residential / Da Zheng Development Group Co. Ltd, Taidong
" Jian Kang DaD i ", Residential / Nan Fang Development Group Co. Ltd, Tainan
" Jing Cheng Shi Jie ", Residential / Jing Cheng Development Group Co. Ltd, Kaohsiung
" Dong Shan He ", Residential / Jian Mei Development Group Co. Ltd, Pingdong
" Jian Mei " , I Hotel / Jian Mei Development Group Co. Ltd, Pingdong
" Yuan Shan Zhu Ren ", Residential Complexes /Rui Tang Development Group Co. Ltd, Kaohsiung
" Kun Lun Commercial District ", Harbin
" Li Jing Tian Di, Mei Jing Tian Di " , Residential Complexes / SUNBETTER GROUP, Harbin
" Rong Yao Tian Di, Rong Yao Shi Ji " , Residential Complexes / SUNBETTER GROUP, Harbin
" Hong Jing Tian Di, Shi Ji Guang Chang " , Residential Complexes / SUNBETTER GROUP, Harbin
" De Yi Ju " , Residential Complexes / SUNBETTER GROUP, Harbin
" Yang Ming " Villa & Club / Yang Ming Develpment Group, Shanghai
" Xing Hai Guo Bao ", Residential Complexes / SUNBETTER GROUP, Dalian
" Xing Hai Jig Zuan " Commercial / SUNBETTER GROUP, Harbin
" Hua Feng Shi Mao Guang Chang " , Commercial & Hotel / Hua Feng Development Group, Harbin
" Da Du Hui Mi Lan D.C ", Residential Complexes / SUNBETTER GROUP, Harbin
" Ti Po Wan", Guo Tai Sheng Da Group, Dalian

林怡君
Y.C.LIN

学历

1991年 成功大学建筑学士
1993年 成功大学建筑硕士

经历

张文明建筑师事务所建筑师
沈祖海建筑师事务所设计师

主要作品

台北孙逸仙医院
台大体育馆
台东台湾史前文化博物馆
屏东尖美建设"东山河"集合住宅
高雄树德技术学院"设计大楼"
山集团王象建设"夏都"集合住宅、别墅
南投慈济921援建工程
— 社寮中学
— 社寮小学
— 延平小学
— 中洲小学
大连顺迈集团"星海国宝"商服及住宅
国立海洋生物博物馆『第二研究中心』
鼎宇建设 园邸 店铺集合住宅
鼎宇建设 绿墅 店铺集合住宅
京城建设 京城贵宾 店铺集合住宅
慈济 菲律宾静思堂
慈济 台中清水静思堂
慈济 香港分会
国立海洋生物博物馆 世界水域馆
鼎宇建设 美术之星 集合住宅
宏 建设 凤凰之响曲 集合住宅
茂盈建设 千利修 集合住宅

Education

Cheng Kung University, B. Arch(1991)
Cheng Kung University, M. Arch(1993)

Experience

Architect, W.M.CHANG Associates Architects & Planners
Designer, Haigo Shen & Partners, Architects & Engineers

Major Works

" Sun Yi Xian " Hospital / Taipei
" Turnhall of National Taiwan University " / Taipei
" National Museum of Prehistory " / Taitong
" Dong Shan He ", Residential / Jian Mei Development Group Co. Ltd, Pingdong
" Design Building of Sue-Te University " / Kaohsiung
" Xia Du " , Villa & Residential / Wang Xiang Development Group Co. Ltd
Tzu-Chi Foundation, Nantou
-" She-Liao Middle School"
-" She-Liao Elementary School"
-" Yan-Ping Elementary School"
-" Zhong-Zhou Elementary School "
" Xing Hai Guo Bao ", Residential Complexes / SUNBETTER GROUP, Dalian
" Second Research Building " / National Museum of Marine Biology & Aquarium
" Yuan Di " Ding Yu Development Group Co. Ltd
" Liu Shu " Ding Yu Development Group Co. Ltd
" Jing Cheng Gui Bin " , Residential Complexes / Jing Cheng Development Group Co. Ltd
" Taizhong Manila Subcommission (Philippines)" / Tzu-Chi Foundation
"Taizhong Gingshui Subcommission(Taiwan)" / Tzu-Chi Foundation
"Taizhong Hongkong Subcommission (Hongkong)",/ Tzu-Chi Foundation
" Waters of the Words" / National Museum Of Marine Biology & Aquarium
"Mei Shu Jr Shing" Ding Yu Development Group Co. Ltd
"Feng Huang Jiao Xiang Qu" Hung Ji Development Group Co. Ltd
"Chien Li Hsiu" Mao Ying Development Group Co. Ltd

映時設計
TIME MIRROR international

刘庆介
Brian C. Liu

刘庆介

设计总监

作为一名 Time Mirror international 的设计总监，Mr. Brian Liu 的良好信誉来自于他专业的、广泛的、以服务为主的建筑和室内设计经验。

Mr. Brian Liu 于美国南加大建筑研究所取得建筑硕士学位。十余年美国洛杉矶和中国项目的工作经验，使他涉及各种类型的项目，包括建筑项目及室内设计项目，从销售展厅、餐饮会所、房地产开发、私人住宅到高科技集团办公室大楼设计，皆有他优秀的作品。

Brian C. Liu

DESIGN DIRECTOR

As a design director of **Time Mirror** International, the good reputation of Mr. Brian Liu comes from his professional, extensive, service-oriented experiences in architecture and interior design.

Mr. Brian Liu received Master of Architecture degree from University of Southern California, USA. More than 10 years China projects makes him taking a part in a wide and various design styles, including architectural and interior design projects have various design styles, including architectural project and interior design project. He has very good representative works from the design of sales exhibition hall, dining club, real estate development, and private accommodation to office building design of high-tech group.

(C.J Chen Architects)
陈昭榕建筑师事务所

陈昭榕
C.J Chen

陈昭榕

对陈昭榕先生来说，建筑创意是无国界的。在完成国内的建筑专业教育，取得建筑师以及公务员资格后，毅然前往纽约进修。对于国际大都会纽约不仅仅是艺术创意的殿堂，更累积了国际的设计实力以及眼界之提升。

陈昭榕先生透过与美国纽约设计中心，纽约建筑师 Andrew Macnair 带领的设计团队以及台湾事务所的通力合作，打造优质的建筑团队。

C.J Chen

For Mr. Chen, the architectural creativity is without borders. After completing the professional education on architecture at home and obtaining the qualifications of architect and civil servant, he decided to continue his study in New York. Because this international metropolis is not only a paradise for artistic creativity, but also a place which accumulates the design strength all over the world and which can broaden people's vision.

Through the cooperation with New York Design Center, the design team led by Andrew Macnair, a New York architect and Taiwan Design Firm, Mr. Chen endeavors to build a high-quality architectural design group.

黄山市建筑设计研究院
Huangshan Architectural Design and Research Institute

洪祖根
Zugen Hong

洪祖根

男，1961 年 11 月出生，高级工程师，国家一级注册建筑师，黄山市建筑设计研究院院长、安徽省徽派建筑工程技术研究中心主任，现为安徽省政协委员、黄山市政协常委、安徽省土木建筑学会常务理事、安徽省工程勘察设计协会常务理事、副秘书长、黄山市建筑学会副会长，2004 年荣获中国建筑学会先进工作者称号。

主持和参与的设计获奖项目有：

安徽省徽文化博物馆（省一等奖）、黄山管委会办公楼（省一等奖、部三等奖）、黄山狮林饭店（省二等奖、部鼓励奖）、黄山市老干部活动中心（省三等奖）、黄山市建筑设计研究院办公大楼等，并参与编制国家标准设计《地方传统建筑（徽州地区）图集》。

Zugen Hong

Male, born in November 1961, senior engineer, national first grade registered architect, dean of Architectural Design and Research Institute, Huangshan Construction Engineering Technology Research Center of Anhui School. At present, he is the commissioner of the CPPCC in Anhui Province, the standing member of the CPPCC in Huangshan, managing director of Civil Architectural Institute in Anhui Province, managing director and deputy Secretary General of Engineering Survey and Design Institute in Anhui, vice president of the Architectural Society of Huangshan City. In 2004, he received the title of advanced worker of the Architectural Society in China.

Projects he hosted and participated are: "Cultural Museum in Huizhou, Anhui" (Provincial First Prize, the Third Prize of Ministry of Construction), Huangshan Management Committee Office Building (Provincial First Prize, the Third Prize of Ministry of Construction), Huangshan Lion Forest Hotel (the Provincial Second Prize, Encouragement Prize of Ministry of Construction), Huangshan Exchange Center for Old Cadres (Provincial Third Prize), Office Building of Architectural Design and Research Institute, Huangshan City, etc., and he participates in national architecture standard design---"Local Traditional Architecture Atlas (Huizhou region)".

韩 毅
Yi Han

韩 毅

男，1970 年 11 月出生，国家一级注册建筑师，高级建筑师，黄山市建筑设计研究院总建筑师。

主持和参与的主要设计项目有：

黄山市示范幼儿园（建设部三等奖、省二等奖）、黄山市阜安新村规划设计（省三等奖）、祁门县城总体规划（省鼓励奖）、黄山市黟县水墨宏村（科技部集约型住宅推广奖，全国工商联地产商会第四届"精瑞住宅科学技术奖、建筑文化金奖"）、山东济南外国语学校、景德镇锦绣昌南"中国瓷园"，福州市政府招待所（于山宾馆）、歙县老年活动中心、歙州学校、黄山汤口环球大酒店、黄山市房管局办公大楼、黄山市教委办公大楼等。

Yi Han

Male, born in November 1970, National Class I registered architect, Senior Architect, Chief Architect of Architectural Design and Research Institute, Huangshan.

Projects he hosted and participated are: Huangshan Demonstration Kindergarten (Third Prize of Ministry of Construction, the Provincial Second Prize), planning and design for Fuan Village, Huangshan City(Provincial Third Prize), overall planning for Qimen County (Provincial Encouragement Award), Shuimo Hongcun Village, Yixian County, Huangshan (Promotion Award for Intensive Houses of Ministry of Science and Technology, "Science and Technology Award of Fine Housing" and "Construction Culture Award" given by the Estate Commerce of National Federation of Chamber", Foreign Language School in Jinan, Shandong, "China Porcelain Garden " of Jingdezhen in the splendid Southern area of Jiangxi, Fuzhou Government Hostel (Yushan Hotel), Shexian Activity Center for Elders, Xizhou School, Tangkou Universal Hotel of Huangshan, Office Building of the House Administration in Huangshan City, Office Building of Board of Education in Huangshan City.

HFUT Institute of Architectural Design
合肥工业大学建筑设计研究院

祁小洁
Xiaojie Qi

祁小洁

合肥工业大学建筑设计研究院副总建筑师
高级建筑师
国家一级注册建筑师

主要作品
合肥市南园新村蝶形住宅
荣获 建设部建筑设计二等奖
合肥工业大学翡翠湖校区学生公寓
荣获 安徽省建筑设计三等奖
泾县榔桥镇黄田古村落历史文化民村保护规划
荣获 安徽省规划设计一等奖
世纪阳光花园会所及小学会所
郑州高等专科学校新校区规划及单体建筑

Xiaojie Qi

Vice Chief Architect of Architectural Design Academy of Hefei University of Technology
Senior Architect
National class I Registered Architect

Major Works
Nanyuan new butterfly residential quarters, Hefei
Award: The Second Prize of Architectural Design issued by Ministry of Construction
Student housing, Emerald Lakes Campus of Hefei University of Technology
Award: The Third Prize of Anhui Province architectural design
Protection planning for Huangtian historic and cultural village, Liangqiao town, Jing County, Anhui Province
Award: First Prize of Anhui Province planning and design
Century Sunshine Garden Club and Elementary School Club
The new campus planning and single buildings of Zhengzhou junior colleges

查 昕
Xin Zha

查 昕

合肥工业大学建筑设计研究院副总建筑师
合肥工业大学建筑设计研究院教育建筑设计院院长
高级工程师
国家一级注册建筑师
建筑学硕士

主要作品
合肥工业大学学科楼群
荣获 中国勘察设计协会优秀勘察设计三等奖
中科院合肥物质科学研究院研究生教育基地
荣获 安徽省勘察设计协会优秀勘察设计二等奖
安庆师范学院新校区办公楼
荣获 安徽省勘察设计协会优秀勘察设计三等奖

Xin Zha

Vice chief Architect of Architectural Design Academy of Hefei University of Technology
Dean of Education & Architectural Design Institute of Architectural Design Academy of Hefei University of Technology
Senior Engineer
National Class I Registered Architect
Master of Architecture

Major Works
Academic building complex of Hefei University of Technology
Award: The Third Prize of Outstanding National Survey and Design issued by China Investigation & Design Association
Education base for graduate students of Hefei Institute of Physical Science
Award: The Second Prize of Anhui Outstanding Survey and Design issued by Anhui Investigation &Design Association
The new campus office building of Anqing Teachers College
Award: The Third Prize of Anhui Outstanding Survey and Design issued by Anhui Investigation &Design Association

葛 军
Jun Ge

葛 军
合肥工业大学建筑设计研究院副总建筑师
合肥工业大学建筑设计研究院房产设计院院长
高级建筑师
国家一级注册建筑师
工商管理硕士

主要作品
安徽广播影视学院新校区规划
安徽金孔雀温泉度假别墅区规划
蚌埠兰亭书苑小区
科大创新广场
安科生物科研研发大厦
合肥京东方生活配套区

Jun Ge

Vice chief Architect of Architectural Design Academy of Hefei University of Technology
Dean of Housing Design Institute of Architectural Design Academy of Hefei University of Technology
Senoir Architect
National First-Class Certified Architect
MBA degree

Major Works
Planning and design for the new campus of Anhui Broadcasting Movie and Television College
Planning and design for Villa District of Anhui Golden Peacock Spa Tourism Resort
Residence community design for Bengbu Lanting Reading Garden
The High Technology and Innovation Square
The research and development block of Anhui Anke Biotechnology (Group) Co., Ltd
The Living Supporting Area of Hefei BOE

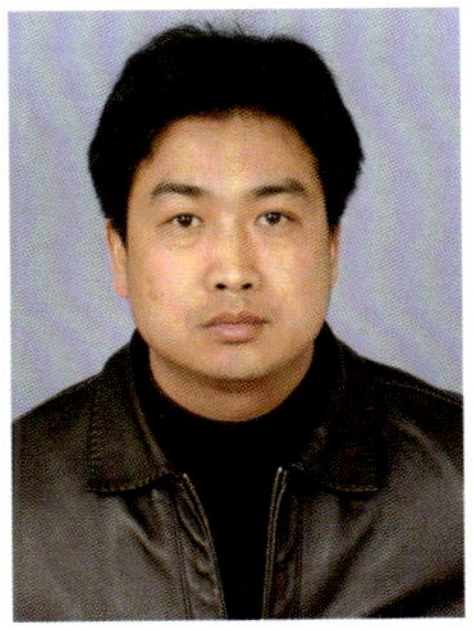

张一敢
Yigan Zhang

张一敢

合肥工业大学建筑设计研究院副总建筑师
合肥工业大学建筑设计研究院医疗建筑设计院院长
高级建筑师
国家一级注册建筑师

主要作品
繁昌县行政中心
芜湖技师学院新校区
信地城市广场综合体
沃野花园小区

Yigan Zhang

Vice chief Architect of Architectural Design Academy of Hefei University of Technology
Dean of Medical & Architectural Design Institute of Architectural Design Academy of Hefei University of Technology
Senoir Architect
National Class I Certified Architect

Major Works
The administrative center of Fanchang County
The new campus of Wuhu Technician College
The Xindi City Plaza Complex
Residence Community of Woye Garden

中科院建筑设计研究院

Institute of Architecture Design and Research, Chinese Academy of Sciences

崔 彤

Tong Cui

崔 彤

1962 年 10 月出生，清华大学建筑系毕业获建筑学硕士学位。国家优秀设计金奖获得者、全球华人青年建筑师奖获得者、享受国务院专家政府津贴、现任中科院建筑设计研究院副院长、总建筑师。

在中科院设计院工作期间，多次担任国家级及中科院重大项目的设计主持人和负责人工作，并获中国科学院创新先进个人。

多次在重大的建筑设计竞赛中获奖，并应邀参加国内外学术交流的讲演，多次参加国内重大项目的评审。其撰写的论文多次发表于国内外著名杂志及出版物。

主要获奖

获中国建筑学会建筑创作大奖	2009 年
获全球华人青年建筑师奖	2008 年
获科学中国人年度大奖	2007 年
获中国建筑学会优秀建筑创作奖	2006 年
获全国优秀工程设计金质奖	2005 年
获建设部优秀工程建筑设计一等奖	2004 年
获北京市优秀工程建筑设计一等奖	2004 年
获首都建筑设计汇报展建筑创作优秀奖及十佳建筑奖	

最新完成作品

中国科学院研究生院新校区规划及设计	设计／建成：2008/2010 年
建造地点：北京雁栖湖	建筑面积：310 000 平方米
中国国家开发银行	设计／建成：2006/2009 年
建造地点：北京长安街	建筑面积：151 000 平方米
中国人民银行重点库（721 国家重点工程）	设计／建成：2006/2009 年
建造地点：北京	建筑面积：73 500 平方米
光大国际中心	设计／建成：2004/2007 年
建造地点：北京	建筑面积：120 000 平方米
歌华大厦	设计／建成：2004/2006 年
建造地点：北京	建筑面积：108 500 平方米
泰国曼谷中国文化中心	设计／建成：2008/2010 年
建造地点：泰国曼谷	

Tong Cui

Born in October 1962, and received Master Degree of Architecture from Department of Architecture, Tsinghua University. Cui is the winner of National Excellent Design Award and get the Global Chinese Young Architects Award, enjoying the expert subsidies from State Department. Currently Cui is the vice president and Chief Architect of Architectural Design & Research Institute, Chinese Academy of Sciences.

When he worked in the CAS design institute, Cui undertook the position of design supervisor or manager for major projects both of the national level and the CAS level, won the title of innovative advanced individual of Chinese Academy of Sciences.

Cui has won prizes in major architectural design competitions for several times, and been invited to participate in lectures of academic exchanges both at home and abroad, as well as in the review of national major projects. Papers about his work have been published in famous magazines and publications, both domestic and international.

Main Awards

Architecture Design Award by Architectural Society of China 2009
Global Chinese Young Architects Award 2008
Annual Award of Chinese Science 2007
Excellent Architecture Design Award by Architectural Society of China 2006
the gold prize of National Excellent Engineering Design 2005
the first prize of Excellent Architectural Design by Ministry of Construction 2004
the first prize of Beijing Excellent Architectural Design 2004
the Excellent Award of Architectural Design and the Top Ten Architectures of the Capital Architectural Design Exhibition

The latest completed projects

- The planning and design of the new campus for Graduate School of Chinese Academy of Sciences Design / Completion: 2008/2010; Construction Location: Yanxi Lake, Beijing, Floor Area: 310,000 m²
- China Development Bank Design / Completion: 2006/2009; Construction Location: Chang An Avenue, Beijing, Floor Area: 151,000 m²
- Key library of People's Bank of China (721 national key project) Design / Completion: 2006/2009; Construction Location: Beijing, Floor Area: 73,500 m²
- Everbright International Center Design / Completion: 2004/2007; Construction Location: Xicheng District, Beijing, Floor Area: 120,000 m²
- Song Wah Building Design / Completion: 2004/2006; Construction Location: Beijing, Floor Area: 108,500 m²
- Chinese Cultural Centre in Bangkok, Thailand Design / Completion: 2008/2010; Construction Location: Bangkok, Thailand

张 京

Jing Zhang

张 京

现为中科院建筑设计研究院有限公司副院长兼常务副总建筑师，

多年来主持了近 60 项工程设计项目，包括国家重大科学工程 LAMOST 项目；中科院数学院、声学所、理论所、生物物理所科研建筑及公共建筑多项；上海万科、沈阳万科等住宅区规划及单体；中科院研究生院研究生公寓等工程。累计完成建筑面积近百万平方米。获省部级优秀设计奖三项。在建筑设计与规划、室内设计等方面有较好的专业素质及丰富的实际工程经验。对建筑节能及生态建筑有一定研究，发表论文多篇，主编国家标准图集《实验室建筑设备》。多次赴美国、加拿大、欧洲、日本、韩国、巴西、智利、澳大利亚等地考察调研。于 2006 年参加西班牙著名建筑事务所研修班，现为中科院翻译工作者协会会员。

主要获奖

1998 年　上海万科城市花园新十二区获上海市优秀规划设计二等奖
1998 年　上海万科城市花园“海棠苑”欧陆式多层住宅获上海市优秀住宅单体设计二等奖
1999 年　上海万科城市花园“海棠苑”中科院优秀设计一等奖
2007 年　主编国家标准图集《实验室建筑设备》
2008 年　沈阳万科城一期获国家科技部 第五届精瑞住宅科学技术奖——住区规划设计优秀奖
2009 年　建国六十周年国庆天安门彩车展示 LAMOST 天文望远镜项目照片

Jing Zhang

Now, he is the Vice President and Deputy Chief Architect of Institute of Architectural Design in Chinese Academy of Sciences Co., Ltd..

For many years Jing Zhang has presided over nearly 60 design projects, which includes the major national scientific project LAMOST; the scientific research and public buildings such as CAS Mathematics Institution, Acoustics Institution, theory Institution and the Biophysics Institution; the residential planning and monomer such as Shanghai Vanke and Shenyang Vanke, and projects such as the graduate student apartment of Graduate School, Chinese Academy of Sciences. The total floor area covers approximately one million square meters. Zhang has won three Outstanding Design Awards of the provincial or ministry level. He has excellent professional quality and rich practical experience in the architectural design and planning, the interior design, etc. Zhang also makes certain contributions to the building energy conservation and ecological construction, and has published many papers. He edited the national standard atlas "Laboratory Construction Equipment". For many times Zhang went to the United States, Canada, Europe, Japan, Korea, Brazil, Chile and Australia, etc. to carry out investigations and researches. In 2006 he participated in the training classes of a famous architectural firm in Spain, and is now a member of Translators Association of Chinese Academy of Sciences.

Main Awards

1998 the new 12 district of Shanghai Vanke City Garden won the second prize of Shanghai Outstanding Planning and Design
1998 the European-style multi-storey residential dwelling "Begonia Garden" of Shanghai Vanke City Garden won the second prize of Shanghai Excellent Residential Architectural Design
1999 "Begonia Garden" of Shanghai Vanke City Garden was awarded the first prize of Good Design by Chinese Academy of Sciences
2007 edited the national standard atlas "Laboratory Construction Equipment"
2008 the first period construction of Shenyang Vanke City was awarded Excellence Award of Residential Planning and Design in the Fifth Glitter Residential Science and Technology Award by the State Ministry of Science and Technology
2009 displayed photos of LAMOST telescope project on the Tiananmen Square floats on the sixtieth anniversary of the founding of China

刘 峰

Feng Liu

刘 峰

副院长，副总建筑师，国家一级注册建筑师
1997 年　毕业于同济大学建筑城规学院，获建筑学硕士学位
1997 年至今　中科院建筑设计研究院历任建筑师、所长、副总建筑师

主要获奖

海南三亚红树林五星级度假酒店，61 770 平方米，2005 年建成
天津西青假日风景花园，300 000 平方米，2007 年建成
荣获“全国双节双优杯住宅方案竞赛金奖”及“詹天佑大奖优秀住宅小区金奖”
中国农业大学生命科学楼，32 236 平方米，2008 年建成
郑州隆福国际商住综合体，180 000 平方米，2010 年建成
荣获全国人居经典建筑规划设计方案竞赛建筑金奖
石家庄国际会展中心，380 000 平方米，国际竞标优秀方案
中科院地球化学所（贵阳）金阳新所园区，140 000 平方米，2012 年建成

Feng Liu

Vice President, Deputy Chief Architect, National First-Class Certified Architect

In 1997, graduated from College of Architecture and Urban regulations, Tongji University, master degree of architecture

Since 1997, worked as Architect, Director, Deputy Chief Architect in Institute of Architecture Design and Research, Chinese Academy of Sciences (ADCAS)

Main Awards

Sanya Mangrove Resort Hotel in Hainan
Holiday Landscape Garden in Xiqing, Tianjin
Life Sciences Building in China Agricultural University
Longfu Residential Area in Zhengzhou
Shijiazhuang International Convention & Exhibition Center
Institute Of Geochemistry,Chinese Academy of Sciences,Guiyang

世纪安泰建筑工程设计有限公司
SJAT Architecture & Engineering Design Co.,Ltd.

刘 焱
Yan Liu

谢 忱
Chen Xie

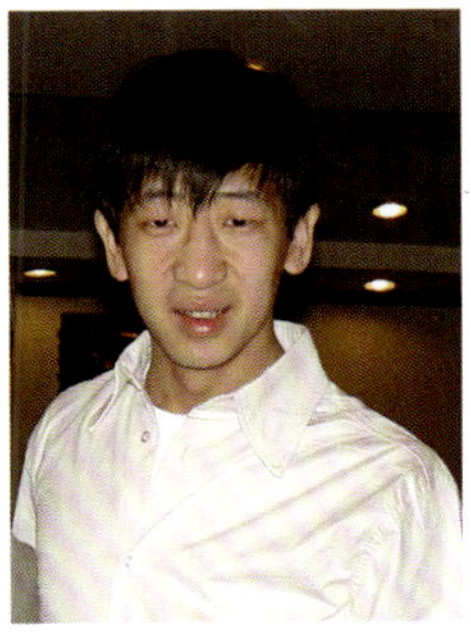
谷 川
Chuan Gu

王 非
Fei Wang

刘 焱

副总经理
国家一级注册建筑师
高级建筑师
教育背景：北京建筑工程学院建筑系毕业
工作经历：曾在北方工业大学和北京工业大学建筑系任教并从事建筑设计工作
曾任北京京澳凯芬斯设计有限公司副总建筑师

部分代表作品
北沙滩住宅小区
北京电信住宅小区
佳兴园住宅小区5、6号楼
唐山华联商务会馆
唐山机场新区住宅
大兴西红门住宅

谢 忱

1994年7月–1999年7月 北京建筑工程学院 建筑学学士学位
1999年7月–2002年4月 中国建筑设计研究院第二建筑设计所 助理建筑师
2002年9月–2004年1月 英国诺丁汉大学建筑学（The University of Nottingham） 获建筑学硕士学位（Master of Architecture）
2004年2月–2005年6月 华成国际建筑工程有限公司 建筑师
2005年7月–2006年7月 德国欧博迈亚工程咨询有限公司（OBERMEYER Engineering Consulting Co.,Ltd.） 建筑师
2006年8月至今 北京世纪安泰建筑工程设计有限公司

部分代表作品
凤凰卫视影视中心（深圳）
获国际竞赛第二名
英国诺丁汉大学体育中心方案设计
获2003CORUS国际学生建筑设计竞赛推荐奖
北京昌平伯朗峰酒店
香河–运河国际生态城总体规划
中国邮政北京邮件处理中心

谷 川

副总建筑师
国家一级注册建筑师

1992–1997年 北京建筑工程学院建筑系 建筑学学士
1997–2000年 北京华特建筑设计顾问公司
2001年至今 北京世纪安泰建筑工程设计有限公司

部分代表作品
部分代表作品
北京西成忆树项目
北京西贸中心项目
北京联盟天地广场项目
北京新天地住宅项目
唐山市机场新区（梧桐大道）项目
安阳市引领示范区商业中心项目

王 非

建筑专业负责人

2000年毕业于北京建筑工程学院建筑学专业
2000年7月–2001年3月 北京华特建筑顾问有限责任公司任建筑师
2001年4月至今 北京世纪安泰建筑工程设计有限公司

部分代表作品
山水文园二期
山水文园E区
朝来绿色家园来春园
北京邮件处理综合中心
西红门项目
京客隆燕尚园店

Yan Liu

Deputy General Manager
National Class-A Registered architect
Senior Architect
Education: Graduated from Beijing Institute of Architectural Engineering, Department of architecture
Work experience: Once taught in department of architecture in North China University of Technology and Beijing University of Technology, served as architectural designer
Former vice General Architect of Beijing-O Kai Fensi Design Co., Ltd.

Representative works
Beishatan Residential
Beijing Telecom Residential
Jiaxing Park Residential Building No.5 and 6
Tangshan Hualian Business Center
Tangshan Airport New Area Residential
Xihongmen residential in Daxing District

Chen Xie

General Creative Architect

1994.7-1999.7 Beijing Institute of Architecture, Bachelor of Architecture
1999.7-2002.4 China Architecture Design Institute, Second Architectural Design Agency, Assistant Architect
2002.9-2004.1 The University of Nottingham, School of Architecture, Master of Architecture
2004.2-2005.6 Huacheng International Construction Co., Ltd. Architect
2005.7-2006.7 German OBERMEYER Engineering Consulting Co., Ltd. Architect
2006.8 hitherto, Beijing Century Aetna Building Engineering Co., Ltd.

Representative works
Phoenix Satellite Television Center (Shenzhen)
Awarded: second prize for international competition
University of Nottingham Sports Centre Design
Awarded: A Commendation in the National Competition for the Corus Undergraduate Design Awards 2003 for Architecture Design
Brown Peak Hotel at Changping, Beijing
General Planning for XIANGHE - Canal International Eco-City
Beijing Mail Processing Center, China Post

Chuan Gu

Deputy Chief Architect
National Class-A Registered architect

1992 to 1997, Beijing Institute of Architecture Bachelor of Architecture
1997 to 2000, Beijing Walter Architectural Design Consultant Company
2001 hitherto, Beijing Century Aetna Building Engineering Co., Ltd.

Representative works
Beijing Xi-Cheng Yishu project
Beijing West Trade Center Project
World Plaza Beijing Union
Beijing New World residential
Tangshan Airport District (Indus Boulevard) Project
Business Center of Anyang new area for model leading

Fei Wang

Architecture Professional Leader
2000 Graduated from Beijing Institute of Architecture
July 2000 - March 2001 Beijing Walter Construction Consultants LLC, architect
April 2001 hitherto, Beijing Century Aetna Building Engineering Co., Ltd.

Representative works
Landscape Culture Garden Phase II
Landscape Culture Garden Area
Chaolai Green Home Garden, Spring Garden
Beijing Integrated Mail Processing Center
Xihongmen Project
Jingkelong Supermarket YanShangyuan Shop

MCC 中冶京诚 北京京诚华宇建筑设计研究院有限公司 CERI ARCHITECTS & ENGINEERS

冯 杰
Jie Feng

冯 杰

高级工程师 一级注册建筑师
1987 年 7 月毕业于河北建筑工程学院建筑工程系工业与民用建筑工程专业
副总建筑师、建筑所副所长

主要业绩

河北电力开发有限公司主厂房、综合楼
国家物资储备局广西 479 处安全改造工程（国家保密工程）
国家物资储备局山西 475 处安全改造工程（国家保密工程）
国家物资储备局湖北 270 处安全改造工程（国家保密工程）
国家物资储备局云南 774 处安全改造工程（国家保密工程）
国家物资储备局河北 133 处安全改造工程（国家保密工程）
国家物资储备局内蒙古 136 处、836 处安全改造工程（国家保密工程）
福地家园住宅小区
寺河 120MW 煤层气发电厂（亚洲最大的煤层气发电环保项目）
中国科学院高能物理研究所科研配套综合楼
京汉香城住宅小区
中冶正兴置业南京河西莲花住宅小区
山西大同凯德．世嘉住宅小区

Jie Feng

Senior engineer First-level registered architect
In July 1987, she graduated from the Architectural Engineering Department of Hebei Architecture Engineering Institute
Assistant chief architect, deputy director of the construction institute

Main Achievements

The integrated building of Hebei Electric Power Development Co., Ltd.
Guangxi 479 Safety Improvement Project of State Reserve Bureau (the national confidential project)
Shanxi 475 Safety Improvement Project of State Reserve Bureau (the national confidential project)
Hubei 270 Safety Improvement Project of State Reserve Bureau (the national confidential project)
Yunnan 774 Safety Improvement Project of State Reserve Bureau (the national confidential project)
Hebei 133 Safety Improvement Project of State Reserves Bureau (the national confidential project)
Inner Mongolia 136, 836 Safety Improvement Project of State Reserve Bureau (the national confidential project)
Blessed Home
120MW Coal and Methane Power Plant (The project is the largest coal-bed methane power plant in Asia)
The comprehensive R & D building of the Institute of High Energy physics
Jinghan Xiangcheng residential buildings
Nanjing Hexi Lotus Residential community of China Metallurgical Zhengxing Real Estate
Cade• Seg a residential district in Shanxi Datong

谢 明
Ming Xie

谢 明

副总建筑师 一级注册建筑师
1987 年毕业于华中理工大学建筑学系

主要业绩

北京东单游泳馆
北京大兴工业区凯驰工业园
北京大兴工业区奥亚大楼
北京浩庭花园住宅二期工程
中冶京诚工程技术有限公司亦庄研发与产业基地生产实验楼
沈阳第一城高层住宅区
太钢集团临汾钢铁有限公司高层住宅区
承德建龙钢铁有限公司住宅小区
烟台世茂海湾 1 号工程

Ming Xie

Deputy Chief Architect First-level registered architect
Graduated from Department of Architecture, Huazhong University of Science and Technology in 1987

Main Achievements

Beijing Dongdan Natatorium
Karcher Industrial Park of Daxing Industrial Zone, Beijing
Aoya Building of Daxing Industrial Zone, Beijing
Beijing Haoting Garden (Phase II)
Production Laboratory Building of Yizhuang R & D and Industrial Base of Capital Engineering & Research Incorporation Limited
The high-rise residential quarter of Shenyang No.1 City
The high-rise residential quarter of TISCO Linfen Iron & Steel Co., Ltd.
The residential Quarter of Chengde Jianlong Iron & Steel Co., Ltd.
Shimao•No.1 The Harbor, Yantai

李越莺
Li Yueying

李越莺

高级建筑师 一级注册建筑师
毕业于武汉理工大学建筑系

主要业绩

沈阳浦和新城住宅安置小区，7 万平方米，该项目荣获 2009 年度全国冶金行业优秀工程设计三等奖
唐山建龙实业公司行政小区，7 万平方米
火神庙商业中心，商业综合体 36 万平方米
天津于家堡金融起步区 03-21 地块，5A 级金融写字楼 13 万平方米
营口北海新区回迁安置小区，70 万平方米
今典集团青岛红树林酒店，旅游综合体 85 万平方米

Yueying Li

Senior Architect First-level registered architect
Graduated from Department of Architecture of Wuhan University of Technology

Main Achievements

Residential District of Puhe new city in Shenyang, 7 million square meters. This project won the third prize for the outstanding engineering design of national metallurgy industry of year 2009
Administrative Area of Tangshan Jianlong Industrial Company Limited, 7 million square meters
Temple of Hephaistos Business Center, commercial complex, 36 million square meters
Yujiabao financial starting area, plot No.03-21, financial office building (Grade 5A), Tianjin, 13 million square meters
Beihai new area move back settlement district, Yingkou,70 million square meters
Qingdao Mangrove Hotel of Jindian Group, tourism complex, 85 million square meters

刘景波
Liu Jingbo

刘景波

主任建筑师
教授级高级工程师 高级建筑师
刘景波具有 30 年的工程设计、管理、项目主持的工作经验，多年来从事了大量工业建筑工程的设计项目。

主要业绩

北京复盛机械有限公司新厂
华特科园超硬材料厂房及办公楼
宏大研究院纺织新材料实验基地
DANIELI 冶金设备有限公司
北京国际信息园标准厂房（二期）
安泰科技永丰新材料产业基地
钢铁研究总院永丰基地 2 号、3 号 A、3 号 B 项目
钢铁研究总院涿州新材料产业园一期工程 A、B 座生产车间
北京安泰钢研超硬材料有限责任公司金刚石制品基地二期工程
安泰科技涿州产业园万吨级非晶带材及制品项目一期、二期工程
北京飞行博达电子有限公司（798 厂）新建生产基地项目
北京东郊农工商联合公司生产基地（物美超市商品配送中心）
安泰科技空港基地特种合金制品项目
钢铁研究总院涿州新材料产业园总体规划

Jingbo Liu

chief architect
Professor of Engineering Senior Architect
Jingbo Liu has engaged in a great many of design works of industrial construction projects for many years and has accumulated 30 years'working experiences of project deisign , management, and project conducting.

Main Achievements

The new plant of Beijing Fusheng Machine Co. Ltd.
The super hard material factory building and office building of Huate Science Park
The Experimental Base of new textile material , Hongda Research Institute
DANIELI Metallurgical Equipment Co. Ltd.
The standard workshop building of Beijing International Information Park(the second phase)
Yongfeng Industrial Base of new material, Antai Scientific Co. Ltd
2#,3#A,3#B projects, Yongfeng Base, Central Iron and Steel Research Institute
Manufacturing shops of building A and building B, Xunzhou Industry Park of new material(project of the first phase), Central Iron and Steel Research Institute
The diamond products base (project of the second phase), Beijing Antai Steel Research of Super Hard Material Co. Ltd.
Xunzhou Industry Park of 10000-ton class amorphous strip and products (project of the first phase and the second phase), Antai Scientific Co. Ltd.
The project of new built manufacturing Beijing Flight Boda Electronic Co. Ltd. (798 factory)
The manufacturing base of Beijing Eastern-suburb Agri-industrical Integrated Company,(goods distribution center of WUMART)
The project of special alloy, Antai Scientific Airport Base
The overall plan of Zhuzhou Industry Park of new Airport Base material, Central Iron and Steel Research Institute

中冶京诚 北京京诚华宇建筑设计研究院有限公司
CERI ARCHITECTS & ENGINEERS

王大治
dazhi Wang

王大治

1993年7月毕业于河北煤炭建筑工程学院 建筑学学士
2009年聘为建筑教授级高级工程师 高级建筑师

主要业绩
百盛二期方案及初步设计
深圳监狱
人民大会堂宴会厅改造
外交部国际会议厅改造
武警水电指挥培训中心大楼建筑设计
北京红居南区危改住宅工程
安泰科技股份有限公司北京空港新材料产业基地
武警招待所二期综合楼工程
铁科院科技创新基地
钢研院永丰基地3、4号厂房
北京14中学校综合楼工程
财富二期写字楼工程

dazhi Wang

Graduated from Hebei Coal Architecture Engineering College in July, 1993.
He was named professor of Engineering and Senior Architect in 2009

Main Achievements
The scheme and preliminary design of Baisheng
Shengzhen prison
The reformation of the Banqueting Hall of Great Hall Of the People
The reformation of International Conference Hall of the Ministry of Foreign Affairs
The design of the Military Police Water and Electricity Command Training Center Building
The reformation of dilapidated residential building of Hongju southern district in Beijing
Beijing Airport New Material Manufacturing Base of Antai Technology Joint Stock Limited Company
The second phase project of multi-functional building of military police rest house
The Technological Innovation Base of the Railway Scientific Institute
No.3 and No.4 Plant building of Yongfeng Base in the Steel Research Institute
The project of Multi-functional Building in Beijing No. 14 Middle School
The second phase project of super high-rise Fortune Office Building

吴国庆
Guoqing Wu

吴国庆

2001年 毕业于清华大学 获建筑学学士
2004年 毕业于米兰理工大学 获建筑学硕士
曾工作于中国建筑设计研究院和中国建筑科学研究院

主要业绩
中关村软件园R17/R18地块
北京金融街F10办公楼
帕尔马某学校竞赛
北京交通大学软件工程楼
天津海益国际中心
陕西西安中华城项目
作品曾获2006年全国人居经典规划设计方案竞赛金奖

Guoqing Wu

Graduated from Tsinghua University and got BA Degree of Architecture in 2001
Graduated from Polytechnic University of Milan and got MA Degree of Architecture in 2004
Has worked at CADRG and CABR

Main Achievements
Block R17&R18 in ZhongGuanCun Software Park
The F10 Office Building in the Finance Street, Beijing
A Primary School, Provincia di Parma, Italy
Building of Software Engineering of Beijing Jiaotong University
Tianjin Haiyi International Center
Chunghwa Town Residential District, Xi'an
The works have been awarded for the Golden Prize of National Habitat Construction Planning and Deigning Competitions.

林鑫钫
Xinfang Lin

林鑫钫

二所方案室副主任
2004年毕业于重庆建筑大学 建筑学学士
国家一级注册建筑师 工程师

主要业绩
北京财富中心写字楼设计
北京798展览中心方案设计
当代MOMA太原样板间设计
北京市第十四中学新址工程项目设计
烟台世茂海湾1号项目设计
邯郸百家村生活区改造工程
唐山建龙公司办公宿舍区规划设计方案

Xinfang Lin

The deputy director of Programming Office of the Second Research Institution.
Graduated from Chongqing Jianzhu University and got BA Degree of Architecture
First grade registered architect Engineer

Main Achievements
The designing of Beijing Wealth Center Office Building
The designing of Beijing 798 Exhibition Center
The designing of model room of Taiyuan Contemporary MOMA
The project designing of the new site of Beijing No. 14 Middle School
The designing of No.1 project Yantai Shimao No.1 the Harbor
The living quarters renovation of Baijiacun, Handan
The designing and planning of working and dormitory region of Tangshan Jianlong Company

陈 星
Xing Chen

陈 星

一所方案室副主任
2006年毕业于重庆大学建筑城规学院 建筑学学士

主要业绩
北京华彬婚礼堂
南京下码头住宅项目
长春新奥蓝城2、3、4期住宅项目
秦皇岛迎宾路住宅项目
中冶京诚北京亦庄第二办公区
中冶京诚营口中试基地

Xing Chen

The deputy director of program room of No.1 Institute
Graduated from the College of Architecture and Urban Planning of Chongqing University and got BA Degree of Architecture in 2006

Main Achievements
Beijing Huabin Wedding Hall
Housing project of Nanjing Xiamatou
Phase 2, 3, 4 housing project of New Aolan city, Changchun
Housing project of Yingbin Road, Qinhuangdao
No.2 Office area in Beijing Yi zhuang of Capital Engineering & Research Incorporation Limited
Yingkou pilot project base of Capital Engineering & Research Incorporation Limited

MCC 中冶京诚 北京京诚华宇建筑设计研究院有限公司 CERI ARCHITECTS & ENGINEERS

梁嘉樑
Jialiang Liang

梁嘉樑

规划所所长
清华大学建筑学学士　城市规划硕士
擅长大型、精品型规划项目，思路清晰、逻辑严密

主要业绩

天津宝坻京津新城规划（2010年）
北京姚家园三才会所建筑方案设计（2009年）
廊坊开发区中心区城市设计（2009年）
中国科学院研究生院怀柔新园区规划设计（2008年）
北京理工大学北院居住区设计（2007年）
北京房山区长阳镇佛满村、赵庄村新农村规划（2007年）
河南郑汴新区城市设计（2007年）

Jialiang Liang

Head of the Planning Institute
Bachelor of Science in Architecture of Tsinghua University
He is good at broad scale and boutique-type planned projects, has clear thought and meticulous logical thinking.

Main Achievements

Beijing and Tianjin New Town Planning, Baodi District, Tianjin (2009)
Conceptual design for Sancai Club of Yaojia Park, Beijing (2009)
Urban design for development zone center area of Langfang (2009)
Planning and design for Huairou new park of Graduate University of Chinese Academy of Sciences (2008)
Design for the north school residential district of Beijing Institute of Technology (2007)
New Countryside Planning for Foman village and Zhaozhuang village, Changyang town, Fangshan District, Bejing (2007)
Urban design for Zhengbian new zone, Henan province (2007)

甘燕琼
Yanqiong Gan

甘燕琼

1976年3月出生，1998年毕业于北方工业大学建筑学院

主要业绩

中日友好医院质子治疗中心	1.1万	90床
深圳龙珠医院	6.5万	600床
绵阳市中心医院	10万	1000床
廊坊市传染病医院	1.3万	190床
天津市 CDC 120 急救中心 第一人民医院 CDC		
总3.9万医院部分	1.2万	100床
云南传染病医院	2.4万	300床
甘肃省 CDC	0.78万	
安徽国际护理学院模拟医院教学楼	1.1万	
厦门第二人民医院	2.9万	250床
厦门圣玛利妇幼保健医院	3.3万	200床
济南妇幼保健医院	2.5万	350床
北京市丰台区医院	3.2万	200床
唐山唐海县人民医院	3.5万	300床
国家康复辅具研究中心	3.3万	150床
钦州市第二人民医院	4.2万	500床
鞍山市烧伤整形医院	10.5万	850床

Yanqiong Gan

Born in March 1976,Architecture College of North China University in 1998

Main Achievements

Proton Therapy Center of Sino-Japanese Friendship Hospital	11,000	90 beds
Shenzhen Pearl Hospital	65,000	600 beds
Mianyang Central Hospital	100,000	1000 beds
Langfang Hospital for Infectious Diseases	13,000	190 beds
Tianjin CDC 120 Emergency Center & First People's Hospital CDC		
39,000 in total for hospital	12,000	100 beds
Yunnan Hospital for Infectious Diseases	24,000	300 beds
Gansu CDC	7,800	
Simulated-hospital Teaching Building of Anhui International College of Nursing	11,000	
Xiamen Second People's Hospital	29,000	250 beds
Saint Mary's Maternal and Child Health Care Hospital in Xiamen	33,000	200 beds
Jinan Maternal and Child Health Care Hospital	25,000	350 beds
Fengtai District Hospital in Beijing	32,000	200 beds
People's Hospital in Tanghai County, Tangshan	35,000	300 beds
National Research Center for Rehabilitation Aids	33,000	150 beds
Qinzhou Second People's Hospital	42,000	500 beds
Anshan Burn-plastic Surgery Hospital	105,000	850 beds

宋志强
Zhiqiang Song

宋志强

2001年毕业于天津大学建筑学院
设计中注重方案的可实施性，主张适度超前及"利于营造"。
尤其是在住宅项目设计中，力争实现市场效益与社会效益的有机结合。

Zhiqiang Song

Graduated from School of Architecture, Tianjin University in 2001
Focus on the feasibility of the program; maintain going further moderately and "good for construction"
Especially in the residential program design, try to achieve the organic combination of market and social benefits.

习 皓
Hao Xi

习 皓

2001年毕业于天津大学建筑学院
主持设计多项工程，尤以商业建筑、办公建筑等见长。

主要业绩

国家康复辅具研究中心
山西烟草大厦
天津万科新城会所

Hao Xi

Education background: graduated from Architecture College of Tianjin University
Designing for many projects, and especially good at commercial buildings and office buildings

Main Achievements

the Researching Center of National Treatment Assistant
Shanxi Tobacco Building
Tianjin Wankexincheng Club

CSA 中天伟业

做建筑设计专家

王志强
Zhiqiang Wang

张建英
Jianying Zhang

侯立超
Lichao Hou

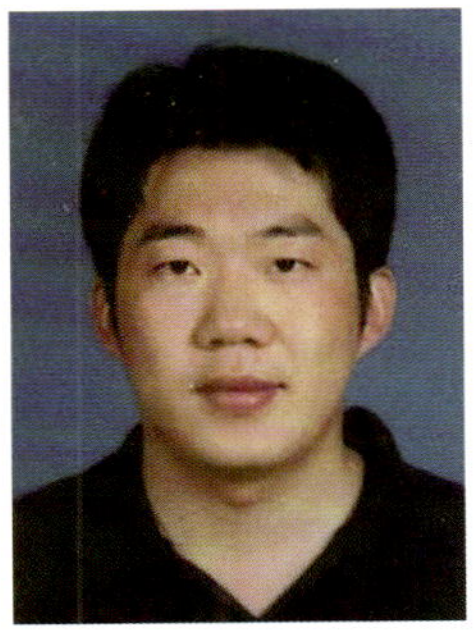
孙 杰
Jie Sun

王志强

中天伟业（北京）建筑设计集团 设计部经理
建筑学硕士
高级建筑师
国家一级注册建筑师

执业经历
1990–1994年 青岛建筑工程学院 建筑系学士
1994–1997年 湖南大学建筑系 建筑学硕士
曾在中国电子工程设计院、北京华特建筑设计顾问有限公司工作
2008年加入中天伟业（北京）建筑设计集团

主要作品
北京奥运媒体村
北京凤凰置地广场
邯郸华信时代广场
山西朔州古城改造
北京中关村生命科学园医疗园A5地块
北京强佑清河新城

张建英

中天伟业（北京）建筑设计集团 副总建筑师
建筑学学士
高级建筑师

执业经历
2001–2009年 中国建筑设计研究院联安国际建筑设计有限公司
2009年加入中天伟业（北京）建筑设计集团

主要作品
北京中关村CADENCE软件学院　合作设计
北京官园桥锦官苑小区1号3号楼　合作设计
北京西城区三里河一区办公楼AB座　合作设计
北京西城区三里河一区办公楼CD座　合作设计
北京西城区三里河三区12号地办公楼　合作设计
北京西城区国家物资储备局办公楼
北京公安交通管理局京顺车管所办公楼
石家庄南二环南王地块住宅区规划设计
北京东城区幸福三村高档住宅区
新疆库尔勒市和合家园小区
海南陵水度假酒店区规划

侯立超

中天伟业（北京）建筑设计集团 副总建筑师
建筑学学士
高级建筑师

执业经历
1995–2000年 吉林建筑工程学院 建筑系学士
曾在中国建筑设计研究院、主题工作室工作
2008年加入中天伟业（北京）建筑设计集团

主要作品
北京通州永顺西街商住项目
鑫苑中国郑州世纪东城
鑫苑中国郑州中央华府
北京凤图腾二期
凤凰岭度假酒店
北京燕栖开发区综合楼
陕西光华集团办公楼

孙 杰

中天伟业（北京）建筑设计集团 主创建筑师
建筑学学士

执业经历
2002–2007年 河北工业大学建筑学院 建筑学学士
曾在天津市建筑设计研究院、北京五合国际
2010年加入中天伟业（北京）建筑设计集团

主要作品
北京西直门中材大厦
昆明和合家园永升广场（160米双子塔）
奥北住宅区规划方案
深圳生命保险大厦竞标方案（200米）
台湾苗北音乐厅
台湾同湖13层新概念办公楼

Zhiqiang Wang

China Sky (Beijing) Architecture Group　Design department manager
Master of Architecture
Senior Architect
National first class registered architect

Working experience
1990-1994 Qingdao Institute of Architecture and Engineering Bachelor of Architecture
1994-1997 Architecture department of Hunan University Master of Architecture
He had worked in China electronics engineering design institute and Beijing Huate Architectural Design and Consulting Co., Ltd.
In 2008, he joined China Sky (Beijing) Architecture Group.

Major works
Olympic Media Village in Beijing
Phoenix place in Beijing
Huaxin Time Square in Handan
Renovation of The ancient city of Shuozhou in Shanxi
Zhong Guancun Life Science Park Medical campus A5 plot
Qiangyouhe New Town in Beijing

Jianying Zhang

China Sky (Beijing) Architecture Group　Vice Chief Architect
Bachelor of Architecture
Senior Architect

Working experience
2001-2009 China Architecture Design & Research Group Lianan architecture design company
In 2009, Jianying Zhang joined China sky (Beijing) architecture group Creation Architect

Major works
Zhongguancun-Cadence Institute of Software Technology　Cooperative design
No.1 and No.3 building Guanjinyuan residential quarter in Guanyuan bridge Cooperative design
The No.1 division office building tower A&B in Sanlihe , Xicheng district Cooperative design
The No.1 division office building tower C&D in Sanlihe , Xicheng district Cooperative design
The No.3 division office building No.1 and No.2 plots in Sanlihe, Xicheng district Cooperative design
Office boiling of National Material Reserves bureau in Xicheng district
Beijing Traffic Management Bureau Department of Motor Vehicles of Jingshun
Planning and design of Nanwang plot residence area in South 2nd ring road in Shijiazhuang
Xingfu San Cun High-end residential quarters in Dongcheng district, Beijing
Korla Hehe home in Xinjiang
Hainan Lingshui resort hotel planning

Lichao Hou

China Sky (Beijing) Architecture Group　Vice Chief Architect
Bachelor of Architecture
Senior Architect

Working experience
1995-2000 Jilin Architectural and Civil Engineering　Bachelor of architecture
He had worked in China Architecture Design &Research group and Idea studio
In 2008, Lichao Hou joined China sky (Beijing) architecture group Creation Architect

Major works
Yongshun West Street in Tongzhou commercial residence project
Xiyuan (China) century Dongcheng in Zhengzhou
Xiyuan (China) central mansion in Zhengzhou
Beijing phoenix totem phase 2
Resort Hotel of Fenghuangling
Beijing Yanqi Economic Development Zone comprehensive building
Guanghua group in Shanxi

Jie Sun

China Sky (Beijing) Architecture Group Creation Architect
Bachelor of Architecture
Creation Architect

Working experience
2002-2007 Hebei University of Technology　school of architecture & art design
Bachelor of architecture
He had worked in Tianjin architecture design & research institute, Beijing 5+1 Werkhart International, Beijing CCI design, Taiwan Jinding architecture design and consulting company Beijing branch.
In 2010, Jie Sun joined China sky (Beijing) architecture group Creation Architect.

Major works
Xizhimen Zhongcai Tower in Beijing
Kunming Hehe Home Yongsheng Square (160-meter petronas twin tower)
Aobei residence area planning scheme
Shenzhen Life Insurance building biding scheme (200 m)
Taiwan Miaobei Concert Hall
Tonghu 13-storey New Concept Office Building in Taiwan

北京清城华筑建筑设计研究院有限公司
THCA Architectural Design & Consulting Institute
(北京清华城市规划设计研究院建筑分院)
Beijing Tsinghua Urban Planning & Design Institute Architectural Department

朱显泽
Xianze Zhu

朱显泽

北京清华城市规划设计研究院　副总建筑师
北京清城华筑建筑设计研究院有限公司　副院长 总建筑师
全国工程建设标准设计专家委员会　委员
高级建筑师
国家一级注册建筑师

工作业绩

《住宅设计规范》编委
《全国民用建筑工程设计技术措施——规划建筑景观》编委
《21世纪高品质高层住宅》副主编
国家标准图集《外装修》设计负责人，获中国勘察设计协会一等奖
全国城镇商品化住宅设计竞赛，获国家建设部一等奖
北京棕榈泉国际公寓，获全国纺织行业优秀工程设计一等奖
北京回龙观后期48班中学，获全国纺织行业优秀工程设计一等奖
北京东方太阳城，获北京市优秀工程设计二等奖
北京望京花园东区，获全国纺织行业优秀工程设计二等奖
北京紫南家园，获北京市优秀工程设计三等奖

Xianze Zhu

Beijing Tsinghua Urban Planning & Design Institute　Vice-chief Architect
Beijing Qingchenghuazhu planning & design institute Co.,Ltd.　Vice-president chief architect
National construction design standard expert committee member
Chief Architect
National Class I registered architect

Research Projects

1 Member of editorial board of *Residential design Standard*
2 Member of editorial board of *National architecture construction design measurement – planning landscape architecture*
3 Vice-chief editor of *21th century high quality high rise residence*
4 Chief designer of *external decoration* of national standards
First Prize awarded of national investigating design associate
5 National city & town commodity residence designing competition
First Prize awarded of national ministry of construction
6 Zonglvquan international apartment of Beijing
First Prize awarded of national textile industry outstanding construction design
7 Huilongguan 48 Classes middle school of Beijing
First Prize awarded of national textile industry outstanding construction design
8 East sun city of Beijing
Second Prize awarded of Beijing outstanding construction design
9 East district of Wangjing garden of Beijing
Second Prize awarded of national textile industry outstanding construction design
10 Zinan residence of Beijing
Third Prize awarded of Beijing outstanding construction design

陈首春
Shouchun Chen

陈首春

1993年毕业于清华大学建筑学院
北京清城华筑建筑设计研究院有限公司　副院长
北京清华城市规划设计研究院　住区规划设计研究所　副所长
中国建筑学会建筑师分会建筑技术委员会　副秘书长
中国绿色建筑与节能专业委员会青年委员会　委员

Shouchun Chen

Graducated from School of Architecture , Tsinghua University in 1993
Beijing Qingchenghuazhu architectural design & research institute Co.,Ltd. Vice President
Beijing Qingchenghuazhu Planning & Design Institute　Settlements in the planning and design institute　Deputy Director
Architectural Society of China architects Architecture Technology committee, Deputy Secretary General
Youth Committee of China Green Building Council　Committee member

北京清城华筑建筑设计研究院有限公司
THCA Architectural Design & Consulting Institute
（北京清华城市规划设计研究院建筑分院）
Beijing Tsinghua Urban Planning & Design Institute Architectural Department

宋晔皓
Yehao Song

宋晔皓

清华大学建筑学院　教授　博士生导师
清华大学建筑学院建筑与城市研究所　所长
北京清城华筑建筑设计研究院有限公司　副总建筑师
中国建筑学会建筑师分会　理事
中国建筑学会建筑师分会建筑技术委员会　副主任
中国绿色建筑与节能委员会绿色建筑理论与实践学组　秘书长
《生态城市与绿色建筑》杂志　副主编
清华大学工学博士
德国慕尼黑工业大学访问学者
国家一级注册建筑师

工作业绩

中国建筑学会，获第五届建筑创作奖佳作奖、2006年第六届青年建筑师奖、2006年第四届建筑创作奖优秀奖
德阳表面工程研究所，获教育部优秀建筑设计二等奖、海星杯第五届威海中国国际建筑设计大赛银奖
常熟图书馆，获2005年建设部优秀建筑设计二等奖、2005年教育部优秀建筑设计一等奖、2004年“WA中国建筑优胜奖”、海星杯第五届威海中国国际建筑设计大赛优秀奖
中国建筑学会主办，“首届中国绿色住宅设赛”“绿色更新：苏南传统水乡地区生态住宅研究”，获第一名，二等奖
博士论文，“结合自然 整体设计：注重生态的建筑设计研究”获“清华大学优秀博士论文”
匈牙利Graphicsoft和世界建筑杂志社共同主办“绿色家园设计竞赛”，获一等奖

Yehao Song

Tsinghua University　School of Architecture of Tsinghua University
Professor　doctorial tutor
Tsinghua University school of Architecture building & urban planning research department　Director
Beijing THCA Architecture Design & Consulting Institute　Vice-chief Architect
Chinese architecture associates architect branch　Director
Chinese architecture associate architect branch architectural techniques committe　Vice-director
Chinese green building & energy saving committee green building theory and practice group　Head secretary
Magazine *Eco-city & Green architecture*　Vice-chief editor
Doctor of Technical Science in Tsinghua University
Visiting scholar of Technical University of Munich
National First-Class certified architect

Research Projects

Architectural Society of China, won Excellent Work Prize in the 5th Architecture Creation Prize, 6th Young Architect Prize in 2006, Honorable Mention in 4th Architecture Creation Prize in 2006
Deyang Surface Engineering Institute, won Second Prize of Excellent Architectural Design of Ministry of Education, Silver Prize of Haixing Cup 5th Weihai China International Architectural Design Megament
Changshu Library, won Second Prize of Excellent Architectural Design of Ministry of Construction in 2005, First Prize of Excellent Architectural Design of Ministry of Education, “WA China Architecture Winning Prize” in 2004, Honorable Mention in Haixing Cup 5th Weihai China International Architectural Design Megagame
“Green Updating: Ecological Residence Study in Sunan Traditional Region of Rivers”, won first place and Second Prize in “First China Green Residence Design Competition” held by Architectural Society of China
Doctoral dissertation “Combining the Nature, Designing entirely: architectural design research concerning ecology”, won “Excerpts of Doctoral Dissertation in Tsinghua University”
Won First Prize in “Green Home Design Competition” jointly held by Hungarian Graphicsoft and World Architecture Periodical Office

王昌兴
Changxing Wang

王昌兴

教授级高级工程师
国家一级注册结构工程师（第一批）
国家注册咨询工程师（第一批）
全国一、二级注册结构工程师执业资格考试命题专家组成员
建设部　绿色建筑评价标识专家委员会委员（节材）
《绿色办公建筑评价标准》起草人之一
《绿色建筑评价标准》补充说明起草人之一
《绿色建筑评价技术指南》编委会成员
众多开发商、设计院、监理单位顾问，培训机构特聘教授
北京清华城市规划设计研究院　结构总工程师
北京清城华筑建筑设计研究院有限公司　结构总工程师

Changxing Wang

Professor level senior engineer
National Class I registered structural engineer (the 1st batch)
National registered consulting engineer (the 1st batch)
Member of experts setting the operation qualifying examination for National first and second class registered structural engineer
Member of experts committee of green architecture evaluation and identification of Ministry of Construction (Nodal wood)
One of the draftsmen of “Green Office Building Evaluation Criterion”
One of the draftsmen of the additional remarks of “Green Architecture Evaluation Criterion”
Member of editorial committee of “Green Architecture Evaluation Technical Manual”
Consultant of many developers, designing institutes, supervision units, distinguished professor of training institutions
Structural chief engineer of Beijing Tsinghua Urban Planning and Design Academy
Structural chief engineer of Beijing Qingchenghuazhu Architectural Design Academy Limited Company

俞伯华
Bohua Yu

俞伯华

北京清城华筑建筑设计研究院有限公司　设计部部长

工作业绩

北京天伦锦城住宅小区
北京回龙观文化居住区F04区住宅
北京天通苑北苑住宅小区
王府井国际商场
兴隆财满街三期商业办公楼
中国驻塞拉里昂使馆经商处
北师大二附中
香山六艺（别墅）
北清嘉园（Townhouse）
香山81号院（Townhouse）
北京金长安商业大堂室内设计
乐成国际学校室内（与澳大利亚DesignInc合作）

Bohua Yu

Beijing Qingchenghuazhu Architectural Design & Research Institute Co.,Ltd.
Minister of Design

Research Projects

Tianlunjincheng Residence Beijing
F04 Huilongguan Residence Beijing
North Tiantongyuan Residence Beijing
International Shopping Mall in Wangfujing
Xinglongcaiman Street Business office building
Business Center Chinese Embassy in Sierra Leone
Branch middle school of North Teacher University
Liuyi villa in Xiangshang
Beiqingjiayuan Townhouse
No. 81 Townhouse in Xiangshan
Inner design of Jinchangan business center hall in Beijing
Inner design of Yuecheng international school (Cooperated with DesignInc)

北京炎黄联合国际工程设计有限公司
Beijing Yanhuang United International Engineering Co.,Ltd.

周 楠
Nan Zhou

周 楠

男，1978年出生
2002年毕业于深圳大学建筑学专业，获得建筑学学士学位。
2003年进入北京炎黄联合国际工程设计有限公司工作，主要从事建筑设计工作，任主任建筑师及项目经理。
2007年通过考试获得国家一级注册建筑师职业资格。在中外建公司七年的工作当中参加并完成了许多各个类型的建筑及规划方案工作。

主要业绩

1. 北京恒奥中心写字楼和全国政协综合楼
2. 北京太阳中心城市综合体
3. 北京龙熙温泉度假酒店
4. 包头名流尚都
5. 河北中石油华北油田第一机械厂活动中心 项目在建
6. 浙江温州雁荡山高尔夫酒店
7. 北京城南故里住宅
8. 惠州西湖山庄
9. 中坤集团北京门头沟杨家峪酒店及康西草原酒店
10. 北京防震减灾中心
11. 贵州保利高尔夫 深化设计阶段
12. 北京延庆辉煌云上别墅区 项目在建

Nan Zhou

Male, born in 1978
2002 Graduated from Shenzhen University, faculty of architecture, Bachelor of Architecture.
In 2003, served at Beijing Yanhuang United International Engineering Co., Ltd., mainly engaged in the architectural design, being as director of architect and project manager.
In 2007, qualified with the national Class A registered architect. During his seven years work at CCI Architectural Co., Ltd., he has participated and completed many different types of construction and planning tasks.

Main works

1. Beijing Heng'ao Centre office building and the CPPCC National Committee integrated building. Completed
2. Beijing Sun City center complex
3. Beijing Longxi Spa Resort
4. Baotou Celebrities Shangdu
5. PetroChina Huabei Oilfield, the 1st activity center of Hebei Machinery Factory. project under construction
6. Yandang Wenzhou Golf Resort, Zhejiang province
7. Beijing south hometown residential
8. Huizhou West Lake Hills
9. Beijing Mentougou Yangjiayu hotel and Hotel of Kangxi Plain, hold by zhongkun Group
10. Beijing Earthquake Disaster Mitigation Center
11. Guizhou Poly Golf. Deepen design stage
12. Beijing Yanqing County Brilliant Cloud Villa. Under construction

安 晨
Chen An

安 晨

男，1979年出生
2002年毕业于西安建筑科技大学建筑学专业。
2007年进入北京炎黄联合国际工程设计有限公司工作，主要从事于建筑方案设计工作。在公司三年的工作当中参加并完成了许多各个类型的建筑及规划方案工作。

主要业绩

1. 三亚美丽之冠七星级酒店项目
2. 郭庄子三限房规划方案投标
3. 天津宝坻橄榄树住宅小区
4. 山西临汾长途汽车站方案
5. 吕梁财政大厦
6. 邢台世贸商住项目
7. 苏丹五星级酒店项目立面改造
8. 北京国际世纪艺术城

Chen An

Male, born in 1979
In 2002, graduated from the Xi'an University of Architecture and Technology Architecture, major in Architecture.
In 2007, serves at Beijing Yanhuang United International Engineering Co., Ltd., mainly engaged in the architectural design work. During the past three years in the company, he has participated in and completed a number of different types of construction and planning works.

Main works

1. Beautiful Crown - seven-star hotel project in Sanya
2. Gouzhuangzi three limitations Housing Planning tender
3. Tianjin Baodi Olive Residential Community
4. Shanxi Linfen coach station program
5. Lu Liang Treasury Building
6. Xingtai Trade commercial and residential project
7. Elevation renovation of five-star hotel project in Sudan
8. Beijing International Century Art City

张凌燕
Lingyan Zhang

张凌燕

女，1981年出生
毕业于山西大学城市设计专业，
2007年初进入北京炎黄联合国际工程设计有限公司工作，主要从事于城市规划和建筑规划设计工作，并着重研究高尔夫社区主题规划，完成了多个高尔夫社区的总体规划方案。

主要业绩

1. 贵阳保利生态体育旅游基地及配套设施综合项目规划设计
2. 潍坊白浪河上游湿地体育公园控制性详细规划
3. 北京市门头沟区古村落保护修缮利用
4. 碛口古镇风貌恢复——五里长街设计方案
5. "滇池西岸"国际高尔夫旅游度假城总体规划设计
6. 江西龙虎山休闲度假中心
7. 昆明呈贡国际社区规划设计方案
8. 江苏镇江征润州概念性总体规划
9. 北京密云人居博物馆控制性规划方案
10. 鹿角湾——中坤岳阳水上国际养生度假中心
11. 海南东郊椰林小镇"梦想庄园"总体规划设计
12. "恒溪谷"度假区总体规划
13. 榆林国际塞上风规划设计
14. 福鼎游艇俱乐部规划方案

Lingyan Zhang

Female, born in 1981
Graduated from Shanxi University,
Urban Design Dept., she has been serving at Beijing Yanhuang United International Engineering Co., Ltd. since the early of 2007, mainly engaged in urban planning and architectural planning and design work, who also focusing on the theme planning of golf community, and has completed a number of golf community overall plan.

Main works

1. Poly eco-sports tourism base and its matching facilities integrated project planning and design in Guiyang
2. Brownsea River upstream wetlands Sports Park Regulatory detailed Plan in Weifang
3. Preservation and Restoration of Ancient Villages in Mentougou District, Beijing
4. Qikou ancient town Recovery – Five Miles Long Street design
5. Planning and Design for "Dianchi Lake West Bank" International City Golf Resort
6. Lung Fu Shan leisure resort in Jiangxi
7. Planning and Design for Chenggong International Community in Kunming
8. Conceptual overall plan for Zhengrunzhou, Zhen Jiang, Jiangsu province
9. Controlling Planning for Miyun Habitat Museum, Beijing
10. Antlers Bay – Zhongkun Yueyang Water international health resort
11. The overall planning for Eastern outskirts of the coconut town "Dream Park", in Hainan
12. "Hengxi Valley" Resort overall Plan
13. Yulin International Frontier Wind Design and Plan
14. Fuding Yacht Club planning program

陈德鹏
Depeng Chen

陈德鹏

男，1975年出生
1998年毕业于清华大学建筑学院，本科学士学位。
2007年进入北京炎黄联合国际工程设计有限公司工作，主要从事建筑专业的设计和管理工作。近三年中参与了各种类型建筑项目的设计工作。

主要业绩

1. 北京大兴网球学校改造项目
2. 北京大兴西红门回迁住宅项目
3. 北京门头新村配套商业项目
4. 山西吕梁大厦
5. 内蒙古包头宝隆阳光尚品住宅小区
6. 郭庄子三限房规划方案

Depeng Chen

Male, born in 1975
In 1998, graduated from Tsinghua University, Faculty of Architecture, Bachelor degree.
In 2007 served at Beijing Yanhuang United International Engineering Co., Ltd., mainly engaged in professional Architectural design and management. For the past three years, he has involved in various types of construction projects either large or little in scale.

Main works

1. Beijing Daxing tennis school improvement project
2. Beijing Daxing Xihongmen move back residential project
3. Beijing Mentou New Village matching commercial project
4. Shanxi Luliang Building
5. Baolong Sunshine Superior Quality Residential in Baotou, Inner Mongolia
6. Guozhuangzi three limitations Housing Planning

北京市古代建筑设计研究所
BEIJING TRADITIONAL CHINESE ARCHITECTURAL DESIGN AND RESEARCH INSTITUTE

徐新伟
Xinwei Xu

徐新伟

副所长
国家一级注册建筑师

部分代表作品
北京市西客站古建部分
北京八宝山告别大厅新建工程
北京古观象台抢险修复
交通银行内外装饰
福佑寺修缮设计
辽宁鞍山玉佛苑玉器市场
德内大街御海王朝四合院

近期项目
辽宁广佑寺
辽宁阜新普安寺
鸦儿胡同四合院
东方太阳城四合院
西海鱼生
山东济南宏济堂博物馆
河北临漳邺城博物馆
内蒙古巴林左旗上京楼

Xinwei Xu

Deputy Director
National Class A Registered Architect

Typical design projects
Chinese Ancient architecture design project of Beijing west railway station
Construction of Beijing Babaoshan farewell hall
Urgent restoration of antique observatory in Beijing
Exterior & Interior Decoration for Bank of Communications
Renovation design of Fuyou temple
Jade market in Anshan jade Buddha Park,Liaoning
Yuhai royal court courtyard dwellings Te-Nai Street

The latest projects
Guangyou temple in Liaoning
Puan temple in Fuxin, Liaoning
Ya'er Hutong courtyard dwellings
Oriental Sun-Town courtyard dwellings
Xihai fresh fish
Hongjitang museum in Jinan, Shandong
Linzhang Yecheng Museum in Hebei province
Shangjing tower in Bairin left Banner, Inner Mongolia

张丽鲜
Lixian Zhang

张丽鲜

总工程师
高级工程师
国家一级注册建筑师
北京市评标专家库评标专家
北京市设计协会理事会理事

部分代表作品
中央电视台无锡影视基地《三国城》工程、《水浒城》工程
北京宛平城东西大街改造工程
北京大学国际数学中心工程
北京大学国家发展研究院工程
山东惠民府衙复原工程
浙江温州安福寺工程

Lixian Zhang

Chief engineer
Senior engineer
National class A registered architect
Tender evaluation expert of Beijing tender evaluation experts data base
Director of Beijing design association administrative committee

Typical design projects
Three kingdoms city and Water Margin City project in Wuxi Movie and TV Base of CCTV
Renovation of East and West Street in Wanping town, Beijing
Beijing international center for mathematical research
National School of Development at Peking University
Reconstruction of Shandong Huimin official residence in feudal China
Wenzhou Anfu temple in Zhejiang

张 越
Yue Zhang

张 越

副所长
国家一级注册建筑师
1999年毕业于中国矿业大学建筑学专业 本科

部分代表作品
澳门渔人码头中国城建筑设计
历代帝王庙修缮
中央戏剧学院外立面装修改造工程
中山公园西门改造设计
前门大街A9–A13新建和改造工程
三里河清真寺修缮工程

近期项目
山东桓台王渔洋故居修缮、复建、新建工程
忠勤祠修缮、复建工程
北京第三十五中学迁址建设工程（古建部分）
台湾饭店四合院

Yue Zhang

Deputy Director
National class A registered architect
Graduated from China University of mining and technology in 1999, bachelor of architecture

Typical design projects
Architecture design of China town in Macau fisherman's wharf
Temple of ancient monarchs renovation
Facade renovation and design of the central academy of drama building
Renovation and design for the west entrance of Zhongshan Park
New construction and renovation of the A9-A13 of Qianmen Street.
Restoration of Sanlihe mosque

Latest projects
Renovation and reconstruction of Wangyuyang former residence in Huantai, Shandong
Renovation and new construction of Zhongqin ancestral temple
Relocation and construction of Beijing No.35 middle school (Chinese ancient architecture part)
Taiwan restaurant quadrangle courtyard

合道设计集团
HORDOR
DESIGN GROUP

范尚华
Shanghua Fan

李　明
Ming Li

赵建群
Jianqun Zhao

林苏挺
Suting Lin

范尚华

职　务：总建筑师
出生日期：1963年
毕业院校：同济大学建筑设计硕士

1963年出生，福州市人。1983年毕业于同济大学建筑系，1991年同济大学建筑城规学院建筑设计专业硕士毕业。现任厦门合道工程设计集团有限公司总建筑师，福州分公司副总经理。

代表作品
芳沁园1号楼
民航培训中心（长乐）
中福西湖花园
福州奥林匹克花园

Shanghua Fan

Position: Chief Architect
Date of Birth: 1963
Graduate College: Tongji University,Master of Architectural Design

Mr. Fan was born in 1963 in Fuzhou. He graduated from the Department of Architecture, Tongji University in 1983. In 1991 he received Master's Degree of Architectural Design in College of Architecture and Urban Planning, Tongji University. Currently he is the chief architect of Xiamen Hordor Engineering Design Group Co., Ltd. and the Deputy General Manager of Fuzhou branch.

Representative Designs
Building 1# of Fang Qin Park
Civil Aviation Training Centre (Changle)
Zhongfu•West Lake Garden
Fuzhou Olympic Garden

李　明

工作背景
2004年毕业于国立华侨大学建筑系，同年进入厦门市建筑设计院有限公司土建一所工作至今。

工作业绩
如厦门湖边花园C区保障性住房方案设计、厦门金林湾花园A、B区方案设计、江苏盐城宝龙城市广场、青岛李沧宝龙城市广场、江苏常州宝龙城市广场、江苏宿迁宝龙城市广场、河南新乡宝龙城市广场、厦门湖里万达城市广场等、厦门市救助管理站改扩建工程、厦门观音山中小学、厦门东海一中一小，厦门高林居住区及相关配套项目等，曾获省级优秀设计奖。

Ming Li

Background
In 2004 Mr. Li graduated from Department of Architecture, national Huaqiao University; since then, he has been working in Brach office for Civil Engineering of Xiamen Architectural Design Institute Co., Ltd.

Performance
As the project design for affordable housing of District C, Xiamen Lakeside Garden; project design for District A and B of Xiamen Jinlinwan Garden, Baorlong Urban Plaza of Yancheng, Jiangsu, Baorlong Urban Plaza of Licang, Qingdao, Baorlong Urban Plaza of Changzhou, Jiangsu, Baorlong Urban Plaza of Suqian, Jiangsu, Baorlong Urban Plaza of Xinxiang, Henan, Wanda Urban Plaza of Huli, Xiamen, the reconstruction and expansion project of Xiamen Salvation Management Station, Xiamen Guanyinshan Primary and Secondary School, First Primary School of Xiamen Donghai First Middle School; Xiamen Gaolin residential area and the related facilities, etc. Mr. Li has received the provincial excellent design award.

赵建群

职　务：副总建筑师
出生日期：1966年8月
毕业院校：苏州城建环保学院

1966年出生，福建厦门人，1990年毕业于苏州科技学院（原苏州城建环保学院）。现任厦门合道工程设计集团有限公司副总建筑师，土建二院院长。参与主持设计多项大型建筑，如厦门海沧行政服务中心、厦门国贸蓝海、国贸春天、国贸金海岸、国贸阳光、南昌朝阳绿洲等项目。

代表作品
国贸花园、国贸广场、国贸金海岸、信隆城二期、海沧行政服务中心、怡祥大厦、国贸阳光、国贸春天、海沧区文化中心、厦门科技创业广场、国贸蓝海、X2004G01-2＃,3＃,4＃、南昌朝阳绿洲、合肥国建新城三期、国贸2007G28,G29地块

Jianqun Zhao

Position: Deputy Chief Architect
Date of Birth: August 1966
Graduate College: Suzhou Institute of Urban Construction & Environmental Protection

Mr. Zhao was born in 1966 in Xiamen, Fujian and graduated in 1990 from Suzhou University of Science and Technology (the former Suzhou Institute of Urban Construction and Environmental Protection). Currently he is Deputy Chief Architect of Xiamen Hordor Engineering Design Group Co., Ltd. and the president of Civil Engineering Design Institute Ⅱ. Mr. Zhao has presided over the design of many large-scale projects, including Xiamen Haicang Administrative Service Center; Xiamen ITG•Coast of Azure; ITG•Spring; ITG•Golden Coast; ITG•Sunshine; Nanchang Chaoyang Oasis and other projects.

Representative Designs
ITG•Garden, ITG Plaza, ITG•Golden Coast, Xinlong City, Phase Ⅱ, Haicang Administrative Service Center, Yixiang Building, ITG•Sunshine, ITG•Spring Haicang Cultural Center, Xiamen Technological Venture Plaza, ITG•Coast of Azure, X2004G01-2#,3#,4#, Nanchang Chaoyang Oasis, Hefei Guojian New City, Phase Ⅲ, ITG Block 2007G28,G29

林苏挺

工作背景
2001年厦门大学建筑学学士,现任厦门合道设计集团有限公司土建一院主任建筑师

参与各种类别工程项目的规划和设计，涉及教育建筑、住宅建筑、商业建筑及办公建筑、公共综合建筑等。兴趣在于智能写字楼、城市综合体、商业酒店、文化建筑等各类建筑的规划设计。参与设计多项综合性建筑,如厦门理工学院，厦门软件园二期，厦门集美大学规划，厦门五缘运营中心等。

Suting Lin

Working Background
Mr. Lin received Bachelor's Degree of Architecture in Xiamen University in 2001. Currently he is the director architect in Civil Engineering Institute Ⅰ of Xiamen Hordor Engineering Design Group Co., Ltd.
Mr. Lin has participated into the planning and design of various types of projects, involving the educational buildings, the residential buildings, the commercial buildings, the office buildings and the public integrated buildings, etc. He is especially interested in the planning and design of intelligent office building, urban complexe, commercial hotel and cultural building, etc. He has taken part in designs of many integrated construction, such as Xiamen University of Technology; Xiamen Software Park, Phase II; the planning of Jimei University in Xiamen; Xiamen Wuyuan Operation Center, etc.

兰州市城市建设设计院

金光辉
Guanghui Jin

高亚平
Yaping Gao

刘元珍
Yuanzhen Liu

金光辉

男　汉族
毕业于兰州理工大学
教授级高级建筑师，国家一级注册建筑师
兰州市第四、五批专业技术拔尖人才
现任兰州城市建筑设计院副院长

参加工作以来，一直从事城市规划、建筑工程的研究、规划及设计工作，共参与或主持完成了各类工程规划、设计100余项，担任项目总负责主持完成了近60项大中型工程的规划和设计工作，为兰州市城市建设做出了积极贡献。参与或主持完成的1个项目获省优秀设计一等奖，1个项目获省优秀设计二等奖，4个项目设计获省优秀设计三等奖，1项科研项目获省科技进步二等奖，市科技进步一等奖。此外还主持审定了《甘肃省工程建设标准设计图集》3本，主持编制完成了《甘肃康居农宅设计图集》2本。

在甘肃省的规划、建筑设计业内具有较高的知名度，有较高规划、建筑专业理论基础，具有主持大型规划民用设计的前期研究规划和设计协调等能力，具有较高的实践经验，尤其在规划方面具有较深造诣，曾主持了近10项大型住宅小区的规划设计工作。

Guanghui Jin

Male　Han nationality
Graduated from Lanzhou scientific and industrial university
senior architect (professor level), Class 1 registered architect.
the fourth&The fifth batch of top-notch professional and technical talents in Lanzhou City.
he is the assistant president of Lanzhou Urban Construction Design Institute
During these years of working, he has engaged in urban planing ,architecture and engineering researching ,planing and design . For these years, he has participated in and taken charge of more than 100 different types of projects planing and designing , making active contribution to Lanzhou city construction . the projects he participated in and took charge of has won many awards, including one first prize of provincial excellent design , one second prize of provincial excellent design, four third prizes of provincial excellent design.one science and technology research project has won the second prize of provincial technology progress prizes and the first prize of municipal technology progress prizes.In the meanwhile, he has conducted and examined 3 volumes of Gansu projects construction standard design collection , taken charge of and composed two volumes of Gansu new rural residential construction design collection
He has good reputation in Lanzhou architecture planing and design area ,specialised in architecture and designing, he has the capacity in taking charge of large scale research planing and design coordination. with great many of practical experience, he made remarkable achievements in planning.,and has taken charge of the planning and design work of almost 10 large-scale Residence Quarters.

高亚平

女　汉族
毕业于西安冶金学院
高级建筑师，国家二级注册建筑师
兰州市培养造就跨世纪学术技术带头人“222工程”

参加工作以来，一直从事城市规划、建筑工程的研究、规划及设计工作，参与或主持完成的主要项目有金源大厦工程（获2000年甘肃省优秀住宅工程设计一等奖）、中国水电四局高层住宅（获2002年甘肃省优秀工程设计二等奖）、水车博览园（国家4A级旅游景点）。

Yaping Gao

Female　Han nationality
Graduated from Xi'an Institute of Metallurgy
senior Architect, Class-B National Registered Architect
A member of Lanzhou "222 Project"—century academic technology leaders.
Ever since her devotion to work, she has been engaged in urban planning, construction research, planning and design. The projects she coducted and participated in iclude:Jinyuan Mansion (awarded the first prize of Gansu Excellent Housing Design,2000); The high-rise residential building of China Hydropower Bureau Ⅳ completed (awarded the second prize of Gansu Excellent Housing Design,2002); Waterwheel Exposition Park completed in 2004 composes three parts: Waterwheel Park(the National 4A Grade Tourist Attractions).

刘元珍

男　汉族
毕业于西北农林科技大学
高级建筑师，国家一级注册建筑师
兰州市“151人才工程”第二批入选人员

参加工作以来，一直从事城市规划、建筑工程设计工作，参与或主持完成的主要项目有兰州市泰生大厦、珠海青木工业园区、珠海江村花园、兰州市西部广场、白银公安局快速反应指挥指挥中心、兰州山水名庭、参与编制完成的02系列《木门》图集（获2005年度省优秀设计二等奖）。

Yuanzhen Liu

Male　Han nationality
graduated from Northwest A&F University
Senior Architect and National level 1 registered Architect
He has ranked among the the second batch of Lanzhou "151 talents project" personel.
Since he joined work, Yuanzhen Liu has devoted in the urban planning and architecure projects designing work. The projects he coducted and participated in iclude Taisheng plaza in Lanzhou, Qingmu Industrial Park in Zhuhai, Jiangcun Garden in Zhuhai, Lanzhou west square, the fast response control center of police station in Baiyin city, Shanshuimingting in Lanzhou. At the same time he took part in and compiled the 02 series of graphic collection wooden door(Award the sencond prize of the excllent design awards in Lanzhou Province,2005).

蘭州市城市建设设计院

刘宏明
Hongming Liu

李雅琴
Yaqin Li

张　绚
Xuan Zhang

宋天乙
Tianyi Song

刘宏明

男　汉族
高级建筑师
1991年7月毕业于甘肃工业大学建筑工程系建筑学专业。
现任兰州市城市建设设计院主任建筑师。
自参加工作以来一直从事城市规划、建筑工程设计工作。

Hongming Liu

Male　Han nationality
Senior Architect
In July 1991, Liu graduated from the Architecture Department of Gansu University of Technology, majored in Architecture.
Urban Construction Design Institute, Lanzhou the director of the architect office.
Since he joined work, Yuanzhen Liu has devoted in the urban planning and architecure projects designing work.

李雅琴

女　汉族
高级建筑师，国家二级注册建筑师。
1986年7月毕业于重庆建筑工程学院城镇建设专业。
现任兰州市城市建设设计院主任建筑师。
自参加工作以来一直从事城市规划、建筑工程设计工作。

Yaqin Li

Female　Han nationality
Senior Architect, National Registered Architect Class-B.
She graduated as an Urban Construction major from Vocational College of Architectural Engineering, Chongqing in July 1986;
Urban Construction Design Institute, Lanzhou the director of the architect office.
Since he joined work, Yuanzhen Liu has devoted in the urban planning and architecure projects designing work.

张　绚

男　汉族
建筑师
1997年毕业于西安建筑科技大学建筑学专业。
现任兰州市城市建设设计院副所长。
自参加工作以来一直从事建筑工程设计工作。

Xuan Zhang

Male　Han nationality
Architect
In 1997, he graduated from the Xi'an University of Architecture and Technology, majored in Architecture
Deputy Director, Urban Construction Design Institute, Lanzhou
Since he joined work, Yuanzhen Liu has devoted in architecure projects designing work.

宋天乙

男　汉族
建筑师
1999年毕业于辽宁工程技术大学建筑系。
现工作于兰州市城市建设设计院建筑一所。
自参加工作以来一直从事建筑工程设计工作。

Tianyi Song

Male　Han nationality
Architect
In 1999,he graduated from Liaoning Technical University, Department of Architecture.
works at Lanzhou Urban Construction and Design Institute A
Since he joined work, Yuanzhen Liu has devoted in architecure projects designing work.

方略建筑设计有限责任公司
Strategy Architectural Design Co.,Ltd.

丘 志
Zhi Qiu

丘 志

1987年	毕业于重庆建筑大学建筑学专业
1994年	毕业于德国慕尼黑艺术学院建筑专业（硕士）
1990–1998年	工作于德国慕尼黑STEIDLE+PARTNER建筑师事务所
1999至今	方略建筑设计有限责任公司（原九三建筑设计研究院）

Zhi Qiu

1987, graduated from School of Architecture, Chongqing University
1994, graduated from the Architecture major of the Art Institute in Munich, Germany (Master)
1990-1998, working in STEIDLE + PARTNER Architects firm in Munich, German
1999 – Present, Strategy Architectural Design Co., Ltd. (former Jiusan Architectural Design and Research Institute)

杨 楠
Nan Yang

杨 楠

1987年	毕业于重庆建筑大学建筑学专业
1988年	工作于建设部北京建筑设计事务所
1991年	工作于德国慕尼黑PETER LANZ ARCHITECTS建筑师事务所
1995年	毕业于加拿大麦吉尔大学城市规划专业（硕士）
1996年	工作于加拿大BRYDEN MARTEL ARCHITECTS建筑师事务所
1999至今	方略建筑设计有限责任公司（原 九三建筑设计研究院）

Nan Yang

1987, graduated from School of Architecture, Chongqing University
1988, working in the Beijing architectural design group Ministry of Construction
1991, working in PETER LANZ ARCHITECTS Architects Group Munich, Germany
1995, graduated from McGill University in Canada, majored in Urban Planning (Master)
1996, working in BRYDEN MARTEL ARCHITECTS Architects firm in Canada
1999 – Present, Strategy Architectural Design Co., Ltd. (former Jiusan Architectural Design and Research Institute)

广西华蓝设计（集团）有限公司
Guangxi Hualan Design & Consulting Group
（原广西建筑综合设计研究院）

庞 波
Bo Pang

庞 波

广西华蓝设计（集团）有限公司执行总建筑师，1991年毕业于广西大学建筑学专业，高级建筑师，国家一级注册建筑师。

1991年开始从事建筑设计至今，先后担任主创和负责设计完成的大中型项目数十项，具有丰富的设计实践经验，特别是在学校建筑、体育建筑、办公建筑等公共建筑及居住建筑设计方面有自己的创作理念，并有多次获得建设部、广西优秀设计奖。

Bo Pang

Bo Pang is a senior architect, a PRC Class I Registered Architect, one of chief architects in Guangxi Hualan limited company. He graduated from Architecture Department of Guangxi University in 1991.
From 1991 to now, he has been in charge of 10 large and medium-sized projects. As for designs of public building and resident building, he has rich experience and special design concept, especially on schools, sports centers and offices. Among his designs, many won Outstanding Design Award issued by Ministry of Construction and Guangxi Province.

徐欢澜
Huanlan Xu

徐欢澜

广西华蓝设计（集团）有限公司副总建筑师。1999年毕业于广西大学，城规与建设专业，研究生，教授级高级建筑师，国家一级注册建筑师。广西建筑协会常务理事，入选广西十佳建筑师，荣获"全国五一巾帼标兵""全国三八红旗手"第五届"广西十大女杰""广西优秀专家"称号。

Huanlan Xu

Huanlan Xu is a senior architect, a PRC Class I Registered Architect and a deputy chief architect in Guangxi Hualan limited Company. She received postgraduate degree from architecture department of Guangxi University in 1991. She is a standing director of Guangxi Architect Commission, one of top ten Outstanding Architects. She is awarded the titles of "National May 1st Female Pacesetter", "March 8th Red Banner holder", "Top Ten Guangxi Female Figures" and "Guangxi Excellent Expert".

广西华蓝设计（集团）有限公司
Guangxi Hualan Design & Consulting Group
（原广西建筑综合设计研究院）

禤晓林
Xiaolin Xuan

禤晓林

广西华蓝设计（集团）有限公司建筑设计院副总建筑师，1991年毕业于华南理工大学，城市规划专业。2007年毕业于华中科技大学，建筑与土木工程专业，获工程硕士学位。教授级高级工程师、国家一级注册建筑师。

主要建筑作品

广西区人民医院门诊综合楼（建设部优秀勘察设计三等奖、广西优秀工程设计一等奖）、广西医科大学第一附属医院内科大楼（全国医院建筑优秀设计三等奖、广西优秀工程设计一等奖）、广西中医学院一附院住院综合楼（全国优秀工程勘察设计行业奖建筑工程三等奖、广西优秀工程设计一等奖）、南宁三中实验综合楼、艺术楼（全国优秀工程勘察设计行业奖建筑工程三等奖、广西优秀工程设计一等奖）、广西化学中毒医疗救治基地（广西优秀工程设计三等奖）南宁市兴宁区人民政府机关综合楼礼堂及服务中心（广西优秀工程设计三等奖），广州珠江医院规划设计、广西区人民医院一分院心理康复中心、河池地区人民医院住院大楼、广西区人民医院肿瘤大楼、广西区人民医院内科大楼改造、广西医科大学肿瘤医院改造、广西江滨医院改造、柳州市工人医院、广西区人民医院一分院高层住宅楼、广西区人民医院高层住宅综合楼、文华明苑、广西财经学院学生生活综合楼。

Xiaolin Xuan

Xiaolin Xuan is a senior architect, a deputy chief architect in Guangxi Hua Lan limited company and PRC Class I Registered Architect.. He received bachelor degree from Urban Planning department of South China University in 1991. He received postgraduate degree from Civil Engineer department of Huazhong University of Science and Technology.

Main Works

Outpatient Complex Building of the People's Hospital of Guangxi Zhuang Autonomous Region (It won the Third Prize of Outstanding Investigation and Design issued by Ministry of Construction and the first prize of Excellent Engineering Design in Guangxi Province),the Medicine Building of First Affiliated Hospital of Guangxi Medical University (It won the Third Outstanding Design on Hospital and the First Prize of Excellent Engineering Design in Guangxi Province),the Inpatient Building of First Affiliated Hospital of Guangxi College of Traditional Chinese Medicine (It won the Third Prize of State Outstanding Investigation and Design on Construction Project and the first prize of Excellent Engineering Design in Guangxi Province),the complex building and art building of Nanning No.3 Middle school (It won the third prize of State Outstanding Investigation and Design on Construction Project and the First Prize of Excellent Engineering Design in Guangxi Province), the Base of Treatment of Chemical Poisoning in Guangxi Province (It won the Third Prize of Excellent Engineering Design in Guangxi Province),the complex building, hall and service center of Nanning city Xingning district (It won the Third Prize of Excellent Engineering Design in Guangxi Province),the project design of Guangzhou Zhujiang Hospital, the Psychology Rehabilitation Center of the People's Hospital of Guangxi Zhuang Autonomous Region, the Inpatient Building of Chihe People's Hospital, Tumors Building of Guangxi People's Hospital, the renovation of the Medicine Building of Guangxi People's Hospital, the renovation of the Tumour Hospital Guangxi Medical University, the renovation of Guangxi Jiangbin Hospital, Liuzhou Worker's Hospital, high-rise apartment for Guangxi People's Hospital, Wenhuaming Uptown and student apartment for Guangxi Finance and Economic College.

徐洪涛
Hongtao Xu

徐洪涛

广西华蓝设计（集团）有限公司华蓝研究院总建筑师。国家一级注册建筑师。1997年东南大学建筑系本科毕业，2005年清华大学建筑学院硕士毕业。热爱建筑创作，积极参加重点项目的方案投标将项目策划、城市设计等与建筑创作紧密结合。

近3年来主要参与的项目：南宁博物馆、广西美术馆、百色市文化科技中心、涠洲岛火山博物馆、华蓝奕园、南宁市青少年活动中心等文化建筑；广西体育中心体育场、越南外交部大楼、广西政协、河池机场等大型公共建筑；泛北论坛会议酒店、北海冠岭山庄国宾馆、涠洲岛度假酒店等酒店类建筑；北海蓝海银湾、广西丽原大厦等居住及综合商业建筑的设计工作。另外还参与了南宁创意产业园、华蓝弈园等项目策划工作，百色市龙景区中心轴线城市设计，南宁市民歌湖等城市设计项目工作。

Hongtao Xu

Hongtao Xu is a chief architect in Guangxi Hualan limited company and a PRC Class I Registered Architect. In 1997 he received bachelor degree from Architecture department of Southeast University. In 2005, he received postgraduate degree in Tsinghua University. Cherishing deep love for architectural creation, he actively participants into bidding for key project, and connects project planning, urban design with architectural creation.

In recent three years, he has participate following projects: culture building including Nanning Museum, Guangxi Art Museum, Baise City Cultural and Technology Center, Wenzhou Island Volcano Museum, Hualan Chess Park, Nanning Teenager Activity Center; public building including The Main Stadium of Guangxi Sports Center, the building of the Vietnamese Ministry of Foreign Affairs, Guangxi People's Political Consultative Building, Chihe Airport; hotels including Fanbei Forum Hotel, Beihai Crown Ridge Resort, Wenzhou Island Resort; residence or commercial building including Beihai Lanhai Silver Bay, Guangxi Liyuan Building. Besides, he participates in many projects planning such as Creative Industrial Park and Nanning Hualan Chess Park, the spool thread with center in Baise City Longjing District and renovation of Nanning Mingge Lake.

周　叱
Chi Zhou

周　叱

广西华蓝设计集团有限公司建筑设计院副总建筑师。1996年毕业于清华大学建筑学院，入选中法总统合作项目〝100名中国建筑师留学法国〞，2003年获法国里尔建筑学院建筑与城市设计硕士学位。国家一级注册建筑师、注册规划师、香港注册规划师、广西建筑学会副秘书长。

主要建筑作品

广西·南宁荔园山庄（建国六十周年中国建筑学会建筑创作大奖、建设部优秀勘察设计三等奖、第四届中国建筑学会建筑创作佳作奖、全国人居经典建筑规划设计方案竞赛 建筑、环境双金奖、广西优秀工程设计一等奖）、南宁国际会展中心一期工程（建国六十周年中国建筑学会建筑创作大奖、第五届詹天佑土木工程大奖、广西优秀工程设计一等奖），百里柳江景观控制性规划（广西优秀规划设计二等奖）。广西区人民政府副省级干部住宅（广西优秀工程设计三等奖）。

主要参编标准图集

《全国小城镇住宅通用（示范）设计——广西南宁地区》（05SJ917-9），荣获2007年度广西优秀标准设计一等奖。

Chi Zhou

Chi Zhou is a deputy chief architect in Guangxi Hua Lan limited company. He received bachelor degree from Architecture Department of Tsinghua University, and was chosen to be a member of a cooperative program named "100 Chinese Architects Study in France". In 2003, he received postgraduate degree from Urban Design department from Ecole Nationale Superieure D'Architecture De Bretagne. He is PRC Class I Registered Architect Qualification, PRC Registered Planner and Hongkong Registered Plaaner and Deputy Secretary-General of Guangxi Architect Commission.

Main Works

Guangxi Nanning Liyuan Villa ((It won the Outstanding Design issued by Architect Commission on the 60th anniversary of the founding of PRC, the third prize of State Outstanding Investigation and Design on Construction Project, Outstanding Design issued by the 4th China Architect Commission The State Classical Design of Double Gold Metal for Architecture and Environment and the first prize of Excellent Engineering Design in Guangxi Province), the Phase I Project of Nanning Interational Exhibition Center (It won the Outstanding Design issued by Architect Commission on the 60th anniversary of the founding of PRC, The 5th China Civil Engineering Zhantianyou Prize and the First Prize of Excellent Engineering Design in Guangxi Province), the Regulatory Planning of Baili Liujiang Landscape (It won the Second Prize of Excellent Engineering Design in Guangxi Province),residence for deputy provincial cadre in Guangixi government (It won the Third Prize of Excellent Engineering Design in Guangxi Province)

Participates to edit atlas

The Standard Design of Residence in Towns---Nanning Guangxi Province (Gui 05SJ917-9). It won the first prize of Excellent Standard Design in Guangxi Province.

雅克设计机构
APCE DESIGN GROUP
ARCHITECTURE • PLANNING • CONSULTANTS • ENGINEERING

侯彦良
Yanliang Hou

侯彦良

1996年至今　雅克设计有限公司　副总建筑师

主要作品
海口西秀海岸带状公园“城廓远眺”广场设计
获省级建筑设计二等奖
济南政务中心区设计
三亚丽景海湾酒店
海峡度假村（五星级酒店）
海南征稽花园
广西崇左市公务员小区
山东寿光圣城中央花园小区
西安交大苏州研究生学院
中石化海南公司办公楼
河南郑州郑东新区民航四季苑小区
海南绿色佳园天上人间小区
湖州仁皇山新区1号地块小区
海口阳光西海岸三期南区
海口东海岸南方电网花园度假小区
万宁宝安新城
海口市青年路廉租房
临高县财政局办公大楼
琼海市京博雅居
中信泰富万宁神州半岛高尔夫球会所及酒店
中信泰富万宁神州喜来登公寓
海口金和信花园
琼海观泉苑小区
海南大学社科实验室与行政办公楼
海南经济学院新校区教学楼群
海南陵水糖厂安置区

Yanliang Hou

From 1996 to now, Deputy Chief Architect in APCE Design Co., Ltd.

Major Works
Design of "City Overlooking" Square of Xixiu Coast belt park, Haikou
The second prize of the provincial excellent architectural design
Design of Jinan Governmental Affairs Center
Landscape Beach Hotel, Sanya
Strait Resort (five-star hotel)
Hainan Levy Garden
Civil Servant Residential Community of Chongzuo, Guangxi Province
Central Garden Residential Community of Holy City Street, Shouguang, Shandong Province
Graduate School of Xi'an Jiaotong University in Suzhou
Office building of Sinopec Hainan Branch
Civil Aviation Seasons Garden in Zhengdong New District, Zhengzhou, Henan Province
Green Garden•Heaven On Earth Community, Hainan
Block No.1 Community of Huangrenshan New Area, Huzhou
South District of Phase Ⅲ of Sunshine West Coast Community, Haikou
Southern Power Grid Garden Resort on the east coast of Haikou
Baoan New Town, Wanning
Low-rent housing of Youth Road, Haikou
Office building of Lingao Finance Bureau
Jingboyaju Residential Community, Qionghai
CITIC Pacific Golf Club and Hotel of Shenzhou Peninsula, Wanning
CITIC Pacific Sheraton Apartment Project of Shenzhou Peninsula, Wanning
Golden Hesine Garden, Haikou
Guanquanyuan Residential Community, Qionghai
Social Science Laboratory and Administration Building of Hainan University
Teaching building group in the new campus of Hainan College of Economics
Resettlement area of Lingshui Sugar Mill, Hainan Province

王忠好
Zhonghao Wang

王忠好

雅克设计有限公司杭州分公司
副总建筑师
建筑二所所长
高级建筑师
国家一级注册建筑师
浙江省建筑学会会员
杭州市建委设计评审专家
1986年毕业于同济大学建筑系
1986年9月—2003年5月　中国新型建筑杭州设计研究院
1997年1月—1998年12月　浙江省建筑科学研究院（借调）
2001年3月—2003年5月　杭州名都建筑设计事务所（借调）
2003年6月至今　雅克设计有限公司

主要作品
杭州市瓶窑大厦
杭州市燃气大厦
杭州红楼饭店
浙江省金华市中级人民法院审判办公大楼
浙江省丽水市中级人民法院审判办公大楼
浙江省东阳市公安技术大楼
中国人民银行淮安中心支行办公大楼
杭州杭政储43号地块多蓝水岸居住小区
湖北省随州香港街
杭州市蒋村商业综合楼
江苏省淮安经济开发区城东花园
江苏省淮安经济开发区徐杨小区
浙江省义乌商城-嘉和广场
浙江省绍兴润和-南岸花城
获得全国人居经典综合大奖
江苏省昆山曼氏商业街

Zhonghao Wang

Hangzhou Branch of APCE Design Co., Ltd.
Deputy Chief Architect
Director of Architectural Institute Ⅱ
Senior Architect
National Class-A Registered Architect
Member of Zhejiang Institute of Architects
Design evaluation expert of Hangzhou Municipal Construction Committee
In 1986, Mr. Wang graduated from Department of Architecture, Tongji University
1986.9-2003.5　CNBM Hangzhou Design & Research Institute
1997.1-1998.12　Zhejiang Academy of Building Research (transference)
2001.3-2003.5　Hangzhou Mingdu Architectural Design Firm (transference)
2003.6-present　APCE Design Co., Ltd.

Major Works
Pingyao Building, Hangzhou
Ranqi Building, Hangzhou
Honglou Hotel, Hangzhou
Trial & office building of Jinhua Intermediate People's Court, Zhejiang Province
Trial & office building of Lishui Intermediate People's Court, Zhejiang Province
Technology building of Dongyang Public Security Bureau, Zhejiang Province
Office building of Huaian Center Branch of People's Bank of China
Blue Waterfront Residential Community of H.Z.C.C. Block 43 #, Hangzhou
Hong Kong Street, Suizhou, Hubei Province
Commercial Complex Building of Jiang Village, Hangzhou
Chengdong Garden in Huaian Economic Development Zone, Jiangsu Province
Xuyang Residential Community in Huaian Economic Development Zone, Jiangsu Province
Yiwu Mall—Jiahe Square, Zhejiang Province
Runhe—South Bank Flower City, Shaoxing, Zhejiang Province
The Comprehensive Award of National Habitat Classics
Maxim Commercial Street, Kunshan, Jiangsu Province

雅克设计机构
APCE DESIGN GROUP
ARCHITECTURE · PLANNING · CONSULTANTS · ENGINEERING

傅培儒
Peiru Fu

傅培儒

雅克设计有限公司 副总建筑师
建筑设计三所 所长
1995年毕业于厦门大学建筑系
1995年7月—1996年8月　厦门大学建筑设计研究院
1996年8月—2006年3月　新加坡著名建筑设计公司
2006年3月至今　雅克设计有限公司

主要作品
马来西亚滨海度假村规划
马来西亚半山公寓规划
武汉盘龙城美式小镇28街
厦门水岸虹桥
海口望海商城二期
海大中日友好交流中心与外国语学院
海南省（海口）青少年活动中心
兴隆南海雨林温泉度假酒店
兴隆明珠温泉花园
兴隆太阳河温泉度假村

Peiru Fu

Deputy Chief Architect of APCE Design Co., Ltd.
Director of Architectural Institute Ⅲ
In 1995, Mr. Fu graduated from Department of Architecture, Xiamen University
1995.7-1996.8 Architectural Design & Research Institute of Xiamen University
1996.8-2006.3 A renowned architectural design firm in Singapore
2006.3-present APCE Design Co., Ltd.

Major Works
Planning of Malaysian Coastal Resort
Planning of Malaysia Banshan Apartment
The 28th Street of American Town in Panlong City Community of Wuhan
Rainbow Shore Community of Xiamen
Haikou Wanghai Mall, Phase Ⅱ
China-Japan Friendship Exchange Center and Foreign Languages College of Hainan University
Hainan (Haikou) Youth Center
SCS (South China Sea) Rainforest Spa Resort Hotel of Xinglong
Pearl Spa Garden of Xinglong
Sun River Spa Resort of Xinglong

游　涛
Tao You

游　涛

雅克设计有限公司厦门分公司 总经理
作品入选《厦门十大青年设计师作品选》
1993年毕业于武汉城市建设学院（现武汉科技大学城市建设学院）规划与建筑系
1995年3月—2004年2月　现代集团上海建筑设计研究院
2004年3月至今　雅克设计有限公司

主要作品
厦门市行政中心改，扩建
厦门市华侨博物院扩建
厦门海沧国家税务局
厦门台商会馆
龙岩市体育公园体育馆
安溪明园大酒店
中国（安溪）特产城
冠豸山温泉酒店及度假村
崇武半月湾度假酒店
许昌健发御园
许昌香格里拉大酒店

Tao You

Xiamen Branch of APCE Design Co., Ltd.
General Manager
His design was selected into Design Collection of Xiamen Top Ten Young Architects
In 1993, Mr. You graduated from Department of Planning and Architecture, Wuhan Urban Construction College (now Urban Construction College of Wuhan University of Technology)
1995.3-2004.2 Xian Dai Group: Shanghai Institute of Architectural Design & Research
2004.3-present APCE Design Co., Ltd.

Major Works
Reconstruction and expansion of Xiamen Administrative Center
Expansion of Xiamen Huaqiao Museum
Haicang State Administration of Taxation, Xiamen
Xiamen Taiwan Businessmen Guildhall
Stadium of Longyan Sports Park
Anxi Mingyuan Hotel
China (Anxi) Specialty Town
Spa Hotel and Resort Village of Guanzai Mountain
Half Moon Bay Resort Hotel of Chongwu
Jianfa Royal Park of Xuchang
Xuchang Shangri-la Hotel

周忠云
Zhongyun Zhou

周忠云

2000年毕业于长春工程学院
2000年6月—2001年9月　北京市香港中厦设计院
2001年9月—2003年6月　澳在利亚柏涛(PTW)设计有限公司
2003年6月—2006年9月　阿奇贝斯(加拿大)设计公司
2006年至今　雅克设计有限公司　副总建筑师
建筑创作中心主任

主要作品
海口海岸·赛拉维住宅小区
海口保税区联检大楼
海南保亭万平家园二期
海南大学社会科学学科群实验室与行政办公楼
三亚海棠湾皇圃大酒店
海口玉沙村南区建筑设计
海口蓝城商务大厦建筑设计
海口医学城规划设计
海南医学院西海岸校区规划设计
海南洋浦半岛体育公园规划
海南广播电视大学建筑设计
海南洋浦开发区滨海生活服务区总体规划
峨嵋天颐国际会议中心及酒店建筑设计
南宁中国东盟国际商贸城规划设计
广西南宁新闻出版城建筑设计
北京师范大学新校区规划设计
北京理工大学规划与景观设计
西安西部通讯产业基地

Zhongyun Zhou

In 2000, Mr. Zhou graduated from Changchun Institute of Technology
2000.6-2001.9 Hongkong Cathay Design Institute in Beijing
2001.9-2003.6 Peddle Thorp (PTW) Design Co., Ltd. in Australia
2003.6-2006.9 Archispace (Canada) Design Corporation
2006-present works as Deputy Chief Architect and Director of Architectural Design Cente in APCE Design Co., Ltd.

Major Works
Seacoast· C'est la Vie Residential Community of Haikou
Combined Detection Building of Haikou Free Trade Zone
Wangping Homeland Phase Ⅱ of Baoting, Hainan
Laboratory and Administration Building of Social science group in Hainan University
Huang Pu Hotel of Begonia Bay, Sanya
Architectural design for South District of Yusha Village, Haikou
Architectural design of Haikou Blue City Commercial Building
Planning and design of Haikou Medical City
Planning and design for West Coast Campus of Hainan Medical College
Planning for Sports Park of Yangpu Peninsula, Hainan Province
Architectural design of Hainan Radio & TV University
General planning for Coastal Life Service Area of Yangpu Development Zone, Hainan Province
Architectural design of Mt. Emei Paradise International Conference Centre and Hotel
Planning and design of Nanning China-ASEAN International Trade City
Architectural design of Guangxi Nanning Publishing House
Planning and design for the new campus of Beijing Normal University
Planning and landscape design of Beijing Institute of Technology
Xi'an Western Communication Industry Base

机械工业第六设计研究院

NO.6 INSTITUTE OF PROJECT PLANNING & RESEARCH OF MACHINERY INDUSTRY

武浩杰
Haojie Wu

武浩杰

机械工业第六设计研究院研究员级高级工程师、院副总工程师，1975 年生，河南洛阳人，1998 年 6 月毕业于西安冶金建筑学院（现西安建筑科技大学）建筑系，建筑学学士，国家一级注册建筑师。

主要作品
湖北中烟工业园区总体规划及武汉卷烟厂易地技改项目
杭州卷烟厂"十一五"易地技术改造项目
厦门卷烟厂"十五"技术改造项目
呼和浩特卷烟厂易地技术改造项目
深圳卷烟厂易地技术改造工程
宝鸡烟厂"十五"技术改造项目
北京卷烟厂易地技术改造项目
武汉重型机床集团有限公司整体搬迁改造项目
宁夏小巨人机床有限公司
中华人民共和国郑州海关
中华人民共和国商丘海关
郑州光大银行
鹤壁夏威夷俱乐部

Haojie Wu

Born in 1975 at Henan Luoyang, he is senior engineer & vice chief engineer of No.6 Institute of Project Planning & Research of Machinery Industry. He was graduated from Architecture Department of Xi'an University of Architecture & Technology in Jun. 1998, Bachelor of Architecture & national Class I certified architect.

Main Works
Removing & Technical Innovation Project of Hubei China Tobacco Industrial Park General Planning & Wuhan Tobacco Company
"10th Five-year Plan" Removing & Technical Innovation Project of Hangzhou Tobacco Company
"10th Five-year Plan" Technical Innovation Project of Xiamen Tobacco Company
Removing & Technical Innovation Project of Huhehaote Tobacco Company
Removing & Technical Innovation Project of Shenzhen Tobacco Company
"10th Five-year Plan" Technical Innovation Project of Baoji Tobacco Company
Removing & Technical Innovation Project of Beijing Tobacco Company
Removing & Transforming Project of Wuhan Heavy-duty Machine Tool Group Co., Ltd.
Ningxia Little Giant MAZAK Co., Ltd.
Zhengzhou Customs of P.R.C
Shangqiu Customs of P.R.C
China Ever bright Bank Zhengzhou Branch
Hebi Hawaii Club

龙文新
Wenxin Long

龙文新

机械工业第六设计研究院研究员级高级工程师、院副总工程师，1967 年生，湖南衡阳人，1989 年毕业于合肥工业大学建筑学专业，建筑学学士，国家一级注册建筑师。

主要作品
郑州国际会展中心
郑州英协路东地块高层住宅小区
郑州市华亚科技广场
郑州市中医院高层病房楼
河南省新华书店中原图书城
河南科技市场数码港
河南建业联盟新城
郑州森林半岛

Wenxin Long

Born in 1967 at Hunan Hengyang, he is senior engineer & Vice Chief Engineer of No.6 Institute of Project Planning & Research of Machinery Industry. He was graduated from Architecture Department of Hefei University of Technology in 1989, Bachelor of Architecture & national Class I certified architect.

Main Works
Zhengzhou International Conference & Exhibition Center
Zhengzhou Yingxie Road High-rise Residential District
Zhengzhou Huaya Scien-tech Square
High-rise Ward Building of Zhengzhou Hospital of Traditional Chinese Medicine
Henan Xinhua Book Shop Zhongyuan Book City
Henan Scien-tech Market Cyberport
Henan Jianye Lianmengxincheng Residential District
Zhengzhou Forest Peninsula Residential District

李文东
Wendong Li

李文东

机械工业第六设计研究院高级工程师、院长助理，1973 年生，天津市人，1996 年 6 月毕业于西安冶金建筑学院（现西安建筑科技大学）建筑系，建筑学学士，国家注册规划师。

主要作品
郑州国际会展中心
郑州紫荆广场
郑州市人大办公楼
郑州烟草研究院异地建院工程科研实验楼
天瑞集团铸造有限公司
顺弛中央特区项目
龙腾盛世苑小区
焦作建业森林半岛
河南检察职业学院新校建设工程
濮阳建业城项目
新乡金龙建业森林半岛
河南源升香榭丽舍住宅小区
郑州鑫苑名家住宅小区
沈阳机床银丰铸造有限公司

Wendong Li

Born in 1973 at Tianjin, he is senior engineer & president assistant of No.6 Institute of Project Planning & Research of Machinery Industry. He was graduated from Architecture Department of Xi'an University of Architecture & Technology in Jun. 1996, Bachelor of Architecture & National Certified Planner.

Main Works
Zhengzhou International Conference & Exhibition Center
Zhengzhou Zijing Square
Office Building of the NPC Zhengzhou Committee
Scientific Research & Experiment Building of Zhengzhou Institute of Tobacco Research
Shenyang Machine Tool Yinfeng Casting Co., Ltd.
Tianrui Group Casting Co., Ltd.
Sunco Zhongyangtequ Residential District
Longteng Shengshiyuan Residential District
Jiaozuo Jianye Forest Peninsula Residential District
New School Construction Project of Henan Procurator Professional School
Puyang Jianye City Project
Xinxiang Jinlong Jianye Forest Peninsula Residential District
Henan Yuansheng Champs Elysees Residential District
Zhengzhou Xinyuan Splendid Residential District
Shenyang Machine Tool Yinfeng Foundry Co.,Ltd.

牛　飚
Biao Niu

牛　飚

机械工业第六设计研究院教授级高级工程师、院副总工程师，1985 年毕业于湖南大学建筑系，建筑学学士，国家一级注册建筑师。

主要作品
盛润国际空间
郑东新区金融大厦
河南省中级人民法院
郑州凯旋门工程
运河上郡
四季花城
河南科技大学新校区
黄河科技大学图书馆
郑州友谊医院

Biao Niu

Senior engineer at professor level & Vice Chief Engineer of No.6 Institute of Project Planning & Research of Machinery Industry. He was graduated from Architecture Department of Hunan University, Bachelor of Architecture & National Certified Planner.

Main Works
Shengrun International Space Mansion
East Zhengzhou New Area Financial Mansion
Henan Intermediate People's Court
Zhengzhou Triumph Arch Project
Yunheshangjun Residential District
Sijihuacheng Residential District
New School of Henan University of Science & Technology
Library of Huanghe University of Science & Technology
Zhengzhou Youyi Hospital

机械工业第六设计研究院

NO.6 INSTITUTE OF PROJECT PLANNING & RESEARCH OF MACHINERY INDUSTRY

张海燕
Haiyan Zhang

张海燕

机械工业第六设计研究院高级工程师、院副总工程师，郑州市首届优秀建筑师。1971 年生，河南省太康县人，1993 年毕业于郑州大学建筑学专业，建筑学学士，国家一级注册建筑师。

主要作品
郑州国际会展中心
中义 · 阿卡迪亚
郑州大上海城商业步行街一期
富田 · 太阳城居住区
顺弛中央特区一、二期
青岛百丽广场
青岛颐和国际
郑州市小李庄城中村改造项目
郑州市中原区小岗刘城中村改造项目

Haiyan Zhang

Born in 1971 at Taikang County Henan Province, senior engineer & Vice Chief Engineer of No.6 Institute of Project Planning & Research of Machinery Industry. She was graduated from Architecture Department of Zhengzhou University in 1993, Bachelor of Architecture & National Class I Certified Architect.

Main Works
Zhengzhou International Conference & Exhibition Center
Zhongyi Acadia Residential District
Zhengzhou Grand Shanghai Shopping Mall Phase I
Futian Sun City Residential District
Sunco Zhongyangtequ Residential District Phase I & II
Qingdao Baili Square
Qingdao Yihe International Mansion
Zhengzhou Xiaolizhuang Urban Village Reconstruction
Zhengzhou Xiaogangliu Urban Village Reconstruction

毛卫东
Weidong Mao

毛卫东

机械工业第六设计研究院院长助理，高级工程师，1971 年生，河南周口人，1994 年 6 月毕业于合肥工业大学建筑系，建筑学学士，国家注册建筑师，方案主创建筑师。

主要作品
周口市中心医院新建病房大楼
河南省中医院综合楼
中州大学新校区学生活动中心
河南永达国际大厦
河南省郑东试验学校新校区
河南省农科院综合实验楼
郑州中原区白庄城中村整体改造项目
郑东新区王鼎国际大厦
安阳市中医药学校新校区
飘安住宅小区

Weidong Mao

Born in 1971 at Zhoukou Henan Province, senior engineer & President Assistant of No.6 Institute of Project Planning & Research of Machinery Industry. He was graduated from Architecture Department of Hefei University of Technology in Jun. 1994, Bachelor of Architecture & National Certified Architect.

Main Works
New-built Ward Building of Zhoukou Central Hospital
Comprehensive Building of Henan Hospital of Traditional Chinese Medicine
Activation Centre of New School of Zhongzhou University
Henan Yongda International Mansion
East Zhengzhou Experiment School
Comprehensive Experiment Building of Henan Institute of Agriculture Science
Zhengzhou Baizhuang Urban Village Reconstruction
East Zhengzhou Wangding International Mansion
Anyang New School of Trational Chinese Medicine
Piao'an Residential District

郭芳慧
Fanghui Guo

郭芳慧

机械工业第六设计研究院高级工程师、副总建筑师，1972 年生，1996 年毕业于郑州工业大学建筑学专业，建筑学硕士，国家一级注册建筑师。

主要作品
兰州移动通信枢纽楼工程
拉萨移动通信枢纽及调度指挥中心
吉林市电信枢纽工程
河南济源市电信综合生产楼
河南移动信阳分公司生产楼
长春卷烟厂"十一五"易地技术改造项目
清水苑住宅小区
橄榄城住宅小区
观屿国际住宅小区
郑州市档案馆

Fanghui Guo

Born in 1972, senior engineer & Vice Chief Architect of No.6 Institute of Project Planning & Research of Machinery Industry. She was graduated from Architecture Department of Zhengzhou University of Technology in 1996, Master of Architecture & National Class I Certified Architect.

Main Works
Lanzhou Mobile Communication Building
Command Center of Lasa Mobile Co., Ltd.
Jilin Telecommunication Hub Project
Comprehensive Production Building of Henan Jiyuan Telecommunication
Production Building of Henan Mobile Xinyang Branch
"11th Five-year Plan" Removing & Technical Innovation Project of Changchun Tobacco Company
Qingshuiyuan Residential District
Ganlancheng Residential District
Guanyu International Residential District
Zhengzhou Archives

包昌亮
Changliang Bao

包昌亮

机械工业第六设计研究院研究员级高级工程师、院副总工程师，享受国家政府特殊津贴。毕业于扬州大学工学院，国家一级注册建筑师。

主要作品
河南省人大办公楼
郑州市人大办公楼
世界客属文化中心
郑州国际会展中心
郑东新区大厦
郑州温哥华大厦
郑州海联国际交流中心大厦

Changliang Bao

Senior engineer at researcher level & Vice Chief Engineer of No.6 Institute of Project Planning & Research of Machinery Industry enjoying special subsidy from state government. He was graduated from Technology Department of Yangzhou University and is National Class I Certified Architect.

Main Works
Office Building of the NPC Henan Committee
Office Building of the NPC Zhengzhou Committee
World HAKKA Cultural Center
Zhengzhou International Conference & Exhibition Center
East Zhengzhou New Area Mansion
Zhengzhou Vancouver Mansion
Zhengzhou Hailian International Communication Center

Annual Review of Chinese Architectural Design Works

2009-2010

中国建筑设计作品年鉴

〇下

《中国建筑设计作品年鉴》编委会 编

華中科技大學出版社
http://www.hustpas.com

图书在版编目（CIP）数据

2009—2010中国建筑设计作品年鉴（上、下册）／《中国建筑设计作品年鉴》编委会 编.
—武汉：华中科技大学出版社，2010.11
ISBN 978-7-5609-6592-5

Ⅰ.①2… Ⅱ.①中… Ⅲ.①建筑设计-作品集-中国-2009—2010-年鉴 Ⅳ.①TU206-54

中国版本图书馆CIP数据核字（2010）第184116号

2009—2010中国建筑设计作品年鉴（上、下册）　　《中国建筑设计作品年鉴》编委会 编

出版发行：华中科技大学出版社（中国·武汉）
地　　址：武汉市武昌珞喻路1037号（邮编：430074）
出 版 人：阮海洪
责任编辑：姚　璐 余航帆
责任监印：秦　英
美术设计：郝泽星 彭　丹 张　琼 胡培瑜
印　　刷：北京信彩瑞禾印刷厂
开　　本：889 mm×1194 mm　1/12
印　　张：82（上：40，下：42）
字　　数：560千字
版　　次：2010年11月第1版
印　　次：2010年11月第1次印刷
书　　号：ISBN 978-7-5609-6592-5 / TU·951
定　　价：960.00元（上、下册）

投稿热线：(010)64155588-8009　　hzjztg@163.com
销售电话：(022)60266190，(022)60266199（兼传真）
　　　　　(010)64155566（兼传真）
网　　址：www.hustpas.com　　www.hustp.com

主　　管：中华人民共和国住房和城乡建设部
主　　办：中国建筑文化中心
编　　辑：《中国建筑设计作品年鉴》编委会
　　　　　北京主语空间文化发展有限公司
主　　编：陈建为
执行主编：肖　峰
编辑部主任：何建国
编　　辑：陈　石 刘笑艳 王　伟 冷娜英
　　　　　代　君 张　澜 时亦飞
编辑部地址：北京市海淀区三里河路13号
　　　　　中国建筑文化中心712室
邮　　编：100037
电　　话：+86-10-88151966/13910120811
传　　真：+86-10-88151958
网　　址：www.ccdu.com.cn
E-mail：zgjsmail@126.com

谨以此年鉴献给为中国建筑设计行业发展而辛勤付出的设计师们!

2009-2010
中国建筑设计作品年鉴

《2009-2010中国建筑设计作品年鉴》(以下简称《年鉴》)
收录了新中国成立以来，重点是近几年的1900多件优秀建筑设计作品，
这些不同类型、不同设计理念的作品基本反映了当前建筑设计行业多元化的发展状况和实践成果，
从中可以直观地领略到中国建筑设计行业融汇中西、贯通古今的设计思想。
对于各地建设主管部门和从事城市规划、建筑设计、景观设计、工程建设、房地产开发人员以及相关科研教育机构掌握建筑设计行业发展趋向、了解新的设计理念，《年鉴》会有很好的参考、借鉴作用。
《年鉴》在编辑过程中，得到了各地建设主管部门和众多设计院所、设计师的大力支持和帮助，
在此，谨向为《年鉴》提供帮助和支持的单位和个人表示衷心的感谢!
《年鉴》在国内公开发行，并委托中国图书进出口(集团)总公司在国外和港、澳、台地区发行。
《年鉴》编辑过程中，不免会有疏漏或错误之处，敬请读者指正，并提出宝贵意见，
以便我们不断提高《年鉴》的编辑水平，满足读者的需求。

《中国建筑设计作品年鉴》编辑部
2010年9月

宗白華先生言，一切藝術綜合于建築，而禮樂詩歌舞劇之表演，亦與建築背景協調成為一片美的生活，以每一文化的強盛時代莫不有偉大的建築計劃，以容納和表現這一豐富之生命。先生此言發表半个多世紀以前，今日我們更應致力于這一和諧的美的實現早日來臨

吴良镛

Saving our environment is the most vital issue that humankind must address today leading into our fears that this millennium may be our last. An ecological approach to our businesses and design is ultimately about environmental integration. Ecodesign is designing for bio-integration.

建筑师在本世纪所担负最大的课题就是用生态设计的方法去解决我们人类社会所面临的环境问题。我们的终极目标就是实现建筑与环境的和谐共生。

杨经文

希望"年鉴"成为全面总结和推动中国建筑创作发展的优秀典籍

何镜堂

贺"中国建筑设计年鉴"出版

作为中国建筑设计历程的记录，终将成为世界建筑史的辉煌篇章

程泰宁

愿《中国建筑设计作品年鉴》成为中国设计力量走向世界设计主流崛起的推动力！

[illegible]

办年鉴，年年办鉴，推精品，精益求精，树大师，不遗余力。

[illegible]

为了中国建筑设计的崛起，为了本土建筑师的声望与力量，请有志同仁加入到"新本土"建筑设计运动中，也请《中国建筑设计作品年鉴》作为此方面的引导

崔愷

推陈出新

把祖国建设得更美好

钟训正

希望"年鉴"成为高品质的传世年鉴。

孟建民

贺《中国建筑设计作品年鉴》出版发行

追求卓越 宁静致远

[illegible]

梁应添

祝贺《中国建筑设计作品年鉴》发行

良师益友

历史年轮 时代鉴证

刘景樑

设计，经营说法，都是一种选择：在喧嚣中选择宁静，在盲目中选择理性，在守旧中选择创意，在浮华中选择真诚。

庄惟敏

希望本年鉴不只是建筑师或设计单位的"目录"，而是大家彼此观摩学习的园地。

潘冀

2009-2010

中国建筑设计作品年鉴

建筑文化是一个最能感染其他的文化，而建筑设计，是一个最没有国界的产品。

中国建筑设计的发展，从过去20多年来，由抄袭西方，模仿西方，到最近两年来的探索中国建筑文化，可以说是慢慢的找到了"自我"。但我很希望中国设计师们，不要重复"义和团"的"自大"。

的确，如何"古为今用"，"洋为中用"，进而"中为西传"，是我们设计人员，未来的重要工作。共勉之。

愿《中国建筑设计作品年鉴》在中国设计界更有权威性、代表性。

潜心研究市场需求
专注挖掘项目价值

认真钻研建筑功能
努力创造优秀建筑

杜怀文

立足澳门 背靠祖国
面向世界 共创繁荣

希望中国建筑师的职业地位进一步提升；希望中国建筑师有更多更好的相互交流的平台；希望中国建筑师所面对的设计市场环境更开放、更公平。

佳作 拙作 都写作
前鉴 后鉴 都年鉴

AIM 陈晓宇

建筑设计能改变人的生活方式并影响人的生活质量，因此，科学设计是我们的责任。

追求品质
倡导原创
宏扬个性
力争创新

建筑文化源于生活，高于生活

愿《中国建筑设计作品年鉴》在中国建筑界树起中国建筑文化的大旗。

愿《中国建筑设计年鉴》成为东西方建筑文化交流的平台。

绘画三江
筆染四海

顺德建筑设计院

回顾新中国60年的建筑设计历程，努力进取，迎接中国建筑创作更加美好的未来。

浓缩精华 展示精品
愿2009年"年鉴"成为建筑师的良师益友

杨键

Epigraphy Cmooittee 编委题词

特邀编委

（按姓氏首字母排序）

目录 上卷

目录 下卷

特邀编委索引 下卷

（按姓氏首字母排序）

设计师索引 下卷

（按所在机构首字母排序）

设计作品分类索引 下卷

设计作品分类索引 下卷

宾馆、酒店、餐厅、招待所

医院、医疗建筑

设计作品分类索引 下卷

设计作品分类索引 下卷

园林、公园、规划设计

其他

Architectural

设计作品
Design Works

重庆世贸中心
重庆国贸大厦
重庆国贸大厦
重庆万豪酒店
重庆新东方女人广场
重庆邹容广场
重庆黄花园立交
重庆时代豪苑
重庆谊德大厦
重庆邹容广场
重庆奎星楼建筑群
重庆群鹰商场
重庆都市广场（重庆大世界）
重庆华侨宾馆
重庆裕轮大厦
重庆聚融广场
重庆韩光百货
重庆扬子岛酒店
重庆商业大厦
长江文具店
重庆美美百货
重庆纽约.纽约
重庆时代豪苑
重庆江北滨江路
重庆地王广场
重庆洲际酒店
重庆帝都广场
重庆建设银行大厦
重庆南国丽景
重庆（和记）商务大厦
重庆东方曼哈顿
重庆御景江山
重庆海逸酒店
重庆民族广场
重庆大都会广场

CQADI

ChongQing Architectural Design Institute of China

中国·重庆市设计院

重庆市设计院成立于1950年，是国家建设部、国家发改委批准的综合甲级勘察设计和工程咨询单位，拥有建筑工程、市政公用、智能化设计、工程勘察、工程咨询、工程造价咨询、施工图审查等国家甲级资质，经国家外经贸委批准，享有对外经营权，并通过ISO 9001：2000质量体系认证，2006年被评为建设部"十五"全国建筑技术创新先进企业，一直以来多次位列全国勘察设计前100强。

目前重庆市设计院下设9个建筑所、6个建筑工作室、3个市政设计所、城市规划研究中心、施工图审定中心等20个生产部门和办公室、组织有人事部、财务部、技术质量部、计划经营部等5个管理部门，以及后勤服务中心及从事建筑勘察设计延伸业务的迪赛实业（集团）有限公司。设计院现有在职职工675人，拥有重庆市设计大师4名、享受国务院政府津贴专家12名，拥有教授级高级建筑工程师46人，高级建筑工程师159人，建筑工程师149人；拥有注册建筑师、注册结构工程师、注册岩土工程师、注册规划师、注册电气工程师、注册公用设备工程师、注册监理工程师、注册造价工程师、注册估价师等注册人员共186人。

重庆市设计院始终坚持"精心设计，求实创新，诚信服务，顾客满意"的质量方针，以优质的作品和高度的责任心竭诚为广大客户提供满意的服务，建院以来，设计院从事的工程涉及众多行业，遍及国内10多个省、市、自治区，在全国各地设计了众多具有影响力的项目，建院以来共完成勘察设计工程项目4000多项，先后有重庆大都会广场、重庆都市广场、四川美术学院综合教学楼等近二百个项目荣获了国家、部、省、市级优秀设计和科技成果奖。

Chongqing Architectural Design Institute ("CQADI"), which was founded in 1950, is an institute with Class A certificate conferred by the Ministry of Construction and the State Development and Reform Commission in geotechnical survey & design and engineering consultation. It has the national Class A qualification in construction engineering, municipal works, design of intelligent property management, geotechnical survey & design, engineering consultation, project cost consultation, blueprint verification,etc. CQADI, authorized by the National Foreign Economic Relations and Trade Commission, is a legal entity to undertake foreign operations. It has acquired the ISO9001: 2000 China Quality Certification as well. Since 1997, it has been one of the Top 100 Best Engineering Design and Geotechnical Survey & Design Institutes of P.R.China for seven consecutive years

Chongqing Architectural Design Institute adheres to the quality guideline of "elaborate design, realistic and innovative approach, considerate services, and good customer satisfaction". It wholeheartedly and dutifully provides its clients with high-quality products to meet their requirements. Over the past 57 years, CQADI has accomplished more than 4000 geotechnical design projects, and about 200 engineering projects undertaken by CQADI have been awarded with Excellent Design and Scientific & Technological Achievements prizes at national, ministerial, provincial or municipal levels including Chongqing Mitroplitan Plaza, the Teaching and Administration Building of Sichuan Fine Arts Institute.

ChongQing

National comprehensive A grade prospecting and design institute

Architectural

National Top100 prospecting and design institute lasting 7 years

Design

Experienced over 40000 architecture project

Institute

Gain over 200 prizes

地址：重庆市渝中区人和街31号　电话：+86-23-63854124　传真：+86-23-63856935　邮件：cqadinet@cta.cq.cn　网址：www.cqadi.com.cn

ECD Architects (Int'l) Ltd.

卓创国际工程设计有限公司
International Creative Engineering & Design Co.,Ltd

卓创国际工程设计有限公司（ECD）是一个具有国际化背景并经国家建设部批准的甲级建筑设计机构，是一家具有独立法人资格的从事民用、工业建筑工程设计甲级资质的设计机构。

公司引进国际职业化建筑师事务所的设计与管理经验，在中国设置上海、北京、深圳、重庆等分支机构。公司在城市规划、城市设计、城市综合体、办公建筑、商业建筑、居住区规划与设计等众多建筑专业领域取得了卓越的成绩。客户包括中国著名的房地产商、大型企业以及政府部门等。

ECD 始终以积极、诚恳的态度面对客户，力求为客户提供最准确的工程设计和综合解决方案。

ECD 与世界各大设计机构有着广泛而深入的联系，Dans.Hanganv Architects、USAstantec、Tang Design Studio 等都是我们良好的战略合作伙伴。

公司现已拥有高素质的各类专业设计人员 200 余人，以规划、公建、住宅、环境景观设计为主，提供前期咨询、方案、设计发展、施工图、施工合同管理等全方位的专业服务，并得到广泛的认可，具有很高的社会信誉，一直与甲方、合作方保持着良好的关系。

公司以完善的现代企业管理制度、雄厚的人力资源、先进的技术设备、独特的设计理念、丰富的实践经验，始终如一地为客户提供一流的服务。多年以来，公司承担的各类工程设计得到了众多房地产开发商和建设施工单位的好评，并多次在全国及省市设计评比中获奖。

卓创国际的核心优势来自于经验丰富的建筑师团队，多元化的国际建筑师群体以及互动的工作方式，开放自由的设计思想，完善的质量管理体系，保证了设计的深层品质。在卓创国际的每一个设计中，都贯穿了对生活、环境、可持续发展及文化传统的关注和思考。二十一世纪充满着机遇与挑战，我们将继续通过提供高品质的产品、高质量的服务来创造更多的新的生活空间。

ECD, an independent legal entity, is an international architectural design institution with national Class A architectural design certificate approved by the Ministry of Construction in PRC and specialized in structural design for civil and industrial architecture.
Through introducing advanced and professional experience in engineering, ECD has set up branches in Chinese mainland like in Shanghai, Beijing, Shenzhen, Chongqing and other cities. The company has made outstanding achievements in such professional fields as urban planning, urban design and urban complex design, office architecture,commercial architecture, residential quarter planning and design.
ECD has large numbers of customers in Chinese mainland, including prominent real estate partners, large-scale enterprises and government departments.
ECD insists on treating each customer in line with principle of sincerity and honesty and providing customers with the precise engineering designs and a set of well-integrated solutions.This company, meanwhile, has also deeply cooperated with major designing institutions all over the world. Dans. Hanganv Architects USA, Stantec, Tang Design Studio and so on are our strategic partners.
At present, ECD has a high-qualified staff team made up of over 200 professional designers in various fields. The business scope of this company mainly covers planning and programming, public construction, construction of housing and environmental landscape design. And this company can supply comprehensive professional services, including preliminary consultation, conceptual design, building drawing, and building management. Through years of hard working and persistence, ECD has been widely acknowledged, gained high credibility and kept good relationship with owners and its partners.
Being recognized by construction units and real estate developers, ECD has undertaken various types of engineering design over these years. What is more, this company has won a number of prizes in national and provincial designing competitions. All these achievements can be due to the modern enterprise management system, abundant human resources, advanced technical equipment, unique design ideas, rich experience in practice and constant first-class service.
Core competencies of this company can be attributed to experienced architect team, which consists of architects from all over the world. Liberal-minded arrangements and the excellent quality management system assure the high-quality designs. ECD's designs embody the specialists' awareness and thoughts of cultural traditions and sustainable development between people's life and environment. In the 21st century, which is a century full of opportunities and challenges, this company will continue to provide products and service of superior quality and create new living space.

地址：重庆江北区金源路 7 号君豪 B 区商务楼 7F
邮编：400020
电话：+86–23–65479362
邮箱：ecd2008@126.com
网站：www.ecdart.com
联系人：左黎

Address: 7F Commercial Building, Junhao Area B, No.7 Jinyuan Road, Jiangbei District, Chongqing, China
P.C.: 400020
Tel: +86-23-65479362
E-mail: ecd2008@126.com
Http: //www.ecdart.com
Contact: Li Zuo

卓创国际工程设计有限公司
International Creative Engineering & Design Co.,Ltd

公建类

2–3 东和冉家坝城市广场
建设地点：重庆
建筑面积：260 000 平方米
设计时间：2010 年

本案是通过一“心”，即由超高层构成的城市地标中心；三“轴”，即由城市景观大道统领城市景观主轴，以及两条副轴——休闲空间轴和商业空间轴——再结合不同功能和流线需要的下沉广场、商业中庭、各种出入口等构成了本案的规划布局。

2-3 Donghe Ranjiaba City Square
Construction Location: Chongqing
Building Area: 260,000 m²
Design Time: 2010

The design of this case is mainly based upon the principle of "one center and three axes". "One center" is the center of the city landmark. The principal axis of "three axe" is the urban landscape principal axis dominated by landscape roads. Two auxiliary axes are space axes for relaxation and commerce. The layout is made up by traffic lines and other functional parts of sinking down plaza, atrium of commercial building and passageways and "one center and three axes".

4–5 昆明红星美凯龙
建设地点：云南 昆明
建筑面积：800 000 平方米
设计时间：2009 年

“红星国际”是以一种崭新的综合体姿态出现在“超市郊化”(Exurbanization) 体系中的。不同于城市中心的 CBD，“红星国际”项目希望通过简洁的设计语言塑造魔幻般的空间形态，来形成强有力的区域吸引力和城市副中心的商业核心。

4-5 Kunming Macalline
Construction Location: Kunming, Yunnan
Building Area: 800,000 m²
Design Time: 2009

Hongxing International appears as a new synthesis in the Exurbanization system. Different from CBD in the downtown, the project of Macalline shapes fantastic spatial pattern through compact design which adds attractiveness of the region and forms commercial core of SCBD.

1 安踏百年树
建设地点：福建 厦门
建筑面积：1 621.3 平方米
设计时间：2006 年

在整体设计上充分利用现有用地，并满足城市规划要求，各部分建筑有机结合，地上地下通盘考虑，合理组织人流、物流，内外交通便捷、畅通，互不干扰，新建筑与周围环境形成有机整体，提高了环境的质量，为城市建设增彩添色。

1 One-hundred-year Anta Construction Project
Construction Location: Xiamen, Fujian
Building Area: 1,621.3 m²
Design Time: 2006

This project fully makes use of the existing sites in the holistic design and meets the demand for urban planning. Each part of construction combines with others organically, taking both overground and underground factors into account. Stream of people and material flow are well organized. Traffic inside and outside, without interfering with each other, becomes smooth and unblocked. New buildings and the surroundings integrated with each other to improve environmental quality and to make the city more colorful.

卓创国际工程设计有限公司
International Creative Engineering & Design Co.,Ltd

6

6 南宁汇东
建设地点：广西 南宁
建筑面积：320 000 平方米
设计时间：2010 年
一个引人注目的、优美的塔楼，可以成为一个地区引以为豪的精神领袖。加以现代手法诠释的现代城市商业空间，使建筑与其所在场地的结合趋于完美。

6 Nanning Huidong
Construction Location: Nanning, Guangxi
Building Area: 320,000 m²
Design Time: 2010
An elegant and exquisite tower can become a spiritual symbol of a region. The construction and its surroundings can be a perfect match through applying modern architectural techniques to illustrate commercial space of the modern city.

7

公建类

7 重庆报业大厦
建设地点：重庆市
建筑面积：100 000 平方米
设计时间：2009 年
建筑的蜂巢状外部结构起到了两个重要作用。首先，作为承重结构使大跨度室内空间得以实现，其次帮助调节光线和热量进入大楼内。通过利用大小不同的六角形窗户的交替穿插，使房间可以获得充足的阳光，同时保持合适的温度，夏天不需要空调，冬天也不会有太多的热量流失。本着统一、和谐的设计思想，方案在每个建筑立面上均采用同样的材质及处理手法，使建筑更加美观。

7 Chongqing Press Plaza
Construction Location: Chongqing
Building Area: 100,000 m²
Design Time: 2009
Honeycomb external structure of the building plays two main roles. One hand, this kind of bearing structure makes it possible for longspan indoor space. On the other hand, it helps adjust light and heat of the building. Rooms have sufficient sunshine and keep the appropriate temperature through alternating different sizes of hexagon windows. Air-conditioners are not necessities in such rooms in summer. And in winter, the structure helps save heat. According to the unitary and harmonious design ideas, the same material and handing method are applied to every elevation of this building. In this way, the building put on its artistic and beautiful appearance.

8 天津红星美凯龙
建设地点：天津
建筑面积：760 000 平方米
设计时间：2010 年
设计的主要目标是创造出一个有活力的商业建筑，重新诠释城市商业空间。设计时利用流体原理，塑造出富有雕塑感的地标性建筑。

8 Tianjin Macalline
Construction Location: Tianjin
Building Area: 760,000 m²
Design Time: 2010
The major objective of the design is to build a dynamic commercial building to newly illuminate urban commercial space. The design takes advantage of principles of fluids to create a landmark with characteristics of sculpture.

9

8

9 重庆新闻出版智能大厦
建设地点：重庆
建筑面积：65 000 平方米
设计时间：2007 年
重庆新闻出版智能大厦是重庆市重点项目，项目建成后，不仅进一步提升了解放碑 CBD 的城市景观，而且极大地拓展了解放碑商务中心的文化功能，成为重庆乃至整个西部地区世贸中心、万豪国际金融中心之后的又一地标建筑，大厦为超高层建筑，地上 54 层为商场、写字楼，地下 4 层为地下商场、机械式停车库以及设备用房，建筑总高度为 187 米。

9 Chongqing News Media Intelligent Building
Construction Location: Chongqing
Building Area: 65,000 m²
Design Time: 2007
Chongqing News Media Intelligent Building is a municipal key project in Chongqing. After the completion of the project, not only the urban scenery is improved in CBD of Jiefangbei, but also functions of culture in commercial center of Jiefangbei are expanded. It has become another landmark in Chongqing and even the southwest area besides World Trade Center and JW Marriott Financial Center. The building is extremely tall, with the total height of 187meters, 54 floors of shopping malls and office premises overground and 4 floors of shopping malls, mechanical parking garage and equipment use area underground.

1

1 晋愉岭海

建设地点：重庆
建筑面积：180 000 平方米
设计时间：2008 年

根据用地的实际情况和建设规模的要求，场地内采用高层住宅围合商业体、多层住宅独立成区的布局方式。商业体与山谷地形契合，使住宅可以最大限度地享受双山公园优美的景观。

1 Jinyu Linghai

Construction Location: Chongqing
Building Area: 180,000 m²
Design Time: 2008

The scale of construction meets the demand for the real situation of the land. The overall arrangement is that high-rise residential buildings become independent to commercial areas and multiple dwellings. The commercial area conforms to the landform of valley and dwellers in the residential area can enjoy the beautiful landscape of Shuangshan Park as much as possible.

2

3

4

2–4 南宁青秀山国际高尔夫球场

建设地点：广西 南宁
建筑面积：150 000 平方米
设计时间：2009 年

项目充分利用坡地地形富于变化的特点，创造出多种形态的空间布局。项目强调产品的多样化，在户型、风格和空间组合上富于变化，在提高项目可识别性的同时提升了别墅作为高端物业的个性特征。

2-4 Nanning Qingxiu Mountain International Golf Course

Construction Location: Nanning, Guangxi
Building Area: 150,000 m²
Design Time: 2009

This project fully makes use of the changeful characteristic of slope landforms and creates variform spatial layouts. Diversification is the distinguishing feature. Various internal layouts, styles and spatial organizations improve the identifiability of this project. Meanwhile, it also embodies idiosyncrasies of high-end property.

居住类

5

6

7

5–7 首金美利山

建设地点：重庆
建筑面积：830 000 平方米
设计时间：2009 年

本地块充分利用基地本身赋予的有利条件及景观要素，着力追求居住的舒适度与高品位，通过建筑与景观的完美融合，形成具有山水特色的高品质小区。

5-7 Shoujinmeili Mountain

Construction Location: Chongqing
Building Area: 830,000 m²
Design Time: 2009

The design of this site takes advantage of the site itself and its landscape. Focusing on comfort and taste, the perfect integration of buildings and scenery creates a high-quality community, in which buildings and the natural view enhance each other's beauty.

卓创国际工程设计有限公司
International Creative Engineering & Design Co.,Ltd

8 成都棠湖
建设地点：四川 成都
建筑面积：190 000 平方米
设计时间：2009 年

规划设计从对内外景观资源的不同需求去创造自然、合谐的氛围开始，营造出独特的居住空间和环境，创造出既有城市文化内涵，又有自然生态、恬静的生活空间。

8 Chengdu Tanghu
Construction Location: Chengdu, Sichuan
Building Area: 190,000 m²
Design Time: 2009

The planning and design is based on different natural landscapes, pursuing the harmony and unity between human beings and nature, creating distinctive living space and environment. The living space there is not only provided with liberty and freedom of urban culture, but with tranquil and ecological characteristics of nature.

8

9

11

居住类

11 东海汤山别院
建设地点：重庆
建筑面积：100 000 平方米
设计时间：2008 年

11 Tangshan Villa of East Sea
Construction Location: Chongqing
Building Area: 100,000 m²
Design Time: 2008

10

9–10 橡树澜湾
建设地点：重庆
建筑面积：1 000 000 平方米
设计时间：2009 年

以创造注重生活环境质量的高品质居住小区为宗旨，使该地段达到功能组织合理、用地配置得当、结构清晰、道路顺畅，配套设施完备等要求，创造出以使用者为中心、尊重环境、舒适优美的生活空间，突出其鲜明的地方特色。

整个小区以规划的城市道路交通系统为出发点，顺应地形建造符合该地段特点的道路交通系统，以脉络状的道路系统将建筑单体有机串联起来，并对建筑与建筑之间的尺度进行合理缩放，使小区内部生活不受干扰，又为小区营造出尺度不同的景观空间。

9-10 Bay of Oak
Construction Location: Chongqing
Building Area: 1,000,000 m²
Design Time: 2009

Based on the principle of building a high-quality residential quarter stressing the living environment, this district meets the requirements for reasonable functionally organized structures, functional space with proper configuration, clear structure, smooth roads, and perfect supporting facilities. This residential quarter is a user centric and enviroment respected area user-centric and environment-respected. with special and distinctive local style and it is very comfortable and pleasant to live.

The planning of this district complies with the road transportation system of this city. Single buildings are in series connection in the shape of network, conforming to landforms and road transportation. Variances of distances between buildings not only make the internal life of this residential quarter tranquil and undisturbed, but also create different scaled landscapes.

居住类

12–14 西安阳光城

建设地点：陕西 西安
建筑面积：290 000 平方米
设计时间：2009 年

设计以人为本，力求建立完善的小区生活系统，创造一种绿意盎然、自然和谐、经典高尚的居住环境。设计充分考虑由于城市发展而造成的居住者生活模式、欣赏格调、品位上的转变，从而规划设计出具有超前意识的现代居住小区。

12-14 Xi'an Sunshine Town

Construction Location: Xi'an, Shaanxi
Building Area: 290,000 m²
Design Time: 2009

The living system of this residential area is refined on the basis of people-oriented. Green is shining through. In harmony with nature, the living condition there is noble and lofty. Taking the development of city into account, the design shows the foresight to build a modern residential quarter, which is adapted to changes of life model, aesthetic needs and tastes. Moreover, the single buildings integrate with the overall planning and buildings into the afforestation and water.

12

13

14

15

15 东方公馆

建设地点：重庆
建筑面积：240 000 平方米
设计时间：2006 年

总平面布局顺应城市的架构，充分照顾到城市道路的关系，结合地形与周边建筑，使建筑与环境协调共生。项目以城市阳台和小区中心绿地为轴线，由两端的建筑组成主要的骨架格局，有机高效地组织了商业、住宅和城市阳台及轻轨站的关系。

15 Oriental Mansion

Construction Location: Chongqing
Building Area: 240,000 m²
Design Time: 2006

The layout complies with the framework of the city. Taking urban roads into account, the overall planning combines terrain, surrounding constructions and landscapes in a fantastic way. The city terrace and the center green in the community are taken as axes and buildings on both sides make up of the skeleton pattern. So the commerce district, residential areas, the city terrace and light rail station are effectively organized.

广州市设计院
GUANGZHOU DESIGN INSTITUTE

法人：赵路
电话：+86-20-87544608
传真：+86-20-87544798
邮编：510620
地址：广州市体育东路体育东横街3号设计大厦
邮箱：yuanzhang@gzdi.com
网址：www.gzdi.com

Legal Person: Lu Zhao
Tel: +86-20-87544608
Fax: +86-20-87544798
P.C.: 510620
Address: Design Mansion, 3 Tiyudong Hengjie, Tiyudong Road, Guangzhou
E-mail: yuanzhang@gzdi.com
Http: //www.gzdi.com

广州市设计院组建于1952年，是国内成立最早的甲级勘察设计单位之一，是国内建筑设计行业中一支重要队伍。如今全院员工达五百多人，其中被授予"全国工程设计大师"称号的设计师一名，经国务院批准享受政府特殊津贴的专家十多名，持有国家注册师资格的员工有一百多名，博士、硕士六十多名，各类高中级技术人才四百多名。

广州市设计院拥有工程勘察设计、工程设计咨询、城市规划、市政设计、工程总承包、工程监理、建筑智能化系统工程、消防、环保、施工图审查等多项资质，并已通过ISO 9001：2000质量体系认证，是全国勘察设计综合实力百强单位、并由国家计委确定为全国旅游旅馆设计四个指导性设计院之一，也是建设部批准的第一批建筑智能化系统工程设计甲级单位并已获得国家授予直接对外经营权。在全国优秀设计院评比中，一直名列前茅，2006年荣获"2006年中国十大建筑设计公司"的称号。

广州市设计院自20世纪50年代以来，尤为注重精品项目的设计及技术的创新，打造出了一批高质量、影响力较强的时代精品，如广州友谊剧院、广州宾馆、广州火车站、白天鹅宾馆、中国大酒店、广州天河体育中心、汕头国际金融大厦、广州市购书中心、广东美术馆、北京广州大厦、广州体育馆、广州大学城中山大学、湖南省政府新机关院办公区、湖南省人民会堂、南沙大酒店、广州天河正佳商业广场、广州发展中心大厦、金海湾花园、广州新电视塔、湖南韶山毛泽东遗物馆、中山图书馆、广州新图书馆、珠江城、山东烟台日报印刷工业园、山东临沂久隆国际城市广场等项目。

广州市设计院已有三百多项（次）工程勘察设计项目获国家、部、省、市级的奖励，其中国家级奖二十多项，部省级奖一百四十多项，市级奖二百项。其中白天鹅宾馆获国家科技进步奖、全国优秀工程设计奖、全国第七次质量月奖；广州天河体育中心获国家科技进步奖、全国优秀工程设计奖、建设部优秀工程设计奖、国际建协IAKS奖；广东美术馆、汕头国际金融大厦、华夏大酒店、广州市购书中心、北京广州大厦等项目获全国优秀工程设计奖；广州大学城中山大学、湖南省政府新机关院办公区、广州体育馆等项目获建设部优秀工程设计奖等。

广州市设计院与国内外众多著名设计公司保持着良好的合作关系，如美国SCM、美国ARQ公司、德国GMP国际建筑设计有限公司、荷兰Information Based Architecture B.V、日本株式会社日建设计、法国巴黎机场公司、ARUP奥雅纳工程顾问、美国TT结构顾问公司、迈进机电顾问公司等。

广州市设计院遵循"精心设计，顾客满意，科学管理，持续改进"的质量方针，坚持走管理创新、质量兴业、科技强院之路，充分发挥自己的品牌优势、人才优势、信誉优势，积极应对各种激烈的挑战，敢于竞争、善于竞争，在竞争中谋求更大更快的发展，提升综合竞争能力。

广州市设计院全体员工贯彻科学发展观，着力培育企业精神，形成强劲的精神动力，为企业的可持续发展提供强大的精神支持。全院员工将更加热诚地为社会各界提供优质服务，用智慧与汗水浇铸一座座城市建设的不朽丰碑。

Guangzhou Design Institute was founded in 1952 as one of the earliest Class A survey and design unit established in China and an important team of architectural design of China. There are more than 500 staff members of the whole institute now, of whom one is awarded with the title of "National Engineering Designer", more than 10 staffs enjoy the special allowance for experts as approved by the State Council, more than 100 hold the qualification of national registered engineers/architects, and there are more than 60 doctors/masters and more than 400 senior and medium level technical staff members of various disciplines.

Guangzhou Design Institute possesses multiple qualifications of engineering survey design, engineering design consultant, city planning, municipal engineering design, project general contracting, project supervision, intelligent building system engineering, fire control, environmental protection, and construction drawing review, etc., and has passed the ISO 9001: 2000 Quality System Certification. Guangzhou Design Institute is one of the 100 top institute in terms of comprehensive ability in survey design, one of the 4 guiding design institutes for the design of tourist hotels of the nation as appointed by the National Planning Commission, and one of the first batch of Class A units for intelligent building system design as approved by the Ministry of Construction. We have been awarded the right of direct foreign business operation by the nation. This institute has always come out top in the national appraisal of excellent design institutes. We were awarded with the title of "One of the Top Ten Architectural Design Companies of China, 2006" in 2006.

Since the 1950s, Guangzhou Design Institute has been paying special attention to the design of competitive products and technical innovation, and has produced a batch of high-quality fine works of the times with quite strong influences, such as Guangzhou Friendship Theatre, Guangzhou Hotel, Guangzhou Railway Station, White Swan Hotel, China Hotel, Guangzhou Tianhe Sports Center, Shantou International Financial Mansion, Guangzhou Book Center, Guangdong Art Gallery, Beijing Guangzhou Mansion, Guangzhou Gymnasium, Guangzhou University City Sun Yat-sen University, Hunan Provincial Government New Office Area Courtyard, Hunan Provincial People's Hall, Nansha Hotel, Guangzhou Tianhe Grand View Mall, Guangzhou Development Center Mansion, Golden Bay Garden, Guangzhou New TV Tower, Hunan Shaoshan Mao Zedong Relic Museum, Sun Yat-sen Library, Guangzhou New Library, Pearl River Tower, Shandong Yantai Daily Printing Industry Garden, Shandong Linyi Jiulong International City Square, and etc.

Guangzhou Design Institute has had more than 300 engineering survey design projects (times) awarded with national, ministerial, provincial and municipal prizes, i.e. more than 20 national prizes, more than 140 ministerial/provincial prizes, and nearly 200 municipal prizes. Among the projects, White Swan Hotel was awarded with National Science and Technology Progress Prize, National Excellent Engineering Design Prize, and the 7th National Quality Month Prize. Guangzhou Tianhe Sports Center was awarded with National Science and Technology Progress Prize, National Excellent Engineering Design Prize, Ministry of Construction Excellent Engineering Design Prize and the International Architecture Association IAKS Prize; Guangdong Art Gallery, Shantou International Financial Mansion, Huaxia Hotel, Guangzhou Bookstores Center, and Beijing Guangzhou Mansion, etc. were awarded with National Excellent Engineering Design Prize; and Guangzhou University City Sun Yat-sen University, and Hunan Provincial Government New Office Area Courtyard, etc. were awarded Excellent Engineering Design Prize by Ministry of Construction, and so on.

Guangzhou Design Institute maintains excellent cooperative relationship with a number of famous domestic and foreign design firms, such as Skidmore, Owings & Merrill, Arquitectonica, GMP International GmbH Architects and Engineers, Information Based Architecture B.V of Holland, Nikken Sekkei Ltd., Paul Andreu Architecte Paris, Ove Arup & Partners Ltd., Thornton-Tomasetti Engineers and Meinhardt (M&E) Ltd., etc.

Guangzhou Design Institute follows the quality policy of "Designing elaborately to the satisfaction of the customers and managing scientifically to make continuous improvement", and insist on the development mode of innovation in management, advancement with high quality, and developing the institute by means of science and technology. We will fully exert our advantages in brand, professionals and reputation, and positively confront ourselves with various severe challenges. We dare to compete, and are good at competition, and try to achieve even bigger and faster development and enhance our comprehensive competitive edge through competition.

The entire staff of Guangzhou Design Institute will carry out the scientific outlook on development, and make an effort to cultivate the enterprise spirit to form a strong spiritual momentum as the powerful guarantee for sustainable development. The whole staff of the institute will provide high-quality services for all walks of life of the society with greater enthusiasm to cast immortal monuments of the construction of the city one after another with our wit and sweat.

白云机场铂尔曼大酒店

南越王宫博物馆

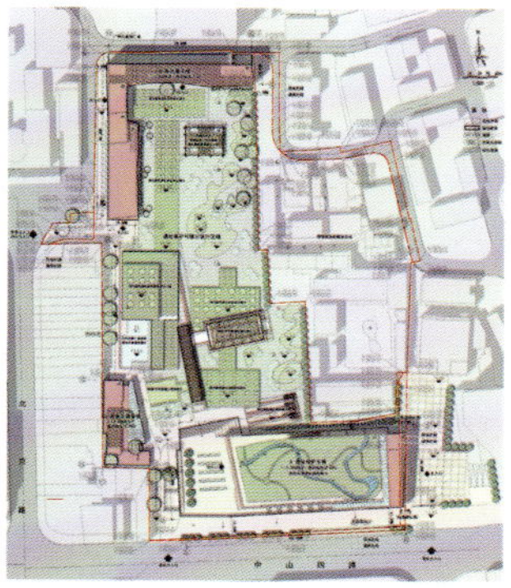

广州科学城综合研发孵化区 A 组团

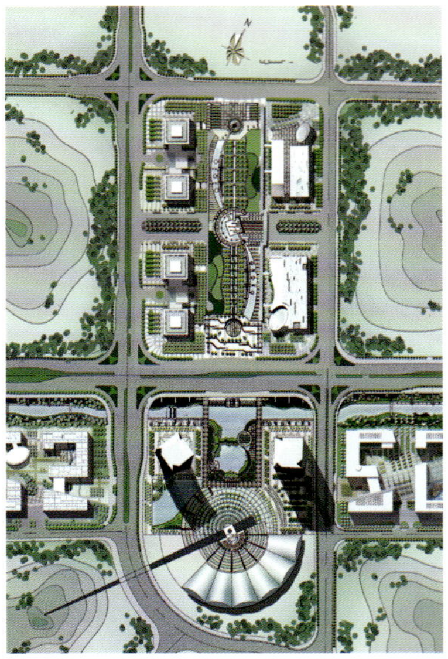

景森設計 JINGSENDESIGN

www.jingsen.cn

广州市景森工程设计顾问有限公司（以下简称景森设计）2001年成立于广州，以“创中国最具影响力的设计企业”为发展目标，不断吸收融汇国际先进技术理念和设计元素并加以创新与运用，为客户提供城市规划、建筑及景观设计、咨询等全程服务。获得国家建筑行业建筑工程设计甲级资质，现为广州市工程勘察设计行业协会理事。

景森设计汇聚大批经验丰富、设计理念先进、思想活跃、精力充沛并且熟悉国内城市规划与建筑设计行业情况的优秀专业人才，先后为500多家知名企业提供服务。目前已在广州、长沙、武汉、昆明、湛江等地设立分公司，为求为客户提供专业、高效、本土化的全程服务。

景森设计专业的设计与服务赢得了客户及业界的广泛好评：

获得万科颁发的“万科深圳区域2007年度设计供方第二名”；

获得2007年度中国房地产最具合作价值专业公司100强——景观设计公司10强；

获得广州勘察设计协会颁发“佛山万科四季花城别墅项目”广州市优秀工程设计二等奖；

获得广州勘察设计协会颁发“佛山万科四季花城小高层项目”广州市优秀工程设计三等奖；

获得2008年度中国房地产最具合作价值专业公司100强——建筑设计公司10强；

中国房地产开发商协会授予景森“2008中国建筑规划.园林景观设计行业广东地区十大最具综合实力品牌机构”；

获得广东省工程勘察设计行业协会颁发“万科四季花城五期别墅项目”2009年度广东省优秀工程设计三等奖。

总部地址：广州市天河区科韵路16号信息港C1栋4楼
4F, Building C1, No.16, Information Harbor, Ke Yun Road, Tian He area, GuangZhou City, Province GuangDong, China
联系电话 Tel: +86-20-22032208　　传真号码 Fax: +86-20-22032200
客服热线 +86-20-22032333　　电子邮件 E-mail: js@jingsen.cn
联　系 QQ: 497283574

· 广州分公司 Filiale in Guangzhou
· 地　址 Address: 广州市天河区科韵路16号广州信息港C1栋4楼
· 联系电话 Tel: +86-20-22032208　　传真号码 Fax: +86-20-22032200
· 联　系 QQ: 1016246065　　电子邮件 E-mail: gz@jingsen.cn

· 长沙分公司 Filiale in Changsha
· 地　址 Address: 长沙市天心区芙蓉南路368号BOBO天下城1栋21001—21007室
· 联系电话 Tel: +86-731-85143456　　传真号码 Fax: +86-731-84114545
· 联　系 QQ: 958516785　　电子邮件 E-mail: cs@jingsen.cn

· 武汉分公司 Filiale in Wuhan
· 地　址 Address: 武汉市江岸区发展大道357号田园商务大厦12楼1201室
· 联系电话 Tel: +86-27-68823507　　传真号码 Fax: +86-27-68823392
· 联　系 QQ: 784572632　　电子邮件 E-mail: wh@jingsen.cn

· 昆明分公司 Filiale in Kunming
· 地　址 Address: 昆明市五一路46号国防大厦14楼C座
· 联系电话 Tel: +86-871-8157858　　传真号码 Fax: +86-871-8157857
· 联　系 QQ: 923941787　　电子邮件 E-mail: km@jingsen.cn

· 湛江分公司 Filiale in Zhanjiang
· 地　址 Address: 湛江市开发区绿华路20号汇景名都3号楼3号商铺之二
· 联系电话 Tel: +86-759-2699888　　传真号码 Fax: +86-759-2699678
· 联　系 QQ: 806351251　　电子邮件 E-mail: zj@jingsen.cn

长城——韶关·长城世家

项目位于韶关市武江区，多为山地丘陵，但水景资源丰富，稍加改造就可以形成层层跌落的流水景观。

规划的核心理念是让环境延伸到人的生活空间，通过景观环境设计实现人与自然的亲近和交流。住区形象充满现代气息，符合本区域特点。住区设计中表现出功能主义的形态构成，简约主义的审美情趣，自然主义的视觉体验，精工唯美的细微环节。居住区完整保留用地原有的〞两山一水〞的空间布局，通过强调两轴——中央景观水轴和连接东西两侧山体的绿轴形成全景的景观脉络，有机地串联起各居住组团。

科学规划 景森设计 JINGSEN DESIGN

越南洽丰——our city

项目位于越南海防市西北边，地块从东到西划为三层。我们对项目的整体规划结合现状，充分利用天然资源，以生态社区为主题，东面结合湿地公园布置高层建筑，内部则引水造湖，建设岛屿，西面沿城市主干道布置风情商业街区、星级酒店和幼儿园等公建。各种住宅拥有最佳的环境，保持私密性的同时更提升用地的土地价值。

小区建筑组团布局西低东高，以减少对内部湖面的压迫感。外围的高层区，主要采用南北向、大间距的围合，形成开阔的组团空间。景观设计将滨水社区引入生活，将闲适氛围融入日常生活中，将水景引入小区环境中，实现〞户户有美景的居住环境〞。

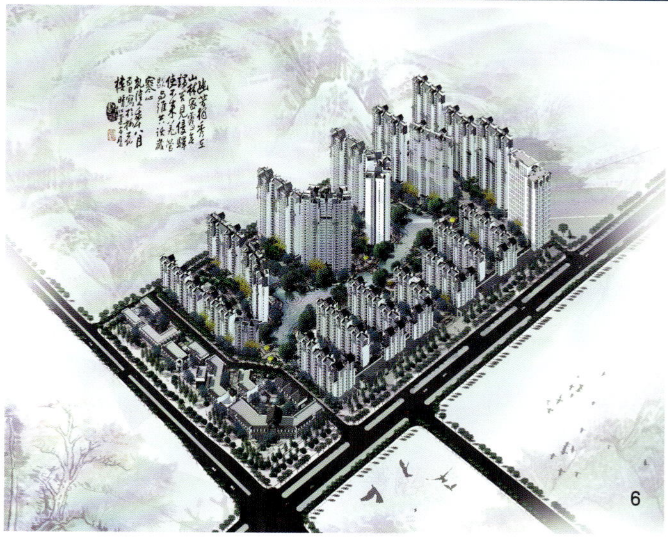

1 美林基业——广州·美林湖国际社区

2 信业——肇庆·珀丽湾

3 逸涛万国——南沙·千山·万国项目

4 珠光——新疆·红雁池水库项目

5 时代——南沙·时代山海湖

6 融侨——西安·曲江观邸

7 中惠——东莞·南城新区

8 凯捷利——连云港·东方之珠

万科——广州·万科云山

项目位于广州市白云区新广从公路以西，坐落在白云风景区内，东侧紧邻南湖风景区，南侧毗邻城市副中心白云新城和白云国际会展中心，东眺连绵的凤凰山脉，南邻面积约8.7公顷的市政规划山体公园，形成"园山中，山在园内"的山景居住模式。云山，不仅将红筒瓦、坡屋顶、拱门、连廊、大草坪等南加州奔放，充满活力的设计元素用于此处，更是将工艺和文化内涵，优雅而精致的生活方式带给居住者。

人 文 建 筑　景 森 设 计　J I N G S E N D E S I G N

珠光——广州·新城御景

项目位于珠江新城N2-1、N2-2地块，处于珠江新城中央商务区的东部。建筑新古典主义的风格奢华而沉稳，加入哥特式的装饰元素，融合古典与现代的建筑精华，延承经典的建筑比例和精致的装饰细节，并与简洁的建筑装饰线条有机结合，使建成项目为现代都市生活注入新的理念。

1

2

3

4

5

6

1 恒信德龙——湖北·摩尔城

2 英惠——昆明·英惠别墅

3 万科——佛山·四季花城

4 云东海花园酒店——三水·山水庄园

5 万科——广州·金域蓝湾

6 珠光——从化·御景花园

7 恒基实业——佛山·恒福国际

8 昆明冶金高等专科学校

8

万达——昆明·滇池卫城G3H4地块

“滇池卫城”项目坐落于昆明滇池国家旅游度假区内，该度假区是昆明市风景优美，环境极佳的高尚旅游、度假、居住社区。

G3地块以泰国北部山地风情为基调，利用该地块存在2米高差的现状，巧妙设计了跌级的毛石砌墙，使流水叠泉与浓郁南亚色彩的花阶等交叠成趣，穿插于建筑之间，运用分层式“台地”园林布局手法解决了园内高差问题，营造出趣味性强的阳光生活社区。H4地块则以泰国南部的临海的地域特征作为设计依据，旨在营造如“芭提雅”蓝天碧海般度假式的生活区。设计中多运用台地、下沉空间、微地形、缓坡等处理手法，丰富场地的节奏感，同时采用浓密的自然植物和粗犷的天然石材，透过特色水景、阳光廊架、风情雕塑小品等元素，营造“一步一景，步移景异”的东南亚园林住宅。

生 态 园 林　景 森 设 计　J I N G S E N D E S I G N

广恒——湛江·君临海岸

项目位于湛江市赤坎区金沙湾观海长廊西边，本方案整体布局以东南亚热带风情为范本，吸取台地园林设计精华，并体现人文文化的内涵，打造出现代与传统融为一体，别具特色风格的小区。设计根据场地高差由东向西设置景点，“热带东南亚风情”的理念一直贯穿设计始终，使每个细节的精雕细琢无不彰显非凡的气派和品位。形式各异的小桥、巴厘岛风情的滨水亭、巧妙设置的水池、亭亭玉立的棕榈科乔木，使小区处处都在诉说着一种浪漫，一种休闲的假日情怀！

1 恒大——广州·御景半岛

2 众一——娄底·众一桂府

3 恒翔——广州·云山雅苑

4 珠光——广州·南沙御景

5 银泰置业——长沙衡阳·银泰红城

6 世纪天乐——昆明·安宁温泉

7 万达——昆明·滇池卫城G2地块

8 恒大——增城·山水城

艺筑建筑设计公司是一个具备雄厚实力的团队。公司专注于房地产建筑设计，能精准把握各项工程的条件和要求，致力于打造出符合客户要求和市场需求的创新精品，向客户提供商业、零售中心、公建、酒店、住宅、规划等多个领域的建筑设计服务及工程与建筑管理服务。我们以专业的眼光和敏锐的市场敏感度，对产品不断求变创新，为客户提供精益求精的设计和优质的服务。诚挚尽责是我公司的座右铭。九成的项目来自既有客户是我公司引以为傲的成绩表。客户对我们的不断支持，使艺筑能够稳步成长。

艺筑设计的广州美林海岸花园、南国奥林匹克花园、云景名都、美林湖畔花园、佛山奥园、合景泰富瑜翠园等项目均取得不俗的销售业绩，近期完成万科花都天景园、南京苏宁天润城、力迅上筑、花都镜湖花园、贵港御临皇府、茂名名雅世家、能兴怡翠世嘉、合景科汇金谷等知名项目设计工作。

"不断创新，打造住宅精品"创造高性能价格比的住宅是本公司不断发展壮大的核心动力。

ART& 艺筑
ARCHITECTURE
广州市艺筑建筑设计有限公司

Art & Architectural Design Company Limited is one of the leading architectural design firms in China since 1998. The firm has an outstanding reputation in meeting clients' needs by delivering creative, value-added,well-designed top-notch projects. These projects include commercial centers, retail centers, public institutions, hotels, residential developments and planning.

We aim consistently to provide a well-balanced living environment by considering the needs of users, developers, environment and socio-economical factors.

Our success is based on recurring business from highly satisfactory clients who awarded projects to our devoted multi-disciplined specialists. They develop and produce solutions for virtually any type of development requested by our clients through reliable services.

地址：广州市越秀区中山二路 3 号粤运大厦 3 楼 A 室
电话：+86–20–61181771
传真：+86–20–37621177
邮箱：winland@win-land.com
网址：www.win-land.com

Address: Room A, 3F, Yueyun Tower, No.3 Zhongshan Road 2th, Guangzhou
Tel: +86-20-61181771
Fax: +86-20-37621177
E-mail: winland@win-land.com
Http: //www.win-land.com

番禺喜盈商业街

工程地点：广州番禺
用地面积：53 419 平方米
建筑面积：41 647 平方米
设计时间：2004 年
开 发 商：广州番禺番发房地产有限公司

设计构思

项目位于广州市番禺区新世纪花园南地块。地块北面是 60 米宽青新路扩展段，地块的对面是新世纪花园。设计采用以白色和暖色为基调的加州风格，体现温馨、活泼的气氛，通过对建筑顶层的艺术处理，丰富了主景观面的整体造型效果。

万科天景花园

设计构思

“居住在风景中”，这是万科天景花园的规划与设计宗旨。该项目位于广州市花都区荔红路西侧，总用地面积六万多平方米，交通方便，主要为居住区，同时设有完善的生活配套设施，如商业设施，包括国际公园、酒店、银行、休闲购物城、餐饮、会所等服务设施。开发商的目标是以丰富的配套资源和区域发展前景，结合多元的景观资源、经典的创意户型和创新规划设计，将万科天景花园打造成一座“都会 • 河畔 • 地标”。

工程地点：广州花都
用地面积：61 325 平方米
建筑面积：167 869 平方米
设计时间：2007 年
开 发 商：万科集团

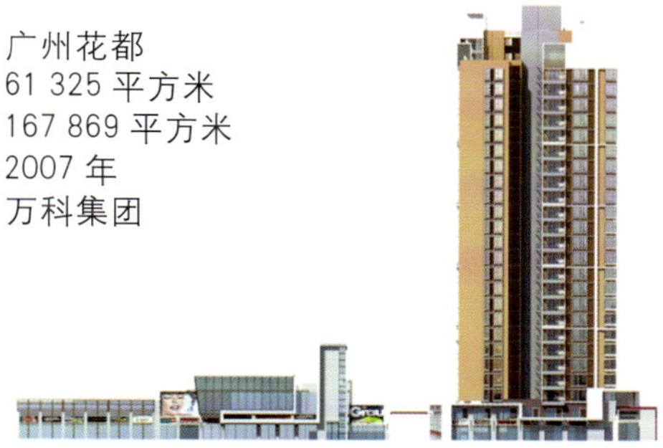

合景泰富 • 瑜翠园

工程地点：广州龙洞
用地面积：44 201 平方米
建筑面积：126 795 平方米
设计时间：2005 年
开 发 商：合景泰富

合景泰富 • 科汇金谷

工程地点：广州萝岗
用地面积：150 082.5 平方米
建筑面积：385 805 平方米
设计时间：2009 年
开 发 商：合景泰富

佛奥广场

工程地点：广州珠江新城
用地面积：6 784 平方米
建筑面积：71 800 平方米
设计时间：2009 年
发 展 商：佛奥集团

顺德·凤城商业办公楼

工程地点：顺德大良
用地面积：6 257 平方米
建筑面积：52 407 平方米
设计时间：2009 年
发展商：广东海骏达置业投资集团

贵港商务区

工程地点：广西贵港
用地面积：98 252.6 平方米
建筑面积：179 508 平方米
设计时间：2009 年
发 展 商：贵港市建设局

设计构思

贯彻“以人为本”的思想，以建设生态型环境为规划目标，创造一个布局合理、交通便捷、绿意盎然、生活方便、富有文化韵味的办公区。

设计延续历史文脉，创造一个内在拥有悠久文化底蕴，外表焕然一新的标志性的办公区。

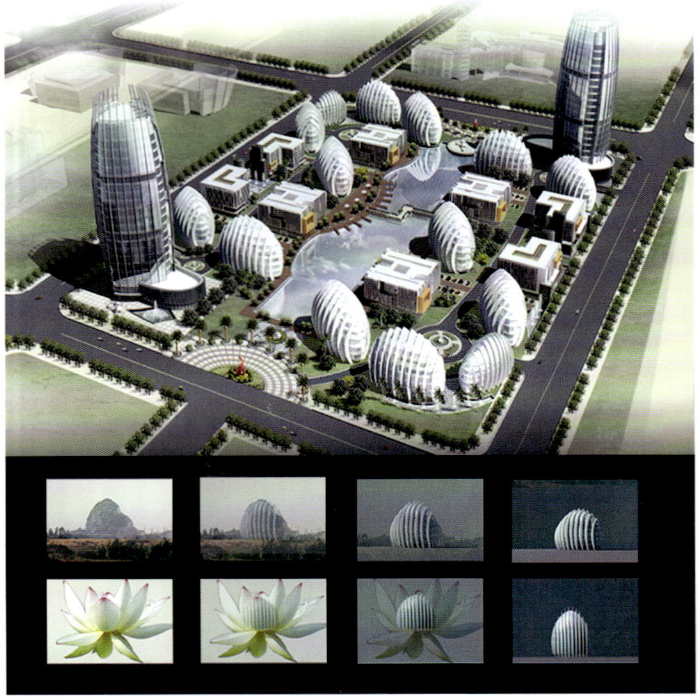

广州天伦林和村改造项目

设计构思

项目位于广州市中心地段，规划设计的小区被定位为高素质，高品位，高档次的超高层（49 层）商业居住小区，将成为广州市未来居住小区的新标杆。本方案最为突出的设计亮点可概括为两区三心四线，规划设计妥善处理了小区内部回迁户及销售户的需要关系并使外部与周边环境自然过渡、自然融合，展示出一种休闲、舒适、尊贵、自然的气质。

工程地点：广州天河
用地面积：64 414 平方米
建筑面积：506 784 平方米
设计时间：2009 年
发 展 商：天伦控股有限公司

1 | 2
3 | 4 | 5

1. 梧州海骏达西堤
2. 佛山季华路项目
3.4. 佛山奥园五期
5. 惠州山水华府四、五期

南沙 08NJY-1 地块商住项目
占地面积：134 760 平方米
建筑面积：307 520 平方米

广东韶关银山高尔夫
占地面积：2 494 222 平方米
建筑面积：3 243 341 平方米

贵港盛世华府
占地面积：51 485 平方米　建筑面积：400 076 平方米

佛山怡翠世嘉
占地面积：92 890.44 平方米　建筑面积：283 840 平方米

茂名名雅世家
占地面积：54 037.07 平方米　建筑面积：317 538.41 平方米

合景泰富 · 科汇金谷
占地面积：150 082 平方米　建筑面积：385 805 平方米

横沙城中村改造
占地面积：64 600 平方米　建筑面积：2 649 572 平方米

珠江村城中村改造
占地面积：340 000 平方米　建筑面积：290 080 平方米

珠江实业集团
PEARL RIVER ENTERPRISES GROUP
广州珠江外资建筑设计院

广州珠江外资建筑设计院

广州珠江外资建筑设计院是华南地区著名的国有设计机构，创建于改革开放之初的1979年，是一所综合性建筑工程甲级设计院。

弹指一挥30年。珠江设计院和翱翔在珠江之滨那只洁白的天鹅——白天鹅宾馆一起诞生，伴随着祖国改革开放的大潮而崛起。我们的作品，是30载艰苦历程的真实记录，也见证了南粤大地建设事业的发展。30年来，珠江设计院坚持以创新为先导、质量为基础、服务为依托，走技术进步、科学管理、繁荣创作的道路，高质量地完成了广州及全国各地面积达上千万平方米的工程设计，项目涵盖公共建筑、居住建筑、景观园林和室内装饰等领域，完成的作品中创新意念贯穿其中，设计风格独树一帜，既有岭南建筑的韵味，又有现代建筑的风采，众多项目荣获国家、部、省、市优秀设计及科技进步奖，赢得了客户及社会各界的广泛赞誉。

这30年来，珠江设计院形态各异、神采飞扬的作品如一颗颗璀璨夺目的明珠装点着繁华都市，似一曲曲灵动流淌的音符谱写着都市的七彩华章。从充满趣味与梦幻童话般意境的广州儿童活动中心到极具强烈现代感的广州世界贸易中心和广州好世界广场，这些作品是20世纪80年代珠江设计院伴随着共和国改革开放的脚步突飞猛进发展的重要标志。进入21世纪，珠江设计院沐浴着祖国敞开国门的和煦春风，立足南粤，走出境外，迎接前所未有的机遇和挑战！与国际建筑界携手共进，开展广泛的技术合作与人才交流。珠江设计院与英国扎哈·哈迪德建筑设计事务所共同设计了广州市标志性文化设施——构思为“圆润双砾”的广州歌剧院；与德国GMP国际建筑设计有限公司合作设计晶莹剔透的珠江边立方体广州电视台新址；在武汉琴台文化艺术中心——琴台大剧院的设计中追求自主创新，大胆采用解构主义象征手法，表达高山流水、金石顿开的深远意境；完成了具有“舰魂铸碑”之意的武汉“中山舰博物馆”以及完全按照澳门当地和欧共体标准进行设计的澳门2005年东亚运动会主体育馆等一个又一个标志性建筑。

30载风风雨雨，30年孜孜以求。珠江设计院的每一件作品，都是建筑师们用智慧和汗水铸就的一座座丰碑，也是建筑师们对日新月异的新理念、新文化、新技术、新材料以及新的审美观的思考和探索。它记载了我们的建筑师、工程师们与珠江设计院一起成长的足迹以及我们进行审美体验和诗意追求的心路历程。

珠江设计院在改革开放大潮中崛起，在30年的艰苦创业中发展壮大，铸精品，树品牌，追求原创，赢得市场，是我们的战略理念。面对未来，我们坚信，秉承求真创新，勇于进取的企业精神，更多异彩纷呈的设计作品一定会呈现在世人的面前！

Selected Works for 30th Anniversary of Guangzhou Pearl River Foreign Investment Architectural Designing Institute.

Guangzhou Pearl River Foreign Investment Architectural Designing Institute is a famous designing institute in South China region, and it was founded in the year of 1979, which was at the very beginning of the reform and development. What's more, it was also the first-class comprehensive architectural designing institute.

Established with White Swan Hotel, which soars up Pearl River, the institute stood out prominently in the spring tide of reform and opening up. Witnessing the construction achievements of South China, our projects are true record of 30-year tough history. Since its establishment, guided by the innovation, based on quality and backed by services, it has insisted on the road of technical progress, scientific management, and prosperous recreation, which made us complete more than 10-million sqm architectural design with high quality in Guangzhou and all over the country. Our projects involve city programs, public buildings, residential buildings, scenery and gardens, and interior decoration, etc. The works our institute have accomplished, which had the notion of innovation through it, have received the widespread praise from clients and all the corners in society at levels of nation, ministry, province and city not only for the taste of Lingnan architecture, but also for the charm of modern architecture, together with the unique design style.

During these 30 years, with various shapes and vivid appearances, the works our institute have accomplished, seem to be dazzling pearls dotted in bustling city, or flowing smart musical notes composing colorful city. From Guangzhou Children Center full of dreamy fairy-tale mood, to Guangzhou International Trade Center and Guangzhou Nice World Square full of modern senses, all of which are important symbols indicating that the institute had made astounding advances with the reform and opening up of our country in the 1980's. Coming into 21st century, bathing in the warm spring breeze brought by opening China, the institute went out of China to meet the unprecedented opportunities and challenges. At the beginning of the new century, our institute cooperated with the international architectural institutes, widely carried out technical cooperation and talent exchange. Our institute cooperated with Zaha Hadid Architectural Designing Institute in Britain for the cultural symbolic facilities in Guangzhou City - design of Guangzhou Opera Theatre. In addition, cooperated with GMP International Architectural Design Co., Ltd in Germany, the institute designed sparkling new Guangzhou Television Station along Pearl River. Pursing independent innovation and boldly adopting symbolism of deconstructionism to express far-reaching atmosphere "High Mountain and flowing water", the institute designed Qintai Grand Theater, which is Qintai Art and Cultural Center. The institute also completed the project of Zhongshan Warship Museum in Wuhan, which implied that warship casts the monument. The Main Gymnasium in Macau for East Asian Games in 2005, in accordance with local and European Community standards, is also designed by our institute.

Having experienced for 30 years, the staff in institute also have pursued for 30 years. Each work is monument created by the wisdom and sweat, and is also the new idea and exploration to this ever-changing world, including new ideas, new culture, new technologies, new materials and new aesthetic taste. It records the developing process of our architects and engineers with the institute, and also records the spiritual course on pursing aesthetic experience.

With our institute standing out prominently in the spring tide of reform and opening up, developing and expanding for 30 years' hard work, we take "making competitive products, setting famous brand, pursing originality, and wining the market" as our strategic principles. On the way to the bright future, all designers of our institute firmly believe that by the long-time formed innovation and dedication spirits, more colorful design projects will be presented in front of the world.

广州歌剧院

武汉琴台大剧院

武汉琴台音乐厅

中山舰博物馆

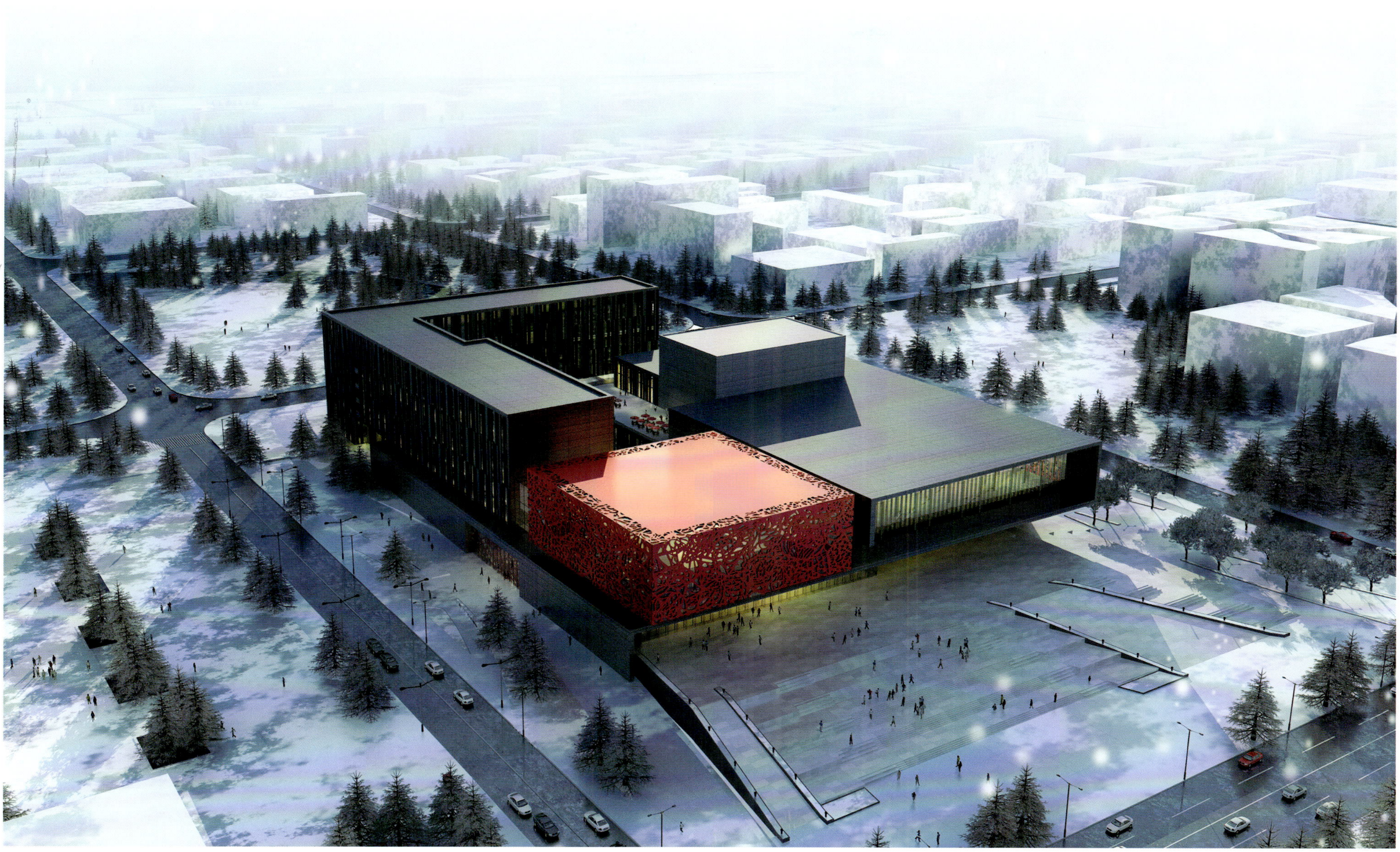

长春保利大剧院

江门演艺中心

汶川博物馆

广州辛亥革命纪念馆

漢森國際設計顧問
International Arch
建築綜合設計 • 漢森(綜合)• 伯盛(甲級)
www.hsarch.com.cn

concept + detail

概念+细节 提升设计附加值

筑 建 卓 越 人 生

拥有260余人的国际化专业团队，竭诚为客户提供规划、建筑、室内、景观一站式国际水准的全专业服务

汉森国际设计顾问伯盛设计（甲级）创立于1993年，技术实力雄厚，专业配套齐全，拥有现代化设计和管理手段，是国内外一批热衷于城市规划和建筑设计的专家组合而成的综合性规划及建筑设计集团。集团推行精英组合，采用国际先进公司管理模式，聚集了近300余名国内外资深的专业设计师。完善及迅捷的电讯、电脑通讯确保广州总部，北京分部，上海分部，成都分部可以协作于同一项目中，共享资源，加快进度。属下机构伯盛设计更是广东省首批获发甲级建筑设计资质的名人事务所之一，并已通过ISO 9001质量体系认证。

公司凭借自身的实力优势，独特新颖的视角开创了新的设计理念，为客户提供规划设计，建筑设计，城市设计，景观设计等领域的一站式服务，以及工业与民用建筑设计和与建筑业有关的各种咨询，技术开发服务。我们的设计方案着重于高水准的设计素质，强调维持高水准的服务水平和合作精神，具有丰富的设计实践和工程经验。公司2004–2007年连续被评为“CIHAF中国房地产二十大品牌影响力规划建筑、景观设计院”；2005年、2006年连续被评为“中国十佳规划建筑设计机构”；2008年被评为“CIHAF中国房地产最佳顾问机构”；2009年更被评为“2009CIHAF中国最具品牌价值设计机构”。

服务的信守承诺，设计的优质超群，使得公司在客户中赢得了广泛信誉。而我们独特的规划、深刻的建筑语言和专业的服务精神，更进一步实现了品质的提升，充分体现设计的附加值。

Founded in 1993, HS Arch International (Class A qualification) has strong technical strength and complete professional equipment as well as modern design and management means, and it is a comprehensive plan and architectural design group staffed by a large number of foreign and domestic experts who keen on urban and architectural planning design. Pursuing elite team and adopting management mode of advanced Hong Kong and international companies, the group has a galaxy of nearly 300 senior foreign and domestic professional designers. With perfect and rapid telecommunication and computer communication, Guangzhou headquarters, Beijing Branch, Shanghai Branch and Chengdu Branch can make cooperative efforts on the same project, and they can share resources as well as speed up the process. Bosheng Design, which is the affiliated institute, is one of the celebrated firms which were granted Class A certificate of architectural design by Guangdong province in the first batch, and it has been also granted the certificate of ISO9001 Quality Management System.

With its own advantages and unique novel perspective, the company creates the new design concept, and provides customers with one-stop service in fields of planning design, architectural design, urban design, interior design, landscape design and so on, as well as industrial and civil design, and various consultation & technology development services on architecture. Focused on high level design quality and emphasized high standard service and team spirit, we have rich experiences and engineering practices in design. From 2004 to 2007 the company has been consecutively ratified as "Top 20 Brand Influence Planning Architecture and Landscape Design Institutes of CIHAF China Real Estate" and "China Top 10 Planning Architectural Design Organizations" in 2005 and 2006, and the company has been also awarded "The Best Advisory Body of CIHAF China Real Estate" in 2008. The best brand value of design organiza.

With credible service and superior design, the company has won the good reputation from our customers, and our unique planning, profound architectural language and professional spirit of service further realize the creativity and originality promotion of the quality and content, and fully reflect the added value of design and commitment to the society.

◎ 通过ISO 9001质量管理体系认证
◎ 2006年《时代楼盘》理事会理事单位
◎ 2006–2007年时代建筑理事会理事单位
◎ 2007年广州建筑装饰行业协会会员单位
◎ 2007中国主流地产金冠奖
——年度最具影响力建筑设计公司
◎ 2008年广州建筑装饰行业协会优秀会员单位
◎ 2008中国主流地产金冠奖
辉煌30年中国建筑人居创新杰出贡献企业、年度最受尊敬建筑设计事务所
◎ 2005–2006年《中国建筑设计作品年鉴》特邀编委单位
◎ 2006–2007年《中国建筑设计作品年鉴》特邀编委单位
◎ 2007–2008年《中国建筑设计作品年鉴》特邀编委单位
◎ 2008–2009年《中国建筑设计作品年鉴》特邀编委单位
◎ 2008–2009年中国最具业主满意度设计机构
◎ 1985–2005年广州地产二十年杰出贡献建筑师事务所（公司）
◎ 2004年度CIHAF中国建筑二十大品牌影响力规划建筑设计事务所（公司）
◎ 2005年度CIHAF中国建筑二十大品牌影响力规划建筑设计景观设计公司
◎ 2005中国10佳规划建筑设计机构奖
◎ 2007年度CIHAF中国最具影响力规划建筑设计机构
◎ 2008年度CIHAF中国房地产最佳顾问机构
◎ 2009年度CIHAF中国最具品牌价值设计机构
◎ 2009广州设计周推荐建筑设计机构
◎ 2005中国地产峰会——中国十佳规划建筑设计机构
◎ 中国建筑文化中心／中国建筑设计作品年鉴编委会
——2007年优秀设计机构
◎ 中国建筑文化中心／中国建筑设计作品年鉴编委会
——2008年优秀设计机构
◎ 第二届中国国际建筑艺术双年展专辑编委会
——优秀设计单位奖
◎ 建设部中国建筑文化中心、中国建筑设计作品年鉴编委会

总部
地址：广州市东风中路448号成悦大厦22楼
电话：+86–20–83390004
传真：+86–20–83365993
邮编：510030
邮箱：hsarch@163.net

北京
地址：北京市朝阳区东三环中路39号建外SOHO小区B座2605室
电话：+86–10–58692111
传真：+86–10–58692990
邮编：100022

工程地点：广东　东莞
建筑面积：66 086 平方米
设计时间：2003 年
竣工时间：2005 年
获得奖项：2005 年广东省优秀设计；
2005 年东莞市优秀工程设计一等奖

Construction Location: Dongguan, Guangdong
Building Area: 66,086 m^2
Design Time: 2003
Completion Time: 2005
Awards:
2005 Excellent Design in Guangdong Province;
No.1 prize for Excellent Architectural Design of Dongguan in 2005

Richwood Garden Hotel　丰泰花园酒店

Langham Place, EDZ, Guangzhou　广州汇裕朗豪酒店

工程地点：广东　广州
用地面积：46 832 平方米
建筑面积：106 337.6 平方米
设计时间：2008 年 4 月

Construction Location: Guangzhou, Guangdong
Site Area: 46,832 m^2
Building Area: 106,337.6 m^2
Design Time: 2008.04

佛山华粤假日酒店
Huayue Holiday Hotel in Foshan

酒店类建筑

佛山三水江南新区酒店
Sanshui Jiangnan New District Hotel

Holiday Hotel Beyond River of Yangshuo　阳朔河畔度假酒店

工程地点：广西　桂林
用地面积：16 800 平方米
建筑面积：36 741 平方米
设计时间：2006 年 7 月
超五星高级渡假酒店，各类型客房 300 余套

Construction Location: Guilin, Guangxi
Site Area: 16,800 m^2
Building Area: 36,741 m^2
Design Time: 2006.07
An extra-five-star Holiday Inn more than 300 rooms

商业类建筑

工程地点：广东 广州
用地面积：17 684.7 平方米
建筑面积：160 000 平方米
设计时间：2005 年 5 月

Construction Location: Guangzhou, Guangdong
Site Area 17,684.7 m^2
Building Area: 160,000 m^2
Design Time: 2005.05

JunWei Commercial Center 骏威商业中心

广州家信商业中心
Joynus Commercial Centre

百信广场
Blossom Plaza

倚能城市之光
Yineng Light of City

上海国际金融大厦
Shanghai International Finance Building

广州富力城商业中心
Business Center in R&F City, Guangzhou

北京富力又一城 C 区
Sector C of R&F Another City, Beijing

东莞星河传说
Dynatown, Dongguan

Silver Beach North Sea 北海银滩

文化类建筑

工程地点：河北 邯郸
用地面积：510 000 平方米
建筑面积：260 000 平方米
设计时间：2006 年 6 月

Construction Location: Handan, Hebei
Site Area: 510,000 m^2
Building Area: 260,000 m^2
Design Time: 2006.06

The New Campus of Handan No,1 Middle School 邯郸市第一中学新校区

Guangdong Experiment Middle School 广东实验中学（高中校区）

工程地点：广东 广州
用地面积：114 826 平方米
建筑面积：80 000 平方米
设计时间：2002 年
竣工时间：2004 年 8 月
获得奖项：2005 年广东省优秀设计

Construction Location: Guangzhou, Guangdong
Site Area: 114,826 m^2
Building Area: 80,000 m^2
Design Time: 2002
Completion Time: 2004.08
Awards: 2005 Excellent Design in Guangdong Province

湛江档案馆
Zhanjiang Archive

临汾一中
Linfen No.1 Middle School

成都龙泉中学
Chengdu Longquan Middle School

番禺展览馆
Panyu Exhibit Center

中国民居风格在高层中的运用

Mountain Poetic House, Qingyuan 清远云山诗意

工程地点：广东 清远
用地面积：200 000 平方米
建筑面积：500 000 平方米
设计时间：2006 年

Construction Location: Qingyuan, Guangdong
Site Area: 200,000 m^2
Building Area: 500,000 m^2
Design Time: 2006

广州云山诗意
Mountain Poetic House

南京紫园
Purple Garden, Nanjing

南沙万国园
Wanguo Garden, Nansha

青岛银岭熙岸
Yinling Xi'an, Qingdao

柳州盛天龙湾
Shengtian Dragon Bay, Liuzhou

山东淄博荷塘月色
Lotus Pond, Shandong

汇东国际花园
Huidong International Garden

东南亚风情体验

工程地点：广东 广州
用地面积：385 559.8 平方米
建筑面积：627 924 平方米
设计时间：2004 年
获得奖项：“广州市 2008 年度优秀工程考察设计评选大会”一等奖

Construction Location: Guangzhou, Guangdong
Site Area: 385,559.8 m^2
Building Area: 627,924 m^2
Design Time: 2004
Awards: The First Prize of 2008 Year Excellent Project Reconnaissance Design of Guangzhou.

R&F City of Guangzhou 广州富力城

番禺信业花园
Xinye Garden, Panyu

广州时代廊桥
Laguna, Guangzhou

成都富力桃园
R&F The Peach Garden, in Chengdu

江门绿荫豪苑
Green Garden, Jiangmen

北海金色海岸
Golden Beach, Beihai

海南祥阁
Harmany Garden, Hainan

海南恩祥紫园
Enxiang Purple Garden, Hainan

Another City of R&F, Beijing 北京富力又一城

工程地点：北京
占地面积：36 000 平方米
建筑面积：1 150 000 平方米
设计时间：2005 年 4 月

Construction Location: Beijing
Site Area: 360,000 m^2
Building Area: 1,150,000 m^2
Design Time: 2005.04

欧陆风情

兰州仁恒国际
Renheng International, Lanzhou

工程地点：甘肃 兰州
占地面积：6 260 平方米
建筑面积：59 437 平方米
设计时间：2006 年 8 月

Construction Location: Lanzhou, Gansu
Site Area: 6,260 m^2
Building Area: 59,437 m^2
Design Time: 2006.08

大连夏家河子
Xiajia Hezi, Dalian

内蒙古奈伦金融住宅小区
Nailun Financial Residence, Inner-Mongolia

新古典主义

临沂尚城沂龙湾
Dragon Bay of Shangcheng, Lin yi,Shandong

惠州万城国际
Wancheng International, Huizhou

顺德龙光水悦城邦
Longguang Shuiyue City, Shunde

顺德康城半岛
Healthy Peninsula, Shunde

三水东海蓝湾
Blue East Bay, Sanshui

YIBO CONCEPTION 益博建築

珠海益博建筑设计工程有限公司

Zhuhai Yibo Architectural Design and Engineering Co.,Ltd.

益博建筑设计工程有限公司，创立于2004年，自创立伊始，便秉承“专注、创新、求精、超越”的设计理念，汇集了一批工程设计经验丰富的资深注册建筑师、工程师以及朝气蓬勃的年轻设计精英。在竞争激烈的国内建筑设计市场中，公司用短短5年时间迅速确立了自己的立足点，并已成长为一支最具设计特色的实战军，我们的创作宗旨就是把中华文化的精髓恰当地融入现代建筑设计之中。

公司联合国内外知名设计团队，发挥各自优势，在珠江三角洲地区创作出很多大型公共及民用建筑，尤其在办公楼设计领域有着丰富的经验，多项设计在实施后，受到了项目委托方的好评。

目前，公司服务项目及作品遍及珠海、深圳、广州、山西等地区，并通过益博建筑设计旗下其他各专业设计团队（景观、规划、室内）为项目委托方提供更多、更好的设计增值服务，不断满足并超越社会与顾客的期望。

Yibo Architectural Design and Engineering Co., Ltd. was established in 2004. Since its foundation, it has been taking "devoting, innovating, improving and surmounting" as the design idea and has collected one group of senior registered architects and engineers with abundant experiences on engineering design, as well as the outstanding young designers full of vigor and vitality. In the intense competition of domestic architecture design market, the company has fixed its own standpoint rapidly by just 5 years and has grown to a combat army with the most design characteristics. The tenet of our creation is to integrate the essence of Chinese culture into the modern architecture design properly.

We unite some domestic and foreign well-known design teams, making use of our respective superiorities, and design well for lots of large-scale public and civil constructions in the Pearl River Delta. We are specialized in the office building design and our designs are highly appreciated by the project consigners. and earn good reputation in this field ofter several implementation of the projects.

At present, our developing market has extended to Zhuhai, Shenzhen, Guangzhou and Shanxi province, etc. Through other subordinate specialized design teams (landscaping, layout, indoor) of Yibo Architecture Design, we supply for project consigners with more and more good value-added services related to design, and satisfy and surpass the expectations from customers and society unceasingly.

地址：珠海市新香洲红山路288号珠海国际科技大厦B座509–511室
电话：+86–756–3229409／2666832／2666831（直线）
传真：+86–756–3229406
邮箱：Arch_yibo@126.com
网址：www.zh–yibo.com

Address: Room 509-511, Block B of Zhuhai International Technology Marsion,
Hongshan Road No.288, Xinxiangzhou, Zhuhai
Tel : +86-756-3229409/2666832/2666831 (Straight Line)
Fax: +86-756-3229406
E-mail: Arch_yibo@126.com
Http: //www.zh-yibo.com

珠海黄金海岸

建筑功能：办公、酒店、商业、住宅
建筑面积：390 000 平方米
设计阶段：方案
业　　主：珠海豪逸实业有限公司

Zhuahai golden coast

Building Function: offices, hotel, commerce and residence
Building Area: 390,000 m^2
Design Stage: Proposal
Proprietor: Haoyi Industrial Corporation Limited in Zhuhai

珠海吉大南山商务综合体

建筑功能：办公、商业、酒店
建筑面积：80 218 平方米
设计阶段：概念方案
业　　主：珠海华业集团

Jida Nanshan Commercial Complex in Zhuhai

Building Function: offices, commerce and hotel
Building Area: 80,218 m^2
Design Stage: concept design proposal
Proprietor: Zhuhai Huaye group

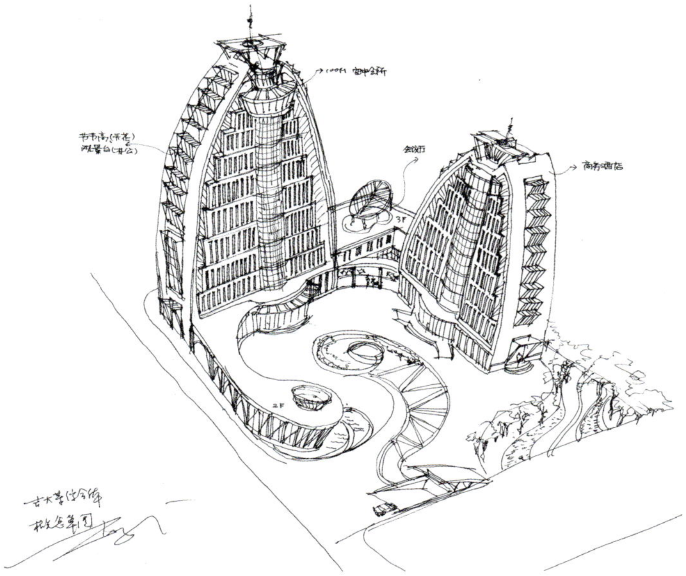

珠海京华假日湾

建筑功能：住宅小区
建筑面积：220 000 平方米
设计阶段：施工图
业　　主：广东家和园房地产开发有限公司

Jinghua holiday bay in Zhuhai

Building Function: residence community
Building Area: 220,000 m^2
Design Stage: construction documents design
Proprietor: Jiaheyuan Real Estate Development Co.,Ltd. in Guangdong

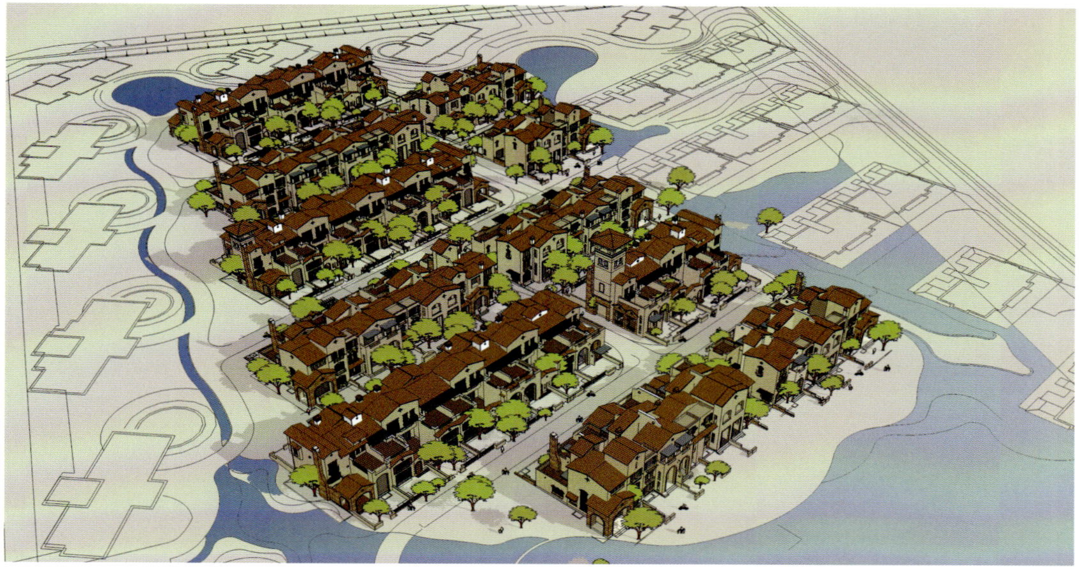

珠海湖畔假日花园

建筑功能：住宅小区
建筑面积：启动区 110 000 平方米
设计阶段：施工图
业　　主：珠海市电力房产开发有限公司

Lakeside holiday garden in Zhuhai

Building Function: Fesidence
Building Area: 110,000 m^2
Design Stage: construction documents design
Proprietor: Power Estate Development Co., Ltd. in Zhuhai

深圳金证科技大厦

总建筑面积：40 000 平方米
设计时间：2005 年
业　　主：深圳市金证科技股份有限公司

Kingdom Technology Plaza

Building Area: 40,000 m^2
Design Time: 2005
Proprietor: Shenzhen Kingdom Technology Co., Ltd.

广东建筑艺术设计院有限公司

Guangdong Architectural Artistic Design Institute Co.,Ltd.

设计构思及设计观念

本项目是我国教育类的高等学府之一，具有自己的文化特征。该学院是教育和培养人才的高等学府。是各优秀人群，深造的场地。因此，设计构思和理念特点：

① 共享与交流：在网络信息时代的今天，人们更需要交流和共享的空间，这需要在校园里创造一个人们交流的场所；

② 高效与秩序：创造高效人流的同时，做到秩序井然，条理清楚，达到最佳“教与学”的效率；

③ 生态与健康：创造一个生态的校园、健康的校园、为国家培养一批批年富力强、充满斗志的科技人才。

功能分区

学院以创新湖为中心，有机的分为五个区。

① 中心区 包括图书馆，交流广场(面积达8 000平方米)，文化中心及国际交流中心(15层高层建筑)；

② 教学区：设有四组教学楼；

③ 实验（训）区：设有两实验楼，四栋实训楼；

④ 体育运动区：设有容纳1 000人的体育馆，一个标准400米跑道的活动场，和一个50米的标准游泳池，另外设有篮球，排球，羽毛球等各类球类运动场所；

⑤ 生活区：有教师住宅，学生宿舍，教职工食堂，学生食堂，学院功能分区明确，互不干扰。

岭南校园特色

校园的造园意旨：学院位于广东省东莞市东莞“可园”为岭南四大名园之一，学院的造园设计承传了岭南园林的文脉，在此基础上创造出具有现代社会特色的岭南园林。

① 校舍建筑的特点创新学院的校舍尊重岭南建筑的特点，以迴廊式空间为主，组合各组建筑物，使建筑空间迂回曲折，别有情趣，配以立面为灰（白）色的墙面，蓝色坡屋面，简洁轻盈，朴素无华

② 校园的景观原则上保留原地形生态环境，校园的低洼地带造人工湖，利用各种景观要素，规划中以人为本，追求美感，以自然为体，使人与自然达到和谐共生。

Design ideas and design concepts

This project is one of higher educational institutions in China with its own cultural identity: "College" is a higher institution for educating and cultivating talents. It is a place for excellent people at all times to study further. Therefore, the features of design ideas and concepts are as follows:

① Sharing and exchange: in the network information times today, people need more space to communicate and share by creating a place where people have met on campus;

② Efficiency and order: create efficient flow of people and be orderly and well organized as well as to achieve the best "teaching and learning" efficiency;

③ Ecology and health: creating an eco-campus and healthy campus, training groups of young and energetic, highly motivated technology professionals for the country.

Functional areas

"College" is organically divided into five areas taking Innovation Lake as the centre:

① Center area: including library, exchange square (an area of 8,000 m^2), Cultural Centre and the International Exchange Center (15-storey high building);

② Teaching area: 4 groups of teaching buildings.

③ Experimental (training) area: there are two experimental buildings and four practical training buildings;

④ Sports Zone: stadium with capacity of 1,000 people; the activity field of a standard 400 m track and a standard 50m swimming pool as well as sports places for all kinds of balls like basketball, volleyball, badminton.

⑤ Living area: teacher's residences; student hostels, faculty dining halls, student canteens.

Functional areas of "College" are clear and there are not interfered each other.

Lingnan Campus features

Garden intention of campus: "College" is located in Dongguan City, Guangdong Province. "Keyuan Garden" in Dongguan is one of the four famous gardens of Lingnan. The garden of "College" is based on the principle of inheriting the context of Lingnan gardens, creating Lingnan gardens with modern social characteristics:

① features of school buildings:

School buildings of "Innovation" college respect the characteristics of Lingnan architectures. They are mainly gallery-style space, integrating each group buildings. This makes building space circuitous, full of special taste together with the gray (white) wall color of facade and blue sloping roof, buildings are concise, light and simple.

② Campus landscape:

In principle, the design retains the original terrain ecology. Low-lying areas on campus are made into artificial lakes. Using a variety of landscape elements, the design is people-oriented and the pursuit of beauty, and takes nature as the body of to make human beings and nature harmonious.

行政办公楼

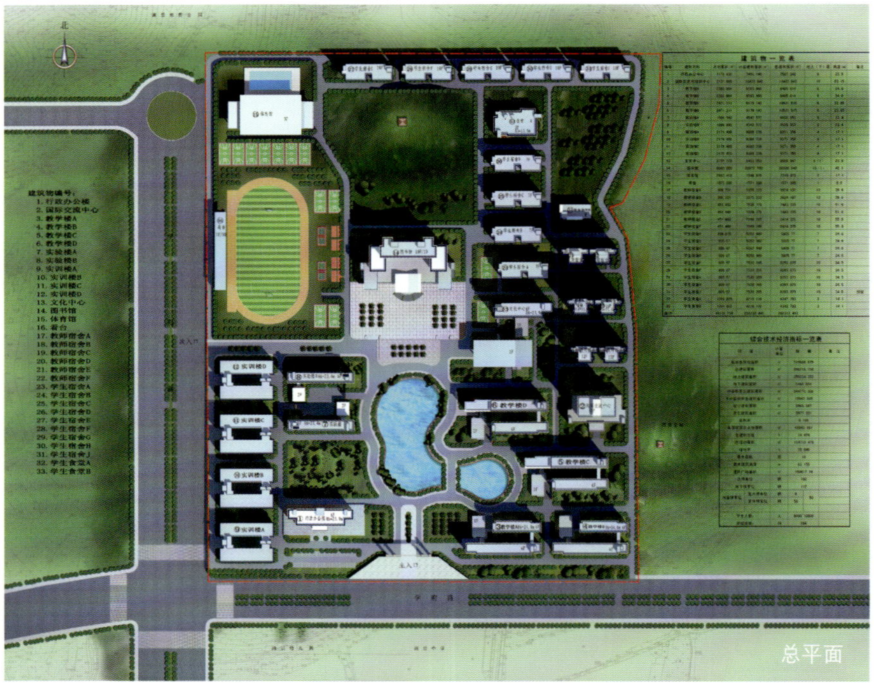

总平面

广东创新科技职业技术学院

鸟瞰图

实验楼

广东创新科技职业技术学院

图书馆

群体教学楼

广东创新科技职业技术学院

广东创新科技职业技术学院

实训楼

学生宿舍楼

教师公寓楼

建设地点：广东 东莞
占地面积：319 986.68 平方米
建筑面积：260 715.16 平方米
建筑密度：14.67%
容 积 率：0.765
学院规模：10 000–12 000 人
建设时间：2010 年 11 月（一期）

参加设计人员
李 坚、邓日新、柯 华
郭 丹、徐明华、尹续福

Construction Location: Dongguan,Guangdong
Site Area: 319,986.68 m^2
Building Area: 260,715.16 m^2
Construction Density: 14.67%
Plot Ratio: 0.765
College Scale: 10,000 - 12,000
Construction Time: completed the Phase Ⅰ in November 2010

Participatants in design
Jian Li, Rixing Deng, Hua Ke,
Dan Guo, Minghua Xu, Xufu Yin

广东创新科技职业技术学院

广州番禺区市桥建筑设计院有限公司

Shiqiao Architecture & Design Institute Co.,Ltd. Panyu Guangzhou

广州番禺区市桥建筑设计院有限公司的前身为广州番禺区市桥建筑设计院，成立于1980年，一直是广州地区一支重要的设计力量，并为全国各地提供优质的服务。

2001年公司转制后，业务迅速扩大，实力也大大增强，形成了拥有建筑工程设计甲级、工程勘察乙级、城市规划丙级、公共市政丙级等资质的勘察设计单位，同时还成立了控股的监理公司，具备招标代理和工程监理乙级资质。公司现有员工150人，其中注册建筑师、注册结构师、注册岩土工程师及规划师共40多人，技术力量雄厚。

近年公司积极参加业内各项设计投标活动，完成了大批工程的设计勘察任务。公司先后在广州、深圳、惠州、中山、韶关、清远、江门、汕尾、肇庆、珠海、长沙、武汉、乌鲁木齐、大同、娄底、无锡、宜兴、桂林、赣州、玉林、郑州等地承接各类设计业务，得到同行的普遍赞誉。

客户的满意是我们最基本的要求，超出客户的满意是我们不懈的追求！公司将一如既往地坚持“精心设计、科学管理、服务至上、开拓进取”的方针提供专业化的服务。

Shiqiao Architecture & Design Institute Co.,Ltd.Panyu, Guangzhou, formerly is Shiqiao Architecture & Design Institute of Panyu was founded in 1980. It has always been an important design strength in Guangzhou, providing high-quality services throughout the country.

After restructuring in 2001, the company's business expands rapidly and its strength is greatly enhanced. The company has become a survey & design unit with qualifications of class-A construction design, class-B engineering survey, class-C urban planning and Class-C public municipal design. It also sets up a holding project management company with qualifications of bidding agent and class-B project supervision. Currently the company has 150 employees and a strong technical force of more than 40 people, including registered architects, registered structural engineers, registered geotechnical engineers and planners.

In recent years, the company actively participates in the design bidding activities and has completed many survey tasks for engineering projects. It has undertaken all kinds of design business in Guangzhou, Shenzhen, Huizhou, Zhongshan, Shaoguan, Qingyuan, Jiangmen, Shanwei, Zhaoqing, Zhuhai, Changsha, Wuhan, Urumqi, Datong, Loudi, Wuxi, Yixing, Guilin, Ganzhou, Yulin, Zhengzhou, etc., receiving universal praises.

To satisfy customers' requirements is our basic purpose, and to exceed their requirements is our relentless pursuit! The company will continue to adhere to the principle of "elaborate design, scientific management, service orientation and pioneering development", providing more professional services!

地址：广州市番禺区市桥清河西路36号
邮编：511400
法人：潘永强
电话：+86–20–84884435
传真：+86–20–84882392
邮箱：shqjzhsjy@21cn.com

Address: No.36, Shiqiao Qinghe Road, Panyu District, Guangzhou
P.C.: 511400
Legal presen: Yongqiang Pan
Tel: +86-20-84884435
Fax: +86-20-84882392
E-mail: shqjzhsjy@21cn.com

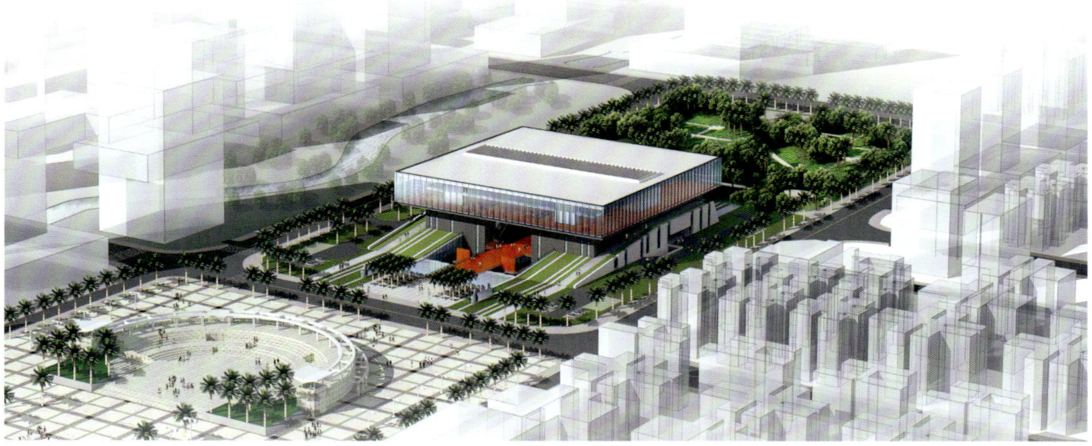

番禺展览中心
用地面积：40 000平方米
建筑面积：50 000平方米
工作内容：入围方案
设 计 师：王畅

Panyu Exhibition Centre
Site Area: 40,000 m^2
Building Area: 50,000 m^2
Work Content: Short-listed options
Designer: Chang Wang

长沙市长江置业沁园春御院
用地面积：100 000平方米
建筑面积：370 000平方米
工作内容：方案设计
设计时间：2005年
项目状态：已竣工

Qin Yuan Chun Yu Yuan of Yangtze River home, Changsha
Site Area: 100,000 m^2
Building Area: 370,000 m^2
Work Content: design program
Design Time: 2005
Project Status: completed

东泓商务大厦
用地面积：6 000 平方米
建筑面积：15 000 平方米
工作内容：投标方案
设 计 师：王畅

East Hung Commercial Building
Site Area: 6,000 m^2
Building Area: 15,000 m^2
Work Content: bidding scheme
Designer: Chang Wang

番禺职业技术学院实训中心
用地面积：8 000 平方米
建筑面积：30 000 平方米
工作内容：中标方案

Training Center of Guangzhou Panyu Polytechnic
Site Area: 8,000 m^2
Building Area: 30,000 m^2
Work Content: successful program

富豪大厦及富豪商业广场
用地面积：16 000 平方米
建筑面积：30 000 平方米
项目状态：建设中

Regal House and Regal Commercial Plaza
Site Area: 16,000 m^2
Building Area: 30,000 m^2
Project Status: Under Construction

海润大厦
用地面积：10 000 平方米
建筑面积：20 000 平方米
工作内容：设计方案

Ocean Mansion
Site Area: 10,000 m^2
Building Area: 20,000 m^2
Work Content: design program

江门市御景酒店
用地面积：50 000 平方米
建筑面积：50 000 平方米
工作内容：中标方案

Scenic Hotel of Jiangmen
Site Area: 50,000 m^2
Building Area: 50,000 m^2
Work Content: successful program

广州市东海酒店
用地面积：20 000 平方米
建筑面积：80 000 平方米
工作内容：中标方案

East China Hotel in Guangzhou
Site Area: 20,000 m^2
Building Area: 80,000 m^2
Work Content: successful program

南沙科技园工业
用地面积：100 000 平方米
建筑面积：250 000 平方米
工作内容：方案设计

Nansha Science & Technology Industrial Park
Site area: 100,000 m^2
Building Area: 250,000 m^2
Work Content: Design program

娄底南苑上和小区
用地面积：100 000 平方米
建筑面积：450 000 平方米
项目状态：建设中

Nan Yuan Shang He Community in Loudi
Site Area: 100,000 m^2
Building Area: 450,000 m^2
Project Status: Under construction

宁乡行政中心由行政大楼
后勤服务中心
会议中心组成
用地面积：200 000 平方米
建筑面积：60 000 平方米
项目状态：已完成
合作单位：长沙市建筑设计院

Ningxiang Administrative Center, including the administration building, the logistics center and the conference center
Site Area: 200,000 m^2
Building Area: 60,000 m^2
Project Status: Completed
Cooperation: Changsha Institute of Architectural Design

南沙腾盈大厦
用地面积：8 000 平方米
建筑面积：30 000 平方米
项目状态：建设中

Teng Ying Building of Nansha
Site Area: 8,000 m^2
Building Area: 30,000 m^2
Project Status: Construction

番禺区石壁行政大楼
用地面积：9 000 平方米
建筑面积：25 000 平方米
工作内容：中标方案

Administration Building, Shek Pik, Panyu
Site Area: 9,000 m^2
Building Area: 25,000 m
Work Content: successful program

桥南中学
用地面积：30 000 平方米
建筑面积：20 000 平方米
工作内容：投标方案

Bridge School
Site Area: 30,000 m^2
Building Area: 20,000 m^2
Work Content: bidding scheme

无锡中堂凯莱大酒店
用地面积：30 000 平方米
建筑面积：150 000 平方米
工作内容：设计方案

Gloria Hotel, Zhongtang Wuxi
Site Area: 30,000 m^2
Building Area: 150,000 m^2
Work Content: design program

宁乡春城批发市场
用地面积：100 000 平方米
建筑面积：180 000 平方米
工作内容：方案设计

Spring City Market of Ningxiang
Site Area: 100,000 m^2
Building Area: 180,000 m^2
Work Content: Design program

石北中学扩建
用地面积：50 000 平方米
建筑面积：30 000 平方米
工作内容：投标方案

The expansion of Rock North School
Site Area: 50,000 m^2
Building Area: 30,000 m^2
Work Content: Bidding program

中山市坦洲锦绣阳光家园
用地面积：110 000 平方米
建筑面积：90 000 平方米
项目状态：已完成

Tan Chau Fairview Sunshine Home, Zhongshan
Site Area: 110,000 m^2
Building Area: 90,000 m^2
Project Status: Completed

韶关中等职业技术学校
用地面积：180 000 平方米
建筑面积：90 000 平方米
项目状态：已完成

Shaoguan Secondary Vocational School
Site Area: 180,000 m^2
Building Area: 90,000 m^2
Project Status: Completed

美国 TEAMZERO 建筑规划设计集团
America Teamzero Design & Planning Group
广州市天作建筑规划设计有限公司
Guangzhou Teamzero Design & Planning Co.,Ltd

美国天作 TEAMZERO 建筑规划设计集团是在建筑设计、城市规划、景观设计方面有较大影响力的综合设计公司。广州市天作建筑规划设计有限公司成立于 2002 年，是美国 TEAMZERO 建筑规划设计集团在亚太地区唯一授权的分支机构。

广州市天作（TEAMZERO）建筑规划设计公司集开发咨询、建筑设计、城市规划、景观设计、旅游规划等专业于一体，公司员工包括博士、硕士在内的技术精英共七十多位，其中国家一级注册建筑师五人，注册规划师七人，以及多位常驻外籍设计师，是具有多元化国际背景的设计公司。

广州市天作（TEAMZERO）公司成立以来完成了政府委托的重点项目以及房地产项目近两百个，其中包括大中型城市的城市发展战略规划、总体规划、控制性详细规划，修建性详细规划以及重点地区的城市设计、建筑设计、景观设计等。并在大型居住区，大学校园、大型商业中心、会展中心、五星级酒店、甲级写字楼等领域取得了丰富的项目经验，成功设计出一批优秀的作品，在业界产生了重大影响。

天作公司长期合作的企业包括万科地产、保利地产、中信集团、美的集团、珠江地产、合生创展集团、阳光 100、银业地产等大中型房地产开发投资公司。设计建成作品包括广东星海演艺集团演艺中心设计、广西桂林阳朔酒吧街规划、东莞世纪新城国际公馆规划与建筑设计、广州万科 · 兰乔圣菲规划与建筑设计、保利圆梦城规划与建筑设计、中信长沙国际公馆规划与建筑设计、广州大学城广州大学新校区建筑设计、宁波国际贸易展览中心城市设计、广州新电视塔广场公园设计及其南广场地下空间建筑设计、珠江罗马家园、珠江骏景高尔夫花园、合生帝景湾、深圳珠江广场、广州国际纺织博览中心等具有影响力的公建和居住区项目。

广州市天作（TEAMZERO）建筑规划设计有限公司多年来始终以追求不断创新为目标，许多作品在国内外频获佳奖。其中天作公司负责的东方夏湾拿花园规划设计获 2005 国际花园社区总决赛“绿色奥斯卡”全球金奖第一名；东莞世纪城国际公馆规划及建筑设计于 2005 年 9 月荣获联合国国际花园社区奖及东莞市金牌户型奖。2006 年，广西阳朔酒吧街新天地项目设计获得了第二届中国国际建筑艺术双年展的最佳创意奖及广东省注册建筑师协会第四届优秀建筑创作奖的佳作奖。2009 年，公司出品的广东东莞骏景高尔夫花园规划与建筑设计以及广州花都银业堡岛规划与建筑设计均获得第六届中国人居典范建筑规划设计“人居典范”设计金奖。同年，广东星海演艺集团（新址）办公楼获得广东省注册建筑师协会第五届优秀创作奖。同时，广州市天作（TEAMZERO）建筑规划设计有限公司于 2005 年被中国住交会组委会授予“中国二十大规划建筑设计事务所（公司）”称号。2007 年获亚洲房地产峰会颁发的亚洲建筑规划设计区域十大品牌机构奖。天作设计总监李少云博士于 2007 年获广州国际设计周全国十大杰出设计师奖。

America Teamzero Design & Planning Group is an influential comprehensive company mainly in architectural design, urban planning and landscape design. Guangzhou Teamzero Design & Planning Co., Ltd, established in 2002, is the only branch authorized by America Teamzero Design & Planning Group in Asia-Pacific area.

Guangzhou Teamzero Design & Planning Co., Ltd is a company mainly in development consultation, architectural design, urban planning, landscape design and tourism planning, with a staff of over 70, including 5 national Class A registered architects, 7 registered planners and a number of outposted designers, some of whom hold Doctor or Master Diploma, thus, it is a design company integrated with multiculture.

Guangzhou Teamzero Design & Planning Co., Ltd has successfully completed nearly 200 key projects commissioned by the government and real estate enterprises since its establishment, including overall urban development strategy planning, overall planning, detailed controlling planning and detailed constructive planning of large and medium cities, and urban design, architectural design, landscape design of key cities. We also have excellent works and rich experiences in large residential areas, university campuses, large business centers, convention centers, five-star hotels and grade A office buildings, which earned good reputation in this field.

It has established long-term cooperation with a number of large & medium real estate enterprises and investment corporations, such as Vanke Real Estate, Poly Real Estate, CITIC Group, Midea Group, Pearl River Real Estate, Hopson Group, Sunshine 100 Group and Silver Industry Real Estate. Completed design works include performance center design of Guangdong Xinghai Performance Group, Guilin Yangshuo Bar Street planning, planning and design of International Residence of Dongguan Century City, Guangzhou Vanke Rancho Santa Fe, Poly Dream City, CITIC International Residence in Changsha, architectural design of new campus of Guangzhou University in Guangzhou University City, Ningbo International Trade and Exhibition Centre, Guangzhou new TV Tower Park and its south Plaza, Pearl River Rome Home and Junjing Garden , Hopson Royal Bay, Shenzhen Pearl River Plaza, Guangzhou International Textile Expo Center and other influential public and residential projects.

Guangzhou Teamzero Design & Planning Co., Ltd always adheres to innovation as the goal, thus, many works won awards at home and abroad.

Among them, planning and design of East Havana Garden won the First Golden Prize of the International Environmentally Sustainable Project Awards 2005 (Global “Green Oscar” Competition); planning and design of International Residence of Dongguan Century City won the International Awards for Livable Communities endorsed by UNEP and the Golden Unit of Dongguan in September 2005; the project of Xintiandi design in Yangshuo won the Best Creative Award in the Second China International Art Biennale Competition and the Fourth Architecture Design Award for excellence held by Guangdong Provincial Association of Registered Architects. In 2009, planning & design of Dongguan Junjing Garden and Planning and Architectural Design of Guangzhou Huadu Silver Castle both won the Gold Design Award of Livable Community in the Sixth China Architectural Planning & Design. In the same year, Office of Guangdong Xinghai Performance Group (new site) received the award of the Fifth Excellent Creations held by Guangdong Provincial Association of Architects.

Meanwhile, Guangzhou Teamzero Design & Planning Co., Ltd was granted with the title of "Top 20 Architecture Planning Firms (Company)" in 2005 by CIHAF. It was awarded Top Ten Brands of Asian Regional Architecture Planning & Design institutions in 2007 at the Asian Real Estate Summit. As the design director in Teamzero, Doctor Li received the Top Ten Outstanding Designers Award in 2007 Guangzhou International Design Week.

广州圣普拉多

Teamzero Design & Planning

深 圳 珠 江 广 场

Teamzero Design & Planning

委 托 方：深圳市珠江房地产开发有限公司
项目地址：广东 深圳
用地面积：61 386.9 平方米
建筑面积：21 000 平方米
完成时间：2008 年
实施情况：国际招标第一名

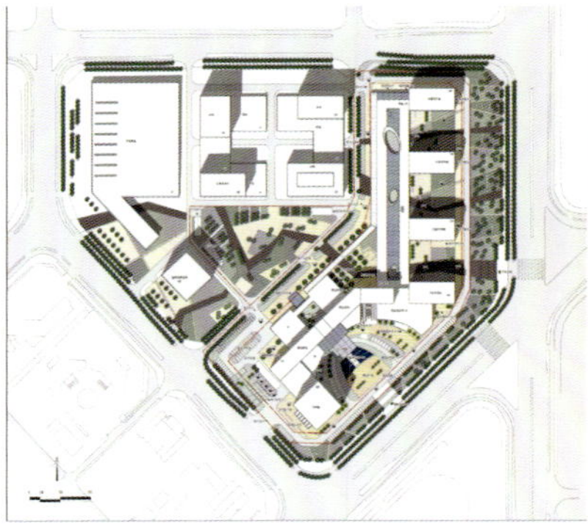

世纪城·国际公馆四期 Ⅰ区、Ⅱ区

Teamzero Design & Planning

建 设 单 位：东莞市世纪城商住开发有限公司
工 程 地 点：广东 东莞
小区总用地面积：207 098 平方米
本期用地面积：122 017 平方米
本期建设总建筑面积：113 486.7 平方米

广东星海演艺集团办公楼

Teamzero Design & Planning

委 托 方：广东星海演艺集团
项目地址：广东 广州
规划面积：7 325 平方米
建筑面积：13 623 平方米
完成时间：2006 年
实施情况：全国投标第一名，在建

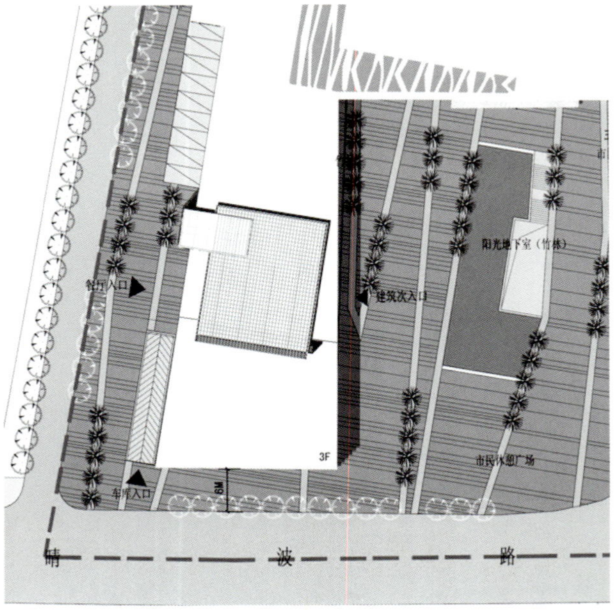

武汉保利圆梦城

Teamzero Design & Planning

申派·国际设计有限公司 Simply Arch International Ltd
广州市申派建筑设计顾问有限公司 Simply Arch Design & Consultant Ltd Guangzhou

公司简介 Profile

申派建筑设计顾问有限公司，成立于 2003 年元月，是具有中国法人地位的本土设计顾问公司。这里是一个不断成长的空间，对建筑艺术的追求以及对创作品质的高度要求使我们凝聚在一起，每个人虽然以不同的方式作出贡献，却都对建筑设计享有共同点信念、相同的热情。申派是一个追求高质量的、卓越的设计团队，公司业务面向全国并提供高质量的设计服务，市场定位于公共建筑，IT 建筑，居住建筑的领域，同时，我们也是一个不断为社会创造高水平作品的团队。通过我们的作品为社会带来更大的收益 。对于我们的事业，我们会要求的更多，因为我们要成为最好。

Simply Arch Design & Consultant Ltd is a local design and consultant company with legal entity in China, is founded in January 2003. Here is a growing space, and we are all pursuing the art of architecture and high quality requirements of the creation, which drove us together. Everyone contributes call the shots in different ways with common beliefs the same passion, on architectural design. It's a pervect design team that provides the clients with high quality and excellert service in the fields of public, IT and residential buildings. At the same time, we are also constantly creating a high level of design work, to bring greater benefits to society. For our business, we ask ourselves for even more, because we want to be the best.

设计思想 Design Philosophy

社会的价值——我们工作和挖掘的核心，
客户的要求——我们工作进步的推动器，
客户的回应——我们设计创作的重要思路，
创造性——广州申派工作的主题，并和客户的精神高度相关。对市场的高度聆听，使我们的创造处于时代发展的前沿。

Social values-the core of our work and excavation,
The requirements of customers-the driving engine of our work and progress,
Customer response-an important creative design idea,
Creativity-the subject of Guangzhou Simply Arch's work,
Highly following the market, keeps our creation at the forefront of the times.

SIMPLY ARCH INTERNATIONAL LTD
Legal Person: Yizhan Li
Address: Floor 14 of Lihua Mansion, No.86 of Donghua West Road, Guangzhou
Tel: +86-20-83838689
Fax: +86-20-82838689-806
Http: //www.simply arch.com

申派建筑设计顾问有限公司
法人：李毅展
地址：广州市东华西路 86 号丽华大夏 14 楼
电话：+86-20-83838689
传真：+86-20-82838689-806
网址：www.simply arch.com

1 武汉 国际机场

建设地点：湖北 武汉
项目类型：机场
占地面积：3 200 000 平方米
设计时间：2007 年
容 积 率：1.2
项目介绍：
　　该项目位于湖北武汉机场，规划总面积 3 200 000 平方米，此项目是与美国 Melenassociates 公司合作的项目。

1 Wuhan International Airport

Construction Location: Wuhan, Hubei
Type: Airport
Site Area: 3,200,000 m^2
Design Time: 2007
Plot ratio: 1.2
Project profile: The project is located in Wuhan Airport with the planned total area of 3,200,000 m^2. This project is with the United States Melenassociates company.

1

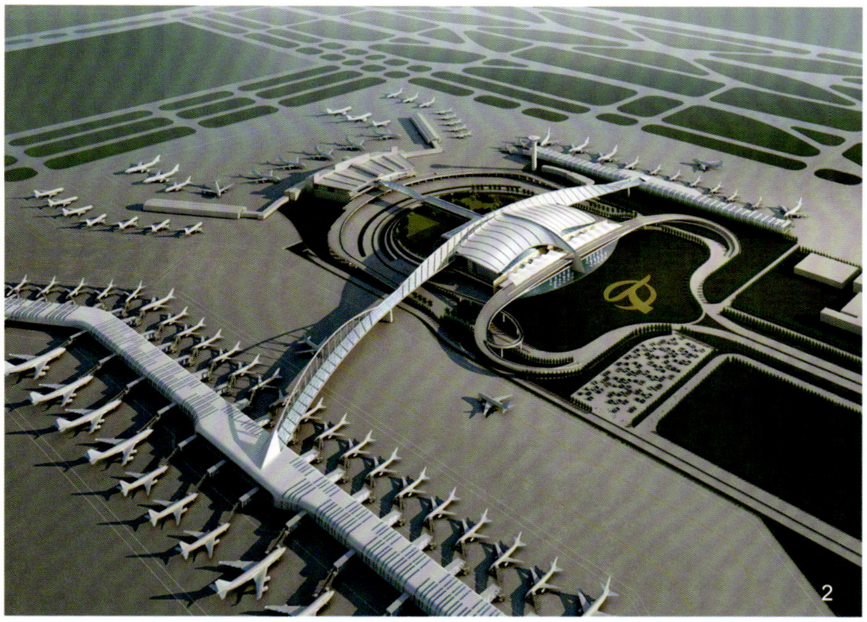
2

2–5 江苏 南京国际机场

建设地点：江苏 南京
项目类型：机场
建筑面积：2 800 000 平方米
占地面积：6 500 000 平方米
设计时间：2009 年
容 积 率：1.3
项目介绍：
　　该项目位于江苏南京国际机场，规划总面积 2 800 000 平方米，其建筑特点在于采用了整体屋顶系统。此项目是与美国 Melenassociates 公司合作的项目。

2-5 Nanjing International Airport

Construction Location: Nanjing, Jiangsu
Type: Airport & Office Building
Building area: 2,800,000 m^2
Site Area: 6,500,000 m^2
Design Time: 2009
Plot ratio: 1.3
Project profile:The project is located in International Airport Nanjing Jiangsu Province. With the planned total area of 2,800,000 m^2, including construction is characterized by an overall roof system. This project With the United States Melenassociates companies.

3

4

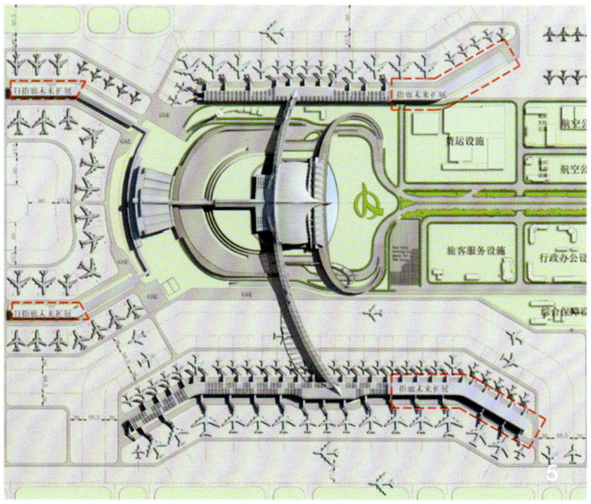
5

6

7

8

6–8 广东 阳江核电专家村

建设地点：广东 阳江
项目类型：居住 商业
建筑面积：76 400 平方米
设计时间：2007 年
容 积 率：0.75

6-8 Expert Village

Construction Location: Yangjiang, Guangdong
Type: Residential & Commercial
Building area: 76,400 m^2
Design Time: 2007
Plot ratio: 0.75

9–12 景德镇陶瓷基地

建设地点：江西 景德镇
项目类型：艺术创造
建筑面积：10 500 平方米
设计时间：2009 年
容 积 率：0.31

9-12 Ceramic Base

Construction Location: Jingdezhen, Jiangxi
Type: Art Creation
Building area: 10,500 m^2
Design Time: 2009
Plot ratio: 0.31

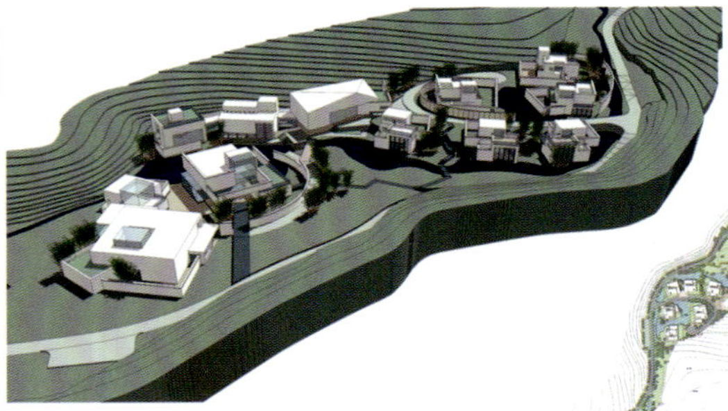
9

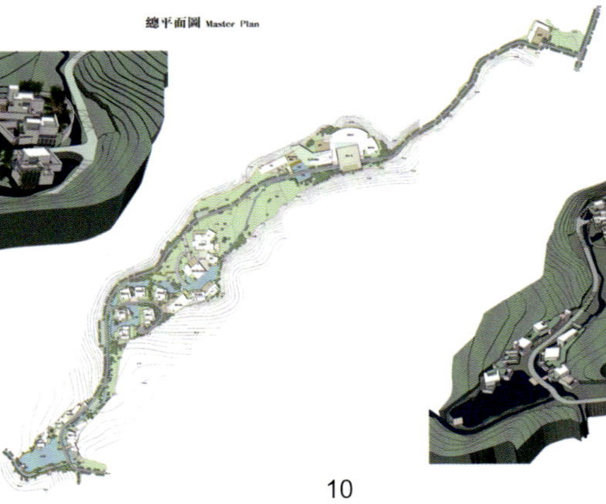
10

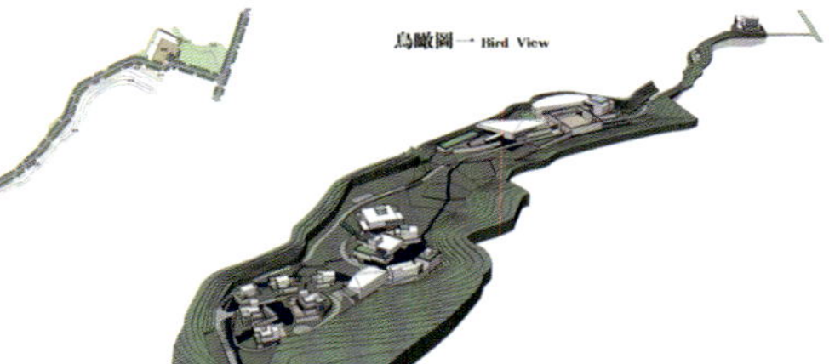
11

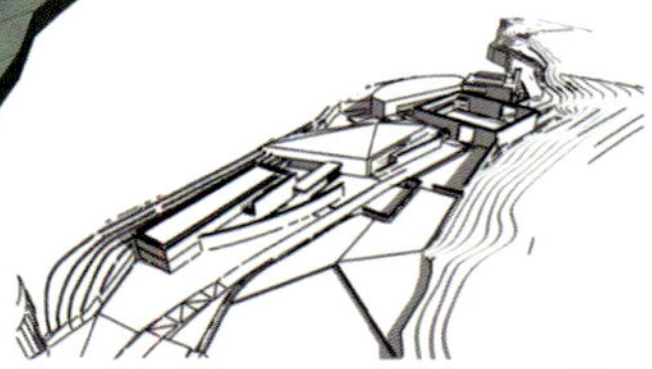
12

13

14

13–14 广东 中信 · 山语湖别墅

建设地点：广东 中山
项目类型：居住
建筑面积：30 000 平方米
设计时间：2009 年
容 积 率：0.8

13-14 Lake Villa language

Construction Location: Zhongshan, Guangdong
Type: Residential
Building area: 30,000 m^2
Design Time: 2009
Plot Ratio: 0.8

15–17 广东 中山 · 怡廷豪园

建设地点：广东 中山
项目类型：居住
建筑面积：65 270 平方米
设计时间：2009 年
容 积 率：1.3

15-17 Verde YiTing

Construction Location: Zhongshan,Guangdong
Type: Residential
Building area: 65,270 m^2
Design Time: 2009
Plot ratio: 1.3

15

16

17

18

19

20

18–22 郴州 · 湘南风情园

建设地点：湖南 郴州
项目类型：居住 商业 办公楼
占地面积：221 000 平方米
设计时间：2009 年

18-22 Shonan style garden

Construction Location: Chenzhou, Hunan
Type: Residential & Commercial
Site Area: 221,000 m^2
Design Time: 2009

23

21

22

23–24 广东 · 阳西御景湾

建设地点：广东 阳西
项目类型：居住 办公楼
占地面积：68 500 平方米
设计时间：2009 年

23-24 Royal Lagoon

Construction Location: Yangxi, Guangdong
Type: Residential & Office Building
Site Area: 68,500 m^2
Design Time: 2009

24

25

26

27

28

29

30

22–30 广东 · 广州益鹏规划

建设地点：广东 广州
项目类型：居住 办公楼
占地面积：55 500 平方米
设计时间：2008 年

22-30 Yi Peng Planning

Construction Location:Guangzhou, Guangdong
Type: Residential & Office Building
Site Area: 55,500 m^2
Design Time: 2008

广州亚泰建筑设计院有限公司

Guangzhou Asiatai Architectural Design Institute CO.,Ltd.

广州亚泰建筑设计院有限公司始建于 1981 年，是国家建设部认定的建筑工程甲级设计单位（证书编号 A144002161）。专业从事各类建筑工程设计及相关的景观设计、工程技术与城市规划咨询等业务。

公司目前人才荟萃，拥有一批正规学院教育出身、工程实践经验丰富的优秀设计师。经过多年的努力，公司主要的工程业绩有华南农业大学、广州天河软件园—IT 商务中心（五星级酒店）、山西世茂大酒店、广州大学城二期工程、大宇宙信息创造（中国）有限公司办公大楼等。

公司设建筑、结构、给排水、电气、暖通、规划、景观、室内等专业，在设计概念与构思方面，亚泰一直尊重传统与历史，融亚洲文化、传统艺术于国际化视野中，并利用自身良好的社会资源，与多所建筑类名校、大设计院、建筑界媒体建立密切合作关系，同时保持一贯塑造精品的发展步伐。亚泰建筑设计以丰富的建筑工程设计经验、杰出的设计作品和设计质量以及良好的服务口碑著称于业内。

在未来，本公司将继续发挥较强的技术力量、先进的技术水平，按照一贯的质量方针和质量标准对外开展业务活动，积极发展同海内外客户的业务往来与合作关系，为客户与社会提供更多、更优秀的作品。

Guangzhou Yatai Architectural Design Institute Co.,Ltd. was established in 1981. It is a Class A construction design unit acknowledged by the National Ministry of Construction (Certificate No.A144002161). The company is engaged in various construction project design and related landscape design, engineering technology and city planning and other services.

Our company has many talent designers who are well-educated and have rich experiences in projects. After years of hard working, we have finished many projects, such as: HuaNan Agricultural University, Guangzhou Tianhe Software District----IT Commercial Center(five-star hotel), Shanxi ShiMao Hotel, Guangzhou College Town 2nd Phase Project, office block of Macrocosms Information Creation (China) Co., Ltd. and so on.

Our company mainly responsible for construction, structure, water supply and drainage, electric, HVAC, planning, landscape and interior design. In the area of design concepts and ideas, Yatai always respects the development of tradition and history, and brings the Asian cultural traditional art into the international view harmony; also we make good use of the social resources, cooperate with several famous architecture schools, design institute and the medium in architecture closely. Asiatai Architecture Design has built a good reputation for its rich architecture design experiences, outstanding design works and design quality and the good services.

The company will take advantage of its technical strength so as to carry out business operation on the basis of its consistent quality policy and objective and actively cooperate with customers at home and abroad.

辛亥革命纪念馆

辛亥革命纪念馆

地 址：广州市海珠区北约新街 5 号 6 楼
邮 编：510310
电 话：+86–20–84204629/84204936
传 真：+86–20–84204936–801
邮 箱：gzatsj@126.com
网 址：www.gzlwsj.cn

Address: The 6th floor, Beiyue New Street, Haizhu District
P.C.: 510310
Tel: +86-20-84204629/84204936
Fax: +86-20-84204936-801
E-mail: gzatsj@126.com
Http: //www.gzlwsj.cn

广州黄埔区南岗镇亨元项目勘察设计（中标）

本项目是广州市政府为解决低收入家庭住房问题而兴建的保障性住宅小区，属于国家重点建设项目。小区包括新建住宅和公建配套项目等，规划用地为 48 105 平方米，建筑面积为 104 336 平方米。

本项目是保障性住宅小区，住宅楼设计是重点，风车型的平面布置形式不但保证建筑每层平面的户数，也确保更多的住户享有更好的建筑朝向和室外景观以及舒适的内部空间。

Hengyuan projects Nangang town in Guangzhou Huangpu District

Our project is ministry of construction engineering key project, and it is the Government of Guangzhou to solve the construction of indemnificatory housing so as to tackle the difficulty in housing for low-income families.Residential housing included public facilities and new phouses planning land 48,105 m^2 and a building area of 104,336 m^2.

This project is an affordable residential district, building design is the key, the wind model not only ensure the graphic form of each layer of the plane, also ensure households have better orientation, landscape and comfortable interior space.

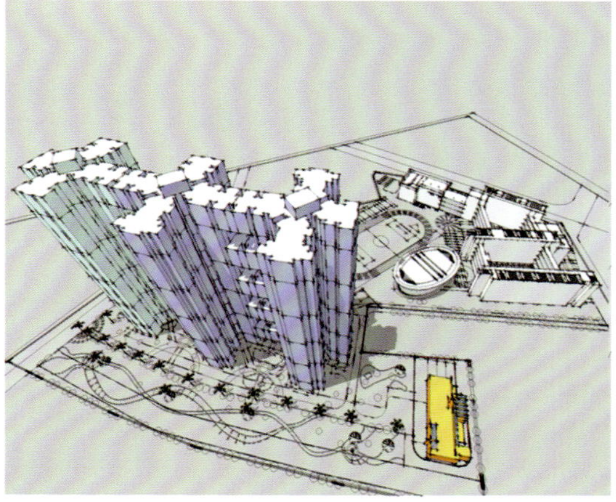

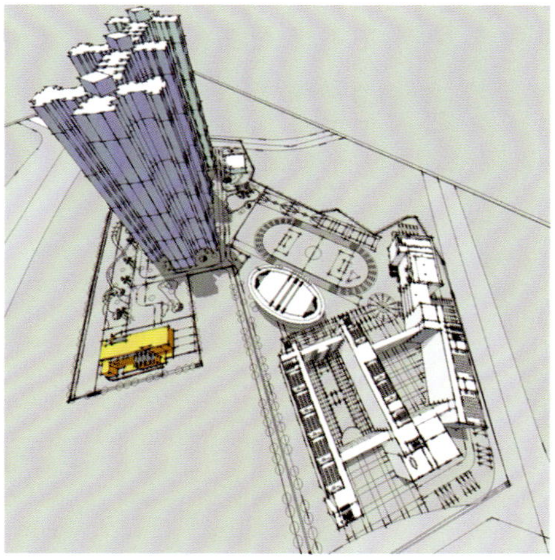

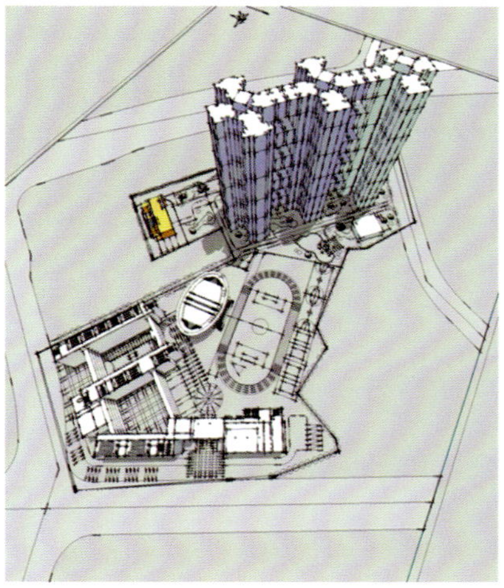

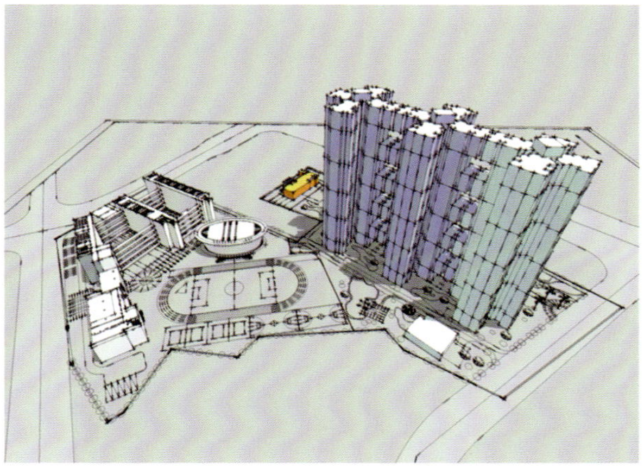

中学总平面图

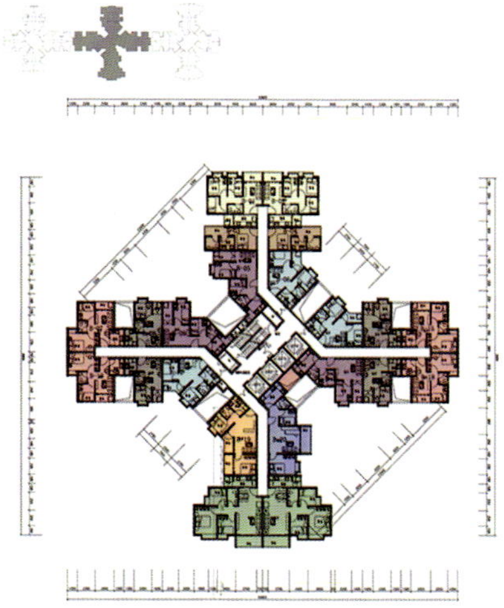

E 栋标准层平面

钦州东方豪庭

本工程位于钦州市扬帆大道西面、市机电工程学校北面、缸瓦窑村附近，规划建设用地面积约为 65 532.23 平方米。项目分两期开发，正在开发建设的为第一期。项目二期的基地被一条城市道路分隔。该居住区呈东向西走向的不规则四边形，东南面为城市主干道，西南面为规划用地，建筑均为连排式高层住宅，另设有一栋 4 层会所和两个地下车库。项目总建筑面积为 250 126 平方米。

QinZhou Oriental Plaza

his project is located in the northern of city electrical engineering school, near the Gangwanyao village, on Yangfan avenue, Qinzhou city, planning construction area about 65,532.23 m^2, constructed in two sections, which is developed as Phase I. The bases of Phase II are separated by a city road. The residential section was the East - West to the irregular quadrangle, southeast of the city main road, south-west for the planned land used, are connected to row-style noble house, while a 4 layers club, build two underground garage. The total construction area is 250,126 m^2.

天河软件园软件产业集中孵化中心

本项目为天河软件园软件产业集中孵化中心及配套工程二期，位于广深高速公路、广汕公路与大观路交汇处。项目总建筑面积为 100 111 平方米，其中计入容积率的建筑总面积为 77 435 平方米，地下建筑面积为 19 930 平方米，架空建筑面积为 2 746 平方米。建筑地上共十层，地下一层，建筑总高为 40 米。

Tianhe Software Park software industry hatching center

Tianhe Software Park software industry hatching center and its 2nd phase project is located in Guangzhou Shenzhen Highway, the connection of Guangzhou-Shantou road and Daguan Road. The overall floorage is 100,111 m^2, and the rate calculation of the floorage is 77,435 m^2, the floorage of the basement is 19,930 m^2, the overhead floorage is 2,746 m^2. The building has 10 floors, one floor underground, ten floors overground, the total height is 40 m.

番禺孵化器办公楼

华天时代科贸楼（在建）

亚泰建筑

建筑之路还很长 ...

GUANGZHOU ASIATAI ARCHITECURAL DESIGN INSTITUTE CO.,LTD.

我们一直在努力 ...

广东工贸职业技术学院教学楼方案

广州软件人才培训中心

金沙洲商场方案

广州亚运体育文化中心

海南思源堂

HSARCHITECTS

华森建筑与工程设计顾问有限公司

1980年，建设部建筑设计院（现中国建筑设计研究院）和香港森洋国际有限公司，在香港和深圳联合创办了合资设计企业——华森建筑与工程设计顾问有限公司。

目前，华森在香港、广州、南京、杭州、苏州分别设立了分支机构，形成以深圳为中心，业务延伸至全国的一体化经营格局。公司业务范围涵盖建筑与工程设计的全部内容，包括建筑与小区规划、结构、给水排水、空调通风、建筑智能化、电力电讯、民用动力、环境景观和建筑消防、人防工程的设计、咨询以及施工图审查。公司是技术实力雄厚、专业配套齐全、拥有现代化设计和管理手段的甲级（综合）设计公司。

基于多年打造的品牌优势、服务各类顾客的丰富经验、业内首屈一指的人才团队、日益完善的经营和服务体系，华森必将成为国内乃至国际一流的设计服务公司，为顾客提供最专业和最优质的服务。

1980, Architectural Design Institute of Ministry of Construction (former China Architectual Design Institute) and Hongkong Senyang International Co.,Ltd., establish the first joint venture design enterprise in flourishing and bustling city Hong Kong and Shenzhen——Huasen Architectural & Engineering Designing Consultant Ltd.

HSA sets up branchs in Hong Kong, Guangzhou, Nanjing, Hangzhou and Suzhou, HSA forms an integrative management pattern, which makes Shenzhen the center, and extends through the country. It's business covers the whole process of architecture and engineering design, including building and community planning, structure, drainage, ventilation, intelligent building, electric power and telecommunications, civil power, environmental landscape and building fire prevention and control, and civil air defense design and consultation etc.It is a top grade company with influential technology, completed professional necessities modern design and management means.

On the basis of brand advantage, service experience, excellence talent group, modern management means, and service system, HSA will be devoted to become "the first-class design company both domestic and abroad, offering the most professional and high-quality service for the customers".

地址：深圳市南山区滨海之窗办公楼六层
邮编：518054
电话：+86-755-86126666
+86-755-86126888
传真：+86-755-86126800
网址：www.huasen.com.cn

Address: 6/F, BinHaiZhiChuang Office Building, Nanshan, Shenzhen, China
P.C.: 518054
Tel: +86-755-86126666
+86-755-86126888
Fax: +86-755-86126800
Http: //www.huasen.com.cn

1

2

华森设计
HSArchitects

超高层

3

1–3 深圳地铁大厦

建设地点：广东 深圳
建筑面积：97 200 平方米
设计时间：2010 年

1-3 Shenzhen Metro Tower

Construction Location: Shenzhen, Guangdong
Building Area: 97,200 m^2
Design Time: 2010

4

5

4–6 东冠大厦

建设地点：浙江 杭州
建筑面积：69 493.63 平方米
设计时间：2005 年

4-6 Dongguan Plaza

Construction Location: Hangzhou, Zhejiang
Building Area: 69,493.63 m^2
Design Time: 2005

6

华森设计 HSARCHITECTS

酒店

7–11 广东从化侨鑫温泉养生谷五星级酒店（合作）

建设地点：广东 从化
建筑面积：140 000 平方米
设计时间：2008 年

7-11 Conghua Qiaoxin Spa Center Five-star Hotel in Guangdong(cooperative project)
Construction Location: Conghua, Guangdong
Building Area: 140,000 m^2
Design Time: 2008

7

8

9

10

11

酒店

华森设计 HSArchitects

12

13

12–14 浙江宁波索菲特万达大饭店

建设地点：浙江 宁波
建筑面积：33 900 平方米
设计时间：2007 年

12-14 Ningbo Sofitel Wanda Hotel, Zhejiang

Construction Location: Ningbo, Zhejiang
Building Area: 33,900 m^2
Design Time: 2007

14

16

15

18

17

15–18 深圳华侨城洲际大酒店（合作）

建设地点：广东 深圳
建筑面积：108 900 平方米
设计时间：2006 年

15-18 Inter-continental Hotel in Shenzhen (cooperative project)

Construction Location: Shenzhen, Guangdong
Building Area: 108,900 m^2
Design Time: 2006

华森设计 HSARCHITECTS

住宅

19–21 洲头咀

建设地点：广东 广州
建筑面积：317 822 平方米
设计时间：2009 年 6 月 –2010 年 3 月

19-21 Zhoutouzui

Construction Location: Guangzhou, Guangdong
Building Area: 317,822 m^2
Design Time: 2009.06-2010.03

20

21

22

22–23 广东珠海方圆云山诗意

建设地点：广东珠海
建筑面积：138 000 平方米
设计时间：2006 年

22-23 Zhuhai Fineland Oriental in Guangdong

Construction Location: Zhuhai, Guangdong
Building Area: 138,000 m^2
Design Time: 2006

23

26

27

住宅

华森设计
HSARCHITECTS

26–31 广州从化明月山溪

建设地点：广东 广州
建筑面积：314 800 平方米
设计时间：2008 年

26-31 Conghua Mingyueshanxi, Guangzhou

Construction Location: Guangzhou, Guangdong
Building Area: 314,800 m^2
Design Time: 2008

28

29

30

31

32

33

34

35

32–35 杭州钱江新城金基晓庐

建设地点：浙江 杭州
建筑面积：297 000 平方米
设计时间：2005—2007 年

32-35 Qianjiang new town Jinjixiaolu in Hangzhou

Construction Location: Hangzhou, Zhejiang
Building Area: 297,000 m^2
Design Time : 2005-2007

SuiADR

深圳大学建筑设计研究院

The Shenzhen University Institute of Architectural Design & Research (SUIADR)

深圳大学建筑设计研究院成立于1984年，是建设部批准的全国综合甲级设计单位(资质编号：190114–SJ)，目前拥有多名全国著名专家学者，是大型甲级综合设计院。在我院完成的几百余项项目中，覆盖了公共建筑、商业建筑、体育建筑、居住建筑、教育建筑及交通建筑等建筑类型。

我院现有各类专业技术及管理人员215人（不含系编教师），国家一级注册建筑师39人，一级注册结构工程师15人，教授级高工18人，高级工程师58人，博士后、博士、硕士、研究生42人，留学归来的设计人员达20多人。

我院每年超过40人次出国进行专业考察并参加国际学术会议。我院参与完成了多项国家、省、市规范、规程、地方标准和设计资料集的编制工作，承担了国家、省、部级科研课题数十项，同时担任了市政府多项大型工程的顾问工作，并与多个国际著名建筑事务所和许多著名房地产开发商建立了全方位的战略合作关系。

建院20多年来，共获国家、部、省、市各级优秀设计和科技奖近百余项，其中包括一项全国优秀结构设计一等奖、两项华夏建设科学技术奖一等奖、四项詹天佑土木工程大奖、两项建筑环境双金奖。深圳大学建筑设计研究院已成为深圳和华南地区最有实力和影响的建筑设计院之一。2006年被国家建设部评定为“十五”全国建筑业技术创新先进企业。

The Shenzhen University Institute of Architectural Design & Research (SUIADR) was established in 1984 and it is authorized, as the national Class A archiectural design company by Ministry of Construction. The main services provided by the institute are town planning, residential planning, architectural design, construction planning, interior design, landscape design, planning and design consulting, etc. Projects finished are located in many cities in China, and the types include public, commercial, sports, academic and education. Projects we have finished are located through the whole country including fields from public architecture, commercial architecture, sports architecture, educational architecture and residential architecture.

After twenty years' growth, we have a strong design team composed by design experts. Currently we have 215 staffs, and among them, 39 are registered architects, 15 are registered structure engineers, 18 are professors, 58 are senior engineers, 42 own master or higher degree, and 20 have overseas education background. We have been emphasizing on combining research, academic and practice, and created a good base. Many of our staffs have been to other countries for international conference and anticipated planning and design projects there. We also published many academic books and they became good reference for the design field in China. Besides above, we have established good cooperation relationship with many real estate developers and archiectural firms, and many of our staff have visited other countries for conference or research.

We have won many awards, including 80 from the city to the national level. Some of the residential design projects won awards. We have become one of the most famous design institutes in China, especially in south east region. We were awarded No. three of the "top 50 design institutes" in Shenzhen in 1995, and in 2000, we passed ISO9001 qualification management system. In 2002 and 2004 we were awarded one of the top ten companies in Shenzhen.

院长：高青
地址：深圳大学校内
邮编：518060
电话：+86–755–26732802 26732803

电话：+86–755–26732802 26732803
传真：+86–755–26732801
邮箱：archjy@szu.edu.cn
网址：www.suiadr.com

惠程电气厂区建筑

建设单位：惠程电气股份有限公司
方案设计：深圳大学QL建筑工作室
施工图设计：国际印象建筑设计有限公司
施工单位：金众集团
建设地点：深圳市龙岗

设计时间：2004年9月
竣工时间：2007年4月（一期）
占地面积：3.01公顷
总建筑面积：39 600平方米
容 积 率：1.2
摄 影：覃力

本项目为国内首座大规模清水混凝土工业建筑，也是深圳市第一座大规模清水混凝土建筑，清水混凝土面积达3万平方米。同时，它也是深圳市具有时代感的一座造型特色突出、简洁明快并与国际设计理念接轨的新型现代工业建筑。

本项目建筑设计强调时代感、现代意识和工业技术，建筑造型体量明晰、简洁大方。厂房中穿插设计庭院景观，营造出轻松活跃的空间氛围。办公楼底层架空，水池绿化伸入建筑内部，使整个厂区的景观环境大为改观。建筑物的外观采用混凝土与花岗岩相结合的处理方式，立面则采用玻璃外挂金属百叶遮阳，既减少了日晒和噪音又极具现代感，再加上绿化景观的辅助，建筑形象个性更显突出。

结合分期建设的要求，将厂前区设置在用地的西北侧，并以此为中心，形成管理办公和休息活动的区域。办公楼布置在用地的西侧，第一期建设的厂房位于用地的中央，东侧为二期开发的用地。厂房与办公楼之间以架空的走廊相连接；机动车交通沿着用地的外围形成环路，便于货物进出；办公楼前的广场禁止车辆进入，形成步行活动的区域。项目空间组织清晰合理、流线便捷、分区明确。

深圳大学师范学院教学实验楼

项目名称：深圳大学师范学院教学实验楼
建设地点：深圳市南山区深圳大学
占地面积：4 700平方米
建筑面积：18 000平方米
设计单位：QL建筑工作室
项目负责人：覃力
建筑设计：覃力、颜奕填、李一凡、杜晓钟
结构设计：张正国、邱松明
电气设计：连建社、刘中平
给排水设计：谢蓉、晏风
暖通设计：郑建超
设计时间：2004年5月
竣工时间：2007年5月
摄　　影：覃力

本项目为教学实验楼，设有普通教室、实验室、办公室及一系列的特殊教室，如音乐教室、美术教室、舞蹈教室、琴房等，此外还有一个黑箱实验剧场和一个小型音乐厅。

项目的设计特点如下：

周边环境的整合。由于师范学院教学实验楼过于接近深圳大学的主校门，因此，我们便采用将建筑化整为零和由南向北退合的方法，来减弱整体造型的厚重感和对校门的压迫感。同时，师范学院的教学综合楼充分利用了南低北高的地形高差，地下室在南面露出地面，作为一层成为直接对外空间，埋在地下的部分还设置了一些下沉庭院，使阳光可以直接照射到地下，增加了空间上的变化。

交往空间的创造。交往空间的创造是本项目的一大特色，在建筑处理上利用了架空层、中庭和自然叠落的屋顶平台，为教师与学生在每一楼层上，提供了一系列变化丰富的公共活动场所。这些处于各个楼层中室内和室外的大尺度的活动空间，与连接着教室的线型走廊交相重叠，产生了十分丰富的视觉感受和空间效果，与人的活动形成互动，是极具活力的动态空间节点。

开放理念的运用。师范学院教学实验楼的另一特征是开放性。设计在几个方向上都利用架空、掏空等手法，打破了建筑内部空间与外部空间之间的界线，使内外空间相互渗透，在建筑内部又以通高的中庭和拔空的庭院，从竖向上将这些不同标高的流动空间连成一体，形成一种上下、内外贯通的开放空间系统。这种开放空间系统，有利于空气流通，使空间和使用者更亲近自然，特别适用于深圳这样的南方地区。

低造价高标准。本工程在设计之初，建设方就提出了控制造价的要求，要求18 000平方米教学楼土建总造价不得超过3 600万元，经过我们的多方努力，在不影响效果的情况下，最终成功地将单方造价控制在2 000元之内。同时，建成后的效果也获得了多方的好评，被评价为深圳大学造价最低、效果最好的建筑。

深圳大学图书馆二期

设计团队：
设计总负责人：杨文焱
建筑设计：杨文焱、成国亭、吕夏青、黄亮、姚小玲
结构设计：张正国、丘松明
水电设计：谢蓉、肖冬开、李月荣
空调设计：郑文国
经济技术指标：
占地面积：21 682.92平方米
总建筑面积：24 544.27平方米
容 积 率：0.85

深圳大学图书馆二期，作为老馆的扩容补充，以理工类藏书为主，选址于深圳大学中心广场南边，原 E 座教学楼旧址。北面隔中心广场与旧馆遥相呼应，南靠秀美的杜鹃山，东望学生活动中心，西临文山湖。设计首先关注校园建筑秩序的延续问题，希望保持中心广场特定的校园格调，同时维护和加强空间形态秩序。通过二进制代码转译环境要素，从中获得形态构成法则，以求得新旧建筑在形式和空间构成上的相似性。布局和材料组织的成分均简化为0与1的单元，依照对信息源的反应进行组织和排列。建筑在面对中心广场的方向上依据其中轴，排列实体（1）与虚空（0），使轴线延伸至杜鹃山顶，维持老馆与杜鹃山之间的视觉联系。建筑体量的水平向依据使用功能排列条状的建筑体量（1）与院子（0），以获得建筑空间层次和气流的变化。竖向上则是架空平台与实体的交错组织，形成内（0）、外（1）的空间编织交错，拓展阅览和交流空间。建筑外观的设计，强调室内图书空间的品质和建筑表达。U型玻璃（1）和白玻璃（0）的交替应用和凹（0）凸（1）的交错排列，形成室内漫射光照，避免大量的直射和眩光，同时获得与老馆外形的某种同质联系，以及对时间变迁的信息反映。

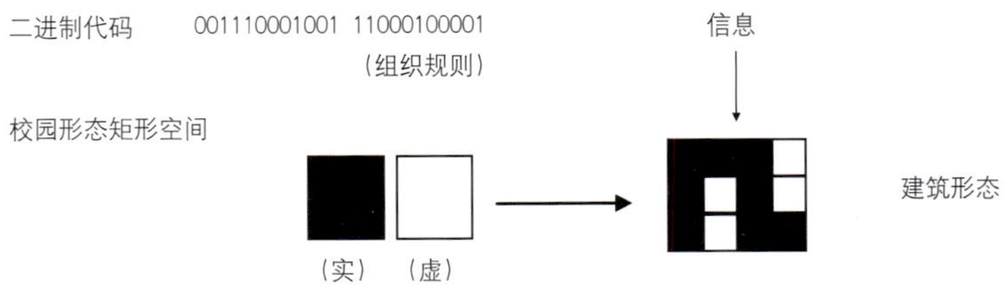

设计体现了以下的先进性：

1. 图书馆二期的建筑形式以一种新的概念和方式进行组织，在延续深圳大学校园环境的氛围方面恰如其分，这既是形态上的和谐与空间秩序的延展，也是优雅的空间格调，为校园注入了具有新时代印记的新元素和活力；

2. 建筑积极回应气候影响。架空建筑接地层，大面积设置水池，组织流动的气流，使建筑处于良好的微气候当中。由此形成的大片室外阴影区域拓展了室内阅读空间，并提供活动场所。设计结合朝向和建筑周边高大树木的位置，设置建筑墙面的开闭，保证室内良好的视野和光照，避免强烈光照对建筑的影响，降低营运能耗；

3. 使用新型材料。大面积采用U型玻璃作为外围护材料，充分发挥其材料的热工特性，获取室内自然漫射光照，营造优雅的建筑空间和形象和良好的室内阅读光照；

4. 采用合理的技术，慎用资源，在投资限额下（3600元/平方米）求得了优良的现代图书馆的室内外空间，较充分地达到了建筑的建设目的，各专业的技术均追求建设与运行消耗的最小化，符合可持续发展之道。

深圳大学艺术村

主要设计单位：深圳大学建筑设计研究院

建设地点：深圳市南山区深圳大学内

主要创作人员：覃力 张道真 骆静静 温雅乔

建设规模：2 400平方米

建成时间：2008年8月

摄　　影：覃力

创作构思说明

深圳大学艺术村地处深圳大学校园内，北面正对着深圳大学的北门，地理位置显要。所谓艺术村，实际上就是深圳大学的美术馆，该基地原为艺术系教师李瑞生先生设计建造的很有特色的“鬼屋”的原址，所以新建的美术馆需要非常引人注目，并肩负着重新振兴该地域建筑环境的作用。

“鬼屋”并非建筑师的李先生自己动手设计建造的，其原是一组建筑，自成一个院落，院内有大片的荔枝林和竹丛，地形东低西高，环境幽美。建筑全部用毛石砌筑，形态随地形自由变化，造型独特，曾经是深圳大学乃至深圳市的一个极具个性的艺术家工作室。后来由于产权问题，“鬼屋”主体建筑被拆除了，学校在原址上临时修建了音乐教室、琴房和简易的展厅，但是，院落里面的几座石砌建筑仍然被保留了下来，新建的美术馆就选在原来主体建筑的位置，面对着北校门，沿着深圳大学校内的环路临路而建。

由于这样特殊的场所条件和极不规则高低起伏的地形特点，我们在设计时便考虑以自由变化的建筑形式来适应环境，用小尺度的形体组合，延续原来“鬼屋”的造型特征。这种不规则的平面和灵活丰富的造型形态，不但可以营造出不同于常态的内部空间效果，还可以增加展示面的长度。以独特的造型去吸引人们的注意力，使这座建筑成为深圳大学北校门内轴线上的独特风景。

在建筑空间组织上，我们利用地形的高差，使建筑空间与环境的结合立体化。正面的主入口设有大台阶，可以直接进入美术馆的二层。一层的主展厅则通过下沉的室外展示场地与建筑围合的内部庭院相连。各展厅之间呈线性连接，流线简明顺畅。序厅和作为展览间歇的休憩空间则面向荔枝林开窗，拥有非常好的景观视野，而展厅的内部却相对封闭，以利于展品的展示。整座建筑空间不仅高下相合，而且大小、明暗组织有序，富于展览建筑空间的特征。

HHD sz

深圳市华汇设计有限公司

SHENZHEN HUAHUI DESIGN CO.,LTD.

华汇工程建筑设计有限公司（HHD）是由中外合资的具有国家建筑甲级和规划乙级设计资质的综合性建筑设计单位，公司通过 ISO 9001：2000 国际质量认证。公司领导层由多名国内著名高等院校毕业的学生、学有专长的工学硕士以及海外归来的资深建筑师组成。公司队伍充满活力，锐意创新。专业人员约 300 人，其中高级职称 80 人、中级职称 98 人、初级职称 40 人。

公司拥有国家一级注册建筑师 32 人、一级结构工程师 17 人、在各方面经验丰富的技术专家多名，还有多名资深外籍建筑师，以及由中国科学院院士担任的高级顾问。

公司创始人周恺先生为中国勘察设计大师，是全国民营设计机构唯一获得官方大师称号的建筑师。

华汇公司的设计群体由具有丰富的工程设计实际经验和较高理论水平的专家、学者、教授以及在设计行业展现才华的中青年建筑师、工程师们组成。

华汇公司给客户提供全面优质的城市、居住区规划，城市设计，各类民用与工业建筑设计以及景观设计服务。

华汇的设计宗旨是坚持最优设计、与客户密切配合，全面质量保证。

本公司自创建以来，在全国各主要城市承接并完成了大量的单体建筑设计、城市设计、小区规划设计及建筑前期策划、可行性研究等工作，项目内容包括住宅综合开发项目、星级酒店、商业建筑、文化建筑、教育建筑、博览建筑、体育建筑等。大量作品获得国家及省市级优秀设计奖，受到业主与同仁的普遍赞誉。

深圳华汇设计有限公司作为华汇设计的重要组成部分，自 2003 年成立以来，已设计完成了许多经济效益与社会效益俱佳的建筑作品，尤其在住宅设计领域，凭借其优质的设计服务，已与多家全国性的品牌发展商结成了长期合作关系，如万科地产、华侨城地产、中海地产、招商局地产、保利地产、龙湖地产、华润置地、佳兆业、花样年、中航城等。

公司的有项目连续获得国家重要设计奖项。公司首席建筑师肖诚先生是 2009 年亚洲建协建筑金奖、2007 全球华人青年建筑师奖、2008 中国建筑学会中国青年建筑师奖、第五届中国精锐科技住宅奖建筑设计金奖 等众多重要奖项的获得者，并作为影响中国建筑设计界的重要人物之一，获邀入选《建筑中国 60 年 · 人物卷》

Huahui Design Co.,Ltd. (HHD) is a joint venture architectural design institute with qualifications of National Architecture Class-A and Planning Class-B, and it has passed through the ISO9001: 2000 International Quality Certification.

The leadership of the company, full of vitality and innovation, is constituted by a number of specialized engineering masters graduated from famous universities and senior architects returning from overseas. There are about 330 professional designers , there are 80 with senior titles, 98 with intermediate titles and 40 with junior titles.

The company has 32 national Class-A registered architects, 17 Class-A structural engineers and many technical experts with rich experience. It also has many senior foreign Architects and senior consultants assumed by academicians from Chinese Academy of Sciences. Mr. Zhou, the founder of the company, is an investigation and design master in China, the only architect receiving the official title of master in the private design institutes at home.

Huahui's design group composes experts, scholars and professors who have rich experience in engineering design and high theoretical foundation, as well as the young architects and engineers displaying their talents in the design industry.

Huahui endeavors to provide customers with high-quality urban and residential planning, urban design, various types of civil and industrial building design and the landscape design.

The design purposes of Huahui are: adherence to the optimum design; close coordination with customers and comprehensive quality assurance.

Since its foundation, Huahui has undertaken and completed a large number of projects in the major cities throughout the country, such as the single building design, urban design, community planning & design, pre-construction planning and the feasibility studies. These project items include the integrated residential development projects, star-level hotels, commercial buildings, cultural buildings, educational buildings, exhibition buildings as well as sports buildings, etc. Many of these designs have won national, provincial and municipal Excellent Design Awards, received universal praises from customers and associates.

As an important component of Huahui Design, Shenzhen Huahui Design Co.,Ltd. has completed the design for many constructions with both excellent economic and social benefits ever since its foundation in 2003. With its high-quality design services, especially in residential design area, Shenzhen Huahui has established a long-term relationship with many national brand developers, such as Vanke Real Estate / OCT property / China Overseas Property / China Merchants Property Development / Poly Real Estate / Lake Properties / China Resources Land/ Kaisa Group / Fantasia / CNAC City.

HHD has been awarded important national design awards for years in succession. Mr. Xiao, the chief architect of the company, is the winner of many important awards: the Gold Prize of Architecture by Asian Architecture Society in 2009; the Global Chinese Young Architects Award in 2007; the Young Architects Award by Architectural Society of China in 2008; the Gold Prize of Architecture Design in the Fifth China's Elite Residential and Technology Award, etc. As an important figure that has greatly influenced China's architectural design, Mr. Xiao has been selected into the book "Construction of China over 60 Years • People Volume".

肖 诚

Cheng Xiao

肖 诚

1972 年生，天津大学建筑学硕士
深圳华汇设计有限公司总经理、总建筑师
国家一级注册建筑师，高级建筑师

曾荣获主要奖项

2009 年 亚洲建协建筑金奖
2007 年 全球华人青年建筑师奖
第七届中国建筑学会青年建筑师奖

Cheng Xiao

Born in 1972, Master of Architecture in Tianjin University
General Manager and Chief Architect of Shenzhen Huahui Design Co.,Ltd.
National Class-A Registered Architect, Senior Architect

Major Awards

2009 The Gold Prize of Architecture by Asian Architecture Society
2007 Global Chinese Young Architects Award
Young Architects Award of the Seventh Architectural Society of China

华侨大学厦门学院图书馆

建设地点：福建 厦门
用地面积：15 000 平方米
总建筑面积：39 280 平方米
容 积 率：2.4
建筑层数：地上 5 层，地下局部 1 层
设计时间：2006 年
施工时间：2007–2008 年

The Library of Xiamen College, Huaqiao University

Construction Location: Xiamen, Fujian
Site Area: 15,000 m^2
Total Building Area: 39,280 m^2
Plot Ratio: 2.4
Building Layers: five floors above ground and one partial floor under ground
Design Time: 2006
Construction Time: 2007-2008

深圳万科金域华府

建设地点：广东 深圳
建筑规模：22 万平方米
设计时间：2006 年
一期竣工时间：2008 年

Shenzhen Vanke Jinyuhuafu Community

Construction Location: Shenzhen, Guangdong
Construction Scale: 220,000 m^2
Design Time: 2006
Completion Time of Project Phase I: 2008

武汉茂园建筑设计

建设地点：湖北 武汉
占地面积：9 000 平方米
建筑面积：2 000 平方米
设计时间：2008 年
竣工时间：2009 年

Architectural Design of Wuhan Mao Park

Construction Location: Wuhan, Hubei
Site Area: 9,000 m^2
Building Area: 2,000 m^2
Design Time: 2008
Completion Time: 2009

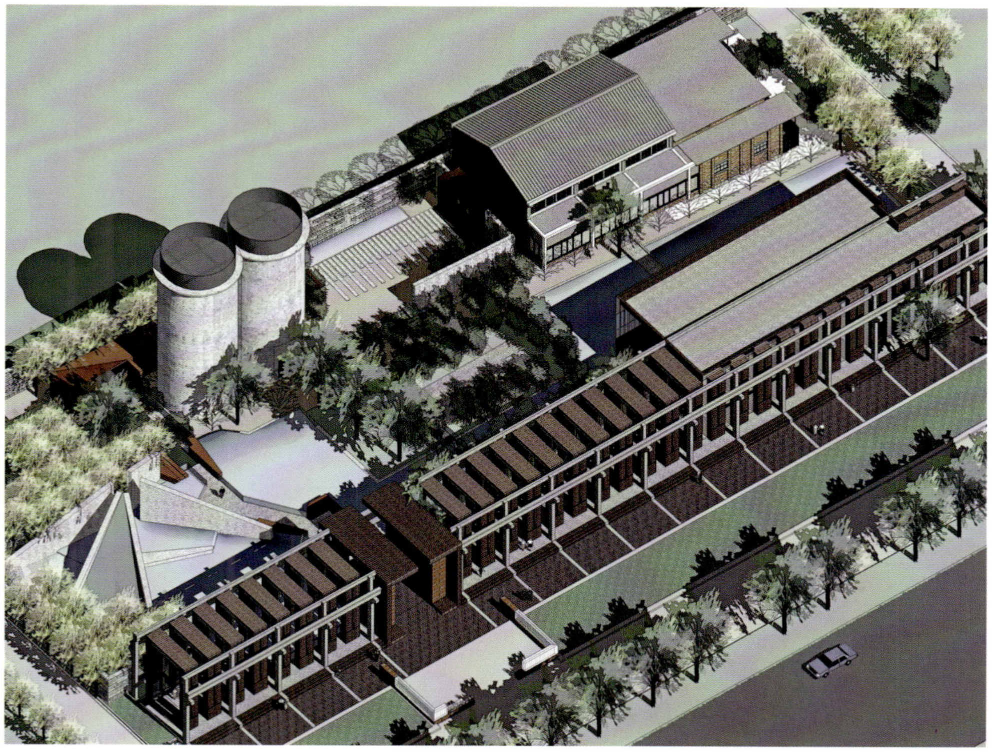

万科 蓝山，广州，中国

建设地点：广东 广州
占地面积：74 034 平方米
建筑面积：148 068 平方米
容 积 率：2.0
绿 地 率：39%
总 户 数：约 1 030 户
开发周期：3 年，分 2 期
设计时间：2004 年 6 月 –2005 年 5 月
竣工时间：2007 年
主要产品：高层、合院情景、联排住宅、叠拼住宅

VANKE•BULE MOUNTAIN, GUANGZHOU, CHINA

Construction Location: Guangzhou, Guangdong
Site Area: 74 034 m²
Building Area: 148 068 m²
Plot Ratio: 2.0
Greening Ratio: 39%
Household Number: about 1 030 in total
Development Phases: two phases in three years
Design Time: 2004.06-2005.05
Completion Time: 2007
Main products: High-rise building, Scene-conditioned courtyard houses, Townhouse, Layered Town

TSINGHUA-YUAN

深圳市清华苑建筑设计有限公司

TSINGHUA YUAN ARCHITECTURAL DESIGN LTD. SHENZHEN

地 址ADD：深 圳 市 南 山 区 桃 园 路 南 苑 新 村 4 号 楼
电话TEL：+86-755-26493788　传真FAX：+86-755-26079280
E-mail：qhy@qhy. sina. net　网址WEBSITE：www.thy.cn

深圳市清华苑建筑设计有限公司的前身是清华大学建筑设计研究院深圳分院，研究院分院成立于1995年4月，于2000年7月经相关部门批准，改制为有限责任公司。公司持有建设部颁发的“建筑工程”甲级资质证书和广东省建设厅颁发的“城乡规划编制”乙级资质证书。公司于2002年通过“ISO 9001：2000质量管理体系”认证。

在清华校训“厚德载物，自强不息”的指引下，公司全体员工努力工作，精心设计，赢得了顾客和社会的支持与信任。公司成立以来，完成各类型工程设计项目830多项（含在建项目），完成规划类用地面积约3.2万公顷，完成建筑设计面积约5 000万平方米，获部、省、市级优秀工程设计奖62项。公司现有九个专业设计室，九个综合设计所、室和一个结构安全诊治中心，共有员工400余人，其中09%以上为工程技术人员，有国家一级注册建筑师、国家一级注册结构工程师、注册规划师五十多人，拥有自建办公楼，配置品质优良的技术装备。2005年公司荣获“深圳市勘察设计行业优秀企业”称号。

目前公司积极尝试绿色建筑设计和太阳能在建筑领域应用技术，争取在绿色建筑、节能环保方面做出成绩。公司将一如既往地遵从校训“严谨、勤奋、求实、创新”的指导，努力提高设计水平和服务质量，刷新品牌，回报社会。

Tsinghua Yuan Architectural Design Limited Shenzhen, (TYAD), formerly Shenzhen Branch of Architectural Design and Research Institute of Tsinghua University(TUADRI), was founded in April 1995, and reorganized in July 2000 into a limited liability company. TYAD has been accredited as a “Class A” Architectural Design Institute by the State Ministry of Housing and Urban-Rural Development, and “Class II” City Plan Design Institute by the Construction Department of Guangdong Province. In addition, TYAD has passed the ISO-9001:2000 Quality Assurance Standard in 2002.

Driven by the guidance “upholding things with large virtue and making unremitting efforts to improve ourselves” of Tsinghua University, TYAD is committed to provide high quality design and customer service. With every staff’s consistent efforts, it has received great reputation from clients and the society. In the past years from 1995, the institute has completed totally more than 830 projects which covers planning site areas over 32,000 ha. and building floor areas up to 50,000,000 sq.m.. 62 of these projects were awarded by the official authorities including State, Province and Municipality Ministry. TYAD is an organization with nine design studios, nine comprehensive design departments and a special group of analysis and treetmant gov structural safety. There are staffs up to 400, 90 percent of whom are engineers and technicians. More than 50 designers own the State professional certificates including architecture, structure and city planning. The institute possesses of its own office building and is well equipped. In 2005, TYAD is awarded the title of “Outsanding Business of Survey and Design Institute of Shenzhen City”.

TYAD will continue to carry on the guidance of school motto “Rigor, Diligence, Truth and Innovation”, and to promote the atmosphere to pursue the high-quality design standard and service, as well as to encourage the practice of the new technology in order to make new achievement in sustainable design and energy-saving.

1

2

公 建 类

1-2 广东医学院附属松山湖医院（东莞市第二人民医院）
用地面积：238 013.86 平方米
总建筑面积：282 800.00 平方米
容 积 率：0.97
建设地点：广东 东莞
建设单位：东莞市广医医疗投资有限公司

1-2 Songshan Lake Hospital affiliated Guangdong Medical College (Dongguan Second People's Hospital)
Site Area: 238,013.86 m^2
Gross Building Area: 282,800.00 m^2
Plot Ratio: 0.97
Construction Location: Dongguan, Guangdong
Construction Unit: Dongguan Guangyi Medical Investment Co.,Ltd.

3-4 深圳创意产业园 3# 厂房绿色设计
改造前建筑面积：16 201.2 平方米
改造后建筑面积：24 260 平方米
建设地点：广东 深圳
建设单位：招商房地产有限公司

3-4 Green design on No.3 plant of Shenzhen Creative Industrial Park
Land area before transformation: 16,201.2 m^2
Land area after transformation: 24,260 m^2
Construction Location: Shenzhen, Guangdong
Construction Unit: Zhaoshang Real Estate Co.,Ltd.

“深圳创意产业园 3 号厂房绿色设计”是针对旧厂房进行绿色节能改造，并赋予其全新办公功能的项目。通过对旧的建筑结构、给排水、电气及暖通的集成设计，使项目达到比同类建筑综合节能 65%，此节能率在 2008 年该公建投入使用后的 8 ～ 9 月份的能耗实测中得到了初步证实。

As the project, green design on No.3 plant of Shenzhen Creative Industrial Park aims at changing the old plant to be green energy-saving and gives it a new office function. According to the integration design of the old building, structure, drainage, electrical and HVAC, this project can save 65% integrated energy compared with the same buildings, and the saving rate has been initially confirmed in energy consumption between August and September when it has been put into use.

3

4

5

8

公 建 类

8 深圳港大铲湾港区集装箱码头一期工程控制大楼
用地面积：10 182.46 平方米
总建筑面积：22 050.10 平方米
建设地点：广东 深圳
建设单位：深圳大铲湾现代港口发展有限公司

8 Control Building of the 1st phase project of Container Terminal in Shenzhen Bay and Da Chan Bay
Site Area: 10,182.46 m^2
Gross Building Area: 22,050.10 m^2
Construction Location: Shenzhen, Guangdong
Construction Unit: Shenzhen Da Chan Bay Modern Port Development Co.,Ltd.

6

5–7 四川省仁寿一中新校区灾后重建工程及仁寿县体育场灾后重建工程
用地面积：23.88 公顷
总建筑面积：69 600 平方米
建设地点：四川 仁寿
建设单位：四川省仁寿第一中学校

5-7 Post-disaster reconstruction on new campus of Renshou No.1 Middle School in Sichuan and Renshou Stadium
Site Area: 23.88 ha
Gross Building Area: 69,600 m^2
Construction Location: Renshou, Sichuan
Construction Unit: Renshou No.1 Middle School in Sichuan

9

7

9 中国人民银行深圳市中心支行
用地面积：3 286 平方米
总建筑面积：23 000 平方米
建设地点：广东 深圳
建设单位：中国人民银行深圳市中心支行

9 Shenzhen Center Branch of The People's Bank of China
Site Area: 3,286 m^2
Gross Building Area: 23,000 m^2
Construction Location: Shenzhen, Guangdong
Construction Unit: Shenzhen Center Branch of The People's Bank of China

10 深圳万科华昱花园
用地面积：62 472 平方米
总建筑面积：96 969 平方米
容 积 率：1.55
建设地点：广东 深圳
建设单位：深圳市万科房地产有限公司

10 Shenzhen Vanke Huayu Garden
Site Area: 62,472 m^2
Gross Building Area: 96,969 m^2
Plot Ratio: 1.55
Project location: Shenzhen, Guangdong
Construction Unit: Shenzhen Vanke Real Estate Co.,Ltd.

10

12

11

13

11 阳光 · 敔山湾
用地面积：237 923 平方米
总建筑面积：240 753.93 平方米
容 积 率：0.75
建设地点：江苏 江阴
建设单位：江苏阳光置业发展有限公司

11 Sunshine•Yushan Bay
Site Area: 237,923 m^2
Gross Building Area: 240,753.93 m^2
Plot Ratio: 0.75
Construction Location: Jiangyin, Jiangsu
Construction Unit: Jiangsu Sunshine Real Estate Development Co.,Ltd.

12–13 荣和 · 山水绿城（B1 地块）
用地面积：84 840.24 平方米
总建筑面积：361 627.8 平方米
容 积 率：3.54
建设地点：广西 南宁
建设单位：广西荣和有限责任公司

12-13 Ronghe•Shanshui Green City(plot B1)
Site Area: 84,840.24 m^2
Gross Building Area: 361,627.8 m^2
Plot Ratio: 3.54
Construction Location: Nanning, Guangxi
Construction Unit: Guangxi Ronghe Co.,Ltd.

14

15

14–15 惠州东江湾．绿景花园
用地面积：498 360 平方米
计容积率建筑面积：596 508.2 平方米
容 积 率：1.20
建设地点：广东 惠州
建设单位：惠州市银泰达实业有限公司

14-15 Huizhou Dongjiang Bay, Lvjing Garden
Site Area: 498,360 m^2
Building area accounting plot ratio: 596,508.2 m^2
Plot Ratio: 1.20
Construction Location: Huizhou, Guangdong
Construction Unit: Huizhou Yintaida Industry Co.,Ltd.

16

16 东莞松山湖金多港片区城市设计
规划面积：267.4 公顷
建设地点：广东 东莞
委托单位：东莞松山湖科技产业园区管委会

16 Urban design of Jinduo Harbor Area of Songshan Lake in Dongguan
Planning Area: 267.4 ha
Construction Location: Dongguan, Guangdong
Entrust organization: Administrative Committee of Dongguan Songshan Lake Science and Technology Industry Park

18 成都三十七度国际社区
用地面积：233 公顷
总建筑面积：1 000 000 平方米
建设地点：四川 成都
建设单位：成都岗宏房地产开发有限公司

18 Chengdu 37 Degree International Community
Site Area: 233 ha
Gross Building Area: 1,000,000 m^2
Construction Location: Chengdu, Sichuan
Construction Unit: Chengdu Ganghong Real Estate Co.,Ltd

17

规 划 类

17 厦门福隆体育公园（国家乒乓球训练基地）
用地面积：11.4 公顷
总建筑面积：80 000 平方米
建设地点：福建 厦门
建设单位：厦门福隆体育产业有限公司

17 Xiamen Fulong Sports Park (National Table Tennis Training Base)
Floor Area: 11.4 ha
Gross Building Area: 80,000 m^2
Construction Location: Xiamen, Fujian
Construction Unit: Xiamen Fulong Sports Industry Co.,Ltd.

18

19

21

22

20

19–20 中兴通讯盐田研发基地
用地面积：110 236.31 平方米
总建筑面积：140 846.67 平方米
容 积 率：1.12
建设地点：广东 深圳
建设单位：深圳市中兴发展有限公司

19-20 ZTE Yantian R & D Base
Floor Area: 110,236.31 m^2
Gross Building Area: 140,846.67 m^2
Plot Ratio: 1.12
Construction Location: Shenzhen, Guangdong
Construction Unit: Shenzhen ZTE Development Co.,Ltd.

21–22 重庆“融侨半岛．左海湾”修建性详细规划
用地面积：96.53 公顷
总建筑面积：2 455 300 平方米
建设地点：重庆
建设单位：融侨长江（重庆）房地产有限公司

21-22 Construction Detailed Planning on Rongqiao Peninsula•Left Bay in Chongqing
Floor Area: 96.53 ha
Gross Building Area: 2,455,300 m^2
Construction Location: Chongqing
Construction Unit: Rongqiao Changjiang(Chongqing)Real Estate Co.,Ltd.

承构设计

美国承构建筑师事务所
深圳市承构建筑咨询有限公司
Made & Make Architects

地址：深圳市福田区深南大道 2008 号中国凤凰大厦 1 号楼 20C
电话：+86–755–33067800
传真：+86–755–33067801
邮箱：made_make@126.com

Address: 20C of Phoenix Building No.1, Shennan Road No.2008, Futian District, Shenzhen
Tel: +86-755-33067800
Fax: +86-755-33067801
E-mail: made_make@126.com

美国承构建筑事务所注册于美国纽约，现有建筑设计师 60 多人，其中多名为美国注册建筑师。深圳市承构建筑咨询有限公司为美国承构在中国的代表机构。作为一个有实力的设计机构，承构建筑公司的作品覆盖全国乃至国外，服务于全国的客户。

承构建筑有限公司自成立以来，已完成数百万平方米的建筑设计，包括住宅、教育建筑、商业建筑、办公综合体建筑、城市设计等。客户群包括香港置地、龙湖地产、中海地产、招商地产、华润置业、中冶集团、中信实业、创维集团、大唐电力集团、奥林匹克花园等众多知名企业。公司完成了许多兼具社会效益和市场效益的优秀的建筑作品。进入 2009 年承构公司与香港 studio off 联合成立室内设计部，同时获得了国内甲级资质，具备开展施工图设计的能力，进一步完善了公司的设计服务系统。

承构建筑的实践包括建筑单体设计，城市设计，城市规划以及室内设计等相关领域。我们认为建筑实践是一种承接的行为，建筑实践与现存的城市地理环境、社会经济条件息息相关。成功的建筑实践不仅满足建筑功能及美学的需求，更应根植于市场。团队的合作努力铸就了每个设计项目的优秀质量，并且通过构造新的空间，延续了城市和文件的整体记忆，并提供给人们一种全新的生活体验。

承构的设计宗旨

关注市场——在产品设计上注意结合市场和领先市场。
关注客户——保证与客户的密切配合和沟通。
关注设计——每一个成功的建筑都应是有灵魂的，很难说清好的建筑应包含哪些方面，但创意与激情、理想与现实、艺术与市场统一的建筑作品是我们始终追求的。
关注建造——在设计阶段就注重建筑所在地的技术及材料条件，避免设计上的浮夸和不切实际，在施工阶段密切配合甲方，全面保证质量。

Made & Make Architects registered in New York, U.S and now it has over 60 architectural designers, including many American. registered architects. Shenzhen Made & Make Architects is the Chinese representative institution of Made & Make Architects. As an institution with enormous potentiality, Made&Make Architects provides services to the clients all over China, and their projects cover the whole country and some other regions in the world.

Since its establishment, Made&Make Architects has completed an architecural design area of over 1,000,000 m^2, including residential, academic and commercial buildings as well as office complex progect, urban design. etc. the clients of Made & Make Architects has lots of clients, including famous companies like Hongkong Land Holdings Limited, Longfor Properties Co. Ltd., Zhonghai Real Estate, Investment Estate, Huanrun Real Estate, China Metallurgical Group Corp, Skyworth Group, China Datang Corporation, Olympic Garden etc. Shenzhen Made & Make Architects has created many architectural works with great social and market benefits. In 2009, Made & Make Architects and Hong Kong Studio Off jointly set up the Department of Interior Design. Meanwhile, the Company purchased national Class A qualification. It has the ability to carry out constructional design and has further improved the company's design service system.

Their architectural design practice covers monomer architectural design, urban design, urban planning, interior design and other relevant areas. We consider architectural practice as a way of preservation, architectural design practice is closely related to the geographic environment and social economical conditions. A successful architectural design work should not only meet the function and aesthetic demand, but also should be based on the market, preserving the record of city and documents by creating new space and providing new life experience as well.

The great team work ensures the solid quality of every design project

Focus on the market – strive to be the leader of the market by intergrating the projects design and the market demands.
Focus on the clients – guarantee the close cooperation and communication with the clients.
Focus on the designing works – each successful architecture has its own spirit, it is difficult to tell which elements should be contained in a great architecture design work. But what we pursue is the creativity and enthusiasm, ideal and reality, the intergration of art and the market.
Focus on the construction – pay great attention on the local technical and material conditions at the design stage, avoiding anything grandiose and unpractical, closely cooperate with the clients at the construction stage to guarantee the quality.

创维总部大厦

本工程为深圳市南山区科技园的一座标志性建筑。建筑建成后创维总部将设在这里，同时建筑的裙房功能还有转化为商业的可能性。本设计采用“躺倒的高层”的概念进行设计，赋予建筑转化成线型商业空间的可能性。“躺倒的高层”以一个“音箱”式的吸入式入口作为高潮点，入口覆以巨大的LED显示屏，具有非常好的商业昭示性。同时又是创维作为电视成像企业的企业标志所在。这个面向城市的入口空间也将成为创维大厦企业文化部分的载体。

Skyworth Headquarter Building

It is a landmark building in the Science and Technology Park, Nanshan District, Shenzhen. When the construction is completed, Skyworth headquarters will be set within this building. Meanwhile the podium of the construction has a possibility to be applied to commercial use. This design adopts the concept of a "lying high-rise", offering possibilities for the building to be converted to linear commercial space. This “lying high-rise” takes a slot-in entrance of the "sound box" shape as its climax point—the entrance is covered by a huge LED screen, which is a very good commercial demonstration and displays the symbol of Skyworth as a TV imaging company. Therefore, this entrance space facing the city has become the carrier of the corporate culture of Skyworth Building.

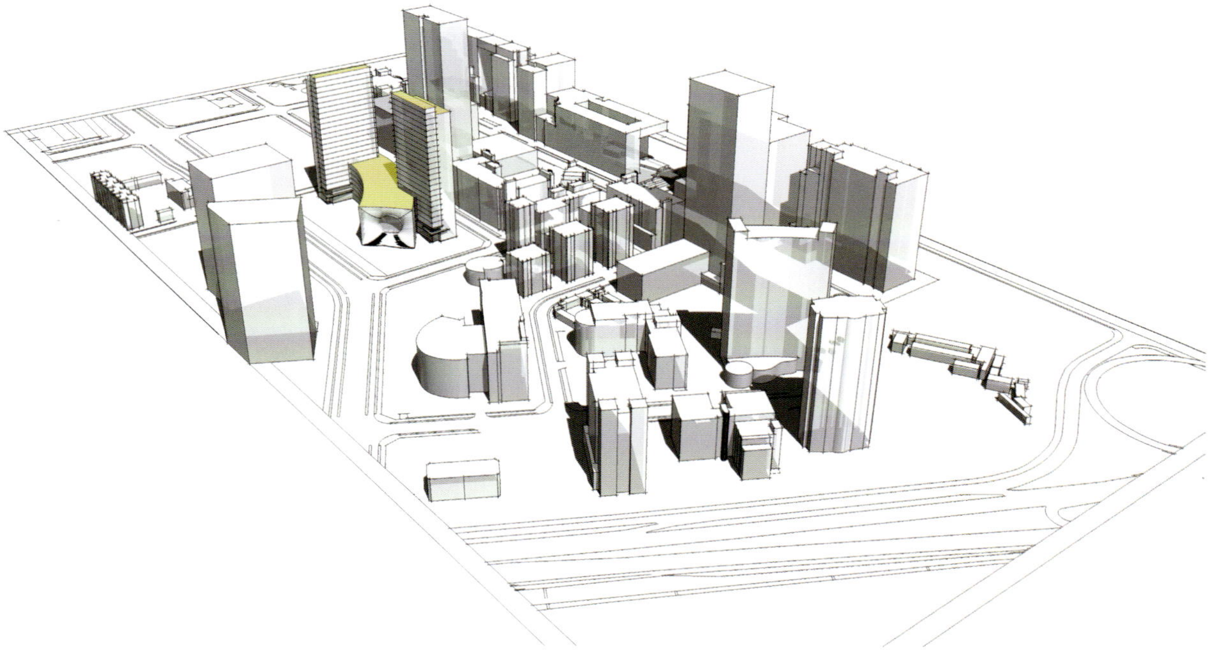

奥誉中心

工程地址：广东 广州
开发公司：广州奥林匹克花园地产有限公司
设计时间：2009 年

奥誉中心是一座现代化的购物中心，规模 30 万平方米，包括四星级酒店、酒店式公寓、大型购物中心及商业步行街、办公楼等，是广州市面向番禺区未来最大型的城市中心商业综合体。

Aoyu Center

Construction Location: Guangzhou, Guangdong
Developer: Guangzhou Olympic Garden Real Estate Co.,Ltd.
Design Time: 2009
Aoyu Center is a modern shopping center whose project scale is 300,000 m^2, including four-star hotels, serviced apartments, large-scale shopping malls, commercial pedestrian streets and office buildings, etc. It will be the largest urban central commercial complex in Panyu District, Guangzhou in the near future.

浙江长兴金郁金香酒店

工程地址：浙江 湖州
开发公司：上海金宏浦投资有限公司
设计时间：2009 年

由荷兰金郁金香酒店管理集团经营的长兴郁金香酒店是一座五星级的现代化度假酒店，酒店拥有 250 间客房及大型度假别墅区。

Changxing Golden Tulip Hotel in Zhejiang

Construction Location: Huzhou, Zhejiang
Developer: Shanghai Jin hongpu Investment Company
Design Time: 2009

Operated by the Netherlands Golden Tulip Group, Changxing Golden Tulip Hotel is a five-star modern Holiday Inn with 250 rooms and a large holiday villa resort.

深圳市石厦村改造项目

工程地址：广东 深圳
开发公司：深圳市新天时代投资有限公司
设计时间：2009 年

深圳市石厦村改造项目可以使这片曾经的城中村复兴成为具有活力的城市综合体，其中包括 180 米的超高层写字楼，两栋 160 米高的酒店式公寓，大型购物中心及住宅社区，总建筑面积达到 50 万平方米。

其中超高层写字楼的高区、中区、低区部分标准层平面层向外错出，形成具有特色的体量造型。同时塔楼的核心筒也随着低区、中区、高区层层退进，使核心筒始终居于标准层的中间，保持了结构的稳定。

Reconstruction Project of Shixia Village, Shenzhen

Construction Location: Shenzhen, Guangdong
Developer: Xintian Times Investment Co.,Ltd., Shenzhen
Design Time: 2009

The reconstruction project of Shixia Village in Shenzhen can revive this once urban village into a vibrant city complex which includes a super high-rise office building of 180 m in height, two serviced apartments of 160 m in height, large-scale shopping malls and residential communities, with a gross floor area up to 500,000 m^2.

Part of the standard storeys at the high area, middle area and low area of the super high-rise office building stagger at each layer, forming a characteristic body-mass mode. At the same time the core tube of the tower building withdraws along with each storey of the low area, middle area and high area, ensuring its middle position at the standard floor to maintain the structural stability.

天津响螺湾国泰大厦

工程地址：天津
开发公司：天津滨海国泰
设计时间：2009 年

国泰大厦楼高 201 米，以方形平面开始向上逐渐收进，最终以一个透明的玻璃叶子结束。造型形体赋予了建筑多角度的体形变幻。

基地上的"Y"字流线不单可将人流方便地引入基地，更可将地块合理的分成三个"Block"，其中一块为超高层写字楼，另外两块则为带中庭的酒店式公寓楼。

Cathay Building, Xiangluowan, Tianjin

Construction Location: Tianjin
Developer: Tianjin Binhai Guotai
Design Time: 2009

Cathay Building is 201-meter-high. Its square plane gradually shrinks upward and is ended with a transparent glass leaf: this body shape of gradual changes imparts to the building a multi-angle shape shift.

The Y-shaped streamline can not only facilitate the introduction of human flow into the base, but also reasonably divides the land into three blocks: one is the super high-rise office building, while the other two are serviced apartment buildings with an atrium.

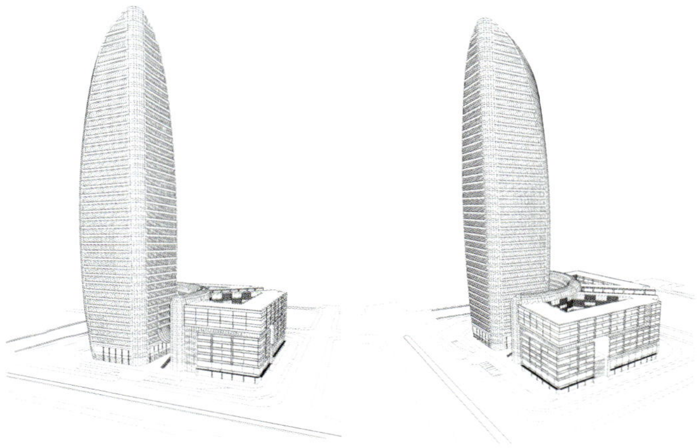

重庆龙湖春森彼岸

工程地址：重庆
开发公司：重庆市龙湖地产
设计时间：2002—2010 年

春森彼岸坐落于长江沿岸，规模达到 60 万平方米，其中包含大型高层建筑、小高层住宅社区、超高层酒店式公寓及办公楼、星级酒店、爬坡商业街和大型滨江商业综合体。

春森彼岸的地形高差极大，南北长达 2 千米，有很长的沿江面。方案充分利用这一特点，制造出了优美起伏的天际线，立面"乱"中取胜，既有统一效果，又诠释出重庆临山临水的城市特点。

Longhu Chunsen Land, Chongqing

Construction Location: Chongqing
Developer: Chongqing Longfor Real Estate
Design Time: 2002 - 2010

Chunsen Land is located along the Yangtze River with the project scale of 600,000 m^2. It includes the large-scale high-rise and small high-rise residential communities, super high-rise serviced apartments and office buildings, star-level hotels, slope-climbing commercial streets and large-scale riverside commercial complex.

The altitude difference of Chunsen Land is very great and the distance from north to south is about 2 km with a long shore line. Taking full advantage of this feature, the program creates a beautiful skyline of fluctuation and projects, the facade from the "chaos", producing a uniform effect while interpreting Chongqing's urban characteristic of mountain & water adjacency.

重庆龙湖观山水

工程地址：重庆
建筑规模：200 000 平方米
开发公司：重庆龙湖地产
设计时间：2005 年
（注：本项目知识产权归柴晟个人所有）

Longfor Landscape, Chongqing

Construction Location: Chongqing
Construction Scale: 200,000 m^2
Developer: Chongqing Longfor Real Estate
Design Time: 2005

重庆龙湖大城小院

工程地址：重庆
开发公司：重庆市龙湖地产
设计时间：2006 年

这个规模达 8 万平方米的住宅社区以其特殊的户型设计——上部花园洋房 + 底部迷你别墅——使我们获得了国家颁发的住宅设计户型专利奖。

（注：本项目知识产权归柴晟个人所有）

Longfor Dachengxiaoyuan Community, Chongqing

Construction Location: Chongqing
Developer: Chongqing Longfor Real Estate
Design Time: 2006

This residential community whose project scale is 80,000 m^2 has been awarded the national patent of residential dwelling design for its uniqueness—the upper garden house + the bottom mini villa.

重庆龙湖睿城

工程地址：重庆
开发公司：重庆市龙湖地产
设计时间：2007 年

睿城是一个采用现代手段演绎的以川西民居风格为设计主题的院落式合院别墅，每个合院形成一个具有“新都市主义”韵味的邻里单元。叠式别墅与联排别墅的组合方式，使建筑群落具有民居群落的起伏感。

（注：本项目知识产权归柴晟个人所有）

Longfor Intelligent City, Chongqing

Construction Location: Chongqing
Developer: Chongqing Longhu Real Estate
Design Time: 2007

Intelligent City is a courtyard-villa community of the compound style which takes the modern interpretation of Western Sichuan residential community as the design theme. Each courtyard forms a neighborhood unit with the flavor of "new urbanism" and the combination of stacked villa and the joint-row villa imparts the construction group a sense of ups and downs of the residential community.

上海龙湖蓝湖郡

工程地址：上海
开发公司：上海龙湖地产
设计时间：2008 年

蓝湖郡采用的十二合院设计，混入了华东地区的地域特色——里弄式住宅——形成了一个“里弄”+“邻里中心”的居住模式。每个合院拥有四条独立连接三栋别墅的里弄，其所并联的十二栋别墅可共享比它们第一层的中心庭院——“邻里中心”。

（注：本项目知识产权归柴晟个人所有）

Longfor•Blue Lake County, Shanghai

Construction Location: Shanghai
Developer: Shanghai Longfor Real Estate
Design Time: 2008

Blue Lake County adopts the design of twelve courtyards. Combined with the geographical characteristics of East China—terrace house, the design produces a residential mode: "lane" + "neighborhood center". Each courtyard has four lanes which independently links three villas, and the twelve villas of parallel connection share the undergrourd central courtyard which sinks one floor underground— “neighborhood center".

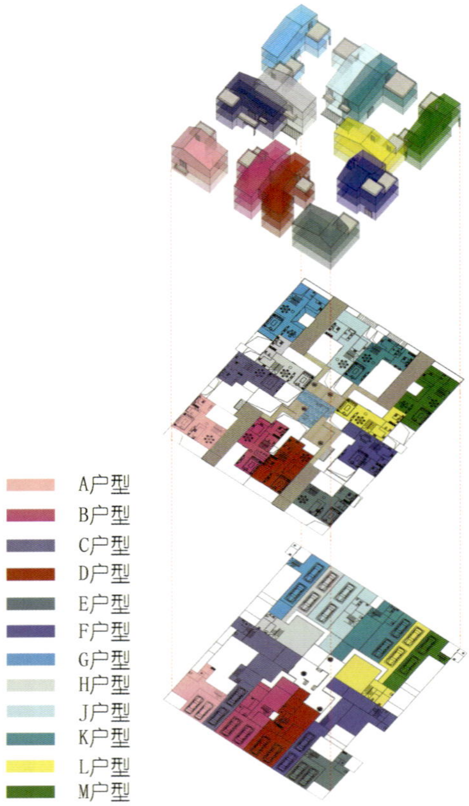

重庆龙湖悠山郡

工程地址：重庆
开发公司：重庆龙湖地产
设计时间：2007 年

悠山郡是我们一系列合院式别墅研究的开始。我们的研究课题——“以标准化的生产模式”产生“非标准化的视觉效果”——在这个项目中有着充分的体现。所以在这个项目里，合院并非主题，而是通过一个标准的“合院组合”，再经过拼接、翻转、生长、形成的具有“非标准化视觉效果”的别墅社区才，这是我们的中心议题。

（注：本项目知识产权归柴晟、许崹所有）

Longfor Youshan County, Chongqing

Construction Location: Chongqing
Developer: Chongqing Longfor Real Estate
Design Time: 2007

Youshan County is the beginning of our systemic research on courtyard villa. Our study subject—to produce non-standardized visual effect by standardized mode of production—has been fully manifested in this project. Therefore in this project design, the courtyard is not our theme. The core issue is the villa community of “non-standardized visual effect” formed through a standard “courtyard combination” and then through the connection, reversal and development.

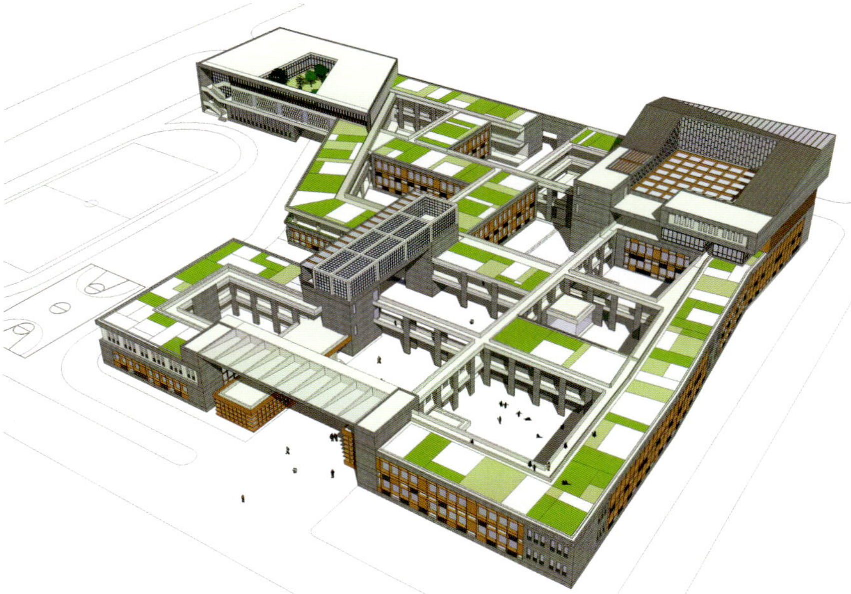

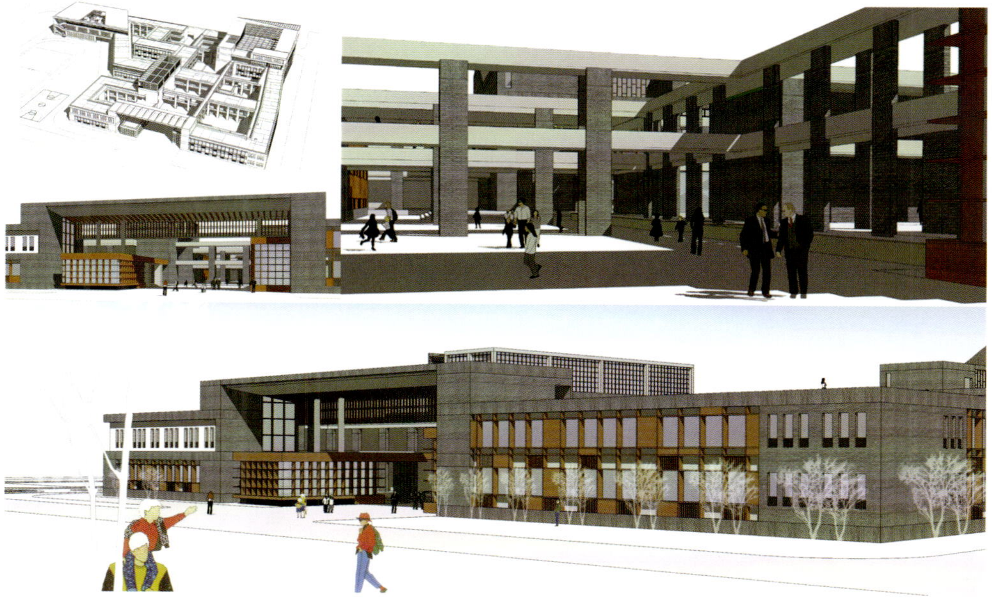

东莞建设小学

工程地址：广东 东莞
开发公司：东莞市政府教育局
设计时间：2006 年

东莞建设小学是由于设计者对之前其身处的教学环境的厌恶感而催生的设计。那个时代的学校建筑一律是五层高、兵营般冰冷的房子，操场和地面的活动场地与教学楼相距遥远。课间十分钟时，学生们只能拥挤在狭小的走廊里……

因此东莞建设小学将有限的地面空间放置于教学区中，形成六个年级各自独立的院子。院子被两层的教学楼所围合，尺度亲切。每个院子内独立的教师办公及交通系统可确保孩子活动的安全与教学的方便。

建筑的屋顶采用绿覆设计。

建筑采用灰、红、白、黑四种东莞民居的典型颜色设计。力图使其具有地域性。

（注：本项目知识产权归柴晟个人所有）

Dongguan Construction Primary School

Construction Location: Dongguan, guangdong

Developer: Education Bureau of Dongguan Municipal Government

Design Time: 2006

The design of Dongguan Construction Primary School is derived from the designers' abhorrence to the former teaching environment. All school buildings of that time are five-storey, barrack-like cold houses; the playground and activity field are so far from the teaching building that during the10 minutes' break, the students are crowded in the narrow corridor ...

Dongguan Construction Primary School places the limited land space into the teaching area, forming a separate yard for each of the six grades. The yard is enclosed by the two-storey teaching building and the scale is moderate. The teachers' office area and the transport system at each yard can ensure the safety of children and the convenience for teaching.

The roof of the building is covered by green.

The construction is designed with four colors of Dongguan civil house: gray, red, white and black, attempting to produce an effect of regional characteristics. (Note: The intellectual property belongs to Sheng Chai.)

江与城天主教堂

工程地址：重庆
开发公司：香港置地 & 龙湖地产
设计时间：2008 年

江与城教堂的设计取材于“摩西带领犹太人出埃及记到摩西与上帝以刻在石头上的‘摩西十诫’为盟到最后寻找到圣地伽兰”的宗教故事。因此，石头、分开的海水在方案中起到象征的意义。从而创造出的宗教的序列感和圣洁感，这也才是建筑师的立意所在。

（注：本项目知识产权归柴晟个人所有）

River & City Catholic Church

Construction Location: Chongqing

Developer: Hong Kong Property & Longfor Real Estate

Design Time: 2008

The design of River & City Catholic Church draw its idea from the religious story about Moses: the Exodus of the Jews, the Ten Commandments on the stone with God and the final arrival at the Holy Land Jialan. Therefore, the stone and separate seawater have symbolic significance in the program. However, the religious sequence and holiness created in the design is the architects' true intention. (Note: The intellectual property belongs to Sheng Chai.)

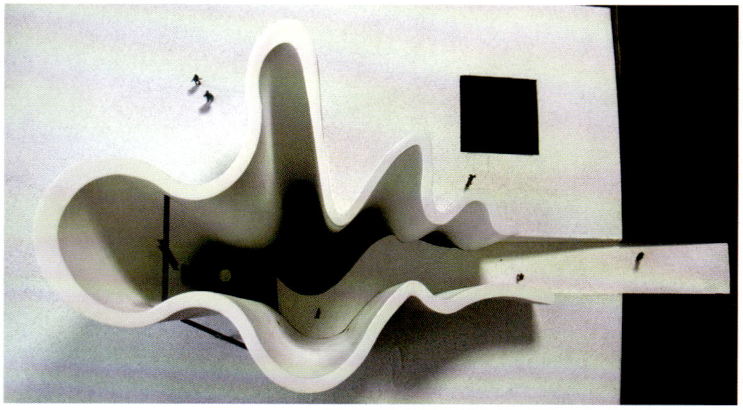

深圳协建建筑设计有限公司

Shenzhen Introduction of Xiejian Architectural Design Co., Ltd.

地址：深圳市福田区益田路江苏大厦B座901-906室
电话：+86-755-82960669
传真：+86-755-82960960
邮箱：szxj@china.com
szxj168@163.com

Address:Room 901-906, block B, Jiang su building, Yitian road, Futian District, Shenzhen
Tel: +86-755-82960669
Fax: +86-755-82960960
E-mail: szxj@china.com
szxj168@163.com

深圳市协建建筑设计有限公司始建于 2000 年，由多名经验丰富且长期从事建筑设计工作的注册建筑师组成。业务范围包括建筑设计、园林设计、项目策划及管理等。公司实行人才资本经营的原则，招贤引才、海纳百川，构成了坚实的设计及管理技术力量。在设计中，一贯注重以理论为基础，所有项目均是通过与业主共同对场地深入研究、反复比较，对不同的城市背景作出的不同解析。

公司成立十年来，伴随着深圳特区的发展而不断成长，公司作品遍及深圳、中山、重庆、成都、西安、长沙、苏州、扬州、张家港、武汉等多个城市。主要作品有深圳尚都、深圳熙龙湾、深圳熙园山院、深圳西岸观邸、深圳家和盛世、西安水岸枫桥、重庆云湖天都、重庆可乐小镇、张家港皇家首座等大型综合社区设计项目。

公司以不断的沟通、周全的策划、有效的协调、精湛的技术、优质的服务为支柱，以人才管理为中心，实行经济效益、社会效益、设计责任直接同设计者结合的经理负责制，其特点为在确保质量前提下的高效、高速。

"勤恳务实、追求卓越"是我们的一贯宗旨，我们希望这种追求完美的精神能够成为我们所有员工的动力，不断地去创造更加美好的生活。

Founded in 2000, Shenzhen Xiejian Architectural Design Co., Ltd. assembles a number of experienced registered architects who have been engaged in the architectural design work for a long time. Our service include architectural design, landscape design, project planning and management, etc. Implementing the principle of human capital management, our company constitutes a stable design and management strength by inviting and absorbing talents. In the design, we emphasize the theoretical foundation and make sure that all the projects are in-depth study and repeated comparison along with the owners, offering different analysis to different city backgrounds.

During the ten years since its foundation, our company grows along with the development of Shenzhen Special Economic Zone and its design has reached many cities in China, such as Shenzhen, Zhongshan, Chongqing, Chengdu, Xi'an, Changsha, Suzhou, Yangzhou, Zhangjiagang and Wuhan, etc. Its major designs include: Shangdu, Shenzhen; Xilongwan, Shenzhen; Xiyuan Shanyuan, Shenzhen; Xi'anguandi, Shenzhen; Jiaheshengshi, Shenzhen; Maple Waterfront, Xi'an; Yunhutiandu, Chongqing; Cola Town, Chongqing; Royal's First, Zhangjiagang and other large-scale Synthesis community design projects.

Our company takes continually communication, sound planning, effective coordination, superb technology and excellent service as foundation pillars and the talent management as its core. It also adopts the manager responsibility system which directly combines the designers with the economic benefit, social benefit and the design responsibility, achieving high-performance, high-speed under the premise of quality assurance.

Our consistent aim is "Diligent, pragmatic and constant pursuit of excellence". We hope that this spirit of striving for perfection will become the driving force of all our employees, encouraging them to create a better life.

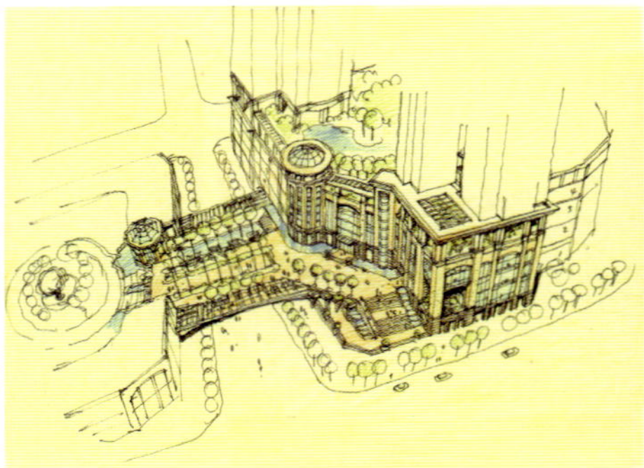

深圳尚都

项目业主：深圳市鸿荣源房地产开发有限公司
建设地点：广东 深圳
用地面积：南区：19 000 平方米　北区：38 000 平方米
建筑面积：南区：110 000 平方米　北区：165 000 平方米
设计时间：2003 年 10 月 –2004 年 8 月
项目状况：已建

Shangdu Community, Shenzhen

Project Owner: Hongrongyuan Real Estate Development Co., Ltd., Shenzhen
Construction Location: Shenzhen, Guangdong
Site Area: South District: 19,000 m^2　North District: 38,000 m^2
Building Area: South District: 110,000 m^2　North District: 165,000 m^2
Design Time: 2003.10-2004.08
Project Status: Completed

深圳北商务中心

项目业主：熙园置业深圳有限公司
建设地点：广东 深圳
用地面积：550 000 平方米
建筑面积：2 050 000 平方米
设计时间：2004 年 1 月 –2005 年 6 月

Northen Shenzhen Business Center

Project Owner: Shenzhen Co., Ltd. of Xiyuan Property
Construction Location: Shenzhen, Guangdong
Site Area: 550,000 m^2
Building Area: 2,050,000 m^2
Design Time: 2004.01-2005.06

重庆茶园子别墅

项目业主：重庆光华控股集团有限公司
建设地点：重庆
用地面积：500 000 平方米
建筑面积：328 000 平方米
设计时间：2009 年 –2010 年 3 月

Chayuanzi Villa, Chongqing

Project Owner: Guanghua Holding Group Co., Ltd., Chongqing
Construction Location: Chongqing
Site Area: 500,000 m^2
Building Area: 328,000 m^2
Design Time: 2009-2010.03

深圳家和盛世

项目业主：深圳市宇宏投资集团有限公司
建设地点：广东 深圳
用地面积：46 000 平方米
建筑面积：164 000 平方米
设计时间：2007 年 4 月
项目状况：已建

Jiaheshengshi Community, Shenzhen

Project Owners: Yuhong Investment Group Co., Ltd., Shenzhen
Construction Location: Shenzhen, Guangdong
Site Area: 46,000 m^2
Building Area: 164,000 m^2
Design Time: 2007.04
Project Status: Completed

深圳熙园山院

项目业主：鸿荣源置业集团深圳有限公司
建设地点：广东 深圳
用地面积：68 000 平方米
建筑面积：106 000 平方米
设计时间：2008 年 10 月 –2009 年 3 月

Xiyuan Shanyuan Community, Shenzhen

Project Owners: Shenzhen Co., Ltd. of Hongrongyuan Real Estate Group
Construction Location: Shenzhen, Guangdong
Site Area: 68,000 m^2
Building Area: 106,000 m^2
Design Time: 2008.10-2009.03

深圳熙龙湾

项目业主：深圳市鸿荣源房地产开发有限公司
建设地点：广东 深圳
用地面积：117 000 平方米
建筑面积：475 000 平方米
设计时间：2005 年 3 月

Xilongwan Community, Shenzhen

Project Owner: Hongrongyuan Real Estate Development Co., Ltd., Shenzhen
Construction Location: Shenzhen, Guangdong
Site Area: 117,000 m^2
Building Area: 475,000 m^2
Design Time: 2005.03

深圳西岸观邸

项目业主：深圳市鸿荣源房地产开发有限公司
建设地点：广东 深圳
用地面积：28 000 平方米
建筑面积：117 000 平方米
设计时间：2005 年 6 月
项目状况：已建

Xi'anguangdi Community, Shenzhen

Project Owner: Hongrongyuan Real Estate Development Co., Ltd., Shenzhen
Construction Location: Shenzhen, Guangdong
Site Area: 28,000 m^2
Building Area: 117,000 m^2
Design Time: 2005.06
Project Status: Completed

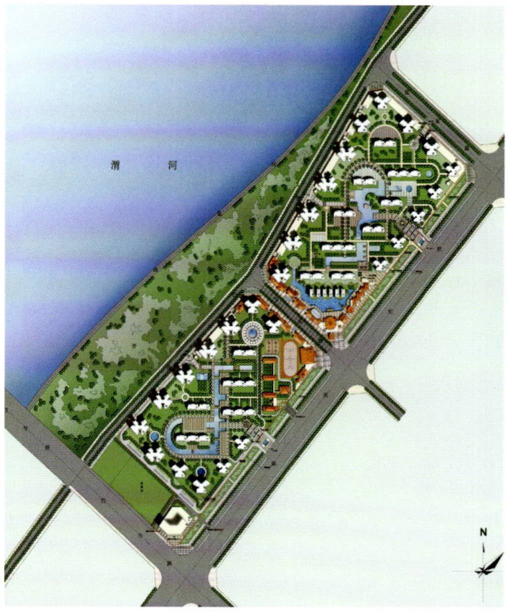

西安枫桥水岸

项目业主：咸阳宇宏房地产发展有限公司
建设地点：西安 咸阳
用地面积：21 000 平方米
建筑面积：780 000 平方米
设计时间：2008 年 8 月

Maple Waterfront, Xi'an

Project Owner: Yuhong Real Estate Development Co.,Ltd., Xianyang
Construction Location: Xianyang, Xi'an
Site Area: 21,000 m^2
Building Area: 780,000 m^2
Design Time: 2008.08

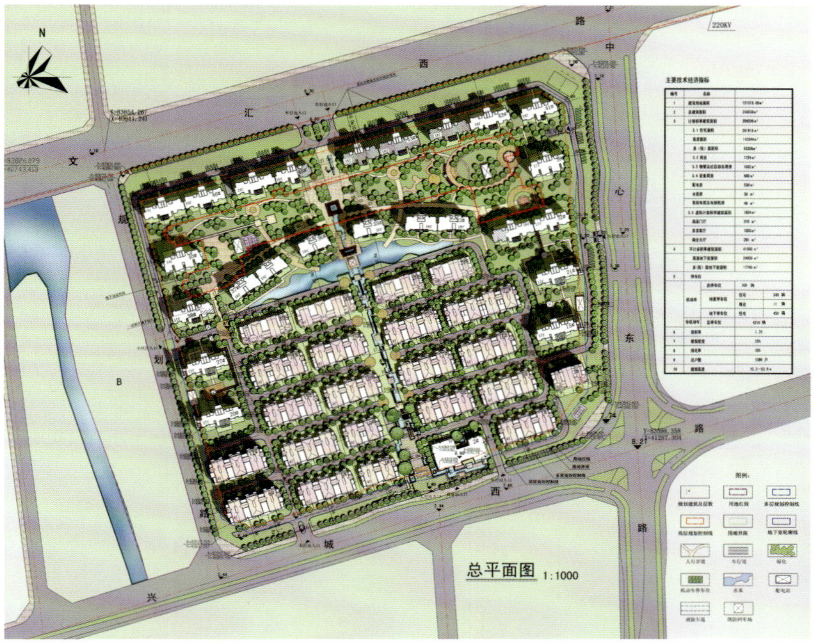

扬州 340 地块规划方案

项目业主：扬州天俊置业有限公司
建设地点：江苏 扬州
用地面积：121 100 平方米
建筑面积：247 600 平方米
设计时间：2009 年 10 月

Planning for Block 340, Yangzhou

Project Owner: Tianjun Property Co., Ltd., Yangzhou
Construction Location: Yangzhou, Jiangsu
Site Area: 121,100 m^2
Building Area: 247,600 m^2
Design Time: 2009.10

深圳依山郡

项目业主：深圳招商房地产有限公司
建设地点：广东 深圳
用地面积：150 000 平方米
建筑面积：300 000 平方米
设计时间：2005 年 3 月

Yishanjun Community, Shenzhen

Project Owner: Shenzhen Investment Real Estate Co., Ltd.
Construction Location: Shenzhen, Guangdong
Site Area: 150,000 m^2
Building Area: 300,000 m^2
Design Time: 2005.03

重庆可乐小镇

项目业主：重庆光华控股集团有限公司
建设地点：重庆 江北大石坝
用地面积：95 000 平方米
建筑面积：539 000 平方米
设计时间：2007 年 11—2008 年 6 月

Cola Town, Chongqing

Project Owner: Guanghua Holding Group Co., Ltd., Chongqing
Construction Location: Dashiba, Jiangbei District, Chongqing
Site Area: 95,000 m^2
Building Area: 539,000 m^2
Design Time: 2007.11-2008.06

重庆云湖天都

项目业主：重庆佳乐九龙房地产开发有限公司
建设地点：重庆
用地面积：175 000 平方米
建筑面积：320 000 平方米
设计时间：2003 年 —2005 年 8 月

Yunhutiandu Community, Chongqing

Project Owner: Jialejiulong Real Estate Development Co., Ltd., Chongqing
Construction Location: Chongqing
Site Area: 175,000 m^2
Building Area: 320,000 m^2
Design Time: 2003-2005.08

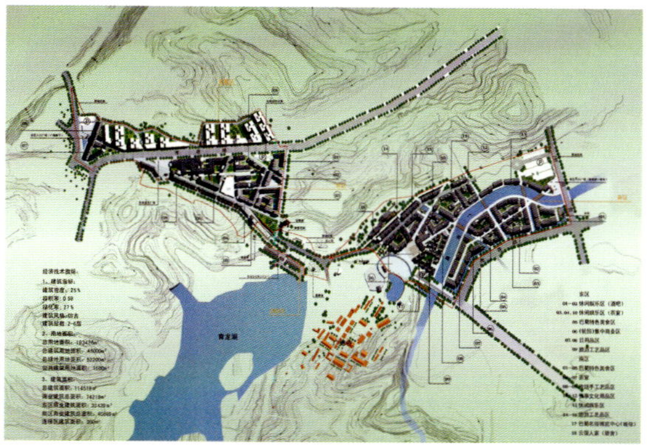

四川广德旅游商业文化街

项目业主：四川遂宏天合旅游发展有限公司
建设地点：四川 遂宁
用地面积：东区：57 000 平方米　南区：99 000 平方米
建筑面积：东区：66 000 平方米　南区：34 000 平方米
设计时间：2004 年 11 月

Guangde Tourism Business Cultural Street, Sichuan

Project Owner: Suihongtianhe Tourism Development Co., Ltd., Sichuan
Construction Location: Suining, Sichuan
Site Area: East District: 57,000 m^2　South District: 99,000 m^2
Building Area: East District: 66,000 m^2　South District: 34,000 m^2
Design Time: 2004.11

重庆合川中心区规划方案

项目业主：重庆兆甲房地产开发有限公司
建设地点：重庆 合川
用地面积：374 100 平方米
建筑面积：1 673 400 平方米
设计时间：2010 年—至今

Planning of Hechuan Central Area, Chongqing

Project Owner: Zhaojia Real Estate Development Co., Ltd., Chongqing
Construction Location: Hechuan, Chongqing
Site Area: 374,100 m^2
Building Area: 1,673,400 m^2
Design Time: 2010 - present

大唐世纪建筑设计事务所
DatangCN Architecture Design Institute

深圳市大唐世纪建筑设计事务所是经国家建设部批准具有建筑设计甲级资质，从事规划、建筑、景观及工程设计和咨询的专业设计公司。作为一支异军突起的建筑设计行业的生力军，公司发展迅速，成果丰硕。公司按国际化模式组建，坚持以研究型、精品化的设计实践作为发展方向，积极拓展与著名设计机构及地产商的合作关系。业务范围包括公共建筑、商业建筑、文教建筑、旅游建筑、居住建筑、景观设计等。

大唐世纪广纳国际及国内高级专业人才，技术力量雄厚，其主要成员已先后承担建筑设计任务百余项，项目包括大型办公商业建筑、居住区、工业园区、物流园区等。特别是在法院建筑和旧城改造项目方面有丰富经验。

我们深谙房地产策划、开发、设计等方面的实践知识和经验，并对房地产运作程序与规律十分熟悉，从而具有较高的产品策划与市场定位能力。事务所倡导与开发商合作发展的经营理念，注重与开发商建立和发展伙伴关系。创新设计让我们相识；优质服务让我们相知；良好的社会与经济效益让我们共同发展。我们丰富的专业经验使我们能充分理解项目的可操作性并确保项目的经济性，为客户创造最大的社会与经济价值。

我们倡导“专业－敬业，作品－精品”的核心价值观和专心创作、精益求精，共同实现“超越期待，设计未来”的企业愿景。我们坚持理论与实践经验、创意与技术、个性与团队、专业与市场合理结合，注重建筑的自身及城市的属性，注重建筑设计中的方法论，并力争凝聚每一个成员的能力，共同营造一个学习、交流、实践、成长的广阔空间。

我们强调企业流程控制和质量管理，注意方案与施工过程中无逢式衔接，使创意在工程中得以完美实现。同时我们还有着强烈的成本控制意识，以及一整套对建筑工程细部节点与材料深入设计的流程。凭借科学的管理体制，创新的设计理念及优质的后期服务，公司业务不断拓展。目前，大唐世纪的足迹已遍布广东、北京、天津、陕西、甘肃、河南、河北、湖南、湖北等省市。

建筑设计是用尽可能完美的手段去解决项目的需求和问题的过程，基于这一立场，大唐世纪建筑事务所推行“重视方案，珍惜施工图”的设计理念，公司将始终如一地尽可能全面、合理地认知每一个项目的属性，专业而又合理的完成项目的设计工作，用每个成果去表述自我服务社会。

翠竹金翠园

建设地点：深圳
建筑面积：74 856.32 平方
设计时间：2008 年 2 月

Construction Location: Shen
Building Area: 74,856.32 m²
Design Time: 2008.2

深圳市公安局龙岗分局三所三队一中心

建设地点：深圳
建筑面积：110 126 平方米
设计时间：2009 年 6 月

Construction Location: Shenzhen
Building Area: 110,126 m²
Design Time: 2009.6

佳兆业沈阳超高层

建设地点：沈阳
建筑面积：210 000 平方米
设计时间：2009 年 12 月

Construction Location: Shenyang
Building Area: 210,000 m²
Design Time: 2009.12

尚品花园

建设地点：深圳
建筑面积：217 973 平方米
设计时间：2007 年 8 月

Construction Location: Shenzhen
Building Area: 217,973 m²
Design Time: 2007.8

大亚湾经济技术开发区综合办公区及石化大厦

建设地点：大亚湾
建筑面积：126 084 平方米
设计时间：2009 年 12 月

Construction Location: Dayawan
Building Area: 126 084 m²
Design Time: 2009.12

大亚湾经济技术开发区科技创新产业园研发综合楼

建设地点：大亚湾
建筑面积：43 500 平方米
设计时间：2010 年 3 月

Construction Location: Dayawan
Building Area: 43,500 m^2
Design Time: 2010.3

深房集团光明项目

建设地点：深圳
建筑面积：181 470 平方米
设计时间：2010 年 4 月

Construction Location: Shenzhen
Building Area: 181,470 m^2
Design Time: 2010.4

烟台东海金域蓝湾住宅小区 B 区

建设地点：烟台
建筑面积：292 679 平方米
设计时间：2010 年 6 月

Construction Location: Yantai
Building Area: 292,679 m^2
Design Time: 2010.6

专业力致敬业　作品力致精品

公司地址：深圳市罗湖区爱国路金通大厦 B 座 14 层
邮编：518003
电话：+86–755–25541187
传真：+86–755–25545548
邮箱：datangcn@126.com
网址：www.datangcn.com.cn
甲级证书号：A144007485

POF 华域普风 深圳市华域普风设计有限公司

Shenzhen Pofart Design Consulting Co., Ltd.

地址：深圳市南山区海德三道海岸城东座 1306，1308，1309
法人：梅坚
电话：+86-755-86290985
传真：+86-755-86290409
邮箱：pofsz@126.com
网址：www.pofart.com.cn

Address: Room 1306, 1308, 1309 of East Block of Coast Building, No.3 Haide Road, Nanshan District, Shenzhen, PRC
Legal Representative: Jian Mei
Tel: +86-755-86290985
Fax: +86-755-86290409
E-mail: pofsz@126.com
Http:// www.pofart.com.cn

深圳市华域普风设计有限公司是近年活跃于深圳的建筑与城市规划设计公司。公司主要合伙人及国外主创人员具有建筑专业留学及工作背景，并拥有在专业领域内多年的设计实践经验。

普风致力于探讨什么是中国所需要的现代建筑，并始终坚持以当代的建筑学思想及朴实的工作态度营造现代中国人的生活环境。在各项建筑活动中，华域普风擅长方案设计尤其是建筑细部设计的控制，从方案创意到现场实施，华域普风均能提供优质服务。

公司具有与政府部门和一线城市运营开发商的合作经验，其近期作品分布于深圳、成都、南京、珠海、佛山、东莞、贵阳、长沙、合肥、海口、三亚、博鳌、绵阳等二十几个城市。

普风始终不断为客户提供跻身于一线品牌的产品设计。公司为客户所提供的建筑作品，不仅取得了良好的经济效益，同时也为客户创造了卓越的品牌社会效益。此外，普风还凭借其团队成员在业内十多年的项目操作经验，坚持将与客户共同确定设计条件作为设计的重要阶段，并以之作为普风实践的根本引导。

Shenzhen Pofart Design Consulting Co., Ltd. is a well-known design company of architectural and urban planning in Shenzhen. Main Leaders and partners of this company are professional and experienced staffs with oversea education and work experience fo architecture.
Pofart dedicated itself to exploring what kind of modern architecture is needed in China. Meanwhile, it insists that modern Chinese living environment be created in line with contemporary architecture philosophy and through down-to earth working attitude. Pofart provides Cexcellent services in various stages of construction activities, from making plans and proposals to putting them into practice. This company boasts conceptual designs, especially architectural detail designs.
Pofart has rich experience in cooperating with government departments and key urban operators. Its recent principal works of architecture are distributed in over twenty cities like Shenzhen, Chengdu, Nanjing, Zhuhai, Foshan, Dongguan, Guiyang, Changsha, Hefei, Haikou, Sanya, Boao, and Mianyang.
Pofart has constantly supplied customers with well-known architectural designs. These works of architecture bring about preferable benefits not only economically but also socially. In addition, through practical experience over a decade in this field, this company has realized that it is a vital phase for design to work together with clients to determine design conditions. Respecting the positions of customers is a guideline for Pofart in practice.

合肥华润置地澜溪镇

用地面积：2 794 平方米
建筑面积：9 197 平方米
建设地点：安徽 合肥
项目用途：住宅用地
开 发 商：华润置地（合肥）有限公司

项目概况

项目位于合肥市高新开发区，周边交通便捷，自然景观资源丰富。建筑整体构思用退台的概念，与山体的自然曲线完美呼应，创造出自然流畅的建筑群衔接面，并提供了面向山水绝好风景的空中平台。

项目采用平层户型与跃层双面户型相结合的布局方式，每户均有大尺度的观景阳台，跃层户型提供屋顶露台，将景观资源最大化利用，体现出豪宅品质，让每位住户从进入建筑的那一刻开始，就有了与自然环境亲密接触的体验。

项目采用现代主义手法，简约大气、高雅时尚。建筑运用了金属板、金属网、钢化清玻等现代化材质，同时将屋顶绿化等自然元素融入其中，体现工业化与自然结合的精神。

Hefei French Annecy of China Resources Land

Site Area: 2,794 m²
Building Area: 9,197 m²
Construction Location: Hefei, Anhui
Project Purpose: Residential Area
Developer: China Resources Land(Hefei) Co., Ltd

Project introduction

The project locates at the High-tech Development Zone of Hefei, around which the transportation is convenient and the natural landscape resources are rich. The integral conception of the architecture applies the concept of set-back floors, perfectly echoing with the natural curve of the massif, creating natural and fluent architectural complex cohesion, and providing an aerial platform facing the excellent landscape of the mountains and water.
The project applies the strategy of combining leveling unit with duplex level. Each unit possesses a large-scale viewing balcony, and the duplex unit provides the roof terrace, in order to maximize the utilization of the landscape resources and to represent the quality of villa, which makes every resident get the experience of intimately contacting the natural environment as soon as he steps into the architecture.
The project utilizes the modernism technique, thus it is concise and liberal, elegant and fashionable. The architecture adopts modernized materials such as metal plate, metal mesh, tempered clear glass etc., and meanwhile integrates natural elements like roof greening, representing the spirit of combining industrialization with the nature.

绵阳长虹仙海湖创新研发园

用地面积：189 292.19 平方米
建筑面积：113 571.49 平方米
建设地点：四川 绵阳
项目用途：住宅用地
开 发 商：四川长虹电子集团有限公司

设计思路

项目地块沿仙海湖展开，内部地形起伏，景观条件优越。建筑总体布局顺应地形地貌，不同类型的住宅布置在形似3个C形的山脊和山坳上，形成一系列尺度合理、富有韵律的住区组团。建筑形式从传统院落形式中捕捉灵感，追求自然、朴实、简约、纯粹的风格。外墙以简单的白色基调为主，局部设计为木色和灰色的墙面。一栋栋尺度宜人的建筑隐身于自然的树林之中，充分展现当代东方人居的魅力。

Mianyang Changhong Xianhai Lake Research & Innovation Park

Site Area: 189,292. 19 m²
Building Area: 113,571. 49 m²
Construetion Location: Mianyang, Sichuan
Project Purpose: Residential Area
Developer: Sichuan Changhong Electronics Group Co., Ltd.

Design Idea

This project is developed in light of the landform of Xianhai Lake, featuring beautiful natural scene. General arrangement of architecture complex conforms to the topography there. Different housing types are arranged on three ridges and level grounds of mountains like shape of C, presenting an appearance of a series of well-structured dwelling zones. The design is inspired by the traditional courtyard which is characterized with being natural, compact, simple and pure. White is the predominant color of the outer wall, and there are wood and grey walls in local design. Buildings hiding in the woods display oriental charming residential areas .

成都华润置地凤凰城销售中心

用地面积：530 平方米
建筑面积：215 平方米
建设地点：四川 成都
项目用途：公共建筑
开 发 商：华润置地（成都）有限公司

设计思路

项目将原有的小书店进行改造。设计上保留其原有的墙体和结构，对原建筑的外部附以新的设计，外立面具有其独立的结构基础。设计运用现代的手法及理念巧妙地用横向及竖向的钢构架构成了富于虚实变化的立面形式。钢件结合花槽设计不但具有十足的现代感，也使得小建筑本身成为周围活动场所的一个景观点，吸引了周围人的眼球，这一点对于开发商来说有一定的现实意义。

Chengdu China Resources Phoenix City Sales Center

Site Area: 530 m²
Building Area : 215 m²
Construetion Location: Chengdu, Sichuan
Project Purpose: Public Building
Developer: China Resources Land (Chengdu) Co., Ltd.

Design Idea

This sales center is transformed from a small book store whose original wall and the structure are retained. Exterior of the original building is on new wrappers. Interlocking transversal and vertical steel structures constitute changeable and dimensional facade. The design of steel chamfer not only has the sense of contemporary, but also makes the sales center a scenic spot of the surrounding. This construction catches people's eyes and also brings about benefits to the developer.

成都华润置地凤凰城 C 区

用地面积：40 428.72 平方米
建筑面积：225 191.36 平方米
建设地点：四川 成都
项目用途：住宅用地
开 发 商：华润置地（成都）有限公司

设计思路

本案采用围合式的规划布局，在充分利用基地周边城市景观的同时，营造大尺度的小区景观庭院。住区内 85% 以上的住户可获得城市公园景观或小区中心花园景观，满足了资源利用最大化与住宅均好的要求。在基地东部临城市滨水公园处，布置十字形点式高层住宅，与相邻地块形成差异化的建筑肌理和可识别的城市形象。

立面设计采用亲和、适度、现代的建筑语言，营造独具特质的住区形象，根据住宅体量的划分设计，将不同的建筑体块分别以横向及竖向立面元素进行设计，形成简明、清晰、强烈的视觉效果。建筑外立面色彩以灰色、白色和赭黄色为主，材料以面砖、金属、玻璃为主，底层部分采用石材，以体现所需具备的质感。

Block C of Chengdu China Resources Land Phoenix City

Site Area: 40,428. 72 m²
Building Area: 225,191. 36 m²
Construetion Location: Chengdu, Sichuan
Project Purpose: Residential Area
Developer: China Resources Land(Chengdu) Co., Ltd.

Design Idea

The encircling style of layout is applied in this case. On basis of fully employing surrounding city landscapes, the design creates large-scale scenic courtyard. The construction makes the best use of landscaps. Over 85% of residents can enjoy landscapes of city parks or landscape of the central park in the neighborhood in their houses. In the east of this area which is next to the city waterfront park, the arrangements of crossing spot tall building, distinction of the textures between buildings and its adjacent lots as well as distinctive city landmarks are elaborately made.

The design of façade is attractive, moderate and modern, constructing an idiosyncratic residential quarter image. Traversal and vertical facades consist different buildings, assuming an aspect of being simple, clear and powerful. Grey, white and ochraceous are main colors of facades. Principle construction materials are face bricks, metals and glass, while the bottom layer is made up of dimension stone to highlight the high quality.

海南博鳌幸福海

用地面积：101 077.77 平方米
建筑面积：92 580.00 平方米
建设地点：海南 琼海
项目用途：住宅用地
开 发 商：幸福联盟投资有限公司

设计思路

项目总体布局顺应城市规划结构，注重海岸、社区及小镇之间的形态关系。地块南端规划为酒店，酒店建筑以感性的形态特征隐喻《博鳌由来》故事中的莲花，并将此主题以"莲花酒店"、"莲花公园"、"莲花广场"、"莲花水池"等建筑景观要素为载体加以充分演绎。

销售中心结合滨海公园统一设计，前期开发力求形成海岛度假特征鲜明并极具视觉冲击力的卖场形象。

Hainan Boao Happiness Sea

Floor Area: 101,077. 77 m²
Building Area: 92,580 m²
Construetion Location: Qionghai,Hainan
Project Purpose: Residential Area
Developer: Happiness Alliance Investment Co., Ltd.

Design Idea

The overall arrangement of this project complies with the urban planning, stressing the relations among the seashore, the community and the village in terms of patterns. In south of the area is the hotel, whose shape indicates lotus flowers in the story of Boao. Taking lotus as the theme of the landscape gardening, there are Lotus Hotel, Lotus Park, Lotus Square and Lotus Pool to illustrate the story.

The design of the sales center will integrate with the coastal park. The construction of sales center is completed prior to that of coastal park. The strong visual impact of sales center in the island resort improves the image of the whole area.

华润置地绵阳项目

用地面积：292 611.11 平方米
建筑面积：996 828.90 平方米
项目位置：四川 绵阳
项目用途：住宅用地
开 发 商：华润置地（绵阳）有限公司

设计思路

社区结构顺应城市规划发展轴线，由西往东沿商务商业中心、商业办公中心及西山绿色核心带发展该片区的形象风貌轴，统领整个片区的建筑空间及景观步行系统结构，营造大批量建设住宅的特定之美。

建筑形式追求现代、自然、朴实的风格，摒弃多余的装饰，强调设计感及品质感，反映建筑与城市总体景观的良好关系。销售卖场结合城市公园统一设计，形成完整、大气并极具视觉冲击力的城市形象。

China Resources Land Mianyang Project

Site Area: 292,611. 11 m²
Building Area: 966,828. 90 m²
Construetion Location: Mianyang, Sichuan
Project Purpose: Residential Area
Developer: China Resources Land(Mianyang) Co., Ltd.

Design Idea

The structure of this community conforms to development axes of urban planning. The townscapes of this area from east to west are commercial center, office buildings and green center of West Mountain. Axes, determining architectural space and walking landscapes networking system of the whole area, construct unique beauty of a large-scale residential quarter.

The form of the architecture gives a style of contemporary, nature and simplicity , discarding redundant decoration, focusing on the sense of designing, which quality, which express the good association of architecure and urban landscape. The design intergrated the sales site and the urban park, creating a complete and grand city image with great visual impact.

海口现代美居物流园

用地面积：486 665 平方米
建筑面积：844 688.8 平方米
项目位置：海南 海口
项目用途：商业、办公、仓储用地
开 发 商：海南现代科技集团

设计思路

本次规划着重合理高效的功能分区，对现有场地进行竖向设计，本着大景观的设计原则，将整个建筑群作为大地景观的一部分进行规划。建筑屋顶经过精心的绿化设计，并成为景观体系中的亮点，与外部自然景观、地面绿化一起，形成多层次的立体景观体系。

单体设计秉持模块化原则，现代简约，符合商业及仓储的经营租赁需求。

Haikou Modern Mercure Logistics Park

Site Area: 486,665 m²
Building Area: 844,688. 8 m²
Construetion Location: Haikou, Hainan
Project Purpose: Commercial, Office & Warehouse Land
Developer: Hainan Modern Technology Group

Design Ideas

This planning emphasizes highly-effective function district. In line with the large landscape's principle of design, the existing location is designed vertically and the entire architectural complex is planned as a part of the large landscape. The roof is a greening design which is a highlight in the landscape designing system. The exterior natural landscape, the afforested ground, as well as greening design of the roof form the multi-level and dimensional landscape system.

The design of single buildings upholds the modular principles, which are modern and brief meeting the rent demand for commercial and the warehousing management.

珠海恒隆御水岸

用地面积：7 394.12 平方米
建筑面积：49 815.98 平方米
建设地点：广东 珠海
项目用途：住宅用地
开 发 商：珠海经济特区恒隆房产开发有限公司

设计思路

本项目地块所处位置地理环境优越，西侧的前山水道为项目提供了高品质的水景资源，周边被多层及小高层住宅包围。

项目以最大化利用城市自然景观资源为设计思路，建筑以两栋塔楼错开布置，巧妙地利用建筑错位关系实现了户户可观江景的目标，并且满足了建筑间距及日照的要求。建筑形式力求体现“当代”“滨水”“高尚”的主题特征，建筑语言简洁而令人印象深刻。主色调以深暖灰色为底色，其表面悬浮灵动而有韵律的白色元素，给人轻盈、清新、雅致之感。

Zhuhai Hallon Yushui Shore

Site Area: 7,394. 12 m²
Building Area: 49,815. 98 m²
Construetion Location: Zhuhai, Guangdong
Project Purpose: Residential Area
Developer: Hallon (Zhuhai Special Economic Zone) Real Estate Development Co., Ltd.

Design Ideas

This project locates in a superior geographical position. The west side of Former Landscape Road provides beautiful waterscape, surrounded by multi-layered and medium housing.

The designing idea aims at making the best use of urban natural landscape. Two turrets are positioned staggered. The ingenious location makes it possible to view the river scenery and meets the request for spacing and lighting. The architectural form makes effort to manifest the theme of “the present age” “water front” “nobility”. The construction style is plain and impressive. The bottom color is dark warm gray, with white element on its surface. And the dominant tone is overall relaxed, fresh and elegant.

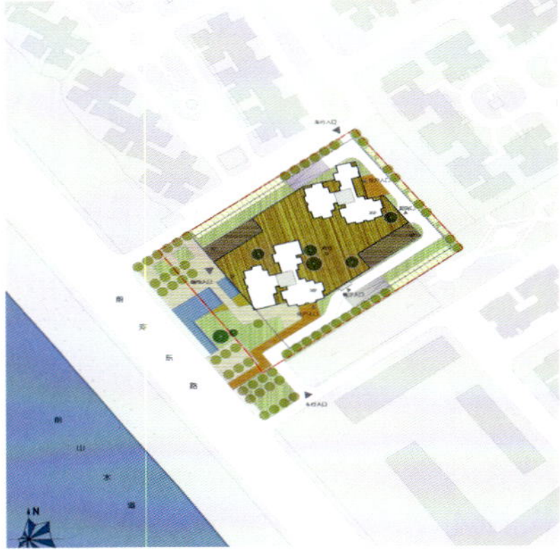

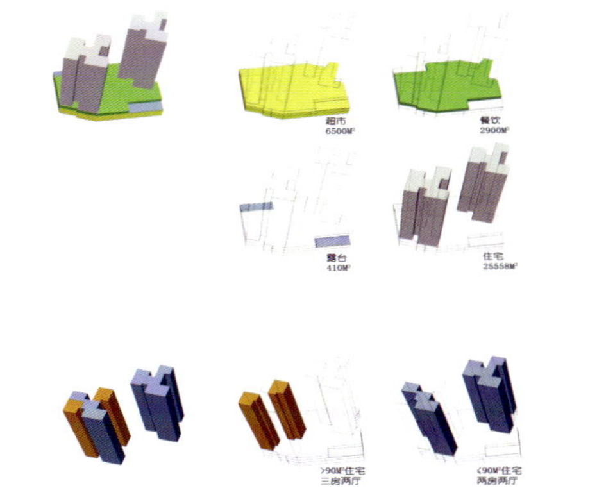

乐山嘉州新城

用地面积：61 970 平方米
建筑面积：107 740 平方米
建设地点：四川 乐山
项目用途：住宅用地
开 发 商：乐山新业置地发展有限公司

设计思路

从总体规划着手，通过对外立面颜色的控制，改变现有建筑给人的单调感，避免小区内部空间同质化，使住宅单体和组团均具备标志性，同时也利于宅间景观的塑造。

对部分可调整构件，如阳台、窗、空调机位等进行简化，将造型语汇梳理成系统，有利于形成住宅简约现代的风格。

Leshan Jiazhou Newtown

Site Area: 61,970 m²
Building Area: 107,740 m²
Construetion Location: Leshan, Sichuan
Project Purpose: Residential Area
Developer: Leshan Xinye Real Estate Development Co., Ltd.

Design Ideas

According to the overall plan, facade colors are controlled by which changes the existing feeling of monotony, avoids homogenization of the inner space, and makes both single buildings and the complex distinctive. In addition, it is good for landscape shaping.

Some adjustable components like the balcony, the window, the air conditioning space are simplified and objects in similar shapes are put into one system to form a modern brief housing style.

合肥天下锦城

用地面积：290 386 平方米
建筑面积：920 000 平方米
建设地点：安徽 合肥
项目用途：住宅用地
开 发 商：安徽皖投置业有限责任公司

设计思路

首期规划以超大中心花园及水系为主要景观轴线，营造出低密度居住社区的氛围和品质。设计将电梯洋房布置在内部景观资源最好的位置；将高层住宅散点式的布置在其余各期的外围，从而形成外部密集内部疏朗的格局，实现了景观资源最大化利用与住宅均好的要求。主会所布置在首期开发区内，并结合一期主入口的统一规划进行设计。

小区的商业设施规划结合底层架空停车，分别布置在小区各期城市道路沿街面。社区的每一组团均有独立出入口，形成各自独立的交通环线。

多层洋房色彩以黄，灰，褐三色为主要元素，给人以亲切自然的居住氛围。高层建筑在形式上力求体现现代、清新、简约、大气的主题特征，建筑语言简洁而令人印象深刻。

Hefei Glourious City Garden

Site Area: 290,386 m²
Building Area: 920,000 m²
Construetion Location: Hefei, Anhui
Project Purpose: Residential Area
Developer: Anhui Wantou Real Estate Co., Ltd.

Design Ideas

In the first phase of the design, the huge central garden and the river system consist the primary landscape axis, which build the atmosphere and the quality of the low density housing community. The elevator and foreign-style houses are arranged in best position of interior landscapes. High-rise buildings are dispersed in periphery of other phases. Thus, the pattern of extensive exterior and sparse interior is formed. In this way, the plan meets requests of scenic maximum use of landscapes and an overall satisfied housing. The main chamber is arranged in the first phase of development, and united with the design of main entry.

The planning of commercial facilities combines with parking facilities of elevated open space on the ground floor, distributed in community roads to urban roads in other phases. Each group of this community has an independent access, forming the independent loop line of transportation.

Dominant colors of multi-layered foreign-style houses are yellow, grey, and brown, making people feel close to nature. The design of high-rise buildings attempts to manifest themes of present age, and being fresh and ease. The style is profound and impressive.

三亚金茂海景花园

用地面积：11 364.06 平方米
建筑面积：90 655.2 平方米
建设地点：海南 三亚
项目用途：住宅用地
开 发 商：广东创茂建设工程有限公司

三亚金茂海景花园项目用地位于三亚市大东海片区。土地资源的稀缺，使得规划的首要原则在于对用地和海景资源利用的最大化。

建筑布局通过运用板点结合的方式，沿东、北、西三面对称布置两栋板式高层住宅，南面结合用地楔形，通过具有针对性的户型平面设计，布置一栋点式高层住宅。

一、二层沿街布置两层通高骑楼商业，塑造宜人的购物氛围。三层为绿化架空层，空间通透开敞，主要设置住宅塔楼的各辅助空间和包括屋顶泳池、咖啡茶座等休闲会所设施。各栋塔楼在架空层通过玻璃连廊连接，人流从各栋独立大堂导向架空层总入户大堂，并与首层入户大堂连接，形成系统的步行景观入户流线。

Jin Mao · Seaview Garden Sanya

Site Area: 11,364.06 m^2
Building Area: 90,655.2 m^2
Constru Ction Location: Sanya, Hainan
Project Purpose: Residential land
Developer: Guangdong Chuangmao Construction Engineering Co., Ltd.

The project site of Jin Mao•Sea-view Garden is located in Dadonghai District, Sanya City. Because of the scarcity of land resources, the primary principle of the planning is to maximize the utilization of land and seaside scenery resources.

The layout of the building adopts the combination of plates and points. In either side of east, north and west, there are two board high-rise buildings in symmetrical arrangement. In the south, the design takes into consideration the wedge-shaped land and deposits there a point mode high-rise building with the targeted unit graphic design.

On the first and second floor, there are two arcade commercial buildings of equal height along the street, creating a pleasant shopping atmosphere. The third floor is the green empty space, open and transparent, which is mainly equipped with the auxiliary space of the residential tower buildings and the recreational facilities such as the roof swimming pool and coffee house, etc. The tower buildings connect with each other through the glass corridor in the empty space. The stream of people is introduced through each independent hall into the main entrance hall of the empty space, which is linked with the lobby on the first floor, forming a systemic walking-home landscape streamline.

廣東華方工程設計有限公司

GUANGDONG HUAFANG ARCHITECTS & ENGINEERS LTD.

广东华方工程设计有限公司最早成立于1992年4月，成立之初就凭借技术服务优势迅速发展壮大，经过多年的发展，现已成为一家综合性的设计服务机构，具备国家建设部颁发的建筑甲级（A144014879）、规划乙级（082037）、人防乙级（A244014876）、消防甲级、工程咨询丙级资质。公司成立十多年来，完成了大量优秀工程项目，业务范围涉及区域规划、住宅、商业、酒店、写字楼、医院、学校、室内外装修等领域，在同行中享有较高的知名度。

公司在东莞、广州、中山、成都、南宁均设有分公司，拥有多名具有海外工作经验的建筑师，高、中级技术人员300多人，专业技术力量雄厚。公司拥有完善的质量管理体系，并通过了ISO 9001：2000质量体系认证，能对设计产品的前期策划至建成竣工验收阶段的全过程提供优质的设计咨询服务。

华方价值观：专业、规范、诚信、进取。
华方理念：客户想到的我们要做到，客户没想到的我们也要做到。
华方宗旨：客户与我共发展。
发展愿景：打造有特色的、极富专业水准的设计企业，努力成为建设全程服务领域的行业能手。

Since its foundation, Guangdong Huafang Architects & Engineers Co., Ltd, certified with Class A Design Certificate and Engneering General Contractor Certificate, has accomplished many outstanding engineering projects with the support of enterprises and construction bureaus of province/city. It gained good reputation in this industry for good quality, elegance, novely and grandeur.

Huafang, with abundant technical power, rich experience and complete service facilities, has an excellent team of design, decoration, fitment and construction. At present, it engages in providing excellent service and effective way to fulfill customers needs, with several departments, suh as scheme dept., construction design dept, engineering design dept, a ecoration dept, etc.

广州地址：
广州市东华南路19号广湾十八商务港6楼
邮编：510100
电话：+86–20–83650298
传真：+86–20–83644106
网址：www.gdhuafang.com
E–mail：walford@163.net

Addressress: 6F Harbour 18 Business tower.19 Donghua South Rd, Guangzhou, China
P.C.: 510100
Tel: +86-20-83650298
Fax: +86-20-83644106
Http: //www.gdhuafang.com
E-mail: walford@163.net

东莞地址：
东莞市南城区元美路华凯广场A栋19–20层
邮编：523070
电话：+86–769–22820769
传真：+86–769–22820756
网址：www.gdhuafang.com
E–mail：winway866@163.net

Address: A19-20 Huakai Plaza, Yuanmei Rd, Nancheng, Dongguan, China
P.C.: 523070
Tel: +86-769-22820769
Fax: +86-769-22820756
Http: //www.gdhuafang.com
E-mail: winway866@163.net

中山地址：
中山市东区起湾北路38号A401
邮编：528403
电话：+86–760–88305009
传真：+86–760–88361065
网址：www.gdhuafang.com
E–mail：h6099@126.com

Address: A401, 38# Qiwan Bei Rd.East district, zhongshan, China
P.C.: 528403
Tel: +86-760-88305009
Fax: +86-760-88361065
Http: //www.gdhuafang.com
E-mail: h6099@126.com

成都地址：
成都市金牛区星辉西路5号水晶花园16楼–137
邮编：610081
电话：+86–28–83221058
传真：+86–28–83222068
网址：www.gdhuafang.com
E–mail：hc.sc@126.com

Address: Room 137.16F, Shuijing Garden, No.5, Road, Xinghui West Rd, Jinniu Sichcan, China
P.C.: 610081
Tel: +86-28-83221058
Fax: +86-28-83222068
Http: //www.gdhuafang.com
E-mail: hc.sc@126.com

南宁地址：
南宁市高新区创新西路9号科铭大厦2楼
邮编：530007
电话：+86–771–3212415
传真：+86–771–3215912
网址：www.gdhuafang.com
E–mail：nnwalford@yahoo.com.cn

Address: 2F, Keming Building, 9# Chuangxin Road West Rd, Nanning, Guangxi China
P.C.: 530007
Tel: +86-771-3212415
Fax: +86-771-3215912
Http: //www.walford.cn
E-mail: nnwalford@yahoo.com.cn

廣東華方工程設計有限公司 • GUANGDONG HUAFANG ARCHITECTS & ENGINEERS LTD.

1 香樟绿洲
2 虎门港职工过渡房及东莞海事局生活配套设施工程设计
3 湖北通山县东山豪园
4 虎门港职工过渡房及东莞海事局生活配套设施工程设计
5 中山市黄圃镇奥城水岸花园
6 半山居空中别墅

GUANGDONG HUAFANG ARCHITECTS & ENGINEERS LTD. • 廣東華方工程設計有限公司 •

7

8

9

10

11

7 阳江酒店
8 大连中拥蓝天下 B 区
9 恒福国际二期
10 橡树林高尔夫会所及酒店式公寓
11 中山眼科中心珠江新城院区
12 深圳市慢性病防治院改扩建工程
13 中拥温泉海景别墅
14 宏远制药厂松山湖项目

12

13

14

GUANGDONG HUAFANG ARCHITECTS & ENGINEERS LTD. • 廣東華方工程設計有限公司 •

15 中山火炬职业技术学院体育馆
16 肇庆市第一人民医院新院设计方案
17 虎门港总部大楼
18 南城区文化体育中心
19 芙蓉和苑
20 虎门客运站

佛山市顺德建筑设计院有限公司

Introduction of Foshan Shunde Architectural Design Institute Co., Ltd.

佛山市顺德建筑设计院有限公司成立于1958年，1993年率先改制组建有限责任公司，是建设部核定的行业甲级建筑设计院，也是中国勘察设计协会的理事单位之一。

佛山市顺德建筑设计院有限公司现有国家级专家、高级工程师、注册建筑师、注册结构师、注册设备师、注册规划师、市政工程师、景观设计师、室内工程师等各类工程设计技术人员三百多人，可承接国内外各类型工业与民用建筑设计、城市规划、市政设计、建筑机电设备安装设计、园林绿化设计及室内设计。

多年来，佛山市顺德建筑设计院有限公司，坚持质量第一，加强技术管理，积极创优创新，全面实施TQC全面质量管理，并于2008年获得ISO 9001国际质量认证。佛山市顺德建筑设计院有限公司以百年企业为目标，致力于全面提升整个设计的每个环节的服务质量，为社会打造更加优秀的设计作品。

Foshan Shunde Architectural Design Institute Co., Ltd was founded in 1958 . In 1993, Foshan Shunde Architectural Design Institute took the lead in restructuring to be a limited company. Foshan Shunde Architectural Design Institute Co., Ltd is a first class architectural design institute which is checked and ratified by national development department and is also a director unit of Chinese reconnaissance design institute.

Foshan Shunde Architectural Design Institute Co., Ltd now has 300 designers, which include many types of construction design engineers, such as, national specialists, high-class engineers, registered architects, registered construction designers, registered equipment designers, registered urban planners, municipal engineers, landscape architects, and interior designers. Foshan Shunde Architectural Design Institute Co., Ltd can take full charge of all types of industrial and civil building design, urban planning, municipal engineering, building equipment installment design, landscape design and interior design.

For the past years, Foshan Shunde Architectural Design Institute Co., Ltd has insisted on quality first, enhanced technical management, got into the swing of innovation, and implemented TQC quality control, and obtained ISO9001 international quality attestation in 2008. Foshan Shunde Architectural Design Institute Co., Ltd aims at being a hundred years' company, commits itself to upgrade servings quality of all parts of the whole design and creates more excellent design work for the society.

主要获奖作品

珠海宾馆获1984年建设部银质奖及广东省优秀设计一等奖
珠海九洲城获1986年广东省优秀设计三等奖
法国巴黎中国城获1995年省优秀设计二等奖
顺德清晖园扩建工程获广东省第九次优秀设计三等奖
顺德宝林寺获广东省第十次优秀设计三等奖
顺德港规划与建筑设计获广东省优秀设计表扬奖
顺德区郑裕彤中学规划与建筑设计获1995年省优三等奖
万家乐综合楼获广东省优秀设计表扬奖及国家“鲁班奖”工程
佛山市顺德区行政服务中心东座与西座，获广东省第12次优秀工程设计二等奖
第五届中国花卉博览会中心展馆，获广东省第12次优秀工程设计三等奖
佛山市顺德一中高中部，获2008年度全国工程勘察设计行业优秀工程勘察设计三等奖，广东省第13次优秀工程设计二等奖
佛山市顺德区一中高中部修建性详细规划，获广东省优秀城市规划二等奖
佛山市顺德区迎接亚洲艺术节和省运会市容整治规划，获广东省优秀城市规划设计表扬奖
美的海岸花园一、二区（美的 依兰尼斯），获2009年度广东省优秀设计三等奖，获2007-2008年度佛山市优秀一等奖及获2008年度顺德区建设工程设计行业优秀建筑设计二等奖。
新乐从中学，获2009年度广东省优秀设计三等奖，获2007-2008年度佛山市优秀二等奖及，获2008年度顺德区建设工程设计行业优秀 建筑设计二等奖。
顺德宜新路综合发展，获2007-2008年度佛山市优秀二等奖。
大良医院，获2007-2008年度佛山市优秀三等奖。
北京加多宝饮料有限公司厂区工程生产办公楼，获2007-2008年度佛山市优秀表扬奖，及获2008年度顺德区建设工程设计行业优秀建筑设计二等奖。
顺德输变电生产综合楼，获2008年度顺德区建设工程设计行业优秀建筑设计三等奖。

Main awarded design work

Zhuhai Hotel – National development department silver prize and 1984 Guangdong Province excellent design first prize
Zhuhai Jiuzhoucheng – 1986 Guangdong Province excellent design third prize
China Town of Paris ,France – 1995 Guangdong Province excellent design second prize
Shunde Qinghuiyuan enlargement project - Guangdong Province the ninth excellent design third prize
Shunde Baolin Monastery - Guangdong Province the tenth excellent design third prize
Shunde port planning and architectural Design - Guangdong province excellent design praise prize
Shunde Zhengyutong Middle School planning and architectural Design - 1995 Guangdong province excellent design third prize
Wanjiale Complex Building - Guangdong Province excellent design praise prize and National luban prize
Foshan Shunde Administration Service Center east wing and west wing - Guangdong Province the twelfth excellent design second prize
China the Fifth Flower Exhibition Center - Guangdong province the twelfth excellent design third prize
Foshan Shunde No.1 Middle School Senior Department – 2008 National reconnaissance design excellent design third prize, Guangdong Province the thirteenth excellent design second prize
Foshan Shunde No.1 Middle School Senior Department construction detail planning - Guangdong Province excellent urban planning second prize
Foshan Shunde planning of renovating city for welcoming Asia Art Festival and Guangdong Province Sports Meeting - Guangdong province excellent urban planning praise prize
MIDEA•ILANES project, was awarded the 3rd prize of 2009 Guangdong Province excellent architecture design award, 1st prize of 2007-2008 Foshan excellent architecture design award.
New Lecong Middle School, was awarded the 3rd prize of 2009 Guangdong Province excellent architecture design award, 2nd prize of 2007-2008 Foshan excellent architecture design award.
New world center, was awarded the 2nd prize of 2007-2008 Foshan excellent architecture design award.
Daliang Hospital, 3rd prize of 2007-2008 Foshan excellent architecture design award.
JDB Beverage Co., Ltd. Beijing Plant Engineering Production Office, was awarded a prize of 2007-2008 Foshan excellent architecture design award, and 2nd prize of 2008 Shunde excellent architecture design award.
Shunde Transformer Substation Production Building, was awarded the 3rd prize of 2008 Shunde excellent architecture design award.

地址：广东省佛山市顺德区大良立田路建筑设计大厦
电话：+86-757-22287320、+86-757-22600168
总机：+86-757-22600333、+86-757-22600666
传真：+86-757-22600338、+86-757-22281470
邮编：528300
网址：www.sdadi.com

Address: Archi-design Bldg., Litian Road, Daliang, Shunde, Foshan, Guangdong Province
Tel: +86-757-22287320 / +86-757-22600168
+86-757-22600333 / +86-757-22600666
Fax: +86-757-22600338 / +86-757-22281470
P.C.: 528300
E-mail: sadi@21cn.com
Http: //www.sdadi.com

设计理念

建筑必须要融入环境，并与本地传统相呼应。本设计的目的是要通过融汇西方医疗设计精髓和本地医疗设计经验，创造一个有亲和力的治疗环境，而达到为本地区服务的目的。

本设计专注于营造一个精密的医疗系统，它具有吸引力，具有宁静、积极、和谐的特点。建筑形式准确地 服从于其运作和功能的需要。这是一个不仅拥有治疗功能而且具有提升精神状态作用的医疗系统。此外，人性化的理念是医疗环境设计中不可缺少的元素。

场地的规划布局既有效率，也具有清晰、导向性强的特点，这样可以使场地未来有更灵活的发展空间。公共区域与私密性区域在场地和医疗规划中都有清晰地区分。

精确、合理的建筑形态通过一条生机勃勃的生态廊体现出来。这条“生态廊”形成一个舒适的室内人行空间，同时联系医院各组成部分。它既是一个可持续发展的元素，又是组织骨架的一个重点。

可持续设计对于人们的健康和设备的持久性有至关重要的影响，好的可持续性设计首先是要最大化利用自然减少能源消耗，而人工的能源只作为辅助能源提高运作效率。

建筑的朝向令建筑获得最多的自然光和最好的通风，减少过多的太阳热量的获取，尽可能利用地下温度和可再生能源。

作为一个公立医院，它以开放的格局面向社会大众，以便将尽量多的空间留作绿化用地及室外花园用地。

佛山市顺德区第一人民医院

建设地点：广东 佛山
占地面积：133 333 平方米
建筑规模：225 000 平方米

FIRST PEOPLE'S HOSPITAL OF SHUNDE

Construction Location: Foshan, Guangdong
Site Area: 133,333 m^2
Building Scale: 225,000 m^2

Design concept

Buildings must be merged into the surroundings and correspond with the local tradition. The objective of the design is to create a likable therapeutic milieu in the service of the community through the integration of the essence of Western medical design and the local medical design experience.

The design focuses on creating a sophisticated medical system, which is attractive, quiet, positive, and harmonious. The architectural form accurately obeys the needs of its operation and function. This is a medical system not only healing the body but also improving the mental state. Besides, the concept of humanity is the indispensable element in the design of medical environment.

The layout of the site is only efficient but also clear and has strong guidance quality, which can make a more flexible development potential for the site in the future. There is a clear distinction of the public area and privacy area in the planning of site and medical treatment.

Accurate and reasonable architectural form is reflected through a vibrant ecological corridor. The “ecological corridor” forms a comfortable indoor pedestrian space, and connects each component of the hospital as well. It is an element of sustainable development as well as a focal point of organizational framework.

Sustainable design has an important impact on people's health and the sustainability of equipment. Good sustainability first means maximizing the utilization of natural resources and reducing energy consumption, while artificial energy source can improve operation efficiency only as the supplementary energy.

The building orientation can let the building get more natural light and the best ventilation, reduce the gain of excessive solar heat, and make as much use of underground temperature and renewable energy as possible.

As a public hospital, it opens to the public by the pattern of opening up, leaving as much space used as the land of greening and outdoor garden as possible.

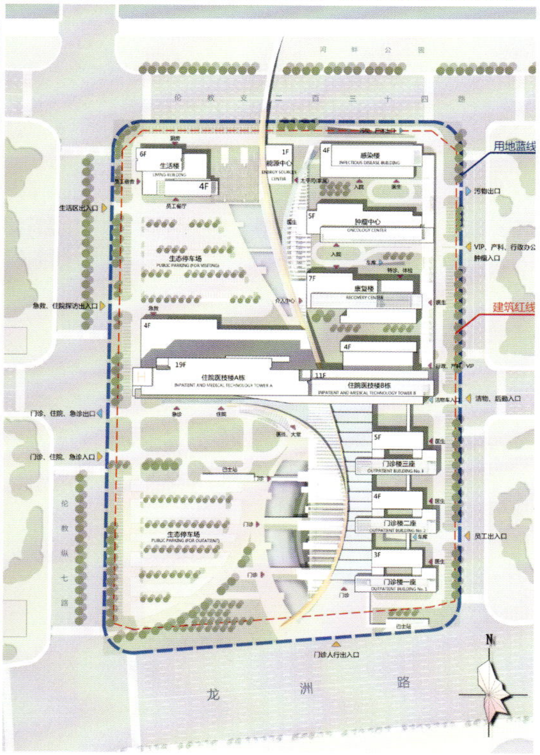

佛山市高明区职业技术学校

占地面积：319 520 平方米
总建筑面积：195 205.4 平方米
容 积 率：0.61

Gaoming District Vocational School

Site Area: 319,520 m^2
Gross Building Area: 195,205.4 m^2
Plot Ratio: 0.61

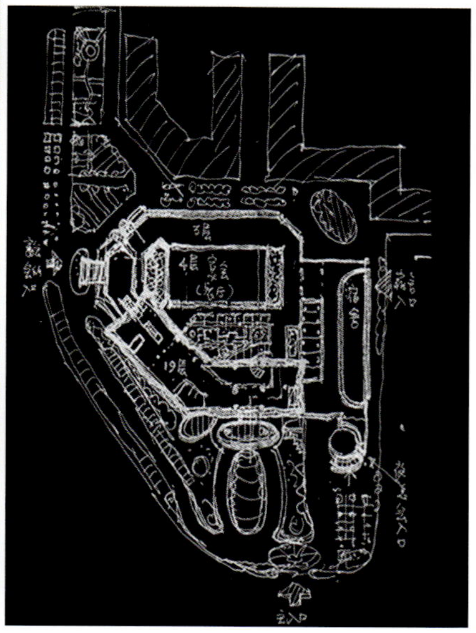

顺德金茂华美达广场酒店

酒店用地面积：26 010 平方米
酒店总建筑面积：60 000 平方米
容 积 率：2.31
建设规模：400 间客房

SHUNDE JINMAO RAMAD PLAZA HOTEL

Site Area of the Hotel: 26,010 m^2
Total Building Area of the Hotel: 60,000 m^2
Plot Ratio: 2.31
Construction Scale: 400 Rooms

顺德喜来登酒店

占地面积：70 000 平方米
建设规模：400 间客房

SHUNDE SHERATON HOTEL

Site Area: 70,000 m^2
Construction Scale: 400 Rooms

Project Introduction

Midea Yuhaidongjun Residential Quarter is located in Ronggui, Shunde District, Guangdong Province. This project has an overlook of Desheng River landscape, and is positioned as an elegant residential area in Ronggui. Its gross land area is 380,000 square meters and the design focuses on the townhouses and the high-rise buildings with a river view.

Design Concept of the Project Planning:

The traffic division of pedestrians and vehicles produces an atmosphere of comfort and safety, meeting the trend of social development; the appropriate application of the terrain creates a landmark luxury community along the river bank. The planning design emphasizes on the maximization of landscape usage and the product planning focuses on the treatment of the differential details; these two combines to form a human community of diverse landscape change and rich cultural connotation.

The architectural design inherits the characteristics of shading and ventilation in Lingnan Region. And together with the combination of resort buildings in Southeast Asia and the modern simple architectural features, this design displays a unique style of building facade with the embracing of the construction and the green space.

The traffic division of pedestrians and vehicles in the transport system, together with the innovative townhouses of people-vehicle-separation, produces an atmosphere of comfort and safety for the residents.

The community is equipped with a central water-system axis, which from east to west connects the whole community together. Taking advantage of the two plots on the back and south side, integrating clubs, commercial centers and other facilities together, the design forms a clear treatment of plot difference, assigns green spaces with various features to products of different levels, and maximally utilizes landscapes inside and outside of the community.

项目概况

美的御海东郡位于广东顺德容桂，坐拥德胜河江景，项目定位为容桂片区高尚住宅区。项目占地共计 38 万平方米，项目建筑以联排低层住宅及高层江景洋房为主。

项目规划设计理念

人车分流，安全舒适，适应社会发展趋势，合理利用地形，打造江岸标志性豪宅社区。规划设计注重景观利用的最大化，产品规划注重差异化细节处理，有利于建立景观层次的变化多样和富有文化内涵的人文社区空间。

建筑设计秉承岭南地区遮阳、透风的建筑特色，把东南亚的度假建筑及现代简约建筑特色相互结合，呈现出特有的建筑立面风格，建筑与绿化空间相互辉映。建立人车分流的道路交通系统，创新形态的人车分流联排低层住宅，为住户提供安全舒适的居住空间。

小区中设计了一条中央水系景观轴，东西贯穿整个小区，利用背面及南面的两片用地的大景观，融合会所、商业及其他配套设施，形成清晰的片区差异化处理，使不同档次的产品获得不同特色的景观绿化空间，并使区内区外的景观得到最大化地利用。

美的御海东郡商住社区

地块一
总用地面积：133 334 平方米
总建筑面积：264 864.11 平方米
建筑密度：24.30%
容 积 率：1.6

地块二
总用地面积：254 630 平方米
总建筑面积：496 275.42 平方米
建筑密度：23.95%
容 积 率：1.6

Plot I
Gross Site Area: 133,334 m^2
Gross Building Area: 264,864.11 m^2
Building Density: 24.30%
Plot Ratio: 1.6

Plot II
Gross Site Area: 254,630 m^2
Gross Building Area: 496,275.42 m^2
Building Density: 23.95%
Plot Ratio: 1.6

肇庆中源名苑工程项目简介

肇庆市中源名苑项目位于肇庆市 110 区黄岗路东侧地块，总用地面积 114 496 平方米，净用地面积 90 433 平方米，总建筑面积 34 610 平方米，建筑密度 32.5%，容积率 3.2。项目位于肇庆市城市新开发区 110 地区，地块周边均为规划的住宅发展用地。地块因南北向规划市政路而分为东西两块，两地块均呈不规则方形，这给小区整体设计提高了难度。

项目以市场为导向，在保证质量的前提下做好成本控制与优化，强调规划设计的精细化与适度创新，同时以户型创意性为基准，将其定位为肇庆市具有"国际感、艺术化、空间观"的创意产品。

项目从设计初期乃至不断的深化设计过程中，委托方与设计方一直紧密联系，不断研讨和分析，最终高质高效地完成了工程设计。

项目规划上布局合理，联系紧凑，空间尺度合理，单体设计上有创新意念，市场反响好，后期施工阶段执行力度强，作品完整度较高。

Introduction of Zhaoqing Zhongyuan Garde

Zhaoqing Zhongyuan Garden is located in the land plot at the east side of Huanggang Road, 110 Block, Zhaoqing city, which covers a total land area of 114,496 m^2, net area of 90,433 m^2, total building area of 34,610 m^2, building density of 32.5% and plot ratio of 3.2. The project is located in 110 Block of the new urban development area of Zhaoqing city, surrounded all by the planned residential development. Because of the north-south design of municipal Road, the plot is divided into east and west, both of which are irregular square. This also causes hardships for the overall design of the residential area.

With the market as its orientation, this project does well in cost control and optimization under the premise of guaranteeing quality, and emphasized the refinement and modest innovation of the plan. At the same time, based on the unit creativity, we will position it as the creative product with "international sense, art and space view".

In the initial design and even the continuous deepening design process of the project, the entrusting party and the design side have been constantly working together to discuss and analyze, and finally completed the design with high quality and efficiency. The project planning has reasonable layout, compact structure, and reasonable spatial scale; the individual design has innovative ideas and a good market response; in the post-construction stage, the enforcement is good and the work has a high degree of integrity.

廣東建聯建築設計有限公司

KINGLINK DESIGN & DEVELOPMENT CONSULTANTS LTD.

地址：广东省汕头市韩江路 17 号华景广场 603
邮编：515041
电话：+86–754–88367095 / 88167784
传真：+86–754–88367095
网址：www.kld–arch.com
邮箱：kld@kld–arch.com

Address: Room 603 Huajing Development Park on No.17 Hanjiang Road Shantou, Guangdong
P.C.: 515041
Tel: +86-754-88367095 / 88167784
Fax: +86-754-88367095
Http: //www.kld-arch.com
E-mail: kld@kld-arch.com

广东建联建筑设计有限公司（简称 KLD）是由一批从事建筑工程设计行业 15 年以上的具有高学历、高技术职称和各专业注册执业资格的以资深设计团队为核心的专业建筑设计公司，拥有国家建筑行业建筑工程甲级设计资质公司。依照市场化的运作和现代企业管理模式向客户提供包括项目总承包、项目策划、总体规划和建筑工程各专业从方案到施工图设计和咨询服务。

公司以专业化、精细化以及优质的服务和良好的信誉立足于市场，注重对市场和消费需求的研究，注重于对各类项目的运作流程和管理模式的理解，整合各方面的智慧和经验，注重于各专业设计的协调和优化，追求总体设计的完美，力求把每个项目都做成精品。

KingLink Design & Development Consultants Ltd(KLD) is a professional architectural designing company composed by a group of experienced designers engaged in architectural design industry for more than 15 years, with high education, high-tech titles and registered professional qualifications of different fields as the core of the company. It has architectural engineering Design Qualification, A Grade of the national construction industry. In accordance with the operation of the market and modern enterprise management mode, the company provides for the clients services including: project contracting, project planning, general plan and services from plan to construction documents design and consultation of various professions of architectural engineering.

The company enjoys high reputation in the market for specialization, refinement, quality service and well-deserved reputation. It focuses on the research of the market and consumption demand, and pays attention to the understanding of the operation process and management mode of various types of project. Integrating of the wisdom and experiences of all aspects, the company pays attention to the co-ordination and optimization of various professional designs, pursues the perfection of the overall design, and strives to do every project into the realm of quality.

公司的主要客户

1. 香港和记黄埔地产有限公司
2. 中信房地产股份有限公司
3. 龙光地产股份有限公司
4. 花园集团有限公司
5. 香港东兴集团有限公司
6. 香港信利电子国际有限公司
7. 广东宜华房产股份有限公司
8. 广东泰安房地产集团有限公司
9. 广东隆泰房地产集团有限公司
10. 广东锦峰集团房地产有限公司
11. 汕头市中环房地产开发有限公司
12. 汕头市侨银房地产开发有限公司
13. 汕头市南信房地产开发有限公司
14. 汕头市长平房地产开发有限公司
15. 汕头市经纬集团华南贸易广场开发有限公司
16. 汕头市海逸投资（集团）有限公司
17. 汕头大洋（集团）公司
18. 西宁安泰房地产开发有限公司

Main clients of the company

1. Hutchison Whampoa Properties Limited of Hongkong
2. CITIC Real Estate Co., Ltd.
3. Longguang Real Estate Co., Ltd.
4. Garden Resorts Co., Ltd.
5. Dongxing Group of Hong Kong Co., Ltd.
6. Xinli Electronic International Ltd. of Hongkong
7. Yihua Real Estate Co., Ltd. of Guangdong
8. Taian Real Estate Group Co., Ltd. of Guangdong
9. Longtai Real Estate Group Co., Ltd. of Guangdong
10. Jinfeng Group Real Estate Co., Ltd. of Guangdong
11. Zhonghuan Real Estate Development Co., Ltd. of Shantou
12. Qiaoyin Real Estate Development Co., Ltd. of Shantou
13. Nanxin Factory Building Development Co., Ltd. of Shantou
14. Changping Real Estate Development Co., Ltd. of Shantou
15. Jingwei Group Southern Trade Plaza Development Co., Ltd. of Shantou
16. Haiyi Investment (Group)Co., Ltd. of Shantou
17. Dayang (Group) Corporation of Shantou
18. Antai Real Estate Development Co., Ltd. of Xining

公司核心团队（个人）获奖项目

1. 金海湾大酒店（五星级）荣获建设部优秀设计二等奖、广东省优秀设计一等奖
2. 帝豪酒店（五星级） 荣获建设部优秀设计三等奖、广东省优秀设计一等奖
3. 汕头市人民政府工作楼 荣获广东省优秀设计二等奖
4. 丽日度假酒店 荣获广东省优秀设计二等奖
5、华景广场 荣获广东省优秀设计三等奖
6、汕融大厦 荣获广东省优秀设计三等奖

Our core team (personal) awards

1. Golden Gulf Hotel (five star) won the second prize (Ministry of Construction P.R.China) and the first prize(Province)
2. Regcency Hotel(five star) won the third prize (Ministry of Construction P.R.China) and the first prize(Province)
3. Shantou Municipal Government office building won the second prize(Province)
4. Korea Day Holiday Hotel Shantou won the second prize of the provincial excellent design awards
5. huajing square won the third prize of the provincial excellent design awards
6. Shanrong tower won the first prize(Province)

海岸明珠：君庭

泰安华庭

KINGLINK DESIGN & DEVELOPMENT CONSULTANTS LTD.

中信凯旋蓝岸花园

中机国际工程设计研究院

地址：湖南省长沙市韶山中路 18 号
电话：+86–731–85383382
传真：+86–731–85383588
邮箱：market@cmiei.com
网址：www.cmiei.com

Address: Shaoshan Middle Road No.18
Changsha, Hunan Province
Tel: +86-731-85383382
Fax: +86-731-85383588
E-mail: market@cmiei.com
Http: //www.cmiei.com

中机国际工程设计研究院（原机械工业部第八设计研究院）创建于 1951 年，是集工程咨询设计、工程总承包、项目管理、工程建设监理、工程造价及技术咨询服务为一体的国家大型甲级综合设计研究院，坐落于风景秀丽的历史文化名城——长沙。

本院现持有机械、兵器、船舶、轻工、建筑、市政、环境保护等行业工程设计甲级资质，持有工程咨询、工程造价咨询、建筑智能化、消防设计及工程建设总承包、工程监理等甲级资质，持有压力容器和压力管道设计资格、建设项目环境评价乙级证书，持有电子、电力、商物粮、化工、石化、医药、城乡规划等行业乙级工程设计证书，并具有国家批准的国外承包工程劳务合作经营许可证和进出口贸易权。研究院还先后通过了 ISO 9001 质量管理体系认证和湖南省高新技术企业认定。

经过近六十年的发展，本院已成为机械、建筑、市政、环保等多个行业领域具有竞争实力和良好声誉的科技型企业。建院以来已累计完成国内外咨询设计、工程总包和监理等项目 13 000 余项，取得科学研究成果 500 余项，科研设计成果遍及全国各地及十余个国家和地区，获得国家、部、省级优秀科技成果、优秀设计奖 180 余项，拥有国家专利和专有技术 20 余项，为国家多项技术标准和规范的主、参编单位。

作为科技型企业，本院拥有一支高素质的科研设计人才队伍，涵盖 30 多个专业门类，现在，中机国际设计人员的足迹已经遍及马尔代夫、斯里兰卡、印度尼西亚、赞比亚、塞内加尔、土耳其、赤道几内亚、蒙古等国家和地区，树立了良好的口碑。中机国际工程设计研究院在不断提高自我、挑战自我的同时，将竭诚为国内外合作伙伴提供优质的产品和满意的服务。

China Machinery International Engineering Design & Research Institute (originally the Eighth Institute of Design & Research of the Ministry of Machine-building Industry) is a large-scale comprehensive design and research institute with national Class A designing certificate, covering such areas as engineering consultation and design, engineering project contracting, project management, engineering construction supervision, engineering cost and technology consultation. The Institute is situated at the picturesque historic city Changsha.

Our Institute is authorized with Class A designing certificate in such areas as machine-building, military projects, shipbuilding, light industry, architecture, municipal services and environmental protection; Grade-A certificate in engineering consultation, engineering cost analyses, intelligent building designing and construction, fire-protection design, engineering project contracting and construction supervision; Grade-B certificate in the design of pressurized vessels and pipelines, environmental evaluation of construction projects; Class B certificate in electronics, power generation, commercialized food circulation and storage, chemical and petrochemical engineering, pharmaceutical industry, rural and urban planning; It is also empowered by the State to project contracting in overseas labour services and to carry our both imports & exports with foreign countries. Has successively passed the Quality Management System Standard Certification ISO9001 and the qualification of high-tech enterprises in Hunan Province.

After its development of almost 60 years, our Institute is now a sci-tech oriented enterprise that is highly developed and competitive both at home and abroad in such fields as machine-building, architecture and architectural engineering, municipal constructions, environmental protection, and so on. Ever since the founding of our Institute, we have successfully accomplished both at home and abroad more than 13,000 projects of engineering consultation and design, turn-key contract and supervision, together with 500-odd sci-tech research projects. Our sci-tech research and engineering designing projects are now all over the country and cover more than 10 other countries and regions. Because of our excellent research and design, we have been honored with more than 150 prizes at the State, ministerial and provincial levels. We now possess 20-odd State patents and patented technology, and are the editor or co-editor of many State technical standards and codes.

As a sci-tech oriented enterprise, our Institute is in possession of a contingent of highly-qualified research and design personnel, covering more than 30 faculties and specialties. Now, the footprint of the staff in China Machinery International Engineering Design & Research Institute has covered many countries and regions worldwide, including Maldives, Sri Lanka, Indonesia, Zambia, Senegal, Turkey, Equatorial Guinea, Mongolia and so on, established a good reputation. With the constant commitment of self improvement and self challenge, China Machinery International Engineering Design & Research Institute is striving for providing the superb quality products and satisfactory service to the cooperative partners.

东方电机有限公司华东新能源制造基地

项目地点：浙江 杭州
新征土地：16.7 公顷
新建面积：90 000 平方米
总投资约 10 亿元。年产 1.5 到 3.5 兆瓦直驱风力发电机组 350 台（套）、潮汐发电机组 10 台套（30 兆瓦）。

西安西玛电机（集团）有限公司整体工程迁建项目

项目地点：陕西 西安
项目征地面积：197 900 平方米
用地面积：172 400 平方米
固定资产投资：25 696 万元
年产各类大、中、小型电机共 860 万千瓦。

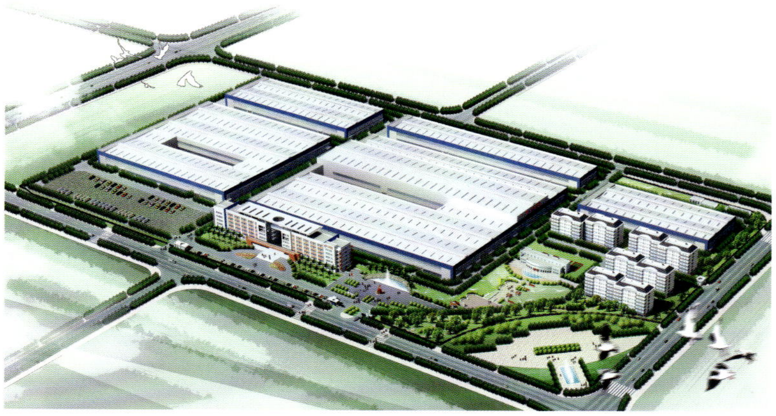

长沙电机厂有限公司整体迁建项目

项目地点：湖南 长沙
征地面积：246 000 平方米
用地面积：223 000 平方米
一期固定资产投资：20 275 万元
项目达产后，具备年产各类大、中、小型电机共 590 万千瓦的生产能力。

南京长风新能源股份有限公司风力发电机项目

项目地点：江苏 南京
征地面积：26 公顷
投资规模：10.5 亿元
生产规模 4 000 台 2 兆瓦风力发电机，新建结构车间、冲压车间、电工装配车间、机加车间，总计建筑面积约 80 000 平方米并引进德国 VEM 公司技术。

江西东元电机有限公司整体迁建项目

总建筑面积：88 000 平方米
工程总投资：3 亿元人民币
固定资产投资：2 亿元
生产大中型电机和水轮发电机组，新建厂房最大跨度为 36 米，最大吊车吨位 75 吨。

山河智能天津工业园

项目地点：天津
占地面积：1 200 000 平方米
总建筑面积：600 000 平方米
一期建筑面积：150 000 平方米
主要生产起重机械、路面机械、建筑机械、环卫机械及其他装备，项目达产后年产各种工程机械产品 10 000 台以上，年生产能力超过 100 亿元。

中联重科泉塘工业园一、二期建设项目

项目地点：湖南 长沙
占地面积：400 000 平方米
总建筑面积：180 000 平方米
主要生产中大吨汽车起重机、全系列全地面起重机，年产中大吨位工程起重机械 4 000 多台，年生产能力超过 60 亿元。

中国南车集团株洲电力机车有限公司城规车辆扩能建设

项目地点：湖南 株洲
新建建筑面积：50 000 多平方米
本项目属扩建改造项目，包括新建车体联合厂房、组装联合厂房扩建、中转库扩建等，实施后新增年产量为 300 辆城轨车辆。

长沙金洲社会主义新农村改革与发展试点项目

项目地点：湖南 长沙
项目状态：一期已建成
设计时间：2009 年
建筑面积：19 000 平方米

香樟 · 兰亭

项目地点：湖南 长沙
项目状态：建设中
设计时间：2009 年
建筑面积：230 000 平方米

永州远志新外滩

项目地点：湖南 永州
项目状态：建成
设计时间：2007 年
建筑面积：193 000 平方米

设计理念——环境［人文 · 生态 · 市场］
　　设计既要满足现实的市场需要，又要满足更高层次的规划要求，使功能与周边环境相契合，使之起到应有的先导性，这就涉及到在多个“维度”上都使建筑能嵌入环境当中的问题。

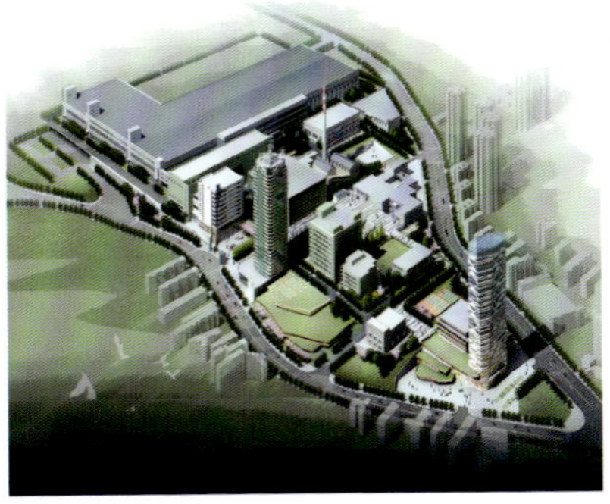

湖南中烟工业有限责任公司技术中心

设计时间：2009 年
建筑面积：49 000 平方米

新城国际公馆综合楼设计

项目地点：湖南 长沙
项目状态：建设中
设计时间：2007 年
建筑面积：280 000 平方米

赫曦台

项目地点：湖南 长沙
项目状态：待建
设计时间：2007 年
建筑面积：6 120 平方米

赫曦台重建原址位于岳麓山风景名胜区麓山景区东北部，建设总用地 6.07 公顷，建筑主体占地 3 577.50 平方米，总建筑面积 6 120 平方米，总投资约 1.5 亿人民币。赫曦台平面呈十字形，三明二暗共五层，通高 47.45 米，坐西南朝东北，倚群山瞰江流，地理地相俱佳。

邵阳医专梅子井校区

项目地点：湖南 邵阳
项目状态：建成
设计时间：2006 年
建筑面积：53 000 平方米

湖南农业大学图书信息中心

项目地点：湖南 长沙
项目状态：建成
设计时间：2006 年
建筑面积：33 000 平方米

和庄（原名：江南丽湾）

项目地点：湖南 长沙
项目状态：一期已建成，二期待建
设计／建成时间：2006 年
建筑面积：226 018 平方米

场地为西偏北 19° 的狭长地带，东西长约 700 米，南北宽约 70 米。地块呈东北角高、西南角低的楔型，地势落差较大，场地限制较多。项目净用地面积 50 127.91 平方米。

宁乡县人民医院

项目地点：湖南 宁乡
项目状态：建成
设计时间：2001 年
建筑面积：90 000 平方米
合作单位：中国中元国际工程公司

本工程各建筑物以主廊、联廊相连，形成高低错落的建筑群，立面造型采用叠翠绿瓦坡屋面与珍珠白墙面、横向轻巧遮阳板、玻璃天窗相结合的处理方式，突出了南方建筑通透轻盈的形象特征。

他城（天鸿国际）

项目地点：湖南 长沙
项目状态：建成
设计时间：2005 年
建筑面积：72 000 平方米

南华大学附属第一人民医院门诊大楼

项目地点：湖南 衡阳
项目状态：建设中
设计时间：2007 年
建筑面积：约 65 000 平方米

建筑的外部造型及立面设计考虑到不同视距及角度的观赏效果。运用大面积的变化及细小的变动，使立面造型简洁而生动，远看有着强烈的节奏感及雕塑感，近观可看到建筑的细部，透过玻璃更可看到室内绿化，绿色无处不在，使得建筑融入自然，建筑与环境相呼应。

怀化市第三人民医院内科医技楼

项目地点：湖南 怀化
项目状态：深化设计阶段
设计时间：2010 年
建筑面积：约 29 000 平方米

永州中心医院

项目地点：湖南 永州
项目状态：建成
设计时间：2006 年
建筑面积：120 000 平方米

本工程区域内道路交通便捷，使病人就医十分方便。基地内现状地势起伏较大，植被茂密，自然环境优美。

全院由综合医疗区、传染病区、教学科研区、办公后勤服务区和生活区五大主要区域组成，并以医院主街为纽带联系起来。建筑围绕院区中心的景观山布置，利用地势高差，将门急诊的主入口架高和永州大道持平，而架空层下设停车场，在院区的西北部开设后勤、货物出入口，院区的东侧规划路则开设传染病区出入口和污物出口。

湖南城市学院规划建筑设计研究院

Urban Planning & Architectural Design Institute Of Hunan City University

湖南城市学院规划建筑设计研究院于 1986 年正式成立，是湖南城市学院下设的二级机构，同时也是湖南省建设厅直属设计单位。研究院拥有城乡规划、建筑工程、工程咨询和科技咨询四项甲级资质，岩土工程勘察、市政设计和工程测量三项乙级资质。研究院下设总工办、综合办公室、计划经营室、规划设计室、建筑设计室、市政设计室、勘察设计室、城市设计研究所、数字城市研究所、测绘队、结构研究所等机构，对外承接各类城镇规划、专项规划以及城市设计、建筑设计、市政设计、岩土工程勘察和工程咨询业务，并主要从事城乡规划、城市设计、数字城市、地籍测绘、结构检测及加固等方面的研究。

Urban Planning & Architectural Design Institute of Hunan City University was officially established in 1986, which is the Class B subsidiary of Hunan City University and is affiliated to Department of Construction of Hunan Province. It has Class A qualification in urban and rural area planning, architecture construction, project consulting and technology consulting areas. Three Class B qualifications of municipal design, geotechnical investigation, project consulting. With the organization of chief engineer offices, general offices, planning and marketing offices, planning and design offices, architecture design offices, municipal design offices, investigation and design offices, urban design research center, digital urban research center, surveying and drawing team, structure research center, their design business covers many areas including urban and town area planning, special project planning, urban design, architecture design, municipal design, geotechnical investigation and project consulting. They are mainly engaged in the research of urban and rural area planning, urban design, digital city, cadastre survey, structure examination and reinforces and other aspects.

地址：湖南省益阳市益阳大道西 238 号
法人：谭献良
邮编：413000
电话：+86–737–4233101
传真：+86–737–4244250
邮箱：cysjy@vip.163.com
网址：www.cysjy.net

Address: Yiyang City, Hunan Province Yiyang Road No. 238
Corporate: Xianliang Tan
P.C.: 413000
Tel: +86-737-4233101
Fax: +86-737-4244250
E-mail: cysjy@vip.163.com
Http: //www.cysjy.net

建 筑 Construction

湖南城市学院规划建筑设计研究院
Urban Planning & Architectural Design Institute Of Hunan City University

1

1—4 龙华 · 世纪城

建设地点：湖南 永州

委 托 方：永州龙华投资置业有限公司

总建筑面积：270 000 平方米

设计时间：2010 年

Longhua Century city

Construction Location: Yongzhou, Hunan

Entrusting Party: Yongzhou Longhua Real Estate Co., Ltd.

Total Building Area: 270,000 m^2

Design Time: 2010

2

3

4

1

2

3

4

1—3 书香名邸二期

建设地点：湖南 益阳

委 托 方：湖南德远置业有限公司

总建筑面积：28 400 平方米

设计时间：2010 年

Phase II of Shuxiang mingdi

Construction Location: Yiyang , Hunan

Entrusting Party: Hunan Deyuan real estate Co., Ltd.

Total Building Area: 28,400 m^2

Design Time: 2010

4 汨罗市人民医院住院大楼

建设地点：湖南 汨罗

委 托 方：汨罗市人民医院

总建筑面积：22 000 平方米

设计时间：2009 年

Warding building of Miluo People's Hospital

Construction Location: Miluo, Hunan

Entrusting Party: Miluo People's Hospital

Total Building Area: 22,000 m^2

Design Time: 2009

建 筑 Construction

湖南城市学院规划建筑设计研究院
Urban Planning & Architectural Design Institute Of Hunan City University

1

2

3

4

1—3 中奇梓湖东城

建设地点：湖南 益阳
委 托 方：益阳中奇房地产开发有限公司
总建筑面积：356 000 平方米
设计时间：2010 年

Zhongqi zihudongcheng
Construction Location: Yiyang , Hunan
Entrusting Party: Zhongqi Real Rstate Development Co., Ltd. in Yiyang
Total Building Area: 356,000 m^2
Design Time: 2010

4 沉江水立方大酒店

建设地点：湖南 沅江
委 托 方：沅江水立方房地产开发有限公司
总建筑面积：27 000 平方米
设计时间：2009 年

Yuanjiang Water Cube Hotel
Construction Location: Yuanjiang, Hunan
Entrusting Party: Yuanjiang Water Cube real estate development Co., Ltd.
Total Building Area: 27,000 m^2
Design Time: 2009

1 益阳市精神卫生疾病防治中心住院大楼
建设地点：湖南 益阳
委 托 方：益阳市精神卫生疾病防治中心
总建筑面积：13 000 平方米
设计时间：2009 年

Ward building of Yiyang Mental Healthy Center
Construction Location: Yiyang, Hunan
Entrusting Party: Yiyang Mental Healthy Center
Total Building Area: 13,000 m^2
Design Time: 2009

2—3 钻石华庭
建设地点：湖南 益阳
委 托 方：南县城镇建设综合开发公司
总建筑面积：22 000 平方米
设计时间：2010 年

Diamond Mansion
Construction Location: Yiyang, Hunan
Entrusting Party: Urban Construction Development Company of NanXian County
Total Building Area: 22,000 m^2
Design Time: 2010

1

2

3

Architectural

设计作品 Design Works

诚士设计

湖南诚士建筑规划设计有限公司

HUNAN CHENGSHI ARCHITECTURAL DESIGN AND CITY PLANNING Co., LTD.

地址：湖南省长沙市人民东路 59 号
电话：+86-731-84718448
传真：+86-731-84718448-8888
邮箱：csadcl@vip.163.com

Address: Renmin East Road No. 59 Changsha, Hunan Province
Tel: +86-731-84718448
Fax: +86-731-84718448-8888
E-mail: csadcl@vip.163.com

本公司系建筑设计甲级、规划设计乙级资质的设计企业。自一九九七年成立以来，经历了创业、发展、壮大的艰难历程，至今已拥有专业技术人员 45 人，其中一级注册建筑师 4 人、注册规划师 3 人、一级注册结构工程师 4 人、注册设备工程师 3 人、注册电气工程 2 人，二级注册建筑、结构师各 1 人，本公司拥有中级技术职称及以上的人员达到 36 人，是一支专业齐全的素质高、能力强、能打硬战，充满活力的技术骨干队伍。

十多年来，我公司奉行＂天道酬诚＂的宗旨，坚持＂质量就是生命＂的管理原则，完成了省内外逾百个项目的设计工作，近五年完成施工图建筑工程 400 多万平方米，完成从总体规划、分区规划到详细规划和城市设计各阶段的规划项目数十个，曾多次获得省、市优秀设计奖，我公司的方案创新，设计质量、施工服务均受到建设单位的好评。

近年来，我公司积极组织和参加绿色建筑、建筑节能、新技术推广的学习，以提高技术人员科技创新和运用新技术、新材料的能力，同时，公司大力推行设计过程的精细化，加强后期服务的主动性，并成立了一个专门的研究中心，使我们的设计水平和服务意识更上一个台阶，为公司的＂可持续＂发展奠定了坚实的基础，也为更好的服务提供了制度保障。

We are a design company with qualifications of architecture Class A and planning Class B. Since 1997, our company has come through the progress of establishment, development and booming. Now there are 45 professional technical talents working in our company, including 4 registered architects, 3 registered planners, 4 first-class registered structural engineers, 3 registered utility engineers, 2 registered electrical engineers; 1 Class 2 registered architect and registered structural engineer and more than 36 staff of medium-grade professional title or higher above. Our company is a dynamic key technical team with high quality personnel who are of high capacity and efficiency

For over 10 years, holding the belief of “success rewards the integrity” and adhering to the management principle of “quality is the life of the enterprise”, our company has finished over 100 design projects in and out of the province. we have worked on construction and architectural projects of over 4,000,000 m^2 and about 10 planning projects of city district planning, detailed planning and urban design. For several times. We have won the provincial and municipal excellent design awards, and have high reputation in construction profession for our innovative scheme, high quality design, and good construction service.

In recent years, our company actively organized and participated the study of green architecture, energy-saving in architecture, the promotion of new technology, to improve the technical staff's capacity of innovation and application of new technique and new materials. Meanwhile, the company puts emphasis on the design process of high precision, giving active support of service in later stage. Besides, we have established a specific research center, improving our design level and service, laying a solid foundation of a sustainable development and well guaranteed a better service.

蒋怀庆作品选

“绿色建筑”——用心去做

Doing with Your Heart

▲▼ 长沙太阳星城绿色市范小区

▲ 长沙市委党校方案设计鸟瞰图

▲ 尽量保留每一棵大树

◀ 办公楼建成实景

◀ 办公楼的门厅实景

有机建筑

生态环境

和谐社会

► 新建办公楼、教学楼前保留的原生态大树

▼ 教学楼建成实景

▲ “731” 新餐厅主入口建成实景

建筑与自然共生　人与人和谐交流

▼ “731” 新餐厅入口门廊实景

▼ 湖南省外侨办 “731” 新餐厅综合楼方案设计鸟瞰图

建筑立面不是一张表皮，而是建筑群与自然共生的复合立面。建筑的主入口处采用钢结构点式玻璃墙的门厅与雨篷、弧形文化墙与柱廊结合的手法，使人与自然共生的主题和建筑群的时代感都得到体现。一、二层错落有致的生态坡屋顶让建筑群的天际线舒展而有层次变化。建筑不用昂贵的材料堆砌，而由生态黏土面砖、环保涂料、天然文化石与生态水泥瓦、透明玻璃、墨绿色喷涂铝合金等普通材料构成。建筑与其中的植物和水景共同形成生态丰富、典雅的立面效果。

设计建筑更是设计生活，倡导一种健康阳光的生活模式。建筑设计时而将自然引入建筑中，时而将建筑延伸入自然环境中，让人与自然和谐交流，使人们回归自然、享受自然。

▼ “731” 新餐厅入口广场水景实景

▲ 钢结构点支式玻璃门厅的对景——与自然共生的大餐厅

▲ 外走廊与小包房相连处形成的天井

◀ 将多功能用房串联起来的生态外廊

室内空间室外化、室外空间室内化，建筑艺术与技术完美结合

◀ 钢结构点支式玻璃门厅

▶ 大餐厅与玻璃门厅相交处的自然环境

▶ 通向康乐区的生态外廊

外走廊将各功能用房联系起来，并形成了各样的天井，其间的植物破土而出向天生长，用生命感动着人们。

▲ 中型会议室建成实景

▼ 庭园木平台休憩区

设计建筑就是设计生活

▼ 大套房客厅前的自然景观

◀ 湖南省外侨办“731”贵宾楼方案设计鸟瞰图

设计从整体规划到建筑单体，从内部空间到外部造型，再到环境设计一气呵成，甚至家具、装饰、雕塑都要统一精心配置。建筑艺术巧妙结合，创造出一个充满文化内涵的生态建筑空间，给人们带来美好的感受。

▲ 南昌科技大学图书馆方案设计

不同的建筑应有不同的性格

建筑与自然共生，给人以丰富的空间、诗一般的意境，不仅满足人们物质功能的需要，更给人以精神的升华，使人与自然，与建筑产生情景对话和心灵的共鸣。

建筑是有生命的，空间是建筑的灵魂，空间不仅是二维、三维（视觉的）、也是四维（时间的），更是五维（心灵的）。

▼ 双溪老年活动中心方案设计

▼ 武陵源国际度假酒店

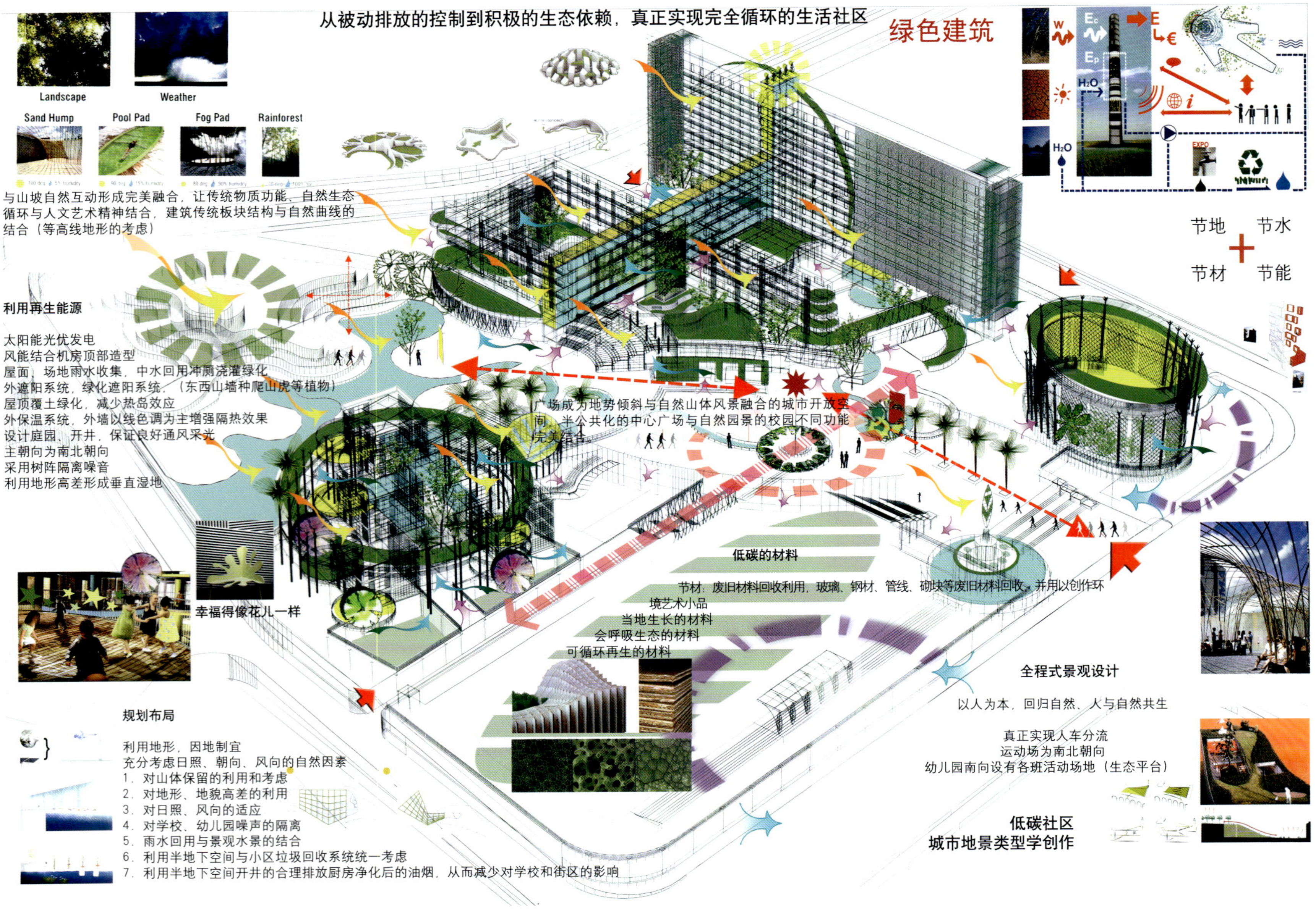

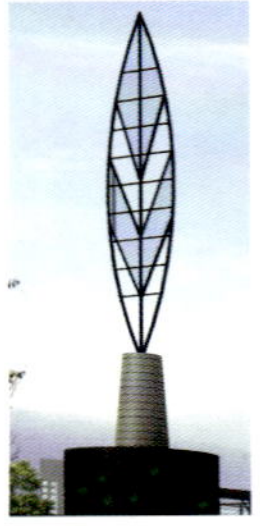

美学艺术来源于自然生活

仿生的设计——提炼自然的"碎片"，在建筑平面构图中，在建筑立面造型中，在景墙、铺地中，在雕塑等景观小品和家具中。

有机建筑

让建筑成为生命的载体
建筑充满自然语的联想和记忆
有别于冰冷工业生产化的住人机器

将花园作为建筑群的中心
将生命作为建筑的中心

曾益海工作室 ZENG YIHAI STUDIO

长沙有色冶金设计研究院 CHANGSHA ENGINEERING AND RESEACH INSTITUTE OF NONFERROUS METALLURGY

Tel: +86-731-84397126
Fax: +86-731-84458515
E-mail: zengyihai@yahoo.cn
www.cinf.com.cn

Yihai Zeng 曾益海

Class 1 registered architect(PrC) 国家一级注册建筑师
State registered urban planner 国家注册规划师
Member of Architectural society of China(AsC) 中国建筑学会会员
Member of the Hong Kong Institute of Architects(hKiA) 香港建筑师学会会员
Deputy Director of Hunan Architect Association 湖南省建筑师学会副理事长
Awarded Hunan Excellent Survey Designer 2008 2008 年被评为“湖南省优秀勘察计师”

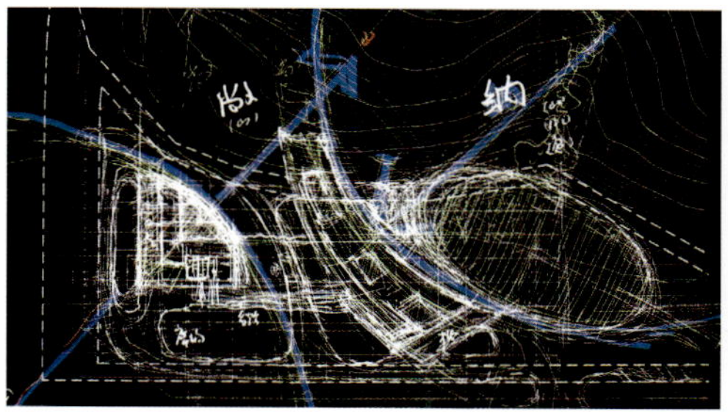

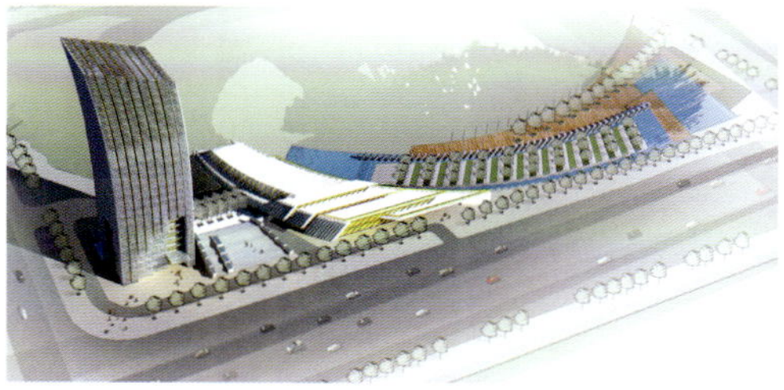

株洲外事大院概念设计

项目位置：湖南 株洲
项目规模：建筑面积 34 100 平方米
设计时间：2009 年

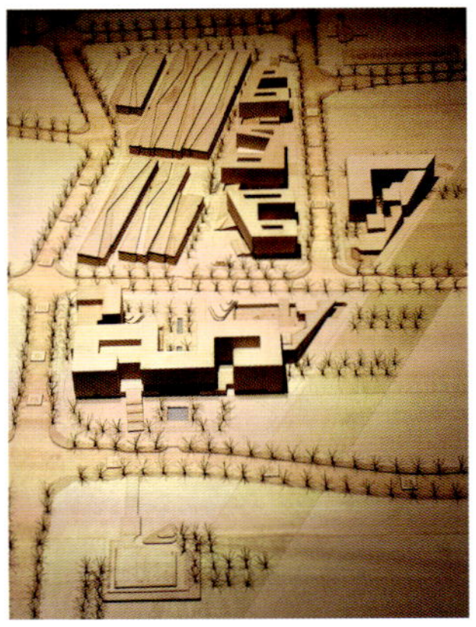

中铝集团科学技术研究院

项目位置：北京

项目规模：规划用地面积 19.6 公顷，建筑面积 300 000 平方米

设计时间：2010 年

湖南省环保厅办公楼

项目位置：湖南 长沙

项目规模：建筑面积 33 254 平方米

设计时间：2005 年

湘江一号住宅区茶室 ↗

项目位置：湖南 长沙
项目规模：175 平方米
设计时间：2009 年

花瑶风景名胜区接待中心 ↘

项目位置：湖南 邵阳
项目规模：2 800 平方米
设计时间：2010 年

现代投资大厦

项目位置：湖南 长沙
项目规模：建筑面积 84 732 平方米
设计时间：2006 年

长沙市国土局办公楼

项目位置：湖南 长沙
项目规模：建筑面积 31 437 平方米
设计时间：2002 年

湘江轮渡码头

项目位置：湖南 长沙
项目规模：建筑面积 8 599 平方米
设计时间：2008 年

武广客运高速长沙车站

项目位置：湖南 长沙
设计时间：2006 年

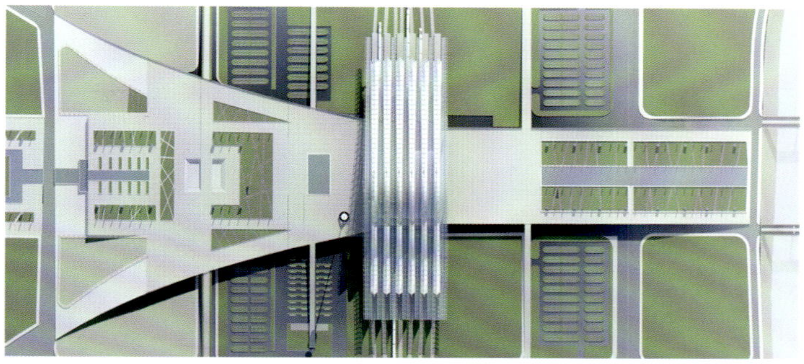

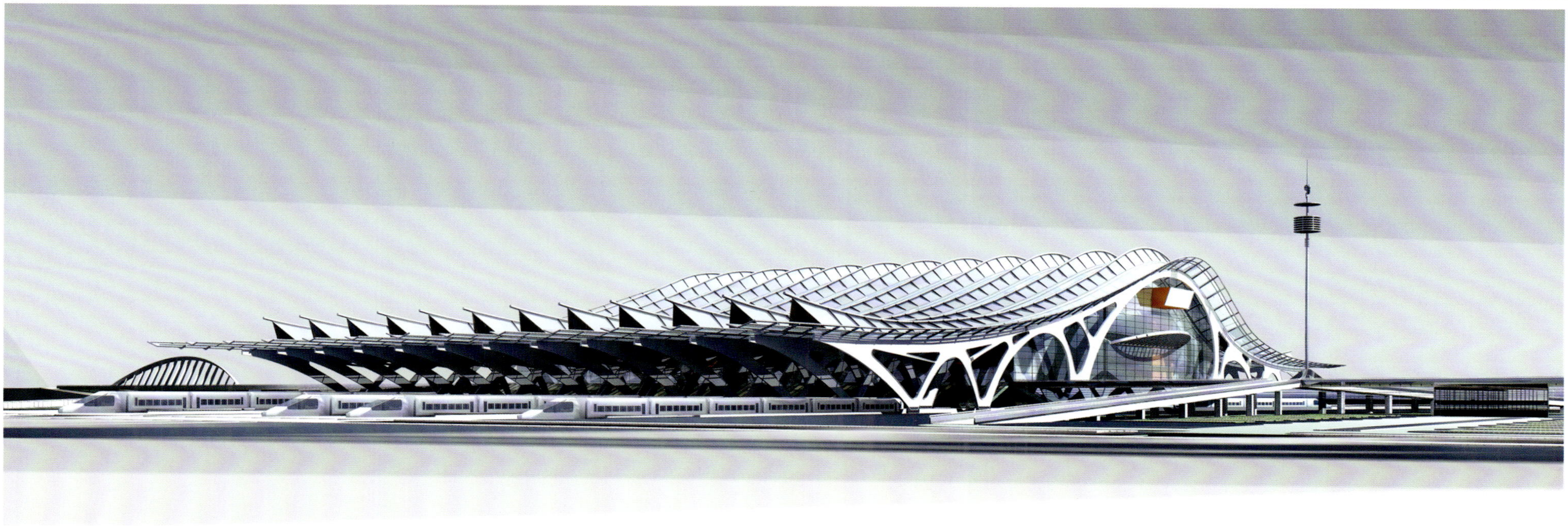

总体鸟瞰

长沙贾谊故居修复设计

贾谊是汉代著名思想家、政治家，其思想在两千年中对中国古代的政治思想产生了重要的影响。贾谊曾任长沙王太傅，其故居在长沙市太平街。贾谊去世后人们将其故居改成纪念贾谊的祠庙，称"贾太傅祠"，并发展为园林。1938 年贾谊故居毁于长沙大火。现长沙市政府决定修复贾谊故居，并委托柳肃教授担任修复设计工作。此项目于 2005—2008 年完成设计，预计 2011 年动工。

Design of Jia Yi's residence repair

Jia Yi is a famous thinker and statesman in Han Dynasty (202 BC -220 AD) and he has a great influence on Chinese traditional political thoughts. Jia was titled "Taifu" in Changsha, Hunan province, which was in charge of the country affairs when the emperor was young. His residence was located in Taiping Street of Changsha, and when Jia passed away, his residence became the memorial temple called "Jia Taifu's Temple". The temple was developed to a garden, but fell into ruins in a fire in 1938. The Changsha municipal government decided to repair his residence and entrusted Professor Liu Su to design it. The design was completed between 2005 and 2008, and it will start in 2011.

南向展开立面

北向展开立面

日本串木野日中文化交流园

两千年前，秦始皇派方士徐福带领三千童男童女和工匠到东海神山寻找长生不老的仙药，徐福一去不返，传说徐福当年就是在今天日本西海岸的串木野登陆的。日本政府认为这是中国文化第一次大规模传入日本，决定在此修建日中文化交流园，园中建造纯中国式园林建筑。经过考察，日方委托柳肃教授进行设计。柳肃教授于 2003 年开始设计，交流园于 2004 年建成。项目受到各方好评，日方还将"设计、监修，中国湖南大学柳肃"的字样刻成石碑嵌在亭子基座上。

Sino-Japanese Communication Garden in Kush Kino

Two thousand years ago, the first emperor of Qin Dynasty (221BC—207BC) sent Xu Fu headed for East Ocean to look for elixir with 3 000 pure youths and maidens and craftsmen, but Xu never returned. It was said that they landed in Kush Kino of Japan; Japanese government considered this the first time for Chinese culture to spread into Japan, so the government decided to establish Sino-Japanese Communication Garden where typical Chinese garden was built. Professor Liu Su was responsible for this design. The design was finished in 2003 and the garden was completed in 2004. The garden was well-praised and the character "Designer and supervisor-Liu Su, Hunan University of China" was carved into a stone tablet on the base of pavilion.

日本现场指导

整体建成

水映亭

LIUSU 柳肃 works

屈子祠外观

岳麓书院屈子祠复原设计

长沙岳麓书院是中国古代四大书院之一，是目前保存下来最大、最完整的书院之一。岳麓书院古代就有屈子祠，用以纪念古代大诗人屈原。由于屈子祠年久失修，加之周边环境的改变，于是研究决定在岳麓书院后部一片山洼中将其复原重建。该项目于 2003—2004 年完成设计，并于 2006 年建成。

Repair of Qu Yuan Temple in Yuelu Academy

Yuelu Academy in Changsha,Hunan Province is one of the biggest Chinese traditional academies and well-preserved ones. Qu Yuan Temple was established in ancient times mourning for the great poet QuYuan. Because it has been out of repair for long years plus changes in surroundings, it was decided to repair the Temple around the hills behind the Academy. The design was completed between 2003 and 2004 and the Temple was established in 2006.

后部鸟瞰

后部回廊

内部小庭院

漏窗

曾国藩墓区修复规划设计

曾国藩墓在望城县坪塘镇，原有御碑亭、墓庐屋、牌楼、神道等建筑。御碑亭和墓庐屋内有当时皇帝和慈禧太后题写的巨幅御碑。“文革”期间地面建筑被毁，但御碑、牌楼和神道的石构件仍有相当部分保存下来，湖南省文物局和长沙市文物局决定将其重新修复，并委托柳肃教授进行规划设计。该项目于2005年完成规划设计方案，并通过专家论证评审，目前正在施工建设之中。

Design of Zeng Guofan's tomb restoration

Zeng Guofan's tomb was situated in Pingtang Town of Wangcheng County, Hunan province, where pavilions housing royal tablets, Mulu houses, memonial arches and other buildings were built. There were large stone tablet inscribed by Emperor and Empress Dowager Cixi inside pavilions and Mulu houses. During the Cultural Revolution, the ground buildings were demolished, but some tablets and memorial arches were preserved. Cultural Relics Bureaus of Hunan Province and Changsha city decided to repair it and entrusted Liu Su to design it. In 2005, the design plan was finished and the project is under construction after the experts' inspection.

曾国藩墓复原规划前

曾国藩墓复原规划后

故居复原建成后实景

田汉故居复原设计

田汉是我国著名作曲家，共和国国歌的作者。其故居在长沙县果园镇，“文革”期间被毁，成为一片田地。2005 年当地政府决定将其复原重建，并委托柳肃教授进行设计。本设计是在对现场进行考古发掘并作了大量调查研究的基础上完成的，该项目于 2006 年建成。

Design of Tian Han's residence repair

Tian Han is a famous Chinese composer and author of the national anthem. His former residence at Guoyuan Town of Changsha County, Hunan province was destroyed during the Cultural Revolution and became a field. In 2005, the local government decided to restore it and entrusted Liu Su to design it. The design was completed after a great amount of archaeological explorations and studies and the project ended in 2006.

修复前近景

修复加固后

望城县惜字塔修复保护设计

望城县茶亭镇惜字塔由于某种特殊原因在塔顶上长出一株大树。大树已有百年以上树龄，成为树塔结合的天下奇观。但树根的生长对塔体结构造成破坏，使塔有倒塌的危险。当地政府委托柳肃教授进行修复保护设计。在进行建筑、结构、植物等大量研究的基础上，2006–2008 年完成设计，并于 2009 年完成施工，待到新叶长出，证明保护修复完全成功。

Design of Xizi Tower's repair and protection

At the top of Xizi Tower in Chating Town of Wangcheng County province grew a tree for some reason. The tree was over one hundred years old and the combination of the tree and tower has become a world wonder. But the growing root damaged the tower and caused the falling anytime, the local government entrusted Liu Su to take the design to repair and protect the tower. After studies on buiding, structure and plant, Liu finished the design between 2006 and 2008 and the project was completed in 2009. The new leaves will prove the success of the protection.

崇圣祠

明伦堂

岳麓书院崇圣祠、明伦堂修复设计

崇圣祠、明伦堂是长沙岳麓书院文庙的重要组成部分，1938 年毁于战火，一直未再重建。2002 年决定重建，委托柳肃教授进行修复设计。该项目于 2003 年完成设计，并于 2004 年建成。

Design of Chongsheng Temple and Minglun Hall repairs inside Yuelu Academy

Chongsheng Temple and Minglun Hall were important parts of the Confucian Temple in Yuelu Academy, and were destroyed in a battle of 1938. In 2002, it was decided to rebuild the temple and the hall and Professor Liu Su was commissioned to design it. The design was accomplished in 2003 and the buildings were completed in 2004.

LIUSU 柳 肃 works

茶陵洣江书院复原重建设计

茶陵洣江书院复原重建设计

湖南茶陵是古代教育发达之地，宋至清代曾出过数十名进士、状元。洣江书院是茶陵古代最著名的书院之一，“文革”中被毁。当地政府决定重建，委托柳肃教授进行设计。该项目于 2009 年完成设计，目前正在施工中。

Design of Mijiang Academy repair in Chaling

The ancient education was well developed in Chaling, Hunan province and scores of excellent scholars were born here from Song to Qing Dynasty. Mijiang Academy was one of well-known Academies in ancient Chaling and was destroyed during Cultural Revolution. Local government decided to rebuild it and entrusted Liu Su to design it. In 2009, the design was finished and the project is under construction at present.

茶陵城楼复原重建

茶陵城楼复原重建

茶陵县古城墙是湖南省内保存最完好的，也是唯一的一段宋代城墙。其城楼在“文革”期间被毁，并在上面建了工厂，对城墙造成破坏。当地政府委托柳肃教授进行复原重建设计。该项目于 2005 年完成设计，并于 2007 年建成。

Restoration and reconstruction of Chaling gate tower

Ancient city wall within Chaling County, Hunan Province, is the best preserved and only section of Song Dynasty (960AD—1279). Its gate tower was destroyed during the Cultural Revolution, and on the top of it built a factory which did damage to the wall. Local Government commissioned Professor Liu Su to take the job of restoration and reconstruction design. In 2005, Liu finished the design, and the construction was completed in 2007.

张国焘故居修复

张国焘是中国共产党早期领导人之一，其故居在江西省上栗县，原是一个大家族聚居的传统民居。因年久失修，大部分已毁，当地政府委托柳肃教授进行修复设计。该项目于 2009 年完成设计，目前正在施工之中。

Repair of Zhang Guotao’s Residence

Zhang Guotao was one of the early leaders of Chinese Coummunist Party, and his residence was located in Shangli County, Jiangxi province, which was a traditional house inhabited by a big family. Because the residence was in bad repair and destroyed mostly, local government commissioned Professor Liu Su to design it. The design was finished in 2009 and the construction is in process now.

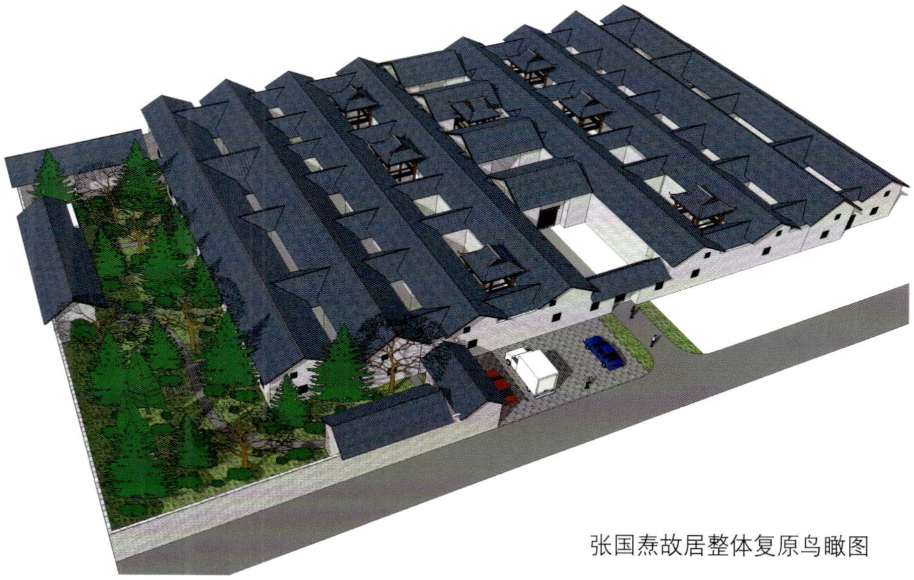
张国焘故居整体复原鸟瞰图

湘潭市建筑设计院

XIANGTAN CITY BUILDING DESIGN INSTITUTE

湘潭市建筑设计院始建于 1980 年，1987 年成为湖南省地市级设计院中首个获得甲级设计资质的设计院，目前拥有工程咨询甲级资质，建筑工程设计甲级资质，市政道路工程甲级设计资质，工程监理甲级资质，城市规划乙级资质，园林景观乙级资质。湘潭市建筑设计院是一个专业技术人才密集的国有科技型企业，拥有一支具有较强竞争力、高素质、专业化的设计人才队伍。设计院现有职工近 300 人，各类技术人员近 200 人，其中国家注册工程师近百人。院下还设有一个城乡规划设计所、两个方案创作室、一个市政景观设计所、四个建筑工程综合设计所，以及重庆分院、广西分院、东莞分院、温州分院、惠州分院、广州分院、华夏监理公司等下属机构。经过 30 年的发展壮大，我院综合实力不断增强，业务范围不断扩大，经济效益日渐递增，2009 年全院湘潭本部合同额近亿元，人均产值跻身于湖南省设计行业前列。

湘潭市建筑设计院以"质量第一、信誉第一、服务至上、力创精品"为宗旨，采用 ISO 9001:2000 质量管理体系，在奉献社会的同时，不断发展自己。近年来我院已与国内外众多知名设计单位和事务所建立了长期良好的合作关系，在全国范围内承接各类设计项目数千项。良好的设计均得到了建设主管部门、业主和广大市民的好评。近几年先后有彭德怀纪念馆、广西钦州文化中心、齐白石纪念馆、重庆市委党校学员综合楼、湘潭市岳塘区行政中心、厦门集美师专教学楼、新疆伊宁大酒店、湘潭市九华经济区办公大楼、广东技师学院教学楼、湖南华菱湘潭钢铁有限公司行政中心、湘潭市体育中心等 30 多个项目获国家或省市优秀设计奖，其中 2009 年度优秀作品数量和奖项在湖南省 200 多家设计单位中排名第二，华菱湘潭钢铁有限公司行政中心更是同时荣获 2009 年全国优秀工程勘察设计行业奖三等奖和湖南省 2009 年度优秀工程设计一等奖，为城市建设增添了一道道靓丽的风景线。

精诚合作是湘潭市建筑设计院一贯的作风，开拓创新是永恒的追求，让客户满意是永远的信念，创建精品设计是坚定的誓言。湘潭市建筑设计院愿意竭诚为广大业主提供优质设计和服务，携手奋进，实现新的跨越！

Xiangtan City Building Design Institute founded in 1980, became the first design institute who received Class A design qualification in Hunan province in 1987. Now it has Grade-A qualification of engineering consulting, Grade-A qualification of architecture and construction design, Grade-A qualification of Municipal Road engineering design, Grade A certificate for construction supervision, Grade-B qualification of urban planning and Class-B qualification of landscape architecture. Xiangtan City Building Design Institute is a professional technicians intensive state-owned science and technology enterprise, with a highly competitive, high-quality, professional design talent team. Now it has nearly 300 employees, including nearly 200 technicians of all kinds, and about 100 national registered engineers. The institute has one urban and rural planning design institute, two program creation workshop, one municipal landscape design institute, and four construction engineering comprehensive design institutes. It also has branch institutes in Chongqing, Guangxi, Dongguan, Wenzhou, Huizhou and Guangzhou, and Huaxia Supervision Company. After 30 years development, with the continuous growth in comprehensive strength, our institute continuously expands its business scope, and dramatically increases its economic efficiency. Xiangtan institute achieved the contract of nearly 100 million yuan and its per capita output value ranked among the best of Hunan design industry in 2009.

In the principle of "quality first, credit first, service-oriented, and making every effort to achieve competitive products", Xiangtan City Building Design Institute applies the ISO9001:2000 quality management system to constantly develop itself while contributing to society. In recent years, our institute has established long-term good cooperative relationship with many domestic and international renowned design units and business quarters. It has undertaken thousands of design items of many kinds throughout the nation and its good design has received a favorable report from the construction administrative department, business proprietor and the wider public. In recent years, there have been more than 30 projects which have won the national or provincial excellent design award, for example, Peng Dehuai memorial hall, Guangxi Qinzhou cultural center, Qi Baishi memorial hall, students' complex building of Chongqing Party School of CPC, administration center of Yuetang District in Xiangtan, classroom building of Xiamen Jimei teachers training school, Xinjiang Yining Hotel, office building of Jiuhua economic region in Xiangtan, classroom building of Guangdong Technician College, administration center of Hualing Xiangtan Steel Ltd. of Hunan province, and Xiangtan sports centre. The number and awards of excellent works in 2009 was ranked second in more than 200 design units in Hunan province. The administration center of Hualing Xiangtan Steel Ltd. was award the third prize of national award for outstanding engineering survey and design industry in 2009 as well as the first prize of excellent engineering design in Hunan province in 2009, which adds a beautiful scene to urban construction.

Sincere cooperation is our persistent style of work; innovation is our eternal pursuit; customer satisfaction is our perpetual belief; creating boutique design is our firm pledge. Xiangtan City Building Design Institute is willing to wholeheartedly provide quality design and service to all our clients, and advanced hand in hand to achieve a new leap!

地址：湖南省湘潭市雨湖区熙春路 300 号
法人：杨晓军（院长）
邮编：411100
电话：+86–731–58223227
传真：+86–731–58223227
邮箱：xtjzsj@xtjzsj.com
网址：www.xtjzsj.com

Address: N0.300 Xichun Road, Yuhu District, Xiangtan city, Hunan province
Legal: Xiaojun Yang (President)
P. C.: 411100
Tel: +86-731-58223227
Fax: +86-731-58223227
E-mail: xtjzsj@xtjzsj.com
Http: //www.xtjzsj.com

华菱湘钢职工体育活动中心
建设地点：湖南 湘潭
竣工时间：2007 年
建筑面积：8 000 平方米
获奖情况：湖南省 2009 年度优秀工程设计二等奖

Staff sports center of Hualing Xianggang
Construction Location: Xiangtan, Hunan
Completion Time: 2007
Building Area: 8,000 m^2
Awards: Second prize of excellent engineering design in Hunan Province in 2009

湖南华菱湘潭钢铁有限公司行政中心
建设地点：湖南 湘潭
竣工时间：2007 年
建筑面积：8 000 平方米
获奖情况：2009 年全国优秀工程勘察设计行业奖三等奖
湖南省 2009 年度优秀工程设计一等奖

Administration Center of Hualing Xiangtan Steel Ltd. of Hunan
Construction Location: Xiangtan, Hunan
Completion time: 2007
Building Area: 8,000 m^2
Awards: Third prize of national award for outstanding engineering survey and design industry in 2009
First prize of excellent engineering design in Hunan Province in 2009

彭德怀纪念馆
建设地点：湖南 湘潭
竣工时间：1998 年
建筑面积：4 000 平方米
获奖情况：湖南省 2000 年度优秀工程设计一等奖

Peng Dehuai memorial hall
Construction Location: Xiangtan, Hunan
Completion time: 1998
Building Area: 4,000 m^2
Awards: First prize of excellent engineering design in Hunan Province in 2000

湖南省湘潭市岳塘区行政中心
建设地点：湖南 湘潭
竣工时间：2007 年
建筑面积：45 000 平方米
获奖情况：湖南省 2008 年度优秀工程设计二等奖

Administration Center of Yuetang District in Xiangtan, Hunan Province
Construction Location: Xiangtan, Hunan
Completion time: 2007
Building Area: 45,000 m^2
Awards: Second prize of excellent engineering design in Hunan Province in 2008

湘潭九华经济区九华大厦
建设地点：湖南 湘潭
竣工时间：2007 年
建筑面积：23 000 平方米
获奖情况：湖南省 2008 年度优秀工程设计三等奖

Jiuhua Building in Jiuhua Economic Region in Xiangtan
Construction Location: Xiangtan, Hunan
Completion time: 2007
Building Area: 23,000 m^2
Awards: Third prize of excellent engineering design in Hunan Province in 2008

湘潭市体育中心
建设地点：湖南 湘潭
竣工时间：2010 年
建筑面积：25 000 平方米

Sports Centre of Xiangtan city
Construction Location: Xiangtan, Hunan
Completion time: 2010
Building Area: 25,000 m^2

湘潭市国土资源交易大厦
建设地点：湖南 湘潭
竣工时间：2010 年
建筑面积：18 000 平方米

Territorial resources exchange building in Xiangtan
Construction Location: Xiangtan, Hunan
Completion time: 2010
Building Area: 18,000 m^2

湘潭市商业银行
建设地点：湖南 湘潭
竣工时间：2006 年
建筑面积：20 000 平方米
获奖情况：湖南省 2008 年度优秀工程设计三等奖

Xiangtan Commercial Bank
Construction Location: Xiangtan, Hunan
Completion time: 2006
Building Area: 20,000 m^2
Awards: Third prize of excellent engineering design in Hunan Province in 2008

湘潭市广电中心
建设地点：湖南 湘潭
竣工时间：2006 年
建筑面积：22 000 平方米
获奖情况：湖南省 2006 年度优秀工程设计二等奖

Xiangtan Broadcast and Television Center
Construction Location: Xiangtan, Hunan
Completion Time: 2006
Building Area: 22,000 m^2
Awards: Second prize of excellent engineering design in Hunan Province in 2006

湘潭市政协办公楼
建设地点：湖南 湘潭
竣工时间：2006 年
建筑面积：9 000 平方米
获奖情况：湖南省 2008 年度优秀工程设计二等奖

CPPCC office building of Xiangtan city
Construction Location: Xiangtan, Hunan
Completion Time: 2006
Building Area: 9,000 m^2
Awards: Second prize of excellent engineering design in Hunan Province in 2008

湘潭市齐白石艺术中心
建设地点：湖南 湘潭
竣工时间：2005 年
建筑面积：10 000 平方米
获奖情况：湖南省 2005 年度优秀工程设计二等奖

Qi Baishi Arts Centre in Xiangtan
Construction Location: Xiangtan, Hunan
Completion Time: 2005
Building Area: 10,000 m^2
Awards: Second prize of excellent engineering design in Hunan Province in 2005

土木風設計
TUMUFENG DESIGN

吉林土木風建築工程設計院有限公司

Ji Lin TMF Architecture & Engineering Design limited Company

土木风设计是由吉林土木风建筑工程设计有限公司和吉林土木风环境艺术设计有限公司组成的集建筑设计、规划设计、景观设计、装饰设计为一体的综合性设计公司。公司持有建设部颁发的建筑工程设计：装饰工程设计甲级资质，规划设计、风景园林设计乙级资质。

公司现有各级专业设计人员 78 人，其中国家一级注册建筑师 6 人，一级注册结构工程师 7 人，国家注册电气工程师、国家注册公用设备 6 人，国家注册规划师 3 人，省级青年设计大师 1 人。公司通过设计实践不断吸收和学习国内外优秀设计公司的成功经验和先进的经营理念，努力将公司建设成为省内一流，在国内具有影响力的集建筑、规划、景观、装饰四位一体的设计公司。

Tumufeng design is composed of Jijlin Tumufeng Architectural Construction-Engineering Design Co., Ltd. and JijlinTumufeng Environmental Art Design Co., Ltd. It is a comprehensive deisign compay which provides architectural design, planning design, landscape design and decoration design. Currently, it has obtained the class A qualification of architectural engineering design and decoration construction design, and landscape architecture deisgn class B qualification.

The company now has 78 professional deisign staff, including 6 national class 1 certified architects, 7 national class 1 registered structural engineers, 6 national registered electricengineers and national registered HVAC engineers, 3 national registered planners and one provincial youngster design master. The company keeps learning from the excellent international design companies, drawing their experiences and advanced management ideas through practice, striving to establish a provincial first roted and national famous deisign company which covers architecture, landscape planning and design and decoration areas.

法定代表人：石铁军
地址：吉林省长春市东民主大街 519 号
电话：+86-431-88591360
传真：+86-431-88591360
邮箱：tmfjz@163.com
网址：www.tumufeng.com

Legal representative: Tiejun Shi
Address: No.519, Minzhu Street, Changchun, Jilin
Tel: +86-431-88591360
Fax: +86-431-88591360
Email: tmfjz@163.com
Http: //www.tumufeng.com

荣大天玺酒店

建设地点：吉林 长春
项目类别：办公楼
建筑面积：53 400 平方米
设计时间：2010 年 1 月

Rongda Royal Seal Hotel

Location: Changchun, Jilin
Project Type: Office building
Building Area: 53,400 m^2
Design Time: 2010.01

安华保险综合楼

建设地点：吉林 长春
项目类别：办公楼
建筑面积：45 000 平方米
设计时间：2009 年 8 月

AnHua Insurance complex building

Location: Changchun, Jilin
Project Type: Office building
Building Area: 45,000 m^2
Design Time: 2009.08

长春市高新北区兴华学校及幼儿园

建设地点：吉林 长春
项目类别：办公楼
建筑面积：21 000 平方米
设计时间：2009 年 6 月

Xinghua school and kindergarden in north of high and new echnologyarea in Changchun

Location: Changchun, Jilin
Project type: office building
Gross flor area: 21,000 m^2
Deisign time: 2009.06

吉林农委办公楼

建设地点：吉林 长春
项目类别：办公楼
建筑面积：40 000 平方米
设计时间：2009 年 12 月

Jinlin municipal commission of rural affairs building office

Location: Changchun, Jilin
Project type: Office building
Building Area: 40,000 m^2
Design Time: 2009.12

长春市社保办公楼

建设地点：吉林 长春
项目类别：办公楼
建筑面积：32 000 平方米
设计时间：2010 年 2 月

Social Security Office building in Changchun city

Location: Changchun, Jilin
Project type: Office building
Building Area: 32,000 m^2
Design Time: 2010.02

长春市兴瑞冲压件有限公司

建设地点：吉林 长春
项目类别：办公楼
建筑面积：19 079 平方米
设计时间：2008 年 10 月

Changchun Xingrui Stamping Co., Ltd.

Location: Changchun, Jilin
Project type: Office building
Building Area: 19,079 m^2
Design Time: 2008.10

敦化市渤海活动中心

建设地点：吉林 敦化
项目类别：体育馆
建筑面积：7 200 平方米
设计时间：2009 年 4 日

Bohai Civic Center in Dunhua

Location: Dunhua, Jilin
Project type: Stadium
Building Area: 7,200 m^2
Design Time: 2009.04

长春天伦中央区规划方案

建设地点：吉林 长春
项目类别：商业综合体
建筑面积：129 079 平方米
设计时间：2008 年 8 月

Planning project of Tianlun Central District in Changchun

Location: Changchun, Jilin
Project type: Commercial building complex
Building Area: 129,079 m^2
Design Time: 2008.08

长春市市委党校

建设地点：吉林 长春
项目类别：办公楼
建筑面积：109 079 平方米
设计时间：2010 年 7 月

Changchun party school

Location: Changchun, Jilin
Project type: Office building
Building Area: 109,079 m^2
Design Time: 2010.07

长春市中医药大学附属医院

建设地点：吉林 长春
项目类别：医院
建筑面积：67 000 平方米
设计时间：2008 年 7 月

Changchun party school

Location: Changchun, Jilin
Project type: Office building
Building Area: 109,079 m^2
Design Time: 2010.07

東南大學建築設計研究院
Architecture Design & Research Institute of Southeast University

地 址：江苏省南京市四牌楼二号
邮 编：210096
电 话：+86-25-83793178
传 真：+86-25-57713341
邮 箱：vinh@seu.edu.cn
网 址：www.adriseu.com

Address: No.2,Sipailou,Nanjing, Jiangsu
P.C.:210096
Tel: +86-25-83793178
Fax: +86-25-57713341
E-mail: vinh@seu.edu.cn
Http: //www.adriseu.com

东南大学建筑设计研究院始建于1965年，属国家教育部领导，是国内一流的著名设计院之一。目前，全院共有399人，其中具有中、高级技术职称的有员工240人，国家一级注册建筑师53人、一级注册结构工程师42人，其他注册工程师、造价师86人。

本院现有建筑行业建筑工程设计甲级、公路行业（公路）甲级、市政行业（道路、桥梁、风景园林）甲级、文物保护规划与设计甲级、智能建筑（系统工程设计）甲级、电力行业（火力发电、新能源发电）乙级、市政公用行业（热力、环境卫生）乙级等工程设计资质，另具建筑咨询甲级、公路咨询甲级、市政公用工程（道路、桥隧、给排水）咨询甲级资质。

自本院建院以来，已在全国各地承接完成了数千项工程设计任务，尤其是改革开放以来，本院承接完成了周恩来纪念馆、吴建雄纪念馆、武汉大学人文科学馆、南京大学——霍普金斯大学中美文化交流中心、中国近代史遗址博物馆文化服务区、南京国际展览中心、南京狮子山阅江楼、金陵图书馆新馆、江苏省档案新馆、南京国投大厦、南京国泰君安金融中心、江苏银行总部办公楼、南京紫金山庄（五星级）、南京电视台演播中心、南京鼓楼医院仙林国际医院、南京大观·天地MALL、东莞人民大会堂、深圳龙岗文化体育中心、太仓市行政文化中心、太仓市传媒中心、东南大学九龙湖新校区、南师大新校区。与境外设计公司合作完成了南京国际金融中心、中国国际艺术实践展项目、南京丁山花园香格里拉酒店（五星级）、苏州吴宫喜来登大酒店（五星级）等数百项大型办公、宾馆、度假村、商业中心、写字楼及文化、体育、医院、纪念建筑、高等院校的建筑、规划设计任务。同时，本院还承接完成了南京市宁溧路快速化改造、南京市玄武大道快速化改造、湖州市南太湖特大桥、江阴市澄南大道、澄东大道、芙蓉大道改建、昆山市绿地大道、中央党校掠雁湖景观桥梁等省市级重点工程、江苏双良科技热电公司热电项目、江苏阳光热电有限公司热电厂、菲律宾大壮钢铁高炉煤气发电工程、印尼SPV(PT.SOUTH PACIFIC VISCOSE)公司热电工程、常州市武进区垃圾(600t/d)焚烧项目、南京华润热电330MW机组供热项目、江苏王子制纸自备发电站、南通联海生物热电公司生物质综合利用热电项目、南京长江二桥公园、南京聚宝山公园、巩义市文化公园、东莞市行政文化中心广场等大量市政、交通、热电和景观工程设计及咨询任务。

本院以"精心设计，勇于创新，讲究信誉，优质服务，持续改进，顾客满意"为质量方针，以ISO 9001–2008标准为管理体系，取得了很好的公司信誉和社会效益。在历届国家和部委、省、市的优秀设计评选中取得了优异的成绩，本院先后荣获各级优秀设计奖四百多项，其中国家级银质奖6项、铜质奖3项，"全国建筑设计行业国庆60周年建筑设计大奖"1项，中国建筑学会授予的建国60周年不同时期我国现代建筑优秀代表性作品的最高奖项"中国建筑学会建筑创作奖"4项，建设部优秀设计奖及由建设部委托全国工程勘察设计行业协会评选的"全国工程勘察设计行业优秀工程勘察设计行业奖"50项，教育部优秀设计奖90项，江苏省优秀设计奖130项，江苏省科技进步奖二等奖三项。2008年获江苏省勘察设计企业综合实力排序（建筑设计）前十名（本院排名第一），2009年获得江苏省勘察设计质量管理先进单位和江苏省工程勘察设计行业"诚信单位"。

本院的企业精神是"始于点划，止于至善"，全院员工在工作实践中将积极创新、追求卓越，努力使客户更满意，让世界更精彩。

Architectural Design & Research Institute of Southeast University was founded in 1965, it is a prestigious first class design institute under the leadership of Ministry of Education. Now it has 399 staffs, including 240 staffs with intermediate and senior technical professional titles, 53 national Class 1 registered architects, 42 Class 1 registered structure engineers, and 86 registered engineers and registration moderators in other relevant areas.

Architectural design & research Institute of Southeast University currently obtains numerous professional qualifications, including class A qualification of architecture construction design, class A qualification of road design, Class A qualification of municipal design(roads, bridges, landscape architecture), historical relic protection and design Class A qualification, intelligent building(system engineering design) design Class A qualification, power industry(fuel-burning power and new energy power) Class B qualification, Class B qualification of municipal public design utility design(heating power and environmental sanitation). Besides, it also has the Class A qualification of architecture consulting, road design consulting and municipal public engineering (roads, tunnels, water supply and drainage).

Since the foundation, the institute has undertaken thousands of national design projects. Especially after the reform and opening up policy implemented in China, the Insitute has undertaken a series of programs including Zhou Enlai Memorial Hall, Wu Jianxiong Memorial Hall, Humanities and Science Museum of Wuhan University, Sino-American Cultural Exchange Center of Nanjing University - Johns Hopkins University, cultural service area of the site of modern Chinese history museum, the Nanjing International Exhibition Center, Yuejiang Tower of Nanjing Lion Rock, Jinling New Library, Archives New Museum in Jiangsu Province, Nanjing National Investment Building, Nanjing Guotai Junan Financial Center, headquarters building of Jiangsu Bank, Nanjing Zijin Resort (five star), television studio in Nanjing, Nanjing Xianlin Drum Tower Hospital, International Hospital, Nanjing Grand World Tiandi MALL, Dongguan Great Hall, Longgang cultural and sports center in Shenzhen, Taicang administrative and cultural center, Media Centre of Taicang City, Jiulonghu new campus of Southeastern University, new campus of Nanjing Normal University, co-complete an international financial center in Nanjing with out-China design companies, China International Art practice development project, Nanjing Dingshan Garden Shangri-La Hotel (five star), Suzhou WuGong Sheraton Hotel (five star) and other hundreds of large-scale programs in functions of offices, hotel, resorts, commercial centers, office buildings and culture, sports, hospitals, monuments, institutions of higher learning in the architecture, planning and design projects. At the same time, it also has undertaken the shorttime Reconstruction of Nanjing NingLi Road, Reconstruction of Nanjing Xuanwu Avenue, Huzhou South Tai Lake Bridge, Chengnan Road and Chengdong Road of Jiangyin City, Furong Avenue, Green Road of Kunshan City, Central School of landscape bridges Lueyan Lake and other provincial and municipal key projects, Jiangsu Shuangliang Technology Thermal Power Corporation Cogeneration, Thermoelectric Power Plant of Jiangsu Sunshine Co., Ltd., Philippines Dazhuang steel blast furnace gas power generation projects in Indonesia SPV (PT.SOUTH PACIFIC VISCOSE) Thermal Engineering Company, Wujin District garbage burning project of Changzhou City (600t / d), Nanjing China Resources Thermal Power 330 MW Unit Heating Project, Jiangsu Oji-owned power plants, Integrated biomass thermal power project Nantong Lianhai thermal utilization company. Nanjing No. 2 Yangtse River Bridge park, Nanjing Jubaoshan park, Gongyi culture park, Gongguan municipal cultural centre square, municipal design, transportation, pyroelectricity and landscape design and consulting task.

With the guide line of aiming at intensive design and innovation, in the pursuit of good credit , excellent quality. Making constant progress, achieve satisfaction of our clients, the institute brings the"ISO9001-2008"standard into administration system, gaining good reputation and social benefit. In the previous national ministries, provincial and municipal excellent design elections, they made outstanding performance, awarded over 400 excellent design awards of all levels, including 6 national silver medals, 3 bronze medals, one national architectural design industry architecture design award of China's 60th Anniversary national day, 4 architectural creation awards of ASC, the highest prize of national architecture design representative works in modern China, which was given by Architectural Society of China. 50 excellent design awards of Construction Ministry and other prizes of China investigation & design industry excellent engineering design awards, as well as 90 excellent design awards of Ministry of education, 130 excellent design awards of Jiangsu province, the second and third prize of technology advancement awards of Jiangsu province. In 2008, it is in the first place of the top 10 best investigation & design enterprise (architecture design) in Jiangsu. In 2009, it is awarded the advanced unit of quality management and the credit enterprise of Jiangsu investigation & design industry.

The design philosophy of the institute is starting from the details and completing by the perfection. All the staff in the institute actively pursue the innovation and outstanding design in their working practice, striving for greater satisfaction of clients and making a wonderful world.

东南大学建筑设计研究院

1 南京南捕厅历史文化风貌保护区四号地块三期工程（南区）

2 淮安电视塔

3 联创科技大厦

东南大学建筑设计研究院

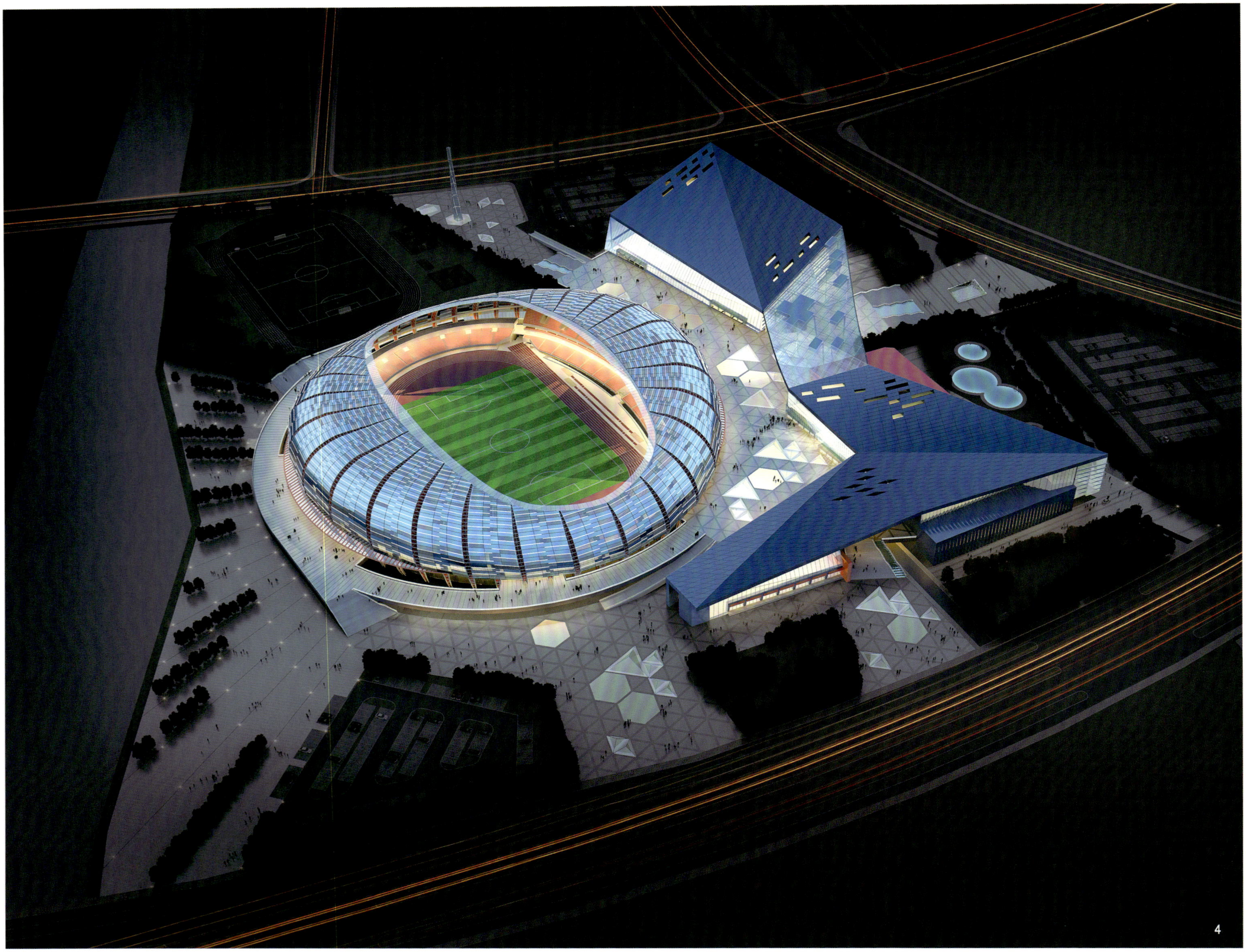

4 靖江体育中心

东南大学建筑设计研究院

5 国家电网公司智能电网科研产业（南京）基地

6 江苏银行总部大楼

7 人民日报社报刊业务综合楼

Architectural

设计作品
Design Works

長江都市
CHANGJIANG URBAN

地址：南京市洪武路 328 号　　邮编：210002
院服务台：+86-25-84567200
市场经营部电话：+86-25-84567208
院长办公室电话：+86-25-84567201
行政部电话：+86-25-84567213
传　真：+86-25-84567300
电子邮箱：jihuake@nanjing-design.com

Address: No 328 Hongwu Road, Nanjing　　P.C.: 210002
Institute Service Desk: +86-25-84567200
Marketing Department Tel: +86-25-84567208
Dean Office Tel: +86-25-84567201
Administration Department Tel: +86-25-84567213
Fax: +86-25-84567300
E-mail: jihuake@nanjing-design.com

南京长江都市建筑设计股份有限公司始创于 1976 年，前身为南京市民用建筑设计研究院，2009 年底由原公司派生分立。自创建至今历经了三十年的发展，形成以城市设计、办公建筑、商业建筑、科研教育建筑、居住建筑设计为特色和主要产品，且具有一定实力的区域性、综合性、有中等规模和鲜明特色的现代建筑设计企业，在江苏省乃至华东地区具较高声誉及竞争力。

公司目前具有国家建筑行业甲级设计资质，设有城市设计、居住区规划、建筑、结构、给排水、电气、智能化、采暖与通风、人防工程、室内装饰、景观园林、计算机应用等专业。能承接各类民用建筑及工业建筑的可行性研究、城市规划、建筑设计、技术咨询以及工程项目管理等相关业务。

公司现有在职员工 287 人，设计技术人员占总人数 90%，其中具有建筑师、工程师中级以上职称的专业技术人员占设计、技术人员总人数的 54%。具有国家注册建筑师、注册结构工程师、注册设备专业取得注册执业资格的员工占设计人员 30%。

我司近几年先后设计了许多功能齐全、技术复杂并且有较大影响的大中型公共建筑，如南京长发中心超高层建筑群、苏宁南京鼓楼科技园超高层办公群、南京总统府历史文化旅游街区（1912 街区）、芜湖侨鸿国际大酒店（五星级）、红太阳华东摩尔、苏宁电器成都商业中心、滨海体育中心、南京航空航天大学江宁校区东区、南京徐庄软件产业基地、南京中山植物园展览温室等。

作为江苏省内规模最大的居住建筑专业化设计企业，我司设计资质齐备、人力资源充足、技术力量雄厚、设计业绩众多。针对房地产市场的迫切需求，率先展开了设计总包一体化全过程设计服务工作。包括前期策划、居住区规划、居住建筑设计、景观园林设计及成品房精装修设计。向客户提供高质量的全方位技术服务，实现居住建筑高品质保证。在互利双赢的合作基础上，我司与万科地产、江苏新城股份公司、金地地产南京公司、碧桂园集团、江苏保利地产、江苏苏宁置业等大客户建立了战略合作伙伴关系。使设计合作更顺畅、配合更密切、效益更大化。

在关系国计民生的政府解决城市中低收入者住房主导产品方面，我司对政府保障性住房设计近年来投入较大力量，也取得了很大成绩，如设计规划了 2010 年度南京市政府重大民生项目——丁家庄创业园保障性住房和西善桥岱山新城（目前全市最大保障性住房项目近 400 万）。在大量设计实践的基础上编撰了《南京市保障性住房设计导则》和《南京市保障性住房小区景观设计导则》，为南京市政府保障性住房建设工作作出贡献，同时成为该领域权威设计机构。

公司除规划设计以外还致力于住宅科技方面的研究，先后主编了江苏省《住宅设计标准》和大型住宅设计类工具书《住宅设计资料集》。同时，我司率先在省内住宅领域研究绿色建筑技术及在实际工程设计中的运用，并取得阶段性成果，在全省住宅设计领域绿色建筑设计和住宅工业化推广上起示范作用。

作为江苏省内规模最大的居住建筑专业化设计企业，率先展开了设计总包一体化全过程设计服务工作，包括前期策划、居住区规划、居住建筑设计、景观园林设计及成品房精装修设计，向客户提供高质量的全方位技术服务，实现居住建筑高品质保证。

公司始终坚持尊重客户、理解客户，持续提供超越客户期望的产品和服务的经营宗旨，“精心设计、科学管理、客户至上、持续改进”的质量方针。高度重视设计创新、技术质量、成本控制、设计服务等环节。近年来，先后有几十项设计工程项目获得建设部、省、市级优秀设计，优秀工程和科技进步奖。2009 年，我司获江苏省勘察设计企业综合实力排序第五名，并被授予江苏省勘察设计质量管理先进单位和诚信单位称号，2010 年获全国工程勘察设计行业优秀民营设计企业称号。

公司注重和培养设计人员的创新能力和技术水平，运用先进的技术设备，发挥自身优势和品牌，提高核心竞争力，进一步开拓设计市场，以“建筑服务社会”的企业使命为社会提供优质设计和服务。

Nanjing Yangtze River Architectural Design Co., Ltd. was founded in 1976, formerly known as Nanjing Institute of Civil Architecture Design, and the company derived from the original institute by the end of 2009. Since its foundation, the company has undergone thirty years of development, and has developed into a regional, comprehensive, medium-sized modern architectural design enterprise with urban design, office buildings, commercial buildings, research and education construction, residential building design as its features and main products. The company has a high reputation and competitiveness in Jiangsu Province, even in East China.

The company currently possesses State Class A design qualification in construction industry, and is engaged in many fields such as urban design, residential planning, architecture, structure, drainage, electrical, intelligent, heating and ventilation, air defense works, interior decoration, landscape gardening, specialized computer applications and so on. The company can undertake feasibility studies, urban planning, architecture design, technical consulting and engineering project management and other related businesses in all types of civil and industrial constructions.

The company now has 287 employees, among whom 90% are designers and technical personnel, 54% designers and technical personnel are architects, engineers with intermediate or advanced professional titles. The registered architects, registered structural engineers and professional equipment engineers with registered qualification accounted for 30% of the designers.

In recent years, our company have designed a number of full-featured, technically complex and influential medium-sized public buildings: Nanjing Changfa Center high-rise buildings, Suning Nanjing Drum Tower Science and Technology Park high-rise office buildings, Nanjing Presidential Palace historical and cultural tourism district (1912 block), Wuhu Qiao Hong International Hotel (five star), Red Sun East China Moore, Suning Appliance Chengdu Commercial Centre, Binhai Sports Center, Nanjing University of Aeronautics and Astronautics Jiangning East Campus, Nanjing Xuzhuang software industry base, Nanjing Zhongshan Botanical Garden Exhibition greenhouses and so on.

As the largest residential building professional design firm in Jiangsu province, our company has comprehensive design qualifications, sufficient human resources, strong technical forces, and large number of outstanding design achievements. Aiming at the urgent need of the real estate market, our company first started the comprehensive design services of design integration of the whole process, including pre-planning, neighborhood planning, residential building design, landscape garden design and existing house refined decoration, to provide customers with a full range of high quality technical services and to achieve high quality assurance in residential buildings. On the basis of mutual benefit and win-win cooperation, our company has established a strategic partnership with Vanke Real Estate Division, Jiangsu Newtown Company, Nanjing golden earth real estate company, Country Garden Group, Jiangsu Baoli real estate, Jiangsu Suning house purchasing and other major customers to ensure more harmonious design cooperation, closer coordination, and to gain more benefits.

With the respect to the housing-led products of government's solving the low-income people's livelihood in the cities, our company have put more energy in the design of indemnificatory housing in recent years, and has made great achievements. We have designed 2010 Nanjing government major livelihood projects—Dingjia Village innovation park affordable housing and Xishanqiao Daishan New Village (the city's largest affordable housing project with nearly 400 million square meters now). Based on a large number of design practice, we have edited *Nanjing indemnificatory Housing Design Guidelines* and *Nanjing indemnificatory Housing Landscape Design Guidelines*, which contribute to the construction of Nanjing affordable housing and also makes our company become an authoritative design agency in this field.

In addition to planning and design, the company is also committed to the research of housing science and technology, and has edited Jiangsu Province Residential Design Standards and the reference book *Residential Design Data Set* for large residential design. Meanwhile, our company takes the lead in studying the green building technology and its application in practical engineering design in residential house field in our province of and obtained phased objectives, and set an example in the green building design and residential industrial promotion in the field of residential design in our province.

As the largest residential building professional design firm in Jiangsu province, our company first started the comprehensive design services of design integration of the whole process, including pre-planning, neighborhood planning, residential building design, landscape garden design and existing house refined decoration, to provide customers with a full range of high quality technical services and to achieve high quality assurance in residential buildings.

We always insist on the management objectives of respecting the customers, understanding the customers, and providing the products and services exceeding the customers' expectations continuously, and the quality policy of meticulous design, scientific management, customer first, and continuous improvement. We attach great importance to design innovation, technical quality, cost control, design services and other sectors. In recent years, there have been dozens of design projects awarded the Ministry of Construction, provincial, municipal excellent design prize, outstanding engineering prize and technology progress prize.

In 2009, in the rank of overall strength of survey and design enterprises in Jiangsu Province, our company ranked the fifth, and was awarded the title of design and survey quality management advanced unit and integrity unit in Jiangsu Province. In 2010, our company was awarded the title of excellent private design firm in national engineering survey and design industry.

The company focuses on and develops the innovative ability and technical level of designers. Using the advanced technology and equipment, and taking advantages of their own and the brand to enhance core competitiveness, and to further develop the design market, the company takes "Construction Serve the Society" as the enterprise mission to provide high-quality design and service for the society.

南京银城大厦

汤山公馆

徐庄软件产业基地

南京岱山新城

无锡金色港湾

翡翠谷

盐城苏宁广场

江北新城总部大厦

苏州城发建筑设计有限公司

Suzhou Institute of Urban Development Architectural Design Co., Ltd.

公司办公大楼
总体鸟瞰效果图

地　址：苏州市东环南路 1 号和诚大厦 6 楼
电　话：+86-512-62873966
传　真：+86-512-67773882
邮　箱：chengfa@126.com
网　址：http://www.szudad.com

Address: The 6th Floor of HeCheng Building, Donghuannan Road No.1, SuZhou
Tel: +86-512-62873966
Fax: +86-512-67773882
E-mail: chengfa@126.com
Http: //www.szudad.com

苏州城发建筑设计院有限公司坐落于美丽、和谐、开放的苏州工业园区，是一家专门致力于城市规划、建筑设计的专业设计机构。设计资质为国家建筑工程甲级资质。

本院的企业精神是“专业、敬业、高效、务实”。“一切为客户着想”的理念贯穿于每一个设计环节。我们一贯认为，对客户负责才是真正地对自己负责。在规划、设计的服务过程中让客户放心、省心、满意是我们所追求的目标，运用专业技能结合我们的区域优势，尽可能多的为客户排忧解难。

我院在苏州设计了一系列较有影响的项目。一些项目无论是规模，还是设计难度都创苏州之最。我们擅长各类民用建筑的设计，熟悉各类工业建筑的工艺要求和各类建筑的法规、法则。与客户有着良好的沟通与交流，特别是在产品的定位、产品的地域性特征及造价控制方面有独特的体会和经验。

我院专业配置齐全，团队作战能力突出，在各专业配合及成本控制方面有较大的优势，在业内及客户中具有良好的口碑。我们熟悉境外设计公司的工作流程并与众多境外知名设计公司有着良好的合作关系。

我院全体设计人员具有良好的教育背景、开阔的思路和很强的市场服务意识，为业主科学合理地降低和控制开发成本是我们一贯的追求。在院领导的高度重视和支持下，我们有决心和信心从方案设计阶段开始就做好设计质量、工程成本控制、设计进度及施工过程中的技术服务等方面的工作，力求给业主一份满意的答卷。

Located in the beautiful harmonious and open Suzhou industrial park, Suzhou Institute of Urban Development Architectural Design Co., Ltd. is a professional design institute dedicated to professional urban planning and architectural design, with national class A qualification of architectural engineering.

Our enterprise spirit is professional, dedicated, efficient and pragmatic. Everything for the customer, This loyalty concept permeates every aspect of the design. We have always believed that being highly responsible for each of our customers is the key to success. In the course of planning and design, quality and customers, satisfaction are the goals we strike to pursue. Combined with our professional skills and advantage of localization, we provide the optimal solutions to our customers.

Our institute has designed numerous influential projects in Suzhou, which reach the highest level in scale and design difficulty in Suzhou city. We are expert in the design of civic use architecture, knowing well of the technology requirement of various industrial architectures and different architectural laws and regulations. We have a good communication with our clients, especially with good understanding of the project position, the fondness of the local inhabitants and experience of budget control.

Our institute has complete professional facilities and competent teams, with large advantage in professional coordination and cost control area, which gain us good fame among the clients in this profession. We have established cooperation relation with many famous overseas design companies and are familiar with their working process.

The staffs in our institute are all with good education background, open mind and strong sense of market services, striving for lowering and controlling the cost of the project development. With the great concern and support of the leaders, we have the determination and confidence to do well in works of design quality, project budget control, design schedule and the service of construction from the scheme design stage, offering the best works to our clients.

宜兴 东城 SIHO
YiXing Dongyu SOHO

建设地点：江苏 宜兴
建筑面积：48 000 平方米
设计时间：2009 年

Construction Location: YiXing, Jiangsu
Building Area: 48,000 m^2
Design Time: 2009

特点介绍

传承与创新、城市形象、古典与现代的对比。

项目位于江苏省宜兴市东部新城，地理位置十分优越。

项目力图打造一个高雅、时尚的多元化服务场所，通过营造高品位的氛围，聚拢城市精英人群，刺激周边消费，带动区域内经济发展。

设计上一方面我们将古典、优雅的酒吧文化镶入现代规划设计中，追求一种形神兼备的现代服务产业，同时通过建筑的宏大视觉效果，塑造东部新城新的轮廓线。从而达到传承和创新的愉悦化人性空间，激发并增强人们记忆和情感的积淀。

西南角人视图

Features Introduction

Inheritance and innovation, urban image, the contrast of classical and modern.

The project is located in the East New Town in Yixing City, Jiangsu Province. The geographic location is very excellent.

The project intends to create an elegant, stylish diversified service place. By creating the atmosphere with high quality, it gathers the urban elite groups to stimulate peripheral consumption and promote economic development in the region.

On one hand, we insert classical, elegant bar culture into the modern planning and design, seeking a modern service industry and culture with form and spirit. Meanwhile, we pursue grand vision of narration by building and create a new contour line of the East New Town to achieve the pleasant human-friendly space of heritage and innovation and to stimulate and enhance the accumulation of people's memories and feelings.

1

2

3

1–3 绵竹江苏工业园孵化基地
建设地点：四川 绵竹
建设时间：2009 年

Mianzhu incubator base of industrial park, Jiangsu
Construction Location: Mianzhu, Sichuang
Construction Time: 2009

特点介绍
复兴与重振绵竹工业，体现恢复、发展和升腾的现代精神；
整体布局序列化，使用空间立体化，建筑形象现代化；
建造与城市和建筑有机融合的主题景观广场；
营造舒适的庭院化办公环境，提倡绿色办公理念。

Features Introduction
Rehabilitating and revitalizing Mianzhu industries, embodying the modern spirit of the recovery, development and ascension;
Serializing the overall layout, using three-dimensional space, the image of modern architecture;
The theme landscape square organically integrating urban and architecture;
Constructing comfortable courtyard office environment and promoting the concept of green office.

4 春菊大厦
建设地点：江苏 苏州
设计时间：2009 年

Spring-Chrysanthemum Mansion
Construction Location: Suzhou, Jiangsu
Design Time: 2009

5 都市 VIP 规划与建筑设计
建设地点：江苏 苏州
设计时间：2010 年

Urban VIP Planning and Architectural Design
Construction Location: Suzhou, Jiangsu
Design Time: 2010

4

北向透视效果图 5

1

1 吴江汾湖购物中心
建设地点：江苏 苏州
设计时间：2009 年

Fenhu Shopping Centre, Wujiang
Construction Location: Suzhou, Jiangsu
Design Time: 2009

2–3 苏州市市区公安监管场所迁扩建工程
建设地点：江苏 苏州
设计时间：2010 年

Removal Expansion Project for the Urban Public Security Supervision and Management Areas, Suzhou
Construction Location: Suzhou, Jiangsu
Design Time: 2010

入口广场效果图 2

总体鸟瞰效果图 3

人民日报社新大楼设计（2009-2012年）

设计主持人：东南大学建筑学院 周琦 教授
建筑设计参与者：钱锋、张力、虞晗、王为、庄昉、孔辉、周文祥、左静楠、曾德萍、朱晓洲、赵然。

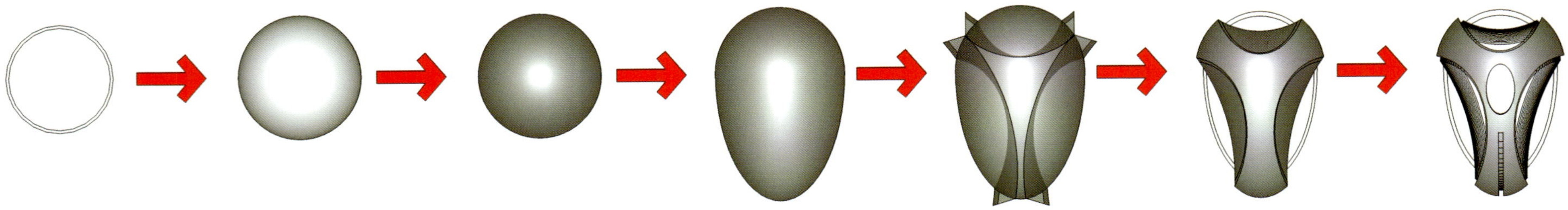

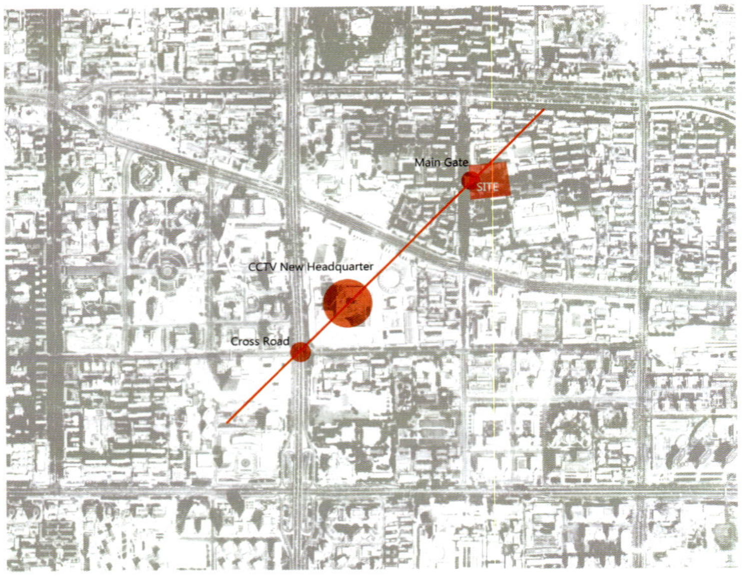

场地环境图

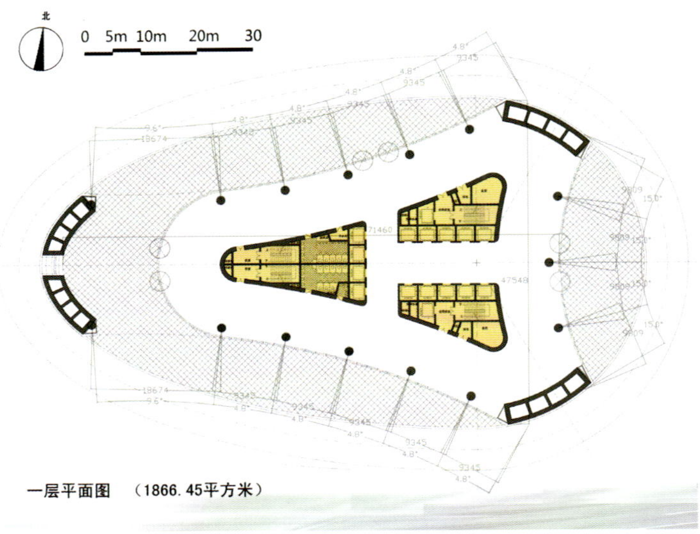

一层平面图

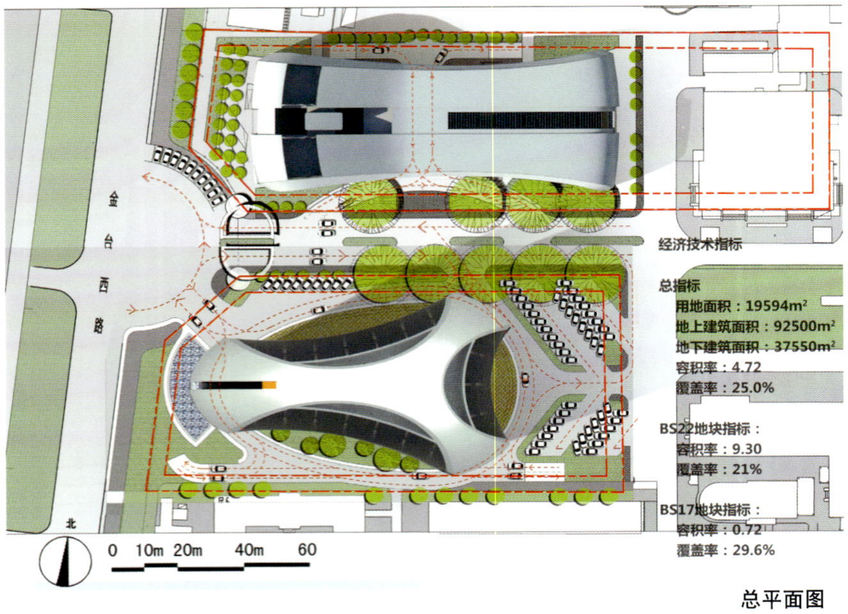

总平面图

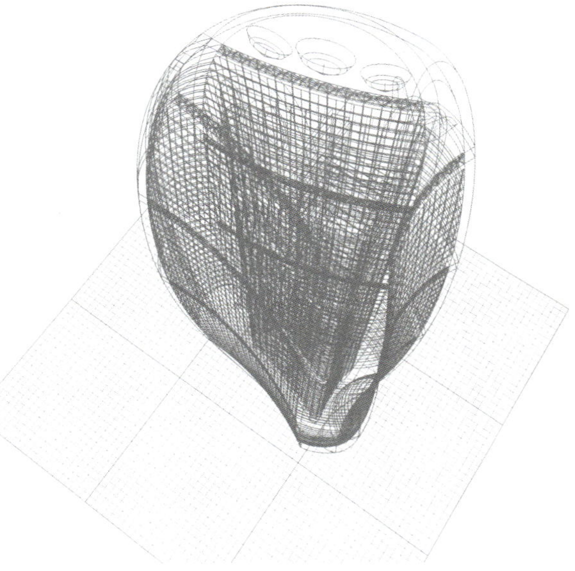

数字化三维模型

人民日报社新大楼是人民日报社建社以来最重大的建设项目之一，由国际会议中心、图书馆和办公楼组成，建筑高度180米，地上建筑面积98 000平方米。本设计方案通过象征的手法与建筑形态上的动感，体现着建筑的标志性和现代感，是时代特色同传统文化内涵的融合。

设计思想包含以下三个方面。

以人为本的设计理念：人字形平面是人民日报社主题的象征，它建构出尺度宜人的空间关系，曲线造型同时使建筑形态自然柔和。

与自然环境和谐统一：对自然环境采取最小干预原则，维持基地原有的良好生态条件，对道路进行退让处理，在视觉上弱化了高层建筑的体量，有效避免了高层建筑对周边建筑日照上的不利影响，使其同城市和谐共生。

积极响应时代精神：独特的建筑造型理性地表达了结构的受力逻辑，产生于稳定结构中的奔腾的动势呼应了平稳快速发展的时代风貌。

设计作品
周琦
Zhou Qi
网址：www.zhoustudio.com

人民日报社效果图

研究与实践方向

建筑形态学设计与研究

Design and Research on Architectural Morphology

形态学是关于形式的科学。建筑形态学致力于创造美的形式，衡量的标准在于它是否具有内在而持久的感染力和生命力，并以贴切的环境、空间、场所作为其得以实现的物质载体。

自然的形式是美的，因为美蕴藏在内部整体的有机联系之中。因此，若要人工介入自然，创造美的形式，就必须通过各种途径找到那种联系，即如同自然一般和谐共生的存在方式，这是形态学研究面临的最大挑战和终极目标。要达到这一目标，只有通过人的主观意识，将社会、文化、功能、地理、气候、技术等诸多因素有机结合并找到彼此间的联系，进行艺术的再创造，才能产生具有生命力的美的形式。

这里从整个建筑形态学研究的范畴中择取三个方面，即基于自然力的结构形态学、影响规划理念及技巧的地理有机形态学以及关乎人们生活常态的居住形态学。建筑形态学就是在这三个方面上进行探索与实践，并在此基础上发展出复合形态学的研究。

中国近代建筑研究与保护

Conservation on China near-Modern Architecture

中国近代建筑研究与保护主要研究自鸦片战争以来中国沿海开埠城市建筑发展与变迁，尤其是在西方建筑文化影响下所产生的综合性文化与历史因素。近20年来，城市更新与发展加快，对这批近代遗产的保护就显得尤为重要。经过我们10多年的研究与实践，在城市关系、历史脉络、保护与利用以及更新与改造的关键技术等众多方面进行了不懈的探索与努力，完成了一批理论研究成果与建筑保护实例。

中西建筑比较与建筑评论

Comparison between Chinese and Western Architecture & Architecture Criticism

建筑批评是理论研究的组成部分。建筑批评的对象是建筑、建筑师和社会环境，通过对建筑师的创作思想，建筑作品的设计、建造和使用过程，社会个体和群体的鉴定和评价，进行研究描述、分析阐释、比较论证、评价判断，评价出建筑以及其建筑师的价值和品质。借助建筑批评的手段，通过高度抽象和理性的分析来剖析经典作品的设计过程——从建筑师的行为特点入手，思考建筑从概念到建成的全部过程的，探寻达到最终结果途径。

随着社会的演化，作为建筑体的人变化，城市环境和自然环境的改变，工程技术的进步，建筑学科将会应对越发复杂的问题。建筑批评的研究，试图从哲学和意识形态、过程和发展、技术、社会历史因素的演变等不同层面寻索了理性因素，并以此作为指导设计的依据和源泉。

广州东莞同沙大桥（2004年）

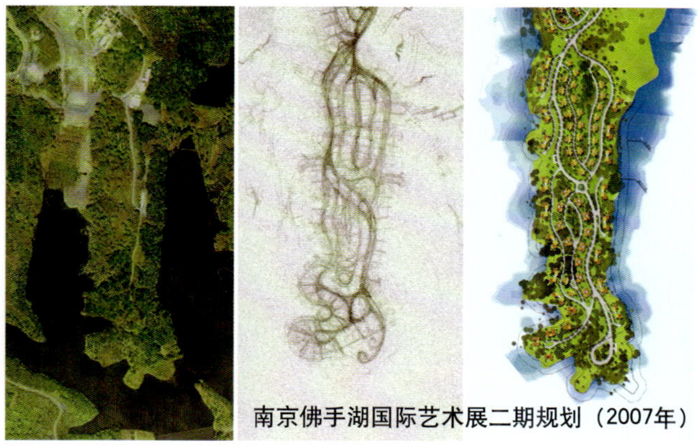

南京佛手湖国际艺术展二期规划（2007年）

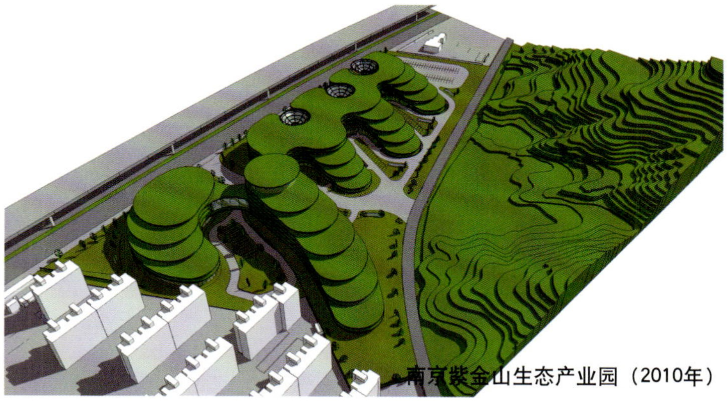

南京紫金山生态产业园（2010年）

单位简介 / Unit Profile

在经历十多年的建筑实践之后，傅筱副教授于 2007 年成立了集筑建筑工作室（integrated architecture studio，IA）。该工作室依托于南京大学建筑与城规学院良好的学术平台和南京大学建筑规划设计研究院良好的技术平台，长期致力于建筑创意和建造品质控制的实践工作。工作室通过集成建筑信息模型技术、生态技术等技术资源，至今已经完成多项高品质的建筑作品，近年来多次获得国家、省、部级建筑设计奖项以及 2008WA 中国建筑奖佳作奖、中国首届 BIM 建筑设计最佳建筑设计二等奖、江苏省三维建筑设计竞赛一等奖、法国罗纳－阿尔卑斯大区生态建筑入围奖、第十二届威尼斯双年展—CA'ASI 中国新锐建筑创作展入围奖等有影响力的奖项。

After more than 10 years of architecture practice, Proffessor Fu has founded the integrated architecture studio (IA) in 2007. With the support of the institute of architecture and urban planning of Nanjing University, and through collaborating with the design & research Institute of Nanjing University, IA studio focus on the creative design and the quality of the construction all the time. Integrating with all kinds of technical resources, such as BIM, ecological technique, IA studio has completed many high quality projects recently and has gained many influential awards: the 2008 WA Chinese Architecture Award, the Second Prize of first Chinese BIM Architecture Design Competition, the First prize of Jiangsu 3D Design Competition, the finalist of Rhone-Alpes Ecological Building Award France, 12th Venice Biennale-2010 CA'ASI creative exhibition of chinese architecture and so on.

集筑建筑工作室

单位法定中文名称：集筑建筑工作室
地址：江苏省南京市汉口路 22 号南京大学费彝民楼 1512
法人：傅筱
传真：+86–25–83621626
邮箱：ia–studio@vip.163.com
网址：www.arch–ia.com

Integrated Architecture Studio, IA
Address: Room 1512, Feiyimin Building Nanjing University, Hankou Road No.22, Nanjing, Jiangsu
Legal person: Xiao Fu
Fax: +86-25-83621626
E-mail: ia-studio@vip.163.com
Http: //www.arch-ia.com

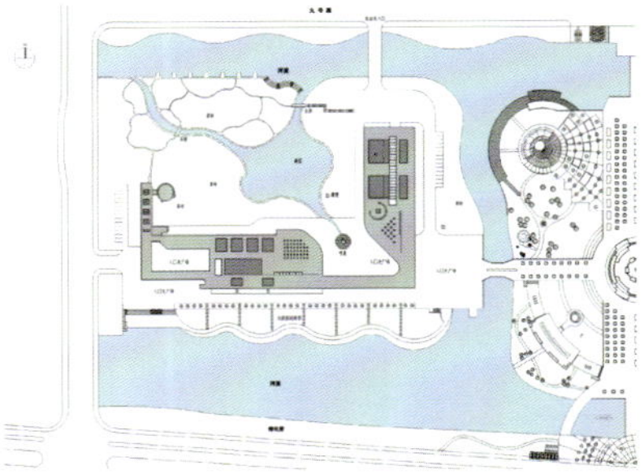

长兴县图书馆、档案馆建筑基地位于城市文化行政新区西侧

占地面积：86 909 平方米
建筑面积：18 500 平方米
容 积 率：0.21
设计时间：2003 年
建成时间：2005 年
获奖情况：全国工程勘察设计行业优秀工程勘察设计（2008 年）二等奖
江苏省城乡建设系统优秀勘察设计奖（2005 年） 一等奖
江苏省第十二届优秀工程设计（2007年） 一等奖

注：此为傅筱在东南大学建筑设计研究院工作期间完成的设计作品。

The construction site of library and archives of Changxing County is located at the west of urban cultural new administrative district.
Site Area: 86,909 m²
Building Area: 18,500 m²
Plot Ratio: 0.21
Design Time: 2003
Complete Time: 2005
Awards: the Second Prize of national excellent engineering investigation and design awards (2008)
The First Prize of excellent engineering investigation and design awards (2005) in Rural-urban Construction system of Jiangsu Province
The First Prize of the 12th Excellent Project Design Awards (2007) in Jiangsu Province

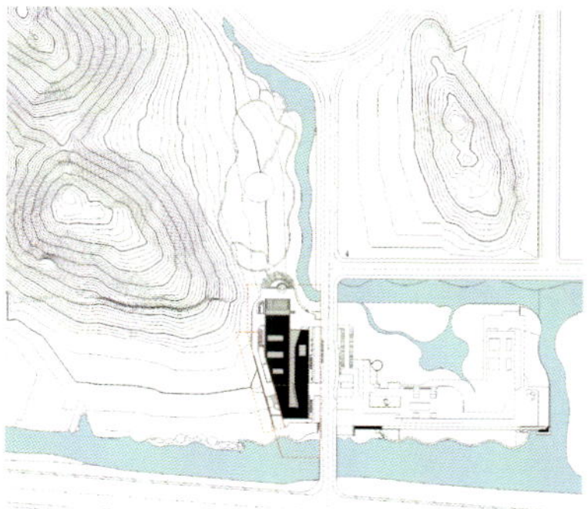

长兴县传媒中心项目，建筑地点为长兴龙山行政新区西侧

我们的设计概念是将空间还给城市市民，建筑需成为一个可以让人自由前往的场所，并可让人自由活动，因此我们采用了三个主要设计手法：

1. 将公共部分的屋顶设计成自由漫步道，从基地北端引导人们缓缓而上，步道串联了图书馆公园、岐山公园和甘家河公园；

2. 建筑北面设计了室外观演广场，将建筑与梅山公园紧密联系起来；

3. 建筑对外大厅设计成多功能厅，人们可以自由前往，人们在这里可以喝咖啡、上网、观看展览、观演等活动，大厅向北与自由漫步道联系起来，在这里可以喝茶并观看梅山美景。此外建筑采用了双层通风幕墙、种植屋面、架空隔热屋面等技术以节约能耗。由于设计率先采用全三维 BIM 技术完成建筑施工图。

该项目获得 2009 中国首届 BIM 建筑设计大赛最佳建筑设计二等奖，并收入第十二届威尼斯双年展 –CA'ASI 中国新锐建筑创作展官方目录。

Media center project in Changxing County is located at the west of the new administrative district. Our design conception is, returning the urban space to the residents of the city, making the architecture a place where people can easily get access to and carry on their activities. We adopt three major design methods: 1. turning the roof in the public section into the promenade, which links the library park, Qishan Park and Ganjiahe Park, leading people from the north of the site base slowly going upward. 2. setting up an outdoor theatrical square in the north of the architecture, connecting the architecture with the Meishan Park. 3. transforming the hall to a multifunctional hall where people can carry on leisure activities such as drinking coffee, going on line, watching performance and exhibitions. As the north part is connected with the promenade, people can see the wonderful view of Meishan and drink tea here. Besides, to save the energy resources, this project well utilizes the technique of double skin, planted roof and insulating board overhead roofing .Being the first to complete the construction working drawing with the three-dimensional BIM technique.

This project has won the Second Prize of the best architecture design in 2009 the first BIM architectural design competition.And beselected in the list of 12th Venice Biennale-2010 CA'ASI creative exhibition of chinese architecture.

占地面积：20 272 平方米
建筑面积：23 660 平方米
容 积 率：0.72
设计时间：2007 年
建成时间：2009 年

注：此为傅筱在东南大学建筑设计研究院工作期间完成的设计。

Site Area: 20,272 m^2
Building Area: 23,660 m^2
Plot Ratio: 0.72
Design Time: 2007
Completion Time:2009

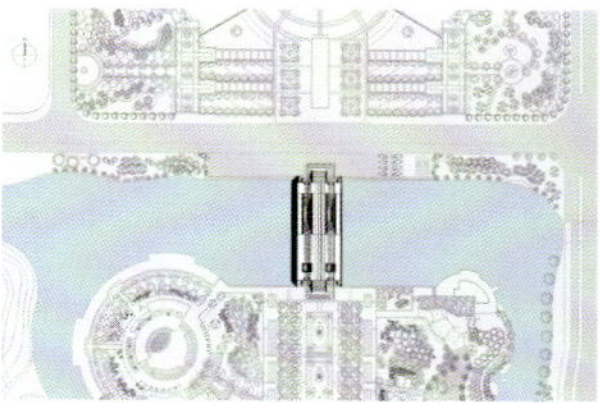

S 景观步行桥位于浙江省长兴县市民广场和行政广场主轴线上，正对政府行政主楼。桥跨度为 54 米，主要建造材料为型钢、玻璃、混凝土、塑胶、合成木板。

概念：
1. 满足轴线上礼仪性的需求，同时又具有自由的景观性；
2. 创造一个可让市民活动的空间。

策略：
1. 桥面处理成左右对称的形式，满足了轴线上对称性的需求和礼仪性活动，而桥侧面和内部则由一系列变化的空间组成，增强了步行桥的景观性。通过桥的空间设计可以欣赏周边景色，桥自身也将成为一个城市景观；
2. 弯曲的钢结构连接南北两岸。结构的弯曲带来了空间竖向上的变化，于是休闲的空间产生了，桥的纯粹通行功能被打破，桥演化为可容纳更多行为活动的空间场所。

反馈：
1. 成为了当地老百姓乐于前往的休闲交流场所；
2. 增强了周边环境的景观性；
3. 打破了国内常见的政府广场形象，创造出开放、民主的城市空间形态。

该项目获得 2008WA 中国建筑奖，并入围第十届威尼斯双年展 –CA'ASI 中国新锐建筑创作展。

S Landscape bridge lays on the main axis of People's Square and Municipal Square,facing the government office main building. The bridge span is 54 m , with building material of merchant steel , glass , concrete, plastic and synthetic wood.

Conception:
1. meeting the needs of etiquette activities on the main axis, offering the unrestrained landscape.
2. creating an activity space for city inhabitants.

Strategy:
1. the bridge floor is bilateral symmetry, meeting the demand of being symmetry and the function of holding ceremony activities; the side and interior of the bridge is composed of diversified space.
It provides a space for people to enjoy the views, and the bridge also become an urban landscape itself.
2.The crooked steel structure links the banks of the north and south, the winding of the structure brings the vertical change of space , it thus creates the leisure space, not only with the traditional function of passage , it also becomes a place for more activities.

Feedback:
1. offering leisure space for local residents
2. highlighting the surrounding environmental landscape
3. changing the old image of Municipal Square, creating an opening and democratic urban space pattern.

This project win honorable mention in 2008WA Chinese Architecture Award and be selected in 12th Venice Biennale-2010CA'ASI creative exhibiton of chinese architecture.

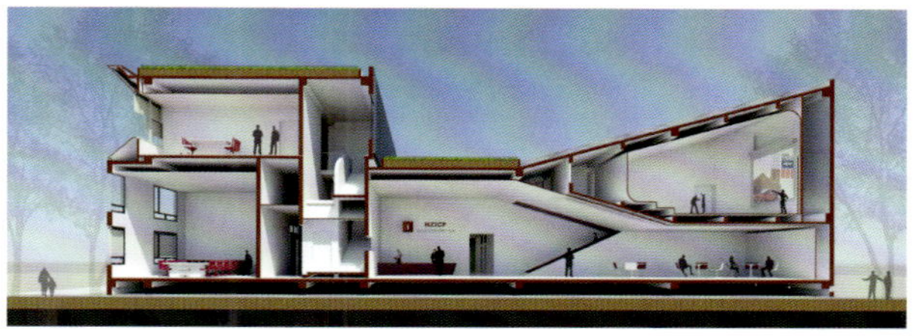

南京紫东国际招商中心办公楼是一个从生态概念入手设计的建筑。作品主要关注了两点，一是如何将传统的生态技术与现代生态技术相结合，强调建筑中生态技术的综合利用；二是将建筑艺术与生态技术相结合，打破"先艺术后生态技术"的创作模式。在这里，我们通过作品来探索的是建筑的形体生成与生态技术措施的运用是紧密关联的，生态技术可以作为一种设计的方法和原创动力，从而设计出令人愉悦的建筑艺术。鉴于设计上的探索性，该项目获得了法国罗阿大区政府和时代建筑联合颁发的生态建筑入围奖，并在上海世博会法国罗阿馆展出。

Nanjing Zidong international Merchants Center office building is an architectural stressing on the ecological conception. This project concentrates on two key points: 1. how to combine the traditional ecological technique with modern ecological technique, emphasize the comprehensive application of ecologic technique in the architecture design. 2. integrating the architectural art with ecologic technique, breaking the mode of art comes before the ecologic technique. Here through the design works, we are exploring: The form and structure of architecture is closely connected with the application of ecological technique measures, as a design method and motive power of creativity, the ecological technique measures can produce pleasant architectural art. For the above discovery, this project has won the outer critics' award issued by France Ah-region government and Time Architecture and be put on display in Shanghai Expo in France Rhône-Alpes site.

设计时间：2010 年 3 月—2010 年 4 月
建成时间：2010 年 12 月
占地面积：6 350 平方米
建筑面积：1 512 平方米
容 积 率：0.24

Design Time: 2010.03-2010.04
Completion Time: 2010.12
Site Area: 6,350 m^2
Building Area: 1,512 m^2
Plot Ratio: 0.24

中国建筑东北设计研究院有限公司

China Northeast Architectural Design & Research Institute Co.,Ltd.

中国建筑东北设计研究院有限公司系国家大型综合建筑勘察设计单位，始建于1952年，隶属于中国建筑工程总公司（世界五百强企业），具有建筑工程甲级设计资质，同时具有岩土勘察、工程咨询、工程监理、装饰工程、施工图审查等甲（壹）级资质。

公司专业配置齐全，包括建筑、规划、园林、结构、给排水、暖通、电气、电讯、动力、建筑经济、建筑装饰、建材、勘察、岩土、监理等。业务涵盖建筑工程、城乡规划、市政工程、风景园林、人防工程、新型建材、装饰工程、岩土勘察、工程咨询、工程监理、工程造价、商物粮等行业。

公司在沈阳总部设有大师工作室、八个综合设计所、国际业务部、规划与景观事业部、基础设施事业部、投资管理与项目咨询事业部、科技研发中心及咨询公司、岩土公司、装饰公司、监理公司和审图公司，在深圳、厦门、福州、大连设有分公司，并在北京、上海、重庆、青岛、长春和营口设有分支机构。公司实行数字化网络管理。

公司现有职工1 119人，其中国家建筑设计大师1人，辽宁省设计大师8人，享受政府特殊津贴专家20人，国家一级注册建筑师和一级注册工程师312人，教授级高级建筑师和教授级高级工程师70人，高级建筑师和高级工程师245人，建筑师和工程师211人。

公司成立五十多年来，共完成20 000多项各类工业与民用建筑设计，工程涉及各个行业，遍及国内27个省、市、自治区及欧洲、亚洲、非洲等十几个国家和地区。先后荣获国家级、省部级优秀设计奖500余项。多年来，主编和参编了几十项国家、行业及省、市技术标准与设计规范。在历届全国勘察设计单位综合实力评比中，均位居百强之列，名列建筑设计单位前茅，在国内外建筑界享有较高的声誉。

近年来，公司连续获中国勘察设计协会“优秀勘察设计院”奖、BCI Asia“TOP10中国十大建筑设计公司”等荣誉称号。

公司在工程设计、技术咨询等方面全面贯彻GB/T 19001—2008、GB/T 24001—2004和GB/T 28001—2001标准，确保向用户提供优质的产品和满意的服务。

China Northeast Architectural Design & Research Institute Co., Ltd. (CNADRI) established in 1952 is a large-scale, comprehensive architectural design company, affiliated to China State Construction Engeering Corporation Ltd (CSCEC) which has been listed as one of the World Top 500 Companies. It possesses National Class A Design Qualification as well as Class A(1) Qualifications in geotechnical investigation, engineering consultation, engineering supervision, decoration engineering and checking of working drawings.

There are all the necessary specialties, including architectural, planning, landscape, structural, water supply and sewerage, HVAC, electrical, telecommunications, power, building economy, building decoration, building materials, exploration, geotechnical, and supervision specialties. And its business scope covers the following fields: building construction, urban and rural planning, municipal works, landscape gardening, civil air defense, new building materials, decoration works, geotechnical investigation, engineering consultation, engineering supervision, engineering cost and commercial, logistics & grain storage.

The CNADRI headquarters in Shenyang consist of Architect Studio, eight integrated design departments, International Business Department, Planning and Landscape Division, Infrastructure Division, Investment Management and Project Consulting Division, Science and Technology Development Department, as well as an engineering consulting company, a geotechnical engineering company, a decoration company, a supervision company and a drawing checking company. There are branch offices in Shenzhen, Xiamen, Fuzhou, Dalian and affiliates in Beijing, Shanghai, Chongqing, Qingdao, Changchun and Yingkou. The whole company adopts digitized network management.

There are 1,119 staff members working for the company now. Among them, one is a national architectural design master; eight are design masters of Liaoning Province, 20 have been awarded special government allowances, 312 are national Class A certified architects or engineers, including certified mechanical engineers; 70 are senior architects or engineers of professor rank, 245 are senior architects or engineers and 211 are architects or engineers.

In more than 50 years since the establishment of the company, 20,000 industrial and civil architectural designs of various kinds have been completed. The projects cover a broad range of fields and spread throughout 27 provinces, cities and autonomous regions in China, as well as more than 10 European, Asian, African countries and regions. CNADRI has won over 500 outstanding design awards of national, provincial and ministerial levels. Over the years, CNADRI has been the editor-in-chief or has participated in the compilation of numerous national provincial, municipal and industrial design codes & technical standards. In every past National Comprehensive Strength Eveluation held for exploration & design companies, CNADRI has been listed as one of the China Top 100 Companies and also been listed at the top of the architectural design companies. Therefore, it enjoys high prestige in the construction industry both at home and abroad.

In recent years, CNADRI has been consecutively awarded the ‘Excellent Exploration & Design Institute’ by China Exploration & Design Association and the “TOP 10 Architects in China’ by Building & Construction Interchange(BCI) Asia.

During the engineering design and technical consultation, CNADRI strictly follows the standards in GB/T 19001-2008, GB/T 24001-2004 and GB/T 28001-2001 so as to provide high-quality products and satisfactory service for the clients.

法人：冯晓明
地址：沈阳市和平区光荣街65号
邮编：110006
电话：+86—24—62123555（总机） +86—24—62123636（市场部）
传真：+86—24—23861440
网址：www.cscecnei.com E—mail：cnadri@cscecnei.com

Legal Person: Xiaoming Feng
Address: 65 Guangrong Street, Shenyang, China
P.C.: 110006
Tel: +86-24-62123555(Switchboard)
+86-24-62123636(Marketing Department)
Fax: +86-24-23861440
Http: //www.cscecnei.com
E-mail: cnadri@cscecnei.com

中国建筑东北设计研究院有限公司主楼

中国人民志愿军烈士墙

建设地点：辽宁 沈阳　　项目类型：纪念建筑

建筑面积：1 000 平方米　　占地面积：3 000 平方米

设计时间：2009 年

项目介绍

项目以庄严、肃穆、沉重为主调，以下沉纪念馆的 18 个乳白色玻璃立方体和坡道墙上黑色磨光花岗岩及上面所镌刻的 18 万志愿军烈士的名字，来承载对牺牲在抗美援朝战场上的英雄儿女的缅怀与敬仰。

Chinese People's Volunteers Martyrs Memorial Wall

Construction Location: Shenyang, Liaoning　　Type: Memorial buildings

Building Area: 1,000 m^2　　Floor Area: 3,000 m^2

Design Time: 2009

Project Introduction

Solemn, solemn, heavy, the following 18 Shen milky glass memorial cubes and ramps polished black granite wall are inscribed the names of 180,000 volunteer soldiers, carrying a pair of sacrifices in the Korean battlefield heroes and heroines of the recall with veneration.

沈阳展览中心

建设地点：辽宁 沈阳

项目类型：会展建筑

建筑面积：168 000 平方米

设计时间：2009 年

竣工时间：2010 年

Shenyang Exhibition Centre

Construction Location: Shenyang, Liaoning

Type: Conference and Exhibition Building

Building Area: 168,000 m^2

Design Time: 2009

Completion Time: 2010

东北大学科技交流中心

建设地点：辽宁 沈阳　　项目类型：办公建筑

建筑面积：21 000 平方米　　设计时间：2009 年

竣工时间：2010 年

项目介绍

东北大学科技交流中心采用相对灵活的布局形式，以 L 形的半围合体量与椭圆形的体量相结合，通过明确的体量关系和富于变化的空间形态，来展现大学校园建筑的简洁、明快，体现时代的特点。

Northeastern University Science and Technology Exchange Center

Construction Location: Shenyang, Liaoning　　Type: Office building

Building Area: 21,000 m^2　　Design Time: 2009

Completion Time: 2010

Project Introduction

Academic Exchange Center, Northeastern University, the relative flexibility of layout. L-shaped semi-confining fit oval body volume and amount of the combination, through a clear relationship between body mass and variegated space form, to show campus buildings clean, crisp, showing characteristics of the times.

鞍山银光购物中心

建设地点：辽宁 鞍山　　项目类型：商业
建筑面积：48 000 平方米　　设计时间：2009 年
竣工时间：2010 年

项目介绍

购物中心以实为主，在建筑的底部与中部采用玻璃幕墙，重点强调出入口空间和上部橱窗展示的空间。扶梯结合建筑层层跌落的形体设置，成为建筑立面上一道活动的风景线，透射出大厦内部的商业活动，同时，也可以将其理解为晶体从岩石中露出，向人们传递着高雅、富贵的商业理念。

Anshan Silver Shopping

Construction Location: Anshan, Liaoning　　Type: Commercial Buildings
Building Area: 48,000 m^2　　Design Time: 2009
Completion Time: 2010

Project Introduction

Shopping by entities based in the construction of the bottom of the glass curtain wall and the central, stressed the import and export of space and the upper window display space, and focused. Escalators, architectural layers of falling body set, a facade a scene of activity, a reflection of commercial activities within a building. At the same time, it can be interpreted from the rock crystal with elegance and business concept.

沈阳西站综合交通枢纽

建设地点：辽宁 沈阳　　项目类型：交通建筑
建筑面积：23 600 平方米　　设计时间：2009 年
竣工时间：2010 年

项目介绍

项目采用流线型屋面造型，充分体现交通建筑的风格特点，流畅且充满动感。整体化的造型，加强了城市界面的连续性。建筑材料以铝板和玻璃幕墙为主，局部配合水平向的百叶装饰，完美的曲线、优雅舒展的体量、晶莹剔透的五个立面，打造出高雅脱俗的综合交通枢纽。

Shenyang West Station integrated transport hub

Construction Location: Shenyang, Liaoning　　Type: Traffic Building
Building Area: 23,600 m^2　　Design Time: 2009
Completion Time: 2010

Project Introduction

Streamlined shape with the roof fully reflects the architectural style of the traffic, fluid and dynamic. The overall shape of the city, strengthens the continuity of the urban interface. Aluminum and glass curtain wall building materials to the main level of the local match to the louver decoration. Perfect curves, elegant stretch of body mass, sparkling five facade, elegant and refined to create a comprehensive transportation hub.

沈阳市政府广场地下商城及地铁车站

建设地点：辽宁 沈阳　　项目类型：交通
建筑面积：150 000 平方米　　设计时间：2009 年
竣工时间：2011 年

项目介绍

地下轨道交通与商业空间的结合是现代城市建筑的趋势，通过修建沈阳地铁二号线和七号线，借助城市轨道交通的发展，适度开发商业，同时改造市府广场景观。项目充分利用下沉广场和景观优势，吸引广场上的人流驻足购物，形成休闲购物、商业、景观与建筑融为一体的现代城市公共空间。

Shenyang City Plaza mall and Subway Station

Construction Location: Liaoning, Shenyang　　Type: Traffic Building
Building Area: 150,000 m^2　　Design Time: 2009
Completion Time: 2011

Project Introduction

Underground rail transit and commercial space, combined with the trend of modern urban architecture, through the construction of Shenyang Metro Line and on the 7th line, with appropriate development of urban rail transport development business, while transforming the City Hall landscape. Make full use of advantages of Sunken Plaza and landscaping to attract passenger traffic shopping plaza, the formation of leisure, commercial, landscape, architecture integration of modern urban design.

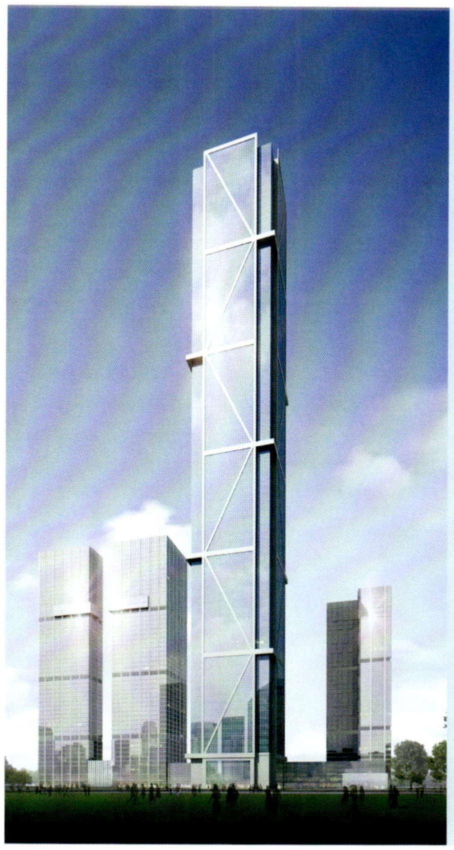

沈阳国际金融中心

建设地点：辽宁 沈阳
项目类型：办公建筑
建筑面积：640 000 平方米
设计时间：2009 年
竣工时间：2012 年

项目介绍

建筑的铝框及玻璃幕墙系统设计以对角线支撑为单元，组成外围框架，形成亮丽的巨塔形象。

外部立面简洁、平滑，局部以悬挑的形式表现，用以营造现代办公建筑的特色。

Shenyang International Finance Centre

Construction Location: Shenyang, Liaoning
Type: Office building
Building Area: 640,000 m^2
Design Time: 2009
Completion Time: 2012

Project Introduction

Mutually supporting in a diagonal position, the aluminum framework and glass curtain wall system form a basic unit, creating the external framework and the beautiful image of the great towers.

The external facade is designed to be simple and smooth, with some parts overhanged, in order to create an office building with modern features.

沈阳盛世长安商业广场

建设地点：辽宁 沈阳
项目类型：商业建筑
建筑面积：268 000 平方米
设计时间：2009 年
竣工时间：2010 年

Cheung Shopping Mall in Shenyang Golden Age

Construction Location: Shenyang, Liaoning
Type: Commercial Buildings
Building Area: 268,000 m^2
Design Time: 2009
Completion Time: 2010

四川奥翔航空产业园飞行模拟训练中心

建设地点：四川 成都
项目类型：工业建筑
建筑面积：19 400 平方米
设计时间：2009 年
竣工时间：2010 年

Sichuan Province Austrian Aerospace Industrial Park

Construction Location: Chengdu, Sichuan
Type: Industrial Buildings
Building Area: 19,400 m^2
Design Time: 2009
Completion time: 2010

福建省平潭新客运站

建设地点：福建 平潭
项目类型：交通建筑
建筑面积：39 900 平方米
设计时间：2009 年
竣工时间：2010 年

项目介绍

建筑立面设计将复杂的多功能体归纳至简洁、明晰的体块组合之下。横向起伏的波浪状幕墙造型，强调一种动感的态势，采用动态平衡的手法，来与客运站的流向、动态、平衡的意境相吻合，同时隐喻海面波浪起伏的意象。

Pingtan new terminal in Fujian Province

Construction Location: Pingtan, Fujian
Building Area: 39,900 m^2
Design Time: 2009
Completion Time: 2010

Project Introduction

In the external facade design, it places the complex multi-function body in the simple and distinct architecture block. The curtain wall is of rolling waves image, highlighting a state of dynamic. It adopts the dynamic equilibrium technique, conforming to the passengers flow, motion and balance in the passenger transportation terminal, offering an image of continuous waves on the sea.

沈阳长白万科城

建设地点：辽宁 沈阳
项目类型：住宅
建筑面积：280 000 平方米
设计时间：2009 年
竣工时间：2010 年

项目介绍

建筑群体的布局综合考虑了土地的充分利用、环境质量的提高，同时兼顾气候、日照、节能等基本要求，把握空间组合及给人的感受，并充分考虑与城市天际线和街景的结合。组团分区既兼顾建筑形态的渐变还考虑了分区和局部区域的形态布局。

Shenyang Changbai Vanke City

Construction Location: Shenyang, Liaoning
Type: Residential Building
Building Area: 280,000 m^2
Design Time: 2009
Completion Time: 2010

Project Introduction

For the basic requirements such as fully utilizing of the land, environmental quality improvement and other elements of climate, sunshine and energy-saving in the planning of the architecture, the designers have all taking these factors into account, in the meanwhile, they also lay stress on the combination of architectural space, the feeling that people get, and the connection on urban skyline and streetscape. It emphasizes the gradual change of architecture patterns, the division of the cluster and the geography layout in partial region as well.

辽宁省建筑设计研究院
LIAONING PROVINCIAL BUILDING DESIGN & RESEARCH INSTITUTE

辽宁省建筑设计研究院（LDI）是国家甲级建筑设计和甲级工程勘察单位，也是国家首批具有建筑智能化专项工程设计资质的单位，是沈阳市高新技术企业。本院创建于1956年1月，由沈阳、大连、锦州三市的设计人才汇集而成，至今已发展成为能够为工程建设提供全方位综合服务，专业设置齐全、设计手段先进、技术力量雄厚、社会信誉卓著的优秀勘察、设计和咨询单位。全院共职工262人，其中教授级高级建筑师和教授级高级工程师25人，高级建筑师和高级工程师67人，建筑师和工程师103人，其中具有国家级执业资格注册人数80人，享受政府特殊津贴专家6人。本院下设4个综合性建筑设计研究所、建筑方案研究所以及林立岩大师工作室、环境艺术研究所、建筑经济所、地源热泵应用技术研究所等部门，院属公司有岩土工程公司、项目管理咨询公司、施工图审查咨询有限公司等实体。我院于2006年成立了大连分院。

建院53年来，我院已经向社会提供了约1.3万项设计成果，工程项目不仅遍布辽宁，而且拓展到北京、上海、广州、深圳、杭州、宁波、无锡、西宁、威海、大庆、吉林、梅河口、通辽、赤峰等城市，以及也门、布基纳法索、俄罗斯、喀麦隆、塞舌尔、巴基斯坦等国家。近年来，我院与德国、澳大利亚、美国、英国、日本、韩国等国家以及香港、台湾等地区的建筑师事务所有着广泛的合作，与他们共同投标、合作设计了国内的多个项目。

20世纪80年代以来，我院共获国家、部、省、市优秀工程设计（勘察）奖213项和科技进步奖78项，并荣获辽宁建设系统质量管理奖、重合同守信用单位、辽宁省工程勘察设计行业CAD应用先进单位、档案管理国家一级单位、辽宁省“八五”期间建设科技工作先进单位、全国建筑节能先进单位、沈阳市节能设计先进集体、全国建设技术创新工作先进单位等奖励和称号。

2000年12月我院总工程师林立岩同志被中华人民共和国建设部授予“中国工程设计大师”称号。2008年院长杨晔、副院长及总工师李庆钢被授予首批“辽宁省工程设计大师”称号；院长杨晔同志1990年获联合国教科文组织奖（UNESCO PRIZE）；1998年获中国建筑学会青年建筑师奖，2004年当选“沈阳市十大中青年建筑师”、“沈阳市十大杰出青年”；2006年被评为“辽宁省劳动模范”、“辽宁省优秀共产党员”，童年被中国勘察设计协会授予“优秀企业家”称号，2008年被授予“辽宁省十大杰出青年”称号并当选2008年奥运火炬手，2009年荣获“全国建筑设计行业国庆60周年管理创新奖”。副院长兼总工程师李庆钢2005年荣获“第五届辽宁青年科技奖十大英才”称号，2010年获“辽宁省首批领军人才”荣誉称号。总建筑师曹辉2007年获“辽宁五一劳动奖章”，2009年获“辽宁省直属机关十大杰出青年”称号。

设计院及所属岩土工程公司、项目管理咨询公司于2000年12月通过ISO 9000（1994版）质量体系认证，2003年通过GB/T 19001－2000 idt ISO 9001:2000标准质量体系认证，2009年通过GB/T 19001—2008版质量管理体系认证。

The institute (ldi) is a national Class A architectural design and project survey firm as one of the intelligent building special item project designing and one of the first group of institutes in the country; it has qualification of intelligent building special item project design. Founded in january, 1956, the institute has collected design talents from Shenyang, Dalian and Jinzhou. Now it has become an excellent building design and exploration firm with complete specialties, advanced design method, great technological power and excellent social reputation. The institute has 270 staffs, including 25 professors of architects and engineers, 67 advanced architects and engineers, 103 architects and engineers, 80 national registered architects and engineers and 6 experts enjoying special state allowance. The institute sets four complex architecture design and research section, master lin liyan studio, environmental art institute and building economy section as well as iat on gshp. Companies affiliated institute include geotechnical engineering company, project managing consultation company, Construction Drawing Review Consultation Co., Ltd. and so on.

More than 50 years since the foundation of the institute, it has provided more than 13,000 achievements to the society. These engineering projects can be found not only in Liaoning but also in many other cities such as Beijing, Shanghai, Guangzhou, Shenzhen, Hangzhou, Ningbo, Wuxi, Xining, Weihai, Daqing, Jilin, Meihekou, Tongliao, Chifeng as well as some foreign countries like Yemen, Burkina Faso, Russia, Cameroon, Seychelles. Recent years, the institute has broad cooperation with some foreign architects firms including Germany, Australia, America, Britain, Japan and Korea. We work together to win the bid and design a number of domestic projects.

Since 1980s, the institute has won 213 excellent project design (exploration) prizes and 78 technological progress prizes with national, ministerial, provincial and municipal level. Moreover, it has been awarded series of honors and titles such as Liaoning Provincial Construction System Quality Management Prize, Unit focusing on Contract and Credit, Liaoning Provincial Exploration and Design Industry CAD Applying Advanced Unit, National Class I File Management Unit, Advanced Unit for Construction Technology Work in Liaoning Provincial “Eighth Five-Year Plan” Period, National Building Energy=Saving Advanced Unit, Shenyang Energy-Saving Design Advanced Group, and National Construction Technology Creation Work Advanced Unit.

In December 2000, Mr. Lin, Chief Engineer of the institute, was awarded the honorary title of "Master of China engineering design" by the Ministry of Construction of People’s Republic of China. In 2008, President Ye Yang and Vice President Qinggang Li were awarded the first batch of title for Master of Liaoning Engineering Design. Mr. Yang Ye, the president, was awarded the UNESO PRIZE in 1990, the Youth Architect Prize of CAA in 1998, and he was also selected into "Top 10 Youth Architects of Shenyang" and "Top 10 Excellent Youths of Shenyang" in 2004. In 2006 he was selected as Liaoning Province Working Model, Liaoning Excellent Communist, and was awarded the title of Excellent Entrepreneur by China Exploration Design Association. In 2008 he was awarded the title of Top ten Youths of Liaoning Province, and was selected as the 2008 Olympic torch bearer. In 2009 Ye Yang was awarded "Franchise Innovation Award of the 60th National Day" by CEDA. Vice President Qinggang Li were awarded "Liaoning Youth Science and Technology Award for Top Ten of excellence" in 2005. Hui Cao, as the chief architect, was awarded May 1st Labor Medal of Liaoning in 2007, and also awarded the title of Top Ten Excellent Youths of directly subordinate organization under Liaoning Province in 2009.

Geological Engineering Co., Ltd, and Project Management Consultant Co., Ltd, subordinated by institute and the institute obtained the certificates of ISO-9000 Quality System approval (1994 version) in December 2000, and GB/T 19001-2000 idt ISO 90001:2000 Standard quality system approval in 2003.

联系方式 / Contact Us

网址/Http://www.lnbdri.com.cn　邮箱/E-mail: lnbdri@vip.163.com

沈阳

地址/Address: 辽宁省沈阳市和平区和平南大街84号

No.84, Heping nan street, Heping district, Shenyang, China

电话/Tel: +86-24-23389443　传真/Fax: +86-24-23389612　邮编/P.C.: 110005

大连

地址/Address: 辽宁省大连市高新园区黄浦路624号

No.624, Huangpu street, Hi-tech Zone, Dalian, China

电话/Tel: +86-411-39750590　传真/Fax: +86-411-39750590　邮编/P.C.: 116023

中海 · 寰宇天下一期工程

建设地点：沈阳
设计时间：2010年
建筑面积：434 190.26平方米

中海城

建设地点：沈阳
占地面积：24 828公顷
设计时间：2010年

以人为本——体现"以人为本"的设计理念，综合分析居民的生活模式和心理需求，塑造处处关怀人的环境，实现人与自然的"天人合一"，实现社会效益、环境效益和经济效益的有机结合。

生态环境——充分利用大自然赋予我们的阳光、空气、水土资源，创造满足生态需求的可持续发展的高品位社区环境，让建筑在自然中生长，让自然融合于建筑，营建出21世纪的"绿色生态家园"。

建筑采用简约的新古典风格，色彩上以米色及暖红色为主，局部点缀少量深灰色。合理的色彩搭配使整个园区显得沉稳、大气，并和周围美好的环境达到一种和谐的平衡。高层住宅沿街布局整齐、合理，具有很强的韵律感，从而形成统一而又富于变化的城市天际线。

LDI

辽宁省建筑设计研究院
LIAONING PROVINCIAL BUILDING DESIGN & RESEARCH INSTITUTE

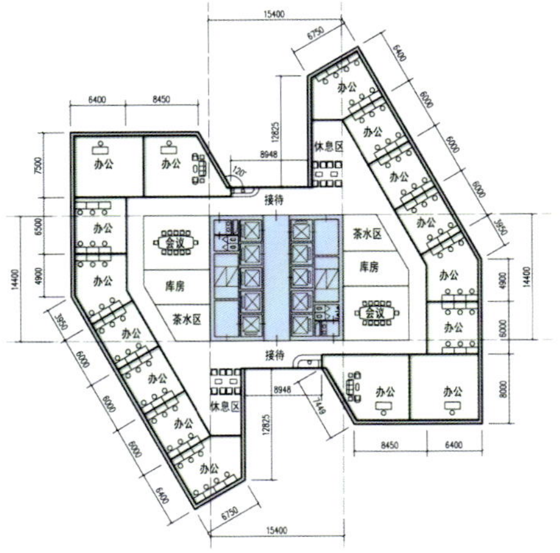

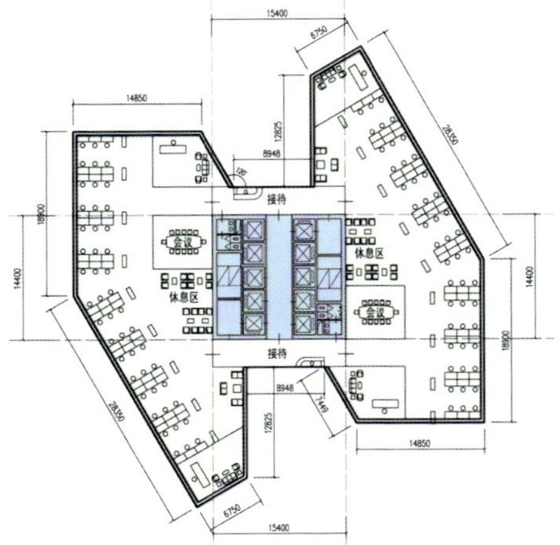

东北传媒大厦

东北传媒大厦用地形状方整，处在整个地块的东南角，面临青年大街和文体路，设计目标是将东北传媒大厦设计成沈阳南大门的地标性建筑，建立起新建筑与周边环境和城市肌理之间有机的联系，将建筑融入城市空间，既尊重城市空间，又丰富了城市空间。

项目名称：东北传媒大厦
合作设计：Beye Scheid Architekten
用地面积：8 412.65平方米
建筑面积：85 520.78平方米
建筑高度/层数：187.50米/43层
建设地点：辽宁 沈阳
设计/建成时间：2007年/2010年

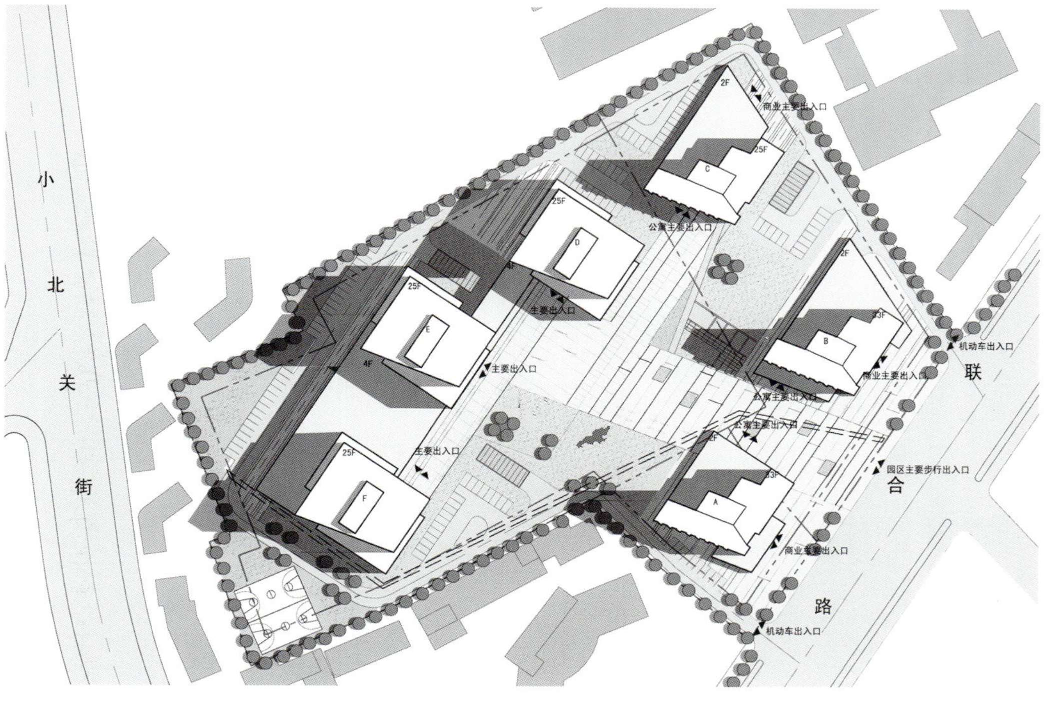

沈阳中韩科技产业团地

整个用地形状呈不规则多边形，分为工业用地和商业用地两部分。工业地块借鉴韩国楼宇化工厂模式，主要功能为产品研发与科技孵化器；商业地块规划为公寓，考虑到与城市道路、可朝向的关系，单体建筑呈角度布置。六栋建筑形成围合空间。人们从园区步行主入口进入基地内部并通向各栋建筑入口，置身于科技与文化的氛围中。

项目名称：沈阳中韩科技产业园地
用地面积：52 534.60平方米
建筑面积：298 503.83平方米
建筑高度/层数：145.20米/33层
建设地点：辽宁 沈阳
设计/建成时间：2009年/2012年

LDI

LDI

沈阳裕景中心

本项目用地西侧的青年大街和北侧的中山路是沈阳市重要的交通性和商业性道路，而南侧的药王庙路和东侧的兴林街则是车流量较小的生活性道路。周边道路向用地内渗透，形成了由西南向东北的分界线，自然地将整个用地划分为居住功能区和商业功能区。设计充分利用了青年大街和中山路的商业功能，强化了地块与外部的联系，同时保证了居住功能区的完整和安静。商业功能区包括两栋超高层塔楼和一处〞小山〞式一层商业建筑。居住功能区包括六栋超高层住宅和两栋超高层公寓，两栋公寓分别位于西南角和东北角，临近城市主干路，六栋住宅位于居住功能区的中心，减少了交通噪声的干扰。

由于建筑的布局与用地呈45度夹角，使得用地周边出现多处三角形空间。这部分空间在地面上成为建筑与街道的缓冲空间，建立了园区与城市之间的和谐关系。地下一层三角形外墙局部回退，为机电专业外管网设计提供了空间条件，同时最大限度地利用了用地的空间资源。

项目名称：沈阳裕景中心
用地面积：84 128平方米
建筑面积：1 288 838平方米
建筑高度/层数：380米/90层
建设地点：辽宁 沈阳
设计/建成时间：2009年/2013年

本溪老干部活动中心

本次基地选址邻于本溪市自然风景优美的望溪公园，北邻望溪公园2号门，南邻望溪公园3号门，选址用地面积0.70公顷，基地现为啤酒屋的小型建筑群落，坐落在基地南北两块台地之上，高差3米。基于道路和台地高差考虑，设计师把基地北侧定位建筑群的主要出入口方向。

该方案把“绿色建筑，健康营造”作为本方案的设计主旨。结合基地周边望溪公园的自然景观，设计师提出使人、建筑与生态环境之间形成一个良性系统的目标。设计保护了原有地貌的特点，并有意识恢复自然山体，使建筑自然“生长”在基地中。设计师从人性—空间、立面—造型两个角度上进行具体分析设计。

工程名称：本溪老干部活动中心
建设地点：辽宁 本溪
建筑面积：4 800平方米
建筑高度：10.5米
层　　数：地上2层
设计时间/建成时间：2010年

潍县战役纪念馆

本工程建设地点为山东省潍坊市东风东街，占地约5公顷。

潍县战役纪念馆位于潍坊市中心，南临东风街，东临虞河，是集纪念参观、革命教育、红色旅游等功能于一体的现代纪念馆。

项目规划设计从城市周边环境及用地分析，建立了有序列关系的室外环境，引导城市中的参观者逐步进入室内展陈空间。设计师在烈士陵园的有利位置上安排了纪念馆建筑，丰富了城市景观。室内部分由一层的展陈空间和二层的全景画空间组成。平面的中心设有楼梯和坡道双向交通路线，防止人流交叉。设计师还考虑了无障碍设计。室外墙面主要采用浅褐色火烧板，与纪念主题相契合。圆柱形主体外圈包围着若干片逐级升高的景观墙，寓意着革命烈士前仆后继，后人继续完成革命事业。

工程名称：潍县战役纪念馆
建设地点：山东 潍坊
建筑面积：3 416.47平方米
建筑高度：24米
层　　数：地上2层
设计时间/建成时间：2009年/在建

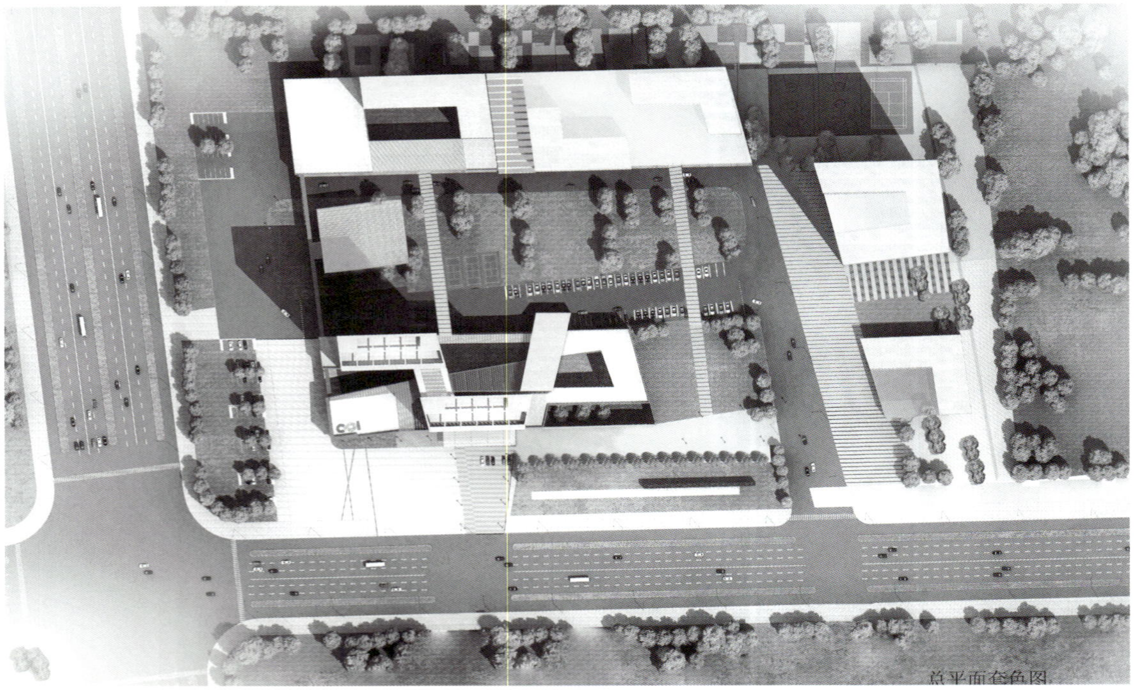

总平面套色图

中电投蒙东能源集团有限责任公司生产调度指挥中心

本工程建设地点为内蒙古通辽市开发区，占地10万顷，分三期建设。生产调度中心于一期建设，西临创业大道，南临清沟大街，并与绿化景观河隔路相望，是集行政办公，科研机构，综合服务等功能于一体的现代的总部大楼。

项目规划设计从城市周边环境及用地分析入手，建立了多层次的空间系统，平衡了道路交通，在场地的有利位置，安排核心建筑，突出企业形象，丰富城市景观。单体建筑，由共享的景观大厅将前后错落的两个主体有效地连接，形成舒展的"Z"型。石材的凝重和玻璃的通透形成鲜明的对比。裙房整体连贯，西侧与北侧屋面向上折起，寓意能源由原始状态转化为可以由人们直接使用的能量，它将会发挥无限的力量，指引企业前进的方向。

工程名称：中电投蒙东能源集团有限责任公司生产调度指挥中心
建设地点：内蒙古 通辽
建筑面积：43 758平方米
建筑高度：78.4米
层　　数：地上19层，地下1层
设计时间/建成时间：2008年/2010年

沈阳铁路机械学校新校区总体规划

所有的功能被有机地连接成一个整体“机械”——一部高效率、有人情味的、有精神内涵的机械。

网格系统继承了城市的脉络，轴网对应四方朝向，自然地与城市融为一体。两重校园大门模糊了城市与校园的界限，模糊“内与外”“有与无”的界限。老教学楼的再生、老校区大树的迁移，将旧校区的场所记忆部分“移植”给新场所，延续记忆，延续历史。中央主题广场位于规划的显要位置，并作为保留空间和聚集场所。每个建筑都拥有由墙体及柱廊围合成的广场或庭院。

简约和机械之美是建筑的主导形式。

工程名称：沈阳铁路机械学校新校区总体规划
建设地点：沈阳经济技术开发区
建筑面积：160 000平方米
设计时间/建成时间：2009年

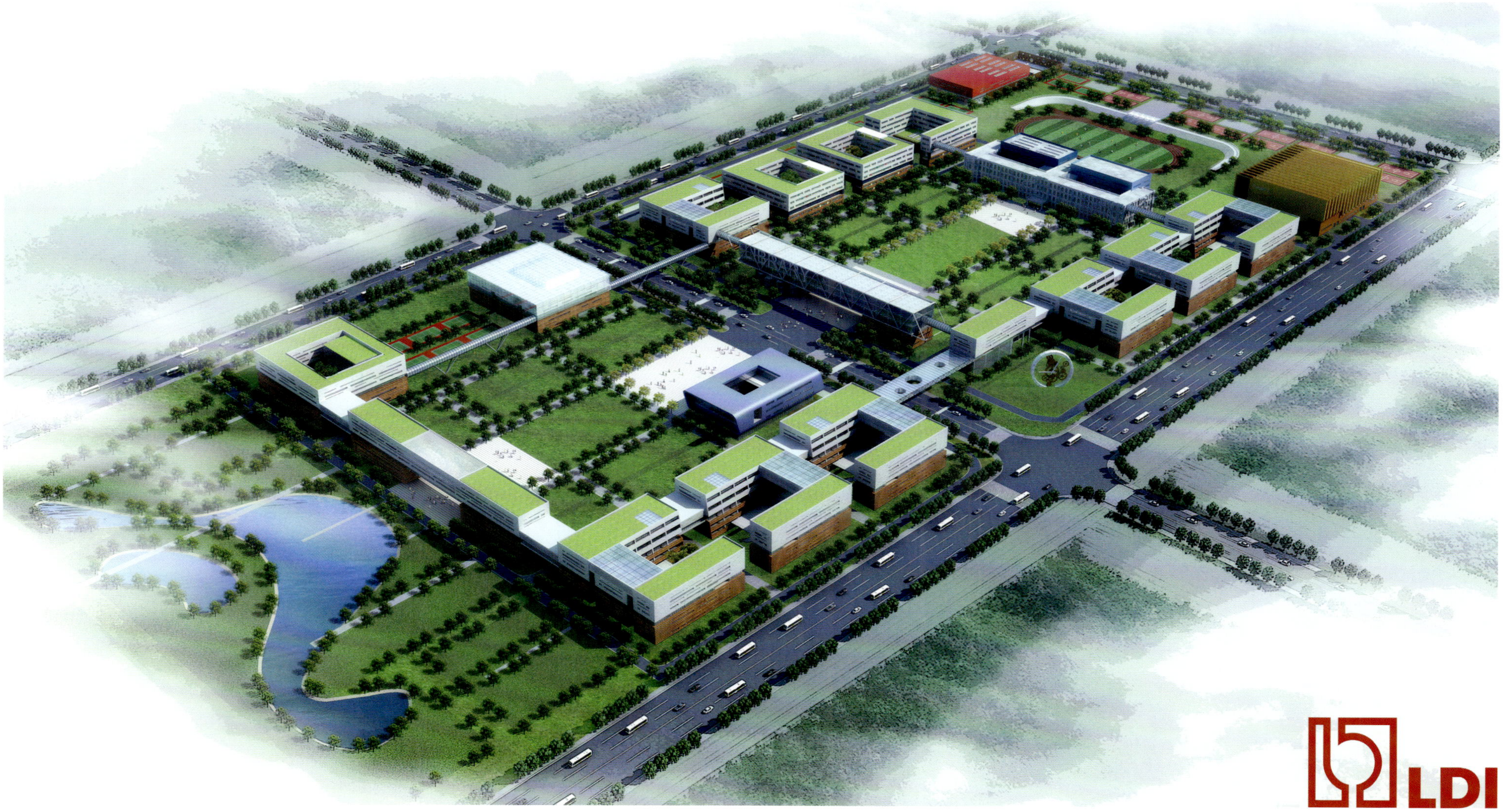

LDI

UDS

U D

UDS 建筑师工作室是大连市建筑设计研究院的先导建筑师团队。工作室以市场化运作为机制，同时借鉴国内成功设计公司的经营管理模式，业务范围涉及酒店、商业、教育、医疗、居住等多个领域，已逐渐形成一支具有专业化建筑设计经验的团队，并提供从项目策划、单体设计到室内设计、景观环境的全程设计服务。

建筑创作是融合的过程，经济条件、技术水平、文化差异等都促进或制约着作品的完成。完美的建筑也许只存在于理想中，但对理想的追求永远是我们前行的发动机。建筑创作又是一种立场，是不可救药的完美主义，是一丝不苟的精益求精，我们就是这样一批充满热情的创作者，绘制生活、实现梦想、创造未来是我们对业主、对社会始终如一的承诺。

As the avant-garde team of Dalian Architectural Design & Research Institute in terms of architectural design, UDS Architecture Studio is guided by market principles in its daily operation. By borrowing operation modes from many successful domestic and foreign colleagues, the studio has expanded its scope of business to cover almost all aspects of architectural design including hotels, commercial buildings, education institutes, medical facilities, residential buildings, etc. With prominent expertise in architecture, the studio as a team provides one-stop service from project planning, single unit design to interior design and landscaping to its clients.

Architectural creation is a course of blending that is promoted or constrained by a number of factors such as economic condition, technology capability, cultural variation...Though they know that perfect architecture may only exist in people's dreams, it is the pursuit for dreams that drives them forward all the way. In their eyes, architectural creation is also a way of life, an incurable perfectionism and the most captious criticism...any word that they could think of to describe their career. Born to be passionate creators, it is their job to paint our life, to make our dreams come true and to create a beautiful future for our world. And that will be their permanent commitment to the clients and the society.

地址：中国大连市西岗区胜利路 102 号 308 室
电话：+86-411-84313789
传真：+86-411-84303187
电子邮箱：ud_studio@126.com

Address: No.102 SHENGLI RD. XIGANG DIST. DALIAN, CHINA
Tel: +86-411-84313789
Fax: +86-411-84303187
E-mail: ud_studio@126.com

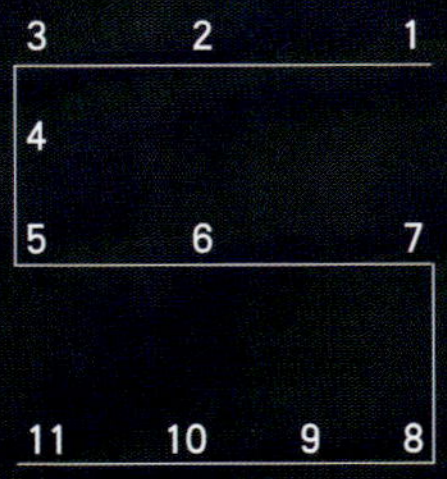

1 大连国际会议中心
澳大利蓝天组合作设计
2 大连开发区文化中心
建设部优秀勘察设计三等奖
3 大连开发区第十高级中学
中国建筑学会建筑创作嘉奖
4 大连大学理学院
5 大连大学文学院
6 大连邢良坤陶艺馆
7 大连金石滩模特影视学校
中国建设部优秀勘察设计三等奖
8 大连外经贸办公楼保护
9 沈阳友谊宾馆
10 大连北方金融中心
中国建设部优秀勘察设计二等奖
11 大连开发区东城国际住宅

沈陽萬宸建築規劃設計有限公司

SHENYANG WANCHEN ARCHITECTURE&PLAN DESIGN CO.,LTD

沈阳万宸建筑规划设计有限公司，成立于2005年9月，是经国家建设部批准成立的具有建筑甲级、规划丙级设计资质的股份制设计公司。公司现有员工75人，公司业务范围包括建筑、规划、园林、效果图制作等。公司自成立至今虽仅四年多的时间，但公司业务发展迅猛，承接的工程项目遍及沈阳、本溪、盘锦、阜新、朝阳、葫芦岛、营口、吉林等地。公司秉承"创新求真、精益求精、创立品牌"的设计理念，竭诚为广大业主提供一流的服务。公司员工正以饱满的创作激情、精诚合作的团队精神，努力争创辽沈地区第一建筑设计品牌！

Shenyang Wanchen Architecture&Plan Design Co., Ltd was founded in September, 2005. It gets the approval of the Ministry of Construction to set up and it's also a company that has qualifications of Class A construction and Class C of planning in design. Currently, the company has 75 employees. The scope of business includes construction, planning, landscape, renderings production and so on. Although only a little more than four years since its inception, the company has a rapid business development, undertaking projects throughout Shenyang, Benxi, Panjin, Fuxin, Chaoyang, Huludao, Yingkou, Jilin and so on. Adhering to the design concept of "truth-seeking and innovation, striving to be perfect, and creating a brand", the company dedicates to providing the best service to the owners. The staffs of the company are now full of creative passion and with a spirit of sincere cooperation, striving to become the first architectural and design brand of Liao-Shen District.

地址：沈阳市和平区南京南街80号美龙家园A座26楼
电话：+86-24-23509455/23509466
传真：+86-24-23508881
网址：www.wcjzsj.com

Address: 26th floor, Building A of Meilong Home Community,
No. 80 of Nanjing south street, Heping District, Shenyang city.
Tel : +86-24-23509455/23509466
Fax :+86-24-23508881
Http//: www.wcjzsj.com

盘锦广厦新城

项目规模60万平方米。诠释高品质现代中式住宅，营造高品位北方益居园区。
设计时间/建成时间：2008年/2010年

Panjin Guangsha New City

The scale of project is 600,000m^2. It is a perfect interpretation of high-quality modern Chinese residence, creating a north high quality eco-city.
Design/Complete Time: 2008/2010

本溪建设大厦、工会大厦、市规划展示馆

建筑群力求营造向上生长的力量感，轮廓清晰挺拔，庄重大方。项目规模5万平方米。
设计时间/建成时间：2006年/2007年

Benxi Construction Building, Union Building, the Urban Planning Exhibition Hall

This set of buildings sought to exhibit a sense of growing-up force, with a clear upright outline and dignified appearance.It is 50,000 m^2.
Design/Complete Time: 2006/2007

葫芦岛山城露园住宅小区

30万平方米的缓坡山地住宅区，其富有张力的入口建筑将商业与景观及交通功能有机地结合起来。
设计时间/建成时间：2008年/2011年

Mountain City and Dew Park Residential community of Huludao

It is a 300,000 m^2 residence architecture building on the gentle slope of mountains. The architecture of the entrance well combines the business function ,landscape function and transportation function.
Design/Complete Time: 2008/2011

鞍山兴隆大家庭商业摩尔

40万平方米的辽宁地区最大的商业综合体，如山石般巨大的体量体现出颇具现代时尚感且气势磅礴的商业建筑。
设计时间/建成时间：2009年/2012年

Anshan Xinlong Happy Family Store

It covers 400,000m^2 and is the largest commercial complex in Laoning.Its large cubic content of building gives an expression of modern, fashion and of great momentum.
Design/Complete Time: 2009/2012

沈阳建筑大学建筑设计研究院

Architectural Design and Research Institute of Shenyan jianzhu University

沈阳建筑大学建筑设计研究院成立于 1979 年，依托大学，技术力量雄厚、装备先进，于 2003 年 10 月 23 日通过 ISO 9001 质量体系认证。现执有建筑行业（建筑工程）甲级，建材行业（水泥工程、新型建筑材料工程）专业乙级，市政行业（给水工程、排水工程、道路工程、桥梁工程、热力工程）专业乙级，风景园林工程设计专项乙级证书。在其所承担的工程设计中有 60 余项获得部、省、市级优秀设计奖，并取得较好的社会效益和经济效益，得到社会各界和建设单位的赞扬和好评。

Architectural Design & Research Institute of Shenyan Jianzhu University was established in 1979. With support from the univeristy, relying on powerful technical resources and the advanced equipments, it has passed the ISO9001 quality system certification in 2003. Now it has obtained professional qulifications in different divisions of architecturd design as below: architecure industy Grade A (architecture and engineering), building materials industry (Cement Engineering, new building material engineering); Municipal Engineering Grade B(water supply system engineering, waste engineering, Road Works, Bridge Engineering, heating engineering); professional Grade B certification of landscape projects design. Among all the projects they have undertaken, 60 projects has won the excellent design awards in municipal, provincial and national construction design competitions. Their design works have made good social and economic benefits, which also offer them good reputaion and praises from the society and the construction companies.

地址：沈阳市浑南新区浑南东路 9 号
法人：孙长征
电话：+86–24–24507799
传真：+86–24–24506333
邮件：sjd24507799@163.com
网址：www.jzsj.sjzu.edu.cn

Address: Hunnan New District In Shenyang City Road On the 9th Hunnan
Legal: Changzheng Sun
Tel: +86-24-24507799
Fax: +86-24-24506333
E-mail: sjd24507799@163.com
Http: //www.jzsj.sjzu.edu.cn

辽宁机电职业技术学院仪器仪表学院

项目概况

辽宁机电职业技术学院新校区位于丹东市浪头开发区仪器仪表产业园。依照丹东浪头开发区总体规划，学院与丹东仪器仪表产业园一期工程共同构成了产业起步区。

设计理念

1. 空间开放化

新校址位于开发区干道与世纪南路的交汇处，校园空间通过50米绿带与开发区城市空间相互交融。校园规划设计以开放性的视角将校园空间和产业园空间连成整体。

2. 专业集成化

主题空间的塑造，着力突出仪器仪表工业的特征。规划设计将工业中心、览展中心作为仪器仪表园起步区的公共要素并沿道路集成设置，通过时光广场、巨型日晷雕塑、机械仪表展示等突出了仪器仪表工业的专业特征。

3. 环境人文化

根据总体规划理念，确立了一轴、两带、三区的规划空间结构：

“一轴”指的是由校园主入口广场－学术广场－生活广场－商业内街等一系列建筑及景观所组成的空间主轴线序列。

“两带”指湿地景观带和运动景观带构成的南北带状娱乐休闲区。

“三区”是指校前区、主教学区、生活服务区所组成的各具特色的建筑空间。

建筑及外部空间的“点”、“线”、“面”的有机结合，以多种方式丰富了空间形式和景观内容，满足了不同层次的校园人文环境需求。

技术经济指标

1. 规划用地面积：约14.56公顷
2. 规划建筑面积：10 1837平方米（地上）

其中：①工业中心：12 581平方米　②行政中心：8 805平方米　③展览馆：10 811平方米　④图书信息中心：7 590平方米　⑤机械馆：12 405平方米　⑥仪器仪表馆：12 405平方米　⑦食宿综合楼：30 880平方米　⑧风雨操场：2 598平方米　⑨架空长廊：3 762平方米　⑩设备用房用地范围：2 469.2平方米（地下）

3. 容积率：0.699
4. 绿地率：30%
5. 建筑密度：22%
6. 地面停车位：136辆
7. 建筑高度：1层不大于18米，2层不大于18米，
3层不大于24米，9层不大于38米。

Instrument and Meter Institute of Liaoning Jidian Polytechni

Project Profile

The new campus of Liaoning Jidian Polytechnic is located in Instrumentation Industrial Park of Langtou Development Zone in Dandong. In accordance with the master plan of Langtou Development Zone, this college in coorperation with the first phase project of Dandong Instrumentation Industrial Park constitute Industry Start Zone.

Design Concept

1. Opening-up of Spaces

The new campus is located at the conjunction of the main line in the development zone and South Century Road and it is integrated with the urban space by a 50-meter green belt. From an open perspective the campus planning connects the campus space and the industrial park space as a whole.

2. Professional Integration

The shape of the theme space focuses on highlighting the characteristics of the instrumentation industry and its planning takes the industrial center and exhibition center as public elements of the start zone, displaying them along the road. The professional features of the instrumentation industry are projected by Time Square, the giant sundial sculpture and the mechanical instrument exhibition, etc.

3. Humanization of Environment

According to the general planning concept, a special structure of "One Axis, Two Belts and Three Zones" is established.

"One Axis" refers to the main spatial axis composed by a series of construction and landscape: the main entrance plaza of the campus—Academic Square—Life Square—commercial streets, etc. "Two Belts" refers to the north-to-south recreational area composed by the wetland zone and the sport landscape. "Three Zones" refers to the distinctive architectural space formed by the college front area, the main teaching area and the living service area.

The organic combination of the "point", "Line" and "face" between the building and the external space enriches the spatial form and the landscape content in many ways, meeting the campus human environmental needs of different levels.

Technical and Economic Indicators

1. Planning Area: about 14.56 ha
2. Planning floor Area: 101,837 m^2 (above ground)

Of which: ① Industrial Center: 12,581 m^2 ② Administration Center: 8,805 m^2 ③ Exhibition: 10,811 m^2 ④ Book Information Center: 7,590 m^2 ⑤ Machinery Hall: 12,405 m^2 ⑥ Instrument and Meter Center: 12,405 m^2 ⑦ Accommodation complex building: 30,880 m^2 ⑧ Gymnasium: 2,598 m^2 ⑨ Overhead gallery: 3,762 m^2 ⑩ Equipment room/land area: 2,469.2 m^2 (underground)

3. Plot Ratio: 0.699
4. Greening Rate: 30%
5. Construction Density: 22%
6. Ground Parking Spaces: 136
7. Building Height: One layer: not more than 18 meters; Two layers: not more than 18 m;
Three layers: not more than 24 meters; Nine layers: not more than 38 m.

辽宁金海建筑设计研究院
ARCHITECTURE

辽宁金海建筑设计研究院有限公司是建筑工程甲级资质设计勘察单位，公司改革转制以来，先后完成了邮电大厦、宽甸中心医院（760个床位）、电视发射观光塔、昆山市富贵大厦、海南省金秋阁、北京潭柘寺生态园、沈阳惠民馨苑等项目的设计任务，先后创造了多项精品设计。此外，公司还先后完成了中国驻朝鲜大使馆、俄罗斯布市及伊尔库司克市五星级宾馆、住宅、公共建筑等项目。

目前公司业务已同国际接轨，今后公司将继续遵循科学发展观的重要思想，努力为社会提供更加优质的服务。

Liaoning Jinhai Architectural Design Research Institute Co., Ltd. is a survey & design company with Class A architecture engineering qualification. Since reformation and restruction, they have finished many projects, including Post Telecom Plaza, the Center Hospital of Kuandian County(760 beds), the TV programming sight-seeing tower, Fugui Plaza of Kunshan City, Jinqiu attic in Hainan Province ,ecological garden of Beijing Tanzhe Temple, Shenyang Huimin Xinyuan and other design Projects, creating numerous classic design works. Besides, they worked on many overseas design projects, such as Embassy of the People's Republic of China in the Democratic People's Republic of Korea, five-star hotels, residence, public architecture projects and other design tasks in Russia.

With its business development in the international market, Liaoning Jinhai Architecture Design Research Institute will continue to follow the significant scientific outlook on development, providing high standard services to the society.

地址：辽宁省丹东市沿江开发区太阳大厦B座126号
电话：+86–415–2814421
传真：+86–415–2814421

Address: No.126, tower B, Taiyang Plaza, Dandong Riverside Development Zone, Liaoning
Tel: +86-415-2814421
Fax: +86-415-2814421

1 丹东市雄业大厦

本工程为丹东市雄业大厦总建筑面积30 000平方米，共28层

Dandong xiongye plaza

This project is Xiongye Plaza in Dandong city. The total gross floor area is 30,000 m^2 with 28 floors.

2 中国宽甸检察院

本工程为中国宽甸检察院总建筑面积8 000平方米，共12层

Kuandian Procuratorate

This project is procuratorate in Kuandian. The total gross floor area is 80,000 m^2 with 12 floors.

3 中国丹东国际人才中心

本工程为丹东国际人才中心总建筑面积10 484平方米，共十四层

Dandong International Talent Center

This project is Dandong international talent center The total gross floor area is 10,484 m^2 with 14 floors.

4

5

6

7

8

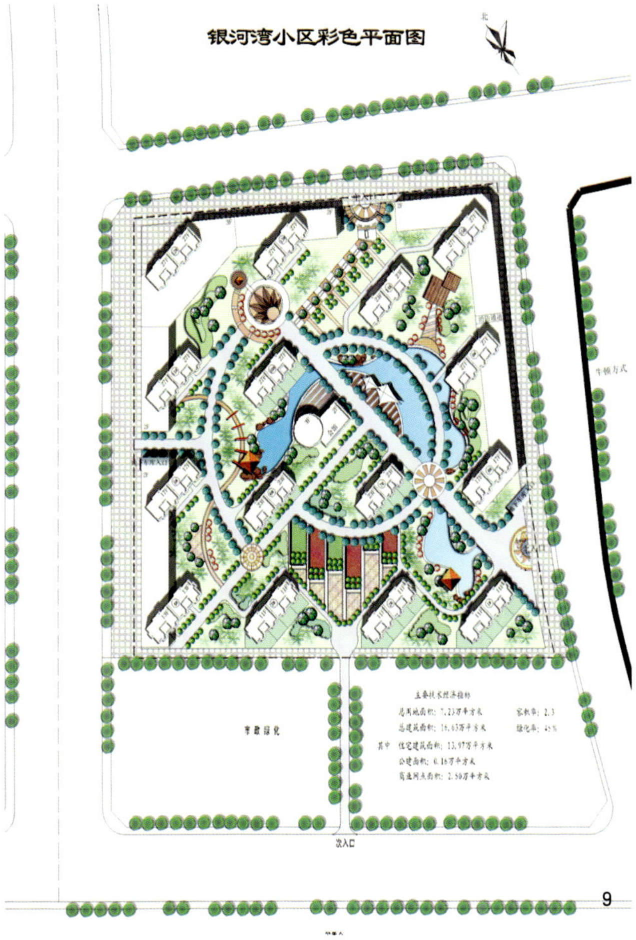

9

4 丹东市新加坡城三、四期设计方案

本工程为丹东市新加坡城三、四期，设计方案总建筑面积为280 000平方米

Singapore town phase 3 and 4 in Dandong city

This project is the design scheme for phase 3 and 4 of Singapore town in Dandong City, the total gross floor area is 280,000 m^2.

5 沈阳市左岸慧晶三期工程设计方案

本工程为沈阳市左岸慧晶三期工程，总建筑面积为150 000平方米。

Phase 3 of Zuo'an Huijing project in Shenyang design scheme

This project is the Phase 3 of Zuo'an Huijing project in Shenyang, the total gross floor area is 150,000 m^2.

6 沈阳惠民馨苑

本工程为沈阳惠民馨苑住宅小区，总建筑面积为194 368.54平方米。

Huimin Xinyuan in Shenyang

This project is Shenyang Huiming Xinyuan, total gross floor area is 194,368.54 m^2.

7 宽甸伟业北方山奇

本工程为中国宽甸伟业北方山奇大市场，总建筑面积为53 807平方米。

Kuandian Weiye Beifang Shanqi

This project is Kuandianweiye Beifangshanqi market, the total gross floor area is 53,807m^2.

8 太阳锦江花园

本工程为丹东市太阳锦江花园小区A，B座，总建筑面积为51 722平方米。

Taiyang Jinjiang Garden

This project is building A & B of Taiyang Jinjiang Garden in Dandong city, the total gross floor area is 51,722 m^2.

9 银河湾小区

本工程为丹东市东方测控银河湾小区，总建筑面积为166 300平方米。

Yinhe wan Community

This project is Yinhe wan Community residence community of Dandong Dongfang Measurement and Control Technology Co., Ltd, the total gross floor area is 166,300 m^2.

工大设计 GRANDDESIGN

内蒙古工大建筑设计有限责任公司

INNER MONGOLIA GRAND ARCHITECTURE DESIGN CO.,LTD.

法人代表、董事长：张鹏举
地址：中国·内蒙古呼和浩特市爱民路49号
邮编：010051
电话：+86–471–6576005/6575347/6577170
传真：+86–471–6576322
邮箱：Gdsijy@yahoo.com.cn
网址：www.Nmggdsj.com.cn

Legal Representative, the Chaiman: Pengju Zhang
Address: No.49 Aimin Road, Hohhot City, Inner Mongolia China
P.C.: 010051
Tel: +86-471-6576005 / 6575347 / 6577170
Fax: +86-471-6576322
E-mail: Gdsijy@yahoo.com.cn
Http://www.Nmggdsj.com.cn

内蒙古工大建筑设计有限责任公司前身为成立于1979年的内蒙古工业大学建筑设计研究院，2002年2月改制为有限责任公司，2004年通过ISO 9001：2000质量体系认证，逐步建立起完善的现代企业制度，经过二十余年的发展，公司取得的成就令人瞩目，现已持有国家授予的建筑工程甲级设计证书。

公司现有职工64人，其中教授4人，副教授8人，高级工程师12人，工程师10人，一级注册建筑师5人，一级注册结构工程师5人，是一支具有高素质的精干的设计队伍，同时公司以高校学术力量为依托，充分发挥广大专业教师的设计潜能，在内蒙古自治区树立了自身独特的行业地位。公司业务范围含概建筑设计、规划设计、室内设计、景观设计、岩土工程地质勘探、工程技术咨询、预应力结构设计及施工等内容。

Inner Mongolia Grand Architecture Design CO.,LTD. was first established as Inner Mongolia University of Technology Architecture Design Institute in the year of 1979, and has been awarded the National Class One Design Qualification Certificate for architecture Design Engineering. The company has finished incorporation in Febuary 2002, and passed the authentication of ISO9001 2000 quality system in December 2004.
The company now has 64 people,including 4 professors, 8 associate professors, 12 senior engineers, 10 engineers, 5 Class One Registered Architect, 5 Class One Registered Structure Engineers, and is a high-efficient design team. By making use of the academic strength of the university, the company has established solid position in the industry in Inner Mongolia. In recent years, the company has worked in Inner Mongolia and other provinces in China, as well as in the Republic of Mongolia, and has won many awards for excellent design works granted by the Ministry of Construction of China and Inner Mongolia, being widely-praised in society.

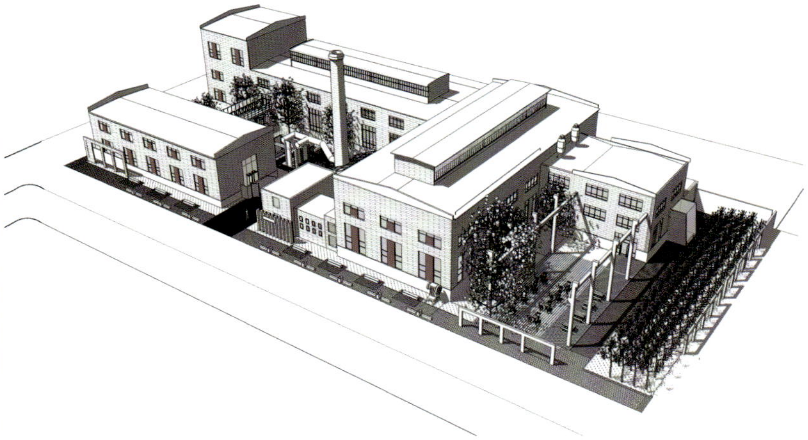

内蒙古工业大学建筑馆

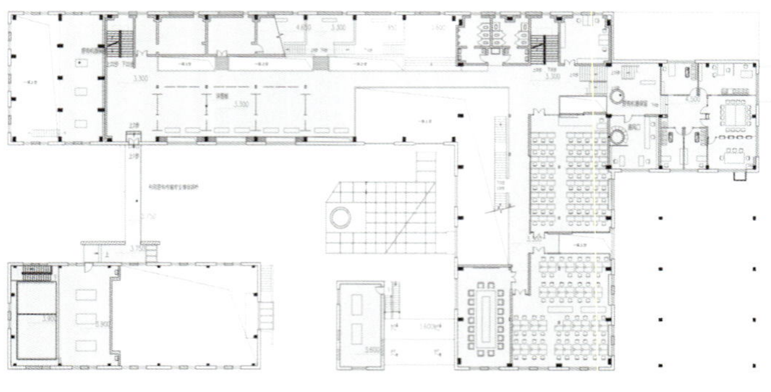

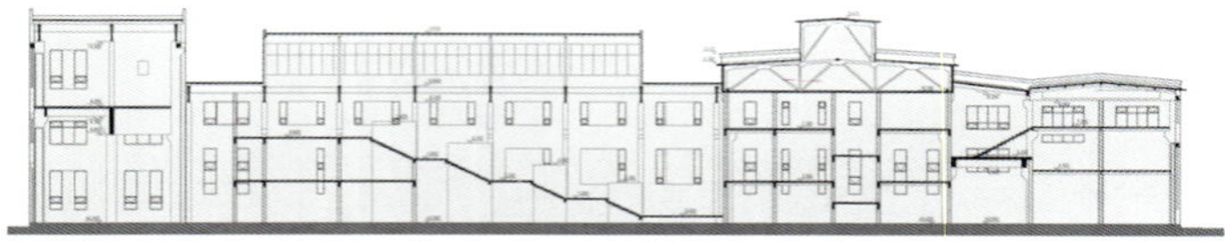

内蒙古工业大学建筑馆，由废弃的工业大学机械厂铸工车间改建而成，原面积2 760平方米，通过在十几米的桁架空间内加层改造，使用面积达到5 910平方米。设计中面临的问题有：(1)在保存历史记忆的前提下，创造适合建筑学教学特点的开放场所空间；(2)在有限资金下进行限额设计。

基于以上难点，设计着重于：(1)在尊重场所感的前提下营造空间，赋予旧材料设备新的功能，将一种特定的历史记忆融在了其中；(2)以最适功能原则、积极主动的原则，从建造现场出发，采用低技的生态策略，综合多效解决技术问题。

Architectural Museum of Inner Mongolia University of Technology: It is converted from the abandoned foundry workshop of the university's machinery factory. The original land area of 2,760 m^2 has been extended to 5,910 m^2 through the story-adding transformation within the truss space of more than 10 meters in height. Difficulties encountered in the design include: 1. create an open space suitable for the teaching features of architecture under the premise of preserving historical memory; 2. carry out the cost-limited design with a limited amount of funds.
Based on the above-mentioned difficulties, the design focuses on: 1. under the premise of respecting the place sense, create space and impart to the old materials and equipment new functions, merging into them a particular historical memory; 2. following the principle of optimal function and positive initiative, starting from the construction site, adopt the ecological strategy of low technology and solve technical problems in a comprehensive and efficient way.

1

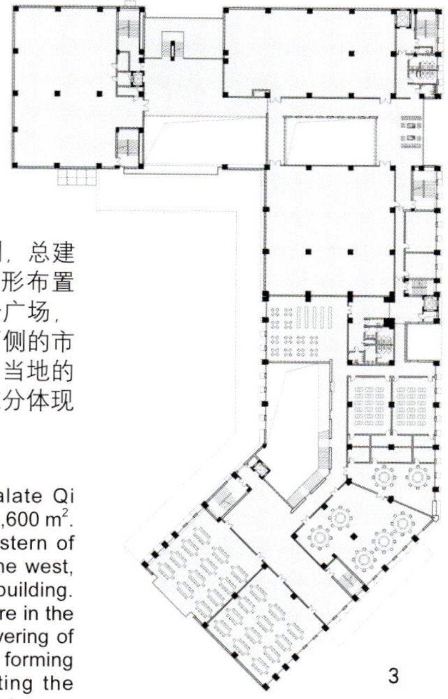
3

达拉特旗博物馆、图书馆

项目位于达拉特旗市政广场的东侧，总建筑面积为20 600平方米。建筑形体成L形布置于基地的东北侧，在西面则围合成一个广场，成为建筑前的过渡空间，同时，也与西侧的市政广场互相呼应，融为一体。建筑采用当地的石材覆面，形成了厚重敦实的体量，充分体现了地方气候和环境特点。

Dalate Qi Museum & Library

This project is located in the east of Dalate Qi Municipal Square, with a total floor area of 20,600 m^2. The L-shaped building lays in the northeastern of the base, while encloses into a square in the west, forming a transitional space in front of the building. Meanwhile, it echoes with the municipal square in the west, integrating with it into a whole. The covering of the building adopts the local stone material, forming a thick and solid body mass, fully reflecting the characteristics of local climate and environment.

1–3 达拉特旗博物馆 图书馆

4–6 达拉特旗影剧院 科技馆 少年宫

2

5

达拉特旗影剧院、科技馆、少年宫

项目位于达拉特旗市政广场的南侧，总建筑面积为23 000平方米。建筑朝向市政广场一侧设置了室外大台阶，将人流引向了二层内院，为不同功能单位提供了多个独立出入口的同时，也形成了具有场所感的室内外空间。建筑采用折板式的屋面和墙体将建筑的各个部分统一为一体，塑造出了自由流畅、简洁现代的建筑形体。

Dalate Qi Cinema, Science and Technology museum & Youth Palace

This project is located in the south of Dalate Qi Municipal Square, with a total floor area of 23,000 m^2. An outdoor stage is installed towards the municipal square, which can introduce the stream of people into the internal court on the second floor. This stage not only provides different functional units with several independent entrances, but also forms indoor and outdoor space with strong sense of place. The building adopts folded-plate roof and walls, which effectively integrate various parts into an organic whole, creating an architectural form of grace, simplicity and modernity.

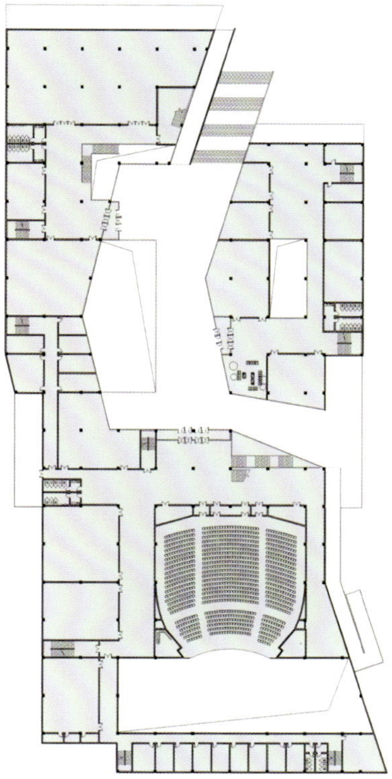
4

6

XY Design 内蒙古新雅建筑设计有限责任公司

Inner Mongolia Guxinya Architectural Design Limited Company

内蒙古新雅建筑设计有限责任公司成立于1999年，现拥有员工68名，是独立资质的国家甲级设计公司，其设计范围涉及公共建筑、居住区建筑设计及城市规划、环境景观设计，并成功地实施了一批建筑作品，包括金融、银行、办公、展览、文化中心、商场、医院、学校、司法、住宅等建筑，曾获得全国优秀勘察设计铜奖一项，建设部二等奖一项，三等奖五项，自治区优秀设计一等奖八项，另外还获得世界华人建筑协会优秀设计奖两项。我公司立足内蒙古、面向全国，并积极拓展国际合作空间，已有数次与国外设计事务所合作的成功经验。

本公司坚持时代性和创新性的设计理念，并充分尊重当地历史文脉、环境和施工技术等条件，做出了大量的因时、因地、得体的设计，这些作品极具有震撼力、感染力，引导和推动了当地建筑创作的潮流。新雅强调精品意识，从每个项目的方案构思到具体设计乃至每个细部都做到精益求精、反复推敲、不断提炼升华，最终塑造出精品。

主要设计作品

包头市信托公司综合楼
呼市回民区祥和小区三区物业管理综合办公楼
内蒙古西部区审计培训中心综合楼
呼和浩特市学府花园住宅小区
呼和浩特市经济技术开发区招商服务中心
呼和浩特市五大班子综合楼
内蒙古出入境检验检疫局综合实验楼
呼市仁和房地产开发有限公司素质拓展训练基地办公楼
阿盟驻呼办事处干部培训中心
蒙古风情园·马文化博物馆
包头青山区阳光大厦
蒙古风情园·成吉思汗纪念堂
呼和浩特市回民区政府办公楼
包头市青山区政府
学府康都
集宁师专图书馆
鄂尔多斯市国宾馆
巨华国际大酒店

Inner Mongolia Guxinya Architectural Design Co., Ltd. founded in 1999, a national Class A corporation with sole investment, has 60 existing employees. This company undertakes the responsibility for designs on public buildings, residential areas, urban planning and environmental landscape. It has managed to put forwards various architecture works including financial building, bank, office, exhibition hall, cultural center, market, hospital and residence. Among these works, one won the Bronze Medal of State Outstanding Investigation and Design, one Second Prize and five Third prizes issued by Ministry of Construction. Eight projects were awarded the First Prize of Outstanding Design in autonomous region. Besides, two won the Outstanding Design issued by World Association of Chinese Architecture. Based on Inner Mongolia, this company extends cooperation with other provinces and other counties. By doing this, it accumulates rich experiences related cooperation with foreign design firm.

While conducting designs, this company adheres to the concept of contemporary spirit and innovation, respects local history and humanity, and considers the environmental conditions and construction technology. Many designs are not only according to specific time and site but in good taste. These works shocks us and arouses resonance with us. They intend to guide and push the trend of local architectural design. Xinya emphasizes top quality consciousness, so to every project, from plan idea to concrete design, architect considers carefully, keeps improving, refines constantly and ensures to create elaborate designs.

Main Works

The Comprehensive Building of Baotou Trust Company
The Comprehensive Building of property management in Block No.3 of Xianghe Community, Hui Autonomous Region, Huhhot
The Comprehensive Building of Audition Training Center in west Inner Mongolia
Huhhot Xuefu Garden Residential Area
Merchants Service Center in Economic and Technological Development Zone in Huhhot
The Comprehensive Building of Huhhot CPC, NPC, Government, CPPCC and Commission for Discipline Inspection
The Integrated Experimental Building of Inner Mongolia Entry-exit Inspection and Quarantine Bureau
The Office Building for quality development training of Huhhot Estate Development Co., Ltd.
Cadre Training Center of League of Arab States Office located in Huhhot
Mongolic Fengqing Park•Horse Cultural Museum
Sunshine Building in Mount Qing District, Baotou
Mongolic Fengqing Park• Genghis Khan Memorial Hall
The Office Building for Government of Hui Autonomous Region, Huhhot
Government of Mount Qing District, Baotou
Xuefu Kangdu
Library of Jining Teachers College
Ordos State Hotel
Juhua International Hotel

地址：内蒙古呼和浩特市金桥开发区金桥路19号
法人：温捷强
邮编：010040
电话：+86-471-4503896
传真：+86-471-4503887
邮箱：xinya_wjq@sina.com

Address: No.19 Jinqiao Road, Jinqiao Development Zone,Huhhot, Inner Mongolia
Legal Person: Jieqiang Wen
P.C.: 010040
Tel: +86-471-4503896
Fax: +86-471-4503887
Email: xinya_wjq@sina.com

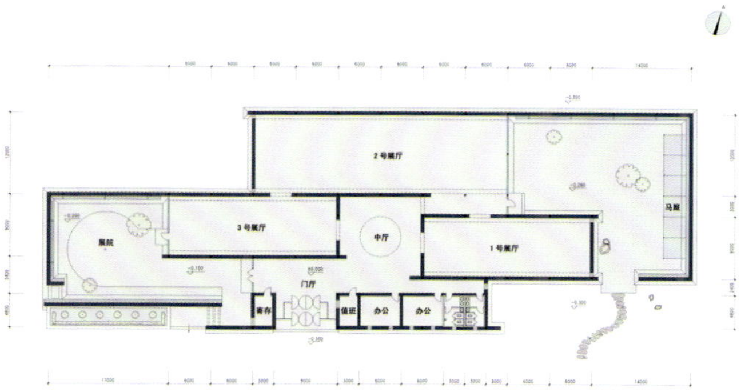

马文化博物馆

建设地点：内蒙古 呼和浩特
总建筑面积：1 300 平方米
完工时间：2006 年 8 月

The Museum of Horse Culture in China

Construction Location: Huhhot, Inner Mongolia
Total Building Area: 1,300 m^2
Completion Time: 2006.08

新雅艺墅 A 座

建设地点：内蒙古 呼和浩特
总建筑面积：19 294.12 平方米
完工时间：2008 年 5 月

Tower A of Xinyayi Villa

Construction Location: Huhhot, Inner Mongolia
Total Building Area: 19,294.12 m^2
Completion Time: 2008.05

成吉思汗纪念堂

建设地点：内蒙古 呼和浩特
总建筑面积：960 平方米
完工时间：2006 年 7 月

Memorial Hall of Genghis Khan

Construction Location: Huhhot, Inner Mongolia
Total Building Area: 960 m^2
Completion Time: 2006 .07

集宁师专图书馆

项目地点：内蒙古 乌兰察布
总建筑面积：15 624 平方米
完工时间：2006 年 11 月

Library of Jining Teachers' College

Construction Location: Ulanqab, Inner Mongolia
Total Building Area: 15,624 m²
Completion Time: 2006.11

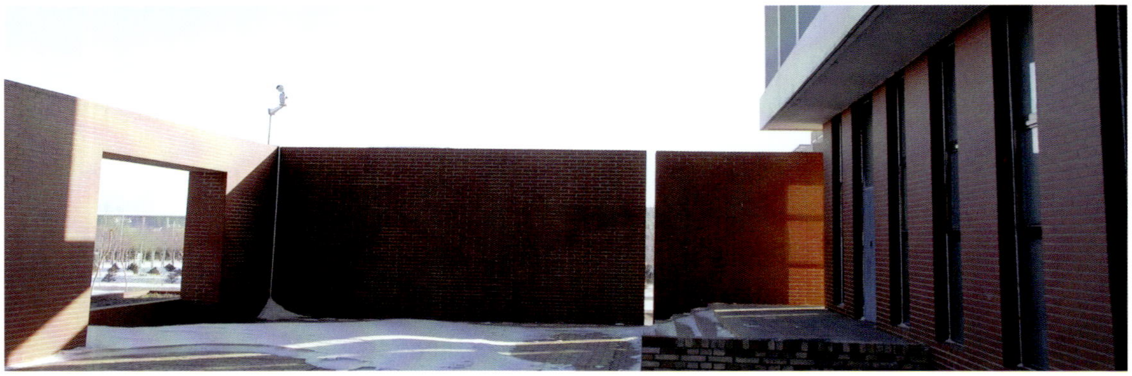

Architectural

设计作品

Design Works

THE 6TH BRANCH

李　鹏
分院院长　高级工程师
1987 年毕业于北京建筑工程学院
1987–1997 年在山东省建筑设计研究院
从事医用工程项目设计工作
1997 年至今，从事医院建筑设计管理工作

『桃李不言，下自成蹊』

企业的声誉是在顾客的千百次检验中建立起来的
但我们深知历史的荣誉只代表昨天的辉煌
光荣与梦想是我们孜孜以求的希望
不断开拓才是前进的力量
我们将以更大的信心和实力把明天做得更好

山东省建筑设计研究院成立于 1953 年，是山东省规模最大、实力最强的建筑设计、 市政设计、 工程勘察甲级设计单位。

山东省建筑设计研究院技术力量雄厚，设计手段先进，质保体系健全。现有在职职工 461 人，其中工程技术人员 414 人。其中工程技术应用研究员 47 人，高级职称 132 人，中级职称 97 人，国家一级注册建筑师 30 人，国家一级注册结构工程师 41 人，其他专业注册 77 人，享受政府特殊津贴的 11 人。

山东省建筑设计研究院第六分院成立于 1995 年，是隶属于山东省建筑设计研究院的一个综合设计分院，主要从事医院建筑设计及其他工业与民用建筑设计、市政工程设计、规划与咨询等设计业务。

山东省建筑设计研究院第六分院专业人才齐全，技术力量雄厚，现有建筑、规划、结构、给排水、暖通空调、强弱电、建筑经济等专业技术人员 36 人。其中应用研究员 4 人，高级工程师（建筑师）13 人，工程师 15 人，助理工程师 4 人；具有注册执业资格的国家一级注册建筑师 5 人，一级注册结构师 6 人。全员配备先进设备，网络化协同工作。

山东省建筑设计研究院第六分院自成立以来，实行全面质量管理，顺利通过 ISO 9001 质量体系认证。重视建筑创作，讲求诚信服务。成立 15 年来，秉承省院“科技立院，改革兴院，文明建院”的治院方针，奉行“质量第一，用户至上，诚信设计”的经营理念；勇于开拓市场，承揽完成了大量的工程设计项目，建筑作品遍布齐鲁大地及江苏、天津、安徽、福建、广东、辽宁、内蒙、河南、河北、青海、甘肃、西藏等国内二十几个省市自治区。近年来，尤其在医院建筑设计方面，六分院取得了突出成绩，仅 2005 年—2009 年度就先后完成了 52 项医院建筑设计项目，设计面积一百多万平方米，其中在广州暨南大学附属第一医院、江苏无锡第二人民医院、东南大学附属中大医院等项目的全国招标中，连续中标，在江苏宜兴中医院新城院区、福建龙海第一医院、广东省中西医医院等异地新建项目中，我分院报送的设计方案以丰富、现代的造型和功能完善的布局赢得评审专家的一致好评。另外，我分院连续中标承接了内蒙古自治区医院内科楼、保健楼、门诊病房综合楼等四项工程设计，其总建筑面积超过 20 万平方米。赢得院方的信任和好评。进入 2010 年以来，设计师们面对每一项工程，无论是投标、中标还是甲方委托的设计项目，都将全力以赴，对设计方案反复调整，悉心比较，保证提交让业主满意的设计成果。在兰州军区兰州总医院、江苏泰州第四人民医院等大型医用工程项目中，我分院设计人员，精心完成设计，报送方案赢得专家及业主高度赞誉并中标实施。这些工程业绩既为我院开拓了医疗建筑设计市场，又为山东省院赢得了社会声誉和经济效益。六分院全体员工在总院质量方针指导下，本着用户第一，质量至上的宗旨，牢固树立品牌意识，力创精品工程，先后有多项工程荣获部、省级优秀设计奖。

山东省建筑设计研究院第六分院全体员工不仅在工程设计中一丝不苟，精益求精，在分院企业文化建设上也走在总院各部门的前列。分院通过不断更新自己的网页与客户及各相关单位进行交流，与国际、国内多家专业医疗建筑协会及杂志社建立了紧密的联系，积极参与卫生建设系统的专业活动，已有多批次专业人员余分赴美、欧、日及港澳地区进行学术交流和参观活动

山东省建筑设计研究院第六分院全体员工一定会团结一致，同心同德，坚持三个文明建设，坚持质量第一，优质服务的一贯方针，在相关领导的关心及各界朋友的支持下共创新的辉煌。

山东省建筑设计研究院第六分院

SHANDONG PROVINCIAL ARCHITECTURAL DESIGN & RESEARCH INSTITUTE–THE SIXTH BRANCH

医用工程专项设计研究所

地址：山东省济南市经四路
小纬四路四号
邮编：250001
电话：+86–531–87913048
+86–531–87913099
邮箱：plee@vip.163.com
网址：www.yysj.com.cn

山东省建筑设计研究院第 6 分院

SHANDONG PROVINCIAL ARCHITECTURAL DESIGN & RESEARCH INSTITUTE—THE 6TH BRANCH

医用工程（专项）设计研究所

1

2

1–2 东南大学附属中大医院（施工中）

总体规划及综合病房楼设计
建筑面积：75 000 平方米
建筑功能：病房、医技
建筑层数：地上 20 层
地下 2 层

江苏大学附属医院（04 年中标方案）

杭州萧山第四人民医院

西宁解放军第四医院 施工中

3

3 宜兴市中医院总体规划及单体设计（施工中）

工程规模：72 000 平方米
工程状态：施工图中
工程特点：本工程位于江苏省宜兴市，为宜兴市中医院在新址上的新建工程，总体规模 800 床，主要功能包括了门急诊、医技、病房、后勤办公用房等。

泰州市第四人民医院 中标工程

鄂尔多斯中医院 中标方案

SHANDONG PROVINCIAL ARCHITECTURAL DESIGN & RESEARCH INSTITUTE--THE 6TH BRANCH

广州中山大学附属第三医院（投标方案）

青岛大学附属第一医院（06年中标方案）

广州医学院第二附属医院（投标方案）

广州中山大学附属第二医院（投标方案）

本溪钢铁集团总医院（中标方案）

兰州军区总医院中标工程

山东省建筑设计研究院第 6 分院

SHANDONG PROVINCIAL ARCHITECTURAL DESIGN & RESEARCH INSTITUTE--THE 6TH BRANCH

—— 医用工程（专项）设计研究所

无锡第二人民医院 施工中

福建龙海第一医院 施工中

4

4–5 暨南大学附属第一医院综合楼（施工中）

建筑面积：72 000 平方米
工程状况：施工图中
建筑功能：病房、医技

5

宜兴中医院 施工中

广东省中西医结合医院 施工中

北京三0九医院 中标方案

6

6 江苏大学附属医院外科大楼

工程规模：50 000 平方米

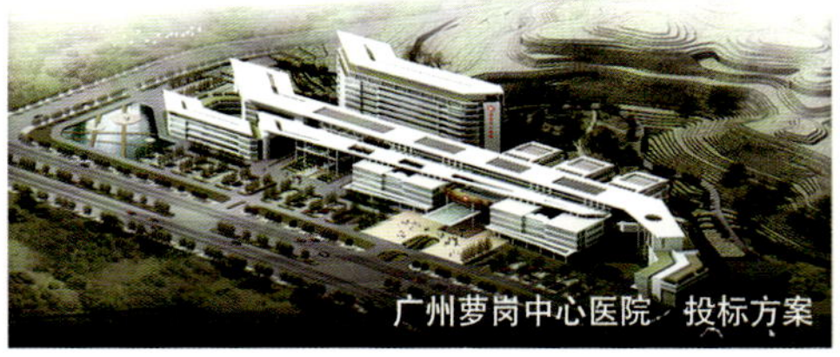
广州萝岗中心医院 投标方案

SHANDONG PROVINCIAL ARCHITECTURAL DESIGN & RESEARCH INSTITUTE--THE 6TH BRANCH

东南大学附属中大医院

内蒙古自治区医院（05 年中标方案）

本溪钢铁集团总医院（中标方案）

山东中医药大学附属第一医院

深圳市妇幼保健院

山东省建筑设计研究院第 6 分院

SHANDONG PROVINCIAL ARCHITECTURAL DESIGN & RESEARCH INSTITUTE--THE 6TH BRANCH

医用工程（专项）设计研究所

7

江苏大学附属医院

陕西省中医院

洛阳市第五医院

8

9

7–8 江苏泰州市第四人民医院（中标工程）

建筑面积：84 000 平方米
建筑规模：600 床
建筑地点：江苏 泰州
内部功能：门诊医技、病房等。

9 辽宁本钢集团总医院外科楼（建成工程）

兰州大学附属第一医院

SHANDONG PROVINCIAL ARCHITECTURAL DESIGN & RESEARCH INSTITUTE--THE 6TH BRANCH

深圳盐田区人民医院

济宁市人民医院

哈尔滨医科大学第二附属医院

常州一院

山东省建筑设计研究院第 6 分院

SHANDONG PROVINCIAL ARCHITECTURAL DESIGN & RESEARCH INSTITUTE—THE 6TH BRANCH

—— 医用工程（专项）设计研究所

10

10 兰州军区总医院医技综合楼（中标工程）

建筑面积：120 000 平方米
建筑地点：甘肃 兰州
内部功能：门诊、医技、病房等

河北滦南县人民医院

辽宁朝阳市中心医院

11

13

12

11–12 内蒙古自治区医院门诊楼（中标工程）

建筑面积：90 000 平方米
建设地点：内蒙古 呼和浩特
内部功能：门诊、医技、病房等

13 天津汉沽医院门诊病房综合楼（建成工程）

山东省中医院

福建泉州方祥医院

SHANDONG PROVINCIAL ARCHITECTURAL DESIGN & RESEARCH INSTITUTE--THE 6TH BRANCH

深圳第三人民医院

淮安市第四医院

湖南省中医院

蓬莱中医院

山东省建筑设计研究院第 6 分院

SHANDONG PROVINCIAL ARCHITECTURAL DESIGN & RESEARCH INSTITUTE—THE 6TH BRANCH

医用工程（专项）设计研究所

14

15

14–15 广东省中西医结合医院总体规划及单体设计（08 中标方案）

工程规模：110 000 平方米
工程状态：施工中
工程特点：本工程位于广东省佛山市，为广东省中西医结合医院，在原址的扩建工程，总体规模 1100 床，其中一期规模 600 床，主要功能包括了门急诊／医技／病房等。

无锡第二人民医院 施工中

福建龙海第一医院 施工中

宜兴中医院 施工中

16

18

广东省中西医结合医院 施工中

北京三0九医院 中标方案

17

16–17 福建龙海市第一医院总体规划及单体设计（施工中）

工程规模：70 000 平方米
工程状态：施工图中
工程特点：本工程为龙海医院的新区规划，一期规模 800 床，主要功能包括了门急诊／医技／病房等 。

18 内蒙古自治区医院内科大楼（建成工程）

工程规模：72 000 平方米

广州萝岗中心医院 投标方案

SHANDONG PROVINCIAL ARCHITECTURAL DESIGN & RESEARCH INSTITUTE--THE 6TH BRANCH

杭州富阳医院

江苏大丰人民医院

山东中医药大学附属第一医院

内蒙古自治区医院干部病房楼

深圳市妇幼保健院

• 公司简介／Company Profile

山东大卫国际建筑设计有限公司，成立于1998年10月，是经国家建设部批准成立的甲级勘察设计股份制公司。公司主要从事民用建筑设计，尤其对房地产开发、工程前期咨询、策划及规划设计有较深入的研究。

公司由建筑师申作伟先生作为主要创办人。申作伟先生从事建筑、规划设计20余年，设计作品获全国性优秀设计奖十几项，获山东省优秀设计奖30余项。申先生历任临沂市规划设计研究院院长、山东省规划设计研究院院长、中国居住区规划专业委员会委员等职，是首批国家一级注册建筑师、国家注册规划师、山东省工程勘察设计大师、山东省优秀建筑师、山东省十佳注册师。2003年申先生被评为中国新本土居住设计名家，2007年，获“中国城市建设卓越人物”称号，2008年3月，被授予“山东省十大杰出青年勘察设计师”，同年8月，被授予“中国建筑设计行业优秀管理人才成就奖”，2009年，被评为“人居十年·2009中国建筑规划十大影响力设计大师”，同年，荣获“2009中国建筑业十大创新建筑设计师”称号，与此同时荣获2009年度“法国金牌建筑设计师”称号，2010年4月被山东省住房和城乡建设厅授予“山东省勘察设计行业管理优秀勘察设计院长”称号。申作伟先生现已成为建筑设计界的学术带头人。

公司从成立之日起即以“设计精品，服务社会”的口号，不断提高设计水平，坚持走品牌设计之路，公司现有设计师130余人，是一支具有高度职业荣誉感和良好职业道德的设计队伍。公司注重建筑设计创新，立足于中国本土文化，融汇国际先进的设计理念和技术，精益求精，完成了一批享誉省内外的精品设计项目，得到了社会各界的一致好评，公司每年均有设计作品在各类评比中获奖。

公司于2003年通过了ISO 9000质量体系认证，设计项目已遍及山东、北京、海南、沈阳、重庆、四川、山西、宁夏、云南、河南、河北等地。公司以创建百年设计公司为奋斗目标，以质量和服务求生存促发展，本着“团结、责任、真诚、创新”的设计理念，在激烈的市场竞争中，赢得了较高的声誉。

我们真诚地期待和社会各界的交流与合作!

Shandong David International Architectural Design Ltd., which was founded in October 1998, is a class A survey and design joint-stock company approved by the Ministry of Construction, mainly engaged in design of civil buildings, especially experienced on real estate development, primary consulting, planning and design of project.

The company was started by the architect Mr. Shen. Since getting occupied with architecture, plan and design for more than 20 years, Mr. Shen's work gained 15 national excellent design prizes, and more than 30 of Shandong Province. He has served as president of Planning & Design Research Institute in Linyi and Shandong Province, Chinese Planning & Design Committee member, also first national class A registered architect, national registered planner, and master of engineering investigation and planning, excellent architect in Shandong Province, Top 10 registered architect of Shandong province. He was recognized as Chinese new local young residential design expert in 2003, was named as "Chinese Remarkable People of City Construction" in 2007, "Top 10 Excellent Young Designers of Shandong Province" in March, 2008, and gained "Chinese Excellent Architecture and Design Manager Achievement Award" in August as well. In 2009, he was elected as "Top 10 Chinese Influential Architectural Design Masters in 2009", "Top 10 Chinese Creative Architect" and "French Prominent Architect". What's more, he was awarded "Excellent President of Investigation and Design Institute in Shandong Province" by Shandong Residential and Urban-Rural Construction Bureau in April 2010. Now Mr. Shen has become an academic leader of architectural design industry.

With the slogan "Provide outstanding works to serve the society" from the first day, the company has kept improving design competence, and maintaining the specialty of the brand. More than 130 architects work here, having high sense of professional honor and professional ethic. The company emphasizes much on creation, at the same time integrates Chinese local culture and international advanced concepts and techniques of design, and always endeavors to do the best. It has accomplished some outstanding projects famous in and outside Shandong Province, and obtained accordant appreciation from people from different circles. Every year, the company has new projects which gain prizes in various appraisements.

The company has obtained the certificate of ISO9000 in 2003, and the design projects cover Shandong, Beijing, Hainan, Shenyang, Chongqing, Sichuan, Shanxi, Ningxia, Yunnan, Henan, Hebei and so on. Having the goal of establishing a centuries-old enterprise, this company tries to survive and develop mainly on quality and service. It gained high fame in the severe market competition using the idea of "cooperative, conscientious, genuine, and creative".

We respectfully await commercial exchanges and cooperation from all circles!

DAWEIDESIGN

山东大卫国际建筑设计有限公司

Shandong David International Architectural Design Co., Ltd.

地址：济南市英雄山路166号
邮编：250002
电话：+86-531-82719857/82719858
传真：+86-531-82719879
网址：www.daweiarch.com
邮箱：dawei_sheji@163.com

Address: No.166 Yingxiongshan Road, Jinan
P.C.: 250002
Tel: +86-531-82719857/82719858
Fax: +86-531-82719879
Http: //www.daweiarch.com
E-mail: dawei_sheji@163.com

• 近期获奖作品

北京优山美地东韵生态住宅小区
获：2006年度全国优秀工程勘察设计金奖
2005年度建设部城乡优秀勘察设计一等奖
2005年山东省优秀工程勘察设计一等奖
2004年“中国人文珍品大院”称号
2008年全国优秀工程勘察设计行业二等奖
2009年第三届全国民营优秀设计华彩奖金奖
首届全国民营设计企业优秀工程设计“华彩奖”银奖
全国人居奖铜奖

济南东环国际广场
获：2005年山东省优秀工程勘察设计一等奖
2007年全国民营设计企业优秀工程设计“华彩奖”银奖

山东电力集团　鲁能中心
获：2003年度山东省优秀工程勘察设计一等奖

北京优山美地东韵生态住宅小区A区
获：2007年建设部优秀建筑设计二等奖
2007年山东省优秀勘查设计一等奖

山师附属中学幸福柳分校
获：2007年度山东省优秀工程勘察设计二等奖
2007年全国民营设计企业优秀工程设计“华彩奖”铜奖
2008年全国优秀工程勘察设计行业三等奖

青岛海情大酒店二期工程
获：2007年度山东省优秀工程勘察设计二等奖

泰安东尊　优山美地度假区
获：2007年度山东省优秀工程勘察设计二等奖
2009年山东泰安东尊　优山美地度假区荣获全国优秀工程勘察设计行业三等奖

• Awarded Works

Beijing Yosemite • Dongyun Residential Quarter
2006 Gold Medal of National Excellent Engineering Investigation and Design
2005 First Prize of Excellent Urban-Rural Investigation and Design, Ministry of Construction
2005 First Prize of Excellent Engineering Investigation and Design, Shandong, China
2004 "Chinese Cultural Treasures Building" Title
2008 Second Prize of National Excellent Engineering Investigation and Design Industry
2009 Gold Medal, the 3rd National Private Excellent Design Huacai Award
Silver Medal, the 1st National Private Excellent Design Huacai Award
Bronze Medal, National Habitat Award

Donghuan International Plaza, Jinan, Shandong, China
2005 First Prize of Excellent Engineering Investigation and Design, Shandong, China
2007 Silver Medal, National Private Excellent Design Huacai Award

Shandong Electric Power Group, Luneng Center
2003 First Prize of Excellent Engineering Investigation and Design, Shandong, China

District A of Beijing Yosemite • Dongyun Residential Quarter
2007 Second Prize of Excellent Architectural Design, Ministry of Construction
2007 First Prize of Excellent Engineering Investigation and Design, Shandong, China

The Middle School affiliated to Shandong Normal University, Xingfuliu Branch Campus
2007 Second Prize of Excellent Engineering Investigation and Design, Shandong, China
2007 Bronze Medal, National Private Excellent Design Huacai Award
2008 Third Prize of National Excellent Engineering Investigation and Design Industry

The second phase project of Qingdao Haiqing Hotel
2007 Second Prize of Excellent Engineering Investigation and Design, Shandong, China

Tai'an Dongzun Yosemite Holiday Resort
2007 Second Prize of Excellent Engineering Investigation and Design, Shandong, China
2009 Third Prize of National Excellent Engineering Investigation and Design Industry

已建成项目

澜泊湾

临沂澜泊湾贵宾楼位于临沂兰山区，南望沂河、地势平坦、交通便利、环境幽雅。本项目是集住宿、餐饮和会议于一体的酒店建筑，总建筑面积为 13 287.59 平方米。

根据现有交通，合理的布置不同的出入口，为建筑的平面争取最大限度的景观视野。主入口设置在滨河大道附近，主入口前设置广场绿化、小品等形成立体景观，与周围景观相融合，并尽量减少与过往人流、车流的交叉。

建筑的平面功能分区明确，互相独立又互相联系，建筑平面采用园林式布局，内部功能围绕庭院布置，做到建筑与环境相结合。基地北侧为人工湖，建筑沿湖面方向展开，争取最大限度的利用现有景观。建筑造型为现代中式风格，利用不同的体块穿插和高低错落的屋顶组合，将中式建筑加以创新，意图通过对中国传统建筑的继承与创新，以及对本土文化的关心与关注，寻求传统文化与现代生活的契合点，形成独特的建筑风格，使整个建筑独具韵味、灵活多变。

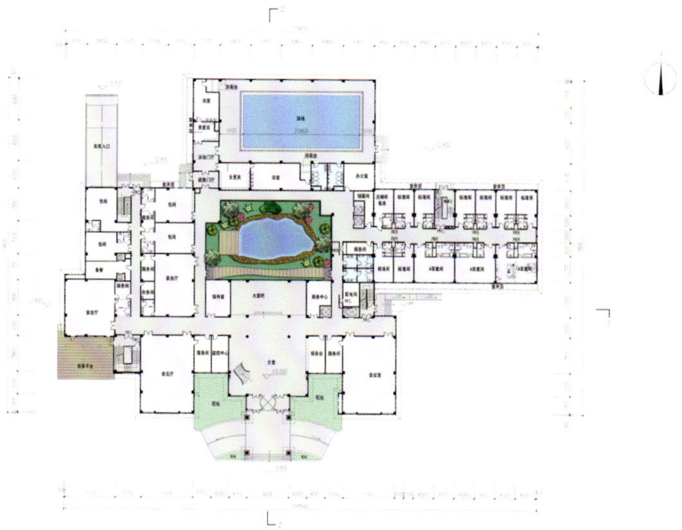

Lanpo Bay

Linyi Lanpo Bay locates in Linyi Lanshan District on the west of Yin River. It is flat, with an elegant environment and easy transportation. This is a multi-functional hotel for lodging, dinning and conference. Its gross building area is 13,287.59 m^2.

Entrances and exit are set differently according to existing traffic system. Architects integrate existing landscape into surface design as much as possible. Main entrance locates near Binhe Road. In front of the main entrance, stereoscopic landscape such as landscaping squares and scenes are set, which not only melts into surrounding environment but also avoid traffic and people stream as much as possible.

Surface functional zones are divided distinctively, and they compose an independent but connected structure. Architectural surface is distributed as a garden. Internal functional zones are set around yard, which combines perfectly with environment. In the west of building site, there is a man-made lake. The layout of architecture is spreading out in front of lake, which will make best use of existing landscape. This architecture should belong to modern Chinese style, but different blocks and uneven roof display innovation of Chinese style. Through inheritance and innovation of Chinese traditional architecture, care and concern to local culture, architects intend to pursuit the correspondency between traditional culture and modern life, put forwards architecture with special character, making architecture charming and flexible.

已建成项目

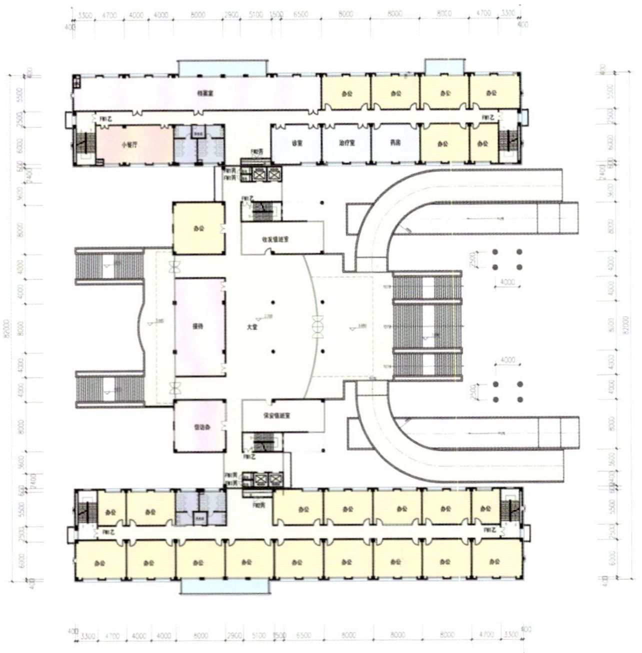

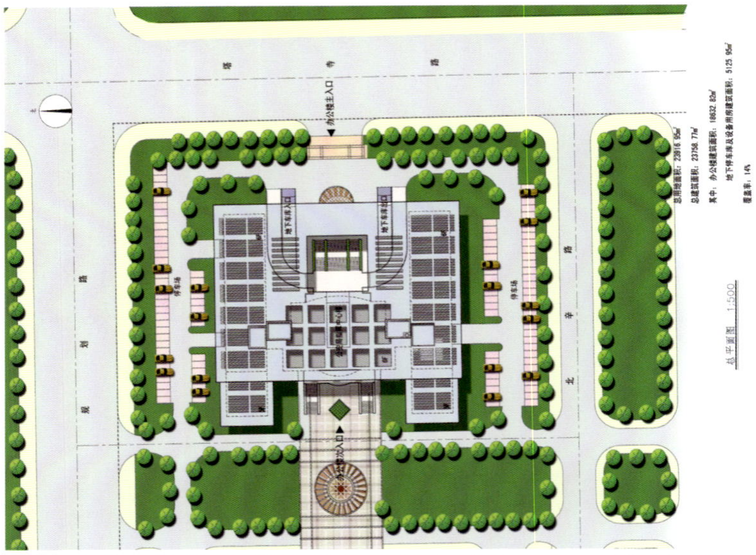

滕州公安局

滕州市公安局指挥中心楼位于滕州市商务中心区东北角，南侧为滕州市会展中心，西侧是新建的市政府大楼，区域位置十分优越，是商务中心区的重要组成部分之一。项目地下一层，地上六层，建筑总面积为 23 758.77 平方米。

根据规划要求，大楼用地为南北长、东西短的矩形，结合地形及对建筑功能的考虑，将整个大楼设计成“H”形平面。“H”形平面可以在用地范围内多布置绿化景观，提升大楼的环境档次，利于功能分区，方便大堂、会议及指挥中心等高大空间的设计，并保证办公空间的品质。

建筑的立面造型力图兼顾公安局大楼所特有的威严与庄重感同时不失现代建筑的新颖别致。巨大的凹入式庭院配合花坛、汽车坡道及大台阶给人以强烈的震撼感。加上入口两侧的八个巨大的束柱和顶板及顶板上巨大的洞口射下的光线，使得建筑气氛十分庄严。大楼整体十分对称、完整，但又不乏丰富的细部变化，建筑的南北两侧的体量强调以实为主，而东西向则以虚为主，虚实对比得当，整体的颜色稳重而又不显压抑。

本项目以合理的功能布局，威严庄重而新颖别致的立面造型深受使用单位的好评，并已成为当地的地标性建筑。

Tengzhou Police Office

The commanding center building of Tengzhou Police Office locates in the northwest corner of Tengzhou Bussiness Center. On the south is Tengzhou Exhibition Center, while west is newly built government building. The district position is very important and it is an important part of business center. With gross building area of 23,758.77 m^2, it has one floor basement and six floors.

According to the requirements of plan, this building is shaped into a rectangle. Architectural north and south is long, while west and east is short. Taking territory and architectural function into consideration, architects shape this building into a letter "H". On the surface of "H", green landscape is set as much as possible, which will upgrade the surrounding environment, distinguish functional zones, make hall, conference center and commanding center convince and keep office space in high quality.

The design of façade intends to display the grandeur of traditional police office and innovation of modern building. Large receding yard, flower bed, automobile slops and big steps strike us. 8 large columns on both sides of entrance, giant roof and light from huge hole on the roof create solemn environment. The whole building is unified and symmetric, but rich in details. Entities on south and north sides emphasize reality, while other two sides emphasize fantasy. Reality and fantasy compare with each other, which makes the color modest but not boring.

This project received the client's high approval by rational distribution of functional zones, grandeur but innovative façade design, and it becomes a landmark building in this area.

近期作品

1–2 青岛二疗城市综合体
建筑面积：33.5 万平方米
建设地点：山东 青岛
设计时间：2009 年 10 月—2010 年 3 月

1-2 Qingdao No.2 Sanatorium Urban Complex
Building Area: 335,000 m²
Construction Location: Qingdao, Shandong
Design Time: 2009.10-2010.03

3–4 胶南茂昌城市广场
用地面积：8.52 公顷
建筑面积：36 万平方米
建设地点：山东 胶南
设计时间：2009 年 10 月

3-4 Maochang Urban Plaza, Jiaonan
Site Area: 8.52 ha
Building Area: 360,000 m²
Construction Location: Jiaonan, Shandong
Design Time: 2009.10

5–6 济南趵突泉南路城市综合体
建筑面积：6.7 万平方米
建设地点：山东 济南
设计时间：2009 年 3 月

5-6 Spouting Spring South Road Urban Complex, Jinan
Building Area: 67,000 m²
Construction Location: Jinan, Shandong
Design Time: 2009.03

7–8 沂沭泗河流域防汛调度楼鸟瞰图
建筑面积：1.4 万平方米
建设地点：江苏 徐州
设计时间：2009 年 6 月

7-8 Aeroview of Sihe basin flood control office building, Yishui
Building Area: 14,000 m²
Construction Location: Xuzhou, Jiangsu
Design Time: 2009.06

9 山东莱钢永锋钢铁有限公司办公楼
建筑面积：2 万平方米
建设地点：山东 齐河
设计时间：2009 年 12 月

9 Office building of Shandong Laigang Yongfeng Steel Corp
Building Area: 20,000 m²
Construction Location: Qihe, Shandong
Design Time: 2009.12

近期作品

10 莒南县城北新区城市设计
用地面积：800 公顷
建设地点：山东 莒南
设计时间：2010 年 3 月

10 Urban design for new area of north Junan County
Site Area: 800 ha
Construction Location: Junan, Shandong
Design Time: 2010.03

11 德州嘉诚 · 名都国际
用地面积：33 公顷
建筑面积：56 万平方米
建设地点：山东 德州
设计时间：2009 年 10 月

11 Jiangcheng Mingdu International, Dezhou
Site Area: 33 ha
Building Area: 560,000 m²
Construction Location: Dezhou, Shandong
Design Time: 2009.10

12 济南国华凤凰路小区规划
用地面积：17.69 公顷
建筑面积：27 万平方米
建设地点：山东 济南
设计时间：2009 年 9 月

12 Planning for Guohua Fenghuanglu Residence Community, Jinan
Site Area: 17.69 ha
Building Area: 270,000 m²
Construction Location: Jinan, Shandong
Design Time: 2009.09

13–14 山东高速雪野国际度假区
建筑面积：5.82 万平方米
建设地点：山东 莱芜
设计时间：2009 年 2 月

13-14 Shangdong Expressway Xueye International Holiday Resort
Building Area: 58,200 m²
Construction Location: Laiwu, Shandong
Design Time: 2009.02

15 淄博中房会所
建筑面积：8000 平方米
建设地点：山东 淄博
设计时间：2009 年 9 月

15 Zibo China Real Estate Group Club
Building Area: 8,000 m²
Construction Location: Zibo, Shandong
Design Time: 2009.09

16 济南函谷十八商住楼
建筑面积：2.58 万平方米
建设地点：山东 济南
设计时间：2010 年 2 月

16 Hangu No.18 business-living building, Jinan
Building Area: 25,800 m²
Construction Location: Jinan, Shandong
Design Time: 2010.02

近期作品

17 城建学院图书信息中心
建筑面积：2 万平方米
建设地点：山东 济南
设计时间：2009 年 4 月

17 Book Info Center of School of City Construction
Building Area: 20,000 m²
Construction Location: Jinan, Shandong
Design Time: 2004.04

18 济南医疗器械产品质量检测中心
建筑面积：2.5 万平方米
建设地点：山东 济南
设计时间：2009 年 10 月

18 Jinan Medical Apparatus and Instruments Quality Detection Center
Building Area: 25,000 m²
Construction Location: Jinan, Shandong
Design Time: 2009.10

19 小清河展览馆
建筑面积：2 800 平方米
建设地点：山东 济南
设计时间：2009 年 7 月

19 Xiaoqing River Exhibition Center
Building Area: 2,800 m²
Construction Location: Jinan, Shandong
Design Time: 2009.07

20–21 济南师范学校图书楼
建筑面积：2.5 万平方米
建设地点：山东 济南
设计时间：2009 年 7 月

20-21 Library of Jinan Normal School
Building Area: 25,000 m²
Construction Location: Jinan, Shandong
Design Time: 2009.07

22–23 济南师范学校艺术楼
建筑面积：1.53 万平方米
建设地点：山东 济南
设计时间：2009 年 7 月

22-23 Art building of Jinan Normal School
Building Area:15,300 m²
Construction Location: Jinan, Shandong
Design Time: 2009.07

近期作品

24

25

26

24 宁夏国际农产品和穆斯林商贸物流中心
用地面积：230 公顷
建筑面积：155.8 万平方米
建设地点：宁夏 银川
设计时间：2009 年 3 月

24 Ningxia International Agricultural Products and Muslim Commercial Trade Center
Site Area: 230
Building Area: 1,558,000 m²
Construction Location: Yinchuan, Ningxia
Design Time: 2009.03

25 淄川服装城改造
建筑面积：改造 4.5 万平方米，新建 8.8 万平方米
建设地点：山东 淄川
设计时间：2009 年 9 月

25 Renovation of Zichuan Clothing City
Building Area:45,000 m² (renovation area); 88,000 m² (newly-built area)
Construction Location: Zichuan, Shandong
Design Time: 2009.09

26 无棣商业街
建筑面积：5.3 万平方米
建设地点：山东 无棣
设计时间：2010 年 4 月

26 Wudi Commercial Street
Building Area: 53,000 m²
Construction Location: Wudi, Shandong
Design Time: 2010.4

27–28 淄博 SHOPPING MALL1
建筑面积：8.39 万平方米
建设地点：山东 淄博
设计时间：2009 年 7 月

27-28 Zibo Shopping Mall
Building Area: 83,900 m²
Construction Location: Zibo, Shandong
Design Time: 2009.07

29–31 山东省检察官培训学院枣庄分院
建筑面积：3.0 万平方米
建设地点：山东 枣庄
设计时间：2009 年 10 月

29-31 Shandong Prosecutors Training Institute (Zaozhuang Campus)
Building Area: 30,000 m²
Construction Location: Zaozhuang, Shandong
Design Time: 2009.10

27

28

29

30

31

近期作品

32 济宁天仙嘉苑小区规划
用地面积：13 公顷
建筑面积：30 万平方米
建设地点：山东 济宁
设计时间：2010 年 1 月

32 Planning for Tianxian Jiayuan Residence Community, Jining
Site Area: 13 ha
Building Area: 300,000 m²
Construction Location: Jining, Shandong
Design Time: 2010.1

33 淄博中房翡翠山居
用地面积：35.5 公顷
建筑面积：30.7 万平方米
建设地点：山东 沂源
设计时间：2009 年 3 月

33 Mountain Jade Residence Community, Zibo China Real Estate Group
Site Area: 35.5 ha
Building Area: 307,000 m²
Construction Location: Yiyuan, Shandong
Design Time: 2009.3

34 沧州美好家园
建筑面积：18 万平方米
建设地点：河北 沧州
设计时间：2010 年 4 月

34 Cangzhou Better Homes & Gardens
Building Area: 180,000 m²
Construction Location: Cangzhou, Hebei
Design Time: 2010.4

35–36 临沂汤泉 · 齐鲁园
用地面积：18.5 公顷
建筑面积：9.7 万平方米
建设地点：山东 临沂
设计时间：2009 年 11 月

35-36 Linyi Hot Spring Qilu Garden
Site Area: 18.5 ha
Building Area: 97,000 m²
Construction Location: Linyi, Shandong
Design Time: 2009.11

37 博山水岸名居
建筑面积：6.0 万平方米
建设地点：山东 博山
设计时间：2009 年 12 月

37 Boshan Shuian Housing
Building Area: 60,000 m²
Construction Location: Boshan, Shandong
Design Time: 2009.12

山东建築大學设计研究院

DESIGN & RESEARCH INSTITUTE OF SHANDONG JIANZHU UNIVERSITY

地址：山东省济南市历山路96号
法人：赵学义
电话：+86-531-86367169
传真：+86-531-86956156
邮箱：SDJDSJY@163.com

Address: No. 96 Lishan Road, Jinan, Shandong Province
Legal Representative: Xueyi Zhao
Tel: +86-531-86367169
Fax: +86-531-86956156
E-mail: SDJDSJY@163.com

山东建筑大学设计研究院始建于1960年，隶属于山东建筑大学，由山东省建设厅领导，是山东省规模较大的综合性建筑设计院之一。

现有建筑行业建筑工程设计甲级、城市规划设计甲级、工程咨询甲级、风景园林专项乙级、市政工程（给水）乙级、工程加固与鉴定、施工图设计审查等工程设计资质。

目前，设计院共有设计人员130余人，其中中、高级职称技术人员80余人，国家一级注册建筑师5人，一级注册结构师9人，国家注册规划师13人，其他注册工程师20人，香港建筑师学会会员4人，香港结构工程师学会会员2人，省“十佳”注册师6人，省优秀建筑规划师2人。

设计院技术力量雄厚，专业技术人员配置齐全，下设3个建筑设计综合所，城市规划所、建筑方案创作室、咨询中心、节能建筑研究所等7个相关专业设计所及施工图审查中心、图文制作中心。

设计院通过严格、科学的设计管理，热情周到的服务，先进的质量管理体系，赢得了良好的信誉与社会效益，2001年通过了ISO 9001质量管理体系认证。近十年，在国家、部、省级的优秀设计评选中，先后获各项优秀设计奖65项，其中国家级奖1项，建设部优秀设计奖6项，山东省优秀设计奖51项，优秀咨询成果奖11项。

设计院将遵循“精心设计、优质服务、确保顾客满意”的质量方针，开拓创新，重视技术进步，注重人才培养，竭诚为社会各界提供满意的服务。

Established in 1960, Design and Research Institute of Shandong Jianzhu University is subject to Shandong Jianzhu University and Construction Department of Shandong Province. It is one of the large-scale comprehensive architectural design institute in Shandong province.

At present, it is equiped with Class A architectural engineering design in architecture industry, Class A construction design, Class A urban planning and design, Class-A consulting engineering, special Class B landscape park architecture ,class B municipal projects (water), project reinforcement and identification, design review of construction plans, and other engineering design qualifications.

There are more than 130 design staffs in the Design Institute, of whom 80 members are technical staffs with medium and senior professional titles, 5 the national Class A registered architects, 9 Class A registered structural designers, 13 national registered planners, and 20 other registered engineers, 4 members of the Hong Kong Institute of Architects, 2 members of Hong Kong Structural Engineers Association, 6 provincial "Top Ten" registered designers, and 4 provincial outstanding construction planning designers .

With strong technical force of Design Institute, and professional and technical personnel fully equipped, it has 3 integrated architectural design institutes, urban planning institute, construction programme workshops, counselling centres, energy-saving building Institute and specialized institutes with the total number of 7 domains seven related speciality Institutes, construction drawing revien center, and graph designing centre.

Through rigorous scientific design and management, warm and thoughtful service, and advanced quality management system, the Design Institute has won a good reputation and social benefits, while it passed the ISO 9001 quality management system certification in 2001. In the past 10 years, it has been successively awarded for various titles 65 times in excellent design contests held by the State, Ministry, and Province, including 1 state-level award, 6 for provincial level excellent design in the Ministry of Construction, 51 for outstanding design award in Shandong Province, and 11 for outstanding advisory achievement awards.

The Design Institute will follow the the quality principle "well-designed, high-quality service and ensuring customers' satisfaction", at the same time, pay attention to pioneer and innovation, attach much importance to technological advances, and pay attention to personnel training and dedicate to providing satisfactory services for our society.

规划作品 Planning Works

博山文姜公园景观设计

用地面积：10 万平方米
设计时间：2009 年 4 月
建设地点：山东 淄博

Landscape design of Boshan wenjiang park

Site Area:100,000 m^2
Design Time: 2009. 04
Construction Location: Zibo, Shandong

海洋亚沙会酒店、配套住宅开发项目规划设计

用地面积：19.46 万平方米
设计时间：2009 年 8 月
建设地点：山东 烟台

Haiyang Asian Beach Games hotel, panning and design for the Residential Development Project

Site Area: 194,600 m^2
Design Time: 2009.08
Construction Location: Yantai, Shandong

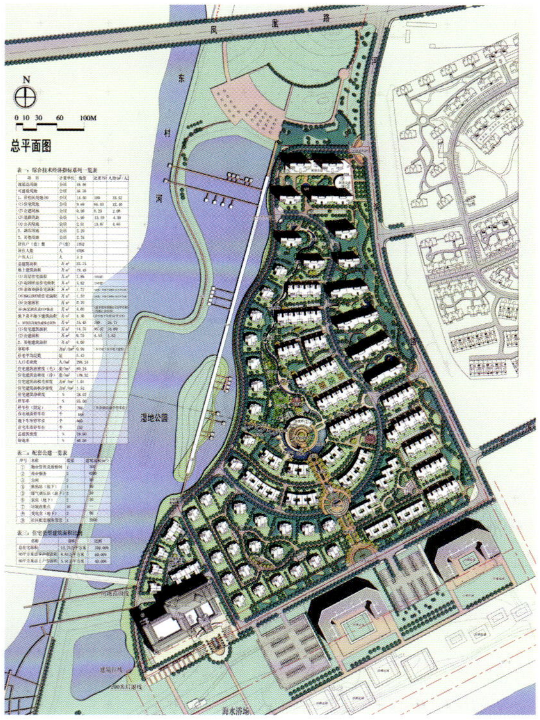

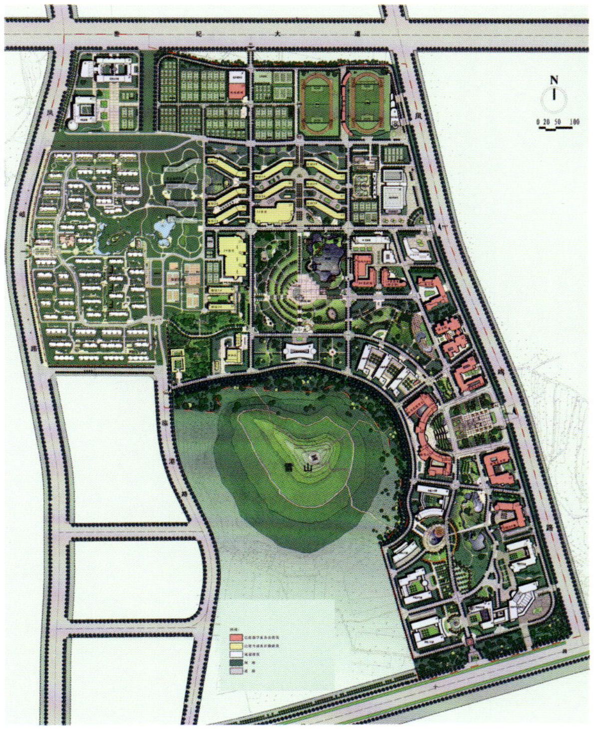

山东建大教授花园规划设计

用地面积：28.71 万平方米
设计时间：2008 年 5 月
建设地点：山东 济南

Planning and design of professor garden of Shandong Jianzhu Universtry in shandong

Site Area: 287,100 m^2
Design Time: 2008.05
Construction Location: Jinan, Shandong

规划作品 Planning Works

第十一届全运会省建比赛训练场馆及配套设施项目规划设计

用地面积：93 万平方米

设计时间：2008 年 2 月

建设地点：山东 济南

Planning and design for the competition training venues and other facilities of the Eleventh National Games

Site Area: 930,000 m^2

Design Time: 2008.02

Construction Location: Jinan, Shandong

淄博高新区副中心（石桥办事处新驻地地块）控制性详细规划设计

用地面积：93 万平方米

设计时间：2008 年 2 月

建设地点：山东 济南

Detailed plan for Sub-Centre of High Tech Industrial Development zore, zibo

Site Area: 930,000 m^2

Design Time: 2008.02

Construction Location: Jinan, Shandong

泰山新湖绿园景观设计

用地面积：12 万平方米

设计时间：2009 年 3 月

建设地点：山东 泰安

Landscape design of Taishan new lake green garden

Site Area: 120,000 m^2

Design Time: 2009.03

Construction Location: Tai'an, Shandong

Works of Architecture 建筑作品

山东城建学院图书馆

建设地点：山东 济南
总建筑面积：20 000 平方米

设计理念："山水之间显书卷，规矩方圆承文脉"

Library of Shandong Urban Construction Vocational College

Construction Location: Jinan, Shandong
Total Building Area: 20,000 m^2

Design Idea: to reveal the sense of civilization in landscape preserve the culture in discipline.

泰安报业文化中心及附属工程

建设地点：山东 泰安
总建筑面积：47 000 平方米
地下建筑面积：11 000 平方米
报业办公：16 000 平方米
四星级酒店：15 400 平方米
艺术馆：4 600 平方米

Tai'an Newspaper Cultural Center and ancillary works

Construction Location: Tai'an, Shandong
Total Building Area: 47,000 m^2
Underground Building Area: 11,000 m^2
Newspaper Office Area: 16,000 m^2
Four-star Hotel Area: 15,400 m^2
Art Museum: 4,600 m^2

设计理念："山"体塑造气势
"梭"形展现动势
"条"肌理表现"报业"文化

Design Idea: The "mountain" form creates momentum
The "Shuttle" shape shows dynamic posture
The "strip" texture expresses the culture of "newspapering"

Works of Architecture 建筑作品

山东大学威海分校科技大厦

建设地点：山东 威海

总建筑面积：25 000 平方米

设计理念

1．"整体性" "地域性"

设计中突出了环境、建筑及内部空间设计的整体性，使建筑与环境相融合，其形体和色彩都采用了与现有建筑相同的设计手法。

2．"开放性" "生长性"

Science Building of Shandong University in Weihai

Construction Location: Weiha, Shandongi

Total Building Area: 25,000 m^2

Design Idea

1. Integrity,regionalism

It focuses on the integrity of environment, architecture and space design. The architectures mix well with environment, it adopts the same design technique with the architecural design in its form and colors.

2. Openness and development

山东泰源中医药研发有限公司

建设地点：山东 泰安

总建筑面积：20 000 平方米

设计理念

因山就势，错落有致，天人合一

形神兼备，寓意其中。

Shandong Taiyuan Chinese Medicine R & D Co., Ltd.

Construction Location: Taian, Shandong

Total Building Area: 20,000 m^2

Design Idea

Conform to the location of mountain, interspersed orderly, harmonious between man and nature, unity of form and spirit, with implied meaning in it.

Works 建筑作品 of Architecture

济南师范艺术楼

建设地点：山东 济南
总建筑面积：14 900 平方米

设计理念

形。坡屋顶、花格窗是形象设计中的主要元素；

神。动静分区，避免相互干扰，下沉演艺广场营造出具有诗意的另行空间；

意。多变的空间体验和可以量度的尺度体验成为建筑设计追求的意境；

境。无障碍设计等细节上的处理为学生提供了舒适的学习环境。

Artistic building of Jinan Teacher Training College

Construction Location: Jinan, Shandong
Total Building Area: 14,900 m^2

Design Idea

Form, pitched roof and lattice window are the main elements in the design. Spirit, lively and quiet places are divided so as to avoid mutual interference. Sink Performing Arts Plaza can create special space which is filled with poetry.

Artistic conception, varied spatial experience and the experience for the measurable space become the artistic conception which the architectural design pursues.

Environment, the detailed design of no-obstacle plan can provide comfortable learning environment for the students.

济南师范图书馆

建设地点：山东 济南
总建筑面积：25 000 平方米

设计理念

立面的处理中强调形体的整体性，突出建筑的体量感，衬托出图书馆的核心地位。立面细节的设计上突出表现了建筑的文化内涵，功能布局中着重强调分区明确合理，高效利用面积，减少使用面积的浪费，交通便捷，人流组织合理，避免相互干扰。

Library of Jinan Teacher-training College

Construction Location: Jinan, Shandong
Total Building Area: 25,000 m^2

Design Idea

The facade design emphasizes the wholeness of the form, highlight the architectural mass of the building and make the core position of the library stand out. The facade design highlights the cultural spirit of the building and layout of functions emphasizes the specified and justified characteristics of the zoning. It can make good use of the land and reduce the waste of the usable floor area; the traffic is convenient, and it has a reasonable arrangement for the traffic flow, so as to avoid interferences.

威海东鲁建筑设计院

获奖情况

- 五洲·太阳城总体规划设计方案——2004 年首届国际人居建筑设计竞赛铜奖
- 槐云西钦村旧村改造规划设计方案——2004 年首届国际人居建筑设计竞赛优秀奖
- 华夏高级技工学校 I 区——2008 年全国优秀工程勘察设计行业建筑工程项目三等奖
  2007 年度山东优秀工程勘察设计行业建筑工程项目二等奖
  2009 年第五届国际人居建筑设计竞赛铜奖
- 威海华夏技术学院教学区——2009 年度山东优秀工程勘察设计行业建筑工程项目二等奖
- 威海职业技术学院教学组团——2009 年度山东优秀工程勘察设计行业建筑工程项目二等奖
- 威海华夏会馆 F 区——2009 年度山东优秀工程勘察设计行业建筑工程项目三等奖
- 威海华夏会馆 M 区——2009 年度山东优秀工程勘察设计行业建筑工程项目三等奖

Awards

- The overall planning and design of Wuzhou Sun City—the Bronze Prize of the First International Habitat Architectural Design Competition in 2004
- The planning and design of Xiqin Village, Huai-Yun in the transformation of old villages—the Award of Excellence in the First International Habitat Architectural Design Competition in 2004
- District I of Huaxia Senior Technical School—winning the third prize of construction project in 2008 National Outstanding Engineering Survey and Design, the second prize of construction project in 2007 Shandong Outstanding Engineering Survey and Design; the Bronze Prize of the Fifth International Habitat Architectural Design Competition in 2009
- The teaching area of Weihai Huaxia Technical College—the second prize of construction project in 2009 Shandong Outstanding Engineering Survey and Design
- The teaching complex of Weihai Vocational College—the second prize of construction project in 2009 Shandong Outstanding Engineering Survey and Design
- District F of Weihai Huaxia Hall—the third prize of construction project in 2009 Shandong Outstanding Engineering Survey and Design
- District M of Weihai Huaxia Hall—the third prize of construction project in 2009 Shandong Outstanding Engineering Survey and Design

1 威海市公安局科技大楼
The Science and Technology Building of Public Security Bureau in Weihai

2 威海迪上尚捷年办公楼
The office building of Weihai Di Shang Shangjienian

地址：威海市环翠区金线顶路28 号 4 楼
邮编：264200
电话：+86–631–3660000
传真：+86–631–3660000
邮箱：donglu@vip.163.com
网址：www.whsjy.com

Address: 4th Floor of Building No. 28 of Jinxianding Road, Huancui District, Weihai
P.C.: 264200
Tel:+86-631-3660000
Fax:+86-631-3660000
E-mail: donglu@vip.163.com
Http: //www.whsjy.com

威海东鲁建筑设计院

3 威海华夏高级技工学校 I 区 –2
District I-2 of Weihai China Senior Technical School

4 威海华夏高级技工学校 I 区
District I of Weihai China Senior Technical School

5 华夏主题公园景区大门
The door of China theme park scenery

6 威海华夏接待中心区
Weihai China reception center

7 主教学楼威海华夏高级技工学校教学区
The main teaching building of the teaching area of Weihai China Senior Technical School

山东圣凯建筑设计咨询有限公司

Shandong Sheng Kai Architecture Design Consulting Co., Ltd.

地址：烟台市建设路 6 号　邮编：264000
电话：+86-535-6256265　6252718　传真：+86-535-6241894
邮箱：zgh001@public.ytptt.sd.cn 或 skjzsj@163.com
网址：www.skjzsj.com 或 www.skjzsj.cn

Address: No.6, Jianshe Road, Yantai city　P.C.: 264000
Tel: +86-535-6256265 6252718　Fax: +86-535-6241894
E-mail: zgh001@public.ytptt.sd.cn or skjzsj@163.com
Http: //www.skjzsj.com or www.skjzsj.cn

山东圣凯建筑设计咨询有限公司，始成立于 1973 年，原为烟台市第二建筑设计院，1998 年 5 月由事业单位改制为有限公司。公司有着浓厚历史传统和文化底蕴，是山东省建委确定扶持的全省建设行业重点单位之一。

公司专业配套齐全，技术力量雄厚，现有员工 190 多人，各类专业技术人员占员工总数的 90％，拥有众多国家一级注册专业技术人才。具有建筑行业建筑工程设计、工程勘察甲级资质，城市规划编制乙级资质及市政工程设计、工程咨询、工程监理、招标代理等多种技术资质，能够胜任工业与民用建筑、城市规划、园林景观项目设计、各种工程技术服务以及工程总承包服务。

公司信守"设计一流产品，造就一流员工，建设一流强院"的企业哲学，坚持"对客户和社会负责，勇于奉献，和谐发展"的核心价值观。多年来，公司设计的一大批项目在烟台地区享誉盛名，使"圣凯"逐步成为烟台地区知名设计品牌。公司近年来设计的重要作品有烟台栈桥、中医药专科学校、文登汤伯温泉旅游度假区、冰轮集团办公楼、东方商务中心、北方月光兰庭、龙海家园、锦绣新天地、南山世纪华府、烟台开发区富士康厂房、观海路控规、海阳市留格庄镇总体规划、烟台夹河岛生态园等。

公司坚持人才强院战略，不断提高设计水平和质量，深化圣凯服务理念，为客户提供优质服务，生产经营业绩呈现跨越发展的良好势头。公司的发展战略目标是建设成为具有工程总承包全过程服务能力的总承包企业和项目管理企业，为社会作出更大的贡献。

Shandong Sheng Kai Architecture Design Consulting Co., Ltd. was founded in 1973, formerly the second Institute of Architectural Design of Yantai City.It was transformed into a limited company from a public institution in May 1998. With strong historical tradition and cultural heritage, the company is one of units of the province's construction industry supported by Shandong Provincial Construction Committee.

The company has complete specialty and strong technical force. It has a staff of more than 190. All kinds of professional and technical personnel account for 90% of the whole staff, including many National First class registered professional and technical talents. The company has the Class-A qualification of construction project design of construction industry and engineering investigation. It has Class-B qualification of urban planning and other technical qualifications such as municipal engineering design, engineering consultation, project supervision, tender agent. The company can operate general contracting services. It is also capable of other services like industrial and civil construct urban planning, landscape garden project design, various engineering technical services, and engineering consultation services.

The company believes the business philosophy of “designing the best products, creating first-class staff, building first-class institution,” and insists on the core value of “responsible for customers and society, brave spirit of sacrifice, harmonious development”. Over the years, a large number of projects designed by Sheng Kai have won good reputation in Yantai area and gradually has made “Sheng Kai” become a well-known brand among architectural design industry. The works in recent years include: Yantai trestle, Traditional Chinese Medicine College, Wendengtangbo Spa Resort, Moon Group Office Building, Orient Business Centre, North Moonlight Blue Court, Longhai Homes, Jinxiu New World, Nanshan Century Buiding, Foxconn Plant in Yantai Development Zone, Control Regulations of Haiguan Road, overall Planning of Liu gezhuang Town in Haiyang City, and Jiahe Island ecological park in Yantai.

Adhering to the strategy of reinvigorating institution through human resource development, the company has continuously been improving the design level and quality and deepening the service concept of Sheng Kai. Providing customers with quality service, production and operating results show a good momentum with leapfrog development. The company's strategic goal is to build a general contracting and project management enterprise which has the capability of the entire process service for a general contracting and to make a greater contribution to society.

烟台栈桥

建设地点：山东 烟台

用地面积：16 014 平方米

Yantai Trestle

Construction Location: Yantai, Shandong

Floor Area: 16,014 m^2

北方月光兰庭
建设地点：山东 烟台
建筑面积：230 000 平方米

Northern Moonlight Court
Construction Location: Yantai, Shandong
Building Area: 230,000 m²

烟台市观海路控制性详细规划及环境整治规划
项目地点：山东 烟台
项目规模：道路长度 11.38 千米，规划范围约 6 平方千米

The detailed regulatory plan and environmental remediation plan for Guanhai Road, Yantai
Construction Location: Yantai, Shandong
Project Scale: The road length is 11.38 km, The planning area is about 6 km²

东方国际商务中心
建设地点：山东 烟台
建筑面积：130 620 平方米

Oriental International Business Center
Construction Location: Yantai, Shandong
Building Area: 130,620 m²

冰轮集团办公楼
建设地点：山东 烟台
建筑面积：42 720 平方米

The office building of Moon Group
Construction Location: Yantai, Shandong
Building Area: 42,720 m²

夹河绿岛
建设地点：山东 烟台
总用地面积：270 000 平方米

Green Island of Jia River
Construction Location: Yantai, Shandong
Gross Floor Area: 270,000 m^2

锦绣新天地
建设地点：山东 烟台
规划面积：2 337 000 平方米

Fairview New World
Construction Location: Yantai, Shandong
Planning Area: 2,337,000 m^2

日报社印刷工业园
建设地点：山东 烟台
建筑面积：40 000 平方米

Daily Printing Industrial Park
Construction Location: Yantai, Shandong
Building Area: 40,000 m^2

龙海家园
建设地点：山东 烟台
总建筑面积：1 227 948 平方米

Longhai Home
Construction Location: Yantai, Shandong
Gross Building Area: 1,227,948 m^2

威海汤伯温泉度假村

建设地点：山东 文登

建筑面积：44 526 平方米

Weihai Tang Bo Spa Resort

Construction Location: Wendeng, Shandong

Building Area: 44,526 m^2

生态度假区

建设地点：烟台市莱山区

规划面积：400 000 平方米

Eco-Resort

Construction Location: Laishan, Yantai

Planning Area: 400,000 m^2

乡村记忆

建设地点：山东 烟台

规划面积：114 100 平方米

Village Life

Construction Location: Yantai, Shandong

Planning Area: 114,100 m^2

中医药专科学校

建设地点：山东 烟台

建筑面积：174 090 平方米

Traditional Chinese Medicine College

Construction Location: Yantai, Shandong

Building Area: 174,090 m^2

青岛时代建筑设计有限公司
Qingdao Times Architectural Design Co., Ltd.

青岛时代建筑设计有限公司，建筑设计甲级资质（同时具备市政公用设计、风景园林设计、规划设计、工程勘察及建设工程监理资质），是集设计、科研、监理、科贸为一体的技术密集性的科技企业。公司本着“设计重质量，以优取胜；服务求完美，以佳取胜；合作讲信誉，以诚取胜”的方针，竭诚欢迎各界朋友共同合作，共展宏图大业。

Qingdao Times Architectural Design Co., Ltd. with the qualification of architectural design of Class A (both has municipal utility design, landscape design, planning and design, engineering investigation and construction supervision qualification) is a technology-intensive, scientific and technological enterprise, which integrates design, research, supervision, and technology. Following the spirit of "design of the quality, win by excellence; service for the perfect, high grade; and cooperation with credit, integrity first", the company sincerely welcomes all friends from all walks of life to cooperate and achieve success together.

法　人：金杰
地　址：青岛经济技术开发区黄浦江路57号建设大厦
电　话：+86-532-86898676　+86-532-86890940
传　真：+86-532-86898676
邮　编：266555
网　址：www.qdsct.cn
邮　件：qdsct@tom.com

Legal Representative: Jie Jin
Address: Jianshe Mansion, No.57 Huangpujiang Road, Qingdao Economic and Technological Development Zone
Tel: +86-532-86898676　+86-532-86890940
Fax: +86-532-86898676
P.C.: 266555
Http: //www.qdsct.cn
E-mail: qdsct@tom.com

青岛开发区维客武家庄购物中心
占地面积：18 624平方米
建筑面积：42 939平方米

Wujia Village Shopping Center of WEEKLY in Qingdao Development zone
Site Area: 18,624 m²
Building Area: 42,939 m²

本项目基地三面临街，交通便利。项目主入口位于南侧，沿街设停车位及地下车库出入口，北侧为货流出入口。人流与货流分设有效地避免了流线的交叉和干扰。建筑形体干净利落，运用现代设计手法和具有对比质感的建筑材料，突出“维客”企业标志，烘托浓郁的商业气氛，营造出生机勃勃的维克企业文化。

This project designs for the Wujia Village Shopping Center of WEEKIY in Qingdao Development Zone. The building site faces to street in three sides and owns convenient transportation. The main entrance of this project is located in the south, the parking spaces and underground garage entrances are along street, and the north entrance is cargo flow. Passenger and cargo flow grouped effectively prevent the cross and interference of flow line. Using modern design techniques and contrast texture building materials, the neat building form highlights "the WEEKIY" logo. The sequencing formed by large-scale advertising contrasts the strong business climate and creates a vibrant enterprise culture of WEEKIY.

孟子文化产业园

孟子文化产业园的整体布局可以概括为“两心，两轴，七个部分”。

两心指人文历史景观核心和自然风光景观核心。两轴指历史发展轴和两孟文化轴。七个部分是指孟子文化产业园的七个组成部分，包括孟子故宅景区、孟子文化广场景区、儒家文化产业园区、旅游服务商业配套区、孟母林景区、孟子故里居民区、圣母中医院。该项目在“蓝星杯”第五届中国威海国际建筑设计大奖赛中被评为优秀奖。

Mencius Cultural Industrial Park

The whole design of Mencius Cultural and Industrial Park can be generalized as "double cores, double axes and seven parts".
Double cores: The core of cultural and historical landscape and natural scene landscape. Double axes: The axis of historical development and Mencius Temple and Mencius Mansion. Seven parts: Scenic spot of Mencius's old house, Mencius cultural square, Confucius cultural industrial park, commercial part for tourist service, scenic spot of Mencius's mother's forest, residential site of Mencius's native village, and Goddess hospital of Chinese medicine.This project had won the excellent awards in the blue star cup--the fifth Weihai International Architecture Design Competition.

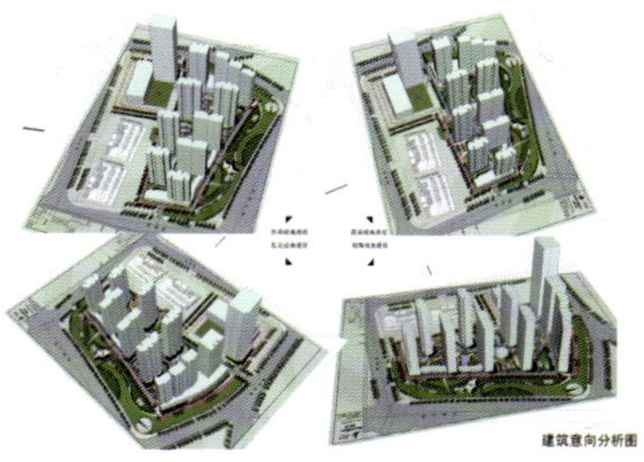

青岛.东方至尊
用地面积：49 986平方米
建筑面积：312 110平方米

Qingdao. Eastern Supreme
Site Area: 49,986 m^2
Building Area: 312,110 m^2

东方至尊项目位于青岛胶南市市政府东侧，地理位置优越，项目定位为集住宅、公寓、办公、商业为一体的综合性高端小区。

设计引入立体化交通，利用现状地形的高差，将高层底部架空处理，将景观步道与人行出入口天桥结合，人流、车流处于不同的标高。此外，公寓楼与经贸大厦相邻布置，高层住宅"板点结合"，错落布局，整体紧凑而局部舒缓，形成"由商到住"的自然过渡。建筑外观为简洁的古典风格，玻璃幕墙与石材墙面刚柔相济，形成了视觉上的虚实对比；低窗台、宽窗户为室内空间提供了广阔的视野。

The Eastern Supreme project of Qingdao is located in the east of municipal government of Jiaonan, and the location is convenient, which positioned as an integrated high-end residential and set of residential, apartment, office and commercial.
The project is designed into three-dimensional transport, using the existing topography of the elevation to overhead process the top bottom and combine with the path of landscape trail and pedestrian bridge entrances. The pedestrian flow and traffic flow are designed at different elevation, completely segregation. In addition, the apartment building and trade building are adjacent to each other. High-rise residential is the integration of board and point, distribution scattered, overall compact and local relief, so as to achieve the natural transition of "live by the business". The appearance treats of simple classical style, the glass curtain wall and stone wall are hardness and softness, forming the actual situation comparison visually. Low windowsill and wide windows provide a broad perspective for the interior space, the processing of jumping horizon line of the top presents the building a kind of calm and motifs personality.

内蒙古兴安盟科尔沁活动中心
用地面积：21 510平方米
建筑面积：17 594平方米

Nei Mongol Xingan League Horqin Activity Center
Site Area: 21,510 m^2
Building Area: 17,594 m^2

挖掘人文、历史精髓，尊重地域特色，强化文脉风格。

发展基地优势，重构基地格局，突显区域中心地位。

时代建筑语汇诠释时代建筑形象。功能合理，流线畅通，经济节能。

Digging the soul of culture and history, respecting regional character and underlining traditional context.
Taking advantage of conditions in building site, redistributing the structure of building site and emphasizing the regional central.
Expressing the architectural image by architectural vocabulary.
Functional zones are reasonable, smooth Outline, and energy saving construction process.

"月华 水畔"居住区
用地面积：562 222平方米
建筑面积：832 088平方米

Residential Site named "Yuehua Shuipan"
Site Area: 562,222 m^2
Building Area: 832,088 m^2

项目位于呼伦贝尔市，设计遵循高品质及现代化的整体构思，强调居住与配套设施的有机联系及独立性，设计师从对用地现状环境仔细分析，结合项目地形的特点，合理安排用地分块及空间形态形成了具有秩序、层次感的空间序列。

项目的整体布局可以概括为"一心 两轴 三区"。

一心是指依托伊敏河河心岛设计的景观式结构核心；两轴是指纵贯南北的内河水轴和中央商业休闲轴；三区：设计中依托规划一路与规划二路将基地划分为中部核心商业区，北部住宅区，南部住宅区三个部分，其中南部住宅区由规划三路分成南北两个组团。

This project is located in the city of Hulunbuir . Architects intend to build a modern and high-quality residential site. This design emphasizes the organic connection and independence between residence and supporting facilities. Based on the regional characters, after fully analyzing existing environment and rationally building blocks, spatial patterns, fuctional zones and traffic systems, an organic order and multilayer space sequence is formed.
The whole layout can be generalized as "one core, two axes and three districts".
One core refers to landscape in Inter-river Island in Yimin River. Two axis refer to the axes of river running from north to south and the central commercial and recreation axes.Three parts: Based on No.1 planning road and No.2 planning road, this district can be divided into mid commercial site, north residential site, and south residential site. The south residential site is divided into two parts by No.3 planning road.

山西省建筑设计研究院

Shanxi Architectural Design and Research Institute (SADI)

山西省建筑设计研究院创建于1953年，是山西的第一家国家综合甲级建筑设计研究院，并获得国际IS 9001：2000、ISO 14001：2004和GB/T28001–2001三个体系认证。研究院现有员工500多人，其中建筑、城市规划、结构、给排水、暖通空调、建筑电气、工程造价、岩土工程地质、企业管理等专业技术人才470余人（各类高级技术人才近200人），国家一级注册建筑师和结构工程师90多人。研究所设综合设计所、建筑方案创作所、综合勘察处、惠州分院、工程造价中心、新型结构研究中心和岩土工程研究所等设计科研机构，另有华翔监理公司、沃尔泰项目管理公司、通建施工图审查中心等下属企业。服务项目有城市规划、工程咨询、工程设计、工程勘察、工程监理、项目管理、施工图审查和工程总承包等内容。半个多世纪的精心设计、锐意创新，为山西的城乡建设作出了突出的贡献，在大型公共建筑、医疗建筑、体育建筑、商业建筑、文化教育建筑、居住建筑和邮电、电力、餐饮、宾馆、工厂等工业与民用建筑工程设计及建筑节能设计中积累了丰富的经验。多年来累计完成工程设计8 000多项，建筑面积达5 000余万平方米，项目遍布全国各地，远及亚洲、非洲、欧洲。作品获国家、部、省级奖励120余项，享有良好的社会声誉。

Shanxi Architectural Design and Research Institute (SADI), founded in 1953, was the first national Grade-A comprehensive design institute established in Shanxi Province. It has been awarded the international system certificates of ISO 9001:2000, ISO 14001:2004 and GB/T28001-2001.

SADI specializes in city planning, engineering design and consultation, geological survey and overall contract management. SADI owns more than 470 architects, engineers and experts in architecture, city planning, building structure, water supply and drainage, air conditioning and heating, power supply, quantity estimate, geological engineering and management. 14 design and research departments and 8 subordinate enterprises are set up under the institute. With more than half century's meticulous design and incessant innovation, the projects designed covered all various trades and occupations including health, education, sports, trade, finance, post and telecommunication, electric power, dinning, hotel, office and housing etc.. 8,000 projects, which covered a total area of 50 million square meters, have been completed. These projects spread all over the country and in the international market of Asia, Africa and Europe. 120 national and provincial rewards were awarded to our works.

To create fine architecture and a wonderful life are our goals. We shall do our best to meet the requirements of the clients, provide best services to them and contribute more good works to the society.

地址：山西省太原市府东街5号
电话：+86–351–3285401/3285361
传真：+86–351–3073613
邮编：030013
网址：www.sxjzsj.com
邮箱：web@sxjzsj.com sadri@public.ty.sx.xn

Address: No.5 of Fudong Street, Taiyuan, Shanxi Province
Tel: +86-351-3285401 / 3285361
Fax: +86-351-3073613
P.C.: 030013
Http: //www.sxjzsj.com
Email: web@sxjzsj.com sadri@public.ty.sx.xn

SADI

1

2

3

■ 1 山西省实验中学学术交流中心
建筑面积：45 500平方米
设计时间：2009年9月

Academic Exchange Center of Shanxi Experimental Secondary School
Building Area: 45,500 m^2
Design Time: 2009. 09

■ 2 海关监控大厦
建筑规模：33 000平方米
设计时间：2009年6月

Customs Control Tower
Construction Scale: 33,000 m^2
Design Time: 2009. 06

■ 3 太原理工大学多学科中心
建筑规模：40 000平方米
设计时间：2010年4月

Multi-disciplinary center of Taiyuan University of Technology
Construction Scale: 40,000 m^2
Design Time: 2010. 04

4

5

6

7

■ 4 朔州市廉政教育基地
建筑面积：22 000平方米
设计时间：2010年5月

Taiyuan Longcheng Secondary School
Building Area: 22,000 m^2
Design Time: 2010. 05

■ 5 鄂尔多斯博物馆
建筑规模：41 610平方米
设计时间：2006年10月

Ordos Museum
Construction Scale: 41,610 m^2
Design Time: 2006. 10

■ 6 河北涉县八路军129师纪念馆
建筑规模：10 000平方米
设计时间：2009年11月

Memorial Hall for 129 Division of Eighth Route Army in She County, Hebei
Construction Scale: 10,000 m^2
Design Time: 2009. 11

■ 7 忻州档案馆
建筑面积：10 000平方米
设计时间：2009年7月

Xinzhou Archives
Building Area: 10,000 m^2
Design Time: 2009. 07

■ 8 山西能源工业国际分销物流港大厦
建筑面积：97 000平方米
设计时间：2009年7月

International Distribution Logistics Centre Building of Energy Industry in Shanxi
Building Area: 97,000 m^2
Design Time: 2009.07

■ 9 临汾市新医院
建筑面积：170 000平方米
设计时间：2009年9月

New hospital in Linfen City
Building Area: 170,000 m^2
Design Time: 2009. 09

■ 10 太原科技大学综合实验教学楼
建筑规模：34 000平方米
设计时间：2009年5月

Comprehensive Experimental and Teaching Building of Heavy Machinery Institute
Construction Scale: 34,000 m^2
Design Time: 2009.05

■ 11 襄汾医院
建筑面积：80 000平方米
设计时间：2009年3月

Xiangfen Hospital
Building Area: 80,000 m^2
Design Time: 2009. 03

■ 11–13 龙城中学

规划用地：116 157平方米
建筑占地：19 425平方米
建筑面积：70 418 平方米
容积率：0.53
绿化率：43.4%

The planning and architecture design project of Longcheng middle school in Longcheng new district

Planning Area: 116,157 m^2
Floor Area: 19,425 m^2
Building Area: 70,418 m^2
Plot Ratio: 0.53
Greening Ratio: 43.4%

规划理念

1.统一规划，合理布局。实现现代化教学模式，具有灵活性、通用性和可调性。
2.功能上分区明确，营造一种理性、典雅、大气、富有文化内涵的氛围，形成个体特点鲜明，总体协调，布局合理，宁静优美的校园环境。
3.塑造高文化品位的建筑艺术形象与环境品质。
4.重视自然资源和环境资源的有效利用，合理布置建筑楼群。

Planning Ideas

1.Unify the planning and establish reasonable layout, achieving morden eduction mode, which is flexible, universal and adjustable.
2.With distinct function division, creat rational, elegant and grand atmosphere , building a well planned campus with tranqulity and serenity.
3.Creat the architecture. with artstic image and high environmental quality.
4.Well intergrate and utilize the natural resources and environmental resources, well locate the layout of th buildings.

14

15

14–17 重庆市北部新区民心佳园——鸳鸯组团公共租赁房及配套廉租住房

项目性质：重庆市公共租赁住房

建设地点：重庆

用地面积：323 774平方米

建筑面积：1 134 577平方米

容积率为：3.79

实施时间：2010年3月

Minxin Garden in new district of Chongqin north——public rental and supporting facilities low rent house

Project Property: Public rental in Chongqing

Construction Location: Chongqing

Site Area: 323,774 m^2

Building Area: 1,134,577 m^2

Plot Ratio: 3.79

Construction Time: 2010. 03

16

17

总体规划以提高环境质量，强调以人为本，以处理人与建筑、人与交通、人与绿化、人与空间以及人与人之间的关系为主，探求更好的规划设计模式，体现新型生态绿色社区的特色。设计中寻求动态的空间对称和相互渗透，形成相互流通的建筑形态，创造21世纪开放的人文空间。

廉租房户型设计考虑在30平方米～50平方米之间；公租房户型设计考虑在50平方米～80平方米之间。最终确定的公租房户数为5 681户，廉租房户数为12 195户。

The overall planning focus on the environmental quality and human. The architects pay great attention to the relationship between people and architecture, people and transportaion, people and planting, people and space. Exploring a better planning design mode, it embodies the new ecological green community characteristic. In the pursuit of the animated symmetric space, combination and interchangeable architecture, creating a 21st century openning human space.

The size of the low rent houses units are mainly from 30 m^2 to 50 m^2, the size of the public rental are ranging from 50 m^2 to 80 m^2. The final public rental units are 5,681 and the low rent house units are 12,195.

Weaison 沃易森(西安)建筑景观设计

(Xi'an) Landscape Architect Co., Ltd

Weaison (China) Landscape Archirects & Planners

地址：陕西省西安市高新区科技二路65号A座C层
电话：+86–29–88335526/88335540
传真：+86–29–88322528
邮箱：weaison@126.com

Address: Floor C, Building A, No.65 Keji 2nd Rd., Hi-and-New Tech Park of Xi'an, Shaanxi
Tel: +86-29-88335526 88335540
Fax: +86-29-88322528
E-mail: weaison@126.com

沃易森（西安）建筑景观设计有限责任公司是中国和澳大利亚合资的设计公司，目前主要针对中国西北部地区，同时也涉及华东、华南地区部分城市的设计市场。

公司主要从事风景旅游、景观建筑与规划、公共艺术设计。公司自成立以来，参与或主持了多项规模不等的景观设计项目以及景观工程建设施工项目。多年的积累，我们得出了“全方位”的设计思想，不断从综合性规划、建筑、人文地理、经济因素等方面来认识景观设计，并以此为基础进行景观规划、建筑、景观建筑以及相应领域的设计工作，如旧城改造设计、景观配套设施设计、城市导示设计，我们试图以景观设计来改善人与环境的关系。

当前，中国的城市化进程迅速，我们将在这一进程中不断地摸索和探求设计的目标与价值，分析已实践和正在实践的项目，从而创造出更有效、更令人愉快的景观环境。

Weaison Architecture Landscape Designing is a Sino-Australia joint uenture in China. It focuses on the design markets in northwest China currently and also involves the design markets in some cities of east China and south China.

Our company is mainly engaging in designing of scenery tourism, architectural landscape and planning and public art. We also have participated in or presided many landscape designing and constructing projects of different scales since the establishment of the company. Through years of accumulation, we have epurated the designing idea of "Omnibearing design" and with comprehensive planning, architecture, civil geography, economic factors in mind to recognize landscape planning, building and landscape building as well as relevant fields such as old city rebuild, landscape completion facilities designing. City guiding designing with responsibility, view of value and sensitivity in order to solve the issue between human and the environment. Cities in China are developing rapidly and greatly. we will grope for and explore goals and values, analyze the projects both practiced and practicing currently, and create more efficient and enjoyable landscape environment.

西安浐霸生态区商务中心
项目地点：陕西 西安
项目规模：110 000平方米
项目类别：景观环境设计

2006年5月，该项目是我公司配合西北建筑设计研究院设计的建筑主体而完成的外部影景观环境设计。项目在研究景观设计的出发点时，充分认识和理解建筑设计思想，并从建筑特点分析、环境分析、主题分析、总体认识上对该项目的景观设计进行思考。充分明确景观设计的宗旨，即“建筑设计是一种秩序，景观设计就是增加另一种秩序”。为此，本项目的设计思想、设计方向、设计步骤、空间配置等都力求与建筑主体共生，力求与环境共生。

目前项目正在建设施工中。

Commercial Center of Xi'an Chanba Ecological Area
Project Location: Xi'an, Shanxi
Project Scale: 110,000 m^2
Project Type: landscape environment design
In May 2005, cooperated with China Northwest Building Design Research Institute, we finish the design of the exterior environment of the main body. In researching the starting point of landscape design, the project has full awareness and understanding of design concept and makes the environmental landscape design concept from architectural features analysis, environment analysis, theme analysis and the overall understanding. It also fully clears the aim of landscape design which is "Architectural design is a kind of order, and landscape design is to add another order", and therefore, the symbiosis with the main building and environment is persevered in design concept, design direction, design steps, and space allocation.
The project is under construction.

水景城
项目地点：浙江 杭州
项目规模：250 000平方米
项目类别：环境景观设计

本项目坐落于浙江杭州，设计更多考虑的是如何恰当地利用水资源和如何突出城市肌理特性的问题。景观设计结合杭州地区人们的生活和审美习惯，尤其注重在设计过程中强调景观带来的情感和空间感受，从而实现项目的最终目标。

项目设计完成于2004年，并获得国际花园小区奖和建设部的中国名盘奖。

Waterscape City
Project Location: Hangzhou, Zhejiang
Project Scale: 250,000 m^2
Project Type: environment landscape design
As the design project located in Hangzhou, Zhejiang, the design gives more consideration on how to make use of rich water and how to distinguish urban texture characteristics. According to the living habit and aesthetic habit of local people, particularly focusing on emotion and space perception brought by the landscape in the design process, the design is to achieve the ultimate goal of the project.
The project design was completed in 2004 and was granted International Garden Community Award and China Noted Building Award by the Ministry of Construction.

西安市大明宫区域重点地段城市设计

项目地点：陕西 西安
项目规模：1 000 000平方米
项目类别：城市设计

2008年6月，受大明宫研究院邀请，我公司与西安建筑科技大学全天景观规划设计研究所联合参与西安市大明宫区域重点地段城市设计方案。其城市设计方案重点在于研究空间功能、空间品质、空间格局以及设计手法，并提出大明宫周边环境的建设思路，以及落实设计的措施等。

本项目完成时间为2008年7月，获得大明宫研究院颁发的二等奖。

Urban design for key sections in Xi'an Daming Palace area

Project Location: Xi'an, Shanxi
Project Scale: 1,000,000 m^2
Project Type: urban design

In 2008, all-weather Landscape Planning and Design Institute of Xi'an University of Architecture and Technology was invited by the Daming Palace Research Institute to participate in designing key sections in Xi'an Daming Palace area. Its urban design program focuses on space function, space quality, spatial pattern and design methods, and also raises the construction idea for the surrounding environment of Daming Palace and the implementation measures.

This project was completed in July 2008, and was awarded the Second Prize by the Daming Palace Research Institute.

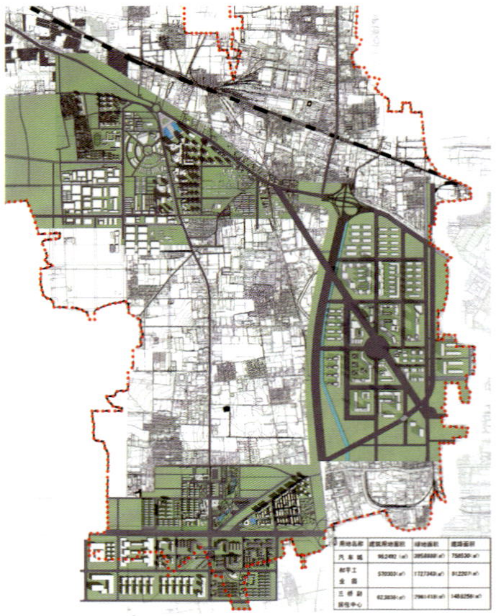

西安未央区·三桥地区空间、重构策略规划

项目地点：陕西 西安
项目规模：31.5平方千米
项目类别：控制性详细规划

受西安规划未央分局委托，我公司对西安市未央区三桥地区进行控制性详细规划设计。本项目要求在该地区现状条件下作空间规划和城市概念设计，包含产业经济空间与布局、城市空间发展、历史文化、景观与绿色规划等方面的内容。

Space Reconstruction Strategic Planning for Xi'an Weiyang District • Sanqiao Area

Project Location: Xi'an, Shanxi
Project Scale 31.5 km^2
Project Type: regulatory detailed planning

Authorized by Weiyang Branch of Xi'an City Planning Bureau, we make the regulatory detailed planning design for Sanqiao area in Weiyang District, Xi'an. This project is characterized by space planning design and urban concept design which are based on current situation of the region. It also includes some other aspects such as industrial economy space and layout, urban space development, historical culture, landscape, green planning, etc.

伊甸园入口

大连金石滩高科技农业观光示范园

项目地点：辽宁 大连
项目规模：2平方千米
项目类别：控制性详细规划及旅游项目规划

2003年2月，我公司与东南大学建筑设计研究院深圳分院共同承担了位于金石滩旅游度假区中心部位的大连金石滩高科技农业观光示范园的设计项目。本项目要求是建立一个凸显农业主题的大型主题公园，同时，在项目中还包括了学校、住宅区块、酒店等的设计内容。

玛雅街市

Dalian Jinshitan Hi-tech Agricultural Tourism Demonstration Park

Project Location: Dalian, Liaoning
Project Scale: 2 km^2
Project Type: regulatory detailed planning and tourism project planning
We cooperate with Shenzhen Branch of Architectural Design & Research Institute of Southeast University in Feb., 2003 to design Dalian Jinshitan Hi-tech Agricultural Tourism Demonstration Park, which is located in central area of Jinshitan Tourism Resort. This project is the large-scale agricultural theme park, and also includes design of school, residential block and hotel, etc.

城市风景 都市印象

项目地点：陕西 西安
项目规模：45 000平方米
项目类别：环境景观设计

2007年，受西安高科房地产委托，我公司与委托方共同制定景观设计策略，明确了景观主题以及副主题，从景观空间的各层次和设计手法上充分体现主题的内涵和副主题的意义，并在设计与实施的过程中，不断地调整和充实设计内容，力图体现出项目的景观特征和风格。

目前，项目工程正在实施过程中。

Urban landscape• Metropolitan Impression

Project Location: Xi’an, Shaanxi
Project Scale: 45,000 m^2
Project Type: environment landscape design
Entrusted by Xi’an Gaoke Real Estate Co., Ltd, we jointly make landscape strategy with the trustor in 2007. Clearing the theme and sub-theme of the landscape, it fully reflects the connotation of the theme and significance of the sub-theme from all levels of landscape space and design methods, and meanwhile, it constantly adjusts and enriches its content in process of design and implementation to embody landscape features and style.
The project is under construction.

上海现代建筑设计（集团）有限公司
Shanghai Xian Dai Architectural Design (Group) Co.,Ltd.

地址：中国上海市石门二路258号　邮编：200041
电话：+86–21–52524567
传真：+86–21–62464000
网址：www.xd–ad.com.cn
邮箱：xiandai@xd–ad.com.cn

Address: No.258 Shimen Er Road, Shanghai, 200041, P. R. China
Tel: +86–21–52524567
Fax: +86–21–62464000
Http: //www.xd-ad.com.cn
E-mail: xiandai@xd-ad.com.cn

上海现代建筑设计(集团)有限公司是一家以建筑设计为主的现代科技型企业，集团旗下拥有华东建筑设计研究院和上海建筑设计研究院等20余家专业公司和机构。公司在住房和城乡建设部历年全国勘察设计单位综合实力测评中跻身于前6名，在2005年中国工程设计企业60强排名中跻身于前3名，2001—2006年连续6年被上海市人民政府授予“上海市跨国经营先进企业”，同时连续6年被美国ENR“工程新闻记录”列入“国际工程设计公司200强”和“全球工程设计公司150强”。

集团现有职工3 000余人，其中包括中国工程院院士2名、国家设计大师5名。各专业技术人员占职工人数的83%，高级工程师占职工人数的24%，技术力量雄厚。

集团业务领域涵盖工程项目建设全过程的咨询服务，包括规划、建筑、结构、机电、建筑声学、市政设计、工程总承包、项目管理、工程监理、环境与装饰设计、工程地质勘察、项目可行性研究、科技咨询、投资开发等内容。2006年全集团营业净收入超过13亿元人民币。

至今集团累计完成了3万余项工程的设计与咨询任务，创作成果遍及全国29个省市及20余个国家和地区，其中800余项工程设计、科研项目和标准设计获国家、省（市）级优秀设计和科研进步奖。

集团已在北京、天津、重庆、武汉、青岛、深圳、西安、海南、厦门、大连、南宁、苏州、香港等地共设有25个分院(公司)或办事处，同时集团已与美国、加拿大、英国、法国、德国、澳大利亚、日本、新加坡、韩国、中国香港、中国台湾等国家和地区的设计机构合作设计的项目达400多个。

集团的专业设计人员以完美的创意和先进的建筑技术，将古典与现代、艺术与商业、舒适度与功能性、美感与科技完美地结合在一起，赋予建筑独特的风格和超凡的魅力，最大限度地满足了客户的需求。

Background of Shanghai Xi'an Dai Architectural Design (Group) Co., Ltd. (herein after referred to as "The Group") is a modern scientific & technological enterprise majoring in architectural design. Over 20 professional companies including East China Architectural Design & Research Institute and Shanghai Institute of Architectural Design & Research are the members of The Group. The Group has been ranked Top 6 of "The Top 100 Nationwide Prospecting & Design Companies in China" in terms of comprehensive competition strength assessed by the State Ministry of Construction in recent years. The Group was ranked Top 3 of "The Top 60 Engineering Design Firms in China" in 2005. The Group was ranked by the People's Government of Shanghai as one of "The Top 20 Trans-national Business Enterprises in Shanghai" from the year 2001 to 2006. Furthermore, The Group has been ranked by US magazine <Engineering News Record> as one of "The Top 200 International Design Firms" and one of " The Top 150 Global Design Firms" from the year 2001 to 2006.

The Group, having a strong technical strength, has a staff of over 3,000 among which 2 are academicians of the Chinese Academy of Engineering, 5 are National-class Design Masters in Architecture, 83% are professionals of different disciplines, 24% are senior architects/engineers.

The service scope of the group covers the whole process of the construction of the project, including urban planning, architectural design, structural design, MEP design, architectural acoustical design, infrastructure design, general contractor, project management, engineering supervision, environmental & interior design, geological exploration, feasibility study, scientific & technological consultant, investment & development, etc. The gross revenue of the group reached the amount of RMB 1,329 million in 2006.

The Group has completed the design and consultation service for more than 30,000 projects which spread across 29 provinces and cities of China as well as more than 20 countries all over the world. Over 800 designs of projects, scientific & technological researches and standardized designs undertaken by the Group have been granted National or Provincial/Municipal Awards for Outstanding Design and Progressive Scientific & Technological Achievements.

The Group has 25 branches (subsidiary companies) or offices in Beijing, Tianjin, Chongqing, Wuhan, Qingdao, Shenzhen, Xi'an, Hainan, Xiamen, Dalian, Nanning, Suzhou, Hong Kong, etc. More than 400 projects have been completed in collaboration with the national and regional design institutions from USA, Canada, UK, France, Germany, Australia, Japan, Singapore, South Korea, Hong Kong, Taiwan, etc.

Combining classic style with modern style, arts with commerce, comfort with functions and sense of beauty with science & technology by means of consummate skill and advanced architectural technique, The Group brings forth the unique style and charm to the architecture so as to fulfill the customers' needs to the utmost.

上海浦东国际机场 II 期

建设性质：交通建筑
建设地点：浦东国际机场
用地面积：1 099 000平方米
建筑面积：480 000平方米
设计时间：2004年8月—2006年1月
主体建筑结构形式：钢结构屋面、钢筋混凝土框架
主要外装修材料：玻璃幕墙、清水混凝土、花岗石幕墙

项目说明

浦东国际机场 II 期工程建设总目标是建设一个枢纽型航空港，以满足2015年4 000万年旅客量（加上 I 期2 000万年旅客量共计可达到6 000万年旅客量）。II 期工程建设分为两个阶段，第一阶段先建设T2航站楼、交通中心、总体道路及相关的配套设施设计，其中航站楼主楼部分可以处理年旅客量为4 000万人次，长廊部分为2 200万人次。II 期一阶段于2007年底完成。

Phase II of Shanghai Pudong International Airport

Construction character: transportation building
Construction Location: Pudong International Airport
Site Area: 1,099,000 m^2
Building Area: 480,000 m^2
Design Time: 2004.08-2006.01
Structure form of main body: steel roof, reinforced concrete frame
Main outside decoration materials: glass curtain wall, exposed concrete, granite curtain wall

Project Description

Phase II of Pudong International Airport aims at establishing "hub model" airport to accommodate 40 million passengers in 2015(plus 20 million passengers of Phase I to reach a total passenger volume of 60 million). Phase II project is divided into two phases, during Phase I, T2 airport building, transportation center, general road and relative supporting facilities are planning to build firstly, and main body of airport building can handle 40 million person times per year, and part of gallery can handle 22 million person times per year. The second stage project in Phase I was completed at the end of 2007.

上海光源工程

上海光源工程总体布局和主体建筑的设计构思源于光源装置所特有的旋转与散射的轨迹，并结合自然界的鹦鹉螺曲线，在总体布局和主要建筑单体上产生空间上的旋转动线，使其产生强烈的动感，反映出光源项目所特有的美感。

Project of Shanghai Light Source

Overall layout of Shanghai Light Source and design concept of main building stem from the unique path of rotation and scattering, combined with the nautilus curve in natural world to create revolving line in upper space on overall layout of the main building and major single building, so as to have a strong sense of movement and reflect the unique beauty of the light project.

上海世博家园

项目总体规划充分考虑了世博会定向安置项目所面临的突出问题——经济投入与高品质的建设要求之间的矛盾。项目属于经济型住宅的开发项目，但无论在总体规划、景观构想，还是在单体设计方面都要求体现“超前性、先进性、整体性、示范性”的原则，创造出能够体现2010年世博会精神的居住区，这是本方案的特殊性和难点所在，也是公司在设计中努力寻求突破的主要着眼点。设计力求在功能布局上符合城区总体规划要求，并优化提升地域环境品质，强化资源共享及可持续发展。该项目的建成将为世博会动迁居民提供一个功能配套完善、生活和谐的具有新上海城市生活特色的大型居住区。

Shanghai World Expo Home

Overall planning of the project gives full consideration on the outstanding problems of resettlement projects - the conflict between economic input and high-quality building requirements. The project belongs to an economic-type residential development project. Both in the overall planning, landscape concept and in the single design, the principle of "foresight, advanced, comprehensive, exemplary " is required to embody to create residential district which can reflect the spirit of the 2010 World Expo. To try to be in accordance with the overall urban planning requirements, optimize the geographical environmental quality and strengthen the sharing of resources as well as sustainable development are the uniqueness and difficulties of the project, which are also the main focus of the company to seek a breakthrough. The project was built for the World Expo relocation residents to provide a new Shanghai's major urban residential district with completed functional system and peaceful living environment.

同济大学建筑设计研究院
Architectural Design & Research Institute of Tongji University

同济大学建筑设计研究院成立于 1958 年，是全国知名的集团化管理的特大型甲级设计单位。研究院持有国家建设部颁发的建筑、市政、桥梁、公路、岩土、地质、风景园林、环境污染防治、人防、文物保护等多项设计资质及国家计委颁发的工程咨询证书，是目前国内设计资质涵盖面最广的设计单位之一。

经过近 50 年的积累和进取，我院拥有了雄厚的设计实力、丰富的人力资源、先进的设计理念与技术。全院现有职工 758 人，一级注册建筑师 98 名，一级注册结构工程师 114 名。1986 年以来我院共有近 200 项设计作品获奖。

作为一所国际著名高校的设计单位，该院非常重视建筑教育，培养了一大批硕士生、博士生。此外，我院还具备丰富的对外交流合作经验，曾成功地与来自美国、加拿大、德国、法国、西班牙等国的著名事务所合作，并互派员工交流学习。

按 ISO 9001 标准建立的质量保证体系通过中国 SAC 和美国 RAB 双重认证。自 2001 年起，我院与中国人民保险公司签订了每年累计赔偿一亿元人民币的工程设计险合同。我院有能力提供顶尖的设计产品和一流的咨询服务。

地址：上海市四平路 1239 号
邮编：200092
电话：+86–21–65987788
传真：+86–21–65985121
网址：www.tjadri.com
邮箱：info@tjadri.com

The Architectural Design and Research Institute of Tongji University, founded in 1958, is one of the leading design groups in China with the most comprehensive range of disciplines available in the country, allowing it to develop many types of design project. It holds design certificates issued by the State Ministry of Construction for China in architecture, municipal engineering, bridge engineering, highway engineering, geotechnical investigation, geology, landscaping, environmental engineering, civil air defence and cultural relic protection. It also holds an Engineering Consulting Certificate issued by the State Planning Committee.

The Design Institute has developed its strength in design, manpower and technology through its 50 years of experience. It employs 758 professionals including 98 National Class A Registered Architects and 114 State First Class Registered Structural Engineers. It has won nearly 200 prizes for its design work since 1986.

Affiliated to a well known university, the Design Institute has a tradition in architectural education. It has cultivated a number of graduate students. In addition, the Design Institute has successfully collaborated with many well known design firms from the United States, Canada, Germany, France and Spain etc. Exchange of staff is also part of the collaboration.

The Design Institute has been approved to the ISO9001 Quality Assurance Standard by Shanghai Audit Centre and RAB, the American ISO assessment organization. In 2001, the Design Institute signed an Engineering Design Insurance Contract with PICC, one of the leading insurance companies in China. This gives total cover of 100 million RMB per annum.

Address: No.1239 Siping Road, Shanghai
P.C.: 200092
Tel: +86-21-65987788
Fax: +86-21-65985121
Http: //www.tjadri.com
E-mail: info@tjadri.com

同济大学汽车学院

建设地点：上海
设计时间：2002—2003 年
建筑面积：584 100 平方米

项目概况

校园共分为六大片区：校前区（行政服务培训区）、公共教学区、二级学院区、研发实训区、生活区和体育运动区。各片区内建筑群体以院落式的方式组合，形成多院落的交往空间，使人与自然环境形成直接的对话。

考虑校园"统一规划，分期实施"的建设原则，校园规划不但考虑一、二期工程完全实施后的整体性，还保证校园仅在一期工程实施后的完整性。组团式建筑布局使各功能建筑分片建设，组团内部建筑可以分期建设，使校园规划形成可持续发展的、有机生长的设计结构。

Automotive Engineering College, Tongji University

Construction Location: Shanghai
Design Time: 2002-2003
Building Area: 584,100 m^2

Project Profile

The campus is divided into six areas including front area (administrative service training area), public teaching area, area of secondary institute, R & D practical training area and living area as well as sports area. Buildings in each area are located with courtyard type to form communication space of multiple courtyards, which makes people directly feel the natural environment.

Considering the principle of "integrated plan, phased implementation", the campus planning not only focuses on the integrity after the implementation of Phase Ⅰ and Ⅱ projects, but also needs to guarantee the completeness of campus after Phase Ⅰ project. The group architectural layout ensures each functional building is built with areas and interior building is constructed in phases, becoming organically growing structure with the sustainable development.

秦皇岛体育场

建设地点：河北 秦皇岛
建筑面积：48 000 平方米

项目概况

体育场观众席内圈椭圆形与外圈椭圆形的中心线东西相距 15 米，这种设计可使视线较好的西看台容纳更多的观众。看台共设有 33 000 个坐席，其中有 2‰的伤残人士坐席，东西看台后侧还分别设有包厢。运动场内设有半径为 36.5 米的 400 米环形跑道标准田径场，跑道内设有 104 米 × 68 米的天然草皮足球场。

Qinhuangdao Stadium

Construction Location: Qinhuangdao, Hebei
Building Area: 48,000 m^2

Project Profile

The central line of inner ellipse spaces out 15 meters apart the outer ellipse from east to west, which can ensure the West Stand with better sight quality accommodate more audience. The stand sets up a total of 33,000 seats, of which 2‰ is for the disabled. Boxes are set up behind the East and West stands. The stadium is equipped with standard track and field with 36.5-meter radius and 400-meter circular track; meanwhile, inside the track there is a football field of 104 m × 68 m with natural sod.

中国银联上海信息处理中心

建设地点：上海
用地面积：20 984 平方米
建筑面积：15 387 平方米

项目概况

中国银联是中国货币与金融电子化的先驱，本次工作目标就是要为之提供一个与其品质相适应的具有典型现代主义特征的空间。

1．理性、可靠、坚固、现代的建筑造型，与中国银联的企业形象相吻合。

2．由于项目地处两条河道的交界处，自然环境优美，故采用园林式的建筑布局，为生产办公提供自然化的景观环境。在景观设计中运用了大量的水景，大面积平静的水面及设置于其上的休憩平台使工作者能拥有平和的心境，置身其间感受建筑与水景的相互辉映，使身体与心灵都融入到这种极富意境的体验中。

Shanghai Information Processing Center of China Union Pay

Construction Location: Shanghai
Site Area: 20,984 m²
Building Area: 15,387 m²

Project Profile

China Union Pay is the pioneer of Chinese currency and financial computerization, and its working goal is to furnish the qualified space with typical modernism characteristics.

1. With the rational, reliable, solid and modern architecture molding, which is matched the enterprise features of China Union Pay.

2. Because of the natural beauty brought by its location at the junction of two rivers, the project adopted garden-style building layout to provide naturalized landscape for the production and office work. In the design, a large amount of waterscape is used . A large area of calm water surface and the over platform are the supporting of the peace mind of staff. The building embraces the waterscape, and the peaceful environment matches the smart mind, which are all highly mood experience.

浙江省公安指挥中心

建设地点：浙江 杭州
建筑面积：32 000 平方米
设计时间：2000 年
竣工年份：2003 年

项目的设计目的是为浙江省公安厅提供全新的建筑设施与功能布局。设计从城市设计入手，尊重总体规划和周围环境，充分考虑西湖风景区的自然景观，将已建公安厅大院内建筑与环境设施进行统筹设计、规划，满足在当前基本构架的基础上建立准确有效的工作模式，体现高效率、简洁的风格。

Zhejiang Police Control Center

Construction Location: Hangzhou, Zhejiang
Building Area: 32,000 m^2
Design Time: 2000
Completion Time: 2003

The project aimed at providing the brand new building facilities and functional layout for Department of Public Security Zhejiang Province at the beginning of the 21st century. Started from urban design and respected the master design and surrounding environment, the project gave full consideration to the visual landscape of the West Lake scenic area, and it made overall plan and design for the existed buildings and environmental facilities inside the compound to establish the accurate and effective working mode on the basis of the current basic framework, reflecting the high efficient and simple style.

PDG INTERNATIONAL

泛太平洋设计集团有限公司（加拿大）

Pan Pacific Design Group Ltd. (Canada)

泛太平洋设计集团有限公司（加拿大）1994年成立于加拿大多伦多，并于1995年在上海成立设计分支机构。公司设计范围包括城市规划及建筑设计，设计范围涉及酒店、办公、商业文化设施、住宅、学校、体育场所等，并重视发展城市景观设计、建筑设计与室内设计的衔接与渗透。

公司依托国际化优势，以国际加本土的专业化团队，形成自己的创作和服务特色。在创作过程中，公司全面认知项目的土地价值、开发定位、文脉延续、艺术感知及心理体验，以开放、理性和积极的姿态，去寻求创作中的每一个机会和最佳答案。公司还重视实施过程的技术控制和提升，逐步发展了一套适应于当代市场多样变化的服务体系。

公司成立至今已完成了上海浦东新区接待中心、九寨天堂国际会议度假中心、德隆集团总部大楼、成都新国际会展中心和上海烟草集团科教中心等国内重大工程的设计，并赢得了建设部全国人居经典综合金奖、建设部建筑形态金质奖、建设部创新风暴中国建筑设计示范住宅单项奖、上海市优秀住宅小区工程设计项目一等奖、上海市优秀勘察设计二等奖、上海市住宅小区优秀规划奖等近三十个设计奖项。2004年12月至2005年12月，公司获得由中国住交委员会等颁发的2004年度中国建筑二十大品牌影响力规划建筑设计事务所（公司）奖及2005年度中国二十大品牌影响力规划建筑、景观设计公司，2006年获得法国建筑设计师协会、国际房地产商协会和亚洲房地产峰会组织委员会等颁发的"2006年度亚洲建筑规划设计机构100强"以及由中国建筑规划设计师协会等颁发的"2007引领中国建筑规划设计十大品牌设计机构"等奖项。

泛太平洋设计集团有限公司（加拿大）于2001年1月成立上海泛太建筑设计有限公司，并于2005年9月成立上海泛巢建筑设计事务所。

上海泛太建筑设计有限公司

SHANGHAI PAN-PACIFIC ARCHITECTURAL DESIGN CO., LTD.

上海泛巢建筑设计事务所

SHANGHAI PAN-PACIFIC SPACE-WORK ASSOCIATES

上海泛太建筑设计有限公司成立于2001年1月，于同年获得上海市首批颁发的建筑专业专项设计资质，2006年改为建筑工程设计乙级资质，上海泛巢建筑设计事务所注册于2005年3月，同年获建设部建筑设计专项甲级资质，均是泛太平洋设计集团有限公司（加拿大）在中国设立的全资控股企业。公司承接从规划、建筑方案到施工图全程所有专业的设计工作。

公司以高标准的设计与服务建立起了市场知名度和信誉度，在此基础上，公司着眼全国市场，在江苏、浙江、四川、山东、吉林等近20个省开展了一系列项目，并在成都成立了分支机构。

目前，公司有设计人员近120名，专业涵盖建筑、结构、机电、室内设计和景观等，其中国家一级注册建筑师25名，外籍建筑师10名，建筑学博士2名，硕士25名，他们具有一流的设计能力和工程协调能力。自2003年起，公司在各年度民用建筑设计市场民营设计企业二十强排行榜中排名的均位于前列，自2001年来，公司每年都有数个设计作品获得国家及地方评比的奖项。

随着社会经济与技术的发展，公司致力于前所未有的人性化的建筑创造，并通过我们的双手实现梦想…

PDG was established in Toronto, Ontario, Canada in 1994. Since the opening of the Shanghai Office in 1995, PPDDG has become one of the most active design firms in the Chinese market.

Inspired by international trends of modern design, PDG offers its expertise in architectural design, urban planning, landscape design and interior design. With over 100 professionals from all over the world, PDG combines Chinese and western philosophy in its commitment to excellence under the company's mission, that is, to maximize the value of the project for the clients, users and society in general through design.

Based on the specific project, our team works closely with the clients to achieve a full understanding of their facility needs and site conditions. The application of creativity to solve challenging problems, the pursuit for excellence and the spirit of the team collaboration are the means for our success.

In 2001, Shanghai Pan-Pacific Architectural Design Co.,Ltd., which was licensed by the Chinese government and in 2005, Shanghai Pan-Pacific Space-work Associates was incorporated in Shanghai and thus became the legal base for PDG to work in China.

On recognition for excellence on both professional design and service, PDG has been awarded numerous with national awards both from the government and professional associations.

加拿大
地址：2403–1250 Hasting Street，Vancouver，BC，Canada V6E 4T7
电话：001–(604) 899 2639
传真：001–(604) 899 2692
上海
地址：上海市浦东新区张江高科技园区伽利略路 11 号伽利略商务公馆 6 单元（201023）
电话：+86–21–31265800
传真：+86–21–50814700
邮箱：ppddg@ppddg.com
网址：www.ppddg.com

Canada
Address: 2403-1250 Hasting Street, Vancouver, BC, Canada V6E 4T7
Tel: 001- (604) 899 2639
Fax: 001- (604) 899 2692
Shanghai
Address: Unit 6 (201023), Galileo Business Residence, No.11 of Galileo Road, Zhangjiang Hi-Tech Park, Pudong New Pistrict, Shanghai
Tel: +86-21-31265800
Fax: +86-21-50814700
E-mail: ppddg@ppddg.com
Http: //www.ppddg.com

PDG INTERNATIONAL

常州奔牛机场

项目地址：江苏 常州
建筑面积：58 000 平方米
设计时间：2008 年
项目状态：方案

Benniu Airport, Changzhou

Project Location: Changzhou, Jiangsu
Building Area: 58,000 m²
Design Time: 2008
Project Status: programming

山东潍坊滨海经济开发区 CBD

项目地址：山东 潍坊
规划面积：540.25 公顷
设计时间：2009 年
项目状态：方案

Binhai Economic Development Zone CBD in Weifang, Shandong

Project Location: Weifang, Shandong
Planning Area: 540.25 ha
Design Time: 2009
Project Status: programming

控制性详细规划范围

港营街以北，海港路以东，白浪河以西，新港街以南。规划总用地面积为 540.25 公顷。

规划理念

本方案以"太阳城"主题为依托，综合规划建筑、绿化、水系、路网。规划设计既强化核心区的中心地位和 CBD 的整体性，又与规划的白浪湾滨海度假区、科教创新区形成呼应，构成了滨海新城新的特色。

规划结构：

中央商务区 CBD 分为南、北两个大小不同的功能区，分别规划不同的功能主题，由北向南分别为北区的金融商务区、南区的行政主题商务区、居住区。

南、北两区各成体系，利于分期建设。按照建设计划，应先开发 2003 山东潍坊滨海经济开发区中央商务区南区，形成规模后再开发北区，最终形成完整的 CBD 城区。

The range of regulatory detailed planning: north areas of Gangying Street, east areas of Haigang Road, west areas of Bailang River, south areas of Newport Street. The overall site area of planning is 540.25 hectares.
Planning concept:
"Sun City" --- the program comprehensively plans architecture, greening, water system, road network on the basis of the "Sun City" theme. The program not only strengthens the central position of the core area and the integrity of the CBD, but also forms the symmetry in the formation with the planned Bailang Bay Binhai Resort, Innovation Zone of Science and Education. These constitute the city texture of Binhai New City CBD with prominent characteristics.
Planning structure: the Central Business District (CBD) is divided into north and south functional areas of different sizes. They are planned with different functional themes. From north to south, they are respectively the North District--- the financial business district, South District --- administrative theme business district and residential area.
The north and south districts have their own system, which is beneficial to be constructed in phases. In accordance with the construction plan, number 03-- South District of regulatory detailed planning for the Central Business District of Binhai Economy Development Zone in Weifang, Shandong will be firstly developed. After south district has developed, the north district will be developed. Finally they form a complete CBD urban area.

沪宁城际铁路常州站

项目地址：江苏 常州
占地面积：119 998 平方米
建筑面积：151 573 平方米
设计时间：2009 年
项目状态：在建

Changzhou Station, Shanghai-Nanjing Railway, Changzhou

Project Location: Changzhou, Jiangsu
Site Area: 119,998 m^2
Building Area: 151,573 m^2
Design Time: 2009
Project Status: under construction

山东潍坊大学城

项目地址：山东 潍坊
规划面积：8 850 平方米
设计时间：2009 年
项目状态：方案

Master Plan of the City of Universities, Weifang

Project Location: Weifang, Shandong
Planning Area: 8,850 m^2
Design Time: 2009
Project Status: programming

本次规划的范围为

港营街以南，海河路以东，海惠路以西，港城街以北。规划总用地面积为1120.13 公顷。

规划目标

建设成“产学研一体化，高科技孵化器”的科教新区，建设成为潍坊市重要的学院建设成集中区，集科教，研发，实训等功能为一体的科教创新园区。

规划理念：

“种子、希望”——大学城“希望之星”的总体意象。

设计体现了“种子萌芽、生命成长、孕育希望、生生不息”的理念。一个从无到有的大学城就犹如一颗刚刚发芽的种子，生机勃勃、面向未来，大学城的规划理念正是吸取了这个理论的精华。

规划结构：

规划根据概念性规划的要求，结合本规划用地的自然条件，以及规划区未来城市空间发展态势，规划形成“一岛，一环，两翼，四角，两边”的空间结构。

The range of this planning: south areas of Gangying Street, east areas of Haihe Road, west area of Haihui Road, north area of Gangcheng Street. The overall floor area of planning is 120.13 hectares.

Planning objectives:

This place will be built into a New Zone of Science and Education of "enterprises-universities-researches integration, high-tech incubator". It will be an important centralized area of colleges in Weifang City. It is an innovation area of science and education, integrating functions of science and education, research and development, training and so on.

Planning concept:

"Seed, hope" --- the overall image of "Star of Hope" of the City of Universities

The design embodies the concept of "seed germination, growing up of life, nurturing hope, persistence". As a City of Universities, it grows out of nothing. It is like a vibrant seed which just germinates and faces future. The planning for the City of Universities absorbs the essence of this kind of concept, which gives people the overall impression of "seed, hope".

Planning Structure:

Under the requirement of conceptual planning, combining with the natural conditions of land use planning, taking the future development trend of urban space in planning area into consideration, the idea forms the spatial structure of "one island, one ring, two wings, four corners, two sides".

成都 IBP 总部公园

项目地址：四川 成都
占地面积：429 000 平方米
建筑面积：528 290 平方米
设计时间：2007 年
项目状态：2010 年竣工

IBP Headquarters Park, Chengdu

Project Location: Chengdu, Sichuan
Site Area: 429,000 m²
Building Area: 528,290 m²
Design Time: 2007
Project Status: completed in 2010

山东潍坊展览馆

项目地址：山东 潍坊
占地面积：15 900 平方米
建筑面积：25 290 平方米
设计时间：2009 年
项目状态：方案

Exhibition Hall in Weifang, Shandong

Project Location: Weifang, Shandong
Site Area: 15,900 m²
Building Area: 25,290 m²
Design Time: 2009
Project Status: programming

建筑语言：

建筑物开敞明亮，表面机理和屋顶天窗形似宝石，色彩夺目。建筑寓意潍坊蕴藏着丰富的资源，具有极大的发展潜力。

狭长的中庭空间，富于光影变化，色彩奇妙绚丽，形态变化万千，是一个美妙的空间，同时兼具通风采光、太阳能发电等多重功能，建筑是可呼吸的生态建筑。

Architectural language: architecture is open and bright. The shapes of the surface mechanism and the roof-lights are like glittering gems, implicating that Weifang has rich resources and has great development potential.

The narrow atrium space is filled with changes of light and shadows with wonderful and gorgeous color and ever-changing forms. It is a wonderful space, combing multiple features like ventilation and lighting, solar power etc, which makes the building a breathing ecological one. Visit and tour become a voyage of discovery.

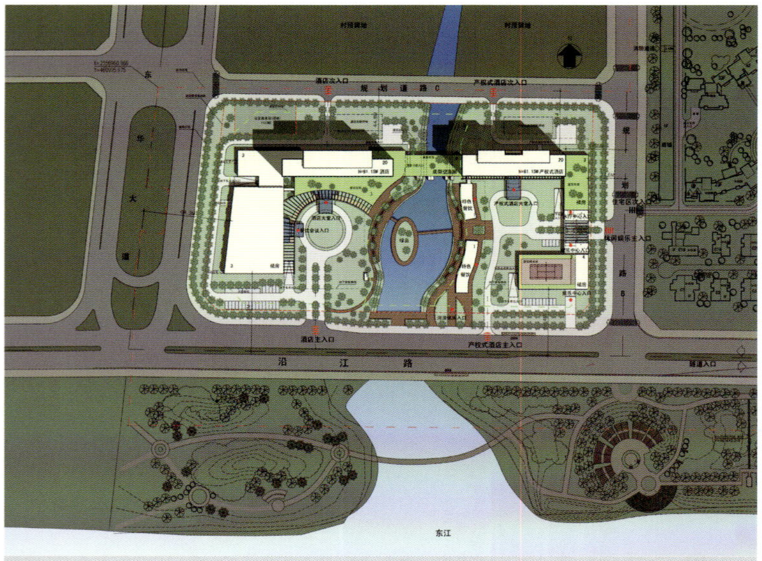

广州增城保利国宾酒店

项目地址：广东 广州
占地面积：34 007 平方米
建筑面积：94 412 平方米
设计时间：2010 年
项目状态：方案

Poly Ambassador Hotel, Zengcheng, Guangzhou

Project Location: Guangzhou, Guangdong
Site Area: 34,007 m^2
Building Area: 94,412 m^2
Design time: 2010
Project Status: programming

武汉市工人文化宫

项目地址：湖北 武汉
占地面积：65 729 平方米
建筑面积：43 510 平方米
设计时间：2009 年
项目状态：方案

Workers Cultural Palace, Wuhan

Project Location: Wuhan, Hubei
Site Area: 65,729 m^2
Building Area: 43,510 m^2
Design Time: 2009
Project Status: programming

本区域未来将成为高档住宅社区，用地周边高楼环绕。现有的规划绿化带无法承担连通长江—沙湖的城市低温廊道作用，用地内建筑本身也应顺应城市低密度空间的发展趋势，因此，采用流线型地景式景观建筑。这种建筑有利于形成自然风道和整体连续的绿化景观带。

建筑布局顺应地形南北向布置，最大限度顺应东南主风向，从而营造宜人小气候。错落有致、富于变化、充满动感的建筑群体，体现江城武汉的独特地域文化内涵。

设计师根据具体建筑功能空间尺度及空间感受对形体进行推敲，设计了简洁、流畅的建筑形态。

In future, the region will be superior residential communities. The land is surrounded by high buildings. Planning greening belt can not afford the role as low-temperature corridor of the city to connect Yangtze River and Shahu Lake. The building itself within land should also comply with spatial alignment of low density band of the city, and streamlined landscape-style landscape architecture, forming a natural wind tunnel and the overall continuous greening landscape belt.

Conforming to the terrain, the architectural layout and constitute is laid out by north and south trending, forming the flowing public space of the construction. The layout follows the southeast prevailing wind, and dredges greening corridors beside rivers to create a pleasant microclimate. Changeable architectural groups arranged at random but regular like hundred dynamic boats compete to win, contain a unique regional culture connotation of the city of the Changjiang River--- Wuhan.

According to functional spatial dimension of specific architectural and spatial feelings, designers consider the shape and determine the simple building shape to create a symbolic image of the group.

四川省美术馆新馆

项目地址：四川 成都
占地面积：6 666 平方米
建筑面积：21 906 平方米
设计时间：2008 年
项目状态：方案

New Museum of Art Museum, Sichuan

Project Location: Chengdu, Sichuan
Site Area: 6,666 m²
Building Area: 21906 m²
Design Time: 2008
Project Status: programming

本项目是对武汉锅炉厂老厂区整体搬迁后留下的 53 公顷用地进行整体改造，需将其规划建设成大型居住社区。用地位于武昌区中心地段，紧邻中南、亚贸商业圈和武昌高校区，交通便利，城市配套设施完善，文化氛围浓郁，是非常适宜居住的城市板块。项目定位为高品质居住社区。

规划设计特点

1. 城市更新：从城市旧区功能改造和环境更新的高度认识基地，弥补城市结构缺陷，改善城市机能，强化项目规划建设对城市更新的积极贡献。

2. 大型社区：定位为大型城市居住社区，集居住功能与城市休闲、商务、文化生活为一体，创造出一个丰富多彩的城市生活新载体。

3. 延续文脉：充分尊重地块现有的肌理和文脉特征，积淀历史遗存，将其变成住区中独特的资源，因地制宜地创造标志性个性形象。

4. 持续发展：按独立组团划分地块，整体规划，分期建设。兼顾整体和组团关系，适应持续发展的市场需求。

顺畅的交通结构、合理的组团划分、切实可行的分期、多元化的住宅产品、完善的社区配套、富有特色的景观系统，以及延续文脉的建筑环境风貌，使本项目在充分融入城市环境的同时凸显出大型社区高品质的生活氛围，达到“闹中取静，别有情趣”的效果，成为城市更新改造中的亮点。

This project is the overall transformation for 53 hectares land left by the overall removal of the old factory of Wuhan Boiler Factory, planning and constructing large residential communities. The land is located in the central area of Wuchang District, close to Central South and Asia Trade Commercial District, High Education Area in Wuchang. There is convenient transportation, perfect urban facilities and rich culture atmosphere. It is a very livable city block. The project is positioned as high quality residential communities.

Design features:

1. Urban renewal: knowing the base from the high position of functional transformation of the old districts and the environment renewal, the design makes up the defects urban structure, improves the urban functions, and strengthens the positive contribution of project planning and construction to urban renewal.

2. Large-scale community: the design is positioned as the large-scale urban residential community. It incorporates the interaction of resident function, the city leisure, business, cultural life to create a rich new medium of urban life.

3. Extending context: the design completely respects the land's texture and context features, accumulating historical remains which are turned into unique resources in residences. The design adjusts measures to local conditions to create the indicative individual image.

4. Sustainable Development: the design divides blocks according to independent groups. The blocks are overall planned and constructed by stages. The design is both covering the whole and the groups in order to meet market demand for sustainable development.

The features such as the smooth traffic structure, the reasonable division of groups, the practical stages, diversified residential products, perfect community facilities, distinctive landscape systems, and the built environment style which extends context, make the project highlight the high-quality life atmosphere of large-scale community as well as fully integrate city environment to achieve the effect of "getting quietness from noise, amazing and attractive landscape", becoming the bright spot in the renovation of the urban renewal.

武锅项目总体规划与建筑方案设计

业主：武汉三方置业有限公司
项目地址：湖北 武汉
占地面积：532 620 平方米
建筑面积：1 065 240 平方米
设计时间：2007 年
项目状态：方案

Overall project planning and architectural design of WBC

Owner: Wuhan tripartite Properties Limited
Project Location: Wuhan, Hubei
Site Area: 532,620 m²
Building Area: 1,065,240 m²
Design Time: 2007
Project Status: programming

无锡九龙仓古运河 71 号地块

项目地址：江苏 无锡
占地面积：131 646 平方米
建筑面积：2 384 646 平方米
设计时间：2008 年
项目状态：在建

Block No. 71 of Canggu Canal in Jiulong, Wuxi

Project Location: Wuxi, Jiangsu
Site Area: 131,646 m²
Building Area: 2,384,646 m²
Design Time: 2008
Project Status: under construction

外滩地区功能发展与综合交通规划

项目地址：上海
占地面积：3.14 平方千米
设计时间：2004 年
项目状态：2010 年竣工
合作单位：上海市工程设计研究院
茂盛（亚洲）工程顾问有限公司
获奖：外滩地区功能发展和综合交通规划方案设计第一名

The Bund District Functional Development and Comprehensive Traffic Planning Shanghai China

Project Location: Shanghai
Site Area: 3.14 km²
Design Time: 2004
Project Status: completed in 2010
Cooperation Unit: Engineering Design Institute, Shanghai
Maunsell (Asia) EngineeringConsultants Co., Ltd.
Awards: the First Prize for program planning of The Bund District Functional Development and Comprehensive Traffic Planning

规划总体目标是协调外滩地区城市发展和交通需求的矛盾，提升城市功能，改善城市环境，实现地区的可持续性发展。

The overall objectives of planning are to coordinate conflicts between the urban development and transportation needs to Bund District, to enhance urban functions, to improve the urban environment and to realize sustainable development of the region.

郑州新客站西广场商业

项目地址：河南 郑州
占地面积：51 620 平方米
建筑面积：310 417 平方米
设计时间：2009 年
项目状态：方案

West Plaza commerce of Passenger Station, Zhengzhou

Project Location: Zhengzhou, Henan
Site Area: 51,620 m^2
Building Area: 31,0417 m^2
Design Time: 2009
Project Status: programming

武当山太极湖生态文化旅游区湖滨区规划

项目地址：湖北 武当山
占地面积：88 200 平方米
建筑面积：20 793 平方米
设计时间：2009 年
项目状态：方案

Riverside Planning of Taijihu Ecological and Cultural Resort, Wudangshan

Project Location: Wudang, Hubei
Site Area: 88,200 m^2
Building Area: 20,793 m^2
Design Time: 2009
Project Status: programming

中国建筑上海设计研究院有限公司——李犁工作室（创作部）

CSCEC Shanghai Design Academy Company Limited—Li Li Studio (Creative Department)

李犁工作室于2001年底成立于上海，已完成了一系列策划、城市设计与规划、建筑、景观、室内、家具的设计项目。工作室的主持人李犁1987年毕业于南京工学院（今东南大学）建筑系建筑学，获学士学位，在其20年的设计生涯中，他历经了90年代初的广东特区建筑热潮，90年代末的浙江迅猛发展期及2000年以来上海的建设风暴，使他对城市、地域重建、文化、平衡、风水、群落、历史、空间等进行了综合思考，并由此形成了李犁工作室"因缘而生"的平衡建筑观。

综合考虑自然与人文因素，努力通过多种手法，建立建筑与自然环境"场"及社会环境"场"的联系，这其中包括融合建筑与其周边环境、当地人文环境的关系，以达到"因缘而生"的有机生态平衡。建筑应是整个自然与社会生态链中的一环，其如同人体器官一样，生来就有其位置和功能，因此有机的、适宜的、可持续发展的建筑，能带来人们需要的工作与生活质量。

2006年，加拿大BSN建筑师事务所来上海设办事处，其绿色生态理念与李犁工作室的"因缘而生"平衡建筑观惊人相似，意气相投，遂与李犁工作室达成战略伙伴关系。李犁同时兼职BSN设计总监，并将乔治·贝尔德教授的生态观引入李犁工作室的设计理念中。

2007年8月，《建筑细部》杂志出版特辑"加拿大BSN建筑师事务所＋李犁工作室"。至此，李犁工作室致力于绿色建筑，先后与上海斯宾茨建筑节能技术有限公司，中国建筑上海设计研究院有限公司节能事业部达成合作关系。

2008年7月，李犁加入中国建筑，并担任中国建筑上海设计研究院有限公司第八设计所所长。2009年12月，李犁作为中国建筑上海设计研究院有限公司创作部总监，任《建筑技艺》杂志理事。

2009年12月28日，中国最大的建筑设计集团——中建设计集团在北京中建商务大厦揭牌成立，其通过组织机构调整的跨越，成为国际知名、国内一流，行业内不可或缺的重要设计企业，实现在重要城市三分天下有其一的态势。

电话：+86–21–52715812
传真：+86–21–62850220

Li Li Studio was founded at the end of 2001 in Shanghai. It has finished a series of design projects of engineering, urban design and planning, architecture, landscape, indoor and furniture. The host Li Li of the studio graduated and got Master's Degree in architecture major, Department of Architecture of Nanjing Engineering Institute (the present Southeast University). During 20 years' design career, he has experienced the architecture upsurge of Guangdong Special Zone in early 1990s, the swift and violent development of Zhejiang in late 1990s as well as the constructing storm of Shanghai since 2000. Therein he has carried out comprehensive consideration on city, regional reconstruction, culture, balance, geomancy, community, history and space, which all form the balance construction view of "to create according to chance".

The studio comprehensively puts both natural and humanity factors into consideration, and endeavors to construct the relationship between architecture and natural environment "field" and social environment "field" through various techniques, including blending the relation between the architecture and its surroundings and the local humanity environment, so as to achieve the organic ecological balance of "to create according to chance". Architecture should be a link within the whole natural and social ecological chain; like an organ of human body, it should be born with its position and function. Organic, suitable and sustainably developing architecture can bring the work and living quality that meets people's requirement.

In 2006, the Canadian BSN Architects Office came to Shanghai to set an agency, and its green ecology concept is amazingly similar with Li Li Studio's balance construction view of "to create according to chance". Therefore they chimed in easily and formed strategic parternership. Li Li meanwhile holds the post of BSN's Design Director, and Professor George Baird's ecological view has also been introduced into Li Li Studio.

In August 2007, the magazine "Details in Architectures" published a Special "Canadian BSN Architects Office + Li Li Studio". Thus Li Li Studio devoted itself to green construction, and has successively achieved strategic cooperation with Shanghai SBC Green Building Energy Service Company Limited and Energy-saving Business Unit of CSCEC Shanghai Design Academy Company Limited.

In July 2008, Li Li joined CSCEC, and took the post of Director of the 8th Design Office in CSCEC Shanghai Design Academy Company Limited. In December 2009, as the Chief Inspector of Creative Department in CSCEC Shanghai Design Academy Company Limited, Li Li held the post of director of the magazine of "Architecture Techniques".

On December 28th, 2009, the biggest architecture design group—CSCEC Design Group was established in Beijing CSCEC Business Building. Through the spanning of the adjustment of institutional framework, it has become an important design enterprise which is famous in the world, first-class in China and indispensable in the industry, and realized the formation of the situation that taking one third of the whole in major cities.

Tel: +86-21-52715812
Fax: +86-21-62850220

1

2

3

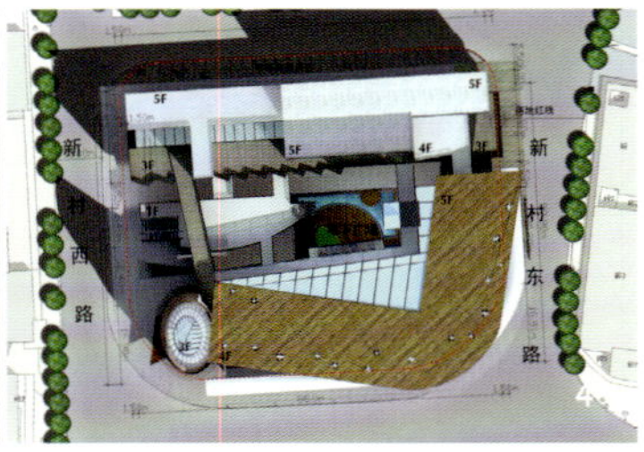
4

5

1–5 景德镇瓷都新天地

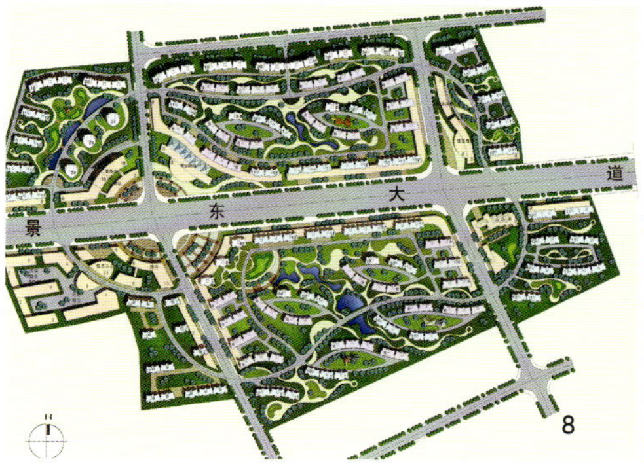

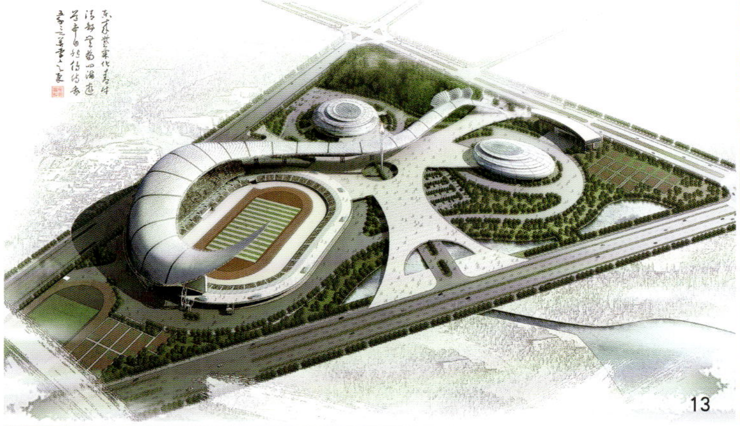

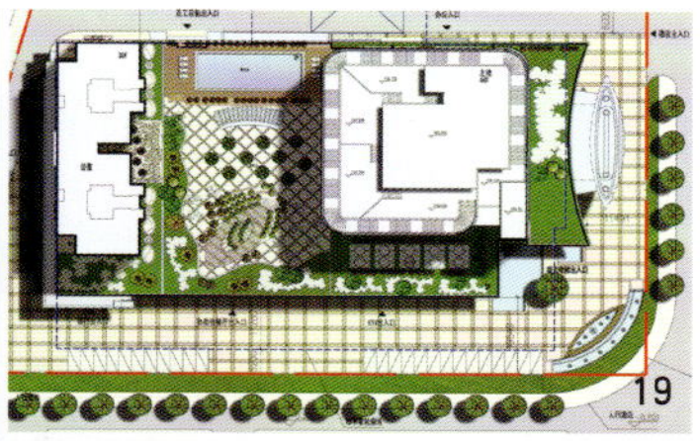

6–9 景德镇景东大道

10–12 上海车市

13–17 宣城体育中心

18–21 宿州中央大厦

22–25 宁夏国宾馆

上海垣恒建筑设计咨询有限公司
Shanghai C@C Architectural Design Co.,Ltd.

上海垣恒建筑设计咨询有限公司在成立之初就坚持走国际化的发展道路并采用先进的经营理念和完善的管理制度。公司与国际建筑市场接轨，实行人性化的管理模式。公司承接的项目以建筑、规划、景观、室内等的方案设计及施工图设计为主，并与国内多家甲级设计院及加拿大、德国的设计公司有良好的合作关系。

公司拥有专业的设计团队和扎实的技术力量，以有着20年国企总工程师工作背景的国家一级注册建筑师王一鸣为核心带头人，他拥有丰富的设计、投标、施工管理、工程技术等方面的经验，承做过许多大型公共建筑、住宅及工业建筑项目，并多次在上海及全国的工程竞标中一举中标，在业内享有良好的声誉。

Shanghai C@C Architectural Design Co.,Ltd. has been sticking to its international development principle, advanced business philosophy and perfect management system, linking itself to the international architectural market and implementing humanistic management mode. The company gives priority to architecture, planning, landscape, indoor scheme design and construction documents design, and has established sound cooperation ties with many domestic Class A design institutes and other countries such as Canada and Germany.

The company boasts a professional design team and huge technical strength. The company, led by Mr. Wang Yiming (a national Class A Registered Architect, with 20 years of experiences in state-owned companies as a chief engineer and with abundant experiences in design, tendering, construction management and engineering), has undertaken the construction of a great many of large-scale public buildings, residential buildings and industrial buildings and has won quite a number of bids for the projects in Shanghai and around China, and the company is broadly recognized in this industry.

>>>

总建筑师：王一鸣
设计团队：乔豪、陈蕾、陈思妍、王秋、卜中勤、盛成涛、李欣波、王佳成

地　址：上海市江场三路228号507室
电　话：+86-21-61077016
传　真：+86-21-61077018
邮　箱：shanghaiyuanheng@126.com

Address: Room 507, No.228 of Jiangchang West Road, Shanghai
Tel: +86-21-61077016
Fax: +86-21-61077018
E-mail: shanghaiyuanheng@126.com

包头市蒙古族中学新校区

在本项目的设计中我们从蒙古族文化中抽取具有符号象征意义的两种器物——马头琴和蒙古弓，作为整个设计规划布局的结构主轴与文化主轴。马头琴和蒙古弓，一文一武，一张一弛，不仅充分体现了蒙古族中学的民族文化特征，又能符合现代教育所倡导的德智体全面发展的教学主旨。

设计中将四栋教学楼设置在校区西北部，与大报告厅、实验楼、办公楼通过环廊相连，共同组成马头琴的头部。将科技楼、会议讲座展览厅、艺术楼、体育馆连为一体，共同组成马头琴的琴身及琴弓，这样的布局形式使得教学区域与文体课外活动区域两大教育组成部分，既相对独立又相互联系，共同组成包头市蒙古族中学的文化主轴。

校区主入口处的弧形大门，与校区前道路共同组成了蒙古弓的形态，从而形成了整个学校的动态主轴。

北京师范大学包头附属学校

项目位于包头市滨河新区东南部，用地面积为148 155平方米，建成后的学校将引入北师大小学部至高中部的全程基础教育配套功能，以省级示范学校为标准建造，其中小学为36个班，初中、高中各为18个班，学校总建筑面积为85 876平方米，其中地上总建筑面积为83 661平方米，地下建筑面积为2 215平方米。设计中充满浓郁的文化气息，有互动的学习环境。学校的规划采用“庭院式布局”手法，以建筑围合为基本手段，院落为表现形式，一组或者多组的建筑围绕一个中心广场，从而构成对称布局的建筑群。

包头机电工业职业学校

项目位于包头市九原区职业教育园区西北部占地面积为138 181平方米，总建筑面积为49 385平方米。

包头铁道职业技术学院

项目位于包头市九原区职业教育园区东北部，占地面积200 093平方米，总建筑面积为91 110平方米。

山东省诸城市碧海云天花园及嘉福悦国际大酒店

项目位于诸城市北部，基地北临北环路，西侧为和平北街，东侧为潍河国家水利风景区，地块总用地面积100 000平方米，依托其东侧的潍河国家水利风景区优美的景观，以及周边区域行政中心和商业配套的发展，加上地块内部嘉福悦国际大酒店和商业的建设，使该住宅项目拥有较大的开发升值前景，周边拥有良好的景观资源，以及交通环境，各个配套设施相对完善。依托区域的发展，项目将建成新的商务CBD中心区，同时该居住用地，也最有机会占据周边同类高端市场的份额，成为高档时尚住宅的价值标杆。

包头市东德新城

项目位于包头市滨河新区核心区域，滨河新区政府东侧。基地北接腾飞大街，东临共青大道，南侧为秋实路，西侧为礼贤路。基地总面积为450 745平方米，规划设计为高层住宅及双拼叠加多层住宅区及其相关配套的沿街商业、学校及公园绿地，总建筑面积为565 662平方米，基地周边为住宅用地，周围还规划有商业、办公等配套设施，区位条件极佳。

休宁横江半塔地块住宅建设项目

项目位于安徽省黄山市休宁县，基地用地面积为135 839平方米，规划设计为高层住宅、小高层住宅、多层住宅、低层住宅及其相关小区配套的管理用房和沿街商业等。

总体设计通过对基地比较全面的分析与理解，以地块中部的原生态树林为空间核心，将开放的空间沿地形依次展开，并与之形成互动，共同构成一轴双环多组团的整体架构。

HSarchitects
上海汉思建筑设计事务所

地址：中国上海长宁路 1018 号
龙之梦大厦 901 室
邮编：200042
电话：86–21–33727290
传真：86–21–33727291
网址：www.HSarchitects.cn

Address: Room 901 No.1018 Changning Road, Cloud Nine Plaza, Shanghai, China
P. C: 200042
Tel: 86-21-33727290
Fax: 86-21-33727291
Http: //www.HSarchitects.cn

上海汉思建筑设计事务所是专业从事建筑室内外工程设计的合伙企业（简称 HSa），HSa 作为一支专业化的设计团队，长期致力于大型城市综合体项目的设计研究和实践，尤其在大型商业、超高层写字楼、豪华酒店和与公共交通设施相关的设计等方面积累了丰富的经验。设计服务范围涵盖了建筑工程设计、园林景观设计、室内装饰设计。事务所以突出的综合优势为客户提供兼具创新理念和商业价值的完美设计，创造美好的城市环境和空间景致。

HSa 具有建筑设计单项甲级资质（A131004549）。

HSarchitects (HSa) is a practice-led corporate of architects, focus on urban and architecture design. Being a dedicated group of professionals, HSa has gained renowned reputation and sound experience in large scale projects over long-term practice, mainly in shopping mall, high-rise office, luxury hotel and related public transport facilities. HSa's services range from architecture, interior and landscape design. We believe, with our broad and varied knowledge and experience, that innovative, sustainable design and commercial success could be achieved as a cohesive whole for both our clients and community, creating an excellent public space and improving the quality of urban environment.

HSa has Architecture Design Certificate Class 1(A131004549).

2

1

3

1 潍坊财富广场

Weifang Fortune Plaza

建设地点：山东 潍坊
建筑规模：250 000 平方米
建筑类型：大型城市综合体
设计日期：2004 年
建设阶段：完工项目

Construction Location: Weifang,Shandong
Construction Scale: 250,000 m^2
Type:Multi-used
Design Date: 2004
Project Status: Completion

2–3 莘庄凯德龙之梦购物广场

Xinzhuang Capitaland Cloud Nine Shopping Mall

建设地点：上海
建筑规模：190 000 平方米
建筑类型：大型城市综合体
设计日期：2006 年
建设阶段：在建项目

Construction Location: Shanghai
Construction Scale: 190,000 m^2
Type: Multi-used
Design Date: 2006
Project Status: Under construction

4

5

4–6 沈阳龙之梦亚太中心

建设地点：辽宁 沈阳
建筑规模：4 370 000 平方米
建筑类型：大型城市综合体
设计日期：2007 年
建设阶段：在建项目

Shenyang Asia Pacific Center

Construction Location: Shenyang, Liaoning
Construction Scale: 4,370,000m^2
Type: Multi-used
Design Date: 2007
Project Status: Under construction

6

7

7 沈阳铁西龙之梦购物中心

建设地点：辽宁 沈阳
建筑规模：300 000 平方米
建筑类型：大型城市综合体
设计日期：2009 年
建设阶段：在建项目

The longemont Shenyang Tiexi Shopping Mall

Construction Location: Shenyang,Liaoning
Construction Scale: 300,000 m^2
Type: Multi-used
Design Date: 2009
Project Status: Under construction

8

10 大同宾馆贵宾楼

建设地点：山西 大同
建筑规模：35 000 平方米
建筑类型：酒店
设计日期：2006—2008 年
建设阶段：完工项目

Daotong Hotel VIP Building

Construction Location:Datong,Shanxi
Construction Scale: 35,000 m^2
Type: Hotel
Design Date: 2006-2008
Project Status: Completion

8–9 宏泉丽笙酒店

建设地点：上海
建筑规模：25 000 平方米
建筑类型：酒店
设计日期：2006 年
建设阶段：完工项目

Hongquan Radisson Hotel

Construction Location: Shanghai
Construction Scale: 25,000 m^2
Type: Hotel
Design Date: 2006
Project Status: Completion

9

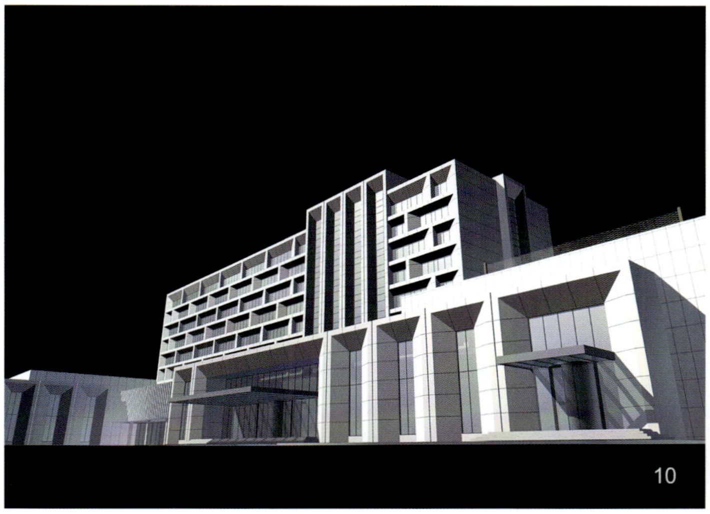
10

11

11 祥和星宇

建设地点：上海
建筑规模：300 000 平方米
建筑类型：居住建筑
设计日期：2003—2005 年
建设阶段：完工项目

Xianghe Stella

Construction Location: Shanghai
Construction Scale: 300,000 m^2
Type: Residential
Design Date: 2003-2005
Project Status: Completion

WEN KUN
上海文坤建筑设计有限公司
Shanghai Wenkun Architectural Design & Research Co.Ltd.

地址：上海市普陀区真北路 915 号绿洲中环商务中心 3 号楼 907 室
邮编：200333
法人：宋文玻
电话：+86–21–32513789
传真：+86–21–32513789
邮箱：sh_wenkun@sina.com

Address: Room 907, NO.3 of Lvzhou Zhouhuan Business Center, NO.915 of Zhenbei Road, Putuo District, Shanhai
P.C.: 200333
Legal: Wenbo Song
Tel: +86-21-32513789
Fax: +86-21-32513789
E-mail: sh_wenkun@sina.com

WENKUN 是一家富有前瞻性的建筑设计及城市规划设计公司，在商业设计领域处于国内领先地位，与国内顶尖的零售业巨头（茂业集团有限公司）等公司长期合作，对于国内商业地产的发展有独到的眼光和深入的研究。公司通过灵活多变的商业规划手段，来满足商业项目在不同区位、不同人文环境的需求，从而打破传统商业在地域方面的局限性。

WENKUN 除上海之外，在深圳、成都、合肥相继设立分支机构，各类设计人员约 100 人，是一家综合性的建筑方案和规划设计公司。

WENKUN 在多种建筑类型上均有实践，特别擅长于集超高层的办公、酒店，娱乐购物为一体的大型城市综合体类建筑，以及高端别墅类、高品质大型居住区、旅游休闲度假类等建筑规划及方案设计。

WENKUN 强调设计质量与服务并重，为客户提供最优的方案，与客户实现共赢。

WENKUN, whose business ranges from architectural design to urban planning, has conducted a prospective observation and profound research on development of domestic commercial real estate. It possesses the top position in sucha field as commercial design. Furthermore, it established long-term cooperation with Maoye Group Co.,Ltd. which is a retail giant in home. Through flexible design method, commercial projects which this company took over broke the limitation of terrain and met the requirements of different region and different cultural environment.

Except for Shanghai, WENKUN has founded the establishment of separate offices in Shenzhen, Chengdu, Hefei. WENKUN is a comprehensive architectural design and research company with 100 designers of various kinds.

With rich experiences on various kinds of architectures, WENKUN is especially good at overall planning of large and complex city with high-rise office buildings, hotels, supermarkets and entertainment settings, and architectural designs of high-end villa, large residence community with high quality and tourism setting.

WENKUN emphasizes high quality and good service, provides clients with best products and achieves win-win situation.

提供服务的企业

大型国有企业：淮南矿业集团；
川投置信房地产开发发展有限公司；
安徽省皖北煤电恒馨房地产开发有限公司。

商业地产领军企业：深圳茂业（集团）股份有限公司；
深国投房地产开发有限公司；
莱蒙置业有限公司。

国内地产领军企业：成都置信实业（集团）有限公司（成都本土最大和最具影响力的房地产开发商）；
沿海绿色家园发展有限公司；
常州新城宏业房地产有限公司（综合实力第一名，中国房地产百强）；
中天城投集团（贵州省最大房地产公司）；
华南国际工业原料城（深圳）有限公司（国内领先的大型综合物流及交易中心开发商和营运商）；
大中华国际实业有限公司。

获奖

成都・置信丽都・花园城 获 2009 中国土木工程詹天佑奖优秀小区金奖

捐赠项目

成都・绵阳小枧沟镇置信小学

项目

深圳——世界金融中心・大中华广场・城建观澜别墅区・东方时代广场
广州——珠江巨眼摩天轮・番禺商业中心
成都——置信丽都花园城（外侧）・置信丽都花园城（内侧）B 区
无锡——茂业城・夹城里 8 号地・太湖新城
常州——茂业泰富城・莱蒙世界之旅・武宜路 1、7 和 8 号地块
海口——仁邑国际社区
遵义——万里路商业中心・凉山片区平庄小区
安阳——彰德府文化商业步行街
合肥——松芝万象城・家天下花园・滨湖商业中心・新地中心
宿州——皖北煤电新居住园区
辽宁——海城裕通商业广场
天津——东方时代广场・海河一号
南京——百盛麒麟东郊小镇邻里中心
大连——经典生活
武汉——沿海商业中心

Companies which Wenkun provides services to

Large-scale state-owned enterprises: Huainan Mining Industry Group
Sichuan Investment Zhixin Real Estate Development Co.Ltd.
Anhui Wanbei Coal & Electricity Hengxin Real Estate Development Co.Ltd.

Leading commercial real estate enterprises: Shenzhen Maoye Group
Szitic Real Estate Development Co.Ltd.
Top Spring Real Estate Co.Ltd.

Domestic leading real estate enterprises: Chengdu Zhixin Industrial Group Co.Ltd.(the largest local real estate developer which has great influence in Chengdu).
Coastal Green Home Development Co.Ltd.
Changzhou Xinchenghongye Real Estate Development Co.Ltd(with the best comprehensive strength, ranking among the top 100 real estate companies)
Zhongtian Urban Development Group(the largest real estate company in Guiyang)
China South International Industrial Materials City(domestic leading large logestic and trading center)
Great China International Investment Group

Awards

Chengdu • Zhixin Lidu • Garden City won the golden medal of China Civil Engineering Zhantianyou Prize, 2009

Contribution Projects

Chengdu • Xiaojiangou Primary school in Mianyang

Projects

Shenzhen —World Financial Center, Great China International Exchange Square, Chengjian Guanlan villas, Oriental Times Square
Guangzhou —— Pearl River Eye in Zhujiang, Fanyu Commercial Center
Chengdu ——Zhixinlidu garden(outside& inside, part B)
Wuxi——Maoye town, No.8 plot of Jiachengli, Taihu New city.
Changzhou ——Maoye Taifu City. Laimeng Street Mall. NO.1, 7, 8 Blocks along Wuyi Road
Haikou ——Renyi International Community
Zunyi ——Wanlilu commercial center, Pingzhuang residence area in Liangshan district.
Anyang ——Zhangdefu culture and commercial walk street
Hefei ——Songzhiwanxiang city, Jiatianxia garden, binhu commercial center, Xindi center
Suzhou —— Anhui Wanbei coal & electricity group new residence quarter
Liaoning ——Haichengyutong Commercial Square
Tianjin ——Oriental times square.the No.1 of the Haihe River
Nanjing —— Baishengqilin Dongjiaotown neighborhood community
Dalian —— Classic Life
Wuhan —— Coastal Business Center

成都置信丽都花园城（外侧）

开发商：川投置信房地产开发有限责任公司
总用地面积：79 900 平方米
总建筑面积：219 300 平方米
容积率：2.46
荣获 2009 中国土木工程詹天佑奖优秀小区金奖

海口仁邑国际社区

开发商：海南联华星河投资有限公司
总用地面积：218 500 平方米
总建筑面积：519 200 平方米
容积率：2.0

在设计中，我们以"观念超前、以人为本、定位恰当、结合市场"为总的设计思想。

整体建筑采用了简洁、明朗的东南亚建筑风格。既传承了古典主义的庄重大气，又摒弃了繁复与浮华。形成与现代人的审美标准相契合的建筑风格，各种建筑语汇和细节的巧妙运用，更加展现了本案的高贵品质和典雅气质。

标准层平面

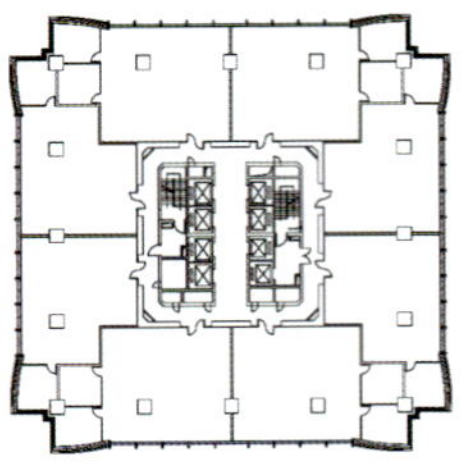

五十层平面

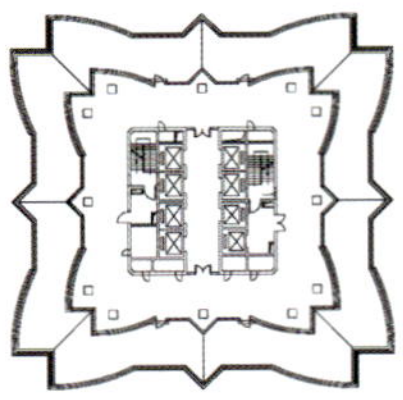

五十五层平面

五十六层平面

建筑平面的生成

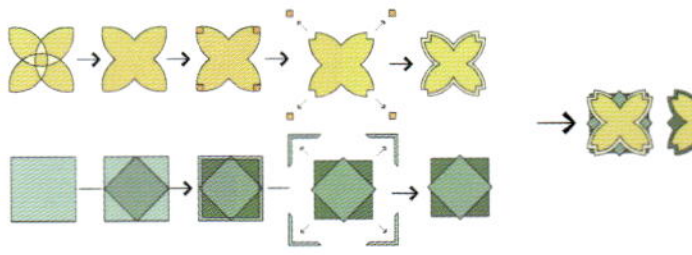

世界金融中心

建设地点：深圳 罗湖
开发商：深圳茂业（集团）股份有限公司
总建筑面积：146 800 平方米

金融中心的设计诠释了生长和收分形式在高层建筑的多重含义。其独特的顶部造型，借鉴了竹笋的生长形式和中国古代密檐塔的收分手法，像一根笔直的"竹笋"节节向上。基于上述的理念和手法，金融中心的造型采用了四个组合在一起的花瓣形体和横向水平窗格层叠向上的形式，在顶部花瓣逐台收分，最终融为一体。这四个花瓣形体与方柱体错位叠合而成为建筑的主体，在阳光角度变化时会产生丰富而微妙的色彩光影变化。两侧的弧形铝板中间的玻璃幕墙角部被切削出竖直凹槽，这不仅保持了自身的独立性而且使建筑更具动感。位于弧形铝板中间的玻璃幕墙采用竖向线条划分空间，夹在其间的是次一级韵律的横向水平窗格，而角部的弧线墙面将这种水平窗格转化为完全展开的横向层次划分，由此形成了楼身纵向与横向线条不断交织又突显出竖向线条的醒目变化。

无锡茂业城

开发商：深圳茂业（集团）股份有限公司
总用地面积：52 000 平方米
总建筑面积：57 000 平方米
容积率：6.5

本案力求形成融特色商业、餐饮、文化、休憩等功能于一体的传统商业中心，成为一个历史与未来、传统与现代相互交织的品牌街区。

常州茂业泰富城

开发商：深圳茂业（集团）股份有限公司
总用地面积：87 600 平方米
总建筑面积：311 400 平方米
容积率：3.0

在本规划中，商业及住宅地块形成双"L"形相咬合，住宅区充分利用了地块的自然景观资源。商业区则利用沿街面与道路转角，实现了客流的最大化和商业利润的最大化。公寓位于两地块中部，这种布局可使本建筑对周边建筑的影响减到最小。

建筑立面统一而富有变化，清晰有力的建筑轮廓，丰富而又协调的建筑表面，简洁生动的建筑语言以及建筑材料的穿插变化，使得本方案大气而又不失细节。现代建筑风格体现得淋漓精致，将使它成为常州市武进区今后的一大亮点。

遵义万里路商业中心

开发商：贵州中天置业房地产有限责任公司
总用地面积：117 000 平方米
总建筑面积：576 700 平方米
容积率：5.05

地块根据建筑功能规划成四种功能分区：广场、商业、住宅、公寓和酒店，四种功能相对完整却又相辅相成。

常州莱蒙都会

开发商：常州莱蒙都会置业有限公司
总用地面积：99 400 平方米
总建筑面积：557 500 平方米

该项目是常州最大的、以商业为主体的城市综合体，对中心城区的商贸业经济发展起着举足轻重的作用，一期工程的建筑群是具有繁荣商贸经济特色、具有新时代商业街景双重功效的标志性建筑。

常州武宜路 1、7 和 8 号地块

开发商：常州新城宏业房地产有限公司
规划总用地：177 800 平方米
总建筑面积：1 040 000 平方米
容积率：4.50

河南安阳彰德府文化商业步行街

开发商：安阳彰德府置业有限公司
总用地面积：43 700 平方米
总建筑面积：140 900 平方米
容积率：1.43

整体商业建筑大致可以分为以下几大版块龙头主力店、风情商业街、高级会所和停车楼。各个版块通过一条内街有机地连接在一起。

松芝・万象城 规划方案

开发商：义福房地产发展有限责任公司
总用地面积：33 900 平方米
总建筑面积：217 900 平方米
容积率：3.28

武汉沿海商业中心

开发商：沿海绿色家园发展有限公司
总用地面积：77 600 平方米
总建筑面积：196 400 平方米
容积率：1.99

都易国际设计与发展公司（DOT.INT）创办于香港，是一家专注于高端住区的专业设计公司。

2004 年，都易国际在上海成立了独立设计咨询机构——上海都易建筑设计有限公司，主要研究创作新古典主义高档住区，将国际先进设计理念与本土文化相结合，不断探索创新，打造精细化产品，为业主提供快速、全面的服务。

都易在新古典主义设计领域取得了全国领先地位，率先在杭州、南京、宁波、重庆、青岛、成都、武汉、郑州、福州等地成功打造了 ART DECO 文化概念的国际化高档社区，帮助客户取得了巨大的商业成功。

新 古 典 主 义 精 品 住 宅

DOT.INT 都易国际

CHANGE THE CITY

CULTURE · SPACE

www.dotint.com.cn

DOT.INT Design and Development Co., Ltd, established in Hong Kong, is a company which concentrated on specialized design of high-end residence communities. Shanghai DOT.INT Architectural Design Co., Ltd, an independent design company founded in Shanghai in 2004, mainly specializes in high-end residence communities design with neoclassicism style, sparing no effort to promote their fast and comprehensive services all-around as well as combining the advanced international design conception and localized perception.

The company obtains the national leading position in neoclassicism design, and creates internationalization high-end residence communities by ART DECO as the subject concept in places such as Hangzhou, Nanjing, Ningbo, Chongqing, Qingdao, Chengdu, Wuhan, Zhengzhou, Fuzhou for the first time, which brings the huge commercial success.

都易国际设计与发展有限公司（香港）
上 海 都 易 建 筑 设 计 有 限 公 司

香港九龙尖沙咀加拿芬道 20 号加拿芬广场 1006 室

Unit 1005, Carnarvon Plaza, 20 Carnarvon Rd.,Tsimshatsui, Kowloon, Hong Kong

T: 00852-27340415　　F: 00852-30117096

上海市中山西路 999 号华闻国际大厦 611 室 / 200051

Unit 610~611, Huawen International Building,
No.999, West Zhongshan Rd., Shanghai

T: 0086-21-32503750　　F: 0086-21-32503950

E: dotint@vip.163.com　　W: www.dotint.com.cn

【杭州 · 钱塘集团 · 梧桐公寓】

HANGZHOU/QIANTANG GROUP/PHOENIX TREE MANSION LA: 20,582m² / GFA: 68,546m²

所获荣誉

2006 年 中国国际花园社区
2006 年 首届杭州“好户型”评比大赛获奖楼盘
2006 年 杭州第六届人居展十大亮点楼盘
2006 年 杭州最值得期待十大楼盘

《82% 杭州人是通过梧桐公寓认识 Art Deco 这个英文单词的》
《杭州日报》 2008 年 7 月 3 日

梧桐公寓的出现，让 Art Deco 在杭州获得了广泛的时尚拥趸。据悉，目前三墩、丁桥、转塘、庆春路等杭州热点区域的楼盘，正在纷纷加入追随梧桐公寓的行列。一时间，Art Deco 在杭州房地产界风起云涌。

【宁波·浙大网新置业·ART 蓝湾】

NINGBO / INSIGMA GROUP/ART BLUE BAY

LA: 62,999m^2 / GFA: 110,247m^2

所获荣誉

中国（浙江）标志楼盘
IACE 2006 年 国际人居范例奖
2007 年 最值得期待楼盘
宁波首届城市规划杯优秀规划人居住宅小区

【宁波·浙大网新置业·ART 蓝海】

NINGBO / INSIGMA GROUP/ART BLUE SEA

LA: 46,478m^2 / GFA: 101,300m^2

DOT.INT 都易国际

【福州·融汇集团·融汇江山】

FUZHOU/RONG HUI GROUP/RONGHUI JIANGSHAN

LA: 32,054m²/ GFA: 117,000m²

用来居住的建筑应该是稳重的，而且是有气质的——高尚是一种生活态度，尊宅是一种居住高度。

建筑采用华贵坚实的外观设计，线条精炼简洁，峻拔中透出典雅稳重的气质，与其主人的坚韧沉稳形成呼应。进口石材干挂并结合高级面砖以及高档铝合金门窗等材质的运用，提升了高品质生活的健康与环保效果，更值得一提的是五星级酒店入户大堂与高档精装地下室大堂、双大堂的设计，既体现了尚品尊宅，又彰显出大宅风范。

【杭州·万科地产·金色家园】

HANGZHOU / VANKE GROUP / GOLDEN HOMESTEAD

LA: 55,576m² / GFA: 124,258m²

DOT.INT 都易国际

【重庆·协信地产·协信公馆】

CHONGQING / SINCERE GROUP / ABOVE THE WORLD

LA: 24,794m² / GFA: 170,986m²

DOT.INT 都易国际

以 ARTDECO 风格树立起城市地标建筑的新形象，是协信集团继阿卡迪亚国际标准住宅区之后在渝中区打造的又一顶级豪宅居住区，而协信公馆顶级会所建成后将成为重庆人的交流中心。

DOT.INT 都易国际

【南京·万科地产·金域缇香】

NANJING / VANKE GROUP / JINYUTIXIANG

LA: 48,937.5m^2 / GFA: 126,115.62m^2

【杭州·万科地产·西溪蝶园·II】

HANGZHOU /VENKE GROUP / XIXIDIEYUAN-II

LA: 105,212m^2 / GFA: 242,258m^2

DOT.INT 都易国际

【合肥 · 钱塘集团 · 金色梧桐 · II】

HANGZHOU / QIANTANG GROUP / GOLDEN PHOENIX TREE-II

LA: 18,647m² / GFA: 184,901m²

【成都 · 华润置地 · 二十四城 · IV】

CHENGDU / CRLAND / TWENTY-FOUR CITY-IV

LA: 38,864m² / GFA: 155,456m²

GN® 栖城
GN INTERNATIONAL DESIGN

GN 栖城简介

GN INTERNATIONAL DESIGN 国际性设计机构，成立于美国特拉华州，同时在中国拥有上海栖城建筑规划设计有限公司和上海栖城景观规划设计有限公司，并在长沙、昆明设有办事处，并拥有中国建设部颁发的建筑设计资质。

GN 在中国拥有由 100 多名来自世界各地的专业人士构成的强大设计团队。我们提供的服务涉及城市规划设计，建筑设计，环境设计，室内设计，机电设计顾问等专业领域。自 2003 年进入中国市场，GN 已经在 30 多个城市完成了数百项设计和顾问服务，并屡获国内外各类专业奖项。我们的客户包括房地产开发商、各类企事业单位、政府机关等部门。我们的设计作品涵盖居住类、商业办公、酒店度假、教育、公共市政等多个类别。

GN 有别于其他国际设计机构的地方在于，我们具有完美的专业技术、全新的设计理念、协调和创新能力，同时拥有完善的开放式管理体系，这些独到之处使得 GN 在快速发展的同时能够保持冷静的思考能力、卓越的设计能力和非凡的预见能力。

GN 在中国一直致力于当代建筑文化的复兴，将推进中国本土建筑的发展创新为使命，以 GN 对社会的思考和 GN 的设计作品为载体，像中国的建筑设计市场推行 GN 的文化价值观。

Established in Delaware (U.S.A), GN INTERNATIONAL DESIGN LLC is an international design team, which has gained the professional design qualification acknowledged by China Ministry of Construction. Its Chinese branches include Shanghai Architectural Planning and Design Company, and Shanghai Architectural Landscape Planning and Design Company, along with two other branch offices in Changsha and Kunming city respectively.

GN International Design LLC boasts a powerful team of more than 100 professional architects from all over the world. The service GN provided involves urban planning, architecture design, landscape design, interior design, consultation for equipment and electricity design, etc. Since entered into the Chinese market in 2003, GN has completed hundreds of design and consulting service for more than 30 cities, which has won various professional awards at home and aboard. GN's clients involve real estate developers, types of enterprises, government organizations, etc. GN's design works cover diversified fields including residential facilities, office buildings, commercial facilities, hotel resorts, municipal public facilities, and so forth.

Different from other international design companies, GN excels at absorbing, coordinating and innovative ability in techniques and concepts; meanwhile, it also possesses impeccable open management system, which ensures imperturbable reckoning, superior design and outstanding prediction no matter how rapid it develops.

Under the commission of promoting the local architectural innovation, GN has been devoting itself to contemporary architectural culture revival in China, and realizing culture and society value through its design works.

GN

上海西站交通大厦

GN 上海栖城 A

项目名称：上海西站交通大厦
项目地址：上海
建筑规模：6 435 平方米
设计时间：2009 年 12 月

项目简介：
项目基地位于西站交通枢纽的核心地段，南临上海西站南广场，北靠沪宁城际高铁及其站房。由于城市设施的功能需求，使得建筑内部需要布置地铁人行出口与地铁出风口。本案以动感的线性空间为设计主题，使其建筑形象与飞驰的城际高铁形成呼应。同时本案以雕塑感的立面形态，成为城市区域中可识别性地标。

Project Name: Shanghai West Railway Station Transportation Mansion
Construction Location: Shanghai
Construction Scale: 6,435 m^2
Design Time: 2009.12

Brief introduction:
The base locates at the core district of the West Railway Station's transportation hinge, facing the South Square of Shanghai West Railway Station in the south, close to HuNing Intercity High-speed Railway and its station building to the north. Subway pedestrian exports and subway air outlets should be assigned within the building. The design theme of this case being dynamic linear space, this building echoes with and is on a par with the galloping Intercity High-speed Railway from afar. Besides, the facade shape of the sculptural look in this case makes it an identifiable landmark of the specific site.

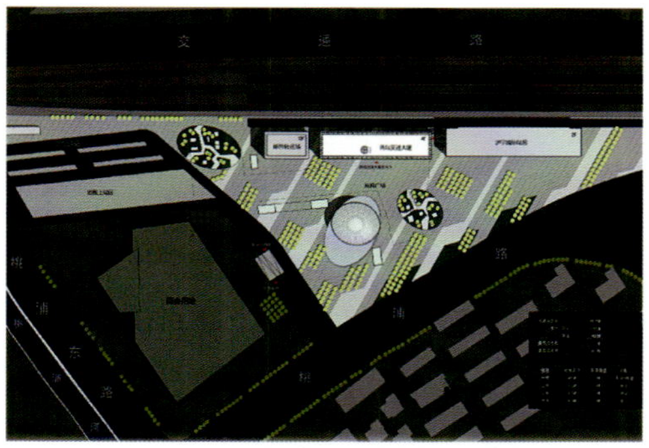

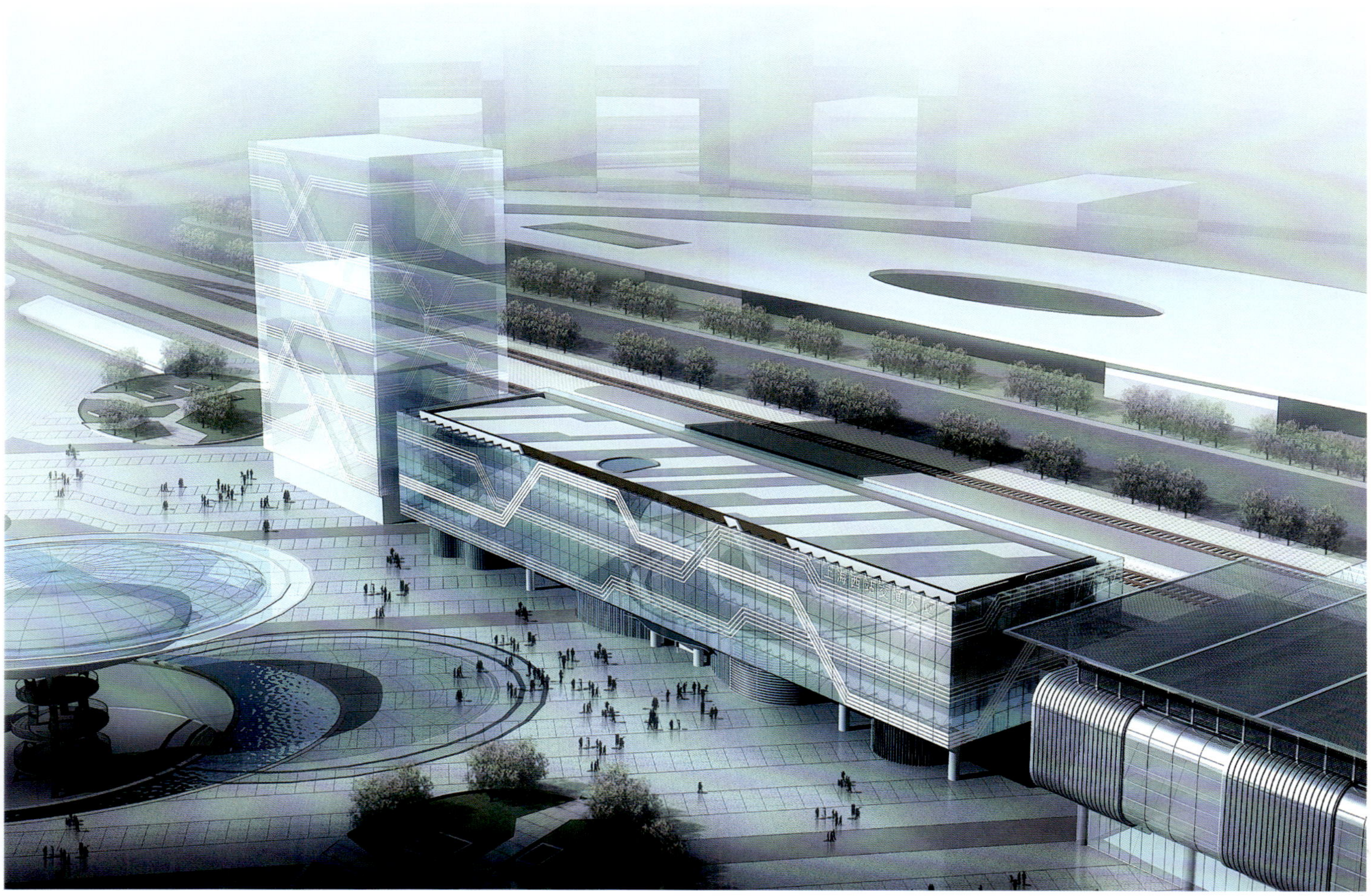

桐乡市文化中心

GN 上海栖城 B

项目名称：桐乡市文化中心
项目地址：浙江 桐乡
占地规模：42 706 平方米
建筑规模：36 507 平方米
设计时间：2008 年 7 月

项目简介：

桐乡市文化中心的设计从桐乡典型的江南水乡民居中汲取灵感，通过对“粉墙黛瓦”的丰富演绎，唤醒人们对地域文化和历史的共鸣。通过一些三角形连续屋面和不规则方窗的设计，以及 U 形玻璃墙材质的运用，使建筑体现出强烈的现代感并被赋予了新时代的建筑内涵。

运用颇具动感的平面布局形式，形成兼具时代感和中国传统精神的庭院空间，既适应于紧张的用地限制，又符合展览流线的需求，使多个不同主题和规模的建筑单体形成独立而又相互联系的建筑群体，每个建筑单体面向中央广场，均有很强的识别性和引导性。

Project: Tongxiang Cultural Centre
Construction Location: Tongxiang, Zhejiang
Site Area : 42,706 m^2
Construction Scale: 36,507 m^2
Time: 2008.07

Brief introduction:

Inspired by Tongxiang's typical Yangtze River Delta dwellings, through rich interpretation of "white walls and black tiles", this project rouses people's resonance about regionalism and sense of history. Applying some designation of triangle continuous roofs, irregular square windows and U-shape glass wall materials, it strongly represents the contemporary sense.

The application of the shape of a plane layout, which is considerably rich in modern fluency, forms the court space which is also rich in Chinese traditional spirit. It not only adapts to the site limit of competition, but also accords with the requirement of streamline. These make many single constructions of various themes and scales form architectural groups both diverse and united, and each single construction bears strong identification and directivity towards the central square.

桐乡振石大酒店

GN 上海栖城 C

项目名称： 桐乡振石大酒店
项目地址： 浙江 桐乡
占地规模： 30 681.70 平方米
建筑规模： 61 492 平方米

项目简介：

酒店主体建筑沿 320 国道和庆丰南路展开，呈"人"字形的带状平面布局，以使建筑获得更多城市界面与城市形象展示面。其中沿 320 国道为酒店主体楼和裙房，沿庆丰南路南侧为办公楼，酒店和办公楼通过一个连廊联系起来，形成一个城市综合体。酒店的主入口位于基地东北角的 320 国道和庆丰南路交叉口，酒店主体适当后退，形成一个气势恢宏的花园广场，并将其作为进入酒店的景观序曲。在西北角的 320 国道和西侧规划道路的交叉口，设置酒店的宴会厅和娱乐中心的出入口。宴会厅主要用于政府及大型会议接待，在形象上相对独立，出入口左右对称，庄重典雅。右则为娱乐中心，主要功能为 KTV、游泳池、健身房及洗浴中心。在西侧规划道路上还设置了酒店的员工和货物出入口，将出入口设于此处是由于考虑到这条道路较为隐蔽，对酒店干扰最小。在庆丰路上还设置一个会议和中餐厅的出入口，方便非住店顾客的使用。

基地东南侧为巨石办公楼，与主体酒店离开适当距离，拥有单独的出入口，同时又与连廊和酒店裙房相联系，在功能上实现一定的联系，在形象上又形成一个整体。

在基地西南侧较为僻静处设置几栋商务别墅，可以灵活经营、分期开发，提升了地块的整体开发价值。几组建筑共同围合出一个内庭院，这个内庭院为生活在都市的客人提供了一个僻静优雅的休闲场所，这种内庭园林对于嗓杂城市来说，可谓一个世外桃源，进一步提升了酒店的居住品质。项目的整体规划简洁合理，收放有致，充分体现了其作为国际五星级宾馆的气质，也充分考虑了桐乡地区的文化特征、建筑功能及风格特色。

交通的合理布置是本案的又一大特色。酒店的室外设置一定量的地上停车位，以满足停车要求，大部分停车位则位于地下车库内，这样就可以腾出大量的地面来布置景观绿化。宾客室外步行区主要集中在内庭花园，商务别墅区与办公楼人流也能顺利通达到花园中，实现景观共享。

基地的四个角分别布置酒店主入口、宴会娱乐入口、办公入口、商务别墅入口。车行交通主要在建筑外围解决。主体宾馆楼周围形成环状的消防通道，环绕酒店四周。酒店背后的道路为 2 米宽硬质铺装人行道路，道路外侧各两米宽度的范围内的地面按照可承载消防车压力的路面设计，平时则植草作为绿化景观，紧急时供消防车通行。

巨石桐乡酒店要求集会议、度假、餐饮、娱乐、休闲、养生和住宿为一体，打造豪华而富有时代感并秉承桐乡地区文化内涵的五星级酒店项目。依托桐乡丰富的自然和文化资源，建立一个休闲、度假、旅游等高生活品位的优越会议度假场所，酒店按五星级标准设计。

Project: Tongxiang Zhenshi Grand Hotel
Construction Location: Tongxiang, Zhejiang
Site Area: 30,681.70 m^2
Construction Scale: 61,492 m^2

Brief introduction:
The major building of the hotel spreads along 320 National Road and Qingfeng South Road in a "Y" ribbon pattern, in order to obtain more urban interface and urban image revealing face. Therein along 320 National Road are the major buildings and podiums of the hotels, and along the south side of Qingfeng South Road is the office building. Connected by a corridor, the hotel and the office building form an urban complex. The main entrance to the hotel locates at the northeast corner of the base, the cross of 320 National Road and Qingfeng South Road. The main body of the hotel draws back properly to form a garden square spraying imposing manner, which forms the ceremony and scenery prelude of the access into the hotel. The banquet hall and entertainment passageways are set at the cross of the 320 National Road and the west planning road in the northwest. The banquet hall is mainly applied for government and bull session reception. It is relatively separated in image, bilateral symmetry in entrance and exit and dignified and elegant in manner. On the right side is the entertaining center, with the main function of KTV, swimming pool, gym and bath center. Considering the west planning road is relatively covert and disturbs the hotel at the slightest, the entrance and exit of the hotel staff and goods are located on this road. A passageway for conferences and Chinese dining room is set on Qingfeng Road as well, in convenience for the non-in-hotel customers' application.
On the southeast side of the base are Jushi Office Buildings etc., which are separated from the main building of the hotel, having special passageways and is connected to the hotel and the annex with corridor. It is relatively related in function and forms a complex in image. Herein the image of the office building could as well be revealed to the outside.
In the quiet places on the southwest side of the base are some business villas which could be flexibly operated and developed by stages, and they promoted the integral developing value of the area. Several buildings jointly enclose an inner yard, which provides for the urban residents a nice landscape and a quiet and elegant leisure place. For the noisy city, such inner yard gardens could be regarded as a Xanadu, which further promotes the habitat quality of the hotel. The integral planning is concise and justice, nice in operating, which fully represents the disposition of an international five-star hotel and considers the cultural feature, function and stylish characteristics of Tongxiang district.
The proper arrangement of the traffic is another characteristic of this program. Several overground parking spaces are set outside the hotel; to meet the parking requirement, most parking spaces are located in the underground garage, and this could makes vast ground to lay out landscape greening. The guests' outdoor pedestrian district is mainly concentrated within the inner yard garden, by which the stream of people in the business villas and office buildings can also get through, and thus the landscape sharing can be realized.
To the satisfaction of one and all, on the four corners of the base locate respectively the main entrance to the hotel, entrance to banquet entertainment, entrance to office, entrance to business villas. The vehicle transportation is mainly settled on the periphery of the building. Circular fire fighting accesses are formed around the main buildings of the hotel, surrounding the hotel. Behind the hotel are 2-meter-wide rigid paved pedestrian roads. The ground within the range of 2 meters outside the roads, according to the pavement design that could bear the fire engine pressure, is used as greening with grass planted at ordinary times, and for fire engine's pass at need. Outdoor fire hydrants are set outside according to regulations.
Jushi Tongxiang Hotel demands to integrate conferences, holiday, repast, entertainment, regimen and lodging as a whole, and to forge a five-star hotel program both luxurious and rich in contemporary sense, and inheriting Tongxiang district culture. Relying on Tongxiang's rich natural and cultural resources, to build an advanced conference and holiday site for entertainment, holiday and tourism with high life quality, the hotel will be designed at five-star standard.

鹰潭市夏埠新区市民广场工程

GN 上海栖城 D

项目名称： 鹰潭市夏埠新区市民广场工程
项目地点： 江西 鹰潭
占地面积： 94 372 平方米
建筑面积： 163 860 平方米

设计理念：

夏埠中心区位于鹰潭城市的北部、信江北岸，与老城区隔江相望，具有十分优越的开发区位条件。依据上位规划，夏埠中心区定位为以文化展示、商务办公、商业金融、旅游服务和生活居住等功能为主导的鹰潭城市新的中心区，并将其作为城市跨江发展的主要启动区块，因此，夏埠中心区的开发建设对于强化鹰潭主城区的中心城市服务职能，提升城市文化品位，推进城市可持续发展都有着十分重要的意义。

以夏埠新区建设“生态之城、休闲胜地、宜居家园、现代都心”的战略目标为指引，规划将夏埠中心区建设成为集商务办公、酒店服务、零售商业、文化娱乐及生态居住等多元功能于一体，服务于鹰潭全市的现代化城市中心区。

Project: Xiafu New Area Civic Plaza Project in Yingtan
Construction Location: Yingtan, Jiangxi Province
Site Area: 94,372 m^2
Total Building Area: 163,860 m^2

Design concept:
Xiafu Central Area locates in the north of the city of Yingtan, on the north bank of Xinjiang River, facing the old city across the river, and it owns fairly advantageous conditions of developing zone. According to superordinate planning, Xiafu central area is positioned as new center of the city of Yingtan with the dominance of functions such as cultural representation, commercial office, business finance, tourism service and living residence etc., and as the main starting section of the city's cross-river development as well. Therefore, the development and construction of Xiafu Central Area is fundamental to the strengthening of Yingtan main urban zone's central city service function, the promotion of the city's cultural taste, and the impelling of the city's sustainable development.
Guided by the strategic goal of "ecological city, leisure resort, livable home, contemporary center" of Xiafu New Area's construction, the planning will construct Xiafu center area into a contemporary urban center combining the multiple functions of commercial office, hotel service, retail commerce, cultural entertainment and ecological residence etc. as a whole and serving the whole city of Yingtan.

Architectural

设计作品
Design Works

上海桑叶建筑设计咨询有限公司
Pdi International Co.,Ltd.

专注于商业地产及城市综合体的开发与研究
——Focus on development and research of commercial real estate and urban complex

Pdi是一家结合众多设计精英及全方位设计咨询理念的公司。

多年来，Pdi植根于中国，一直致力于探索中国各地区城市文明与地域文化的发展与形成，针对地域历史、民风、民俗等方面进行细致的分析和研究，运用国际化城市文明发展专业理念以及国内、外的发展实务经验，结合各区域城市状况，对代表城市文明建筑精品的“城市综合体”等商业建筑进行深度研究，将对城市的理解与希望融入城市建筑之中，使之成为能够充分代表城市历史、文化、发展、时尚的“精品”，从而提升建筑的生命力与活力，造就了每一个“城市生命”设计与实施的完美演绎，并积累了大量的实践经验。

Pdi专业于将战略思维、策略计划、实施设计、跟踪设计有机地整合在一起，针对不同地域的都市结构、不同的开发阶段，Pdi以不同的设计主题构想，为客户提供适时正确的服务要求，坚持从城市文明、地域文化、市场发展、建筑艺术及业主需求的角度出发，特别强调设计的“动态三宜”——即因时、因地、因人制宜。针对关键性问题提出解决策略和项目定位，并不断进行价值探讨，为业主创造更大价值的空间产品，保障整个开发案的有效发展，实施创造新城市文明建筑。

目前，我们已成功地完成了国内多个不同都市和处于各种经济发展阶段的大小城市的商业地产及都市复合型建筑的设计，并得到了客户的一致认可。

Pdi is the company which combines design elites from the Mainland, Taiwan and Hong Kong, and integrates all-around design consultation concepts.

Rooted in China for many years, Pdi has committed itself to exploring the development and formation of urban civilization and regional culture in China, and has made detailed accumulation on geographical history, folk customs and so on. It adopts professional concepts of international urban civilization development and practical experience from home and combines urban status in all regions to carry out in-depth research on "urban complex", which represents outstanding constructions of urban civilization, and other commercial buildings. It integrates understanding and hopes into the urban constructions, and makes them fully represent the urban history, culture, development, and fashion so as to enhance the vitality and vigor of construction, create the perfect combination of design and implementation for each "urban Life", and accumulate a large number of practical experience.

Focused on organically integrating strategic thinking, strategic planning, implementation design and tracking design, Pdi makes different design theme concepts and provides right timely service according to different regions and urban culture as well as various stages of development. It adheres to be based on the perspective of urban civilization, and regional culture, market development, architectural art and owners' demands, and particularly, it puts emphasis on the design of "dynamic three suits ", which includes "adjust measures to suit specified periods, tailored measures to suit local conditions and measures to suit different persons". It provides resolution strategy and project orientation for key issues, and continues to make value inquiry so as to create space products with greater value, ensure the implement of the effective development of the plan and create new urban civilization constructions.

At present, we have successfully completed the design for commercial real estate and urban complex buildings at various urban hierarchies and economic development stages, and have been unanimously ratified by customers.

地址：上海市长宁区新华路543号1号楼4楼F-G座
电话：+86-21-62817700
传真：+86-21-52300755
邮箱：pdi@pdi.net.cn

Address: Building F-G, 4 F, No.1 Building, Xinhua Road No.543, Changning District, Shanghai
Tel: +86-21-62817700
Fax: +86-21-52300755
E-mail: pdi@pdi.net.cn

金海岸国际商业城（C区）

建设地点：江苏 海安
发 展 商：江苏沿海房产开发有限公司
占地面积：86 428平方米
建筑面积：128 778平方米
容 积 率：1.49
绿 化 率：25.1%
设计时间：2008年

Gold Coast International Commercial City(C section)

Location :Haian,Jiangsu
Developer: Jiangsu coastal real estate developing co.,ltd.
Floor Area: 86,428 m^2
Building Area: 128,778 m^2
Plot Ratio: 1.49
Landscaping Ratio: 25.1%
Design Time : 2008

舞钢别墅规划

建设地点：河南 宁德
占地面积：18 000平方米
建筑面积：14 377平方米
容 积 率：0.78
建筑密度：22%
绿 化 率：30%
设计时间：2009 年

Planning of Villas in Wugang
Location: Ningde, Henan
Site Area: 18,000 m^2
Building Area: 14,377 m^2
Plot Ratio: 0.78
Building Density: 22%
Landscaping Ratio: 30%
Design Time: 2009

济宁太平洋广场（概念方案一）

用地面积：32.89公顷
建筑面积：726 597平方米
容 积 率：2.2

Jining Pacific Plaza (Conceptual Scheme one)
Site Area: 32.89 ha
Building Area: 726,597 m^2
Plot Ratio: 2.2

济宁太平洋广场（概念方案二）

用地面积：32.89公顷
建筑面积：726 039平方米
容 积 率：2.2

Jining Pacific Plaza (Conceptual Scheme two)
Site Area: 32.89 ha
Building Area: 726,039 m^2
Plot Ratio: 2.2

内蒙古巴彦淖尔市-直属局地块

用地面积：19 259平方米
建筑面积：143 623平方米
容 积 率：6.0
建筑密度：33.9%
绿 地 率：35%

South plot of Maryland Park, Bayan Nur

Site Area: 19,259 m^2
Building Area: 143,623 m^2
Plot Ratio: 6.0
Building Density: 33.9%
Landscaping Rate: 35%

商丘市行署地块概念方案

建设地点：河南 商丘
占地面积：183 895平方米
建筑面积：595 339平方米
容 积 率：3.24

Shangqiu administrative office land
Location: Shangqiu, Henan
Site Area: 183,895 m^2
Building Area: 595,339 m^2
Plot Ratio: 3.24

霞浦住宅概念规划

建设地点：福建 宁德
占地面积：89 160平方米
建筑面积：194 435平方米
容积率：2.18
建筑密度：30%
绿化率：35%
设计时间：2010年

Residence planning in Xiapu

Location: Ningde Fujian
Site Area: 89,160 m^2
Building Area: 194,435 m^2
Plot Ratio: 2.18
Building Density: 30%
Landscaping Ratio: 35%
Design Time: 2010

神垕钧都–新天地

建设地点：河南 禹州
发 展 商：禹州新天地建设开发有限公司
占地面积：10 123平方米
建筑面积：34 201平方米
容 积 率：1.49
建筑密度：40.4%
设计时间：2008年

Shenhoujundu-New World
Location: Yuzhou, Henan
Developer: New World Real Eastate Development Co.,Ltd. Yuzhou
Site Area: 10,123 m^2
Building Area: 34,201 m^2
Plot Ratio: 1.49
Building Density: 40.4%
Design Time: 2008

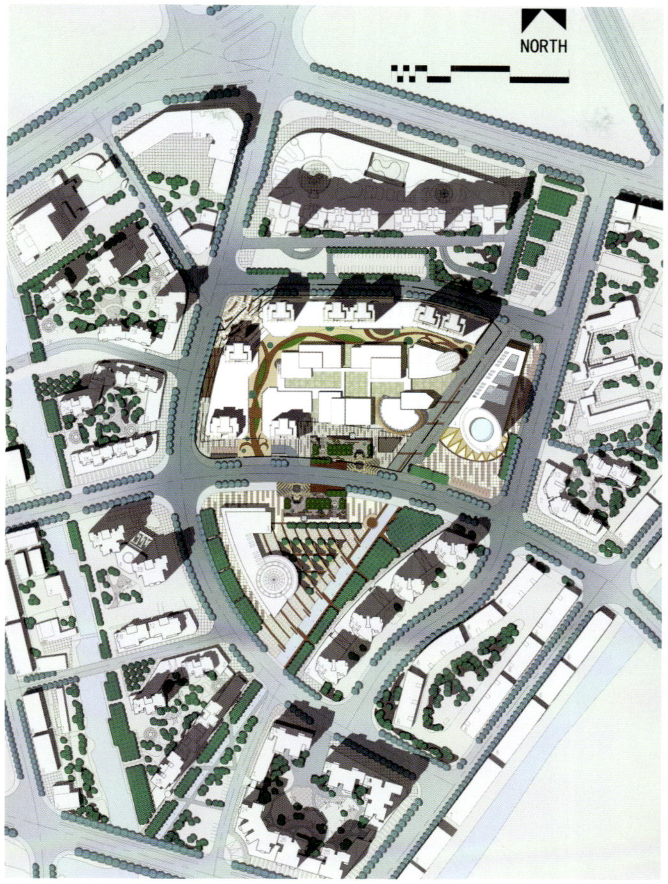

温州龙港商业中心

建设地点：浙江 温州
占地面积：28 355平方米
建筑面积：137 035.6平方米
容 积 率：3.28
绿 化 率：13%
设计时间：2008年

Longgang Commercial Center in Wenzhou

Location: Wenzhou, Zhejiang
Site Area: 28,355 m^2
Building Area: 137,035.6 m^2
Plot Ratio: 3.28
Landscaping Ratio: 13%
Design Time: 2008

蓝岳华府

建设地点：江苏 姜堰
占地面积：63 650平方米
建筑面积：190 950平方米
容 积 率：3
建筑密度：27.6%
绿 化 率：30%
设计时间：2010年

Lanyue mansion
Location: Jangyan, Jiangsu
Site Area: 63,650 m^2
Building Area: 190,950 m^2
Plot Ratio: 3
Building Density: 27.6%
Landscaping Ratio: 30%
Design Time: 2010

内蒙古巴彦淖尔市-美丽园南地块

用地面积：59 493平方米
建筑面积：246 670平方米
容 积 率：4.15
建筑密度：45%
绿 地 率：54.21%

South plot of Maryland Park, Bayan Nur

Site Area: 59,493 m^2
Building Area: 246,670 m^2
Plot Ratio: 4.15
Building Density: 45%
Landscaping Rate: 54.21%

MAP（建格）设计研究中心成立于 2004 年，专业领域涵盖城市规划、城市设计、建筑设计、景观设计，设计作品包括区域发展研究、城市总体规划、商业街区、办公综合体、中小学、居住社区等，各类项目广泛分布于上海、天津、山东、江苏、浙江、福建、陕西、江西、河南、安徽、山西、黑龙江等地的三十余座城市，受到包括政府机构和商业机构在内的各类业主的广泛好评。

MAP（建格）坚持原创精神与专业化服务，拥有多元技术背景组合而成的设计师团队与专业顾问群，熟悉国际先进经验、行业前沿理论与中国发展实际，注重将系统化的理念研究与具体的设计实践相结合，提供具有国际品质的设计成果。

MAP（建格）坚持整体分析和以人为本的设计理念，在特定的区域发展背景下，研究场地属性与潜质，研究人作为空间主体的活动特征，进而寻求具体问题的独特解决方式。

在全球化的浪潮中，MAP（建格）试图通过对中国社会经济发展、城市化进程和公众利益维护的综合思考，发掘城市文化的演进轨迹，探索城市生活的改良道路，实现当代设计师的社会角色。

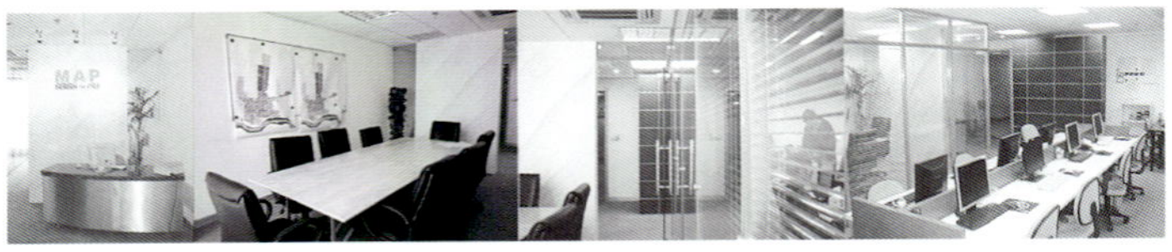

地址：上海市杨浦区大学路 243 号 504 室
邮编：200092
电话：+86–21–55153090
传真：+86–21–55153090–215
网址：www.map2004.com
邮箱：map_ga@163.com

白佛山景观大道

山东 东平
2010 年
总用地面积 214 300 平方米
在建

萌山公园

山东 嘉祥
2009 年
总用地面积 1 726 300 平方米
总建筑面积 125 800 平方米
在建

F 宅

福建 福清
2009 年
总用地面积 3 310 平方米
总建筑面积 2 064 平方米
在建

九龙湾公寓式办公楼

山东 威海
2009 年
总用地面积 9 500 平方米
总建筑面积 49 600 平方米

贵溪洋新城

福建 三明 | 2008 年 | 总用地面积 1 440 800 平方米 |
总建筑面积 1 735 900 平方米

青甸湖区域战略发展研究

浙江 绍兴
2006 年
总用地面积 34 平方千米

MAP Design Office was founded in Shanghai, China, in 2004. We provide professional services including Urban Planning, Urban Design, Architecture and Landscape Architecture with experience in Regional Development Study, Master Plan, Commercial, Office, Education and Residential Area. Our projects spread around more than 30 cities in China and are highly valued by clients including local governments and business companies.

MAP Design Office presents original-design works and professional services. Our professional working team combines multi-background designers and consultants who are familiar with advanced international theories and have extensive experience in China. We focus on both systematic research and design practice to provide international-quality planning and design.

MAP Design Office emphasizes site analysis and human-oriented design. In the specific regional context, we study the potential character of the site with people inside to find the unique approach to the specific issue.

In the wave of globalization, MAP Design Office keeps thinking about social and economic development, urbanization procedure and public interests in order to explore the evolution of Chinese city culture, find the road to improve the quality of city life, and establish the role of contemporary urban planners and architects.

建格設計
Modern Architecture & Planning design office

东平博物馆·东平文化发展中心

山东 · 东平 | 2009 年 | 总建筑面积 13 700 平方米（博物馆）21 800 平方米（文化发展中心）| 在建

这一次文化建筑的实践发生在一座具有悠久历史的城市，直白仿古的手法被摒弃，体量组合、材质对话成为了建筑形象的主角，变幻的光影和适时出现的文化符号成为建筑述说城市文化的理想语言。

建格設計
Modern Architecture & Planning Design Office

记忆 活力 影像
MEMORY VIGOR SCENE

隙
INTERLUDE

宋、郑之间有隙地焉。——《左传 哀公十二年》
蒐于农隙。——《国语 周语上》
三农之隙。——张衡《东京赋》
秋冬之隙。——苏轼《教战守》
隙地未尽辟。——洪亮吉《治平篇》

青浦新城中央公园

上海 青浦 | 2009 年 | 总用地面积 686 000 平方米 | 公开竞赛（入围）

身处拥挤的超级大都市，水乡的传统早就踪迹难觅，视觉层面的形式符号已不足以提供心灵的慰藉，所以，在这一次的尝试中，“隙”成为了主题：寻求时空与心灵的间隙，化解所谓的“中央公园”的中心指向，托起繁杂尘世中的绿色空间，用三维体验释放城市活力，用街巷空间唤起水乡记忆，而这些街巷融入城市路网后，又恰好修补了被汽车化尺度所破坏的城市肌理。

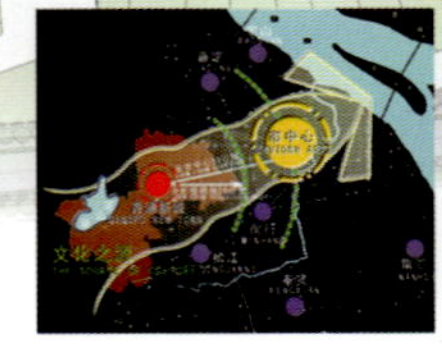
宏观区位 /CITY

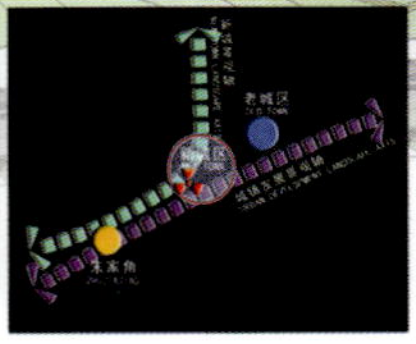
中观区位 /DISTRICT

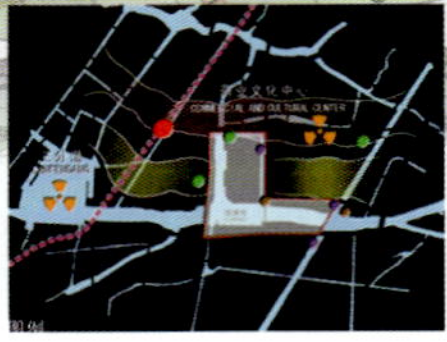
微观区位 /CENTRAL PARK

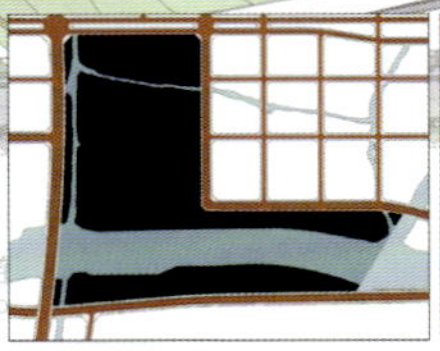
界定——现状网络
LIMIT-PRESENT GRID

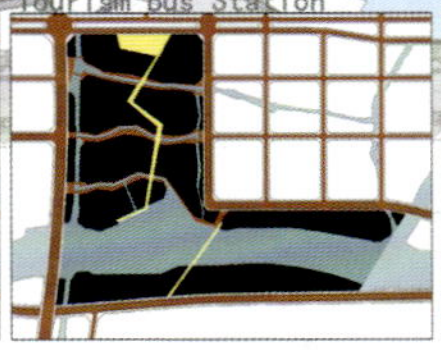
切割——肌理延续
CUT-TEXTTURE CONTINVITY

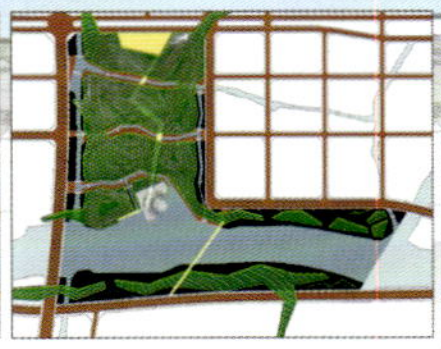
深化——功能完善
DEEPEN-FUNCTION PERFECTION

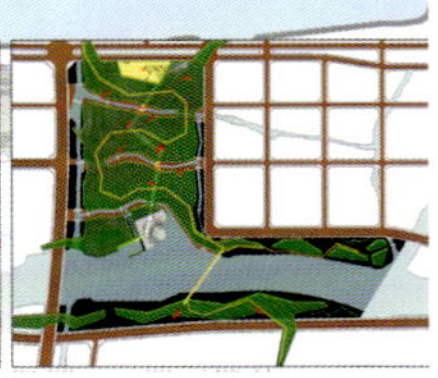
雕琢——品质提升
POLISH-QUALITY IMPROVEMENT

东平城市水景

山东 东平 | 2009 年 –2010 年 | 在建

清月湖

东源湖

东原路水景带

水街

荷花广场

龙王湾

MMA 建格設計
MODERN ARCHITECTURE & PLANNING DESIGN OFFICE

嘉祥一中新校区

山东 嘉祥 | 2008—2009 年 | 总用地面积 180 700 平方米 | 总建筑面积 112 300 平方米 | 在建

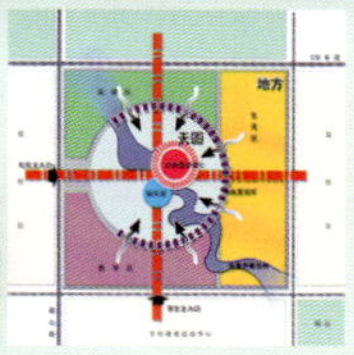

一座封闭管理的学校同时还是城市景观中轴东侧的重要界面、城市生态公园北侧的重要对景，有限的用地给设计带来了挑战，也为整体化的建筑群组合模式创造了可能，自然水系融入“天圆地方”的总体布局。严谨的空间轴网和完善的功能组合联结在一起，形成了校园设计的唯一答案。

金桥出口加工区 16 号街坊

上海 | 2009 年 | 总用地面积 43 500 平方米 | 总建筑面积 178 700 平方米

这一块地只有两面临路，却汇聚着会展中心、甲级写字楼、五星级酒店、总部办公楼四大功能体，整体性的空间组织模式，精心安排的交通流线和优美的滨河绿地成为本项目的突出亮点。

建格設計
MODERN ARCHITECTURE & PLANNING DESIGN OFFICE

上海景源建筑设计事务所

LEAD International Design L.T.D U.S.A

建筑甲级证书编号：090167–sj　Architecture Design Class A, Ref.No.090167-sj

LEAD 国际设计（集团）是成立于美国的专业规划设计公司．在全世界范围内，LEAD 可为客户提供城市规划设计、大型综合区及开发区的规划与设计、各类风景旅游区的规划设计、环境与保护、景观规划与设计、住宅区和商业区规划与设计、教育园区规划设计、各类建筑设计、室内设计以及房地产开发策划与咨询等方面的服务。

为进一步拓宽市场，2005 年 LEAD 国际设计（集团）与上海景源建筑设计事务所合作，成为由 2001 年中国国家建设部核准成立的，具有甲级资质的设计公司合作并授权其为在中国全权代理机构。

目前 LEAD– 景源已具有 60 余人的企业人员规模，由具有海外设计公司工作经历的海归人员以及具有博士学位、硕士学位的规划师、建筑师、景观设计师作为业务骨干，由一批富有职场经验的高级工程师、国家注册建筑师、注册结构工程师作为业务总监，并聘请学校和研究机构的知名学者作为顾问，力求为客户提供高标准、全方位的专业服务，旨在建立既有国际水准又符合本土规范要求的专业团队。

LEAD international design (group), a specialized planning and design company, was established in US. LEAD provides customers with urban planning and design, planning and the design for the large-scale mixed-use districts and development zones, planning and the design for various kinds of scenic spots, environment design and protection, planning and the design for landscapes, planning and the design for residential districts and business districts, planning and the design for campuses, different kinds of architectural design, indoor design as well as services such as offering planning guidance and technical advice on property development all over the world.

Approved by the Ministry of Construction, Shanghai Jingyuang Architectural Design Institute was established in 2001 and it is an Architectural Design Institute with a class-A certificate. To further expand the market, Shanghai Jingyuang Architectural Design Institute was authorized as LEAD's general agency in China in 2005.

At present LEAD- Jingyuan has 60 staffs all together. They are overseas returnees with rich overseas designing experience, planners and architects with doctorate and the master's degrees, as well as the backbone of business--landscape designers. The team is made up by senior engineers with rich work experience, registered architects, and registered structural engineers who are business directors. This institute, in the meanwhile, hires well-known scholars and experts from universities and research institutes as consultants. It makes effort to provide customers with high-standard and omni-directional professional service, and aims to establish a specialized team which is international professional and at the same time meets the domestic standards and requirements.

地址：上海市杨浦区荆州路 334 弄 108 号荣广商务中心 1 号楼 102、2 号楼
电话：+86–21–51807928 51807925 51807926 51807927
邮箱：jingyuan2002@online.sh.cn

Address: No.1 Building of Rongguang Commercial Center, 334 Lane of Jingzhou Road, Yangpu District, Shanghai
Tel: +86-21-51807928 / 51807925 / 51807926 / 51807927
E-mail: jingyuan2002@online.sh.cn

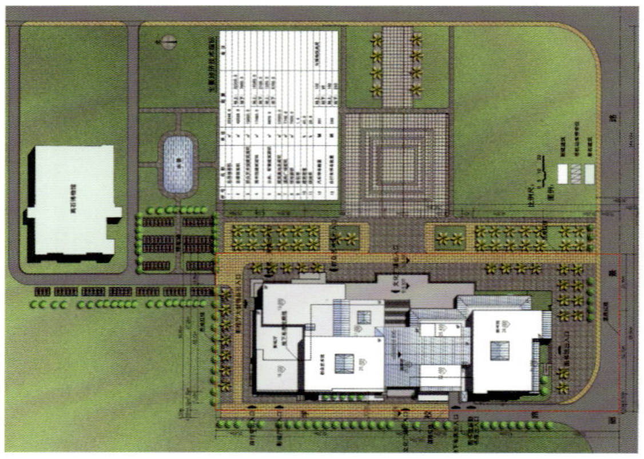

吕梁市文化广场工程

建设地点：山西 吕梁
用地面积：25 346 平方米
建筑面积：40 000 平方米

项目在总体构思上，充分利用地块的有利条件，努力创造一个年轻的社会共享空间，同时也体现了离石区建设功能齐全、形象突出且具有地方特色的文化广场的特点，力求将吕梁市文化广场建设成一个标志性的绿色建筑。

Luliang Culture Square Plaza

Construction Location: Luliang, Shanxi
Site Area: 25,346 m²
Building Area: 40,000 m²

This project makes full use of advantages of the plot to create young communal space in the general plotting. Meanwhile, this project aims to construct Luliang culture square a green iconic building and to embody concepts of buiding Lishi district into a place of complete-function and outstanding image and with with regional features.

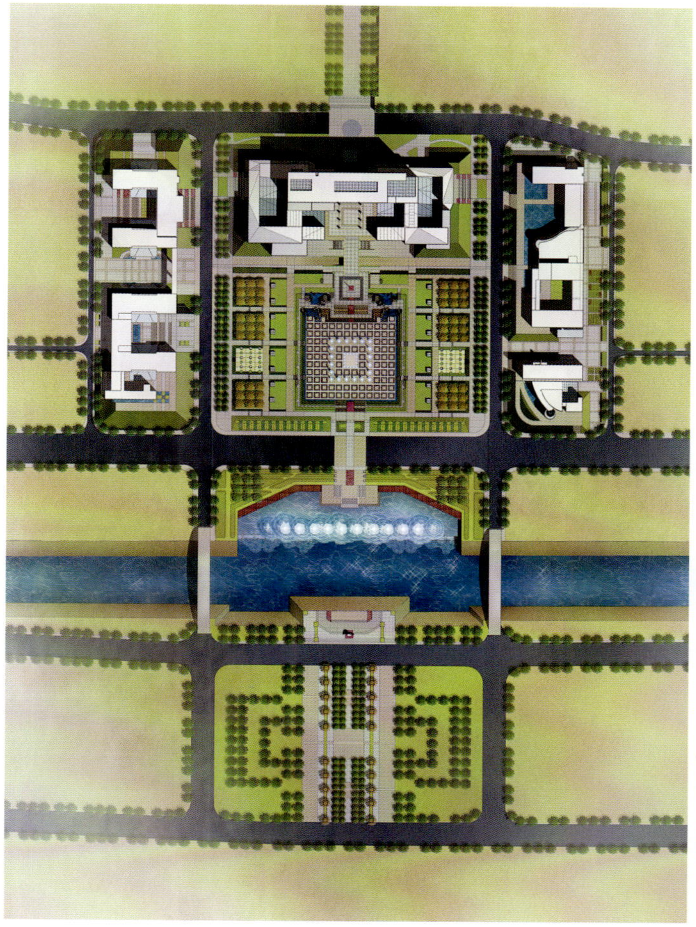

吕梁市离石区东城新区文化商务中心

建设地点：山西 吕梁
占地面积：12.2 公顷

项目地上总建筑面积约 12.7 万平方米，地下建筑面积约 3.3 万平方米，其中商务办公中心用地面积约 3.2 公顷，地上建筑面积约 5 万平方米，地下建筑面积约 1.3 万平方米。

规划定位

项目以“山、水、绿、城”四点元素的融合作为设计的最大特色、以现代化商务与文化功能为主体，组成城市核心区域，使项目成为绿树葱茏、水色掩映，区别于一般城市空间的独特场所。

Culture and commercial Center in New East City of Lishi District, Lvliang

Construction Location: Lvliang, Shanxi
Floor Area: 12.2 ha

Overground gross floor area is about 127,000 m². Underground gross floor area is about 33,000 m². Among them, land area of the business center is 3.2 ha. Overground floor area is about 50,000 m². And underground floor area is about 13,000 m².

Planning Orientation

The most significant feature of this project is integration of mountain, water, green vegetation and the city. Although modern business and cultural functions is a priority, the urban core, which is a different place from average city space, is surrounded by green water and verdant hills.

中国建筑西南设计研究院有限公司

CHINA SOUTHWEST ARCHITECTURAL DESIGN AND RESEARCH INSTITUTE CORP.LTD

中国建筑西南设计研究院有限公司始建于1950年，是中国同行业中成立时间最早的大型综合甲级建筑设计院之一，隶属世界500强企业中国建筑工程总公司。建院50多年来，本院设计完成了近万项工程设计任务，项目遍及我国各省、市、自治区及全球10多个国家和地区，是我国拥有独立涉外经营权并参与众多国外设计任务的大型建筑设计院之一。2004年以来，本院连续四年被亚洲建筑师协会评为“中国十大建筑设计公司”，并获得“全国工程勘察设计百强”企业称号。

目前，本院有各类专业技术人员1 500余人，其中教授级高级建筑师、教授级高级工程师40余人，高级建筑师、高级工程师260余人，国家一级注册建筑师、一级注册结构工程师、注册造价工程师、注册设备工程师、注册电气工程师300余人，培养出中国工程勘察设计大师4人、国家有突出贡献中青年专家1人，建设部有突出贡献中青年专家2人，四川省学术和技术带头人3人，四川省有突出贡献专家2人，享受政府特殊津贴23人，四川省工程设计大师11人。

作为中西部最大的建筑设计院和国家基本建设的重要国有骨干企业，本院以“精心设计，服务社会”为己任，坚持以繁荣建筑创作为宗旨，不断完善创新设计理念，力创建筑设计精品，在工程设计和科研方面获国家级、部级和省级以上优秀奖近500项，并取得了国家优秀设计金质奖四项、银质奖三项、铜质奖五项的佳绩。50多年的设计耕耘，西南院在博览文化建筑、体育建筑、医疗建筑、教育建筑、旅游建筑、居住建筑以及空间结构等设计领域具有独特的设计优势，而严格、规范的ISO 9001质量体系认证管理更使本院的设计质量为业界广泛认同。

在加强生产经营、科技创新和内部管理工作的同时，本院认真贯彻企业党建工作和企业文化建设的各项工作，提倡三个文明建设同步发展、共同进步。本院在改革发展中相继获得建设部“八五期间全国工程建设管理先进单位”、“精神文明建设先进单位”、全国优秀勘察设计院、四川省先进单位、“重点工程建设先进单位”、省级“最佳文明单位”以及“国家‘十五’期间科技进步先进集体”等荣誉称号。

我院的经营范围包括建筑行业建筑工程、人防工程设计及相应的咨询与技术服务，市政公用行业给水、排水、热力、桥梁、隧道、风景园林等工程设计及相应的咨询与技术服务，智能化建筑系统工程设计及相应的咨询与技术服务，商物粮行业、通信铁塔等工程设计及相应的咨询与技术服务，城市规划设计，城市设计，室内外装饰设计，建筑、公用工程科研试验项目，工程总承包及项目管理，工程监理，境外建筑工程的勘测、咨询、设计和监理项目，以及这些项目所需的设备、材料出口。本院可对外派遣上述项目勘测、咨询、设计和监理劳务人员。

Being established in 1950, China Southwest Architectural Design and Research Institute Corp.Ltd, as one of the large-scale, Class A and comprehensive architectural design institutes, is founded among the earliest in the same industry in China. It is subordinated to one of the Fortune Global 500--China State Construction Engineering Corporation. The institute has fulfilled almost 10 thousand design tasks during the 50 years or more since it was founded, which extend all over China and 10 other countries and regions . Boasting independent and foreign related operating authority, it has been awarded by Asia Institute of Architectures as one of "China's Top 10 Architectural Design Firms"for the successive three years since 2004. Moreover, it has been awarded as "one of the Best 100 National Reconnaissance Design Corporations".

At present, there are more than 1,500 professional and technical personnel in the institute, including over 40 senior architects and engineers with the rank of professor, over 260 senior architects and engineers, over 300 national Class 1 registered architects, structural engineers, cost plan engineers, facilities engineers and electrical engineers; 4 national engineering design masters,1 national expert with great contribution in middle and young age, 1 expert with great contribution in middle and young age awarded by ministry of construction, 3 provincial expert with great contribution in middle and young age, 2 provincial academic and technical experts, 23 experts who get special allowance from the government and 11 provincial engineering design masters.

As the biggest architectural design institute in middle west of China, the important and backbone state-owned enterprise in the country's capital construction, the institute takes "Elaborate design and service for society"as the task and persists in the aim of prospering architectural creation. We constantly improve design concepts, bring forth new ideas, and take efforts to produce fine works in architectural design. Thus, we have won over 500 items of national, departmental and provincial award in engineering design and scientific research, including 4 gold, 3 silver and 5 bronze awards of National Excellent Design. More than 50 years' efforts in design make the institute show unique design advantages in international fair culture architecture, sports buildings, medical treatment buildings, educational buildings, tourism buildings, residential buildings and spatial structure design and so on. And the strict and standardized authentication management of the ISO9001 quality system makes the design quality of the institute recognized by the whole industry.

Except for enhancing production and operation, technology innovation and development and interior management, the institute also thoroughly implements party building and spirit building, advocating the simultaneous development of "the Three Civilization Construction". Our institute bears fruits in its reform and development. It successively wins the honors of "Advanced Enterprise of National Engineering Construction and Management during the 8th Five-Year Plan", "the Advanced Enterprise in Spiritual Civilization", "Advanced Enterprise of Key Engineering Construction", "The Best Civilized Enterprise" of the Province, "National Excellent Reconnaissance Design Institute", "Provincial Advanced Enterprise" and "Advanced Group in Science and Technology advancement during the 10th Five-Year Plan".

地址：四川省成都市星辉西路8号　　邮编：610081
院办公室：
电话：+86-28-83233046　传真：+86-28-83231006
经营处：
电话：+86-28-83233031　传真：+86-28-83225229
客户服务：+86-28-83233315
邮箱：xnyyb@vip.163.com
网址：www.xnjz.com

Address: No.8 Xinghui West Rd., Chengdu, Sichuan　P.C.: 610081
Institute office:
Tel: +86-28-83233046 Fax: +86-28-83231006
Business office:
Tel: +86-28-83233031 Fax: +86-28-83225229
Customer Service: +86-28-83233315
E-mail: xnyyb@vip.163.com
Http: //www.xnjz.com

西部国际金融中心

总建筑面积：375 900平方米
完成设计时间：2008年3月
合作设计：法国安德鲁建筑事务所

采用适宜技术造就风格独特的办公建筑，其突出的特征在于花瓣形的曲线体形与树枝图案铝板构成的银色外表墙体。金融中心将成为成都南部副中心最富吸引力的标志性建筑群体。

Chengdu Western International Financial Center
Gross Building Area: 375,900 m^2
Completion Time: March 2008
Co-designer: Andrew Architects, France

The features of the office buildings applied with proper techniques and unique style are the petal shapes and the silver envelopes composed with branch patterned aluminum sheet. Financial Center will be the most attractive and melodious landmark in the vice city center in south Chengdu.

Chengdu Shuangliu International T2 Airport Terminal

Gross Building Area: 305,000 m^2
Completion Time: 2009
The terminal is located at the east side of the runway, connecting with the north end of the original terminal. Because the depth of the construction site is low, the "W"-shaped gallery layout is adopted. There are 23 close plan seats and six far plane seats, which maximize the use of plane parking space.
A wide station square is arranged at the landside of the terminal. There are spring and center green land in the center of the square and car parks with inter-phase greenbelts in other areas, forming a Green Square with smooth flow of traffic.
The hierarchical entry and exit layout is adopted in the new terminal. Through the overpass, the passengers are led to the departure hall on the second floor. The departure hall of 25-meter height is constructed with the hyperbolic space steel truss of 60 m net-cross, bright and spacious. The non-rod intersection line single-layer steel reticulated shells roof system used in the 33-meter wide corridor is a domestic initial technique, and the corridors are used as the waiting place for the departing passengers. Through the large French windows, the passengers can see the operation and movements of the aircraft, so that the anxiety of the waiting passengers can be relived. The first floor is used as the luggage hall and the arrival hall. After the arriving passengers pick up their baggage, they can directly go into the city by all types of vehicles under the viaduct, conveniently and quickly. The equipment rooms are arranged in the underground construction.
In the design of the new terminal, the structural system and the construction image, the image of the architecture and the interior space are organically integrated. By using the new technology steel roof system and the exposed steel structure used in interior ceiling and the outer hood, the rhythm of the new steel structure is full displayed, and combined with semi-hidden frame of the curtain wall, the stretching and unique, concise and generous, bright visual effect is created, forming a gentle and beautiful curve in the appearance. The construction image has a strong uniqueness.

成都双流国际机场T2航站楼

总建筑面积：305 000平方米
完成设计时间：2009年

航站楼位于跑道东侧，与原有航站楼北端相连。根据建设场地进深较浅的状况，采取了“山”字形的走廊式布局，设有23个近机位和6个远机位，最大限度地利用了场地停机。

在航站楼陆侧布置了宽阔的站前广场，广场中部设有旱喷和中心绿地，其余为以绿带相间的停车场，形成流线通畅的绿化交通广场。

新航站楼采用进出港分层布局，由高架桥将旅客引入二层出港大厅，净高25米的大厅采用60米净跨的双曲空间钢桁架，使空间显得气度恢弘、宽敞明亮。33米宽的指廊采用国内首创的无拉杆相贯线单层钢网壳屋盖体系，作为出港旅客候机使用。透过落地大玻璃窗可以看到飞机的起降和运作，减缓了旅客候机的焦急情绪。底层为行李提取厅、进港迎客厅，进港旅客提取行李后能直接通往高架桥下的各类车辆进入市区，疏散快捷。设备机房均布置在地下层。

新航站楼在设计中将结构体系和建筑形象有机地结合在一起，将建筑造型和内部空间有机地结合在一起。建筑设计运用新技术的屋盖钢结构体系，室内天棚和外挑出沿均采取外露钢结构的手法，充分展示新型钢结构的韵律美，结合外墙半隐框幕墙，创造出舒展别致、简洁大方、内外通透的视觉效果，形成了外观曲线柔和优美，具有强烈识别性的建筑形象。

重庆江北国际机场

总建筑面积：约75万平方米

重庆江北国际机场规划定位为亚洲领先、世界一流的大型商业门户中转枢纽——中国第四大复合枢纽机场。作为重庆市对外交往和扩大开放的重要城市窗口，重庆江北国际机场的建设喻示重庆经济文化等各方面的腾飞。设计方案融合"两江汇流"与"比翼神鸟"寓意。"两江汇流"的构思赋予航站楼流畅的构型曲线，强调江水流动意向，也凸显了重庆对多元文化的兼容并蓄；"比翼双飞"延续自巴人的神话传说，航站楼建筑造型从来自两侧走廊的曲面向中央汇聚，强调比翼相连、振翅欲飞的建筑形态，寓意城市和谐发展，走向世界的美好愿望。多种交通形式高度集成，凸显了商业门户中转枢纽机场的特征。贯彻设计始终的可持续发展观，无论在总体规划、航站楼设计，还是景观造型上处处得到了体现。

国资委《央企要情》以"中国建筑扭转由外方主创大型国际空港设计方案局面"为题对本项目进行了报道，高度肯定了本项目中标所取得的非凡意义。

Chongqing Jiangbei International Airport

Gross Building Area: about 750,000 m^2

The plan of Chongqing Jiangbei International Airport is positioned as the transit hub of Asia's leading and world class large commercial portal as well as China's fourth largest complex hub airport. As an important city window for foreign exchange and expanding opening-up of Chongqing City, its construction indicates the rapid development of economy and culture of Chongqing. The design planning mixes the allegory of "two streams confluences" and "a pair of magic birds": the idea of "two streams confluences" gives the terminal smooth configuration curve, emphasize the meaning of the river flow and highlight the full inclusiveness of multiculturalism of Chongqing; "fly wing to wing" lasts the myths of Ba People. The terminal architectural styles are designed in this way: it converges to the center from the curved surface of the corridor of both sides and emphasizes the architectural form of the association of two wings and the flap of the wings, which means the harmonious development of the city and good wishes of opening to the world. The high degree of integration of various traffic patterns highlights the characteristic of commercial portal transit hub airport. The sustainable development, which runs through the whole design process, is reflected from the general plan, terminal design to the landscape modeling.

"Important issues of central enterprises"of SASAC reported the project on the title of "China architecture reverses the tide of design planning of large international air harbor sponsored by foreign side", and highly appreciated the great significance achieved by the bid winning of this project.

成都地铁2号线

春熙路站是地铁2号线和3号线的换乘站，地下建筑面积为24 384平方米，车站总投资约2.2亿。地铁站与红星路综合楼、九龙仓地块等周边物业相连，春熙路站是成都地铁目前为止设计、施工最为复杂的地铁站。未来将成为成都市交通枢纽，是成都跨向繁华都市的桥梁。

东门大桥站位于东门大桥东侧，为地下3层车站，总建筑面积为13 081平方米，投资约1.2亿。该项目与天府时代广场共同设计、开发，力求打造成都市又一地标性建筑，东门大桥站是快速与奢华的结合点，是成都未来的又一购物天堂。

Chengdu Metro Line 2

Chunxi Road Station is the transfer station of Metro Line 2 and Line 3. The underground floor area is 24,384 m^2 and the total investment on this station is about 220 million. This subway station connects with the neighboring estates, such as the integrated building of Red Star Road and the Wharf lot. So far, Chunxi Road Station is the most complex station in Chengdu Metro System in terms of the design and construction. In future it will become the transportation hub of Chengdu, forming a bridge towards the urban prosperity. East Gate Bridge Station is located in the east side of East Gate Bridge. It is a 3-storey underground station, with the gross floor area of 13,081 m^2 and total investment of about 120 million. This station is designed and developed together with Tianfu Time Square, intended to present another landscape building in Chengdu. East Gate Bridge station is a joint point of high speed and luxury, another shopping paradise for the future Chengdu.

常州体育会展中心

总建筑面积：195 198平方米
完成设计时间：2006年11月
2007年中国建筑工程总公司优秀方案设计一等奖

本工程包括3.5万人体育场、6 000座体育馆及1 000个展位的会展中心、2 000座游泳跳水馆和火炬观光塔。设计充分考虑常州市的地域特征，采用大面积浅水池和斜坡绿化，营造浓郁的江南水乡意境。场、馆结合的设计理念，使体育场与游泳跳水馆如漂浮在清水之上的玉兰花。体育馆与会展中心如涟漪之上的花蕾。结构采用了平面管桁架及索承单层网结构等新技术、新工艺。新颖的设计构思，受到了当地领导、群众以及国内专家的高度评价。

Exposition Center of Changzhou Sports Center

Gross Building Area: 195,198 m^2

Completion Time: 2006.11

This project includes a stadium with an audience capacity of 3.5 million, a gymnasium with 6,000 seats, an exhibition center with 1,000 booths, a natatorium with 2,000 seats and a torch sightseeing tower. The design takes into full consideration the geographical characteristics of Changzhou and adopts large areas of shallow pool and green slopes, creating a rich atmosphere of Jiangnan water town. The design concept of a combination of the stadium and the gymnasium makes the stadium and the natatorium look like Magnolia flowers floating above the crystal water, while the gymnasium and the exhibition center are like the buds drifting on the ripple. The structure applies the new technology such as the plane steel-pipe truss and the cable-supported single-layer lattice shell. This novelty of the design concept has received positive evaluation from the local leaders and people, as well as the experts at home and abroad.

重庆市袁家岗体育中心

袁家岗体育中心位于重庆市高新技术开发区，建设规模为60 000座灯光比赛场，建筑面积为62 508平方米，作为2004年亚洲杯足球赛主赛场之一，它是重庆市新世纪大型城市基础设施重点项目。

重庆袁家岗游泳馆、跳水馆位于重庆袁家岗体育中心内，是重庆市新世纪直辖以来的大型城市基础设施项目之一，能承接国际、国内的旅游、跳水、水球、花样旅游和蹼泳等比赛。

Chongqing Yuanjiagang Sports Center

Yuanjianggang Sports Center is located in high and new technology industrial development zones, Chongqing. There are 60,000 lighting competition fields with the construction area of 62,508 m². As the main competition venue of the Asian Football Cup in 2004, it is a major urban infrastructure project for the new century large-scale city. Yuanjianggang Swimming Pool and Diving Hall is located in Yuanjianggang Sports Center. It is one of the large-scale urban infrastructure projects since Chongqing became the municipality directly under the Central Government in the new century. It can undertake the international and domestic tourism, diving, water polo, synchronized swimming, fin swimming and so on.

巴基斯坦人马座

业主名称：巴基斯坦PakGulf公司
建设地点：巴基斯坦 伊斯兰堡
总建筑面积：32 000平方米

本工程位于巴基斯坦首都伊斯兰堡，地处一个具有极其重要商业价值的地段。它由1栋7星级豪华酒店、2栋精装公寓、1栋办公楼和5层购物中心组成。波浪形购物中心的屋顶是连接工程的主要构成部分，反映出城市与喜马拉雅山脉的联系，形成一个辨识度高的外形，同时它的规模和尺寸使它像迪拜的帆船酒店一样，在建成后成为伊斯兰堡的一个关键的国际标志建筑物。

The Centanurus Project, Pakistan

Owners: Pakistan PakGulf
Construction Location: Islamabad, Pakistan
Gross Building Area: 32,000 m^2
Project Overview: This project is located in a block of important commercial value in the capital of Pakistani—Islamabad. It consists of one 7-star luxury hotel, two deluxe apartments, one office building and a 5-storey shopping mall. The wave-shaped roof of the shopping center is the main component for connecting the project, reflecting the background of the Himalayas and forming a clear signature shape. Meanwhile its scale and dimension render the building similar to the Dubai's BurjAl-Arab Hotel. After completion this project will be a key international landmark in Islamabad.

利比亚班加西住宅项目

总建筑面积：1 000 000平方米

建筑师根据规划区所处的位置和对班加西城市传统居住区布局的调查与分析，探讨伊斯兰教城市空间形态产生的素内方外圆的要素，以及中心放射的空间肌理特征。建筑师通过借鉴理想城市设计理念。项目强调公共设施的多元化以及与自然的有机契合，建立起规则与自由形态共生的设计理念。项目中心区围绕居住区，公共空间采用规则的路网，外围居住区则是自由的布局形态，形成外曲内直的规划结构，强调与自然环境及心理需求的结合、规则网络与自由曲线的结合。

Residential project in Benghazi, Liby

Gross Building Area: 1,000,000 m^2
According to the location of the planning area and the research & analysis of the layout of conventional residence sections in Benghazi City, they probe into the spatial tissue characteristic of elements inside and outside the circle and center radiation generated by Islamic urban spatial morphology, and borrow ideas from the design idea of the ideal city to emphasize the diversification of public facilities and its organic juncture with nature. The design idea of the co-existence of rules and free form is established. The architects use the regular network in the central area around the public space of the residential area, while in the peripheral residential quarter the free arrangement form is applied. It forms the planning structure of curved outside and vertical inside. It emphasizes the integration with the natural environment and psychological need, and the integration of regular network and free curve.

西藏博物馆

建设地点：西藏 拉萨
建筑面积：15 000平方米
占地面积：20 000平方米
设计时间：1997年
竣工时间：1999年

馆址位于拉萨罗布林卡（夏宫）东门外附近，北望布达拉宫。原址内多高大杨、柳树，建成后均被保留下来，景观环境较好。

建筑的性质和特殊的地理位置，决定了建筑的造型必须体现西藏独特的民族文化和地域特色，大体量的块状组合、大片收分石墙（均采用当地石材），平顶女儿墙棕色饰带曲尺变形、仿藏式琉璃瓦坡顶的举架出檐组合，以及斗拱、柱、枋和门、窗等细部形式，都借鉴了藏族传统建筑的手法。立面采用左右对称构图，使建筑下简上繁，突出了建筑中部。在出挑的栏板外绘制藏式白底兰花图案，中央为江泽民题写的馆名金字，庄重而凝练，也体现了西藏独特的族风貌，同时，寓改造于借鉴之中，排除西藏传统建筑中常有的宗教气息，对传统形式加以变形和简化，使建筑不乏时代气息，作为博览建筑本身，它也成为一个重要的展品。室内设计也以体现藏文化为主。

Tibet Museum

Construction Location: Lhasa, Tibet
Building Area: 15,000 m^2
Floor Area: 20,000 m^2
Design Time: 1997
Completion Time: 1999
Design overview and main characteristics:
the museum is located nearby the outside of the East Gate of Lhasa Nor-bu gling-ka (Summer Palace), overlooking the Potala Palace in the north. There are lots of tall poplars and willows in-situ, which are retained after the completion for the environment consideration.
The nature and the special geographical location of the construction determine the model to reflect the unique culture of Tibet and its geographical characteristics. with the massive amount of block combinations, and large slice of stone wall (using the local stone), the flat-topped parapet deformation brown colt disppharynx and Tibetan imitation glazed tile over the brim canopies portfolio as well as the brackets, columns, piles and doors, windows and other detail forms, all borrow from the traditional Tibetan way of architecture. The facade uses symmetrical composition, below simple while above complicated, highlighting the central, in the outer steeplechase drawing the Tibetan white background with orchids patterns, and in the centre it is the golden museum name inscribed by Former President Jiang which is solemn and concise, also embodies the Tibetan style. In addition, containing transformation in the drawing, the exclusion of traditional Tibetan buildings religious style, to deform and simplify the traditional forms, so that there is adequate in flavor of the times in the architecture, as the Expo building itself , it has become an important exhibits too. And the interior design aims to mainly reflect the Tibetan culture as well.

昆明春城财富中心

总用地面积：44.78公顷
净用地面积：33.97公顷
总建筑面积：170公顷

春城财富中心项目用地位于呈贡城市中心区南北中央干道彩云南路西端。春城财富中心将购物、餐饮、娱乐、休闲、旅游、社交、商务等功能集于一体，包括28.5万平方米的零售商业、8万平方米的商务办公、3.6万平方米的SOHO公寓、85万平方米的高档住宅等功能，项目目标力求将其打造昆明新城的中央商业商务区。

Spring-city Fortune Center

Gross Land Area: 44.78 ha
Net Land Area: 33.97 ha
Gross Floor Area: 170 ha
The project site of Spring-city Fortune Center is located at the west end of Caiyun South Road—the south-to-north central road of Chenggong urban center. This center integrates functions such as the shopping, dining, entertainment, leisure, tourism, sociality and business into a whole. It encompasses a retail business area of 285,000 m^2, a commercial office area of 80,000 m^2, SOHO apartments of 36,000 square meters and the high-end residential quarter of 850,000 m^2. The design of this project strives to build a central commercial district for Kunming New Town.

基准方中

建筑设计事务所

JZFZ ARCHITECTURAL DESIGN ASSOCIATES

成都 Chengdu
西玉龙街6号
新世纪广场大厦12层
邮政编码：610017
联系电话：028–86582121

重庆 Chongqing
高新区新南路164号
水晶国际大厦10层
邮政编码：401147
联系电话：023–67039777

西安 Xi'an
南关正街88号
长安国际中心A座10层
邮政编码：710068
联系电话：029–87651710

执行合伙人：
钟明 ZhongMing 先生

电子邮件：
yw@jzfz.com.cn
网络主页：
www.jzfz.com.cn

基准方中建筑设计事务所是以创建领先的行业模式为基本目标的设计团队，追求设计与市场、设计与社会的和谐共生，目的是站在可持续发展的高度，创造具有持久价值的设计产品（作品）。"创新、技术、服务"是基准方中的设计理念。基准方中要求其建筑师应以合格的、国际化的职业建筑师为追求目标，弱化团队中的个人英雄主义，在设计合格的优质建筑产品的同时，承担更多的社会责任与历史责任。

基准方中视创新为企业生存和发展的根源和动力，不断地以产品设计生产过程的服务创新、管理创新和技术创新来激发和推动产品创新。同时，基准方中清楚地认识到，优秀的建筑设计企业必然是制度创新、管理创新、技术创新和文化创新发展的实践者、推动者和领先者，也必然是行业产品质量标准的制订者和领导者。事务所在充分吸取和学习国内外成功的大型综合建筑设计企业在制度、管理和技术标准建立方面的成功经验和科学方法的基础上，结合国内外建筑设计业在技术、市场和文化等方面的现实和发展趋势，有目的、有计划地逐步建立市场经济条件下服务于现代技术服务型企业的经营理念、管理制度和企业文化，在未来两到三年内初步建立具有国内领先水平的设计各阶段和设计成果的专业和项目设计全过程的综合技术标准。

基准方中坚定不移地在项目研究、咨询、设计、管理的全过程贯彻以人为本的设计、宏观和微观优化的设计、空间使用品质性价比更高的设计、充满历史和文化意味的设计、项目各相关专业按照产品市场目标协调一致，进而创造出更高价值的设计产品，并在此过程中为客户、社会和相关服务商提供积极健康、高效愉悦的服务体验。

事务所成立七年多以来，在良好的经济和行业发展背景下，取得了快速的发展。员工人数已从2002年的不到30人发展到600余人。专业配备完整，专业化水平、综合服务水平与能力不断提升，基准方中基于更好地服务于市场和客户目标，其制度化建设和产品标准的重新建立已初步完成。

2008年8月和2009年7月，基准方中重庆事务所和西安事务所先后投入了运营，标志着基准方中走向全国，在更广阔和更具挑战的舞台寻求进一步发展。事务所计划在未来3～5年中于全国各省市，特别是中西部各省会城市开设若干区域事务所，使基准方中成长为以全国为市场的行业领先企业。

基准方中长期坚持从国内各大知名高校毕业生中选拔、招聘优秀学生，以充实员工队伍。通过持续进行全方面的技术培训和职业道德教育，初步形成了事务所比较完善的人才梯队。同时，基准方中一直重视并积极吸纳行业内具备丰富技术经验和良好职业道德，认同基准方中企业价值观和设计服务理念的现代高素质技术人才的加盟。

基准方中的中、长期发展目标是成为具有高度社会责任感和历史责任感、市场化的的企业，以创造具有持久价值的建设产品为根本目的，并成为具有强大研究、咨询、设计和管理能力，服务于建设的大型综合技术服务公司。

JZFZ Architectural Design Assosiates was established in Chengdu Nov. 2002. and Class A Engineering Design Certificate by the Ministry of Construction of P.R.China.
With a efficient, pleasant service experience, JZFZ firmly provides designs based on human needs, sustainability, economy, quality, history and culture. Each technical department contributes to make more innovative values by corresponding to the market goal.
As a long term aspiring, we aim to be a large and integrative technical service company with mighty abilities of researching, consultation, designing and managing. We take the creation of architectural products as our basic goal and serve the construction with high sense of socialhistorical responsibility in a market way.

钟明 Ming Zhong

高级建筑师

1982年8月毕业于同济大学建筑系建筑学专业，建筑学学士；
1982年8月至1995年先后在中国建筑西北设计研究院和中国建筑西南设计研究院工作任高级建筑师；
1999年创办四川基准建筑设计有限公司；
2001年与周方中先生共同发起创办成都基准方中建筑设计事务所。现任首席执行官、执行合伙人。

周飚 Fan Zhou

高级建筑师
国家一级注册建筑师

1989年毕业于东南大学建筑系，建筑学学士；
1989年至2002年6月在中国建筑西南设计研究院从事设计工作；
2002年7月加入成都基准方中建筑设计事务所，任总建筑师、总经理、高级合伙人。

龚进 Jin Gong

高级建筑师
注册建筑师

1982年7月毕业于重庆建筑工程学院建筑系，建筑学学士；
1982年8月至2002年7月在中国建筑西南设计研究院从事建筑设计工作；
2002年7月加入成都基准方中建筑设计事务所，任首席总建筑师、高级合伙人。

李强 Qiang Li

高级建筑师
国家一级注册建筑师

1986年华南理工大学建筑学毕业，工学学士；
1986年至2000年，四川省建筑设计院高级建筑师、副总建筑师；
2001年–2002年，在德国多家著名建筑和景观设计事务所研修工作；
2004年2月加入成都基准方中建筑设计事务所，任执行总建筑师、合伙人。

黎亮 Liang Li

高级建筑师
国家一级注册建筑师

1992年苏州科技大学建筑系毕业，城市规划学学士；
1992年至2007年9月在中国建筑西南设计研究院从事设计工作；
2007年10月加入成都基准方中建筑设计事务所，执行总建筑师、合伙人。

中国现代五项赛事中心

建筑面积：33 923.50平方米
建筑密度：10.6%
建筑高度：29.5米
设计时间：2009年
项目进程：建设中

项目包括约3.6万平方米的赛事中心场馆（包含1.03万平方米的体育馆、2.58万平方米的体育场和辅助用房区），以及室外比赛区和赛事后勤区。使得国际现代五项锦标赛在历史上第一次实现了所有5项比赛全部集中于一个赛事中心内举办的情况。建成后，在锦标赛举办期间室内体育馆可容纳3 000名观众，体育场可容纳5 000名观众。锦标赛结束后则可将部分赛事区域改造成为集大型商业、休闲场所和运动会所为一体的多功能商业运动休闲中心。在2009年7月举行的公开设计竞赛中，本方案以科学合理的功能组织，简洁优美的外观造型以及其在材料、结构形式、预算和施工周期控制等方面表现出的良好可实施性，获得了评委和赛事组织者的一致认可，获选为中标实施方案。

广元公共卫生中心

建筑面积：31 395.4平方米
建筑密度：23.32%
建筑高度：65米
设计时间：2008年
项目进程：建设中

广元市公共卫生中心是一个综合建设项目，其子项目内容包括广元市疾病控制中心、妇幼保健院、紧急救援中心、卫生执法监督支队和红十字会5家单位。在提出总体规划布局上考虑统一整合，使各单位相对独立、突出疾病控制主体这三大原则，同时，以"合理功能布局和良好的交通流线组织"为主要设计目标，首先考虑将交通最为复杂的区域进地流线组织，在使用上各单位将拥有自己相对独立的出入口。在用地东北角上，还设置了可供各单位共同使用的功能区，包括职工食堂和大型会议室，其目的在于使各项资源得到合理利用。

棕榈泉国际中心

建筑面积：268 929.23平方米
建筑密度：57.17%
建筑高度：120～180米
设计时间：2009年
项目进程：建设中

成都棕榈泉国际中心项目地处成都新政务中心区，为超大型综合商业体。地面建筑由超高层写字楼和综合商业楼组成，地下4层具备商业（包括超市、临售、餐饮）并与地铁站直接相连。建筑立面以简化的欧式古典风格为主，力图塑打造一个典雅、人文、现代、豪华的高品质建筑，使其既有现代建筑的简洁，又充满精致和人性化的细节。

华润二十四城住宅及商业

建筑面积：约220万平方米
建筑密度：50%
建筑高度：100～180米
设计时间：2006年—2010年
项目进程：建设中，部分已建成

二十四城位于成都市二环路东二段，是一个集住宅、商业、办公等功能为一体的超大型城市综合社区。自2006年开工建设至今，该项目已经成为该区域最有影响力，市场认可度极高的标志性项目。高品质的建筑单体设计成为楼盘品牌的重要支撑和亮点。

城投集团观城项目

建筑面积：255 483平方米
建筑密度：22%
建筑高度：100米
设计时间：2010年
项目进程：建设中

观城项目位于成都市东二环与东三环之间。其周边为成熟的居住社区，是块城市里闹中取静的适居之所。项目定位于打造复合型的高档宜居社区，由6+1、8+1多层电梯洋房、33层高层电梯住宅和商务公寓组成。建筑风格以英式风格为主，结合artdeco的建筑设计元素，既体现了新古典建筑风格特点，又彰显了建筑的高档品质。

中国成达工程有限公司

CHENGDA

CHENGDA ENGINEERING CORPORATION OF CHINA

中国成达工程有限公司是一个全资国有公司，始建于1958年，前身为化工部第八设计院，经过几十年的努力，现已成为以设计为主体，实行工程总承包的国家级综合性大型工程公司，并成为全国首批获得工程设计综合甲级资质的企业。

公司现有在册职工1216人，其中中国工程院院士1人，国家工程设计大师1人，教授级高级工程师111人，高级职称390人，中级职称335人。

中国成达工程有限公司在国内及海外成功执行了几千个项目，并有200余个项目获得国家级奖项，2008年公司营业收入达33.8亿元，其中境外收入19.5亿元。成达公司在2009年度ENR公布的“中国承包商和工程设计企业双60强”的排名如下：ENR中国工程设计企业60强排名第2位、最具成长工程设计企业排名第8位、最具国际拓展力工程设计企业排名第1位；在2008年世界200家最大的国际设计公司排行榜中排名第38位；在2008年全球顶级150家国际工程设计公司排名中排名第65位。

公司承接的民用建筑主要以写字楼、酒店、场馆、商业等大型公共建筑群及住宅为主，以质量第一、用户第一、信誉第一、服务第一为经营宗旨，以不断改进的技术和管理，向国内外提供规范的工程建设的全程服务，确保顾客满意。

CHENGDA ENGINEERING CORPORATION OF CHINA, a wholly state-owned company, founded in 1958, was formerly known as No.8 Institute of the Ministry of Chemical Industry. After decades of efforts, it has become the state-level large integrated full-service engineering company and project turn-key contractor mainly engaged in engineering design, and become one of the enterprises which firstly gained Class-A comprehensive qualifications of engineering design.

There are currently 1,216 registered workers in the company, including 1 academician of Chinese Academy of Engineering, 1 national engineering design master, 111 professor-level senior engineer, 390 staffs with senior titles and 335 staffs with mid-level titles.

CHENGDA ENGINEERING CORPORATION OF CHINA has successfully carried out thousands of projects at home and aboard, and more than 200 projects have won national-level rewards. The company revenue income in 2008 is 3.38 billion yuan in 2008, including 1.95 billion yuan oversea income. The ranks of Chengda corporation in “Top 60 Chinese Contractor and Design Firms” announced by ENR are as follows: Second rank of ENR “Top 60 China Engineering Design Enterprises”, eighth rank of “Engineering Design Enterprises Which Have Most Growth Potential”, first rank of “Engineering Design Enterprises Which Have Most International Expansion Power”, 38th rank of the ranking list of “The World’s 200 Biggest International Design Companies” in 2008, 65th rank of “Global Top 150 International Engineering Design Companies” in 2008.

Main civil architectures undertaken and completed by the company are office buildings, hotels, venues, business and other large public buildings. In the principle of “quality first, customer supreme, credit first and service first”, the technology and management with continual improvement are adopted to provide the normal services for the whole implementation of project and ensure the clients satisfactory.

地址：四川省成都市天府大道中段279号
国际事业部：+86-28-65531428
经营部：+86-28-65537017
传真：+86-28-65530000
+86-28-65530001
网址：www.chengda.com
邮箱：chengda@chengda.com

Address: No.279 of Tianfu Avenue, Chengdu, Sichuan Province
International Division: +86-28-65531428
Business Dept: +86-28-65537017
Fax: +86-28-65530000
+86-28-65530001
Http:// www.chengda.com
E-mail: chengda@chengda.com

中国成达工程公司总部大楼项目
合作单位：上海欧邦建筑设计公司
建设地点：四川 成都
占地面积：13 398.5 平方米
建筑面积：77 209 平方米
使用时间：2008 年

The headquarter building of China Chengda Engineering Corporation
Cooperation: Shanghai Orban Architectural Design Company
Construction Location: Chengdu, Sichuan
Site Area: 13,398.5 m^2
Building Area: 77,209 m^2
Time for Use: 2008

成达佳园
建设地点：四川 成都
建筑面积：230 000 平方米
设计时间：2005 年 9 月

Chengda Jiayuan
Construction Location: Chengdu, Sichuan
Building Area: 230,000 m^2
Design Time: 2005.09

九寨天堂甘海子国际会议中心
建设地点：四川 九寨沟
合作单位：泛太平洋设计与发展有限公司（加拿大）（方案合作）
占地面积：217 978 平方米
建筑面积：139 800 平方米
设计时间：2002 年
竣工时间：2003 年

Ganhaizi International Conference Center of Jiuzhai Paradise
Construction Location: Jiuzhaigou, Sichuan
Cooperation: PAN-PACIFIC DESIGN & DEVELOPMENT GROUP LTD. (CANADA) (program co-operation)
Site Area: 217,978 m^2
Building Area: 139,800 m^2
Design Time: 2002
Completion Time: 2003

项目位于海拔 2 680 米的四川省阿坝藏羌族自治州，距著名的九寨沟风景区 26 千米，地处藏羌民族地区，整个建筑群背靠群山，掩映在原始森林中，建筑将现代文明与原始自然风光融为一体，人与自然、古老与现代、原始与文明在这里得到了完美、和谐的统一。

工程设计新颖独特，结构形式多样，外观美轮美奂，为一环保生态建筑群，是集会议、旅游、度假、娱乐于一体的大型超五星级酒店。通过生态与人文并重、保护与发展兼顾的综合开发方式成为九寨沟国际旅游精品中一大亮点，是四川省标志性建筑之一。

This project is located 2,680 m above the sea level in Abazang Qiang Autonomous Prefecture, Sichuan Province. It has a distance of 26 km from the famous Jiuzhai Valley landscape and located in Tibetan and Qiang ethnic minority areas. The whole building is in front of the mountains and sets off in the forests, integrating the modern civilization and the primary landscapes. It makes a harmonious and perfect unification between man and nature, ancient and modern, as well as primary and civilization.

The design for this project is novel and unique, structural forms are of various forms, and the outlook is beautiful and splendid, so it is a set of eco-buildings. It is a large super five-star hotel that gathers the function of conference, travel, vacation and entertainment. It has become one of the remarkable points in Jiuzhai Valley international tourism and a landmark building of Sichuan Province, by emphasizing a comprehensive exploration method that pays attention to both ecology and culture, as well as protection and development.

九寨天堂甘海子甲蕃古城
建设地点：四川 九寨沟
建筑面积：260 802 平方米
使用时间：2005 年 5 月

The ancient town of Jiafan, Ganhaizi, Jiuzhaigou
Construction Location: Jiuzhaigou, Sichuan
Building Area: 260,802 m^2
Time for Use: 2005.05

项目位于四川省九寨沟县章扎镇甘海子地区，项目设计包括剧场、博物馆、官寨、羌寨等。基地西侧紧临九环线，距九寨沟沟口约 20 千米，交通方便，风景秀丽，场地内有一条河流延伸通过，两侧为山地，地形为高山间河流深切形成的沟各地带，地形平缓开阔，地貌单元属白龙江河漫滩和一级阶地前缘地带。该项目是集旅游、度假、观演、展览等于一体的建筑群，能提供大型演出及展览用地，同时能满足九寨沟日益增长的客房需求，进一步带动该地区经济发展。

This project is located in the Ganhaizi region of Zhangzha Town, Jiuzhaigou County, Sichuan Province. The project designs include: the theater, the museum, Guangzhai and Qiangzhai, etc. The west of the base is in close neighborhood of Jiuzhaigou circular highway, about 20 km from Jiuzhaigou Port. It enjoys convenient transportation and beautiful sceneries, with a river running through the base and mountains perching on both sides. The terrain, flat and open, is the gully area formed by the river between mountains, and the geomorphic unit belongs to the floodplain and Class-one front terrace area. Integrating services for tourism, vacation resorts, performances and exhibitions as a whole, the project is able to provide large-scale performances and exhibitions, meeting the growing room needs of Jiuzhaigou and further promoting the region's economic development.

项目位于成都的核心地区，地理位置显要，是将东大街打造成为金融一条街的重要组成部分。因此，设计旨在追求建筑本身的性格的同时，也注意保持建筑与环境的和谐统一。项目为两幢高层双塔及高层裙房的建筑综合体，建筑功能复杂，含商业、电影院、写字楼并配有地下车库。

This project, located in the core area of Chengdu, is an important component of the East Avenue and the Financial Street. Therefore, while pursuing the building's own characteristic, the design also pays attention to ensure its harmony with the environment. The project is a building complex with two high-rise towers and a high-rise annex. It is equipped with comprehensive functions, including the business center, the movie theater, the office building and the underground garage.

阳光保险大厦
四川省电影大厦
建设地点：四川 成都
占地面积：11 828 平方米
建筑面积：113 920 平方米

Sunshine Insurance Building
Sichuan Movie Bilding
Construction Location: Chengdu, Sichuan
Site Area: 11,828 m^2
Building Area: 113,920 m^2

某酒店灾后重建项目
建设地点：四川 都江堰
用地面积：244 471 平方米
建筑面积：60 644 平方米

The post-disaster reconstruction project of a hotel
Construction Location: Dujiangyan, Sichuan
Floor Area: 244,471 m^2
Building Area: 60,644 m^2

项目位于世界著名的“中国历史文化名城”、“中国优秀旅游城市”都江堰市的东部城乡结合地带，紧邻川苏都江堰科技产业园。项目尊重场地精神，以保护自然生态环境和延续历史文脉为前提，形成天、地、山、水、人紧密融合的，集科研、生产、生活、休闲、体育锻炼、酒店会议、餐饮为一体的地域经济新的有机综合体，力求建筑群体与自然环境相融合，构筑成复合型为测试中心服务的综合旅游项目，实现各功能之间互补共生，最终形成一个完整的经济链系统。

This project is located in the eastern fringe area of Dujiangyan—the city with the title "Chinese historical and cultural city" and "China Excellent Tourism City", which enjoys a worldwide reputation. It is in close neighborhood of Chuansu Science and Technology Industrial Park. Respecting the site spirit and taking the environment protection and history continuation as its premise, this project produces a new organic synthesis of regional economy integrating scientific research, production, living, entertainment, physical exercise, hotels, conference and restaurant as a whole, and with the close combination of heaven, earth, mountain, water and people. It strives to achieve a harmonious felation between the building complex and natural environment. The design produces an integrated tourism project which serves the test center and the functional symbiosis among its components form a complete economic chain.

成都万通成达商务中心
建设地点：四川 成都
建筑面积：300 000 平方米

Wantong Chengda Business Center in Chengdu
Construction Location: Chengdu, Sichuan
Building Area: 300,000 m^2

武汉新城国际博览中心
建设地点：湖北 武汉
建筑面积：1 000 000 平方米

International Expo Centre of Wuhan New City
Construction Location: Wuhan, Hubei
Building Area: 1,000,000 m^2

项目位于武汉新区汉阳四新滨江地带，南临三环线上的白沙洲长江大桥，东侧直面长江，规划用地面积 1 702 400 平方米，总建筑面积 2 400 000 平方米，拥有国际标准展位 7 400 个。项目从西向东，依次分为会展、会议和酒店三大片区。

This project is located in Bingjiang Region, Hanyang, Wuhan, the south of which neighbors Baishazhou Yangtze River Bridge while the east the Yangtze River. The planned land area is 1,702,400 m^2 and the gross floor area 2,400,000 m^2, with 7,400 international standard booths. From west to east, there are mainly three areas, i.e. the exhibition area, the conference area and the hotel area in proper order.

烟草兴业大厦
合作单位：上海欧邦建筑设计公司
建设地点：四川 成都
建筑面积：106 623 平方米
设计时间：2008 年 9 月

Tobacco Industrial Building
Cooperation: Shanghai Orban Architectural Design Company
Construction Location: Chengdu, Sichuan
Building Area: 106,623 m^2
Design Time: 2008

项目位于成都世纪城中心广场西南角的制高点，北面为天鹅湖，地理位置较为显要，是世纪城的城市景观和人文环境的重要组成部分

建筑风格：贯彻简洁、创新的设计思想，力求塑造一个鲜明的标志性建筑物。通过玻璃体量的转折，创造出一个充满雕塑感的建筑形式。

设计构思：最初的灵感源自方形的“门 (mén)”，从一个最简单的立方体出发，通过对体量的切割，在会展中心广场上形成一个“世纪之门”的标志。通过立方体的穿插环抱也寓意团结向上和开放交融的精神内涵，交织重叠的巨门形象，也彰显出世纪城全球化和信息化的特质。

This project is located in the commanding height of the southwest corner of Central Plaza in Chengdu Century City, neighboring Swan Lake to the north. Enjoying a prominent location, it is an important component to the urban landscape and human environment of Century City.
Architectural Style: By implementing the design concept of simplicity and innovation, the design strives to create a distinctive landmark. The turning of the vitreous volume creates an architectural form with a strong sense of sculpture.
Design concept: The initial inspiration comes from the square of the Chinese character "(mén)": starting from the simplest cube, through the cutting of the volume, a sign "Gate of the Century" is created on the Central Plaza. The interspersing and embracing of these cubs embody the spiritual implication of unity and open up, and the overlapping of the giant-door image also highlights the Century City's globalization and informationization.

成都世纪城—假日酒店

成都世纪城—娇子国际会议中心

成都世纪城新国际会展中心
合作单位：泛太平洋设计与发展有限公司（加拿大）
建设地点：四川 成都
建筑面积：700 000 平方米

成都世纪城——四星假日酒店
建筑面积：87 000 平方米
客房数量：780 间

成都世纪城——会议中心
建筑面积：10 600 平方米

成都世纪城——五星洲际酒店
建筑面积：76 000 平方米
客房数量：560 间

New International Convention and Exhibition Center of Chengdu Century City
Cooperation: PAN-PACIFIC DESIGN & DEVELOPMENT GROUP. LTD. (CANADA)
Construction Location: Chengdu, Sichuan
Building Area: 700,000 m^2

Chengdu Century City—the four-star Holiday Hotel
Building Area: 87,000 m^2
Number of Rooms: 780

Chengdu Century City—Convention Center
Building Area:10,600 m^2

Chengdu Century City—the five-star InterContinental Hotel
Building Area: 76,000 m^2
Number of Rooms: 560

成都世纪城—洲际大酒店

四川西南广厦建筑设计院有限责任公司

地址：四川省成都市马鞍南路 110 号马鞍大厦 7、8 楼
电话：+86-28-83383369
邮箱：guangsha@vip.163.com

Address: 7th and 8th floor of Maan Mansion, No.110 South Ma'an Road, Chedu, Sichuan Province
Tel: +86-28-83383369
Email: guangsha@vip.163.com

四川西南广厦建筑设计院有限责任公司是国家批准的建筑甲级设计公司，现有员工均为行业内的中青年技术骨干。公司在建筑、结构、给排水、电气、暖通、概算、室内设计、环境设计、规划设计等各专业上人员配备齐全，并拥有全国同行业内最先进的技术设备和现代化的办公条件，设计、管理全面实现信息化、网络化，技术力量雄厚，有能力承接各类复杂的大型设计项目。

四川西南广厦建筑设计院有限责任公司的前身是成都广厦建筑设计事务所，于 1988 年成立，由四川省土木建筑学会主管，中国建筑西南设计研究院主办。2002 年经股份制改造，省工商局核准更名为四川西南广厦建筑设计院有限责任公司。

改制重组后的四川西南广厦建筑设计院有限责任公司极具活力，人员业精技强，注重科学管理，并特别强化设计质量、服务质量，在为业主提供全面优质服务上具有非常明显的优势和实力。改制完成后，公司已相继完成各类大、中型设计项目 70 余项，并多次在重大方案投标中获奖中标，其中童子街 29 号改造工程、都江堰市人民医院、蓝光金荷花、遵义商业银行、达州市市政中心、棕北国际商务公寓、核工业部 416 医院、四川报业集团印务中心等颇具影响力的项目，因其优秀的设计和完善的服务，获得了业主的一致好评，提升了公司的知名度，取得了良好的社会效益。

四川西南广厦建筑设计院有限责任公司凭借实力、诚信和不懈的追求，必将在建筑设计领域打造出“西南广厦”的知名品牌，和社会各界一起，共创中国建筑设计行业的美好未来。

Sichuan southwest Guangxia Architectural Design Institute Co., Ltd. is grade-A building & design company approved by the state. Almost the existing staffs are middle-aged professional technology backbones. The company has various talents on construction, structure, water and drain system , electricity, HAVC, budgetary estimate, indoor design, environment design, urban planning and so on. It also has the most advanced equipment and modern work condition in the country. It has realized the informatization on design and management. With advanced technique, it has ability to handle all kinds of complex large-scale design project.

The former name of Sichuan Southwest Guangxia Achitectural Design Institute Co., Ltd. is Chengdu Guangxia Institute. It was founded in 1988, managed by Sichuan Institute of Civil Engineering and organized by China Southwest Architectural Design & Research Institute. After shareholding reform, Chengdu Guangxia Institute changed into Sichuan Southwest Guangxia Achitectural Design Institute Co., Ltd., approved by Provincial Commerce and Industry Bureau.

After reform, Sichuan Southwest Guangxia Architectural Design Institute Co., Ltd. becomes more and more vigorous and professional. It emphasizes scientific management, tries to improve the quality of design and service, and intends to provide best and comprehensive services for clients. This company has completed 70 large and medium scale projects, and wins the tender for planning many times. Among these, the rebuilding project for No.29 Tongzi Street, Dujiangyan People's Hospital, Languang Jinhehua, Zuiyi Commercial Bank, Dazhou Government Center, Zongbei International Business Apartment, 416 Hospital invested by Ministry of Nuclear Industry, Print Center of Sichuan News Group and other influential projects. For its outstanding design and consummation service, this company has obtained owner's high praise. In the mean time, the image of company has been promoted, which brings good social benefits.

Relying on strength, honesty and unremitting pursue, Sichuan Southwest Guangxia Architectural Design Institute Co., Ltd. will develop a famous brand of "Southwest Guangxia", and create a bright future for the architecture industry with all sectors of our society.

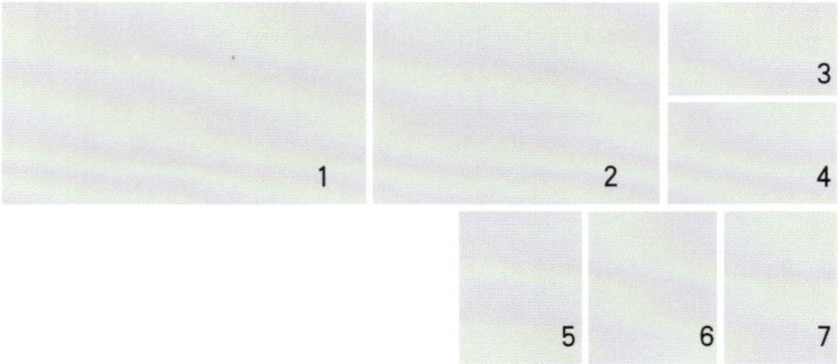

1　巴中医院
2–4　金地新城市花园
5–7　416 医院

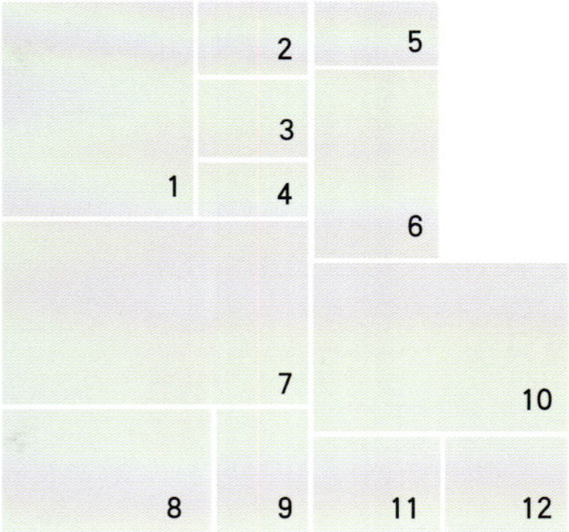

1–4 蓝光金荷花国际时装城
5–6 蓝光紫檀山 4、5 号地
7 尚都服饰广场
8–9 双流体育场
10–12 四川日报报业集团新印务中心

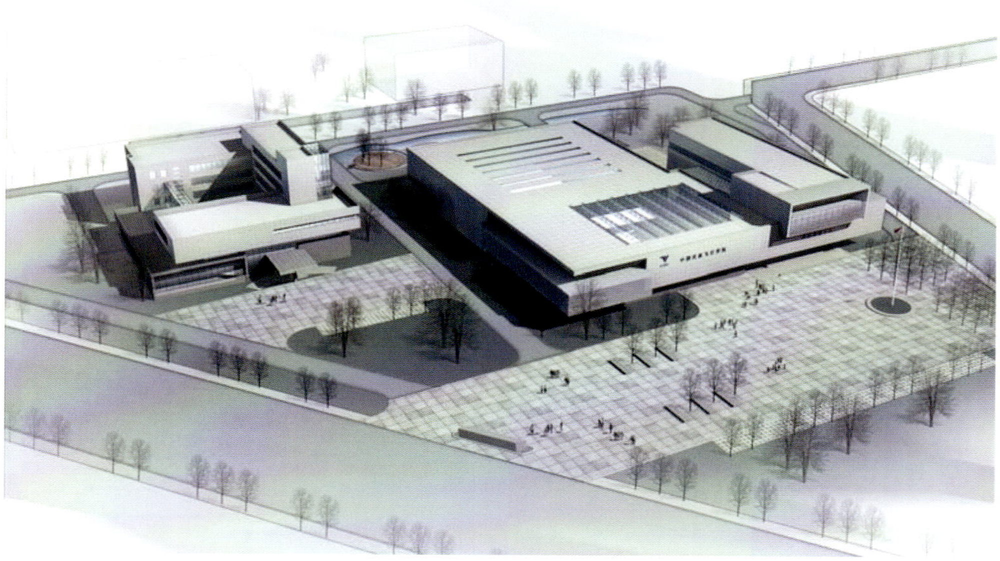

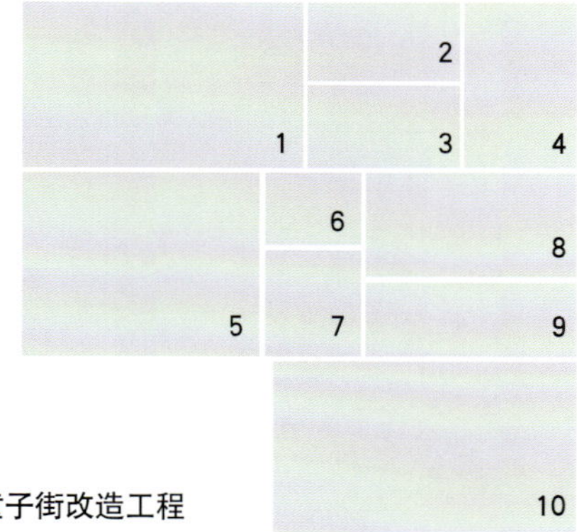

1–3　四川省财校
4　四川省机关事务管理局童子街改造工程
5–7　新都文广大厦
8–10　中国民航飞行学院模拟飞行中心模拟中心

成都长源建筑设计研究院有限公司

地址：成都市石马巷 8 号红鑫大厦四楼
邮编：610017
电话：+86-28-86957595
传真：+86-28-86953456
网址：www.cd-cy.com.cn

Address: F4 Hongxi Building No.8 Shi Maxiang Chengdu
P.C.: 610017
Tel: +86-28-86957595
Fax: +86-28-86953456
Http: //www.cd-cy.com.cn

成都长源建筑设计研究院有限公司是一个涉足项目策划、建筑规划、建筑工程设计的综合性建筑设计研究院。公司成立于 90 年代初期，是成都虹鑫投资集团旗下的独立法人企业。主要领导及骨干均来自中国建筑西南设计研究院。

成都长源建筑设计研究院有限公司现为国家甲级资质的设计公司（A151009731）。现有国家一级注册建筑师五名，国家二级注册建筑师两名，国家一级注册结构工程师三名，国家二级注册结构工程师二名以及近百名设计技术人员，公司在城市规划、商业设施、办公建筑以及住宅小区设计等方面有丰富的实践经验和卓越的设计理念，所参与的工程设计项目涉及各个领域。在十多年的工程设计实践中，设计及竣工面积逾数百万平方米，并与新加坡、加拿大等的国际设计机构有着长期而广泛的合作。

Chengdu Changyuan Architectural Design and Research Institute Co.,Ltd. has a wide range from project and architectural planning to engineering design. This company, founded in early 1990s, is an enterprise with an independent legal entity which is invested by Chengdu Hongxin Group. Most of backbones and leaders are from China Southwest Architecture Design and Research Institute.

Chengdu Changyuan Architectural Design and Research Institute is granted National Grade A qualification (A151009731). It has 5 PRC Class I Registered architect, 2 PRC Class II registered Architect, 3 PRC Class III Registered Structural Engineer, 2 PRC Class Registered Structural Engineer and more than 100 technical staffs. This company has rich experiences and excellent design concepts on urban planning, design of commercial facilities, office buildings and residential quarters. They took part into many engineering design projects covering every field. In over ten years, this company designs and completes more than one million square meters. It also established extensive cooperative relationship with international design institute of Singapore, Canada and etc..

1 中房 星云国际大厦

Zhongfang Xingyun International Mansion

Chengdu Changyuan Architecture Dsdign And Research Institate Co.,Ltd.

2–5 新都 泽瀚帝景
Xindu Zehan Empire Landscape

6–9 中航 云岭
高尔夫别墅
Zhonghang Yunling
Golf Villa Block

10–12 峨眉 阅湖郡
Emei Yuehu County

13–15 广汉 经典上城
Guanghan Classical Uptown

16

17

18

19

20

21

22

23

24

26

25

16–19 卓锦城
Zhuojin City

20–22 恩威 华阳古城村
Enwei Huayang Gucheng Country

23–26 西安北客站站前规划
The planning of area in front of Xi'an North Railway Station

Chengdu Changyuan Architecture Dsdign And Research Institate Co.,Ltd

Chengdu Changyuan Architecture Dsdign And Research Institate Co.,Ltd.

VTR（美国）国际设计研究有限公司（中国机构）
VTR (USA) International Design Institute Ltd (Chinese Organization)

"建筑设计，反映着每个时代科技、经济与文化的内涵。设计的意义也随着时间而演变，成为一种新的国际语言。"

VTR（美国）国际设计研究有限公司（中国机构）（简称：VTR），是一家拥有建筑工程甲级设计证书、甲级工程咨询证书的专业设计单位，并在2008年通过了国家质量管理GB/T19001-2000idt9001：2000标准认证。

VTR作为国际性的规划设计机构，拥有来自各个建筑领域的优秀设计人员及支持人员，形成开放而多元化的设计文化，让前端生活哲学与设计细节同步，无间距地与国内快速提升的价值观和市场相结合。公司现有员工168人，其中教授级高级工程师7人，副教授级高级工程师39人，工程师76人，助理工程师30人。几年来VTR完成了一大批城市标志性建筑，如：战旗文工团西南剧院、阿坝州政府驻成都政务中心、成都送仙桥古玩市场、自贡报业大厦、希望集团·美好国际商城、剑阁政务中心等项目。

现在，VTR已不只是一批专业人士形成的团队，而是在这创新时代，着眼于创造新的价值观与生活艺术的团队，伴随着建筑设计，这必将令设计价值发挥至极致，为客户缔造更大的产业价值。

"Architectural design reflects the science, technology, economy and culture of the times. The aim of design is changing with the time and is becoming a new international language."

VTR (USA) International Design Institute Ltd (Chinese Organization) (VTR for short) specializes in design with Grade A certificates for construction engineering design and engineering consulting service. The company has passed the certification of the national quality management system of GB/T19001-2000idt9001:2000 in 2008.

As an international design and planning company, VTR has a team of excellent designer and supporters in various fields of construction. It has developed an open and diverse design culture, making elegant life philosophy congruent with its design and combining fast growing values with the market. The company has 168 employees, among whom 7 are professor-level senior engineers, 39 are associate professor-level senior engineers, 76 are engineers and 30 are assistant engineers. In recent years VTR has completed a host of urban land mark buildings, including the Zhanqi Troupe Southwest Theatre, Aba Prefecture Government's Administrative Service Center in Chengdu, Chengdu Songxianqiao Antique Market, Zigong Press Building, Hope Group • Meihao International Shopping Center, and the Jiange Administrative Service Center.

Now, VTR is not just a team of excellent professionals, it focuses on exploring new values and new lifestyles in the new age. It will give full play to its design value and create greater industrial value for its customer.

地址：中国成都红星路二段120号亚新大厦
首席执行官：谢绍宁
电话：+86-28-86786662
传真：+86-28-86788919
邮件：vtr@vtr-sc.com
网址：www.vtr-sc.com

Address: Yaxin Building, No.120, Second Section of Hongxing Road, Chengdu, China
CEO: Shaoning Xie
Tel: +86-28-86786662
Fax: +86-28-86788919
Email: vtr@vtr-sc.com
Http: //www.vtr-sc.com

金英汇

占地面积：4 021.17 平方米
建筑面积：106 131.54 平方米
设计时间：2009年4月

本项目位于成都市金牛区抚琴西北街，由1栋35层超高层建筑与周边两栋住宅共同组成。该超高层建筑功能较为复杂，包含四星级酒店及政务中心两大块。在设计中我们合理组织各功能，充分做到了流线相对独立，互不交叉干扰。立面设计采用新古典主义风格，合理使用配合功能的元素和材料，强调比例及细节，打造出挺拔大气而又不失层次变化的全新建筑形象。

Jinyinghui Building

Floor Area: 4,021.17 m^2
Building Area: 106,131.54 m^2
Design Time: 2009.04

Located in the Northwest Fuqin Street in Jinniu District of Chengdu, the project consists of a 35-storey high rise building and two surrounding residence buildings. The high rise building has complicated functions, including two major parts: the four star hotel and the administrative service center. In design we have rationally organized their functions, and have paid full attention to independent streamlines without cross interference. Its façade of neoclassic style adopts matching elements and materials with attention to proportion and details. It appears upright and imposing with multi-layered structures.

西南剧场

占地面积：2 863 平方米
建筑面积：14 341 平方米
设计时间：2007 年 12 月

战旗文工团西南剧场位于成都市正通顺街，为 1200 座甲级剧场。该项目建筑造型独特，在合理布置功能的同时，强调体量的夸张与材质的通透，并注重细部的处理。其柔美的屋顶造型取意于飘扬的战旗，幕墙上挺拔的竖向金属条则象征着一根根拨动优美音符的琴弦。整栋建筑飘逸灵动，充满了张力，为西南地区的部队文艺建设树立了崭新的里程碑。

Southwest Theatre

Floor Area: 2,863 m^2
Building Area: 14,341 m^2
Design Time: 2007.12

Zhanqi Troupe Southwest Theatre is located in Zhengtongshun Street of Chengdu and is a Class A theatre and can seat 1,200 people. The project has a unique shape. and emphasizes exaggeration of space and transparency of materials and pays attention to detail expression while having a rational layout. Its soft roof is shaped like a fluttering flag. The vertical and straight metal strips on the curtain wall symbolize strings playing beautiful music notes. The whole building is grace and elegant with full vitality. It is a milestone for the army culture in the southwest of China.

堰上芙蓉 · 印象

占地面积：43 669.11 平方米
建筑面积：38 038.22 平方米
设计时间：2010 年 2 月

本项目秉承三道堰镇作为千年古镇的精神归属，将“水”作为规划设计中的灵魂与载体，试图诠释新时代背景下中国传统居住理念与现代人居模式之间的共生共振点。项目以水系为媒介建立人与自然的对话平台，规划中分别考虑了公共水景与私人住宅水景，并使两者成为整体，相互连通，建立完整的亲水体系。在完成水系规划的同时，以此为基础形成区域内的交通骨架，保证了水系的决定性位置。在本次设计中，提出了“岛居”概念，将建筑完全布置于由主水系构成的“岛”之上，提高了居住的品质。而由主水系衍生出的支水系延伸至建筑的后院之中，使水与建筑融为一体。在居住建筑的布置方式上，注重通风、采光的要求，取与南北向成 25 度为主要朝向平行布置建筑，建筑间距满足防火及日照要求。在沙三线与项目建筑之间留出 14 米的绿化隔离带处理为滨水景观带，一方面塑造临城市道路的城市公共空间展示面，另一方面提高临道路居住品质。全区内作景观绿化处理，达到人、建筑、自然三者之间的共生和谐。

Yanshang Furong · Image

Floor Area: 43,669.11 m^2
Building Area: 38,038.22 m^2
Design Time: 2010.02

The project, on the basis of the ancient town of Sandaoyan, uses “water” as the soul and carrier in the design to represent the traditional Chinese folk living and modern living pattern against the background of the new time. It is a platform for dialogues between human and nature with water as the media. In the plan, the designers have integrated the public waterscape and private residential waterscape to establish a complete waterscape design. While completing the water system planning, the design also completed the traffic network in the area, guaranteeing the water system’s location. In the design, the “island living” concept has been proposed, placing the architecture on the islands full of water systems to improve living quality. And the water systems will further extend to the architectural yards, integrating water with architecture. For the living architectural design, the ventilation and sunshine are considered, with the 25° in the north-south direction parallel to the architectural design. The architectural distance can also meet the need of the best fireproof and sunshine. A green insulation belt of 14 m between the boundary line and the project architecture should be the waterscape for the urban public space demonstration and roadside living quality. The area is has been landscaped and afforested to provide harmonious existence of human, architecture and nature.

安宁明珠大酒店

占地面积：17 038.7 平方米
建筑面积：37 792.1 平方米
设计时间：2006 年 8 月

安宁明珠大酒店位于米易县新城区中心地区，紧邻安宁河、新政府办公区。交通十分便利，满足现代使用功能要求的目的。打造全国一流五星级酒店，是本次方案设计的根本宗旨。讲究以人为本，天人合一，建筑与城市景观的融合。人与自然的融合，是本次设计的精髓。

Anning Mingzhu Hotel

Floor Area: 17,038.7 m^2
Building Area: 37,792.1 m^2
Design Time: 2006.08

Anning Mingzhu Hotel is located in the center of the new urban area of Miyi County, close to the Anning River and the New Government Office Area, with convenient traffic. The fundamental purpose of the project is to satisfy demand of modern functions and to make a national first-class five star hotel. It emphasizes human orientation and harmonious society, paying special attention to the integration between the architecture and the city and between human and nature.

威远中医院

占地面积：5 389.6 平方米
建筑面积：25 917.32 平方米
设计时间：2009 年 2 月

威远县中医院位于威远新建的内环路侧，交通方便快捷，同时整个用地高差较大，最高处达 16 米，不利于医院的常规建设。

针对整个用地的优劣势，在与院方交流后，我们确定以下的规划设计思路：

（1）发挥优势，将病患出入口设于内环路上，方便快捷；
（2）采用分散式布局，满足院方分期实施的需要；
（3）因地制宜，结合地形的特点，将整个用地分为三个台地，打造具有川西特色的坡地建筑；
（4）整个建筑的展开面平行于道路布置，使其成为新的地标；
（5）整个规划设计满足医院“三区”“四口”“五条流线”规划要求；
（6）在建筑形态方面，考虑到中医所蕴含的中国传统文化内涵，我们借鉴了四川传统民居的结构特点，结合现代建筑的做法，形成别具一格的现代中式风格。通过穿斗式屋架，木质花窗，青砖，青瓦等形成鲜明的地方特色。

Weiyuan TCM Hospital

Floor Area: 5,389.6 m^2
Building Area: 25,917.32 m^2
Design Time: 2009.02

Weiyuan TCM Hospital is located on the side of the newly built Inner Ring Road in Weiyuan, with convenient traffic.The height difference in the project land is relatively big, with difference of 16m at the maximum, so it is unfavorable for the ordinary hospital construction.

On the basis of the project advantage and disadvantage, after consultation with the hospital, we determined following design plan:

1. Displaying the project advantage, setting the patient's entrance and exit on the Inner Ring Road, with convenient traffic.
2. Adopting distributed layout to satisfy the hospital's phased construction.
3. Considering the location, dividing the land into three parts, building the slope architecture peculiar to west Sichuan.
4. The whole complex should be parallel to the road layout to become a new landmark.
5. The whole planning and design should meet the hospital's planning requirements of “three areas”, “four entrances” and “five lines”.
6. Considering the TCM hospital's characteristics in terms of architectural pattern, modeling on the traditional folk house in Sichuan and modern architecture to form the unique Chinese style with column and tie construction, wooden window, blue brick, blue tile, etc.

天津市建筑设计院

TIANJIN ARCHITECTURAL DESIGN INSTITUTE

天津市建筑设计院（简称TADI）创立于1952年，历经五十多年历史沧桑，现已发展成为技术实力雄厚、人才济济的天津地区最大的综合建筑设计院。

TADI现有员工1118人，其中全国工程设计大师3人、国务院批准享受政府特殊津贴专家12人，国家人事部批准有突出贡献的中青年专家2人。天津市中青年授衔专家4人、正高级建筑工程师45人、高级建筑工程师330人、中级建筑工程师257人、并有国家一级注册建筑师69人、国家二级注册建筑师33人、国家一级注册结构工程师76人、国家级注册监理工程师14人、造价工程师13人、注册城市规划师12人、注册岩土工程师3人、注册咨询（投资）工程师18人、注册电气工程师22人、注册公用设备工程师（给排水）26人、注册公用设备工程师（暖通空调）24人。

TADI具有国家住房和城乡建设部颁发的甲级建筑工程设计资质、甲级城乡规划编制资质、甲级风景园林专项工程设计资质、房屋建筑工程监理甲级资质、工程招标代理机构甲级资质、工程造价咨询企业甲级资质、施工图设计文件一类审查机构资质、乙级人防工程设计资质、国家发改委颁发的工程咨询甲级资质、国际建筑工程咨询协会（菲迪克）会员单位。

TADI设有建筑、规划、结构、给排水、暖通空调、电气照明、自动控制、广播通讯、经济技术分析、岩土工程、城镇规划、景观与环境设计和室内装修设计等专业，主要承接民用建筑设计、工业建筑设计、城市规划及居住区、住宅小区规划设计；承接民用与工业建筑项目的可行性研究、工程咨询、建设工程监理、工程项目总承包以及技术合作、技术成果转让等业务，承接境外工程咨询、设计和监理、国外经济技术合作、智能建筑（系统工程）设计等业务。为更好地拓展业务，设计院相继在海南、厦门、广州、上海、重庆等地成立了分院。

TADI工程技术人员积极进取，锐意创新，先后有10项设计获国家优秀设计奖，其中有三项获金质奖，近230项工程设计、科研和标准设计获得了建设部、市级优秀设计奖、优质工程奖及科研成果奖等奖项。其中：瑞景花园住宅小区工程、天津奥林匹克中心体育场工程荣获中国土木工程詹天佑奖，天津意式风貌保护区的保护和整修规划、天津海河两岸城市设计、天津津湾广场修建性详细规划均荣获全国优秀城市规划设计二等奖、天津市优秀城市规划设计一等奖，平津战役纪念馆、天津体育馆、天津第二南开中学、周恩来、邓颖超纪念馆、天津奥林匹克中心体育场、天津站荣获新中国成立60周年建筑创作、建筑设计大奖。近几年设计院完成了天津奥林匹克中心体育场、天津医科大学总医院、中华剧院、泰达国际会展中心、泰达体育场、国际金融中心、天津市迎宾馆四号楼、哈尔滨市第十四中学、天津科技广场、天津港企业文化中心、天津空港国际汽车园、天津市东丽区华明示范小城镇、华夏未来少儿艺术中心——4D影院、天津电视台数字电视大厦、中国航天科工集团第三研究院六六八工程、天津城建管理职业技术学院、天津中新生态城服务中心、中船重工大厦、迁安市人民医院、天津市中心妇产科医院迁址新建工程、天津金融城津湾广场一期工程、天宾商务中心、小白楼音乐厅、天津站交通枢纽工程轨道换乘中心工程停车中心等一批大型重点公建项目。设计院以其全新的设计理念和现代高新技术，展示了TADI的技术实力和创新、创优水平。

TADI于1996年获得ISO9001国际质量体系认证，于2002年4月4日率先实现GB/T19001—2000—ISO9001：2000标准转换。2010年3月取得2008版IDSO9001标准认证证书。

TADI质量方针：为顾客设计好每一平方米的建筑，提供优质的设计全过程服务，科技领先、锐意创新，实现质量管理体系的持续改进，使顾客满意。

Tianjin Architectural Design Institute (TADI) was founded in 1952. After more than 50 years development, it has become a comprehensive architectural design institute with strong technology capacity and abundance of talents.

The total number of TADI staff is 1118, among which there are 3 national engineering design masters, 12 people owning government special allowance confirmed by the State Council, 2 young and middle-aged professionals having outstanding contribution confirmed by the Department of Personnel, 4 young and middle aged titled experts of Tianjin, 45 full senior architects/engineers, 330 associate senior architects/engineers, 257 junior architects/engineers, 69 Class I national certificated architects, 33 Class II national certificated architects, 76 Class I national certificated structural engineers, 14 national certificated supervision engineers, 13 budget engineers, 12 certificated city planners, 3 certificated geotechnical engineers, 18 certificated consultant/investment engineers, 22 certificated electrical engineers, 26 certificated public equipment engineers (water supply and sewerage), and 24 certificated public equipment engineers (heating ventilation air conditioning).

TADI owned qualification of Class-A engineering design from the Ministry of Residence and Urban-Rural Construction, Class-A urban-rural planning, Class-A building engineering supervision, project tender agent, Class-A project budget consultant company, Class-A working drawing censoring institute, Class-B civil air defense work design, Class-A engineering consultant from the National Development and Reform Commission, and has been a member of International Association of Engineering Consultant (FIDIC).

There are majors in TADI including architecture, planning, structure, water supply and sewerage, heating ventilation air conditioning, electrical lighting, automatic control, radio communications, economic and technical analysis, geotechnical engineering, urban planning, landscape and environmental design and interior decoration design etc. TADI mainly undertakes business like civil building design, industrial building design, city planning and residential quarters planning and design, also works on feasibility analysis of civil and industrial building projects, engineering consultant, construction supervision, engineering project contracting, technique cooperation, transfer of technical product and so on. It can carry on engineering consultant, design and supervision abroad, international economic and technological cooperation and intelligent building (systematic engineering) design as well. To extend the business, it successively established branches in Hainan, Xiamen, Guangzhou, Shanghai, and Chongqing.

With aggressive endeavor of engineering technicians in TADI, 10 of their works have won national excellent design awards, three of which are golden prizes. Besides, almost 230 engineering design works, research products, and standard design have gained prizes including municipal and ministerial excellent design awards, high-quality projects awards and research achievement awards. Ruijing Garden Residential Quarter and Tianjin Olympic Center Stadium won Chinese Zhan Tianyou Civil Engineering Award. Protection and renovation planning for Tianjin Italian scenes conservation area, urban design along the Haihe River in Tianjin, detailed reconstructive plan for Tianjin Jin Bay Plaza won the Second Prize of National Excellent Urban Design and First Prize of Tianjin Excellent Urban Planning and Design. Beiping-Tianjin Campaign Memorial Hall, Tianjin Gymnasium, No.2 Nankai Middle School of Tianjin, Zhou Enlai and Deng Yingchao Memorial Hall, Tianjin Olympic Center Stadium, and Tianjin Railway Station were awarded architectural invention and architectural design prizes of the 60th anniversary of PRC. During the recent years, with new design concepts and modern high-tech, TADI completed a series of key public projects, showing high level of technical expertise and capacity of innovation and excellence-creating. Works of TADI include Tianjin Olympic Center Stadium, Central Hospital of Tianjin Medical University, Chinese Theatre, TEDA International Convention and Exhibition Centre, TEDA Stadium, International Finance Center, No. 4 Building of Tianjin Gust House, No. 14 Middle School of Harbin, Tianjin Science and Technology Plaza, Culture Center of Tianjin Port, International Auto Land of Tianjin Airport, Tianjin Dongli District Huaming Model Town, Cathay Future Children' Art Center, including a 4-D cinema, digital TV building of Tianjin TV, 668 Project of the Third Research Institute of China Aerospace Science and Industry Group, Tianjin Urban Construction Management and Vocation Technology College, Service Center of the Zhongxin Ecological City in Tianjin, China Shipbuilding Heavy Industry Building, People's Hospital of Qian'an, relocation project of Tianjin Central Obstetrics and Gynecology Hospital, Phase I of Jin Bay Plaza Project of Tianjin Financial City, Tianbin Business Center, Xiaobailou Music Hall, and parking center of Transfer Center Project of Traffic Junction Project in Tianjin Railway Station.

TADI has gained ISO9001 certification of international quality system in 1996, and firstly achieved GB/T19001-2000-ISO9001: 2000 standard transition on April 4th, 2002. By March 2010, it has got IDSO9001 standard certification 2008 edition.

The guideline of TADI is to do good job on each square meter of architectures for the customers, and to provide high-quality design service during the whole process. Being the lead in technology and encouraging original creation, TADI has tried to keep improving quality management system to further satisfy all the clients.

地址：天津市河西区气象台路95号 **邮编**：300074 电话：+86-22-23543000 **传真**：+86-22-23345260 **网址**：www.tadi.net.cn **邮箱**：taditadi@tadi.net.cn

Address: N.95 of Qixiangtai Road, Hexi District, Tianjin **P.C.**: 300074 **Tel**: +86-22-23543000 **Fax**: +86-22-23345260 **Http**://www.tadi.net.cn **E-mail**: taditadi@tadi.net.cn

天津梅江会展中心

建设地点：天津
用地面积：24.7 公顷

设计说明

项目是世界经济论坛 2010 年新领军者年会（简称"夏季达沃斯论坛"）的主会场。

建筑造型设计受到飞机造型的启发，两个象征翅膀的屋面，舒展、优雅，有机地把会展的功能整合在一起。作为城市新兴的大型公共建筑，它涵盖了一系列必不可少的功能设施，包括了核心功能、交通功能、配套功能以及辅助功能四大类功能设施，同时建筑还与周边及城市其他重要的公共设施进行合理、有机的整合，优化了城市的功能配置，为会展建筑获得良好的经营环境、增强办展水平创造了条件。

项目设计充分体现了节能环保的需求，屋顶部分采用了通透的天窗设计，利用自然光线即可满足室内日间照明。空调系统采用地源热泵等新技术、新工艺，地源热泵污染物排放将比空气源热泵减少能耗 40%以上，与电供暖相比将减少能耗 70%以上，真正实现节能减排。

Meijiang Convention and Exhibition Center Tianjin

Construction Location: Tianjin
Floor Area: 24.7 ha

Design Introduction

This project is the main venue of the Annual Meeting of New Global Leaders in 2010 World Economic Forum (the "Summer Davos").

Inspired by aircraft model, its architectural style is designed with two roofs in symbol of stretch and elegant wings, organically integrating the exhibition functions into one figure. As the large-scale public building nowadays emerging, it not only covers a range of essential functions facilities, including four main categories i.e. the core function, transportation function, supporting function and auxiliary fuction, but also matches with neighboring and other important public facilities in the city reasonably and organically, optimized for urban functions configuration, thus creating good business environment, enhance the conditions for exhibition.

Project design fully reflects the energy-saving and environmental protection, while the roof adopts transparent design, by this way, its indoor lighting can be met making use of natural light; air-conditioning system adopts ground source heat pump and other new technologies, whose pollution emissions decreased by 40% or more over air source heat pump, reduced by 70% or more than electric heating, undoubtedly realizes energy saving.

滨海文化商务中心

建设地点：天津
用地面积：41.1 公顷
建筑面积：495 402 平方米
设计说明

项目作为滨海新区行政、服务、会议、展示、接待、办公等各种活动的综合型商务中心，在功能构成、人流疏散、自然环境等方面进行了综合考虑。

项目布局从整体出发，强调轴线关系与空间序列。总体布局为“和”字形格局，表达追求“和谐”的中国传统人文理念。设计整体风格统一协调、大气庄重、现代明快，突出绿色建筑，节能减排的设计理念。建筑组群和外部环境相互协调，互为呼应。

Binhai Cultural Business Center

Construction Location: Tianjin
Floor Area: 41.1 ha
Building Area: 495,402 m^2
Design Introduction
As the comprehensive business center of Tianjin Binhai New Area for administrative, service, conference, exhibition, reception, office and other activities, this project has taken many aspects into consideration like function composition, crowd dispersal, natural environment, etc. Based on the overall layout, the project stressed the relationship between the axis and spatial sequence. Its overall layout is in shape of Chinese "character", so as to convey the Chinese tradition of humanism that is in pursuit of "harmony". The overall style is harmony, solemn, modern and bright, and its design idea is in highlight of green building, and energy conservation. Architecture Groups echo with the external environment to create coordination and interactive effect.

天津碧桂园凤凰酒店

建设地点：天津
设计时间：2007 年 4 月—2008 年 7 月
占地面积：21 904 平方米
建筑面积：300 646 平方米
设计说明

建筑物布局依据规划要求于用地界线内做了相应的退让，形成不同区域的主题广场，便于人流集散及假日休闲活动。项目与月亮岛一水之隔，北、东两侧滨临海河，与海河北岸的"海河外滩公园"隔河相望，交通便利，是响螺湾商务区内位置极佳的地块。

项目充分利用了得天独厚的景观资源，吸引着大量前来与会、购物、娱乐、住宿的人流，提升了大型建筑综合体的经济价值。设计重视环境的利用及保护，采用环保型建材、屋顶绿地等措施，达到对环境的利用及保护的目的。

Tianjin Bigui Garden Phoenix Hotel

Construction Location: Tianjin
Design time: 2007. 04 – 2008. 07
Floor Area: 21,904 m^2
Building Area: 300,646 m^2
Design Introduction
As planning requires, its building layout makes the corresponding concessions within the boundaries, as a consequence, it forming various subject squares in different regions, to facilitate the flow distribution and holiday recreation. The project next to the Moon Island only across a stream, bordering the Haihe River at its north and east side, overlooking "Haihe Bund Park" at the north shore across the river, which is an excellent block convenient in communication location in Xiangluo Bay business district.
This project takes full advantage of the unique landscape resources, attracting a large number of people come here for gathering, shopping, entertainment, and accommodation, that means great enhance in economic value promoting as a large building complex. In Design, it emphasizes on the usage and protection of the environment, adopting environment-friendly building materials green roofs and other measures to achieve environmental use and protection purposes.

中国水利博物馆

建设地点：浙江 杭州
用地面积：77 000 平方米
建筑面积：32 524 平方米（地上）
设计说明

项目坐落于杭州市萧山区"中国国际水城"南入口处，项目高度为 130.28 米，北侧为钱塘江大堤，南侧为规划道路，东侧为浙江省农业高科技示范园区，西侧为钱江世纪城。项目基地内地势平坦、开阔，基地中部拥有一片大面积人工湖。建筑本身位于人工湖水面南侧，四周无其他建筑物。建筑的使用耐久年限为 100 年。

China Water Conservancy Museum

Construction Location: Hangzhou, Zhejiang
Floor Area: 77,000 m^2
Building Area: 32,524 m^2 (on the ground)
Design Introduction
This Project is located at the South entrance of "China International Water City" in Xiaoshan District, Hangzhou city, in height of 130.28 m. It faces to the Qiantang River levee in the north, the planned road to the south, Zhejiang High-tech Agricultural Demonstration Zone to the east, and the Qianjiang Century City to the west side. This venue is characterized by flat and open, with a large artificial lake in the central space. The building itself is located in the south of artificial lake, no other buildings nearby surrounded. Use durable life spans 100 years.

万丽天津宾馆

建设地点：天津
设计时间：2009 年 2 月—2009 年 7 月
占地面积：15 523.63 平方米
建筑面积：95 534 平方米

设计说明

1．注重总图规划、交通组织、景观生态、消防车道等设计。
2．建筑空间完整、功能分区明确、设施完备、硬件标准一流。
3．设计充分考虑节能、生态、环保问题并合理利用土地。
4．立面造型尊重城市、尊重环境，通过塑造建筑独特的形象来体现建筑自身的性格，提升该地区整体视觉品质，与周边环境相互融合。
5．项目定位为白金五星级酒店，建筑内外品质均达到国内一流。

Renaissance Tianjin Hotel

Construction Location: Tianjin
Design Time: 2009. 02-2009. 07
Floor Area: 15,523.63 m^2
Building Area: 95,534 m^2

Design Introduction

1. Focus on the Overall Planning, traffic, landscape ecology, firefight road design, etc.
2. Complete architectural space, clear functional area, complete facilities, and top standard hardware.
3. Give full consideration to energy-saving design, ecology, environmental protection, and rational use of land.
4. Elevation modeling respects for the city and the environment, to create a unique image of the building to reflect the architectural character of its own, to enhance the overall visual quality of the region, better merging with the surrounding environment.
5. Platinum five-star hotel standard, builds first-class quality brand inside and outside the building.

天津市胸科医院

建设地点：天津
用地面积：91 413 平方米
建筑面积：96 000 平方米
设计时间：2008 年
项目状态：在建
病 床 数：850 床
门诊人次：2500 人次／日门诊

设计说明

1、“以患者为中心”即在设计中体现为“以水平流线为主”，为心胸疾病患者提供最便捷的流线。门诊医技设两层，住院部分为七层。门诊住院围绕医技形成门诊、急诊、住院部包围医技的围合状功能布局，其间通过医院街联系，流线更紧凑救治最便捷住院楼内外科分区设置，使患者与医技之间实现流程简化、时间短、效率高，便于管理，有利患者治疗与康复的目的。

2、“经济、节能”即在设计中体现为降低建筑楼层层数，即裙房两层，主楼六层，建筑结构经济、抗震好、建设周期短，并且利于紧急情况疏散，减少遮挡“模块化单元”与内庭院单元利于诊室等房间采光通风，使单元设备能源使用可独立控制，能耗降低，地下室餐厅等可利用两栋住院楼“V”字形布局进行采光，减少楼横向长度，形成丰富的景观，减少对建筑北向的遮挡，使两侧房间均能采光。

Tianjin Thorax Hospital

Construction Location: Tianjin
Floor Area: 91,413 m^2
Building Area: 96,000 m^2
Design Time: 2008
Status: under construction
Number of beds: 850
Visits: 2500 visits per day

Design Introduction

1. "Patient-oriented" To provide with the most convenient flow line for heart disease patients in essence of "mainly in flow line", the out-patient Medical technology has two storeys, while the hospital apartment occupies seven storeys. In this way, it forms an enclosure-like layout i.e., "hospital, emergency and outpatient around Medical technology", among which connected by Hospital Street, so that the flow line is more compact in the most convenient. The "hospital building with individual Med/Surg" Patients and Medical Technology processes is shorten, efficient, manageable, and favorable for treatment and rehabilitation of patients.

2. "economic& energy-saving" Presentation: "lower layer" two-storey podium, six-story main building, economical structure, better seismic, short construction period, better emergency evacuation, reduce the block between "modular unit" and the units within the courtyard, which contributes to lighting ventilation for consulting room and other rooms. Its unit equipment can be controlled independently to save energy, and the basement restaurants and other departments are with floorlight. The two hospitals is in V - shaped layout, aiming to reduce the horizontal length of the floor, and also form a rich landscape as well, besides, it can also reduce the block to the north, thus both sides in the room gains lighting.

天津数字电视大厦

建设地点：天津
用地面积：24.6 公顷
建筑面积：302 800 平方米

设计说明

项目包括数字电视大厦主楼（高度为 150 米）、演播中心、综合服务中心、多功能剧场、奥运国际电视新闻中心、数字广播大厦和数字网络大厦。

建筑设计针对特定的场地和功能，汲取中国传统城市的营建理论和造园手法，合理有序的规划了各个不同功能区域，强化了“广电城”的设计理念，同时，将对公共建筑与城市生活、科技理念与人文思想、标志性与大众性、务实原则与建筑意向、近期效果与可持续发展等一系列关联性问题进行思考和探索，并将这些思考与探索充分融汇于设计方案之中。

演播中心将各个演播厅平铺布置，通过横纵两条生态长廊联系在一起，功能明确、空间丰富，塑造了独具广电建筑特色、时代精神和地域特征的建筑群体形象。

Tianjin Digital TV Mansion

Construction Location: Tianjin
Floor Area: 24.6 ha
Building Area: 302,800 m^2

Design Notes

The project includes the main body of digital TV Mansion (150 m high), studio center, multi-service centers, multi-functional theater, Olympic international television news centers, digital broadcasting mansion and digital network mansion.

In respect to specific site and function, its architectural design learns from theory and practices of traditional Chinese city garden construction, reasonably and orderly planned different areas with various functions, and strengthen the "Radio City" design concept; meanwhile it also integrates pondering and exploration to a series of relevant relationship between public buildings and urban life, science and technology concepts and human thinking, iconic and popular, practical principles and architectural interest, recent effective and sustainable development issues, etc., taken into fully design consideration.

Tile studio center makes a tile layout for individual studio, which linked by two horizontal and vertical corridors, thus creates definite function, and abundant space as well. Overall it presents an group image of the building with unique R&T architectural features, era spirit along with specific geographical features.

津湾广场

建设地点：天津
用地面积：110 000 平米
建筑面积：600 000 平米
设计时间：2008 年
竣工时间：2009 年（一期竣工）

设计说明

项目是天津市中心城区建设中塑造区域空间景观、带动提升区域综合环境的重要组成部分。津湾广场沿海河由北向南分为 3 个层次塑造城市空间。

第一层次，沿河设多层建筑，建筑高度控制在 21.95 米左右。建筑体量及尺度为解放北路现有建筑的延伸；

第二层次，由 1 幢 18 层的建筑及青年宫组成的，建筑高度在 90 米左右，构成城市中高尺度的天际轮廓线；

第三层次，由 3 幢分别为 180 米、235 米、299.8 米的超高层建筑组成。随着金融城项目的展开，现在的解放北路即将重拾昔日的辉煌。未来的解放北路将成为以金融业、旅游业为主的商务发展区。

Jin Wan Plaza

Construction Location: Tianjin
Floor Area: 110,000 m^2
Building Area: 600,000 m^2
Design Time: 2008
Completion Time: Phase I in 2009

Design Introduction

This project plays an important part in Tianjin Central City construction to shape the landscape of regional space, and promote enhancing the integrated environment of the region. Jin Wan Plaza can be divided into three levels along Haihe River from north to south, to shape the urban space:

the first level: multi-storey buildings along the river, building height within 21.95 m. Its building body and scale is an extension of the existing building, Liberation North Road;

The second level: composed by a 18-storey building and the Youth Center, building height around 90 m, constituting a high-scale skyline for the city;

The third level: composed by three super high-rise buildings respectively in height of 180 m, 235 m, and 299.8 m. With the development of Commerce City Project, the current Liberation North Road, will soon regain its former glory. In the future, Liberation North Road will be an financial industry and tourism-based business development area.

天大建筑

Design changes the world

天津大学建筑设计规划研究总院创建于1958年，是国家甲级建筑设计、甲级城市规划设计、甲级文物保护工程勘察设计、甲级工程咨询单位，研究院依托于国家重点大学人才优势、学科优势和技术优势，凭借先进的技术装备和严谨的工作作风，近年来研究院在全国范围内完成的各种类型、各种规模的建筑设计、城市规划设计屡获各类奖项，受到项目委托单位及社会各界的好评。

研究院现有一线设计人员260余人，其中45人为国家一级注册建筑师，35人为国家一级注册结构师，15人为国家注册城市规划师，拥有硕士学位的员工占全院总人数的40%以上。研究院专业配套齐全，人员素质较高，单位在教育和科研建筑、旅游和疗养建筑、办公和商贸建筑以及大型医院建筑、体育建筑设计和城市规划、古建筑保护规划等方面都有不同的成就。其中“天津美术学院美术馆”、“郑州大学工科园”、“天津市滨海新区响螺湾商务区城市设计”、“深圳欢乐谷”、“宝鸡市青铜器博物馆”、“天津蓟县地质博物馆”等项目在国内外影响深远。近几年来，研究院在天津海河重大开发项目的国际竞赛中，曾力挫群雄，摘取一、二等奖的桂冠，同时研究院在工业建筑设计、古建筑复原设计、设计监理、深基坑支护以及建设项目策划和可行性研究等方面也具有很强的竞争实力，业务范围已遍布全国各地，并已开始走出国门，迈进世界市场，取得了骄人业绩。

“实事求是”是天津大学的校训，它也已经深深地融入本院的企业文化与服务理念之中。特色鲜明、风格独特的天津大学建筑设计规划研究总院，愿为中国的城市建设精心勾绘最具地方特色、最具竞争力、最具时代精神的发展蓝图。

Established in 1958, Research Institute for Planning, Architectural Design of Tianjin University is ranked as national Class A in architectural design, urban planning and design, survey and planning in heritage protection project, and engineering consulting unit. Relying on the advantage of national key university in talent, disciplines and technology, by virtue of advanced techniques, equipment and strict working style, our research institute has been successively awarded many awards in nation-wide various types and scales of architectural design, urban planning and design in recent years, and has won praises from project commissioned units and community.

The academy now owns more than 264 front-line design staffs, of which 45 are registered architects at national Class-A, 35 are registered mechanical engineer at national Class-A, 15 are national registered urban planners, and the master's degree holders occupy more than 40% of the total number. Complete professional composition and high-quality personnel contribute a lot in achievements such as education and scientific research construction, tourism and convalescence construction, office and commercial construction, large hospital buildings, sports architecture design, urban planning as well as protection of ancient buildings, etc. wherein "Gallery of Tianjin Academy of Fine Arts", "Technology and Science Park of Zhengzhou University", "Urban Design for Tianjin Binhai New Area Xiangluo Bay Business District", "Shenzhen Happy Valley", "Chinese Bronzes Museum"," Tianjin Jixian Geological Museum" and other projects has won reputations at home and abroad, prevalent in the world. In the international competition of major projects on developing Haihe River in Tianjin in recent years, it has beat around and crowned the First and Second Prize respectively. It is also strongly competitive in respect of industrial building design, ancient architecture recovery design, design supervision, deep excavation support, construction project planning and feasibility studies, etc. Its business has spread all over the country as far as joining the world market, and has achieved remarkable results.

As the motto of Tianjin University, "Seeking truth from facts" has been deeply integrated into our corporate culture and service philosophy. With distinct features and unique style Academy of Architectural Design & City Planning Tianjin University is willing to provide city-building in China with the carefully drawn development blueprint which has the most local characteristics, the most competitiveness and most of the spirit of the times.

天津大学建筑设计规划研究总院

Academy of Architectural Design & City Planning of Tianjin University

地址：天津市南开区卫津路92号 邮编：300072 电话：+86–22–27404753 传真：+86–22–27401845

Address: Wei Jin Road, No.92, Nankai District, Tianjin P.C.: 300072 Tel : +86-22-27404753 Fax : +86-22-27401845

邮箱：td_design@126.com 网址：www.aaut.net.cn

E-mail: td_design@126.com Http: //www.aaut.net.cn

1–3 天津大学26教学楼

建设地点：天津大学
设计时间：2007年10月
竣工时间：2009年5月
建筑规模：5万平方米
获奖情况：2010年天津市“海河杯”一等奖

由于历史的原因，校园核心空间轴线两侧校区的占地面积、建筑风格、基本色调、墙面肌理都存在较大的反差。我们把26教学楼的设计当作一次寻求对话、整合校园、强化特色、传承文脉的机会，在设计中，通过完整统一的形态、较大的建筑尺度在和棕褐色陶土砖的外墙肌理，力求使其与湖对岸历史建筑在体量上均衡、在视觉上的呼应，从而强化校园核心空间，营造沉稳、质朴的校园整体氛围。

1

2

3

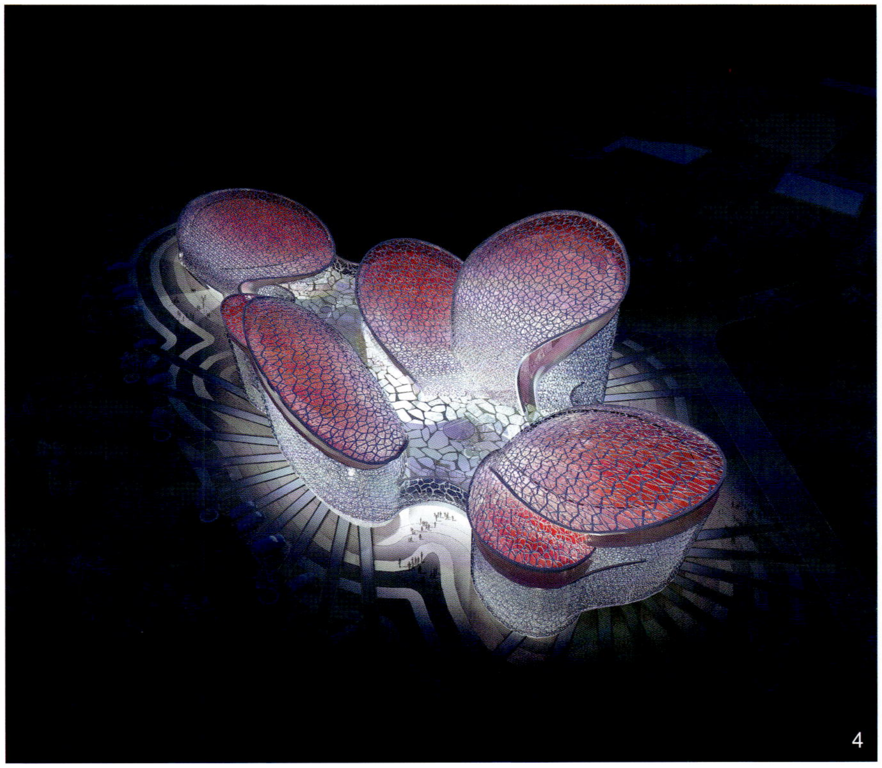
4

5

4–5 重庆巾帼园（妇女文化中心）

建设地点：重庆市
设计时间：2009年
建筑规模：2万平方米
结构/材料：框架结构/彩釉玻璃
中标情况：扎哈　哈迪德、荷兰MVRDV、日本URBAN株式会社、美国柏森设计公司等五家著名境外公司竞标，中标，获得一等奖

“绽放”的核心设计主题

作为重庆妇女的文化中心，如果不能恰当和充分地展现出重庆妇女独有的魅力和风采，那真是最大的遗憾！

重庆女人有着女人共有的善良，温柔，勤劳的共同点，更有着美丽，时尚，自信，泼辣的独特个性。什么样的建筑形态才是最恰当的表现呢？终于，一个充满激情的建筑形态让我们心潮澎湃，这就是“绽放”的形态！“绽放”的美是一种动态的美，自由的美，丰富的美。它不仅仅象征着重庆女人像山茶花一样美丽的外表，更体现出一种时代女性特有的动感。

无论是从外观表情，还是从内心感受，我们相信，“绽放”是每个时代女性都想要诠释，想要表达的主题。

6

6–7 曹妃甸国际论坛永久会址

项目地点：河北省曹妃甸
设计时间：2010年
中标情况：击败英国伟信建筑设计公司、美国东肯国际设计公司一举中标

7

“曹妃甸国际论坛”建设项目位于曹妃甸生态城“龙湾”的“龙头岛”上。

曹妃甸本是一片荒滩湿地，我们希望通过一组充满动感的建筑群体来展示一个从荒滩小鸟身上的羽毛最终变成凤凰羽翼的奇妙的过程，并以此体现曹妃甸从一个海边荒滩变成国际生态城的伟大奇迹，同时体现出曹妃甸国际论坛“可持续发展”的永久主题。

我们希望通过这个形似凤凰羽翼的建筑，明确地体现出“曹妃甸国际论坛”所在地——唐山——的特点，更体现出“曹妃甸国际论坛”建筑区别于国内外所有其他国际论坛建筑的唯一性和标志性。

这个凤羽般的建筑群采用了集中与分散相结合的建筑布局形式。中央是大型会议区，两侧分别是酒店餐饮和高端接待区，高端接待区采用别墅式酒店的形式，以满足元首级贵宾的特殊需要。三个部分相对独立又有联系，随着论坛知名度提高，不同部分还留有扩建和发展的空间。

为了保障工期，我们采用了“铺开工作面”的做法，把造型活泼的低层会议区与标准、简洁的高层酒店结合起来，并采用大量钢结构装饰构件，把施工期间的现场作业与工厂加工结合起来，既缩短工期，又保障质量。

凤羽般的论坛建筑与“龙头岛”的整体自然环境融合在一起，岛上利用中水回用、雨水收集、清洁能源等手段实现岛上自身的循环和平衡，让“龙头岛”成为一个展示生态技术的封闭单元。

灵动的建筑群可以布局在岛上的不同地点，创造出多样的建筑景观，我们最终让凤羽建筑变成龙的眉毛，并与龙的眼睛形成一个整体，突出“龙头岛”的主题，同时在岛的南侧，规划一个大型庆典广场，将为在10月份唐山秋高气爽的季节召开的“曹妃甸国际论坛”提供一处与大自然融合的海边大型集会庆典场所，与论坛主体建筑一起，创造出大气、开放的整体空间景观。

我们希望这个凤羽般的建筑轻盈地落在“龙头岛”上，能与龙湾和谐地融合在一起，给人们带来永久的平安和吉祥，并呈现出极具中国特色的龙飞凤舞，龙凤呈祥的完美格局。

8–9 古雷半岛概念发展规划

建设地点：福建省漳州市
设计时间：2010年4月
建筑规模：68平方千米

规划引入低碳、绿色城市等理念，建立生态安全与新风水格局，营建生态谷，使多元复合的城市功能与自然山海景观共生，古雷半岛与古雷港工业互为支撑，是一个具有海岸风情和高标准环境的生态新城，将成为海峡两岸合作交流的重要平台和示范区。

8

9

10

10–14 漳州元光文化生态园概念规划

建设地点：福建省漳州市芗城区
设计时间：2009年9月
建筑规模：8.34平方千米
中标情况：中标项目

发展理念：全面保护、和谐联动。规划设计全面保护历史遗存，保护山水生态，保护风水格局，同时利用多种界面带状带动关联产业的发展，以旅游带动人气，以人气提升区域价值。

建设理念：综合开发、多向发展。规划设计将历史遗存与现代消费结合，将"朝圣文化、特色旅游、商业地产、新家园"多重叠加，多向发展，以聚集人气，获得全面收益。

景观理念：山水大空间、风水大格局。规划设计尊重地域特色和风水格局，借山理水，以山之自然美、水之龙形气势，烘托圣王墓地的千年风水格局，以显宝地之灵气，汇四方之人气。

文化理念：圣王文化、闽台乡情。规划设计围绕开漳圣王文化，主打文化牌、乡情牌，利用圣王文化，彰显漳台文化交流主线，打造引人入胜的活力地带。

11

12

13

14

15

16

15–16 天津利顺德大饭店修缮改造工程

建设地点：天津市
获奖情况：2010年天津市"海河杯"一等奖

设计师在"保持建筑原形制、原结构、原材料、原工艺的原则基础上进行修缮设计"。缺失部分的修补必须与整体保持和谐，并使修缮不扭曲其艺术或历史见证。

这项改造工程最大限度地保护和延续文物建筑的历史信息，体现对历史的尊重。扩建建筑修缮应充分考虑其对旧址建筑的影响，协调建筑之间的关系，使得建筑整体价值最大化，同时尽量借鉴旧址建筑的立面符号，避免建筑之间形态上的冲突，以达到建筑整体的协调统一。在满足国家关于节能的相关规范和行业标准的基础上，综合考虑先进生态技术的运用，以实现建筑的可持续发展，达到良好的社会效应。

17

18

19

17–19 天津市宝坻低碳工业园

建设地点：天津市宝坻区
设计时间：2009年8月
建筑规模：12.7平方千米

宝坻低碳工业园区是天津市示范产业园区，以国际先进的低碳产业技术研发、孵化、产业化为主导，推广应用最新节能环保技术，实现二氧化碳等温室气体低排放的工业园区。

宝坻低碳工业区位于宝坻城区南部，津蓟高速公路九园出口处的九园工业区内，北侧紧邻京津新城，是宝坻区离天津市最近的工业区。规划范围东至潮白河，西至宝白公路，南至九园公路，北至青龙湾引河，规划总用地面积为12.7平方千米，起步区用地面积2.0平方千米。

目标定位：结合宝坻的交通区位及生态环境优势，将其建设成功能辐射京津，充分展现京津新城全新、生态、环保形象的国家级低碳工业示范园区。

园区主要功能涵盖三大类型：高附加值低碳制造产业，新型研发创制产业，总部经济、会所、公寓等配套服务产业。

总体构思将生态的观念引入城市设计，通过对“中心、组团、廊道、轴线、路径”等主要城市设计要素的提取，宝坻低碳工业园区力图创造一个功能明确、分区合理、自然生态环境优良的高效率的现代工业园区，在此基础上形成的城市设计结构体系可以用“点轴相连、组团分区、生态渗透”12个字来概括。

整体空间布局规划依照重点突出，主次分明的原则形成三个主要的区域功能中心，并同时向周边区域辐射扩散，进而构成中间高、四周低的空间形象。产业岛、高科技中心岛、低碳产业研发岛、总部经济会议岛、蜿蜒曲折的河流、生态的绿地穿插其间，共同组成疏密有致、层次丰富的景观。

20

20–21 泉州台商投资区总体规划

建设地点：福建泉州
设计时间：2009年11月至今
建筑规模：规划用地范围200平方公里，主城区远期用地规模82平方公里
中标情况：中标项目

山海相连，指状生长，百里青山入半城。依托泉州在海西区重要的地理位置和规划区域内优良的山水环境，规划力求打造先进制造业和高端服务业支撑的滨水生态新城，形成泉州湾创新服务平台、东部城市新地标和全国对台经贸合作网络上的重要节点。

21

22 天津大学生体育中心

建设地点：天津工业大学　　建筑规模：3万平方米
设计时间：2010年　　中标情况：中标项目

此方案设计理念与天津工业大学校园文脉相契合

建筑体型方正平直，与校园的其他建筑相呼应。最有特色的外观设计是方形与菱形网络的铝合金方通表层，它恰如其分地展示了服装工艺疏密有致的编织手法，加之以金色为主的外层结构，能给人以耳目一新之感。

力求拟建低碳形“绿色建筑”

本方案设计了柱状天窗，并以此作为自然采光通风构件，根据实际条件，采用地源或水源热泵空调系统，以及雨水，空调热回收技术，同时大量采用 Led光源，不但节约了宝贵的能源，同时在夜间创造出繁星闪烁的景观效果。

设计方案解决了校园建筑与城市重要节点相融的关系。稳重大气的建筑风格在全国的大学体育建筑中独树一帜，建筑提现了天津大学生朝气蓬勃的精神风貌，更展现了天津大都市的魅力与风采。

23–24　1895天大建筑创意大厦

建设地点：天津市南开区　　竣工时间：2010年　　中标情况：中标项目
设计时间：2009年　　建筑规模：3.7万平方米

作为南开区2009年第二批科技发展二十大重点项目之一的天津大学建筑设计规划研究总院新办公大楼——“1895”天大建筑创意大厦——于2009年7月30日举行奠基仪式，设计总院洪再生院长主持仪式，天津大学校领导和南开区领导及有关部门负责人共同为奠基石培土。

1895天大建筑创意大厦位于鞍山西道科贸街东部，南临鞍山西道，东临卫津路，西临天大天财公司，北临天大北五村住宅区，建设规划用地8 400平方米，主体地上10层，地下1层，总建筑面积37 000平方米，于2010年底竣工。大厦建成后将成为天津大学建筑设计规划研究总院的办公大楼和天津大学建筑相关学科知名教授的创业研发及转化基地。大厦的建成改善了设计总院的办公条件，提升了总院形象，并且符合天津市整体发展规划、南开区创意产业的定位，彰显天津大学建筑品牌优势，支撑天津大学建筑学科发展。天津大学建筑学科发现重点在建筑行业再生能源利用、生态建筑技术转化、建筑科技咨询、建筑新技术方面进行研究及推广应用，引进如建筑规划软件开发的IT公司、高端效果图制作公司等创意产业，使建筑设计与创意产业结合，使企业与创意设计结合，直至完成终端产品，引进与建筑规划创意产业相关的正版化设计软件，构建软件服务平台，充分实现网络资源高度共享，在知识产权维权意识不断增强的形势下，为中小企业提供正版资源，节约中小科技创意企业在软件方面的投资，实现包括建筑项目前期策划、立项咨询、科研编制、方案设计、项目代建、工程招标代理、房地产评估、环境评价、节能设计、日照分析、广告设计、动漫研发、专利代理、财务审计等环节在内的建筑创意产业“一条龙”服务，形成高技术产业集群，同时，加强国际交流合作，提高天津建筑水平。此外，大厦中将设有国际开放工作室、教授工作室和艺术沙龙等，使之成为创意产业的孵化器，成为天津大学的重要基地。

25–26　郑州高新区孵化器

设计部门：彭院士设计工作室与设计五所合作
建设地点：郑州市高新区
总体规模：110.245平方米
获奖情况：教育部（2008）优秀勘察设计评选 三等奖

天津市交通建筑设计院

TIANJIN COMMUICATION ARCHITECTURAL DESIGN INSTITUTE

地址：天津市河西区友谊路13号
邮编：300201
电话：+86–22–28138698
传真：+86–22–28138698
邮箱：jtjyzgb@126.com

Address: No.13 Youyi Rd., Hexi District, Tianjin
P.C.: 300201
Tel: +86-22-28138698
Fax: +86-22-28138698
E-mail: jtjyzgb@126.com

建筑　规划　港工　水运　道路　环境　咨询

天津市交通建筑设计院是以建筑工程设计、水运工程设计、规划设计、道路工程设计、工程咨询为主体的甲级设计院。设计院可向国内外提供工业与民用建筑、水运工程（港口及航道）、物流业工程提供项目建议书、可行性研究报告、规划设计、初步设计、施工图设计、技术咨询等服务。

设计院以满足顾客要求为宗旨，强化服务管理和质量管理，多年来我院坚持"规范、先进、安全、经济、美观"的设计原则，竭诚与新老顾客合作，并为顾客提供优质的设计、满意的服务。

企业理念：
以质量求生存、设计创品牌；
以服务赢市场、诚信建和谐；
以产品树形象、创新谋发展；
以团队铸事业、高效展未来。

您的远见，加上我们的智慧，是完成一件旷世作品的前提；
我们的责任就是帮助我们的客户实现他们的梦想；
我们坚持为客户规划出最详尽的宏伟蓝图，用我们的努力给我们的客户提供最大的利益保障，同时通过巧妙的技术、完美的构思和换位的思考来确保您梦想的实现。

Architecture; Planning; Port Engineering; Water transportation; Road; Environment; Consultation

Tianjin Communication Architectural Design Institute is a Class-A design institute mainly engaged in architectural engineering design, water transportation design, planning and design, road engineering design, engineering consultation, etc. We can provide services such as project proposal, feasibility study report, planning and design, preliminary design, construction drawing design and technical consultation for the industrial and civil construction, waterway engineering (port and waterway) and logistics engineering at home and abroad.

Our institute strengthens the service design and quality management, upholding the aim of meeting the customers' requirements. Over these years, we insist the principle of " Standard, advanced, secure, economic and aesthetic" in our design work, striving to provide the best deisgn works and services to cooperating customers .

Our enterprise Idea is :
surviving by quality; creating brand by design;
winning the market by service; building harmony by sincerity;
establishing the image by product; seeking development by innovation;
pursuing career by team cooperation; molding future by efficiency.

Your vision, combined with our intelligence, is the prerequisite for the completion of a masterpiece.
Our responsibility is to help our customers achieve their dreams.
We stick to provide the most detailed and grand blue print for our customers and do our best to guarantee our customers' benefit, realizing their dream by crafty skills, perfect design and the way of thinking what the consumers think.

天津张贵庄出入境边防检查站
建筑面积：7 800 平方米
设计时间：2009 年 5 月

Exit-and-entry Inspection Station of Zhangguizhuang, Tianjin
Building Area: 7,800 m^2
Design Time: 2009. 05

1 天津港公安局东疆公安基地
建筑面积：11 000 平方米
设计时间：2009 年 4 月

Police staion Dongjiang base of Tianjin port
Building Area: 11,000 m^2
Design Time: 2009. 04

2 天津市永悦物流中心
建筑面积：160 000 平方米
设计时间：2009 年 12 月

Tianin Yongyue logistic center
Building Area: 160,000 m^2
Design Time: 2009. 12

3 天津市大寺镇社区文化活动中心
建筑面积：8 800 平方米
设计时间：2009 年 3 月

Community Cultural Activity Center of Dasi town, Tianjin
Building Area: 8,800 m^2
Design Time: 2009. 03

4 南洋电缆厂办公楼
建筑面积：90 000 平方米
设计时间：2009 年 11 月

Building office of Nan Yang Cable
Building Area: 90,000 m^2
Design Time: 2009. 11

5 内蒙古自治区化德县第一小学教学楼
建筑面积：10 000 平方米
设计时间：2009 年 1 月

Dehua No.1 primary school in Inner Mongolia
Building Area: 10,000 m^2
Design Time: 2009. 01

6 天津市镇海县唐官屯加工物流园
建筑面积：10 平方千米
设计时间：2009 年 8 月

Tangguantun processing and logistics park in Zhenhai county ,Tianjin
Building Area: 10 km^2
Design Time: 2009. 08

7 宁河新世纪广场
建筑面积：90 000 平方米
设计时间：2009 年 12 月

Ninghe New Century Square
Building Area: 90,000 m^2
Design Time: 2009. 12

天津天咨拓维建筑设计有限公司

TIANJIN TIANZI TUOWEI ARCHITECTURAL DESIGN CO.,LTD

天津天咨拓维建筑设计有限公司于1999年成立于天津，公司设计范围涵盖建筑设计和城市规划，包括酒店、办公、住宅、商业、学校、文化设施等内容，并重视发展城市景观设计、建筑设计与室内设计的衔接与渗透。在建筑设计日益专业化、开放化、市场化的背景下，公司于2004年进行了企业改革，并以现代企业制度的要求做到了公司的运作规范化、管理先进化、模式专业化。公司设计资质为建筑综合甲级资质。

公司拥有一支由60多人组成的优秀的设计团队，其中设计骨干多毕业于清华大学、天津大学等知名学府。通过流畅高效的团队合作，设计师将卓越的创作实力和丰富的工程经验融入公司的项目之中。设计团队的哲学是建立在理解客户需要、要求和目的的宗旨上的。每个精心为业主营造的项目，都体现了设计者对项目现实背景和文化环境的思索，对地域物理环境的尊重，以及对社会经济、艺术审美等方面的综合考虑。因为公司以此作为设计的出发点，所以公司的设计卓有成效。一流卓越的技术，已经建立的在合理预算基础上的完美运作，以及与多方面长期愉快的合作关系是公司追求的设计目标。

成立十年间，公司通过对先进设计理念和建筑技术的整合，博采众长、精心研究，已完成各类重大工程项目百余项。建成项目多次获得天津市优秀建筑设计奖。

团队精神

方案创作强调以设计主创人为中心，设计组随时研讨，根据分工，同步协调展开设计工作，形成互动式的工作状态。工程设计强调以工程主持人为中心，主创人对整体空间处理及细部效果可根据工程设计进度同步把握。

创新意识

从机械的工作中跳出来，去体验活生生的、具有创造性的现实生活。我们主张多视角、多思维，不排斥急功近利的消费文化，推崇具有个性的原创文化。

服务意识

在建筑创作活动中，我们学会了换位思考，更多地去倾听、分析和满足业主的需求，建立诚信的合作关系。

强调行为自律和职业道德。在创精品与追求经济利益之间，强调道德与良好习惯的平衡。

The team spirit

For the schematic creation, the main designers are centered with coordination of the project designers.

For the project design, the project designer in charge is to control the space design and detail impression according to the design process.

Creation consciousness

We expect to extricate ourselves from the mechanical working and experience lively and creative life.

We maintain the consumption culture with quick result and instant profit and assert multi-views, multi-thinking and . And we appreciate the original creative culture with special characteristics.

Service consciousness

During the architectural creation, we have understood how to listen, analyze and meet the requirements of the clients based on their conditions and to establish honest cooperation.

We emphasize in corruptive behavior and professional moral. Both of the moral and the sound habit should be balanced during creating the excellent works and concentrating on the economic benefit.

天津市北辰教育中心
主持建筑师：方轶 徐珩
建设地点：天津
建筑面积：70 000 平方米
设计—竣工时间：2008 年—2009 年

项目建成后将是北辰区重要建筑节点之一，其 94 米的高度也将成为天津市教育建筑中的第一高楼。规划设计强调为学校提供具有完备功能设施的校园规划，强调形式多样化和个性化的校园空间。教学区具有严谨、理性又不失活泼和多样化的特点。生活区建筑体形自由、灵活，在强调创造生活氛围的同时又能与教学区的建筑很好地融合在一起。

Beichen Education
Principal in Charge: Yi Fang, Xu Heng
Construction Location: Tianjin
Building Area: 70,000 m^2
Design / Completion Time: 2008-2009

This project after completion will become one of the important focal-point constructions in Beichen District, and its height of 94 meters will make it the highest building in Tianjin's educational architecture. The design focuses on a campus planning which can provide comprehensive facilities for the school and a campus space of diversity and personality. The teaching area is rigorous and rational, yet still lively and diverse. The building shape of the living area is free and flexible, creating an atmosphere of life and integrating perfectly with the buildings in the teaching area.

天津红桥区红星职专

主持建筑师：宝建华 张杰

建设地点：天津

建筑面积：48 000 平方米

设计／竣工时间：2007 年 9 月—2009 年 10 月

本项目规划从学生心理特点出发来进行空间规划。校园构图规整、结构理性、形象现代严谨，这种严谨、理性的空间形象体现了现代教育的整体风向。

由功能结构自然分为三个区域：教学区，生活区，高层办公和运动区。在每个区域之间都设有一个主通道，与校园的前广场相连。弧形的入口大厅，将三个功能区域连接起来。通过入口大厅，全校师生可以方便地到达校内各个功能区域。通高的玻璃幕墙与宽大的前广场交相呼应，形成建筑与景观以及周边环境的有机结合。

Red Star Secondary Vocational School in Hongqiao District, Tianjin

Principal in Charge: Bao Jianhua, Zhang Jie

Construction Location: Tianjin

Building Area: 48,000 m^2

Design/Completion Time: 2007.9-2009.10

The space planning of this project is carried out based on the psychological characteristics of students at school. The campus is regular in composition, rational in structure and modern, rigorous in image. This rigorous, rational spatial image shows the overall direction of modern education.

The campus is naturally divided into three area by its functional structure: the teaching area, the living area and the high-rise office building & sports area. There is a main passageway between each region, which leads to the square in the front of the campus. The curved entrance hall connects these three functional areas together, through which teachers and students can easily reach anywhere in the campus. The high glass curtain wall echoes with the spacious square, producing an organic integration of the architecture, landscape and the surrounding environment.

邢台市中兴东大街片区项目
主持建筑师：方轶 张阳 李雅娟
建设地点：河北 邢台
建筑面积：500 000 平方米
设计时间：2010 年

Zhongxing East Avenue Area Project in Xingtai
Principal in Charge: Fang Yi, Zhang Yang, Li Yajuan
Construction Location: Xingtai,Hebei
Building Area: 500,000 m^2
Design Time: 2010

本方案力图将建筑塑造为一个融入城市环境的有机体，避免形成城市建筑的不连续性与随机性。商业外部形象大气、稳重，注重与周边地块建筑形态和色彩的协调统一。商业内部形象则以创造出全天候热闹的节日气氛为目的，多方面地运用了丰富的色彩搭配组合与精致的夜景灯光等设计手段。

This project is to create the construction into an organic integrity which is in harmony with the urban environment, avoiding the discontinuity and randomness in urban architecture. The external business image, gracious and staid, focuses on the coordination in form and color with buildings upon its surrounding land. The internal business image is to create a festive atmosphere around the clock by various design means such as rich color combination and exquisite night landscape lighting.

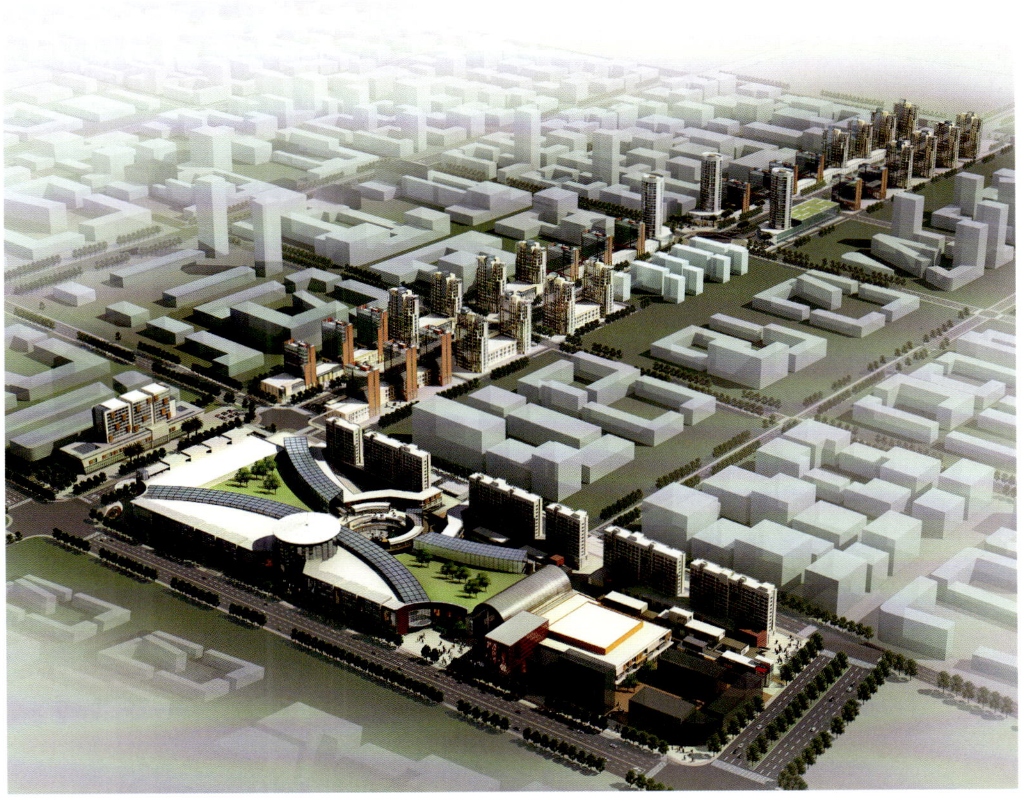

1

2

3

4

5

1．大港港东新城海滨园
建设地点：天津
建筑面积：77 000 平方米

2．天津雅安道体育公园
建设地点：天津
建筑面积：100 000 平方米

3．银川·青年 SOHO 规划设计
建设地点：宁夏 银川
建筑面积：160 000 平方米

4．北闸口示范小城镇城市中心区
建设地点：天津
建筑面积：669 000 平方米

5．汇淼温泉花园
建设地点：河北 保定
建筑面积：68 000 平方米

纵观以上五个项目，其商业或住宅的鲜明特色和独特品质见证了多年来我公司积累的丰富的住宅与商业建筑相结合的设计经验。对于住宅建筑而言，虽然技术难点相对较少，但是为了在市场上占有一定份额并满足业主的要求，公司针对市场特点研究了一套对住宅建筑设计项目行之有效的商业化运作方法和措施，而在商业建筑方面我公司专门设立了商业建筑专业化设计团队，每年组织专业的国内外考察和学术交流，参与大型综合商业项目的设计工作。在今后的工作中，公司有实力也有信心承接更有挑战性的商业住宅项目，通过各团队的合作创作出更满足市场要求的具有更高档次的优秀作品。

Seashore Park of Gangdong New City, Dagang
Construction Location: Tianjin
Building Area: 77,000 m^2

Tianjin Sports Park of Yaan Road
Construction Location: Tianjin
Building Area: 100,000 m^2

Planning and Design of Yinchuan • Youth SOHO
Construction Location: Yinchuan,Ningxia
Building Area: 160,000 m^2

The urban center of Beizhakou Demonstration Town
Construction Location: Tianjin
Building Area: 669,000 m^2

Huimiao Spa Garden
Construction Location: Baoding,Hebei
Building Area: 68,000 m^2

For these five projects, their distinctive characteristics and unique qualities of either commercial or residential construction have witnessed our company's rich design experiences in integrating the residential and commercial construction together, which is accumulated over the years. As far as the residential construction is concerned, although there are relatively few technical difficulties, in order to hold a certain market share and to meet the customers' requirements, our company come up with a set of methods and measures of commercial operation specifically dealing with residential construction projects according to the market characteristics. As for the commercial construction, our company establishes a professional design team of commercial construction, organizes researches and academic exchange at home and aboard for a specific subject each year and participates into the design operation for large-scale integrated commercial projects. In the future work, our company has the strength and confidence to undertake commercial and residential projects of more challenge, creating through team cooperation outstanding designs of higher grade which better meet the market requirements.

TSDI

铁道第三勘察设计院集团有限公司

TSDI

铁道第三勘察设计院成立于 1953 年，是以铁路、公路、城市轨道交通等勘察设计为主的跨行业科技型国家甲级大型综合设计院，是信用等级 AAA 企业，综合实力在全国勘察设计百强企业中名列前茅，拥有对外经营权，通过英国劳埃德质量认证公司颁发的 ISO 9001 质量体系认证，是全国勘察设计单位推行全面质量管理的第一个达标单位，在城市轨道交通、高速铁路、重载铁路、铁路客运专线、磁悬浮铁路等高科技领域也保有领先的技术储备和科研成果。

Established in 1953, the Third Railway Survey and Design Institute is considered as a large-scale National A-level comprehensive Design Institute, which majors in the multi-industrial Survey and Design for the projects such as railway, highway and urban track construction. Credited as an AAA-level institute, it is also listed as one of the Top 100 National Survey and Design Institutes, and has passed the ISO 9001 Quality management System, owning the foreign managerial authority. It is the first enterprise that reaches the comprehensive quality standards set by the National Survey and Design Institutes. The Institute also enjoys much advanced scientific and technical research achievements in the high technological fields such as urban track traffic, high speed railway, heavy loading railway, railway for passenger transport, and magnetic suspension railway construction.

法人姓名：王洪宇 / 地址：天津市河北区中山路 10 号 / 邮编：300142 / 电话：+86–22–26326134 / 传真：+86–22–26326134 / 网址：www.tsy.com.cn

设计思路

北京南站地处北京市崇文区西南角与丰台区右安门地域的交界处，前身为永定门火车站。在北京铁路枢纽客运系统“四主两辅”的规划格局中，北京南站作为主要客站，承担着京沪高速和京津城际始发终到列车作业。新建北京南站拥有 24 条到发线，13 座客运站台，最高聚集人数为 10 500 人，建筑面积超过 30 万平方米，建成时，为亚洲规模最大的铁路客运站。

作为现代大型交通枢纽站，北京南站是由国有铁路、市郊铁路、地铁等轨道交通与公交、出租车、社会小汽车等城市交通体系综合而成的一个高效、有机、和谐的系统。

北京南站的建筑形态与城市规划紧密结合，为了顺应北京市方格网的城市格局，站房采用椭圆形的建筑形态，受地域文化和周边建设条件的影响，车站整体造型设计来源于椭圆形态和天坛概念的引申。把天坛圆形平面的三重檐运用到椭圆的平面上，最高的屋檐变形成高架候车厅的弧形屋盖。车站两翼的雨篷，采用轻巧的悬垂梁结构，恰好对应着两重屋檐的变化形式。

北京南站坚持以以人为本的设计理念，紧凑合理的功能布局，为旅客提供便利的换乘条件。主要功能空间分为高架层、地面层、地下层 3 个主要的空间层次。高架层为铁路旅客进站层；地面层为铁路站台层，也是公交车旅客进站层，简单的南北进站厅，起到联系上下各层的交通功能；地下一层跨越整个车场的地下部分形成车站的交通换乘空间。两侧为国铁出站厅及快速进站厅。两个地铁站的共用付费区布置在大厅的中央部位，地铁 4 号线站台位于地下二层，地铁 14 号线站台位于地下三层。出租车载客层和小汽车的地下停车库布置在东西两侧地下空间内，有专用通道与交通换乘大厅相连。交通换乘大厅南北两侧直接与南北广场地下空间相连，最大限度地方便地铁客流的换乘。

北京南站地处北京城市中心区，因此减少占用城市土地，节约社会资源是南站设计的重要的原则。在有限的城市空间里，车站从规划上将城市交通、广场、景观纳入车站建筑的内部空间。将小汽车的高架进站道路直接引到高架候车厅入口处，利用环行车道与城市各方向道路衔接。大规模开发利用地下空间，布局紧凑的南北广场采用立体广场的形式，与客流组织分层接驳，同时地下汽车库为城市提供约 7 万平方米的停车面积。车站设计中采用太阳能发电技术，利用新型的薄膜光电池特性，将其设置于屋面采光带中，兼顾遮阳功能。同时，热电冷三联供等技术的应用，使北京南站在铁路项目在节能环保方面起到了示范作用。

北京南站采用先进的设计方法，首先是房桥结构的一体化设计。对于车流交通、客流组织、消防疏散等设计内容进行性能化研究，采用计算机模型技术，通过对真实场景的模拟，来辅助完成设计方案的确定和服务等级的评价。

立面创意说明

独特的地理位置和深远的文化背景，注定了北京南站的建筑品格。屋面造型设计汲取天坛的造型元素，中心主体建筑突出，两侧逐渐跌落的雨棚与整体建筑和谐统一，在城市中形成优美的轮廓线。建筑整体通透流畅，充分体现交通建筑的特点，直接反映出内部功能及空间组成。利用现代的技术手段，实现天坛的建筑形象，为昔日至高无上的皇家穹宇加入大众化和现代化的特色，使北京南站成为同时具有文化性和时代感的公共建筑。

先进技术的引用

北京南站在设计中运用了多项国内先进技术，如热电冷三联供加污水源热泵技术、太阳能光伏发电技术、现代化信息技术及综合调度系统、大跨度悬垂梁结构体系、现代化屋面构造系统、真空排污系统、消防性能化设计技术等。

竣工时间：2008 年 7 月

TSDL 铁道第三勘察设计院集团有限公司

主要技术指标

站房最高聚集人数：10 500 人
站房总建筑面积：25.4 万平方米
雨篷建筑面积：7.1 万平方米
雨篷最低点高度：16.5 米，最高点：30 米
高架环形桥面积：2.3 万平方米
地下车库停车泊位：909 辆
建筑平面尺寸：北京南站站房外形为椭圆形建筑，椭圆长轴 500 米，短轴 350 米，总高度：40 米
建筑层数：站房地上二层，地下三层，局部设有夹层（地下二层和三层分别为北京地铁 4 号和 14 号线的站台层）。

北京南站

设计思路

天津站作为我国第一条客运专线——京津城际的起始站，同时还将承担津秦客运专线、京沪高速的旅客发送的功能。面对新世纪的机遇和挑战，当代铁路客站设计最首要的问题是如何由传统客站模式向适应新时期要求的现代化铁路客站转变的问题。天津站的设计正是在经过不断创新不断探索的基础上产生的作品。

现有天津站站房建成于 1988 年，位于天津市中心，河北区与河东区交界处，接驳地铁 2、3、9 号线。站场部分地上高架层为候车厅，地面层为站台层，地下一层为地下出站大厅。根据"人性化、捷运化、信息化、生态化"的交通发展战略目标，实现车站广场交通、集散、信息、景观及休闲的功能集约设计。充分考虑车站内外各种交通方式的衔接，以满足旅客换乘方便、快捷的需求。交通组织方案，以实施公共交通优先、人车分流、长效管理的原则。天津站坐落于海河北侧，是城市最主要的交通枢纽。站房分为主站房、子站房和高架候车厅三部分，采取高架进站、地下出站、南北进出的布局模式。

天津站最高聚集人数为 7 000 人，建设内容包括新建北站房、地下进出站大厅、高架候车厅、地下出站地道、站台无柱雨篷和既有南站房改造，总建筑面积为 18.7 万平方米。北站房地上二层，局部为四层，地下四层，其中地下二至四层分别为天津地铁通过区域，交通组织明确了地区道路的功能，避免过境交通与进出枢纽交通产生冲突，并为公交车、出租车、自备车的集散规划了各自的路线。

立面创意说明

由于南站房及其钟塔已成为天津市的标志之一，本着尊重历史、延续历史的观点，我们保留了既有站房的整体风格，对其外檐进行重新装修。利用石材来表现建筑的厚重，利用大面积玻璃幕墙体现建筑的亲和力。

配合建筑的狭长形用地，我们将北站房设计为弧形，意在与后广场的下沉景观广场呼应并呈微微环抱之势。建筑立面处理简洁流畅，其边柱形式、站房高度及石材与玻璃幕墙结合，并与现有南站房相呼应，且不失现代感，使建筑显得流畅、舒展、通透、精细，富于秩序。

结合北站房的有机曲线造型，建筑的平面设计酷似"鸟翼"，而且翅膀具有力量、速度及运动的象征，这正体现了铁路建筑高效的内在本质。设计将北站房及高架候车厅屋面整体考虑，北站房上方舒展的两翼，沿着铁路法线对称布置，屋面曲线由中央向两侧微微弯曲，寓意为"时空之翼"。由于高架候车厅宽度较大，我们结合两侧的进站通廊，将此部分屋面局部下沉，以打破对"鸟翼"的体量压迫，同时，屋面边线处理为曲线造型。整个屋面在接近南站房的一段采取下沉的处理手法，体现对现有站房的尊重和呼应。

先进技术的引用

在客流资料的基础上建立客流模拟体系，模拟中采用各种空间物体图例表示实体在空间内运行及其路线信息，应用"最小成本"原理来对行人的活动进行仿真，以得到最优方案。最后，通过分析结果指导车站方案的平面及流线布局。

通过对车站室内外声音的产生、传播和接收的研究，及对人体产生的生理、心理效应的分析，提出改善和控制站房声环境质量的技术和管理措施。利用隔声、吸声、消声、减振设备材料以及释放等方式给旅客提供一个良好的室内声环境。

从自然采光和人工照明两方面进行研究，通过视环境评价方法、光源显色性评价方法、建筑色彩体系和色卡的研究，确定合理的站房采光设计方案。

竣工时间：2008 年 8 月

TSDL 铁道第三勘察设计院集团有限公司

主要技术指标
站房最高聚集人数：7 000 人
站房总建筑面积：18.7 万平方米
雨篷建筑面积：约 8 万平方米
建筑平面尺寸：新建北站房长 242 米，进深 42 米；已有南站房长 208 米，每部分进深 22.3 米。
新建北站房檐口高度：29 米
建筑层数：北站房地上二层，局部为四层；地下四层，其中地下二至四层分别为天津地铁 2、3、9 号线。
站场部分地上高架层为候车厅，地面层为站台层，地下一层为地下出站大厅。

天津站

设计思路

丹东站设计工程为改造工程。在站区范围内，除丹东站站房和必要的广场道路外，设计基本根据基地的地形条件，最大限度的增加了站区范围内的绿化面积，体现绿色、环保的设计理念。

本工程站房采用双层候车的模式，采取简洁的"下进下出"的旅客进出站流线，一层建筑面积为 7 500 平方米，主要设计布置 2、3 站台及远期四站台候车室、售票厅、出站厅等。二层建筑面积 5 500 平方米，主要布置国际及基本站台候车室、贵宾室、设备管理用房等，功能分区明确，流线清晰。

主入口处大厅设计为通高的共享空间，通透明亮，不失大气。候车室采用白色垂片吊顶与筒灯相结合，结构梁均匀布置在垂片间，在保证候车室净空的前提下，为乘客打造优美舒适的候车空间。二层回马廊玻璃隔断与不锈钢的结合使整个空间更具通透感。在候车厅、进站厅、站台及地下空间的设计中，更加注重和关心旅客在使用上的感受，通过开敞自然空间元素的运用、室内温度环境、室内导识系统及自然光的研究及控制，提升车站室内空间品质，体现以人为本的设计理念。

设计中合理确定各空间大小，创造开敞通透的换乘空间；合理设置内部功能及流线组织，使旅客通过清晰便捷的流线快速乘降。加强车站的导向性，创造出舒适、安全、节能的候车环境。

站房雨棚采用无站台柱雨棚，雨棚柱立于股道间。轻巧的结构形式、为旅客提供了宽敞、通透的进出站空间。站台铺装采用灰色烧毛面花岗岩，装修风格简洁明了，墙面与地面采用米黄色石材，顶棚采用石膏板造型，空间错落有致，简洁精致。

立面创意说明

丹东站立面造型简洁流畅，条形窗具有浓郁的地方特色。建筑整体造型庄重大气，充分体现口岸交通建筑的特点并反映内部功能及空间组成，在各个部分协调统一的前提下进行细部划分。站房实体为大片石材墙面与竖向、水平条窗结合，形成虚实对比和过渡的立面效果。建成后的丹东站成为丹东区域交通的重要组成部分，它将是现代化铁路枢纽的代表，在现代化铁路的发展进程中发挥巨大的作用。

竣工时间：2008 年 8 月

主要技术指标
站房最高聚集人数：3 000 人
站房总建筑面积：1.53 万平方米
雨篷建筑面积：3 万平方米
建筑平面尺寸：站房东西向长 147 米，南北向进深 58 米
总 高 度：16.58 米
建筑层数：站房地上二层，局部三层；地下二层，局部设设备架层。

丹东站

设计思路

铜陵市位于安徽省南部，地处长江中下游南岸，西距安庆市、东距芜湖市均在 100 千米左右。城市面积 1 113 平方千米。

铜陵东站是线下式站房。站房采用简洁的上进下出的客流组织模式。站场结合城际站台，二层为基本站台旅客发送层，三层为中间站台的旅客发送层。

站前区地形相对平坦，铜九线站台设计标高 21.01 米，站房设计标高定于 21.16 米。站前广场高度为 15.36 米，除去站房及 12 米宽的汽车匝道外，公交、长途、社会车、出租车场等均位于 15.51 米标高处，以方便各个车场与城市干路的衔接。

站房采用双层候车的模式。二层作为远期城际候车室（近期作为普通、软席候车室），以满足乘客快速进站的需求；三层候车室远期作普通候车室，功能分区明确、流线清晰。一、二站台出站地道净宽为 6.0 米，出站地道出入口净宽为 4.5 米，两端均为台阶。为了给远期城际站台预留出站地道，地道设在站舍内东侧，只设一个出口，地道净宽为 4.0 米，内设台阶。旅客天桥净宽为 8.0 米，出入口净宽为 4.5 米。行包地道净宽为 5.2 米，净高为 3.0 米。

设计充分利用用地的进深和宽度，形成"四场一轴"的总图格局。在站房中心视线上布置反映城市文化主题的景观广场，广场北侧为长途车场、公交车场，南侧为社会车、出租车及预留车场。四个广场通过一条轴线式的贯穿南北的步行道相连，使旅客在各种交通设施间高效、安全、方便的换乘。在中轴广场两侧是由站前路引向站房的汽车匝道，社会车和出租车在这里行驶到站前平台，落客后离开。中间景观广场，为车站建筑提供了良好的绿化环境。

作为车站人车流集散的广场，我们不仅考虑了其辐射城市的交通效应，也考虑了它作为城市对外窗口的景观效应。沿站房中轴线对应于城市进站主干道的区域，我们设计了与红星河呼应的水轴线广场，广场充分注重与北京东路及沿新大道的对景关系。广场上的铜文化雕塑走廊，在体现铜文化与城市历史文脉的联系的同时，使旅客到达和离开铜陵时都能看到这样一个有着浓郁地域文化与优雅景观的城市广场，给旅客留下深刻的城市印象，同时停车场周围也布置了大量的绿化与广场景观相呼应。站前广场绿化率达到 40% 以上，是绿色的生态景观广场。

结合站前广场设计，我们充分理解"铜文化"对铜陵市的重要意义，结合规划设计，我们在远、中、近三个层面分别规划了以铜为主题的雕塑景观轴线，以体现铜陵作为铜都的文化内涵。

立面创意说明

车站周围自然景色优美，群山环抱，树木与水资源丰富，佳景天成。为了不破坏自然景色的和谐与优美，从深层次反映绿色铜陵的精神所在，设计采用与群山相呼应的曲线型屋面，屋面所形成的柔和而富于变化的天际线与周围自然景色完美地融合在一起；墙面主要采用镀膜与透明 LOW – E 玻璃相结合的处理方式，使之与周围景色互为映衬、互相渗透；入口处设计了一个铁路与城市共享的"灰"空间，功能上起到联系上、下二层交通与入口大厅过渡空间的作用。由于站房的主立面向西，为避免西晒，特意采用大屋面悬挑和金属百叶遮挡的设计，玻璃则采用了镀膜与低辐射玻璃。

车站的基座处理较为厚重，与顶部飘浮的屋面及通透的玻璃材质形成强烈的对比，由此寓意今日铜陵市的快速发展与腾飞来自于城市几千年文化的沉淀与积累，世代传承的铜采掘与冶炼文化是城市腾飞的根基。

造型设计紧扣城市"古朴厚重，融旧铸新，自强不息，敢为人先的城市精神，成为铜陵城市精神的完整表达。铜陵东站不仅是铜九铁路的标志性建筑，更将成为铜陵城市形象新的载体。

先进技术的引用

墙面采用镀膜与透明 LOW – E 玻璃相结合的处理方式，使之与周围景色互为映衬。

竣工时间：2009 年 9 月

TSDL 铁道第三勘察设计院集团有限公司

主要技术指标

站房最高聚集人数：1 200 人

站房总建筑面积：站房总建筑面积为 20 890 平方米，其中铁路客运用房面积为 13 417 平方米，长途汽车 5 576 平方米，邮政 674 平方米，商业 1 223 平方米。

建筑平面尺寸：站房南北方向长约 209.4 米，进深 39 米

总高度：28 米

建筑层数：从广场方向看均为三层，局部设夹层。

地面层层高 5.5 米，候车厅一层层高 8 米，二层网架下部最低处净高 6.3 米，售票用房、售票厅层高为 8 米，车站办公层高为 4 米，公安办公区及设备用房一、二层层高均为 5 米，贵宾候车层高为 8 米。

从站场方向看新建站房中央最高处屋面檐口高度 22.50 米，两侧屋面檐口高度为 15.5 米；从广场方向看新建站房中央最高处屋面檐口高度为 28 米，两侧屋面檐口高度为 21 米。

铜陵东站

Architectural

设计作品
Design Works

天津华厦建筑设计有限公司

TIANJIN HUAXIA ARCHITECTURAL DESIGN CO.,LTD

天津华厦建筑设计有限公司始创于 1992 年，是以现代企业制度而组建的股份制民营设计单位，公司注册资本 1 000 万元，具有国家建筑工程设计甲级资质，城乡规划编制乙级资质，市政行业乙级资质，其业务范围涉及建筑设计、市政设计、规划设计等综合性服务。

董事长兼总经理刘存发本着以诚为本、锐意进取、举贤纳才的理念，汇聚了一批资深专家和业界精英，其中管理团队精诚干练、高效务实，设计团队创新求异、精益求精，开发团队熟韵市场、善于开拓，是一支充满活力、开放而多元的团队。公司在学校、医院、宾馆、展厅、商厦、办公等综合性的公共建筑及民用住宅、住宅小区和各类厂区、车间、大跨度钢结构工程等建筑设计、规划设计等方面经验丰富。公司成立十几年，认真执行国家有关设计规范和规定，严把设计质量关，精心设计，严格管理。公司先后承担了 1 400 余项大中型设计工程，设计规模约 1 200 万平方米，设计项目遍及京、津、东北三省及冀、鲁、豫、晋和内蒙古、福建、新疆等地，其中多项设计工程获奖。

公司在狠抓设计业务的同时，十分重视软件正版化的推广工作，并于 2009 年荣获“中国勘察设计协会颁发的全国工程勘察设计行业国庆 60 周年国产 CAD 软件应用先进单位”。

公司已通过 GB/T19001－2008 质量管理体系认证，工程项目设计从方案、设计招投标、设计合同、施工图到项目管理、图纸档案、现场服务等均满足体系规定的标准。

公司遵循精心设计、优质服务、科技领先、锐意创新的宗旨，实现质量管理的持续改进，力求达到顾客满意，以“爱岗敬业，忠诚守信，团结协作，以德兴企”的企业精神，努力为业主提供满意、安全、经济、适用的设计工程和技术服务。

Tianjin Huasha Architectural Design Co., Ltd. was found in 1992. It is a privately run design enterprise of joint-stock system built up under the modern enterprise system. The company owns a registered capital of 10 million yuan, and possesses the Class A certificate of constructional engineering design, Class B aptitude of town and country planning compilation and B class aptitude of municipal trades in national construction trades. Its scope of business involves comprehensive services such as architectural design, municipal design and planning design etc.

The chairman and general manager Cunfa Liu, who is honest-orentied and determined to forge ahead, values a lot the right and capable people, and thus collects a group of senior experts and industry elites, among whom the management team is faithful and capable, effective and practical; the design team is innovative and always seeks for greater perfection; the development team, an energetic, open and polybasic fighting group, is familiar with the market and good at extension. The company has gained outstanding achievements in aspects like constructional design and planning design of comprehensive public constructions such as schools, hospitals, hotels, exhibition halls, commercial buildings and office buildings, and civil residences, residential communities as well as various plant areas, workshops and large-span steel structure projects etc. During more than ten years after the company was founded, it has seriously carried out relevant design specification and relative provisions of the nation, strictly controlled the quality standard of the design, and been meticulous in design and strict in management. It successively undertook more than 1,400 design projects of large and medium size. The design scale reaches around 12 million square meters, and the design programs cover places of Beijing, Tianjin, Three Provinces in Northeast China, Hebei, Shandong, Henan, Shanxi, Inner Mongolia, Fujian, Xinjiang etc., among which many design programs has won awards.

In the meantime of emphasizing design business, the company attaches great importance to the promotion of the software authorization, and in 2009 it won the award of "Domestic CAD Software Application Advanced Unit in 60th Anniversary of the Nation in National Projects Survey and Design Industry issued by China Survey and Design Association".

The company has passed GB/T19001-2008 Quality Management System Certification. The designs of the construction projects are in high-level controlled condition under the specified standard of system from programs, design biddings, design contracts, working drawings to project managements, files of drawing, field service etc.

Abiding by the purpose of "meticulous design, excellent service, leading in technology, keen in innovation", the company realizes the continuous improvements of the quality managements to meet the customers' satisfaction. With the enterprise spirit of "passionate in job, faithful and honest, united and cooperative, promoting the enterprise with morality", the company makes great efforts in providing the business owners satisfied, safe, economical and applicative design projects and technical services.

地址：天津市南开区华苑产业园区榕苑路 16 号鑫茂科技园中心楼三层
电话：+86-22-58693336-41
传真：+86-22-58598916
邮箱：tjhxjzsjgs@126.com

Address: 3rd floor of the Xinmao Technopark Central Building, Rongyuan Road No.16, Huayuan Industry Park, Nankai District, Tianjin
Tel: +86-22-58693336-41
Fax: +86-22-58598916
E-mail: tjhxjzsjgs@126.com

1 唐山海港汽车客运站
建设规模：22 000 平方米
建设地点：天津

Haigang Bus Station in Tangshan
Construction Scale: 22,000 m^2
Construction Location: Tianjin

2 天山水榭花都怡泽轩
建设规模：45 073 平方米
建设地点：天津

Tianshanhuixiehuadu yizexuan
Construction Scale: 45,073 m^2
Construction Location: Tianjin

3 中惠团泊湾工程
建设规模：292 588 平方米
建设地点：天津

Zhonghuituanbo Bay project
Construction Scale: 292,588 m^2
Construction Location: Tianjin

1 新华英郡
建设规模：72 600 平方米
建设地点：天津

Xinhuaying county
Construction Scale: 72,600 m^2
Construction Location: Tianjin

2 弘盛王朝住宅小区
建设规模：49 550 平方米
建设地点：河南 商丘

Hongshengwangchao residence quarter
Construction Scale: 49,550 m^2
Construction Location: Shangqiu, Henan

3 武警天津市总队后勤部新建指挥中心
新建指挥中心
建设规模：44 000 平方米
建设地点：天津

New command center
Construction Scale: 44,000 m^2
Construction Location: Tianjin

1

2

1 天津市静海县医院
建设规模：29 998.18 平方米
建设地点：天津

The Surgery Complex of Jinghai Hospital, Tanjin
Construction Scale: 29,998.18 m^2
Construction Location: Tianjin

2 青海西宁长乐港湾
建设规模：85 015.4 平方米
建设地点：青海 西宁

Changle Harbor in Xi'ning, Qinghai
Construction Scale: 85,015.4 m2
Construction Location: Xi'ning, Qinghai

3 嫩江公园道 1 号
建设规模：246 494.33 平方米
建设地点：黑龙江 黑河

Nenjiang River Park Way No.1 residential quarter
Construction Scale: 246,494.33 m^2
Construction Location: Heihe, Heilongjiang

4 静海县人民检察院办案和专业技术用房
建设规模：9 673.35 平方米
建设地点：天津

Houses for Case and Technique Works
Construction Scale: 9,673.35 m^2
Construction Location: Tianjin

3

4

新疆印象建设规划设计研究院有限公司

新疆印象建设规划设计研究院有限公司即原是新疆建设规划设计研究院，是一个致力于打造独具“新疆印象，不同创想”的，集文化思考、城市导演、工程设计于一体的创新型特色设计研究机构，公司设有建筑、规划、景观、装饰、旅游五个设计所及新疆印象建设沙龙、陆易农教授工作室、王洋创作事务所、徐乃钦方案室、影视动漫工作室、阎顺研究员工作室、倪泽玄、伊莎雕塑家工作室等机构。现拥有教授级高级工程师和一、二级注册建筑师、结构师、注册城市规划师、注册环境艺术师、雕塑家等各类专业人才 80 余人，其中包括享受国务院政府津贴的、自治区人民政府城乡规划管理委员会专家委员陆易农，还有中国策划 20 年十大策划专家、中国环境艺术委员会专家委员会委员、新疆优秀中青年勘察设计工作者王元新，一级注册结构师、总工程师张晓民，一级注册建筑师、总建筑师苗新育，一级注册建筑师、国家注册规划师、副院长张世鹏，国家特许注册规划师、总规划师陆易农等多位资深专家。

近几年设计院有 30 余项规划、建筑、景观设计成果获国家建设部、自治区、兵团、乌鲁木齐、昌吉等优秀设计奖。其中达坂城王洛宾音乐艺术城获 2008 年中国策划 20 年经典策划金奖、2009 年昌吉回民小吃街和莎车十二木卡姆故乡园两项工程设计获得的中国人居典范规划设计方案金奖。

院长王元新一直强调“创新与研究是设计院的第一生产力”，强调“德业双臻，高度决定视野、人品决定产品”，强调”七分产品研究，三分作品设计”，强调“减法做大，加法做强”的设计理念。公司还出版了新疆第一部规划专著《论城市的有机属性》（陆易农著），并在国内外学术刊物中发表论文 80 余篇。

Xinjiang Image Construct Planning Design Institute Co., Ltd, formerly known as Xinjiang Construct Planning Design Institute, is striving for establishing a new characteristic design institute which is of Xinjiang characteristic and various creative ideas, fully integrating the culture speculation, city planning and architecture design.

Xinjiang Image Construct Planning Design Institute Co., Ltd has five sub-design institutes of architecture, planning, landscape, decoration, and tourism. Besides, it has seven working studios including Xinjiang image construct salon, Lu Yinong studio, Wang Yang workshop, Xu Naiqin program office, animation work studio, Yanshun researching office and sculptor's studios of Ni Zexuan, Yisha. There are more than 80 professional staff in this institute , including senior Engineers, national class one and two registered architects and registered structure designers, registered urban planners, registered environment artists , sculptors and other professional talents , among them, the committee member of the government of Xinjiang Uygur autonomous region urban - rural planning committee Yinong Lu who is an expert enjoys government allowance financed by the state council; one of the top 10 expert planner in 20 years- Yuanxin Wang who is the committee member of China environment and art committee and excels in investigation and design area; Besides, national registered class one structure designer, chief engineer Xiaomin Zhang; national class one registered architect, chief architect Xinyu Miao; national class one registered architect, national registered planner, vice president Shipeng Zhang and other experts are working in the institute.

In recent years, there are more than 30 planning, architecture, and landscape design projects winning excellent design awards in competitions organized by Ministry of Construction, Autonomous Regions, Corps, Urumqi and Changji region. Among them, Dabancheng Wangluobin Music City has awarded Gold prize in China planning classic cases for 20 years in 2008, both Jichang Hui People Snacks Street and Shache 12 Muqam Home Park have won the gold metal of habitat model planning award.

The president Yuanxin Wang has always emphasized four aspects related with design: "Innovation and research are the primary productivity of our institute"; "Designers should have both integrity and capability, because the vision is broaden by the capability, and the quality of works depends on the integrity"; "Designer should put 70% effort on research and 30% on designing"; "Focusing on details can make our institute stronger, if not the institute just have more orders". Mr. Wang published the first monograph on design in Xinjiang named The Study on Organic Property of Urbans(Yinong Lu) , and published 80 thesis's.

董事长：王元新
地址：新疆乌鲁木齐苏州东街 568 号金邦大厦 15 楼
电话：+86—991—7819196
传真：+86—991—7819189
邮箱：dongfang923334@vip.com
网址：www.xjxcpd.com

Chairman :Yuanxin Wang
Address: 15th floor, 568 Jinbang Building, Suzhou East Street, Urumqi, Xinjiang
Tel: +86-991-7819196
Fax: +86-991-7819189
E-mail: dongfang923334@vip.com
Http: //www.xjxcpd.com

1 昌吉科技馆政务中心大厅
项目地点：新疆 昌吉
创作设计：王元新 徐乃钦 王雪丰

1 Government Afairs Center Hall in Changji Science & Technology Museum
Project Location: Changji Xinjiang
Major Designers: Yuanxin Wang, Naiqin Xu, Xuefeng Wang

2 新疆沙湾大盘美食城
建筑面积：5 951.06 平方米
项目地点：新疆 沙湾
创作设计：王元新 王洋 冯爱疆 谢强
参与设计：冯雪 张世鹏 韦世武 张晓民 张学萍 刘阳 蒋万成 赛福拉 宗江 侯小磊 展宗芳

2 Xinjiang Shawan Dapan Food Town
Building Area: 5,951.06 m^2
Project Location: Shawan, Xinjiang
Major Designers: Yuanxin Wang, Yang Wang, Aijiang Feng, Qiang Xie
Other Designers: Xue Feng, Shipeng Zhang, Shiwu Wei, Xiaomin Zhang, Xueping Zhang, Yang Liu, Wancheng Jiang, Fula Sai, Jiang Zong, Xiaolei Hou, Zongfang Zhan

3 库尔勒梨园民街
项目地点：新疆 库尔勒
创作设计：王元新 徐乃钦 张娟 陆易农

3 Kuerle Pear Liyuanmin Street
Project Location: Kuerle, Xinjiang
Major Designers: Yuanxin Wang, Naiqin Xu, Juan Zhang, Yinong Lu

4 吉木萨尔县博物馆
项目地点：新疆 吉木萨尔
创作设计：王洋

4 Museum of Jimsar County
Project Location: Jimsar, Xinjiang
MajorDesigner: Yang Wang

5 五家渠沁园生态区别墅规划
项目地点：新疆 五家渠
创作设计：徐乃钦 王元新 黄宣烨
参与设计：冯雪 张晓民 张雪萍
王鹰 时卫国 展宗芳 刘阳

5 Qin Gargen Ecological District Villa Planning in Wujiaqu
Project Location: Wujiaqu, Xinjiang
Major Designers: Naiqin Xu, Yuanxin Wang,Xuanye Huang
Other Designers: Xue Feng, Xiaomin Zhang, Xueping Zhang, Ying Wang, Weiguo Shi, Zongfang Zhan, Yang Liu

6 维斯特温泉假日酒店
项目地点：新疆 乌鲁木齐
建筑设计：宫立新

6 West SPA Holiday Inn
Project Location: Urumqi, Xinjiang
Architecture Designer: Lixin Gong

7 亚中小区花园洋房
建筑面积：74 945.8 平方米
项目地点：新疆 昌吉
创作设计：王洋 罗莎

7 Yazhong Residence Garden Villa
Building Area: 74,945.8 m^2
Project Location: Changji, Xinjiang
Designers:Yang Wang, Sha Luo

8 农八师中心法院中庭
项目地点：新疆 石河子
创作设计：刘魁杰

8 Nongbashi Intermediate Court Central Hall
Project Location: Shihezi, Xinjiang
Designer: Kuijie Liu

9

10

11

12

9 天和印象建筑、景观、及室内设计方案
建筑面积：45 976.6 平方米
创作设计：王洋
参与设计：戴金梅 王红霞 罗莎

9 Architecure Design, Landscape Design and Interior Design of Tianhe Image
Building Area: 45,976.6m^2
Major designer:Yang Wang
Other Designers: Jinmei Dai, Hongxia Wang, Sha Luo

10 五彩湾温泉会所
建筑面积：9 000 平方米
项目地点：新疆 吉木萨尔
设计时间：2010 年
创作设计：黄康 胡世亮
参与设计：胡世亮 颜辉捷 严志鹏 乔迁

10 SPA Club of Wucaiwan
Building Area: 9,000 m^2
Project Location: Jimusaer, Xinjiang
Design time: 2010
Major desingers: Kang Huang , Shiliang Hu
Designers: Shiliang Hu, Huijie Yan, Zhipeng Yan, Qian Qiao

11 野生动物园
项目地点：新疆 天山
创作设计：徐乃钦 时卫国

11 Wild Animal Park
Project Location: Tianshan, Xinjiang
Major Designers: Naiqin Xu, Weiguo Shi

12 亚中花园洋房
建筑面积：74 945.8 平方米
项目地点：新疆 昌吉
创作设计：王洋 罗莎

12 Yazhong Residence Garden Villa
Building Area: 74,945.8 m^2
Project Location: Changji, Xinjiang
Major Designers: Yang Wang, Sha Luo

13 喀什莎车公安局
项目地点：新疆 莎车
创作设计：徐乃钦 王洋
赵清华 姚芝

13 Shache Public Security Bureau
Project Location: Shache, Xinjiang
Building Designers: Naiqin Xu, Yang Wang, Qinghua Zhao, Zhi Yao

14 喀什远方国际物流港
项目地点：新疆 喀什市
创作设计：王元新 王洋 赛福拉
苗青梅 罗莎 张娟

14 Friendship Village
Project Location: Kashi, Xinjiang
Building Designers: Yuanxin Wang, Yang Wang, Fula Sai, Qingmei Miao, Sha Luo, Juan Zhang

15 库尔勒孔雀河二期“百果园”
建筑面积：85 000 平方米
项目地点：新疆 库尔勒
创作设计：侯菲菲 王元新

15 Kongquehe Phase 2 Baiguoyuan in Kuerle
Building Area: 85,000 m^2
Project Location: Kuerle, Xinjiang
Building Designers: Feifei Hou, Yuanxin Wang

13

14

15

16 亚中玫瑰花园小区
建筑面积：74 945.8 平方米
项目地点：新疆　昌吉
创作设计：王洋　罗莎　赵清华

16 Yazhong Rose Garden Residence
Building Area: 74,945.8 m^2
Project Location: Changji, Xinjiang
Major Designers: Yang Wang, Sha Luo,Qinghua Zhao

17 焉耆人民广场
用地面积：70 000 平方米
项目地点：新疆　焉耆
创作设计：王元新　戴金梅　伊莎
参与设计：蒋万成　时卫国　张晓民

17 Yanqi People Squre
Building Area: 70,000 m^2
Project Location: Yanqi, Xinjiang
Major Designers: Yuanxin Wang ,Jinmei Dai,Sha Yi
Other Designer: Wancheng Jiang,Weiguo Shi, Xiaomin Zhang

18 焉耆人民广场实景
建筑面积：70 000 平方米
项目地点：新疆　焉耆
创作设计：王元新　戴金梅　伊莎
参与设计：蒋万成　时卫国　张晓民

18 Yanqi People Square Outdoor Scene
Building Area: 70,000 m^2
Project Location: Yanqi, Xinjiang
Major Designers: Yuanxin Wang, Jinmei Dai,Sha Yi
Other Designers: Wancheng Jiang, Weiguo Shi, Xiaomin Zhang

19 华彩春成花园
建筑面积：964 921 平方米
项目地点：新疆　乌鲁木齐
创作设计：王洋　罗莎
参与设计：赵清华

19 Huacaichuncheng Garden
Building Area: 96,4921 m^2
Project Location: Urumqi, Xinjiang
Building Designers: Yang Wang, Sha Luo
Other Designer: Qinghua Zhao

20 新疆福寿园
项目地点：新疆　乌鲁木齐
首席创作：王元新
参与创作：袁多　伊莎

20 Xinjiang Fushou Garden
Project Location: Urumqi, Xinjiang
Key designer: Yuanxin Wang
Other Designers: Duo Yuan,Sha Yi

21 阜康畅岁园
用地面积：60 000 平方米
项目地点：新疆　阜康
创作设计：王元新　戴金梅　苗青梅

21 Changsui Garden in Fukang
Building Area: 60,000 m^2
Project Location: Fukang, Xinjiang
Building Designers: Yuanxin Wang, Jinmei Dai, Qingmei Miao

22 农八师 143 团部（花园镇）总体规划
项目地点：新疆农八师 143 团
创作设计：王元新　陆易农　戴金梅　王冀东

22 No.143 Regiment of Science and Technology Commission of 8th Division of Xinjiang Crops
Project Location: No.143 regiment of Science and Technology Commission of 8th Division of Xinjiang Crops
Building Designers: Yuanxin Wang, Yinong Lu, Jinmei Dai, Jidong Wang

23 哈密市领先花园二期设计方案
项目地点：新疆　哈密
创作设计：孙艳　胡睿
参与设计：刘红娟　王维　葛塔生　叶志鹏

23 Design of Hami Lingxian Garden Phase 2
Project Location: Hami, Xinjiang
Building Designer: Yan Sun, Rui Hu
Other Designers: Hongjuan Liu, Wei Wang, Tasheng Ge, Zhipeng Ye

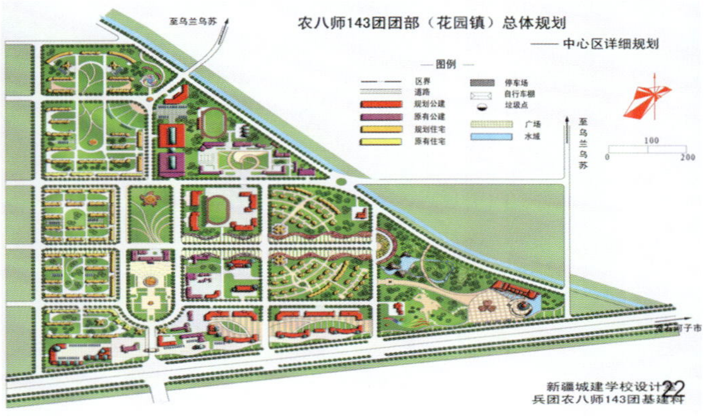

24

25

26

24 阜康滨河园林景观设计方案
占地面积：200 000 平方米
项目地点：新疆　阜康
创作设计：王元新　洪枫　李超峰

24 Landscape Design Project for The Riverside in Fukang
Building Area: 200,000 m^2
Project Location: Fukang, Xinjiang
Major Designers: Yuanxin Wang, Feng Hong, Chaofeng Li

25 吉木萨尔高层住宅
项目地点：新疆　吉木萨尔县
创作设计：王洋　黄俊华

25 High Rise Residence in Jimusar
Project Location: Jimusar, Xinjiang
Major Designers: Yang Wang, Junhua Huang

26 亚心之窗国际之都—乌鲁木齐南旅基地广场
建筑面积：385 000 平方米
项目地点：新疆　乌鲁木齐
策划主创：戴金梅　王元新
参与设计：伊莎　徐乃钦　倪泽玄　陆易农　阎顺　吴建伟　王红霞　黄海峰　董芳　王冀东

26 Yaxin International City—Urumqi Nanlv Jidi Square
Building Area: 385,000 m^2
Project Location: Urumqi, Xinjiang
Major DesigneRS: Naiqin Xu,Yuanxin Wang
Other Designers: Sha Yi, Naiqin Xu, Zexuan Ni, Yinong Lu, Shun Yan, Jianwei Wu, Hongxia Wang, Haifeng Huang, Fang Dong, Jidong Wang

27 北庭园浮雕墙
项目地点：新疆　吉木萨尔
总体设计：王元新
雕塑设计制作：伊莎

27 BeiTingyuan County Relief Wall
Project Location: Jimusaer, Xinjiang
Key Designer:Yuanxin Wang
Designer and manufacurer: Yi Sha

28 和静文化体育活动中心
项目地点：新疆　和静
创作设计：苗新育

28 Hejing Culture & Sports Center
Project Location: Hejing, Xinjiang
Major Designer: Xinyu Miao

29 昌吉农展馆
项目地点：新疆　昌吉
创作设计：王元新　徐乃钦　杨国洪

29 Agriculture Exhibition Center in Changji
Project Location: Changji, Xinjiang
Major Designers: Yuanxin Wang, Naiqin Xu, Hongguo Yang

27

28

29

30 建设学院教学楼
项目地点：新疆　乌鲁木齐
创作设计：王元新　钟光明

30 School Teaching Building Construction
Project Location: Urumqi, Xinjiang
Major Designers: Yuanxin Wang, Guangming Zhong

31 莎车县木卡姆故乡园
项目地点：新疆　莎车县
创作设计：王元新　王海霞

31 Muqam Gome Park in Shache
Project Location: Shache, Xinjiang
Major Designers: Yuanxin Wang, Haixia Wang

30

31

32 库车体育馆
项目地点：新疆　库车
创作设计：徐乃钦　赛福拉　罗莎

33 昌吉花儿大剧院
建筑面积：17 000 平方米
项目地点：新疆　昌吉
创作设计：郑玲　成江华　朱兴国

34 和田金地华府
项目地点：新疆　和田
创作设计：苗新育

35 巩乃斯避暑山庄
项目地点：新疆　巩乃斯国家森林公园
创作设计：王元新　徐乃钦
参与设计：时卫国　冯雪　赛福拉　黄海峰　张杰
张学萍　展宗芳　王鹰　张晓民

36 奇台一中天和分校
占地面积：117 700 平方米
项目地点：新疆　奇台
创作设计：王元新　苗青梅　王洋

37 策勒县体育馆
建筑面积：12 000 平方米
项目地点：新疆　策勒
创作设计：郑玲　成江华　朱兴国

38 和静县财政局希望投资公司办公楼建筑设计
建筑面积：新疆　和静
创作设计：李扬　苗新育

39 哈巴河县双语幼儿园
项目地点：新疆　哈巴河
创作设计：孙艳

32 Kuqa Coliseum
Project Location: Kashi, Xinjiang
Major Designers: Naiqin Xu, Fula Sai, Sha Luo

33 Changji Huaer Theater
Building Area: 17,000 m^2
Project Location: Changji, Xinjiang
Major Designers: Ling Zheng, Jianghua Cheng, Xingguo Zhu

34 Hetian Jindi Mansion
Project Location: Hetian, Xinjiang
Major Designer: Xinyu Miao

35 Gongnaisi Mountain Resort
Project Location: Xinjiang kunas national forest park
Major Designers: Yuanxin Wang, Naiqin Xu
Other Designers: Weiguo Shi, Xue Feng, Fula Sai, Haifeng Huang, Jie Zhang, Xueping Zhang, Zongfang Zhan, Yingwang, Xiaomin Zhang

36 Qitai County No. one Middle School Tianhe Division
Building Area:117,700 m^2
Project Location: Qitai, Xinjiang
Major Designers: Yuanxin Wang, Qingmei Miao, Yang Wang

37 Cele Stadium
Building Area: 12,000 m^2
Project Location: Cele, Xinjiang
Major Designers: Ling Zheng, Jianghua Cheng, Xingguo Zhu

38 The Bureau of Finance in Hejing Hope Investment Company Office Building Architecure Design
Project Location: Hejing, Xinjiang
Major Designer: Yang Li, Xinyu Miao

39 Biliingual Kindergarden in Habahe County
Project Location: Habahe, Xinjiang
Major Designer: Yan Sun

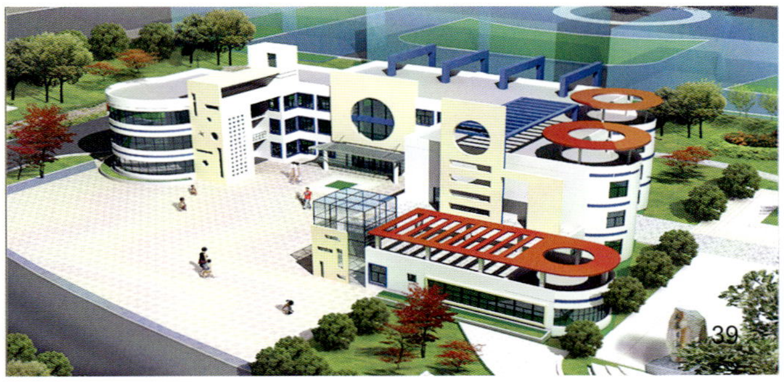

40 新欧国际家居博览中心
创作设计：王洋　赵德昌

40 Xin'ou International Furniture Exhibition Center
Major Designers: Yang Wang, Dechang Zhao

41 和静县黄府花园
项目地点：新疆　和静
创作设计：郑睿媛　苗新育

41 Huangfu Garden in Hejing County
Project Location: Hejing, Xinjiang
Major Designers: Ruiyuan Zheng, Xinyu Miao

42 哈密悦隆广场
项目地点：新疆　哈密
创作设计：孙艳 刘红娟 王维
葛塔生 叶志鹏 胡睿

42 Hami Ruilong Square
Project Location: Hami, Xinjiang
Major Designer: Yan Sun, Hongjuan Liu, Wei Wang, Tasheng Ge, Zhipeng Ye, Rui Hu

43 亚心之窗国际之都
——乌鲁木齐南旅基地办公楼
建筑面积：385 000 平方米
项目地点：新疆　乌鲁木齐
创作设计：徐乃钦　王元新
参与设计：李祉祯

43 Yaxin International City –Urumqi Nanla Base Office Building
Building Area: 385,000 m^2
Project Location: Urumqi, Xinjiang
Major Designers: Naiqin XU, Yuanxin Wang
Other Designers: Zhizhen LI

44 和静县石林宾馆
项目地点：新疆 和静
创作设计：罗岚 苗新育

44 Hejing Shilin Hotel
Project Location: Hejing, Xinjiang
Major Designers: Lan Luo, Xinyu Miao

45 沙湾大盘美食文化城
占地面积：44 508.3 平方米
项目地点：新疆 沙湾
主创设计：王元新 冯爱疆
李天鹏 谢强
参与设计：韦世武 赛福拉 张晓民
刘阳 宗江 将万成
侯小磊 王鹰 吴建伟
冯雪 展宗芳

45 Xinjiang Shawan Dapan Food Town
Building Area: 44,508.3 m^2
Project Location: Shawan, Xinjiang
Key Designer: Yuanxin Wang,Aijiang Feng, Tianpeng Li, Qiang Xie
Other Designers: Shiwu Wei, Fula Sai, Xiaomin Zhang, Yang Liu, Jiang Zong, Wancheng Jiang, Xiaolei Hou, Ying Wang, Jianwei Wu, Xue Feng, Zongfang Zhan

46 巩乃斯宾馆
创作设计：王小燕　苗新育

46 Gongnaisi Hotel
Major Designers: Xiaoyan Wang, Xinyu Miao

47 那拉提景区管委会接待中心
项目地点：新疆　新源
创作设计：杨充

47 Nalati Scenic Spot Administration Reception Center
Project Location: Xinyuan, Xinjiang
Major Designer: Chong Yang

48

49

50

51

52

53

54

48 奎屯天北新区街景设计
项目地点：新疆　奎屯市
创作设计：王元新　戴金梅

48 Street Landscape Design of Tianbei New District in Kuitun
Project Location: Kuitun, Xinjiang
Major Designers: Yuanxin Wang, Jinmei Dai

49 库尉之窗
占地面积：18.5 公顷
项目地点：新疆　库尔勒
创作设计：王元新　侯菲菲
　　　　　田中奎　刘鸿　张杰

49 Windows of Kuwei
Building Area: 18.5 ha
Project Location: Kuerle, Xinjiang
Major Designers: Yuanxin Wang, Feifei Hou, Zhongkui Tian, Hong Liu, Jie Zhang

50 伊河山庄
项目地点：新疆　伊宁市
创作设计：王元新　徐乃钦
　　　　　刘魁杰　文军

50 Yihe Village
Project Location: Yining, Xinjiang
Major Designers: Yuanxin Wang, Naiqin Xu, Kuijie Liu, Jun Wen

51 维斯特二期住宅
项目地点：新疆　乌鲁木齐
创作设计：钟香萍　杨金风　张丽敏
　　　　　宫立新

51 West Residence Phase 2
Project Location: Urumqi, Xinjiang
Major Designers: Xiaoping Zhong, Jinfeng Yang, Limin Zhang,
Lixin Gong

52 天山野生动物园科普教育馆
项目地点：新疆　达坂城
创作设计：徐乃钦

52 Tianshan Wild Animal Park Science and Education Hall
Project Location: Dabancheng, Xinjiang
Major Designer: Naiqin Xu

53 库尔勒梨园民街
项目地点：新疆　库尔勒
创作设计：王元新

53 Kuerle Pear Garden Min Street
Project Location: Kuerle, Xinjiang
Major Designers: Yuanxin Wang

54 亚洲地理中心规划方案
项目地点：新疆　乌鲁木齐
创作设计：王元新　高波

54 Asia Geography Center Planing Project
Project Location: Urumqi, Xinjiang
Designers: Yuanxin Wang, Bo Gao

55 中哈霍尔果斯国际合作中心连接通道标志塔
项目地点：新疆　伊犁
创作设计：王元新
参与设计：买买提江　将万成　王红霞　吴建伟

55 Horgos International Trading Center Linking Passagelandmark Tower
Project Location: Yili, Xinjiang
Key Designer: Yuanxin Wang
Other Designers: Maimaitijiang,Wancheng Jiang, Hongxia Wang, Jianwei Wu

56 中哈霍尔果斯合作中心广场
项目地点：新疆　霍尔果斯
创作设计：王元新　王冀东

56 Horgos Cooperation Center Square
Project Location: Horgos, Xinjiang
Major Designers: Yuanxin Wang, Jidong Wang

57 哈什市阔乃则尔巴格路街景整治设计
项目地点：新疆　喀什
创作设计：王元新　徐乃钦
参与设计：赛福拉　买买提江　姚芝

57 Planning Design of Street Landscape of Kuonaizeerbage Road in Kashi
Project Location: Kashi, Xinjiang
Major Designers: Yuanxin Wang, Naiqin Xu
Other Designers: Fula Sai, Maimaitijiang, Zhi Yao

58 霍尔果斯国际贸易中心
项目地点：新疆　伊犁
创作设计：仲新荣　王元新

58 Horgos International Trading Center
Project Location: Yili, Xinjiang
Major Designers: Xinrong Zhong, Yuanxin Wang

59 喀什莎车人民法院
项目地点：新疆　喀什
创作设计：徐乃钦　王洋
建筑表现：赛福拉　罗莎　姚芝

59 Kashi Biliingual Kindergarden in Habahe County
Project Location: Kashi, Xinjiang
Major Designers: Naiqin Xu, Yang Wang
Architects: Fula Sai, Sha Luo, Zhi Yao

60 杜氏旅游
项目地点：新疆　昌吉
创作设计：郭立阳

60 Dushi Travels
Project Location: Changji, Xinjiang
Major Designers: Liyang Guo

61

62

63

64

65

61 “金沙湾”主雕
项目地点：新疆　沙湾
创作设计：王元新　伊莎　李羽

61 Jinshawan Key Sculpture
Project Location: Shawan, Xinjiang
Major Designers: Yuanxin Wang, Sha yi

62 莎车十二木卡姆故乡园雕塑墙
项目地点：新疆　莎车
策划设计：王铁民　王元新
设计制作：伊莎　陈述

62 Shache Uyghur 12 Muqami Home Land Park Sculpture
Project Location: Shache, Xinjiang
Planed and Designed by: Tiemin Wang, Yuanxin Wang
Designed and manufactured by: Shu Chen, Sha yi

63 张骞出使西域
项目地点：新疆　鄯善
创作设计：李羽

63 Zhang Qian Seeks Xiyu
Project Location: Shanshan, Xinjiang
Major Designers: Yu Li

64 师徒西行
项目地点：新疆　鄯善
创作设计：李羽

64 Journey to the West
Project Location: Shanshan Xinjiang
Major Designers: Yu Li

65 鄯善沙雕公园
项目地点：新疆　鄯善
创作设计：李羽

65 Shanshan Sand Sculpture Park
Project Location: Shanshan Xinjiang
Major Designers: Yu Li

66 中哈霍尔果斯国际边境合作中心概念规划
项目地点：新疆　霍尔果斯
创作设计：王元新　李雄飞　陆易农　王冀东　苗青梅

66 Conception design of China and Kazakhstan Horgos International Border Cooperation Center
Project Location: Horgos Xinjiang
Major Designers: Yuanxin Wang, Xiongfei Li, Yinong Lu, Jidong Wang, Qingmei Miao

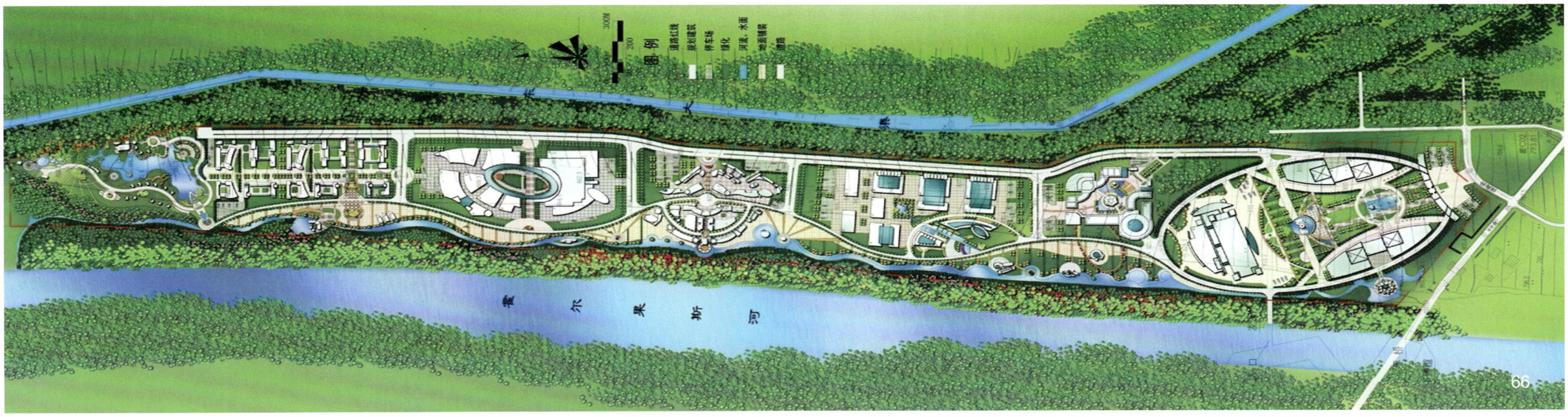
66

67 鄯善驼铃天下
设计制作：李羽
效果表现：魏小未

68 库尔勒市河北社会福利院
项目地点：新疆　库尔勒
创作设计：徐乃钦　冯雪　塞福拉

69 团结门主雕
项目地点：新疆　莎车
创作设计：倪泽玄　王铁民

70 丝绸之路雕塑
项目地点：新疆　乌鲁木齐
创作设计：倪泽玄　王元新

71 沙湾大盘美食文化城
规划面积：44 000 平方米
项目地点：新疆　沙湾
创作设计：王元新　李天鹏　冯爱疆

72 天路雕塑
项目地点：新疆　乌鲁木齐
创作设计：倪泽玄

67 Tuolingtianxia Shanshan
Designed by: Yu Li
Graphic: Xiaowei Wei

68 Korla City, Hebei Social Charity House
Project Location: Korla, Xinjiang
Major Designers: Naiqin Xu, Xue Feng, Fula Sai

69 Main body of TUANJIEMEN
Project Location: Shache, Xinjiang
Major Designers: Zexuan Ni, Tiemin Wang

70 The Silk Road Sculpture
Project Location: Urumqi, Xinjiang
Major Designers: Zexuan Ni, Yuanxin Wang

71 Xinjiang Shawan Dapan Food Town
Floor Area: 44,000 m^2
Project Location: Shawan, Xinjiang
Major Designers: Yuanxin Wang, Tianpeng Li, Aijiang Feng

72 TIANLU Sculpture
Project Location: Urumqi, Xinjiang
Major Designers: Zexuan Ni

Architectural

设计作品
Design Works

新疆四方建築設計院有限公司
XINJIANG SI FANG INSTITUTE OF ARCHITECTURAL DESIGN

公司简介 \ Company Introduction

“新疆四方建筑设计院有限公司”成立于1993年1月1日，其前身为“乌鲁木齐经济技术开发区建筑勘察设计院有限责任公司”，2001年5月1日与“乌鲁木齐建筑设计院”合并成立“乌鲁木齐建筑设计研究院有限责任公司”，2009年11月2日重新组建为“新疆四方建筑设计院有限公司”。

我院是一所具有建筑工程设计甲级、规划设计乙级、岩土勘察设计及施工乙级、市政设计丙级、咨询设计丙级等设计资质的综合性设计院。我院现有职工115人，其中国家各类注册师21人、高级职称36人，许多优秀设计师曾获得国家、建设部、自治区优秀设计大奖，具有较高的高端人才优势。

自建院以来，我院本着“顾客第一，信誉第一，质量第一，服务第一”的原则取得较好的社会效应。设计作品涉及办公、宾馆、医院、体育建筑、幼儿园、学校建筑、纪念性建筑、住宅、居住区规划、残疾人建筑等各个领域，涌现了许多优秀作品，例如，新疆屯河大厦、独山子大酒店、宾尤大酒店、独山子游泳馆、兵团二钢幼儿园、新疆康达小区、新疆军区医院门诊病房楼、新疆维吾尔医医院病房楼，乌鲁木齐市烈士陵园陈列馆、米东区公务员小区（用地90公顷，建筑面积152万平方米）华源．博瑞新村居住小区、市残联残疾人服务中心等项目。

随着国家西部大开发政策的实施，我院将抓住这一历史性机遇，强化具有地方、地域特色的精品设计理念，闯出一条具有新疆地方文化，地域特色的原生态建筑理论和实践之路。

Xinjiang Sifang Architecture Design Institute was established on January 1st, 1993, organically was called Urumqi economic development zone Architecture Survey and Design company.On May 1st, 2001, combined with Urumqi Architecture Design Institute, the company changed name as Urumqi Architecture Design and Research Institute. In November 2nd, it was transformed as Xinjiang Sifang Architecture Design Institute.

Our institute is a comprehensive design institute with the Class A qualification of architectural design, Class B qualification of planning design, the second class qualification of geotechnical investigation design and construction, the third class qualification of municipal design and consulting design. It now has 115 staff, including 21 registered masters, 36 people with senior professional titles; among them, many excellent designers has won outstanding design awards of national, construction of ministry and autonomous region. We have advantage in the high level talents and professionals.

Since the establishment of the institute, holding the principle of “the clients, credit, quality and the service come first.”, we have received good social feedback. Our design projects cover office buildings, hotels, hospitals, sports architecture, kinder garden, schools, memorial architecture, residence, housing planning, architecture for the disabled and other areas, creating numerous outstanding works, such as Xinjiang Tunhe Plaza, Dushanzi Hotel, Binyou Hotel, Dushanzl swimming pool, kinder garden of Crops Ergang, Kangda residence quarter in Xinjiang, Xinjiang Military Command Hospital outpatient and warding building, The Uyghur hospital warding building, the exhibition hall of Urumqi Memorial Manorial Park, residence district for public officials in Midong (site area: 90 ha, building area: 1,520,000 m^2) Huayuanborui new village, Disabled Persons' Federation service center, and so on.

With the carrying out of the west development policy, we will grasp; this historical opportunity, promoting the fine design idea of regional and local features; starting a new road of Xinjiang culture and characteristic, with ecological theory and practice.

质量方针：原创设计　追求卓越　立足诚信　服务一流

质量目标：设计产品合格率达100%，优质率达90%，确保两年内获自治区以上优秀设计两项，院创优项目五项

Quality principle: Original design, Commitment to excellence, Credit-oriented, Superb service

Quality target: The passing rate of the design products is 100%, rate of excellence is 90%, making sure to get two excellent prizes of provincial or above and five excellent projects.

联系方式 \ Contact Information

单　位：新疆四方建筑设计院有限公司
地　址：新疆乌鲁木齐市水磨沟区安居南路70号中国万向招商大厦13、14、19层
联系人：朱飞
电　话：+86-991-4614563/4881093
传　真：+86-991-4677309/4697637
邮　箱：xjsf_d@sina.cn

Company name: Xinjiang Sifang Architecture Design Institute
Address: The 13th, 14th, 19th floor of China Wanxiang Merchants Building, No.70 of Anjun South Road, Shuimogou district, Urumqi, Xinjiang
Contact: Fei Zhu
Tel: +86-991-4614563/4881093
Fax: +86-991-4677309/4697637
E-mail: xjsf_d@sina.cn

新疆四方建築設計院 有限公司
XINJIANG SI FANG INSTITUTE OF ARCHITECTURAL DESIGN

乌鲁木齐市烈士陵园烈士事迹陈列馆

透视效果

建筑面积：3 900 平方米
建设地点：新疆 乌鲁木齐
设计时间：2008 年
主设计师：朱飞　赵天兵　宁杰
设计理念：设计采用中轴对称的手法，利用建筑单体向两边延伸，形成视觉层次，符合纪念建筑庄重的建筑属性。

报告厅

一楼展厅

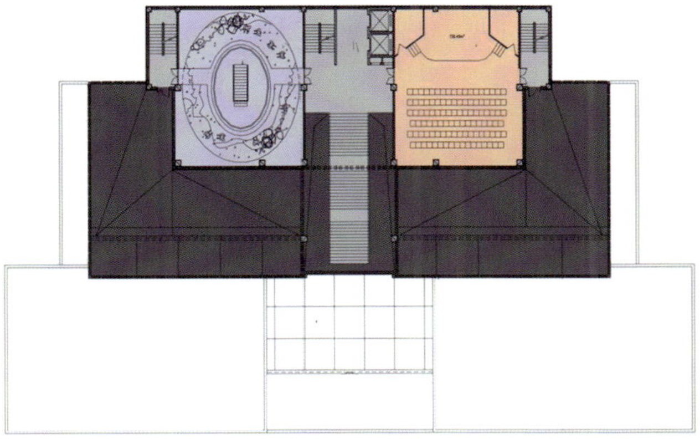

二楼展厅

入口门厅

缅怀厅

烈士厅

半影画厅

新疆四方建筑设计院有限公司
XINJIANG SI FANG INSTITUTE OF ARCHITECTURAL DESIGN

新疆维吾尔自治区维吾尔医医院综合门诊病房楼

建筑面积：37 200 平方米
建设地点：新疆 乌鲁木齐
设计时间：2009 年
主设计师：付晨 宁杰
设计理念：设计师对地方民族特色准确把握，利用抽象的手法处理民族符号和色彩，充分体现新疆的地域文化。

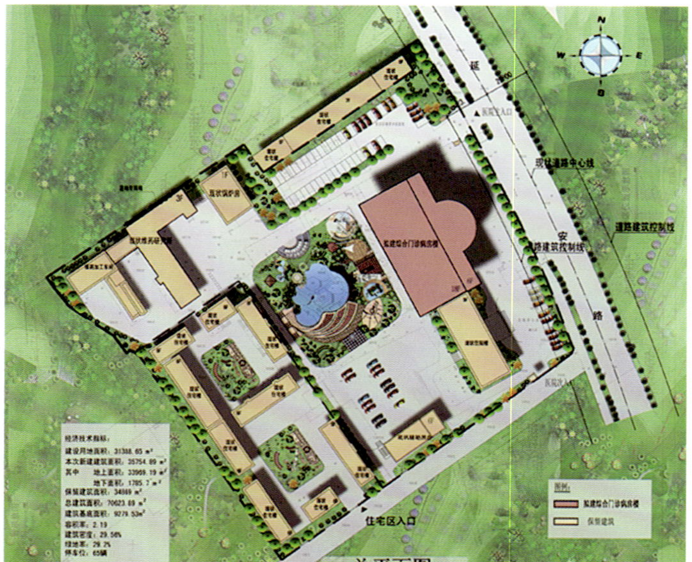

总平面图

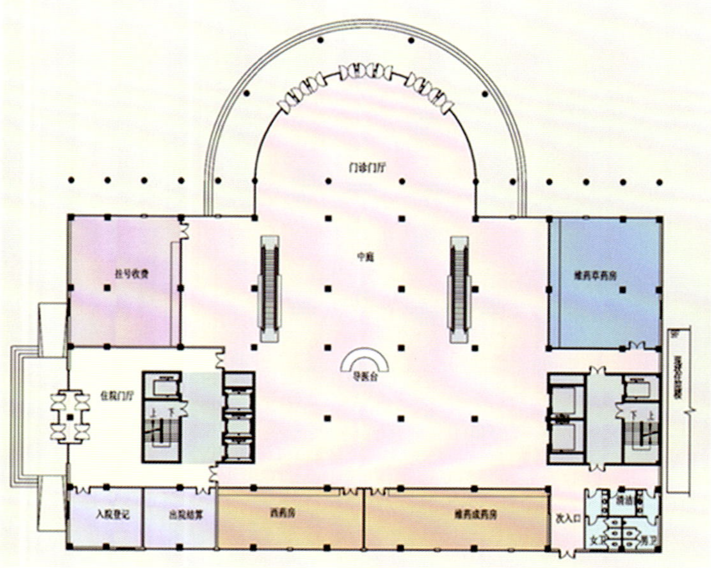

库尔勒鹭洲国际小区

建筑面积：205 000 平方米
开 发 商：巴州昌和房地产公司
设计时间：2010 年
主创人员：刘豫 付晨 陈治刚 毛文全
工程情况：在建

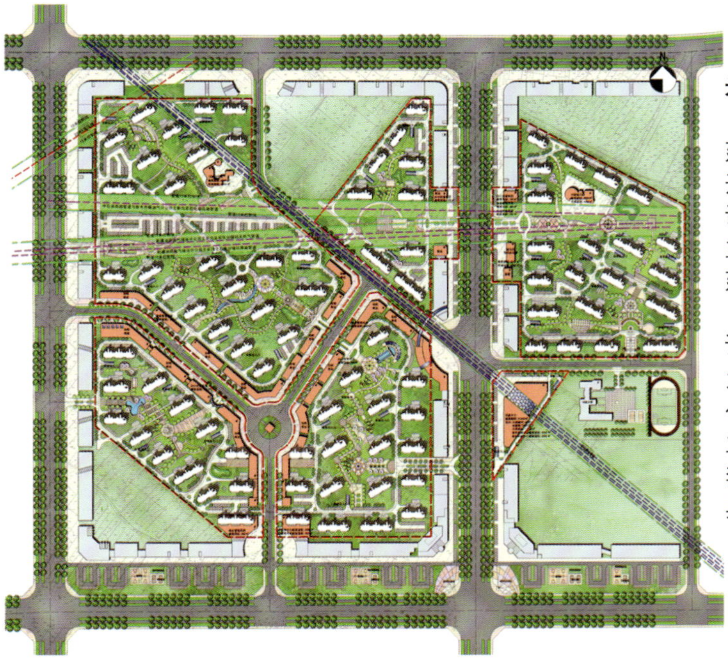

米东公务员小区

建筑面积：1 030 000 平方米
开 发 商：乌鲁木齐市米东区政府
设计时间：2010 年
主创人员：朱飞 刘豫 陈志刚
建设情况：在建

项目应用社区与生态景观设计理念，强调以公务员人群特色为本的设计思路，强化南部体育公园，北部中心区区间性步行商业街，中部商业与医疗等服务配套中心等级差异化布局，营造各具特色的各级社区空间。

新疆四方建築設計院有限公司
XINJIANG SI FANG INSTITUTE OF ARCHITECTURAL DESIGN

规划项目

华凌·河南路项目

建筑面积：374 000 平方米
开 发 商：新疆华凌集团
设计时间：2010 年
主创人员：朱飞 刘豫 曹雪鼎
工程情况：在建

项目应用绿色节能设计理念，强调以古树保护带作为高层区与多层区的空间分割，在强化服务配套核心化（邻里中心）的基础上，打造新型社区化的城市次干道商业空间。

珠光·鸿雁池国际滨水度假项目

建筑面积：658 000 平方米
开 发 商：广州珠光集团
设计时间：2010 年
主创人员：朱飞 刘豫 曹雪鼎 陈志刚
工程情况：在建

项目应用滨水度假设计理念，强调以水域保护带规划新疆水生动植物园、新疆军垦文化公园、新疆海洋地质公园、以水库水面规划新疆水上竞技运动中心。强调低碳的可持续规划的实施。

新疆四方建筑设计院有限公司
XINJIANG SI FANG INSTITUTE OF ARCHITECTURAL DESIGN

规划项目

华源·博瑞新村

建筑面积：356 000 平方米
开 发 商：新疆华源集团
设计时间：2007 年
主创人员：朱飞 刘豫
工程情况：获建设部 2008 年绿色节能示范工程
获建设部 2008 年广厦奖

项目应用绿色节能设计理念，建设太阳能一体化楼宇系统，同时引入同层排水技术与中水系统相配套。设计依托本区建设经验，建立新疆 65% 住宅节能标准。

特变·阳光绿景

建筑面积：182 000 平方米
开 发 商：新疆特变电集团
设计时间：2007 年
主要创作人员：朱飞 刘豫
获奖情况：建设部 A 级住宅认证项目

项目应用绿色节能设计理念，强调以枯山水景观为主景观体系的多样化再造。建筑布局以弧形为主，强化了人视点的控制设计。在入口设置纵向商业街，满足景观与服务需求。

新疆民用建筑设计院有限公司

XIN JIANG CIVIL BUILDING DESIGN INSTITUTE CO.,LTD.

新疆民用建筑设计院有限公司前身为新疆机械电子工业设计院，成立于 1985 年 6 月。公司原隶属于新疆机械电子工业厅，受行业滑坡影响，到 1997 年陷入困境。1997 年底，以史耀华同志为核心的新的领导班子成立后，坚持推进改革，转换公司内部经营机制，积极开拓民用建筑市场，使得单位综合效益有了极大的提高。1998 年起经营收入持续高速增长，预计 2010 年经营收入是 1997 年的 60 倍。公司于 2000 年 3 月改制成功，成为新疆勘察设计行业事业单位改企转制的典范，2000 年成为“新疆勘察设计行业综合实力 20 强单位”。建筑设计资质自 1999 年升为乙级，到 2001 年升为甲级，由此成为新疆第三家建筑设计甲级资质拥有者，实现了快速跨越，并由原来单一的行业设计发展成为拥有建筑设计甲级、工程咨询甲级、城市规划乙级、工程勘察乙级、市政、水利设计丙级等多项资质的综合实力较强的单位。企业注册资本已增至 500.7 万元。随着企业综合实力和社会信誉的整体提升，所占市场份额不断扩张，工程规模不断扩大，逐步形成了“立足新疆，辐射内地，面向中亚”的市场格局。由该院作为投资主体、具备独立法人资格和规划设计资质的西安创元建筑设计院有限公司也已展开服务。

地址：新疆乌鲁木齐市新医路 2 号汇文大厦 8 楼
邮编：830054
电话：+86-991-4338259
传真：+86-991-4335977
邮箱：xjmyjzsjy@126.com
网址：www.mysjy.com.cn

新疆民用建筑设计院有限公司
XIN JIANG CIVIL BUILDING DESIGN INSTITUTE CO.,LTD.

1 新疆天山房产“雅南高第”居住小区规划与建筑设计
2 新疆喀什日报社（丰材）居住小区规划与建筑设计
3 新疆沙湾县润泽房地产开发有限公司高层建筑
4 新疆驰达房地产开发有限公司驰达大厦
5 新疆兵团机关“南湖明珠”小区规划与建筑设计
6 新疆塔城地区建设局高层住宅
7 新疆福瑞嘉幸房地产开发有限公司“嘉幸 尚品”高层住宅
8 新疆塔城地区闻琴路步行街高层建筑

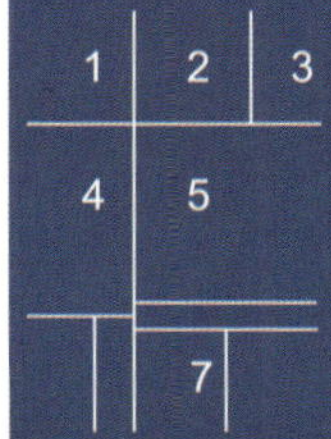

雲南省設計院

YUNNAN DESIGN INSTITUTE

地址：中国 · 昆明 · 新闻路 259 号
电话：+86-871-4142994
传真：+86-871-4141085
邮编：650032
网址：www.ydi.cn
邮箱：ydi@ydi.cn

Address: 259Xinwen Road · Kunming · China
Tel: +86-871-4142994
Fax: +86-871-4141085
P.C.: 650032
Http: //www.ydi.cn
E-mail: ydi@ydi.cn

云南省设计院诞生于 1952 年。作为从事工业与民用建筑、城市规划、轻工、市政工程、医药、建材等行业全国知名的大型综合性省级勘察设计企业，多年来秉承“精心设计、严格管理、满意服务”的执业宗旨，为国家经济建设做出了重大的贡献，一直受到社会各方的广泛赞誉。我院现有员工 687 人，技术人员占全院总人数的 90%，其中国家级设计大师 1 人，国家注册建筑师 50 余人，注册结构工程师 80 余人，注册造价师 4 人。具备高级技术职称人员 70 余人，中级技术职称人员 220 余人。设计院拥有规划、总图、建筑、景观、结构、给排水、电气、智能、暖通、轻工工艺、医药食品、环境工程、建材、设备、概预算、岩土工程勘察等二十余个专业，另有各类特种技术专业设计如大跨度预应力结构、网架、特殊桩基础、特种结构加固、空间膜屋面、建筑声学等。我院下设建筑设计研究分院、结构设计研究分院、规划设计研究分院、建筑设备设计研究分院、电气设计研究分院、市政设计研究分院、工程咨询室、勘察分院、工程监理公司和工程承包公司等专业设计分院。

设计院在省内率先采用设计与管理计算机局域网络系统，并与因特网连接，配备各类专业设计软件和先进的电脑外围设备，使我院的设计与管理水平处于国内领先地位。云南省设计院有着 50 余年的光辉历史，共承担各类设计项目 2 万余项，建筑面积 6 900 万平方米，设计成果遍及国内外，先后共获得国家级、部级、省级优秀设计奖及科技进步奖等 200 余项，并于 1999 年通过了 ISO 9001 国际质量体系认证。我院坚持“质量第一、顾客至上”的企业理念，以精湛的技术、优良的质量竭诚为客户提供工程设计、技术咨询等优质服务，愿与海内外各界人士携手合作、共谋发展。

Established in 1952, Design Institute of Yunnan Province has been engaging in many fields including industrial and civil buildings, urban planning light industry, municipal engineering, medicine and building materials industries, etc. As a well-known large-scale comprehensive provincial survey and design enterprise, it has been adhering to the operation principle - "well-designed, rigorous management, satisfactory service" for many years, which also makes significant contribution for nation-wide economic development, winning widely praises by all the parties of society. Currently there are 687 employees at our design institute, wherein technical staffs account for 90% of the total number of staff, including 1 state-grade designer, over 50 national registered architects, over 80 registered structural engineers, and 4 certified budgeting personnel. More than 70 staffs are qualified with senior professional titles and over 220 staffs with intermediate professional titles. The majors of our designers span across more than 20 fields including plan, general plan, architecture, landscape, structure, drainage, electrical, intelligent, HVAC, light industrial products, medicine, food, environmental engineering, construction materials, equipment, budget, geotechnical engineering, etc., in company with all kinds of special technical professional design, while long-span prestressed structure, grid, special foundation base, special structural reinforcement, space membrane roofing, building acoustics, etc. Our professional design branches including Architectural Design & Research Branch, Structural Design & Research Branch, Planning & Design Branch, Construction Equipment Design Branch, Electrical Design & Research Branch, Urban Design & Research Branch, Engineering Consulting Center, Survey Branch, Engineering Supervision & Management Company, Engineering Contractor Company and so on.

Being the first Institute in the province to adopt design and management of computer local network systems, we are equipped with various professional design softwares and advanced computer peripherals equipment, so that our design and management level tops the domestic leading position. The Design Institute of Yunnan Province claims 50 years of glorious history, shouldering over 20,000 of all kinds of design projects, covering construction area of 69 million square meters. Its design results throughout home and abroad, successively winning national, ministerial, provincial outstanding design awards and Technology Progress Award for over 200 times, and it has passed the ISO 9001 international quality system certification since 1999. Our Institute clings to the business philosophy of "quality first, customer supreme", providing clients with sophisticated technology and excellent quality in engineering design as well as technical consulting services, etc., which is also willing to work together with people of all circles at home and abroad for mutual development.

昆明野鸭湖度假住宅区

主创建筑师：张军　王昆军　朱力　吕建芸　张薇　张丹
杨媚　李雪涛　张羽　冯高山　潘建云

项目概况

项目位于昆明空港经济区西北侧，基地地处野鸭湖旅游风景区（以下简称野鸭湖）内，远离城市喧嚣，被群山环抱，风景区内数百亩水面被青山绿树所环绕，生态环境良好，是云南省首批旅游小镇开发建设项目之一。其二期包括C、D、E、F、G共五区。

规划目标

（1）尊重野鸭湖生态林区空间结构和形态，整合地域环境，保持山势形态的整体连续性。

（2）以体现“生长”为设计核心，重复发掘温和地区环境的地域特征，创造丰富宜人的山野环境。

（3）把握野鸭湖生态环境特征，塑造具有温和地域特征的“有机体”，重建城市生活场所及邻里关系的空间组织方式，并有效传承山居聚落的传统文化特征。

Kunming Yeyahu Resort Residential District

Chief Architects: Jun Zhang, Kunjun Wang, Li Zhu, Jianyun Lv, Wei Zhang, Dan Zhang, Mei Yang, Xuetao Li, Yu Zhang, Gaoshan Feng, Jianyun Pan

Project Description

Kunming Yeyahu Landscape Holiday City is located in the northwest of Kunming Airport Economic Zone. The site lays within the Yeyahu tourism scenic region (hereinafter Yeyahu for short), far away from the city bustle and circled by mountains. Several hundred acres of water in the scenic area is surrounded by trees and mountains, forming a good ecological environment; therefore it becomes one of the first developed projects for tourism towns in Yunnan Province. The project phase 2 includes five block C, D, E, F and G.

Planning Objectives

(1) Respect the spatial structure and form of Yeyahu ecological forest, integrate with the regional environment and maintain the overall continuity of the mountain form

(2) Taking "growth" as the core of the design, repeatedly explore the geographical features of the mild local environment and create a rich and pleasant mountain ambience

(3) Grasp the Yeyahu ecological characteristics, shape an “organism” of the mild regional features, reconstruct the spatial organization of urban living spirit and neighborhood relation and effectively inherit the traditional culture of the mountain village.

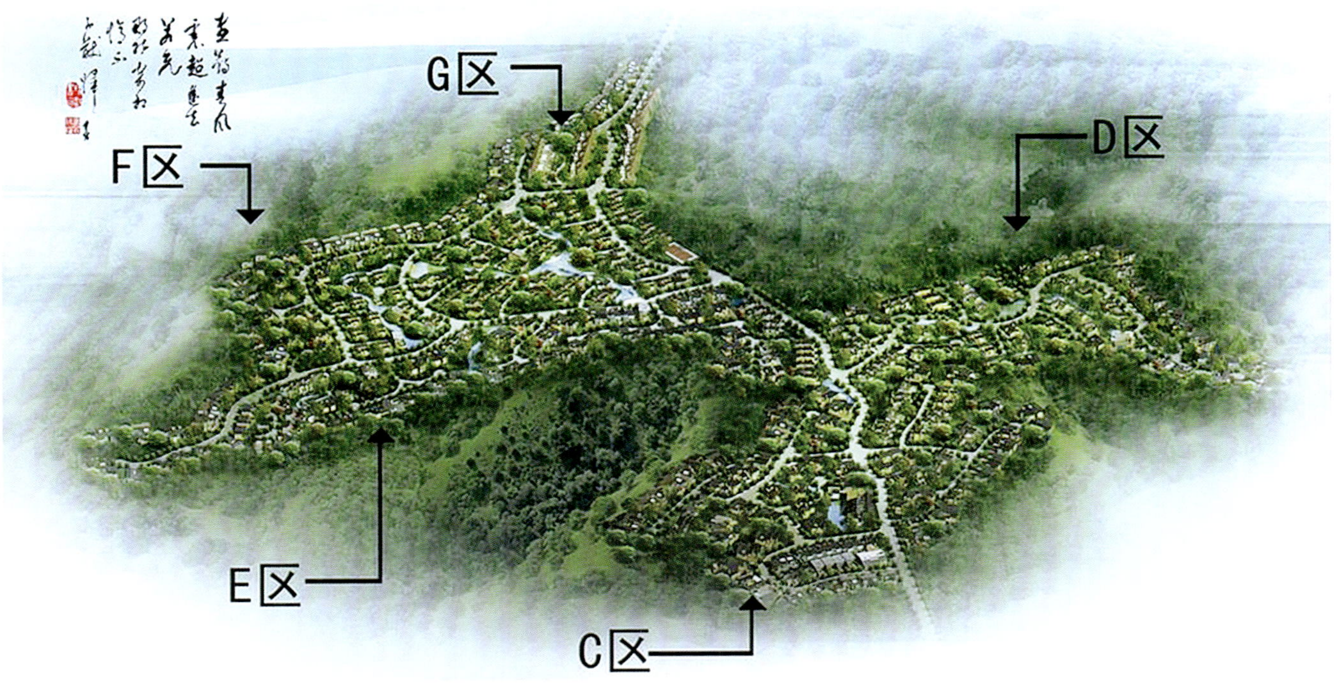

昆明市野鸭湖风景区

规划用地：31.7 公顷
建筑面积：130 000 平方米

设计思路

项目位于昆明市野鸭湖风景区，北纬25°独特的地理条件决定了建筑独特的红土高原风格面貌，乡土特有的气息在这里更加浓郁。顺山而建，应时而变，更突显了建筑的唯一性。

Yeyahu scenic region in Kunming

Planned area: 31.7 ha
Building Area: 130,000 m^2

Design Idea

This project is located in Yeyahu scenic region in Kunming. Its geographical condition of Latitude 25° determines for the construction the peculiar red-soil highland style, enriching the unique atmosphere of ramming soil. By building along the mountains and varying according to the terrain, the building appears to be more distinct and singular.

海埂会议中心——会堂
建筑面积：73 000 平方米
设计时间：2006 年 12 月—2007 年 8 月
项目状态：在建
主创建筑师：张军　张羽　周贻亮　何茂林　杨媚　万进　胡志杰　潘建云

项目简介

项目肩负着云南省重要会议接待、高规格接待、文化艺术活动中心的重要任务，向全国乃至向世界传达着云南特有的政治、经济、文化的信息。因此，具有云南特色的地域文化、民族文化以及时代感成为这里最重要的关键词语，同时，在设计的手法上也全面传达了具有云南特色的建筑语言及风貌。

Haigeng Convention Center—Auditorium
Building Area: 73,000 m^2
Design Time: 2006.12-2007.8
Status: under construction
Chief Architects: Zhang Jun, Zhang Yu, Zhou Yiliang, He Maolin, Yang Mei, Wan Jin, Hu Zhijie, Pan Jianyun

Project Description

Haigeng Auditorium undertakes the important tasks of major meeting reception, high-level reception and cultural arts center, conveying messages about politics, economy and culture of Yunnan to the nation, even to the world. Therefore, the local culture, national culture and senses of the times with Yunnan characteristics become the most essential key words in the design. At the same time, its design approach fully conveys Yunnan's representative architectural language and scenery.

昆明 · 铭真 · 运动休闲会所
建筑面积：8 500 平方米
设计时间：2008 年 10 月—2009 年 5 月
项目状态：在建
主创建筑师：张军　朱力　刘美逢　施鸿麟

项目简介

项目为高尔夫会所，选址位于呈贡大渔乡，紧邻滇池，环境优美。建筑功能主要包括员工办公及会员接待两个功能。其中会员接待功能包含了休息、会议及宴会功能，而办公区则是高尔夫工作人员办公的区域。

Kunming • Mingzhen • Sports & Leisure Club
Building Area: 8,500 m^2
Design Time: 2008.10-2009.05
Status: under construction
Chief Architects: Zhang Jun, Zhu Li, Liu Meifeng, Shi Honglin

Project Description

The project is a golf club located in Dayu Town in Chenggong county. It is adjacent to the Dianchi Lake and enjoys a beautiful environment. Its architectural function mainly includes staff office and membership reception: the former is the office region for the golf staff while the latter undertakes functions for rest, conference and banquet.

丽江悦榕酒店
建筑面积：18 000 平方米
占地面积：10 公顷
设计时间：2004 年
竣工时间：2006 年
获奖情况：2008 年国优银奖
2008 年中国勘察设计协会建筑工程优秀设计一等奖
主创建筑师：李光熙 徐锋 方泰生 罗文兵 王莉 王宏伟 陈妤婕 刘建 王茂珍 陈文斌 李兵

项目简介

设计中着重考虑了地域文化的体现和环境的融合，体现传统的回归，重视文化的传承，符号语言的重构，注重人文关怀，从细节入手进行建筑环境的重塑，尊重自然生态环境和绿色建筑设计，从而体现出以人为本、尊贵豪华、景观优先、亲和自然的设计原则。以精细化的设计，严谨的施工控制，最终形成高完成度建筑群落，使建成后的丽江悦榕度假酒店成为丽江乃至全国均无先例，独具唯一性的豪华旅游休闲度假酒店。

Lijiang Yuerong Hotel
Building Area: 18,000 m^2
Land Area: 10 ha
Design Time: 2004
Completion Time: 2006
Awards: National Silver Award in 2008
The first prize of Excellent Architectural Design in China Association of Construction Survey and Design in 2008
Chief Architects: Li Guangxi, Xu Feng, Fang Taisheng, Luo Wenbing, Wang Li, Wang Hongwei, Chen Yujie, Liu Jian, Wang Maozhen, Chen Wenbin, Li Bing

Project Description

The project design lays stress on the local culture reflection and environment integration; embodies the return of the tradition, focuses on the inheritance of culture and the reconstruction of symbolic language; pays attention to humanities and reshapes the construction environment from the details; respects the natural environment and emphasizes the green-oriented architectural design, thus reflecting the design principle of people orientation, exclusive luxury, landscape priority and natural affinity. Its refined design and strict construction control eventually produces a building community with perfect completion, making the completed Lijiang Yuerong Resort Hotel the unique luxury tourism & leisure resort hotel, unprecedented in Lijiang, even across the whole country.

昆明市建筑设计研究院有限责任公司

昆明市建筑设计研究院有限责任公司，是由 1964 年成立的昆明市建筑设计研究院在 2005 年通过产权制度改革，改制而成的。现有国家甲级设计资质：建筑行业建筑工程设计甲级（包含工业与民用建筑工程设计、建筑装饰工程设计、消防设施工程设计、轻型房屋钢结构工程设计、建筑幕墙工程设计、建筑智能化系统工程设计、照明工程设计），并可从事资质证书许可范围内相应的建设工程总承包业务以及项目管理和相关的技术与管理服务；城乡规划编制甲级；工程咨询甲级；工程勘察专业岩土工程甲级。现有国家乙级设计资质：市政公用行业的道路工程设计；给水工程设计；排水工程设计；风景园林工程专项设计。同时拥有国家商务部核发的对外经营资格证书，云南省建设厅颁发的建筑企业资质证书并通过了 ISO 9001 质量管理体系认证。

公司人才济济，除了有一批行业内各专业的正高级工程师、高级工程师和云南省及昆明市勘察设计咨询行业的专家、学术带头人之外，还有各类专业的注册工程师 86 人。由他们所获得的荣誉称号有全国建设行业劳动模范、享受云南省政府特殊津贴的专业技术人员、云南省勘察设计大师、云南省有突出贡献优秀专业技术人员、昆明市有突出贡献优秀专业技术人员、昆明市优秀科技工作者、昆明市青年科技奖、昆明市政协委员、昆明市人大代表、昆明市劳动模范、昆明市委“优秀党务工作者”等。

“一主两翼”的《发展战略规划》为公司确立了发展目标与方向。其中包括作为公司主业发展的建筑工程勘察、设计及相应的工程技术咨询、城乡规划编制、工程前期策划、建设项目可行性研究、市政工程及园林景观绿化设计、建筑智能化系统工程设计、承包境外建筑工程的勘测、咨询、设计和监理项目，以及上述境外项目所需的设备、材料出口、对外派遣劳务人员以及作为两翼发展的工程项目代建管理、房地产开发、设备安装、工程监理、桩基测试、岩土工程施工、园林绿化施工等。目前公司正向着做强主业，做大两翼的战略方向发展，并已具备了为投资主体全过程服务的能力。公司的经营市场也从本省扩展到了西南地区，乃至东南亚、西欧等国家。

公司勘察设计业务硕果累累，优秀作品遍布国内外，其中获得部优设计的项目有 12 项，省优勘察设计项目有 115 项，市优勘察设计项目有 58 项。公司连续十五年获昆明市“重合同守信用”单位的荣誉，获得云南省“重合同守信用”单位的荣誉证书，为昆明市和云南省的城镇建设谱写了美好篇章、作出了贡献。公司编制的《成果集》纪录了我们的发展史，从 20 世纪六七十年代昆明标志性建筑，到八九十年代省内各地、州、市星罗棋布的居住小区、金融、商务写字楼、体育文化、工业等建筑，再到 21 世纪的别墅区域、高楼大厦、城乡规划等项目，无不呈现我们的发展轨迹，体现我们的实力所在。

公司设计的昆明市中级人民法院审判庭享有“中国第一庭”的美誉，以其为代表的省内一批公检法建筑，以其洗练的个性，深厚的内涵，推进着公检法建筑创作思维的新模式。公司近年设计的项目延安医院住院综合楼、云南农业大学科技综合楼、昆明公交集团多层车库及运营调度指挥中心在 2009 年获得国家工程建设质量银质奖、省优设计一等奖，同时瑞丽市景成温泉旅游度假村、云南映象（杨丽萍）大剧院二个设计方案获得 IUAES 第十六届世界大会中国民居建筑文化专题展组委会和中国民族建筑研究会联合颁发的“中国建筑文化经典示范工程”奖，彰显了公司设计质量水平跨上新高度。而公司呈现在世人面前的亮丽璀璨的现代建筑，则各具特色，更显妖娆。我们实现了连续七年的快速发展，公司步入了健康稳定的可持续发展的轨道，跟上了时代发展的步伐。

在企业文化建设中，我们始终将“敬业奉献，求实创新，忠诚客户，追求卓越”作为企业文化的宗旨；我们始终坚持“三个一”的敬业精神和工作态度，坚持“三个共同”的人生价值观；我们始终以“关注顾客，竭诚服务；敬业奉献，恪尽职守；持续改进，追求卓越”作为公司质量管理体系的质量方针；我们始终坚持“优质服务，诚信经营”的经营之道；我们始终以最优质的产品和高度的责任心竭诚为广大客户提供满意的服务。近年来，企业文化建设和精神文明建设硕果累累，被评为“昆明市劳动关系和谐企业”，获得云南省委和昆明市委“云岭先锋”工程先进基层党组织，工会工作连年获得昆明市城建工会考核一等奖。

公司经过四十五年的不懈努力，积四十五年经验与深厚的文化底蕴，目前已成为实力雄厚，业绩卓著的规划、勘察、设计、咨询科技型企业。在实施“三个管理年”与“精细化管理”之后，2010 年提出了“管理创新，科技领先，依法经营，品牌立院”的“创新发展”工作方针，这是我们今后一段时期实现“一主两翼”《发展战略规划》的指导原则与奋斗目标。我们愿与各行各业携手合作，大展宏图。我们坚信在上级主管部门的领导下，通过不断努力，坚持科学发展观，解放思想，开拓创新，振奋精神，励精图治，一定能将企业做大做强，保持企业稳定、协调、可持续地发展，为城乡建设作出更大贡献。

1

2

3

4

5

6

7

8

9

10

11

12

13

14

1 昆明市中级人民法院审判厅　　2–8 云南省科技馆　　9–14 瑞丽景成地海温泉度假区

1

2

3

4

5

1–5 大理西洱河旅游文化长廊

6–11 昆明市儿童福利院家庭资源中心

12–14 云南省文山体育馆

15–20 云南省第一人民医院住院综合楼

6

7

8

9

10

11

12

13

14

15

16

17

18

19

20

1–8　公交营运调度指挥中心
9–10　中央丽城
11–17　云南映象（杨丽萍）大剧院
18–22　时代人才市场

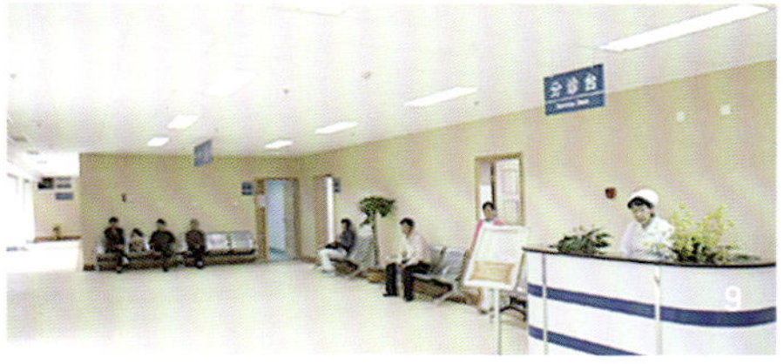

1–3　东川清真寺
4–9　昆明市延安医院
10–13　云南农业大学图书馆
14–19　鼎业束河名人酒吧街

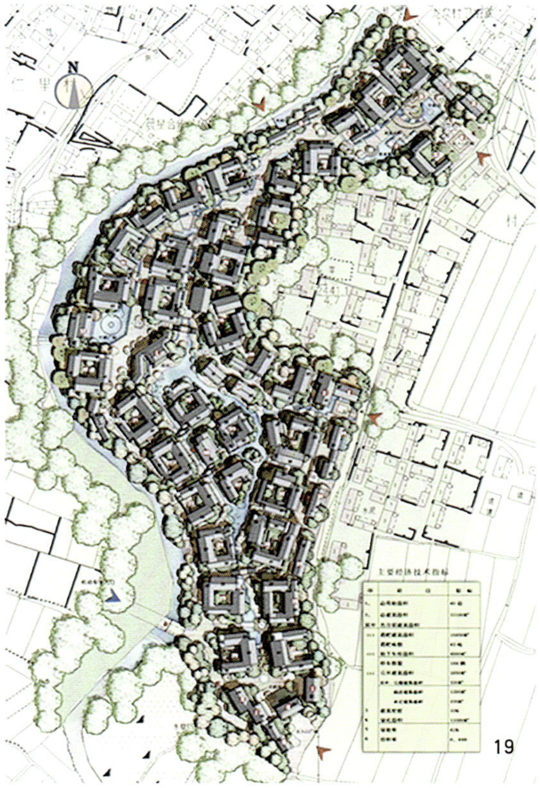

泛亞設計 PAN-ASIA DESIGN

电话：+86–871–6169777
传真：+86–871–6169222
网址：http://www.fysj.cc/
邮箱：fy@fysj.cc
地址：云南省昆明市北辰大道丽景园东路 122 号
邮编：650224

Tel: +86-871-6169777
Fax: +86-871-6169222
Http: //www.fysj.cc/
E-mail: fy@fysj.cc
Address: Kunming, Yunnan, North Star Avenue, Regency Park, 122 East Road
P.C.: 650224

云南泛亚工程设计院有限公司成立于 1993 年，现为国家甲级建筑工程设计单位。2003 年我院完成了改企转制工作，现有员工 70 余名，其中各类注册工程师比例已近技术人员的三分之一，同时我院建立了一支集建筑、规划、结构、设备、策划咨询、室内、环艺、图像等多专业在内的高度协作、高能高效的专业设计团队，为泛亚的长远发展奠定了坚实的基础。

我院设计的作品始终坚持“民族传统、地方特色和时代精神”的创作方向，在探索云南多民族建筑文化的继承和创新、城市改造与可持续发展方面作了大量的研究和设计实践，设计出“瑞丽电影院”（傣族风格）、“丽江客运站”（纳西风格）、“新平商城”（彝族土掌房风格）、“玉溪小庙街商贸城”和“石屏 · 州衙老街”（汉文化）、“爨园小区”（爨文化）、“昆明东部客运站”（石林风格）等一批风格鲜明的作品，并使这些作品成为地方标志性建筑。

我院一直致力于大中型房地产特色项目的策划、规划、设计和研究，涌现出春之城、丽水天锦 · 翡翠湾、公园道一号、茉莉唐朝、明日城市、新域盛景、水木江南、鑫城国际等佳作。

我院积极参与行业建设，现是云南唯一建筑学刊物《云南建筑》的联合主办单位、《时代建筑》理事单位、中国勘察设计协会民营设计分会、省建筑设计协会常务理事单位。为加强学术交流，推动建筑教育事业的发展，我院与昆明理工大学建筑工程学院共同设立“泛亚建筑论坛”每年邀请国内外知名专家学者莅临讲学，并在我院建立了昆明理工大学教学科研实习基地。

改制以来，依托“机制为根、人才为本，技术为先，品牌为源”的二次创业理念，全院目标明确，把握机遇，积极进取，综合实力逐年增强，并涌现了一批获奖作品，受到社会各界的高度赞誉。2010 年我院被评为全国勘察设计行业“优秀民营设计企业”。

Yunnan Pan-Asian Engineering Design Institute Co., Ltd was established in 1993. Now, it is the national Class-A design unit. Our institute completed the work of enterprises-reform and mechanism-establishing system in 2003. There are more than 70 employees, among whom the proportion of various registered engineers has taken up almost one-third of technical personnel. And we have established a highly collaborative, highly energetic and highly efficient professional design team which is consisted of multi-disciplinary majors including architecture, planning, structure, equipment, planning consultation, interior, environment art, images and etc, which has laid a solid foundation for the Pan-Asian's long-term development.

Our design works always insist on the creative direction of “national traditions, local characteristics and spirit of the times". We have done extensive research and design practice on exploring the inheritance and innovation of multi-ethnic architectural culture in Yunnan, on the planning of characteristic real estate projects, on urban renewal and development. We have designed a number of works with distinctive styles such as Rayli Cinema (Dai nationality), Lijiang Bus Station (Naxi style), Xinping Mall (house of soil structure of Yi nationality), and “Old Government Street, Shiping County" (Chinese culture), "Cuan Park Community" (Cuan Culture), "Kunming East Bus Station" (Stone Forest style), and has become local landmarks.

Our institute has committed to project planning, planning, design and research of characteristic projects of medium-sized real estate, emerging masterpieces like Spring City, Lishui Tianjin Green Bay, 1st Park Road, Jasmine Tang Dynasty, Tomorrow City, New Domain Spectacular, Shuimu Jiangnan and Xincheng International.

Our institute actively involves in industry construction. The institute is the co-sponsor of *Yunnan Architecture* which is the only architectural publication in Yunnan, a member of the council of "Time Architecture", and a member of council of Private Design Branch of Survey& Design Society in China and Provincial Architectural Design Association. To strengthen academic exchanges and promote the development of architectural education, our institute co-establishes "Pan-Asian Architecture Forum" with the College of Architectural Engineering of Kunming University of Science and Technology. We annually invite international and domestic famous experts and scholars to give lectures, and establish research practice base for Kunming University of Science and Technology in our institute.

Since restructuring our institute's system, relying on the second undertaking idea of “mechanism-rooted, talents-oriented, technology first, the brand is the source", the whole institute has clear goals and seize opportunities to make progress, so that comprehensive strength gradually increases and a number of award-winning works emerge, receiving highly praises from the society. In 2010, our institute was awarded “outstanding private design firm" of national survey and design industry.

昆明东部客运站

工程规模：40 830 平方米
建设地点：云南 昆明
设计 / 建成：2008 年 /2009 年

东部客运枢纽是未来昆明市主城区公路设施体系——"七大客运站"的重要组成部分。东部客运站主要经营途经昆石公路进入昆明主城的滇东、滇南方向客运，这一地区是中国喀斯特地貌最典型的聚集区，石林更是云南和昆明的旅游名片。建筑造型的设计灵感来源于那有着"大自然雕塑博物馆"美誉的喀斯特地貌和以"雄、奇、险、旷"著称的景观，建筑细部的溶蚀肌理、浮雕沟纹，形成在形似与神似之间，建筑融于石中的造型特点。结合城市景观规划要求，建筑构思新颖、个性鲜明、简洁大方，它将成为昆明又一高品质城市节点的标志性建筑。

Kunming East Bus Station

Project Scale: 40,830 m^2
Construction Location: Kunming,Yunnan
Design/Completion: 2008/2009

The east passenger hub is an important component of "Seven Bus Stations"---the future road facilities in main urban zone in Kunming. It is the bus station of the main access to the east urban zone of Kunming. East Bus Station entries into passenger transport of main city in Kunming in the direction of eastern and southern Yunnan by the way of Yunshi main road. This area is the most typical region of Karst landscape, and Stone Forest is the travel card of Yunnan and Kunming. Building form comes from Karst landform with "natural sculpture museum" reputation and landscape famous for "powerful, singular, precipitous, spacious", as well as corrode texture of details and relief grooves. In the resemblance between shape and spirit, the building fuses into stones. Combining requirements of urban landscape planning, the idea of building is novel, distinctive, simple and generous. It would create another high-quality node logo of Kunming.

丽江客运站

工程规模：9 000 平方米
建设地点：云南 丽江
设计 / 建成：2005 年 /2006 年
获奖：省优三等奖、
中国建筑文化经典示范工程

Lijiang Bus Station

Project Scale: 9,000 m^2
Construction Location: Lijiang,Yunnan
Design/Completion: 2005/2006
Award: the third prize in the provincial excellent projects;
the classic demonstration project of Chinese architectural culture

本规划用地是丽江古城风貌保护地带，是古城的有机组成部分。"以城市发展规划为基础"、"以人为中心"是总体规划的出发点，也是用地片区能否和谐融入城市的重要保证。

客运站的设计不是古城民居的"克隆"，更不是现代建筑的大杂烩。它应该是在民族风格基础上寻求突破与创新，吸取纳西民居中的精华部分，将其尽量凝练、简化，在保留民居中刚劲、挺拔、洒脱敦厚的风格基础上，使建筑造型更为简洁、明快、灵秀，使新客运站的形象在诗意环境中更有生命力。

建筑形象受古城四方街科贡楼的启发，逐级内收抬升的钟楼使规划限高的建筑形象得以突破，并使该建筑进一步确立了"控制性"地位。候车厅主入口前的拱桥造型，展现给旅客流畅、气势恢宏的感受。传统的纳西民居符号结合现代材料和设计手法，给人以耳目一新的震撼感。

The planning land is the control area of Lijiang ancient city and it is an integral part of the ancient city. "Based on urban development planning" and "human-centered" are the starting point for the overall planning, and an important guarantee for whether land area can be integrated into urban.

Bus Station is neither the "cloning" of ancient residential area nor a hodgepodge of modern architecture. It seeks breakthroughs and innovation based on the national style, and adopts the essence of the Naxi Houses. It is given conciseness and simplification as far as possible while retaining bold, upright, honest, free and easy style, it is more concise, lively, and witty, which makes the image of the new bus station full of vitality in poetic environment.

Building image is inspired by Kegong building in Square Street in ancient city. The sequentially rising bell tower makes the building image with limited height get a breakthrough and further establish "controlling" status of the construction. The arch shape in front of the main entrance of waiting room displays fluency, magnificent momentum. Traditional Naxi houses symbols combines modern materials and modern design techniques to give people a fresh shock.

云南农村干部学院

工程规模：47 000 平方米
建设地点：云南 昆明
设计／建成：2009 年／2010 年

Yunnan Rural Cadres College

Project Scale: 47,000 m²
Construction Location: Kunming,Yunnan
Design/Completion: 2009/2010

云南农村干部学院平面设计为 13 层“S”形主楼和南北各一个的圆形附楼。南侧圆形附楼为主门厅，北侧圆形附楼为商务配套功能楼。立面呈现一主两附的形态特征，南门厅的“树阵”造型展现了“植根农村”和学员“广泛的群众基础”，倒圆锥形的造型则展示了学院“开放、民主”的氛围，北侧的 4 层圆体建筑则是一个“果实”的造型，寓意云南农村干部学院和云南农业大学的农学特性和硕果累累的学术成果。

建筑组群构图均衡，一气呵成的曲线造型为城市和校园增添了丰富的轮廓线，产生了动人心魄的艺术效果。

The graphic design of Yunnan Rural Cadres College includes an S-shaped main building of 13 floors and a circular annex building on both sides of north and south. The circular annex on the south is the main hall and the north performs the business-matching functions. The elevation displays a mode of “one core, two annexes”; the "tree array" of the south lobby embodies the concept of "rooted in the rural villages" and the students' “broad mass fundation"; the inverted cone shape provides an atmosphere of "liberty and democracy"; the 4-layer park body on the north is of the shape of "fruit", implying the agricultural features and rich fruits of Yunnan Rural Cadres Collage and Yunnan Agricultural University.
The graphic design of the building complex is balanced, and its smooth curve adds to the rich contours of the campus as well as the whole city, producing an inspiring artistic effect.

鑫城国际

工程规模：320 000 平方米
建设地点：云南 陆良
设计／建成：2009 年／2010 年
获奖：中国领袖地产十大国际品质楼盘金奖

项目综合分析地块与城市、周边环境及项目定位关系，使总体规划在空间布局和平面形态上呈现“一轴、两片、三核”和“立体空间平台体系”的特色，充分展现滇东区域城市综合体的风采。

商务双子楼力图体现城市主轴线上“城市之门”的意念，双塔在顶部相连成为一个整体，建筑的立面中央形成一个宽 20 米，高 60 米的巨大门洞，寓意“开放的陆良欢迎您”。“城市之门”也将成为城市的标志性建筑。

Xincheng International City

Project Scale: 320,000 m²
Construction Location: Luliang,Yunnan
Design/Completion: 2009/2010
Award: the Gold Prize of Top Ten Real Estate of International Quality • Chinese Leading Real Estate

After a comprehensive analysis of the relationship between the land and city, the surrounding environment and project orientation, the overall planning shows the feature of "one axis,two pieces,three cores" as well as that of "three-dimensional spatial platform system" in spatial distribution and the plane pattern, fully demonstrating the characteristics of urban complex of the whole Eastern Yunnan.
The Business Twin Towers endeavor to embody the idea of "City Gate" along the axis line: the twin towers connect into a whole at the top, and the central part forms a huge doorway of 20 meters in width, 60 meters in height, meaning "An open Luliang welcomes you." The surface of "City Gate" will be the city's landmark.

▲ 石屏州衙老街

工程规模：12 000 平方米
建设地点：石屏古城
设计 / 建成：2007 年 /2008 年
获奖：中国建筑文化经典示范工程

Old Government Street, Shiping County
Project Scale: 12,000 m^2
Construction Location: Shiping County
Design / Completion: 2007/2008
Award: the classic demonstration project of Chinese architectural culture

石屏古城至今已有上千年历史。古城布局极具特色，东西南北两条主街在州衙前交汇，若干条巷道在密密麻麻的屋宇间纵横交错，且无街不弯，无巷不曲，形成石屏古城特有的建筑格局。弯街曲巷把城区隔离成大小不等的十多块，恰如龟身裂纹，隐含“九宫八卦”的阵势布局，故而石屏古城又称为“神龟托起的古城”。

石屏古城建城时按风水选择地形，将城区划为“九铺一中心”。本规划所涉及的正是这“一中心”（古州衙中心地段）街区的风貌保护与更新。

Shiping County has been a history of thousands of years up to now. The layout of this ancient city has unique features. Two main streets of all directions intersect in front of government office in feudal China. Certain laneways interweave between the closely-packed houses. Moreover, streets are curved, and lanes are bent. All these form the unique architectural pattern of Shiping County. Curved Streets and bent lanes divide the city area into more than 10 blocks, just like the crack on turtle body, hiding the layout of " Eight-diagram Tactics of Nine Directions". Therefore, Shiping County is also called "the ancient city held up by the turtle."

The terrain is chosen in terms of geomancy when Shiping County is established. The city area is divided into "Nine paves and a center". This plan is right involved in the environment protection and restoration of "a center" area (the central area of the ancient government office).

“三纵三横一核心”的棋盘式布局再现了传统的街巷体系肌理，错位的主街中部自然形成了戏台前的文化广场，成为规划设计的核心部分，既汇聚了人气又展示了“花灯之乡”的地方文化特色。

规划的最大亮点不是戏台核心的布局，而在于找到了文庙文星阁与戏台之间，历史文化轴线的关系，并使两者之间形成对话，促进了融合，进而充分挖掘玉溪城市的文化底蕴，极大提升了片区价值。历史遗迹的“一颗印”民居经过精心地规划后，也融入了新的街巷体系，并产生了较强的场所感和认知感。

The chessboard layout of "three vertical streets and three horizontal streets with one core" reproduces traditional texture of streets and alleys system. The middle part of main malposed street naturally forms a cultural plaza in front of the stage, which becomes a core part of the planning and design. It does not only bring popularity together, but also shows local culture of "Lantern town ".

The chessboard layout of "three vertical streets and three horizontal streets with one core" reproduces traditional texture of streets and alleys system. The middle part of main malposed street naturally forms a cultural plaza in front of the stage.

▼ 小庙街商贸城

工程规模：46 000 平方米
建设地点：玉溪 小庙街
设计 / 建成：2002 年 /2003 年
获奖情况：2005 年省优二等奖

Xiaomiao Street Mall
Project Scale: 46,000 m^2
Construction Location: Xiaomiao Street, Yuxi
Design / Completion: 2002/2003
Awards: the second prize of provincial excellent projects in 2005

▲ 春之城
总建筑面积：216 000 平方米
建设地点：云南 昆明
设计 / 建成：2003 年 /2005 年
获奖：省优二等奖

The Spring City
Gross Building Area: 216,000 m²
Construction Location: Kunming,Yunnan
Design / Completion: 2003/2005
Award: the second prize in the provincial excellent projects

春之城被誉为首席江景美宅。总体规划结合地形和周边环境进行设计，板式高层采取“南低北高”、“南透北围”的手法，以确保最大限度争取南向阳光和江景。在桥头堡的三角形位置规划一栋 30 层圆形点式高层，规划布局“板点辉映”并呈“二龙戏珠”之势。主导风向的西南侧为六层住宅，空间变化丰富，阳光、轻风在内庭中均可享受。

The Spring City is known as the supreme house of beauty with a river view. Its planning combines the terrain and the surrounding environment, and the plane-typed high-rise residential buildings adopt the mode of "south low while north high," "south windows while north walls", maximally displaying the sunshine to the south and the river view. A 30-storey round-point high-rise residential building is planned at the triangular position of bridgehead and the planning layout shows the feature of "plane-point co-echoing" and the trend of "Erlongxizhu". The southwest side along the leading wind direction stands a six-storey residential, rich in spatial changes, with sunshine and breezes dancing freely in the chamber.

▼ 明日城市花园
工程规模：253 000 平方米
建设地点：云南 昆明
设计 / 建成：2005 年 /2007 年
合作单位：澳洲 U/A/ 香港中元
获奖：省优一等奖

Tomorrow City Garden
Project Scale: 253,000 m²
Construction Location: Kunming, Yunnan
Design / Completion: 2005/2007
Cooperation unit: Australia U / A / Hong Kong Zhongyuan
Award: the first prize in the provincial excellent projects

“绿色、生态、可持续发展”是当今无可争辩之时代主题。本项目在户型、交通、道路、娱乐、休闲、水系、绿化、小品等诸多因素的处理上，始终遵循“尊重自然，以人为本”的原则，同时强调对室外环境，尤其是山地、绿化、水系、小品的营造，再加上建筑形态空间序列的变化，各步道铺地等景观轴，把小区之间，组团、院落之间连成一体，使园林绿化与建筑造型、空间形态紧密结合，使其相得益彰。最终描绘出一幅现代感强、人文气息浓厚，具有完善生活配套设施的现代山地田园城市生活的社区图景。

"Green, ecology, sustainable development" is the indisputable theme of the times. The project always follow the principle of " respect for nature, people-oriented" to dealing with many factors such as house type, transportation, roads, entertainment, leisure, waters, green, building pieces, while stressing to construct the outdoor environment, especially for mountains, green, waters, building pieces. With changes of spatial sequence of construction pattern,landscape-axis of trails connects communities, groups and courtyards together, which unite green of garden and park to architectural form and spatial form closely. Eventually it paints a modern community's picture in which urban life with perfect supporting facilities is melting into the mountains and gardens, where a strong sense of times and human breath can be truely felt.

浙江省建筑设计研究院

Zhejiang Province Institute of Architectural Design and Research

法人代表、院长：施祖元
地址：中国杭州安吉路 18 号
邮编：310006
电话：+86-571-85154691
传真：+86-571-85151540
电子信箱：zadri@mail.hz.zj.cn
网址：www.ziad.cn

Legal Representative/President: Zuyuan Shi
Address: No.18 Anji Lu, Hangzhou, Zhejiang, P.R. of China
P.C.: 310006
Tel: +86-571-85154691
Fax: +86-571-85151540
E-mail: zadri@mail.hz.zj.cn
http: //www.ziad.cn

浙江省建筑设计研究院是中国勘察设计单位综合实力百强之一，国家工商总局首批命名的"重合同守信用企业"单位，1997 年通过 ISO 9001 质量体系认证，是建筑设计、市政公用设计（包括给排水、园林设计等）、智能建筑设计、工程咨询、基坑支护和玻璃幕墙专项设计、工程总承包等国家甲级资质设计院，以及城市规划设计、人防工程设计等乙级资质设计院，并具有国家授予的独立对外经营权。院下辖十个设计所、咨询所和上海分院、市政分院、景观分院、装饰分院及浙江华丽装饰工程公司、浙江华智建筑技术推广中心、浙江华之建筑设计有限公司、浙江建院工程咨询有限公司等。

本院专业齐全、设备先进、技术力量雄厚，现有职工近 500 人，其中国家级工程设计大师 2 名，教授级高级工程师、教授级高级建筑师 40 余名，国家一级注册建筑师、一级注册结构工程师 140 余名，高级工程师、高级建筑师 110 余名，工程师、建筑师 200 余名。本院自 1952 年建院以来共承担国内外大中型工业与民用建筑项目设计一万余项，有 400 余项分别获国家、部、省级优秀设计奖，100 余项科研项目分别获国家、部、省级科技成果奖、科技进步奖、全国计算机优秀软件奖和国家专利等。1997 年被国家建设部授予"精神文明建设先进单位"，2000 年被国家科学技术部授予"全国 CAD 应用工程示范企业"，2006 年获建设部"'十五'全国建设科技进步先进集体"称号、中国勘察协会"优秀勘察设计院"荣誉称号及浙江省省级文明单位荣誉称号。

本院以高素质的人才，科学的管理，先进的技术和手段形成可靠的资源，以"设计创新，质量创优，诚信求实，团结敬业，发展争先"为宗旨，为用户提供百分之百的合格设计产品。

通过深化改革和开拓创新，本院设计水平不断提高。在完成的国家和省市重点工程中，设计精品迭出并屡获殊荣，如上海国际会议中心、浙江世界贸易中心、上海金钟广场、黄龙体育中心、上海对外贸易学院、上海东方绿舟、杭州萧山国际机场、浙江图书馆、杭州游泳馆、西藏体育馆、山东大厦等一大批体现科技进步、反映时代精神的优秀建筑作品，深得社会各界的广泛好评与赞誉。

Zhejiang Province Institute of Architectural Design and Research(ZIADR) ranks the Top 100 engineering survey & design enterprises with comprehensive power in China. It is also one of the first to be awarded the title of Contract-following and Credit-keeping Enterprise by the State General Administration for Industry and Commerce. Since 1997 ZIADR has been assessed and registered against the quality management standard of ISO 9001. It has been identified with Class A Certificates involving in the architectural design, municipal works design (including water supply & drainage and landscaping design), intelligent building design, engineering consultation, specialized designs such as foundation pit support and curtain wall and turnkey contract, etc. It also has Class B Certificates for the urban master planning design and air-raid shelter design. ZIADR has an independent foreign trade right authorized by the state. It has ten departments working on the design and consultation, such as Shanghai Branch, Zhejiang Huali Decoration Engineering Company, Zhejiang Huazhi Building Technology Promotion Center, Zhejiang Huazhi Architectural Design Co., Ltd. and Zhejiang Jianyuan Engineering Consultation Co. Ltd., etc.

ZIADR has all kinds of disciplines, state-of-the-art equipment and rich technical resources. It has approximate 500 staffs including two national design masters, over 40 professor -grade senior architects and engineers, over 140 Grade-1 national registered architects and engineers, over 110 senior architects and engineers and more than 200 architects and engineers. Since its founding in 1952, ZIADR has worked on the design for over 10 000 large or medium-sized civil and industrial projects in domestic and foreign market. Over 400 projects have been awarded the national, ministerial or provincial honors for their excellent designs and more than 100 hi-tech programs have won the national, ministerial or provincial prize for the achievement and progress in science and technology, advanced computer software prize and national patents. In 1997 ZIADR was granted the honor of "Role Model in Ethical & Cultural Progress" by the Ministry of Construction. In 2000 it was given the honor of "National Model Enterprise in CAD Application" by the Ministry of Science and Technology. In 2006, it had the honor of "Model Collective in Tenth-five National Construction Science & Technology Progress "by the Ministry of Construction, "Model Survey & Design Institute"by the China Exploration and Design Association and province-level model enterprise in ethical & cultural progress.

Depending on the high-qualified talents, scientific management, advanced technology and means to create the reliable resources, aiming at design innovation, quality excellence, credibility and truth seeking, solidarity and dedication, developing to take the lead, ZIADR provides its clients with complete qualified and satisfied products.

With reform deepening, pioneering and innovating, ZIADR keeps on improving its design standard. For those completed national, provincial and municipal key projects, there are many outstanding designs which have been honored many times, including Shanghai International Convention Center, Zhejiang World Trade Center, Shanghai Golden Bell Plaza, Huanglong Sports Center, Shanghai Foreign Trade College, Shanghai Oriental Green Boat, Hangzhou Xiaoshan International Airport Terminal, Zhejiang Library, Hangzhou Natatorium, Tibet Gymnasium and Shandong Building, etc., reflecting the development in science and conveying the sense of times , and winning high reputation and favorites from the people of all circles.

2

1. 波特菲诺中心
总建筑面积：298 792 平方米
设 计 年 份：2008 年

2. 湖州东部新区城市设计
总建筑面积：830 000 平方米
设 计 年 份：2009 年

2

1. 恩泽医疗中心
总建筑面积：224 000 平方米
设计 / 竣工年份：2006 年 /2009 年
合 作 单 位 ：株式会社日本设计

2. 阿里巴巴杭州软件生产基地
总建筑面积：156 000 平方米
设计 / 竣工年份：2006 年 /2009 年
合 作 单 位 ：澳大利亚 HASSELL 设计公司

3. 江苏东海水晶博物馆
总建筑面积：25 000 平方米
设计 / 竣工年份：2008 年 /2009 年

4. 上海电影博物馆暨电影艺术研究所业务大楼
总建筑面积：103 515 平方米
设 计 年 份 ：2008 年
合 作 单 位 ：美国威尔考特事务所

5. 永康市总部中心商务办公楼群
总建筑面积：320 000 平方米
设 计 年 份 ：2009 年

3

4

5

1

2

3

1. 杭州萧山国际机场二期（国内航站楼）
总建筑面积：167 900 平方米
设 计 年 份 ：2009 年
合 作 设 计 ：华东建筑设计有限公司

2. 杭州武林门旅游客运中心
总建筑面积：120 000 平方米
设计／竣工年份：2003 年／2010 年
合 作 单 位 ：德国 GMP 国际建筑设计有限公司

3. 杭州地铁滨康站及汽车南站城市综合体
总建筑面积：680 000 平方米
设 计 年 份 ：2009 年
合 作 单 位 ：澳大利亚爱伦建筑设计事务所

4. 温州永强机场新航站楼室内装饰
总建筑面积：22 730 平方米
设计／竣工年份：2006 年／2008 年

1. 银泰·海威国际
总建筑面积：120 000 平方米
设计／竣工年份：2005 年／2009 年

2. 杭州亚包大厦
总建筑面积：241 524 平方米
设 计 年 份 ：2007 年

3. 杭州六合天寓
总建筑面积：250 000 平方米
设计／竣工年份：2003 年／2006 年

4. 杭州同人·春江时代
总建筑面积：230 000 平方米
设计／竣工年份：2003 年／2008 年

1

2

1. 上海香溢花城售楼中心装修设计
设计年份：2009 年

2. 杭州东方金座
总建筑面积：53 000 平方米
设计／竣工年份：2002 年／2007 年

3. 上海新弘国际城售楼中心装修设计
设计年份：2009 年

3

1. 杭州西湖文化广场
总建筑面积：370 000 平方米
设计／竣工年份：2003 年 /2010 年

2. 上虞百官广场
总建筑面积：128 000 平方米
设 计 年 份 ：2006 年

3. 萧山云石山庄
总建筑面积：13 000 平方米
设 计 年 份 ：2009 年

4. 台州明和雅苑小区
总建筑面积：200 000 平方米
设 计 年 份 ：2009 年

1

2

3

4

5

1．临海市体育中心体育馆
总建筑面积：36 000 平方米
设计／竣工年份：2001 年／2007 年

2．台州体育中心主体育场
总建筑面积：49 090 平方米
设 计 年 份：1999 年

3．浙江省委党校、浙江行政学院、省社会主义学院新校园建设
总建筑面积：117 600 平方米
设计／竣工年份：2005 年／2008 年

4．青川中学
总建筑面积：47 081 平方米
设 计 年 份：2009 年

5．上海师范大学奉贤校区南区
总建筑面积：117 000 平方米
设计／竣工年份：2002 年／2008 年

ZUADR

浙江大学建筑设计研究院

Architectural Design and Research Institute of Zhejiang University

单位规模

目前有 8 个综合设计所，9 个建筑专业工作室，2 个专业设计所，以及岩土工程、市政交通、风景园林、建筑装饰、建筑幕墙 5 个专业分院组成。全院职工 400 余名，其中中国工程设计大师 1 名，中国建筑学会青年建筑师奖获得者 3 名，一级注册建筑师 71 名，一级注册结构工程师 55 名。

历史沿革、专业特长、工作业绩

浙江大学建筑设计研究院始建于 1953 年，是国家重点高校中最早成立的甲级设计研究院之一，至今已有五十多年历史。现执有国家建筑行业工程设计（含建筑装饰工程设计、建筑幕墙工程设计、轻型钢结构工程设计、照明工程设计和消防设施工程设计）、市政公用行业（排水、道路、桥梁、城市隧道）、工程咨询（规划咨询、编制项目建议书、编制项目可行性研究报告、项目申请报告、评估咨询、工程设计资金申请报告）、建筑智能化系统专项工程设计、工程勘察专业类岩土工程（勘察、咨询、监理）和风景园林工程设计专项等甲级证书，施工图设计文件审查房建、市政（道桥、给排水）一类证书；城乡规划编制、公路行业（公路）、市政行业（给水、环境卫生）、文物保护工程勘察设计（古建筑维修保护、近现代建筑维修保护、文物保护规划）、工程勘察（工程勘察专业类岩土工程）等乙级证书，并具有对外承包工程资格。业务范围涉及高层、超高层的大型办公、宾馆、商业综合体、行政办公楼，学校校园规划与设计，影剧院、图书馆、博物馆等文化建筑，居住区规划与设计，体育建筑，医院类建筑，城市设计，智能建筑设计、室内设计，风景园林与景观设计，市政公用工程，岩土工程，幕墙设计，古建筑和近现代建筑的维修保护、文物保护规划等。

坚持"营造和谐、放眼国际、产学研创、高精专强"的办院方针，2000 年 1 月 16 日通过 ISO 9001 质量体系认证。建院以来坚持设计、教学、科研相结合，依托浙大，定聘中国工程院院士、长江学者等高科技人才作为技术支撑，繁荣建筑创作，积极参与市场竞争，并广泛开展国际学术交流与工程联合设计，与日本、美国的 HOK、德国的 GMP、澳大利亚的 COX、PTW 等国际知名的设计公司进行合作。在历届优秀建筑设计评选中获得 460 余项国家、部、省级优秀设计奖、优质工程奖及科技成果奖，取得了较好的社会声誉和经济效益，得到社会各界和建设单位的赞扬和好评。

Profile

At present Architectural Design and Research Institute of Zhejiang University has 8 comprehensive design institutes, 9 professional studios, 2 professional design institutes, 6 campuses including Geotechnical, M. Infrastructure, Landscape Architecture, Architecture Design, Curtain Wall. The Institute has over 400 staffs including one master of China Engineering Design, 3 winners of Architectural Society of China Young Architect Award, 71 national Class 1 Registered Architects and national 55 Class 1 Registered Structural Engineers.

History, specialization and performance

Founded more than half a century ago in 1953, the Architectural Design and Research Institute of Zhejiang University has been one of the earliest Class-A design institutes established among national key universities. Today, the Institute has been granted with some Class-A certificates of national architecture engineering design (including architecture decoration engineering design, curtain walling engineering design, light-weight steel structure engineering design, illumination engineering design and fire fighting engineering design), municipal public industry (water drainage, road, bridge and urban tunnel), engineering consultation (planning consultation, project compiling proposal, feasibility research report of project compiling, project application report, evaluation consultation and application report of engineering design fund), special architecture intelligent system engineering design, engineering survey specialized in geotechnical engineering (survey, consultation and supervision) and landscape architecture engineering design, first class certificates of construction drawing design document examination, house building and municipal administration (road, bridge and water supply and drainage), some Grade-B certificates of urban-rural planning, roadway industry (roadway), municipal public (water supply, and environmental hygiene), culture relics protection engineering survey & design (ancient architecture maintenance and protection, modern architecture maintenance and protection, culture relics protection planning), engineering survey (engineering survey specialized in geotechnical engineering), etc. It also has the qualification of foreign contract engineering. The business covers high-rise or super high-rise large office, hotel, business complex and administrative office building; planning and design of campus; cinema, library, museum and other cultural buildings; residential community planning and design; construction of sports facilities; construction of hospital; urban design; intelligent architectural design, interior design; landscape architecture and landscape design; municipal public engineering; geotechnical engineering; curtain walling design; maintenance and protection of ancient building and modern building, planning of cultural relics protection and etc.

Under the principle of "Building harmonious atmosphere; taking the world in view; industry-study-research-innovation; stronger, higher standard, more professional", the institute passed the ISO9001 quality system certification on January 16, 2000. Ever since the establishment, the Institute has consistently integrated design, education and research, depending on Zhejiang University, technically relying on high-tech talents of hired Academician of Chinese Academy of Engineering, Cheung Kong Scholar and etc, vividly enhancing architecture creation, actively participating in market competition, widely carrying out international academic communication and joint project design. In addition, the Institute has also carried out cooperation with internationally-recognized design companies including HOK (US), GMP (Germany), COX (Australia), PTW (Australia). More than 460 best design projects of state, ministry or provincial level, high-quality engineering and technical achievements have been awarded in previous high-quality architecture design appraisals, good social reputation and economic benefit have been achieved, and praise from all works of life and construction companies has been obtained.

单位地址：中国杭州浙大路 38 号
法人姓名：董丹申　院长兼总建筑师、研究员、一级注册建筑师
联系电话：+86–571–87951034
传　　真：+86–571–87951725
网　　址：www.zuadr.com

Address: No.38, Zheda Road, Hangzhou, China
Legal person: Danshen Dong, President of Institute and Chief architect, Researcher, National Class 1 Registered Architect
Tel: +86-571-87951034
Fax: +86-571-87951725
Http: //www.zuadr.com

安源大剧院

建设地点：江西 萍乡
建筑面积：21 692 平方米
占地面积：17 500 平方米
设计时间：2009 年
建筑高度：30.40 米
容 积 率：1.04

设计特点

安源大剧院选址于江西省萍乡市新城区，包含 1 230 座剧场、400 座多功能厅以及配套商业服务等功能。

设计表现了清新的艺术氛围，寻求地方文脉，创造和谐的人文场所精神，使安源大剧院成为具有地域特征和高雅艺术品格的标志性文化建筑。设计以种子发出新芽为立意，将其神韵赋予建筑形态之中，使得安源的文化和历史在设计中得以经典重现，达到形式与内涵的完美统一。

整个建筑以平缓的曲线、简洁的体块、灰白的主调色彩来表现建筑的沉着、柔和，将大剧院素雅的外形与室内外错落有致的树木、水面映衬，在设计意图上，突出表现建筑丰富的体形特征和个性特征。

大剧院位于优美的自然环境之中，体型圆润、祥和，灵动秀美，充分显示其特性与本质之美，成为城市舞台中的珍品。

Anyuan Theater

Construction Location: Pingxiang, Jiangxi
Building Area: 21,692 m^2
Site Area: 17,500 m^2
Design Time: 2009
Construction Height: 30.40 m
Plot Ratio: 1.04

Design Features

Anyuan Theater is located at new district of Pingxiang City, Jiangxi, including the theater with 1,230 seats, multifunctional hall with 400 seats and equipped with supporting commercial service, etc.

The design aims to present a fresh artistic atmosphere, chase the local history and culture, and create harmonious cultural site sprit, making the Anyuan Theater a landmark cultural building of geographic features and elegant artistic characters. The design takes the new sprouting of the seeds as its theme and endows the architectural form with its verve, so as to recur the culture and history of Anyuan in the design and realize the perfect unity of form and connotation.

The whole building interprets the stability and tenderness of architecture with gentle curves, simple blocks and main color white and gray, and relieves the plain and graceful appearance of the theater against well-proportioned woods and water surface indoor and outdoor. The design intent intends to show the diversified shape characters and individual features of the architecture.

The Theater is surrounded by beautiful natural sceneries, the rounded and beautiful shape fully demonstrates its characters and the beauty of nature, and becomes a treasure of the city stage.

孙武文化园

建设地点：江苏 苏州
建筑面积：23 360 平方米
占地面积：36.738 公顷
设计时间：2007 年

Sun Wu Cultural Park

Construction Location: Suzhou, Jiangsu
Building Area: 23,360 m^2
Site Area: 36.738 ha
Design Time: 2007

设计特点

穹窿山风景区位于苏州市中心西南部 20 千米处，孙武文化园位于穹窿山主峰下的茅蓬坞与仓坞境内，基地三面环山，东临苏福公路。项目交通便利，景观资源丰富。文化园区跨越两个山坞，园区共占地 36.7 公顷。

根据孙武文化园景区自然山水分布特点，设计合理安排各大景区，科学组织游览观光路线，使整个文化园形成“一湖、一轴、一线、三区”的结构形态。战鼓广场，孙武书院及野外拓展营共同形成孙武文化园区的“寻源”主题游线。

设计通过模拟春秋时期吴国的建筑形态，调整自然与人工的比例，控制建筑物、构筑物体量，来重现春秋时期的景观，将主要人工场所集中布置在主轴线上，周边环以地形起伏的自然山地和曲折跌落的溪涧，创造出一系列对景、借景、框景的视觉引导，有意使游人的视线停留在传统的、典型的景观组合构图上，以唤起游人对春秋时期吴国宫舍和穹隆山水特征的追忆与联想。

Design Features

Qionglong Mountain scenic area is located 20 miles from the southwest of the downtown of Suzhou, and Sun Wu Cultural Park is located between Maopeng Wu and Cang Wu which are under the main peak of Qionglong Mountain. With the convenient transportation and rich landscape resources, the park is surrounded by mountains and is west to Sufu Road. The Park crosses two cols, covering a total area of 36.7 ha.

The scenic areas are arranged reasonably and the sightseeing routes are organized scientifically according to the natural landscape distribution characteristics of Sun Wu Cultural Park scenic area, forming a structural form of “one lake, one axis, one line and three zones” for the entire cultural park. War drum Square, Sun Wu College and outward bound camp together comprise the “origin tracing” thematic tourist line of Sun Wu Cultural Park scenic area.

The design reproduces the drizzly landscape of Spring and Autumn Period by imitating the architectural form of Wu State in Spring and Autumn Period, adjusting the proportion of nature and manual work and controlling the architectural volumes of building and structure. It arranges the main artificial areas to be centralized on the main axis and surrounded by undulating natural mountain and winding mountain stream in ring, which creates visual guide of a series of scenic focal point, borrowed scenery and enframed scenery, and intentionally highlights the traditional typical landscape portfolio composition for the tourists, so as to raise the tourists' retrospection and association of characteristics of Wu architectures and Qionglong landscapes in Chunqiu Period.

湘庄、荷庄

建设地点：浙江 萧山
建筑面积：7 500 平方米
占地面积：44 110 平方米
设计时间：2007 年
建筑高度：最高 11.5 米
容 积 率：0.17

Xiang Manor, Lotus Manor

Construction Location: Xiaoshan, Zhejiang
Building Area: 7,500 m^2
Site Area: 44,110 m^2
Design Time: 2007
Construction Height: max. 11.5 m
Plot Ratio: 0.17

设计特点

材料用旧，则手法从新，反之材料用新，则手法从旧，设计变法之道，执其两端而用其中。
浙江民居用材简朴但空间丰富，手法古朴而趣味盎然，设计取法于此，“何陋之有”。

Design Features

Utilizing modern techniques when using old materials, or traditional techniques when using new materials, is considered an appropriate approach to architectural design.
Using simple material to create rich spaces, Zhejiang local housing has made great achievements, and as a result architectural design is of a high quality.

杭州师范大学仓前新校区一期工程教学区

建设地点：浙江 杭州
建筑面积：147 700 平方米
占地面积：81 700 平方米
设计时间：2009 年
建筑高度：12 ～ 30 米
容 积 率：1.533

The Phase I Project Teaching Area of Hangzhou Normal University Cangqian New Campus

Construction Location: Hangzhou, Zhejiang
Building Area: 147,700 m^2
Site Area: 81,700 m^2
Design Time: 2009
Construction Height: 12-30 m
Plot Ratio: 1.533

设计特点

教学区以湿地书院为设计理念，从地域化和场所感入手，创造独具特色的建筑群体。

总图布局的核心思路就是运用张弛、疏密、刚柔的设计手法，塑造分散聚落状的组合空间，以此来呼应建筑所处的湿地环境。

设计采用江南传统的"水街"、"聚落"形式，通过建筑的自由围合营造内聚性的教学和研究空间，使校园建筑回归传统的"书院"形态。

设计运用杭州的青砖、墙门、花窗等细部元素，用现代构造技术演绎传统建筑意向，以水墨画般黑、白、灰色为基调，呈现素雅、和谐的景致。

设计将人流量小的学院办公室和研究用房沿核心湿地布置，从而提高景观的利用率，将人流量大的教室沿主要道路周边布置，以利于校园学生的集中疏散。

利用湿地和水街，构成教学区的景观骨架。依托湿地环境，主体建筑采用低层高密度建筑模式，并与环境紧密相融。建筑 3 ～ 7 层不等，设计利用建筑的高低错落，营造出开阔、疏朗的校园环境。

Design Features

The teaching area takes Wetland College as its design concept, and creates characteristic building clusters from localization and sense of place.

The core idea of the whole plan layout is to resolve and analyze the architecture, and shapes a scattered and gathered combination space by using the design methods of flexible, density and "hard and soft", so as to match the wetland environment in which the architecture is located.

The area adopts the traditional "Water Street" and "Settlement" forms of south of Yangtze River, builds a cohesive teaching and research space through the free enclosing of the architecture, and takes the campus architecture back to the traditional form of "Chinese college".

The Area interprets the traditional architecture intentions with modern building technologies and add the detailed elements, such as flashed bricks, wall door and decorative open window and others of Hangzhou style. It is finally to present a group atmosphere of simple and elegant, harmonious and powerful which is based on three colors of plain color series of Chinese painting style: black, white and gray.

The college offices and research room of small flow are arranged along the core of wetland so as to improve the utilization of landscape. The classrooms of large flow are arranged along around the main roads so as to be convenient for the evacuation of campus students.

The wetland and water street consist the landscape framework of the area. The principle part of the architecture adopts low high-density architectural mode based on the wetland environment and is closely blend with the environment. The building ranges from 3-7 floors, and forms a open and bright campus environment by using the scattering of the building.

临安博物馆

建设地点：浙江 临安
建筑面积：7 445 平方米
占地面积：39 306 平方米
设计时间：2009 年
建筑高度：18.6 米

设计特点

临安博物馆是为展现临安的自然地理、人文历史而建立的，选址位于临安市西区块，西靠玲珑山，南为待建的烈士公园。

建筑从临安极具地域代表性的钱王砖石遗址中吸取灵感，将建筑体量纵横搭砌，营造地下宫殿一隅被发掘出来的景象，喻义吴越博大幽远的文化历史。部分建筑嵌入大地中，南部以草坪缓坡的方式将建筑与周边的绿化自然相接，使整个建筑看似从自然山体中生长出来一般，与玲珑山浑然一体。建筑与环境一气呵成，含蓄又不失个性。设计通过层层院落的组织，将自然山水引入建筑内部，使人为空间与自然环境交相呼应，这一处理成功避免了地下建筑自然采光、通风不足的缺陷。自上而下的公共流线组织强化了神秘探寻的主题，参观者在行进过程中，慢慢远离尘世喧闹，逐渐沉浸到历史氛围中。

在博物馆形体处理上，充分考虑它与纪念馆的关系，两者如同散落在主轴两侧的古器，体现吴越的历史文化。

Lin'an Museum

Construction Location: Lin'an, Zhejiang
Building Area: 7,445 m^2
Site Area: 39,306 m^2
Design Time: 2009
Construction Height: 18.6 m

Design Features

Lin'an Museum is established for presenting the physical geography and culture and history, it is located at the west area of Lin'an, west to Linglongshan, and south to martyr park.

The building absorbs the inspiration from Empire Qian Masonry Site, the representative architecture in Lin'an. It builds the architecture freely and forms an imaginary of one corner of underground palace is explored, which is just like the exploiting of profound culture and history of Dynasty Wu and Yue, so as to embed the part of the building into the earth. The south part freely connects the building with surrounding greening in the form of lawn slope, making the building just like it grows from the natural mountain and blends into Linglongshan. The building is coherently combined with environment, implicit but unique. Meanwhile, it introduces the natural scenery into the interior of the building by the layout of courtyards, in this way the artificial space is consistent with natural environment, and this arrangement has successfully fix the defect in the lacking of natural lighting and inadequate ventilation of the underground building. The up-to-down public stream line arrangement strengths the theme of mysterious exploring, making the visitors gradually immerses into the atmosphere of history and far away from the reality during the course of visiting.

When dealing with the physical form of the building, the relationship of which with memorial museum is taken into an acount, and the both two as the ancient artifacts scattered on the two sides of the principle axis imply the characteristic traditional wisdom of Wu-Yue Civilization: Yin and Yang, tenon-and-mortise work.

三门剧院

建设地点：浙江 三门

建筑面积：13 967.9 平方米

用地面积：18 125.5 平方米

设计时间：2010 年

竣工时间：在建

建筑高度：29.45 米

容 积 率：0.589

Sanmen Theater

Construction Location: Sanmen, Zhejiang

Building Area: 13,967.9 m^2

Floor Area: 18,125.5 m^2

Design Time: 2010

Completion Time: under construction

Construction Height: 29.45 m

Plot Ratio: 0.589

设计特点

1. 分散性——独立开放、经营的可能性

建筑的综合功能，将按各自的独立性布局，它具体由大剧院、电影超市、文化展示组成。剧院和文化展示往往是在不同时间进行的，而电影超市却有持续性的营业要求。独立经营的方式可以节省资源，实现经济与形式的统一。馆与馆之间通过绿色中庭间隔，通过大平台连接，节能并不一定体现在技术上。

2. 完整性——它需要极具视觉冲击力的建筑形象

我们坚持建筑造型的完整性。

这不仅仅因为它是一个文化共同体，也是因为它兼具城市轴线与景观大道作用，所以它要作为一个更有力量的建筑物出现于此。

3. 导向性——城市轴线与景观轴线的连接点

在本案中，我们在滨海大道方向上设置大型的入口空间。三门剧院不仅仅是一个功能建筑，更是有效连接中轴线与景观广场"中转装置"。

Design Features

1.Dispensability- the possibility of independently open and operation

The comprehensive building function will be divided according to their own individuality, which is made of the theater, film market and cultural exhibition. The operating of the theater and cultural exhibition is unmeant and at different times, while the film market requires a continous operation. The disperse organization enables the realization of developing based on its own and the unity of economy and form by independently operation to save the resources.

The halls are separated by green atrium and are linked by a big platform which is taken as an open space. It shows that the energy saving is not necessarily reflected on the technology.

2. Integrity- it requires a architecture image with strong visual impact

Although the operation and function of the building is dispersive, we insist the integrity on the modeling of the building.

This is not only because it appears as a cultural community, but also its special positioning of locating at junction of city axis and landscape avenue as well as the requirements of building a strong architecture here.

3.Orientation-the junction linking city axis and landscape axis

In this case, we put a large entrance gray space at the direction of Binhai Avenue. This gray space is open and consists with greening landscape square to the east through the large platform, and opens to middle axis to the north through large punch. Therefore, Sanmen Theater is not only a functional architecture, but a "transfer device" linking middle axis and landscape square effectively.

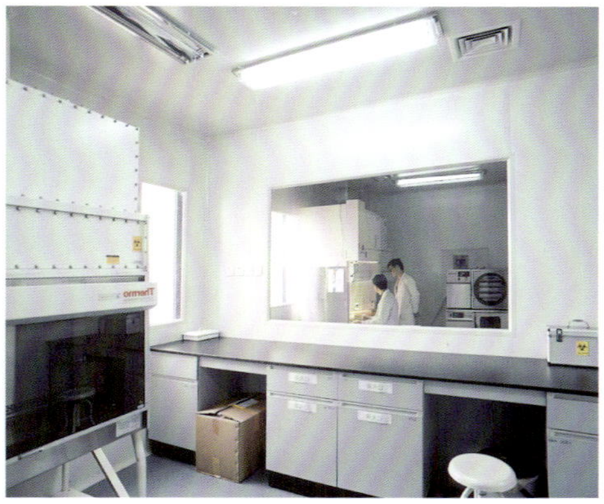

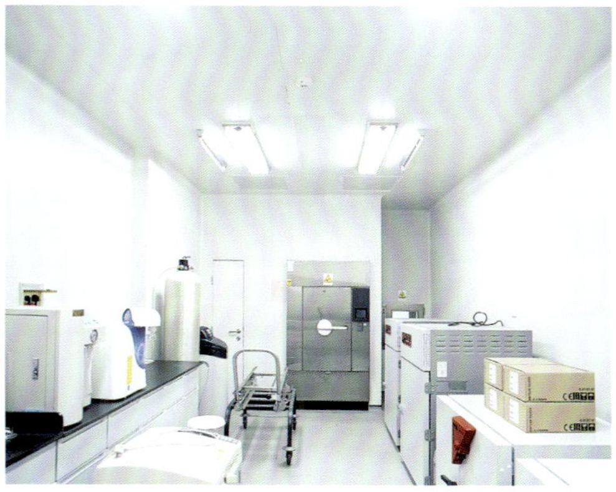

浙江省疾病预防控制中心

建设地点：浙江 杭州
建筑面积：44 986 平方米
用地面积：26 666 平方米
设计时间：2004—2005 年
竣工时间：2008 年
建筑高度：63.85 米
容 积 率：1.47

Zhejiang CDC

Construction Location: Hangzhou, Zhejiang
Building Area: 44,986 m^2
Land Area: 26,666 m^2
Design Time: 2004-2005
Completion Time: 2008
Building Height: 63.85 m
Plot Ratio: 1.47

设计特点

2003 年 sars 疫情爆发后，国家对疾控项目进行重点投入。浙江省疾病预防控制中心是浙江省 2004 年五大百亿工程之一，选址位于杭州市滨江区内，主要由理化实验用房、微生物实验用房、综合业务用房、地下车库等组成。

设计强调节地节能、环境友好的原则

1. 充分考虑城市景观与环保的要求，在总体布局中采用南低北高的布局方式，实验用房布置在基地南侧，位于地块的下风向，避免了废气对整个地块的污染。办公主楼布置在地块西北侧，靠近城市干道交叉口。

2. 各楼的建筑朝向除考虑尽可能避免东西朝向外，还注意利用周围的景观资源。

3、实验楼、综合业务楼均采用大进深平面布局形式。实验楼的柱网尺寸充分考虑不同实验室的模数要求，从而实现合理的空间组合，并为未来的升级改造提供可能。实验楼内的特殊用途实验室，如空气模拟仓、二级生物安全实验室、三级生物安全实验室、洁净动物房、聚合酶链式反应实验室和二恶英实验室等，其布局与空间尺寸均满足了工艺要求。两个三级生物安全实验室、洁净动物房均已通过了国家相关部门的认证和验收，是浙江省规模最大的高等级生物安全实验室。

4、建筑造型与功能紧密结合。

Design Features

After the SARS outbreak in 2008, disease control projects became our country's key investment projects. Zhejiang Disease Control and Prevention Center is one of Zhejiang's five biggest 10 billion RMB projects in 2004, which is located in Binjiang District of Hangzhou. It mainly consists of physical and chemical experiments rooms, microbiological experiment rooms, comprehensive business rooms, underground garage, etc.

The principle of land and energy saving and environmental protection is emphasized in its design.

1. The requirements of urban landscape and environmental protection shall be considered, and the layout way of low south and high south shall be adopted in overall layout. Laboratory rooms shall be set at the south side of the base, and shall be positioned at the below wind direction, so as to avoid the pollution of exhaust gas to the whole land. The main office building shall be set at the northwest side of the plot, which is near the crossing of urban main streets, and is favorable for forming the urban landscape.

2. The orientation of all buildings should avoid the east-west orientation, and should also pay attention to the utilization of the near landscape resources. Qiantang River Landscape, urban CBD landscape, the fourth bridge landscape, etc. can all be introduced into the building.

3. Both laboratory building and comprehensive business building adopt large depth plane, which is beneficial for saving land and energy. The column grid size of laboratory building fully considers different modulus requirements of different labs, which could realize the reasonable space composition, and provide the possibility for the upgrading and alteration in the future. The layout and space size of special labs in laboratory building such as air simulation module, second level biological safety laboratory, third level biological safety laboratory, clean animal room, polymerase chain reaction laboratory, and dioxin laboratory well meet the process requirements. Two third level biological safety laboratories and clean animal room have passed the certification and examination of relevant departments of the country, which are the largest advanced biological safety laboratory in Zhejiang Province.

4. Building model and function are combined closely.

HSS 嘉和设计 ARCHITECTS & PLANNERS ®

法定中文名称：浙江嘉和建筑设计事务所有限公司
法人名称：赵曼
网　　址：Http://www.hss.cc
联系电话：+86-571-85865351
传　　真：+86-571-85865250
电子邮件：mail@hss.cc
通信地址：浙江省杭州市建国北路586号
嘉联华铭座14楼
邮政编码：310004

“中国最具创新力建筑设计事务所”

HSS嘉和设计来自中国创意之都浙江杭州，是一家专业优势突出的特色公司。公司一贯主张“用专业的方法做成功的项目”，同时HSS擅长把客户初期模糊的想法变成符合需求的成功产品。

HSS的优势不在规模，在于其拥有多元优势突出的专业团队，专而精、精而强，多元而互补地贯穿项目全过程，同时也在于其创新实用的设计服务模式，接轨国际，强调精品意识和服务意识。作为浙江省首家甲级建筑设计事务所，HSS是走专业化道路的成功实践者。

HSS嘉和设计成立于2000年，以建筑设计为核心，涵盖规划、结构、景观、机电等专业，拥有甲级设计资质。公司还投资参股了国内最早提出“房地产全程价值整合”理念的机构——动力地产咨询代理公司——架构起崭新的以市场需求为导向的设计服务模式。公司员工总人数200余人，其中HSS核心部有40余人。

HSS嘉和设计以“精品、人和”为价值主张，多年来已积累了六大项目设计领域的成功经验包括房地产住宅，风情小城及地块园区规划，商业市场及酒店建筑，科研、文化及教育建筑，钢结构及工业厂房，景观环境。

近年来，通过创新以及和国内外的合作，HSS的业务发展兴旺，赢得了广泛的声誉和众多成功的客户。客户中不乏知名浙商、上市公司和世界500强公司。获得过的国内外奖项包括“中国最具创新力建筑设计事务所”、“建设部优秀设计单位”、“当代中华建设名家”、“浙江省重点建设项目工程设计咨询推荐单位”等荣誉称号，我们设计的项目还获得过“联合国人居署迪拜国际最佳范例（中国）推动奖”、“全国人居经典建筑规划双金奖”、“中国最佳山水别墅大奖”、“中国最佳创新户型楼盘”等。HSS的项目作品还被选入“中国建筑设计作品年鉴—建国六十周年优秀设计成果专集”。

HSS嘉和设计热诚地欢迎与客户建立密切的伙伴关系，并积极寻求能帮助其项目更加成功的设计和服务。

"The most innovative architectural design associates in China"

HSS Jia He design that comes from Hangzhou - the city of originality of China, is a particular company with outstanding professional predominance. "Do successful projects with professional methods", HSS is good at turning clients' early abstract ideas into the successful product that accords with the requirements.

The predominance of HSS lies not only in its scale, but in its professional team with many outstanding advantages, which are professional, refined, and powerful. HSS has many advantages and those advantages are complementary and run through the whole process; it lies in its innovative and practical design service mode, which meets the tendency of the international modes and emphasizes quality and service consciousness. As the first level A architectural design associates in Zhejiang province, HSS is the successful practicer that goes by the professional road.

HSS Jia He Design was found in 2000, and it takes planning architecture as its core, which covers specialties of layout, structure, landscape, mechanism and electricity, and it has level A design qualifications. The company has invested and shared the national institution that first put forward the idea of "Whole Value Integrated for real estate" --dynamical estate consultation and agency, which put up design service mode that takes the brand-new market requirements as the orientation. It has totally about 200 employees, and about 40 in the core department of HSS.

HSS Jia He design takes "high-quality product, harmony of people" as the key value claim, and it has accumulated the successful experiences for the projects in six fields for years: real estate inhabitancy; amorous little city and land garden layout; commercial market, and hotel building; scientific research, cultural and educational building; steel structure and industrial workshop; sight environment.

In recent years, through the positive creation and the broad cooperation with national and international enterprises, HSS has developed and flourished continuously, which has won broad praise and got many successful clients. The clients are Zhejiang famous companies, listed companies and world top 500 companies. It has won national and abroad awards includes "The most creative architecture and design office of China", "excellent design unit of architecture department", "celebrated present China architecture office", "Recommending Unit for the Design and Consultation of Zhejiang vital construction project"; our designed projects have won many national and international awards such as "United Nations Inhabitancy Dupai international best example (China) promoting award", "Double Awards for National Inhabitancy Classic Architecture Planning", "China Best Natural Villa award", "China best creative house type and building groups", etc. The projects of HSS have been selected in "the architectural design product annual----special collection of excellent design products for China's 60th Anniversary".

HSS Jia He design heartily invites the clients to build close fellowship, and positively seek the design and service that can help its projects be more successful.

安吉九龙逸品住宅设计
设计时间：2010年4月
设计规模：120 000平方米
设计深度：方案设计、初步设计、施工图设计

"低调的奢华，安吉精品豪宅的价值方向"
项目坐落在安吉灵峰山脚，浒溪水畔。以111 322平方米的总占地面积，120 000余建筑面积，直承安吉珍贵山水地脉。以2–4层为主的低密度精品建筑，特色庭院景观与地块周边茂密清幽的树林相结合，运用现代建筑理念，以一种低调、沉静而不失华贵地散发着优雅的气息，筑就安吉最具价值的精品豪宅，让人得到精神与物质的双重享受。

Design for Kowloon Elegance Residence Community, Angie
Design Time: 2010.4
Scale: 120,000 m²
Designed Depth: Conceptual design, preliminary design, construction drawing design

"Modest luxury, objective of Boutique Apartment in Angie"
This project is situated at the foot of Mount Lingfeng and beside Xuxi bank. Its total floor area is 111,322 m² and area of structure over 120,000 m². It is mainly made up of two-to-four-storied fine architectures. Meanwhile, featuring sceneries of courtyard, this project integrates with the lush forests in the surrounding. Modern construction ideas are applied to design, calmness, elegance and luxury are its keynotes. The goal of this project is to build the most valuable elite housing in Angie and to make residents here enjoy material and spiritual comfort.

湖州南浔天都置业住宅设计
设计时间：2009年12月—2010年8月
设计规模：100 000平方米
设计深度：方案设计、初步设计、施工图设计

"现代、自然、亲和——回归的心灵住宅"
本项目以生态、自然的规划手法，结合回归的现代生活理念，追求现代、自然、亲和的居住空间，巧妙地将现代建筑设计与南浔地域文化相结合，形成一个以低层住宅和高层住宅为主要建筑形态，尺度宜人、环境优美的高品质社区。

Design for Nanxun Tiandu Properties, Huzhou
Design Time: 2009.12-2010.8
Scale: 100,000 m²
Designed Depth: Conceptual design, preliminary design, construction drawing design

"Modernity, nature, affinity---Resident of returning of the soul"
The project makes use of the planning approaches of ecology and nature, integrates with the modern living concept and pursues the living space modernity, nature and affinity. With modern architectural design and Nanxun local culture skillfully combined, this project forms a noble community with the low and high residential housing as the main architectural form, pleasant and the environment is beautiful.

浙江烟草集团香溢会所

设计时间： 2009—2010年
设计规模： 4 200平方米
设计深度： 规划设计、方案设计、初步设计、施工图设计

本项目根据业主的需求，以高品质休闲度假基地为设计的出发点，满足特定客户的细致需求。方案在建筑布局与地势利用、平面功能、建筑造型等方面都进行了多种尝试，最终确立了本项目的三大设计特色。

1. 以景观资源为导向的建筑布局设计；
2. 以精致高档餐饮、度假为核心的功能设计；
3. 融合休闲、优雅、尊贵气质的造型设计。

Design for Xiangyi Club of Zhejiang Tobacco Group

Design Time: 2009-2010
Scale: 4,200 m²
Designed Depth: Planning programming,conceptual design, preliminary design, construction drawing design

According to the needs of the project owners, this project makes the high quality leisure base as the starting point, and meets the detail needs of the specific customers. This project makes many attempts in the architectural layout and the utilization of terrain, flat functions and architectural image. Finally, the following three main design features are established.

1. The view resource-oriented architectural layout design;
2. The fine, high-end dinning and holiday resort centered functional design;
3. Modeling design with the features of leisure, elegance and noble qualities integrated together.

连云港义乌中国小商品城

设计时间： 2009—2010年
设计规模： 170 000平方米
设计深度： 方案设计、初步设计

"义乌小商品城在连云港的成功再现"

本项目位于连云港海州区，主要设计产品为大集中市场、风情街和小专业市场。项目充分整合周边资源，以大集中市场为核心发动机，带动风情街区与小专业市场共赢发展，并通过波浪形的造型符号和风格混搭的手法，体现连云港山海融汇的地域文化和大型小商品城的特色。

Design for Lianyungang Yiwu China Commodity City, Jiangsu

Design Time: 2009-2010
Scale: 170,000 m²
Designed Depth: Conceptual design, preliminary design

"Reappearance of Yiwu China Commodity City in Lianyungang"

This project is located in Haizhou District, Lianyungang. Main design products are large-scale concentrated markets, characterized streets, specialized markets. Surrounding resources are integrated into this project. Centering large-scale concentrated markets, which promotes win-win developments of characterized streets and specialized markets, this design, through mixing of symbols and patterns of waves embodies the regional culture mingling mountains with oceans in Lianyungang and characteristics of small commodities markets.

湖南省烟草公司邵阳市公司办公楼及后勤基地

设计时间：2009—2010年
设计规模：80 000平方米
设计深度：方案设计、初步设计、施工图设计

“根据地形营造的山环水绕”

本方案是为邵阳烟草局量身定做，解决了三大设计难点，提供了一个高起点、高品位、高效率的建筑方案。方案设计时充分结合了山丘地形，为办公楼和住宅营造“山环水绕、坐北朝南”的地势境界做了充分的准备。同时注重内部环境与外部环境的紧密联系，将阳光、空气、水和绿化与各种空间进行一体化设计，营造私密性、半私密性等不同的庭院空间，创造了造型错落有致，立面不奢华张扬但大气、亲切、有品位的建筑物和优美的环境。

Design for office building and support base of Shaoyang Branch, Hunan Tobacco Company

Design Time: 2009-2010
Scale: 800,000 m²
Designed Depth: Conceptual design, preliminary design, construction drawing design

“In line with the terrain, the construction is surrounded by water and mountains.”
It is a custom-made architecture for Tabacco Monopoly Bureau of Shaoyang. This design solved three problems, which makes the building scheme a high-starting point, high-grade and efficient one. Based on local landforms of hills, office buildings and housing are orientated north and south and surrounded by water and hills. Meanwhile, the design stresses connection between inner environment and natural environment, which embodies integrated design of sunshine, air, water, vegetation and space. Enclosed and semi-enclosed courtyards are well-proportioned. Facades of buildings is of grandness, high taste and elegance rather than luxury.

湖南机场后勤基地别墅设计
设计时间：2009年7月
设计规模：65 000平方米　199栋别墅
设计深度：规划设计、方案设计、初步设计

本方案结合地形地势，控制建造成本，是为机场集团及长沙别墅市场量身设计。项目在地势利用、户型平面、外观造型、组团构图四个关键部分都进行了突破性的创新，创造了本项目的四大特色。

立体全景观围合花园；
性价比清晰匹配的户型设计；
简约西班牙和法式风格；
低成本多样化的和美总体建筑氛围。

Design for support base villas of Changsha Airport
Design Time: 2009.7
Scale: 65,000 m² / 199 villas
Designed Depth: Planning programming, conceptual design, preliminary design

This design, which customizes villas for Airport Group and for Changsha market, is based on local landforms and terrains and controls the construction cost. The project is known for its ground-breaking innovations: applications of landscapes, room type, shape and appearance, and residential groups, which also consist four distinctive characteristics of this project:
It is a panorama enclosed garden. Cost performance of house pattern designs is clear and reasonable. Villas are a compact Spanish and French style. The overall construction environment is elegant and delicate, while it is diversified and low-cost.

ZMA

浙江省现代建筑设计研究院

浙江现代投资项目管理咨询有限公司

Zhejiang Modern Architectural Design & Research Institute

浙江省现代建筑设计研究院创建于1979年，主要从事各类大中型民用建筑设计及相关工程咨询业务，持有国家建设部颁发的建筑工程甲级设计资质证书、国家发展改革委员会颁发的工程咨询甲级资格证书，并通过了ISO9001：2008质量管理体系认证。

该院现有各类专业技术人员100余名，其中中高级职称专业技术人员60余名，拥有国家一级注册建筑师、一级注册结构工程师、注册电气工程师、注册公用设备工程师、注册咨询工程师及注册造价师等30余名。技术力量雄厚，专业配备齐全，技术装备先进。研究院下设建筑设计研究创作所（医疗建筑设计研究所），建筑设计一所、二所、环境与室内艺术设计所、建筑智能工程设计所等生产部门，同时，为拓展设计与咨询市场，扩大经营范围，便于向业主提供多方位咨询与技术服务，研究院还组建了浙江现代投资项目管理咨询有限公司，主要从事项目管理和工程咨询业务。我院以“精益求精创一流设计，顾客至上争一流服务”为质量方针，以ISO 9001：2008标准为质量管理体系，建院三十多年来，在全国各地承接完成了数百项各类工业与民用建筑工程设计和咨询业务，业务范围遍及浙江、福建、江苏、安徽、江西、河南、湖南、广东、山东、河南、甘肃、陕西等省市，并取得20余项省级优秀设计奖。在对外合作与交流上，我院与德国建筑学会及其成员机构建立了广泛的合作与交流关系，同时，还与美国、澳大利亚等国的多家工程咨询设计机构建立了友好交流关系。

浙江省现代建筑设计研究院是一个学习型、创新型、合作型，不断追求进步、超越自我的团队。面对激烈的市场竞争，面对挑战与机遇，着眼未来，追求持续发展，将更精、更美、更新作为不懈的追求目标，充分发挥自身技术、知识密集的优势，逐步形成为固定资产投资活动提供全程技术性、管理性服务的咨询设计服务体系，力争以优秀的咨询设计和良好的技术服务取得信誉，为国家建设事业做出应有的贡献。

Incorporated in 1979, Zhejiang Modern Architectural Design & Research Institute (ZMA) mainly specializes in the design of various types of large and medium-scale civil constructions and relevant project counseling. It is certified as a Class A architectural design contractor by the Ministry of Construction as well as a Class A engineering counseling service provider and government-funded project contractor by the State Development and Reform Commission. It has also passed ISO9001:2008 certification.

ZMA now has 100 plus professionals and technicians of various types, including 60 plus professionals and technicians with a medium or senior occupational title as well as 30 plus Class A state-level registered architects, Class A registered structural engineers, registered electric engineers, registered public utility engineers, registered counseling engineers and registered cost appraisers. With such a strong technical prowess and well-established advanced technical and professional equipment, ZMA has such subsidiaries and departments as Architectural Design and Research Innovation Office (Medical Building Design and Research Office), No. 1 and No. 2 Architectural Design Offices, Xu Chuang Architectural Design Office, Environment and Indoor Artistic Design Institute, Building Intelligent Engineering and Design Office, etc. Meanwhile, in order to develop its businesses in the design and counseling market, expand its scope of businesses and offer all-around counseling and technical services to customers, ZMA has also established Zhejiang Modern Investment Project Management and Counseling Co., Ltd that specializes in project management and government-funded project undertakings. Cherishing the quality tenet of “better for the best, client first and topnotch service” and having passed ISO9001-2008 certification, ZMA has provided design and counseling services for several hundreds of industrial and civil construction projects across China in the past 30 years since its incorporation, with businesses covering Zhejiang, Fujian, Jiangsu, Anhui, Jiangxi, Henan, Hunan, Guangdong, Shandong and other provinces and municipalities. It has also been awarded 20 plus province-level trophies for excellent designs. In terms of foreign cooperation and exchanges, ZMA has also established extensive cooperation and exchanges with German Architectural Association and its membership organs. Meanwhile, it also enjoys cordial relationship with many project counseling and design organs in the US, Australia and other countries.

ZMA has a hardworking and innovative team that emphasizes on cooperation and keeps going ahead. Confronted with fierce market competition as well as challenges and opportunities, ZMA is casting its eye to the future and aims to excel in all aspects. ZMA will fully capitalize on its technological and knowledge advantages in order to establish a counseling and design service system that provides total-process technical and management services for fixed asset investment activities, hence winning market reputation with its excellent counseling, design and technical services and making contributions to social development.

1

2

3

4

6

1–2 广东省中山市中医院

3–6 河南省平顶山市新城区医院

7

8

9

10

11

12

13

7–8 江苏省扬州市南方协和医院
9–13 瑞安人民医院病房大楼

14

15

16

17

18

19

14–15 浙江省人民医院综合病房楼
16–19 陕西省榆林市第二医院迁建工程

20

21

22

23

20–23 山东省潍坊市第一人民医院

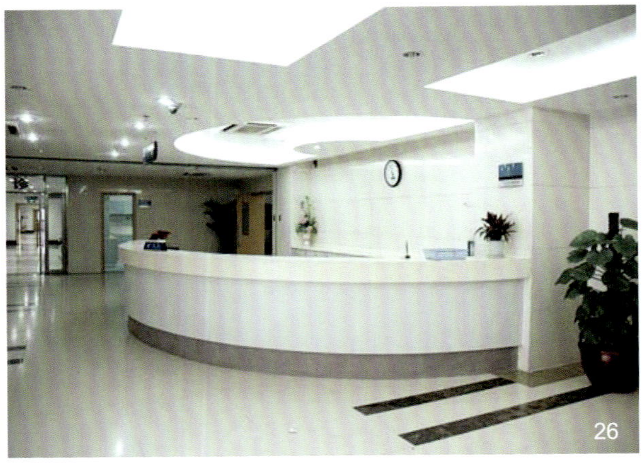

24–29 宁波市鄞州第二医院

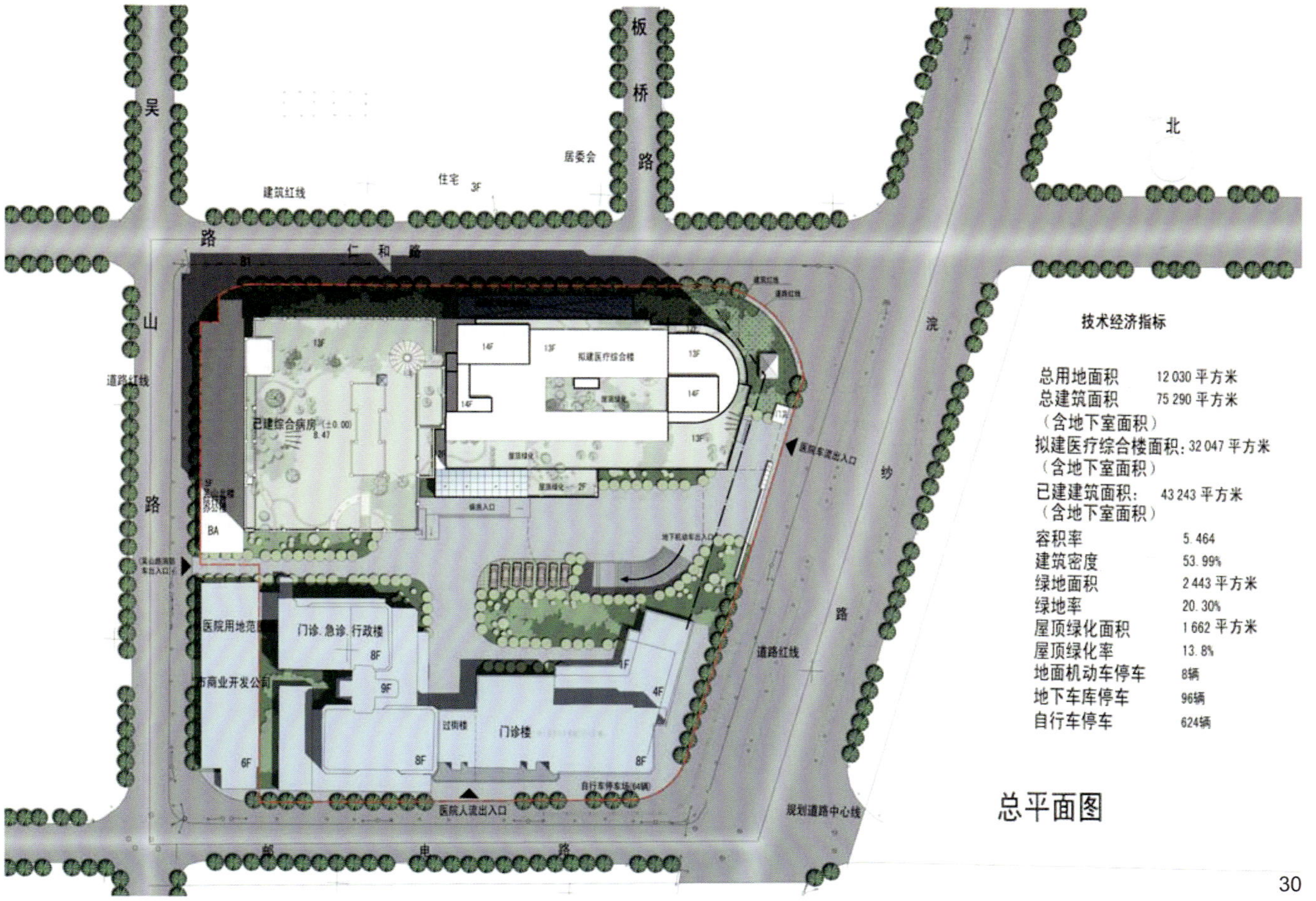

30

35

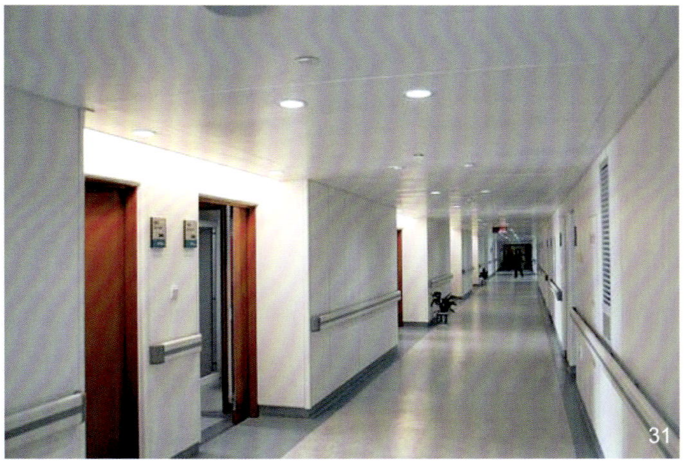
31

33

36

32

34

30–36 浙江省中医院

禾泽都林设计机构

HESOMS DESIGN INSTITUTE

"诠释理想城市，打造商业地产设计第一品牌"是禾泽都林设计机构设计机构的追求。公司由掌握创新思维和先进技术的设计师、具备国际视野与先进理念的技术新秀、具有专业执业资格的高学历人才以及资深文学、商业、经济等方面的专家所组成，百余人的强大设计团队为禾泽都林成为国内领先的设计机构插上了腾飞的翅膀。"坦诚做人，真诚做事"的企业文化给禾泽都林跨越式的发展注入了生机和活力。

禾泽都林关注城市，研究文化。公司秉承"东方文化现代化"的观点，顺应"大规划、大建筑、大景观"的设计理念，把工程学、美学、哲学三者有机结合，探索了当今及未来城市的发展道路，力求使设计城市体现和谐之美。

禾泽都林关注结果，尊重客户、帮助客户实现理想、为客户提供超值服务、为社会提供传世精品是禾泽都林人的奋斗目标。完成一个项目，成为一方地标，引领一个城市的发展，用凝固的作品获得社会的认可，赢得各界的好评是禾泽都林的前进动力。

禾泽都林关注市场，追求质量。公司以中国长三角洲为中心，向其他城市不断扩张，业绩现已遍布北京、天津、上海、杭州、昆明等三十余座城市。我们的作品大气睿智、浑然天成；我们的质量精雕细刻、精益求精。近年设计作品获得了全国人居经典综合大奖、IACE2006 国际人居建筑金奖、中国国际建筑艺术双年展建筑设计一等奖等十五个国家级大奖。

禾泽都林当选为中国商业地产最佳设计机构，立足中国建筑市场，与大学、行业内高端的协会、媒体、企业建立长期战略伙伴关系，多次参与中国国际建筑文化交流、生态建设高端论坛活动。

禾泽都林人将快乐设计、用心工作，力求传承东方文化，打造高品质的禾泽都林品牌，服务客户、奉献社会。

2010 年禾泽都林的发展主题为国际化，我们将以更规范、更高的标准、更严格的要求为社会提供国际化的服务。

It is Hesoms' pursuit to interpret ideal city and to forge the first brand of real estate design. Designers that master innovative thinking and technology, technological youngsters with international outlook and advanced idea, powerful design team made up of more than a hundred high-educated talents with professional practice qualification as well as senior experts in aspects of literary, commerce and economy etc, all provide soaring wings for Hesoms to become domestically leading internalized design institution. The enterprise culture of "Being honest in life, being sincere in work" has transfused energy and vigor into Hesoms' leapfrog development.

Hesoms pays close attention to city and researches into culture. Taking orders of the point of view of "Oriental culture modernization" and complying with the design concept of "Great planning, great construction, great landscape", it has organically combined engineering, aesthetics and philosophy together, explores the urban developing road of the present and of the future, and designs the city's harmonious beauty.

Hesoms pays close attention to the result and respects the customers. It is Hesoms staff's fighting goal to help the customers' realize their ideals, to offer them overflowing service and to provide the society with precious works. Accomplishing a project, becoming a local landmark, leading a city's development, Hesoms gains the society's applauses and good reputation from all walks of life with concretionary works.

Hesoms pays close attention to the market and pursues the quality. Centered with Yangtze River Delta of China, it spreads continuously to other cities, and its achievements have been all over more than thirty cities as Beijing, Tianjin, Shanghai, Hangzhou and Kunming etc. Our works are generous and sagacious, like nature itself; our quality is precise and fine, always seeking for greater perfection. In recent years we have won fifteen awards of national grade such as Comprehensive Award of National Habitation Classics, IACE2006 International Habitation Construction Golden Prize, First Prize of Architecture Design in Biennial of China International Architecture Art, etc.

Elected as the Best Design Institution of China Commercial Real Estate, Hesoms keeps a foothold of China architecture market, sets up a long-term strategic partnership with universities, top associations in the industry, media and enterprises, and for many times takes part in cultural communication of China international architecture and high-end forum activities of ecological construction.

The Hesoms staff will design joy, work with concentration, inherit the oriental culture, forge the Hesoms Brand of high quality, serve the customers and devote to the society.

In 2010 Hesoms' developing theme is internalization. We will serve the society under more standardized and higher criterion and stricter demand.

地址：杭州市文二路 391 号西湖国际科技大厦 B2–8 层
邮编：310003
电话：+86–571–87219177
传真：+86–571–87215877
电子邮箱：webmaster@hesoms.com
网址：www.hesoms.com

Address: 8th Floor of West Lake International Technology Mansion B2, No. 391 of Wen'er Road, Hangzhou
P.C.: 310003
Tel: +86-571-87219177
Fax: +86-571-87215877
E-mail: webmaster@hesoms.com
Http: //www.hesoms.com

项目地点：浙江 嘉兴
项目规模：700 000 平方米
建筑类别：商铺、多层、联排、独栋、小高层、高层
项目进程：已建成

2004 年中国优秀规划设计楼盘
2005 年度浙江省住宅产业十大杰出企业
2007 年度嘉兴市楼盘销售冠军
2007 年度十大消费者协会推荐楼盘首位
嘉兴市南湖新区 2008 年度十佳特色地产

巴黎都市是目前嘉兴市最大的综合房产项目。项目由亲水别墅、联排、景观公寓、生态商住住宅、单身公寓、购物中心、商业休闲步行街、实验小学及幼儿园等组成。设计充分挖掘用地特征，形成贯通南北的中央景观大道和穿越东西的塞纳河休闲带，并将整个用地有机分割成各个组团，组团与组团之间通过道路及景观环境串联起来，形成封闭的组团与开放的小区，有利于增进小区住户的邻里关系。项目现已建设成为引领嘉兴居住文化的风向标，在建其间一度供不应求，达到销售高峰。

地理环境

规划地块位于嘉兴市秀城区，整个项目位于中环东路西侧、中环南路北侧，西侧为中港河，北面有百盛花园。此项目处于南湖区与嘉兴市政府之间，周边有体育中心、森林公园、嘉兴一中、医院等完善的配套设施，地理位置十分优秀，是未来嘉兴市经济、政治、文化发展的中心，具有得天独厚的交通及地理区位优势。

人文意境

一个城市的聚居区，要成为适宜人聚居的区域，还要有一定的人文环境。"封闭的组团，开放的社区"，是禾泽都林在设计中遵循的规划原则。"巴黎都市"小区划分了10个组团，中间是开放的广场，小区的配套商铺、会所也都是开放性设计。任何行人和附近的居民都可以进入小区的底层商铺购物、消费，甚至是在河道景观区域拍摄婚纱照、在会所里举行婚宴，小区的街道变成了城市的一部分。而对于小区的住户来说，每十幢左右的高层住宅围合成的组团具有私密性。在相对小型的社区空间中，人会更有安全感和归属感。"巴黎都市"的居民在回家的路上可经过丰富的空间，从开放的城市街道进入社区，再进入组团，安全感渐进形成。进入组团后，人们会发现所有景观、步道的尺寸都是适合人的尺度。社区内人车分流，所有汽车通过小区的街道直接驶入各组团的地下车库。从人的心理学角度分析，窗下安静的步行街，会让居民感觉舒服。

巴黎都市项目设计特色主要体现在四大方面：运用巴黎的浪漫休闲文化，创造完善的城市配套，创造多样的建筑类型，创造优美的住宅环境。禾泽都林的设计师从四个方面诠释了其专业的设计理念。

"分"——采用周边城市环状道路，使环路内部纯步行化，整个小区具有宽而连续的漫步路径。

"绿"——"小区—院落"二级结构使小区绿地更为宽敞，设计强调绿色对自然生态的促进及其对邻里交往的积极影响力，构筑巴黎都市符合生态发展的景观活动网络。

"透"——社区住宅由小高层、高层、多层住宅及排屋组成，局部架空使院落与中央步行绿地、院落之间视觉空间更为流畅，有助于漫游形式的实现和小区微气候的改善。

"文"——延续巴黎城市格局、商业文化脉络与现代居住理念相结合的总体布置形式，使小区更具识别性与个性特色。

室内设计说明

生存于钢筋水泥的丛林中，人们渴望质朴、休闲的家居环境。巴黎都市希望在追求休闲的同时体现文化的深度，以感性的手法诠释了休闲概念，表现了质朴、自然、精致的休闲文化。简单的造型，在关键部位体现精彩，大气而流畅。设计中尽量多地运用竹、藤、原木等自然材料，并控制使用油漆、胶合板等存在有害物质的材料数量。装修后空气环境质量令人非常满意，既营造了闲逸氛围又保证了环保性能，可谓一举两得。

一进门的装修就十分引人入胜，一个全是玻璃材质做成的地台直入眼帘，其透明的竹竿使人与环境亲近了许多。客厅餐厅间用藤编制做屏障，保持了良好的连贯性。浅棕色肌理的墙面、透光的石柱，为客厅营造了温馨的氛围。雕刻了篆书文字的杉木火烧板点明了餐厅的主题，使人仿佛回到兰亭感受畅饮的酣畅淋漓。门高达 2.3 米，深色调彰显高雅格调。

茶室是一派休闲的南洋风情，竹帘、藤编、植物花草，乱中有序。旋转陈列柜，增加了空间的灵活性，可以在此放置主人喜爱的羽球拍。主卧融合了三个空间：更衣空间、睡眠空间、休闲阳台。主卫塑造了概念性的卫浴空间，顶部排列的杉木、竹竿与马赛克、亚细亚世纪石相配，极具休闲风韵。

九龙一号

项目地址：浙江 杭州
项目类别：排屋
用地规模：100 350 平方米
建筑规模：103 223 平方米
绿 化 率：30%
容 积 率：1.0
设计时间：2006 年
获奖情况：2006 年全国人居经典综合大奖楼盘；IACE2006 国际人居建筑金奖

项目介绍

九龙一号位于银湖板块杭州野生动物世界旁，背靠富春江山麓，南依九龙大道，总占地 10 公顷，建有联体排屋和少量独立别墅。

九龙一号采用市场中少有的中式设计风格，建筑古朴隽永，院落布局考究，白墙青瓦下描绘出独有的中国气质。九龙一号景观设计的目标就是体现中国古典园林意境，打造一个与现代生活方式相融合的中式庭院空间。

从九龙一号中，我们可以看到一个集东方古典园林意境与现代建筑风格为一体的能“藏风”“聚气”的私家院落，古典建筑的意境体现在建筑上，景观设计更体现了设计对中国传统文化的尊重。

在小区公共景观方面，设计了一条自然的山谷水溪，流淌于小区内部。小溪随地形起伏由北向南流淌，穿梭于各组团之间，并与地势结合，形成“三湖三涧”的优美景观效果，使整个小区山与水紧密结合，构筑出错落有致的天际线和怡人的景观效果。

在主入口景观方面，设计采用退台处理的方式，形成豪华的入口景观广场，并于其间设置会所，并设置景观湖。建筑隐于入口景观广场后，形成私密性，可谓“花间隐榭，水际安亭”。

佳源·凤凰广场（商业、别墅）

项目地点：江苏 泰兴
项目规模：218 608 平方米
项目进程：方案阶段

云南璀璨明珠城（城市居住）

项目地点：云南 腾冲
项目规模：546 694 平方米
项目进程：施工图阶段

宣城·大唐国际城（城市商业）

项目地点：安徽 宣城
项目规模：111.57 公顷
项目进程：建设中

藝天建築

浙江大学建筑系高裕江建筑工作室

地址：浙江省杭州市朝晖路 182# 国都发展大厦 A 座 120H
邮编：310014
电话：+86-571-85803763　13705712111
传真：+86-571-85803763
邮箱：YTJZ2008@sina.com
网址：www.hzytjz.com

艺天建筑——浙江大学建筑系高裕江建筑工作室，由原南京市建筑设计研究院总建筑师、现任教于浙江大学建筑系的教授级高级建筑师高裕江先生主持，是由多位具有国内外著名高校专业学历背景的学者教授、博士组成的设计创作和研究团队。工作室依托浙江大学建筑系、设计研究院、南京市建筑设计研究院等较强科研生产学术机构，开展建筑创作的研究、教学和设计实践工作。工作室以创作建筑艺术精品、形成品牌、服务社会为目标，以浙江大学的产、学、研相融汇的办学指导思想为原则，同时兼有理论和实践相结合的优势，真正体现出建筑学科的实践型学术特点。工作室已成一支有国际竞争力的实战型设计研究团队。近年来以本工作室核心成员为创作主体的建筑设计作品，获得多项国家级、省部级优秀设计奖，并在多项国际国内竞赛、竞标中荣获第一名或中标。工作室在大型公建如现代高层办公楼、博物馆、图书馆、度假酒店、星级宾馆等类型创作中，具有较丰富的研究设计经验。

中国 天津　　天津图书馆

中国 舟山　　浙江海洋学院

中国 宁波　　云龙镇文化中心

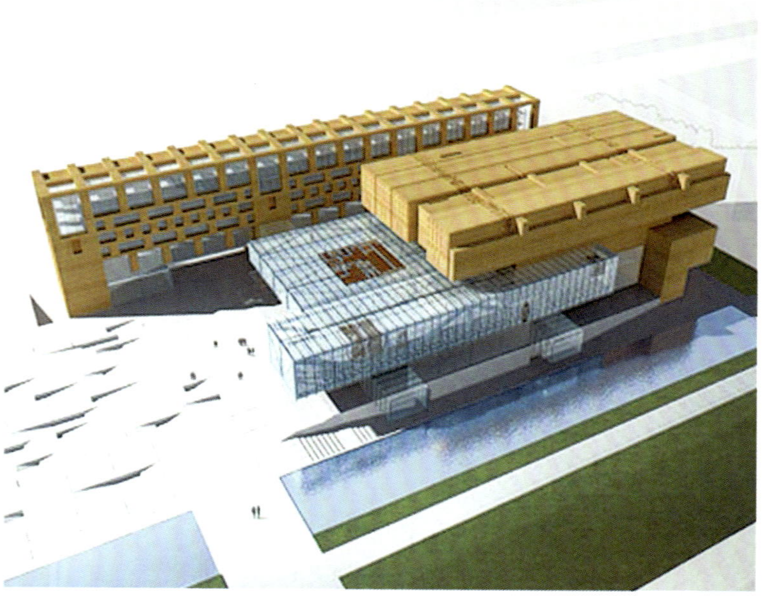

中国 南京　　江苏省档案馆

中国 宁波　　宁波鄞州银座

Address: 120H, Block A of Guodu Development Building,
No.182 of Zhaohui Road, Hangzhou, Zhejiang
P.C.: 310014
Tel : +86-571-85803763 13705712111
Fax : +86-571-85803763
E-mail: YTJZ2008@sina.com
Http:// www.hzytjz.com

Yitian Architectural Studio--GaoYujiang Architecture Studio of Architecture Department of Zhenjing University, is Presided over by Gao Yujiang, former chief archited of Nanjing Architectual Design & Research Institute and curreutly teaching in Architecture Department of Zhejiang University, together with scholars and doctors from renowned universities home and abroad. Supprted by Architecture Department of Zhejiang University, Nanjing Architectual Design & Research Institute and other scientific research and academic institutes, the studio focuses on architeltural creation research, teaehing and design practice work. Ainling to creat excellent works, it is based on the guidelines of integration of creation, study and research. With the advantage of corubining practice and theory, the studio truely manifests the characters of practice oriented architectural science and has become a research and design team with international competitiveness in actual practices. Over the years, the studio has won many national and provincial awards and won the first prizes as well as bids in international and domestic competitions. It has rich research and design experiences in large-scale public buildings like hign-level modern buildings, museum, library, hotels and so on.

中国 宁波　布利杰总部大楼

中国 杭州　浙江省档案馆

中国 宁波　姜山文化体育中心

中国 南京　南京博物院概念设计

中国 南京　中国银行南京运营中心

宁波中鼎建筑设计研究院
Ningbo Zhongding Architectural Design & Research Institute

地址：浙江省宁波市环城西路南段 695 号
邮编：315000
电话：+86-574-87495078
传真：+86-574-87497098
网址：www.zddesign.cn
E-mail：zddesign@zddesign.cn

宁波中鼎建筑设计研究院是一家以建筑设计与研究为主的股份合作制企业，于 2002 年成立，为国家甲级建筑设计院，旨在培养与造就一支高素质的建筑设计队伍，并潜心于城市与建筑的研究与创作，不断为社会奉献精品工程。在工程设计方面充分发挥技术优势，专长于建筑功能、形式与经济三方面的综合效益分析与研究。在科题研究方面能积极探索具有学术前瞻性的基础理论，推进建筑艺术与技术的进步。

中鼎建筑设计研究院致力于应用先进的理念、技术和设备进行工程设计研究，依靠技术人员的创造力、丰富的实践经验和现代信息技术等手段，来应对各种复杂的建设条件和不同客户的要求，并提供先进可靠、经济安全的解决方案。注重形式与空间设计、合理配置投资、善于利用自然资源、追求人与环境的协调发展。

研究院将秉承不断为社会奉献精品工程、推进建筑技术进步为宗旨，坚持“诚信为本，追求卓越”的质量方针，为客户提供先进的设计技术和优质的工程服务。

一、古建筑创新研究领先省内外

天居寺：舍宅为寺是众多历史人物的功德行为，陈武帝在后半生将老宫殿改为天居寺，与其说是寺，不如说是宫殿，宫殿就拥有奢华与雄伟，这即是我们的设计思路。

陈武帝故居：陈武帝本为渔民，后从戎投军，并建南朝。其故居设计则应合乎其出身背景，草亭以喻草笠生活特征，民居简朴以喻故居之原本，院落架空于水上，以解渔民之印象，突出南方干栏式建筑特色。

二、商业广场设计探索了新模式

余姚四明广场：让市民拥有更多的城市空间，让建筑在传统与时尚的对比中呈现恒久之美，让环境在都市文化的场景中焕发活力。

昆山商业广场：简洁而富有特色的建筑形态，在滨江环境中展现都市的魅力，内部的购物空间与休闲空间融为一体，让人的行为在广场中多样化，这也是城市设计的理念—建筑不能游离于环境。

丹阳商业广场：在 5.87 公顷土地中，规划酒店、百货、公寓、商业广场等近 30 万平方米的建筑规模是一件不容易的工作。在合理组织交通的前提下，创造性的营造江景休闲一条街、风情一条街、天幕广场、高端公寓区以及具有丹阳市地标特征的 146 米高的五星级酒店。

Address: No.695, south part, West HuangchengROAD, NingBo, Zhejiang
P.C.: 315000
Tel: +86-574-87495078
Fax: +574-87497098
Http: //www.zddesign.cn
E-mail: zddesign@zddesign.cn

Founded in 2002, as a Class-A national architectural design and research institute, Ningbo Zhongding Architectural Design Research Institute is a joint-stock corporation and has been developing a high quality architectural force team, focusing on architectural design and research. We have intended to create and study urban and architecture as well as devoting excellent projects. In design, the institute exerts fully technical superiority, specializes in providing a combination of synthesis benefits among architectural function, form and economy by analysing and researching. In research, the institute zealously explores fundamental theory toward academic perspective.

Ningbo Zhongding Architecture Design Research Institute dedicates to applying advanced conceptions, technologies, equipments to research of project design. Depending on talent of technicians' creation, abundant practical experience and modern informational techniques, the institute could not only deal with complicated constructional troubles or satisfy different clients' demand, but could also provide safe and economical settlements, which pay attention to the design between structure and spacetomake use of natural resource to pursuit harmony between human and environment.

By constantly offering excellent projects and promoting architectural technology as the principle, the institute insists on a quality policy of pursuiting honesty and excellence in order to offering advanced design technology and prominent project service.

三、文化建筑研究初显影响力

镇海中学改造研究：特别强调建筑与镇海中学校园内的文庙历史建筑相协调，旨在打造与镇海中学超一流教学质量相匹配的建筑文化特色。

四、创意建筑不断出现，让城市更加美好

梅山岛旅游服务区城市设计：作为现代物流与仓储保税功能区，同样应为人们提供一个充满海景风情与诗意般生活的服务区，让小岛拥有城市滨水生活特质。

五、重大项目专业综合能力不断提高

镇海新城（东区）整合设计：六栋商务楼通过公共空间的设置与街面设计，为办公商务人员提供能共享都市文化的公共空间，配置宜人的休闲环境。

横溪小学：我院为横溪小学提供了一套既充满地方建筑文化特色又具一定时代气息的设计方案。我们认为，校园建筑应该更多在地方文化特色方面做出探索，方可避免千篇一律的乏味建筑设计。

渔轮厂地块改造设计：让历史在前卫的环境中存在，勾起历史记忆，激发审美热情，让建筑成为城市雕塑，以象征或隐喻的手法将宁波作为滨海城市的特征要素：贝壳、沙滩或海洋动物。设计更加强调优美的城市江景，美轮美奂！

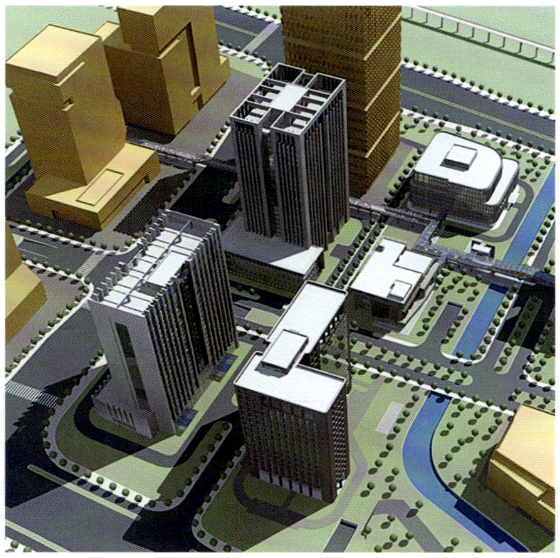

南部商务区（二期）：充分整合地块间的公共空间，综合利用地下室空间整合环境设施，使其一体化，力求拥有更为动人的公共环境效果。

Editor,Designer

编委、设计师

卓创国际工程设计有限公司
International Creative Engineering & Design Co.,Ltd

柏疆红
Jianghong Bai

柏疆红

教授级高级建筑师
建筑学博士
1996 年毕业于西安科技大学
现任卓创国际工程设计有限公司董事长 总建筑师

部分作品
重庆亚太商谷
首金美利山
南宁中铁 凤岭山语城
重庆同天绿岸
南宁青秀山高尔夫球场
成都棠湖光电所牧场及高尔夫别墅
东海汤山别院
金科东方王榭
西安阳光城别墅
融科 橡树澜湾别墅
协信 彩云湖别墅
三亚东和福湾酒店、别墅、洋房
查氏集团重庆国际金融中心
中投远达

Jianghong Bai

Senior professor level architect
Doctor of Architecture
Graduated from Xi'an University of Science and Technology in 1996
Bai is now the chairman and chief architect of ECD Engineering Design Co., Ltd.

Major Designs
Chongqing APEV
Shoujinmeili Mountain
Nanning China Railway Fenglingshanyu Town
Chongqing Tongtian Green Shore
Nanning Qingxiu Mountain International Golf Course
Chengdu Tanghu Photoelectricity Pasture &Golf Villas
Tangshan Villa of East Sea
Vision Tech Oriental Wangtse
Xi'an Sunshine Villa Town
Raycom Bay of Oak Villas
Association Letter Caiyun Lake Villas
Sanya Donghe Blessing Bay Hotel, Villas & Western-style Housing
Zha Group Chongqing International Finance Center
China Investment Corp Yuanda

梁育诚
Yucheng Liang

梁育诚

美国加州注册建筑师
现任 ECD 卓创国际上海区域设计总监。
具有近 20 年海内外的建筑设计经历，历任首席设计师主持并完成过多项具挑战性的建筑项目，梁先生在高端居住社区、商业中心、办公建筑等领域获得较高的造诣与声望。

部分作品
加州首府凯悦花园别墅 112 栋
美国硅谷 HP 计算机公司的办公大楼
上海虹桥 236 栋高尔夫花园别墅
美国洛杉矶市中心办公楼及仓库
美国硅谷医疗及办公大楼
美国永和超市集团商场
北京市办公大楼及住宅大楼设计案
加州首府太平洋广场

Yucheng Liang

Registered architect of California
Currently Liang is the design director of ECD Shanghai Branch.
Liang has over 20 years experience for architectural design at home and aboard and having presided over and accomplished a number of challenging projects, Mr. Liang is prestigious in fields of high-end community, commercial center and office building, etc.

Major Designs
Hyatt Garden Villa Building 112, Capital of California State
HP Offic Building, Silcon Valley,USA
Shanghai Hongqiao Golf Garden Villas Building 236
Los Angles Center Office Building & Public Store, USA
Silcon Valley Medical &Office Building, USA
Shopping Mall of Marina Supermarket Group, USA
Beijing Office Block and Residential Buildings
Pacific Plaza, Capital of California State

景森設計 JINGSENDESIGN 广州市景森工程设计顾问有限公司

肖 至
Zhi Xiao

肖 至

董事、总裁
国家一级注册建筑师

主要作品
贵州都匀布依族苗族民俗博物馆
获广州市2004年度优秀工程设计二等奖
保利–广州 · 百合花园
获广州市2004年度优秀工程设计表扬奖
西安财经学院
广州三银大厦
华信–湛江 · 城市假日
获湛江市第九次优秀设计一等奖（第一名）
万科–广州 · 四季花城 · 别墅项目
获广州市2008年优秀工程设计二等奖
小高层项目
获广州市2008年优秀工程设计三等奖
云东海花园–三水 · 山水庄园
获广州市2008度优秀设计奖三等奖
珠光 · 广州 · 新城御景
万科–广州 · 万科云山
万科–广州 · 金域蓝湾
珠光 · 流溪御景山庄

Zhi Xiao

Director, President
National I Class Registered Architect

Major Works
The Folk Custom Museum of the Miao and Bouyei Nationalities in Duyun, Guizhou
Awarded the Second Prize for Excellent Engineering Design in Guangzhou, 2004
Poly-Guangzhou • Lily Garden
Awarded the Mention Prize for Excellent Engineering Design in Guangzhou, 2004
Xi'An University of Finance and Economics
Guangzhou Sanyin Plaza
Huaxin-Zhanjiang • Urban Holiday
Awarded the ninth First Prize for Excellent Design of Zhanjiang(Number One)
Vanke• Guangzhou Wonder Land, villa project
Awarded the Second Prize for Excellent Engineering Design in Guangzhou, 2008
The small high-rise residential projects
Awarded the Third Prize for Excellent Engineering Design in Guangzhou, 2008
Yundonghai Garden-Sanshui • Royal Manor
Awarded the Third Prize for Excellent Design in Guangzhou, 2008
Zhuguang • Guangzhou • Luxury Extravagance
Vanke-Guangzhou • Vanke Yunshan
Vanke- Guangzhou • Paradiso
Zhuguang • River Flows Royal Landscape

益博建築 YIBO CONCEPTION

张 益
Yi Zhang

张 益

1966年生	国家一级注册建筑师
1987年7月	毕业于太原理工大学　建筑系
1995年9月–1996年7月	清华大学建筑学院
1987年7月–1997年	化工部第二设计院
1998年–1999年	惠州大学土木系讲师
1999年–2000年	分别就职于深圳陈世民建筑事务所、法国欧博城市规划设计公司
2000年至今	中美合资珠海艺蓁工程设计有限公司　副总经理
其中2004年创建	珠海益博建筑设计工程有限公司总经理

近年主要设计作品
1996年两项作品收录于《全国中小型民用建筑图集》

珠海金山软件大厦	2.6万平方米	已建成
珠海星城大酒店	1.7万平方米	已建成
深圳金证科技大厦	3.5万平方米	已建成
珠海翠华城市风景	10万平方米	已建成
珠海京华奥园	7万平方米	已建成
珠海华业临海花园	9万平方米	施工中
珠海财富商务大厦	2.8万平方米	施工中
珠海翠菁苑、澳景湾花园		施工中
柬埔寨金边玫瑰滨江园	8万平方米	施工中
珠海安广大厦	2.6万平方米	设计中

Yi Zhang

Born in 1966, national First-grade Registered Architect
In July 1987 graduated from Taiyuan University of Technology, Department of Architecture
1995.9–1996.7 Tsinghua University School of Architecture
1987.7–1997 Second Design Institute of Ministry of Chemical Industry
1998–1999 Lecturer for Civil Engineering Faculty at Huizhou University
1999–2000 Respectively, in Shenzhen Chen Shimin Construction Firm, and French O'Brien Urban Planning and Design Company
From 2000 to present, Deputy General Manager of Sino-US joint venture Zhuhai Yi Zhen Engineering Design Co., Ltd.
During this period, in 2004 established Zhuhai YiBo Architectural Design Engineering Co., Ltd. as General Manager

Main design works in recent years
Two works of 96 years included in the *National Atlas of Small and Medium-sized Civil Construction*
Zhuhai Kingsoft buildings, 26,000 m^2, completed
Zhuhai Star City Hotel, 17,000 m^2, completed
Shenzhen Jinzheng Science and Technology Building, 35,000 m^2, completed
Zhuhai Cuihua Urban Landscape, 100,000 m^2, completed
Zhuhai Jinghua Aoyuan, 70,000 m^2, completed
Zhuhai Huaye Shore Garden, 90,000 m^2, under construction
Zhuhai Fortune Commercial Building, 28,000 m^2, under construction
Zhuhai-ching Tsui Court, Australian Bay Garden, under construction
Cambodia Phnom Penh Rose Riverside Garden, 80,000 m^2, under construction
Zhuhai, An Guang Building, 26,000 m^2, in design

盛宇宏先生从业二十余年，拥有着丰富的海外和国内著名事务所工作经验，主持了大量大型设计项目，获奖无数。他强调设计应以提升生活的品质、营造生活的感受为目标，强调多专业国际整合（规划、建筑、室内、景观），强调概念创新为先、细节保证为本，讲究创意、本原、互动，以客户的成功为己任，带领着整个团队快速发展，进入一个全新的领域。已建成的设计作品包括大量优秀的住宅区、酒店、学校和商业项目。他亲自经历了国内房地产开发商业化的全过程，在通过设计品质和设计与市场的结合提升项目附加值等方面卓有建树。

总裁

近年获奖项目

一、公建类

2008年广州市优秀工程设计二等奖
——富力城商业中心
2006年广东省优秀工程设计二等奖
2007年国家鲁班奖
——广东丰泰花园酒店（五星级）
建设部.中国建筑文化中心2006年第二届中国建筑艺术双年展最佳创意奖
——广西阳朔河畔酒店（五星级）
广东省第十二次优秀工程设计三等奖
2005年度优秀建筑设计项目一等奖
——东莞丰泰花园酒店主楼
广东省优秀设计二等奖
——广东实验中学

二、住宅类

2009年全国人居经典建筑规划设计方案竞赛"规划金奖"
——天庆中央公园
2009年全国人居经典建筑规划设计方案竞赛"综合大奖"
——佛山汇银奥林匹克花园
2009年度广东省优秀工程设计一等奖
——广州富力城
2009年度广东省优秀工程设计二等奖
——从化夏日港湾
——佛山汇银奥林匹克花园
2009年度广东省优秀工程设计三等奖
——临沂沂龙湾
——广州城市假日园
全国优秀工程勘察设计行业奖三等奖
——兰州仁恒国际
2008年广州市优秀工程设计一等奖
——广州富力城
——佛山奥林匹克花园
2008年广州市优秀工程设计二等奖
——汇侨新城项目
——从化夏日港湾
——山东临沂尚城
——广州市城市假日园二期
——锦东花园
2007年度广东省优秀工程设计二等奖
——时代廊桥
2007年度广东省优秀工程设计一等奖
——广州云山诗意人家
2007年中国居住创新典范最具品牌经典价值示范楼盘
——北京富力又一城
2006年度优秀工程设计三等奖
——兰州仁恒会所
2006年度优秀工程设计二等奖
——时代廊桥
——现代广场G1～G6栋
——兰州仁恒A1～A2栋
2007年度广东省优秀工程设计二等奖
——兰州仁恒A1～A2栋
建设部.中国建筑文化中心2006年第二届中国建筑艺术双年展建筑设计优秀奖
——广州富力广场
中国最具影响力中国名盘
2006年广州市优秀工程设计一等奖
——广州云山诗意人家
2002年广州新人居最佳建筑设计金奖
2002年广州新人居十佳精品小区
2002年广州新人居最佳园林景观金奖
——时代花园

三、室内类

2009年第三届广州建筑装饰设计大赛银奖
——佛山依云水岸中、欧式样板房
2009年第三届广州建筑装饰设计大赛优胜奖
——方圆公园一号售楼部

沙励励女士从创业之日起，致力于研究中国传统文化和将西方先进管理理念与公司实际相结合。在经营管理的二十年中，她凭着丰富的海外和国内知名公司管理经验，秉承"顾客至上、真情服务"的宗旨，不断更新管理模式，并辅之以"高效、准确、稳健"的指导方针，积极推动建筑艺术家创造出极富内涵的作品如中式徽派风格、岭南亚热带风情，以及东南亚特色等精品建筑，这些建筑既简约时尚，又融合了中西方文化精髓，得到发展商和同行的高度认可，使汉森实现真正质的飞跃。

董事长

广东建筑艺术设计院有限公司
Guangdong Architectural Artistic Design Institute Co.,Ltd

邓日新
Rixin Deng

邓日新

1965年毕业于华南理工大学建筑学专业，从事建筑设计四十余年，资深高级建筑师和一级注册建筑师。

主创代表项目

1.丽雅苑
工程项目地点：广东南海桂城
工程设计面积：180 000平方米
工程性质：多高层小区

2.云景花园
工程项目地点：广州同和
工程设计面积：130 000平方米
工程性质：高层小区

3.兰海城
工程项目地点：广州新港中路
工程设计面积：100 000平方米
工程性质：高层小区

4.东莞光大花园
工程项目地点：东莞新城区
工程设计面积：300 000平方米
工程性质：高层商住楼

5.东方花园
工程项目地点：广州天河东圃
工程设计面积：85 000平方米
工程性质：高层小区

6.旭景家园
工程项目地点：广州车坡路口
工程设计面积：210 000平方米
工程性质：高层小区

7.新世纪豪庭
工程项目地点：东莞市横沥镇
工程设计面积：110 000平方米
工程性质：多层建筑、连体别墅、小高层、会所、幼儿园

Rixin Deng

Rixin Deng, graduated from South China University of Technology in1965, engaged in Architectural design over 40 years, Senior class 1 Registered Architect.

Represent Projects

1. Liya Court
Location: Nanhaiguicheng in Guangdong
Area: 180,000 m^2
Purpose: High-storey Residence

2. Yunjing Garden
Location: Tonghe in Guangzhou
Area: 130,000 m^2
Purpose: High-storey Residence

3. Lanhai Town
Location: Xinggangzhong road in Guangzhou
Area: 100,000 m^2
Purpose: High-storey Residence

4. Dongguan Guangda Garden
Location: New town district in Dongguan
Area: 300,000 m^2
Purpose: High-storey Residence and office building

5. East Garden
Location: Tianhe Dongpu in Guangzhou
Area: 85,000 m^2
Purpose: High-storey Residence

6. Xujing Garden
Location: ChepoLukou in Guangzhou
Area: 210,000 m^2
Purpose: High-storey Residence

7. Xinshiji Court
Location: Hengli town in Dongguan
Area: 110,000 m^2
Purpose: Multi-storey building, townhouse, high-rise building, club, Kindergarten

广州亚泰建筑设计院有限公司
Guangzhou Asiatai Architecture Design Institute CO.,Ltd.

王 成
cheng Wang

王　成（院长）

职　　称：高级建筑师
毕业院校：1994年毕业于 武汉工业大学建筑系
　　　　　1999年华南理工大学获城市规划硕士学位

工程业绩
华南农业大学跃进南区教学楼、宿舍楼、生活区配套
广州天河软件园——IT商务中心（五星级酒店）
广州天河软件园孵化中心
山西花园国际大酒店（五星级）
山西世茂广场太钢不锈钢本部写字楼
广州市广仁大厦
广州医学院微创外科大楼
广州大学城二期工程
大宇宙信息创造（中国）有限公司办公大楼
钦州东方豪庭
华天时代科贸楼
亨元保障性住房项目勘察设计

Cheng Wang (Director)

Title: Senior Architect
Graduation School: Graduated from the Department of Architecture of Wuhan University in 1994
Received Master degree of the Urban Planning of Huanan University in 1999

Projects
Teaching building, dormitory, and living area of Leap forward in the Southern District of South China Agricultural University
Guangzhou Tianhe Software Park——IT Business Centre (five-star hotel)
Incubation Center of Guangzhou Tianhe Software Park
Shanxi Garden International Hotel (five-star)
Headquarter Office of Taiyuan Iron and Stainless Steel in Shanxi Shimao Plaza
Guangzhou Guangren Building
Minimally Invasive Surgery Building of Guangzhou Medical College
Second Phase Project of Guangzhou University City
The Office Building of Big Universe Information Creation (China) Co., Ltd.
Qinzhou Oriental Palace
Huatian Era Technology & Trade Building
Heng Yuan survey and design of affordable housing project

湖南城市学院规划建筑设计研究院
Urban Planning & Architectural Design Institute Of Hunan City University

谭献良
Xianliang Tan

谭献良

1971年11月出生，湖南益阳人，中共党员。

湖南城市学院规划建筑设计研究院院长，硕士；高级工程师、国家注册结构工程师；湖南省勘察设计协会常务理事、湖南省城乡规划协会理事、益阳市勘察设计协会副理事长、湖南省房屋建筑和市政基础设施工程施工评委库专家。

Xianliang Tan

Tan was born in Yiyang of Hunan province in November, 1971; he is the party member of Chinese Communist Party.

He is the president of Urban Planning & Architectural Design Institute of Hunan City University , Being a senior engineer of master's degree , he is also a national registered structural engineer, holding the titles of Hunan investigation and design association board member and Hunan urban and rural area Planning association board member, vice director of Yiyang investigation and design association, Hunan housing architecture and municipal Infrastructure project construction judge expert.

HSARCHITECTS

华森建筑与工程设计顾问有限公司

肖 蓝
Lan Xiao

肖 蓝

职务：副总经理、副总建筑师
1995年毕业于天津大学建筑系，由建设部建筑设计院派至深圳华森
建筑与工程设计顾问有限公司，工作至今。

主要作品
成都美林湾
建筑面积：480 000平方米
设计时间：2006年

惠阳东方新城
建筑面积：280 000平方米
设计时间：2005年

杭州钱江新城·晓庐
建筑面积：300 000平方米
设计时间：2005年

惠阳半岛一号
建筑面积：130 000平方米
设计时间：2005年

广州岭南新世界二期
建筑面积：350 000平方米
设计时间：2004年

杭州青枫墅园
建筑面积：550 000平方米
设计时间：2004年

深圳市中心区北中轴深圳书城
建筑面积：83 000平方米
设计时间：2003年

深圳万科十七英里
建筑面积：50 000平方米
设计时间：2003年

杭州兰庭国际住宅小区
建筑面积：100 000平方米
设计时间：2003年

杭州中浙太阳·国际公寓
建筑面积：198 000平方米
设计时间：2002年
获奖情况：2003年全国人居建筑规划设计方案竞赛综合大奖
建筑师奖

深圳中华商贸城桃源居三、四、五区规划设计，生态公园规划设计
建筑面积：240 000平方米
设计时间：2001年

深圳华为总部单身公寓小区
建筑面积：116 000平方米
设计时间：1999年
获奖情况：2004年获深圳市第十一届优秀规划设计三等奖

深圳万科城市花园
建筑面积：61 000平方米
设计时间：1997年
获奖情况：1999年获全国第八届优秀工程设计金奖
1998年获深圳市建筑设计金牛奖

Lan Xiao

Position: Deputy Manager Deputy General Manager
Graduated from the Department of Architecture in Tianjin University in 1995, she has been sent by Building Design Institute of Ministry of Construction to Shenzhen Huasen Architectural & Engineering Designing Consultant Co., Ltd. and has been working so far.

Major Works
Mayland Bay, Chengdu
Building area: 480,000 m^2
Design Time: 2006

Oriental New City, Huiyang
Building area: 280,000 m^2
Design Time: 2005

Qianjiang New City • Xiaolu, Hangzhou
Building area: 300,000 m^2
Design Time: 2005

Top Peninsula Life, Huiyang
Building area: 130,000 m^2
Design Time: 2005

Phase II of Lingnan New World, Guangzhou
Building area: 350,000 m^2
Design Time: 2004

Green Villa Garden, Hangzhou
Building area: 550,000 m^2
Design Time: 2004

Shenzhen Books City in central axis of north central district, Shenzhen
Building area: 83,000 m^2
Design Time: 2003

Vanke 17 Miles, Shenzhen
Building area: 50,000 m^2
Design Time: 2003

Lanting International Housing Estate, Hangzhou
Building area: 100,000 m^2
Design Time: 2003

Zhongzhe Sun • International Apartment, Hangzhou
Building area: 198,000 m^2
Design Time: 2002
Award:
Comprehensive Prize for China Habitat Architecture Design Plan Competition in 2003
Architect Prize

The Third, Fourth and Fifth Stage Design&Plan for China Commercial Town Taoyuan Residence, Shenzhen
Building area: 240,000 m^2
Design Time: 2001

Bachelor Apartment District of Huawei Headquarters, Shenzhen
Building area: 116,000 m^2
Design Time: 1999
Award:
The Third Prize in the 11th Shenzhen Excellent Design & Plan Competition

Vanke City Garden, Shenzhen
Building area: 61,000 m^2
Design Time: 1997
Award:
The Golden Prize for the 8th National Excellent Project Design Competition in 1999
The Golden Cow Prize for Shenzhen Architecture Design Competition in 1998

TSINGHUA-YUAN
深圳市清华苑建筑设计有限公司
TSINGHUA YUAN ARCHITECTURAL DESIGN LTD SHENZHEN
地址ADD：深圳市南山区桃园路南苑新村4号楼
电话TEL：0755-26493788　传真FAX：0755-26079280
E-mail：qhy@qhy.sina.net　网址WEBSITE：www.thy.cn

李念中
Nianzhong Li

李念中

公司董事、总经理、副总建筑师
国家一级注册建筑师、高级工程师

教育背景
1984年　清华大学建筑系建筑学　工学学士
2005年　北京大学　高级工商管理硕士

工作经历
1984–1989年　深圳大学建筑系建筑设计研究院
1989–1991年　澳大利亚访问学者
1991–2000年　深圳大学建筑系建筑设计研究院
2000–2003年　深圳大学城规划建设处
2004–2010年　深圳市清华苑建筑设计有限公司

主要业绩
深圳大学二次规划设计
深圳市龙岗区片区规划设计
深圳大学城规划设计
深圳市坂田万科四季花城
深圳万科金色家园
深圳翠海花园
深圳市罗湖区黄贝岭村旧村改造
招商三洋3#厂房改造
招商海月华府
义务水木清华
镇江红豆江南府邸

Nianzhong Li

Company Director, General Manager and Deputy Chief Architect
National level 1 registered Architect, senior engineer

Education Background
1984 School of Architecture in Tsinghua University　bachelor of engineering (BE)
2005 Peking University　Executive Master of Business Administration (EMBA)

Working experiences
1984—1989 The Institute of Architecture Design&Research, Shenzhen University Architecture Department
1989—1991 Australia visiting scholar
1991—2000 The Institute of Architecture Design&Research, Shenzhen University Architecture Department
2000—2003 Planning & Construction division of Shenzhen University
2004—2010 Tsinghua Yuan Architectural Design Limited, Shenzhen

Major accomplishments
The second planning and design for Shenzhen University
Planning and design of Longgang district in Shenzhen
Planning and design of Shenzhen college town
Vanke siji huacheng in Bantian of Shenzhen
Vanke Golden Garden in Shenzhen
Cuihai Garden in Shenzhen
Transformation project of old Huangbeiling village in Luohu,Shenzhen
Zhaoshang Sanyang NO.3 warehouse reconstruction
Zhaoshang Haiyue Huafu
Yiwu Shuimu Tsinghua
Zhenjiang Hongdoujiangnan Mansion

POF 华域普风 深圳市华域普风设计有限公司
Shenzhen Pofart Design Consulting Co., Ltd

梅　坚
Jian Mei

梅　坚

法国巴黎拉维莱特建筑学院建筑学硕士
国家一级注册建筑师
曾实践于欧洲多个建筑设计事务所，现主持深圳市华域普风设计有限公司
从业二十年来，致力于城市规划设计、建筑与景观设计

主要作品

成都　华润置地凤凰城C区
　　　成都华润置地凤凰城销售中心
佛山　万科金域蓝湾
海口　现代美居物流园
合肥　华润置地澜溪镇
南京　金地自在城
绵阳　长虹仙海湖创新研发园
长沙　五矿龙湾国际社区
成都　华润置地翡翠城二期
深圳　鸿新花园二期
三亚　金茂海景花园
合肥　天下锦城
珠海　恒隆御水岸项目
乐山　嘉州新城一期

Jian Mei

Mr. Mei studied at Villette School of Architecture (Master of Architecture) in Paris, France
He has got PRA class I registered architect qualification.
He worked in many architectural design offices.
He is taking charge of Shenzhen Pofart Design Consulting Co., Ltd. now.
Having been an architect for over 20 years, Mr. Mei devotes himself to urban planning study, building design and landscape design.

Major works

China Resources Land Phoenix City Area C &Phoenix City Sales Center of China Resources Land, Chengdu
Vanke Kingsway & Blue Bay, Foshan
Modern Logistic Park, Haikou
China Resources French Annecy, Hefei
Golden &Comfortable City, Nanjing
Changhong Xianhai Lake Research & Innovation Park, Mianyang
Mintals Dragon Bay International Community, Changsha
China Resources Land Jady city (Phase II), Chengdu
Hongxin Garden (Phase II), Shenzhen
Jinmao Sea View Garden, Sanya
Glourious City Garden, Hefei
Hallon Yushui Shore, Zhuhai
Jiazhou Newtown (Phase I), Leshan

美国承构建筑师事务所
深圳市承构建筑咨询有限公司
Made&Make Architects

柴　晟
Sheng Chai

柴　晟
董事总经理

教育背景
1990—1995年　天津大学建筑系学士
1995—1998年　天津大学建筑系硕士
1999—2001年　英国格林尼治大学建筑学硕士

资历
2001—2001年　中建（国际）方案创作室主任
2002—2003年　中外建（深圳）建筑设计公司副总经理
2003—2008年　深圳市华汇设计有限公司董事总监
2008年至今　深圳市承构建筑咨询有限公司董事总经理

主要作品
公共建筑部分
深圳创维总部大厦
广州番禺奥誉商业综合体
天津国泰大厦（200米）
天津华贸大厦（180米）
浙江长兴金郁金香度假酒店（五星级）
上海新江湾城23–5地块商业综合体项目
天津宁发国际花园酒店
重庆春森彼岸（获詹天佑大奖中的最佳建筑设计奖，55万平方米综合社区、超高层写字楼、酒店、公寓、滨江商业）
深圳市联通总部大厦
中山歌剧院
深圳市高交会馆
住宅部分
重庆龙湖大城小院（获住宅户型专利奖）
上海龙湖蓝湖郡（获住宅户型专利及上海优秀住宅小区奖）
重庆龙湖观山水
重庆龙湖悠山郡
重庆龙湖江北体育公园项目
重庆龙湖睿城
香港置地及重庆龙湖江与城原山
同景国际新城
重庆龙湖大学城木鱼石项目
北京首驰禧福会项目
成都龙湖三千城项目
重庆中冶林荫大道项目
西安龙湖大兴项目
苏州中海国际社区项目
教育类
深圳市第二外国语中学
东莞建设小学
华侨大学厦门分院人文教育学院
山东工商学院

Sheng Chai
Managing Director

Education Background
1990-1995　Bachelor of Architecture at Tianjin University
1995-1998　Master of Architecture at Tianjin University
1999-2001　Master of Architecture at Greenwich University, UK

Qualifications
2001-2001　Zhongjian (International) Program Creation Studio　Director
2002-2003　Sino-foreign Construction (Shenzhen) Architectural Design Corporation　Deputy General Manager
2003-2008　Huahui Design Co., Ltd., Shenzhen　Supervising Director
2008-present　Shenzhen Made&Make Architects Architectural Consulting Co., Ltd., Managing Director

Major Design
Public Construction
Skyworth Headquarter Building, Shenzhen
Aoyu Commercial Complex, Panyu, Guangzhou
Tianjin Cathay Building (200M)
Tianjin Huamao Building (180M)
Golden Tulip Resort Hotel, Changxing, Zhejiang (five star)
Commercial Complex Project of Block 23-5, New Jiangwan City, Shanghai
Ningfa International Garden Hotel, Tianjin
Cunsen Land, Chongqing (the Best Architectural Design of Zhan Tianyou Awards; 550,000 square meters of integrated community, super high-rise office buildings, hotels, apartments and riverside business area)
Unicom Headquarters Building, Shenzhen
Zhongshan Opera Theater
Shenzhen gao jiao hall
Residential Community
Longfor•Dachengxiaoyuan Community, Chongqing (Patent of Residential Dwelling Design)
Longfor•Blue Lake County, Shanghai (Patent of Residential Dwelling Design and the award of Excellent Residential Quarter in Shanghai)
Longfor Landscape, Chongqing
Longfor•Youshan County, Chongqing
Chongqing long lake Jiangbei Sports Park Project
Longfor•Rui City, Chongqing
Hong Kong Property and Longhu River & City•Yuanshan, Chongqing
Tongjing International New Town
Wood•Fish•Stone Project of Longfor University City, Chongqing
Hantop Xifuhui Project, Beijing
Longfor•3000 City Project, Chengdu
MCC Boulevard Project, Chongqing
Longfor•Daxing Project, Xi'an
China Overseas International Community Suzhou
Educational Construction
Shenzhen No.2 Foreign Studies Middle School
Dongguan Construction Primary School
Humanistic Education College of Xiamen branch of Huaqiao University
Shandong Institute of Business

许　峰
Feng Xu

许　峰
董事总监

教育背景
1990—1995年　天津大学建筑系学士
1997—1999年　美国莱斯大学建筑设计硕士
2003—2005年　美国哈佛大学城市设计专业硕士学位

资历
1999—2003年　美国SBA建筑师事务所
2005—2007年　美国KPF建筑师事务所
2007—2008年　万科集团总部任总建筑师
2008年至今　深圳市承构建筑咨询有限公司任董事总监

主要作品
深圳市石厦村商务中心区改造（180米超高层办公及酒店式公寓，大型综合商业）
成都龙湖世纪城（150米超高层住宅）
西安龙湖盛景别墅设计
中海独墅湖别墅区
韩国宋都新城D20，D21居住区设计
上海嘉里中心高层办公楼设计
中国沈阳恒隆广场高层办公楼设计
美国麻省理工大学研究生宿舍
马塞诸塞州伊斯兰文化中心
广州大学城中心区规划
广州大学学生宿舍

Feng Xu
Supervising Director

Education Background
1990-1995　Bachelor of Architecture at Tianjin University
1997-1999　Master of Architecture at Rice University
2003-2005　Master of Urban Design at Harvard University

Qualifications
1999-2003　SBA Architect Firm, U.S.
2005-2007　KPF Architect Firm, U.S.
2007-2008　Vanke Group headquarters　Chief Architect
2008-present　Shenzhen Made&Make Architects Architectural Consulting Co., Ltd. Supervising Director

Major Design
Restriction of Central Business District, Shixia Village, Shenzhen (including a super high-rise office building of 180m in height, hotel apartments and large-scale business complex)
Longfor•Century City, Chengdu (a high-rise residential community of 150m in height)
Design for Longfor•Grand View Villa, Xi'an
Dushu Lake Villa of China Overseas
Design for Residential District D20 and D21 of Songdu New Town, Korean
Design for the high-rise office building of Jiali Center, Shanghai
Design for the high-rise office building of Hang Lung Plaza, Shenyang, China
Graduate students' dormitory of MIT
Islamic Cultural Centre, Massachusetts
Planning for the central district of Guangzhou University Town
Dormitory building of Guangzhou University

佛山市顺德建筑设计院有限公司

Introduction of Foshan Shunde Architectural Design Institute Co., Ltd

梁昆浩
Kunhao Liang

梁昆浩

职务：董事长、院长
职称：一级注册建筑师、高级建筑师、高级规划师

梁昆浩，从事建筑设计行业近50年，现任顺德建筑设计院有限公司董事长、院长。在建筑创作上成就卓著，曾获全国总工会授予“五一”劳动奖章、国务院授予全国劳动模范称号、1991年6月开始享受国务院特殊津贴、荣获广东省优秀中青年专家、全国优秀设计院长、第八届全国人大代表。

在从事建筑设计工作期间，梁昆浩潜心研究仿古建筑、中国庭院、园林建筑，并且用自己多年以来积累的经验加以消化吸收，逐步形成具有自己独特的“岭南建筑新体系”的创作风格。其中珠海“九州城”集城郭之雄、园之美于一体，成为珠海游览胜景之一。“珠海宾馆”则采用并列式庭院组合的处理手法，使山水楼宇与绿树空间错落有致，更兼饰以夏、商、周等图案，惟妙惟肖地体现了岭南庭院的艺术风格。

顺德“仙泉酒店”坐落在树木葱茏的顺峰山麓，水光山色，相互辉映，酒店主楼点缀其间，气势雄伟，独得临湖依山之景．内部空间组织的动静结合，形色多样，变化合理，极具情趣。“珠海宾馆”设计的成功，使得不少海外人慕名前来，也使其有机会承接法国巴黎“中国城”、丹麦哥本哈根的“中丹经济文化中心”和“波兰分拨中心”项目等，欧洲各大报纸多年来对巴黎中国城作了详细报道。

《南方日报》1994年7月8日以“中国城”拓展西欧市场的窗口为题，报道了广东贸易投资展会的情况，论述了中国城作为广东省乃至全国拓展西欧贸易的窗口所起的作用，据1993年4月24日《参考消息》报道，“法国巴黎中国城”已于1993年3月被联合国教科书组织宣布为世界文物重点保护区之一。

1995年，该项目获全国优秀设计银质奖，对欧洲政治和文化艺术都产生了较大的影响，为弘扬我国传统文化艺术起到积极作用。梁昆浩同志大胆改革，勇于创新的精神还表现在企业的管理上。1993年底，顺德建筑设计院在全国同行业中率先组建有限责任公司，内部管理也不断改革创新，连续多年保持良好的经济效益，受到建设部的高度重视。2009年获中国勘察设计协会颁布的全国建筑设计行业国庆60周年“改制创新奖”。

主要获奖作品

1.珠海宾馆获1984年建设部银质奖及广东省优秀设计一等奖
2.珠海九洲城获1986年广东省优秀设计三等奖
3.法国巴黎中国城获1995年省优秀设计二等奖
4.顺德清晖园扩建工程获广东省第九次优秀设计三等奖
5.顺德宝林寺获广东省第十次优秀设计三等奖
6.顺德港规划与建筑设计获广东省优秀设计表扬奖
7.顺德区郑裕彤中学规划与建筑设计 获1995年省优三等奖
8.万家乐综合楼获广东省优秀设计表扬奖及国家“鲁班奖”工程
9.佛山市顺德区行政服务中心东座（办证中心）与西座（报批中心），获广东省第12次优秀工程设计二等奖
10.第五届中国花卉博览会中心展馆，获广东省第12次优秀工程设计三等奖
11.佛山市顺德一中高中部，获2008年度全国工程勘察设计行业优秀工程勘察设计三等奖，广东省第13次优秀工程设计二等奖。
修建性详细规划，获广东省优秀城市规划二等奖
12.佛山市顺德区一中高中部修建性详细规划，获广东省优秀城市规划二等奖
13.佛山市顺德区迎接亚洲艺术节和省运会市容整治规划，获广东省优秀城市规划设计表扬奖

佛山市顺德建筑设计院有限公司
Introduction of Foshan Shunde Architectural Design Institute Co., Ltd

梁昆浩
Kunhao Liang

Kunhao Liang

Position: Board chairman Institute president
Professional title: First-class registered architect,Senior architect,Senior urban planner

Liangkun Hao, who has engaged in architectural design for nearly 50 years, is the incumbent president of Shunde Architectural Design Institute Co., Ltd. Due to his distinguished creative achievements in architecture, he has been awarded the May 1st Labor Medal by National Federation of Trade Union, the title of the National Labor Model by the the State Council, as well as the title of Outstanding Young and Middle-age Expert in Guangdong Province, the title of Fine President of the architectural institute in China. And he has also been honored to receive special government allowance granted by the State Council and was elected as deputy to the Eighth National People's Congress.
While Dedicating to architectural design,Liang Kunhao concentrates his efforts on study of antique buildings, the Chinese-style courtyard and the garden architecture.He assimilates these knowledge well on the basis of his expertise accumulated for years , and gradually shapes up a distinctive "new building system of Lingnan " in his design. Here are some examples. Jiuzhou City in Zhuhai is a combination of the majesty of the outer wall and the beauty of the inside gardens, becoming one of the tourist interests in Zhuhai. And in Zhuhai Hotel ,the landscape ,the building and trees,which are spaced in picturesque disorder by paralleling the courtyard compound and decorated with the Xia, Shang, Zhou and other dynasty designs, giving a vivid expression to the artistic style of the Lingnan Garden.
Shunde Immortal Spring Hotel is nestled at the foot of the the verdant Shun Peak with the rivers and mountains enhancing each other 's beauty. Its main building exclusively enjoys the picturesque scenery near the mountain and by the lake. In return, it also embellishes such beautiful scenery with its magnificence. The internal structure of the main building is really fanciful with association of the static and dynamic aspect of the design, diverse shape and color and proper change. Such a successful masterpiece attracts numerous people overseas. Therefore, it is blessed with the great opportunity to serve as the Paris Chinatown which was reported in detail by the European major newspapers, the Sino-Denmark Economic and Cultural Centre ,Poland Distribution Centre and etc..
The South Daily reported the trade and investment exhibition in Guangdong with the title of "China town-a window to expand the west Europe market" on July 8th ,1994 ,and also illustrated its role as a window to expand the business of Guangdong ,even the whole China in west Europe. According to the report of Reference on April 24th, 1993, Paris Chinatown has been listed as one of the UNESCO World Heritage sites in March,1993.
This project won National Fine Design Silver Medal Award in 1995. It has not only exerted tremendous influence on the politics,culture and art in Europe, but also plays a positive role in carrying forward our traditional fine culture and art. Mr Liang Kunhao has also managed the company with such bold reform and innovation. At the end of 1993, Shunde Architecture Design Institute took a leadership in setting up a limited liability company. This , along with the reforming and innovation in its internal management, contributes to the maintainance of sound economic efficiency for years. Therefore, it received great attention of the National Construction Department. In 2009, the institute won the "Restructuring and Innovation Award" in the national architectural design field on the 60th anniversary of the founding of New China rewarded by the China Survey and Design Association.

Main awarded design works

1. Zhuhai Hotel – National development department silver prize and 1984 Guangdong Province excellent design first prize
2. Zhuhai Jiuzhoucheng – 1986 Guangdong Province excellent design third prize
3. China Town of Paris, France – 1995 Guangdong Province excellent design second prize
4. Shunde Qinghuiyuan Enlargement Project - Guangdong Province the ninth excellent design third prize
5. Shunde Baolin Monastery - Guangdong Province the tenth excellent design third prize
6. Shunde Port Planning and Architectural Design - Guangdong province excellent design praise prize
7. Shunde Zhengyutong Middle School Planning and Architectural Design - 1995 Guangdong Province excellent design third prize
8. The comprehensive building of Wanjiale was awarded the encouragement prize of the Excellent Design in Guangdong Province and the National Luban Award
9. The east wing(the certificates issuing centre) and the west wing(the reporting and approving centre) of the Administration Service Centre Shende District, Fushan was awarded the second prize of the 12th Excellent Design in Guangdong Province
10. The exhibition hall for the 5th China Flower Expo was awarded the third prize of the 12th Excellent Design in Guangdong Province
11. Buildings for the Senior- high- school Department of Sunde No.1 Middle School in Foshan was award the third prize of thet Survey and Design for Excellent Project and the second prize of the 13th Excellent Design in Guangdong Province
12. The detailed constructing plan for the Senior- high- school Department of Sunde No.1 Middle School in Foshan was awarded the second prize of The Excellent City Plan in Guangdong Province
13. The renovation plan of the Shunde District, Foshan in preparation for the Asia Art Festival and Guangdong Sports Game was awarded the encouragement prize of The Excellent City Plan in Guangdong Province

中机国际工程设计研究院

易 凡
Fan Yi

易 凡

男，工业电气自动化专业，本科学历。
1992年，被评为机械电子工业部优秀科技青年
1997年取得高级工程师职称。
历任专业主任工程师、专业组长、设备自动化所所长、中机国际工程设计研究院院长助理、副院长等职。
2009年4月起任中机国际工程设计研究院院长。

易凡同志长期从事非标设备研制及自动化专业的工程咨询、设计、科研及管理工作，主持过工程咨询、设计、科研等项目数十项，特别是在电机试验站微机控制系统的研究与设计工作中取得了突出的业绩，为我国电机试验站的技术发展和行业整体水平的提高做出了突出贡献。其承担的自动跳槽式绕线机微机控制系统的设计与研究项目，成果通过了原机械电子工业部鉴定，达到当时国内领先水平，并获得国家专利；其主持的电机、汽轮机、发电设备等多家国家大型骨干企业的电机试验站设计与实施工作，其试验技术、试验线、控制系统等一直处于国内领先水平。

担任院长以来，易凡积极推动理念创新、制度创新和机制创新工作，在面临全球性经济危机的关键时刻，坚定信心、勇挑重担，并通过改进经营工作和加强制度化建设，带领企业和全体员工克服困难，稳步前进，取得了突出的成绩。

主要学术成果

1.承担的上海电气集团临港重装备基地百万千瓦级核电核岛、常规岛及重型装备自主化制造工程项目获得2009年机械工业优秀工程咨询勘察设计一等奖。

2.承担的350MW级空冷气轮发电机型式试验测控系统项目获得中国机械工业集团公司机械工业科学进步二等奖。

3.承担的上海电气临港重型机械装备有限公司1800MW级汽轮发电机试验站技术专题研究报告获得2007年中国工程咨询协会全国优秀工程咨询成果一等奖与2007年机械工业优秀咨询成果一等奖。

Fan Yi

Male, majored in Industrial electric automatization, Bachelor degree.
He was rated as Excellent Science and Technology Youth of Mechano-electronic Industry Ministry in 1992.
Senior Engineer in 1997.
He has successively held the posts of professional engineer in charge, professional group leader, director of Equipment Automatization Institute, Assistant to President and vice-president of China Machinery International Design Institute, etc and so on.
Since April 2009, he has been president of China Machinery International Design Institute.
For long term Fan Yi has been engaged in work of engineering consultation, design, scientific research and management on development of non-standard equipments as well as on major of equipments and automatization. He has taken charge of tens of engineering consultation, design and scientific research projects, and acquired especial achievements on research and design work of Control System of Microcomputer in Electric Machine Experimental Station, which contributes greatly to the technology development of the domestic electric machine experimental station and the improvement of the whole industry level. The result of the design and research project of Control System of Microcomputer of automatic trench-hopping coiling machine that he took charge in has passed the identification of the former Mechano-electronic Industry Ministry, reached the domestic leading level at that time, and acquired the national patent; he has presided over the design and implementing work of the electric machine experimental station of many national large-sized backbone enterprises in the electric engineering industry, such as electric machine, steam turbine, generating equipment etc., of which the experimental technology, experimental line and controlling system have always been on the leading level in China.
Since he held the post of the president of the Institute, he actively promotes the innovation in idea, system and mechanism. In the crucial moment of confronting the global economic crisis, he was determined to take the heavy responsibilities; and through improving the operation and strengthening the systematized construction, he lead the enterprise and all the staff to overcome the difficulties, to make steady progress, and to obtain outstanding achievements.

Major academic achievements

1. He took charge in the manufacturing engineering project of million kilowatt-level nuclear power island, conventional island and weight equipment autonomation in the Harbor-side Weight Equipment Base of Shanghai Electric Group, which won First Prize of Excellent Engineering Consultation, Survey and Design of Engineering Industry of 2009.
2. He took charge in the project of Type Tests Measurement and Control System of 350MW-level air cooled turbo-generator, which won Second Prize of Engineering Industry Scientific Progress of China National Machinery Industry Corporation.
3. He took charge in the technological monographic study report of 1800MW-level turbonator experimental station of Shanghai Electric Harbor-side Weight Mechanized Equipment Limited Company, which won First Prize of National Excellent Engineering Consultation Achievements of China Engineering Consultation Society of 2007 as well as First Prize of Engineering Industry Excellent Consultation Achievements of 2007.

辽宁省建筑设计研究院
LIAONING PROVINCIAL BUILDING DESIGN & RESEARCH INSTITUTE

杨 晔
Ye Yang

院长杨晔，1968年7月出生于沈阳，1990年毕业于西安冶金建筑学院（现名为西安建筑科技大学）建筑系，2007年赴新加坡南洋理工大学（NTU）人文及社会科学学院学习公共管理专业，获公共管理硕士学位（MPA），教授级高级建筑师、国家一级注册建筑师，中国建筑学会会员、香港建筑师学会会员，中国建筑学会建筑分会理事、辽宁省土木建筑学会副理事长、辽宁省勘察设计协会副理事长、辽宁省建筑师协会副理事长，辽宁省注册建筑师管理委员会委员、国家二级注册建筑师考试命题专家组成员，沈阳建筑大学硕士导师，辽宁省直青联常委、辽宁省青联常委，沈阳市十大中青年建筑师、沈阳市十大杰出青年、辽宁省劳动模范、辽宁省优秀共产党员、辽宁省享受政府特殊津贴专家、中国勘察设计协会优秀企业家、第十六届辽宁省十大杰出青年、辽宁省首批工程设计大师。

建筑观

在技术材料发展迅速、思想意识呈现多元化、个性需要追求多样的今天，建筑师应努力寻求建筑中蕴涵的情感，寻求建筑随时间变化而在空间上与人形成的互动关系与影响。

人的生命在不同的建筑中度过并消耗，建筑却在这个过程中获得了生命与情感，虽然建筑的类型、规模、功能、对象、地域、环境等各有不同，但都会与其中的使用者——人，——产生各种各样的影响与关系。一个有社会责任感的建筑师就是要利用各种可能的和必要的技术手段和空间方式发现并提供积极的影响。

主要作品

1991–1993年，中辽国际大厦、沈阳高登大酒店、沈阳华阳国际大厦、营口港务局综合楼；1994年–1998年，辽宁电视台彩电中心、鲁迅美术学院教学综合楼、葫芦岛电视大楼、葫芦岛辽西大厦、沈阳反贪局综合楼、辽宁工商行政管理局工商干校、铁法商业大厦。

1999–2001年，鲁迅美术学院高层学生公寓综合楼、通辽新世纪大酒店、通辽地税局、中辽国际塞舌尔公司基地综合楼、喀麦隆雅温德市政府广场改造、沈阳宫后里商业中心（兴隆大家庭）；2002年，鲁迅美术学院校园系列工程、辽沈战役纪念馆园区改扩建工程、沈阳监狱迁建工程；2003年，本溪银泽家园小区、沈阳市技术监督局高新技术产品检测中心；2004，俄罗斯伊尔库斯克办公楼、沈阳市图书馆、儿童活动中心、宁波科技园区I–3地块、鲁美大连校区、沈阳曼哈顿商业广场；2005年，沈阳市检察院、沈阳人防指挥中心、沈阳医学院国际护理综合楼、营口高新技术开发区中心医院、宁波华茂才源大厦、大连天一国际软件大厦；2006年，沈阳城市规划展示馆、沈阳新世纪大厦、平顶山惨案遗址纪念馆、东北电网调度交易中心、鲁迅美术学院文体馆；2007年，沈阳新世界酒店二期、东北传媒文化广场；2008年，蒋少武摄影博物馆、格林创意大厦；2009年，沈北新区财落中心镇规划、沈北新区清水台镇规划、沈阳北海乐购超市、辽宁体育训练中心（第十二届全运会比赛训练场馆）；2010年，沈阳裕景中心、厦门裕景中心、沈阳中韩科技产业园地。

主要获奖

1990年，第17届UIA加拿大蒙特利尔大会世界建筑大学设计竞赛第一名、联合国教科文组织奖（UNESCO PRIZE）；1992年，沈阳宫后里商业城（现名沈阳东亚商业广场）方案设计竞赛二等奖（无一等奖，中标实施方案）；1997年，沈阳市、优秀勘察设计一、三等及表扬奖、沈阳市建筑节能与墙体革新工作先进个人；1998年，辽宁省文化艺术中心、省博物馆新馆工程二等奖（第二名）；中国建筑学会青年建筑师奖、优秀奖，辽宁省"九五"城乡住宅设计方案竞赛获辽宁省、优秀工程勘察设计二等奖；1999年，沈阳高登大酒店获辽宁省优秀工程勘察设计一等奖、鲁迅美术学院综合教学楼获辽宁省优秀工程勘察设计二等奖；2001年，辽宁电视台彩电中心获辽宁省优秀工程勘察设计一等奖；2004年，沈阳市图书馆、儿童活动中心国际建筑设计竞赛第一名，沈阳市十大中青年建筑师、沈阳市十大杰出青年、辽宁省新世纪百千万人才工程"百人层次"；2005年，第五届辽宁省优秀科技工作者、辽宁省享受政府特殊津贴专家、辽宁省建设行业优秀工程技术人才、第四批辽宁省省级优秀专家、辽宁省省直机关优秀共产党员；2006年，辽宁省劳动模范、辽宁省优秀共产党员、中国勘察设计协会优秀企业家、沈阳市图书馆、儿童活动中心获全国绿色建筑创新奖综合奖--公建 建设部三等奖；2008年，第十六届辽宁省十大杰出青年、奥运火炬手，辽宁省首批工程设计大师；2009年，辽宁省杰出青年志愿者、辽宁省首批工程设计大师；2010年，全国建筑设计行业国庆60周年管理创新奖、"抚顺市平顶山惨案遗址纪念馆"获中国勘察设计协会优秀建筑工程设计二等奖、"辽宁东北抗联史实陈列馆工程" 获中国勘察设计协会优秀建筑工程设计三等奖。

山东省建筑设计研究院第 6 分院
SHANDONG PROVINCIAL ARCHITECTURAL DESIGN & RESEARCH INSTITUTE—THE 6TH BRANCH
— 医用工程（专项）设计研究所

王 岗
Gang Wang

王 岗

职务：六分院副院长
职称：研究员
国家一级注册建筑师

经 历

1991年毕业于北京建筑工程学院建筑系，学习建筑学专业，获得学士学位。

毕业后分配到山东省建筑设计研究院，一直从事医疗卫生建筑设计工作至今。有着二十年设计医院的经验，先后以项目总负责人的身份设计建成了众多的医院设计案例，近年来，不断参加国内大型综合医院的设计投标工程，多个有影响项目获得中标并建成，投入使用，获得医院方的高度评价。

设计医院感悟

难！医疗卫生建筑因其功能复杂、科室众多、流线交叉，在满足医疗卫生规范的同时，还要符合建筑设计规范，这众多各种因素交织混杂在一起，要设计好医院项目确实很难！

易！医疗卫生建筑也有规律可循，只要掌握了医院设计的纲领主线，围绕着主线展开设计和深入发展，医院建筑项目设计起来又会相对容易。这个纲领主线就是无论是医院规划设计还是单体设计都要遵循"洁污分流明确，避免交叉感染"。

业 绩

1.江苏大学附属医院项目负责人，项目总建筑面积5万平方。
2.东南大学附属中大医院总体规划及病房大楼项目负责人，项目总建筑面积7.4万平方米。
3.广州暨南大学附属医院项目负责人，项目总建筑面积7.2万平方米。
4.广东省中西医结合医院门诊医技综合楼项目负责人，项目总建筑面积11万平方米。
5.江苏大学附属医院总体规划及病房大楼项目负责人，项目规划面积25公顷，总建筑面积5万平方米。
6.江苏仪征市人民医院总体规划及病房楼、门诊楼、科技综合楼项目负责人。
7.山东省威海经区医院综合病房楼项目负责人。
8.内蒙古自治区医院内科综合楼项目负责人，项目总建筑面积7万平方米。
9.辽宁本溪钢铁集团总医院外科综合楼项目负责人。

Gang Wang

Post: Vice president of the sixth subsidiary
Professional title: Researcher
National Class I registered architect

Experience

He graduated from architecture department in Beijing Institute of Civil Engineering and Architecture in 1991, obtained the bachelor degree of architecture.

After graduation, Gang Wang joined Shandong provincial architectural design institute and engaged in the hospital architecture design works. With 20 years of hospital architecture design experiences, he participated and took charge of numerous hospital design projects as the project leader and designer. In recent years, he has attended lots of national hospital architecture design project biddings. with many awarded tenders projects come into service, he has gained great reputation in our clients in medical area.

Enlightenment of hospital architecture design

It is tough! With the characteristic of complex function division, numerous offices and overlapping streamline of hospital architecture, the design need to meet both the medical regulation requirement and the architectural design criterion. With such different elements mixed together, we can say, hospital and medical architecture design project is no easy task.

It is simple as well --- hospital and medical architectural design has its law. As long as the designers master the mainline of the hospital architecture design, thoroughly promoting the design, the hospital architecture projects turn to be relatively easy. The mainline is, no matter in hospital master plan or single building design, we should follow the rule of "Clearly divide the clean and filthy split flow, prevent the cross infection."

Achievement

1. Design director and project leader of Jiangsu University affiliated hospital ward building project. Total gross floor area is 50,000 m^2
2. Design director and project leader of Zhongda Hospital Southeast University master planning and ward building design project, Total gross floor area is 74,000 m^2
3. Design director and project leader of Jinan University affiliate hospital project, total gross floor area is 72,000 m^2
4. Guangdong integrated western and Chinese medicine hospital building project leader and design director ,total gross floor area is 110,000 m^2
5. Design director and project leader of Jiangsu University Affiliated Hospital master planning and ward building design project. Project planning area: 25 ha , total gross floor area is 50,000 m^2
6. Jiangsu Yizheng People's hospital master planning, ward building, out-patient building and technology comprehensive building design project leader and design director
7. Weihai Jingqu hospital comprehensive ward building design project leader and design director
8. Inner Mongolia people's hospital building project leader and design director. Total gross floor area is 70,000 m^2
9. Design director and project leader of General Hospital in Bengang comprehensive building project.

山东大卫国际建筑设计有限公司
Shandong David InternationalArchitectural Design Co.,Ltd.

申作伟
Zuowei Shen

申作伟

学位：硕士
专业：城市规划
职称：教授级高级建筑师
国家一级注册建筑师
国家注册规划师
香港注册建筑师

社会兼职及荣誉

中国居住区规划专业委员会委员（2003 年）
建设部优秀勘察设计评选委员会委员
中国新本土居住设计名家
山东省优秀建筑师、山东省十佳注册建筑师

个人经历

1985 年 7 月 –1997 年 6 月
临沂市规划设计研究院，院长兼临沂市规划局副局长
1997 年 6 月 –1998 年 10 月
山东省城乡规划设计研究院，院长
1998 年 10 月至今
山东大卫国际建筑设计有限公司　董事长

申先生自从业以来，曾获得全国优秀设计奖八项，获山东省优秀设计奖三十余项，并有十几篇论文在《建筑学报》等全国性学术刊物上发表。

"建筑师应该是未来生活的设计者和创造者，引导人们的生活水准向更高层次和更高标准迈进。因此，尽最大努力满足现代人们的生活需求是建筑师的生存方式。"

"建筑设计的发展趋势应该是把传统建筑的精髓与现代生活方式相结合，用现代技术、现代材料和现代的审美观去设计有文脉的建筑。"

北京优山美地别墅区涵盖的"新本土"理念，崇尚亲情感、邻里感，围合了自然、安逸、私密的空间，使人们回归至居住本性。

Zuowei Sheng

Degree: master
Major: City Planning
Professional title: Professor of Senior Architect
National Class I Registered Architect
National Registered Planner
Hong Kong Registered Architect

Social positions and honors

1. The member of Chinese professional residence planning committee (2003)
2. Commissary of Excellence Reconnaissance and Design Committee of Ministry of Construction
3. Chinese famous new local residence designer.
4. The excellent architect and one of the top ten registered architect of Shandong province

Personal Experience

Form July 1985 to June 1997
He was working in City Planning and Design Institute of Linyi City, and took the position as director of the institute, at the same time he also took the position of deputy director for the Planning Bureau .
From June 1998 to October 1998
He was working as a director of Urban and Rural Planning Design Institute of Shandong Province.
From October 1998 until now
He worked as the chairman of the board of Shandong David International Architectural Design and Consultant Co.,Ltd.
Since the work began, he has got 8 awards of national excellent design prizes, more than 30 awards of excellent design prizes in Shandong Province, and he has published more than ten papers on the national academic journals such as Architectural Journal etc.
"The architect should be a designer and creator for the future lives who can lead people's living standard to achieve the advanced level. Therefore, trying their best to satisfy the needs of modern people is architects' survival duty".
"The trend of development of architecture and design should be the combination of traditional constructions and modern living styles. It should use modern materials and the conceptions of beauty to design the buildings which have literary venations".
The concept of "new native land" for Beijing Villa Area of beautiful mountains and lands advocates the feelings of family love and neighborhood; it encloses a secret space of nature and comfort. It makes people return to their residential natures.

山东建筑大学设计研究院
DESIGN & RESEARCH INSTITUTE OF SHANDONG JIANZHU UNIVERSITY

赵学义
Xueyi Zhao

赵学义

山东建筑大学设计研究院院长兼总建筑师、教授、硕士生导师
山东建筑大学第四、七批专业技术拔尖人才
山东建筑学会建筑创作委员会副主任委员
国家一级注册建筑师
国家注册规划师
国家注册咨询（投资）工程师
山东省优秀建筑师
山东省十佳优秀执业(建筑)师

部分获奖工程

山东省实验小学逸夫教学楼
获教育部逸夫第十一批捐资工程优秀项目二等奖
济南逸夫职业教育培训中心综合楼
获山东省优秀勘查设计二等奖国家教委逸夫捐资工程项目优秀工程一等奖
山东建筑工程学院图书馆
获建设部优秀勘查设计二等奖山东省优秀勘查设计一等奖
中国威海首届人居建筑设计大奖赛参赛作品
获大赛金奖
山东省建筑工程学院新校区详细规划
获建设部优秀设计二等奖
山东省建筑工程学院新校区图书信息中心
获中国建筑学会建筑创作佳作奖山东省优秀勘察设计一等奖

Xueyi Zhao

President and general architect Design Research Institute of Shandong Jianzhu University
Professor, Master's Supervisor
The fourth and seventh batch of top-notch professional and technical personnel of Shandong Jianzhu University
Vice director of Architectural Creation Committee of Shandong Architecture Institute
National Class 1 registered architect
National registered planner
National registered consulting (investment) engineer
Excellent architect in Shandong Province
One of Top Ten excellent architects in Shandong province

Part of his rewarded projects

Yifu Academic Building in Shandong Experimental Primary School
Awarded the second prize for the excellent projects of the 11th Yifu Donating projects by Ministry of Education
Multi-use building of Yifu Vocational Education Training Center in Jinan
Awarded the second prize for Shandong excellent surveying and designing
Awarded the first prize for Yifu Donating projects by Ministry of Education
Library of Shandong Jianzhu University
Awarded the second prize for excellent surveying and designing by Ministry of Architecture
Awarded the first prize for Shandong excellent surveying and designing
Awarded the golden prize for the first Weihai Human Settlements Architectural Design Competition
Detailed planning for new campus of Shandong Jianzhu University
Awarded the second prize for excellent design by Ministry of Architecture
Book Information Center in new campus of Shandong Architecture University
Awarded the excellent work for architectural creation of Architectural Society of China, ASC
Awarded the first prize for Shandong excellent survey and design

湖南诚士建筑规划设计有限公司
诚士设计
HUNAN CHENGSHI ARCHITECTURAL DESIGN AND CITY PLANNING CO.LTD.

蒋怀庆
Huaiqing Jiang

蒋怀庆

湖南诚士建筑规划设计有限公司
总建筑师
首席设计师
国家一级注册建筑师
湖南省建筑师学会理事

从业十九年，一直在追寻，追寻一个梦，在不同的项目实践中反复坚持着一个理念：
建筑是有生命的，建筑是有性格的，空间是建筑的灵魂，
空间不仅是"二维"、"三维"的空间（视觉冲击与身体感知），空间也是"四维"的场所（随时化而生长），空间更是"五维"的境界（随心灵升华而感动）。
设计不仅是设计建筑，而是设计生活，倡导一种健康人性化的生活模式。
生活是一种智慧，智慧是一种真爱——生态环境，有机建筑，和谐社会。
人与自然共生，建筑与自然共生。
让建筑成为大自然交响乐中的一部分，或是序曲、或是高潮、或是背景、虽各得其所，却同样可以感动人们。
这种感动，让上帝所创造的人与自然从效果图中的配景，变为真正的主角；
这种感动，以人为本，让人们回归自然，享受自然，
这种感动，尊重自然，
——保护水体、植被采用再生能源让建筑与自然生态体系有机循环共生。

审视这些年的创作实践，真可谓不敢有半点懈怠，虽备受磨砺却执著坚持，虽满含艰辛却充满激情。其中之过程如十月怀胎，孕育每一个建筑生命都很不容易——从规划到单体，从内部空间到外部造型，再到环境景观一气呵成，甚至家具装饰、雕塑……也要量身定做，在生态、可持续发展的指导思想下，因地制宜，节能、节地、节水、节材，努力研究各种当地生态的材料和构造，寻求又经济又方便的施工途径，减少过多的装饰和堆砌，强调内空间与外部造型的完美统一，以空间和功能结构本身的规律和变化创造出美，让建筑之艺术与技术巧妙结合，最终看见孩子落地才有了欣喜的微笑。
母不嫌子丑，现将他呈现给大家……

Huaiqing Jiang

Hunan Chengshi Architectural Design and City Planning Co., Ltd.
Senior Architect
Chief Designer
Class A Registered Architect
Director of Hunan Architect Association
Dedicated in this industry for 19 years, he has been running for a dream, a dream that he has been insisted by and by in different project works:
Architecture has life,
Architecture has characters,
Space is the soul of architecture,
Space is not only a " two vector ", " three vector " space (visual impact and physical sense)
Space is also a "four vector " place (growing all the time)
Space is even a "five vector " realm (heart touching as sublimation)
As to design, it is not only on buildings, but the design on life, a lifestyle promoted with healthy and humane.
Life is nothing but wisdom, wisdom embraces love - the ecological environment, organic architecture, and harmonious society.
Human being is in symbiosis with nature, while architecture and nature coexist.
May architecture be a part of nature symphony, no matter to be overture, climax, or background, it can always get people moved for its role playing.
Such touching, enables the God-created man and nature to turn to a true hero that formerly a objective view in effect maps;
Such touching is people-oriented, allowing whom to return to nature, and enjoying nature,
Such touching, shows honor to the nature,
–Protect water bodies and vegetation, adopt renewable energy to realize the symbiosis cycle between architecture and natural organic ecological systems.
Look inwardly for those years of creative practice, it indeed can be claimed to no daring to any slack off, insisting all the way in regardless how tempered, with difficulty time and time but ample of passion. This process can be compared with a mother in way for ten months, i.e., every building work is not easy - from planning to monomer, from interior to exterior styling, along with the environmental landscape at one go, and even furniture, sculpture, etc...It also requires specific tailoring, under the guiding ideology of ecology and sustainable development, based on local conditions, reducing consumption in energy, land, water and materials, to study various local ecological materials and structures, in pursuit of an economical and convenient way for construction so as to reduce the excessive decoration and piling. Meanwhile we focus on the perfect integration of interior space and exterior styling, taking advantage of rules and changes between space and functional structure in their own to create beauty, integrated the art of architecture technically and tacitly. So she smiles eventually on seeing the baby born.
A mother never thinks her son ugly, now she presented it to all. . .

山西省建筑设计研究院
Shanxi Architectural Design and Research Institute (SADI)

张学锋
Xuefeng Zhang

张学锋

院长
教授级高级工程师
国家一级注册建筑师
中国建筑学会建筑师分会人居环境专业委员会理事
山西省勘察设计协会建筑设计委员会秘书长
山西省建筑师学会副理事长
山西省建工学院兼职教授
山西省太原市规划委员会规划咨询专家

1987年7月毕业于太原工业大学土木工程学院建筑专业，同年分配到山西省建筑设计研究院从事建筑设计工作。工作期间成绩突出，1994年被派往日本研修一年。

张学锋作为一名优秀设计师，对山西建筑创作的发展及山西新地域文化的形成等方面进行了积极的探索与思考，为政府在城市建设中的科学决策提供了有力的技术支持和科学依据，同时他入选了山西省“新世纪学术技术带头333人才工程”省级人选。亲自主持与参加完成大中型或国家及省重点工程项目百余项，其中多项工程获省级优秀奖，为山西重点工程建设项目实施做出了突出贡献。

代表工程
太原邮电业务综合楼
山西大学图书馆
太原工业大学科学楼
山西省人大常委会办公楼
太原机场改扩建工程
山西晋城学院
山西大剧院

个人荣誉
2003年
荣获山西省建设行业第二届青年科技奖
优秀青年设计专家
山西省劳动竞赛一等功
2005年
荣获中国勘察设计协会先进工作者
山西省勘察设计协会先进个人
2006年
荣获山西省首届优秀青年建筑师
2007年
荣获山西省直机关劳动模范
2008年
荣获山西省五一劳动奖章

Xuefeng Zhang

President
Professor of senior engineer
National 1st Class A Registered Architect
Member of Professional Committee of Human Environment, Architects Association Branch of Architectural Society of China
Secretary General of Architectural Design Committee, Shanxi Exploration & Design Association
Vice president of Shanxi Institute of Architects
Adjunct Professor of Shanxi Construction Engineering College
Planning Consultants of Taiyuan Municipal Commission of Urban Planning, Shanxi Province

In July 1987 he graduated from Department of Civil Engineering in Taiyuan University of Technology in architecture major, and in the same year he was assigned to Architectural Design & Research Institute of Shanxi Province and engaged in architectural design. Because of his outstanding achievements during the working period, he was sent to Japan for advanced study and training in 1994.

As a good designer, he makes positively exploration and thinking on development of Shanxi architectural creation and formation of new Shanxi geographical culture, providing strong technical support and scientific basis for government's scientific decision on urban construction. He was selected as the provincial candidate for “New Centurial Leading Academic Technology A333 Talent Project “of Shanxi province. He was responsible for large and medium-sized projects or key projects in national and provincial level over a hundred items, many of which were awarded the provincial excellent prizes, making outstanding contributions to the implementation of Shanxi key engineering construction projects.

Representative Projects
Complex building for Taiyuan Posts and Telecommunications
Library in Shanxi University
Science Building in Taiyuan University of Technology
Office building for People's Congress Standing Committee of Shanxi Province
Expansion project for Taiyuan Airport
Shanxi Jincheng College
Shanxi Grand Theatre

Awards
In 2003, was awarded the second “Youth Science and Technology Prize” in Shanxi Architecture Industry; was selected as Excellent Youth Design Expert; was awarded the Top-grade Merit Citation in Shanxi Labor Competition
In 2005, was awarded the title of “Advanced Worker” by China Exploration & Design Association; was awarded the title of “Advanced Individual” by Shanxi Exploration & Design Association
In 2006, was awarded the title of first outstanding young architect in Shanxi Province
In 2007, was awarded the title of working model in Shanxi Provincial Authorities
In 2008, was awarded 5·1 Labor Medal of Shanxi Province

WEN KUN
上海文坤建筑设计有限公司
Shanghai Wenkun Architectural Design & Research Co.Ltd.

宋文玻
Wenbo Song

宋文玻

总建筑设计师
西安建筑科技大学建筑学

主要设计作品
深圳世界金融中心
深圳南中国一号
无锡茂业城
常州茂业　泰富城
常州武宜路1、7和8号地块规划
常州莱蒙世界之旅
成都置信花园城
深圳城建观澜别墅区
遵义万里路商业中心
合肥松芝万象城
无锡太湖新城
安阳彰德府文化商业步行街
成都金履国际商贸城
佛山九鼎国际城南区
无锡新世界国际纺织城（住宅）
海口仁邑国际社区

Wenbo Song

Chief Architect
Xi'an University of Architecture and Technology Architecture

Main Works
World Financial Center, Shenzhen
South China Number 1, Shenzhen
Maoye City, Wuxi
Maoye Taifu City, Changzhou
Planning of Block 1, 7 and 8 in Wuyi Road, Changzhou
Street Mall named Laimeng World Tour, Changzhou
Zhixin Garden City in Chengdu
Guanlan Villas Area of Urban Construction, Shenzhen
Commercial Centre in Wanli Road, Zunyi
Songzhi • Wanxiang City, Hefei
Taihu Lake New City, Wuxi
Zhangdefu Culture and Commercial Pedestrian Street, Anyang
Jinlu International Trade City, Chengdu
South District of Jiudin International City of Foshan
Wuxi New World International Textile City (residential)
Renyi International Community, Haikou

GN® 栖城
GN INTERNATIONAL DESIGN

沈利江
Kahn Shen

沈利江

GN 栖城执行董事

世界华人建筑师协会会员
中国建筑学会会员
法国建筑设计师协会会员
2000 年 美国达拉斯国际青年建筑师大赛一等奖
2005 年 DI · 设计新潮——年度最佳建筑人获得者
2006 年 “中国建筑规划设计师杰出 100 位”

Kahn Shen

GN Xicheng Inc. executive director

WACA membership
Membership for Architectural Society of China
Membership for French Association of Architects and Designers
USA Dallas International Newby Architects Competition, 2000, First Prize
DI—Architect of the Year, 2005
China Top 100 Outstanding Architectural Planners & Designers, 2006

中国建筑西南设计研究院有限公司
CHINA SOUTHWEST ARCHITECTURAL DESIGN AND RESEARCH INSTITUTE CORP.LTD

龙卫国
Weiguo Long

龙卫国

中国建筑西南设计研究院有限公司院长，执行董事，总经理，教授级高级工程师，清华大学工学硕士。

中国勘察设计协会常务理事，中国勘察设计协会建筑设计分会副会长，四川省勘察设计协会副理事长，四川省土木建筑学会副理事长，成都市土木建筑学会副理事长，木材及复合材结构专业委员会主任委员。

代表作品

川综商场
王子大酒店
霍英东蒲洲港国际商贸广场
新世纪电脑城
重庆时代广场
南桥花苑
四季花城
成都双楠广场
北斗星大厦
重庆龙湖"观山水"

主编规范和手册

《木结构设计手册》（第三版）
《木结构设计规范》
《木骨架组合墙体技术规范》

学术论文著作

《直接焊接连接钢管节点静承载力分析》
《管节点始屈荷载有限元分析》
《深层搅拌桩在高层建筑中的应用》
《异形柱受力性能及结构设计有关问题探讨》
《混凝土结构耐久性设计建议》
《规格材机械分级方法研究》
《中国木结构需要开展的研究工作》

Weiguo Long

He is executive director, general manager of CSWADI. He is also a Master of Engineering graduating from Tsinghua University and a Professor of Engineering. In addition, He is a standing director of China Survey & Design Association, Vice Chairman of Architectural Design Branch of China Survey & Design Association, Vice Chairman of Sichuan Investigation &Design Association, vice chairman of Sichuan Civil Engineering Society, Vice Chairman of Chengdu Civil Engineering Society, and Chairman of Material and Multiple Structure Professional Committee.

Representative Works

Chuanzhong Mall
Prince Hotel
Henry Fok Puzhou Port International Trade Square
New Century Computer City
Chongqing Time Square
South Bridge Garden
Four Season Flower City
Chengdu Shuangnan Square
Big Dipper Edifice
Chongqing Longhu Landscape

Edited Standards and Manuals

"A Handbook for Design of Wood Structure" (the 3rd edition)
"Code for Design of Timber Structure"
"Technical Specification for Timber Framing Composite Wall"

Academic Publications

"Research on Static Bearing Capability of Directly Wielded Joints of Steel Tubes"
"Finite Element Analysis of Bearing Capability of Joints"
"Application of Deep Mixing File on High-rise Building"
"Research on Mechanical Behavior and Structural Design of Special Column "
"Suggestions related with Duality Concrete Structure"
"Classification Method of Dimension Stock and Machine"
"The Necessary Research on Chinese Wood Structure"

天津市建筑设计院
TIANJIN ARCHITECTURE DESIGN INSTITUTE

刘 军
Jun Liu

刘 军

院长
总建筑师
国家一级注册建筑师
正高级建筑师
中国城市规划协会理事
中国工程咨询协会理事
中国建筑学会资深会员
天津市城市规划协会副理事长
天津市建筑学会理事
天津市工程咨询协会副会长

个人荣誉

2004年天津市"十五"立功先进个人
2006年天津市优秀共产党员
2008年规划建设系统支援抗震救灾优秀共产党员
2010年天津市"五一"劳动奖章先进个人
天津市规划局授衔专家
天津市第十届青年科技奖
天津市规划局学科（专业）技术带头人

代表作品

天津海河两岸城市设计
荣获 全国优秀城市规划设计二等奖
天津市优秀城市规划设计一等奖 2009年
天津津湾广场修建性详细规划
荣获 全国优秀城市规划设计二等奖
天津市优秀城市规划设计一等奖 2009年
天津市天钢柳林城市副中心城市设计
荣获 全国优秀城市规划设计三等奖
津市优秀城市规划设计一等奖 2009年
政协俱乐部扩建工程
荣获 优秀工程勘察设计行业二等奖
天津市优秀勘察设计二等奖 2009年
天津意式风貌保护区的保护和整修规划
荣获 全国优秀城市规划设计二等奖
天津市优秀城市规划设计一等奖 2007年
瑞景居住区17号地项目
荣获 国家优秀工程设计 金奖
泉州邮电技工培训学校
荣获 福建省优秀工程奖
天津丽苑钢结构试点工程
荣获 天津市优秀工程设计一等奖

Jun Liu

Head of Institute
Chief Architect
National Class 1 Reqistered Architect
Senior Professional Architect Title
Director of China Urban Planning Association
Director of China Engineering Consulting Association
Senior member of Architectural Society of China
Vice Chairman of Tianjin Urban Planning
Director of Tianjin Institute of Architects
Vice President of Tianjin Association of Engineering Consultants

Personal Collection

In 2004, he was awarded as Meritorious Individuals for the tenth five-year plan in Tianjin
The outstanding party member of Tianjin, 2006
Outstanding party member of planning and construction system to support earthquake relief, 2008
Tianjin "5.1" Labor Medal advanced individuals, 2010
Endowed expert in Tianjin City Planning Bureau
The Tenth Tianjin Youth Science and Technology Award
Disciplines (professional) technology leader of Tianjin Planning Bureau

Representative works

Urban design ashore the Haihe River in Tianjin
Won: The Second Prize for National Outstanding Urban Planning and Design
The First Prize for Excellent Urban Planning and Design in Tianjin 2009
Tianjin Jinwan Plaza site plan
Won:The Second Prize for National Outstanding Urban Planning and Design
The First Prize for Tianjin Excellent Urban Planning and Design 2009
Tianjin Urban Design of Tiangang Liulin City sub-center
Won: The Third Prize for National Outstanding Urban Planning and Design
The First Prize for Tianjin Excellent Urban Planning and Design 2009
Political Consultative Conference (CPPCC) Club Expansion project
Won:The Second Prize for excellent engineering survey and design
The Second Prize for Excellent Engineering Design in Tianjin 2009
Tianjin Italian Conservation Area protection and renovation planning
Won:The Second Prize for National Outstanding Urban Planning and Design
The First Prize for Tianjin Excellent Urban Planning and Design 2007
Ruijing residential area Plot No.17
Won: Gold Award for National Excellent Engineering Design
Telecommunications technician training school in Quanzhou
Won:Fujian Outstanding Project Award
Pilot project of Tianjin Liyuan Steel-work
Won:The First Prize for Tianjin Excellent Engineering Design

天津大学建筑设计规划研究总院
Academy Of Architectural Design & City Planning Tianjin University

洪再生
Zaisheng Hong

洪再生

专　　业：城市规划与设计
研究方向：城市规划与城市设计、城市更新与城市再开发、历史文化名城保护规划等
学　　位：日本神户大学工学博士
职　　称：教授、博士生导师
行政职务：天津大学建筑设计规划研究总院院长
天津大学城市规划设计研究院院长

主要学习工作简历

1984年，毕业于天津大学建筑学建筑专业
1989年，获天津大学建筑系城市规划与设计专业硕士学位
1996年，获日本神户大学工学博士学位
1996年10月归国后曾任天津大学建筑学院城规系副主任、天津大学城市规划设计研究院院长
2005年5月至今任天津大学建筑设计规划研究总院院长

社会活动

中国城市规划学会历史文化名城学术委员会委员
中国城市科学研究会　理事
天津市人民政府规划委员会委员
天津市工程评标专家
天津市城市规划协会、城市规划协会学会副理事长
中国勘察设计协会高等院校勘察设计分会副理事长

论文

《建造符合老人心理、生理特点的老人建筑——日本老人院建筑的构成与设计初探》
《儿童游戏行为与空间的对应关系》
《岳阳楼景区规划设计初探》
《天津传统文化商贸区规划设计探讨》
2005年的论文《烟台城市色彩规划探讨》
2005年的论文《山海风情・人间仙境"——烟台整体城市设计研究》
2006年的论文《转型期我国特大城市规划编制体系的创新实践比较》
译著：《世界城市住宅小区设计 日本卷》
译著：国家教委《社区居住的可持续发展研究》
著作《天津建筑风格》（副主编、执行主编）
入选《Dumon+78位优秀设计师》一书

主要获奖

"天津城市形态的演变与城市风貌的未来"科研项目，获得2002年天津市科学技术进步　二等奖
"北洋大学纪念公园—天津北洋园"规划设计，获2001年度全国优秀规划设计　三等奖
天津梅江居住区规划设计竞赛　一等奖
天津春华里居住区规划设计竞赛　二等奖
"烟台整体城市设计"获省部级　二等奖
"北京亦庄经济技术开发区发展规划"获省部级　二等奖
"响螺湾商务区城市设计"获天津优秀设计　二等奖
天大25教学楼
获　教育部2005年优秀设计　一等奖
建设部2008年优秀设计　二等奖
全国优秀工程勘察设计　铜奖
潍坊中医院获2008年全国勘察设计优秀设计　三等奖

Zaisheng Hong

Major: Urban Planning and Design
Research direction: urban planning and urban design, urban renewal and urban re-development, the protection planning of historical and cultural city, etc.
Degree: Kobe University in Japan, Dr. Engineering
Title: Professor and PhD supervisor
Administrative service:
Dean of the general academy of architectural design planning research of Tianjin University
Dean of Academy of Urban Planning and Design of Tianjin University

personal profile

In 1984, graduated from Tianjin University, Architecture Department, and majored in Architecture
In 1989, obtained the master's degree for urban planning and design, Department of Architecture, Tianjin University
In 1996, obtained D. Engineering at Kobe University in Japan
In Oct 1996, he returned to China and once served as deputy director of the Urban Planning Department of School of Architecture of Tianjin University; Dean of Academy of Urban Planning and Design of Tianjin University
Since May 2005, he has been the dean of Academy of Architectural Design & City Planning Tianjin University

Social activities

Committeeman of famous historical and cultural cities Academic Committee of China Urban Planning Society
Director of Chinese city science research association
Committeeman of Tianjin Municipal People's Government Planning Committee
Tianjin project evaluation expert
Vice Chairman of Tianjin Urban Planning Association, Urban Planning Association & Institute
Vice Chairman of higher education institutions for survey and Design branch of China Survey and Design Association

Thesis

"To Make the Construction in Line with the Psychological and Physiological Characteristics of the Old - the Composition and Design Study in Building Homes for the Aged in Japan"
"Corresponding Relations of children's Playing Behavior and Space"
"Design for the Yueyanglou Tower and Its Environs"
"A Plan for the Tianjin traditional Cultural and Commercial District"
In 2005, "Color Planning of Yantai"
In 2005, "Urban Design Study of Yantai"
In 2006, "A Comparative Study on the Practical Creativity in Metropolitan Urban Planning Drafting System in China in the Transitional Period"
Translation work: World urban residential quarters—Japan Volume
Translation work: Study on ecological sustainable-development community in residential area for state education committee
Work Tianjin Architecture Style (associate editor, executive editor)
Be absorbed into the book Dumon +78 talented designers

Main Awards

"The evolution of the Tianjin urban form and future urban character " project, was awarded the Second Prize for 2002 Tianjin scientific and technological progress
The planning and design "Memorial Park of Northern University - Tianjin Northern Garden", was awarded the Third Prize for 2001 National Excellent Planning and Design
The First Prize for Tianjin Meijiang Living Area Planning Design Competition
The second prize for Tianjin Chunhua Lane residential area planning and designing competition
The Second Prize at provincial and ministerial-level for "Yantai overall urban design"
The Second Prize at provincial and ministerial-level for "Beijing Yizhuang Economic and Technological Development Zone Development Plan"
"Urban Design for Xiangluo Bay Business District" was awarded the Second Outstanding Design Award in Tianjin
The building 25 of Tianjin University
In 2005, was awarded the First Prize for outstanding design by Ministry of Education in 2008, the Second Prize for outstanding design by Ministry of Construction National and the Bronze Prize for National Excellent Engineering Survey and Design
Weifang Chinese medicine hospital was awarded the Third Prize for 2008 National Outstanding Survey and Design

天津大学建筑设计规划研究总院
Academy Of Architectural Design & City Planning Tianjin University

彭一刚
Yigang Peng

彭一刚

教授
1932年9月生于安徽合肥市

博士生导师
中国科学院院士
国家勘察设计大师
天津大学建筑设计规划研究总院名誉院长

1953年毕业于天津大学土木工程系，国务院学位委员会第四届学科评议组建筑学学科召集人。

经彭一刚教授设计的建筑有天津大学建筑馆、天津水上公园熊猫馆、山东刘公岛甲午海战馆、平度市现河公园、伦敦中国城等。山东甲午海战馆个性鲜明，并富有深刻的历史文化内涵，深受广大群众的喜爱，得到建筑界的广泛赞誉。彭一刚教授在尽心教书育人、设计创作的同时，还致力于建筑理论的研究，由他撰写的《建筑空间组合论》获得国内外专家学者的高度评价，此外他还撰写了《中国古典园林分析》、《传统村镇聚落景观分析》、《创意与表现》等6部专著，先后获得首届全国优秀建筑图书一等奖、全国优秀科技图书二等奖、国家教委科技进步二等奖和北方十省市科技图书一等奖。在国内外曾发表学术论文60余篇，被学术界认为处于建筑理论研究的前沿。

Yigang Peng

Professor
He was born in September 1932 in Hefei, Anhui province

Doctoral supervisor
CAS academician
National Survey and Design Expert
Honorary president of academy of architectural design & city planning, Tianjin University

As architecture subject convener of the 4th Subject Assessing Group affiliated with State Council academic Degree Committee, he graduated from department of Civil Engineering in Tianjin University in 1953.
Professor Peng Yigang has designed a number of buildings such as Architecture Museum of Tianjin University, Panda House in Tianjin Water Park, Jiawu Sea Battle Museum in Shandong Liugong Island, Xianhe Park in Pingdu city, London Chinatown and so on. With the unique personality and profound historical and cultural connotations, Sea Battle Museun in Shandong is favored by lots of people and has been praised by the architectural field. Professor Peng devotes to teaching and educating people as well as design and creativity. Meanwhile, he has also committed to researching the architectural theory. "Architectural Spatial Combination Theory" wrote by him is highly evaluated by experts and scholars at home and abroad. Moreover, 6 works are written including Analysis of Chinese Classical Gardens, Landscape Analysis of Traditional Village Settlement, Creation and Representation and so on. They are awarded the First Prize for the 1st national excellent architectural book, the Second Prize for national excellent scientific & technological books, the Second Prize of scientific and technological advancement by national education committee and the second prize of scientific & technological book for Northern Ten provinces. He has published academic papers of more than 60 at home and abroad, and his academic research is considered being in the forefront of architectural theory.

TSDI
铁道第三勘察设计院集团有限公司

周铁征
Tiezheng Zhou

周铁征

1989年毕业于天津大学建筑系，国家一级注册建筑师，历任铁三院建筑院工程师、高级工程师、副总工程师、总建筑师。

荣誉

获得茅以升铁道工程师奖，获得2006年度天津市“五一”劳动奖章和天津市劳动模范称号，
2007年获得全国知识型职工称号，
2009获全国“五一”劳动奖章。

设计经验

承担青藏铁路拉萨河大桥建筑风格的设计，并主持了北京南站、天津火车站、上海虹桥站、天津西站等一批重大的铁路建筑设计项目。

Tiezheng Zhou

National Class 1 Registered Architect
He graduated from Department of Architecture of Tianjin Universityin 1989.
He successively holds positions as engineer, senior engineer, vice-chief engineer and chief engineer in No.3 Iron Architectural Institute

Awards

Mao Yisheng Railroad Engineer Award
Tianjin May 1 Labor Medal, 2006
The honorary title of Tianjin labour model, 2006
The honorary title of National Knowledgeable Worker, 2007
National May 1 Labor Medal, 2009

Working Experience

He designs for Lhasa River Bridge of Qinghai-Tibet Railway
He organizes many large-scale projects for railway architectural design including Beijing South Railway Station, Tianjin Station, Shanghai Hongqiao Station and Tianjin West Railway Station.

新疆印象建设规划设计研究院有限公司
（原新疆建设规划设计研究院）

王元新
Yuanxin Wang

王元新

高级建筑师
全国注册高级环境艺术师
1958年3月生于新疆奎屯

王元新曾就读于西安建筑科技大学建筑学专业和兵团教育学院艺术设计专业，历任新疆奎屯五五新镇规划办主任、兵团农七师设计院建筑室主任、乌鲁木齐开发区设计院副院长兼总建筑师。

2001年至今，任新疆建设规划设计研究院院长、新疆印象建设规划设计研究院董事长、新疆建设职业技术学院学术委员、新疆城市规划协会常务理事、新疆勘察设计协会理事、中国环境艺术委员会专家委员会委员、喀什莎车县和阜康市人民政府顾问、中国旅游电视协会新疆分会成员、新疆印象建设沙龙发起人。

正如王元新本人在1988年中国人民大学出版社出版的《名人与普通人的格言》一书中发表的格言那样"建筑创作的灵魂在于求新，不要重复他人，也不要重复自己，不怕有缺点，就怕无特点"。多年来他一直在自己喜爱并从事的项目策划、城市规划、建筑设计、风景园林、旅游规划、艺术创作等复合性设计创作中执著地追求着、探索着。曾被奎屯市首届科学大会评为"优秀科技工作者"，1999年新疆建设厅授予其"新疆优秀中青年勘察设计工作者"称号(全疆首届评选共十八位获此殊荣)。2008年1月12日王元新在北京人民大会堂荣获"中国策划20年十大策划专家"称号，其代表作"达坂城王洛宾音乐艺术城"设计案例获"中国策划20年经典案例金奖"，主创的昌吉回民小吃街和莎车十二木卡姆故乡园两个项目获2009中国人居典范规划设计金奖。

由王元新主持设计的多项规划设计成果，获全国优秀设计奖三项、全国三等奖一项，获新疆三等奖三项。获兵团新农村方案竞赛一、二、三等奖共九奖的前八项奖，获乌鲁木齐新农村方案竞赛二、三等奖共八项。建筑设计作品分获兵团一、二等奖，新疆三等奖共三项。1997年设计的"库尔勒巴音小区规划与设计方案"，获全国"1997跨世纪住宅小区方案竞赛优秀奖"，方案中将太阳能在住宅坡屋顶上的应用，这属全国首创，其全套方案入选《全国跨世纪住宅小区方案图集》和《新世纪畅销住宅图集》中。

王元新一贯主张用东方智慧做人，用西方理念做事的观点，强调地域性、文化性、时代性，强调"七分产品研究，三分作品设计""新疆印象，不同创想"的设计观。曾先后发表学术论文近三十篇，其设计代表作有世界首个跨国合作区中哈霍尔果斯国际边境合作中心、昌吉科技馆、昌吉博物馆、达坂城王洛宾音乐艺术城、中哈两国通道标志塔、喀什莎车十二木卡姆故乡园、昌吉回民小吃街、阜康天池景区瑶池苑、沙湾大盘美食文化城、库尔勒梨园民街等。

Yuanxin Wang

Senior architect
Registered environmental artist
He was born in Kuitun Xinjiang province in March, 1958.

He studied architecture and art design in Xi'an University of Architecture and Technology and Bingtuan Education Institute. He has been the director of planning office in Wuwu new town in Kuitun Xinjiang Province, director of Research Institute in 7th division of Corps, the vice president and chief architect of Xinjiang Image Construct Planning Design Institute. From 2001, he holds many significant positions ,such as the president and the executive director of Xinjiang Image Construct Planning Design Institute , the academic committee of Xingjiang Vocation & Technical College of Construction, the director of Xinjiang Urban Planning Association, the director of Investigation and Design Association, a consultant of People's Government of Fukang city and Shache county. He is also a member of China Environmental Art and Design Alliance and Xinjiang branch of China Tourist TV association and the sponsor of Xinjiang Image Construct Salon.

Yuanxin Wang closely follows the motto declared in his book The Mottos of Celebrities and Common People, which is said as "the soul of innovate design relies on the pursuit of innovation; we should neither repeat the design of others nor ourselves; we should not fear the flaws, but fear the loss of characters." Over these years, he engaged in his favorite works of project planning, urban planning, architecture design, landscape architecture, travel planning, artistic creation and other composite design works. He was entitled as "the outstanding worker of science and technology" on the first science conference. In 1999, he was given the title of "the outstanding young worker in Xinjiang Investigation and Design"(only 18 people was honored in the first year) . On January 12th, 2008, Wang was awarded as the "top ten planning experts of China for 20 years" in great hall of the people. His representative work "Dabancheng WangLuobin Music City" won the gold metal of "Classical Design for 20 Years". The "Jichang Hui People Snacks Street" and "Shache 12 Muqam home park"designed chiefly by Wang won the gold metal of "China habitat model planning awards". The projects designed by Yuanxin Wang has won many prizes, including three national outstanding awards, one national third prize ,three third prizes of xinjiang autonomous region, the top eight awards out of nine in new rural planning competition organized by Corps, eight awards of the second and third prizes in planning competition organized by Urumqi. In 1997, the "Planning and Design of Korla Bayan community" won "the Excellent Prize on the Cross-century Habitation". The design initiatively provided that the solar energy can be installed on the roof of house. The design scheme was recorded into the Collective Drawings of Cross-century Habitations and Collective Drawings of New Century popular Habitations.

Yuanxin Wang holds a view that behaving should follow oriental wisdom while working should follow western ideology. He emphasizes that the design should consider three factors including the regions, cultures and spirit of times. He promotes principles of design such as "designer should pay 70% effort on research and 30% on designing", "Impression of Xinjiang, Difference in Creation". He has published 30 academic theses. His representative works include the first transnational corporative center---the cooperative center on Korgas China-Kazakhstan border, Changji science and technology centers, Changji museum, Dabancheng Wang Luobin Music City, the symbolic tower on the China-Kazakhstan channel, Shache Twelve Muqam home park, Jichang Hui People Snacks Street, Fukang Heavenly Lake Scenic and Jade Pool Court, Shawan Dapan food culture townand Korla Liyuan man street.

新疆四方建筑设计院有限公司
XINJIANG SI FANG INSTITUTE OF ARCHITECTURAL DESIGN

朱 飞
Fei Zhu

朱 飞

董事长
高级建筑师 一级注册建筑师
1978—1982年毕业于新疆农业大学水利工程系
1987—1988年毕业于苏州城建环保学院建筑系

主要建筑设计作品
人民广场联合办公楼（省级优秀设计一等奖，建设部二等奖）
新疆石化花展厅（兵团优秀设计二等奖，农业部三等奖）
华源博瑞新村（建设部广厦奖，建设部绿色2A示范工程）
新疆烈士陵园
新疆军区医院病房楼
新疆屯河大厦

Fei Zhu

Chairman of the Board
Senior architect first class registered architect
From 1978 -1982, he studied in and graduated from Department of Water Conservancy engineering in Xinjiang Agricultural University
From 1987-1988, he studied in and graduated from Architecture Department in Suzhou Institute of Urban Construction and Environmental Protection

Major architecture design works
People's Square united office building (won the First Prize of Provincial Excellent Design, the Second Prize of Ministry of construction)
Xinjiang Shihuhua exhibition hall (won the Second Prize of Corps' Excellent Design, the Third Prize of the Ministry of Agriculture excellent design)
Huayuanborui new village (won the Guangsha Prize of Ministry of construction, and the green 2 A model project of Ministry of construction)
Xinjiang Memorial Manorial Park
Xinjiang Military Command Hospital
Xinjiang Tunhe Plaza

刘 豫
Yu Liu

刘 豫

院总规划师
重庆大学建筑城规学院 城市设计专业硕士
从业以来先后主持设计数十个大中型项目，其中主要有
喀什老城区旧城改造与抗震防灾规划
库尔勒经济技术开发区总体规划（200平方公里）
乌鲁木齐经济技术开发区甘泉堡片区控规（28.3平方公里）
湖北永清片区城市设计国际咨询案
遵义红花岗片区城市设计（1.2平方公里）
昌吉西陇房产华尔兹音乐城项目规划（建设部地源热泵与太阳能示范工程）
南湖二期安居工程（建设部金奖小区）
特变阳光绿景小区（建设部A级住宅认证）
华源博瑞新村小区（建设部绿色2A节能示范工程，广厦奖）
五家渠沁园旅游度假区（省级优秀设计表扬奖）
米东公务员小区
康公务员小区

Yu Liu

Principal Planner of Xinjiang Sifang Architecture Design Institute
Chongqing Jianzhu University Master of urban design
He has conducted over 10 large and medium sized projects, including
renovation of the old town of Kashi City and the earthquake disaster reduction and prevention planning
Master plan of Korla economic development zone(200 km^2)
Regulatory plan of Ganquanpu District in Urumqi Economic Development zone (28.3 km^2)
International consulting case of urban design for Yongqing region in Hubei
urban design for Honghuagang region in Zunyi(1.2 km^2)
Planning of Waltz music town of Changji Xilong Real estate (Ground Source Heat Pump and solar energy model project of Ministry of construction)
Nanhu phase 2 comfortable housing Project (Gold Prize residence quarter of construction ministry)
Tebianyangguanglvjing residence district (A-class residence certificate)
Huayuanborui new village (won theGuang Sha Prize of Ministry of construction, and the green 2 A model project of Ministry of construction)
Qinyuan tourist & holiday resort in Wujiaqu(won the Honor Prize of Provincial Excellent Design)
Residence district for public officials in Midong
Residence district for public officials in Fukang

昆明市建筑设计研究院有限责任公司

简宇航
Yuhang Jian

简宇航

总工程师兼总建筑师
教授级高级工程师
国家一级注册建筑师
中国建筑学会资深会员
世界华人建筑师协会会员
香港建筑师学会会员
获全国建设系统劳动模范荣誉称号
云南省消防协会理事
昆明市建筑学会理事
昆明市政府规划局城市规划建设专家
昆明理工大学在职硕士研究生校外导师

经　历

1962年9月，出生于昆明市
1979年7月–1983年7月，就读于华南工学院（现华南理工大学）建筑学系
1983年9月至今，昆明市建筑设计研究院有限责任公司工作

工作经验

参加工作至今，先后完成各类规划、设计项目近百项，其中云南省中医学院附属医院门诊综合楼、中国人民银行云南省分行业务楼（金融大楼）、昆明供电局电力调度通信楼等均获得云南省优秀设计奖并有多篇论文在《云南建筑》、《建筑意匠》等学术杂志上发表，在学术会议上交流。

主要作品

云南省中医学院附属医院门诊综合楼
中国人民银行云南省分行业务楼（金融大楼）
昆明供电局电力调度通信楼
玉溪供电局电力调度通信楼
昆明市延安医院住院大楼、门诊大楼
滇池世界花苑居住区
和信花园

Yuhang Jian

Chief Engineer and Chief Architect
Senior Engineer of professor level
National Class A Registered Architect
Senior member of Architectural Society of China
Member of World Association of Chinese Architects
Member of Hong Kong Institute of Architects
Won the honorary title of National Model Worker of the Constructing System
Council member of Fire Protection Association of Yunnan Province
Council member of Kunming Academy of Architecture
Urban planning Construction Expert of Kunming Municipal Planning Bureau
Out-of-school Supervisor of on-the-job graduate students for Master's Degree in Kunming University of Science and Technology

Experience

September 1962, born in Kunming.
July, 1979-July, 1983, studying at Department of Architecture in South China Engineering College (the present South China University of Technology)
September, 1983 to present, working in Kunming Architectural Design Research Institute Limited Company

Working experience

Since beginning to work until present, he successively accomplished nearly a hundred programming and designing projects of all classes, among which the Outpatient Service Building of Affiliated Hospital of Yunnan College of Traditional Chinese Medicine, Business Building of Yunnan Branch of People's Bank of China (Finance Building), Power Switching Communication Building of Kunming Power Supply Bureau, and etc have all won Excellent Design Award of Yunnan Province. He has published many papers in academic journals such as "Yunnan Architecture", "Architecture Conception" etc., which have been interchanged in academic conferences.

Representative works

1.Outpatient Service Building of Affiliated Hospital of Yunnan College of Traditional Chinese Medicine
2.Business Building of Yunnan Branch of People's Bank of China (Finance Building)
3.Power Switching Communication Building of Kunming Power Supply Bureau
4.Power Switching Communication Building of Yuxi Power Supply Bureau
5.Inpatient Building and Outpatient Building of Yan'an Hospital of Kunming
6.Huayuan Residential District of Dianchi World
7.Hexin Garden

泛亞設計
PAN-ASIA DESIGN
Introduciton to Shiqiao Architecture & Design Institute Ltd.Co.,Panyu Guangzhou

李 伟
Wei Li

李 伟

国家一级注册建筑师、高级建筑师。1972年2月生于云南昆明。1994年毕业于云南工学院（现昆明理工大学）建筑学系，获学士学位。现为云南泛亚工程设计院有限公司（甲级）院长、设计总监。

李伟感言他有幸成长于云南这片经济相对落后但少数民族建筑文化富饶的沃土，这样的成长经历使他的设计作品始终坚持“民族传统，地方特色和时代精神”的创作方向。他主持并主创的项目在探索云南多民族建筑文化的继承和创新，特色房地产项目的策划、规划及城市更新与发展方面作了大量的研究和设计实践。瑞丽电影院（傣族）、丽江客运站（纳西族）、新平商城（彝族）、春之城、昆明东部客运站等多个项目等都成为地域特色鲜明的建筑作品，并历获各级设计奖项。

由他主创的小庙街商贸城和丽江客运站被收录到《中国当代优秀青年建筑师作品集》，并荣获“中国当代优秀青年建筑师”称号，在IUAES第十六届世界大会上荣获“中国建筑文化杰出贡献奖”，现任《云南建筑》编委、《时代建筑》、中国勘察设计协会民营设计分会、省勘察设计行业协会及省土木建筑学会理事等社会职务。

设计理念：

感悟生活，享受创作，品味建筑，快乐设计。

建筑创作是一个需要认清民情、分析地情、体会人情，再注入激情的过程。

主创设计项目：

昆明春之城
（200 000平方米，2005年建成，省优设计二等奖）
玉溪小庙街商贸城
（46 000平方米，2003年建成，省优设计二等奖）
丽江客运站
（9 000平方米，2006年建成，中国建筑文化经典示范工程）
水木江南
（16 000平方米，中国最佳宜居示范工程）
鑫城国际
（320 000平方米，在建，中国领袖地产十大国际品质楼盘金奖）
瑞丽市电影院
（2001年建成，省优设计三等奖）
玉溪文庙恢复重建
（2005年建成，省优设计三等奖）
新平商城
（53 000平方米，2004年建成，省优设计表扬奖）
石屏州衙老街
（13 000平方米，2008年建成，中国建筑文化经典示范工程）
昆明东大门—昆石主线收费站
（2001年建成）
昆明南大门—昆玉主线收费站
（1999年建成）
丽水天锦.翡翠湾
（180 000平方米，在建）
昆明东部客运站
（41 000平方米，在建）
金领时代
（170 000平方米，在建）

Wei Li

The national Class A registered architect, senior architect, was born in Kunming, Yunnan in February 1972. In 1994, he graduated from Department of Architecture in the Yunnan Institute of Technology (now Kunming University of Science and Technology), and granted bachelor's degree. At present, he is dean and design director of Pan-Asia Institute of Engineering Design Co., Ltd. Yunnan (Class-A).

He is fortunate enough to grow up in Yunnan---this piece of fertile land where economy is relatively backward, but architectural culture of minority is rich. Design works always insist on the creative direction of "national traditions, local characteristics and spirit of the times,". Projects which he hosts and mainly creates, has done extensive research and design practice on exploring the inheritance and innovation of multi-ethnic culture buildings in Yunnan, on the planning of characteristic real estate projects, on urban renewal and development. Rayli cinema (Dai nationality), Lijiang Bus Station (Naxi nationality), Xinping Mall (Yi nationality), Spring City, Kunming East Bus Station and other projects have become construction works with the distinctive regional characteristics, and have won design awards at all levels. Xiaomiao Street Hall and Lijiang Bus Station of which he is the principal designer are listed in *Works of Outstanding Young Contemporary Chinese Architects* and he won the "Outstanding Young Contemporary Chinese Architects". In the Sixteenth World Congress in IUAES, he won the "Outstanding Contribution Award for China Architectural Culture". Currently, he is editor of *YunNan Architecture*, *Time Architecture*, Design Survey and Design Institute of China branch of the private, Yunnan Provincial Survey and Design Association and director of provincial Civil Engineering Society and other social functions.

Design Concepts

Percept life, enjoy creation, taste architecture, design happily.

Architectural creation is a process which needs to recognize situation of the people, to analysis the geography situation, to understand human feelings, and then to infuse the passion.

Principal design projects

Spring City, Kunming
(200,000 m^2, completed in 2005, the second prize of provincial excellent projects)
Xiaomiao Street Hall, Yuxi
(46,000 m^2, completed in 2003, the second prize of provincial excellent projects)
Lijiang Bus Station
(9,000 m^2, completed in 2006, the classic demonstration project of Chinese architectural culture)
Shuimu Jiangnan
(16,000 m^2, the best livable demonstration project in China)
Xincheng International
(320,000 m^2, under construction, the Gold Prize for top ten international quality of China's leaders' real estate)
Rayli City Cinema
(completed in, the third prize of provincial excellent projects)
Restoration of Confucius Temple, Yuxi
(completed in 2005, the third prize of provincial excellent projects)
Xinping Mall
(53,000 m^2, completed in 2004, the praise award of provincial excellent projects)
Old Government Street, Shiping County
(13,000 m^2, completed in 2008, the classic demonstration project of Chinese architectural culture)
Kunming East door - Kunshi mainline toll station
(completed in 2001)
Kunming South door – Kunyu mainline toll station
(completed in 1999)
Lishui Green Bay,
(180,000 m^2, under construction)
Kunming East Bus Station
(41,000 m^2, under construction)
Gold-collar age
(170,000 m^2, under construction)

浙江省建筑设计研究院
Zhejiang Province Institute of Architectural Design and Research

施祖元
Zuyuan Shi

施祖元

院长（法人代表）
教授级高级工程师
国家一级注册结构工程师

浙江大学土木工程学系岩土工程专业博士
国务院政府特殊津贴专家
全国建设科技进步先进个人
浙江省有突出贡献中青年专家

中国建筑学会抗震防灾分会理事
中国建筑学会建筑结构分会理事
中国建筑学会建筑结构分会高层建筑结构专业委员会委员
中国建筑学会建筑施工学术委员会基坑工程专业委员会委员
《建筑结构》杂志编委
浙江省科学技术协会第八届委员会常委
浙江省土木建筑学会秘书长、常务理事
浙江省土木建筑学会建筑结构分会会长
浙江省土木建筑学会土力学与地基处理专业委员会委员
浙江省力学学会常务理事
浙江省力学学会岩土力学与工程专业委员会副主任委员
中国城市科学研究会绿色建筑与节能专业委员会委员

主要作品
1. 上海青少年素质教育基地——东方绿舟
2. 浙江省桐乡市邮电局综合大楼
3. 杭州西湖文化广场
4. 浙江省国际金融中心
5. 杭州剧院改扩建工程和西湖文化广场（基坑支护设计）

获奖情况
1. "高层建筑钢结构设计技术"2007年获浙江省科技奖二等奖
2. 浙江省标准《建筑地基基础设计规范》2006年获浙江省科技奖三等奖
3. "带撑双排桩式围护结构的性状研究"2008年获浙江省科技奖三等奖
4. "预制钢筋混凝土槽板组合H型桩应用研究"1997年获浙江省科技进步奖三等奖
5. "深基坑支护设计计算研究"1999年获浙江省科技进步奖二等奖
6. 浙江省标准《建筑基坑工程技术规程》2000年获浙江省科技进步奖三等奖
7. "建筑工程中应用地下连续墙的设计计算理论及施工技术研究"2004年获浙江省科技奖二等奖

Zuyuan Shi

Director (Legal Representative)
Professor-Level senior engineer
National 1st class registered construction engineer

Dr. of geotechnical engineering in Department of Civil Engineering, Zhejiang University,
Expert benefiting from the special allowance of Government of the State Council
Advanced Individual of Science and Technology progress of National Construction
Young expert with outstanding contributions in Zhejiang Province

Director of Architectural society of China SRDR branch
Director of Structure Branch of Architectural Society of China
Member of the Professional Committee for high-rise building structure of Architectural Society of China
Member of Professional committee of the foundation engineering of Construction Engineering Academic Committee of Architectural Society of China
"Building Structure" Journal Editorial Board
Member of the Standing Committee of the Eighth Commission of Zhejiang Science and Technology Association
The secretary-general and executive director of Institute of Civil Engineering in Zhejiang
President of Construction Structure Branch of Institute of Civil Engineering Structure in Zhejiang
Member of Soil Mechanics and Foundation Professional Committee of Institute of Civil Engineering in Zhejiang
Executive director of Institute of Mechanics in Zhejiang
Vice Chairman of Geotechnical Mechanics and Mechanical Engineering Committee of Institute of Mechanics in Zhejiang
Member of Green Building and Energy professional Committee of China Urban Studies Society

Major works
1. Quality Education Base for Shanghai youth - Oriental Land
2. Post Office Complex in Tongxiang City, Zhejiang Province
3. Hangzhou West Lake Cultural Square
4. Zhejiang International Financial Centre
5. The extension Project of Hangzhou Theater and the West Lake Cultural Square (excavation design)

Awards
1. "Design technology of Steel structure of high-rise buildings" won the second prize of Zhejiang Science and Technology Award in 2007
2. The standard of "Building Foundation Design Code" won the third prize of Zhejiang Science and Technology Award in 2006
3. "Traits studies of the bracing structure of double row pile with bolsters" won the third prize of Zhejiang Science and Technology Award in 2008
4. "Application research of H-piles of precast reinforced concrete slot board composition" won the third prize of Zhejiang Science and Technology Progress in 1997
5. "Design and calculation research of Deep Foundation" won the second prize of Zhejiang Science and Technology Progress in 1999
6. The standard of "Technical Specification for Building Foundation" won the third prize of Zhejiang Science and Technology Progress in 2000
7. "Design calculating theory of applying underground continuous wall and construction technology research in Construction" won Zhejiang Science and Technology Award in 2004

浙江大学建筑设计研究院
ZUADR Architectural Design and Research Institute of Zhejiang University

董丹申
Danshen Dong

董丹申

职务职称

设计院院长、总建筑师
建筑工程学院副院长
研究员、硕士生导师
国家一级注册建筑师

主要作品和获奖情况

1.浙江省电力生产调度大楼
中国土木工程詹天佑奖 (2009)
建设部优秀建筑工程设计二等奖 (2008)
中国建筑学会建筑创作大奖 (2009)
浙江省优秀工程设计一等奖 (2008)
2.中国井冈山干部学院
中国建筑学会建筑创作大奖 (2009)
浙江省优秀工程设计一等奖 (2006)
中国威海国际建筑设计大奖赛银奖
3.中国禄丰侏罗纪世界遗址馆
建设部优秀建筑工程设计二等奖 (2009)
中国建筑学会建筑创作大奖 (2009)
教育部优秀建筑设计一等奖 (2009)
4.象山丹城基督教堂
中国建筑学会建筑创作大奖 (2009)
5.浙江大学医药学院组团
建设部优秀建筑工程设计三等奖 (2008)
中国建筑学会建筑创作佳作奖 (2008)
教育部优秀建筑设计一等奖 (2007)
6.浙江大学紫金港校区实验中心
建设部优秀勘察设计三等奖 (2003)
中国建筑学会创作佳作奖 (2004)
教育部优秀建筑设计一等奖 (2003)
7.中国轻纺城职业技术训练中心
建设部优秀勘察设计三等奖 (1998)
教育部优秀建筑设计一等奖 (1998)
浙江杯优秀工程设计金奖 (1997)
8.上海大学图书馆
建设部优秀勘察设计三等奖 (2001)
教育部优秀设计一等奖 (2001)
9.绍兴县行政中心
浙江省优秀工程设计一等奖 (2004)
10.丽水电力生产调度中心
建设部优秀建筑工程设计三等奖 (2008)
浙江省优秀工程设计一等奖 (2007)
11.上海宝安大厦
浙江省优秀工程设计一等奖 (2005)
12.良渚文化村白鹭湾君澜度假酒店
浙江省优秀设计一等奖 (2009)
13.浙江省疾病预防控制中心
浙江省优秀设计一等奖 (2010)
14.杭州山水人家住宅小区
詹天佑大奖优秀住宅小区金奖 (2005)
15.杭州亲亲家园住宅小区（B区）
詹天佑大奖优秀住宅小区金奖 (2006)

Danshen Dong

Job title

President/chief architect
Vice-president of College of Civil Engineering and Architecture, Zhejiang University
Researcher/ supervisor of graduate student
National Class 1 Registered Architect

Main works and Awards

1. Zhejiang Electric Power Generation and Distribution Building
Tien-yow Jeme Civil Engineering Prize (2009)
Second Prize of Best Architecture Engineering Design by Ministry of Construction (2008)
Architecture Creation Award of Architectural Society of China (2009)
First Prize of Best Engineering Design of Zhejiang(2008)
2. China Executive Leadership Academy Jinggangshan
Architecture Creation Award of Architectural Society of China (2009)
First Prize of Best Engineering Design of Zhejiang(2006)
Silver Prize of China Weihai International Architecture Design Prix(2006)
3. The Chinese Lufeng Ruins Museum of Jurassic World
Second Prize of Outstanding Architecture Engineering Design of Ministry of Construction (2009)
Architecture Creation Award of the Architectural Society of China (2009)
First Prize of Best Architectural Design by Ministry of Education (2009)
4. Dancheng Church, Xiangshan
Architecture Creation Award of the Architectural Society of China (2009)
5. Architectures of Medical College of Zhejiang University
Third Prize of Best Architecture Engineering Design by Ministry of Construction (2008)
Best Architectural Creation of the Architectural Society of China(2008)
First Prize of Best Architectural Design by Ministry of Education(2007)
6. Laboratory Center of Zijingang Campus of Zhejiang University
Third Prize of Best Investigation & Design by Ministry of Construction (2003)
Best Architectural Creation of the Architectural Society of China(2004)
First Prize of Best Architectural Design by Ministry of Education(2003)
7. Professional Training Center of China Light & Textile Industrial City Group Co., Ltd.
Third Prize of Best Investigation & Design by Ministry of Construction (1998)
First Prize of Best Architectural Design by Ministry of Education(1998)
Gold Prize of Zhejiang Cup Best Engineering Design(1997)
8. Library of Shanghai University
Third Prize of Best Investigation & Design by Ministry of Construction (2001)
First Prize of Best Architectural Design by Ministry of Education(2001)
9. Administrative Center of Shaoxing County
First Prize of Best Engineering Design of Zhejiang(2004)
10. Lishui Electric Power Generation and Distribution Center
Third Prize of Best Investigation & Design by Ministry of Construction (2008)
First Prize of Best Engineering Design of Zhejiang(2007)
11. Shanghai Baoan Building
First Prize of Best Engineering Design by Zhejiang(2005)
12. Jun Lan Egret Bay Resort Hotel Hangzhou, Liangzhu Culture Residence
First Prize of Best Design of Zhejiang(2009)
13. Zhejiang Provincial Center for Disease Control and Prevention (CDC)
First Prize of Best Design of Zhejiang(2010)
14. Hangzhou Poetic Home Residential Community
Gold Prize of Best Residential Community of Tien-yow Jeme Award(2005)
15. Hangzhou Warm Town Residential Community (Area B)
Gold Prize of Best Residential Community of Tien-yow Jeme Award(2006)

HSS
嘉和设计
ARCHITECTS & PLANNERS

赵 曼
Mandy Zhao

赵 曼

总经理 设计总监
国家一级注册建筑师
高级建筑师
浙江省建筑师协会会员
中国勘察设计协会民营分会常务理事
"中华建设名家"称号获得者

赵曼毕业于浙江工业大学土木专业，在继续读取了重庆大学建筑系硕士学位后，出国深造于英国利物浦大学建筑学院，并最终获得了副博士学位。作为嘉和设计的创始人之一，赵曼主持设计了公司不同发展时期的多项设计作品。现担任公司的设计总监并负责统筹公司的发展方向。

赵曼在风情小城及地块园区规划、大型景观房产、商业酒店建筑、教育公共建筑等领域有二十年的实践经验。中都·青山湖畔、太湖旅游度假区规划、余杭图书馆、芙蓉国·沁园等均为其不同时期的代表作品。同时，定期在国内国际学习、工作和交流的经历中，也磨炼出其在建筑、规划、景观等领域可以信赖的专业眼光。赵曼以"自然、精致、简约"为设计风格，将西方先进规划设计理念和中国的具体实践相结合，其作品常常经过深层面的思考，追求卓越及可操作性。

赵曼善于敏锐地、创造性地把握设计方向，能够挖掘客户的真实需求，并将其模糊的思想创造细化成为符合相关需求的产品。同时，多年的设计经验与市场嗅觉，以及对复杂设计问题的全面分析能力，使其善于为各类商业开发的建设项目寻求实现价值的最佳途径。尤其是在各类投资项目的设计、策划、运营、管理方面颇有建树。

同时，赵曼极富探索精神，以智慧的言语和充满哲理的思想引领公司设计团队实现战略成长。对历史和社会、文化和艺术、经济活动和管理实践的深入研究和综合把握，使其对设计业务拥有独特的视角，令设计作品和相关工作在文化与市场的背景下，取得超过客户预期的价值。

Mandy Zhao

General Manager Chief Designing Officer
National Class I Certified Architect
Senior Architect
Member of Zhe Jiang Achitect Association
Executive Director of Private Branch of China Survey and Design Association
Winner for the title of "The Celebrated Chinese Architect"

Mandy Zhao graduated from Civil Engineering faculty of Zhe Jiang University of Technology, and then got Master's degree in Architecture faculty of Chongqing University, and then took further study in the Architecture School of British Liverpool University, and finally got Vice-doctor's Degree. As one of the founder of Jia He Design, Engineer Zhao has held and designed many design works in the different development periods of the company, which has realized periodic development. She is the chief designing officer now and is responsible for making overall plan for the development direction of the company.

Engineer Zhao has twenty years' practical experience in the amorous little town and massif garden plan, large-scale landscape estate, commercial hotel architecture, academic and public architecture, etc. At the same time, the experience of periodic national and international study, work and communication has made her the professional insight that can be trusted in the field of architecture, planning and landscape, etc. Engineer Zhao takes "natural, exquisite, concision" as her design style, and she has combined the western advanced layout design idea and national specific practice, and her work often has had deep consideration, seeking excellence and operability.

Mandy Zhao is keen on incisively and creatively grasping the design method and she can find the real requirements of the clients, and then create and refine its abstract idea into the products that accords with related requirements. At the same time, her years' design experience and marketing investigation, and the complete analysis ability for complicated design problems, makes her good at seeking the best way for the architecture project of every kind of commercial development to realize its value. She has especially great achievement in the design, planning, operation, management of every kind of investment project.

At the same time, Mandy Zhao is full of exploratory spirit and she has led the design team of the company to realize strategic growth with her wisdom words and philosophic idea. The deep study and comprehensive understanding of history and society, culture and art, economic activity and management practice makes her the special insight for design operation. All these elements have made her design works and related work, under the culture and market background achieve the value that beyond clients' expectation.

浙江省现代建筑设计研究院
浙江现代投资项目管理咨询有限公司
ZMA
Zhejiang Modern Architectural Design & Research Institute

李 晨
Chen Li

李 晨

男 1972年出生 杭州人
国家一级注册建筑师 高级工程师
1995年毕业于华中科技大学建筑学专业
现任浙江省现代建筑设计研究院有限公司董事长、院长
多年来从事医疗建筑设计工作，具有丰富的设计经验和良好的工程业绩，多个项目获得过钱江杯等各类优秀设计奖项，曾在各种专业杂志上发表了多篇论文。

主要作品

浙江大学附属妇产科医院病房大楼
浙江省新昌县急救中心
浙江省宁波市鄞州人民医院新建工程
福建省闽东医院病房大楼
浙江大学附属第二医院脑科中心
国际保健中心
浙江省温州市瑞安人民医院病房楼
山东省潍坊市公共卫生服务中心总体规划及单体
浙江省人民医院病房综合楼
江苏省泰兴人民医院门诊病房综合楼
河南省洛阳正骨医院综合大楼
河南省妇保病房综合大楼

Chen Li

Male,was born in 1972 in Hangzhou
National Class I Registered Architect and Senior Engineer
In 1995, he graduated from Architecture Department of the Huazhong University of Science and Technology
He is the chairman and president of Zhejiang Xiandai Architectural Design & Research Co., Ltd.
Having engaged in design for medical building for many years, he has accumulated rich design experience and made good performance.
Among these, several projects won prizes of Outstanding Design in "Qianjiang Cup" competition. Mr. Li also published many theses on professional magazines.

Representative works

Ward building of Affiliated Obstetrics and Gynecology Hospital of Zhejiang University
Emergency center in Xinchang of Zhejiang Province
Zhejiang Ningbo Yinzhou People's Hospital
Ward building of Fujian Mindong Hospital
Cerebral department of No.2 Affiliated Hospital of Zhejiang University
International Health Care Center
Ward building of Zhejiang Wenzhou Ruian People's Hospital
Overall planning of Shandong Weifang public health service center and its entites
Rebuilding project for comprehensive ward building of Zhejiang People's Hospital
Comprehensive outpatient and ward building of Jiangdu Tsizhou People's Hospital
Comprehensive building of Henan Luoyang Hospital of Bone-setting
Comprehensive ward building of Henan Maternal and Child Hygiene Hospital and etc.

浙江省建筑设计研究院
Zhejiang Province Institute of Architectural Design and Research

黄山农
Shannong Huang

吕正中
Zhengzhong Lv

魏　强
Qiang Wei

姚之瑜
Zhiyu Yao

黄山农

主任建筑师
教授级高级建筑师
高级规划师
国家一级注册建筑师
主要作品
杭州东方金座
杭州吴山商城
杭州紫庭花园
天台中央花园
杭州西湖国贸中心
厦门聚泰商业城
海宁行政中心

吕正中

教授级高级工程师
高级规划师
国家一级注册建筑师
主任建筑师
主要作品
1.上海九海广场（瑞安大厦）（已建）
2.上海徐汇体育活动中心（斯波特酒店）（已建）
3.上海市普陀区区级机关综合办公楼群
4.浙江工商大学下沙校区
5.上海浦东机场国航工作区
6.杭州六合房产天寓项目
7.厦门观音山国际商务运营中心启动区（已竣工）
8.上海普陀区图书信息中心（图书馆）（已竣工）
9.亚洲包装中心杭州亚包大厦（在建）
10.中华航空大厦（在建）

魏　强

建筑学硕士
高级建筑师、高级规划师
国家一级注册建筑师
主要作品
1. 西湖文化广场
2. 中国杭州黄龙饭店改扩建工程
3. 湖州市规划大楼
4. 中国工商银行湖州市分行
5. 湖州市双子大厦
6. 海宁体育馆
7. 舟山市体育馆
8. 四川省青川县 青川中学
9. 浙江同济科技职业学院
10. 银泰海威国际住宅

姚之瑜

同济大学硕士
国家一级注册建筑师
主任建筑师
主要作品
1.绍兴市图书馆
2.嘉兴市图书馆
3.山东淄博图书馆
4.杭州地铁滨康站综合体（与澳大利亚爱伦事务所合作）
5.阿里巴巴总部办公楼（与澳大利亚Hassell事务所合作）
6.台州市中心医院
7.恩泽医疗中心（与日本设计合作）
8.温岭市医疗中心
9.嘉兴中港名都大酒店

Shannong Huang

Director Architect
Professor-level Senior Architect
Senior Planner
National 1st class registered architect
Major works
Oriental Gold Tower, Hangzhou
Wushan Mall, Hangzhou
Purple Court Garden, Hangzhou
Tiantai Central Garden
West Lake International Trade Center, Hangzhou
Juthai Commercial City, Xiamen
Haining Administrative Center

Zhengzhong Lv

Professor-level senior engineer
Senior Planner
National 1st Class Registered Architect
Director Architect
Major works
1. Shanghai Nine Seas Plaza (Ryan Mansion) (built)
2. Shanghai Xuhui Sports Centre (SPOT Hotel) (built)
3. Administrative Office Buildings in Putuo District, Shanghai
4. Xiasha Campus, Zhejiang University of Commerce and Industry
5. Air China work area of Pudong Airport in Shanghai
6. Tianyu project of Liuhe Real Estate, Hangzhou
7. Starting area for International Business Operation Center of Guanyin Mountain, Xiamen (completed)
8. Book Information Center in Putuo District, Shanghai (Library) (completed)
9. Hangzhou Yabao building of Asian Packaging Center (under construction)
10. China Air Mansion (under construction)

Qiang Wei

Master of Architecture
Senior Architect/ Senior Planner
National 1st class registered architect
Major works
1. Culture Square of Westlake
2. The extension project of the Dragon Hotel in Hangzhou, China
3. Planning building in Huzhou City
4. Huzhou Branch of Industrial and Commercial Bank of China
5. Twin Towers in Huzhou City
6. Stadium of Haining
7. Stadium of Zhoushan
8. Qingchuan high school in Qing chuan County, Sichuan
9. Tongji Vocational College of Science and Technology, Zhejiang
10. Yintai Hiwell International Residential Community

Zhiyu Yao

Master of Architecture of Tongji University
National 1st Class Registered Architect
Director Architect
Major works
Shaoxing Library
Jiaxing Library
Zibo Library, Shandong
Building complex of Binkang Metro Station, Hangzhou (Cooperation with ArenQuest-QUT, Australia)
Headquarter Office Buildings of Alibaba (Cooperation with Hassell firm, Australia)
Center Hospital in Taizhou,Zhejiang
Enze Medical Center ,Zhejiang (Cooperation with NihonSekkei, Japan)
Medical Center in Wenling,Zhejiang
Zhonggang Mingdu Grand Hotel, Jiaxing

卓创国际工程设计有限公司
International Creative Engineering & Design Co.,Ltd

卢 菁
Jing Lu

卢 菁
建筑硕士
英国特许建造师
高级工程师

部分作品
安踏百年树
东和冉家坝城市广场
昆明红星美凯龙
南宁汇东一号（超高层）
天津红星美凯龙
重庆新闻出版智能大厦（超高层）
南方君临酒店
重庆江区行政中心
重庆巴蜀鲁能中学
东方公馆
晋愉岭海
橡树澜湾
光华可乐小镇

Jing Lu
Master of Architecture
UK Chartered Builder
Senior Engineer

Major Designs
Anta Century Tree
Ranjiaba Urban Plaza, Donghe
Kunming Meikailong
Nanning Huidong No.1 (super high-rise)
Tianjin Meikailong
Intelligent Building of Chongqing Press and Publication (super high-rise)
Carlton Hotel
Chongqing Jiangjin Administrative Center
Bashu Luneng Middle School, Chongqing
Oriental Mansion
Forest Sea
The Oak's Villa
Guanghua Cola Town

梁 华
Hua Liang

梁 华
一级注册建筑师
注册规划师
管理学博士
现任卓创国际工程设计有限公司常务副总经理

部分作品
重庆东和春天
重庆同天绿岸
重庆广厦城
南宁中铁凤岭山语城
重庆维丰长寿湖
重庆融汇半岛
东和冉家坝城市广场
海南博鳌美丽熙海岸
南宁金源城
重庆合川芙蓉园

Hua Liang
Class-A Registered Architect
Registered planner
Doctor of Management
Liang Hua is the Deputy General Manager of ECD Engineering Design Co., Ltd.

Major Designs
Chongqing Donghe Spring
Chongqing Tongtian Green Shore
Chongqing Guansha City
Nanning China Railway Fenglingshanyu Town
Chongqing Weifeng Lake Longevity
Chongqing Ronghui Peninsula
Donghe Ranjiaba City Square
Hainan Boao Beautiful &Sunny Seashore
Nanning Jinyuan City
Chongqing Hechuan Furong Garden

董高翔
Gaoxiang Dong

董高翔
建筑学硕士
现任卓创国际工程设计有限公司上海区域执行总经理、设计总监

部分作品
武汉百瑞景中央生活区
成都龙湖三千里
上海玫瑰苑联排别墅
南昌金融街紫金园
上海湖北大厦办公酒店（125 米高）
上海佘山东紫园别墅
上海合生城邦别墅及商业（全国优秀工程华彩奖）
上海锦宝大厦办公楼及商业（100 米高）
美国加州大学教学楼抗震改造

Gaoxiang Dong
Master of Architecture
Dong is the executive manager and chief designer of Shanghai branch of ECD Engineering Design Co., Ltd.

Major Designs
Wuhan Bridge Living Capital
Chengdu Longhu Sanqianli
Shanghai Roes Garden Townhouse
Nanchang Finacial Street Violet Gold Garden
Shanghai Hubei Hotel & Office Building (125 m H)
Shanghai East Sheshan Violet Villas
Shanghai City-states Villas & Commercial Buildings(Huacai Award for National Excellent Projects)
Shanghai Jinbao Office & Commercial Building(100mH)
Earthquake resistant engineering and retrofitting of Academic Building of University of California

广州市景森工程设计顾问有限公司

林少斌
Shaobin Lin

林少斌

董事长、建筑事业部总经理
建筑设计高级工程师

主要作品
东莞移动通信综合楼
南宁永新区办公区
湖南省政府新办公区
获广东省优秀设计一等奖
广东省电视中心二期
湘天-长沙·奥林匹克花园
广东恒基-佛山·恒福新城
瑞盛-长沙·云栖谷
湛江中国城大酒店
广东恒基-佛山·恒福国际
衡阳 银泰红城（首期）
三一重工-长沙·榔梨培训学校
三一重工-深圳产业园
三一重工-昆山工业园

Shaobin Lin

Director, General Manager of Building Division
Senior Engineer on Architectural Design

Major Works
Multiple-use Building of Dongguan Branch of China Mobile
Working area in Yongxin District, Nanning
New working area of Hunan Provincial Government
Was awarded the First Prize for Excellent Design of Guangzhou
The second phase of Guangdong TV Center
Xiangtian- Changsha•Olympic Garden
Guangdong Heng Ji- Foshan•Hengfu New City
Ruisheng•Changsha•Yunqi Valley
Zhanjiang China City Hotel
Guangdong Heng Ji- Foshan•Hengfu International
Hengyang•Yintai Red City (the first phase)
Sany Heavy Industry – Changsha•Langli Training School
Sany Heavy Industry – Shenzhen Industrial Park
Sany Heavy Industry – Kunshan Industrial Park

侯则林
Zelin Hou

侯则林

董事、副总裁
建筑设计高级工程师
国家一级注册建筑师

主要作品
新世界芳村逸彩庭园
获2003年广州市优秀工程设计二等奖
南海国土局地籍资料库
获2005年广东省优秀工程设计三等奖
东莞愉景花园高层商住楼
获2008年广州市优秀工程设计一等奖
高明市地税局办公大楼
获2005年广东省优秀工程设计二等奖
珠江新城I7-4地块保利大厦
获2008年广州市优秀工程设计一等奖
广州深凌宇厂房一期
获2005年广州市优秀工程设计三等奖
珠江新城I2-5地块保利大厦
获2007年广州市优秀工程设计二等奖
保利琶洲国际广场
获2008年广州市优秀工程设计一等奖

Zelin Hou

Director, Vice President
Senior Engineer on Architectural Design
National 1st Class Registered Engineer

Major Works
New World Yicai Garden in Fang Village
Awarded the Second Prize for Excellent Engineering Design of Guangzhou in 2003
Cadastral Database of Nanhai Bureau of Land and Resources
Awarded the Third Prize for Excellent Engineering Design of Guangdong in 2005
High-rise business–living building of Yujing Garden in Dongguan
Awarded the First Prize for Excellent Engineering Design of Guangzhou in 2008
Office building of Gaoming Municipal Local Taxation Bureau
Awarded the Second Prize for Excellent Engineering Design of Guangdong in 2005
Poly Plaza on I7-4 Block of the Pearl River New City
Awarded the First Prize for Excellent Engineering Design of Guangzhou in 2008
The first phase of factory building for Guangzhou Shen Lingyu
Awarded the Third Prize for Excellent Engineering Design of Guangzhou in 2005
Poly Plaza on I2-5 Block of the Pearl River New City
Awarded the Second Prize for Excellent Engineering Design of Guangzhou in 2007
Poly Pazhou International Plaza
Awarded the First Prize for Excellent Engineering Design of Guangzhou in 2008

贾 涛
Tao Jia

贾 涛

董事
长沙公司总经理

主要作品
三湘置业-长沙·国会山观邸
致盛愿景-长沙·山水未来城
湘天-长沙·奥林匹克花园
瑞盛-长沙·云栖谷
双盈-长沙·卧龙湾
银泰置业-衡阳·银泰·红城（一期）
娄底众一-众一桂府
衡阳合创-珠江·愉景湾
湖南北美-北美·阳光城
德利华-长沙·德润园
景鹏-长沙·湘域西岸
亦成投资-长沙·青竹湖度假酒店
鑫远投资-湘府·文苑
鼎丰置业-新田东湖
汇一置业-长沙·紫晶城
泰禹实业-长沙·泰禹家园

Tao Jia

General Manager of Shangsha Company

Major Works
Sanxiang Property-Changsha•Capital Hill Residence
Zhisheng Vision-Changsha•Shanshui Future City
Xiangtian-Changsha•Olympic Garden
Ruisheng-Changsha•Yunqi Valley
Shuangying-Changsha•Wolong Bay
Yintai Property-Hengyang•Yintai•Red City (the first phase)
Loudi Zhongyi-Zhongyi Guifu Residence
Hengyang Hechuang- the Pearl River•Yujing Bay
Hunan Beimei-Beimei•Sun Town
Delihua-Changsha•Derun Garden
Jingpeng-Changsha•Xiangyu West Bank
Yicheng Investment- Changsha•Qingzhu Lake Holiday Hotel
Xinyuan Investment Group-Xiangfu Residence•Wenyuan Garden
Dingfeng Property-New Tiandong Lake
Huiyi Property-Changsha•Amethyst Town
Taiyu Industry Group-Changhsha•Taiyu Homestead

广州市景森工程设计顾问有限公司

袁 颖
Ying Yuan

何志坚
Zhijian He

张云峰
Yunfeng Zhang

袁 颖

董事、昆明公司总经理
建筑设计工程师
国家一级注册建筑师

主要作品
保利–广州 · 百合花园
获广州市2004年度优秀工程设计表扬奖
贵州都匀布依族苗族民俗博物馆
获广州市2004年度优秀工程设计二等奖
昆明冶金高等专科学校
花都机场花园
云南天乐时光园
西安财经学院新校区
南宁桃花源
云南天晖 · 嵩明商业城
广州三银大厦
华信–湛江 · 城市假日
番禺置业–金海岸花园
万科–广州 · 四季花城（二期）
湘天–长沙 · 奥林匹克花园
南宁罗马假日
云南澄江酒店
巨人地产–云南 · 财富中心

Ying Yuan

Director, General Manager of Kunming Company
Engineer on Architectural Design
National 1st Class Registered Architect

Major Works
Poly-Guangzhou·Lily Garden
Awarded the Mention Prize for Excellent Engineering Design of Guangzhou in 2004
The Folk Custom Museum of the Miao and Bouyei Nationalities in Duyun, Guizhou
Awarded the Second Prize for Excellent Engineering Design of Guangzhou in 2004
Kunming Metallurgy College
Huadu Airport Garden
Tianle Time Park in Yunnan
New campus of Xi'An University of Finance and Economics
Nanning Peach Fairyland
Yunnan Tianhui•Songming Commercial City
Guangzhou Sanyin Plaza
Huaxin-Zhanjiang•Urban Holiday
Panyu Property-Golden Coast Garden
Vanke-Guangzhou•Wonder Land (the second phase)
Xiangtian-Changsha•Olympic Garden
Nanning Roman Holiday
Yunnan Chengjiang Hotel
Ju Ren Estate-Yunnan•Fortune Center

何志坚

董事、建筑设计工程师
国家一级注册建筑师

主要作品
南海丽雅苑住宅小区
南沙国际公寓H、J、K栋（公寓）
保利百合花园（住宅小区）
获广州市2004年度优秀工程设计表扬奖
冼村村民住宅综合楼（高层住宅）
万科四季花城五期Ａ、Ｂ、Ｃ区（住宅小区）
别墅项目
获广州市2008年优秀工程设计二等奖
小高层项目
获广州市2008年优秀工程设计三等奖
万科四季花城六期Ａ、Ｂ区（住宅小区）
南沙富佳花园（住宅小区）
北海“黄金海岸”（别墅区）

Zhijian He

Director, Engineer on Architectural Design
National 1st Class Registered Architect

Major Works
Residential community of Liya Garden in Nanhai
Buildings H, J, K of Nansha International Apartment(apartment)
Poly Lily Garden (residential community)
Awarded the Mention Prize for Excellent Engineering Design of Guangzhou in 2004
Residential Mutiple-use building for Xian Villagers (high-rise residence)
The fifth phase of Districts A, B, C of Vanke Wonder Land(residential community), Villa project
Awarded the Second Prize for Excellent Engineering Design of Guangzhou in 2008
The small high-rise residential projects
Awarded the Third Prize for Excellent Engineering Design of Guangzhou in 2008
The sixth phase of Districts A, B of Vanke Wonder Land
Nansha Fujia Garden (residential community)
"Gold Coast" of the North Sea (Villa Area)

张云峰

园景事业部总经理

主要作品
大学城广东外语外贸大学
万科–广州 · 蓝山花园
保利–广州 · 紫微花园
广州科学城中心区
方圆–广州 · 公园一号公寓
恒大–广州 · 御景半岛
娄底众一–众一桂府
万达–昆明 · 滇池卫城
银泰置业–衡阳 · 银泰 · 红城
广西 · 贵港东湖岚湖嘉荷
云南 · 天乐时光园
湛江 · 骏景花园

Yunfeng Zhang

General Manager of Landscape Division

Major Works
Guangdong University of Foreign Studies in University Town
Vanke-Guangzhou•Blue Mountain Garden
Poly-Guangzhou•Ziwei Garden
Central Area of Guangzhou Academic City
Fangyuan-Guangzhou•No.1 Park Apartment
Hengda-Guangzhou•Royal Landscape Peninsula
Loudi Zhongyi-Zhongyi Guifu Residence
Wanda-Kunming•The Dianchi Lake Acropolis
Yintai Property-Hengyang•Yintai•Red City
Guangxi•Residential area of Lanhu Jiahe of East Lake in Guigang
Yunnan•Tianle Time Park
Zhanjiang•Junjing Garden

ART&ARCHITECTURE
广州艺筑建筑设计有限公司

范文峰
Wenfeng Fan

范文峰

总建筑师
1989年毕业于华南理工大学建筑学系。从业二十多年来，以其务实、专注的作风在行内获得无数好评。“设计无限可能”是他对建筑设计的基本理念和不变的追求，尤其擅长满足客户和使用者的需要，以灵活变通的设计手法，全方位满足业主期望。

主要建筑设计作品
美林湖畔花园
佛山奥林匹克花园
力迅上筑
万科天景花园
南京苏宁天润城

Wenfeng Fan

Chief Architect
He graduated from Department of Architecture of South China University of Technology in 1989. Having worked more than 20 years, he has won numerous praises, thanks to his pragmatic, focused style. "Endless possibilities of design" is his basic philosophy of architectural design and constant pursuit. In particular, he is good at meeting the needs of clients and users as well as the expectations of the owners by flexible design techniques.

Main architectural works
Merlill Lake Park
Foshan Olympic Garden
Lixun Building
Vanke Sky-view Garden
Suning Tianrun City, Nanjing

陈玉慧
Yuhui Chen

陈玉慧

华南建设学院　　建筑学

主要建筑设计作品
美林海岸
合景泰富瑜翠园
东莞世纪滨江豪园
佛山奥园

Yuhui Chen

South China School of Construction Architecture

Main architectural works
Merlill Coast
KWG Yucui Park
Century Riverside Villa, Dongguan
Foshan aoyuan

胡庆山
Qingshan Hu

胡庆山

西安建筑科技大学　　建筑学

主要建筑设计作品
合景泰富·科汇金谷
茂名名雅世家
万科天景花园
花都山水豪庭

Qingshan Hu

Xi'an University of Architecture and Technology School of Architecture

Main architectural works
KWG • International Creative Valley
Maoming Elegant Hereditary House
Vanke tianjing garden
Huadu shanshui court

徐海纯
Haichun Xu

徐海纯

广州市建筑工程学校　　建筑学

主要建筑设计作品
南海怡翠馨园
惠州山水华府
佛奥．棕榈园
佛山奥园

Haichun Xu

School of Architecture and Engineering, Guangzhou School of Architecture

Main architectural works
Nanhai yicui garden
Huizhou Sunford
Fo'ao Palm Garden
Foshan aoyuan

广州亚泰建筑设计院有限公司
Guangzhou Asiatai Architecture Design Institute CO.,Ltd

张鸿儒
Hongru Zhang

张鸿儒

职　　称：国家一级注册结构工程师
毕业院校：1986年毕业于华南理工大学结构专业
1997年12月被评为高级工程师

工程业绩
汕头市建设银行办公大楼
北京陶然宾馆
开平工业大厦
中山嘉路华花园高层商住楼
广州市传染病医院住院手术楼
广州光大银行大厦

Hongru Zhang

Title: National Class I registered structural engineer
Graduated from structure specialty of South China University in 1986; awarded senior engineer in December 1997

Projects
China Construction Bank Office Building in Shantou
Beijing Tao Ran Hotel
Kaiping Industrial Building
High-rise commercial and living building in Ka Rover Garden Zhongshan Road
Inpatient and Surgeon Building of Infectious Disease Hospital in Guangzhou
China Everbright Bank Building in Guangzhou

汪振扬
Zhenyang Wang

汪振扬

职　　称：国家一级注册建筑工程师
毕业院校：1991年毕业于华侨大学建筑学专业
2002年12月被评为高级建筑师
工程业绩
惠州市人民银行
广州国际电子大厦(中外合作项目)
珠海市体育中心训练馆
广州国际科贸大厦
封开县国家税务局办公楼

Zhenyang Wang

Title: National Class I registered architectural engineer
Graduated from Huaqiao University, majored in Architecture in 1991
Awarded senior architect in December 2002

Projects
People's Bank Building in Huizhou
Guangzhou International Electronics Tower (Sino-foreign cooperation program)
Zhuhai Sports Center Training Center
Guangzhou International Technology & Trade Building
State tax bureau building in Fengkai County

袁栩丰
Xufeng Yuan

袁栩丰

职　　称：高级建筑师
毕业院校：2001年毕业于广东工业大学城镇建设专业
2006年被评为中级建筑师
工程业绩
广园东香江花园（翡翠绿洲）规划设计
万科四季花城建筑方案设计
万科云山建筑设计
北海黄金海岸花园建筑设计
佛山恒福新城建筑设计

Xufeng Yuan

Title: Senior Architect
Graduated from Guangdong University of Technology, cities and towns major in 2001
Awarded Intermediate Architects in 2006

Projects
Planning and design of Guangyuandong Rivera Garden (Emerald Oasis)
Architecture Design program for Vanke Wonderland Garden
Architecture design for Vanke Yunshan
Architecture design for North Sea Gold Coast Garden
Architecture design for Foshan Hengfu New City

易　芹
Qin Yi

易　芹

职　　称：高级建筑师
毕业院校：2004年毕业于湖南大学建筑系　硕士学位

工程业绩
招商金山谷二期
招商金山谷四期合
江西余江活动中心
广东某大学图书馆投标
江西吉安行政中心

Qin Yi

Title: Senior Architect
Graduated from Hunan University in 2004　　Master of Architecture

Projects
The second phase of Investment Jinshan Valley
The fourth phase of Investment Jinshan Valley
Yu Jiang Activity Centre, Jiangxi
Tender for a Library of a certain university, Guangdong
Ji'an administrative center, Jiangxi

TSINGHUA-YUAN
深圳市清华苑建筑设计有限公司
TSINGHUA YUAN ARCHITECTURAL DESIGN LTD SHENZHEN
地 址ADD：深圳市南山区桃园路南苑新村4号楼
电话TEL：0755-26493788 传真FAX：0755-26079280
E-mail：qhy@qhy. sina. net 网址WEBSITE：www.thy.cn

唐志华
Zhihua Tang

唐志华

公司董事、总建筑师、副总经理
国家一级注册建筑师、高级工程师

教育背景
1984年 清华大学建筑系建筑学 工学学士
1989年 清华大学建筑系建筑学 工学硕士

工作经历
1989—2002年 深圳市建筑设计研究总院
2002—2010年 深圳市清华苑建筑设计有限公司
2007—2010年 中央美术学院客座教授

Zhihua Tang

Company Director, Chief Architect, Deputy General Manager
National 1st Class registered architect, Senior Engineer

Education background
In 1984, Department of Architecture in Tsinghua University, Bachelor of Engineering
In 1989, Department of Architecture in Tsinghua University, Master of Engineering

Work experience
1989-2002, Shenzhen General Institute of Architectural Design and Research
2002-2010, Tsinghua Yuan Architectural Design Ltd. Shenzhen
2007-2010, Guest Professor of Central Academy of Fine Art

林彬海
Binhai Lin

林彬海

公司董事、副总建筑师、所长
国家一级注册建筑师、高级工程师

教育背景
1988年 清华大学建筑系建筑学 工学学士
1990年 清华大学建筑系城市规划与设计 工学硕士

工作经历
1990—1998年 深圳市建筑设计研究总院第一设计院
1998—2010年 深圳市清华苑建筑设计有限公司
2009年4月 荣获〝深圳市勘察设计行业首届十佳提名青年建筑师〞

Binhai Lin

Company Director, Deputy Chief Architect, Director-General
National 1st Class registered architect, Senior Engineer

Education background
In 1988, Department of Architecture in Tsinghua University, Bachelor of Engineering
In 1990, urban design and plan in department of Architecture in Tsinghua University, Master of Engineering

Work experience
1990-1998, the 1st designing institute of Shenzhen General Institute of Architectural Design and Research
1998-2010, Tsinghua Yuan Architectural Design Ltd. Shenzhen
In Aprial 2009, he was awarded the 1st Nominated Top Ten Young Architects in Shenzhen Exploring and Planning Industry.

徐金荣
Jinrong Xu

徐金荣

副总建筑师、室主任
国家一级注册建筑师

教育背景
1985年 华中科技大学建筑系建筑学 工学学士
1990年 清华大学建筑系建筑学 工学硕士
2007年 中欧国际工商学院 工商管理硕士

工作经历
1995—1998年 中建国际（深圳）设计顾问有限公司
1999—2005年 深圳市建筑设计研究总院第三设计院
2005—2010年 深圳市清华苑建筑设计有限公司

Jinrong Xu

Deputy Chief Architect, Director
National 1st Class registered architect

Education background
In 1985, Department of Architecture in Huazhong University of Science and Technology, Bachelor of Engineering
In 1990, Department of Architecture in Tsinghua University, Master of Engineering
In 2007, China Europe International Business School, MBA

Work experience
1995-1998, CSCEC International (Shenzhen) Design Consultants Co., Ltd
1999-2005, the 3rd designing institute of Shenzhen General Institute of Architectural Design and Research
2005-2010, Tsinghua Yuan Architectural Design Ltd. Shenzhen

TSINGHUA-YUAN
深圳市清华苑建筑设计有限公司
TSINGHUA YUAN ARCHITECTURAL DESIGN LTD. SHENZHEN
地址ADD：深圳市南山区桃园路南苑新村4号楼
电话TEL：0755-26493788 传真FAX：0755-26079280
E-mail：qhy@qhy.sina.net 网址WEBSITE：www.thy.cn

张 曙
Shu Zhang

张 曙

总规划师、室主任
国家注册城市规划师

教育背景
1990年 重庆建筑工程学院建筑系城市规划 工学学士

工作经历
1990–1994年 无锡市规划设计院
1995–2005年 深圳市规划设计院
2006–2010年 深圳市清华苑建筑设计有限公司

Shu Zhang

Chief Planner, Director
National 1st Class registered urban planner

Education background
In 1990, urban design in Department of Architecture in Chongqing Institute of Architectural Engineering, Bachelor of Engineering

Work experience
1990-1994, Wuxi Institute of Planning and Design
1995-2005, Shenzhen Municipal Planning Design Institute
2006-2010, Tsinghua Yuan Architectural Design Ltd. Shenzhen

张旭奎
Xukui Zhang

张旭奎

副总规划师、所长
国家注册城市规划师、高级规划师

教育背景
1992年 重庆建筑工程学院风景园林 工学学士
1994年 重庆建筑工程学院风景园林规划与设计 工学硕士

工作经历
1994–1997年 北京中国城市规划设计研究院
1997–1999年 中国城市规划设计研究院深圳分院
1999–2010年 深圳市清华苑建筑设计有限公司

Xukui Zhang

Deputy Chief Planner, Director-General
National 1st Class registered urban planner, Senior Planner

Education background
In 1992, Landscape Architecture in Chongqing Institute of Architectural Engineering, Bachelor of Engineering
1994, Landscape Planning and Design in Chongqing Institute of Architectural Engineering, Bachelor of Engineering, Master of Engineering

Work experience
1994-1997, China Beijing Academy of Urban Planning Design
1997-1999, Shenzhen Branch of China Academy of Urban Planning Design
1999-2010, Tsinghua Yuan Architectural Design Ltd. Shenzhen

佛山市顺德建筑设计院有限公司
Introduction of Foshan Shunde Architectural Design Institute Co., Ltd

陈霖峰
Linfeng Chen

张小明
Xiaoming Zhang

潘 晓
Xiao Pan

陈霖峰

职务：副董事长、副院长、总经理
职称：一级注册建筑师、高级建筑师、高级规划师、香港建筑师学会会员

华南理工大学建筑设计专业研究生毕业，工学硕士，一级注册建筑师、高级建筑师、高级规划师，2005年获得第二届内地与香港建筑师互认资格。现任佛山市顺德建筑设计院有限公司副董事长、总经理，佛山市顺德区第十一届、第十二届政协委员，佛山市顺德区土木建筑学会副理事长，香港建筑师学会会员。1989年6月至今在佛山市顺德建筑设计院有限公司工作，1989年7月至1991年4月在设计院方案室工作，为方案主创人员，1999年开始任职副院长，2003年兼任总经理，全面负责公司的经营管理工作。多年主持建筑创作参加评比获奖颇丰，其主持设计的顺德一中高中部获2008年全国行业优秀工程设计三等奖。作为方案主创人员，顺德区郑裕彤中学获1995年省优三等奖。主设计第五届中国花卉博览会中心展馆及主入口"欢乐的蝴蝶"，获省优秀设计三等奖。在省优秀城乡规划设计项目评比中，"佛山市顺德一中高中部修建性详细规划"被评为"省优二等奖"。"佛山市顺德区迎接亚洲艺术节和省运会市容整治规划"被授予"表扬奖"。"佛山市顺德区行政服务中心"获"省优二等奖"和"市优一等奖"。近期他与美国 HMC Architects组成设计联合体设计的顺德区第一人民医院项目在国际竞标中获优胜奖并成功中标。积极推进企业改革创新，充分调动多方积极性，为公司的业务发展、质量管理和技术创新提供了有力支撑。2008年获中南建筑设计行业管理人才成就奖。

Linfeng Chen

Position: Assistant Chairman, Assistant Institute President,General Manager
Professional title:First-class registered architect, Senior architect Senior urban planner, Member of Hong Kong Architectural Institute

Graduated from South China University of Technology, major in architectural design, master of engineering, grade A registered architect, senior architect, senior urban designer, got the architect reciprocal qualification between inland and HongKong. At present hold the position of vice-chairman of directors, general manager of Shunde Architect Designing Institute Co. Ltd. in Foshan city, the 11th and 12th secession of Chinese political advisor of Shunde district, Foshan city, vice-president of Civil Construction Institute of Shunde District, Foshan city, the member of Hong Kong Institute of architect. From June, 1989 till now, he has been working in Shunde Architect Designing Institute Co. Ltd. in Foshan city; from July, 1989 to April, 1991, he worked as the chief programmer in program office of the Designing Institute; he became the vice-president in 1999, and held a current post of general manager, overall in charge of management and administration work of the company in 2003. He resided the architect designing to participate in the evaluations and rewarded a lot in over the years, Senior high school of Shude No.1 Middle School which he designed achieved the third prize of the National Engineering Design Competition; as a chief programmer, the design of Zhengyutong Middle school achieved the provincial third prize; he designed the central Exhibition Hall of the Fifth National Flower Show in China and main entrance "Happy Butterfly" as a chief designer, which win the provincial third prize of designing. In the Provincial Urban and Rural Planning designing evaluation, "the site plan of Shunde No.1 Middle School in Foshan city" is rated "the provincial second prize". "The control and planning of city appearance for the Festival of Asia Arts and Provincial Sport Meeting" is honored with "special mentions". "Administration Service Center of Shunde District in Foshan city" gained "the second provincial prize" and "the first civil prize". Recently, the project of "the First People's Hospital in Shunde District" which cooperated with American HMC Architects as a designing unity achieved the winning prize and won the bidding successfully in the international competitive bidding. He made vigorous effect to encourage the innovation of the company, fully aroused the mass's enthusiasm, and provided effective support for business development quality control and technical innovation of the company. He achieved the artistic achievement award of personnel management in Southern of China's Architect Designing Industry in 2008.

张小明

职务：一室主任
职称：一级注册结构师、二级注册建筑师、高级结构工程师

1989年6月至今在佛山市顺德建筑设计院有限公司工作，2001年开始任一室主任，同年年兼任总工程师。2006年考取华南理工大学博士研究生。

1997年参加全国一级注册结构工程师考试获通过，成为首批获得一级注册资质的结构工程师，其后通过了建筑结构高级工程师的职称评审，并于2008年获得第四届内地与香港结构工程师互认资格。参加工作以来，完成了多项顺德地区的标志性建筑的结构设计，有大跨度的空间结构、高层结构等。如顺德展览中心、陈村花卉世界、顺德彩虹桥、顺峰山公园主入口牌坊、喜来登酒店及海景公寓等。获得过国家优秀设计奖，省、市优秀设计奖，市科技进部奖多项，包括顺德一中高中部国家优秀设计叁等奖、第五届花卉博览会主展馆广东优秀设计叁等奖、顺德展览中心顺德科技进步奖叁等奖。在国家级杂志及会议上发表了多篇论文。

Xiaoming Zhang

Position Director of the program office 1
Professional title
Grade 1 registered structures engineer grade 2 certified architect senior structural engineer

Zhang Xiaoming worked in Shunde Architect Designing Institute Co. Ltd. in Foshan city from June, 1989 to now. He worked as the director of the program office 1 from 2001 held the office of chief engineer in 2006, and was admitted to South China University of Technology to get a doctor degree in the same year.
He passed the Grade 1 registered structures engineer examination in 1997 and became the first group of structural engineers who have the Grade 1 registered structures engineer certification. And then he passed the evaluation commendation of senior structural engineer. He got the fourth secession of architect reciprocal qualification between inland and HongKong. Since he took the job, he has accomplished many structural designing of symbolic buildings in Shunde area, including apace structures which have long span and high-rise structures, such as Shunde Exhibition Center Chencun Flower World,Shunde rainbow bridge the memorial archway of the main entrance of Shunfengshan Park Sheraton Hotel and Haijing department etc. He achieved many national provincial and civil designing prizes and civil technology advancement awards, including Senior High School Campus of Shunde No.1 Middle School which achieved the third prize of the National Engineering Design Competition the third prize of designing, the main exhibition hall of the Fifth National Flower Show the third prize of Shunde technology advancement award of Shunde exhibition center. Besides, he published many theses on national magazines and conferences.

潘 晓

职务：方案一室主任
职称：建筑师

资深建筑设计师，从事建筑设计专业多年，特别是对中国古园林建筑和岭南园林建筑都有长时间的研究和设计实践。尤擅长手绘。近三年参与规划和设计的代表性工程项目有其伦教文化广场、伦教文化大楼的建筑方案与建筑设计。参与大良河西岭南文化风情食街的规划与建筑方案设计、福建双岩寺的方案规划与设计。

Xiao Pan

Position:First scheme design office Director
Professional title:Architect

As a senior designer, he has worked on architectural design for many years, with emphasises on archaized building, Chinese courtyard, landscape building, and he was especially good at hand drawing. His representative works are: Lunjiao Culture Plaza; Lunjiao Culture Building; Daliang Culture Food Street; Fujian Shuangyan Temple.

佛山市顺德建筑设计院有限公司

Introduction of Foshan Shunde Architectural Design Institute Co., Ltd

蔡文锋
Wenfeng Cai

蔡文锋

职务：方案一室/副主任
职称：一级注册建筑师/建筑师

重庆建筑大学建筑学毕业，资深建筑设计师，特别在多层、高层民用建筑、大跨度公共建筑等建筑范围内积累了较多的设计经验。近三年设计的代表工程项目有顺德区龙江医院、顺德区大良医院、顺德区杏坛医院、顺德区联塑员工村、山东青岛澳龙国际钢铁物流城、广东南海穆天子山庄（四期）、广东同江医院立案设计。

Wenfeng Cai

Position:First scheme design office Assistant Director
Professional title:First-class registered architect

Graduated from Chongqing Architectural University, Senior designer, especially good at designing multi-storey building, and big span commonality building. His representative works are: Shunde Longjiang Hospital, Shunde Daliang Hospital, Shunde Xingtan Hospital, Shunde Liansu Employee Residential Building, Shandong National Steel City, Nanhai Mutainzi Villa District, and Guangdong Tongjiang Hospital.

林文津
Wenjin Lin

林文津

职称：一级注册建筑师、建筑师

华南理工大学建筑学毕业，资深建筑设计师，擅长大型商住综合项目、高标准酒店、高级居住小区的设计。近年具代表性的项目有顺德陈村花卉世界主展馆及主入口设计（获省优秀设计三等奖）、顺德新世界宜新路综合发展项目（由万怡酒店、新世界商业中心、及五栋近100米高的住宅楼组成）、顺德金茂华美达酒店（按五星标准设计）、龙江奥斯登堡豪庭住宅小区（19万多平方米）、乐从南北钢地块商住小区（14万平方米）、翠怡花园住宅小区、御祥豪庭、均安顺峰小学等。

Wenjin Lin

Professional title:First-class registered architect

Graduated from South China Science University Architectural Department, Senior designer, especially good at designing big commercial-residential project, high-class hotel, high-class residential district.
His representative works are China the Fifth Flower Exposition Exhibition Center - Guangdong Province the twelfth excellent design third prize, Shunde Xinshijie development project (include: Wanyi Hotel, Xinshijie Commercial Center and five residential buildings), Shunde Jinmao Huameida· Hotel, Longjiang Aosidengbao Residential District, Lecong Nan bei gang Residential District, Yi cui Residential District. Junan Shunfeng Elementary School.

罗振韶
Zhenshao Luo

罗振韶

职务：方案二室主任
职称：建筑师

近期主要业绩

华南理工大学建筑学硕士，资深建筑设计师。曾参与多个大型公建、学校及高级住区设计，积累了丰富的设计经验，对校园规划及住宅区规划设计有系统性研究与实践。

近期主要成果有顺德第一中学高中部规划建筑设计、高明职业技术学院校园规划建筑设计、顺德杏坛中心小学规划建筑设计、陈村花卉世界中心展馆建筑设计、湖北李宁工业园制衣及制鞋厂区规划建筑设计、顺德美的御海东郡住区规划建筑设计、顺德佛罗伦萨花园二期规划建筑设计、论文研究《基于价值分析的住区规划思路》。

Zhenshao Luo

Position:Second scheme design office director
Professional title:Architect

Brief introduction

Graduated from Wuxi Light Industry University Landscape Design Department, Bachelor Degree, Senior designer.
His representative works are The construction designing of Senior high school campus of Shunde No.1 Middle School; The construction designing of the Gaoming Professional Technology College campus; The construction designing of Xingtan Central Primary School campus of Shunde; The construction designing of Chencun flower world central exhibition hall; The construction planning and designing of dress making and shoes making site area of Lining Industrial Park in Hubei Province; The construction designing of Meidi Yunhaidongjun Residential Area of Shunde; The construction designing of the Second Phase of Fuluolunsi Park of Shunde; Thesis of "Thinking of Residential Planning Based on Value Analysis".

朱剑伐
Jianfa Zhu

朱剑伐

职务：方案三室主任
职称：建筑师

华南理工大学建筑设计专业研究生，资深建筑设计师。作为主创设计师，在住宅区规划设计、学校、会展中心等公共建筑设计方面积累了丰富的经验，擅长处理功能复杂的综合项目设计，能将功能、空间、形式与项目发展目标诸多关系进行有效平衡。设计作品得到市场的肯定，并多次获得嘉奖，其中省优秀设计三等奖2项，市优秀设计一等奖2项，二等奖3项。近期主持项目有阳江上东新城二期住宅区规划及建筑设计、肇庆中源名苑住宅区规划及建筑设计、新中源商务大厦、英德英州明珠广场项目、锦力办公大楼、乐从沙滘小学。

Jianfa Zhu

Position:Third scheme design office director
Professional title:Architect

Master' degree of architectural design of South China Science University, Senior designer. Especially good at designing residential district, school, exhibition center, and can balance the function, space, and appearance of the project. He has obtained two Guangdong province's third prizes, two Foshan City's first prizes, and three Foshan City's second prizes. His representative works are: Yangjiang Shangdongxincheng Residential District, Zhaoqing Zhongyuanmingdu Residential District, Xinzhongyuan Commercial Building, Yingde Yingzhou Mingzhu Plaza, Jinli Office Building, and Lecong Shajiao Elementary School.

中机国际工程设计研究院

罗　劲
Jin Luo

罗　劲

1990年毕业于湖南大学建筑系建筑学专业，现任院总建筑师。
罗劲工作室首席建筑师，中国建筑学会会员，湖南省土木建筑协会理事，湖南省建筑师学会理事，中国国机集团青年联合委员会委员，中华人民共和国住宅和城乡建设部优秀期刊《中外建筑》 杂志副主编，高级工程师，国家一级注册建筑师。2006年入选湖南美术出版社《德式门——建筑·景观·室内》设计新锐榜。

主要作品
长沙职教城机电职业学院规划及单体建筑
湖南省邵阳医专梅子井校区规划及单体建筑设计
百年名校长沙市明德中学新址搬迁项目
百年名校雅礼中学新建教学楼
岳阳纸业集团长沙总部规划及单体建筑设计
湖南天心实业发展有限公司新城国际公馆综合小区
五矿集团，湘潭嘉盛地产开发有限公司尊城小区项目
娄底市政府搬迁工程规划及单体建筑设计
湖南中烟工业有限责任公司技术中心大楼方案设计
武广客运高速长沙车站建筑概念设计

Jin Luo

In 1990, she graduated from Hunan University, Department of Architecture, majored in Architecture. Currently she serves as chief architect of Insititute, chief architect of Jin Luo studio, membership of China Architectural Society, Association of Civil Engineering of Hunan Province, Hunan Institute of Architects, membership of China State Machine Group Youth Joint Committee, associate editor of excellent journal " Architecture World" magazine hold by Ministry of Housing and Urban of The People's Republic of China, senior engineer, national Class I registered architect. in 2006, his design works are on the cutting-edge billboard of "German-style doors - Building. Landscape. Interior" published by Hunan Fine Arts Publishing House.

Main works
Planning and single building design of Changsha Electromechanical Vocational College
Planning and Design of single building of PlumWell Campus at Shaoyang Medical College
New site relocation Project of Century elite Mingde secondary school in Changsha
New Academic Building of Century prestigious Yali School
Changsha Headquarters planning and design of single building of Yueyang Paper Group
New city international Mansion residence of Hunan Tianxin Industrial Development Co., Ltd.
Respective City Community Project of Five Minmetals Group, Xiangtan Jiasheng Real Estate Development Co., Ltd.
Planning and design of single building of gov. elocation project, Loudi City
China Tobacco Hunan Industrial Co., Ltd. Technical Center Building Design
Concept design for Wuhan-Guangzhou high-speed passenger station building at Changsha station

吴　卓
Zhuo Wu

吴　卓

1992年毕业于湖南大学建筑系。
先后担任院副总建筑师，工程二所所长，2010年成立吴卓建筑工作室。主要专注于规划、建筑及城市设计领域的设计与研究。高级工程师、国家一级注册建筑师。

主要作品
凯达·香格里嘉园
中铁·水映加州团购别墅区
中地·凯旋城二、三期工程
益阳滨江·财富中心
长沙市总工会办公大楼
岳阳市国际会展中心
长沙黎托新城

Zhuo Wu

He graduated from Hunan University in 1992, Department of Architecture, and has served as vice general architect of the institute, then as head of No.2 Engineering Institute. In 2010, Zhuo Wu set up Architecture Studio; mainly focused on design and study of planning, architectural and urban design. He is Senior engineer, national registered architect.

Main works
Kaida • Shangri Greenville
China Railway • Group purchase villas of Shuiying Jiazhou
Phase II,III China Real Estate • Triumph city
Yiyang Riverside • Fortune Center
Union Office Building in Changsha
International Convention and Exhibition Center in Yueyang City
Changsha Lituo Metro

谭迎风
Yingfeng Tan

谭迎风

1989年毕业于武汉城市建设学院(现华中科技大学)。
现为院建筑一所建筑部部长，主要进行医院建筑的设计。国家二级注册建筑师。

主要作品
长沙市第八人民医院医疗综合楼
怀化市第一人民医院住院大楼
涟源市一人民医院门急诊综合楼
岳阳市一人民医院、珍珠山分院医疗综合楼一期、二期工程
安化县人民医院医疗综合楼
宁乡县人民医院办公楼、4号病房楼
湖南省亚华置业有限公司东一商业广场
湖南农业大学留学生公寓、第九教学楼、图书馆
怀化华天大酒店
星沙华天LOHO小镇
怀化市第二人民医院续建病房楼
平江县人民医院医疗综合楼

Yingfeng Tan

In 1989, She graduated from the Wuhan Urban Construction Institute (now Huazhong University of Science and Technology). Currently She serves as construction minister of architecture Institute. She mainly shoulders for the design of hospital buildings. She is national Class 2 registered architect.

Main works
Medical Building of People's Hospital VIII, Changsha City
In-patientdepartment Building of People's Hospital I, Huaihua City
Complex building of People's Hospital Emergency Department, Lianyuan City
Phase II of People's Hospital of Yueyang City, Medical Building 1 of Pearl Hill Branch
People's Hospital Medical Building, Anhua County
People's Hospital Office Building, Ward Building No.4, Ningxiang County
Commercial plaza, Hunan Province Yahua Properties Limited East
abroad student apartments of Hunan Agricultural University, ninth teaching building, library
Huaihua Huatian Hotel
Xingsha Huatian LOHO town
Continued construction Ward Building of Second People's Hospital, Huaihua City
People's Hospital Medical Building, Pingjiang County

唐　昊
Hao Tang

唐　昊

1995年毕业于长沙理工大学建筑学专业。
现为院建筑二所副所长。专长于医疗建筑与住宅建筑设计及项目的整体运作与实施。国家一级注册建筑师。

主要作品
南华大学附属第一医院外科住院楼
南华大学附属第二医院住院大楼、
南华大学附属第三医院门诊大楼
邵东县人民医院住院楼
洪江市人民医院整体设计
衡阳县人民医院住院楼
鑫天鑫城住宅小区 御景湾住宅小区 山城·明珠住宅小区
华润置地
张家界天利佳园住宅小区
一心花苑住宅小区

Hao Tang

In 1995, he graduated from the Changsha Institute of Technology, majored in architecture. Currently serves as the Deputy Director of second Institute of Architecture and specializes in medical buildings and residential architectural design, as well as the overall operation and implementation of the project. Registered architect at the national A level.

Main works
Surgery In-patient department Building, First Affiliated Hospital of Nanhua University
In-patient department Building, Second Affiliated Hospital of Nanhua University
Clinic Building, Third Affiliated Hospital of Nanhua University
In-patient department Building of People's Hospital at Shaodong County
The overall design of People's Hospital of Hongjiang City
In-patient department Building of People's Hospital of Hengyang County
XinTianxincheng residential area Royal Lagoon residential area Mountain City • Pearl residential district
China Resources Land
TianLi Homeland residential area in Zhangjiajie
Yixin Garden Residential District

中机国际工程设计研究院

曹 罡
Gang Cao

胡清波
Qingbo Hu

李晓蕾
Xiaolei Li

张先政
Xianzheng Zhang

曹罡

1999年从武汉城市建设学院（现华中科技大学）规划建筑系建筑学专业毕业，从事工业与民用建筑设计工作。
2005年担任建筑一所建筑部副部长，2010年担任建筑一所所长助理。国家一级注册建筑师。

主要作品
香格里·嘉园
维一星城·国际
中天广场
长沙市第八医院综合医疗楼
德普·美洲故事
中城·丽景香山二期
托斯卡纳二期
上海城二期
湘江雅颂居
中粮北纬28度

胡清波

2001年毕业于合肥工业大学建筑学系，建筑学学士。
2006年毕业于深圳大学建筑学院，工学硕士。方案主创建筑师。

主要作品
湖南建设银行档案馆
怀化、邵阳、岳阳、郴州雷达站
湖南理工学院综合实验楼
益阳华天大酒店贵宾楼
东莞松山湖益普生科技研发园
湖南省委党校学员住宿楼

李晓蕾

1997年获得浙江大学建筑学学士学位。
2005年获得日本横滨国立大学硕士学位。国家一级注册建筑师。
曾先后工作于江苏省建筑设计研究院，株式会社佐藤综合设计，株式会社日本设计，2009年进入中机国际工程设计研究院江苏分院，现为江苏分院总建筑师兼建筑所所长。

主要作品
响水绿岛小夜曲项目
深圳市龙岗中心医院门急诊综合楼
金鹰国际花园高层公寓
淮阴师范学院王营新校区规划
中海盐仓桥项目方案
日本霞关大厦广场改扩建
江苏省委党校新校区规划
淮阴师范学院理工实验楼
泰州经济开发区外商服务中心
南京地铁线网控制中心
淮安市淮海东路项目
金轮新都汇项目

张先政

1996年毕业，分配到长沙有色冶金设计研究院，先后在深圳福田设计院、深圳设计总院、戚杨设计师事务所等单位任职。2008年来到中机国际，奉行在实践中学习、在学习中提高、在提高中超越，做一个纯粹的建筑行者的理念。国家一级注册建筑师。

主要作品
深圳科技创新海岸服务中心
第一湾住宅小区
神龙湾酒店
中山市实验中学
深圳德福花园
深圳福永酒店
深圳星河Mall
长沙城南郡
水韵花都
先锋东外滩

Gang Cao

After graduation from Wuhan Urban Construction Institute (Huazhong University of Science & Technology), he has been engaged in industrial and civic architecture design works.
In 2005, he held the post of vice director of the architecture department in No. 1 institute and in 2010, he became the assistant of No. 1 institute chairman. He is national class I registered architect.

Main works
Shangri Jiayuan
Weiyixingcheng International
Zhontian square
Changsha No.8 Hospital Medical Building
Depu. America story
phase II of Zhongcheng lijingxiangshan
Toscana phase II
Shanghai town phase II
Xiangjiang yasong house
COFCO 28 degree north

Qingbo Hu

After obtained the bachelor degree of architecture in HeFei University of Technology in 2001.
he got the master degree of enginnering in The College of Architecture & Civil Engineering Shenzhen university in 2006 . Now he is the project creation architect.

Main works
China construction bank Archives in Hunan
Radar station of Huaihua, Shaoyang , Yueyang and Chenzhou
Integrated laboratory building in Hunan Institute of Science and Technology
VIP Building of Yiyang Huatian Hotel
Songshanhu Yipusheng science & technology research park in Dongguan
Party school of the provincial Party Committee dormitory building

Xiaolei Li

After obtained bachelor degree of architecture from Zhejiang University,Li got the master degree in Yokohama National University.Now Li is a national class I registered architect.
Having worked in companies of Jiangsu Architecture Design & Research Institute , Ingerosec Corporation and NIHON SEKKEI, Inc., Li joined the Jiangsu branch of China Machinery International Engineering Design & Research Institute in 2009. Now Li is the chief architect and architecture department director of Jiangsu branch.

Main works
Xiangshui green island serenade project
Outpatient and emergency building for Longgang center hospital in Shenzhen
Golden eagle international garden high-rise apartment
Huaiyin Teachers College Wangying new campus planning project
Zhonghai yancang bridge project scheme
Kasumigaseki Building Square Expansion project
New campus planning of Jiangsu Party School
Laboratory Building of Huaiyin Teachers College
Service Center for Foreign Investors in Taizhou economic development zone
Nanjing Metro System Control Center
Huaihai esat road project in Huaihai city
Jinlun Xinduhui project

Xianzheng Zhang

After graduation in 1996, Xianzheng Zhang was assigned to Changsha engineering and research institute of nonferrous metallurgy , then he worked in Futian design institute, Shenzhen General Institute of Architectural Design and Research and Qiyang design work studio. In 2008, he joined China Machinery International Engineering Design & Research Institute. He adheres to the philosophy of learning from the practice, improving in the study, surpassing in the improvement , to be a solid architecture practitioner.
He is national Class I registered architect.

Main works
Shenzhen science and technology new costal service center
The first Bay Residence
Shenlongwan Hotel
Experimental Middle School in Zhongshan City
Shenzhen Defu garden
Shenzhen Fuyong Hotel
Shenzhen XingHe Mall
Changshachengnan Mansion
Shuiyunhuadu
Xianfeng east bund

湖南城市学院规划建筑设计研究院
Urban Planning & Architectural Design Institute Of Hunan City University

陈咏明
Yongming Chen

曹 文
Wen Cao

龚树松
Shusong Gong

唐生伟
Shengwei Tang

陈咏明

学历：中南大学　本 科
职称：高级工程师
职务：湖南城市学院规划建筑设计研究院总工程师

Yongming Chen

Educational background: Central South University　Bachelor degree
Professional title: senior engineer
Post: Chief engineer in Urban Planning & Architectural Design Institute of Hunan City University

曹 文

学历：湖南大学　本 科
职称：高级建筑师
职务：湖南城市学院规划建筑设计研究院建筑室主任

Wen Cao

Educational background: Hunan University　Bachelor degree
Professional title: senior architect
Post: Director of architecture department in Urban Planning & Architectural Design Institute of Hunan City University

龚树松

学历：湖南城市学院　本 科
职称：工程师、国家二级注册建筑师
职务：湖南城市学院规划建筑设计研究院建筑室副主任

Shusong Gong

Educational background: Hunan City University　Bachelor's degree
Professional title: engineer, national Class 2 registered architect
Post: Vice director of architecture department in Urban Planning & Architectural Design Institute of Hunan City University

唐生伟

学历：湖南城市学院　本 科
职称：工程师、国家二级注册建筑师

Shengwei Tang

Educational background: Hunan City University　Bachelor degree
Professional title: engineer, national class 2 registered architect

新疆四方建筑设计院有限公司
XINJIANG SI FANG INSTITUTE OF ARCHITECTURAL DESIGN

付 晨
Chen Fu

付 晨

1961年3月出生新疆察布查尔县
职称：高级建筑师 一级注册建筑师
职务：新疆四方建筑设计院总建筑师
1983年毕业于新疆大学工民建专业，西安建筑科技大学建筑学进修，同济大学城市设计专业进修
1983–2004年新疆建筑设计研究院副总建筑师
2004–2008年乌鲁木齐建筑设计院副院长兼副总建筑师
2008至今新疆四方建筑设计院总建筑师
新疆医科大学一附院门诊楼（自治区优秀设计二等奖）
巴州人民医院急救中心（自治区优秀设计二等奖）
新疆师范大学体育馆（自治区优秀设计二等奖）

Chen Fu

Fu was born in Chabuchar,Xinjiang in March, 1961
Professional title: Senior architect Class I registered architect
Post: Chief architect of Xinjiang Sifang architecture design institute
Graduated from Xinjiang university, civil engineering major, got the get further education in architecture in Xi'an University of Architecture & Technology and urban design in Tongji University.
In 1983 he became the Vice Chief architect of Xinjiang Architecture Design Institute
2004-2008 Vice President and Vice Chief architect of Urumqi Architecture Design Institute.
From 2008, he took charge of the chief Architect of Xinjiang Sifang architecture design institute.
Outpatient building of the first teaching hospital of Xinjiang medical university (the second prize of excellent design in Xinjiang autonomous region)
BaZhou People's Hospital emergency center (the second prize of excellent design in Xinjiang autonomous region)
Gymnasium of Xinjiang Normal University (the second prize of excellent design in Xinjiang autonomous region)

曾益海工作室
长沙有色冶金设计研究院

曾益海
Yihai Zeng

曾益海

1989年毕业于重庆工程学院建筑学专业
1989至今在长沙有色冶金设计研究院工作
现任院总建筑师
国家一级注册建筑师
国家注册规划师
中国建筑学学会会员
香港建筑学学会会员
湖南省建筑学学会副理事长

主要奖项
湖南省2006年优秀工程设计一等奖
湖南省2006年优秀工程设计三等奖
湖南省2007年优秀工程设计一等奖
2008年被湖南省人事厅、建设厅评为“湖南省优秀勘查设计师”
湖南省2009年优秀工程设计三等奖

Yihai Zeng

1989 Bachelor in Architecture Department, ChongQing Architecture and Civil Engineering Institute
Since 1989, has worked in Changsha Engineering and Research Institute of Nonferrous Metallurgy
General Engineer of the Institute
National Class 1 registered architect(PRC)
National registered urban planner
Member of Architectural Society of China(ASC)
Member of the Hong Kong Institute of Architects(HKIA)
Deputy Director of Hunan Architect Association

Major Awards
The First Prize of Hunan Province Excellent Project Design 2006
The Third Prize of Hunan Province Excellent Project Design 2006
The First Prize of Hunan Province Excellent Project Design 2007
Awarded Hunan Excellent Survey Designer 2008
The Third Prize of Hunan Province Excellent Project Design 2009

东南大学周琦工作室

周 琦
Qi Zhou

周 琦

周琦1957年生，1982年获得南京工学院建筑结构工程学士学位，1990年获得东南大学建筑学硕士学位，2002年获得美国伊利诺理工学院建筑学博士学位。其间，在大连，南京，美国芝加哥从事建筑设计与研究工作，2003年回国。现为东南大学建筑学院教授，博士生导师，建筑历史与理论研究所所长。一级注册建筑师，中国建筑学会会员，世界遗产保护协会会员，国际建筑科学院（IAA）会员。

Qi Zhou

Dircetor of Research Institute of Architectural History and Theory, SEU

B. S.(NIT,China)1982;
M. Arch. (SEU,China)1990;
Ph. D. Arch.(IIT,USA)2002;
CAA; Ass. AIA; ICOMOS, IAA

東南大學建築設計研究院
Architecture Design & Research Institute of Southeast University

马晓东
Xiaodong Ma

马晓东

1982年9月—1986年6月 金陵职业大学系建筑专业毕业
1986年7月—1989年8月 金陵职业大学土木系助教
1989年9月—1992年2月 东南大学建筑系研究生毕业，获硕士学位
1992年3月至今 东南大学建筑设计研究院、高级建筑师、国家一级注册建筑师
现任：东南大学建筑设计研究院副总建筑师、硕士生导师、城市建筑工作室设计主持

Xiaodong Ma

September 1982-June 1986 graduated from Jinling Institute of Technology Bachelor of Architecture
July 1986-August 1989 Department of civil engineering in Jinling Institute of Technology teaching assistant
September 1989-Feburary 1992 Southeast University School of Architecture master of Architecture
From March 1992 Architectural design & research institute of Southeast University. Senior architect, national Class I registered architect
He currently is the chief architect in Architectural Design & Research Institute of Southeast University, Master student supervisor, Urban architectural lab design director.

曹伟
Wei Cao

曹伟

1989年9月—1993年6月 东南大学建筑系建筑学本科毕业
1993年9月—1996年4月 东南大学建筑系研究生毕业，获硕士学位
1996年4月至今 东南大学建筑设计研究院、高级建筑师、国家一级注册建筑师
现任：东南大学建筑设计研究院建筑创作研究所所长、副总建筑师

Wei Cao

September 1989-June 1993 Southeast University School of Architecture, bachelor of architecture
September 1993-April 1996 Southeast University School of Architecture, Master of Architecture
From April 1996 Architectural design & research institute of Southeast University, senior architect, national Class I registered architect
He currently is the director of building design creation institute in Architectural Design & Research Institute of Southeast University, Associate chief architect

高崧
Song Gao

高崧

1985年 毕业于东南大学建筑系，建筑学专业
1985—1996年 东南大学建筑设计研究院，建筑师
1996年 获东南大学建筑系硕士学位
1997—2000年 新加坡康建筑与都市设计事务所，高级建筑师
2000年至今 东南大学建筑设计研究院，高级建筑师、国家一级注册建筑师
现任：东南大学建筑设计研究院副院长、硕士生导师、城市建筑工作室设计主持、副总建筑师

Song Gao

1985 Southeast University School of Architecture, Bachelor of Architecture
1985-1996 Architectural Design & Research Institute of Southeast University, architect
1996 Southeast University School of Architecture, Master of Architecture
1997-2000 Kang Architecture and Urban Design Institute in Singapore, Senior architect
From2000 Architectural design & research institute of Southeast University.Senior architect, national Class I registered architect
He currently is the vice president of Architectural Design & Research Institute of Southeast University. Master student supervisor, Urban architectural lab design director, Associate chief architect

韩冬青
Dongqing Han

韩冬青

1980年9月—1984年6月 南京工学院建筑学专业本科，获工学学士学位
1984年7月—1988年8月 扬州工学院城镇建设系任教
1988年9月—1991年6月 东南大学建筑设计及其理论专业研究生，获工学硕士学位
1991年9月—1994年7月 东南大学建筑设计及其理论专业研究生，获工学博士学位
1994年9月—1996年12月 同济大学建筑与城市规划学院博士后流动站
1997年1月至今 东南大学建筑学院任副教授、教授
东南大学建筑设计研究院国家一级注册建筑师
现任：东南大学建筑学院教授、博士生导师
东南大学建筑设计研究院总建筑师、城市建筑工作室设计主持

Dongqing Han

September 1980-June 1984 Nanjing institute of Technology major in Architecture bachelor of engineering
July 1984-August 1988 taught in the department of cities and towns construction in Yangzhou Institute of Technology
September 1988-June 1991 Architecture design and the theory Master Candidate in Southeast University master of engineering
September 1991-July 1994 Architecture design and the theory ph. d. candidate in Southeast University doctor of engineering
September 1994-December 1996 Tongji University School of architecture and urban planning mobile postdoctoral centers
From 1997 Januray School of Architecture Southeast University associate professor, professor.
Architectural Design & Research Institute of Southeast University national Class I registered architect
He currently is the professor and PhD candidate supervisor in School of Architecture Southeast University.
Chief architect, urban architectural lab design director in Architectural Design & Research Institute of Southeast University

辽宁省建筑设计研究院
LIAONING PROVINCIAL BUILDING DESIGN & RESEARCH INSTITUTE

曹辉 总建筑师
Hui Cao Chief architect

1993年毕业于武汉工业大学建筑系，现任辽宁省建筑设计研究院总建筑师、教授级高级建筑师、国家一级注册建筑师，全国注册建筑师考试命题专家组成员，辽宁省土木建筑学会理事、建筑师分会常务理事、绿色建筑委员会常务委员。2007年被辽宁省总工会授予职业道德建设十佳标兵，获得辽宁省"五一"劳动奖章；2009年被授予"辽宁省直机关十大杰出青年"称号。

代表作品：沈阳市人民检察院综合业务楼、沈阳市人防应急指挥中心、鲁迅美术学院大连校区艺术博物馆、东北金融管理中心、四川安县文体中心、抚顺浙商国际商贸城、沈阳中海国际社区等多项大型工程项目。

杨欣刚 常务副总建筑师
Xingang Yang Deputy chief architect

1989年毕业于哈尔滨建筑工程学院建筑系建筑学专业。现任辽宁省建筑设计研究院执行总建筑师，国家一级注册建筑师，香港建筑师学会会员，辽宁省土木工程学会建筑师分会委员，辽宁省建筑标准设计委员会委员。

主要项目：鲁迅美术学院大连校区、营口万都大酒店、沈阳邮件处理中心、沈阳海关设施楼、沈阳光达大厦、锦州元都大酒店、沈阳万和阁国际酒店及公寓、营口万和阁大厦、辽阳市中级人民法院综合审判楼、法库客运站工程、铁岭人寿保险公司办公楼、辽宁省国家安全厅办公楼、辽宁工业展览馆改造等工程。

郭红晶 副总建筑师
Hongjing Guo Deputy chief architect

1987年毕业于合肥工业大学建筑系建筑学专业，获工学学士学位。现任辽宁省建筑设计研究院副总建筑师、国家一级注册建筑师、教授研究员级高级建筑师、中国建筑学会会员、辽宁省土木建筑学会建筑师分会理事、香港建筑师学会会员。

主要项目：沈阳新世界百货二期工程、2006沈阳世界园艺博览会百合塔、沈阳曼哈顿广场、沈阳河畔新城精装住宅、沈阳河畔公馆、中共辽宁省委党校学员大厦及学员宿舍、营口鲅鱼圈"昌宇星河湾"规划及建筑设计、营口鲅鱼圈"海港花园"规划及建筑设计、沈阳市监狱城迁建工程（沈阳第一监狱、第二监狱、女子监狱、新收犯监狱、监狱医院、武警营房等）、沈阳空港来登酒店、沈阳故宫博物院文物库等。

王红玮 副总建筑师
Hongwei Wang Deputy chief architect

1988年毕业于沈阳建筑工程学院建筑系建筑学专业（现沈阳建筑大学建筑与规划学院），同年7月到辽宁省建筑设计研究院参加工作。现任院副总建筑师、高级建筑师、国家一级注册建筑师。

主要项目：辽宁大学图书馆，获1997年度沈阳市优秀工程勘察设计三等奖、1998年度全国优秀教育建筑设计优秀奖；沈阳亿达·唯美品格住宅小区设计，获2003–2004年度辽宁省优秀工程勘察设计一等奖；何氏眼科沈阳绿谷生物技术观光园"孵化器"工程，获"沈阳市城乡建设委员会2007年度优秀工程设计勘察一等奖"；"2005–2006年辽宁省优秀工程勘察设计三等奖"等。

张卓明 副总建筑师
Zhuoming Zhang Deputy chief architect

1996年毕业于沈阳建筑工程学院建筑系，现任第二设计研究所所长、院副总建筑师，国家一级注册建筑师，高级建筑师。

主要设计项目：中国医科大学风雨操场、营口经济技术开发区中心医院（辽宁省优秀设计一等奖、沈阳市优秀设计一等奖）等十余项在建及建成医院工程，龙腾大厦（现日神广场）、沈阳华丽山庄（现曼哈顿庄园）、中国人民银行辽宁重点库、中润科技城、海韵大厦（现海润国际）、宁波科技园区I–3地块工程、御龙逸城等工程。

朱士壮 第四设计研究所副所长
Shizhuang Zhu Deputy director

2003年毕业于哈尔滨工业大学建筑学院，获建筑学学士学位。同年考取哈尔滨工业大学研究生，2005年硕士研究生毕业，获建筑学硕士学位。现任第四设计研究所副所长。

感言：

我喜欢具有雕塑感的建筑，也喜欢用创作雕塑的方式去创作建筑。如果建筑形态能够恰如其分地表达出内部空间的原则和秩序，并能有机地生长在基地环境之中，这就是一个优秀的建筑，至少可以说是合格的建筑。

建筑设计研究所
Architectural design institute

建筑设计研究所现有建筑设计师13人，规划设计2师人。该所主要负责大型公共建筑的方案投标以及施工图设计工作。该设计团队年轻、富有朝气，他们方案，概念构思新颖、设计的现代他们通过与国外知名建筑事务所合作投标中总结了一套与国外设计团队合作的经验，在多次大型建设项目投标中获得了优异成绩。

建筑方案所已承揽了省内外多项大型公共建筑的方案和施工图设计，其中建成的工程遍布辽宁各地，也有多个工程获得省、市优秀设计奖，其中，辽宁东北抗联史实陈列馆获得辽宁省及沈阳市优秀设计奖，鲁迅美术学院（沈阳校区）体育馆荣获辽宁省及沈阳市优秀设计奖。

自主设计的项目有东北大学自动化产业园、沈阳音乐学院教学楼、辽宁省城市建设学院、辽宁省孤儿学校、朝鲜族第六中学、抚顺平顶山惨案纪念馆、沈阳格林豪森创意大厦、沈阳市文化艺术中心、中国人民银行、中国防空博览园、辽宁体育训练中心（柏叶基地）等。其中规划项目有沈阳财落中心镇规划、沈阳兴隆台中心镇规划、沈阳清水台中心镇规划以及一些住宅和旅游度假村规划等。

建筑设计研究所通过对建筑文化的理解，认为建筑应该是很朴实的，是完全的几何形体，是具有空间和功能的。公共建筑没有花边，不搞怪异和时髦，但是必须尊重当地的文化。对文化的妥协是建筑与文化的碰撞，不是对商业的妥协。建筑设计讲求在多样中达到统一，在统一中创造多样。

孙皓 第四设计研究所所长助理
Hao Su Director assistant

1995年毕业于沈阳建筑工程学院，获建筑学学士学位。现任第四设计研究所所长助理。

感言：

建筑依赖于自己的时代，它是时代内在结构的结晶，显示出时代的面貌。这就是技术与建筑紧密结合的原因，我们的真正希望是二者结合在一起，到有一天，其中之一就是另一个的表现，只要到那时，我们才有值得称之为建筑的建筑，建筑才成为我们时代真正的标志。——摘自《建筑与技术》（密斯·凡·德·罗）

沈力源 建筑师
Liyuan Shen Director architect

2005年毕业于沈阳建筑大学，获筑学学士，同年考取西安建筑科技大学研究生，2008年获建筑学硕士学位。现任第四设计研究所建筑师。

感言：

arch（弧）+tech（技术）+ture（真实的）=architecture（建筑学），古意的建筑学是砌筑拱的技术，在我看来，现在的建筑学就是砌筑更美好生活的技术。

山东省建筑设计研究院第 6 分院

SHANDONG PROVINCIAL ARCHITECTURAL DESIGN & RESEARCH INSTITUTE—THE 6TH BRANCH

— 医用工程（专项）设计研究所

钱宁亚
Ningya Qian

钱宁亚

职务：六分院建筑总工
职称：高级工程师
一级注册建筑师

1996年东南大学建筑系本科毕业之后，就进入山东省建筑设计研究院工作，一直从事医疗建筑设计工作至今。先后参加了多个全国医疗建筑设计的投标，并在多省市取得中标项目。

曾参与完成的医院项目部分如下

江苏省人民医院外科病房大楼方案及施工图设计（该工程获得2005山东省优秀设计二等奖）；
江苏省人民医院传染楼方案及施工图设计；
泰安医学院附属医院门诊医技大楼方案及施工图设计；
宁波市医疗中心李惠医院住院楼方案及施工图设计（该工程获得2005山东省优秀设计一等奖，2006建设部优秀设计二等奖）；
河北乐亭县医院病房楼方案及施工图设计；
江苏省常熟市第二人民医院病房综合大楼方案及施工图设计；
安徽省望江县医院整体迁建工程方案及施工图设计；
山东省东营市人民医院门诊病房综合楼方案及施工图设计；
济南军区第401医院病房综合楼方案及施工图设计；
济南军区青岛第二疗养院康复大楼方案及施工图设计；
山东省陵县人民医院门诊病房综合楼方案及施工图设计；
山东省淄博昌国医院迁建工程方案及施工图设计；
山东省济南市第三人民医院病房楼设计。

Ningya Qian

Post: Senior engineer
Professional title: Senior engineer
National Class I registered architect

After graduation from architecture department of southeast university in 1996, Ningya Qian joined the Shandong provincial architectural design institute and engaged in the hospital architecture design works. He has attended lots of national hospital architecture design project biddings and awarded many tenders in different provinces and cities.

The hospital architecture projects he has worked on include

Jiangsu provincial hospital surgical ward building project and construction documents design (this project won the Second Prize in excellent design projects of Shandong province 2005
Jiangsu Provincial People's Hospital infectious disease building project and construction documents design
Hospital building of Taian medical university affiliated hospital project and construction documents design
Ningbo Medical Treatment Center lihuili hospital project and construction Documents design (this project has won the First Prize of Shandong Excellent Design Competition 2005, and the Second Prize of 2006 Excellent Design Competition of Construction Ministry)
Hebei Leting hospital ward building scheme and construction documents design.
Changshu No.2 people's hospital comprehensive ward building project and construction documents design
Anhui Wangjiang Hospital relocation project scheme and construction documents design
Shandong Dongying People's Hospital Out-patient Building Scheme and Construction documents Design
NO.401 hospital of Jinan military command scheme and construction documents design
Jinan military command the second convalescent hospital building project and construction documents design
Shangdong Ling county people's hospital out- patient building project and construction documents design
Shandong Zibo Changguo hospital relocation project and construction documents design.
Shandong jinan the third hospital ward hospital design

程　超
Chao Cheng

程　超

职务：六分院建筑部主任
职称：工程师

2004年本科毕业之后，就进入山东省建筑设计研究院工作，一直从事医疗建筑设计工作至今。先后参加了多个全国医疗建筑设计的投标，并在多省市取得中标项目。

曾参与完成中标的医院项目如下

辽宁本钢总医院外科综合楼方案及施工图设计；
内蒙古自治区医院内科大楼方案设计；
广州暨南大学附属第一医院病房楼方案及初步设计；
解放军总医院附属第二医院设计方案；
辽宁朝阳中心医院病房大楼；
解放军第四医院设计方案；
山东省中医药大学附属医院规划设计方案；
福建龙海市第一医院总体规划及单体设计方案；
烟台蓬莱中医院整体新建工程设计方案；
烟台开发区医院设计方案；
内蒙古鄂尔多斯中医院规划设计方案；
兰州军区总医院医技综合楼工程设计方案。

Chao Cheng

Post: Director of architecture department in Shandong Provincial Architectural Design Institute
Professional title: Engineer

After graduation in 2004, Chao Cheng joined the Shandong Provincial Architectural Design institute and engaged in the hospital architecture design works. He has attended lots of national hospital architecture design project biddings and awarded many tenders in different provinces and cities.

The hospital architecture projects he has worked on include

General Hospital in Bengang comprehensive building scheme and construction documents design
Inner Mongolia people's hospital building scheme design
Scheme and preliminary design of Guangzhou Jinan University No.1 affiliate hospital ward building
Second Affiliated Hospital of Chinese PLA General Hospital design
Ward building of Chaoyang central hospital in Liaoning
Design project of No.4 Military Hospital
Planning and design scheme of Shandong University of Traditional Chinese Medicine affiliated hospital
Master planning and single building design project of Longhai No.1 hospital in Fujiang Province
Reconstruction project of Fenglai traditional Chinese Medical Hospital
Scheme design of Yantai development district hospital
Erdos traditional Chinese medical hospital planning and design scheme
Comprehensive building project design scheme of general hospital of Lanzhou military command

山东省建筑设计研究院第 6 分院
SHANDONG PROVINCIAL ARCHITECTURAL DESIGN & RESEARCH INSTITUTE--THE 6TH BRANCH
— 医用工程（专项）设计研究所

徐 备
Bei Xu

徐 备

2003年本科毕业之后，就进入山东省建筑设计研究院工作，一直从事医疗建筑设计工作至今。先后参加了多个全国医疗建筑设计的投标，并在多省市取得中标项目。

曾参与完成的医院项目如下

江苏大学附属医院病房大楼方案及施工图设计，该工程获得全国十佳医院称号；
江苏省大丰市人民医院病房大楼方案及施工图设计；
无锡市第二人民医院病房大楼方案及施工图设计；
宜兴市中医医院城东新院方案设计；
江苏省江都市妇幼保健院保健楼方案设计；
山东省文登市中心医院门诊综合楼方案设计；
深圳市盐田区人民医院整体改造工程方案设计；
湖北省宜城市人民医院整体规划设计；
河北省保定市脑血管病医院门诊病房综合楼方案及初步图纸设计；
河南省精神病医院整体规划设计；
广东省中西医结合医院施工图设计；
河北省沧州市中西医结合医院病房大楼方案设计；
泰州市第四人民医院整体搬迁工程方案设计及初步图纸设计。

Bei Xu

After graduation in 2003, Bei Xu joined the Shandong provincial architectural design institute and engaged in the hospital architecture design works. He has attended lots of national hospital architecture design project biddings and awarded many tenders in different provinces and cities.

The hospital architecture projects he has worked on include:

Ward building of Jiangsu University Affiliated Hospital project and construction documents design (this project won the top ten best hospital awards).
Ward building of Jiangsu Dafeng People's Hospital project and construction documents design.
Wuxi No.2 People's Hospital ward building scheme and construction documents design.
Scheme design of Yixing traditional Chinese Medicine Hospital new site in the east of the city.
Scheme design of Jiangdu Maternal and Child Care Service Centre In-patient building.
Scheme design of Shandong Wendeng center Hospital Out-patient building.
Innovation and reconstruction project design of Yantian People's Hospital Shenzhen.
Planning and design of Yicheng People's Hospital in Hubei.
Hebei Baoding Naoxueguan Hospital Comprehensive Out-patient and In-patient building project and preliminary design.
Planning and design of Henan mental hospital.
Guangdong Integrated Western and Chinese Medicine Hospital construction documents design.
Hebei Cangzhou Integrated Western and Chinese Medicine Hospital ward building scheme design.
Scheme primary design and relocation project of Taizhou No.4 People's Hospital.

王 健
Jian Wang

王 健

2004年本科毕业之后，就进入山东省建筑设计研究院，一直从事医疗建筑设计方面的工作。先后参加了全国多个医疗建筑的投标，并在多省市取得中标项目。

曾参与完成中标的医院项目如下

深圳市盐田区人民医院整体改造工程方案设计；
广州暨南大学附属第一医院病房楼方案及施工图设计 ；
解放军总医院第二附属医院综结核病研究所工程；
广东省中西医结合医院总体规划及单体设计方案；
山东省中医药大学附属医院规划设计方案；
平顶山市第一人民医院病房综合楼设计方案；
河北省枣强县人民医院整体迁建工程设计方案；
内蒙古自治区包头市肿瘤医院施工图设计；
内蒙古自治区包头市包钢医院门诊医技综合楼设计方案；
内蒙古自治区鄂尔多斯中医院规划设计方案；
内蒙古自治区医院门诊医技综合楼设计方案；
内蒙古自治区赤峰市巴林左旗医院方案及施工图设计。

Jian Wang

After graduation in 2004, Jian Wang joined the Shandong provincial architectural design institute and engaged in the hospital architecture design works. He has attended lots of national hospital architecture design project biddings and awarded many tenders in different provinces and cities.

The hospital architecture projects he has worked on include:

Innovation and reconstruction project design of Yantian People's Hospital in Shenzhen
Guangzhou Jinan University No.1 affiliate hospital project and construction documents design
General Hospital of PLA No.2 Affiliate Hospital Research Institute for Tuberculosis control project
Guangdong Integrated Western and Chinese Medicine Hospital master plan and single building design.
Planning and design scheme of Shandong University of Traditional Chinese Medicine affiliated hospital
Pingdingshan No.1 People's Hospital Ward Building design scheme.
Hebei Zaoqiang people's hospital relocation project scheme
Inner Mongolia Tumor hospital of Baotou city construction documents design
Baogang hospital medical-technique integrated building design scheme
Inner Mongolia Erdos cCentral hospital planning and design scheme
Inner Mongolia People's hospital medical-technique integrated building design scheme
Inner Mongolia Chifeng Balinzuo Banner hospital project and construction documents design

DAWEIDESIGN

山东大卫国际建筑设计有限公司
Shandong David International Architectural Design Co.,Ltd.

毕金良
Jinliang Bi

毕金良

个人经历
1994–1998 年　辽宁工程技术大学建筑学专业
1998–2001 年　江苏省第一工业设计院
2001 年至今　山东大卫国际建筑设计有限公司二分院院长 建筑师

Jinliang Bi

Personal Experience
The president and architect of the second branch institute of Shandong David International Architectural Design Co., Ltd
1994-1998 Liaoning Technical University Major in Architecture
1998-2001 The First Industry Design Institute of Jiangsu province
2001-now Shandong David International Architectural Design Co., Ltd

王海峰
Haifeng Wang

王海峰

个人经历
同济大学城市规划学硕士 城市规划
1999 年至今　山东大卫国际建筑设计有限公司
现任第三分院院长
工程师、全国一级注册城市规划师

Haifeng Wang

Personal Experience
Education master's degree in urban planning of Tongji University, Urban Planning
Work unit: Shandong Dawei International Architecture & Design Co., Ltd
Professional title: architect, national first-level registered urban designer

张 冰
Bing Zhang

张冰

个人经历
1994–1999 年　毕业于山东建筑工程学院建筑学专业
1999 年至今　在山东大卫建筑设计有限公司
副总经理，创作中心主任

Bing Zhang

Personal Experience
1994-1999 Graduated from architecture department of Shandong Institute of Architecture and Engineering, major in architecture.
He has been working in Shandong David Architectural Design Co., Ltd. from 1999 until now.
Position: deputy general manager and director of creation center.

赵晓东
Xiaodong Zhao

赵晓东

工程师，建筑师，室内设计师
2001 年 6 月　毕业于山东工艺美术学院，环境艺术设计学专业
现任山东大卫国际建筑设计有限公司创研室副主任

Xiaodong Zhao

Engineer, architect, Interior designer
2001.June Graduated from Shandong University of Art & Design, majoring in environment art design
Nowadays works as the deputy director of Innovation Research Office in Shandong David International Architecture Design Co., Ltd.

DAWEIDESIGN

山东大卫国际建筑设计有限公司
Shandong David International Architectural Design Co.,Ltd.

邱英林
Yinglin Qiu

王 妍
Yan Wang

HSS
嘉和设计
ARCHITECTS & PLANNERS

应业飞
Yefei Ying

王 欣
Xin Wang

邱英林

国家一级注册建筑师，高级工程师
1995年7月 毕业于山东建筑工程学院建筑系，建筑学专业
现任山东大卫国际建筑设计有限公司一分院院长、建筑主任工程师

Yinglin Qiu

National First Grade Registered Architect
Senior Engineer
In July 1995 Graduated from the Architecture Department of Shandong Institute of Architecture and Engineering major in Architecture
Now works in the first branch institute of Shandong David International Architectural Design Co., Ltd as the president, and chief architect

王 妍

国家一级注册建筑师，工程师
1999年7月 毕业于山东建筑工程学院建筑系建筑学专业
现任山东大卫国际建筑设计有限公司三分院副院长

Yan Wang

National Class A Registered Architect, Engineer
July 1999, graduated from Shandong Institute of Architecture and Engineering, Faculty of Architecture, Major in Architecture.
Currently serve as Vice President of the third Institute of Shandong Dawei International Architecture Design Co., Ltd.

应业飞

副总经理
设计二部设计总监
国家一级注册建筑师
2000年加入嘉和设计

应业飞1997年毕业于青岛建筑工程学院建筑学专业。

应业飞从事建筑设计十余年。他的设计风格自然、休闲，对形式与造型的美学感觉敏锐，对建筑材料的选用灵活多变；对建筑设计的理解具有鲜明的个人特点。从业至今，他参与和主持了数十个项目的设计和管理工作，总建筑面积已逾260万平方米。应业飞在发掘客户真实需求及为客户解决实际操作问题等方面都能给予专业的意见和指导。他曾经与多位境外知名建筑师共事，在和他们长期密切合作的过程中，总结出更符合国内项目开发需求的设计、服务、理念和方法。

应业飞在住宅区规划、办公建筑、商业建筑领域都有其代表性作品。他在2007年主持设计了山东临沂日月城项目，总建筑面积达70万平方米。该项目荣获2008联合国人居署迪拜国际最佳范例（中国）推动奖。

Yefei Ying

Vice manager
Chief Designing Officer for design bipartite
National First-class Registered Architect
Attend Jia He Design in 2000

Yefei Ying graduated from Qingdao Architecture and Engineering College, majoring in Architecture.
Engineer Ying has engaged in architecture for more than ten years. He has natural and leisured design style and sensitive feeling for the aesthetic form and modeling; active and various selection for architecture material; obvious characteristics for the understanding of architecture design. From work till now, he has attended and held the design and management for tens of projects, and the general area has reached 2,600,000 km^2. Engineer Ying can give professional advice and guidance in finding clients' real requirements and in solving practical operating problems for clients. He has worked with many abroad famous architects,and in the long-term process of cooperation with them, engineer Ying has concluded the required design, service, idea and method that are more suitable for national project development.
Yefei Ying has representative works in the fields of residential district planning, office architecture, commercial architecture, etc. He has held and designed Shan Dong Lin Yi Sun and Moon project in 2007, and the overall area has reached 700,000 km^2. This project has won 2008 United Nations Inhabitancy Administration Dubai International Best Example (China) Advancement Prize.

王 欣

产品总监
国家一级注册建筑师
高级建筑师
"建设科技部专家库人才"称号获得者
2002年加入嘉和设计

王欣1994年毕业于浙江大学建筑学专业。毕业后进入杭州市建筑设计院，担任建筑设计工作。2002年获得国家一级注册建筑师资质，2008年成为建设科技部专家库人才。

王欣曾长期在大型国有设计企业工作，具有丰富的设计和项目管理经验。十五年的建筑设计生涯中，他参与和主持了数十个有影响的公建设计项目，包括早年的杭州萧山新世纪广场、杭州市第一人民医院医疗综合楼以及近年的杭州余杭图书馆、浙江财经学院学术交流中心、行政办公中心等。王欣能带领设计团队与各方紧密配合，在组织、协调和专业知识等方面都具有相当丰富的实践经验。他长期深入施工现场第一线，经常敏锐地发现设计与施工不协调的结点，并且提出完善的解决方案，获得了业主的广泛好评。同时，理性、简洁的思维使王欣成为一名优秀的设计组织管理者，在他身上集中体现了设计师创造性思维和管理者理性思维的有效结合。

王欣在公共建筑设计方面也有深入的研究，尤其在大型专业市场的设计中赢得赞誉。2007年起，他担任总投资20亿元的台州国际塑料城项目的主持建筑师，该项目是目前全国规模最大、业态最新、配套功能最全的国际化塑料产业交易服务平台，并于2009年下半年建成开业。

Xin Wang

Product Director
National first-class Registered Architect
Senior Architect
Winner of "Expert Tank Talent of Architecture and Technology Department"
Attend Jia He Design in 2002

Xin Wang graduated from Zhejiang University in 1994, and majored in architecture. After graduation, he worked in Hang Zhou Architecture and Technology Institute and held the architecture and design work. in 2002, he won national class I registered architect, and then in 2008, he became the expert tank talent of architecture and technology department.
Engineer Wang has been worked in large state-owned designing enterprise, and has rich design and project management experience. In the architecture and design career for 15 years, he has attended and held tens of influential public architectural design projects, including the earlier Hang Zhou new century square, Medical Synthesis Building of Hang Zhou No.1 Public Hospital, and recent Hangzhou Yuhang Library, Scientific Exchanges Center in Zhe Jiang College of Finance & Economics, municipal office center, etc. Engineer Wang can lead the design team to have close cooperation with other parties, and thus has rich practical experience in organization, coordination and professional knowledge. He has been in the construction locale forefront for a long time, and he often find the incompatible node between design and construction, and then put forward perfect solutions, which wins the praise of the owners. At the same time, rational and concise thinking has made Engineer Wang a excellent manager in design and organization, and it shows the effective combination of the designer's creative thinking and the manager's rational thinking on him.
Xin Wang has deeply studied in public architecture and design aspect, and he has won the praise especially in the design of large-scaled professional market. From 2007, he was the officiating architect in the project of Taizhou international plastic city, which is investment of 2 billion yuan . And this project that is the international plastic industry business and service platform with the national largest scale, newest industry status and most complete equipments, will be completed and start practice in the second half of the year 2009.

HSS
嘉和设计
ARCHITECTS & PLANNERS

蒋树勇
Shuyong Jiang

蒋树勇

技术总监
国家一级注册建筑师
高级建筑师
浙江省建筑师协会会员
2002年加入嘉和设计

蒋树勇1991年毕业于西安冶金建筑学院建筑系城市规划专业。毕业后进入浙江省建筑设计研究院，担任建筑设计工作。1998年获得国家一级注册建筑师资质，2001年获取高级建筑师资质，2003年起兼任省中级职称评委专家。

蒋树勇从事建筑设计将近二十年，积累了丰富的经验，具有较强的项目前期创造拓展能力和项目后期的实施支持能力。他熟悉建筑与房地产各类产品设计与技术的发展趋势，对房地产发展有全面的认识和感悟。他主持完成过多个规模在100 000平方米以上的大型住宅区设计，如杭州金都新城、湖畔花园、安吉新嘉园、苏州国香园等。参与并顾问了多个大型的别墅区的规划设计，如长沙机场后勤基地别墅群规划、临安官塘湖别墅区规划等。

蒋树勇对项目运作的以下环节充满信心。

项目前期规划、完善产品策略；在项目市场定位的基础上，综合策划构思，有针对性地提出规划设计；负责设计全过程控制管理，包括规划方案前期、设计、审核、协调、扩初、施工图、审图以及后期实施现场关键效果的控制和协调等；对项目的建筑设计全过程及施工提供专项技术支持。

Shuyong Jiang

Chief Technology Officer
National first-class Registered Architect
Senior Architect
Member of Zhe Jiang Architect Association
Attended Jia He design in 2002

Shuyong Jiang graduated from the architecture faculty of Xi'an Metallurgy and Architecture College in 1991,and majored in urban planning. After graduation, he came to join in Zhe Jiang Architecture and Design Institute, and held architecture and design work. In 1998, he won the national Class I registered architect, and then in 2001, he got the senior architect title, and he also too k charge of judging expert for provincial middle rank technical job title since 2003.

Engineer Jiang has worked for architecture and design for about 20 years, and he has had rich experience and great project previous creation and development ability and project later operation and support ability. He is familiar with the developing tendency of each kind of product design and technology for architecture and estate, and has complete knowledge and realization for real estate development. Engineer Jiang has held and completed multiple residential districts design with the scale above 100 000 m^2, such as the Hangzhou Jindu New City, Lakefront garden, An Ji Xin Jia Garden, Su Zhou Guo Xiang Garden. He has participated in and consulted the planning and design of multiple large-scaled villa districts, such as the layout of the villa clusters in the logistics base of Chang Sha Aerodrome, the layout of villa districts in Lin'an Guantang Lake.

Shuyong Jiang has full confidence for the following steps of project operation:

The project previous planning, perfect product strategy; under the basis of project market orientation, combing the planning idea, put forward pertinent layout and design; take charge to design the control management of the whole process, including the prophase of the planning blue print, design, auditing, coordination, development, shop drawing, drawing inspection, and the key control and coordination for later construction locale; offer special technology support for the whole process of project architecture and design and construction.

石　磊
Lei Shi

石　磊

设计一部副总监
主持建筑师

石磊1999年毕业于浙江大学建筑学专业。毕业后进入浙江城建设计院，担任建筑设计工作。

石磊在十年的建筑设计历程中，一直致力于建筑设计方案的创作和施工图设计，长期处在建筑设计的最前沿，主持和参与设计了逾2 000 000平方米的住宅项目及多个公建项目。他长期担任设计负责人工作，非常擅长与业主进行沟通和协调。在项目的设计全程及实时操作过程中，石磊能主动的竭尽所能为业主方考虑项目运行的种种难点，并用各种合理有效的方法为业主解决设计和操作的问题。

石磊主持设计了温州大自然住宅小区、杭州锦绣江南小区、平湖市湖滨花园等，以及公建类的杭州萧山国税局办公大楼（钢结构）、慈溪市人民政府办公大楼、上海杉达大学光彪学院等，获得了业主方的较高评价。

Lei Shi

Vice Director of No. 1 Design Department
General Architect

Lei Shi graduated from Zhejiang University in 1999, and majored in Architecture. After graduation, he came to join in ZheJiang Architecture and Design Institute, and held architecture and design work. Engineer Shi has bended himself to the creation of architectural design plan and shop drawing design in his ten years' architectural design, he has been in the most front of architectural design, and he has held and designed more than 2 000 000 m^2 residence projects and many public architecture projects. He has served as the design principal, and he is keen on the communication and harmony with the owners. In the whole design process and practical operating process of the project, engineer Shi can actively try his best to consider the difficulties in the operation of the project for the owners, and he would use various rational and effective methods to solve the problems in design and operation for the owners.

Engineer Shi has held and designed Wenzhou natural residence community, Hangzhou Jin Xiu Jiangnan Community, Hubin Garden in Pinghu City, etc. and public architecture like the office building of Hangzhou Xiaoshan National Taxation Office (steel structure), the office building of the People's Government in Cixi city, Guang Biao institute of Shanghai Shanda University, etc., which have obtained higher evaluation of the owners.

方　敏
Min Fang

方　敏

主持建筑师

方敏1999年毕业于浙江大学建筑学专业。毕业后先后在杭州市建筑设计院和浙江城建建筑设计院担任建筑设计工作。

方敏工作十年以来，长期在大型建筑设计院工作。她具有较强的方案设计能力、敏锐的市场观察力和丰富的专业知识。她能敏锐把握客户的需求，并主持和带领设计团队始终冲在设计的最前线。"百分百的设计热情、专业的设计技术水平和认真负责的工作态度"是她带领团队所坚持的工作准则。她独具特色的设计理念、丰富的建筑及规划设计经验、饱满的设计激情以及精益求精的设计态度，都促使她力争把每一个项目都设计成建筑精品。

方敏主持了数个大型公建与住宅项目，包括主持设计淮阴区政府、宁波北仑区行政中心以及参与设计杭州下沙高教园区后勤设施公建规划设计（670 000平方米）等，并获得不少的荣誉。2000年参加建设部村镇办发起的"迈向21世纪中国村镇住宅设计竞赛"，获得了三等奖。2001年主持设计的慈溪市行政中心党政机关办公大楼项目（53 000平方米），荣获"钱江杯三等奖"。2003年主持设计的宁波经济技术开发区北仑区行政中心项目（约110 000平方米），荣获"钱江杯二等奖"。

Min Fang

General Architect

Min Fang graduated from Zhejiang University in 1999, and majored in Architecture. After graduation, he successively came to join in Zhe Jiang Architecture and Design Institute, and Zhejiang architectural design office for urban construction, serving as the architecture and design work.

For the ten years of work, engineer Fang has worked in large-scaled architectural design office. She has comparatively strong plan design ability, acute market observation and abundant professional knowledge. She can see the clients' requirements acutely, and hold and lead the design team to be in the design forefront. "100% design ardor, professional design level and responsible working attitude" is her working rule for leading the team. Her particular design idea, abundant architecture and plan design experience, full design enthusiasm and excelsior design attitude all urge her to exert her efforts to design every project to be the exquisite architecture projects.

Min Fang has held several large-scaled public architecture and residence projects, including that she has held and designed the Huaiyin district government, Ningbo Beilun administration center, and she participated in the design of logistic faculty public architecture plan of Hangzhou Xiasha Gaoshu Park (670 000 m^2) etc., and she has obtained many awards. In 2000, she attended the "Chinese village and town residence design competition towards the 21st century", and won the third award. In 2001, she held and designed the office building of party and government organizations of the administration center in Cixi city (53 000 m^2) and won "the Third Award of Qian Jiang cup". In 2003, she held and designed the Ningbo Economic Development Area---Beilun administration center project (about 110 000 m^2) and won "the second award of Qian Jiang cup".

广州番禺区市桥建筑设计院有限公司
Shiqiao Architecture & Design Institute Co.,Ltd. Panyu Guangzhou

赵应武
Yingwu Zhao

林敢新
Ganxin Lin

蔡明峰
Mingfeng Cai

赵应武

董事，副院长，一级注册建筑师，副总建筑师， 高级工程师
1996年毕业于合肥工业大学建筑学系

主持设计项目
宁乡行政中心总体规划与设计
广州富豪山庄居住小区
长沙富豪山庄居住小区
广州亚运配套酒店−莲花山粤海度假村酒店
广州南沙腾盈大厦
无锡中堂凯莱大酒店

论文发表
《与环境共生——番禺职业技术学院综合楼楼创作体验》2004年1月发表于《华中建筑 》
《环境效益与建筑创作的有机结合——潭山中学规划设计创作体验》2006年1月发表于《中外建筑》
《秩序是建筑生命延续的根本——从番禺区象圣中学扩建工程投标设计中谈起》2008年4月发表于《建材与装饰》
《从异域风情中提炼商业气氛——宁乡春城商业广场侧记》2008年8月发表于《建筑技术》

Yingwu Zhao

Board director; Vice President; Class-A registered architect; Deputy chief architect; Senior engineer
Graduated in 1996 from Department of Architecture, Hefei University of Technology

The design projects
The overall planning and design for Ningxiang Administration Center
Residential quarter of Guangzhou Regal House
Residential quarter of Changsha Regal House
Guangzhou Asian-Game matching hotel —Guangdong Lotus Hill Resort Hotel
Teng Ying Building of Nansha, Guangzhou
Gloria Hotel of Zhongtang Wuxi

Published Works
"Coexistence with the Environment —Building Design Experience of the Comprehensive Building in Panyu Polytechnic" published on "Huazhong Architecture" in January 2004
"Combination of Environmental Benefits and Architectural Creation—Planning and Design Experience of Mountain Lake Secondary School" published on "Chinese and Overseas Architecture" in January 2006
"Order is fundamental to the continuation of the building life—from the Bidding Design for Extension Project of Panyu Xiang Sheng Secondary School" published on "Building Materials and Decoration" in April 2008
"Extracting Business from the Exotic—Notes of Ningxiang Spring City Business Plaza" published on the "Construction Technology" in August 2008

林敢新

建筑室主任，一级注册建筑师，副总建筑师
1984年毕业于华南理工大学建筑学

主持设计项目
吴川市国税大楼
吴川建设局大楼
广州番禺新造中学
广州益欣公司新厂区
广州昌顺物业管理有限公司厂房

获奖作品
1992年获湛江市优秀住宅建筑设计竞赛二等奖
1997年获湛江市第四次优秀工程勘察设计评选的建筑设计三等奖
1999年获湛江市第五次优秀工程勘察设计评选的建筑设计三等奖

Ganxin Lin

Director of Construction Office; Class-A registered architect; Deputy Chief Architect
Graduated in 1984 from Department of Architecture, South China University of Technology

The design projects
Wuchuan Tax Building
Buildings of Wuchuan Construction Bureau
Xin Zao secondary school, Panyu Guangzhou
The new plant of Guangzhou Yixin Company
The factory building of Guangzhou Changshun Property Management Co., Ltd.

Awards
1992, Zhanjiang Excellent Housing Design Competition, the second prize
1997, Zhanjiang Excellent engineering survey and design of the fourth selection, the third prize for architectural design
1999, Zhanjiang Excellent engineering survey and design of the fifth selection, the third prize for architectural design

蔡明峰

方案创作室主任，主创建筑师
2001年毕业于湖南大学建筑学系

主持设计项目
韶关职业技术学校
广州东海大酒店
广州南沙科技工业园
长沙长江置业沁园春御院
娄底南苑上和小区
江门御景酒店
广州海润大厦

Mingfeng Cai

Director of Program Creation Office; Chief creative architect
Graduated in 2001 from Department of Architecture, Hunan University

The design projects
Shaoguan Vocational School
East China Hotel in Guangzhou
Nansha Science and Technology Industrial Park, Guangzhou
Qin Yuan Chun Yu Yuan of Yangtze River home, Changsha
Nan Yuan Shang He Community in Loudi
Scenic Hotel of Jiangmen
Guangzhou Ocean Mansion

山东建筑大学设计研究院
DESIGN & RESEARCH INSTITUTE OF SHANDONG JIANZHU UNIVERSITY

刘 甦
Su Liu

刘 甦

山东建筑大学建筑城规学院院长、教授、全国优秀教师
山东省有突出贡献中青年专家
国家一级注册建筑师、注册规划师
主持山东建筑工程学院图书馆项目
运用形态“楔入”手法，将新馆自然融入于原有建筑群中，形成具有标志性特征的建筑群。
该项目获得2002年度建设部优秀建筑设计二等奖、山东省优秀勘察设计一等奖。

Su Liu

President and Professor of Architecture and Urban Planning Department of Shandong Jianzhu University
He is a nationwide prominent teacher, young and middle-aged experts with outstanding contribution in Shandong Province, National Class A Certified Architect and registered planners.
He has been a director of the library project in Shandong Architectural Engineering Institute.
He used the form "wedge" tactics to make the new library be in perfect harmony with the old buildings, so as to form a group of buildings with significant characteristics.
This project won the Second Prize of "Ministry of Construction Excellent Architectural Design of the Year 2002" and the First Prize of "Jiangsu Excellent Survey and Design".

王 乾
Qian Wang

王 乾

山东建筑大学设计研究院建筑创作研究所主任工程师
山东土木建筑学会会员

部分获奖工程

山东省工会干部学院图书馆
获2009年山东省优秀建筑设计方案一等奖
山东城市建设职业学院图书信息中心
获2009年山东省优秀建筑设计方案三等奖

Qian Wang

Director engineer of Architectural Creation Institute of Design & Research Institute of Shandong Jianzhu University
Member of Shandong Civil Engineering Institute

Award-winning Projects

The library of cadre academy of Shandong Trade Union
This project won the First Prize of "Shandong Excellent Architectural Design Plan of Year 2009"
Book Information Centre of Shandong Urban Construction Training College
This project won the Third Prize of "Shandong Excellent Architectural Design Plan of the Year 2009"

丁 磊
Lei Ding

丁 磊

山东建筑大学设计研究院一分院 主任建筑师
国家一级注册建筑师
国家注册规划师

部分获奖工程

“节能、舒适与经济－多层住宅顶置太阳能分散供热水系统方案”获山东省“力诺瑞特杯”太阳能建筑一体化住宅建筑设计方案竞赛一等奖
“山东省益都卫生学院”被评为2009年度山东省优秀建筑设计方案二等奖
“东营市科达小学” 被评为2009年度山东省优秀建筑设计方案三等奖
“山东中医药大学长清校区办公、实验组团”
获2009年度山东省优秀工程勘察设计二等奖
“山东大学西校区护理学院教学楼”
获2009年度山东省优秀工程勘察设计三等奖

Lei Ding

Chief architect of Shandong Jianzhu University Design and Research Institute No.1 Branch office
National Class A Certified Architect
National Registered Planner

Award-winning Projects

"Energy-saving, comfortable and economical-multi-stories dwelling Building Overhead Solar individual hot water supply system plan" The project won the First Prize of "Power Norit Cup" Residential Building Design Plan for the Integration of Solar and Architecture Contest of Shandong Province
"Shandong Yidu Health Institute" The project won "Shandong Excellent Architectural Design Plan of the Year 2009" the Second prize
"Dongying Keda Primary School" The project won "Shandong Excellent Architectural Design Plan of the Year 2009" the Third Prize
"Changqing campus office and experimental cluster of Shandong Traditional Chinese Medicine University" The project won the Second Prize of "Shandong Excellent Engineering Survey and Design of the Year 2009"
"The teaching building of Nursing and Health School in west campus of Shandong University" The project won the Third Prize of "Shandong Excellent Engineering Survey and Design of the Year 2009"

刘明超
Ming Chao Liu

刘明超

山东建大建筑规划设计研究院规划分院副院长
国家注册规划师
2000年毕业于山东建筑工程学院建筑系城市规划专业

部分获奖工程

山东建筑工程学院新校区修建性详细规划
获山东省优秀城市规划设计一等奖、建设部优秀城市规划设计二等奖
山东建筑工程学院新校区生态廊道绿地景观系统规划
获山东省优秀城市规划设计一等奖
建大花园居住小区修建性详细规划
获山东省优秀城市规划设计一等奖
菏泽学院校本部修建性详细规划
获山东省优秀城市规划设计二等奖
聊城市火车站站前区城市设计
获山东省优秀城市规划设计二等奖

Mingchao Liu

Vice President of Shandong Jianzhu University Designing Research Institute Planning Branch
National Registered planner
He was graduated from urban planning major of Department of Architecture of Shandong Architectural Engineering Institute in 2000

Award-winning Projects

Constructive Detailed Planning for the new campus of Shandong Architectural Engineering Institute
The project won the First Prize of "Shandong Excellent Urban Planning and Design" and won the Second Prize of "Ministry of Construction Excellent Urban Planning and Design"
System planning for Ecological Corridors green space landscape of the new campus of Shandong Architectural Engineering Institute
The project won the First Prize of "Shandong Excellent Urban Planning and Design"
Constructive Detailed Planning for garden residential quarter of Shandong Jianzhu University
The project won the First Prize of "Shandong Excellent Urban Planning and Design"
Constructive Detailed Planning for the main campus of Heze Institute
The project won the Second Prize of "Shandong Excellent Urban Planning and Design"
Urban design for the front area of Liaocheng Railway Station
The project won the Second Prize of "Shandong Excellent Urban Planning and Design"

WEN KUN
上海文坤建筑设计有限公司
Shanghai Wenkun Architectural Design &Research Co.Ltd.

戴东辉
Donghui Dai

戴东辉

重庆大学建筑学　学士

设计作品
大中华太古城商业单体设计
大中华乐山CBD规划方案设计
深圳市龙岗区榭丽花园规划及建筑设计
安徽省淮南市煤电大厦规划及建筑设计
天津市海河一号写字楼建筑设计
安徽省淮南市谢二西村商业设计
安徽省淮南市新社西村商业设计
深圳金光华春华四季园
山东威海环翠区旧村改造
江苏无锡新世界纺织服装城国际公寓
安徽淮南云景华城小区
安徽淮南绿城小区
湖北武汉.沿海商业规划及建筑单体方案设计
常州　茂业泰富城
广东清远联泰别墅区
美源游艇会
浙江物产良渚花苑
海南龙沐湾国际旅游度假区

Donghui Dai

Chongqing University　Bachelor of Architecture

Design works
Dazhonghua Taikoo City Commercial Single Building Design
Dazhonghua Leshan CBD Planning Design
Planning and Architectural Design for Sheli Garden in Longgang District, Shenzhen
Planning and Architectural Design for Coal Power House in Huainan, Anhui Province
Architectural Design for the 1st Office Building of Haihe River, Tianjin
Commercial Design of Xie'r Western Village in Huainan, Anhui Province
Commercial Design of Xinshe Western Village in Huainan, Anhui Province
Chun-hua Four Seasons of Golden Hua, Shenzhen
Renovating old villages of Huancui District in Weihai, Shandong
International Apartment of New World Textile and Garment City in Wuxi, Jiangsu Province
Cloud Scene Huacheng Area in Huainan, Anhui Province
Green City Area in Huainan, Anhui Province
Coastal Business Planning and Single Building Project Design in Wuhan, Hubei Province
Maoye Taifu City, Changzhou
Qingyuan Luen Thai Villa Area, Guangdong
Meiyuan Marina Hall
Materials Liangzhu Garden, Zhejiang
Property Hotel　Sports Apartments
Longmu Bay National Tourist Vacation Resorts, Hainan

张　昆
Kun Zhang

张　昆

西安建筑科技大学建筑学　学士

设计作品
上海莱蒙商业中心
安徽宿州千亩园规划设计
常州茂业　泰富城
合肥松芝万象城
遵义万里路C区商贸区项目
安阳彰德府文化商业步行街
佛山九鼎国际城南区
成都成金发展总部基地项目
东莞塘厦信成　幸福里
常州武宜路1、7和8号地块规划

Kun Zhang

Xi'an University of Architecture and Technology　Bachelor of Architecture,

Design works
Laimeng Commercial Centre, Shanghai
Planning Design for Suzhou Qianmu Park, Anhui
Maoye Taifu City, Changzhou
Songzhi • Wanxiang City, Hefei
Business District of C-Block of Project in Wanli Road, Zunyi
Zhangde Cultural and Commercial Pedestrian Street, Anyang
South District of Jiuding International City in Foshan
Base Project of Chengjin Development Headquarters, Chengdu
Xincheng Happiness Village in Tangxia, Dongguan
Planning of Block No.1, 7 and 8 in Wuyi Road, Cangzhou

王　磊
Lei Wang

王　磊

西安建筑科技大学建筑学　学士

设计作品
成都金履国际商贸城
成都西部鞋都D区
安徽六安天街立面改造
辽宁 • 海城裕通商业广场
海口 • 仁邑国际社区
广东博罗 • 一品中央
安徽宿州大观园 • 家天下D区
安徽宿州灵璧石市场
合肥家天下
成都复合工业新城 • 链生河城国际—双羽湾项目

Lei Wang

Xi'an University of Architecture and Technology　Bachelor of Architecture

Design works
Jinlv International Trade City, Chengdu
D-Block of West Shoes Captial, Chengdu
Facade reconstruction of Lu'an Heaven Street, Anhui Province
Yutong Shopping Mall in Haicheng, Liaoning Province
Renyi International Community, Haikou Province
Yipin Centre in Boluo, Guangdong Province
Grand View Park • D-Block of Home World in Suzhou, Anhui
Lingbi Stone Market in Suzhou, Anhui
Hefei Home World
Chengdu Compound Industrial New Park • Liansheng River City International-- Double Feather Bay Project

王　鹏
Peng Wang

王　鹏

重庆后勤工程学院建筑学　学士

设计作品
云南保山奥新体育城别墅2期
成都金牛宾馆新建政务中心接待楼
深圳大梅沙俊城都会轩（别墅）
重庆建桥工业园服务设施概念设计
绵阳市游仙区小枧沟镇置信小学
广州天河区珠江新城B1—1地块概念方案设计
常州武宜路1、7和8号地块规划

Peng Wang

Chongqing Logistical Engineering University　Bachelor of Architecture

Design works
Phase 2 of Villas of Ao'xin Sports City in Baoshan, Yunnan Province
New-built Administrative Centre Reception Building of Golden Bull Hotel, Chengdu
Dameisha Juncheng Metropolis Building (villa), Shenzhen
Service Faculty Concept Design for Chongqing Jianqiao Industrial Park
Zhixin Primary School of Xiaojiangou Town in Youxian District of Mianyang City
B1-1 Block Concept Program Design of Pearl River New City in Tianhe District, Guangzhou Province
Planning of Block No.1,7 and 8 in Wuyi Road, Cangzhou

设计总监 / 总经理　杨锋

重庆大学　　建筑学学士
B.A. of Chongqing University
北京大学　　EDP
EDP of Peking University
1998–2001年　上海城乡建筑设计院有限公司
2001–2004年　上海天华建筑设计有限公司
　　　　　　主创建筑设计师
　　　　　　北京分公司总经理
2004年至今　都易国际设计与发展有限公司(香港)
　　　　　　上海都易建筑设计有限公司
　　　　　　设计总监
　　　　　　总经理

在都易之前的代表作品为:

上海三湘世纪花城　荣获全国"创新风暴"　综合金奖、智能化小区金奖。上海市优秀主宅小区房型创新奖、上海绿波花园别墅。荣获2002年度上海住宅设计小区创优项目优良奖。
上海南方城小区　荣获2003年度上海市优秀住宅规划设计三等奖。
上海绿洲雅宾利
上海中星海上名门
北京威尔逊大道
上海金桥碧云花园南三街坊
重庆天江鼎城　荣获中国国际建筑艺术双年展国际居住建筑设计艺术金奖

DOT.INT 都易国际
CHANGE THE CITY
CULTURE·SPACE
www.dotint.com.cn

重庆大学景观建筑学学士/上海交通大学工商管理硕士
B.LA. of Chongqing University / MBA of Shanghai Jiaotong University

副总经理　刘念

重庆大学　城市规划　学士
B.UP. of Chongqing University

副总经理　齐丹丹

重庆大学　建筑学　学士
B.A. of Chongqing University

高级项目经理　徐钦勇

都易国际的项目经理

服 务 特 点　SERVICE ADVANTAGES

快速反应，贴身服务
提供系列化成熟产品
为客户创造价值，为客户品牌增值。
Quick response,considerate service.
Provide serialized mature products.
Create value to customers and add value to brands.

发 展 目 标　PERSPECITIVE

中国新古典主义高档住区设计领域的领跑者
To become a pacemaker in Chinese neoclassic high-grade community design field.

南京 · 万科 · 金域缇香
南京 · 万科 · 金域蓝湾
杭州 · 万科 · 西溪蝶园 · II
杭州 · 万科 · 金色家园
上海 · 万科 · 五街坊
常州 · 万科 · 大观楼项目
青岛 · 万科 · 南王项目

昆山 · 绿地 · 白市浦河项目
上海 · 绿地 · 嘉定 E24/25 地块
上海 · 绿地 · 昌吉路项目
宁波 · 浙大网新 · ART 蓝湾
宁波 · 浙大网新 · ART 蓝海
宁波 · 浙大网新 · 泰州路项目
福州 · 融汇 · 融汇江山

杭州 · 钱塘 · 梧桐公寓
杭州 · 钱塘 · 沁园
杭州 · 钱塘 · 蓝鲸国际
合肥 · 钱塘 · 金色梧桐 · II
成都 · 华润 · 二十四城 · IV
福州 · 融桥 · 江南水都
重庆 · 协信 · 协信公馆

郑州 · 建业 · 壹号城邦
武汉 · 中联 · 新华府
青岛 · 盾安 · 蓝山湾
台州 · 地之韵 · 比华利别墅
宜昌 · 华泰 · 江山风华
重庆 · 中关村 · 天江鼎城
TO BE CONTINUED ...

GN® 栖城
GN INTERNATIONAL DESIGN

周 峻
James Zhou

周 峻

GN 栖城副总经理、总建筑师
毕业于同济大学建筑系。历任同济大学建筑设计研究院同远分院主任建筑师、副院长
中国一级注册建筑师、上海市建筑学会会员
1994 年全国大学生建筑设计竞赛二等奖
2000 年建设部优秀勘察设计二等奖
2000 年教育部优秀勘察设计一等奖
2001 年第九届全国优秀工程设计铜奖
2003 年第三届上海国际青年建筑师设计作品展一等奖
2006 年第五届上海国际青年建筑师设计作品展二等奖
2007 年第六届上海国际青年建筑师设计作品展一等奖
2007 年上海市优秀勘察设计二等奖
2007 年首届 40 位小于 40 岁的华人建筑师作品展参展建筑师
2008 年上海市勘察设计行业协会授予“20 位最具大师潜力青年建筑师”
2008 年全国优秀工程勘察设计行业奖（原建设部奖）二等奖
2009 年第十三届全国优秀工程设计铜质奖
2009 年上海市优秀勘察设计二等奖
2009 第八届上海（国际）青年建筑师设计作品展——上海市勘察设计行业协会授予“10位最具大师潜质青年建筑师”

James Zhou

GN Xicheng Inc. Deputy General Manager, Chief Architect
Graduated from TongJi University Architecture Department
Tongyuan Branch of TongJi University Architecture Design Institute, Principal Architect, Subdecanal
PRC Class I Registered Architect
Shanghai IAA member
Second Prize for National College Student Architectural Design Competition, 1994
Second Prize for Outstanding Survey Design by Ministry of Construction, 2000
First Prize for Outstanding Survey Design by Ministry of Education, 2000
Bronze Prize for the 9th National Excellent Project Design, 2001
First Prize for the Third Shanghai International Exhibition of Young Architects Design, 2003
Second Prize for the Fifth Shanghai International Exhibition of Young Architects Design, 2006
First Prize for the Sixth Shanghai International Exhibition of Young Architects Design, 2007
Second Prize for Shanghai Outstanding Survey and Design , 2007.
Exhibited architect for the first Chinese exhibition among 40 architects less than 40 years old, 2007
"the 20 Most Potential Young Master Architect", awarded by Shanghai Association of Survey and Design Industry, 2008
Second Prize for the National Outstanding Project Survey & Design Industry (the former Ministry of Construction Award), 2008
Bronze Award for 13th National Excellent Engineering Design, 2009
Second Prize for Shanghai Excellent Survey and Design, 2009
the Eighth Shanghai (International) Young Architects Design Exhibition, 2009 - the "Top 10 Most Master Potential Young Architects" awarded by Shanghai Investigation and Design Association

管轶群
Yiqun Guan

管轶群

GN 栖城发展与研究中心总监
毕业于德国柏林工业大学建筑学专业 德国建筑学硕士
历任德国阿斯曼 萨洛蒙建筑师事务所中国区合伙人
上海柏尔摩建筑设计与规划有限公司合伙人
复旦大学上海视觉艺术学院外聘讲师
柏林 Nikolai 教堂内部改造方案收录于 2004 年柏林博物馆年鉴
作品“next next next”入选 2002 年威尼斯双年展暨第八届国际建筑展（德国馆）

Yiqun Guan

Chief Inspector of GN Xicheng Research Center for Development
Graduated from the Technical University of Berlin, Germany, majored in Architecture, Master of Architecture
Former partner of Germany Assmann Salomon Architect Agency, China Branch.
Architectural design and planning on Haibaiermo Ltd. Partner
Fudan University, Shanghai Institute of Visual Arts, Adjunct Lecturer
Interior reform program for Berlin Nikolai Church, included in the Yearbook of Berlin Museum, 2004
Work "next next next" was selected into 2002 Venice Biennale Exhibition cum the Eighth International Architecture Exhibition (Germany Pavilion)

吴优德
Youde Wu

吴优德

GN 栖城景观设计总监
曾赴日本留学研修
2001 年任职于日本 M.A.O 一级建筑师事务所
擅长手绘，方案创作。
作品有上海奥林匹克花园二期景观设计，连云港胡沟水库公园景观规划与设计等。

Youde Wu

Chief Inspector of GN Xicheng Landscape Design
Once studied abroad in Japan.
Served in M.A.O, the Class- A Architect Agency in Japan, 2001
Specializes in hand-painted, program creation.
Works including landscape design for Shanghai Olympic Garden Phase II, landscape planning and design of Lianyungang Hugou Reservoir Park, etc.

Filip Eric VANDYCKE
（比利时）

Filip Eric VANDYCKE

GN 栖城高级项目园景师
主要工作经历：
Reichen et Robert & Associes Team manager
Project for the development of the Bouregreg Valley in RABAT (Morocco):
General development and infrastructure plan of the Bouregreg-Sale region. Design of master plan for the first phase.
Project for the development of the Faliro Waterfront in ATHENS (Greece):
Master plan of the Beach Volleyball and Tae kwon do site for the 2004 Olympic Games.
Cite Universitaire Internationale in PARIS (France) :

Filip Eric VANDYCKE

Development strategy and master plan for the university campus.
Grand Projet de ville Herouville Saint-Clair (France):
Masterplan for the rehabilitaition of residential neighborhood.
Waterfront of MOHAMMEDIA (Morocco):
Development strategy and program for coast line with marina.
HENIN-CARVIN in France:
Cite Universitaire Internationale in PARIS (France) :
Development strategy and master plan for the university campus.
Grand Projet de ville Herouville Saint-Clair (France):
Masterplan for the rehabilitaition of residential neighborhood.
Waterfront of MOHAMMEDIA (Morocco):
Development strategy and program for coast line with marina.
HENIN-CARVIN in France:

中国建筑西南设计研究院有限公司
CHINA SOUTHWEST ARCHITECTURAL DESIGN AND RESEARCH INSTITUTE CORP.LTD

钱　方
Fang Qian

钱　方

中国建筑西南设计研究院有限公司总建筑师
国家一级注册建筑师
中国建筑学会建筑师分会理事
中国建筑学会建筑师分会副理事长
四川省建筑师学会常务理事
四川省土木建筑学会常务理事
四川省学术和技术带头人
四川省工程设计大师

代表作品
都江堰拉法基古窑博物馆
重庆袁家岗游泳跳水馆
成都高新商务广场
成都天府科技园
成都高新西区技术创新组团
成都市南部副中心科创中心
四川省广电中心等
设计作品获奖39项

设计理念
设计是没有边界的审美过程，它需要理性游戏的态度，技术无法万能，却是实现设计理念的手段。建筑师通过建筑及其设计理解世界，于是建筑有了思想，设计成了审美的体验。

Fang Qian

Chief Architect of Chinese Southwest Architectural Design Institute Co., Ltd.
PRC Class I Registered Architect,
Director and Vice President of Architect Branch of Architectural Society of China
Executive director of Sichuan Architectural Society
Executive director of Sichuan Civil Engineering Society
A leader in academic and technical fields
A master of engineering design

Selected Projects
Dujiangyan Lafaji Acient Kiln Museum
Chongqing Yuanjiagang Swimming & Diving Pool
Chengdu High-tech Commercial Square
Chengdu Tianfu Technological Park
Technological and Innovative Cluster in Chengdu West High-tech Park
Technological and Creative Center in Chengdu South Vice Center
Sichuan BCTV Center
39 items of awarded projects

Design Concept
Design is a boundless aesthetic process. It is like playing games, but needs rational attitude. Although technology is not universal, it is a necessary method to express design concept. Architects use architecture and its design to understand the world, so the world has thought and the design becomes an experience.

冯　远
Yuan Feng

冯　远

中国建筑西南设计研究院有限公司总工程师
国家一级注册结构工程师
教授级高级工程师，建设部专家
国务院颁发的政府特殊津贴获得者
四川省学术和技术带头人
四川省工程设计大师
四川省政协委员

擅长大跨度空间结构设计，高层建筑结构设计，高层钢结构设计。现任中国土木工程学会空间结构委员会副主任委员，中国工程建设标准化委员会混凝土结构标准技术委员会委员，中国土木工程学会混凝土及预应力混凝土分会理事，中国钢结构协会第四届理事会理事，全国工程建设标准设计专家委员会委员，全国超限高层建筑工程抗震设防审查专家委员会委员，中国钢结构协会专家委员会委员，IASS国际薄壳与空间结构协会会员，四川省土木建筑学会常务理事，《建筑结构》、《四川建筑》、《空间结构》杂志编委。

代表作品
德阳市体育馆
重庆市袁家岗游泳、跳水馆
天府广场
成都高新科技商务广场
常州体育会展中心体育馆

主持的科研项目
高层钢结构框架节点的试验研究与应用
大悬挑、中部抽空、三边支承网壳结构的优化设计与应用
大跨度椭球形索承单层网壳结构设计研究与应用

设计理念
用合理的结构表现建筑的美

Yuan Feng

Chief engineer of Chinese Southwest Architectural Design Institute Co., Ltd
State Class Ⅰ Registered Achitect
Professional Senior Engineer, Professor in Ministry of Construction
A winner of special government allowance granted by State Council
An academic and technological leader in Sichuan
A master of engineering design in Sichuan
A member of Sichuan People's Political Consultative Conference
She is good at designs for spatial structure, high-rise building and high steel structure. She possesses positions as a vice-chairman in Spatial Structure Committee of Chinese Institute of Civil Engineer, a member of Chinese Engineer and Building Standard Committee and Concrete Structure Standard Technical Committee, a director of Concrete and Prestressed Concrete Branch of Chinese Institute of Civil Engineer, a director of the Fourth Executive Council of Chinese Institute of Steel Construction, a member of National Engineering Construction Standard Committee of Experts, a member of National Ant-seismic investigation Committee of experts for buildings over height limitation., a member of Chinese Institute of Steel Structure, a member of IASS International Thin Shells and Spatial Structure Committee, a standing director of Sichuan Institute of Civil Construction, editor of Architectural Structure, Sichuan Architecture, Spatial Structure.

Representative Works
Deyang Stadium
Chongqing Yuen Kong Natatorium
Tianfu Square
Chengdu High-tech Business Centre
Changzhou Sports Convention & Exhibition Center

Presided Research Project
Experimental studies and applications on Frame nodes of high-rise steel structure
Optimized design and applications on overhang, central section evacuator, and three-edged supported Latticed Shell Structures
Design Studies and applications on long-span ellipsoidal cable-supported lattice shell Structures

Design Ideas
Applying appropriate economic structures to display the beauty of the architecture

中国建筑西南设计研究院有限公司
CHINA SOUTHWEST ARCHITECTURAL DESIGN AND RESEARCH INSTITUTE CORP.LTD

戎向阳
Xiangyang Rong

戎向阳

中国建筑西南设计研究院有限公司院副总工程师
国家注册公用设备工程师
中国勘察设计协会建筑环境与设备专业委员会副理事长、四川省制冷学会常务理事兼空调热泵专业委员会副主任委员、全国ATHIC联盟技术委员会专家委员、全国建筑环境与设备（暖通空调）优秀工程设计奖评审委员会委员、《暖通空调》杂志编委。

代表作品

中国建筑西南设计研究院第二办公区办公楼暖通设计
中国人民银行成都分行国家金库暖通设计
三峡大厦暖通设计
成都电力生产调度基地暖通设计
成都双流国际机场T2航站楼暖通设计
铁狮门东大街项目暖通设计
西藏军区暖通与节能设计

Xiangyang Rong

Rong is an assistant chief engineer of CSWADI. He is national a registered public utility engineer and a senior engineer.
Rong is appointed as Vice Chairman of Architectural Environment &Facilities Professional Committee, a branch of China Investigation &Design Association. He is also nominated to a Standing Director of Sichuan Refrigeration Institute and an associate commissioner of Heat Pump Professional Commission, to an expert commissioner of ATHIC technical committee, and to a member of Review and Appraisal Board of Excellent Engineering Design Prizes for National Architectural Environment &Facilities (Heating Ventilating &Air Conditioning). In addition, Rong is an editorial board member of Heating Ventilating &Air Conditioning.

Representative Works

Heating and ventilation design for No. 2 Office Area of CSWADI
Heating and ventilation design for Vault of People's Bank of China Chengdu Branch
Heating and ventilation design for Three Gorges Edifice
Heating and ventilation design for Chengdu Power Generation and Dispatching Base
Heating and ventilation design for Terminal T2 of Chengdu Shuangliu International Airport
Heating and ventilation design for Tishman Speyer East Street Project
HVAC & energy conservation design for Tibet Military Area

邱小勇
Xiaoyong Qiu

邱小勇

中国建筑西南设计研究院有限公司院副总建筑师
香港建筑师学会会员（资格）
四川省建筑师学会理事

创作理念

建筑设计是一个苦中找乐的过程，需要设计者通过观察、思考、分析找出建筑基地及其周边环境所隐含的"基地"特征。用适当的工程技术语言来实现建筑与环境的对话。限制与制约条件是我思考的出发点，建筑与环境共融、共生是我的设计目标。

Xioayong Qiu

Vice Chief Architect of China Southwest Architectural Design & Research Co., Ltd.
A member of Hong Kong Institute of Architects
Director of Sichuan Institute of Architects

Design Concept

Architectural design is a process of seeking happiness from painfulness. It needs designers to find out characters implied in the architectural bases and surrounding environment by observing, thinking and analyzing. It needs suitable technologic language to complete the dialogue between architecture and environment. The architects always consider from limited conditions. How to integrate with environment is his design purpose.

郑 勇
Yong Zheng

郑 勇

中国建筑西南设计研究院有限公司院副总建筑师
成都市建筑师学会理事
四川省企业青年联合会第一届委员会委员
四川省有突出贡献的优秀专家称号

主持国内外各类大中型建筑工程的方案创作及设计工作，并为业主提供优质的策划和咨询服务。尤其在博物馆、教育建筑、酒店、写字楼及大型住宅小区设计领域有独到的见解、突出的创作能力和丰富的设计实践经验，勇于探索、敢于创新。多项设计工作成果先后多次获得省部级奖励。

Yong Zheng

Vice Chief Architect of China Southwest Architectural Design & Research Co., Ltd.
A director in Chengdu Institute of Architects
A member of the First Committee of Sichuan Enterprises & Youth Union
Granted as an "Excellent Expert with Great Contribution in Sichuan"

He has held various kinds of large and medium size projects, and provided outstanding design and consulting service for clients. He has particular understanding, outstanding creative capability and rich experiences on designs for museums, educational buildings, hotels, office buildings and large residential site. He is bold in researching and practicing new ideas. Many projects have received prizes on provincial level.

中国建筑西南设计研究院有限公司
CHINA SOUTHWEST ARCHITECTURAL DESIGN AND RESEARCH INSTITUTE CORP.LTD

刘 艺
Yi Liu

刘 刚
Gang Liu

藝天建築
浙江大学建筑系高裕江建筑工作室

高裕江
Yu Jiang Gao

刘 艺

四川省建筑师学会理事兼副秘书长

2003年获中国建筑学会颁发的青年建筑师奖“设计竞赛优秀奖”
2006年获中国建筑学会颁发的第六届“中国青年建筑师奖”
2007年获中国建筑总公司“中国建筑青年才俊”称号
2008年获四川省国资委“十大杰出青年”提名奖

Yi Liu

Vice Director and vice secretary of Sichuan Institute of Architects

He received “Outstanding Prize in Design Competition” for young architect issued by Architectural Society of China in 2003
He received “Chinese Young Architect Prize” in the sixth competition hold by Architectural Society of China in 2006
He was granted as “Chinese Outstanding Young Architect” in Chinese Architectural Company in 2007
He has received the nomination as “Top Ten Outstanding Youths” issued by Sichuan SASAC in 2008

刘 刚

中国建筑西南设计研究院有限公司城市设计研究中心总建筑师
国家一级注册建筑师

主要主持的作品
锦江大礼堂省人大常委会议厅装修工程
三六三医院建设工程
四川省锦江宾馆新馆
雅砻江流域集控中心大楼
成都市人民南路城市设计
成都市东大街城市设计
成都市指挥街地块（节点）城市设计

Gang Liu

The Chief Architect in Urban Planning & Research Center of China Southwest Architectural Design & Research Co., Ltd
National Class Ⅰ Registered Architect

Decoration project for conference room for Provincial People's Congress Standing Committee in Jinjiang Grand Hall
Construction project for 363 Hospital
New Jinjiang Hotel in Sichuan province
Office building of Centralized control center for Yalong River Basin
Urban design for Renmin South Road in Chengdu
Urban design for East Street in Chengdu
Urban design for Zhihui Street Block(nodal point) in Chengdu

高裕江

教授级高级建筑师　特聘总建筑师
工作室主持人　国家一级注册建筑师　博士

主要获奖
国家优秀建筑设计奖两项（主要设计人之一）
建设部建筑创作奖两项（主要设计人之一）
中国青年建筑师奖
上海国际建筑设计青年建筑师作品二等奖（第一作者）
大连国际建筑设计作品展二等奖（第一作者）
全国文化馆竞赛优秀奖（独立）
建设部优秀设计二等奖三项（第一作者及主要设计人之一）
建设部优秀设计三等奖两项（第一作者）
全国建筑画大赛优秀奖（第一作者）
江苏省图书馆（南京图书馆）省优一等奖（第一作者）
加纳国家剧场省优一等奖，建筑创作特别奖（主要设计人之一，主持程泰宁院士）
马里议会大厦，省优一等奖，建筑创作奖（主要设计人之一，主持程泰宁院士）
海南建设银行大厦（海南第一幢超高层），省优一等奖（主要设计人之一，主持姚振生）
南京玄武湖隧道控制中心，省优二等奖，市优一等奖（第一作者）
绍兴鲁迅电影院工程，省优二等奖，建筑创作奖，省优质工程奖（第一作者）
全国“八五”新住宅设计，省优二等奖（第一作者）
杭州五洲大酒店工程（省重点工程、首家新建五星级宾馆，二十四层高层），省优三等奖，市优二等奖（第一作者）
文化部嵊县越剧博物馆工程（全国首家戏剧类专业博物馆、文化部重点工程），省建筑创作奖，省优三等奖，市优二等奖（第一作者）
杭州九十年代住宅设计方案之一获市优二等奖
杭州九十年代住宅设计方案之二获市优二等奖

Yu Jiang Gao

Professor & Senior Architect; Specially-engaged Chief Architect; President of the studio; National Class I Registered Architect; Doctor

Major Awards
Two National Excellent Architectural Design Awards (as one of the chief designers);
Two Architectural Design Awards by Ministry of Construction (as one of the chief designers);
China Young Architects Award;
The second Prize of Designs of Young Architects in Shanghai International Architectural Design (the first author);
The second Prize in Dalian International Architectural Design Exhibition (the first author);
Excellence Award in National Cultural Center Competition (Independent);
Three Second Prizes of Excellent Design by Ministry of Construction (the first author and one of the chief designers);
Two Third Prizes of Excellent Design by Ministry of Construction (the first author);
Excellence Award in National Architectural Drawing Competition (the first author);
Jiangsu Library (Nanjing Library): the First Prize of provincial excellence (the first author)
Ghana National Theatre: the First Prize of Provincial Excellence; Special Award of architecture design (as one of the chief designers, hosted by Academician Cheng Taining)
Mali Parliament Building: the First Prize of Provincial Excellence; Architecture Design Award (as one of the chief designers, hosted by Academician Cheng Taining)
Hainan Construction Bank Building (Hainan No. 1 super high-rise building): the first prize of provincial excellence (as one of the chief designers, hosted by Yao Zhensheng)
Xuanwu Lake Tunnel Control Center, Nanjing: the Second Prize of provincial excellence; the First Prize of Municipal Excellence (the first author)
Shaoxing Lu Xun Cinema: the Second Prize of Provincial Excellence; Architectural Design Award; Provincial Quality Engineering Award (the first author)
New residential design of National "Eighth Five-Year Plan": the Second Prize of Provincial Excellence (the first author)
Hangzhou Continental Grand Hotel (the provincial key project; the first newly-built five-star hotel; 24-story high-rise building): the third prize of provincial excellence; the second prize of municipal excellence (the first author)
Shaoxing Opera Museum of Sheng County by the Ministry of Culture (the First professional museum for drama in China; the key project of the Ministry of Culture): Provincial Architectural Design Prize; the Third Prize of Provincial Excellence; the second prize of municipal excellence (the first author)
1990s Hangzhou Residential Design Ⅰ: the Second Prize of municipal excellence
1990s Hangzhou Residential Design Ⅱ: the Second Prize of municipal excellence

天大建筑

天津大学建筑设计规划研究总院

Academy Of Architectural Design & City Planning Tianjin University

于敬海

Jinghai Yu

于敬海

党委书记，副总工程师，研究员，天津大学硕士研究生导师，国家一级注册结构工程师

天津市超限高层建筑工程抗震设防审查专家委员会委员，天津市工程评标专家，天津市建筑学会建筑专业委员会主任，天津市钢结构学会副会长。

学历及任职

1987年7月毕业于山东建筑工程学院工业与民用建筑专业

1990年3月毕业于天津大学土木工程系结构工程专业，获工学硕士学位

2008年6月毕业于天津大学土木工程系结构工程专业，获工学博士学位

工程设计业绩

主持设计及审核审定项目40余项。

青岛海上皇宫

河南教育学院主楼

天津罗马花园，高层住宅项目

天津环渤海经贸大厦钢结构设计项目

天津医科大学附属肿瘤医院住院楼扩建

天津响锣湾五矿大厦

山东茌平四馆一场项目

科研业绩

1.共撰写学术论文约30篇，其中"重要期刊"论文9篇，《灰砂蒸压加气混凝土砌体剪压复合受力性能试验》及《蒸压灰砂加气混凝土砌体通缝抗剪强度的试验研究》两篇论文被EI核心检索收入。

2.鉴定完成天津市青年科技基金、河北省科技厅及天津市科委等科技课题4项。

1990–2001年项目名称，提高高层及超高层建筑抗震能力的新技术研究，天津市青年科学基金项目，第五完成人，鉴定验收。

2000–2002年项目名称，砌体建筑物裂缝的防治与研究，天津市科委科技攻关项目，第五完成人，鉴定验收。

2003–2005年项目名称，住宅建筑中结构构建裂缝的成因与预防措施研究，河北省科技攻关项目，第二完成人，鉴定验收。

2003–2007年项目名称，蒸压加气混凝土承重砌块多层节能住宅体系的试验研究，天津市科技攻关项目，第二完成人，鉴定验收。

3.参加编写国家建筑标准设计"混凝土异形柱结构构造(一)"06SG331-1和天津市地方设计标准"天津市空间网格结构技术规程"DB29-140-2005。

工程设计及科研获奖

1.青岛前海海上酒楼，获1998年度教育部优秀设计二等奖；

2.石家庄石门安居小区，获2000年度教育部优秀设计三等奖；

3.砌体建筑物裂缝的防治与研究，获2001年度天津市科技进步三等奖；

4.天津市电力公司供用电综合楼，获2006年度天津市建委优秀设计一等奖；

5.潍坊市中医院综合楼，获2006年度天津市建委优秀设计二等奖；

6.东方地球物理公司职工活动中心。获得2008年度天津市"海河杯"优秀勘察设计（建筑类）二等奖；

7.天津万科东丽湖滑水会所，获得2008年度天津市"海河杯"优秀勘察设计（结构类）二等奖；

Jinghai Yu

Secretary of the Party committee, assistant chief engineer, researcher, master student supervisor of Tianjin University, National Class I Registered Structural Engineer.

Committee Member of Examine Experts on Earthquake Fortification of the Out-of-codes High Rise Buildings in Tianjin, project bidding expert of Tianjin, director of Architecture Major Committee of Tianjin Architecture Academy, vice chairman of Tianjin Steel Structure Academy.

Education background and office holding

In July 1987, graduated from Major of Industrial and Civil Architecture in Shandong Institute of Civil Engineering;

In March 1990, graduated from Major of Structural engineering in Department of Civil Engineering of Tianjin University and got Degree of Master of Engineering;

In June 2008, graduated from Major of Structural engineering in Department of Civil Engineering of Tianjin University and got Degree of Doctor of Engineering.

Achievements of project design

Host in design and examine and verify more than 40 programs.

Qingdao Sea Palace

Main building of Henan Educational College

High-rise residence project of Tianjin Roman Garden

Steel structure design project of Tianjin Circum-Bohai Economy and Trade Building

Extension of the inpatient building of Tumor Hospital affiliated to Tianjin Medical University

Wukuang Mansion of Tianjin Xiangluowan

Four stadiums and one field project in Chiping, Shandong

Scientific Research Achievements

1. Totally composed more than 30 academic papers, among which 9 are published on "Significant Journals", and "Composite stress performance testing of dust autoclaved aerated concrete masonry shear" and "Experimental research on continuous seam shear strength of autoclaved dust aerated concrete masonry" have been selected into EI core retrieval.

2. Identified and accomplished 4 technological subjects of Tianjin Youth Technology Fund, Hebei Technology Office and Tianjin Science Commission.

In 1990-2001, project name: New technical research on the improvement of the earthquake fortification capacity of high-rise and super-tall buildings, project of Tianjin Youth Technology Fund, fifth person to finish, examined and accepted;

In 2000-2002, project name: Prevention and research of masonry construction fissures, Science and technology critical project of Tianjin Science Commission, fifth person to finish, examined and accepted;

In 2003-2005, project name: Cause and prevention measures of structural building fissures of residential constructions, Science and technology critical project of Hebei Province, second person to finish, examined and accepted;

In 2003-2007, project name: Experimental Study of autoclaved aerated concrete bearing block multilayer energy-saving residential system, Science and technology critical project of Tianjin Science Commission, second person to finish, examined and accepted;

3. Took part in composing National Architecture Standard Design "Composition of concrete special-shaped column structures (A)" O6SG331-1 and Tianjin local design standard "Tianjin space grid structure technical regulations" DB29-140-2005.

Project and scientific research awards

1. Qingdao Qianhai Sea Restaurant, won Second Prize of Excellent Design of Education Ministry in 1998

2. Shimen'anju Community in Shijiazhuang, won Third Prize of Excellent Design of Education Ministry in 2000

3. Prevention and research of masonry construction fissures, won Third Prize of science and technology progress of Tianjin in 2001

4. Power supply Building of Tianjin Electric Power Company, won First Prize of Excellent Design of Tianjin Construction Committee

5. Comprehensive building of Weifang hospital of traditional Chinese medicine, won Second Prize of Excellent Design of Tianjin Construction Committee

6. Staff action centre of Eastern Geophysics Company, won Second Prize of Tianjin "Haihe Cup" Excellent Survey and Design (Architecture)

7. Dongli Lake Water Ski Chamber of Tianjin Micronics, won Second Prize of Tianjin "Haihe Cup" Excellent Survey and Design (Structure)

蔡　节

Jie Cai

蔡　节

院执行总建筑师，研究员

从业经历

1982年2月毕业于天津大学建筑系建筑学专业，从事建筑设计余三十年。

1995年晋升为高级建筑师

2002年晋升为研究员

1998年注册为国家一级注册建筑师

代表作品

国家物探局办公大楼

国家消防培训基地总体规划及单体设计

张家口师范专科学院计算中心

大庆石油学院秦皇岛分院教学主楼

河南省农业大学教学主楼

天津大学综合实验楼

天津公安警官学院图书信息中心

河北科技大学国际交流中心

茌平市体育馆、图书馆、博物馆、体育场工程

天津大港文化艺术中心

保定电信枢纽大楼

保定信用社大厦

Jie Cai

Executive Chief Architect of the Institution, researcher

Working experiences

In February 1982, graduated from Major of Architecture, Department of Architecture, Tianjin University, and have been engaged in architecture design for more than 30 years.

In 1995, promoted to be Senior Architect.

In 2002, promoted to be Researcher.

In 1998, registered as National Class I Registered Architect.

Representative works

Office building of National Geophysics Exploration Bureau

Master plan and simplex design of National Fire Control Training Base

Computation Centre of Zhangjiakou Normal Specialized Postsecondary College

Main Teaching Building of Qinhuangdao Branch Institute of Daqing Petroleum Institute

Main Teaching Building of Henan Agricultural University

Comprehensive laboratory building of Tianjin University

Book information centre of Tianjin Public Security Officer Academy

International Communication Centre of Hebei University of Science and Technology

Gymnasium, library, museum and stadium projects of Chiping City

Tianjin Dagang Culture and Art Centre

Baoding Telecommunication Hub Building

Baoding Credit Cooperative Mansion

天津大学建筑设计规划研究总院
Academy Of Architectural Design & City Planning Tianjin University

谌 谦
Qian Chen

谌 谦

副院长，副总建筑师

毕业于天津大学建筑系建筑学专业，建筑学学士，国家一级注册建筑师。曾任职于天津市房地产发展经营集团，1992年进入天津大学建筑设计研究院，现任天津大学建筑设计规划研究总院副院长，副总建筑师。谌谦先生一直从事建筑设计，具有丰富的设计经验，其设计和主持的项目涵盖了教育、办公、公共、居住类等多个领域，并相继获得过教育部等部门颁发的重要奖项。

Qian Chen

Vice president, vice chief architect
Graduated from Department of Architecture in Tianjin University, and majored in architecture, Bachelor of Economy, National Class I registered architect
Once worked in Tianjin Real Estate Development Operation Co., Ltd.
In 1992, he joined in Architectural Design & Research Institute of Tianjin University, and currently works as the Vice President and Vice Chief Architect in Architectural Design & Research Institute of Tianjin University.
Having engaged in architectural design for long time, he has rich design experience, and the projects designed and led by him cover various fields including education, office work, public building, and residence and so on. He has won many awards granted by Ministry of Education and other ministries.

张繁维
Fanwei Zhang

张繁维

天津大学建筑设计规划研究总院 执行总建筑师
1982年毕业于天津大学建筑系
研究员，国家一级注册建筑师
天津市建筑设计专家委员会委员

Fanwei Zhang

Executive chief architect in Architectural Design & Research Institute of Tianjin University
In 1982, he graduated from Department of Architecture in Tianjin University
Researcher, national Class I registered architect
Member of Tianjin Experts Committee of Architectural Design

吴 放
Fang Wu

吴 放

47岁，天津大学硕士研究生，国家一级注册建筑师，高级工程师。天津大学建筑设计研究院副总建筑师、设计四所所长，作为项目负责人，参加设计的郑州大学工科园获得天津市优秀设计一等奖，作为项目设计人参加设计的天津利顺德大饭店修缮改造工程获得天津市优秀设计一等奖，作为项目负责人参加设计的天津大学新园村二期住宅工程获得天津市优秀设计三等奖，作为建筑专业负责人参加设计的天津美术学院美术馆获得国家优秀设计银奖。

Fang Wu

47, Master Degree candidate of Tianjin University, National Class I Registered Architect, senior engineer. Assistant chief architect of Architectural Design Academy and director of 4th Design Institute of Tianjin University. Zhengzhou University Industrial and technological Park that he took part in design as the project principal won First Prize of Tianjin Excellent Design. Renovation and modification works of Tianjin Lishunde Hotel that he took part in design as the project designer won First Prize of Tianjin Excellent Design. Second phase residential project of Tianjin University Xinyuan Village that he took part in design as the project principal won Third Prize of Tianjin Excellent Design. Gallery of Tianjin Academy of Fine Arts that he took part in design as architecture major principal won Silver Prize of National Excellent Design.

张 华
Hua Zhang

张 华

天津大学建筑设计研究院 副总建筑师

毕业于天津大学建筑学院，建筑学硕士，研究员，国家一级注册建筑师。现任天津大学建筑设计研究院副总建筑师，教授创作室主任，从事建筑设计30多年，在博物馆、学校、住宅、高层等方面均具有丰富的设计经验，各项设计多次获得建设部、教育部颁发的各种奖项。他本着做中国的原创建筑的理念，在建筑设计中不断探索，并于2008年在中国建筑工业出版社出版个人建筑作品集《璞思·百筑——原创建筑作品集》。

Hua Zhang

Architectural Design & Research Institute of Tianjin University
Vice Chief Architect
Graduated from Department of Architecture in Tianjin University, Master of Architecture, researcher, national Class I registered architect.
At present, he holds the position of vice chief architect and professor of creative director in Architecture Design & Research Institute of Tianjin University. Having engaged in architectural design for more than 30 years, he has a wealth of design experience in museums, schools, residence, high-rise buildings and so on. His design has been awarded by Ministry of Construction and Ministry of Education for many times. Adhering to the concept of developing original Chinese architecture, he continuously makes the exploration in architectural design, and published his own individual architectural works Original Idea • Hard Practice–Collection of Original Architectural Works by China Architecture & Building Press in 2008.

天津天咨拓维建筑设计有限公司
TIANJIN TIANZITUOWEI ARCHITECTURAL DESIGN CO.,LTD.

宝建华
Jianhua Bao

宝建华

公司合伙人建筑师
1984–1988年　天津城建学院建筑学学士学位
1999–2004年　天津市建工设计院建筑二所所长
2004年7月至今　天津天咨拓维建筑设计有限公司合伙人建筑师
2006年　入选《城市环境设计》青年建筑师栏目
2006年　设计作品天津红桥区民族中学
　　获"天津市优秀设计二等奖"
2008年　设计作品天津津南艺术文化中心
　　获"天津市优秀设计三等奖"
2007–2009年　中国建筑设计作品年鉴特约编委

近期主要建筑作品

天津红桥区民族中学
天津津南艺术文化中心
万隆优联第六城商业街区
天津北辰区普育学校
天津红桥区红星职业学校
天津市农业博览馆

Jianhua Bao

Co-partner architect
From 1984 to 1988, he received Bachelor Degree of Architecture from Tianjin Institute of Urban Construction
From 1999 to 2004, he worked as president of Second Institute of Tianjin Architectural Engineering Design Institute
From Jul. 2004 to present, he works as the co-partner architect of Tianjin Tianzi Tuowei Architectura Design Co., Ltd
In 2006, he was selected into column of the Young Architect in Periodical Urban Space Design
In 2006, his design work "Secondary School for Nationalities" was awarded the Second Prize for Excellent Design of Tianjin
In 2008, design work "Jinnan Art and Culture Centre" was awarded the Third Prize for Excellent Design of Tianjin
From 2007 to 2009, he works as the Special Editorial Board Member of Annals of China's Architectural Design Works

Recent main architectural works

Nationalities Middle School in Hongqiao District, Tianjin
Jinnan Art and Culture Centre in Tianjin
The Six Town commercial street block of Wanlong Youlian
Puyu School in Beichen District, Tianjin
Red Star Vocational School in Hongqiao District, Tianjin
Tianjin Agricultural Museum

郭　健
Jian Guo

郭　健

公司合伙人建筑师
1984–1988年　天津城建学院建筑学学士学位
1999–2004年　天津市建工设计院建筑二所副所长
2004年7月至今　天津天咨拓维建筑设计有限公司合伙人建筑师
2006年　入选《城市环境设计》青年建筑师栏目
2007–2009年　中国建筑设计作品年鉴特约编委

近期主要建筑作品

小韩村教堂
美华大厦
洛卡小镇二期
大场商业会所
渤海明珠超高层酒店及公寓
雅宣住宅项目
海河下河圈公园

Jian Guo

The Partner Architect of the Company
From 1984 to 1988, he graduated from Tianjin Institute of Urban Construction, and received Bachelor's Degree in Architecture
From 1999 to 2004, he worked as vice president of the Second Institute of Architecture, Tianjin Institute of Architectural Engineering
From Jul.2004 to present, he works as the partner architect of Tianjin Tianzi Tuowei Architectural Design Co., Ltd
In 2006, he was selected into Column of the Young Architect in Periodical Urban Space Design
From 2007 to 2009, he works as the Special Editorial Board Member of Annals of China's Architectural Design Works

Recent Principal Works of Architecture

Xiaohan Town Church
Mei Wah Mansion
Luka Town (Phrase II)
Dachang Chamber of Commerce
Bohai Pearl Super-high Hotel Apartment
Grand Residence Project
Lower Reaches Park of the Hai River

方　轶
Yi Fang

方　轶

公司合伙人建筑师
1996–2001年　天津大学建筑学学士学位
2001–2004年　天津大学建筑学硕士学位
2004年7月至今　天津天咨拓维建筑设计有限公司合伙人建筑师
2006年　入选《城市环境设计》青年建筑师栏目
2008年　设计作品天津津南艺术文化中心
　　获"天津市优秀设计三等奖"
2007–2009年　中国建筑设计作品年鉴特约编委

近期主要建筑作品

龙顺庄园中式度假村
天津津南艺术文化中心
渤海明珠超高层酒店及公寓
天津北辰区普育学校
天津北辰区成职教中心综合体
天津市农业博览馆
邢台市中兴东大街片区项目
银川·青年 SOHO 规划设计
汇淼温泉花园

Yi Fang

Co-partner architect
From 1996 to 2001, he received Bachelor Degree of Architecture from Tianjin University
From 2001 to 2004, he received Master Degree of Architecture from Tianjin University
From Jul. 2004 to present, he works as the co-partner architect of Tianjin Tianzi Tuowei Architectural Design Co., Ltd
In 2006, he was selected into column of the Young Architect in Periodical Urban Space Design
In 2008, his design work "Jinnan Art and Culture Centre" was awarded the Third Prize for Excellent Design of Tianjin
From 2007 to 2009, he works as the Special Editorial Board Member of Annals of China's Architectural Design Works

Recent Principal Works of Architecture

Longshun Chinese-style Holiday Manor
Tianjin Jinnan Art and Culture Center
Bohai Pearl Super-high Hotel Apartment
Tianjin Beichen Puyu School
Tianjin Beichen Adult Vocational Education Center
Tianjin Agricultural Expo Museum
Zhongxing East Avenue Area Project in Xingtai
Planning and Design of Yinchuan • Youth SOHO
Huimiao Spa Garden

TSDI
铁道第三勘察设计院集团有限公司

李 京
Jing Li

李 京

1989年毕业于天津大学建筑系，国家一级注册建筑师，历任铁三院建筑院工程师、高级工程师、副总工程师、总建筑师。

设计经验
中标并组织了上海虹桥站、济南西站、天津西站等大型客运站的设计全过程，正在组织实施大连北站、宁波站、沈阳南站等大型客运站的设计工作。

Jing Li

National Class1 Registered Architect
Graduated from Department of Architecture of Tianjin Universityin 1989
He has been on positions of engineer, senior engineer, vice-chief engineer and chief engineer in No.3 Iron Architectural Institute

Working Experience
He has won the bid and organized large-scale projects of railway architectural design including Shanghai hongqiao Station, Jinan west station, Tianjin west Station, and now he is on the design work of large-scale railway architecture such as Dalian North Staion , Ningbo Station, Shenyang South Station .

卢 瑛
Ying Lu

卢 瑛

1989年毕业于天津大学建筑系，国家一级注册建筑师，历任铁三院建筑院工程师、高级工程师、副总工程师。

设计经验
作为建筑专业总工组织了北京南站、丹东站、铜陵东站、京沪高速中间站等大中型客运站的设计全过程。

Ying Lu

National Class 1 Registered Architect
Graduated from Department of Architecture of Tianjin University in 1989.
He has been on the positions of engineer, senior engineer and vice-chief engineer in No.3 Iron Architectural Institute

Working Experience
As the architecture chief engineer,he has orgnized the design process of large-scale railway architecture projects of Beijing South Railway Station, Dandong station, Tongling east station, Beijing-Shanghai express railway intermediate station.

高 虹
Hong Gao

高 虹

1988年毕业于天津城市建设学院建筑学专业，国家一级注册建筑师，历任铁三院建筑院工程师、高级工程师、副总工程师。

设计经验
主持完成天津站改扩建项目建筑专业的全过程设计，主持并基本完成哈大线、沈大段八个车站及所有配套用房的建筑设计，中标并主持了地铁北京南站建筑专业的全过程设计、完成地铁虹桥站（无线换乘）建筑专业的全过程设计，正在组织实施海拉尔站、沈阳北站等大型客运站的设计工作。

Hong Gao

National Class 1 Registered Architect
Graduated from Department of Architecture of Tianjin Institute of Urban Construction in 1988.
He has been on the positions of engineer, senior engineer and vice-chief engineer in No.3 Iron Architectural Institute

Working Experience
He has hosted the reconstruction and expansion project of Tianjin Railway Station,the Shenyang-Dalian part of Haerbin-Dalian line which was almost completed. He has won the bid and taken charge of the architecture design projects of Beijing South Railway Station, Hongqiao Underground Station (Zero-Transfer) and now he is on the design work of Hailar station, Shenyang North Station and other large scale railway projects design.

陶 然
Ran Tao

陶 然

1997年毕业于北方交通大学建筑系，国家一级注册建筑师，历任铁三院建筑院工程师、高级工程师、副总工程师。

设计经验
大连北站、哈尔滨西站设计方案得以中标，并负责天津站建筑专业设计的全过程，正在负责济南西站的建筑设计、哈尔滨站改造及哈尔滨西站工程的设计总体控制。

Ran Tao

National Class 1 Registered Architect
Graduated from Department of Architecture of Northern Jiaotong University in 1997.
He has been on the positions of engineer, senior engineer,and vice-chief engineer in No.3 Iron Architectural Institute

Working Experience
He has won the bid for Dalian north station, Haerbin west station, and taken charge of the architecture design of Tianjin Railway Station, and now he is working on architecture design of Jinan West Station, the reconstruction of Haerbin station, the project design and control of Haerbin west station.

昆明市建筑设计研究院有限责任公司

曹 理
Li Cao

许 峻
Jun Xu

李浩冰
Haobing Li

向剑凛
Jianlin Xiang

曹 理

总建筑师
国家注册规划师
高级工程师

1987年毕业于重庆建筑工程学院建筑系
1997年至今在云南省昆明市建筑设计研究院有限责任公司工作，任总建筑师

多年来在各种重要大型工程项目竞标中多次中标，并多次在国家及省级方案竞赛中获奖。

工程涉及范围包括教育、 医疗、政府、居住、体育、商业、规划、景观多个领域，曾多次荣获云南省、昆明市优秀工程设计一等奖。

许 峻

副总建筑师兼建筑方案创作室主任
国家一级注册建筑师
高级建筑师

1994年毕业于重庆建筑大学建筑系
1994年至今在云南省昆明市建筑设计研究院有限责任公司工作历，任主任建筑师、副所长、副总建筑师

多年来以设计总负责人的身份主持并主创多项大型工程项目，工程涉及范围包括教育、医疗、政府、居住、体育、商业、规划、景观多个领域。

获奖
云南省2003、2005、2007年度优秀工程设计二等奖
昆明市2002、2003、2004、2005年度优秀工程设计一等奖及中共昆明市建设局嘉奖
2010中国勘察设计协会、欧特克软件(中国)有限公司——建筑信息模型(BIM)设计大赛：最佳(BIM)工程设计三等奖。

李浩冰

技术发展中心主任建筑师
高级建筑师

1998年毕业于云南工业大学建筑系
现于昆明理工大学建筑系工程学硕士在读
1998年至今在云南省昆明市建筑设计研究院有限责任公司工作，任主任建筑师

多年来在各种重要大型工程项目竞标中多次中标，并多次在国家及省级方案竞赛中获奖。

工程涉及范围包括教育、 医疗、政府、居住、体育、商业、规划、景观多个领域，曾多次荣获云南省、昆明市优秀工程设计一等奖。

向剑凛

总规划师
国家注册规划师
高级工程师

1996年毕业于武汉城市建设学院，城市规划与建筑系城市规划专业
现任昆明市建筑设计研究院有限责任公司总规划师

在长期的规划设计中，逐渐探索出城市规划与城市文化、民族文化、地域文化结合的理论与方法。力图通过设计结合自然，发现城市的历史文脉和精神气质，找到适合本地发展的建设道路，同时构筑"和而不同"的人文环境、生态环境与发展环境。

Li Cao

Chief Architect
National Registered Planner
Senior Engineer

Graduated from Department of Architecture of Chongqing Institute of Architectural Engineering in 1987
Working as Chief Architect in Kunming Architectural Design Research Institute Limited Company, Yunnan from 1997 to present
Having won the bidding in various of competitions of important large scale projects over the years, and won the awards in project competitions of national and provincial level for many times
The projects range over the scopes of domains such as education, medicine, government, residence, sports, commerce, planning and landscape, and has won First Prize of Excellent Project Design in Yunnan Province and City of Kunming for many times.

Jun Xu

Deputy Chief Architect
Director of Architectural Programming Workshop
National first-class Registered Architect
Senior Engineer

Graduated from Department of Architecture of Chongqing Architecture University in 1994
Working in Kunming Architectural Design Research Institute Limited Company, Yunnan from 1994 to present, as principal architect, deputy director and deputy chief architect
Having managed and created in many large scale projects as the chief designer over the years. The projects range over the scopes of domains such as education, medicine, government, residence, sports, commerce, planning and landscape.
Awards:
Second Prize of Yearly Excellent Project Design of Yunnan Province in years of 2003, 2005 and 2007
First Prize of Yearly Excellent Project Design of City of Kunming in years of 2002, 2003, 2004, 2005 as well as Commendation of Kunming Municipal Construction Bureau
In 2010 won the Third Prize of Best (BIM) Project Design in Design Competition of Building Information Model (BIM) held by China Investigation and Design Association and Autodesk Software (China) Ltd.

Haobing Li

Director Architect of the Technology Development Center
Senior Engineer

Graduated from Department of Architecture of Yunnan Industrial University in 1998
Currently studying for Master's Degree of Engineering in Department of Architecture of Kunming University of Science and Technology
Working as Director Architect in Kunming Architectural Design Research Institute Limited Company, Yunnan from 1998 to present
Having won the bidding in various of competitions of important large scale projects over the years, and won the awards in project competitions of national and provincial level for many times.
The projects range over the scopes of domains such as education, medicine, government, residence, sports, commerce, planning and landscape, and has won First Prize of Excellent Project Design in Yunnan Province and City of Kunming for many times.

Jianlin Xiang

Chief Planner
National Registered Planner
Senior Engineer

Graduated from Major of Urban Planning in Department of Urban Planning and Architecture of Wuhan Urban Construction Institute
Currently working as Chief Planner in Kunming Architectural Design Research Institute Limited Company
During long-term planning and design, he has gradually explored theory and methodology combining city planning with urban culture, national culture and regional culture. He attempts to discover city's historical fabrics and spiritual temperament in order to find out the construction way suitable for the local development through the combination of design and nature. He tries to construct "harmonious yet different" human environment, ecological environment and developing environment.

泛亞設計 PAN-ASIA DESIGN

Introduciton to Shiqiao Architecture & Design Institute Ltd.Co.,Panyu Guangzhou

徐 捷
Jie Xu

徐 捷

总建筑设计师、高级建筑师、国家一级注册建筑师

1968年07月24日生，1986年至1989年毕业于重庆建工学院昆明分院建筑系建筑专业，1989年至2003年在昆明铁路局勘察设计院工作。2004年至今在云南泛亚工程设计院有限公司工作。擅长于教育建筑、城市住宅综合体、山地建筑、高级酒店、步行商业街等公共建筑设计。

代表项目：

昆明铁路局第三中学综合教学楼
重庆九龙坡铁中综和改造工程
春之城
公园道一号
明日城市
丽水天锦 · 翡翠湾
新平商城
玉溪小庙街商贸城
湖泉酒店
腾冲云峰国际温泉养生度假村

Jie Xu

Chief architectural designer, senior architect, national Class A registered Architect of Pan-Asian Engineering

He was born on July 24, 1968. From 1986 to 1989, he studied in architecture department of Kunming Branch of Chongqing Institute of Architecture and majored in Architecture. From 1989 to 2003, he worked in Kunming Railway Survey and Design Institute. From 2004 to present, he works in Pan-Asia Engineering Design Institute Co., Ltd. in Yunnan. He specializes in educational architectures, urban housing complexes, mountain buildings, luxury hotels, pedestrian malls and other public building designs.

Representative Projects

Comprehensive Building of No.3 Middle School of Kunming Railway Bureau
Comprehensive Reconstruction Project of Iron middle school of Jiulongpo in Chongqing
Spring City
Park Road No. 1
Tomorrow City
Lishui Tianjin · Green Bay
Xinping Mall
Xiaomiao Street Mall in Yuxi
Lake Spring Hotel
Tengchong Yunfeng International Hot Spring Resort

蒋 勉
Mian Jiang

蒋 勉

副总建筑设计师、国家一级注册建筑师

1969年12月21日生，1988年至1992年毕业于云南工学院建筑学系建筑专业，1992年至2003年在中国有色十四治勘察设计工程公司工作。2003年至2009年在昆明官房建筑设计有限公司工作。2009年至今在云南泛亚工程设计院有限公司工作。擅长于星级酒店、医院、高级院校、商业步行街、大剧院等公共建筑设计。

代表项目：

建水医院门诊医技综合楼
蒙自新天地商业步行街
蒙自第一高级中学
蒙自高级职业中学
昆明市官渡区高级职业中学
昆明市官渡区云秀学校
红河州大剧院
云南农村干部学院综合楼及云南农业大学田家炳楼

Jiang Mian

Deputy chief architectural designer, National Class A registered Architect of Pan-Asian Engineering

He was born in December 21, 1969. From 1988 to 1992, he studied architecture in Architectural department of Yunnan Technology School, receiving bachelor's degree. From 1992 to 2003, he worked in Survey and design engineering company of the Fourteenth China's nonferrous metallurgy. From 2003 to 2009, he worked in Kunming Official Housing Design Co., Ltd. From 2009 to present, he works in Pan-Asia Engineering Design Institute Co., Ltd. in Yunnan. He specializes in star hotels, hospitals, academic institutions, commercial pedestrian street, the grand theatre and other public building designs.

Representative Projects

Complex building of Outpatient Medical Technology of Jianshui Hospital
Xintiandi Commercial Pedestrian Street in Mengzi
No.1 Senior High School in Mengzi
Senior Vocational School in Mengzi
Senior Vocational School in Guandu, Kunming
Yunxiu School in Guandu, Kunming
The Grand Theatre of Honghe State
Complex Building of Rural Cadre College in Yunnan and Tian Jiabing Building of Yunnan Agricultural University

丁 凡
Fan Ding

丁 凡

副总建筑设计师、创作中心主任、国家一级注册建筑师

1970年8月1日生，1990年至1994年就学于云南工学院建筑学系建筑专业。1994年至1998年在云南省城乡规划设计院进行建筑设计工作。1998年至2002年昆明本土建筑设计事务所从事旅游规划和建筑设计。2003年至2006年在昆明官房建筑设计有限公司工作。2007年至今在云南泛亚工程设计院有限公司工作，任创作中心主任。对历史文化深感兴趣，追求在宏观运作中融入细节。至今已经做了为数较多的房地产项目、仿古商业街区、大中型公共建筑、区域旅游规划等设计。

代表项目：

昆明金色年华商住楼
红河州工商局办公大楼
昆明故城商业街
丽江体育馆
大理惠丰新城
石屏州衙老街
文山供电局营销调度办公楼
南方电网公司云南培训中心

Fan Ding

Deputy chief architectural designer, director of creative centre,National Class A registered Architect of Pan-Asia Engineering

He was born in August 1, 1970. From 1990 to 1994, he studied Architecture in Architectural department of Yunnan Technology School, receiving bachelor's degree. From 1994 to 1998, he worked in Urban and Rural Planning Design Institute in charge of architectural design. From 1998 to 2002, he worked in Yunnan local architectural design firm in charge of tourism planning and architecture design. From 2003 to 2006, he worked in Kunming official Housing Design Co., Ltd. From 2007 to present, he works in Pan-Asian Engineering Design Institute Co., Ltd. in Yunnan, undertakes the director of creative centre. He is deeply interested in history and culture, in pursuit of integrating the details into the macro operation. So far, he has done relatively large number of designs of real estate projects, antique commercial district, medium and large public buildings, regional tourism planning.

Representative Projects

Commercial and Residential Building of Golden Age in Kunming
Office Building of Trade and Industry Bureau in Red River State
Commercial Street of Old City in Kunming
Lijiang Stadium
Huifeng New Town in Dali
Shiping Arcient Government Street
Marketing and Dispatching Office Building of Power Supply Bureau in Wenshan
Training Center in Yunnan of Southern Power Grid Corporation

浙江省现代建筑设计研究院
浙江现代投资项目管理咨询有限公司
ZMA
Zhejiang Modern Architectural Design & Research Institute

王国栋
Guodong Wang

王国栋

国家一级注册建筑师、高级建筑师

现任浙江省现代建筑设计研究院有限公司副总建筑师、院长助理。1990年毕业于华南理工大学建筑系，一直从事建筑设计及管理工作。

从业20年，有丰富的各类公共建筑与其他民用建筑设计实践经验，曾多次获省市级优秀设计奖项。尤其在医疗建筑设计领域，能全方位地清晰把握项目定位，对项目从前期策划、建筑设计到实施的全程操作有独到见解。在理论方面，曾多次在国内多家核心期刊、报纸上发表论文。

主要作品

杭州市第十人民医院新建项目（63 000平方米）
内蒙古赤峰市儿童福利院项目（3 000平方米，获省优秀设计奖）
河南省平顶山市新城区医院新建项目（172 000平方米）
浙江金成集团江南春 闲林山水居住区项目(460 000平方米)
浙江医院医疗综合楼项目（4 500平方米）
山东省泰安中医院病房综合楼项目（50 000平方米）
陕西省榆林市第二医院迁建项目（189 000平方米）

发表学术论文

《关于医院建筑人性化设计的思考》一文，最初发表于《健康报》，后经充实修改发表于《中国医院管理》杂志2007.11期，并转载于《医用工程》杂志2008.4期。
《潍坊市公共卫生服务中心规划方案设计浅析》，发表于《中国医院建筑与装备》2007.9期。（与人合作）
《日本、韩国医院建设给我们的启示》，发表于《中国卫生工程》2007.11期。
《谈医院建设选址》，收录入《2009中国上海医院建筑设计及装备国际研讨会暨展示会论文集》。

Guodong Wang

Guodong Wang, as a National Class Ⅰ Registered Architect and Senior Architect
He is vice chief architect and president assistant of Zhejiang Xiandai Architectural Design & Research Institute Co., Ltd. He graduated from Department of Architecture, South China University of Technology in 1990. He has engaged in architectural design and management until now.
Having engaged for 20 years, he accumulated rich experience on construction for public and civil use. Many designs won local and provincial outstanding design awards. Especially for the design of the medical building, he can grasp the project as a whole. From the pre-planning, building design and final operating, he has unique perspectives. In theory, he published papers on domestic core journals and newspaper many times.

Representative works

Hangzhou No.10 People's Hospital (63,000 m^2)
Nei Mongol Chifeng Welfare Home for Children (3,000 m^2 It won the Outstanding Design on provincial level)
Hospital in New City, Pingdingshan Shi, Henan Province (172,000 m^2)
Jiangnanchun Xianlin Shanshui Residential Site invested by Zhejiang Jincheng Group (460,000 m^2)
Comprehensive Medical Building of Zhejiang Hospital (4,500 m^2)
Comprehensive ward building of Shandong Taian Hospital of Chinese Medicine (50,000 m^2)
Rebuilding project for Shanxi Yulin No.2 Hospital (189,000 m^2)

Academic Theses

"Human Nature Design of Hospital Buildings" was firstly published in Health Newspaper, then in 11th issue of Chinese Hospital Management, 2007 after enriching and revising, and finally reprinted in the fourth issue of Medical Engineer, 2008
"Brief Analysis of Planning of Weifang Public Health Service Center" was published in the ninth episode of Architecture and Equipment for Chinese Hospital, 2007. (Cooperated with others)
"Enlightenment from Japanese and South Korean Hospital" was published in 11th issue of Chinese Hygiene Projects, 2007
"Discussion of Site Selection of Hospital Building" was collected in Essays Published in International Symposium on Hospital Architecture and Equipment in Shanghai, China, 2009

谢作产
Zuochan Xie

谢作产

医疗建筑设计研究所副所长、主任建筑师
国家一级注册建筑师
2001年毕业于华中科技大学

主要业绩

浙江省永嘉县卫生局大楼（20 000平方米）
浙江省奉化人民医院门诊综合楼（35 000平方米）
山东省潍坊市妇幼保健医院（37 000平方米）
浙江大学附属妇产科医院科研综合楼（40 000平方米）
江苏省常州市新北人民医院（128 000平方米）
浙江省淳安县人民医院（53 000平方米）
福建省顺昌县人民医院（52 000平方米）
浙江省嘉善县人民医院（79 000平方米）
江苏省南通市妇幼保健医院（57 000平方米）
河南省清丰县人民医院（55 000平方米）

Zuochan Xie

Deputy Prisident of medical building design institute, Chief architect
National Class-1 registered architect Graduated from Huazhong Uniuersity of Science & technology.

Representative works

Building of Zhejaing Yongjia Public Health Bureau (20,000 m^2)
Comprehensive outpatient building of Zhejiang Fenghua People's Hospital (35,000 m^2)
Shandong Weifang Maternal and Child Hygiene Hospital (37,000 m^2)
Research building of Affiliated Obstetrics and Gynecology Hospital of Zhejiang University (40,000 m^2)
Suzhou Cangzhou Xinbei People's Hospital (128,000 m^2)
Zhejiang Chunan People's Hospital (53,000 m^2)
Fujiang Shunchang People's Hospital (52,000 m^2)
Zhejiang Jiaxian People's Hospital (79,000 m^2)
Jiangsu Nantong Maternal and Child Hygiene Hospital (57,000 m^2)
Henan Qingfeng People's Hospital (55,000 m^2)

浙江省现代建筑设计研究院
浙江现代投资项目管理咨询有限公司
Zhejiang Modern Architectural Design & Research Institute

封素芬
Sufeng Feng

蒋德利
Deli Jiang

曾国良
Guoliang Zeng

封素芬

女 1964年6月出生
国家一级注册建筑师、高级工程师
毕业于浙江大学建筑系，获学士学位
现任职于浙江省现代建筑设计研究院有限公司综合建筑设计二所副所长、主任建筑师

主要作品
宁波市鄞州人民医院病房楼
获2003年浙江省“钱江杯”优秀设计三等奖
获2009年全国医院建筑优秀设计三等奖
宁波市鄞州第二医院迁建工程
获2007年浙江省“钱江杯”优秀设计三等奖
获2009年全国医院建筑优秀设计三等奖
浙江省中医院病房楼
获2008年浙江省“钱江杯”优秀设计二等奖
获2009年全国医院建筑优秀设计三等奖
新昌县人民医院病房楼
获2001年浙江省“钱江杯”优秀设计二等奖
浙江大学医学院附属第二医院急救中心（已建）
浙江省人民医院医疗综合楼（已建）
浙江大学学生公寓（已建）
浙江省长兴县中医院迁建工程（已建）
武警浙江省总队杭州医院病房楼（在建）
杭州市下沙医院（在建）
慈溪市人民医院病房楼（在建）
山东潍坊妇幼保健院（在建）

Sufeng Feng

Ms Song was born in June, 1964
She is a national Class Ⅰ registered architect and a senior engineer
She received bachelor degree from Department of Architecture of Zhejiang University.
She takes positions as deputy director and chief architect in No.2 Comprehensive Architectural Institute of Zhejiang Xiandai Architectural Design & Research Institute Co., Ltd.

Representative works
1. Ward building of Ningbo Yinzhou People's Hospital
Won the third prize of Outstanding Design in “Qianjiang Cup” competition in Zhejiang Province, 2003 and the third prize of National Outstanding Design for hospitals, 2009.
2. Rebuilding Project for No.2 Ningbo Yinzhou Hospital
Won the third prize of Outstanding Design in “Qianjiang Cup” competition in Zhejiang Province, 2007 and the third prize of National Outstanding Design for hospitals, 2009.
3. Ward building of Zhejiang Hospital of Chinese Medicine
Won the second prize of Outstanding Design in “Qianjiang Cup” competition in Zhejiang Province, 2008 and the third prize of National Outstanding Design for hospitals, 2009.
4. Ward building of Xinchang People's Hospital
Won the second prize of Outstanding Design in “Qianjiang Cup” competition in Zhejiang Province, 2001
5. Emergency Center of No.2 Affiliated Hospital of Zhejiang University School of Medicine (established)
6. Medical Synthesis Building of Zhejiang People's Hospital (established)
7. Students' Apartment of Zhejiang University (established)
8. Rebuilding project for Zhejiang Changxing Hospital of Chinese Medicine (established)
9. Ward building of Hangzhou Hospital of Zhejiang Armed Police Force (on construction)
10. Hangzhou Xiasha Hospital (on construction)
11. Ward building of Cixi People's Hospital (on construction)
12. Shandong Weifang Maternal and Child Health Hospital (on construction)

蒋德利

国家一级注册建筑师
2000年毕业于华东交通大学建筑学专业
毕业后进入浙江省现代建筑设计研究院有限公司，担任建筑设计工作，长期处在建筑设计的最前沿，2008年获得国家一级注册建筑师。从业至今，主持并设计完成了三十余个医疗项目工程，项目涉及浙江、江苏、福建、山东、江西、河南、河北、陕西等多个省市和地区，其中的浙江省立同德医院医疗科研综合楼、浙江省台州医院恩泽纪念楼、台州母子医院等多个项目产生了一定的影响力。能带领设计团队与各方紧密合作，在组织、协调和专业知识方面都具有相当丰富的实践经验，获得了业主方的较高评价。

Deli Jiang

National Class Ⅰ registered architect
After graduating from Architecture Department in East China Jiao Tong University, he took a job as a designer in Zhejiang Xiandai Architectural Design & Research Institute Co., Ltd. For a long time, he has been pursuing the most advanced technology for architectural design. He was granted as national Class Ⅰ Registered Architect in 2008. Up to now , he has organized and completed 30 medical projects which located in many places including Zhejiang, Jiangsu, Shandong, Jiangxi, Henan, Hebei, Shanxi and etc. Among these projects, medical research comprehensive building of Zhejiang Shengli Tongde Hospital, Enze monumental building of Zhejiang Taizhou Hospital, Taizhou Maternity & Child Hospital and etc exert great influence. He leads his team, corporates with each party, calculates rich practical experience on organization, cooperation and professional knowledge, and finally receives high approval from clients.

曾国良

建筑设计一所所长、主任建筑师
国家一级注册建筑师、高级建筑师
1996年毕业于苏州城建学院

主要业绩
江苏姜堰市人民医院病房综合楼
获省优秀设计三等奖
浙江省富阳市人民医院病房综合楼
获省优秀设计三等奖
浙江省富阳骨伤科医院迁建工程
广东乐从医院新建工程
浙江台州恩泽楼
获省优秀设计三等奖
浙江省上虞市人民医院迁建工程
获省优秀设计二等奖
浙江省嘉兴市第一医院迁建工程

Guoliang Zeng

Director of No.1 Architectural Design Institute, Chief Architect
National Class Ⅰ Registered Architect, Senior Architect
He graduated from Suzhou Institute of Urban Construction in 1996

Representative works
Comprehensive ward building of Jiangsu Jiangyan People's Hospital
The third prize of outstanding design on provincial level
Comprehensive ward building of Zhejiang Fuyang People's Hospital
The third prize of outstanding design on provincial level
Rebuilding project for Zhejiang Fuyang Hospital of Orthopedics
Guangzhou Lecong Hospital
Zhejiang Taizhou Enze Building
The third prize of outstanding design on provincial level
Rebuilding Project for Zhejiang Shangyu People's Hospital
The second prize of outstanding design on provincial level
Rebuilding project for Zhejiang Jiaxing No.1 Hospital

浙江大学建筑设计研究院
ZUADR Architectural Design and Research Institute of Zhejiang University

殷 农
Nong Yin

李 宁
Ning Li

劳燕青
Lao Yanqing

柳 青
Qing Liu

殷 农

职务职称
副总建筑师
高级建筑师
国家一级注册建筑师

主要作品
1.杭州亲亲家园住宅小区
2.宝鸡职业技术学院
3.宁波鄞州外贸大厦
4.杭州电子科技大学下沙校区图书馆
5.舟山临城舟山医院病房楼
6.福建省直机关干部住宅小区
7.南通好一家军山小镇商业办公楼群
8.浙江传媒学院多层学生（女生）公寓
9.绍兴亲亲家园
10.河南南阳职业技术学院

Nong Yin

Job title
Vice chief architect
Senior architect
National Class 1 Registered Architect

Main works
1. Hangzhou Warm Town Residential Community
2. Baoji Vocational Technology College
3. Ningbo Yinzhou Trade Building
4. Library of Xiasha Campus of Hangzhou Dianzi University
5. Ward Building of Zhoushan Hospital of Zhoushan Lincheng
6. Residential Community of Fujian Provincial Government Officials
7. Junshan Village Business Office Building Group of Nantong Haoyijia
8. Multi-story Student (Female) Apartment of Zhejiang University of Media and Communications
9. Shaoxing Warm Town Residential Community
10. Henan Nanyang Vocational Technology College

李 宁

职务职称
咨询中心副总建筑师
研究员
国家一级注册建筑师

主要作品
1.萧山红山文化活动中心
2.杭州市下城区房屋经营公司江山弄7号楼
3.上海浦东之江大厦
4.余杭粮食局大楼
5.上海新兴大厦
6.浙江瑞安中学
7.绍兴托普信息技术学院
8.绍兴日报住宅小区
9.杭州山水人家住宅小区
10.台州书画院

Ning Li

Job title
Vice-principal Architect of the Consultation Center
Researcher
National Class 1 Registered Architect (PRC)

Main works
1. Xiaoshan Hongshan Cultural Functional Center
2. No. 7 Building in Jiangshannong of Hangzhou Xiacheng Real Estate Co.,
3. Shanghai Pudong Zhijiang Building
4. Building of Yuhang Foodstuff Provision Bureau
5. Shanghai Xinxing Duilding
6. Zhejiang Ruian Middle School
7. Shaoxing TOP Information & Technology College
8. Shaoxing Daily Residence
9. Hangzhou Poetic Home Residence
10. Taizhou Academy of Chinese Painting & Calligraphy

劳燕青

职务职称：
第三建筑设计研究所主任建筑师
高级建筑师
国家一级注册建筑师

主要作品
1.浙江瑞安中学体育馆
2.舟山文化艺术中心
3.浙江衢州市行政中心
4.杭州金成·竹海水韵住宅小区
5.中国井冈山干部学院
6.杭州CBD—尊宝大厦
7.浙江加州国际纳米技术研究院大楼
8.杭州市科技馆方案设计
9.山东潍坊市市委党校
10.云南大理·苍洱映象住宅小区

Yanqing Lao

Job title
Principal architect
Senior Architect
National Class 1 Registered Architect (PRC)

Main works
1. Gymnasium of Zhejiang Ruian Middle School
2. Zhoushan Culture and Art Center
3. Zhejiang Quzhou Municipal Administration Center
4. Hangzhou Jincheng•Zhuhaishuiyun Residential Community
5. China Executive Leadership Academy. Jinggangshan
6. Hangzhou CBD•Zunbao Building
7. Zhejiang Jiazhou International Nano Technology Research Institute
8. Plan design of Hangzhou Municipal Technology Museum
9. Party School of CPC Shandong Weifang Municipal Committee
10. Yunnan Dali•Cangeryinxiang Residential Community

柳 青

职务职称
第六建筑设计研究所主任建筑师
高级建筑师
国家一级注册建筑师

主要作品
1.浙江理工大学下沙新校区综合教学楼、综合实验楼、行政大楼、服装艺术学院和会展中心剧场
2.南昌航空学院新校区图书馆、风雨操场、学生活动中心
3.杭州高级中学行政楼国际交流中心体育健身中心
4.杭州华业南岸晶都住宅小区
5.常州世茂香槟湖一期住宅小区
6.富阳中大“西郊半岛”住宅小区
7.绍兴天马商务大楼
8.浙江汇信机电公司
9.杭州西湖区疾病控制中心
10.诸暨建银大厦

Qing Liu

Job title
Principal Architect of the Sixth Architecture Design Research Institute
Senior Architect
National Class 1 Registered Architect (PRC)

Main works
1. Xiasha New Campus of Zhejiang University of Science and Technology Comprehensive Teaching Building, Comprehensive Laboratory Building, Administrative Building, Garment Art College and Exhibition Center Theater
2. New Campus of Nanchang Hangkong University Library,Fengyu Playground,Student Activity Center
3. Administrative Building, International Communication Center and Sports Center of Hangzhou High School
4. Hangzhou Huaye Nan'anjindu Residential Community
5. Changzhou Shimao Xiangbinhu Phase I Residential Community
6. Fuyang Zhongda "Xijiao Peninsula" Residential Community
7. Shaoxing Tianma Business Building
8. Zhejiang Huixin Electromechanical Company
9. Disease Control Center, Xihu District, Hangzhou
10. Zhuji Jianyin Building